CONVERSION FACTORS

Length

1 in. = 2.54 cm
1 ft = 0.3048 m
1 mi = 5280 ft = 1.609 km
1 m = 3.281 ft
1 km = 0.6214 mi
1 angstrom (Å) = 10^{-10} m

Mass

1 slug = 14.59 kg
1 kg = 1000 grams = 6.852×10^{-2} slug
1 atomic mass unit (u) = 1.6605×10^{-27} kg
(1 kg has a weight of 2.205 lb where the
   acceleration due to gravity is 32.174 ft/s²)

Time

1 day = 24 h = 1.44×10^3 min = 8.64×10^4 s
1 yr = 365.24 days = 3.156×10^7 s

Speed

1 mi/h = 1.609 km/h = 1.467 ft/s = 0.4470 m/s
1 km/h = 0.6214 mi/h = 0.2778 m/s = 0.9113 ft/s

Force

1 lb = 4.448 N
1 N = 10^5 dynes = 0.2248 lb

Work and Energy

1 J = 0.7376 ft·lb = 10^7 ergs
1 kcal = 4186 J
1 Btu = 1055 J
1 kWh = 3.600×10^6 J
1 eV = 1.602×10^{-19} J

Power

1 hp = 550 ft·lb/s = 745.7 W
1 W = 0.7376 ft·lb/s

Pressure

1 Pa = 1 N/m² = 1.450×10^{-4} lb/in.²
1 lb/in.² = 6.895×10^3 Pa
1 atm = 1.013×10^5 Pa = 1.013 bar =
   14.70 lb/in.² = 760 torr

Volume

1 liter = 10^{-3} m³ = 1000 cm³ = 0.03531 ft³
1 ft³ = 0.02832 m³ = 7.481 U.S. gallons
1 U.S. gallon = 3.785×10^{-3} m³ = 0.1337 ft³

Angle

1 radian = 57.30°
1° = 0.01745 radian

STANDARD PREFIXES USED TO DENOTE MULTIPLES OF TEN

Prefix	Symbol	Factor
Tera	T	10^{12}
Giga	G	10^{9}
Mega	M	10^{6}
Kilo	k	10^{3}
Hecto	h	10^{2}
Deka	da	10^{1}
Deci	d	10^{-1}
Centi	c	10^{-2}
Milli	m	10^{-3}
Micro	μ	10^{-6}
Nano	n	10^{-9}
Pico	p	10^{-12}
Femto	f	10^{-15}

BASIC MATHEMATICAL FORMULAE

Area of a circle = πr^2

Circumference of a circle = $2\pi r$

Surface area of a sphere = $4\pi r^2$

Volume of a sphere = $\frac{4}{3}\pi r^3$

Pythagorean theorem: $h^2 = h_o^2 + h_a^2$

Sine of an angle: $\sin \theta = h_o/h$

Cosine of an angle: $\cos \theta = h_a/h$

Tangent of an angle: $\tan \theta = h_o/h_a$

Law of cosines: $c^2 = a^2 + b^2 - 2ab \cos \gamma$

Law of sines: $a/\sin \alpha = b/\sin \beta = c/\sin \gamma$

Quadratic formula:
   If $ax^2 + bx + c = 0$, then $x = (-b \pm \sqrt{b^2 - 4ac})/(2a)$

PHYSICS

JOHN WILEY & SONS, INC.

NEW YORK   CHICHESTER
BRISBANE   TORONTO
SINGAPORE

JOHN D. CUTNELL · KENNETH W. JOHNSON
SOUTHERN ILLINOIS UNIVERSITY AT CARBONDALE

PHYSICS

SECOND EDITION

Acquisitions Editor	Cliff Mills
Developmental Editor	Barbara Heaney
Marketing Manager	Catherine Faduska
Production Manager	Joe Ford
Text/Cover Designer	Madelyn Lesure
Cover Photograph	Allsport/Vandystadt
Production Supervisor	Lucille Buonocore and Hudson River Studio
Manufacturing Manager	Lorraine Fumoso
Copy Editing	Deborah Herbert
Photo Researcher	Charles Hamilton
Photo Research Manager	Stella Kupferberg
Illustration Coordinator	John Balbalis

Library of Congress Cataloging in Publication Data:

Cutnell, John D.
    Physics / John D. Cutnell, Kenneth W. Johnson. — 2nd ed.
        p.   cm.
    Includes index.
    ISBN 0-471-52919-2
    1. Physics.   I. Johnson, Kenneth W.   II. Title.
QC23.C985   1992
    530 — dc20   91-35770
                                                                          CIP

Printed in the United States of America

10 9 8 7 6 5 4 3 2 1

PREFACE

This text is designed for a one-year course in college physics that depends on algebra and trigonometry. In writing this text, we have focused on the needs of students and teachers alike. Students want to learn, and teachers want to teach; yet, each has only so much time every day. This new edition deals with important aspects of learning and teaching physics to facilitate the work of both students and teachers. We were most gratified by the success of the first edition, and have now expanded our goals and features to produce what we believe is a better book. These goals and features are discussed below, and those that are new to the second edition are preceded with a star (★).

GOALS AND FEATURES OF THE SECOND EDITION

GOAL I: To help students see that physics is a wonderfully integrated body of knowledge. In support of this goal, the book includes the following key features:

★ **1. Integration of Concepts.** Whenever anyone is learning something, it can be difficult to grasp the overall picture while absorbing a welter of new ideas. In physics, the overall picture is a marvelous one, in which a surprisingly small number of fundamental ideas are unified into a coherent view of the physical world. To help convey the unity of physics, we have included at the end of most chapters a section entitled INTEGRATION OF CONCEPTS. These sections explore the common ground between fundamental ideas in the current chapter and fundamental ideas from previous chapters. The intent is to help students see that physics is an integrated body of knowledge, each idea taking into account what has come before it.

★ **2. Intersection Essays.** These essays, contributed by Professor Neil Comins, explore the ways in which physics intersects with other disciplines. Written in a narrative style, these essays give a flavor of how physics is used in law enforcement, geology, music, environmental science, space science, and medicine. They show that the use of physics extends far beyond the boundaries of physics laboratories and the interests of professional physicists.

3. Reviews of Previous Material. Since reviewing is an essential step in the learning process, a detailed summary of the material is provided at the end of each chapter. These summaries are condensed but thorough expositions of the material presented in the chapters and are not just compilations of equations. Equations are included, however, unless prohibited by complexity, in which case a reference to the text equation is given.

Reviewing is not something to be done only once, after a chapter is read. Reviewing should be an ongoing process. We have tried to encourage reviewing in the worked-out examples in the text and in the homework questions and problems at the ends of the chapters. In these places, we have taken special care to include situations that combine current chapter material with previous chapter material. Thus, students see that material studied early in the course is connected to material studied later. For purposes of frequent reviewing, we have also provided references in the text to the locations of earlier material when it is needed.

GOAL II: To help students develop problem-solving skills. A number of important features in the book are oriented toward this goal.

★ **1. Reasoning—The Cornerstone of Problem Solving.** Often students get the impression that solving problems is like a magic show, in which equations are pulled out of thin air and miraculously give the right answer. Physicists, however, know that careful reasoning, not magic, is the cornerstone of problem solving and must occur before pencil is ever put to paper. Therefore, we have modified the format in which examples are worked out in the text. Each example now includes a "reasoning" step before the numerical solution is carried out. In this step, we explain what motivates our procedure for solving the problem. We believe that students will benefit from seeing the "reasoning" stated explicitly.

★ **2. Problem Solving Insights.** Seeing how problem-solving techniques are applied in worked-out examples helps students to learn that technique. To reinforce these techniques, we have included short statements in the margins, identified by the label PROBLEM SOLVING INSIGHT. For instance, this PROBLEM SOLVING INSIGHT occurs in Chapter 4: "Applications of Newton's second law always involve the net force, which is the vector sum of all the forces that act on an object." The reinforcement provided by these Insights will be useful to students and supplement what teachers stress in class.

3. Examples Worked Out with Great Care. We have tried hard to provide students with good problem-solving models for their own work. Each of the 308 calculational examples in the text is thoroughly worked out. A three-part format is used (i) Problem statement, (ii) Reasoning, and (iii) Solution. The explicit reasoning step is included before the solution to emphasize that careful thinking should precede any numerical calculations. In general, calculations begin with an algebraic solution for the unknown quantity. Then, known information is substituted into the algebraic result to obtain the numerical answer.

4. Free-Body Diagrams. Teachers are familiar with the importance of free-body diagrams as a valuable problem-solving tool. And all students will learn about them as they study Newton's laws of motion. We use free-body diagrams throughout this text, not just in the early chapters where Newton's second law is introduced and applied. For instance, in Chapter 11, when the relation between pressure and depth in a fluid is developed, a free-body diagram clarifies the discussion considerably. Free-body diagrams are also used in worked-out examples, where appropriate, as in Example 5 in Chapter 18 when we calculate the electrostatic forces that electric charges exert on each other.

5. Solved Problems. Another way in which we hope to build problem-solving skills is with a feature called SOLVED PROBLEMS. These innovative illustrative calculations are included at the end of certain chapters, between the chapter summary and the homework material. They differ from the standard numbered examples that occur within the text material of the chapter in three ways. (i) They deal with concepts that are more difficult and elaborate than those treated in the numbered examples. (ii) At the end of each one, there is a summary of the important points that have been illustrated. (iii) The Solved Problems are intended for use in conjunction with homework assignments. Therefore, each one includes a reference list that identifies three to five homework problems dealing with the same general

concepts as the Solved Problem. These associated homework problems have a high level of difficulty and are not simple repetitions of the Solved Problem with only the data changed. Each of the associated homework problems includes a phrase such as "See Solved Problem 2 for a related problem." Thus, once the basic material of the chapter has been discussed, teachers can focus on the concepts in the solved problems by assigning the associated homework problems.

6. Homework Problems with Structured Levels of Difficulty. Building problem-solving skills involves the use of homework problems that progress from relatively easy, to moderate, to challenging levels of difficulty. In this spirit, we have ranked the homework problems according to difficulty. The most difficult problems are marked with a double asterisk (**), while problems of intermediate difficulty are marked with a single asterisk (*). The easiest problems are unmarked. Some of the problems are organized by section, facilitating easy reference of the material, whereas others are grouped without reference to any particular section under the heading "Additional Problems." A number of the first-edition homework problems have been changed and new problems have been added, with a view toward achieving a smoother transition, or ramping, in difficulty level from the easiest to the most difficult problems.

7. Multiple-Method Calculations. In building problem-solving skills, students benefit from seeing calculations done in more than one fashion. Therefore, the worked-out examples include a number of multiple-method calculations.

GOAL III: To show students that physics principles come into play over and over again in their lives. In working toward this goal, we have incorporated the following features:

★ **1. The Physics of . . .** To help students understand just how prevalent physics is in their lives, we have included a large number of explanations of how physics principles play a role in the operation of devices and techniques. These devices and techniques have been chosen from a wide variety of areas and include medical applications (e. g., cavitron ultrasonic surgical aspirator), automobile features (e. g., inertial seat belt mechanism), transportation (e. g., magnetically levitated trains), home entertainment (e. g., compact disc technology), household applications (e. g., plumbing), information processing (e. g., laser printers), detection devices (e. g., bomb detector), camera technology (e. g., autofocusing mechanism), satellite technology (e. g., geosynchronous satellite), and many more. To highlight the discussions of how physics principles are applied, each application is identified in the margin with the label THE PHYSICS OF. . . . A list of the applications can be found following the table of contents and it includes many that are not found in other texts.

2. Physics and Human Physiology. We have also included discussions and examples that focus on human physiology. Among these are muscle forces, blood pressure, blood flow, breathing, the detection of sound by the ear, the refraction of light by the eye, and the physiological effects of radioactivity. Such topics have been selected because of the straightforward connection they have to physics principles.

3. Worked-out Examples and Homework Material. The emphasis on physics in daily life is carried over into the worked-out examples in the text. For instance, Example 4 in Chapter 14 shows how to determine how long a SCUBA diver can stay under water on a tank of air, and Example 9 in Chapter 22 illustrates how to calculate the voltage produced by a bicycle generator. In a similar way, illustrations of everyday physics pervade the homework questions and problems. Students who use this text will realize that physics does indeed permeate their lives.

DESIGN OF THE SECOND EDITION

The layout and design of a book can contribute significantly to the learning process. We have paid special attention to enhancing and strengthening these elements in the second edition, as indicated below.

★ **1. Full-Color Reproduction.** The second edition uses full color reproduction for all figures and photographs, allowing a more realistic portrayal of physical situations. In addition, we have used color in a consistent way to denote vector concepts:

- displacement ▬▬▬
- velocity ▭▭▭ or ▭▭
- acceleration ▬▬▬
- force ▬▬▬
- electric and magnetic fields ▬▬▬

Such consistency provides an added clarity that facilitates the learning process.

★ **2. Photographs.** We have greatly expanded the use of photographs, and there are now more than 230 of them in the book. Many are included to complement the line art, so as to strengthen the ideas being discussed. Others are included to show that physics applies to a wide variety of situations in the real world.

3. Artwork. To optimize the effectiveness of the artwork, we have focused on only one main idea per picture, whenever possible. Therefore, the figures in the text are often divided into two or more parts, to avoid the confusion of superimposing several ideas on the same drawing. The use of airbrushing and three-dimensional rendering for the line art has been greatly expanded in the second edition. These enhancements make the presentation of physics ideas seem more realistic and less abstract.

4. Important Basic Concepts. One of the tasks that face students as they read an introductory physics book is to distinguish between important basic concepts and other related, but less fundamental ideas. To identify important basic concepts, we have enclosed them within a box headed by a prominent colored band. Since applying these concepts entails using correct units, the appropriate SI units have also been included within the box. The boxes are used sparingly so that they can serve effectively as a guide to the truly basic concepts.

5. Presentation of Equations. We have presented equations in a style that provides for maximum clarity and encourages correct usage. First, we have tried to anticipate the common mistakes that students make. Consequently, the book is liberally sprinkled with explanations and cautionary notes to clarify the meanings of difficult concepts and the conditions under which the concepts can be applied. For additional reinforcement, these conditions are included on the right-hand side of numbered equations as, for example, in Equation 8.9 ($v_T = r\omega$). This equation relates the angular speed ω to the tangential speed v_T and can be applied only if angles are measured in radians, not in degrees. Second, many equations, such as Equation 14.7, have a label on the left side of the page that helps to identify the situation for which the equation is applicable. Lastly, we have written some equations, such as Equation 6.7a, with bracketed labels that explain the physical meaning of each term within the equation.

6. Worked-Out Examples. We have increased the number of worked-out examples in the text by 15 percent, so there are now 308 examples in the book. Real-world situations have been used in the examples wherever possible. As mentioned above, each example is now divided into three parts: Problem statement, Reasoning, and Solution.

7. Homework Problems and Questions. The second edition contains 2428 problems and questions for assignment as homework. This represents an increase of 16 percent in the homework problems. Of the total, approximately 470 are in the form of qualitative questions. Whenever possible, we have used real-world situations with realistic data. Solutions to all problems are available for teachers.

8. Organization. Both learning and teaching are easier when the material is arranged in a logical order. Often, however, logic is in the eye of the beholder, and there is more than one sequence for the material. Therefore, we do not claim that our organization is unique, only that we have tried to achieve maximum clarity. Consistent with clarity, we have chosen section and subsection headings in a way that will lend itself to judicious editing. Material that is a likely candidate for omission is typically located in a subsection at the end of a main section or in a separate section near the end of a chapter. Sections marked with an asterisk can be omitted with little or no impact on the overall development of the material. Certain chapters from the first edition have been combined, with the result that the second edition contains 32 chapters. We feel that this number allows instructors to cover the material at an average rate of about one chapter per week.

9. Significant Figures. Standard procedures for significant figures are followed throughout this text. They are not just introduced at the beginning of the book and then ignored. A review of these procedures is given in Appendix B.

We have tried to produce an error-free book, but no doubt some errors still remain, all of which are solely our responsibility. Please feel free to let us know of any errors that you find.

We hope that this text is useful to both students and teachers and look forward to hearing about your experiences with it. We also hope that our efforts will make your lives easier and your work more enjoyable.

Carbondale, Illinois   *John D. Cutnell*
1992   *Kenneth W. Johnson*

SUPPLEMENTS

An innovative package of supplements to accompany *Physics*, 2nd edition, is available to assist both the teacher and the student.

1. Study Guide with Selected Solutions, prepared by Charles R. McKenzie and Andrew J. Pica (both of Salisbury State College), John D. Cutnell, and Kenneth W. Johnson. The Guide encourages and motivates students with chapter objectives and outlines, explanations of commonly misunderstood topics, sample worked-out problems, solutions to selected textbook problems, and chapter quizzes.

2. Solutions Manual, for instructors only, prepared by Charles R. McKenzie and Andrew J. Pica, co-authors of the **Study Guide.** The Manual contains detailed solutions to all homework problems in the text.

3. Solutions Disk, a computer disk version of the **Solutions Manual,** available in Word Perfect or Microsoft Word formats, for instructors only.

4. Homework Disk, for instructors only. Teachers of large classes often use a computer-graded, multiple-choice homework format and spend considerable time modifying textbook homework problems to suit their own needs. As part of our computerized **Test Bank,** Mark Comella of Duquesne University has converted over 1000 of the chapter-ending problems into a multiple-choice format with reasonable answers, so teachers can generate their homework assignments in a convenient and effective way.

5. Instructor's Resource Manual, prepared by Robert L. Kernall of Old Dominion University. The Manual contains teaching suggestions, lecture notes, demonstration suggestions, alternative syllabi for courses of different lengths and emphases, strategies for incorporating supplements and materials from other texts, as well as conversion notes allowing the instructor to use class notes from other texts.

6. Test Bank, prepared by Mark Comella of Duquesne University. This completely new **Test Bank** contains more than 1500 short-answer questions and problems.

7. Computerized Test Bank. IBM, Apple II, and Macintosh versions of the entire **Test Bank** are available with full editing features to help you customize tests.

8. Four-color, Overhead Transparencies. More than 200 four-color illustrations from the text are provided in a form suitable for projection in the classroom.

ACKNOWLEDGMENTS

One of the most gratifying tasks for authors is thanking those who have so generously contributed their talents, insights, and concerns. Foremost among them are our students, for they have given much to this book by shaping our teaching philosophies and techniques.

We are especially grateful to our editor and friend, Cliff Mills. A patient listener through the good and hard times, Cliff has been a source of guidance and encouragement. He provided a wealth of ideas along the way, many of which are now found in the book.

To Barbara Heaney, our developmental editor, we owe a special debt of gratitude. She has had a significant impact on the quality of the book. We have admired her concern and dedication throughout the project, and cannot imagine what the end result would have been like without her.

To Ed and Lorraine Burke of Hudson River Studio, we are deeply appreciative for their efforts during the production and layout stages. They graciously accepted our countless telephone calls and answered our cries for help with sage advice and good humor. The clean, open layout of the book is testimony to their many talents.

Cathy Donovan at John Wiley, Inc., has become a very special person to us, always willing to help, always doing the job right, and always doing it with a smile and sense of humor. It was a genuine pleasure working with Cathy.

We also wish to acknowledge the help and support of Catherine Faduska. She has worn two hats on this project, first as physics editor during the latter stages of the first edition, and now as marketing manager. Her interest in the book is much appreciated.

The superb photographs in the book are a tribute to the research skills of Charles Hamilton and Stella Kupferberg. It is astonishing how much time and effort goes into finding just the "right" pictures. We are indebted to them for a first-rate job.

Our thanks go to Maddy Lesure for the outstanding design of the book and cover, to John Balbalis for assisting with the illustration program, to Deborah Herbert and Virginia Dunn for copyediting the manuscript, to Lucille Buonocore and Joe Ford for coordinating production, and to Lillian Brady for a great job in proofreading the galleys. We also acknowledge the efforts of Joan Kalkut in coordinating a first-rate supplements package.

During the preparation of the first edition we were fortunate to have Robert A. McConnin as our editor. He was an outstanding editor, and much of the flavor of the book is due to his foresight and leadership. It was a pleasure to work with him.

We also express our thanks to Rose Mary Cutnell for her assistance in typing the manuscript.

It has been most rewarding to work with the physicists who reviewed the manuscript and offered many suggestions for improving our writing style and removing ambiguities and inaccuracies. These individuals are an invaluable resource to the physics community, and we applaud them. They devote their time and effort out of dedication to and love of physics. Our thanks go to each one of them:

Joseph Alward
University of the Pacific
Chi Kwan Au
University of South Carolina
William A. Barker
Santa Clara University
Edward E. Beasley
Gallaudet University
Edward R. Borchardt
Mankato State University
Robert Brehme
Wake Forest University
Michael Bretz
University of Michigan at Ann Arbor
Carl Bromberg
Michigan State University
Michael E. Browne
University of Idaho
Ronald W. Canterna
University of Wyoming
Marvin Chester
University of California at Los Angeles
William S. Chow
University of Cincinnati
Albert C. Claus
Loyola University of Chicago
Thomas Berry Cobb
Bowling Green State University
Lawrence Coleman
University of California at Davis

Lattie F. Collins
East Tennessee State University

Mark Comella
Duquesne University

Henry L. Cote
Catonsville Community College

James E. Dixon
Iowa State University

Duane Doty
California State University at Northridge

Miles J. Dresser
Washington State University

Dewey Dykstra
Boise State University

Robert J. Endorf
University of Cincinnati

Roger Freedman
University of California at Santa Barbara

Robert J. Friauf
University of Kansas

C. Sherman Frye, Jr.
Northern Virginia Community College

John Gagliardi
Rutgers University

Simon George
California State University at Long Beach

John Gieniec
Central Missouri State University

D. Wayne Green
Knox College

Grant Hart
Brigham Young University

Lawrence A. Hitchingham
Jackson Community College

Paul R. Holody
Henry Ford Community College

Darrell O. Huwe
Ohio University

David A. Jerde
St. Cloud State University

R. Lee Kernell
Old Dominion University

Gary Kessler
Illinois Wesleyan University

I. K. Kothari
Tuskegee Institute

Robert A. Kromhout
Florida State University

Theodore Kruse
Rutgers University

Pradeep Kumar
University of Florida at Gainesville

Rubin H. Landau
Oregon State University

Christopher P. Landee
Clark University

Daines Lund
Utah State University

R. Wayne Major
University of Richmond

A. John Mallinckrodt
California State Polytechnic University, Pomona

B. Wieb Van Der Meer
Western Kentucky University

Donald D. Miller
Central Missouri State University

Richard A. Morrow
University of Maine

Kenneth Mucker
Bowling Green State University

David Newton
DeAnza College

R. Chris Olsen
University of Alabama

Robert F. Petry
University of Oklahoma

Peter John Polito
Springfield College

Jon Pumplin
Michigan State University

Loren E. Radford
Baptist College at Charleston

Talat Rahman
Kansas State University

Michael Ram
State University of New York at Buffalo

Jacobo Rapaport
Ohio University

Wayne W. Repko
Michigan State University

Harold Romero
University of Southern Mississippi

Larry Rowan
University of North Carolina

O.M.P. Rustgi
State University of New York at Buffalo

Charles Scherr
University of Texas at Austin

John J. Sinai
University of Louisville

Michael Swift
Catonsville Community College

Ronald G. Tabak
Youngstown State University

Howard G. Voss
Arizona State University

James M. Wallace
Jackson Community College

Walter G. Wesley
Moorhead State University

Henry White
University of Missouri

Jerry H. Wilson
Metropolitan State College

Jerry Wilson
Lander College

There are several reviewers who deserve special mention. Professor Mario Iona (Emeritus, University of Denver) provided a line-by-line review of the manuscript for the first edition. His review exceeded all of our expectations, and it was a remarkable tour de force of careful attention to detail and unflagging insistence on correct physics. We came to respect his work greatly, and it continues to have a marked influence on the book. Professor Edward Borchardt of Mankato State University also provided a careful reading of the entire manuscript. His comments were always carefully thought out and insightful. Many of the positive changes in the second edition were due to his influence. Professor Michael Bretz of the University of Michigan at Ann Arbor was a rich source of stimulating and useful suggestions. We are most grateful for his efforts.

J.D.C.
K.W.J.

Features of the Book . . .

THE PHYSICS OF . . .

a giant roller coaster.

Figure 6.16 The Magnum XL-200 roller coaster; one of the fastest roller coasters in the world, includes a vertical drop of 59.3 m.

Example 9   The Magnum XL-200

One of the fastest roller coasters in the world is the Magnum XL-200 at Cedar Point Park in Sandusky, Ohio (Figure 6.16). This ride includes a vertical drop of 59.3 m. Assume that the roller coaster has a speed of nearly zero as it crests the top of the hill. Neglect friction and find the speed of the riders at the bottom of the hill.

REASONING   Since we are neglecting friction, we may set the work done by the frictional force equal to zero. A normal force acts on each rider, but this force is perpendicular to the motion, so it does not do any work. Thus, the work done by nonconservative forces is zero, and we may use the principle of conservation of mechanical energy to find the speed of the riders at the bottom of the hill.

SOLUTION   The principle of conservation of mechanical energy states that

$$\underbrace{\tfrac{1}{2}mv_f^2 + mgh_f}_{E_f} = \underbrace{\tfrac{1}{2}mv_0^2 + mgh_0}_{E_0} \qquad (6.9)$$

The mass m of the rider appears as a factor in every term in this equation and can be eliminated algebraically. Solving for the final speed gives

$$v_f = \sqrt{v_0^2 + 2g(h_0 - h_f)}$$

The initial speed of the roller coaster is assumed to be zero, $v_0 = 0$, and the vertical height of the hill is $h_0 - h_f = 59.3$ m:

$$v_f = \sqrt{2(9.80 \text{ m/s}^2)(59.3 \text{ m})} = \boxed{34.1 \text{ m/s (about 76 mph)}}$$

6.6 NONCONSERVATIVE FORCES AND THE WORK–ENERGY THEOREM

Most moving objects experience one or more nonconservative forces, such as friction, air resistance, and propulsive forces. The work W_{nc} done by the net nonconservative force is not zero, and, consequently, the total mechanical energy of the object is not conserved. In these situations, the difference between the final and initial total mechanical energies is equal to W_{nc}, as expressed by Equation 6.8:

$$W_{nc} = E_f - E_0$$
$$= (\tfrac{1}{2}mv_f^2 + mgh_f) - (\tfrac{1}{2}mv_0^2 + mgh_0)$$

The next two examples illustrate how Equation 6.8 is used when nonconservative forces are present.

Example 10   The Magnum XL-200, Revisited

In Example 9, we ignored friction. In reality, however, friction is present when the Magnum XL-200 roller coaster descends the hill. The actual speed of the riders at the bottom is 32.2 m/s, which is less than that determined in Example 9. How much work is

---

REAL-LIFE EXAMPLES

Many examples are taken from real-life situations. Such examples help to show students that physics principles come into play over and over again in their lives.

REASONING—THE CORNERSTONE OF PROBLEM SOLVING

The examples in the text have an explicit reasoning step that precedes the numerical solution. In this step, the pertinent physics principles and the reasons for choosing a particular method for solving a problem are discussed.

TECHNIQUES FOR PROBLEM SOLVING

Good techniques make it easier for students to find errors in problem solving. First, the unknown variable in an equation is obtained algebraically in terms of the known variables. Then, numbers are substituted for the known variables.

Table 9.2 Analogies between Rotational and Translational Concepts

Physical Concept	Rotational	Translational
Displacement	θ	s
Velocity	ω	v
Acceleration	α	a
The cause of acceleration	Torque τ	Force F
Inertia	Moment of inertia I	Mass m
Newton's second law	$\Sigma\tau = I\alpha$	$\Sigma F = ma$
Work	$\tau\theta$	Fs
Kinetic energy	$\frac{1}{2}I\omega^2$	$\frac{1}{2}mv^2$
Momentum	$L = I\omega$	$p = mv$

We have seen that Newton's second law for rotational motion, $\Sigma\tau = I\alpha$, has the same form as that for translational motion, $\Sigma F = ma$, so each rotational variable has a translational analog: torque τ and force F are analogous quantities, as are moment of inertia I and mass m, and angular acceleration α and linear acceleration a. The other physical concepts developed for studying translational motion, such as kinetic energy and momentum, also have rotational analogs. For future reference, Table 9.2 itemizes these concepts and their rotational analogs.

9.5 ROTATIONAL WORK AND ENERGY

ROTATIONAL WORK

Work and energy are among the most fundamental and useful concepts in physics. Chapter 6 discusses their application to translational motion. These concepts are equally useful for rotational motion, provided they are expressed in terms of angular variables.

The work W done by a constant force that points in the same direction as the displacement is $W = Fs$ (Equation 6.1), where F and s are the magnitudes of the force and displacement, respectively. In Figure 9.18 a rope is wrapped around a wheel and is under a constant tension F. If the rope is pulled out a distance s, the wheel rotates through an angle $\theta = s/r$ (Equation 8.1), where r is the radius of the wheel and θ is in radians. The work done by the tension force in turning the wheel is $W = Fs = Fr\theta$. But Fr is the torque τ applied to the wheel by the tension, so the rotational work can be written in angular variables as follows:

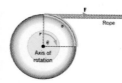

Figure 9.18 The force **F** does work in rotating the wheel through the angle θ.

Definition of Rotational Work

The rotational work W_R done by a constant torque τ in turning an object through an angle θ is

$$W_R = \tau\theta \qquad (9.8)$$

Requirement: θ must be expressed in radians.

SI Unit of Rotational Work: joule (J)

DEFINITIONS AND LAWS

Key definitions and laws are highlighted by enclosing them in a "box." This helps students to identify important concepts quickly.

CAUTIONARY NOTES

Short notes are added to emphasize the condition under which a relation is valid. They help students to apply the relation correctly.

SI UNITS

The SI units of newly defined quantities are clearly identified. Units are an important part of any problem-solving strategy.

when the tape leaves the vicinity of the recording head and, thus, provides a means for storing audio information. Audio information is retained, because at any instant in time the way in which the tape is magnetized depends on the amount and direction of current in the recording head. The current, in turn, depends on the sound intensity picked up by the microphone, so that changes in the sound intensity that occur from moment to moment are preserved as changes in the tape's induced magnetization.

MAGLEV TRAINS

A magnetically levitated train — or maglev, for short — uses forces that arise from induced magnetism to levitate or float above a guideway. Since it rides a few centimeters above the guideway, a maglev does not need wheels. Freed from friction with the guideway, the train can achieve significantly greater speeds than do conventional trains. For example, the Transrapid maglev in Figure 21.36a has achieved speeds of 110 m/s.

Figure 21.36a shows that the Transrapid maglev achieves levitation with electromagnets mounted on arms that extend around and under the guideway. When a current is sent to an electromagnet, the resulting magnetic field creates induced magnetism in a rail mounted in the guideway. The upward attractive force from the induced magnetism is balanced by the weight of the train, so the train moves without touching the rail or the guideway.

Magnetic levitation only lifts the train and does not move it forward. Figure 21.36b illustrates how magnetic propulsion is achieved. In addition to the levitation electromagnets, propulsion electromagnets are also placed underneath the train and along the guideway. By controlling the direction of the currents in the

THE PHYSICS OF . . .

a magnetically levitated train.

THE PHYSICS OF . . .

The physics of interesting, and often high-tech, applications is integrated into the text. These applications show the relevancy of physics to everyday life.

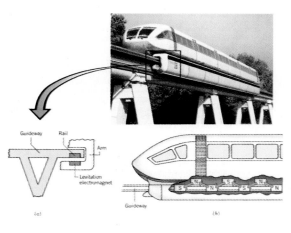

Figure 21.36 (a) The Transrapid maglev (a German train) has achieved speeds of 110 m/s (250 mph). The levitation electromagnets are drawn up toward the rail in the guideway, levitating the train. (b) The magnetic propulsion system.

FREE-BODY DIAGRAMS

Free-body diagrams are used throughout the text. They are a valuable problem-solving aid when Newton's second law is being applied to equilibrium and nonequilibrium situations.

---

106   CHAPTER 4/FORCES AND NEWTON'S LAWS OF MOTION

PROBLEM SOLVING INSIGHT

A moving object may be in equilibrium. It is in equilibrium if it moves with a constant velocity; then, its acceleration is zero. A zero acceleration is the fundamental characteristic of an object in equilibrium.

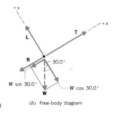

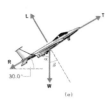

(b) Free-body diagram

(c)

Figure 4.29 (a) A plane moves with a constant velocity at an angle of 30.0° above the horizontal due to the action of four forces, the weight **W**, the lift **L**, the engine thrust **T**, and the air resistance **R**. (b) The free-body diagram for the plane. (c) This geometry occurs often in physics.

REASONING Figure 4.29b shows the free-body diagram of the plane, including the forces **W**, **L**, **T**, and **R**. Since the plane is not accelerating, it is in equilibrium, and the sum of the x components and the sum of the y components of these forces must be zero. To calculate the components, we have chosen axes in the free-body diagram that are rotated by 30.0° from their usual horizontal–vertical positions. This has been done purely for convenience, since the weight **W** is then the only force that does not lie along either axis.

SOLUTION Before determining the components of the weight, it is necessary to realize that the angle β in Figure 4.29a is equal to 30.0°. Part c of the drawing focuses attention on the geometry that is responsible for this fact. There it can be seen that $\alpha + \beta = 90°$ and $\alpha + 30° = 90°$, with the result that $\beta = 30°$. Geometry similar to that in Figure 4.29c occurs often in physics.

The table below lists the components of the forces that act on the jet.

Force	x Component	y Component
W	$-(86\ 500\text{ N})\sin 30.0°$	$-(86\ 500\text{ N})\cos 30.0°$
L	0	$+L$
T	$+103\ 000\text{ N}$	0
R	$-R$	0

Setting the sum of the x components and the sum of the y components of the forces equal to zero yields

$$\Sigma F_x = -(86\ 500\text{ N})\sin 30.0° + 103\ 000\text{ N} - R = 0$$
$$\Sigma F_y = -(86\ 500\text{ N})\cos 30.0° + L = 0$$

These equations can be solved to show that $\boxed{R = 59\ 800\text{ N}}$ and $\boxed{L = 74\ 900\text{ N}}$.

Static friction is a force that sometimes plays a role in keeping an object at rest, that is, in equilibrium. The next example illustrates one way in which information about the static friction force can be obtained.

Example 14   *Measuring the Coefficient of Static Friction*

A block of mass m rests on a hinged board whose angle of elevation is adjustable, as in Figure 4.30a. When the right end of the board is raised, the block remains at rest until a maximum angle θ is reached. If the angle is increased beyond θ, the block breaks loose and slides down the board. Obtain an equation that relates the coefficient of static friction μ_s to the angle θ.

REASONING There are three forces that act on the block, the weight **W** ($W = mg$), the normal force $\mathbf{F_N}$, and the maximum force of static friction $\mathbf{f_s^{MAX}}$. The magnitude of the maximum static frictional force is given by Equation 4.7 as $f_s^{MAX} = \mu_s F_N$. These forces keep the block in equilibrium, and, therefore, they must balance, the sum of the x components and the sum of the y components each being zero. To determine the components, we use the x, y axes shown in Figure 4.30b along with the free-body diagram for the block. The geometry in this diagram is the same as that in Figure 4.29b.

SOLUTION The x and y components of the forces are as follows:

**PROBLEM SOLVING
INSIGHT**

Brief comments in the margin
reinforce important aspects of
problem solving that might
otherwise go unnoticed.

---

180   CHAPTER 7/IMPULSE AND MOMENTUM

Example 3   Assembling a Freight Train

A freight train is being assembled in a switching yard, and Figure 7.3 shows two boxcars in the process of being coupled together. Car 1 has a mass of $m_1 = 65 \times 10^3$ kg and moves at a velocity of $v_{01} = +0.80$ m/s. Car 2, with a mass of $m_2 = 92 \times 10^3$ kg and a velocity of $v_{02} = +1.2$ m/s, overtakes car 1 and couples to it. Neglecting friction, find the common velocity v_f of the two cars after they become coupled.

REASONING   The two boxcars constitute the system. The sum of the external forces acting on the system is zero, because the weight of each car is balanced by a corresponding normal force, and friction is being neglected. Thus, the system is isolated, and the principle of conservation of linear momentum applies. The coupling forces that each car exerts on the other are internal forces and do not affect the applicability of this principle.

SOLUTION   Momentum conservation indicates that

$$\underbrace{(m_1 + m_2)v_f}_{\substack{\text{Total momentum}\\\text{after collision}}} = \underbrace{m_1 v_{01} + m_2 v_{02}}_{\substack{\text{Total momentum}\\\text{before collision}}}$$

This equation can be solved for v_f, the common velocity of the two cars after the collision:

$$v_f = \frac{m_1 v_{01} + m_2 v_{02}}{m_1 + m_2}$$

$$= \frac{(65 \times 10^3 \text{ kg})(0.80 \text{ m/s}) + (92 \times 10^3 \text{ kg})(1.2 \text{ m/s})}{(65 \times 10^3 \text{ kg} + 92 \times 10^3 \text{ kg})}$$

$$= \boxed{+1.0 \text{ m/s}}$$

PROBLEM SOLVING INSIGHT

The conservation of linear momentum is applicable only when the net external force acting on the system is zero. Therefore, the first step in applying momentum conservation to problem solving is to be sure that the net external force is zero.

**PRESENTATION OF
EQUATIONS**

Equations often have bracketed
labels that explain the physical
meaning of each term within
the equation.

Figure 7.3   (a) One boxcar eventually catches up with the other and couples to it. (b) The coupled cars move together with a common velocity after the collision.

   In the previous example it can be seen that the velocity of car 1 increases, while the velocity of car 2 decreases as a result of the collision. The acceleration and deceleration arise at the moment the cars become coupled, because the cars exert internal forces on each other. These forces are equal in magnitude and opposite in direction, in accord with Newton's third law. The powerful feature of the momentum conservation principle is that it allows us to determine the changes in velocity without knowing what the internal forces are. Example 4 further illustrates this feature.

**INTEGRATION OF
CONCEPTS**

These sections explore the common ground between the fundamental ideas in the current chapter and ideas from earlier chapters. The emphasis is on the unity of physics as an integrated body of knowledge built on fundamental concepts.

INTEGRATION OF CONCEPTS

FIELDS AND FORCES

The concept of an electric field is introduced in Chapter 18. An electric field is produced by one or more charged objects and exists in the region around them. Electric field lines are often drawn as an aid in visualizing the magnitude and direction of the electric field within the region. At any given location, the electric field exerts an electric force on a charged object placed there, the force being the product of the charge and the electric field at that point. The direction of the force is either parallel or antiparallel to the electric field, depending on whether the charge is positive or negative, respectively. In the present chapter, we see that a magnetic field is produced by permanent magnets or moving charges, such as an electric current, and exists in the region around them. Magnetic field lines are also drawn as an aid in visualizing the magnitude and direction of the magnetic field. As can an electric field, a magnetic field can exert a magnetic force on a charged object within it, but only if the object is moving and has a velocity component that is perpendicular to the magnetic field. The direction of the magnetic force is perpendicular to the plane defined by the velocity of the object and the magnetic field. Thus, the concept of a field is very useful, for it can be used to describe the electric and magnetic forces that are exerted on charged objects.

THE MAGNETIC FORCE AND NATURE'S FUNDAMENTAL FORCES

There are four fundamental forces in nature, fundamental in the sense that all other forces can be understood as manifestations of one or more of the four. Tension, friction, and the elastic force of a spring, for example, are not fundamental forces, but the gravitational force is. Another force that we have encountered is the force that one electrically charged particle exerts on another charged particle. This force is one part of a fundamental force called the electromagnetic force. The electromagnetic force contains two parts, an electric part and a magnetic part. Both parts, however, derive from the same source, the electric charge carried by the particles. Whether or not the particles are moving, they exert on each other the electric force specified in Chapter 18 by Coulomb's law. When the particles move, the other part of the electromagnetic force also appears, the part that we have called the magnetic force in the present chapter.

SUMMARY

A thorough, but concise, review of the ideas discussed is presented at the end of each chapter. Important concepts are highlighted in boldface type.

SUMMARY

A magnet has a north pole and a south pole. The north pole is the end that points toward the north magnetic pole of the earth when the magnet is freely suspended. **Like poles repel each other and unlike poles attract each other.**

A **magnetic field** exists in the space around a magnet.

The magnetic field is a vector whose direction at any point is the direction indicated by the north pole of a small compass needle placed at that point. The magnitude B of the magnetic field at any point in space is defined as $B = F/(q_0 v \sin \theta)$, where F is the magnitude of the magnetic force that acts on a charge q_0 whose

SOLVED PROBLEMS

These problems are grouped near the end of a chapter and illustrate more sophisticated problem-solving concepts.

REFERENCE LIST

This list identifies homework problems that deal with the same general concepts as the Solved Problem.

SUMMARY OF IMPORTANT POINTS

At the end of each Solved Problem is a summary of the important points that have been discussed.

SOLVED PROBLEMS 51

SOLVED PROBLEMS

Solved Problem 1   A Bus Trip
Related Problems: *7 **8 *59 *61 *64

During a trip, a bus travels 11 km with an average velocity of 21 m/s, but then travels in the same direction for the next 1.0 km at a smaller average velocity of 4.2 m/s, due to the presence of highway construction crews (see the dra Determine the average velocity of the bus for the entir

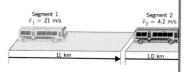

Segment 1
$\bar{v}_1 = 21$ m/s

Segment 2
$\bar{v}_2 = 4.2$ m/s

11 km

1.0 km

REASONING It is important to realize that the avera locity of the bus is *not* obtained by simply adding th velocities and dividing the sum by 2. This method is inc because it does not take into account that most of the tri place at the greater velocity, while only a small part take at the smaller velocity. The correct procedure for calc the average velocity is to use Equation 2.2:

$$\bar{v} = \frac{\text{Displacement}}{\text{Elapsed time}}$$

where the values for the displacement and elapsed tim be those for the entire trip.

SOLUTION The displacement for the entire trip is 1 The elapsed time is the sum of two times, t_1 and t_2. The for the bus to travel 11 km at an average velocity of 21

$$t_1 = \frac{11 \times 10^3 \text{ m}}{21 \text{ m/s}} = 520 \text{ s}$$

Likewise, the time t_2 for the bus to travel the remaining at a velocity of 4.2 m/s is

$$t_2 = \frac{1.0 \times 10^3 \text{ m}}{4.2 \text{ m/s}} = 240 \text{ s}$$

Hence, the elapsed time for the trip is $t_1 + t_2 = 760$ average velocity for the entire trip is, then,

$$\bar{v} = \frac{12 \times 10^3 \text{ m}}{760 \text{ s}} = \boxed{16 \text{ m/s}}$$

A calculator gives $\bar{v} = 15.789$ m/s, but this value m

rounded to $\bar{v} = 16$ m/s, since the data are accurate to only two significant figures.

SUMMARY OF IMPORTANT POINTS The average velocity of a segmented trip is obtained by dividing the total displacement of the trip (which equals the vector sum of the displacements for each segment) by the total time of the trip (which equals the sum of the times for each segment). The average velocity *cannot* be calculated by adding the velocities

ADDITIONAL PROBLEMS

These problems are not keyed to a particular section, and they often combine concepts from different sections. This problem set challenges students to use the material from the entire chapter.

---

278   CHAPTER 10/ELASTICITY AND SIMPLE HARMONIC MOTION

what angular frequency (in rad/s) would the object oscillate on the spring?

*47. A 14.6-kg block and a 29.2-kg block are resting on a horizontal frictionless surface. Between the two is squeezed a spring (spring constant = 1170 N/m). The spring is compressed by 0.152 m from its unstrained length. With what speed does each block move away when the mechanism keeping the spring squeezed is released?

*48. A 1.1-kg object is suspended from a vertical spring whose spring constant is 120 N/m. (a) Find the amount by which the spring is stretched from its unstrained length. (b) The object is pulled straight down by an additional distance of 0.20 m and released from rest. Find the speed with which the object passes through its original position on the way up.

**49. A 70.0-kg circus performer is fired from a cannon that is elevated at an angle of 45.0° above the horizontal. The cannon uses strong elastic bands to propel the performer, much in the same way that a slingshot fires a stone. Setting up for this stunt involves stretching the bands by 3.00 m from their unstrained length. At the point where the performer flies free of the bands, his height above the floor is the same as that of the net into which he is shot. He takes 4.00 s to travel the horizontal distance of 50.0 m between this point and the net. Ignore friction and air resistance and determine the effective spring constant of the firing mechanism.

**50. A spring is mounted vertically on the floor. The mass of the spring is negligible. A certain object is placed on the spring to compress it. When the object is pushed down further by just a bit and then released, one up/down oscillation cycle occurs in 0.250 s. However, when the object is pushed down by 5.00×10^{-2} m to point P and then released, the object flies entirely off the spring. To what height above point P does the object rise in the absence of air resistance?

**51. A 1.00×10^{-2}-kg bullet is fired horizontally into a 2.50-kg wooden block attached to one end of a massless, horizontal spring ($k = 845$ N/m). The other end of the spring is fixed in place, and the spring is unstrained initially. The block rests on a horizontal, frictionless surface. The bullet strikes the block perpendicularly and quickly comes to a halt within it. As a result of this completely inelastic collision, the spring is compressed along its axis and causes the block/bullet to oscillate with an amplitude of 0.200 m. What is the speed of the bullet?

Section 10.6 The Pendulum

52. If the period of a simple pendulum is to be 2.0 s, what should be its length?

53. A grandfather clock can be approximated as a simple pendulum of length 1.00 m and keeps accurate time at a location where $g = 9.83$ m/s². In a location where $g = 9.78$ m/s²,

what must be the new length of the pendulum, such that the clock continues to keep accurate time?

54. Astronauts on a distant planet set up a simple pendulum of length 1.2 m. The pendulum executes simple harmonic motion and makes 100 complete vibrations in 280 s. What is the acceleration due to gravity?

55. A wrecking ball is hanging at the end of a long cable on a crane. A bright student wants to estimate the length of the cable and, therefore, improvises by using a simple pendulum made from a 0.500-m length of string and a stone. The student observes that, in swinging back and forth over a small amplitude, the wrecking ball makes one complete oscillation cycle in the time it takes the stone to complete five cycles. What is the length of the cable?

*56. The period of a simple pendulum is 0.200% longer at location A than it is at location B. Find the ratio g_A/g_B of the acceleration due to gravity at these two locations.

*57. Pendulum A is a physical pendulum made from a thin, rigid, and uniform rod whose length is 1.00 m. One end of this rod is attached to the ceiling by a frictionless hinge, so the rod is free to swing back and forth. Pendulum B is a simple pendulum whose length is also 1.00 m. Obtain the ratio T_A/T_B of their periods for small-angle oscillations.

**58. A point on the surface of a solid sphere (radius = R) is attached directly to a pivot on the ceiling. The sphere swings back and forth as a physical pendulum with a small amplitude. What is the length of a simple pendulum that has the same period as this physical pendulum? Give your answer in terms of R.

ADDITIONAL PROBLEMS

59. The drawing shows how a piston in an automobile engine is attached to the crankshaft, which is rotating with an angular speed of $\omega = 126$ rad/s. If the shadow of point P could be projected onto a screen, the shadow would move in simple harmonic motion. Find (a) the amplitude, (b) the period, and (c) the maximum speed of the motion.

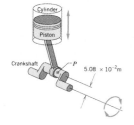

Cylinder

Piston

Crankshaft P 5.08×10^{-2} m

PROBLEMS KEYED TO SECTIONS

Students can review the material in a particular section while they are solving these problems. Instructors can assign problems from different sections to ensure the desired coverage of the material in the chapter.

BRIEF CONTENTS

CONTENTS

INTERSECTION

Physics & Law Enforcement

PART TWO

THERMAL PHYSICS

INTERSECTION

Physics & the Environment

PART FIVE

LIGHT AND OPTICS

INTERSECTION

Physics & Space Science

INTERSECTION

Physics & Medicine

THE PHYSICS OF . . .

APPLICATIONS OF PHYSICS PRINCIPLES

To show students that physics has a widespread impact on their lives, we have included a large number of applications of physics principles. Many of these applications are not found in other texts. The most important ones are listed below along with the page number locating the corresponding discussion. They are identified in the margin of the page on which they occur with a red arrow and the title "THE PHYSICS OF . . ." The discussions are integrated into the text, so that they occur as a natural part of the physics being presented. It should be noted that the list is not a complete list of all the applications of physics principles to be found in the text. There are many additional applications that are discussed only briefly or occur in the homework questions and problems.

PART ONE

MECHANICS

Physics is a science that deals with the fundamental principles that govern the behavior of the physical universe. It is a science that plays a role in nearly every aspect of our lives. Considering the tremendous variety of natural phenomena, you might think that physics consists of an enormous number of concepts and principles. Actually, however, the opposite is true. Physics deals with a relatively small number of fundamental ideas in building a coherent view of the physical universe.

In the first part of this text, we discuss the branch of physics called mechanics. Mechanics is divided into kinematics and dynamics. Kinematics consists of the ideas needed to describe motion. For example, we will see how to describe the motion of a jet plane that starts from rest and eventually becomes airborne. The concept of acceleration will be used to describe how the motion of the plane changes, that is, how it becomes faster and faster. Dynamics focuses on what causes the motion to become faster and faster. In other words, dynamics deals with the way in which forces (such as those produced by the plane's engines) cause the motion to change.

In mechanics, then, we begin by de-scribing motion with the aid of the concept of acceleration. Then, we explore and build upon the relationship between acceleration and force. This building on the common ground between fundamental ideas pervades all of physics.

After discussing acceleration and force, we consider work and energy and will find that the relationship between these two new concepts leads to the principle of conservation of energy. This principle states that energy is conserved, in the sense that it is the same at all points along the motion. In a similar fashion, we will find that the relationship between the ideas of impulse and momentum leads to the principle of conservation of momentum. The idea that certain quantities are conserved in nature is one of the great insights of physics.

The interrelationships between ideas in physics are important not only because they can lead to keen insights, but also because they provide a kind of road map to guide us. To help clarify the interrelationships, we have included at the ends of most chapters a section entitled "Integration of Concepts." These sections explore the common ground between important concepts in the current chapter and concepts discussed previously.

CHAPTER 1

INTRODUCTION AND MATHEMATICAL CONCEPTS

"Who framed Roger Rabbit?" This whimsical movie contains a unique blend of real-life and animated characters that closely interact with each other. Animating a movie requires an enormous amount of artistic work and, as with many modern graphical techniques, relies heavily on computers to aid in the rendering process. All the curved lines, the coloring, the shading, and three-dimensional perspectives of the animation are stored as mathematical relations within the computer. The animator tells the computer how to use these relations in drawing the animation on the screen.

The laws of physics are also described by mathematical relations. With the aid of these relations, phenomena in the physical world can be described in a compact and precise form, a form that biologists, chemists, geologists, and engineers, as well physicists, find indispensable in their work. This chapter introduces some mathematical concepts—like trigonometry and vectors—that will be useful in dealing with the laws of physics as we encounter them throughout this book.

1.1 THE NATURE OF PHYSICS

The science of physics has developed out of the efforts of men and women to describe how and why our physical environment behaves as it does. These efforts have been so successful that today physics encompasses remarkably diverse phenomena. Physics relates to the planets orbiting the sun, a jetliner streaking through the sky, transistors working in stereo and computer systems, lasers being used in eye surgery, and much, much more.

The laws of physics are equally remarkable for their scope. They describe the behavior of particles many times smaller than an atom and objects many times larger than our sun. The same laws apply to the heat generated by a burning match and the heat generated by a rocket engine. The same laws guide an astronomer in using the light from a distant star to determine how fast the star is moving and a police officer in using radar to catch a speeder. Physics can be applied fruitfully to objects as different as subatomic particles, distant stars, or speeding automobiles because it focuses on issues that are truly basic to the way nature works.

The key to understanding the strength of physics is to recognize that its laws are based on experimental fact. This is not to say that intuition and educated guesses are unimportant in physics. The great creative geniuses in science, as in art, work in leaps and bounds that no one can fully understand. However, in physics a flash of insight never becomes accepted law unless its implications can be verified by experiment. Without such verification, a flash of insight provides at best only a hypothesis, often one among many. This insistence on experimental verification has enabled physicists to build a rational and coherent understanding of nature.

The exciting feature of physics is its capacity for predicting how nature will behave in one situation on the basis of experimental data obtained in another situation. Such predictions place physics at the heart of modern technology and, therefore, can have a tremendous impact on our lives. Rocketry and the development of space travel have their roots firmly planted in the physical laws of Galileo Galilei (1564–1642) and Isaac Newton (1642–1727). The transportation industry relies heavily on physics in the development of engines and the design of aerodynamic vehicles. Entire electronics and computer industries owe their existence to the impetus provided by the invention of the transistor, which grew directly out of physical laws describing the electrical behavior of solids. The telecommunications industry depends extensively on electromagnetic waves, whose existence was predicted by James Clerk Maxwell (1831–1879) in his theory of electricity and magnetism. The medical profession uses X-ray, ultrasonic, and magnetic resonance methods for obtaining images of the interior of the human body, and physics lies at the core of all these. Perhaps the most widespread impact in modern technology is that due to the laser. Fields ranging from space exploration to medicine benefit from this incredible device, which is a direct application of the principles of atomic physics.

Because physics is so fundamental, it is a required course for students in a wide range of major areas. We welcome you to the study of this fascinating topic. You will learn how to see the world through the "eyes" of physics, and, in so doing, you will learn how to apply physics principles to a wide range of problems. We hope that you will come to recognize that physics has important things to say about your environment.

1.2 UNITS

SYSTEMS OF UNITS

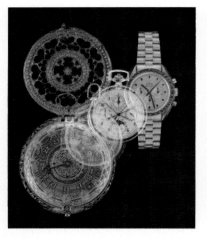

These timepieces, new and old, are attempts to measure time in standard units.

Physics experiments involve the measurement of a variety of quantities, and a great deal of effort goes into making these measurements as accurate and reproducible as possible. The first step toward ensuring accuracy and reproducibility is defining the units in which the measurements are made.

In this text, we will stress the system of units known according to the French phrase "Le Système International d'Unités," referred to simply as *SI units.* This system, by international agreement, employs the *meter* (m) as the unit of length, the *kilogram* (kg) as the unit of mass, and the *second* (s) as the unit of time. Two other systems of units are worth mentioning. The CGS system utilizes the centimeter (cm), the gram (g), and the second for length, mass, and time, respectively, whereas the BE or British Engineering system (the gravitational version) uses the foot (ft), the slug (sl), and the second. Table 1.1 summarizes the units used for length, mass, and time in the three systems.

Table 1.1   Units of Measurement

	System		
	SI	CGS	BE
Length	meter (m)	centimeter (cm)	foot (ft)
Mass	kilogram (kg)	gram (g)	slug (sl)
Time	second (s)	second (s)	second (s)

DEFINITION OF STANDARD UNITS

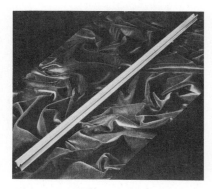

Figure 1.1   The standard platinum-iridium meter bar.

Originally, the meter as a unit of length was defined in terms of the distance measured along the earth's surface between the north pole and the equator. A more accurate measurement standard was eventually agreed upon internationally, and the meter was redefined as the distance between two marks on a bar of platinum–iridium alloy (see Figure 1.1) kept at a temperature of 0 °C. Today, the meter is defined as the distance that light travels in a vacuum in a time of 1/299 792 458 second. This definition arises because the speed of light is a universal constant that is defined to be 299 792 458 m/s.

The definition of a kilogram as a unit of mass has also undergone changes over the years. As Chapter 4 discusses, the mass of an object indicates the tendency of the object to continue in motion with a constant velocity. Originally, the kilogram was defined by use of a specific amount of water. Today, one kilogram is defined to be the mass of a standard cylinder of platinum–iridium alloy, like that in Figure 1.2.

Figure 1.2 The standard platinum-iridium kilogram is kept at the International Bureau of Weights and Measures in Sèvres, France.

Figure 1.3 A cesium atomic clock.

Table 1.2 Standard Prefixes Used to Denote Multiples of Ten

Prefix	Symbol	Factor[a]
Tera	T	10^{12}
Giga[b]	G	10^{9}
Mega	M	10^{6}
Kilo	k	10^{3}
Hecto	h	10^{2}
Deka	da	10^{1}
Deci	d	10^{-1}
Centi	c	10^{-2}
Milli	m	10^{-3}
Micro	μ	10^{-6}
Nano	n	10^{-9}
Pico	p	10^{-12}
Femto	f	10^{-15}

[a] Appendix A contains a discussion of powers of ten and scientific notation.
[b] Pronounced jig'a.

As with the units for length and mass, the present definition of the second as a unit of time is different from the original definition. Originally, the second was defined according to the average time for the earth to rotate once about its axis, one day being set equal to 86 400 seconds. Now, however, the second is defined in terms of a cesium atomic clock like that in Figure 1.3.

BASE UNITS AND DERIVED UNITS

The units for length, mass, and time, along with a few other units that will arise later, are regarded as *base* SI units. The word "base" refers to the fact that these units are used along with various laws to define additional units for other important physical quantities, such as force and energy. The units for these other physical quantities are referred to as *derived* units, since they are combinations of the base units. Derived units will be introduced as they arise naturally along with the related physical laws.

The value of a quantity in terms of base or derived units is sometimes a very large or very small number. In such cases, it is convenient to introduce larger or smaller units that are related to the normal units by multiples of ten. Table 1.2 summarizes the standard prefixes that are used to denote multiples of ten. For example, 1000 or 10^3 meters are referred to as 1 kilometer (km), and 0.001 or 10^{-3} meter is called 1 millimeter (mm). Similarly, 1000 grams and 0.001 gram are referred to as 1 kilogram (kg) and 1 milligram (mg), respectively. Appendix A contains a discussion of scientific notation and powers of ten, such as 10^3 and 10^{-3}.

1.3 THE ROLE OF UNITS IN PROBLEM SOLVING

THE CONVERSION OF UNITS

Since any quantity, such as length, can be measured in several different units, it is important to know how to convert from one unit to another. For instance, the foot

The moon and the planet Venus move across the night sky and are captured in this photograph at 10-minute intervals.

Tailoring Robert Wadlow's clothes presented quite a challenge.

can be used to express the distance between the two marks on the standard platinum–iridium meter bar. There are 3.28 feet in one meter, and this number can be used to convert from meters to feet, as the following example demonstrates.

Example 1   The World's Tallest Man

The tallest man on record was Robert Wadlow, who had a height of 2.72 m. Express his height in feet.

REASONING AND SOLUTION   Since 3.28 feet = 1 meter, it follows that (3.28 feet)/(1 meter) = 1. Multiplying by a factor of 1 does not alter an equation. Therefore,

$$\text{Length} = (2.72\ \text{meters})(1) = (2.72\ \text{meters})\left(\frac{3.28\ \text{feet}}{1\ \text{meter}}\right) = \boxed{8.92\ \text{feet}}$$

A calculator gives this answer as 8.9216 feet. Standard procedures for significant figures, however, indicate that the answer should be rounded off to three significant figures, since the value of 2.72 meters is accurate to only three significant figures. In this regard, the "1 meter" in the denominator does not limit the significant figures of the answer, because this number is precisely one meter by definition of the conversion factor. Appendix B contains a review of significant figures.

When converting between units, it is useful to write down the units explicitly. The units are treated like any algebraic quantity; in Example 1 the colored lines show how the "meter" cancels when the multiplication is performed, leaving only the desired unit of "feet" to describe the answer. In other words, *if the units do not combine algebraically to give the desired result, the conversion has not been carried out properly.* The next example also stresses the importance of writing down the units and illustrates a typical situation in which several conversions are required.

Example 2   Interstate Speed Limit

Express the speed limit of 65 miles/hour in terms of meters/second.

REASONING   This problem requires changing miles to meters and hours to seconds. Commonly used relationships are 5280 feet = 1 mile and 3600 seconds = 1 hour. As a result, (5280 feet)/(1 mile) = 1 and (3600 seconds)/(1 hour) = 1. Multiplying and dividing by these factors of unity does not alter an equation, a fact that will aid us in the conversions.

SOLUTION   By multiplying and dividing by factors of unity, we can find the speed limit in meters per second as shown below:

$$\text{Speed} = \left(65\ \frac{\text{miles}}{\text{hour}}\right)\frac{(1)}{(1)} = \left(65\ \frac{\text{miles}}{\text{hour}}\right)\frac{\left(\dfrac{5280\ \text{feet}}{1\ \text{mile}}\right)}{\left(\dfrac{3600\ \text{seconds}}{1\ \text{hour}}\right)} = 95\ \frac{\text{feet}}{\text{second}}$$

To convert feet into meters we use the fact that (3.28 feet)/(1 meter) = 1:

$$\text{Speed} = \frac{\left(95\,\dfrac{\text{feet}}{\text{second}}\right)}{(1)} = \frac{\left(95\,\dfrac{\cancel{\text{feet}}}{\text{second}}\right)}{\left(\dfrac{3.28\,\cancel{\text{feet}}}{1\,\text{meter}}\right)} = \boxed{29\,\dfrac{\text{meters}}{\text{second}}}$$

In Example 2, it is necessary to know whether a conversion factor such as (5280 feet)/(1 mile) goes into the numerator or the denominator. In general, you can check to see if your choice is correct by verifying that the units combine algebraically to give the desired result for the answer. A collection of useful conversion factors is given on the page facing the inside of the front cover.

UNITS AS A PROBLEM-SOLVING AID

In addition to their role in guiding the use of conversion factors, units serve a useful purpose in solving problems. They can provide an internal check to eliminate certain kinds of errors, if they are carried along during each step of a calculation and treated like any algebraic factor.

Suppose, for instance, that the tank of a car contains 2.0 gallons of gas to start with and that gas is added at a rate of 7.0 gallons/minute. The total amount of gas in the tank 96 seconds later can be obtained by adding the amount put into the tank to the amount present initially. The amount put in can be calculated by multiplying the filling rate by the time the gas pump is on. But if we don't pay attention to the units in the calculation, we can obtain an erroneous result, as the following example shows.

$$\begin{aligned}\text{Total amount} \atop \text{of gas} &= {\text{Gas initially} \atop \text{present}} + {\text{Gas} \atop \text{added}} \\[6pt] &= 2.0\ \text{gallons} + \left(7.0\,\frac{\text{gallons}}{\text{minute}}\right)(96\ \text{seconds}) \\[6pt] &= 2.0\ \text{gallons} + 672\,\frac{\text{gallons} \cdot \text{seconds}}{\text{minute}}\end{aligned}$$

The answer cannot be $2.0 + 672 = 674$, because the units for the two added terms are not the same. *Only quantities that have exactly the same units can be added (or subtracted).* With the filling rate expressed as 7.0 gallons/minute, the correct answer can be obtained only if the time of 96 seconds is converted into minutes:

PROBLEM SOLVING INSIGHT

$$\text{Time} = (96\ \cancel{\text{seconds}})\left(\frac{1\ \text{minute}}{60\ \cancel{\text{seconds}}}\right) = 1.6\ \text{minutes}$$

$$\begin{aligned}{\text{Total amount} \atop \text{of gas}} &= 2.0\ \text{gallons} + \left(7.0\,\frac{\text{gallons}}{\cancel{\text{minute}}}\right)(1.6\ \cancel{\text{minutes}}) \\[6pt] &= 2.0\ \text{gallons} + 11\ \text{gallons} = 13\ \text{gallons}\end{aligned}$$

As indicated by the colored lines, the units of time now cancel algebraically when the multiplication is carried out, leaving only the desired unit of gallons. The procedure of "carrying along the units" serves as an automatic reminder to convert all data used in a calculation into a consistent set of units.

DIMENSIONAL ANALYSIS

We have seen that many quantities are denoted by specifying both a number and a unit. For example, the distance to the nearest telephone may be 8 meters, or the speed of a car might be 25 meters/second. Each quantity, according to its physical nature, requires a certain *type* of unit, regardless of the system of measurement used. Distance must be measured in a length unit such as meters, feet, or miles, and a time unit will not do. Likewise, the speed of an object must be specified as a length unit divided by a time unit. In physics, the term **dimension** is used to refer to the physical nature of a quantity and the type of unit used to specify it. Distance has the dimension of length (symbolized as [L]), while speed has the dimensions of length [L] divided by time [T], or [L/T]. The dimensions of any physical quantity can be expressed as some combination of fundamental dimensions, like length [L], time [T], and mass [M].

Dimensional analysis is used to check mathematical relations for the consistency of their dimensions. As an illustration, consider a car that starts from rest and accelerates to a speed v in a time t. Suppose we wish to calculate the distance x traveled by the car, but are not sure whether the correct relation is $x = \frac{1}{2}vt^2$ or $x = \frac{1}{2}vt$. We can decide by checking the quantities on both sides of the equals sign to see if they have the same dimensions or not. If the dimensions are not the same, the relation is incorrect. For $x = \frac{1}{2}vt^2$, we write the dimensions as follows, using the dimensions for distance [L], time [T], and speed [L/T]:

$$x = \tfrac{1}{2}vt^2$$

Dimensions:
$$[L] \stackrel{?}{=} \left[\frac{L}{T}\right][T]^2$$

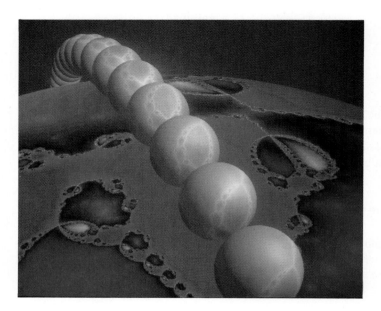

The mathematics of fractals is widely used to create interesting pictures, like the one here. It is also used in physics to explain certain phenomena.

Pure numerical factors like $\frac{1}{2}$ have no dimensions, so they can be ignored. Since dimensions cancel just like algebraic quantities, we have

Dimensions: $[L] \overset{?}{=} \left[\dfrac{L}{T}\right][T]^2 = [L][T]$

The dimension on the left of the equals sign does not match those on the right, so the relation $x = \frac{1}{2}vt^2$ cannot be correct. On the other hand, applying dimensional analysis to $x = \frac{1}{2}vt$, we find that

$$x = \tfrac{1}{2}vt$$

Dimensions: $[L] \overset{?}{=} \left[\dfrac{L}{T}\right][T] = [L]$

The dimension on the left of the equals sign matches that on the right, so this relation is dimensionally correct. If we know that one of our two choices is the right one, then $x = \frac{1}{2}vt$ is it. However, even if a dimensional check does come out correct, it does not guarantee that the relation itself is correct, since dimensional analysis does not account for numerical factors like $\frac{1}{2}$. Nonetheless, dimensional analysis is useful for checking internal consistency, since it will reveal errors in the dimensions of any term.

1.4 TRIGONOMETRY

BASIC TRIGONOMETRIC FUNCTIONS

Scientists use mathematics to help them describe how the physical universe works, and trigonometry is an important branch of mathematics. There are three trigonometric functions that are utilized throughout this text. They are the sine, the cosine, and the tangent of the angle θ (Greek theta), abbreviated as sin θ, cos θ, and tan θ, respectively. These functions are defined below in terms of the symbols given along with the right triangle in Figure 1.4.

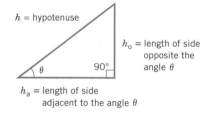

Figure 1.4   A right triangle.

Definition of Sin θ, Cos θ, and Tan θ	
$\sin \theta = \dfrac{h_o}{h}$	(1.1)
$\cos \theta = \dfrac{h_a}{h}$	(1.2)
$\tan \theta = \dfrac{h_o}{h_a}$	(1.3)

h = length of the **hypotenuse** of a right triangle
h_o = length of the side **opposite** the angle θ
h_a = length of the side **adjacent** to the angle θ

The sine, cosine, and tangent are numbers without units, because each is expressed as the ratio of the lengths of two sides of a right triangle. Example 3 illustrates a typical application of Equation 1.3.

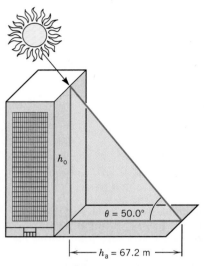

Figure 1.5 From a value for the angle θ and the length h_a of the shadow, the height h_o of the building can be found using trigonometry.

Example 3   Casting a Shadow

On a sunny day, a tall building casts a shadow that is 67.2 m long. The angle between the sun's rays and the ground is $\theta = 50.0°$, as Figure 1.5 shows. Determine the height of the building.

REASONING   In the right triangle in Figure 1.5, the height of the building is the side h_o opposite the angle θ, and the length of the shadow is the side h_a adjacent to the angle. The ratio of the length of the opposite side to the length of the adjacent side is the tangent function, which can be used to find the height of the building.

SOLUTION   We use the tangent function in the following way, with $\theta = 50.0°$ and $h_a = 67.2$ m:

$$\tan \theta = \frac{h_o}{h_a} \tag{1.3}$$

$$h_o = h_a \tan \theta = (67.2 \text{ m})(\tan 50.0°) = (67.2 \text{ m})(1.19) = \boxed{80.0 \text{ m}}$$

The value of $\tan 50.0°$ is found by using a calculator.

Either the sine, cosine, or tangent may be used in calculations such as that in Example 3, depending on which side of the triangle has a known value and which side is asked for. However, *the choice of which side of the triangle to label h_o (opposite) and which to label h_a (adjacent) can be made only after the angle θ is identified.*

Often the values for two sides of the right triangle in Figure 1.4 are available, and the value of the angle θ is unknown. The concept of *inverse trigonometric functions* plays an important role in such situations. Equations 1.4–1.6 give the inverse sine, inverse cosine, and inverse tangent in terms of the symbols used in the drawing. For instance, Equation 1.4 is read as "θ equals the angle whose sine is h_o/h."

$$\theta = \sin^{-1}\left(\frac{h_o}{h}\right) \tag{1.4}$$

$$\theta = \cos^{-1}\left(\frac{h_a}{h}\right) \tag{1.5}$$

$$\theta = \tan^{-1}\left(\frac{h_o}{h_a}\right) \tag{1.6}$$

The use of "-1" as an exponent in Equations 1.4–1.6 *does not mean* "take the reciprocal." For instance, $\tan^{-1}(h_o/h_a)$ does not equal $1/\tan(h_o/h_a)$. Another way to express the inverse trigonometric functions is to use arc sin, arc cos, and arc tan instead of $\sin^{-1}$, $\cos^{-1}$, and $\tan^{-1}$. Example 4 illustrates the use of an inverse trigonometric function.

Example 4   Finding the Depth of a Lake

A lakefront beach drops off gradually at an angle θ, as Figure 1.6 indicates. For safety reasons, it is necessary to know how deep the lake is at various distances from the shore. To provide some information about the depth, a lifeguard rows straight out from the shore a distance of 14.0 m and drops a weighted fishing line. By measuring the length of the line, the lifeguard determines the depth to be 2.25 m. (a) What is the value of θ? (b) What would be the depth d of the lake at a distance of 22.0 m from the shore?

REASONING   Figure 1.6 identifies the sides of the right triangle that are opposite and adjacent to the angle θ; they are $h_o = 2.25$ m and $h_a = 14.0$ m, respectively. We can use the tangent function to find the angle θ in part (a). Then, we can use the value of the angle to find the depth in part (b).

SOLUTION
(a) Using Equation 1.3, we find that

$$\tan \theta = \frac{h_o}{h_a} = \frac{2.25 \text{ m}}{14.0 \text{ m}} = 0.161$$

Now that the value of $\tan \theta$ is known, the angle θ can be obtained by using the inverse tangent:

$$\theta = \tan^{-1}(0.161) = \boxed{9.15°}$$

(b)   Farther from the shore, the sides of the right triangle opposite and adjacent to the angle θ are $h_o = d$ and $h_a = 22.0$ m. Since $\theta = 9.15°$, the tangent function can be used to find the unknown depth; $\tan \theta = h_o/h_a$:

$$h_o = h_a \tan \theta$$

$$d = (22.0 \text{ m})(\tan 9.15°) = \boxed{3.54 \text{ m}}$$

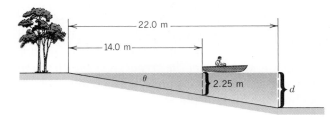

Figure 1.6 If the distance from the shore and the depth of the water at any one point is known, the angle θ can be found with the aid of trigonometry. Knowing the value of θ is useful, because then the depth d at another point can be determined.

THE PYTHAGOREAN THEOREM

The right triangle in Figure 1.4 provides the basis for defining the various trigonometric functions according to Equations 1.1–1.3. These functions always involve an angle and two sides of the triangle. There is also a relationship among the lengths of the three sides of a right triangle. This relationship is known as the *Pythagorean theorem* and is used often in this text.

Pythagorean Theorem

The square of the length of the hypotenuse of a right triangle is equal to the sum of the squares of the lengths of the other two sides:

$$h^2 = h_o{}^2 + h_a{}^2 \qquad (1.7)$$

1.5 THE NATURE OF PHYSICAL QUANTITIES: SCALARS AND VECTORS

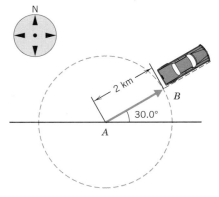

Figure 1.7   A vector quantity has a magnitude and a direction. The arrow in this drawing represents a displacement vector.

The velocity of the cyclists is another example of a vector quantity.

SCALARS

The volume of water in a swimming pool might be 50 cubic meters, or the winning time of a race could be 11.3 seconds. In cases like these, only the magnitude of the numbers matters. In other words, *how much* volume or time is there? The "50" specifies the amount of water in units of cubic meters, while the "11.3" specifies the amount of time in seconds. Volume and time are examples of scalar quantities. A *scalar quantity* is one that can be described by a single number (including any units) giving its magnitude. Some other common scalars are temperature (e.g., 20 °C) and mass (e.g., 85 kg).

VECTORS

While many quantities in physics are scalars, there are also many that are not scalars, quantities for which magnitude tells only part of the story. Consider Figure 1.7, in which a car has moved 2 km along a straight line from point *A* to point *B*. When describing how the car moved, it is incomplete to say that "the car moved a distance of 2 km." This statement would indicate only that the car ends up somewhere on a circle whose center is point *A* and whose radius is 2 km. A complete description would include the direction along with the distance, as in the statement "the car moved a distance of 2 km in a direction 30° north of east." A quantity that deals inherently with both magnitude and direction is called a *vector quantity.* Because direction is an important characteristic of vectors, arrows are used to represent them; *the direction of the arrow gives the direction of the vector.* The colored arrow in Figure 1.7, for example, is called the displacement vector, because it shows how the car is displaced from point *A*. Chapter 2 discusses this particular vector.

It is logical to use the length of the arrow in Figure 1.7 to represent the magnitude of the displacement vector. If the car had moved 4 km instead of 2 km from the starting point, the arrow would have been drawn twice as long. *The length of a vector arrow is proportional to the magnitude of the vector.*

The practice of using the length of an arrow to represent the magnitude of a vector applies to any kind of vector. And in physics there are many important vectors, in addition to the displacement vector. All forces, for instance, are vectors. A force is a push or a pull, and the direction in which a force acts is just as important as the strength or magnitude of the force. The magnitude of a force is measured in SI units called newtons (N). An arrow representing a force of 20 newtons is drawn twice as long as one representing a force of 10 newtons.

SYMBOLS USED FOR SCALARS AND VECTORS

Often, for the sake of convenience, quantities such as volume, time, displacement, and force are represented by symbols. This text follows the usual practice of writing vectors in boldface symbols* **(this is boldface)** and writing scalars in italic symbols *(this is italic)*. Thus, a displacement vector is written as "**s** = 750 m, due east," where the **s** is a boldface symbol. By itself, however, separated from the direction, the magnitude of this vector is a scalar quantity. Therefore, the magnitude is written as "s = 750 m," where the s is an italic symbol.

1.6 VECTOR ADDITION AND SUBTRACTION

ADDITION OF COLINEAR VECTORS

Often it is necessary to add one vector quantity to another, and the process of addition must take into account both the magnitude and the direction of the vectors. The simplest situation occurs when the vectors point along the same direction, that is, when they are colinear, as in Figure 1.8. Here, a car moves 275 m due east along a straight line from point A to point B and then moves 125 m in the same direction from point B to point C. The corresponding two displacements are $\mathbf{s}_{AB}$ and $\mathbf{s}_{BC}$. As usual, boldface symbols denote the vectors. These two separate vectors add to give the total displacement vector $\mathbf{s}_{AC}$, which would apply if the car had moved directly from A to C. With the tail of the second arrow located at the head of the first arrow, the two lengths simply add to give the length of the total displacement. This kind of vector addition is identical to the familiar addition of two scalar numbers (2 + 3 = 5), *and can be carried out here only because the vectors point along the same direction*. In such cases, then, we add the individual magnitudes to get the magnitude of the total, knowing in advance what the direction must be. Formally, the addition is written as follows for the data in Figure 1.8:

$$\mathbf{s}_{AC} = \mathbf{s}_{AB} + \mathbf{s}_{BC}$$

$$\mathbf{s}_{AC} = 275 \text{ m, due east} + 125 \text{ m, due east}$$

$$= 400 \text{ m, due east}$$

The total vector $\mathbf{s}_{AC}$ that results from the addition is referred to as the ***resultant vector.***

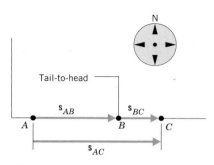

Figure 1.8 Two colinear displacement vectors, $\mathbf{s}_{AB}$ and $\mathbf{s}_{BC}$, add to give the resultant displacement vector $\mathbf{s}_{AC}$.

ADDITION OF PERPENDICULAR VECTORS

Perpendicular vectors are frequently encountered, and Figure 1.9 indicates how they can be added. This figure applies to a car (not shown) that travels due east, from A to B, a distance of 275 m. The displacement vector is $\mathbf{s}_{AB}$. The car then travels due north, from B to C, a distance of 125 m. The second displacement vector is $\mathbf{s}_{BC}$. The resultant displacement vector of the car relative to its starting point is $\mathbf{s}_{AC}$. Once again, the vectors to be added are arranged in a tail-to-head fashion, and the resultant vector points from the tail of the first to the head of the last vector added. The resultant displacement is given by the vector equation

$$\mathbf{s}_{AC} = \mathbf{s}_{AB} + \mathbf{s}_{BC}$$

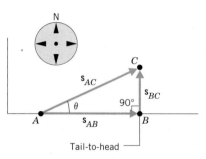

Figure 1.9 The addition of two perpendicular displacement vectors $\mathbf{s}_{AB}$ and $\mathbf{s}_{BC}$ gives the resultant vector $\mathbf{s}_{AC}$.

---

* A vector quantity can also be represented without boldface symbols, by including an arrow above the symbol, e.g., $\vec{s}$.

The addition in this equation cannot be carried out by writing $s_{AC} = 275$ m + 125 m, because the vectors have different directions. Instead, we take advantage of the fact that the triangle in Figure 1.9 is a right triangle and use the Pythagorean theorem (Equation 1.7). According to this theorem, the magnitude of $\mathbf{s}_{AC}$ is

$$s_{AC} = \sqrt{(275 \text{ m})^2 + (125 \text{ m})^2} = 302 \text{ m}$$

The angle θ in Figure 1.9 gives the direction of the resultant vector. Since the lengths of all three sides of the right triangle are now known, either sin θ, cos θ, or tan θ can be used to determine θ:

$$\tan \theta = \frac{s_{BC}}{s_{AB}} = \frac{125 \text{ m}}{275 \text{ m}} = 0.455$$

$$\theta = \tan^{-1}(0.455) = 24.5°$$

Thus, relative to the starting point, the resultant displacement of the car has a magnitude of 302 m and points north of east at an angle of 24.5°.

ADDITION OF VECTORS THAT ARE NEITHER COLINEAR NOR PERPENDICULAR

When two vectors to be added are not perpendicular, the tail-to-head arrangement does not lead to a right triangle, and the Pythagorean theorem cannot be used. Figure 1.10a illustrates such a case for a car that moves 275 m due east from A to B, and then moves 125 m in a direction 55.0° north of west, from B to C. The corresponding displacement vectors are $\mathbf{s}_{AB}$ and $\mathbf{s}_{BC}$. As usual, the resultant displacement vector $\mathbf{s}_{AC}$ is directed from the tail of the first to the head of the last vector added. The vector addition is still given according to

$$\mathbf{s}_{AC} = \mathbf{s}_{AB} + \mathbf{s}_{BC}$$

However, since the triangle in the drawing is not a right triangle, some means other than the Pythagorean theorem must be used to find the magnitude and direction of the resultant vector.

One approach uses a graphical technique. In this method, a diagram is constructed in which the arrows are drawn tail to head. The lengths of the vector arrows are drawn to scale and the angles are drawn accurately (with a protractor, perhaps). Then, the length of the arrow representing the resultant vector is measured with a ruler. This length is converted into the magnitude of the resultant vector by using the scale factor with which the drawing is constructed. In Figure 1.10, for example, a scale of one centimeter of arrow length for each 10.0 m of displacement is used. It can be seen in part b of the drawing that the length of the arrow representing $\mathbf{s}_{AC}$ is 22.8 cm. Since each centimeter corresponds to 10.0 m of displacement, the magnitude of $\mathbf{s}_{AC}$ is 228 m. The angle θ, which gives the direction of $\mathbf{s}_{AC}$, can be measured with a protractor to be $\theta = 26.7°$.

SUBTRACTION OF VECTORS

Suppose that two vectors, $\mathbf{A}$ and $\mathbf{B}$, add together to give a third vector $\mathbf{C}$, according to $\mathbf{C} = \mathbf{A} + \mathbf{B}$. Figure 1.11a shows these vectors. If values for the vectors $\mathbf{C}$ and $\mathbf{B}$ are available, we can calculate vector $\mathbf{A}$ as $\mathbf{A} = \mathbf{C} - \mathbf{B}$. This result is an example of vector subtraction, and it is important to know how to carry out such

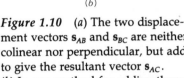

Figure 1.10 (a) The two displacement vectors $\mathbf{s}_{AB}$ and $\mathbf{s}_{BC}$ are neither colinear nor perpendicular, but add to give the resultant vector $\mathbf{s}_{AC}$. (b) In one method for adding them together, a graphical technique is used.

an operation. In practice, vector subtraction is performed exactly as vector addition, except that one of the vectors added (**B** in this case) is multiplied by a scalar factor of -1. To see why, rewrite the equation for vector **A** as follows: $\mathbf{A} = \mathbf{C} + (-\mathbf{B})$. *When a vector is multiplied by -1, the magnitude of the vector remains the same, but the direction of the vector is reversed.* With this in mind, Figure 1.11*b* shows how to calculate vector **A** from the vectors **C** and $-\mathbf{B}$. Notice that vectors **C** and $-\mathbf{B}$ are arranged tail to head and that any suitable method of vector addition can be employed to determine **A**.

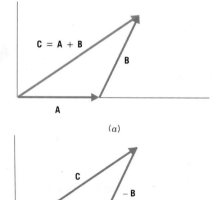

(*a*)

1.7 VECTOR COMPONENTS

THE MEANING OF VECTOR COMPONENTS

Suppose a car moves along a straight line from *A* to *B*. In Figure 1.12 the magnitude and direction of the displacement vector **s** give the distance and direction traveled by the car along the line *AB*. However, the car could also have arrived at *B* by first moving due east, turning through 90°, and then moving due north. This alternative path is shown in color in the drawing and is associated with the two displacement vectors **x** and **y**. The vectors **x** and **y** are called the *x* component and the *y* component of the vector **s**.

The concept of vector components is very important in physics, and two basic features of them are apparent in Figure 1.12. One is that the components add together to equal the original vector, as expressed by the following vector equation:

$$\mathbf{s} = \mathbf{x} + \mathbf{y}$$

In other words, the two components **x** and **y**, when added vectorially, convey exactly the same meaning as does the original vector **s**; that is, they indicate how point *B* is displaced relative to point *A*. Thus, *the components of a vector can be used in place of the vector itself in any calculation where it is convenient to do so.* It will be convenient to use the components of a vector, rather than the vector itself, many times in this text. The other feature of vector components that is apparent in Figure 1.12 is that the components **x** and **y** are not just any two vectors that add together to give the original vector **s**; they are perpendicular vectors.* This perpendicularity is a valuable characteristic of vector components, as we will soon see.

Any vector may be expressed in terms of its components, in a way similar to that illustrated for the displacement vector in Figure 1.12. Figure 1.13 shows an arbitrary vector **A** and its components $\mathbf{A}_x$ and $\mathbf{A}_y$. The components are drawn parallel to convenient *x* and *y* axes and are perpendicular. They add vectorially to equal the original vector **A**:

$$\mathbf{A} = \mathbf{A}_x + \mathbf{A}_y$$

Boldface symbols are used for vector **A** and its components $\mathbf{A}_x$ and $\mathbf{A}_y$. However, the magnitudes of **A**, $\mathbf{A}_x$, and $\mathbf{A}_y$ are scalar quantities and, therefore, are represented by the italic symbols A, A_x, and A_y.

---

* It is possible to introduce vector components that are not perpendicular, but, in general, they are not as useful as those introduced here.

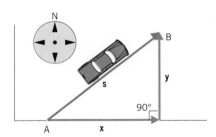

(*b*)

Figure 1.11 (*a*) Vector addition according to $\mathbf{C} = \mathbf{A} + \mathbf{B}$. (*b*) Vector subtraction according to $\mathbf{A} = \mathbf{C} - \mathbf{B} = \mathbf{C} + (-\mathbf{B})$.

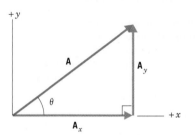

Figure 1.12 The displacement vector **s** and its vector components **x** and **y**.

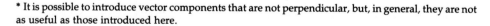

Figure 1.13 An arbitrary vector **A** and its vector components $\mathbf{A}_x$ and $\mathbf{A}_y$.

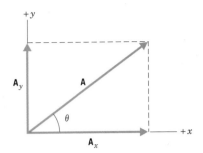

Figure 1.14   This alternative way of drawing the vector **A** and its components is completely equivalent to that shown in Figure 1.13.

There are times when a drawing such as Figure 1.13 is not the most convenient way to represent vector components. Figure 1.14 presents an alternative method. The disadvantage of Figure 1.14 is that the tail-to-head arrangement of A_x and A_y is missing, an arrangement that is a nice reminder that A_x and A_y add together to equal **A**.

The definition given below summarizes the meaning of vector components:

Definition of Vector Components

In two dimensions, the components of a vector **A** are two perpendicular vectors A_x and A_y that are parallel to the x and y axes, respectively, and add together vectorially so that $\mathbf{A} = \mathbf{A}_x + \mathbf{A}_y$.

RESOLVING A VECTOR INTO ITS COMPONENTS

If the magnitude and direction of a vector are known, it is possible to find the components of the vector. The process of finding the components is called "resolving the vector into its components." This process can be carried out with the aid of trigonometry, because the two perpendicular components and the original vector form a right triangle, as Figure 1.13 indicates. Example 5 shows how trigonometry is used to find the components.

Example 5   Finding Vector Components

A displacement vector **s** has a magnitude of $s = 175$ m and points at an angle of $50.0°$ relative to the x axis in Figure 1.15. Find the x and y components of this vector.

REASONING AND SOLUTION 1   The magnitude of the y component can be obtained using the $50.0°$ angle and Equation 1.1, $\sin \theta = y/s$:

$$y = s \sin \theta = (175 \text{ m})(\sin 50.0°) = \boxed{134 \text{ m}}$$

In a similar fashion the magnitude of the x component can be obtained using the $50.0°$ angle and Equation 1.2, $\cos \theta = x/s$:

$$x = s \cos \theta = (175 \text{ m})(\cos 50.0°) = \boxed{112 \text{ m}}$$

REASONING AND SOLUTION 2   It is not necessary to use the $50.0°$ angle to find the x and y components. The angle α in Figure 1.15 can also be used. Since $\alpha + 50.0° = 90.0°$, it follows that $\alpha = 40.0°$. The solution using α yields the same answers as in Solution 1:

$$\cos \alpha = \frac{y}{s}$$

$$y = s \cos \alpha = (175 \text{ m})(\cos 40.0°) = \boxed{134 \text{ m}}$$

$$\sin \alpha = \frac{x}{s}$$

$$x = s \sin \alpha = (175 \text{ m})(\sin 40.0°) = \boxed{112 \text{ m}}$$

PROBLEM SOLVING INSIGHT

When finding the components of a vector, either acute angle of a right triangle can be used to determine the magnitude of the components. The choice of angle is a matter of convenience.

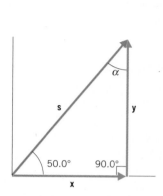

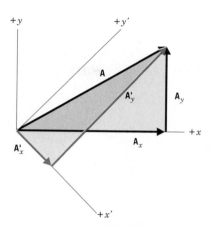

Figure 1.15 The x and y components of the displacement vector **s** can be found using trigonometry.

Figure 1.16 The components of the vector depend on the orientation of the axes used as a reference.

In Example 5 it does not matter whether the 50.0° angle or the 40.0° angle is used in the trigonometric calculation. Both angles lead to the same answers, and the choice is a matter of convenience. In any event, it is possible to check the validity of the answers. Since the components and the original vector form a right triangle, the Pythagorean theorem can be applied to verify that the magnitude of the original vector is indeed 175 m, as given initially:

$$s = \sqrt{(112 \text{ m})^2 + (134 \text{ m})^2} = 175 \text{ m}$$

The values calculated for vector components depend on the orientation of the vector relative to the axes used as a reference. Figure 1.16 illustrates this fact for a vector **A**, by showing two sets of axes, one set being rotated clockwise relative to the other. With respect to the black axes, vector **A** has perpendicular components A_x and A_y; with respect to the colored rotated axes, vector **A** has different components A_x' and A_y'. The choice of which set of components to use is purely a matter of convenience.

> **PROBLEM SOLVING INSIGHT**
>
> When a vector is resolved into components, one can check to see if they are correct; substitute the components into the Pythagorean theorem and verify that the result is the magnitude of the original vector.

VECTORS THAT HAVE ZERO COMPONENTS

Depending on the orientation of the axes used as a reference, it is possible that one of the components of a vector can be zero. Figure 1.17 shows an example of this situation to emphasize that a vector is not zero merely because one of its components is zero. In this drawing, the y component is itself the vector **A**, the x component being zero. Vector **A** would be expressed as the sum of its components according to the following vector equation: $\mathbf{A} = 0 + \mathbf{A}_y$.

For a vector to be zero, every component must individually be zero. Thus, in two dimensions, saying that $\mathbf{A} = 0$ is equivalent to saying that $A_x = 0$ and $A_y = 0$. This seemingly trivial fact plays an important role in physics. In particular, it will be used in Chapter 4 when we describe the equilibrium of an object by saying that the net force acting on the object is zero.

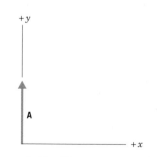

Figure 1.17 The x component of the vector **A** is zero, although the vector itself is not zero.

In a migrating flock of geese, the velocity vectors of each bird are nearly equal.

VECTORS THAT ARE EQUAL

Two vectors are equal if, and only if, they have the same magnitude and direction. Thus, if one displacement vector points east and another points north, they are *not* equal, even if each has the same magnitude of 480 m. In terms of components, two vectors, **A** and **B**, are equal if, and only if, each component of one is equal to the corresponding component of the other. In two dimensions, if **A** = **B**, then $A_x = B_x$ and $A_y = B_y$.

1.8 ADDITION OF VECTORS BY MEANS OF VECTOR COMPONENTS

Vector components provide the most convenient and accurate way of adding (or subtracting) any number of vectors. For example, suppose that vector **A** is added to vector **B**. The resultant vector is **C**, where **C** = **A** + **B**. Figure 1.18*a* illustrates this vector addition, along with the *x* and *y* components of vectors **A** and **B**. Part *b* of the drawing shows the components arranged in a tail-to-head manner. The bottom part of the drawing indicates that the *x* component of **A** and the *x* component of **B** are colinear vectors that add together to give the *x* component of the resultant vector **C**:

$$C_x = A_x + B_x$$

Likewise, the left side of the drawing shows that the *y* components of **A** and **B** are colinear vectors that add together to give the *y* component of **C**:

$$C_y = A_y + B_y$$

The components C_x and C_y of the resultant vector form the sides of the right

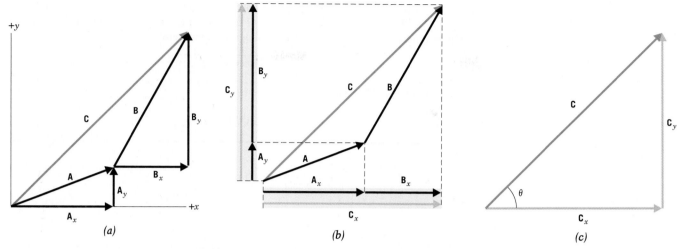

Figure 1.18   (*a*) The vectors **A** and **B** add together to give the resultant vector **C**. The *x* and *y* components of **A** and **B** are also shown. (*b*) The drawing illustrates that $C_x = A_x + B_x$ and $C_y = A_y + B_y$. (*c*) Vector **C** and its components form a right triangle.

triangle shown in Figure 1.18*c*. Thus, we can find the magnitude of **C** by using the Pythagorean theorem:

$$C = \sqrt{C_x^2 + C_y^2}$$

The angle θ that **C** makes with the *x* axis is given by $\theta = \tan^{-1}(C_y/C_x)$. Example 6 illustrates how to add several vectors using the component method.

Example 6   *The Component Method of Vector Addition*

In Figure 1.19*a* vector **A** has a magnitude of 145 m and is oriented 70.0° above the horizontal. Vector **B** has a magnitude of 105 m and points 35.0° below the horizontal. Determine the magnitude and direction of the resultant vector **C** by means of vector components.

REASONING   We resolve the vectors **A** and **B** into their *x* and *y* components. Then the *x* component of **A** is added to the *x* component of **B** to get the *x* component of **C**. In a similar fashion, we obtain the *y* component of **C**. The magnitude and direction of the resultant vector **C** is obtained from its components by using the Pythagorean theorem and the tangent function.

SOLUTION   The first two rows of the table below give the *x* and *y* components of the vectors **A** and **B**. Note that the component B_y is negative, because it points downward, in the negative *y* direction in the drawing.

Vector	x Component	y Component
A	$A_x = (145 \text{ m})\cos 70.0° = 49.6 \text{ m}$	$A_y = (145 \text{ m})\sin 70.0° = 136 \text{ m}$
B	$B_x = (105 \text{ m})\cos 35.0° = 86.0 \text{ m}$	$B_y = -(105 \text{ m})\sin 35.0° = -60.2 \text{ m}$
C	$C_x = A_x + B_x = 135.6 \text{ m}$	$C_y = A_y + B_y = 76 \text{ m}$

The third row in the table gives the x and y components of the resultant vector **C**: $C_x = A_x + B_x$ and $C_y = A_y + B_y$. Part b of the drawing shows **C** and its components. The magnitude of **C** is given by the Pythagorean theorem as

$$C = \sqrt{C_x^2 + C_y^2} = \sqrt{(135.6 \text{ m})^2 + (76 \text{ m})^2} = \boxed{155 \text{ m}}$$

The angle θ that **C** makes with the x axis is

$$\theta = \tan^{-1}\left(\frac{C_y}{C_x}\right) = \tan^{-1}\left(\frac{76 \text{ m}}{135.6 \text{ m}}\right) = \boxed{29°}$$

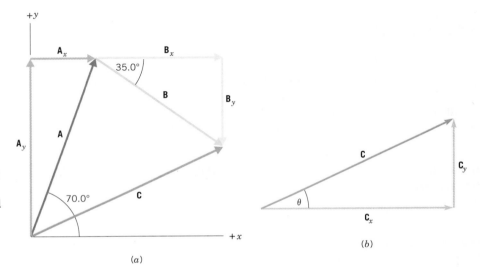

Figure 1.19 (a) The vectors **A** and **B** add together to give the resultant vector **C**. The components of **A** and **B** are also shown. (b) The resultant vector **C** can be obtained once its components have been found.

SUMMARY

Physics is an experimental science that uses precisely defined **units of measurement.** This text emphasizes SI (Système International) units, a system that includes the meter (m), the kilogram (kg), and the second (s) as base units for length, mass, and time, respectively. Units play an important role in solving problems, because the units on the left side of an equation must match the units on the right side. If the units on both sides do not match, either the equation is written incorrectly or the variables and constants in the equation are not expressed in a consistent set of units.

Trigonometry is used throughout physics. Particularly important are the sine, cosine, and tangent functions of an angle θ. These functions can be defined in terms of a right triangle that contains θ. The side of the triangle opposite θ is h_o, the side adjacent to θ is h_a, and the hypotenuse is h. In terms of these quantities $\sin \theta = h_o/h$, $\cos \theta = h_a/h$, and $\tan \theta = h_o/h_a$. Once the value of the sine, cosine, or tangent is known, the angle itself can be obtained using inverse trigonometric functions. The Pythagorean theorem, $h^2 = h_o^2 + h_a^2$, is useful when dealing with the sides of a right triangle.

Two kinds of physical quantities are important, scalars and vectors. A **scalar quantity** is described completely by its magnitude. For a **vector quantity,** however, both magnitude and direction must be specified. Vectors are often represented by arrows, with the length of the arrow being proportional to the magnitude of the vector and the direction of the arrow indicating the direction of the vector. The **addition of vectors** to give a resultant vector must account for both magnitude and direction. When the vectors are all colinear, the addition proceeds in the same way as the simple addition of scalar quantities. When the vectors are not colinear, one procedure for addition utilizes a graphical technique, in which the vectors to be added are arranged in a tail-to-head fashion. The subtraction of a vector is treated as the addition of a vector that has been multiplied by a scalar factor of -1. Multiplying a vector by -1 reverses the direction of the vector.

In two dimensions, the **vector components** of a vector $\mathbf{A}$ are two perpendicular vectors $\mathbf{A}_x$ and $\mathbf{A}_y$ that are parallel to the x and y axes, respectively, and add together vectorially so that $\mathbf{A} = \mathbf{A}_x + \mathbf{A}_y$. Vector compo-

nents provide the best way of adding any number of vectors. A vector is zero if, and only if, each of its components is zero. Two vectors are equal in two dimensions if, and only if, the x components of each are equal and the y components of each are equal.

SOLVED PROBLEMS

Solved Problem 1
Related Problems: *40   *41   **42   **43

Vector $\mathbf{A}$ has a magnitude of 188 units and points 30.0° north of west. Vector $\mathbf{B}$ points 50.0° east of north. Vector $\mathbf{C}$ points 20.0° west of south. These three vectors add to give a resultant vector that is zero. Find the magnitudes of vectors $\mathbf{B}$ and $\mathbf{C}$ by using components.

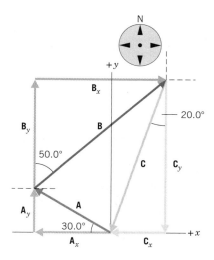

SOLUTION The components of each vector are listed below:

Vector	x Component
A	$-188 \cos 30.0° = -163$
B	$+B \sin 50.0° = +B(0.766)$
C	$-C \sin 20.0° = -C(0.342)$
Total	$-163 + B(0.766) - C(0.342)$

Vector	y Component
A	$+188 \sin 30.0° = +94.0$
B	$+B \cos 50.0° = +B(0.643)$
C	$-C \cos 20.0° = -C(0.940)$
Total	$+94.0 + B(0.643) - C(0.940)$

In these tables the plus and minus signs denote whether the components point along the plus or minus axes. Setting the total x and y components separately equal to zero gives the following equations:

$$-163 + B(0.766) - C(0.342) = 0$$
$$+94.0 + B(0.643) - C(0.940) = 0$$

These equations can be solved simultaneously (see Appendix C for an algebra review) for the unknown quantities B and C:

$\boxed{B = 371 \text{ units}}$ and $\boxed{C = 354 \text{ units}}$.

REASONING In the drawing, vectors $\mathbf{A}$, $\mathbf{B}$, and $\mathbf{C}$ are arranged in a tail-to-head fashion. The fact that the resultant vector is zero $(\mathbf{A} + \mathbf{B} + \mathbf{C} = 0)$ means that the head of vector $\mathbf{C}$ is located at the tail of vector $\mathbf{A}$, thus forming a closed triangle. Since the resultant vector is zero, each of its components must be zero. Therefore, we add the three vectors by the component method and set the total of the x components and the total of the y components separately equal to zero. The result will be two equations that contain the unknown magnitudes of vectors $\mathbf{B}$ and $\mathbf{C}$.

SUMMARY OF IMPORTANT POINTS Whenever a number of vectors add to give a resultant vector that is zero, the components of the resultant vector must separately be zero. Therefore, in two dimensions, the x components of the individual vectors may be added as colinear vectors and the resulting sum set equal to zero. The y components of the individual vectors may be treated in the same way. This approach leads to two separate equations, and provided enough initial information is available, these equations can be solved to yield useful results concerning the original vectors.

QUESTIONS

1. The table below lists four variables along with their units:

Variable	Units
x	meters (m)
v	meters per second (m/s)
t	seconds (s)
a	meters per second squared (m/s²)

These variables appear in the following equations, along with a few numbers that have no units. In which of the equations are the units on the left side of the equals sign consistent with the units on the right side?

(a) $x = vt$

(b) $x = vt + \frac{1}{2}at^2$

(c) $v = at$

(d) $v = at + \frac{1}{2}at^3$

(e) $v^3 = 2ax^2$

(f) $t = \sqrt{\dfrac{2x}{a}}$

2. The variables x and v have the units shown in the table that accompanies question 1. Is it possible for x and v to be related to an angle θ according to $\tan \theta = x/v$? Account for your answer.

3. The variables x, v, and a have the units shown in the table that accompanies question 1. These variables are related by an equation that has the form $v^n = 2ax$, where n is an integer constant (1, 2, 3, etc.) without units. What must be the value of n, so that both sides of the equation have the same units? Explain your reasoning.

4. In the following equation the units of the variables x, v, and t are those shown in the table that accompanies question 1: $v = \frac{1}{3}zxt^2$. What must be the units of the variable z, such that both sides of the equation have the same units? Show how you determined your answer.

5. Using your calculator or a table of trigonometric values, verify that $\sin \theta$ divided by $\cos \theta$ is equal to $\tan \theta$, for any angle θ. Try 30°, for example. Prove that this result is true in general by using the definitions for $\sin \theta$, $\cos \theta$, and $\tan \theta$ given in Equations 1.1–1.3.

6. Sin θ and cos θ are called sinusoidal functions of the angle θ. The way in which these functions change as θ changes leads to a characteristic pattern when they are graphed. This pattern arises many times in physics. (a) To familiarize yourself with the sinusoidal pattern, use a calculator or table of values and construct a graph, with sin θ plotted on the vertical axis and θ on the horizontal axis. Use 15° increments for θ between 0° and 720°. (b) Repeat for cos θ.

7. Which of the following quantities (if any) can be consid-

ered a vector: (a) the number of people attending a football game, (b) the number of days in a month, and (c) the number of pages in a book? Explain your reasoning.

8. Which of the following displacement vectors (if any) are equal? Explain your reasoning.

Vector	Magnitude	Direction
A	100 m	30° north of east
B	100 m	30° south of west
C	50 m	30° south of west
D	100 m	60° east of north

9. Which of the following displacement vectors would yield zero when added together? Account for your answer(s).

Vector	Magnitude	Direction
A	50 m	20° west of north
B	25 m	20° west of north
C	50 m	70° north of west
D	50 m	20° east of south

10. Can two vectors with unequal magnitudes be added together so their sum is zero? Justify your answer.

11. Can two nonzero perpendicular vectors be added together so their sum is zero? Explain.

12. Can three or more vectors with unequal magnitudes be added together so their sum is zero? If so, show by means of a tail-to-head arrangement of the vectors how this could occur.

13. Vectors **A** and **B** satisfy the vector equation $\mathbf{A} + \mathbf{B} = 0$. (a) How does the magnitude of **B** compare with the magnitude of **A**? (b) How does the direction of **B** compare with the direction of **A**? Give your reasoning.

14. Vectors **A**, **B**, and **C** satisfy the vector equation $\mathbf{A} + \mathbf{B} = \mathbf{C}$, and their magnitudes are related by the scalar equation $A^2 + B^2 = C^2$. How is vector **A** oriented with respect to vector **B**? Account for your answer.

15. Vectors **A**, **B**, and **C** satisfy the vector equation $\mathbf{A} + \mathbf{B} = \mathbf{C}$, and their magnitudes are related by the scalar equation $A + B = C$. How is vector **A** oriented with respect to vector **B**? Explain your reasoning.

16. A vector has a component of zero along the x axis of a certain axes system. Does this vector necessarily have a component of zero along the x axis of another (rotated) axes system? Use a drawing to justify your answer.

PROBLEMS

Problems that are not marked with a star are considered the easiest to solve. Problems that are marked with a single star (*) are more difficult, while those marked with a double star (**) are the most difficult.

Section 1.3 The Role of Units in Problem Solving

1. How many seconds are there in (a) one hour and thirty-five minutes and (b) one day?

2. The distance of the Boston marathon is 26 miles, 385 yards. What is the length of this race in meters?

3. Sometimes, highway signs indicate distances to the upcoming exits in both miles and kilometers. One such distance is given as 17.0 miles. What is this distance in kilometers?

4. A 747 jetliner is cruising at a speed of 520 miles per hour. What is its speed in kilometers per hour?

5. One acre contains 43 560 ft². How many square meters (m²) are in one acre?

6. The following are dimensions of various physical parameters that will be discussed later on in the text. Here [L], [T], and [M] denote, respectively, dimensions of length, time, and mass.

	Dimension
Mass (m)	$[M]$
Speed (v)	$[L]/[T]$
Acceleration (a)	$[L]/[T]^2$
Force (F)	$[M][L]/[T]^2$
Energy (E)	$[M][L]^2/[T]^2$

Which of the following equations are dimensionally correct?

(a) $F = ma$
(b) $x = \frac{1}{2}at^3$
(c) $E = \frac{1}{2}mv$
(d) $E = max$
(e) $v = \sqrt{Fx/m}$

*7. Acceleration is the rate of change of the velocity. The acceleration of a car can be expressed in units of miles/hour/hour (or mi/hr²). Another unit for acceleration is meters/second/second (m/s²). Convert an acceleration of 85 mi/hr² to units of m/s².

Section 1.4 Trigonometry

8. The gondola ski lift at Keystone, Colorado, is 2830 m long. On average, the ski lift rises 14.6° above the horizontal. How high is the top of the ski lift relative to the base?

9. A highway is to be built between two towns, one of which lies 35.0 km south and 72.0 km west of the other. What is the shortest length of highway that can be built between the two towns, and at what angle would this highway be directed with respect to due west?

10. A hiker sees a mountain in the distance, the peak of which is known to be 2900 m above sea level. The hiker estimates her line of sight with the top of the peak to be 11° above the horizontal. Assuming the hiker is at sea level and ignoring her height, find the horizontal distance to the point directly under the mountaintop.

11. The corners of a square lie on a circle whose radius is 0.500 m. What is the length of a side of the square?

12. The silhouette of a Christmas tree is an isosceles triangle. The angle at the top of the triangle is 30.0°, and the base measures 2.00 m across. How tall is the tree?

*13. Three buildings, A, B, and C, form the corners of a triangle. Building B is located 210 m from A at an angle of 41° east of north. Building C is located 320 m from A at an angle of 62° east of south. What is the distance between B and C? (*Hint: Consider the law of cosines given in Appendix E.*)

*14. The drawing shows sodium and chlorine ions positioned at the corners of a cube that is part of the crystal structure of sodium chloride (common table salt). The edge of the cube is 0.281 nm (1 nm = 1 nanometer = 10^{-9} m) in length. Find the distance (in nanometers) between the sodium ion located at one corner of the cube and the chlorine ion located on the diagonal at the opposite corner.

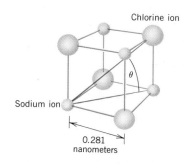

Chlorine ion

θ

Sodium ion

0.281 nanometers

*15. What is the value of the angle θ in the drawing that accompanies problem 14?

**16. The height H of a regular tetrahedron is the perpendicular distance from one corner to the center of the opposite triangular base. The faces of such a tetrahedron are equilateral triangles, and all its edges are equal in length. Show that the ratio between H and the length L of an edge of the tetrahedron is $H/L = \sqrt{2/3}$.

Section 1.6 Vector Addition and Subtraction

17. Two displacement vectors s_1 and s_2 each point 35.0° north of west. One has a magnitude of 4.60 km, while the other has a magnitude of 3.20 km. What are the magnitude and direction of the resultant vector $s_1 + s_2$?

18. The drawing shows a triple jump on a checkerboard, starting at the center of square A and ending on the center of square B. Each side of a square measures 4.0 cm. What is the magnitude of the displacement of the colored checker during the triple jump?

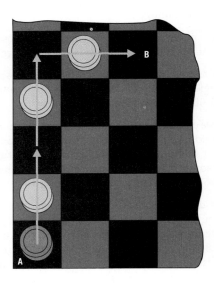

19. Two ropes are attached to a heavy box to pull it along the floor. One rope applies a force of 475 newtons in a direction due west; the other applies a force of 315 newtons in a direction due south. As we will see later in the text, force is a vector quantity. How much force should be applied by a single rope, and in what direction, if it is to accomplish the same effect as the two forces added together?

20. A jogger travels due south and in the process his displacement vector has a magnitude of 4.68 km. He then jogs due west. (a) What is the magnitude of his displacement vector in the due west direction, if the magnitude of his total displacement vector is 7.41 km? (b) What is the direction of his total displacement vector with respect to due south?

21. One displacement vector s_A has a magnitude of 2.43 km and points due north. A second displacement vector s_B has a magnitude of 7.74 km and also points due north. (a) Find the magnitude and direction of $s_A - s_B$. (b) Find the magnitude and direction of $s_B - s_A$.

22. Vector **A** has a magnitude of 48.0 units and points due west, while vector **B** has the same magnitude but points due south. Determine the magnitude and direction of (a) **A** + **B** and (b) **A** − **B**.

23. A boat travels due east for a distance of 5.74 km and then travels 25.0° south of east for a distance of 6.28 km. Use the graphical technique to find the magnitude and direction of the total displacement vector of the boat.

*24. A car is being pulled out of the mud by two forces that are applied by the two ropes shown in the drawing. The dashed line in the drawing bisects the 30.0° angle. The magnitude of the force applied by each rope is 2900 newtons. Arrange the force vectors tail to head and use the graphical technique to answer the following questions. (a) How much force would a single rope need to apply to accomplish the same effect as the two forces added together? (b) How would the single rope be directed relative to the dashed line?

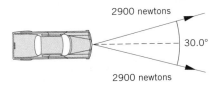

*25. Vector **A** has a magnitude of 8.00 units and points due west. Vector **B** points due north. (a) What is the magnitude of **B** if **A** + **B** has a magnitude of 10.00 units? (b) What is the direction of **A** + **B** relative to due west? (c) What is the magnitude of **B** if **A** − **B** has a magnitude of 10.00 units? (d) What is the direction of **A** − **B** relative to due west?

Section 1.7 Vector Components

26. Vector **A** points along the $+y$ axis and has a magnitude of 100.0 units. Vector **B** points at an angle of 60.0° above the $+x$ axis and has a magnitude of 200.0 units. Vector **C** points along the $+x$ axis and has a magnitude of 87.0 units. Which vector has (a) the largest x component and (b) the largest y component?

27. A displacement vector has a magnitude of 145 m and points at an angle of 28.0° above the negative x axis. What is the magnitude and direction of (a) the x component and (b) the y component of the vector?

28. A bicyclist is headed due east. A 5.00-m/s wind is blowing partially into the rider's face and is coming from a direction that is 35.0° south of east. The speed of the wind and the direction constitute a vector quantity known as the velocity. In effect, then, the rider must "pump" against a component of the wind's velocity vector. What is the magnitude of this component?

29. An ocean liner leaves New York City and travels 18.0° north of east for 155 km. How far east and how far north has it gone? In other words, what are the magnitudes of the components of the ship's displacement vector in the directions due east and due north?

30. A helicopter is traveling with a speed of 67.0 m/s toward a point that is located 38.0° south of east. The speed of the helicopter and the direction constitute a vector quantity known as the velocity. Obtain the magnitude of the velocity component that is directed (a) due south and (b) due east.

31. On takeoff, an airplane climbs with a speed of 180 m/s at an angle of 34° above the horizontal. The speed and direction of the airplane constitute a vector quantity known as the velocity. The sun is shining directly overhead. How fast is the shadow of the plane moving along the ground? (That is, what is the horizontal component of the plane's velocity?)

*32. The magnitude of the force vector **F** is 280 newtons. The x component of this vector is directed along the $+x$ axis and has a magnitude of 150 newtons. The y component points along the $+y$ axis. (a) Find the direction of **F** relative to the $+x$ axis. (b) Find the component of **F** along the $+y$ axis.

*33. The vector **A** in the drawing has a magnitude of 750 units. Determine the magnitude and direction of the x and y components of the vector **A**, relative to (a) the black axes and (b) the colored axes.

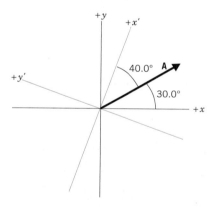

Section 1.8 Addition of Vectors by Means of Vector Components

34. The force vector $\mathbf{F}_A$ has a magnitude of 45.0 newtons and points 30.0° north of east. The force vector $\mathbf{F}_B$ has a magnitude of 75.0 newtons and points due north. Find the magnitude and direction of the resultant $\mathbf{F}_A + \mathbf{F}_B$ by using the component method.

35. A sailboat travels due east for a distance of 1.60 km and then heads 35.0° north of east for another 3.40 km. Using the method of vector components, calculate the vector sum (both magnitude and direction) of these two displacement vectors. Express the direction relative to due east.

36. A golfer, putting on a green, requires three strokes to

"hole the ball." During the first putt, the ball rolls 5.0 m due east. For the second putt, the ball travels 2.1 m at an angle of 20.0° north of east. The third putt is 0.50 m due north. What displacement (magnitude and direction relative to due east) would have been needed to "hole the ball" on the very first putt?

37. A pilot flies from point A to point B to point C, in two straight line segments. The displacement vectors $\mathbf{s}_{AB}$ for the first leg has a magnitude of 243 km and a direction 50.0° north of east. The displacement vector $\mathbf{s}_{BC}$ for the second leg has a magnitude of 57.0 km and a direction 20.0° south of east. The resultant displacement vector is $\mathbf{s}_{AC} = \mathbf{s}_{AB} + \mathbf{s}_{BC}$. What are the magnitude and direction of $\mathbf{s}_{AC}$? Use the component method.

38. Starting from point A, a football player runs the pattern given in the drawing by the three displacement vectors $\mathbf{s}_1$, $\mathbf{s}_2$, and $\mathbf{s}_3$. The magnitudes of these vectors are $s_1 = 5.00$ m, $s_2 = 15.0$ m, and $s_3 = 18.0$ m. Using the component method, find the magnitude and direction θ of the resultant vector $\mathbf{s}_1 + \mathbf{s}_2 + \mathbf{s}_3$.

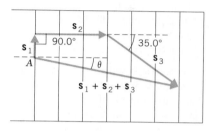

39. You are on a treasure hunt and your map says "Walk due west for 52 paces, then walk 30.0° north of west for 42 paces, and then walk due north for 25 paces." How far north and how far west of your starting point is the treasure?

*40. Three vectors, **A**, **B**, and **C**, add together so that the resultant is zero. Vector **A** points 30.0° north of east. Vector **B** points due north. Vector **C** points 50.0° south of west and has a magnitude of 225 units. By using components, find (a) the magnitude of **A** and (b) the magnitude of **B**. (*See Solved Problem 1 for a related problem.*)

*41. Three vectors **A**, **B**, and **C** add together so that the resultant vector is zero. Vector **A** points 75.0° north of east. Vector **B** points due west. Vector **C** points due south and has a magnitude of 185 units. By using components, find (a) the magnitude of **A** and (b) the magnitude of **B**. (*See Solved Problem 1 for a related problem.*)

**42. What are the x and y components of the vector that must be added to the following three vectors, so that the sum of the four vectors is zero? (*See Solved Problem 1 for a related problem.*)

$$\mathbf{A} = 113 \text{ units, } 60.0° \text{ south of west}$$
$$\mathbf{B} = 222 \text{ units, } 35.0° \text{ south of east}$$
$$\mathbf{C} = 177 \text{ units, } 23.0° \text{ north of east}$$

**43. A sailboat race course consists of four legs, defined by the displacement vectors s_A, s_B, s_C, and s_D, as the drawing indicates. The magnitudes of the first three vectors are $s_A = 3.20$ km, $s_B = 5.10$ km, and $s_C = 4.80$ km. The finish line of the course coincides with the starting line. Using the data in the drawing, find the distance of the fourth leg and the angle θ. *(See Solved Problem 1 for a related problem.)*

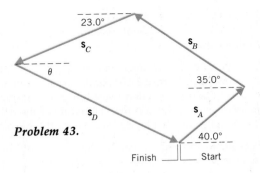

Problem 43.

ADDITIONAL PROBLEMS

44. A displacement vector s_A has a magnitude of 1.62 km and points due north. Another displacement vector s_B has a magnitude of 2.48 km and points due east. Determine the magnitude and direction of (a) $s_A + s_B$ and (b) $s_A - s_B$.

45. An observer, whose eyes are 1.83 m above the ground, is standing 32.0 m away from a tree. The ground is level, and the tree is growing perpendicular to it. The observer's line of sight with the treetop makes an angle of 20.0° above the horizontal. How tall is the tree?

46. The x component of a displacement vector s has a magnitude of 125 m and points along the negative x axis. The y component has a magnitude of 184 m and points along the negative y axis. Find the magnitude and direction of s. Specify the direction with respect to the x axis.

47. A swimming pool has a volume of 4050 ft³. What is the volume in cubic meters (m³)?

48. A frog hops four times: twice forward, once to the right, and once forward again. Each hop covers a distance of 28 cm. What is the magnitude of the frog's displacement?

49. Find the resultant of the three displacement vectors in the drawing by means of the component method. The magnitudes of the vectors are $s_1 = 5.00$ m, $s_2 = 5.00$ m, and $s_3 = 4.00$ m.

50. A student going to class leaves her apartment and walks due north for 1.8 km. She then walks due east for another 1.3 km. (a) What is the shortest distance between her apartment and the classroom? (b) What angle does the shortest distance make with respect to due north?

*51. Consider the two vectors s_1 and s_2 in the drawing for

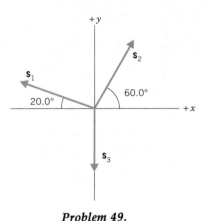

Problem 49.

problem 49. Determine the vector sum $s_1 + s_2$ and the vector difference $s_1 - s_2$ using the method of components. In each case, specify both magnitude and direction.

*52. What is the value of each of the angles of a triangle whose sides are 95, 150, and 190 cm in length? *(Hint: Consider using the law of cosines given in Appendix E.)*

*53. The force vector F_A has a magnitude of 90.0 newtons and points 30.0° north of east. The force vector F_B has a magnitude of 150 newtons and points due north. Use the graphical method and find the magnitude and direction of (a) $F_A - F_B$ and (b) $F_B - F_A$.

**54. Vector A has a magnitude of 6.00 units and points due east. Vector B points due north. (a) What is the magnitude of B, if the vector $A + B$ points 60.0° north of east? (b) Find the magnitude of $A + B$.

CHAPTER 2

KINEMATICS IN ONE DIMENSION

A moving object may be of ordinary size and travel at a low speed, like a sprinter in a track meet. A moving object may also be extremely large and travel at a high speed, like the earth in its journey around the sun or a rocket acquiring sufficient speed for a successful launch. No matter what the example, there are always two aspects to any motion. First, in a purely descriptive sense, there is the movement itself. A moving object does not remain in the same place; it travels from one location to another, perhaps quickly, perhaps slowly. Second, there is the important issue of what causes the motion or what changes it, once it has begun. In this second aspect of motion, forces come into play. A rocket moves upward because of the enormous force generated by its engines, while a touchdown pass is sent on its way by the strong arm of the quarterback. And both the rocket and the football continually experience the force of gravity. The motion of an object can be affected significantly by the forces that act on the object.

Mechanics is the branch of physics that deals with the motion of objects and the forces that change it. Generally speaking, mechanics is divided into two areas: kinematics and dynamics. Kinematics describes the motion of objects without explicit reference to any forces. Dynamics, on the other hand, deals with forces and their effect on motion.

This chapter discusses the fundamental concepts of kinematics and applies them to the description of motion in one dimension. The concepts include displacement, velocity, and acceleration, and Chapter 3 applies them to the description of motion in two dimensions. The study of dynamics depends centrally on three laws of motion discovered by Isaac Newton in the seventeenth century, and Chapter 4 presents these laws.

2.1 DISPLACEMENT

To describe the motion of an object, we must be able to specify the location of the object at all times. Figure 2.1 shows one way of accomplishing this for one-dimensional motion, such as a car traveling along a straight road. Suppose that the initial position of the car is indicated by the vector labeled s_0. As the drawing shows, the length of s_0 is the distance of the car from an arbitrarily chosen origin. At a later time the car has moved to a new position that is indicated by the vector s. The *displacement* of the car Δs (read as "delta s" or "the change in s") is a vector drawn from the initial position to the final position. Displacement is a vector quantity in the sense discussed in Section 1.5, for it conveys both a magnitude (the distance between the initial and final positions) and a direction. The displacement can be related to s_0 and s by noting from the drawing that

$$s_0 + \Delta s = s \quad \text{or} \quad \Delta s = s - s_0$$

Thus, the displacement Δs is the difference between s and s_0, and the Greek letter delta (Δ) is used to denote this difference. It is important to note that the change in any variable is always the final value minus the initial value.

Definition of Displacement

The displacement of an object is the vector whose magnitude is the shortest distance between the initial and final positions of the motion and whose direction points from the initial to the final position.

SI Unit of Displacement: meter (m)

The SI unit for displacement is the meter (m), but there are other units as well, such as the inch and the centimeter. When converting between centimeters (cm) and inches (in.), remember that 2.54 cm = 1 in.

Often, we will have to deal with motion along a straight line. In such a case, a displacement that points in one direction along the line is assigned a positive value, and a displacement pointing in the opposite direction is assigned a negative value. For instance, assume that a car is moving along an east/west direction and that a positive (+) sign is used to denote a direction due east. Then, $\Delta s = +500$ m represents a displacement that points to the east and has a magnitude of 500 meters. Conversely, $\Delta s = -500$ m is a displacement that has the same magnitude, but points in the opposite direction, due west.

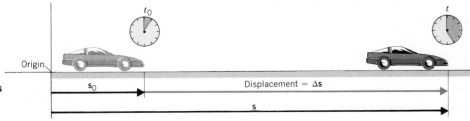

Figure 2.1 The displacement Δs is a vector that points from the initial position to the final position.

2.2 SPEED AND VELOCITY

AVERAGE SPEED

One of the most obvious features of an object in motion is how fast it is moving. If a car travels 200 meters in 10 seconds, we say its average speed is 20 meters per second, the *average speed* being the distance traveled divided by the time required to cover the distance:

$$\text{Average speed} = \frac{\text{Distance}}{\text{Elapsed time}} \qquad (2.1)$$

Equation 2.1 indicates that the unit for average speed is distance divided by time, or meters per second (m/s) in SI units. The next two examples illustrate how the idea of average speed is used.

The average speed of a jogger is the distance run divided by the time of the run.

Example 1   Distance Run by a Jogger

How far does a jogger run in 1.5 hours (5400 s) if his average speed is 2.22 m/s?

REASONING AND SOLUTION   To find the distance run, we rewrite Equation 2.1 as Distance = (Average speed)(Elapsed time):

$$\text{Distance} = (2.22 \text{ m/s})(5400 \text{ s}) = \boxed{12\ 000 \text{ m}}$$

Example 2   A Diving Falcon

With an average speed of 67 m/s, how long does it take a falcon (Figure 2.2) to dive to the ground along a 150-m path?

REASONING AND SOLUTION   To find the time of flight, we rewrite Equation 2.1 as Elapsed time = Distance/Average speed:

$$\text{Elapsed time} = \frac{150 \text{ m}}{67 \text{ m/s}} = \boxed{2.2 \text{ s}}$$

AVERAGE VELOCITY

As useful as it is, the average speed of an object does not reveal anything about the direction of the motion. To take into account the direction, the vector concept of velocity is needed. In Figure 2.1, suppose that the car's initial position is s_0 when the time is t_0. A little later the car arrives at the final position s at the time t. The time for the car to travel between these two positions is the *difference* between t and t_0, which is denoted by the symbol Δt:

$$\text{Elapsed time} = t - t_0 = \Delta t$$

Dividing the displacement Δs of the car by the elapsed time Δt gives the *average velocity* of the car. It is customary to denote the average value of a quantity by placing a horizontal bar above the symbol representing the quantity. The average velocity, then, is written as $\overline{\mathbf{v}}$.

Figure 2.2   If we know the average speed of a diving falcon and the distance of the dive, we can determine the time it takes for the falcon to make the dive.

In this time-lapse photo of cars on a freeway, the velocity of the cars in the left lanes (white headlights) is opposite to that of the cars in the right lanes (red taillights).

Definition of Average Velocity

$$\text{Average velocity} = \frac{\text{Displacement}}{\text{Elapsed time}}$$

$$\overline{\mathbf{v}} = \frac{\mathbf{s} - \mathbf{s}_0}{t - t_0} = \frac{\Delta \mathbf{s}}{\Delta t} \qquad (2.2)$$

SI Unit of Average Velocity: meter per second (m/s)

Equation 2.2 indicates that the unit for average velocity is length divided by time, or meters per second (m/s) in SI units. Velocity can also be expressed in other units, such as kilometers per hour (km/h) or miles per hour (mi/h).

Average velocity is a vector that points in the same direction as the displacement in Equation 2.2. As with displacement, plus and minus signs indicate the two possible directions of the velocity along a straight line. If the displacement points in the positive direction, the average velocity is positive. Conversely, if the displacement points in the negative direction, the average velocity is negative. Example 3 illustrates these features of average velocity.

Example 3   The World's Fastest Jet-Engined Car

A world record of 274 m/s (613 mi/h) for the fastest jet-engined car was set in 1965 by Craig Breedlove in the car *Spirit of America.* Such a measurement is made by calculating the average velocity of the car over a measured course. The driver makes two runs through the course, one in each direction, to nullify any wind effects. Figure 2.3*a* shows that the car first travels from left to right and covers a distance of 604 m in 2.19 s. Figure 2.3*b* shows that in the reverse direction, the car covers the same distance in 2.22 s. From these data, determine the average velocity for each run.

REASONING   During the first run, the car travels to the right, which is the positive direction. Thus, the displacement of the car is $\Delta \mathbf{s} = +604$ m. The corresponding elapsed time is $\Delta t = 2.19$ s. During the second run, the car travels to the left, which is the negative direction. Then, the displacement and elapsed time are $\Delta \mathbf{s} = -604$ m and $\Delta t = 2.22$ s.

SOLUTION   According to Equation 2.2, the average velocities are

Run 1
$$\overline{\mathbf{v}} = \frac{\Delta \mathbf{s}}{\Delta t} = \frac{+604 \text{ m}}{2.19 \text{ s}} = \boxed{+276 \text{ m/s}}$$

Run 2
$$\overline{\mathbf{v}} = \frac{\Delta \mathbf{s}}{\Delta t} = \frac{-604 \text{ m}}{2.22 \text{ s}} = \boxed{-272 \text{ m/s}}$$

The minus sign in the answer for run 2 indicates that the average velocity, like the displacement, points to the left in Figure 2.3*b*. The magnitudes of the average velocities are 276 m/s and 272 m/s. The average of these numbers is 274 m/s and is recorded in the record book.

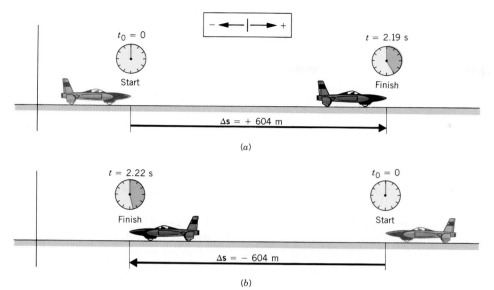

(a)

(b)

Figure 2.3 The arrows in the box at the top of the drawing indicate the positive and negative directions for the displacements of the car, as explained in Example 3.

INSTANTANEOUS VELOCITY

Suppose the magnitude of your average velocity for a long trip was 20 m/s. This value, being an average, does not convey any information about how fast you were moving at any instant during the trip. Surely there were times when the car traveled faster than 20 m/s and times when it traveled slower. The *instantaneous velocity* **v** of the car indicates how fast the car moves and the direction of the motion at each instant of time. The magnitude of the instantaneous velocity is called the *instantaneous speed,* and it is the number (with units) indicated by the speedometer.

The instantaneous velocity at any point during a trip can be obtained by measuring the time Δt for the car to travel a *very small* displacement Δs centered on the point of interest. We can then compute the average velocity over this interval. If the time Δt is small enough, the instantaneous velocity does not change much during the measurement. Then, the instantaneous velocity **v** at the point of interest is approximately equal to ($\approx$) the average velocity $\overline{\textbf{v}}$ computed over the interval, or $\textbf{v} \approx \overline{\textbf{v}} = \Delta \textbf{s}/\Delta t$ (for sufficiently small Δt). In fact, in the limit that Δt becomes infinitesimally small, the instantaneous velocity and the average velocity become equal, so that

$$\textbf{v} = \lim_{\Delta t \to 0} \frac{\Delta \textbf{s}}{\Delta t} \qquad (2.3)$$

The notation $\lim_{\Delta t \to 0} (\Delta \textbf{s}/\Delta t)$ means that the ratio $\Delta \textbf{s}/\Delta t$ is defined by a limiting process in which smaller and smaller values of Δt are used, so small that they approach zero. As smaller values of Δt are used, $\Delta \textbf{s}$ also becomes smaller. However, the ratio $\Delta \textbf{s}/\Delta t$ does *not* become zero but, rather, approaches the value of the instantaneous velocity.

For brevity, we will use the word *velocity* to mean "instantaneous velocity" and *speed* to mean "instantaneous speed." In a wide range of motions, the velocity changes from moment to moment. To describe the manner in which it changes, however, the concept of acceleration is needed, as the next section discusses.

2.3 ACCELERATION

Whenever the velocity of an object is changing, we say that the object is "accelerating." A car, temporarily stopped at a traffic signal, accelerates when the light turns green and the driver steps on the gas; or a car accelerates to a greater velocity to pass another car. While these examples involve an increase in velocity, there are also many examples of acceleration where the velocity decreases, such as a bicycle slowing down when the breaks are applied.

The meaning of *average acceleration* can be illustrated by considering a plane during takeoff. Figure 2.4 focuses attention on how the velocity of the plane changes as the plane moves down the runway. During an elapsed time interval $\Delta t = t - t_0$, the velocity changes from an initial value of $\mathbf{v}_0$ to a final value of $\mathbf{v}$, the change in velocity being $\Delta \mathbf{v} = \mathbf{v} - \mathbf{v}_0$. The average acceleration is defined in the following manner, to provide a measure of how much the velocity changes per unit of elapsed time.

A jet-skier accelerates down the front of a monster wave in the "Pipeline" in Hawaii.

Definition of Average Acceleration

$$\text{Average acceleration} = \frac{\text{Change in velocity}}{\text{Elapsed time}}$$

$$\bar{\mathbf{a}} = \frac{\mathbf{v} - \mathbf{v}_0}{t - t_0} = \frac{\Delta \mathbf{v}}{\Delta t} \tag{2.4}$$

SI Unit of Average Acceleration: meter per second squared (m/s²)

The average acceleration $\bar{\mathbf{a}}$ is a vector that points in the same direction as $\Delta \mathbf{v}$, the change in the velocity. Following the usual custom, plus and minus signs indicate the two possible directions for the acceleration vector when the motion is along a straight line.

We are often interested in an object's acceleration at a particular instant of time. The *instantaneous acceleration* **a** can be defined by analogy with the procedure used in Section 2.2 for instantaneous velocity:

$$\mathbf{a} = \lim_{\Delta t \to 0} \frac{\Delta \mathbf{v}}{\Delta t} \tag{2.5}$$

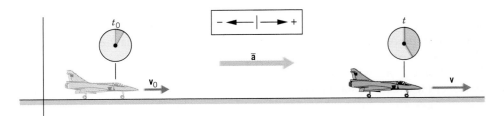

Figure 2.4 During takeoff, the plane accelerates from an initial velocity $\mathbf{v}_0$ to a final velocity $\mathbf{v}$ during the time interval $\Delta t = t - t_0$.

Equation 2.5 indicates that the instantaneous acceleration is a limiting case of the average acceleration. When the time interval Δt for measuring the acceleration becomes extremely small (approaching zero in the limit), the average acceleration and the instantaneous acceleration become equal. Moreover, in many situations the acceleration is constant, so the acceleration has the same value at any instant of time. In the future, we will use the word *acceleration* to mean "instantaneous acceleration." Examples 4 and 5 illustrate these features of acceleration.

Example 4   The Acceleration of a Jet Plane

Suppose the plane in Figure 2.4 starts from rest ($v_0 = 0$) when $t_0 = 0$. The plane accelerates down the runway and at $t = 29$ s attains a velocity of $v = +260$ km/h, where the plus sign indicates the velocity points to the right. Determine the average acceleration of the plane.

REASONING AND SOLUTION   The average acceleration of the plane can be found from Equation 2.4 as

$$\overline{a} = \frac{v - v_0}{t - t_0} = \frac{260 \text{ km/h} - 0 \text{ km/h}}{29 \text{ s} - 0 \text{ s}} = \boxed{+9.0 \; \frac{\text{km/h}}{\text{s}}}$$

PROBLEM SOLVING INSIGHT

The change in any variable is the final value minus the initial value: e.g., the change in velocity is $\Delta v = v - v_0$, and the change in time is $\Delta t = t - t_0$.

The average acceleration calculated in Example 4 is read as "nine kilometers per hour per second." Assuming the acceleration of the plane is constant, an acceleration of $9.0 \frac{\text{(km/h)}}{\text{s}}$ means the velocity changes by 9 km/h during each second of the motion. During the first second, the velocity increases from 0 to 9 km/h; during the next second, the velocity increases by another 9 km/h to 18 km/h, and so on. Figure 2.5 illustrates how the velocity changes during the first three seconds. By the end of the 29th second, the velocity is 260 km/h.

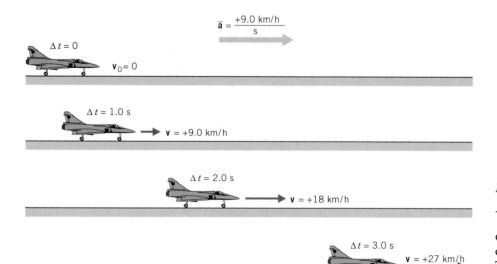

Figure 2.5   An acceleration of $+9 \frac{\text{km/h}}{\text{s}}$ means that the velocity of the plane changes by $+9$ km/h during each second of the motion. The "+" direction for **a** and **v** is to the right.

It is customary to express the units for acceleration solely in terms of SI units. One way to obtain SI units for the acceleration in Example 4 is to convert the velocity units from km/h to m/s:

$$260 \ \frac{km}{h} \left(\frac{\frac{1000 \ m}{1 \ km}}{\frac{3600 \ s}{1 \ h}} \right) = 72 \ m/s$$

The average acceleration then becomes

$$\bar{a} = \frac{72 \ m/s - 0 \ m/s}{29 \ s - 0 \ s} = +2.5 \ m/s^2$$

where we have used $2.5 \ \frac{m/s}{s} = 2.5 \ \frac{m}{s \cdot s} = 2.5 \ \frac{m}{s^2}$. An acceleration of $2.5 \ \frac{m}{s^2}$ is read as "2.5 meters per second per second" (or "2.5 meters per second squared") and means that the velocity changes by 2.5 m/s during each second of the motion.

Example 5 deals with a case where the motion becomes slower as time passes.

Example 5   The Acceleration of a Drag Racer

A drag racer crosses the finish line, and the driver applies the brakes to slow down, as Figure 2.6 illustrates. The brakes are applied initially when $t_0 = 9.0$ s and the car's velocity is $v_0 = +28$ m/s. When $t = 12.0$ s, the velocity has been reduced to $v = +13$ m/s. What is the average acceleration of the dragster?

REASONING AND SOLUTION   According to Equation 2.4, the average acceleration is

$$\bar{a} = \frac{v - v_0}{t - t_0} = \frac{13 \ m/s - 28 \ m/s}{12.0 \ s - 9.0 \ s} = \boxed{-5.0 \ m/s^2}$$

Figure 2.7 shows how the velocity of the dragster changes during the braking, assuming that the acceleration is constant throughout the motion. The acceleration calculated in Example 5 is negative, indicating that the acceleration points to the left in the drawing. As a result, the acceleration and the velocity point in

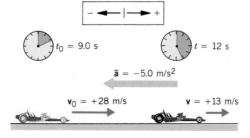

Figure 2.6   The velocity of the car is decreasing, giving rise to an average acceleration that points opposite to the velocity.

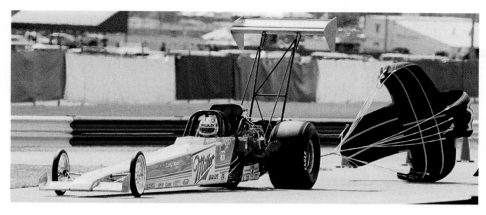

The parachute decelerates the dragster at the end of the race.

opposite directions. **Whenever the acceleration and velocity vectors have opposite directions, the object slows down and is said to be "decelerating."** In contrast, the acceleration and velocity vectors in Figure 2.5 point in the *same* direction, and the object speeds up.

PROBLEM SOLVING INSIGHT

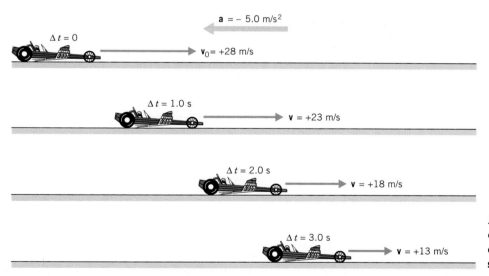

Figure 2.7 Here, an acceleration of -5.0 m/s^2 means the velocity decreases by 5.0 m/s during each second of elapsed time.

2.4 EQUATIONS OF KINEMATICS FOR CONSTANT ACCELERATION

Now that we have the concepts of displacement, velocity, and acceleration to use, we can describe the motion of an object traveling with constant acceleration along a straight line. For convenience, assume that the object is located at the origin $s_0 = 0$ at the time $t_0 = 0$. With this assumption, the displacement $\Delta s = s - s_0$ becomes $\Delta s = s$. It is convenient and customary to dispense with the use of boldface symbols for the displacement, velocity, and acceleration vectors in the equations that follow.

Suppose, for example, that a motorcycle has an initial velocity of $v_0 = +17$ m/s and moves for 4.0 s with a constant acceleration of $a = +3.0$ m/s². For a complete description of the motion, it is also necessary to know the velocity and displacement at the end of the 4.0-s interval. The final velocity v can be obtained from Equation 2.4:

$$\bar{a} = a = \frac{v - v_0}{t}$$

Solving for v yields

$$v = v_0 + at = 17 \text{ m/s} + (3.0 \text{ m/s}^2)(4.0 \text{ s}) = +29 \text{ m/s}$$

The displacement of the motorcycle after 4.0 s can be obtained from Equation 2.2, if a value for the average velocity $\bar{v}$ can be determined:

$$\bar{v} = \frac{s - s_0}{t - t_0} = \frac{s}{t} \quad \text{or} \quad s = \bar{v}t$$

Because the acceleration is constant, the velocity increases at a constant rate. Thus, the average velocity $\bar{v}$ is midway between the initial and final velocities:

$$\bar{v} = \tfrac{1}{2}(v_0 + v) \quad \text{(constant acceleration)} \tag{2.6}$$

Equation 2.6 applies only if the acceleration is constant and cannot be used when the acceleration is changing. The average velocity of the motorcycle is, then,

$$\bar{v} = \tfrac{1}{2}(17 \text{ m/s} + 29 \text{ m/s}) = +23 \text{ m/s}$$

The displacement of the motorcycle can now be determined:

$$s = \bar{v}t = (23 \text{ m/s})(4.0 \text{ s}) = +92 \text{ m}$$

The preceding calculation for the displacement can be summarized algebraically as

$$s = \bar{v}t = \tfrac{1}{2}(v_0 + v)t \quad \text{(constant acceleration)} \tag{2.7}$$

By using Equations 2.4 and 2.7, we were able to complete the description of the motorcycle's motion:

$$v = v_0 + at \quad (2.4) \qquad s = \tfrac{1}{2}(v_0 + v)t \quad (2.7)$$

Notice that there are five kinematic variables in these two equations:

1. $s = $ displacement
2. $a = \bar{a} = $ acceleration (constant)
3. $v = $ final velocity at time t
4. $v_0 = $ initial velocity at time $t_0 = 0$
5. $t = $ time elapsed since $t_0 = 0$

Each of the two equations contains four of these variables, so if three of them are known, the fourth variable can always be found. Example 6 further illustrates how Equations 2.4 and 2.7 are used to describe the motion of an object.

Example 6   The Displacement of a Speedboat

The speedboat in Figure 2.8 has a constant acceleration of $+2.0$ m/s². If the initial velocity of the boat is $+6.0$ m/s, find its displacement after 8.0 seconds.

<u>REASONING</u>   The three known variables are listed in the table:

Speedboat Data

s	a	v	v_0	t
?	$+2.0$ m/s²		$+6.0$ m/s	8.0 s

We can use $s = \frac{1}{2}(v_0 + v)t$ to find the displacement of the boat if a value for the final velocity v can be found. To find the final velocity, it is necessary to use the value given for the acceleration, because it tells us how the velocity changes, according to $v = v_0 + at$.

<u>SOLUTION</u>   The final velocity is

$$v = v_0 + at = 6.0 \text{ m/s} + (2.0 \text{ m/s}^2)(8.0 \text{ s}) = +22 \text{ m/s} \tag{2.4}$$

The displacement of the boat can now be obtained:

$$s = \tfrac{1}{2}(v_0 + v)t \tag{2.7}$$

$$= \tfrac{1}{2}(6.0 \text{ m/s} + 22 \text{ m/s})(8.0 \text{ s}) = \boxed{+110 \text{ m}}$$

A calculator would give the answer as 112 m, but this number must be rounded to 110 m, since the data are accurate to only two significant figures.

The displacement of a speedboat can be determined if its acceleration, initial velocity, and time of travel are known.

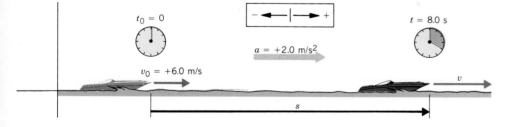

Figure 2.8   An accelerating speedboat.

The solution to Example 6 involved two steps: finding v and then calculating s. It would be helpful if we could find an equation that allows us to determine the displacement in a single step. Using Example 6 as a guide, we can obtain such an equation by substituting v from Equation 2.4 ($v = v_0 + at$) for the v that appears in Equation 2.7 [$s = \frac{1}{2}(v_0 + v)t$]:

$$s = \tfrac{1}{2}(v_0 + v)t = \tfrac{1}{2}(v_0 + \boxed{v_0 + at})t = \tfrac{1}{2}(2v_0 t + at^2)$$

$$s = v_0 t + \tfrac{1}{2}at^2 \quad \text{(constant acceleration)} \tag{2.8}$$

You can verify that Equation 2.8 gives the displacement of the speedboat directly without the intermediate step of determining the final velocity. The first term ($v_0 t$)

on the right side of Equation 2.8 represents the displacement that would result if the acceleration were zero, and the velocity remained constant at its initial value of v_0. The second term on the right ($\frac{1}{2}at^2$) gives the additional displacement that arises because the acceleration changes the velocity to values that are different from its initial value. We now turn to another illustration of accelerated motion in the next example.

THE PHYSICS OF . . .

catapulting a jet from an aircraft carrier.

Example 7   *Catapulting a Jet*

A jet is taking off from the deck of an aircraft carrier, as Figure 2.9 shows. Starting from rest, the jet is catapulted with a constant acceleration of $+31$ m/s² along a straight line and reaches a velocity of $+62$ m/s. Find the displacement of the jet.

REASONING   The data are as follows:

Jet Data

s	a	v	v_0	t
?	$+31$ m/s²	$+62$ m/s	0	

The initial velocity v_0 is zero, since the jet starts from rest. The displacement s of the aircraft can be obtained from $s = \frac{1}{2}(v_0 + v)t$, if we can determine the time t during which the plane is being accelerated. But t is controlled by the value of the acceleration. With larger accelerations, the jet reaches its final velocity in shorter times, as can be seen by solving $v = v_0 + at$ for t.

SOLUTION   Solving Equation 2.4 for t, we find

$$t = \frac{v - v_0}{a} = \frac{62 \text{ m/s} - 0}{31 \text{ m/s}^2} = 2.0 \text{ s}$$

Since the time is known, the displacement can be found:

$$s = \tfrac{1}{2}(v_0 + v)t = \tfrac{1}{2}(0 + 62 \text{ m/s})(2.0 \text{ s}) = \boxed{+62 \text{ m}} \qquad (2.7)$$

PROBLEM SOLVING INSIGHT

"Implied data" is important. For instance, in Example 7 the phrase "starts from rest" means that the initial velocity is zero ($v_0 = 0$).

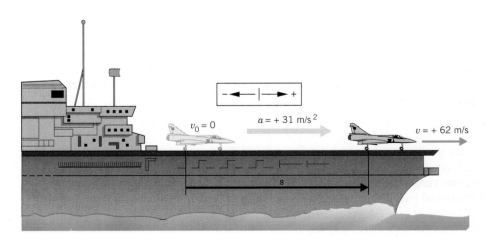

Figure 2.9   A plane is being launched from the deck of an aircraft carrier.

During a launch, like this one from the USS *Saratoga*, a catapult accelerates the jet down the deck of the aircraft carrier.

It is possible to derive a single equation for the displacement s when a, v, and v_0 are known, but the time t is not known, as in Example 7. Solving Equation 2.4 for the time $[t = (v - v_0)/a]$, and then substituting into Equation 2.7 reveals that

$$s = \tfrac{1}{2}(v_0 + v)t = \tfrac{1}{2}(v_0 + v)\boxed{\frac{v - v_0}{a}} = \frac{v^2 - v_0^2}{2a}$$

Solving for v^2 shows that

$$v^2 = v_0^2 + 2as \qquad \text{(constant acceleration)} \qquad (2.9)$$

It is a straightforward exercise to verify that Equation 2.9 can be used to find the displacement of the jet in Example 7 without having to solve first for the time. Table 2.1 presents a summary of the equations that we have been considering. These equations are called the *equations of kinematics.* Each equation contains four variables, as indicated by the check marks (✔) in the table. The next section shows how to apply the equations of kinematics.

2.5 APPLICATIONS OF THE EQUATIONS OF KINEMATICS

The equations of kinematics can be used for any moving object, as long as the acceleration of the object is constant. However, to avoid errors when using these equations, it helps to follow a few sensible guidelines and to be alert for a few situations that can arise during your calculations.

Before attempting to solve a problem, verify that the given information contains values for at least three of the five kinematic variables (s, a, v, v_0, t). A glance back at the data boxes in Examples 6 and 7 will show which three variables were given for these problems. Only when values for at least three variables are known can the equations listed in Table 2.1 be used to find the values for the fourth and fifth variables.

Decide at the start which directions are to be called positive (+) and negative (−) relative to a conveniently chosen coordinate origin. While this decision is

Table 2.1   Equations of Kinematics for Constant Acceleration

Equation Number	Equation	Variables				
		s	a	v	v_0	t
(2.4)	$v = v_0 + at$		✓	✓	✓	✓
(2.7)	$s = \frac{1}{2}(v_0 + v)t$	✓		✓	✓	✓
(2.8)	$s = v_0 t + \frac{1}{2}at^2$	✓	✓		✓	✓
(2.9)	$v^2 = v_0^2 + 2as$	✓	✓	✓	✓	

arbitrary, it is nonetheless an important one. Displacement, velocity, and acceleration are vectors, and their directions must always be taken into account. In the examples that follow, the positive and negative directions will be shown in the drawings that accompany the problems. It does not matter which direction is chosen to be positive. However, once the choice has been made, it should not be changed during the course of the calculation. Example 8 illustrates these important issues.

Example 8   Switching Positive and Negative Directions

Repeat Example 6 using the same data, but now assume the negative direction is to the right, rather than to the left (see Figure 2.10).

REASONING   The three known variables are as follows:

Speedboat Data

s	a	v	v_0	t
?	-2.0 m/s^2		-6.0 m/s	8.0 s

Note that a and v_0 are now negative numbers, consistent with our new choice for the negative direction. The displacement of the boat can be found from Equation 2.8 in Table 2.1. This equation is selected because it contains the four variables of interest, s, a, v_0, and t, with s being the only unknown.

SOLUTION

$$s = v_0 t + \tfrac{1}{2}at^2 = (-6.0 \text{ m/s})(8.0 \text{ s}) + \tfrac{1}{2}(-2.0 \text{ m/s}^2)(8.0 \text{ s})^2 = \boxed{-110 \text{ m}}$$

The magnitude of the displacement is 110 m, the same as in Example 6. Now, however, the displacement is in the negative direction. Since the negative direction now points to the right, the meaning of this answer is exactly the same as it is in Example 6, that is, the boat moves to the right. The physical meaning of the answers in both examples is the same when interpreted in the context of the initial choice for the positive and negative directions.

PROBLEM SOLVING INSIGHT

Sometimes there are two possible answers to a kinematics problem, each answer corresponding to a different situation. Example 9 discusses one such case.

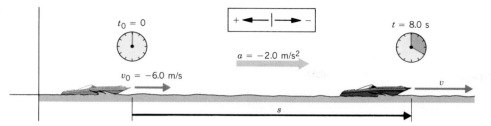

Figure 2.10 This drawing is the same as Figure 2.8, except the direction to the right is chosen to be negative.

THE PHYSICS OF . . .

the acceleration caused by a retrorocket.

Example 9   An Accelerating Spacecraft

The spacecraft shown in Figure 2.11*a* is traveling with a velocity of $+3250$ m/s. Suddenly the retrorocket is fired, and the spacecraft begins to slow down with an acceleration whose magnitude is 10.0 m/s². What is the velocity of the spacecraft when the displacement of the craft is $+215$ km, relative to the point where the retrorocket began firing?

REASONING   Since the spacecraft is slowing down at this stage of the motion, the acceleration must be opposite to the velocity. The velocity points to the right in the drawing, so the acceleration must point to the left, which is the negative direction; thus, $a = -10.0$ m/s². The three known variables are listed below:

Spacecraft Data

s	a	v	v_0	t
$+215\ 000$ m	-10.0 m/s²	?	$+3250$ m/s	

The final velocity v of the spacecraft can be calculated using Equation 2.9, since it contains the four pertinent variables.

SOLUTION

$$v^2 = v_0^2 + 2as = (3250 \text{ m/s})^2 + 2(-10.0 \text{ m/s}^2)(215\ 000 \text{ m}) = 6.3 \times 10^6 \text{ m}^2/\text{s}^2$$

The two possible solutions are

$$v = +2.5 \times 10^3 \text{ m/s} \quad \text{or} \quad v = -2.5 \times 10^3 \text{ m/s}$$

Both of these answers correspond to the *same* displacement ($s = +215$ km), but each arises in a different part of the motion. The answer $v = +2500$ m/s corresponds to the situation in Figure 2.11*a*. Here the spacecraft has slowed to a speed of 2500 m/s, but is still traveling to the right. The answer $v = -2500$ m/s arises because the retrorocket eventually brings the spacecraft to a momentary halt and causes it to reverse its direction of travel, after which it moves toward the left. As the craft moves toward the left, its speed increases due to the continually firing rocket. After a time, the velocity of the craft becomes $v = -2500$ m/s, giving rise to the situation shown in part *b* of the drawing. In both parts of the drawing the spacecraft has the same displacement, but a greater travel time is required in part *b* compared to part *a*. The reader can verify that it takes 75 s for the spacecraft in part *a*, but it takes 575 s in part *b*, where the spacecraft travels a greater distance.

The motion of two objects may be interrelated, so they share a common variable. The fact that the motions are interrelated is an important piece of information. In such cases, data for only two variables need be specified for each

PROBLEM SOLVING INSIGHT

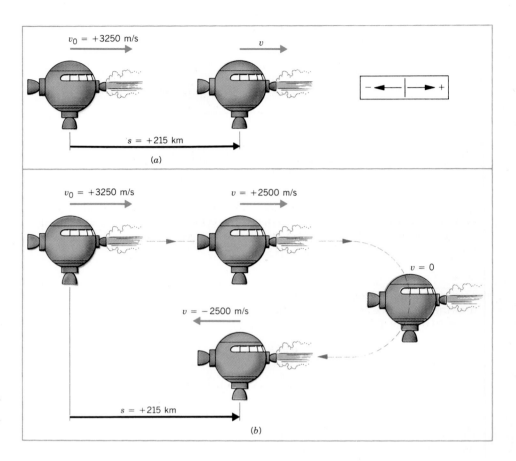

$v_0 = +3250$ m/s

v

$s = +215$ km

(a)

$v_0 = +3250$ m/s

$v = +2500$ m/s

$v = 0$

$v = -2500$ m/s

$s = +215$ km

(b)

Figure 2.11 (a) Because of an acceleration of -10.0 m/s², the spacecraft changes its velocity from v_0 to v. (b) Continued firing of the retrorocket changes the direction of the craft's motion.

object. Example 10 involves the interrelationship between two moving objects, a person and a bus. At first glance it appears that the problem cannot be solved, because only two kinematic variables are specified for each object. However, the time t is the same for the person and the bus. Even though no value is given for the common variable, the problem can be solved. The answer turns out to be interesting.

Example 10   Catching a Bus

In Figure 2.12 a bus has stopped to pick up riders. A woman is running at a constant velocity of $+5.0$ m/s in an attempt to catch the bus. When she is 11 m from the bus, it pulls away with a constant acceleration of $+1.0$ m/s². From this point, how much time does it take her to reach the bus if she keeps running with the same velocity?

REASONING   To catch the bus, the woman must run not only 11 m but also the distance traveled by the moving bus, s_{bus}. Figure 2.12 shows the corresponding displacement vectors, and it follows that

$$s_{bus} + 11 \text{ m} = s_{woman}$$

This equation is the starting point for our analysis.

SOLUTION   Assuming that the clock is zeroed at the instant when the woman is

11 m from the bus, her displacement is $s_{woman} = (5.0 \text{ m/s})t$, since she runs with a constant velocity. Therefore,

$$s_{bus} + 11 \text{ m} = (5.0 \text{ m/s})t$$

This result reveals that we need a value for s_{bus} to calculate t. The known variables for the bus are summarized below:

Bus Data

s_{bus}	a	v	v_0	t
?	$+1.0 \text{ m/s}^2$		0	✓

A check mark has been placed in the box for the time t. Although we do not have an explicit value for t, it is the same for both the bus and the woman. In other words, the time is a "common" variable, and the check mark reminds us of this fact. Equation 2.8 can be used to relate the displacement of the bus and the time:

$$s_{bus} = v_0 t + \tfrac{1}{2}at^2 = \tfrac{1}{2}(1.0 \text{ m/s}^2)t^2$$

This result can be substituted into the expression obtained earlier to give an equation containing only one unknown, the time t:

$$s_{bus} + 11 \text{ m} = (5.0 \text{ m/s})t$$
$$\tfrac{1}{2}(1.0 \text{ m/s}^2)t^2 + 11 \text{ m} = (5.0 \text{ m/s})t$$

or

$$(0.50 \text{ m/s}^2)t^2 - (5.0 \text{ m/s})t + 11 \text{ m} = 0$$

Solving this quadratic equation for t by using the quadratic formula (see Appendix C) reveals that there are two solutions:

$$\boxed{t = 3.3 \text{ s} \quad \text{or} \quad t = 6.7 \text{ s}}$$

Evidently there are two times when the woman can catch the bus. When she catches the bus for the first time at $t = 3.3$ s, she has a *different* velocity than the bus. In fact, the woman has a greater velocity, because the bus started from rest and in 3.3 s has not developed much velocity. If she fails to get the attention of the driver when she reaches the bus, she will actually run past it, since she has the greater velocity. Soon, however, the bus will catch up with her, because it is accelerating while she is maintaining a constant velocity. The two will meet for the second time when $t = 6.7$ s. If the driver still does not notice her, the bus will pull ahead of the woman and leave her behind forever.

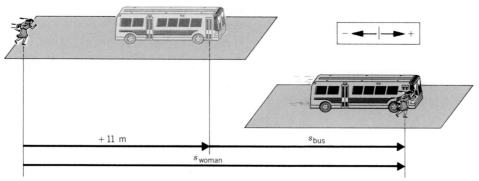

Figure 2.12 A person attempting to catch a bus.

2.6 FREELY FALLING BODIES

FREELY FALLING BODIES AND THE EQUATIONS OF KINEMATICS

Everyone has observed the effect of gravity as it causes objects to fall downward. In the absence of air resistance, it is found that all bodies at the same location above the earth fall vertically with the same acceleration. Furthermore, if the distance of the fall is small compared to the radius of the earth, the acceleration remains essentially constant throughout the fall. This idealized motion, in which air resistance is neglected and the acceleration is nearly constant, is known as *free-fall*. Since the acceleration is constant in free-fall, the equations of kinematics can be used.

The acceleration of a freely falling body is called the *acceleration due to gravity*, and its magnitude is denoted by the symbol g. The acceleration due to gravity is directed downward, toward the center of the earth. Near the earth's surface g is approximately

$$g = 9.80 \text{ m/s}^2 \quad \text{or} \quad 32.2 \text{ ft/s}^2$$

Unless circumstances warrant otherwise, we will use either of these values for g in subsequent calculations. In reality, however, g decreases with increasing altitude and varies slightly with latitude.

Figure 2.13a shows the well-known phenomenon of a rock falling faster than a sheet of paper. The effect of air resistance is responsible for the slower fall of the paper. It is *not* true that gravity makes heavy objects fall faster than light objects. Figure 2.13b, in contrast, illustrates free-fall motion. When the air is removed from the tube, the rock and paper fall with exactly the same acceleration. Free-fall is closely approximated for objects falling near the surface of the moon, where there is no air to retard the motion. A nice demonstration of lunar free-fall was performed in 1971 by astronaut David Scott who dropped a hammer and a feather simultaneously from the same height. Both experienced the same acceleration due to lunar gravity and consequently hit the ground at the same time. The acceleration due to gravity near the surface of the moon is approximately one-sixth as large as that on the earth.

We now turn our attention to several examples that illustrate how the equations of kinematics are applied to freely falling bodies.

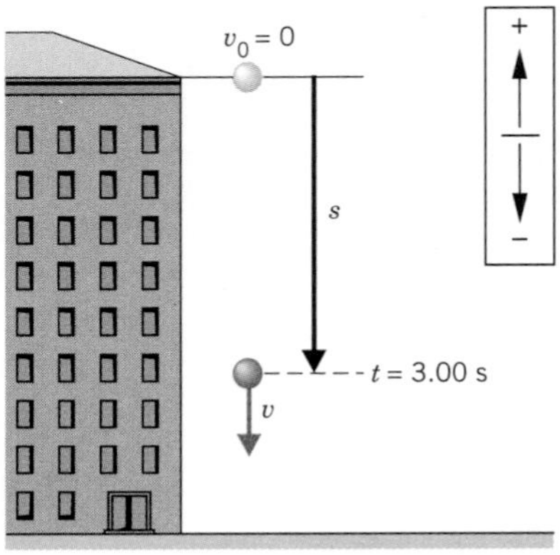

Figure 2.13 (a) In the presence of air resistance, the acceleration of the rock is greater than that of the paper. (b) In the absence of air resistance, both the rock and the paper have the same acceleration.

Figure 2.14 The stone, starting with zero velocity at the top of the building, is accelerated downward by gravity.

Example 11   A Falling Stone

A stone is dropped from rest from the top of a tall building, as Figure 2.14 indicates. After 3.00 s of free-fall, what is the displacement s of the stone?

REASONING   The upward direction is chosen as the positive direction. The three known variables are shown in the box below. The initial velocity v_0 of the stone is zero, because the stone is dropped from rest. The acceleration due to gravity is negative, since it points downward in the negative direction.

Stone Data

s	a	v	v_0	t
?	-9.80 m/s^2		0	3.00 s

Equation 2.8 contains the appropriate variables and offers a direct solution to the problem.

SOLUTION   Using Equation 2.8, we find that

$$s = v_0t + \tfrac{1}{2}at^2 = \tfrac{1}{2}(-9.80 \text{ m/s}^2)(3.00 \text{ s})^2 = \boxed{-44.1 \text{ m}}$$

The minus sign in the answer for s indicates that the displacement vector points in the negative direction, toward the earth.

Example 12   *The Velocity of a Falling Stone*

After 3.00 s of free-fall, what is the velocity of the stone in Figure 2.14?

REASONING   Because of the acceleration due to gravity, the magnitude of the stone's downward velocity increases by 9.80 m/s during each second of free-fall. The data for the stone are the same as in Example 11, and Equation 2.4 offers a direct solution for the final velocity.

SOLUTION   Using Equation 2.4, we obtain

$$v = v_0 + at = (-9.80 \text{ m/s}^2)(3.00 \text{ s}) = \boxed{-29.4 \text{ m/s}}$$

The velocity is negative, indicating that the stone is moving downward.

Examples 11 and 12 reveal an interesting feature of uniformly accelerated motion when a body starts from rest. Equations 2.4 and 2.8 predict that the velocity and displacement at any time during the motion are

$$\left.\begin{array}{l} v = at \\ s = \tfrac{1}{2}at^2 \end{array}\right\} \text{ if } v_0 = 0$$

It is clear from these equations that v is proportional to t, while s is proportional to t^2. For example, if the time is tripled, the velocity is also tripled, but the displacement is increased by a factor of nine. Figure 2.15 shows the velocity and displace-

The cliff diver accelerates downward because of gravity.

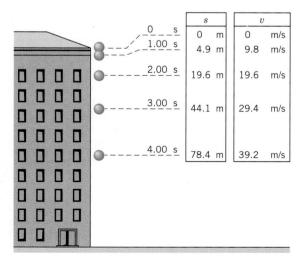

	s	v
0 s	0 m	0 m/s
1.00 s	4.9 m	9.8 m/s
2.00 s	19.6 m	19.6 m/s
3.00 s	44.1 m	29.4 m/s
4.00 s	78.4 m	39.2 m/s

Figure 2.15   The displacement s and velocity v of a freely falling body are illustrated for the first four seconds of fall. The body is dropped from rest.

ment for the first 4 seconds of free-fall, starting from rest. Notice that the object is traveling at a speed of 29.4 m/s (65.7 mi/h) after only 3 s! Very few cars can accelerate from 0 to 29.4 m/s in 3 s, a fact that shows how large the acceleration due to gravity really is.

The acceleration due to gravity is always a downward-pointing vector and acts to increase the speed of an object falling downward. This same acceleration also acts to decrease the speed of an object moving upward, eventually bringing it to a momentary halt and causing it to fall back to earth. Examples 13 and 14 show how the equations of kinematics are applied to the latter situation.

Example 13   How High Does It Go?

A ball is thrown vertically upward with an initial speed of 30.0 m/s. In the absence of air resistance, how high does the ball go?

REASONING   The ball is given an upward initial velocity, as in Figure 2.16. But the acceleration due to gravity points downward. Since the velocity and acceleration point in opposite directions, the ball slows down as it moves upward. Eventually, the velocity of the ball becomes $v = 0$ at the highest point. Assuming that the upward direction is positive, the data can be summarized as shown below:

Ball Data

s	a	v	v_0	t
?	-9.80 m/s^2	0	$+30.0 \text{ m/s}$	

With these data, we can use Equation 2.9 ($v^2 = v_0^2 + 2as$) to find the maximum height s.

SOLUTION   Rearranging Equation 2.9, we find that the maximum height is

$$s = \frac{v^2 - v_0^2}{2a} = \frac{-(30.0 \text{ m/s})^2}{2(-9.80 \text{ m/s}^2)} = \boxed{+45.9 \text{ m}}$$

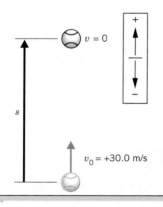

Figure 2.16   The ball is thrown upward with an initial velocity of $v_0 = +30.0$ m/s. The velocity of the ball is momentarily zero when the ball reaches its maximum height.

Example 14   How Long Is It in the Air?

In Figure 2.16, what is the total time the ball is in the air before returning to the ground?

REASONING: METHOD 1   During the time the ball travels upward, gravity causes the speed of the ball to decrease to zero. On the way down, gravity causes the ball to regain the lost speed. Thus, the time for the ball to go up is equal to the time for it to come down. In other words, the total time in the air is twice the time for the upward motion. The data for the ball during the upward trip are the same as in Example 13. With these data, we can use Equation 2.4 ($v = v_0 + at$) to find the upward travel time.

SOLUTION: METHOD 1   Rearranging Equation 2.4, we find that

$$t = \frac{v - v_0}{a} = \frac{-30.0 \text{ m/s}}{-9.80 \text{ m/s}^2} = 3.06 \text{ s}$$

The total flight time is twice this value, or $\boxed{6.12 \text{ s}}$.

REASONING: METHOD 2   It is possible to determine the total time by another method—a method that is not so obvious. When the ball is thrown into the air and returns, the displacement for the *entire trip* is $s = 0$, because the ball returns to its original position. Substituting the data below into Equation 2.8, we can find the time for the entire trip:

Ball Data for Entire Trip

s	a	v	v_0	t
0	-9.80 m/s^2		$+30.0 \text{ m/s}$	?

SOLUTION: METHOD 2   Using Equation 2.8, we obtain

$$s = v_0 t + \tfrac{1}{2}at^2 = (v_0 + \tfrac{1}{2}at)(t)$$

$$0 = [30.0 \text{ m/s} + \tfrac{1}{2}(-9.80 \text{ m/s}^2)t](t)$$

There are two solutions to this equation: $t = 0$ and $t = 6.12$ s. The solution $t = 0$ corresponds to the situation where the ball has not left the ground, so its displacement is zero. The solution $\boxed{t = 6.12 \text{ s}}$ gives the time for a complete up-and-down trip, the displacement also being zero.

Examples 13 and 14 illustrate that the expression "freely falling" does not necessarily mean an object is falling down. A freely falling object is any object moving freely, either upward or downward, under the influence of gravity. In either case, the object always experiences the same *downward acceleration* due to gravity.

SYMMETRY IN THE MOTION OF FREELY FALLING BODIES

The motion of an object that is thrown upward and eventually returns to earth contains a symmetry that is useful to keep in mind from the point of view of problem solving. The calculations just completed indicate that a time symmetry exists in free-fall motion, in the sense that the time required for the object to reach maximum height equals the time for it to return to its starting point.

Another type of symmetry involving the speed also exists. Figure 2.17 shows the ball considered in Examples 13 and 14. At any displacement s above the ground, the speed of the ball during the upward trip equals the speed at the same point during the downward trip. For instance, when $s = +28.0$ m, Equation 2.9 gives two possible values for v, assuming that the initial velocity of the ball is $v_0 = +30.0$ m/s:

$$v^2 = v_0^2 + 2as = (30.0 \text{ m/s})^2 + 2(-9.80 \text{ m/s}^2)(28.0 \text{ m}) = 351 \text{ m}^2/\text{s}^2$$

$$v = \pm 18.7 \text{ m/s}$$

The value $v = \pm 18.7$ m/s is the velocity of the ball on the upward trip, while $v = -18.7$ m/s is the velocity on the downward trip. The speed in both cases is identical and equals 18.7 m/s. Likewise, the speed of the ball just before the ball strikes the ground is equal to its initial speed of 30.0 m/s. This symmetry involving the speed arises because the ball loses 9.80 m/s in speed each second on the way up and gains back the same amount each second on the way down. Thus, the upward speed of the ball equals the downward speed at a given height above the ground.

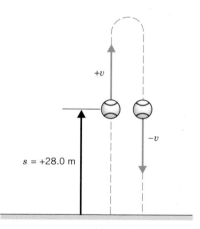

Figure 2.17   For a given displacement along the ball's path, the upward speed of the ball is equal to its downward speed, but the two velocities point in opposite directions.

PROBLEM SOLVING INSIGHT

For an object falling freely near the earth's surface, the acceleration due to gravity is $a = -9.80 \text{ m/s}^2$ (assuming downward to be the negative direction), whether the object is moving upward or downward.

2.7 GRAPHICAL ANALYSIS OF VELOCITY AND ACCELERATION FOR LINEAR MOTION

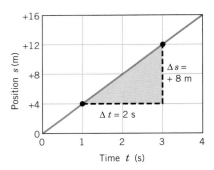

Figure 2.18 A graph of position vs. time for an object moving with a constant velocity of $v = \Delta s/\Delta t = +4$ m/s.

Graphical techniques are helpful in understanding the concepts of velocity and acceleration. Suppose a bicyclist is riding with a constant velocity of $v = +4$ m/s. The position s of the bicycle can be plotted along the vertical axis of a graph, while the time t is plotted along the horizontal axis. Since the position of the bike increases by 4 m every second, the graph of s versus t is a straight line. Furthermore, if the bike is assumed to be at $s = 0$ when $t = 0$, the straight line passes through the origin, as Figure 2.18 shows. Each point on this line gives the position of the bike at a particular time. For instance, at $t = 1$ s the position is 4 m, while at $t = 3$ s the position is 12 m.

In constructing the graph in Figure 2.18, we used the fact that the velocity was $+4$ m/s. Suppose, however, that we were given this graph, but did not have any prior knowledge of the velocity. The velocity could be determined by considering what happens to the bike between the times of 1 and 3 s, for instance. The change in time is $\Delta t = 2$ s. During this time interval, the position of the bike changes from $+4$ to $+12$ m, and the change in position is $\Delta s = +8$ m. The ratio $\Delta s/\Delta t$ is called the *slope* of the straight line:

$$\text{Slope} = \frac{\Delta s}{\Delta t} = \frac{+8 \text{ m}}{2 \text{ s}} = +4 \text{ m/s}$$

Notice that the slope is equal to the velocity of the bike. This result is no accident, because $\Delta s/\Delta t$ is the definition of average velocity (see Equation 2.2). Thus, for an object moving with a constant velocity, the slope of the straight line in a position–time graph gives the value of the velocity. Since the position–time graph is a straight line, any time interval Δt can be chosen to calculate the velocity. Choosing a different Δt will yield a different Δs, but the velocity $\Delta s/\Delta t$ will not change. In the real world, objects rarely move with a constant velocity at all times, as the next example illustrates.

Example 15   A Bicycle Trip

A bicyclist maintains a constant velocity on the outgoing leg of a journey, zero velocity while stopped for lunch, and another constant velocity on the way back. Figure 2.19 shows the position–time graph for such a trip. Using the time and position intervals indicated in the drawing, obtain the velocities for each segment of the trip.

REASONING AND SOLUTION   Using Equation 2.2, we find the following velocities:

Segment 1
$$\bar{v} = \frac{\Delta s}{\Delta t} = \frac{+400 \text{ m}}{200 \text{ s}} = +2 \text{ m/s}$$

Segment 2
$$\bar{v} = \frac{\Delta s}{\Delta t} = \frac{0 \text{ m}}{400 \text{ s}} = 0 \text{ m/s}$$

Segment 3
$$\bar{v} = \frac{\Delta s}{\Delta t} = \frac{-400 \text{ m}}{400 \text{ s}} = -1 \text{ m/s}$$

In the second segment of the journey the velocity is zero, reflecting the fact that the bike is stationary. Since the position of the bike does not change, segment 2 is a horizontal line that has a zero slope. In the third part of the motion the velocity is negative, because the position of the bike decreases from $s = +800$ m to $s = +400$ m during the 400-s interval shown in the graph. As a result, segment 3 has a negative slope, and the velocity is negative.

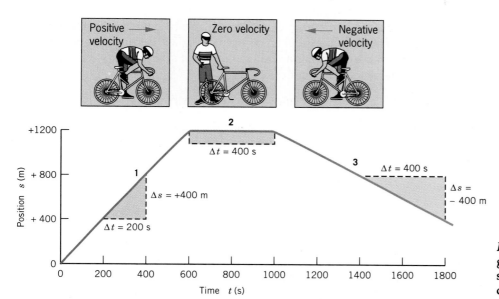

Figure 2.19 This position vs. time graph consists of three straight line segments, each corresponding to a different constant velocity.

If the object is accelerating, its velocity is changing. When the velocity is changing, the s versus t graph is not a straight line, but is a curve, perhaps like that in Figure 2.20. This curve was drawn using Equation 2.8 ($s = v_0 t + \frac{1}{2}at^2$), assuming an acceleration of $a = 0.26$ m/s^2 and an initial velocity of $v_0 = 0$. The velocity at any instant of time can be determined by measuring the slope of the curve at that instant. The slope at any point along the curve is defined to be the slope of the tangent line drawn to the curve at that point. In Figure 2.20, a tangent line is drawn at $t = 20.0$ s. To determine the slope of the tangent line, a triangle is constructed using an arbitrarily chosen time interval of $\Delta t = 5.0$ s. The change in

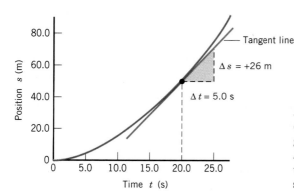

Figure 2.20 When the velocity is changing, the position vs. time graph is a curved line. The slope $\Delta s/\Delta t$ of the tangent line drawn to the curve at a given time is the instantaneous velocity at that time.

s associated with this time interval can be read from the tangent line as $\Delta s = +26$ m. Therefore,

$$\text{Slope of tangent line} = \frac{\Delta s}{\Delta t} = \frac{+26 \text{ m}}{5.0 \text{ s}} = +5.2 \text{ m/s}$$

The slope of the tangent line is the instantaneous velocity, which in this case is $v = +5.2$ m/s. This graphical result can be verified by using Equation 2.4 with $v_0 = 0$: $v = at = (+0.26 \text{ m/s}^2)(20.0 \text{ s}) = +5.2$ m/s.

Insight into the meaning of acceleration can also be gained with the aid of a graphical representation. Consider an object moving with a constant acceleration of $a = +6$ m/s². If the object has an initial velocity of $v_0 = +5$ m/s, its velocity at any time is represented by Equation 2.4 as

$$v = v_0 + at = (5 \text{ m/s}) + (6 \text{ m/s}^2)t$$

This relation is plotted as the velocity versus time graph in Figure 2.21. The graph of v versus t is a straight line that intercepts the vertical axis at $v_0 = 5$ m/s. The slope of this straight line can be calculated from the data shown in the drawing:

$$\text{Slope} = \frac{\Delta v}{\Delta t} = \frac{+12 \text{ m/s}}{2 \text{ s}} = +6 \text{ m/s}^2$$

The ratio $\Delta v/\Delta t$ is, by definition, equal to the average acceleration (Equation 2.4), so the slope of the straight line in a velocity–time graph is the average acceleration.

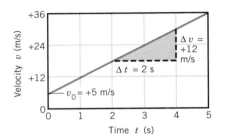

Figure 2.21 A velocity vs. time graph that represents an object with an acceleration of $\Delta v/\Delta t = +6$ m/s². The initial velocity is $v_0 = +5$ m/s when $t = 0$.

SUMMARY

Displacement is a vector that points from an object's initial position to its final position. The magnitude of the displacement is the shortest distance between the two positions.

The **average speed** of an object is the distance traveled by the object divided by the time required to cover the distance: Average speed = (Distance)/(Elapsed time).

The **average velocity** $\overline{v}$ of an object is defined as the object's displacement $\Delta \mathbf{s}$ divided by the elapsed time Δt: $\overline{\mathbf{v}} = \Delta \mathbf{s}/\Delta t$. Average velocity is a vector that has the same direction as the displacement. When the elapsed time Δt is infinitesimally small, the average velocity becomes equal to the **instantaneous velocity v,** the velocity at an instant of time.

Average acceleration $\overline{\mathbf{a}}$ is a vector that is equal to the change in the velocity $\Delta \mathbf{v}$ divided by the elapsed time Δt, the "change in velocity" being the difference between the final and initial velocities: $\overline{\mathbf{a}} = \Delta \mathbf{v}/\Delta t$. When Δt is infinitesimally small, the average acceleration be-

comes equal to the **instantaneous acceleration a.** Acceleration is the rate at which the velocity is changing.

When an object moves with a constant acceleration along a straight line, its displacement s, final velocity v, initial velocity v_0, acceleration a, and the elapsed time t are related by the following equations, assuming that $s = 0$ at $t = 0$:

$$s = \overline{v}t = \tfrac{1}{2}(v_0 + v)t \quad \text{and} \quad v = v_0 + at$$

These two equations can be combined algebraically to give two additional equations that are listed in Table 2.1. The equations in this table are known as the **equations of kinematics.**

In **free-fall** motion, an object experiences a constant acceleration due to gravity and negligible air resistance. All objects at the same location above the earth have the same acceleration due to gravity. The acceleration due to gravity is directed toward the center of the earth and has a magnitude of approximately 9.80 m/s² near the earth's surface.

SOLVED PROBLEMS

Solved Problem 1 A Bus Trip

Related Problems: *7 **8 *59 *61 *64

During a trip, a bus travels 11 km with an average velocity of 21 m/s, but then travels in the same direction for the next 1.0 km at a smaller average velocity of 4.2 m/s, due to the presence of highway construction crews (see the drawing). Determine the average velocity of the bus for the entire trip.

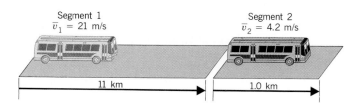

Segment 1
$\bar{v}_1 = 21$ m/s

Segment 2
$\bar{v}_2 = 4.2$ m/s

11 km

1.0 km

REASONING It is important to realize that the average velocity of the bus is *not* obtained by simply adding the two velocities and dividing the sum by 2. This method is incorrect, because it does not take into account that most of the trip takes place at the greater velocity, while only a small part takes place at the smaller velocity. The correct procedure for calculating the average velocity is to use Equation 2.2:

$$\bar{v} = \frac{\text{Displacement}}{\text{Elapsed time}}$$

where the values for the displacement and elapsed time must be those for the entire trip.

SOLUTION The displacement for the entire trip is 12 km. The elapsed time is the sum of two times, t_1 and t_2. The time t_1 for the bus to travel 11 km at an average velocity of 21 m/s is

$$t_1 = \frac{11 \times 10^3 \text{ m}}{21 \text{ m/s}} = 520 \text{ s}$$

Likewise, the time t_2 for the bus to travel the remaining 1.0 km at a velocity of 4.2 m/s is

$$t_2 = \frac{1.0 \times 10^3 \text{ m}}{4.2 \text{ m/s}} = 240 \text{ s}$$

Hence, the elapsed time for the trip is $t_1 + t_2 = 760$ s. The average velocity for the entire trip is, then,

$$\bar{v} = \frac{12 \times 10^3 \text{ m}}{760 \text{ s}} = \boxed{16 \text{ m/s}}$$

A calculator gives $\bar{v} = 15.789$ m/s, but this value must be rounded to $\bar{v} = 16$ m/s, since the data are accurate to only two significant figures.

SUMMARY OF IMPORTANT POINTS The average velocity of a segmented trip is obtained by dividing the total displacement of the trip (which equals the vector sum of the displacements for each segment) by the total time of the trip (which equals the sum of the times for each segment). The average velocity *cannot* be calculated by adding the velocities for the individual segments and dividing the sum by the number of segments.

Solved Problem 2 A Motorcycle Ride

Related Problems: *25 *26 *43

A motorcycle, starting from rest, has an acceleration of $+2.6$ m/s². After the motorcycle has traveled a distance of 120 m, it slows down with an acceleration of -1.5 m/s² until its velocity is $+12$ m/s (see the drawing). What is the total displacement of the motorcycle?

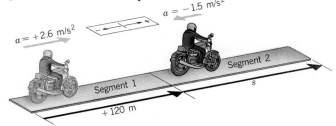

$a = -1.5$ m/s²

$a = +2.6$ m/s²

Segment 2

Segment 1

s

$+120$ m

REASONING The total displacement is the sum of the displacements for the first ("speeding up") and second ("slowing down") segments. The displacement for the first segment is $+120$ m. The displacement for the second segment can be found if the initial velocity for this segment can be determined, since values for two other variables are already known. The key point here is that the initial velocity for the second segment is the final velocity of the first segment.

SOLUTION We can determine the final velocity of the first segment from the given data, recognizing that the motorcycle starts from rest ($v_0 = 0$):

Segment 1 Data

s	a	v	v_0	t
$+120$ m	$+2.6$ m/s^2	?	0	

The final velocity v of the first segment can be calculated from Equation 2.9 ($v^2 = v_0^2 + 2as$):

$$v = \sqrt{v_0^2 + 2as} = \sqrt{2(2.6 \text{ m/s}^2)(120 \text{ m})} = +25 \text{ m/s}$$

We can now use $v = +25$ m/s as the initial velocity for the second segment, along with the remaining data listed below:

Segment 2 Data

s	a	v	v_0	t
?	-1.5 m/s^2	$+12$ m/s	$+25$ m/s	

The displacement can be determined for segment 2 by solving $v^2 = v_0^2 + 2as$ for s:

$$s = \frac{v^2 - v_0^2}{2a} = \frac{(12 \text{ m/s})^2 - (25 \text{ m/s})^2}{2(-1.5 \text{ m/s}^2)} = +160 \text{ m}$$

The total displacement of the motorcycle is 120 m + 160 m = $\boxed{280 \text{ m}}$.

SUMMARY OF IMPORTANT POINTS Often the motion of an object is divided into "segments," each with a different acceleration. When solving such problems, it is important to realize that the final velocity for one segment becomes the initial velocity for the next segment.

QUESTIONS

1. A honeybee leaves the hive and travels 2 km before returning. Is the displacement for the trip the same as the distance traveled? If not, why not?

2. Two buses depart from Chicago, one going to New York and one to San Francisco. Each bus travels at a speed of 30 m/s. Do they have equal velocities? Explain.

3. Often, traffic lights are timed so that if you travel at a certain constant speed, you can avoid all red lights. Discuss how the timing of the lights is determined, considering that the distance between them varies from one light to the next.

4. Give an example from your own experience in which the velocity of an object is zero for just an instant of time, but its acceleration is not zero.

5. At a given instant of time, a car and a truck are traveling side by side in adjacent lanes of a highway. The car has a greater velocity than the truck. Does the car necessarily have a greater acceleration? Explain.

6. The average velocity for a trip has a positive value. Is it possible for the instantaneous velocity at any point during the trip to have a negative value? Justify your answer.

7. An experimental vehicle slows down and comes to a halt with an acceleration whose magnitude is 9.80 m/s^2. After re-versing direction in a negligible amount of time, the vehicle speeds up with an acceleration of 9.80 m/s^2. Other than being horizontal, how is this motion different, if at all, from the motion of a ball that is thrown straight upward, comes to a halt, and falls back to earth?

8. A person standing on a bridge fires a rifle bullet straight up, and then fires another bullet straight down. Neglecting air resistance, which bullet, if either, strikes the water with a greater velocity? Provide a reason for your answer.

9. A ball thrown into the air reaches a maximum height and begins to fall down. Does the acceleration of gravity change, either in magnitude or in direction, when the velocity of the ball changes direction? Explain.

10. For each equation below, show that the units of the left side are the same as the units of each term on the right side:

$$s = v_0 t + \tfrac{1}{2}at^2 \quad \text{and} \quad v^2 = v_0^2 + 2as$$

11. The muzzle velocity of a gun is the velocity of the bullet when it leaves the barrel. The muzzle velocity of one rifle with a short barrel is greater than the muzzle velocity of another rifle that has a longer barrel. In which rifle is the acceleration of the bullet larger? Explain your reasoning.

PROBLEMS

Section 2.2 Speed and Velocity

1. Sound travels at a constant speed of 343 m/s in air. Approximately how much time (in seconds) does it take for the sound of thunder to travel 1609 m (one mile)?

2. A car is traveling at a constant speed of 27 m/s. The driver looks away from the road for 2.0 s to tune in a station on the radio. How far does the car go during this time?

3. In 1985, Said Aouita set the world record for the 1500-m race in a time of 3 minutes, 29.46 seconds. What was his average speed?

4. A plane is sitting on a runway, awaiting takeoff. On an adjacent parallel runway, another plane lands and passes the stationary plane at a speed of 45 m/s. The arriving plane has a length of 36 m. By looking out of a window (very narrow), a passenger on the stationary plane can see the moving plane. For how long a time is the moving plane visible?

5. An 18-year-old runner can complete a 10.0-km course with an average speed of 4.38 m/s. A 50-year-old runner can cover the same distance with an average speed of 4.27 m/s. How much later should the younger runner start in order to finish the course *at the same time* as the older runner?

*6. The three-toed sloth is the slowest moving land mammal. On the ground, the sloth moves at an average speed of 0.037 m/s, considerably slower than the giant tortoise, which walks at 0.076 m/s. After 12 minutes of walking, how much further would the tortoise have gone relative to the sloth?

*7. A sky diver, with parachute unopened, falls 625 m in 15.0 s. Then she opens her parachute and falls another 356 m in 142 s. What is her average velocity (both magnitude and direction) for the entire fall? *(See Solved Problem 1 for a related problem.)*

**8. A car makes a 60.0-km trip with an average velocity of 40.0 km/h in a direction due north. The trip consists of three parts. The car moves with a constant velocity of 25 km/h due north for the first 15 km, and 62 km/h due north for the next 32 km. With what constant velocity does the car travel for the last 13-km segment of the trip? *(See Solved Problem 1 for a related problem.)*

Section 2.3 Acceleration

9. An airplane, starting from rest, moves south and attains a lift-off speed of 56.9 m/s in 18.0 s. What is the magnitude and direction of its average acceleration?

10. A sprinter explodes out of the starting block with an acceleration of 2.3 m/s², which she sustains for 2.0 s. Then, her acceleration drops to zero for the rest of the race. What is her speed at (a) $t = 2.0$ s and (b) the end of the race?

11. If a sports car can go from rest to 27 m/s in 9.0 s, what is the magnitude of its average acceleration?

12. The velocity of a train is +26.4 m/s. At an average acceleration of −1.50 m/s², how much time is required for the train to decrease its velocity to +9.72 m/s?

13. A runner accelerates to a velocity of 5.36 m/s due west in 3.00 s. His average acceleration is 0.640 m/s², also directed due west. What was his velocity when he began accelerating?

*14. A particle is moving with an initial velocity of +112 m/s at $t = 0$. The particle has no acceleration until $t = 3.0$ s, after which it has an acceleration of −4.0 m/s². What is the velocity at $t = 16.0$ s?

**15. Two motorcycles are traveling due east with different velocities. However, four seconds later, they have the same velocity. During this four-second interval, motorcycle A has an average acceleration of 2.0 m/s² due east, while motorcycle B has an average acceleration of 4.0 m/s² due east. By how much did the speeds *differ* at the beginning of the four-second interval, and which motorcycle was moving faster?

Section 2.4 Equations of Kinematics for Constant Acceleration, Section 2.5 Applications of the Equations of Kinematics

16. A soccer player runs at a constant speed of 2.6 m/s. For the next 18 m, he speeds up with an acceleration of 0.45 m/s². What is his speed at the end of the run?

17. A cheetah, the fastest of all land animals over a short distance, accelerates from rest to 26 m/s. Assuming that the acceleration is constant, find the average speed of the cheetah.

18. In getting ready to slam-dunk the ball, a basketball player starts from rest and sprints to a speed of 6.0 m/s in 1.5 s. Assuming that the player accelerates uniformly, determine the distance he runs.

19. A jetliner, traveling northward, is landing with a speed of 69 m/s. Once the jet touches down, it has 750 m of runway in which to reduce its speed to 6.1 m/s. Compute the average acceleration (magnitude and direction) of the plane during landing.

20. A skier, starting from rest, accelerates down a slope at 1.6 m/s². How far has she gone at the end of 5.0 seconds?

21. A truck, traveling at a velocity of 33 m/s due east, comes to a halt by decelerating at 11 m/s². How far does the truck travel in the process of stopping?

22. The length of the barrel of a primitive blowgun is 1.2 m. Upon leaving the barrel, a dart has a speed of 14 m/s. Assuming that the dart is uniformly accelerated, how long does it take for the dart to travel the length of the barrel?

23. With the plane standing on the runway, the pilot brings the engines to full thrust before releasing the brakes. The aircraft accelerates at 2.9 m/s² and reaches a takeoff speed of 58 m/s. (a) Find the time from rest to takeoff. (b) Determine the displacement of the plane.

***24.** A drag racer, starting from rest, speeds up for 402 m with an acceleration of +17.0 m/s². A parachute then opens, slowing the car down with an acceleration of −6.10 m/s². How fast is the racer moving 3.50 × 10² m after the parachute opens?

***25.** Suppose a car is traveling at 12.0 m/s, and the driver sees a traffic light turn red. After 0.510 s has elapsed (the reaction time), the driver applies the brakes, and the car decelerates at 6.20 m/s². What is the stopping distance of the car, as measured from the point where the driver first notices the red light? *(See Solved Problem 2 for a related problem.)*

***26.** A speedboat starts from rest and accelerates at +2.01 m/s² for 7.00 s. At the end of this time, the boat continues for an additional 6.00 s with an acceleration of +0.518 m/s². Following this, the boat accelerates at −1.49 m/s² for 8.00 s. (a) What is the velocity of the boat at $t = 21.0$ s? (b) Find the total displacement of the boat. *(See Solved Problem 2 for a related problem.)*

***27.** A race driver has made a pit stop to refuel. After refueling, he leaves the pit area with an acceleration whose magnitude is 6.0 m/s², and after 4.0 s he enters the main speedway. At the same instant, another race car that is on the speedway and traveling at a constant speed of 70.0 m/s overtakes and passes the entering car. If the entering car maintains its acceleration, how much time is required for it to catch the other car?

***28.** Rederive the equations of kinematics presented in Table 2.1, assuming the initial position is s_0, instead of $s_0 = 0$ as was used in the text.

****29.** A locomotive is accelerating at 1.6 m/s². It passes through a 20.0-m-wide crossing in a time of 2.4 s. After the locomotive leaves the crossing, how much time is required until its speed reaches 32 m/s?

****30.** In the one-hundred-meter dash a sprinter accelerates from rest to a top speed with an acceleration whose magnitude is 2.68 m/s². After achieving top speed, he runs the remainder of the race without speeding up or slowing down. If the total race is run in 12.0 s, how far does he run during the acceleration phase?

Section 2.6 Freely Falling Bodies

31. A penny is dropped from rest from the top of the Sears Tower in Chicago. Considering that the height of the building is 427 m and ignoring air resistance, find the speed of the penny when the penny strikes the ground.

32. From the top of a cliff, a person uses a slingshot to fire a pebble straight downward with an initial speed of 9.0 m/s. After 0.50 s, how far beneath the cliff-top is the pebble?

33. A baseball is thrown upward with an initial speed of 35.0 m/s. What is its speed at $t = 2.00$ s?

34. An arrow is fired straight upward with an initial speed of 15 m/s. How long is the arrow in the air before it strikes the ground?

35. A golf ball rebounds from the floor and travels straight upward with a speed of 5.0 m/s. To what maximum height does the ball rise?

36. Suppose you are visiting a planet in a distant part of the galaxy. To determine the acceleration due to gravity on this planet, you drop a rock from a height of 55 m. The rock strikes the ground 1.9 s later. How many times greater is the acceleration due to gravity on this planet than that on earth?

37. A rifle bullet is shot vertically upward. Twenty-three seconds later the bullet has a velocity of 72.0 m/s, downward. What is the velocity of the bullet when the bullet leaves the rifle?

38. With what initial speed must an arrow be fired straight upward to attain a height of 110 m in 5.4 s?

39. Suppose a ball is thrown vertically upward. Eight seconds later it returns to its point of release. What is the initial velocity of the ball?

40. A diver springs upward with an initial speed of 1.8 m/s from a 3.0-m board. (a) Find the velocity with which he strikes the water. *[Hint: When the diver reaches the water, his displacement is s = −3.0 m (measured from the board), assuming the downward direction is chosen as the negative direction.]* (b) What is the highest point he reaches above the water?

***41.** Four-tenths of a second after bouncing on a trampoline, a gymnast is moving upward with a speed of 6.0 m/s. To what height above the trampoline does the gymnast rise before falling back down?

***42.** A cement block accidentally falls from rest from the ledge of a 53.0-m-high building. When the block is 14.0 m above the ground, a man, 2.00 m tall, looks up and notices that the block is directly above him. How much time, at most, does the man have to get out of the way?

***43.** A rocket is launched from rest with an acceleration of 20.0 m/s², upward. At an altitude of 415 m the engines shut off, but the rocket continues to coast upward. Find the total time that the rocket is in the air, from lift-off until it strikes the ground. *(See Solved Problem 2 for a related problem.)*

*44. A log is floating on swiftly moving water. A stone is dropped from rest from a 75-m-high bridge and lands on the log as it passes under the bridge. If the log moves with a constant speed of 5.0 m/s, what is the horizontal distance between the log and the bridge when the stone is released?

*45. A spelunker (cave explorer) drops a stone from rest into a hole. The speed of sound is 343 m/s in air, and the sound of the stone striking the bottom is heard 1.50 s after the stone is dropped. How deep is the hole?

*46. A roof tile falls from rest from the top of a building. An observer inside the building notices that it takes 0.20 s for the tile to pass her window, whose height is 1.6 m. How far above the top of this window is the roof?

**47. A ball is thrown upward from the top of a 25.0-m-tall building. The ball's initial speed is 12.0 m/s. At the same instant, a person is running on the ground at a distance of 31.0 m from the building. What must be the average speed of the person if he is to catch the ball at the bottom of the building?

Section 2.7 Graphical Analysis of Velocity and Acceleration for Linear Motion

48. A bus makes a trip according to the position–time graph shown in the illustration. What is the average velocity (magnitude and direction) of the bus during each of the segments labeled A, B, and C?

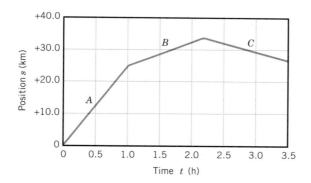

49. A snowmobile moves according to the velocity–time graph shown in the drawing. What is the snowmobile's average acceleration during each of the segments A, B, and C?

50. A person who walks for exercise produces the position–time graph given with this problem. (a) Without doing any calculations, decide which segments of the graph (A, B, C, or D) indicate positive, negative, and zero average velocities. (b) Calculate the average velocity for each segment to verify your answers to part (a).

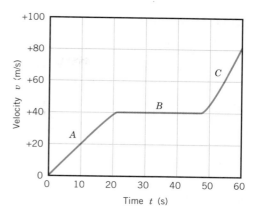

Problem 49

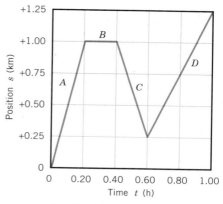

Problem 50

*51. Using the position–time graph that accompanies this problem, draw the corresponding velocity–time graph.

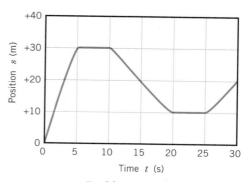

Problem 51

ADDITIONAL PROBLEMS

52. A bicycle racer is moving at a speed of 14 m/s. To pass another cyclist, the racer speeds up with an acceleration of 1.2 m/s². At the end of 6.0 s, how fast is the racer moving?

53. A jogger accelerates from rest to 3.0 m/s in 2.0 s. A car accelerates from 38 to 41 m/s also in 2.0 s. (a) Find the acceleration (magnitude only) of the jogger. (b) Determine the acceleration (magnitude only) of the car. (c) Does the car travel further than the jogger during the 2.0 s? If so, how much further?

54. Two runners in a 1609-m race (one mile) finish with times of 3:52.46 (3 minutes and 52.46 seconds) and 3:52.72. Assuming that both run at their average speeds during the entire time, what distance separates them at the end of the race?

55. From her bedroom window a girl drops a water-filled balloon to the ground, 6.0 m below. If the balloon is released from rest, how long is it in the air?

56. A motorcycle has a constant acceleration of 2.5 m/s². Both the velocity and acceleration of the motorcycle point in the same direction. How much time is required for the motorcycle to change its speed (a) from 21 m/s to 31 m/s, and (b) from 51 to 61 m/s?

57. En route to a Hawaiian vacation, a traveler arrives late at the airport at 1:08 pm. His plane is scheduled to depart at 1:22 pm. To catch the flight, he must run 2.1 km to the gate. What must be his minimum average running speed (in m/s)?

58. An arrow is shot straight up with an initial speed of 50.0 m/s. After reaching its maximum height, the arrow starts down. On the descent, a slight breeze blows the arrow laterally, and it strikes a tree limb that is 30.0 m above the ground. (a) Determine the velocity of the arrow just before the arrow strikes the limb. (b) Find the time the arrow is in the air.

*59. A bicyclist makes a trip that consists of three parts, each in the same direction (due north) along a straight road. During the first part, she rides for 22 minutes at an average speed of 7.2 m/s. During the second part, she rides for 36 minutes at an average speed of 5.1 m/s. Finally, during the third part, she rides for 8.0 minutes at an average speed of 13 m/s. (a) How far has the bicyclist traveled during the entire trip? (b) What is the average velocity of the bicyclist for the trip? (*See Solved Problem 1 for a related problem.*)

*60. A sports car, picking up speed, passes between two markers in a time of 4.1 s. The markers are separated by 120 m. All the while, the car has an acceleration of 1.8 m/s². What is its speed at the second marker?

*61. Suppose that the first one-fourth of the distance between two points is covered with an average velocity of +18 m/s. The average velocity for the remainder of the trip is +51 m/s. What is the average velocity for the entire trip? (*See Solved Problem 1 for a related problem.*)

*62. A woman on a bridge 90.0 m high sees a log floating at a constant speed on the river below. She drops a stone from rest in an attempt to hit the log. The stone is released when the log has 6.00 m more to travel before passing under the bridge. The stone hits the water 2.00 m in front of the log. Find the speed of the log.

*63. A speed trap is set up with two pressure-activated strips placed across a highway, 110 m apart. A car is speeding along at 33 m/s, while the speed limit is 21 m/s. At the instant the car activates the first strip, the driver begins slowing down. What deceleration is needed in order that the average speed of the car is within the speed limit by the time the car crosses the second marker?

*64. In reaching her destination, a backpacker walks with an average velocity of 1.34 m/s, due west. This average velocity results, because she hikes for 6.44 km with an average velocity of 2.68 m/s, due west, turns around, and hikes with an average velocity of 0.447 m/s, due east. How far did she walk while moving east? (*See Solved Problem 1 for a related problem.*)

**65. A ball is thrown downward with an initial speed of 25 m/s from the top of a 210-m-tall building. At the same time, another ball is thrown upward from ground level with a speed of 25 m/s. At what distance from the bottom do the two balls pass each other?

**66. A book accidentally falls from a shelf 4.2 m high. A librarian is standing nearby and moves 0.80 m, starting from rest, to catch the book. What must be his average acceleration if he catches the book when it is 1.8 m above the floor?

**67. A football player, starting from rest at the line of scrimmage, accelerates along a straight line for a time of 3.0 s. Then, during a negligible amount of time, he changes the magnitude of his acceleration to a value of 1.1 m/s². With this acceleration, he continues in the same direction for another 2.0 s, until he reaches a speed of 6.4 m/s. What is the value of his acceleration (assumed to be constant) during the initial 3.0-s period?

CHAPTER 3

KINEMATICS IN TWO DIMENSIONS

Black panthers follow a characteristic curved path when they leap, and so do small house cats, as well as other animals. In fact, there are many examples of this type of motion, as when a dancer leaps through the air, a circus performer is shot from a cannon, or a home run is hit in baseball. In all these cases, the vertical and horizontal locations of the object change simultaneously as time passes. Hence, the resulting motion is two-dimensional and follows the familiar arc-shaped path. In this chapter, we will see that this particular path shape occurs, because once the object is launched, the same acceleration due to gravity influences the moving object, whether it is a panther, a dancer, a circus performer, or a baseball. We will find that no new basic concepts will be needed to understand the details of this kind of two-dimensional motion.

3.1 DISPLACEMENT, VELOCITY, AND ACCELERATION

In Chapter 2 the concepts of displacement, velocity, and acceleration are used to describe an object moving along a horizontal or a vertical straight line. There are also situations in which the motion is along a curved line, as in the famous Indianapolis 500 race where the cars move on an oval track. This type of "two-dimensional" motion can also be described using the same concepts.

Figure 3.1 shows a race car at two different positions along the track. These positions are identified by the vectors $\mathbf{s}_0$ and $\mathbf{s}$ that are drawn from an arbitrary coordinate origin. The *displacement* $\Delta\mathbf{s}$ of the car is the vector drawn from the

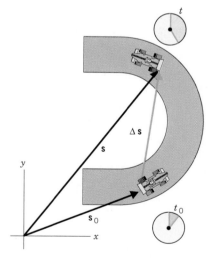

Figure 3.1 The displacement $\Delta\mathbf{s}$ of the car is a vector that points from the initial position of the car to the final position. The magnitude of $\Delta\mathbf{s}$ is the shortest distance between the two positions.

initial position of the car at time t_0 to the final position at time t, the magnitude of Δs being the shortest distance between the two positions. From the drawing it is evident that s is the vector sum of s_0 and Δs, so $s = s_0 + \Delta s$, or

$$\text{Displacement} = \Delta s = s - s_0$$

The displacement here is defined as it is in Chapter 2, except now the displacement vector can lie anywhere in a plane, rather than just along a straight line.

The average velocity $\overline{v}$ of the car between two positions is defined in Equation 2.2 as the displacement, $\Delta s = s - s_0$, divided by the elapsed time $\Delta t = t - t_0$:

$$\overline{v} = \frac{s - s_0}{t - t_0} = \frac{\Delta s}{\Delta t} \tag{3.1}$$

Since both sides of Equation 3.1 must agree in direction, the average velocity vector has the same direction as the displacement. The velocity of the car at an instant of time is its *instantaneous velocity* v. The average velocity becomes equal to the instantaneous velocity v in the limit as Δt becomes infinitesimally small:

$$v = \lim_{\Delta t \to 0} \frac{\Delta s}{\Delta t}$$

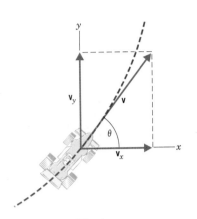

Figure 3.2 The instantaneous velocity v and its two components v_x and v_y.

Figure 3.2 shows the velocity components v_x and v_y, which are parallel to the x and y axes, respectively. Using the components of a vector is advantageous when describing two-dimensional motion.

The *average acceleration* $\overline{a}$ is defined in the same manner as that for one-dimensional motion, namely as the change in the velocity, $\Delta v = v - v_0$, divided by the elapsed time Δt:

$$\overline{a} = \frac{v - v_0}{t - t_0} = \frac{\Delta v}{\Delta t} \tag{3.2}$$

The average acceleration vector has the same direction as the change in velocity. In the limit that the elapsed time becomes infinitesimally small, the average acceleration becomes equal to the *instantaneous acceleration* a:

$$a = \lim_{\Delta t \to 0} \frac{\Delta v}{\Delta t}$$

The acceleration has a component a_x along the x direction and a component a_y along the y direction.

3.2 EQUATIONS OF KINEMATICS IN TWO DIMENSIONS

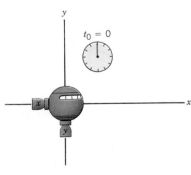

Figure 3.3 At $t_0 = 0$, the spacecraft is assumed to be at the coordinate origin, so $s_0 = 0$.

To understand how displacement, velocity, and acceleration are applied to two-dimensional motion, consider a spacecraft equipped with two engines that are mounted perpendicular to each other. The craft is far from other bodies, so the only forces acting on it are those produced by its own engines. As Figure 3.3 indicates, the spacecraft is assumed to be at the coordinate origin when $t_0 = 0$, so that $s_0 = 0$, and the displacement is $\Delta s = s - s_0 = s$. The components of the displacement vector s along the x and y axes are x and y, respectively.

In Figure 3.4 only the engine oriented along the x direction is firing, and the

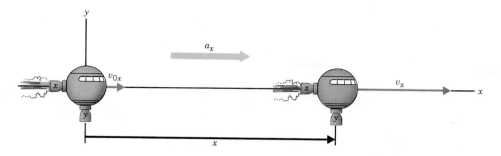

vehicle accelerates along this direction. It is assumed that the velocity in the y direction is zero and that it remains zero, since the y engine is turned off. The motion of the spacecraft along the x direction is described by the five kinematic variables x, a_x, v_x, v_{0x}, and t. Here the symbol "x" reminds us that we are dealing with the x components of the displacement, velocity, and acceleration vectors. If the spacecraft has a constant acceleration along the x direction, the motion is exactly like that described in Chapter 2, and the equations of kinematics can be used. For convenience, these equations are written in the left column of Table 3.1.

Figure 3.5 is analogous to Figure 3.4, except that now only the y engine is firing, and the spacecraft accelerates along the y direction. Such a motion can be described in terms of the kinematic variables y, a_y, v_y, v_{0y}, and t. And if the acceleration along the y direction is constant, these variables are related by the equations of kinematics, as written in the right column of Table 3.1.

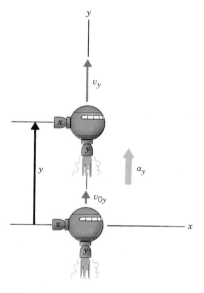

Figure 3.5 The spacecraft is moving with a constant acceleration a_y parallel to the y axis. There is no motion in the x direction, and the x engine is turned off.

Table 3.1 Equations of Kinematics for Constant Acceleration in Two-Dimensional Motion

x Component		Variable	y Component	
x		Displacement	y	
a_x		Acceleration	a_y	
v_x		Final velocity	v_y	
v_{0x}		Initial velocity	v_{0y}	
t		Elapsed time	t	
$v_x = v_{0x} + a_x t$	(3.3a)		$v_y = v_{0y} + a_y t$	(3.3b)
$x = \frac{1}{2}(v_{0x} + v_x)t$	(3.4a)		$y = \frac{1}{2}(v_{0y} + v_y)t$	(3.4b)
$x = v_{0x}t + \frac{1}{2}a_x t^2$	(3.5a)		$y = v_{0y}t + \frac{1}{2}a_y t^2$	(3.5b)
$v_x^2 = v_{0x}^2 + 2a_x x$	(3.6a)		$v_y^2 = v_{0y}^2 + 2a_y y$	(3.6b)

If both engines of the spacecraft are firing *at the same time,* the resulting motion takes place in part along the x axis and in part along the y axis, as Figure 3.6 illustrates. The thrust of each engine gives the vehicle a corresponding acceleration component. The x engine accelerates the ship in the x direction and causes a change in the x component of the velocity. Likewise, the y engine causes a change in the y component of the velocity. *It is important to realize that the x part of the motion occurs exactly as it would if the y part did not occur at all. Similarly, the y part of the motion occurs exactly as it would if the x part of the motion did not exist.* In other words, the x and y motions are independent of each other, and a problem dealing with two-dimensional motion can be considered as two one-dimensional problems. Example 1 illustrates this point.

PROBLEM SOLVING INSIGHT

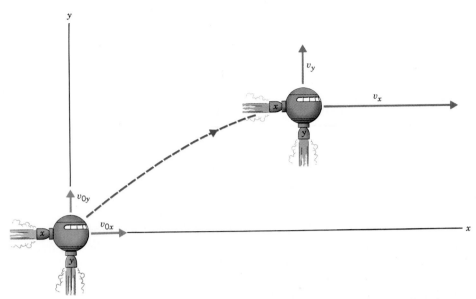

Figure 3.6 The two-dimensional motion of the spacecraft can be viewed as the combination of the separate x and y motions.

Example 1   A Moving Spacecraft

In the x direction, the spacecraft in Figure 3.6 has an initial velocity of $v_{0x} = +22$ m/s and an acceleration of $a_x = +24$ m/s². In the y direction, the analogous quantities are $v_{0y} = +14$ m/s and $a_y = +12$ m/s². After a time of 7.0 s, find (a) x and v_x, (b) y and v_y, and (c) the final velocity (magnitude and direction) of the spacecraft.

REASONING   The motion in the x direction and the motion in the y direction can be treated separately, each as a one-dimensional motion. We will follow this approach in parts (a) and (b) to obtain the location and separate velocity components of the spacecraft. Then, in part (c) the separate velocity components will be combined to give the final velocity.

SOLUTION

(a) The x component of the spacecraft's displacement can be obtained from the data below and Equation 3.5a.

x-Direction Data

x	a_x	v_x	v_{0x}	t
?	+24 m/s²	?	+22 m/s	7.0 s

$$x = v_{0x}t + \tfrac{1}{2}a_x t^2 = (22 \text{ m/s})(7.0 \text{ s}) + \tfrac{1}{2}(24 \text{ m/s}^2)(7.0 \text{ s})^2 = \boxed{+740 \text{ m}}$$

The velocity component v_x can be calculated using Equation 3.3a:

$$v_x = v_{0x} + a_x t = (22 \text{ m/s}) + (24 \text{ m/s}^2)(7.0 \text{ s}) = \boxed{+190 \text{ m/s}}$$

PROBLEM SOLVING INSIGHT

When the motion of an object is two-dimensional, the time variable t has the same value for both the x and y directions of the motion.

(b) The data for the motion in the y direction are listed below.

y-Direction Data

y	a_y	v_y	v_{0y}	t
?	$+12$ m/s^2	?	$+14$ m/s	7.0 s

Proceeding in the same manner as in part (a), we find that

$$y = +390 \text{ m} \qquad \text{and} \qquad v_y = +98 \text{ m/s}$$

(c) Figure 3.7 shows that the velocity of the vehicle is the vector sum of its x and y components. The magnitude v of the velocity can be found by using the Pythagorean theorem:

$$v = \sqrt{v_x^2 + v_y^2} = \sqrt{(190 \text{ m/s})^2 + (98 \text{ m/s})^2} = \boxed{210 \text{ m/s}}$$

The direction of the velocity vector is given by the angle θ in the drawing:

$$\tan \theta = \frac{v_y}{v_x} = \frac{98 \text{ m/s}}{190 \text{ m/s}} = 0.52$$

$$\theta = \tan^{-1} 0.52 = \boxed{27°}$$

After 7.0 s, the spacecraft has a speed of 210 m/s and a velocity vector that points 27° above the positive x axis. At this time the spacecraft is located at a point like that in Figure 3.6, 740 m to the right of and 390 m above the origin.

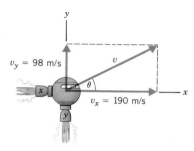

Figure 3.7 The magnitude of the velocity vector gives the speed of the spacecraft, and the angle θ gives the direction of travel relative to the positive x direction.

3.3 PROJECTILE MOTION

The biggest thrill in baseball is a home run. The motion of the ball on its curving path into the stands is a common type of two-dimensional motion called "projectile motion." A good description of projectile motion can often be obtained with the assumption that the moving object (the projectile) experiences only the acceleration due to gravity. This assumption implies that air resistance, which can slow down the projectile, is negligible. In the absence of air resistance, there is no acceleration in the horizontal or x direction, so that $a_x = 0$, and the x component of the velocity remains the same as its initial value ($v_x = v_{0x}$). For our purposes, then, the phrase "projectile motion" means that $a_x = 0$ and $a_y =$ acceleration due to gravity. If the trajectory of the projectile is near the surface of the earth, then the acceleration due to gravity has a magnitude of 9.80 m/s^2. Example 2 illustrates how projectile motion can be described by the equations of kinematics.

Example 2 A Falling Care Package

Figure 3.8 shows an airplane moving horizontally with a constant velocity of $+115$ m/s at an altitude of 1050 m. The directions to the right and upward have been chosen as the positive directions. The plane releases a "care package" that falls to the ground along a curved trajectory. Ignoring air resistance, determine the time required for the package to hit the ground.

Once these Atlantic Bottleneck dolphins leave the water, their motions approximate projectile motion.

REASONING   In falling to the ground, the package moves both downward and to the right. But we can treat the vertical and horizontal parts of the motion separately. The initial velocity of the package in the y direction is zero at the moment the package is released from the plane, so $v_{0y} = 0$. The package is moving at the instant of release, but only in the x direction, not in the y direction. Furthermore, when the package hits the ground, the y component of the package's displacement is $y = -1050$ m, as the drawing shows. The acceleration is that due to gravity, so $a_y = -9.80$ m/s². These data are summarized below:

y-Direction Data

y	a_y	v_y	v_{0y}	t
−1050 m	−9.80 m/s²		0	?

With these data, Equation 3.5b ($y = v_{0y}t + \frac{1}{2}a_y t^2$) can be employed to calculate the time for the package to strike the ground.

SOLUTION   Since $v_{0y} = 0$, it follows from Equation 3.5b that $y = \frac{1}{2}a_y t^2$ and

$$t = \sqrt{\frac{2y}{a_y}} = \sqrt{\frac{2(-1050 \text{ m})}{-9.80 \text{ m/s}^2}} = \boxed{14.6 \text{ s}}$$

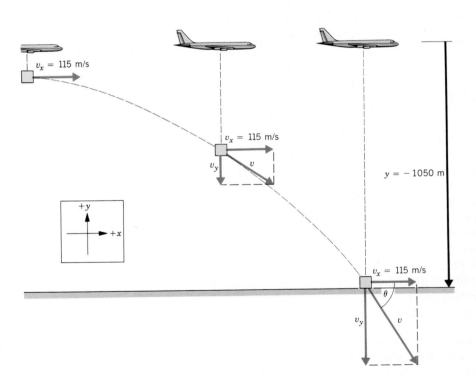

Figure 3.8   The package falling from a plane is an example of projectile motion, as Examples 2 and 3 discuss.

In Example 2, the initial velocity of $v_{0x} = +115$ m/s in the horizontal direction plays no role in determining the time required for the package to reach the ground. In fact, for all types of projectile motion, the time of flight depends only

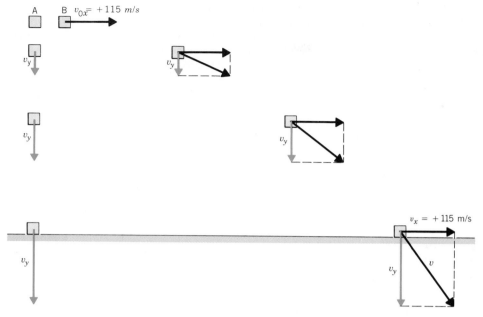

Figure 3.9 Package A and package B are released at the same height and strike the ground at the same time, because their y variables (y, a_y, and v_{0y}) are the same.

on the y variables, and the x variables do not come into play. To emphasize this point, Figure 3.9 illustrates what happens to two packages that are released simultaneously from the same height. Package B is given an initial velocity of $v_{0x} = +115$ m/s in the horizontal direction, as in Example 2, and the package follows the curved path shown in the figure. Package A, on the other hand, is dropped from a stationary balloon and falls directly toward the ground, since $v_{0x} = 0$. Both packages hit the ground at the same time; the time of fall depends only on the y variables (y, a_y, and v_{0y}), and these variables are the same for both packages.

The freely falling package in Example 2 picks up vertical speed on the way down. But the horizontal component of the velocity retains its initial value of $v_{0x} = +115$ m/s throughout the entire descent. Since the plane also travels at a constant horizontal velocity of $+115$ m/s, it remains directly above the falling package. The pilot always sees the package directly beneath the plane, as the dashed vertical lines in Figure 3.8 show. This result is a direct consequence of the fact that the box has no acceleration in the horizontal direction. In reality, air resistance would slow down the package, and it would not remain directly beneath the plane during the descent.

Not only do the packages in Figure 3.9 reach the ground at the same time, but the y components of their velocities are also equal at all points on the way down. However, package B does hit the ground with a greater speed than package A. Remember, speed is the magnitude of the velocity vector, and the velocity of B has an x component, whereas the velocity of A does not. The magnitude and direction of the velocity vector for package B at the instant just before the package hits the ground is computed in Example 3.

"Fido" learns about projectile motion.

Example 3 The Velocity of the Care Package

For the situation shown in Figure 3.8, find the speed of the package and the direction of the velocity vector just before the package hits the ground.

REASONING Since the speed v of the package is given by $v = \sqrt{v_x^2 + v_y^2}$, it is necessary to know values for v_x and v_y at the instant before impact. The component v_x is constant with a magnitude of 115 m/s. The component v_y can be determined by using Equation 3.3b and the data from Example 2 ($a_y = -9.80$ m/s², $v_{0y} = 0$, $t = 14.6$ s).

SOLUTION

$$v_y = v_{0y} + a_y t = (-9.80 \text{ m/s}^2)(14.6 \text{ s}) = -143 \text{ m/s}$$

The speed of the package at the instant before impact is

$$v = \sqrt{(115 \text{ m/s})^2 + (-143 \text{ m/s})^2} = \boxed{184 \text{ m/s}}$$

The velocity vector makes an angle θ with the horizontal, as Figure 3.8 indicates:

$$\cos \theta = \frac{v_x}{v} = \frac{115 \text{ m/s}}{184 \text{ m/s}} = 0.625$$

$$\theta = \cos^{-1} 0.625 = \boxed{51.3°}$$

PROBLEM SOLVING INSIGHT

The speed of a projectile at any location along its path is the magnitude v of its velocity at that location: $v = \sqrt{v_x^2 + v_y^2}$. Thus, both the horizontal and vertical velocity components contribute to the speed.

Often projectiles, like footballs and baseballs, are sent into the air at an angle with respect to the ground. From a knowledge of the projectile's initial velocity, a wealth of information can be obtained about the motion. For instance, Example 4 demonstrates how to calculate the maximum height reached by the projectile.

Example 4 The Height of a Kickoff

A place-kicker kicks a football at an angle of $\theta = 40.0°$ above the horizontal axis, as Figure 3.10 shows. The initial speed of the ball is $v_0 = 22$ m/s. Ignore air resistance and find the maximum height H that the ball attains.

REASONING The maximum height reached by the ball is a characteristic of the vertical part of the motion. The vertical part of the motion can be treated separately from the horizontal part. Making use of this fact, we calculate the vertical component of the initial velocity:

$$v_{0y} = v_0 \sin \theta = (22 \text{ m/s}) \sin 40.0° = +14 \text{ m/s}$$

The maximum height H can be determined by noting that the y component of the velocity v_y decreases as the ball moves upward. Eventually, $v_y = 0$ at the top of the trajectory. The data below can be used in Equation 3.6b ($v_y^2 = v_{0y}^2 + 2a_y y$) to find the maximum height:

PROBLEM SOLVING INSIGHT

When a projectile in two-dimensional motion reaches maximum height, the vertical component of its velocity is momentarily zero ($v_y = 0$). However, the horizontal component of its velocity is not zero.

y-Direction Data

y	a_y	v_y	v_{0y}	t
$H = ?$	-9.80 m/s²	0	$+14$ m/s	

SOLUTION From Equation 3.6b, we find that

$$y = H = \frac{v_y^2 - v_{0y}^2}{2a_y} = \frac{-(14 \text{ m/s})^2}{2(-9.80 \text{ m/s}^2)} = \boxed{+10 \text{ m}}$$

The height H depends only on the y variables; the same height would have been reached had the ball been thrown *straight up* with an initial velocity of $v_{0y} = +14$ m/s.

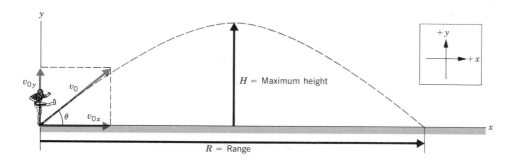

Figure 3.10 A football is kicked with an initial speed of v_0 at an angle of θ above the ground. The ball attains a maximum height H and a range R.

It is also possible to find the total time during which the football in Figure 3.10 is in the air before returning to the ground. Example 5 shows how to determine this time.

Example 5   The Time of Flight of a Kickoff

THE PHYSICS OF . . .

the "hang time" of a football.

For the motion illustrated in Figure 3.10, ignore air resistance and determine the time of flight between kickoff and landing.

REASONING   Given the initial velocity, it is the acceleration due to gravity that determines how long the ball stays in the air. Thus, to find the time of flight we deal with the vertical part of the motion. Since the ball starts at ground level and returns to ground level, the displacement in the y direction is zero. The initial velocity in the y direction is the same as that in Example 4, i.e., $v_{0y} = +14$ m/s. Therefore, we have

y-Direction Data

y	a_y	v_y	v_{0y}	t
0	-9.80 m/s^2		$+14$ m/s	?

The time of flight can be determined from Equation 3.5b.

SOLUTION   Using Equation 3.5b, we find

$$y = v_{0y}t + \tfrac{1}{2}a_y t^2$$
$$0 = [(14 \text{ m/s}) + \tfrac{1}{2}(-9.80 \text{ m/s}^2)t]t$$
$$\boxed{t = 2.9 \text{ s}}$$

The second solution, $t = 0$, is discarded, because it represents the situation where the ball has not yet begun its trip.

An alternative way to compute the total time of flight in a range problem is to calculate the time for the projectile to reach maximum height and then multiply this time by two.

The height reached by this football, kicked by a Pittsburgh Steeler, will depend on the initial speed of the football and the angle of the kick.

Another important feature of projectile motion is called the "range." The range, as Figure 3.10 shows, is the horizontal distance traveled between launching and landing, assuming the projectile returns to the *same vertical level* at which it was fired. Example 6 shows how to obtain the range.

Example 6   The Range of a Kickoff

For the motion shown in Figure 3.10, ignore air resistance and calculate the range R of the projectile.

<u>REASONING</u>   The range is a characteristic of the horizontal part of the motion. Thus, our starting point is to determine the horizontal component of the initial velocity:

$$v_{0x} = v_0 \cos \theta = (22 \text{ m/s}) \cos 40.0° = +17 \text{ m/s}$$

Recall from Example 5 that the time of flight is $t = 2.9$ s. Since there is no acceleration in the x direction, v_{0x} remains constant, and the range is simply the product of v_{0x} and the time.

<u>SOLUTION</u>   The range is

$$x = R = v_{0x}t = (17 \text{ m/s})(2.9 \text{ s}) = \boxed{+49 \text{ m}}$$

The range in the previous example depends on the angle θ at which the projectile is fired above the horizontal. When air resistance is absent, the maximum range results when $\theta = 45°$. Moreover, in all the examples in this section, the projectiles follow a curved trajectory. In general, if the only acceleration is that due to gravity, the shape of the path can be shown to be a *parabola*.

Section 2.6 points out that certain types of symmetry with respect to time and velocity are present for freely falling bodies. These symmetries are also found in projectile motion, since projectiles are falling freely in the vertical direction. In particular, the time required for a projectile to reach its maximum height H is equal to the time spent returning to the ground. In addition, Figure 3.11 shows that the speed of the object at any height above the ground on the upward part of the trajectory is equal to the speed at the same height on the downward part. Although the two speeds are the same, the velocities are different, because they point in different directions.

Another type of symmetry also exists. If you have ever experimented with a stream of water from a hose, you might be aware that there are two possible angles at which to point the nozzle such that the range of the water is the same. Figure 3.12 illustrates three trajectories that have initial angles of 20°, 45°, and

Figure 3.11   The speed of a projectile at a given height above the ground is the same on the upward and downward parts of the trajectory. The velocities are different, however, since they point in different directions.

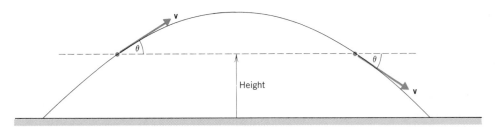

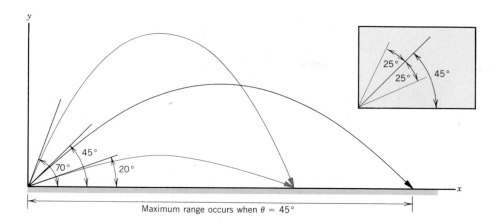

Maximum range occurs when $\theta = 45°$

Figure 3.12 The 20° trajectory and the 70° trajectory have identical ranges. Note from the insert that the two angles are 25° on either side of the 45° line.

70°. The 45° angle gives rise to the maximum range, while the 20° and 70° angles give rise to *identical ranges* that are less than the maximum. There are always two angles giving the same range; one is less than 45° and the other is greater than 45°. The drawing indicates that the two angles are symmetrically placed about the 45° line. In other words, the 20° angle is 25° below the 45° angle, while the 70° angle is 25° above the 45° angle, as the insert in the picture emphasizes.

*3.4 RELATIVE VELOCITY

RELATIVE VELOCITY IN ONE DIMENSION

To someone hitchhiking along a highway, two cars speeding by in adjacent lanes seem like a blur. But, if the cars have the same velocity, each driver sees the other remaining in place, one lane away. The hitchhiker observes a velocity of perhaps 30 m/s. But each driver observes the velocity of the other car to be zero. Clearly, the velocity of an object is relative to the observer who is making the measurement.

Figure 3.13 illustrates the concept of relative velocity by showing a passenger walking toward the front of a moving train. The people sitting on the train see the passenger walking with a velocity of +2.0 m/s, where the plus sign denotes a direction to the right. Suppose the train is moving with a velocity of +9.0 m/s relative to an observer standing on the ground. Then the ground-based observer would see the passenger moving with a velocity of +11 m/s, due in part to the

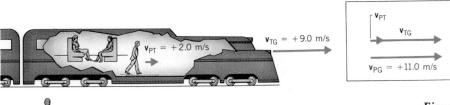

Ground-based observer

Figure 3.13 The velocity $\mathbf{v}_{PG}$ of the passenger relative to the ground-based observer is the vector sum of the velocity $\mathbf{v}_{PT}$ of the passenger relative to the train and the velocity $\mathbf{v}_{TG}$ of the train relative to the ground: $\mathbf{v}_{PG} = \mathbf{v}_{PT} + \mathbf{v}_{TG}$.

An example of relative velocity from the movie *Indiana Jones and the Last Crusade*.

walking motion and in part to the train's motion. As an aid in describing relative velocity, let us define the following symbols:

$\mathbf{v}_{\boxed{PT}}$ = velocity of the $\boxed{\text{Passenger}}$ relative to the $\boxed{\text{Train}}$ = +2.0 m/s

$\mathbf{v}_{\boxed{TG}}$ = velocity of the $\boxed{\text{Train}}$ relative to the $\boxed{\text{Ground}}$ = +9.0 m/s

$\mathbf{v}_{\boxed{PG}}$ = velocity of the $\boxed{\text{Passenger}}$ relative to the $\boxed{\text{Ground}}$ = +11 m/s

In terms of these symbols, the situation in Figure 3.13 is summarized as follows:

$$\mathbf{v}_{PG} = \mathbf{v}_{PT} + \mathbf{v}_{TG} \qquad (3.7)$$

or

$$\mathbf{v}_{PG} = (2.0 \text{ m/s}) + (9.0 \text{ m/s}) = +11 \text{ m/s}$$

According to Equation 3.7, $\mathbf{v}_{PG}$ is the vector sum of $\mathbf{v}_{PT}$ and $\mathbf{v}_{TG}$, and this sum is shown in the drawing. Had the passenger been walking toward the rear of the train, rather than the front, the velocity relative to the ground-based observer would have been $\mathbf{v}_{PG} = (-2.0 \text{ m/s}) + (9.0 \text{ m/s}) = +7.0 \text{ m/s}$.

Each velocity symbol in Equation 3.7 contains a two-letter subscript. The first letter in the subscript refers to the body that is moving, while the second letter indicates the object relative to which the velocity is measured. For example, $\mathbf{v}_{TG}$ and $\mathbf{v}_{PG}$ are the velocities of the Train and Passenger measured relative to the Ground. Similarly, $\mathbf{v}_{PT}$ is the velocity of the Passenger measured by an observer sitting on the Train.

The ordering of the subscript symbols in Equation 3.7 follows a definite pattern. The first subscript (P) on the left side of the equation is also the first subscript on the right side of the equation. Likewise, the last subscript (G) on the left side of the equation is also the last subscript on the right side of the equation. The third subscript (T) appears only on the right side of the equation as the two "inner" subscripts. The colored boxes below emphasize the pattern of the symbols in the subscripts:

$$\mathbf{v}_{\boxed{PG}} = \mathbf{v}_{\boxed{P}\boxed{T}} + \mathbf{v}_{\boxed{T}\boxed{G}}$$

In other situations, the subscripts will not necessarily be P, G, and T, but will be compatible with the names of the objects involved in the motion.

RELATIVE VELOCITY IN TWO DIMENSIONS

Figure 3.14 depicts a common situation that deals with relative velocity in two dimensions. Part *a* of the drawing shows a boat being carried downstream by a river; the engine of the boat is turned off. In part *b*, the engine has been turned on, and now the boat moves across the river in a diagonal fashion because of the combined motion produced by the current and the engine. The list below gives the velocities for this type of motion and the objects relative to which they are measured:

$$\mathbf{v}_{\boxed{BW}} = \text{velocity of the } \boxed{\text{Boat}} \text{ relative to the } \boxed{\text{Water}}$$

$$\mathbf{v}_{\boxed{WS}} = \text{velocity of the } \boxed{\text{Water}} \text{ relative to the } \boxed{\text{Shore}}$$

$$\mathbf{v}_{\boxed{BS}} = \text{velocity of the } \boxed{\text{Boat}} \text{ relative to the } \boxed{\text{Shore}}$$

The velocity $\mathbf{v}_{BW}$ of the boat relative to the water is the velocity measured by an observer who, for instance, is floating on an inner tube and drifting downstream with the current. When the engine is turned off, the boat also drifts downstream with the current and $\mathbf{v}_{BW}$ is zero. When the engine is turned on, however, the boat can move relative to the water, and $\mathbf{v}_{BW}$ is no longer zero. The velocity $\mathbf{v}_{WS}$ of the water relative to the shore is the velocity of the current measured by an observer on the shore. The velocity $\mathbf{v}_{BS}$ of the boat relative to the shore is due to the combined motion of the boat relative to the water and the motion of the water relative to the shore. In symbols,

$$\mathbf{v}_{\boxed{BS}} = \mathbf{v}_{\boxed{B}\boxed{W}} + \mathbf{v}_{W\boxed{S}}$$

The ordering of the subscripts in this equation is identical to that in Equation 3.7, although the letters have been changed to reflect a different physical situation. Example 7 illustrates the concept of relative velocity in two dimensions.

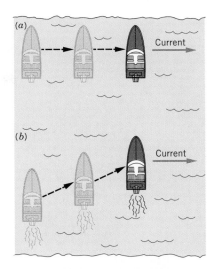

Figure 3.14 (a) A boat with its engine turned off is carried along by the current. (b) With the engine turned on, the boat moves across the river in a diagonal fashion.

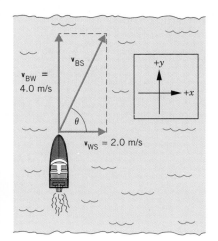

Figure 3.15 The velocity $\mathbf{v}_{BS}$ of the boat relative to the shore is the vector sum of the velocity $\mathbf{v}_{BW}$ of the boat relative to the water and the velocity $\mathbf{v}_{WS}$ of the water relative to the shore: $\mathbf{v}_{BS} = \mathbf{v}_{BW} + \mathbf{v}_{WS}$.

Example 7   Crossing a River

The engine of a boat drives it across a river that is 1800 m wide. The velocity $\mathbf{v}_{BW}$ of the boat relative to the water is 4.0 m/s, directed perpendicular to the current (see Figure 3.15). The velocity $\mathbf{v}_{WS}$ of the water relative to the shore is 2.0 m/s. (a) What is the velocity $\mathbf{v}_{BS}$ of the boat relative to the shore? (b) How long does it take for the boat to cross the river?

REASONING AND SOLUTION

(a) The three velocity vectors are related by $\mathbf{v}_{BS} = \mathbf{v}_{BW} + \mathbf{v}_{WS}$. Since the vectors $\mathbf{v}_{BW}$ and

$\mathbf{v}_{WS}$ are perpendicular, the magnitude of $\mathbf{v}_{BS}$ can be determined from the Pythagorean theorem:

$$v_{BS} = \sqrt{v_{BW}^2 + v_{WS}^2} = \sqrt{(4.0 \text{ m/s})^2 + (2.0 \text{ m/s})^2} = \boxed{4.5 \text{ m/s}}$$

Thus, the boat moves at a speed of 4.5 m/s with respect to an observer on the shore. The direction of the boat relative to the shoreline is given by the angle θ in the drawing:

$$\tan \theta = \frac{v_{BW}}{v_{WS}} = \frac{4.0 \text{ m/s}}{2.0 \text{ m/s}} = 2.0$$

$$\theta = \tan^{-1} 2.0 = \boxed{63°}$$

(b) The component of $\mathbf{v}_{BS}$ that is parallel to the width of the river (see Figure 3.15) indicates how fast the boat is moving across the river; the magnitude of this parallel component is $v_{BS} \sin \theta = v_{BW} = 4.0$ m/s. The time for the boat to cross the river is equal to the width of the river divided by the magnitude of the boat's velocity component parallel to the width:

$$t = \frac{\text{Width}}{v_{BS} \sin \theta} = \frac{1800 \text{ m}}{4.0 \text{ m/s}} = \boxed{450 \text{ s}}$$

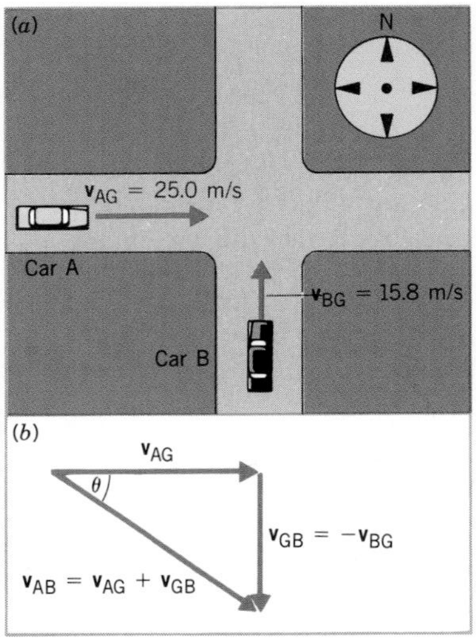

Occasionally, situations arise when two vehicles are in relative motion, and it is useful to know the relative velocity of one with respect to the other. Example 8 considers this type of relative motion.

(a)

(b)

Figure 3.16 Two cars are approaching an intersection along perpendicular roads.

$\mathbf{v}_{AG} = 25.0$ m/s

Car A

$\mathbf{v}_{BG} = 15.8$ m/s

Car B

$\mathbf{v}_{AG}$

θ

$\mathbf{v}_{GB} = -\mathbf{v}_{BG}$

$\mathbf{v}_{AB} = \mathbf{v}_{AG} + \mathbf{v}_{GB}$

Example 8 Approaching an Intersection

Figure 3.16a shows two cars approaching an intersection along perpendicular roads. The cars have the following velocities:

$\mathbf{v}_{\boxed{AG}}$ = velocity of $\boxed{\text{car A}}$ relative to the $\boxed{\text{Ground}}$ = 25.0 m/s, eastward

$\mathbf{v}_{\boxed{BG}}$ = velocity of $\boxed{\text{car B}}$ relative to the $\boxed{\text{Ground}}$ = 15.8 m/s, northward

Find the magnitude and direction of $\mathbf{v}_{AB}$, where

$\mathbf{v}_{\boxed{AB}}$ = velocity of $\boxed{\text{car A}}$ as measured by a passenger in $\boxed{\text{car B}}$

REASONING To find $\mathbf{v}_{AB}$, we would use an equation whose subscripts followed the order outlined earlier. Thus,

$$\mathbf{v}_{\boxed{AB}} = \mathbf{v}_{\boxed{A}G} + \mathbf{v}_{G\boxed{B}}$$

The second term on the right side of this equation involves $\mathbf{v}_{GB}$, the velocity of the ground relative to a passenger in car B, rather than $\mathbf{v}_{BG}$, which is given as 15.8 m/s, northward. In other words, the subscripts are reversed. However, $\mathbf{v}_{GB}$ is related to $\mathbf{v}_{BG}$ according to

$$\mathbf{v}_{GB} = -\mathbf{v}_{BG}$$

This relationship reflects the fact that a passenger in car B, moving northward relative to the ground, looks out the car window and sees objects on the ground moving southward, that is, in the opposite direction. Therefore, the equation $\mathbf{v}_{AB} = \mathbf{v}_{AG} + \mathbf{v}_{GB}$ may be used to find $\mathbf{v}_{AB}$, provided we recognize $\mathbf{v}_{GB}$ as a vector that points opposite to the given velocity $\mathbf{v}_{BG}$. With this in mind, Figure 3.16b illustrates how $\mathbf{v}_{AG}$ and $\mathbf{v}_{GB}$ are added vectorially to give $\mathbf{v}_{AB}$.

PROBLEM SOLVING INSIGHT

In general, the velocity of object R relative to object S is always the negative of the velocity of object S relative to R: $\mathbf{v}_{RS} = -\mathbf{v}_{SR}$.

SOLUTION   From the vector triangle shown in Figure 3.15b, the magnitude and direction of $\mathbf{v}_{AB}$ can be calculated as

$$v_{AB} = \sqrt{v_{AG}^2 + v_{GB}^2} = \sqrt{(25.0 \text{ m/s})^2 + (15.8 \text{ m/s})^2} = \boxed{29.6 \text{ m/s}}$$

and

$$\cos \theta = \frac{v_{AG}}{v_{AB}} = \frac{25.0 \text{ m/s}}{29.6 \text{ m/s}} = 0.845$$

$$\theta = \cos^{-1} 0.845 = \boxed{32.3°}$$

SUMMARY

Motion that occurs in two dimensions can be described in terms of the time t and the x and y components of four vectors: the displacement, the acceleration, the final velocity, and the initial velocity. The motion can be analyzed by treating the x and y components separately. When the acceleration vector is constant, the x components of these vectors (x, a_x, v_x, and v_{0x}) are related by the equations of kinematics, as are the y components (y, a_y, v_y, and v_{0y}). Table 3.1 summarizes the equations of kinematics for two-dimensional motion.

Projectile motion — that of an object (the projectile) moving through the air — is one particular kind of motion in two dimensions. If air resistance can be neglected, the projectile experiences only the acceleration due to gravity. Then, the horizontal component of the projectile's velocity stays constant at all times, while the vertical component changes because of the acceleration due to gravity.

SOLVED PROBLEMS

Solved Problem 1   A Falling Meteorite
Related Problems: *30   *31   *32

A meteorite is being tracked by radar as it falls to earth. When its altitude is 3.00×10^4 m, the radar screen shows that the meteorite has a velocity of 583 m/s at an angle of 28.3° below the horizontal (see the drawing). If the effect of air resistance could be ignored, how much time would elapse before the meteorite strikes the earth, and what would be its velocity (magnitude and direction) just before impact?

REASONING   This problem is similar to Examples 2 and 3 in Section 3.3. Here, however, the meteorite is initially traveling downward at an angle, whereas in Examples 2 and 3 the package is moving horizontally to begin with. As Section 3.3 discusses, the time that a projectile is in the air is governed by the values of the y variables of the motion. The vertical component of the initial velocity of the meteorite is $v_{0y} = -(583 \text{ m/s}) \sin 28.3° = -276$ m/s, where the minus sign indicates the downward direction. These variables are summarized in the following box. From these data and Equation 3.5b, the time can be found.

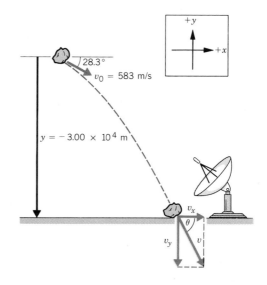

y-Direction Data

y	a_y	v_y	v_{0y}	t
-3.00×10^4 m	-9.80 m/s²		-276 m/s	?

SOLUTION From Equation 3.5b, we obtain

$$y = v_{0y}t + \tfrac{1}{2}a_y t^2$$
$$-3.00 \times 10^4 \text{ m} = (-276 \text{ m/s})t + \tfrac{1}{2}(-9.80 \text{ m/s}^2)t^2$$

Solving this quadratic equation for the time t yields
$\boxed{t = 55.0 \text{ s}}$. The second solution to the quadratic equation,
$t = -111$ s, is not physically meaningful.

The speed v of the meteorite just before impact is given by
the Pythagorean theorem as $v = \sqrt{v_x^2 + v_y^2}$ (see the drawing).
Since $a_x = 0$ in projectile motion, v_x remains constant at all
times, so

$$v_x = v_{0x} = (583 \text{ m/s}) \cos 28.3° = 513 \text{ m/s}$$

The value of v_y can be obtained with the aid of Equation 3.3b:

$$v_y = v_{0y} + a_y t = (-276 \text{ m/s}) + (-9.80 \text{ m/s}^2)(55.0 \text{ s})$$
$$= -815 \text{ m/s}$$

The speed just before impact is

$$v = \sqrt{v_x^2 + v_y^2} = \sqrt{(513 \text{ m/s})^2 + (-815 \text{ m/s})^2} = \boxed{963 \text{ m/s}}$$

From the drawing, it can be seen that the angle θ is given by

$$\cos \theta = \frac{v_x}{v} = \frac{513 \text{ m/s}}{963 \text{ m/s}} = 0.533$$

$$\theta = \cos^{-1} 0.533 = \boxed{57.8°}$$

SUMMARY OF IMPORTANT POINTS The time that a
projectile spends in the air is governed by the y variables of the
motion; the time does not depend on the x variables. However,
once the time has been determined, it can be used to find
kinematic variables in either the y or x direction.

Solved Problem 2   A Home Run

Related Problems: *33   *34   *35

A baseball player hits a home run, and the ball lands in the
left-field seats, 7.6 m above the point at which the ball was hit.

The ball lands with a velocity of 49 m/s at an angle of 31° to
the horizontal (see the drawing below). What is the initial
velocity of the ball when the ball leaves the bat?

REASONING To find the initial velocity of the ball, we must
determine the initial speed v_0 and the angle θ in the drawing.
These quantities are related to the magnitudes of the horizontal and vertical components of the initial velocity (v_{0x} and v_{0y})
by the relations

$$v_0 = \sqrt{v_{0x}^2 + v_{0y}^2} \quad \text{and} \quad \tan \theta = \frac{v_{0y}}{v_{0x}}$$

Therefore, it is necessary to find v_{0x} and v_{0y}, which we do with
the equations of kinematics.

SOLUTION Since $a_x = 0$ in projectile motion, v_x remains
constant throughout the motion, so

$$v_{0x} = v_x = (49 \text{ m/s}) \cos 31° = 42 \text{ m/s}$$

The value for v_{0y} can be obtained from Equation 3.6b ($v_y^2 = v_{0y}^2 + 2a_y y$) and the data displayed below (see drawing for sign
convention):

y-Component Data

y	a_y	v_y	v_{0y}	t
7.6 m	-9.80 m/s²	$(-49 \sin 31°)$ m/s	?	

$$v_{0y}^2 = [(-49 \sin 31°) \text{ m/s}]^2 - 2(-9.80 \text{ m/s}^2)(7.6 \text{ m})$$
$$v_{0y} = 28 \text{ m/s}$$

The initial speed v_0 and angle θ of the baseball are

$$v_0 = \sqrt{v_{0x}^2 + v_{0y}^2} = \sqrt{(42 \text{ m/s})^2 + (28 \text{ m/s})^2} = \boxed{50 \text{ m/s}}$$

$$\tan \theta = \frac{v_{0y}}{v_{0x}} = \frac{28 \text{ m/s}}{42 \text{ m/s}} = 0.67$$

$$\theta = \tan^{-1} 0.67 = \boxed{34°}$$

SUMMARY OF IMPORTANT POINTS If the final param-
eters of the motion of a projectile are known (y, v_x, and v_y in
this problem), the equations of kinematics allow the initial
parameters to be calculated (v_0 and θ in this problem).

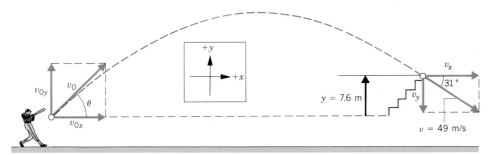

QUESTIONS

1. Suppose an object could move in three dimensions. What additions to the equations of kinematics in Table 3.1 would be necessary to describe three-dimensional motion?

2. At a given time, can an object be moving horizontally and yet have a vertical acceleration? If so, give an example.

3. If an object has a negative acceleration, does it necessarily mean that the object is slowing down? If not, why not?

4. A little league baseball player and a major league baseball player each hit a fly ball to center field. Once in flight, which ball, if either, has the greater acceleration? Provide a reason for your answer.

5. A tennis ball is hit into the air and moves along an arc. Neglecting air resistance, where along the arc is the speed of the ball a minimum? Where along the arc is the speed a maximum? Justify your answers.

6. A wrench is accidentally dropped from the top of the mast on a sailboat. Will the wrench hit at the same place on the deck whether the sailboat is at rest or moving with a constant velocity? Justify your answer.

7. A rifle, at a height H above the ground, fires a bullet parallel to the ground. At the same instant and at the same height, a second bullet is dropped from rest. In the absence of air resistance, which bullet strikes the ground first? Explain.

8. A rock is thrown upward from the surface of the earth. Another rock is thrown upward from the surface of the moon and has the same initial velocity as the rock on earth. Which rock has the greater range, and which attains the greater height? Why?

9. A railroad flatcar is equipped with a cannon that points straight up. The train is moving due east at a constant velocity, and the cannon is fired. Neglecting air resistance, discuss where the shell will strike when it returns to earth.

10. A football quarterback throws a pass on the run and then keeps running without changing his velocity. Is there any way for the quarterback to throw the pass and then catch it himself? Give your reasoning.

11. On a riverboat cruise, a plastic bottle is accidentally dropped overboard. A passenger on the boat estimates that the boat pulls ahead of the bottle by 5 meters each second. Is it possible to conclude that the boat is moving at 5 m/s with respect to the shore? Give a reason.

12. A plane takes off at St. Louis, flies straight to Denver, and then returns the same way. The plane flies at the same speed with respect to the ground during the entire flight, and there are no head winds or tail winds. Since the earth revolves around its axis once a day, you might expect that the times for the outbound trip and the return trip differ, depending on whether the plane flies against the earth's rotation or with it. However, under the conditions given, the two flight times are identical. Explain why.

13. Three swimmers can swim equally fast. They have a race to see who can swim across a river in the least time. Swimmer A swims perpendicular to the current and lands on the far shore downstream, because the current has swept him in that direction. Swimmer B swims upstream at an angle to the current and lands on the far shore directly opposite the starting point. Swimmer C swims downstream at an angle to the current in an attempt to take advantage of the current. Who crosses the river in the least time? Justify your answer.

PROBLEMS

Section 3.1 Displacement, Velocity, and Acceleration

1. A student rides up a dormitory elevator for a distance of 36 m. She then walks 19 m along a straight hallway to her room. What is the magnitude of her displacement from the base of the elevator to her room?

2. A radar antenna is tracking a satellite orbiting the earth. At a certain time, the radar screen shows the satellite to be 162 km away. The radar antenna is pointing upward at an angle of 62.3° from the ground. Find the x and y components (in km) of the position of the satellite.

3. A mountain-climbing expedition establishes two intermediate camps, labeled A and B in the drawing, above the base camp. What is the magnitude s of the displacement between camp A and camp B?

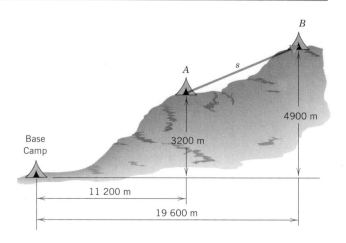

4. A person stands 20.0 m in front of a tall building and looks up toward a window. The person's line of sight makes an angle of 55.0° with the horizontal. How far above the person's eyes is the window?

5. The altitude of a hang glider is increasing at a rate of 6.80 m/s. At the same time, the shadow of the glider moves along the ground at a speed of 15.5 m/s when the sun is directly overhead. Find the magnitude of the glider's velocity.

6. A jetliner is moving at a speed of 245 m/s. The vertical component of the plane's velocity is 40.6 m/s. Determine the magnitude of the horizontal component of the plane's velocity.

7. A dart is thrown upward at an angle of 25° from the horizontal. The vertical component of the dart's velocity is $v_y = +2.2$ m/s. Determine the x component of the velocity.

8. A Formula-one race car is heading due north with a speed of 78.0 m/s. In a time of 3.30 s the car speeds up to 96.0 m/s. What is the average acceleration of the car?

***9.** A sailboat in a race travels the course shown in the drawing. (a) Determine the displacement (magnitude and direction) of the sailboat between points A and B. (b) If the time to move from A to B is 1100 s, what is the average velocity (magnitude and direction) of the boat?

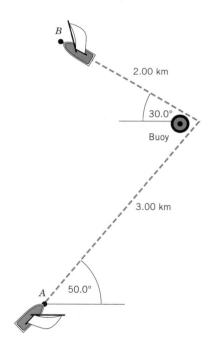

Section 3.2 Equations of Kinematics in Two Dimensions, Section 3.3 Projectile Motion

10. A spacecraft is traveling along the $+x$ direction with a speed of $v_{0x} = 5480$ m/s. Two engines are turned on for a time of 842 s. One engine gives the spacecraft an acceleration in the $+x$ direction of $a_x = 1.20$ m/s², while the other gives it an acceleration in the $+y$ direction of $a_y = 8.40$ m/s². At the end of the firing, find (a) v_x and (b) v_y.

11. The punter on a football team tries to kick a football so that it stays in the air for a long "hang time." If the ball is kicked with an initial velocity of 25.0 m/s at an angle of 60.0° above the ground, what is the "hang time"?

12. A rock climber throws a small first aid kit to another climber who is higher up the mountain. The initial velocity of the kit is 11 m/s at an angle of 65° above the horizontal. At the instant when the kit is caught, it is traveling horizontally, so its vertical speed is zero. What is the vertical height between the two climbers?

13. With a particular club, the maximum speed that a golfer can impart to a ball is 30.3 m/s. (a) What is the longest hole in one that the golfer can make, if the ball does not roll when it hits the green? (b) How much time does the ball spend in the air?

14. During a baseball game a fly ball is hit to center field and is caught 115 m from home plate. Simultaneously, a runner on third base takes off for home, and the center fielder throws the ball to the catcher standing on home plate. The runner takes 3.50 s to reach home, while the baseball is thrown with a velocity whose *horizontal* component is 41 m/s. Which reaches home first, the runner or the ball, and by how much time?

15. A bullet is fired from a rifle that is held 1.6 m above the ground in a horizontal position. The initial speed of the bullet is 1100 m/s. Find (a) the time it takes for the bullet to strike the ground and (b) the horizontal distance traveled by the bullet.

16. A quarterback throws a pass to a receiver, who catches it at the same height as the pass is thrown. The initial velocity of the ball is 15.0 m/s, at an angle of 25.0° above the horizontal. What is the horizontal component of the ball's velocity when the receiver catches it?

17. If a projectile has a launching angle of 52.0° above the horizontal and an initial speed of 18.0 m/s, what is the highest barrier that the projectile can clear?

18. A major league pitcher can throw a baseball in excess of 41 m/s. If a ball is thrown horizontally at this speed, how much can it be expected to drop due to gravity by the time it reaches a catcher who is 17 m away from the point of release? (Pitcher's mounds are raised to compensate for this drop.)

19. A car drives straight off the edge of a cliff that is 54 m high. The police at the scene of the accident note that the point of impact is 130 m from the base of the cliff. How fast was the car traveling when it went over the cliff?

20. A jet fighter is traveling horizontally with a speed of 111 m/s at an altitude of 3.00×10^2 m, when the pilot accidentally releases an outboard fuel tank. (a) How much time elapses before the tank hits the ground? (b) What is the speed of the tank just before it hits the ground?

21. An archer is standing inside a building whose ceiling is 11 m high. An arrow is shot from ground level at an initial speed of 62 m/s. Calculate the angle of firing (above the horizontal) that gives the greatest possible range inside the building.

22. Suppose the water at the top of Niagara Falls has a horizontal speed of 2.7 m/s just before it cascades over the edge of the falls. The height of the falls is 59 m. What is the magnitude of the water's velocity just before the water strikes the bottom? Treat the water particles as if they are in free-fall.

23. A tennis ball is struck such that it leaves the racket horizontally with a speed of 28.0 m/s. The ball hits the court at a horizontal distance of 19.6 m from the racket. What is the height of the tennis ball when it leaves the racket?

24. A motorcycle daredevil is attempting to jump across as many buses as possible (see the drawing). The takeoff ramp makes an angle of 18.0° above the horizontal, and the landing ramp is identical to the takeoff ramp. The buses are parked side by side, and each bus is 2.74 m wide. The cyclist leaves the ramp with a speed of 33.5 m/s. What is the maximum number of buses over which the cyclist can jump?

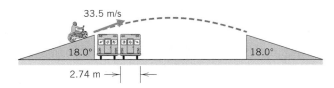

25. A horizontal rifle is fired at a bull's-eye. The muzzle speed of the bullet is 670 m/s. The bullet strikes the target 0.025 m below the center of the bull's-eye. What is the horizontal distance between the end of the rifle and the bull's-eye?

26. A criminal is escaping across a rooftop and runs off the roof horizontally, landing on the roof of an adjacent building. The horizontal distance between the two buildings is 3.4 m, and the roof of the adjacent building is 2.0 m below the jumping off point. What would be the minimum speed needed by the criminal?

27. During a fireworks display, a rocket is launched with an initial velocity of 35 m/s at an angle of 75° above the ground. The rocket explodes 3.7 s later. What is the height of the rocket when it explodes?

*28. A rock, thrown horizontally from the top of a lighthouse, strikes the water 2.6 s later. A straight line is drawn from the top of the lighthouse to the point where the rock strikes the water. This line makes an angle of 35° with respect to the lighthouse. Calculate the initial speed of the rock.

*29. From the edge of a 60.0-m cliff, a small rocket is launched upward with an initial velocity of 23.0 m/s at an angle of 50.0° with respect to the horizontal. At what point

above the ground does the rocket strike the wall of a vertical cliff located 20.0 m away?

*30. A golfer is standing on a fairway and hits a shot to a green that is elevated 6.0 m above the point where she is standing. If the ball leaves her club with a velocity of 43 m/s at an angle of 40.0° to the ground, find (a) the time for the ball to come down on the green, and (b) the speed of the ball just before impact. *(See Solved Problem 1 for a related problem.)*

*31. An airplane is flying with a speed of 240 m/s at an angle of 30.0° with the horizontal, as the drawing shows. When the altitude of the plane is 2.4 km, a flare is released from the plane. The flare hits the target on the ground. What is the angle θ? *(See Solved Problem 1 for a related problem.)*

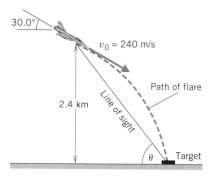

*32. An airplane, with a speed of 97.5 m/s, is climbing upward at an angle of 50.0° with respect to the horizontal. When the plane's altitude is 732 m, the pilot releases a package. (a) Calculate the distance along the ground, measured from a point directly beneath the plane, to the point where the package hits the earth. (b) Relative to the ground, determine the angle of the velocity vector of the package just before impact. *(See Solved Problem 1 for a related problem.)*

*33. A diver springs upward from a board that is three meters above the water. At the instant she contacts the water her speed is 8.90 m/s and her body makes an angle of 75.0° with respect to the surface of the water. Determine her initial velocity, both magnitude and direction. *(See Solved Problem 2 for a related problem.)*

*34. After leaving the end of a ski ramp, a ski jumper lands downhill at a point that is displaced 55 m horizontally from the end of the ramp. His velocity, just before landing, is 25 m/s and points in a direction 38° below the horizontal. Neglecting air resistance and any lift that he experiences while airborne, find his initial velocity (magnitude and direction) when he left the end of the ramp. *(See Solved Problem 2 for a related problem.)*

*35. A golf ball is driven from a level fairway. At a time of 5.10 s later, the ball is traveling downward with a velocity of 48.6 m/s at an angle of 22.2° below the horizontal. Calculate the initial velocity (magnitude and direction) of the golf ball. *(See Solved Problem 2 for a related problem.)*

*36. Stones are thrown horizontally with the same velocity from the tops of two different buildings. One stone lands twice as far from the base of the building from which it was thrown as does the other stone. Compare the heights of the buildings.

**37. A garden hose, pointed at an angle of 25° above the horizontal, splashes water on a sunbather lying on the ground 4.4 m away in the horizontal direction. If the hose is held 1.4 m above the ground, at what speed does the water leave the nozzle?

**38. If the *maximum* horizontal distance ($\theta = 45°$) that a ball can be thrown is 47.0 m, how high can it be thrown straight upward, assuming the same throwing speed in each case?

**39. The drawing shows an exaggerated view of a rifle that has been "sighted in" for a 91.4-meter target. If the muzzle speed of the bullet is $v_0 = 427$ m/s, what are the two possible angles θ_1 and θ_2 between the rifle barrel and the horizontal such that the bullet will hit the target? One of these angles is so large that it is never used in target shooting. (*Hint: The following trigonometric identity may be useful:* $2 \sin \theta \cos \theta = \sin 2\theta$.)

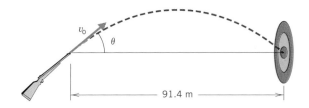

**40. A placekicker is about to kick a field goal. The ball is 27.4 m from the goal post. The ball is kicked with an initial velocity of 19.8 m/s at an angle θ above the ground. Between what two angles, θ_1 and θ_2, will the ball clear the 2.74-meter-high crossbar? (*Hint: The following trigonometric identities may be useful:* $\sec \theta = 1/\cos \theta$ *and* $\sec^2 \theta = 1 + \tan^2 \theta$.)

**41. A small can is hanging from the ceiling by an electromagnet. A rifle is aimed directly at the can, as the figure illustrates. At the instant the gun is fired, the can is released. Ignore air resistance and show that the bullet will always strike the can, regardless of the initial speed of the bullet. Assume that the bullet strikes the can before the can reaches the ground.

Section 3.4 Relative Velocity

42. At some airports there are speedramps to help passengers in getting from one place to another. A speedramp is a moving conveyor belt that you can either stand or walk on. Suppose a speedramp has a length of 105 m and is moving at a speed of 2.0 m/s relative to the ground. In addition, suppose you can cover this distance in 75 s when walking on the ground. If you walk at the same rate with respect to the speedramp that you walk with respect to the ground, how long does it take for you to travel the 105 m using the speedramp?

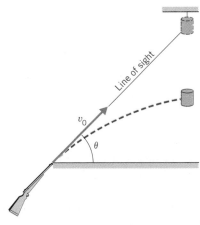

Problem 41

43. A bus has a velocity of 25 m/s due south. A passenger walks to the back of the bus with a speed of 1.0 m/s relative to the bus. What is the velocity (magnitude and direction) of the passenger relative to a person standing on the ground?

44. Two cars, A and B, are traveling in the same direction, although car B is 186 m behind car A. The speed of B is 24.4 m/s and the speed of A is 18.6 m/s. How much time does it take for B to catch A?

45. Two passenger trains are passing each other on adjacent tracks. Train A is moving east with a speed of 13 m/s, and train B is traveling west with a speed of 28 m/s. (a) What is the velocity (magnitude and direction) of train A as seen by the passengers in train B? (b) What is the velocity (magnitude and direction) of train B as seen by the passengers in train A?

46. A swimmer, capable of swimming at a speed of 1.4 m/s in still water (i.e., the swimmer can swim with a speed of 1.4 m/s relative to the water), starts to swim directly across a 2.8-km-wide river. However, the current is 0.91 m/s, and it carries the swimmer downstream. (a) How long does it take the swimmer to cross the river? (b) How far downstream will the swimmer be upon reaching the other side of the river?

47. A remote-controlled model airplane is flying due east in still air. The airplane travels with a speed of 22.6 m/s relative to the air. A wind suddenly begins to blow from the north toward the south with a speed of 8.70 m/s. Find the velocity (magnitude and direction) of the airplane as seen by the controller who is standing on the ground.

48. The captain of an airliner wishes to proceed due west. The cruising speed of the plane is 245 m/s relative to the air. A weather report indicates that a 38.0-m/s wind is blowing from the south to the north. In what direction, relative to due west, should the pilot head the plane?

*49. A ferry boat is traveling in a direction 35.1° north of east with a speed of 5.12 m/s relative to the water. A passenger is walking with a velocity of 2.71 m/s due east relative to the boat. What is the velocity (magnitude and direction) of the passenger with respect to the water?

*50. An oceanliner is heading due north with a speed of 8.5 m/s relative to the water. A small sailboat is heading 45° east of north with a speed of 1.0 m/s relative to the water. Find the relative velocity (magnitude and direction) of the sailboat as observed by the passengers on the oceanliner.

*51. A yacht can travel at a speed of 3.33 m/s relative to the water. The captain wishes to reach a marina that is 3.00 km *directly across* the river from his location. The water is flowing at a speed of 1.11 m/s. (a) At what angle (measured *upstream* relative to the perpendicular line crossing the river) must the captain steer the boat to reach his destination? (b) How much time is required for the yacht to reach the marina?

ADDITIONAL PROBLEMS

52. A baseball player hits a triple and ends up on third base. A baseball "diamond" is a square, each side of length 27.4 m, with home plate and the three bases on the four corners. What is the magnitude of his displacement?

53. What is the smallest muzzle velocity that a bullet can have, if a horizontally fired bullet is to hit a 0.0254-meter-diameter target located 26.2 m away? Assume that the center of the target is on the same horizontal line as is the barrel of the rifle.

54. A diver runs horizontally with a speed of 1.20 m/s off a platform that is 10.0 m above the water. What is his speed just before striking the water?

55. Two cars with different velocities are approaching an intersection. Car A, traveling due east, has a speed of 15 m/s. Car B, traveling due north, has a speed of 21 m/s. What is the velocity (magnitude and direction) of B as seen by the passengers in A?

56. At a certain point along its trajectory, a golf ball has the following velocity components: $v_x = +14.0$ m/s and $v_y = +18.9$ m/s, with upward and to the right being positive. (a) What is the speed at this point? (b) At what angle does the velocity vector point relative to the horizontal direction?

57. A hot-air balloon is rising straight up with a speed of 3.0 m/s. A ballast bag is released from rest relative to the balloon when it is 9.5 m above the ground. How much time elapses before the ballast bag hits the ground?

58. A box of .22-caliber bullets has the following message written on it: "Warning! Range, 2.0 km." Assume that the muzzle speed of a bullet is 340 m/s and calculate the maximum range ($\theta = 45°$). Compare your answer with the "warn-

ing." (Of course, air resistance will slow down the bullet, so its actual range will be substantially less than that calculated.)

*59. A remote-controlled "target" plane is flying horizontally at an altitude of 4.20 km with a speed of 225 m/s. When the plane is directly overhead, a projectile is fired at an angle θ with the ground and has an initial speed of 389 m/s, as the diagram shows. The projectile hits the plane. Find the angle θ.

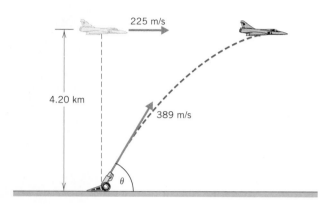

*60. A small aircraft is headed due south with a speed of 57.8 m/s with respect to still air. Then, for 9.00×10^2 s a wind blows the plane so that it moves in a direction 45.0° west of south, even though the plane continues to point due south. The plane travels 81.0 km with respect to the ground in this time. Determine the velocity (magnitude and direction) of the wind with respect to the ground.

*61. A projectile is fired at an angle θ above the horizontal. Prove that the time for the projectile to travel from the ground to its maximum height is equal to the time to travel from its maximum height back to the ground.

*62. A soccer player kicks the ball toward a goal that is 29.0 m in front of him. The ball leaves his foot at a speed of 19.0 m/s and an angle of 32.0° above the ground. Find the speed of the ball when the goalie catches it in front of the net.

**63. A jetliner can fly 6.00 hours on a full load of fuel. Without any wind it flies at a speed of 2.40×10^2 m/s. The plane is to make a round-trip by heading due west for a certain distance, turning around, and then heading due east for the return trip. During the entire flight, however, the plane encounters a 57.8 m/s wind from the jet stream, which blows from west to east. What is the maximum distance that the plane can travel due west and just be able to return home?

**64. A baseball is hit into the air at an initial speed of 36.6 m/s and an angle of 50.0° above the horizontal. At the same time, the center fielder starts running away from the batter and catches the ball 0.914 m above the level at which it was hit. If the center fielder is initially 1.10×10^2 m from home plate, what must be his average speed?

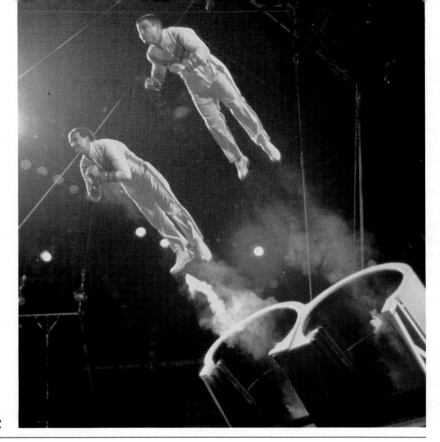

CHAPTER 4

FORCES AND NEWTON'S LAWS OF MOTION

Forces are an important part of our lives. Just ask these human cannonballs who are being shot from cannons at the Ringling Brothers Circus. They are propelled outward by forces within the cannons and, once airborne, are eventually brought back to earth by the force due to gravity. All of us are continually being acted upon by forces, whether we are sitting, running, or skiing down the side of a mountain. Forces cause a car to speed up when the driver steps on the accelerator and to slow down when the brakes are applied. Forces also play a central role in the design of buildings and bridges, and they cause airplanes to fly. In this chapter we will investigate some familiar types of forces, such as the force due to gravity, friction, and the tension in a rope. We will also explore how forces can change the motion of an object or keep an object at rest. Three laws discovered by Isaac Newton are central to understanding the relation between forces and motion, and we will show how useful these laws are in understanding physical phenomena.

4.1 THE CONCEPTS OF FORCE AND MASS

Force is a common word and usually means a pull or a push. A tow truck pulls a stalled car by applying a force to the car. The force of gravity continually pulls all

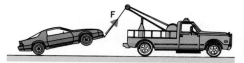

Figure 4.1   The arrow labeled **F** represents the force acting on each object.

objects downward and, thereby, causes a fly ball in a baseball game to return to the earth. And a bulldozer pushes over the wall of a dilapidated building by exerting a force on the wall. Figure 4.1 illustrates each of these examples and uses arrows to represent the forces. It is appropriate to use arrows, because a force is a vector quantity and has both a magnitude and a direction. The direction of the arrow gives the direction of the force, and the length is proportional to the strength, or magnitude, of the force.

The word *mass* also occurs in our vocabulary. A massive supertanker, for instance, is one that contains an enormous amount of mass. As we will see in the next section, such massive objects are difficult to get moving and are hard to stop once they are in motion. In comparison, a penny does not contain much mass. The emphasis here is on the amount of mass, and the idea of direction is of no concern. Therefore, mass is a scalar quantity.

During the seventeenth century, Isaac Newton, starting with the work of Galileo, developed three important laws that deal with force and mass. They are collectively called "Newton's laws of motion" and provide the basis for understanding the effect that forces have on an object. Because of the importance of these laws, a separate section will be devoted to each one.

4.2 NEWTON'S FIRST LAW OF MOTION

THE FIRST LAW

To gain some insight into Newton's first law, think about the game of ice hockey (Figure 4.2). If a player does not hit a stationary puck, it will remain at rest on the ice. After the puck is struck, however, it coasts on its own across the ice, slowing down only slightly because of friction. Since ice is very slippery, there is only a relatively small amount of friction to slow down the puck. In fact, if it were possible to remove all friction and wind resistance, and if the rink were infinitely large, the puck would coast forever in a straight line at a constant speed. Left on its own, the puck would lose none of the velocity imparted to it at the time it was struck. Here we have the essence of Newton's first law of motion:

Newton's First Law of Motion

An object continues in a state of rest or in a state of motion at a constant speed along a straight line, unless compelled to change that state by a net force.

The first law uses the phrase "net force." Often, several forces act simultaneously on a body, and *the net force is the vector sum of all of them.* Individual forces matter only to the extent that they contribute to the total. For instance, if friction and other opposing forces were absent, a car could travel forever at 30 m/s in a straight line, without using any gas after it has come up to speed. In reality, of course, gas is needed, but only so that the engine can produce the

Figure 4.2   The game of ice hockey can give some insight into Newton's laws of motion.

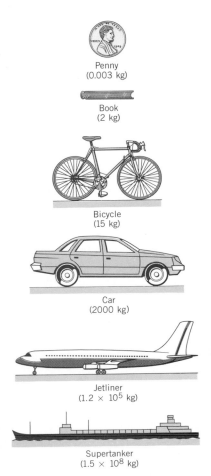

Penny
(0.003 kg)

Book
(2 kg)

Bicycle
(15 kg)

Car
(2000 kg)

Jetliner
(1.2×10^5 kg)

Supertanker
(1.5×10^8 kg)

Figure 4.3   The masses of various objects.

THE PHYSICS OF . . .

seat belts.

THE PHYSICS OF . . .

automobile navigation
systems.

necessary driving force to cancel opposing forces such as friction. This cancellation ensures that there is no net force to change the state of motion of the car.

When an object moves at a constant speed along a straight line, its velocity is constant. Newton's first law indicates that a state of rest (zero velocity) and a state of constant velocity are completely equivalent, in the sense that neither one requires the application of a net force to sustain it. The purpose served when a net force acts on an object is not to sustain the velocity of the object, but, rather, to *change the velocity*.

INERTIA AND MASS

A greater net force is required to change the velocity of some objects than of others. For instance, a net force may cause a bicycle to pick up speed quickly. But when the same force is applied to a freight train, any resulting change in the motion is imperceptible. Accordingly, we say that the train has more *inertia* than the bicycle. Quantitatively, the inertia of an object is measured by its *mass*, a large mass, such as that of a freight train, indicating a large inertia. The following definition of inertia and mass indicates why Newton's first law is sometimes called the law of inertia:

Definition of Inertia and Mass

Inertia is the natural tendency of an object to remain at rest or in motion at a constant speed along a straight line. The mass of an object is a quantitative measure of inertia.

SI Unit of Inertia and Mass: kilogram (kg)

The SI unit for mass is the kilogram (kg), whereas the units in the CGS system and the BE system are the gram (g) and the slug (sl), respectively. Conversion factors between these units are given on the page facing the inside of the front cover. Figure 4.3 gives the masses of various objects, ranging from a penny to a supertanker. The larger the mass, the greater is the inertia. Often the words "mass" and "weight" are used interchangeably, but they should not be. Mass and weight are different concepts, and Section 4.7 will discuss the distinction between them.

Figure 4.4 shows a useful application of inertia. Seat belts unwind freely when pulled gently, so they can be buckled. But in an accident, they hold you safely in place. One mechanism used for seat belts consists of a ratchet wheel, a locking bar, and a pendulum. The seat belt is wound around a spool mounted on the ratchet wheel. While the car is at rest or moving at a constant velocity, the pendulum hangs straight down, and the locking bar rests horizontally, as the black-lined part of the drawing shows. Consequently, there is nothing to prevent the ratchet wheel from turning, and the seat belt can be pulled out easily. When the car suddenly slows down in an accident, however, the relatively massive, lower part of the pendulum keeps moving forward because of its inertia. The pendulum swings on its pivot into the position shown in color and causes the locking bar to block the rotation of the ratchet wheel, thus preventing the seat belt from unwinding.

Inertia also lies at the heart of one type of electronic navigation system being developed for automobiles. Sensors mounted on the car measure the distance traveled and changes in the direction of the motion. A computer uses this information to mark the current location of the car on a map displayed on a dashboard viewing screen, as in Figure 4.5a. So, as the driver proceeds to his destination, he can watch his location marker move on the screen. A main element in this system is the sensor that detects changes in the car's direction. The sensor contains two heated wires surrounded by a gas, as Figure 4.5b shows. When the car moves straight ahead (constant velocity), the gas is uniformly distributed and cools each wire equally. Thus, each wire has the same temperature. However, when the car changes direction, inertia keeps the gas moving straight ahead. If the car turns right, for instance, as in part c of the drawing, the gas is then on the left side of the sensor and cools the left wire more than the right wire. From the difference in temperatures of the wires, the computer can tell which way the car has turned.

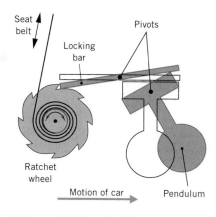

Figure 4.4 Inertia plays a central role in one seat belt mechanism. The colored parts of the drawing show what happens when the car suddenly slows down, as in an accident.

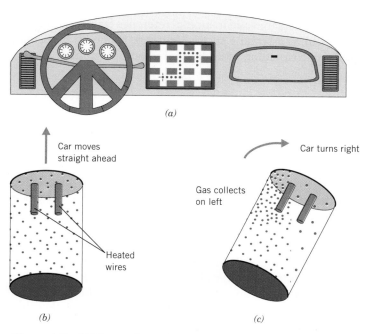

Figure 4.5 (a) Electronic navigation systems for cars will use a dashboard viewing screen to display a map. (b) The temperatures of the heated wires in a gas-filled sensor are the same when the car travels straight ahead. (c) The temperature of the left wire becomes lower than that of the right wire when the car turns right.

AN INERTIAL REFERENCE FRAME

Sometimes Newton's first law (and also the second law) can appear to be invalid to certain observers. Suppose, for instance, that you are a passenger riding in a friend's car. While the car moves at a constant speed along a straight line, you do not feel the seat pushing against your back to any unusual extent. This experience is consistent with the first law, which indicates that in the absence of a net force you should move with a constant velocity. Suddenly the driver floors the gas pedal. Immediately you feel the seat pressing against your back as the car accelerates. Therefore, you sense that a force is being applied to you. The first law leads you to believe that your motion should change, and, relative to the ground outside, your motion does change. But *relative to the car*, you can see that your motion does *not* change, because you remain stationary with respect to the car.

Clearly, Newton's first law does not hold for observers who use the accelerating car as a frame of reference. As a result, such a reference frame is said to be noninertial. All accelerating reference frames are noninertial. In contrast, observers for whom the law of inertia is valid are said to be using *inertial reference frames* for their observations, as defined below:

Definition of an Inertial Reference Frame

An inertial reference frame is one in which Newton's law of inertia is valid.

All of Newton's laws of motion are valid in inertial reference frames, and when we apply these laws, we will be assuming such a reference frame. In particular, the earth itself is a good approximation of an inertial reference frame.

4.3 NEWTON'S SECOND LAW OF MOTION

THE SECOND LAW

Newton's first law indicates that if no net force acts on an object, then the velocity of the object remains unchanged. The second law deals with what happens when there is a net force acting. Consider a hockey puck once again. When a player strikes a stationary puck, he causes the velocity of the puck to change. In other words, he makes the puck accelerate. The cause of the acceleration is the force that the hockey stick applies. As long as this force acts, the velocity increases, and the puck accelerates. Now, suppose another player strikes the puck and applies twice as much force as the first player does. The greater force produces a greater acceleration. In fact, if the friction between the puck and the ice is negligible, and if there is no wind resistance, the acceleration of the puck is directly proportional to the force. Twice the force produces twice the acceleration. Moreover, the acceleration is a vector quantity, just as the force is, and points in the same direction as the force.

Often, several forces act on an object simultaneously. Friction and wind resistance, for instance, do have some effect on a hockey puck. In such cases, it is the net force, or the vector sum of all the forces acting, that is important. Mathematically, the net force is written as $\Sigma\mathbf{F}$, where the Greek capital letter Σ (sigma) denotes the vector sum. Newton's second law states that the acceleration is proportional to the net force acting on the object.

In Newton's second law, the net force is only one of two factors that determine the acceleration. The other factor is the inertia or mass of the object. After all, the same net force that imparts an appreciable acceleration to the hockey puck (small mass) will impart very little acceleration to a semitrailer truck (large mass). Newton's second law states that for a given net force, the magnitude of the acceleration is inversely proportional to the mass. Twice the mass means one-half the acceleration, if the same net force acts on both objects. Thus, the second law shows how the acceleration depends on both the net force and the mass, as summarized below in Equation 4.1.

Newton's Second Law of Motion

When a net force $\Sigma\mathbf{F}$ acts on an object of mass m, the acceleration $\mathbf{a}$ that results is directly proportional to the net force and has a magnitude that is inversely proportional to the mass. The direction of the acceleration is the same as the direction of the net force.

$$\mathbf{a} = \frac{\Sigma\mathbf{F}}{m} \quad \text{or} \quad \Sigma\mathbf{F} = m\mathbf{a} \tag{4.1}$$

SI Unit of Force: $kg\cdot m/s^2$ = newton (N)

According to Equation 4.1, the SI unit for force is the unit for mass (kg) times the unit for acceleration (m/s^2), or

$$\text{SI unit for force} = (kg)\left(\frac{m}{s^2}\right) = \frac{kg\cdot m}{s^2}$$

The combination of $kg\cdot m/s^2$ is called a *newton* (N) and is a derived SI unit, not a base unit; 1 newton = 1 N = 1 $kg\cdot m/s^2$.

In the CGS system, the procedure for establishing the units is the same as with SI units, except that mass is expressed in grams (g) and acceleration in cm/s^2. The resulting unit for force is the *dyne*; 1 dyne = 1 $g\cdot cm/s^2$.

In the BE system, the unit for force is defined to be the pound (lb),* while the unit for acceleration is ft/s^2. With this procedure, Newton's second law can then be used to obtain the unit for mass:

$$\text{BE unit for force} = lb = m\left(\frac{ft}{s^2}\right)$$

$$m = \frac{lb\cdot s^2}{ft}$$

The combination of $lb\cdot s^2/ft$ is the unit for mass in the BE system and is called the *slug* (sl); 1 slug = 1 sl = 1 $lb\cdot s^2/ft$.

Table 4.1 summarizes the various units for force, mass, and acceleration. Conversion factors between force units from different systems are provided on the page facing the inside of the front cover.

Table 4.1 Units for Mass, Acceleration, and Force

System	Mass	Acceleration	Force
SI	kilogram (kg)	meter/second² (m/s^2)	newton (N)
CGS	gram (g)	centimeter/second² (cm/s^2)	dyne (dyn)
BE	slug (sl)	foot/second² (ft/s^2)	pound (lb)

* We refer here to the gravitational version of the BE system, in which a force of one pound is defined to be the pull of the earth on a certain standard body at a location where the acceleration due to gravity is 32.174 ft/s^2.

FREE-BODY DIAGRAMS AND THE SECOND LAW

When using the second law to calculate the acceleration, it is necessary to determine the net force that acts on the object. In this determination a *free-body diagram* helps enormously. A free-body diagram is a diagram that represents the object and the forces that act on it. Only the forces that *act on the object* appear in a free-body diagram. Forces that the object exerts on its environment are not included. Example 1 illustrates the use of a free-body diagram.

Example 1   Pushing a Stalled Car

Two people are pushing a stalled car, as Figure 4.6*a* indicates. The mass of the car is 1850 kg. One person applies a force of 275 N to the car, while the other applies a force of 395 N. Both forces act in the same direction. A third force of 560 N also acts on the car, but in a direction opposite to that in which the people are pushing. This force arises because of friction and the extent to which the pavement opposes the motion of the tires. Find the acceleration of the car.

REASONING   Before Newton's second law can be used to find the acceleration, the net force acting on the car must be determined. Figure 4.6*b* shows the free-body diagram for the car, in which the car is represented as a dot and the motion of the car is chosen to be along the *x* axis. The diagram makes it clear that the forces all act along one direction. Therefore, they can be added as colinear vectors to obtain the net force.

SOLUTION   The net force is

$$\Sigma F = +275 \text{ N} + 395 \text{ N} - 560 \text{ N} = +110 \text{ N}$$

The acceleration can now be obtained:

$$a = \frac{\Sigma F}{m} = \frac{+110 \text{ N}}{1850 \text{ kg}} = \boxed{+0.059 \text{ m/s}^2} \tag{4.1}$$

The plus sign indicates that the acceleration points along the +*x* axis, in the same direction as the net force.

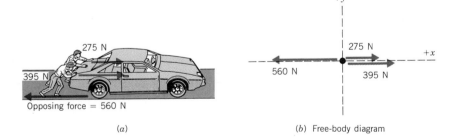

Figure 4.6   *(a)* Two people push a stalled car, in opposition to a force created by friction and the pavement. *(b)* A free-body diagram shows the forces acting on the car.

NEWTON'S FIRST LAW AS A SPECIAL CASE OF THE SECOND LAW

We have been discussing Newton's first and second laws as if they were independent. In fact, the first law can be considered as a special case of the more general second law. When the net force acting on an object is zero, $\Sigma F = 0$. Substituting this value into Newton's second law, $\Sigma F = m\mathbf{a}$, reveals that the acceleration $\mathbf{a}$ is also zero. A zero acceleration means that the object travels with a constant velocity, that is, with a constant speed along a straight line, which is exactly the kind of motion dealt with by the first law.

4.4 THE VECTOR NATURE OF NEWTON'S SECOND LAW OF MOTION

When a football player throws a pass, the direction of the force he applies to the ball is important. Both the force and the resulting acceleration of the ball are vector quantities, as are all forces and accelerations. The directions of these vectors can be taken into account in two dimensions by using x and y components. The net force $\Sigma \mathbf{F}$ in Newton's second law has components ΣF_x and ΣF_y, while the acceleration $\mathbf{a}$ has components a_x and a_y. Consequently, Newton's second law, as expressed in Equation 4.1, can be written in an equivalent form as two equations, one for the x components and one for the y components:

$$\Sigma F_x = ma_x \tag{4.2a}$$

$$\Sigma F_y = ma_y \tag{4.2b}$$

This procedure is similar to that employed in Chapter 3 for the equations of two-dimensional kinematics (see Table 3.1). The vector components themselves in Equations 4.2a and 4.2b will be either positive or negative numbers, depending on whether they point along the positive or negative x or y axis. The remainder of this section deals with examples that show how these equations are used.

A quarterback in action; both the force he applies to the ball and the resulting acceleration are vector quantities.

Example 2   An Accelerating Block

In Figure 4.7 a 15.0-kg block moves on a flat, friction-free, horizontal surface. At the instant shown in part a, two forces, $\mathbf{F}_1$ and $\mathbf{F}_2$, begin acting on the block. $\mathbf{F}_1$ has a magnitude of 35.0 N and acts at an angle of 60.0° with respect to the $+x$ axis, while $\mathbf{F}_2$ has a magnitude of 10.0 N and acts along the $+x$ axis. Find the x and y components of the resulting acceleration.

REASONING   Knowing the forces and the mass of the block, we can use Newton's second law to determine the acceleration components. However, it is the net force that appears in the second law. So our first step is to find the components of the net force. We do this by using individual force components to obtain the x component ΣF_x and the y component ΣF_y of the net force.

SOLUTION   Figure 4.7a and the table below show the force components.

Force	x Component
$\mathbf{F}_1$	$+(35.0\ \text{N})\cos 60.0° = +17.5\ \text{N}$
$\mathbf{F}_2$	$+10.0\ \text{N}$
	$\Sigma F_x = +17.5\ \text{N} + 10.0\ \text{N} = +27.5\ \text{N}$

Force	y Component
$\mathbf{F}_1$	$+(35.0\ \text{N})\sin 60.0° = +30.3\ \text{N}$
$\mathbf{F}_2$	0
	$\Sigma F_y = +30.3\ \text{N}$

PROBLEM SOLVING INSIGHT

Applications of Newton's second law always involve the net force, which is the vector sum of all the forces that act on an object.

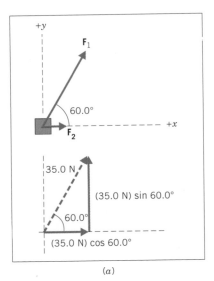

(a)

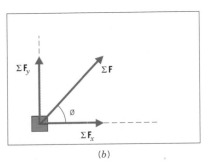

(b)

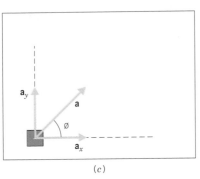

(c)

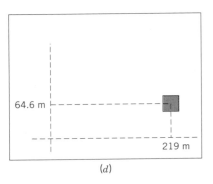

(d)

The plus signs indicate that these components point in the direction of the $+x$ and $+y$ axes, as part b of the drawing shows. The x and y components of the acceleration can now be obtained:

$$a_x = \frac{\Sigma F_x}{m} = \frac{+27.5 \text{ N}}{15.0 \text{ kg}} = \boxed{+1.83 \text{ m/s}^2} \qquad (4.2a)$$

$$a_y = \frac{\Sigma F_y}{m} = \frac{+30.3 \text{ N}}{15.0 \text{ kg}} = \boxed{+2.02 \text{ m/s}^2} \qquad (4.2b)$$

Figure 4.7c shows these components of the acceleration.

Example 3 The Displacement of an Accelerating Block

When the forces begin acting in Figure 4.7a, the block has a velocity of 20.0 m/s directed along the $+x$ axis. Use the acceleration components found in Example 2 and determine the displacement components of the block after 8.00 s has elapsed.

REASONING   Once the forces acting on an object and its mass have been used in Newton's second law to obtain the acceleration, it becomes possible to describe the resulting motion with the aid of the equations of kinematics. Here, we find the x and y components of the block's displacement at a time of $t = 8.00$ s.

SOLUTION   From Example 2 we know that $a_x = +1.83$ m/s^2. We also know that the initial value of the x component of the velocity is $v_{0x} = +20.0$ m/s. Therefore, the x component of the displacement can be obtained from Equation 3.5a:

$$x = v_{0x}t + \tfrac{1}{2}a_x t^2$$
$$= (20.0 \text{ m/s})(8.00 \text{ s}) + \tfrac{1}{2}(1.83 \text{ m/s}^2)(8.00 \text{ s})^2$$
$$= \boxed{+219 \text{ m}}$$

In a similar fashion the y component of the displacement can be obtained from Equation 3.5b, since $a_y = +2.02$ m/s^2 and the initial value of the y component of the velocity is $v_{0y} = 0$: $\boxed{y = +64.6 \text{ m}}$. Figure 4.7d shows the final location of the block.

4.5 NEWTON'S THIRD LAW OF MOTION

Imagine you are in a football game. You line up facing your opponent, the ball is snapped, and the two of you crash together. No doubt, you feel a force. But think about your opponent. He too feels something, for while he is applying a force to you, you are applying a force to him. In other words, there isn't just one force on

Figure 4.7   (a) In Examples 2 and 3, two forces, $\mathbf{F_1}$ and $\mathbf{F_2}$, act on a block. (b) The x and y components of the net force can be used with Newton's second law to find (c) the x and y components of the acceleration. (d) The equations of kinematics can be used to determine the displacement of the block at a later time.

the line of scrimmage; there is a pair of forces. Newton was the first to realize that all forces occur in pairs and there is no such thing as an isolated force, existing all by itself. His third law of motion deals with this fundamental characteristic of forces.

The two teams exert action and reaction forces on each other.

Newton's Third Law of Motion

Whenever one body exerts a force on a second body, the second body exerts an oppositely directed force of equal magnitude on the first body.

The third law is often called the "action–reaction" law, for it is sometimes quoted as follows: "For every action (force) there is an equal, but opposite, reaction."

Figure 4.8 illustrates how the third law applies to an astronaut who is drifting just outside a spacecraft and who pushes on the spacecraft with a force **F**. According to the third law, the spacecraft pushes back on the astronaut with a force −**F** that is equal in magnitude, but opposite in direction. In Example 4, we examine the accelerations produced by each of these forces.

Example 4   The Effects of Action and Reaction

Suppose that the mass of the spacecraft in Figure 4.8 is $m_S = 11\ 000$ kg and that the mass of the astronaut is $m_A = 92$ kg. In addition, assume that the astronaut exerts a force of **F** = +36 N on the spacecraft. Find the accelerations of the spacecraft and the astronaut.

REASONING   According to Newton's third law, when the astronaut applies the force **F** = +36 N to the spacecraft, the spacecraft applies a reaction force −**F** = −36 N to the astronaut. As a result, the spacecraft and the astronaut accelerate in opposite directions. While the action and reaction forces have the same magnitude, they do not create accelerations of the same magnitude. The reason is that the spacecraft and the astronaut have different masses. According to Newton's second law, the astronaut, having a much smaller mass, will experience a much larger acceleration.

SOLUTION   The acceleration of the spacecraft is

$$\mathbf{a}_S = \frac{\mathbf{F}}{m_S} = \frac{+36\ \text{N}}{11\ 000\ \text{kg}} = \boxed{+0.0033\ \text{m/s}^2}$$

The acceleration of the astronaut is

$$\mathbf{a}_A = \frac{-\mathbf{F}}{m_A} = \frac{+36\ \text{N}}{92\ \text{kg}} = \boxed{-0.39\ \text{m/s}^2}$$

Figure 4.8   The astronaut pushes on the spacecraft with a force +**F**. According to Newton's third law, the spacecraft simultaneously pushes back on the astronaut with a force −**F**.

PROBLEM SOLVING INSIGHT

Even though the magnitudes of the action and reaction forces are always equal, these forces do not necessarily produce accelerations that have equal magnitudes, since each force acts on a different object.

Action–reaction situations occur all the time. Figure 4.9 illustrates the process of walking, for example. When you walk, your shoe exerts a force on the earth due to friction. The earth accelerates under the application of this force, but the acceleration is imperceptibly small, since the earth is so massive (approximately 6×10^{24} kg). According to Newton's third law, the earth exerts a reaction force of equal magnitude on the shoe. This reaction force causes you to accelerate forward and, hence, walk. It is difficult to walk on ice, because ice, with its slippery surface,

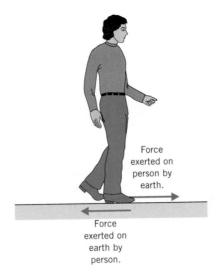

Figure 4.9   To walk, a person exerts a force on the earth. Consistent with Newton's third law, the earth exerts an oppositely directed force of equal magnitude on the person's foot and causes the person to accelerate forward.

Force exerted on person by earth.

Force exerted on earth by person.

THE PHYSICS OF . . .

automatic trailer brakes.

does not permit you to exert a large frictional force on the earth. Correspondingly, the reaction force exerted by the earth is small, and you do not accelerate forward very much.

There is a clever application of Newton's third law in some rental trailers. As Figure 4.10 illustrates, the tow bar connecting the trailer to the rear bumper of a car contains a mechanism that can automatically actuate brakes on the trailer wheels. This mechanism works without the need for electrical connections between the car and the trailer. When the driver applies the car brakes, the car slows down. Because of inertia, however, the trailer continues to roll forward and begins pushing against the bumper. In reaction, the bumper pushes back on the tow bar. The reaction force is used by the mechanism in the tow bar to "push the brake pedal" for the trailer.

Newton's three laws of motion make it very clear that forces play a central role in determining the motion of an object. To predict the motion accurately, we must know all the forces that act on the object. In nature there are four fundamental types of forces, and these will be encountered at various places in this text. They are "fundamental" in the sense that all forces are manifestations of one or more of these four. One of these fundamental forces is the gravitational force, which is the topic of the next section.

Mechanism for actuating trailer brakes

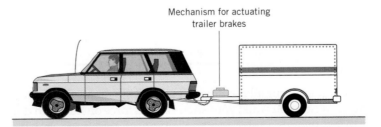

Figure 4.10   Some rental trailers include an automatic brake-actuating mechanism.

4.6 THE GRAVITATIONAL FORCE

NEWTON'S LAW OF UNIVERSAL GRAVITATION

Objects fall downward because of gravity, and Chapters 2 and 3 discuss how to describe the effects of gravity by using a value of $g = 9.80$ m/s² for the downward acceleration it causes. However, nothing has been said about why g is 9.80 m/s². The present section sheds some light on this matter.

The acceleration due to gravity is like any other acceleration, and Newton's second law indicates that it must be caused by a net force. In addition to his famous three laws of motion, Newton also provided a coherent understanding of the *gravitational force.* His "law of universal gravitation" is stated below.

In an avalanche, the gravitational force accelerates a large mass of snow.

Newton's Law of Universal Gravitation

Every particle in the universe exerts an attractive force on every other particle. A particle is a piece of matter, small enough in size to be regarded as a mathematical point. For two particles, which have masses m_1 and m_2 and are separated by a distance r, the force that each exerts on the other is directed along the line joining the particles and has a magnitude given by

$$F = G \frac{m_1 m_2}{r^2} \qquad (4.3)$$

The symbol G denotes the universal gravitational constant, whose value is found experimentally to be

$$G = 6.67(3) \times 10^{-11} \text{ N} \cdot \text{m}^2/\text{kg}^2$$

The constant G that appears in Equation 4.3 is called the ***universal gravitational constant,*** because it has the same value for all pairs of particles anywhere in the universe, no matter what their separation. The value for G was first measured in an experiment by the English scientist Henry Cavendish (1731–1810), more than a century after Newton proposed his law of universal gravitation.

To see the main features of Newton's law of universal gravitation, look at the two particles in Figure 4.11. They have masses m_1 and m_2 and are separated by a distance r. In the picture, it is assumed that a force pointing to the right is positive. The gravitational forces point along the line joining the particles and are

+**F**, the gravitational force exerted on m_1 by m_2
−**F**, the gravitational force exerted on m_2 by m_1

Figure 4.11 The two particles, whose masses are m_1 and m_2, are attracted by gravitational forces +**F** and −**F**.

These two forces have equal magnitudes and opposite directions. They act on different bodies, causing them to be mutually attracted. In fact, these forces are the action–reaction pair required by Newton's third law. Example 5 shows that the magnitude of the gravitational force is extremely small when the masses have ordinary values.

Moon
M_M +F   −F
Earth
M_E

r

M_M +F   −F M_E

Figure 4.12   The gravitational force between two uniform spheres of matter is the same as if each sphere were a particle with its mass concentrated at its center. The earth (mass M_E) and the moon (mass M_M) approximate such uniform spheres.

A satellite view of earth.

Example 5   The Attraction of Gravity

What is the magnitude of the gravitational force that acts on each particle in Figure 4.11, assuming $m_1 = 12$ kg (approximately the mass of a bicycle), $m_2 = 25$ kg, and $r = 1.2$ m?

REASONING AND SOLUTION   The magnitude of the gravitational force can be found using Equation 4.3:

$$F = G\frac{m_1 m_2}{r^2} = (6.67 \times 10^{-11} \text{ N·m}^2/\text{kg}^2)\frac{(12.0 \text{ kg})(25.0 \text{ kg})}{(1.20 \text{ m})^2} = \boxed{1.4 \times 10^{-8} \text{ N}}$$

For comparison, you exert a force of about 1 N when pushing a doorbell, so that the gravitational force is exceedingly small in circumstances such as those here. This result is due to the fact that G itself is very small. However, if one of the bodies has a large mass, like that of the earth (5.98×10^{24} kg), the gravitational force can be large.

As expressed by Equation 4.3, Newton's law of gravitation applies only to particles. However, most familiar objects are too large to be considered particles. Nevertheless, the law of universal gravitation can be applied to such objects with the aid of calculus. Newton was able to prove that an object of finite size can be considered to be a particle for purposes of using the gravitation law, provided the mass of the object is distributed with spherical symmetry about its center. Thus, Equation 4.3 can be applied when each object is a sphere whose mass is spread uniformly over its entire volume. Figure 4.12 shows this kind of application, assuming that the earth and the moon are such uniform spheres of matter. In this case, r is the distance *between the centers of the spheres* and not the distance between the outer surfaces. The gravitational forces that the spheres exert on each other are the same as if the entire mass of each was concentrated at its center. Even if the objects are not uniform spheres, Equation 4.3 can be used to a good degree of approximation if the sizes of the objects are small relative to the distance of separation r.

THE MASS OF THE EARTH

Once the value of the gravitational constant G is known, it becomes possible to determine the mass of the earth M_E. For an object of mass m falling freely near the surface of the earth, the acceleration due to gravity is $g = 9.80$ m/s², if the effects of air resistance are negligible. According to Newton's second law, this accelerated motion is produced by a force whose magnitude is given by the product of the mass and the acceleration, $F = mg$. Since the earth's gravitational attraction provides this force according to Equation 4.3, we find that

$$F = mg = G\frac{M_E m}{R_E^2}$$

The radius R_E of the earth has been substituted for the distance r in Equation 4.3, since the value $g = 9.80$ m/s² applies near the earth's surface. Eliminating the mass m algebraically from both sides of this equation and solving for M_E shows that

$$M_E = \frac{gR_E^2}{G}$$

This remarkable result gives the mass of the earth in terms of parameters (g, R_E, and G) that can be experimentally measured. If we consider the earth to be a perfect sphere of radius $R_E = 6.38 \times 10^6$ m, the mass of the earth is

$$M_E = \frac{gR_E^2}{G} = \frac{(9.80 \text{ m/s}^2)(6.38 \times 10^6 \text{ m})^2}{6.67 \times 10^{-11} \text{ N}\cdot\text{m}^2/\text{kg}^2} = 5.98 \times 10^{24} \text{ kg}$$

4.7 WEIGHT

THE WEIGHT OF AN OBJECT

Newton's law of universal gravitation can be used to express the weight an object has due to the gravitational pull of the earth.

Definition of Weight

The weight of an object on the earth is the gravitational force that the earth exerts on the object. The weight always acts downward, toward the center of the earth. On another astronomical body, the weight is the gravitational force exerted on the object by that body.

   SI Unit of Weight: newton (N)

Using W for the magnitude of the weight,* m for the mass of the object, and M_E for the mass of the earth, it follows from Equation 4.3 that

$$W = G\frac{M_E m}{r^2} \qquad (4.4)$$

Equation 4.4 and Figure 4.13 both emphasize that an object has weight whether or not it is resting on the earth's surface, because the gravitational force is acting even when the distance r is not equal to the radius R_E of the earth. The gravitational force becomes weaker as r increases, since r is in the denominator of Equation 4.4. Example 6 demonstrates how the weight changes, depending on where an object is located with respect to the center of the earth.

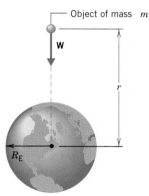

Object of mass m

Mass of Earth = M_E

Figure 4.13 On earth, the weight **W** of an object is the gravitational force exerted on the object by the earth.

---

* Often, the word "weight" and the phrase "magnitude of the weight" are used interchangeably, even though weight is a vector. Generally, the context makes it clear when the direction of the weight vector must be taken into account.

Example 6   The Hubble Space Telescope

The mass of the Hubble space telescope is 11 600 kg. Determine the weight of the telescope when it is (a) resting on the earth and (b) in orbit 596 km above the earth's surface.

REASONING AND SOLUTION
(a) The weight of the telescope on the surface of the earth can be obtained from Equation 4.4 by letting $r = 6.38 \times 10^6$ m, which is the radius of the earth:

$$W = G\frac{M_E m}{r^2} = \frac{(6.67 \times 10^{-11}\ \text{N}\cdot\text{m}^2/\text{kg}^2)(5.98 \times 10^{24}\ \text{kg})(11\ 600\ \text{kg})}{(6.38 \times 10^6\ \text{m})^2}$$

$$\boxed{W = 1.14 \times 10^5\ \text{N}}$$

(b) When the telescope is 596 km above the surface, its distance from the center of the earth is

$$r = 6.38 \times 10^6\ \text{m} + 596 \times 10^3\ \text{m} = 6.98 \times 10^6\ \text{m}$$

The weight now can be calculated as in part (a), except the new value of r must be used: $\boxed{W = 0.950 \times 10^5\ \text{N}}$. As expected, the weight has decreased.

The space age has forced us to broaden our ideas about weight. For instance, an astronaut weighs only about one-sixth as much on the moon as on the earth. To obtain the weight of the astronaut on the moon from Equation 4.4, it is only necessary to replace M_E by M_M (the mass of the moon) and let $r = R_M$ (the radius of the moon). The ratio of the weight W_M of the astronaut on the moon to the weight W_E on the earth is

$$\frac{W_M}{W_E} = \frac{G\dfrac{M_M m}{R_M^2}}{G\dfrac{M_E m}{R_E^2}} = \frac{M_M R_E^2}{M_E R_M^2}$$

Substituting the known values for the masses and radii of the earth and moon reveals that

$$\frac{W_M}{W_E} = \frac{(7.35 \times 10^{22}\ \text{kg})(6.38 \times 10^6\ \text{m})^2}{(5.98 \times 10^{24}\ \text{kg})(1.74 \times 10^6\ \text{m})^2} = 0.165 \approx \frac{1}{6}$$

RELATION BETWEEN MASS AND WEIGHT

Although massive objects weigh a lot on the earth, mass and weight are not the same quantity. As Section 4.2 discusses, mass is a quantitative measure of inertia. As such, mass is an intrinsic property of matter and does not change as an object is moved from one location to another. Weight, on the other hand, is the gravitational force acting on the object and can vary, depending on how far the object is above the earth's surface or whether the object is located near another body such as the moon.

The relation between weight W and mass m can be written in one of two ways:

PROBLEM SOLVING INSIGHT

Mass and weight are different quantities. They cannot be interchanged when solving problems.

$$W = \boxed{G\frac{M_E}{r^2}}\,m \tag{4.4}$$

$$W = m\boxed{g} \tag{4.5}$$

The first of these is Newton's law of universal gravitation, while the second is Newton's second law incorporating the acceleration g due to gravity. These expressions make the distinction between mass and weight stand out. The weight of an object whose mass is m depends on the values for the universal gravitation constant G, the mass M_E of the earth, and the distance r. These three parameters together determine the acceleration g due to gravity. The specific value of $g = 9.80 \text{ m/s}^2$ applies only when r equals the radius R_E of the earth. For larger values of r, as would be the case on top of a mountain, the effective value of g is less than 9.80 m/s^2. The fact that g decreases as the distance r increases means that the weight likewise decreases. The mass of the object, however, does not depend on these effects and does not change.

APPARENT WEIGHT

Usually, the weight of an object can be determined with the aid of a scale. However, even though a scale is working properly, there are situations in which it does not give the correct weight. In such situations, the reading on the scale is called the "apparent" weight to distinguish it from the gravitational force or "true" weight.

To see the discrepancies that can arise between true weight and apparent weight, consider the scale in the elevator in Figure 4.14. The reasons for the discrepancies will be explained shortly. A person whose true weight is 700 N steps on the scale. If the elevator is at rest or moving with a constant velocity (either upward or downward), the scale registers the true weight, as Figure 4.14a illustrates.

If the elevator is accelerating, the apparent weight and the true weight are not equal. When the elevator accelerates upward, the apparent weight is greater than the true weight, as Figure 4.14b shows. Conversely, if the elevator accelerates downward, as in part c, the apparent weight is less than the true weight. In fact, if the elevator falls freely, so its acceleration is equal to the acceleration due to gravity, the apparent weight becomes zero, as part d indicates. In a situation such as this, where the apparent weight is zero, the person is said to be "weightless." The apparent weight, then, does not equal the true weight if the scale and the person on it are accelerating.

Elevators at the Marriott Hotel in Atlanta, Georgia; when an elevator is accelerating, the apparent weight and true weight of the passengers are not equal.

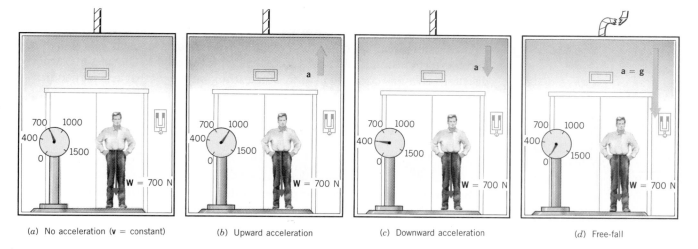

(a) No acceleration (v = constant)   (b) Upward acceleration   (c) Downward acceleration   (d) Free-fall

Figure 4.14 (a) When the elevator is not accelerating, the scale registers the true weight (W = 700 N) of the person. (b) When the elevator accelerates upward, the apparent weight exceeds the true weight. (c) When the elevator accelerates downward, the apparent weight is less than the true weight. (d) The apparent weight is zero if the elevator falls freely, that is, if it falls with the acceleration of gravity.

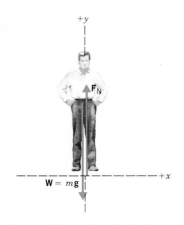

Figure 4.15 A free-body diagram showing the forces acting on the person riding in the elevator of Figure 4.14. **W** is the true weight and **F**$_N$ is the normal force exerted on the person by the platform of the scale.

The discrepancies between true weight and apparent weight can be understood with the aid of Newton's second law. Figure 4.15 shows a free-body diagram of the person in the elevator. The two forces that act on him are the true weight **W** = *m***g** and the normal force **F**$_N$ exerted by the platform of the scale. Applying Newton's second law in the vertical direction gives

$$\Sigma F_y = +F_N - mg = m(\pm a)$$

where $+a$ is used if the person's acceleration points upward and $-a$ is used if the acceleration points downward. Solving for F_N shows that

$$F_N = \underbrace{mg}_{\substack{\text{True} \\ \text{weight}}} + m\,(\pm a) \qquad (4.6)$$

$$\underbrace{}_{\substack{\text{Apparent} \\ \text{weight}}}$$

In Equation 4.6, F_N is the magnitude of the normal force exerted on the person by the platform of the scale. But in accord with Newton's third law, F_N is also the magnitude of the downward force that the person exerts on the scale, namely, the apparent weight.

Equation 4.6 contains all the features shown in Figure 4.14. If the elevator is not accelerating, $a = 0$, and the apparent weight equals the true weight. If the elevator accelerates upward (acceleration $= +a$), the equation shows that the apparent weight is greater than the true weight. If the elevator accelerates downward (acceleration $= -a$), the apparent weight is less than the true weight. If the elevator falls freely (acceleration $= -a = -g$), the apparent weight is zero. The scale registers an apparent weight of zero, because when both the person and the scale fall freely, they cannot push against one another. In this text, when the weight of an object is given, it is assumed to be the true weight, unless stated otherwise.

4.8 THE NORMAL FORCE

In many situations, an object is in contact with a surface, such as a tabletop. Because of the contact, there is a force acting on the object. The present section discusses only one component of this force, the component that acts perpendicular to the surface. The next section discusses the component that acts parallel to the surface. The perpendicular component is called the *normal force*, where the word "normal" is used as a synonym for the word "perpendicular."

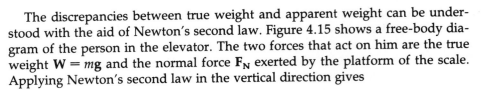

Definition of the Normal Force
The normal force **F**$_N$ is one component of the force that a surface exerts on an object with which it is in contact, namely, the component that is perpendicular to the surface.

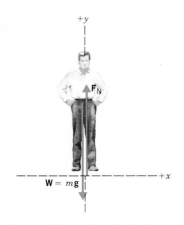

Figure 4.16 Two forces act on the block, its weight **W** and the normal force **F**$_N$ exerted by the surface of the table.

Figure 4.16 shows a block resting on a horizontal table and identifies the two forces that act on the block, the gravitational force or weight **W** and the normal force **F**$_N$. To understand how an inanimate object, such as a tabletop, can exert the normal force, think about what happens when you sit on a mattress. Your weight causes the springs in the mattress to compress. As a result, the compressed springs exert an upward force (the normal force) on you. In a similar manner, the weight of the block causes invisible "atomic springs" in the surface of the table to compress, thus producing a normal force on the block. The normal force is not one of nature's fundamental forces, in the sense that the gravitational force is. Instead, the normal force arises at the atomic level as a manifestation of the fundamental electric force between electrically charged particles within the atoms of the block and the table.

Newton's third law always comes into play in connection with the normal force. In Figure 4.16, for instance, the block exerts a force on the table. Consistent with the third law, the table exerts an oppositely directed force of equal magnitude on the block. This reaction force is the normal force. The magnitude of the normal force indicates how hard the two objects are pressing against each other.

If an object is resting on a horizontal surface and there are no vertically acting forces except the object's weight and the normal force, the magnitudes of these two forces are equal, i.e., $F_N = W$. This is the situation in Figure 4.16. The weight must be balanced by the normal force for the object to remain at rest on the table. If the magnitudes of these forces were not equal, there would be a net force acting on the block, and the block would accelerate either upward or downward, in accord with the second law.

If other forces in addition to W and F_N act in the vertical direction, the magnitudes of the normal force and the weight are no longer equal. In Figure 4.17a, for instance, a box whose weight is 15 N is being pushed downward against a table. The pushing force has a magnitude of 11 N. Thus, the total downward force exerted on the box is 26 N, and this must be balanced by the upward-acting normal force if the box is to remain at rest. In this situation, then, the normal force is 26 N, which is considerably larger than the weight of the box.

Figure 4.17b illustrates a different situation. Here, the box is being pulled upward by a rope that applies a force of 11 N. The net force acting on the box due to its weight and the rope is only 4 N, downward. To balance this force, the normal force needs to be only 4 N. It is not hard to imagine what would happen if the force applied by the rope were increased to 15 N—exactly equal to the weight of the box. In this situation, the normal force would become zero. In fact, the table could be removed, since the block would be supported entirely by the rope. The situations in Figure 4.17 are consistent with the idea that the magnitude of the normal force indicates how hard two objects are pressing against each other. Clearly, the box and the table are pressing against each other harder in part a of the picture than in part b.

Like the box and the table in Figure 4.17, various parts of the human body press against one another and exert normal forces. Example 7 illustrates the remarkable ability of the human skeleton to withstand a wide range of normal forces.

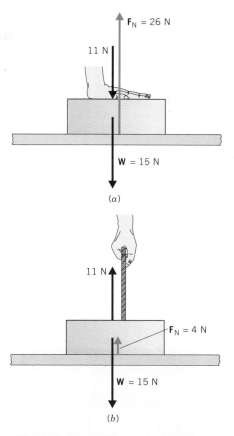

Figure 4.17   (a) The normal force is greater than the weight of the box, because the box is being pressed downward with an 11-N force. (b) The normal force is smaller than the weight, because the rope supplies an upward force of 11 N that partially supports the box.

Example 7   A Balancing Act

In a circus balancing act, a woman performs a headstand on top of a man's head, as Figure 4.18a illustrates. The woman weighs 490 N, and the man's head and neck weigh 50 N. It is primarily the seventh cervical vertebra in the spine that supports all the weight above the shoulders. What is the normal force that this vertebra exerts on the neck and head of the man (a) before the act and (b) during the act?

REASONING AND SOLUTION
(a) Figure 4.18*b* shows the free-body diagram for the man's body above the shoulders. The only forces acting are the normal force $\mathbf{F_N}$ and the weight $\mathbf{W} = 50$ N. These two forces must balance for the man's head and neck to remain at rest. Therefore, the seventh cervical vertebra exerts a normal force of $\boxed{F_N = 50 \text{ N}}$.

(b) Figure 4.18*c* shows the free-body diagram that applies during the act. Now, the total downward force exerted on the man's head and neck is 50 N + 490 N = 540 N, which must be balanced by the upward normal force provided by the vertebra: $\boxed{F_N = 540 \text{ N}}$.

In summary, the normal force does not necessarily have the same magnitude as the weight of the object. The value of the normal force depends on what other forces are present. Later sections in this chapter will show how to calculate the normal force under a variety of situations, including inclined surfaces and accelerating objects.

(*a*)

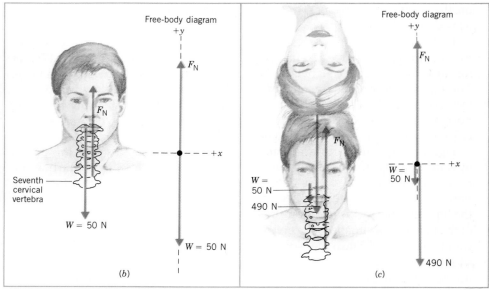

Figure 4.18 (*a*) A balancing act and free-body diagrams for the man's body above the shoulders (*b*) before the act and (*c*) during the act. For convenience, the scales used for the vectors in parts *b* and *c* are different.

4.9 STATIC AND KINETIC FRICTIONAL FORCES

When an object is in contact with a surface, there is a force acting on the object, and the previous section discusses the component of this force perpendicular to the surface. When the object moves or attempts to move along the surface, there is also a component of the force that is parallel to the surface. This parallel force component is called *friction.* Like the normal force, frictional forces are not fundamental forces in the sense that the gravitational force is.

In many situations considerable engineering effort is expended trying to reduce friction. For example, oil is used to reduce the friction that causes wear and tear in the pistons and cylinder walls of an automobile engine. Sometimes, however, friction is absolutely necessary. Without friction, car tires could not provide the traction needed to move the car. In fact, the raised tread on a tire is designed to maintain friction. On a wet road, the spaces in the tread pattern (see Figure 4.19) provide places for water to collect without coming between the tire surface and the road surface, where it would reduce friction and allow the tire to slip. Thus, friction can be either detrimental or absolutely essential.

When the surface of one object slides over the surface of another, *each object* exerts a frictional force on the other. This frictional force is called the **kinetic*** or **sliding frictional force** and, for example, comes into play when a baseball player slides into home plate. Another frictional force, called the **static frictional force**, acts even when there is no relative motion between the surfaces of two objects. The static frictional force is what makes it so difficult to start a heavy box moving across a rough floor.

Surfaces that appear to be highly polished can actually look quite rough when examined under a microscope. Such an examination reveals that two surfaces in contact touch only at relatively few spots, as Figure 4.20 illustrates. The microscopic area of contact for these spots is substantially less than the apparent macroscopic area of contact between the surfaces—perhaps thousands of times less. At these contact points the molecules of the different bodies are close enough together to exert strong attractive intermolecular forces on one another, leading to what are known as "cold welds." Frictional forces are associated with these welded spots, but the exact details of how frictional forces arise are not well understood. However, some empirical relations have been developed that allow us to account for the effects of friction.

▶ **THE PHYSICS OF . . .**

tire treads.

Figure 4.19   The tread design on a tire helps to preserve friction between the tire surface and the wet road surface.

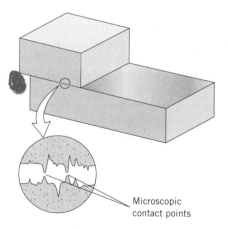

Microscopic
contact points

Figure 4.20   Even when two highly polished surfaces are in contact, they touch only at a relatively few points.

* The word kinetic is derived from the Greek word *kinetikos,* meaning "of motion."

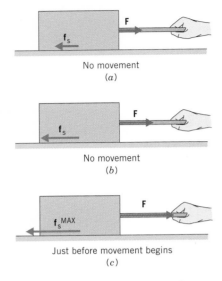

Figure 4.21 Applying a small force **F** to the block, as in parts *a* and *b*, produces no movement, because the static frictional force **f$_s$** exactly balances the applied force. (*c*) The block just begins to move when the applied force is slightly greater than the maximum static frictional force **f$_s$^{MAX}**.

Figure 4.22 The maximum static frictional force **f$_s$^{MAX}** would be the same, no matter which side of the block is in contact with the table.

Figure 4.21 helps to explain the main features of static friction. The block in this drawing is initially at rest, and as long as there is no attempt to move the block, there is no static frictional force. Then, a horizontal force **F** is applied to the block by means of a rope. If **F** is small, as in part *a*, experience tells us that the block still does not move. Why? It does not move because the static frictional force **f$_s$** exactly cancels the effect of the applied force. The direction of **f$_s$** is opposite to that of **F**, and the magnitude of **f$_s$** equals the magnitude of the applied force, $f_s = F$.

Increasing the applied force in Figure 4.21 by a small amount still does not cause the block to move. There is no movement because the static frictional force also increases, by an amount that cancels out the increase in the applied force (see part *b* of the drawing). If the applied force continues to increase, however, there comes a point when the block finally "breaks away" and begins to slide. This breakaway force represents the *maximum static frictional force* **f$_s$^{MAX}** that the table can exert on the block (see part *c* of the drawing). Any applied force that is greater than **f$_s$^{MAX}** cannot be balanced by static friction, and the resulting net force accelerates the block to the right. Thus, the magnitude f_s of the static frictional force equals the magnitude of the applied force and can assume any value from zero up to a maximum of f_s^{MAX}, depending on the applied force:

$$f_s \leq f_s^{MAX}$$

The symbol "≤" is read as "less than or equal to," and the equality holds only when f_s attains its maximum value.

Experimental evidence shows that, to a good degree of approximation, the maximum static frictional force between a pair of dry, unlubricated surfaces has two main characteristics. It is independent of the apparent macroscopic area of contact between the objects. For instance, in Figure 4.22 the maximum static frictional force that the surface of the table can exert on the block is the same, whether the block is resting on its largest side or its smallest side. The other main characteristic of **f$_s$^{MAX}** is that its magnitude is proportional to the magnitude of the normal force **F$_N$**. As Section 4.8 points out, the magnitude of the normal force indicates how hard the two surfaces are being pressed together. The harder the surfaces are pressed together, the larger is f_s^{MAX}, presumably because the number of "cold-welded," microscopic contact points is increased. Equation 4.7 expresses the proportionality between f_s^{MAX} and F_N with the aid of a proportionality constant μ_s, which is called the *coefficient of static friction.*

Static Frictional Force

The magnitude f_s of the static frictional force can have any value from zero up to a maximum value of f_s^{MAX}, depending on the applied force. In other words, $f_s \leq f_s^{MAX}$, where

$$f_s^{MAX} = \mu_s F_N \qquad (4.7)$$

In Equation 4.7, μ_s is the coefficient of static friction, and F_N is the magnitude of the normal force.

It should be emphasized that Equation 4.7 relates only the magnitudes of **f$_s$^{MAX}** and **F$_N$**, *not the vectors themselves.* This equation does not imply that the directions of the vectors are the same. In fact, **f$_s$^{MAX}** is parallel to the surface, while **F$_N$** is perpendicular to the surface.

The coefficient of static friction, being the ratio of the magnitudes of two forces ($\mu_s = f_s^{MAX}/F_N$), is a unitless number. It depends on the type of material from which each surface is made (steel on wood, rubber on concrete, etc.), the condition of the surfaces (polished, rough, lubricated, etc.), and other variables such as temperature. Typical values for μ_s range from about 0.01 for smooth surfaces to about 1.5 for rough surfaces.

Once two surfaces begin sliding over one another, the static frictional force is no longer of any concern. Instead, the kinetic frictional force comes into play and opposes the relative sliding motion. If you have ever pushed an object across a floor, you may have noticed that it takes less force to keep the object sliding than it takes to get it going in the first place. In other words, the kinetic frictional force is usually less than the static frictional force.

Experimental evidence indicates that the kinetic frictional force $\mathbf{f_k}$ has three main characteristics, to a good degree of approximation. It is independent of the apparent area of contact between the surfaces (see Figure 4.22). It is independent of the speed of the sliding motion, if the speed is small. And lastly, the magnitude of the kinetic frictional force is proportional to the magnitude of the normal force. Equation 4.8 expresses this proportionality with the aid of a proportionality constant μ_k, which is called the **coefficient of kinetic friction.**

Kinetic Frictional Force

The magnitude f_k of the kinetic frictional force is given by

$$f_k = \mu_k F_N \tag{4.8}$$

In Equation 4.8, μ_k is the coefficient of kinetic friction, and F_N is the magnitude of the normal force.

Equation 4.8, like Equation 4.7, is a relationship between only the magnitudes of the frictional and normal forces. The directions of these forces are perpendicular. Moreover, like the coefficient of static friction, the coefficient of kinetic friction is a unitless number and depends on the type and condition of the two surfaces that are in contact. Values for μ_k are typically less than those for μ_s, reflecting the fact that kinetic friction is generally less than static friction. The next two examples illustrate the effects of kinetic and static friction.

Example 8   Sled Riding

A sled, traveling at 4.00 m/s, enters a horizontal stretch of snow, as Figure 4.23a illustrates. The sled and its rider have a total mass of 38.0 kg. The coefficient of kinetic friction is $\mu_k = 0.0500$. How far does the sled go before stopping?

REASONING   The sled comes to a halt because the kinetic frictional force opposes the motion and causes the sled to slow down. Therefore, we will determine the kinetic frictional force and use it in Newton's second law to find the acceleration of the sled. Knowing the acceleration, we can determine the stopping distance by employing the appropriate equation of kinematics, as discussed in Chapter 3.

SOLUTION   To determine the magnitude of the kinetic force f_k, it is necessary to know the magnitude of the normal force F_N, since $f_k = \mu_k F_N$. Part b of Figure 4.23

shows the free-body diagram for the sled. Since the sled does not accelerate in the vertical direction, there can be no net force acting vertically on the sled. As a result, the normal force and the weight **W** must balance, so the magnitude of the normal force is $F_N = mg = (38.0 \text{ kg})(9.80 \text{ m/s}^2) = 372$ N. The magnitude of the kinetic frictional force is

$$f_k = \mu_k F_N = (0.0500)(372 \text{ N}) = 18.6 \text{ N} \qquad (4.8)$$

The kinetic frictional force opposes the sliding of the sled and is directed to the left in Figure 4.23.

Now that the frictional force is known, Newton's second law gives the acceleration of the sled as

$$a_x = \frac{F_x}{m} = \frac{-18.6 \text{ N}}{38.0 \text{ kg}} = -0.489 \text{ m/s}^2 \qquad (4.2a)$$

The minus sign appears because the frictional force points along the $-x$ axis in the free-body diagram, and, therefore, so does the acceleration. The stopping distance x can be obtained with the aid of Equation 3.6a from the equations of kinematics ($v_x^2 = v_{0x}^2 + 2a_x x$), where the initial velocity is $v_{0x} = 4.00$ m/s and the final velocity is $v_x = 0$:

$$x = \frac{v_x^2 - v_{0x}^2}{2a_x} = \frac{-(4.00 \text{ m/s})^2}{2(-0.489 \text{ m/s}^2)} = \boxed{16.4 \text{ m}}$$

Figure 4.23 (a) The moving sled decelerates because of the kinetic frictional force. (b) Three forces act on the moving sled, its weight **W**, the normal force $\mathbf{F_N}$, and the kinetic frictional force $\mathbf{f_k}$. The free-body diagram for the sled shows these forces.

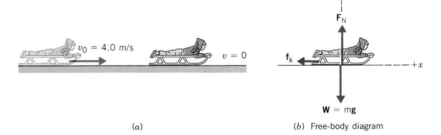

(a)

(b) Free-body diagram

Example 9   The Force Needed to Start a Sled Moving

For the sled in Example 8, the coefficient of static friction is $\mu_s = 0.350$. Determine the horizontal force needed to get the sled barely moving again after it has stopped.

REASONING AND SOLUTION   To set the sled into motion again, sufficient force is needed to overcome the maximum force of static friction. The magnitude of the necessary force must be greater than

$$f_s^{\text{MAX}} = \mu_s F_N = (0.350)(372 \text{ N}) = \boxed{130 \text{ N}} \qquad (4.7)$$

Static friction opposes the impending relative motion between two objects, while kinetic friction opposes the relative sliding motion that actually does occur. In either case, *relative motion* is opposed. However, this opposition to relative motion does not mean that friction prevents or works against the motion of *all* objects. In Figure 4.9, for instance, the foot of a person walking exerts a force on the earth, and the earth exerts a reaction force on the foot. This reaction force is a static frictional force, and it opposes the impending backward motion of the foot, propelling the person forward in the process. Kinetic friction can also cause an object to move, all the while opposing relative motion, as it does in Example 8. In this example the kinetic frictional force acts on the sled and opposes the relative motion of the sled and the earth. Newton's third law indicates, however, that if the earth exerts the kinetic frictional force on the sled, the sled must exert a reaction force on the earth. In response, the earth accelerates, but because of the earth's huge mass, the motion is too slight to be noticed.

4.10 THE TENSION FORCE

Forces are often applied by means of cables or ropes that are used to pull on an object. For instance, Figure 4.24a shows a force **T** being applied to the right end of a rope that is attached to a box. Each particle in the rope, in turn, applies a force to its neighbor. In the process the force is transmitted along the rope and ultimately is applied to the box at the other end, as part b of the drawing shows.

In situations such as that in Figure 4.24, it is often said that "the tension force in the rope is responsible for applying the force **T** to the box." This statement means that both the tension in the rope and the force applied to the box have the same magnitude. However, the common usage of the word "tension" is in the following context: "the tension is the tendency of the rope to be pulled apart." To see the relationship between these two uses of the word "tension," consider Figure 4.24c. The left end of the rope applies the force **T** to the box, and, in accordance with Newton's third law, the box applies a reaction force to the rope. The reaction force has the same magnitude as **T** but is oppositely directed. In other words, a force − **T** acts on the left end of the rope. Thus, forces of equal magnitude act on both ends of the rope and tend to pull the rope apart. The tension force, like the normal and the frictional forces, is not one of nature's fundamental forces. Tension is merely a manifestation of the fundamental forces that exist at the atomic level.

In the discussion above we have used the concept of a "massless" rope without saying so. A massless rope has zero mass, and, of course, such ropes do not exist. The advantage of introducing such an idealization can be understood with the aid of Newton's second law, which indicates that whenever an object has mass, a net force is required to accelerate it. By assuming a massless rope, we are able to set the mass m equal to zero in the second law, with the result that $\Sigma \mathbf{F} = ma = 0$. Therefore, no net force is needed to accelerate a massless rope. To put it simply, we can ignore the rope and assume that a force **T** applied to one end is transmitted undiminished to the object attached at the other end.* In contrast, if the rope in Figure 4.24 had mass, part of the force **T** applied on the right in part a would have to be used to accelerate the rope. In part b, then, the force applied to the box would not be **T**, but some diminished value. In this text we will assume that a rope connecting one object to another is massless, unless stated otherwise. The ability of a massless rope to transmit a force undiminished from one end to the other is not affected when the rope passes around objects such as the pulley (assumed to be massless and frictionless) in Figure 4.25.

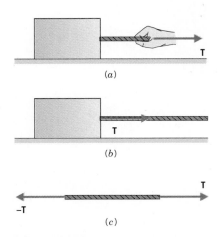

Figure 4.24 (a) A force **T** is being applied to the right end of the rope. (b) The force is transmitted to the box. (c) Forces are applied to both ends of the rope. These forces have equal magnitudes and opposite directions.

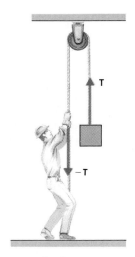

Figure 4.25 The force **T** applied at one end of a massless rope is transmitted undiminished to the other end, even when the rope bends around a pulley, provided the pulley is also massless and there is no friction.

* If a rope is not accelerating, **a** is zero in the second law and $\Sigma \mathbf{F} = ma = 0$, regardless of the mass of the rope. Then, the rope can be ignored, no matter what mass it has.

4.11 EQUILIBRIUM APPLICATIONS OF NEWTON'S LAWS OF MOTION

The Chinese acrobats in this incredible balancing act are in equilibrium.

Have you ever been so upset that it took days to recover your "equilibrium?" In this context, the word "equilibrium" refers to a balanced state of mind, one that is not changing wildly. In physics, the word "equilibrium" also refers to a lack of change, but in the sense that the velocity of an object isn't changing. If its velocity doesn't change, an object is not accelerating. Our definition of equilibrium, then, is as follows:

> **Definition of Equilibrium**
>
> An object is in equilibrium when it has zero acceleration.

According to Newton's second law, this definition implies that the net force acting on an object in equilibrium is zero. To put it another way, the forces acting on an object in equilibrium must balance. If the net force is zero, then in two dimensions the x component and the y component of the net force must each be zero. Thus, the equilibrium condition can be expressed by two equations:

$$\Sigma F_x = 0 \tag{4.9a}$$

$$\Sigma F_y = 0 \tag{4.9b}$$

This section deals with the application of these equations to various equilibrium situations, and there are five steps that are followed in each case.

Step 1. Select the object (often called the "system") to which Equations 4.9a and 4.9b are to be applied. Generally, this will be the object about which the most information is known. It may be that two or more objects are connected by means of a rope or a cable. Then, it may be necessary to treat each object separately according to the following steps.

Step 2. Draw a "free-body" diagram for each object chosen above. As Section 4.3 discusses, a free-body diagram is a drawing that represents the object and shows the forces that act on it, each force with its proper direction. Be sure to include only forces that act on the object. Do not include forces that the object exerts on its environment.

Step 3. Choose a convenient set of x, y axes for each object and resolve all forces in the free-body diagram into components that point along these axes. The emphasis here is on the word "convenient," because the axes are typically selected so that as many forces as possible point directly along the x axis or the y axis. Such a choice minimizes the number of calculations needed to determine the force components.

Step 4. Apply Equations 4.9a and 4.9b by setting the sum of the x components of the forces equal to zero and the sum of the y components of the forces equal to zero.

Step 5. Solve the two equations obtained in Step 4 for the desired unknown quantities, remembering that two equations can yield answers for only two unknowns at most.

Example 10 is a straightforward illustration of how these steps are followed, because only two forces act together to establish the equilibrium.

PROBLEM SOLVING INSIGHT

When an object is in equilibrium, the net force that acts on the object is zero. As discussed here, there are five steps to follow when using this fact to solve problems.

Example 10   Traction for the Neck

THE PHYSICS OF . . .

traction for a neck injury.

During recuperation from a neck injury, the cervical vertebrae are kept under tension by means of a traction device, as Figure 4.26a illustrates. The device creates tension in the vertebrae by pulling to the left on the head with a force **T**, which, in effect, is applied to the first vertebra at the top of the spine. This vertebra remains in equilibrium, because it is simultaneously pulled to the right by a force **F** that is supplied by the next vertebra in line. The force **F** comes about in reaction to the pulling effect of force **T**, in accord with Newton's third law. If it is desired that **F** have a magnitude of 34 N, how much mass m should be suspended from the rope?

REASONING AND SOLUTION   Since the forces **T** and **F** act on the first cervical vertebra, we choose it as the object for analysis. In Figure 4.26b, the free-body diagram for this vertebra shows only the two forces **T** and **F**. Friction between the head and the table is assumed to be negligible, since the head rests on a small rolling platform. At equilibrium the net force must be zero, so that $F = T$. But $T = mg$, so that the necessary mass is

$$m = \frac{F}{g} = \frac{34 \text{ N}}{9.80 \text{ m/s}^2} = \boxed{3.5 \text{ kg}}$$

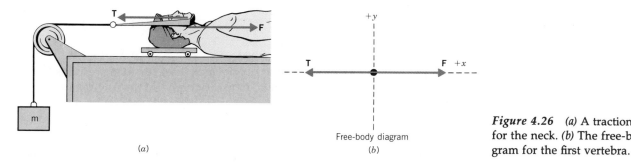

Free-body diagram

(a)   (b)

Figure 4.26   (a) A traction device for the neck. (b) The free-body diagram for the first vertebra.

The next example also deals with a traction device, but now three forces act together to bring about the equilibrium.

Example 11   Traction for the Foot

THE PHYSICS OF . . .

traction for a foot injury.

Figure 4.27a shows a traction device used with a foot injury. The weight of the 2.2-kg mass creates a tension in the rope that passes around the pulleys. Therefore, tension forces T_1 and T_2 are applied to the pulley on the foot, and they have the same magnitude T. It may seem surprising that the rope applies a force to either side of the foot pulley. A similar effect occurs when you place a finger inside a rubber band and push downward. You can feel each side of the rubber band pulling upward on the finger. The foot pulley is kept in equilibrium, because the foot also applies a force **F** to it. This force arises in reaction (Newton's third law) to the pulling effect of the forces T_1 and T_2. Find the magnitude of **F**, ignoring the weight of the foot.

REASONING   The forces T_1, T_2, and **F** keep the pulley on the foot at rest. The pulley, therefore, is in equilibrium, and the sum of the x components and the sum of the y components of the three forces must each be zero. Figure 4.27b shows the free-body diagram of the pulley on the foot. The x axis is chosen to be along the direction of force **F**, and the components of the forces are indicated in the drawing. (See Section 1.7 for a review of vector components.)

SOLUTION Since the sum of the x components of the forces is zero, it follows that

$$\Sigma F_x = T_1 \cos 35° + T_2 \cos 35° - F = 0$$

This equation can be used to calculate F, once the magnitudes of the tension forces are known. The tension in the rope is determined by the weight of the 2.2-kg mass; $T = mg$. Therefore, $T = (2.2 \text{ kg})(9.80 \text{ m/s}^2) = 22 \text{ N}$, and $T_1 = T_2 = 22 \text{ N}$. As a result,

$$F = 2(22 \text{ N}) \cos 35° = \boxed{36 \text{ N}}$$

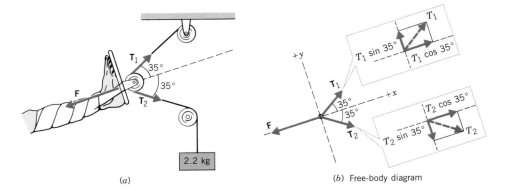

Figure 4.27 (a) A traction device for the foot. (b) The free-body diagram for the pulley on the foot.

(a)

(b) Free-body diagram

Example 12 presents another situation in which three forces are responsible for the equilibrium of an object. However, in this example all the forces have different magnitudes.

Example 12   Replacing an Engine

An automobile engine (weight **W** = 3150 N) is being positioned above an engine compartment, as Figure 4.28a illustrates. To position the engine, a worker is using a rope. Find the tension $\mathbf{T}_1$ in the supporting cable and the tension $\mathbf{T}_2$ in the positioning rope.

REASONING Under the influence of the forces **W**, $\mathbf{T}_1$, and $\mathbf{T}_2$ the ring is at rest and therefore in equilibrium. Consequently, the sum of the x components and the sum of the y components of these forces must each be zero. Figure 4.28b shows the free-body diagram of the ring and the force components for a suitable x, y axes system.

SOLUTION The free-body diagram shows the components for each of the three forces, and the components are listed in the table below.

Force	x Component	y Component
$\mathbf{T}_1$	$-T_1 \sin 10.0°$	$+T_1 \cos 10.0°$
$\mathbf{T}_2$	$+T_2 \sin 80.0°$	$-T_2 \cos 80.0°$
W	0	-3150 N

The plus signs in the table denote components that point along the positive axes, while the minus signs denote components that point along the negative axes. Setting the sum of the x components and the sum of the y components equal to zero leads to the following two equations:

$$\Sigma F_x = -T_1 \sin 10.0° + T_2 \sin 80.0° = 0$$
$$\Sigma F_y = +T_1 \cos 10.0° - T_2 \cos 80.0° - 3150 \text{ N} = 0$$

Solving the first of these equations for T_1 shows that

$$T_1 = \left(\frac{\sin 80.0°}{\sin 10.0°}\right) T_2 = 5.67\ T_2$$

Substituting this expression for T_1 into the second equation gives

$$(5.67\ T_2) \cos 10.0° - T_2 \cos 80.0° - 3150 \text{ N} = 0$$

which can be solved to show that $\boxed{T_2 = 582 \text{ N}}$. Since $T_1 = 5.67\ T_2$, it follows that $\boxed{T_1 = 3.30 \times 10^3 \text{ N}}$. The equations solved above are simultaneous equations with two unknown quantities. Appendix C contains a review of the procedure used to solve such equations.

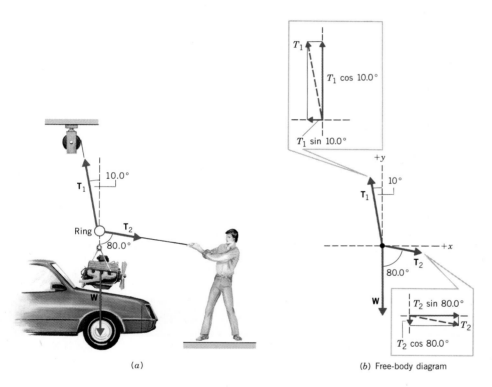

(a)   *(b)* Free-body diagram

Figure 4.28 *(a)* The ring is in equilibrium because of the three forces $\mathbf{T}_1$ (the tension force in the supporting cable), $\mathbf{T}_2$ (the tension force in the positioning rope), and $\mathbf{W}$ (the weight of the engine). *(b)* The free-body diagram for the ring.

An object can be moving and still be in equilibrium, provided there is no acceleration. Example 13 illustrates such a case, and the solution is again obtained using the five steps summarized at the beginning of the section.

Example 13   A Climbing Jet

A jet plane is flying with a constant speed along a straight line, at an angle of 30.0° above the horizontal, as Figure 4.29a indicates. The plane has a weight $\mathbf{W}$ of 86 500 N, and its engines provide a forward thrust $\mathbf{T}$ of 103 000 N. In addition, the lift force $\mathbf{L}$ (directed perpendicular to the wings) and the force $\mathbf{R}$ of air resistance (directed opposite to the motion) act on the plane. Find $\mathbf{L}$ and $\mathbf{R}$.

REASONING   Figure 4.29*b* shows the free-body diagram of the plane, including the forces **W**, **L**, **T**, and **R**. Since the plane is not accelerating, it is in equilibrium, and the sum of the *x* components and the sum of the *y* components of these forces must be zero. To calculate the components, we have chosen axes in the free-body diagram that are rotated by 30.0° from their usual horizontal–vertical positions. This has been done purely for convenience, since the weight **W** is then the only force that does not lie along either axis.

SOLUTION   Before determining the components of the weight, it is necessary to realize that the angle β in Figure 4.29*a* is equal to 30.0°. Part *c* of the drawing focuses attention on the geometry that is responsible for this fact. There it can be seen that $\alpha + \beta = 90°$ and $\alpha + 30° = 90°$, with the result that $\beta = 30°$. Geometry similar to that in Figure 4.29*c* occurs often in physics.

The table below lists the components of the forces that act on the jet.

Force	x Component	y Component
W	$-(86\ 500\ \text{N})\sin 30.0°$	$-(86\ 500\ \text{N})\cos 30.0°$
L	0	$+L$
T	$+103\ 000\ \text{N}$	0
R	$-R$	0

Setting the sum of the *x* components and the sum of the *y* components of the forces equal to zero yields

$$\Sigma F_x = -(86\ 500\ \text{N})\sin 30.0° + 103\ 000\ \text{N} - R = 0$$

$$\Sigma F_y = -(86\ 500\ \text{N})\cos 30.0° + L = 0$$

These equations can be solved to show that $\boxed{R = 59\ 800\ \text{N}}$ and $\boxed{L = 74\ 900\ \text{N}}$.

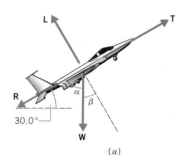

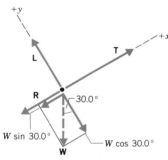

(b) Free-body diagram

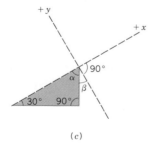

(c)

Figure 4.29   (a) A plane moves with a constant velocity at an angle of 30.0° above the horizontal due to the action of four forces, the weight **W**, the lift **L**, the engine thrust **T**, and the air resistance **R**. (b) The free-body diagram for the plane. (c) This geometry occurs often in physics.

Static friction is a force that sometimes plays a role in keeping an object at rest, that is, in equilibrium. The next example illustrates one way in which information about the static friction force can be obtained.

Example 14   Measuring the Coefficient of Static Friction

A block of mass *m* rests on a hinged board whose angle of elevation is adjustable, as in Figure 4.30*a*. When the right end of the board is raised, the block remains at rest until a maximum angle θ is reached. If the angle is increased beyond θ, the block breaks loose and slides down the board. Obtain an equation that relates the coefficient of static friction μ_s to the angle θ.

REASONING   There are three forces that act on the block, the weight **W** ($W = mg$), the normal force $\mathbf{F_N}$, and the maximum force of static friction $\mathbf{f_s}^{\text{MAX}}$. The magnitude of the maximum static frictional force is given by Equation 4.7 as $f_s^{\text{MAX}} = \mu_s F_N$. These forces keep the block in equilibrium, and, therefore, they must balance, the sum of the *x* components and the sum of the *y* components each being zero. To determine the components, we use the *x*, *y* axes shown in Figure 4.30*b* along with the free-body diagram for the block. The geometry in this diagram is the same as that in Figure 4.29*b*.

SOLUTION   The *x* and *y* components of the forces are as follows:

Force	x Component	y Component
W	$-mg \sin \theta$	$-mg \cos \theta$
$\mathbf{F_N}$	0	$+F_N$
$\mathbf{f_s}^{MAX}$	$+\mu_s F_N$	0

Since the sum of the x components and the sum of the y components must each be zero, it follows that

$$\Sigma F_x = -mg \sin \theta + \mu_s F_N = 0$$
$$\Sigma F_y = -mg \cos \theta + F_N = 0$$

The second equation reveals that $F_N = mg \cos \theta$. Substituting this relation for F_N into the first equation yields

$$-mg \sin \theta + \mu_s (mg \cos \theta) = 0$$

Solving for μ_s gives

$$\mu_s = \frac{\sin \theta}{\cos \theta} \quad \text{or} \quad \boxed{\mu_s = \tan \theta}$$

Thus, from a measured value for the maximum angle θ, the coefficient of static friction can be obtained.

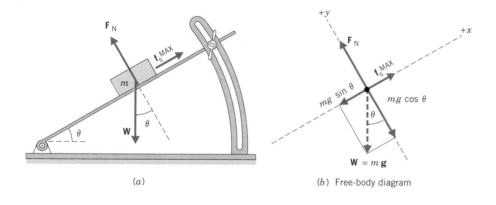

(a)   (b) Free-body diagram

Figure 4.30 (a) A block resting on an inclined board. The maximum angle of the incline, just before the block begins to slip, is given by θ. The forces acting on the block are its weight **W**, the normal force $\mathbf{F_N}$, and the maximum force of static friction $\mathbf{f_s}^{MAX}$. (b) The free-body diagram for the block.

4.12 NONEQUILIBRIUM APPLICATIONS OF NEWTON'S LAWS OF MOTION

When an object is accelerating, it is not in equilibrium. The forces acting on it are not balanced, so the net force is not zero in Newton's second law. However, with one exception, the steps followed in solving nonequilibrium problems are identical to those used in equilibrium situations. The exception occurs in Step 4 of the five steps outlined at the beginning of the last section. Since the object is now accelerating, the representation of Newton's second law in Equations 4.2a and 4.2b applies. These equations are repeated here for convenience:

$$\Sigma F_x = ma_x \quad (4.2a) \quad \text{and} \quad \Sigma F_y = ma_y \quad (4.2b)$$

Example 15 uses these equations in a situation where the forces are applied in directions similar to those in Example 11, except that now an acceleration is present.

Example 15   Towing a Supertanker

A supertanker (mass $= 1.50 \times 10^8$ kg) is being towed by two tugboats, as in Figure 4.31a. The tensions in the towing cables apply the forces $\mathbf{T}_1$ and $\mathbf{T}_2$ at equal angles of 30.0° with respect to the tanker's axis. In addition, the tanker's engines produce a forward drive force $\mathbf{D}$, whose magnitude is 75.0×10^3 N. Moreover, the water applies an opposing force $\mathbf{R}$, whose magnitude is 40.0×10^3 N. The tanker moves forward with an acceleration that points along the tanker's axis and has a magnitude of 2.00×10^{-3} m/s². Find the magnitudes of the tensions $\mathbf{T}_1$ and $\mathbf{T}_2$ in the towing cables.

REASONING   The unknown forces $\mathbf{T}_1$ and $\mathbf{T}_2$ contribute to the net force that accelerates the tanker. An essential step in determining $\mathbf{T}_1$ and $\mathbf{T}_2$, therefore, is to analyze the net force, which we will do using components. The various force components can be found readily by referring to the free-body diagram for the tanker in Figure 4.31b, where the ship's axis is chosen as the x axis.

SOLUTION   The individual force components are summarized below:

Force	x Component	y Component
$\mathbf{T}_1$	$+T_1 \cos 30.0°$	$+T_1 \sin 30.0°$
$\mathbf{T}_2$	$+T_2 \cos 30.0°$	$-T_2 \sin 30.0°$
$\mathbf{D}$	$+D$	0
$\mathbf{R}$	$-R$	0

Since the acceleration points along the x axis, there is no y component of the acceleration. Consequently, the sum of the y components of the forces must be zero:

$$\Sigma F_y = +T_1 \sin 30.0° - T_2 \sin 30.0° = 0$$

This result shows that the magnitudes of the tensions in the cables are equal, $T_1 = T_2$. Since the ship accelerates along the x direction, the sum of the x components of the forces is not zero. The second law indicates that

$$\Sigma F_x = T_1 \cos 30.0° + T_2 \cos 30.0° + D - R = ma_x$$

Using the fact that $T_1 = T_2 = T$ and the given values for D, R, m, and a_x, it can be seen that

$$2T \cos 30.0° + 75.0 \times 10^3 \text{ N} - 40.0 \times 10^3 \text{ N} = (1.50 \times 10^8 \text{ kg})(2.00 \times 10^{-3} \text{ m/s}^2)$$

$$\boxed{T = 1.53 \times 10^5 \text{ N}}$$

It often happens that two objects are connected somehow, perhaps by a draw-bar like that used when a truck pulls a trailer. If the tension in the connecting device is of no interest, the objects can be treated as a single composite object when applying Newton's second law. However, if it is necessary to find the tension, as in the next example, then the second law must be applied separately to at least one of the objects.

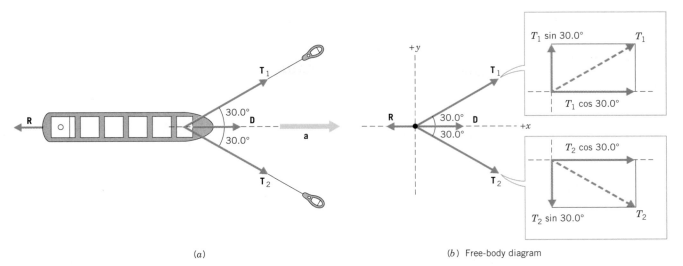

(a)   (b) Free-body diagram

Figure 4.31 (*a*) Four forces act on a supertanker: T_1 and T_2 are the tension forces due to the towing cables, D is the forward drive force produced by the tanker's engines, and R is the force with which the water opposes the tanker's motion. (*b*) The free-body diagram for the tanker.

Example 16   *Hauling a Trailer*

An 8500-kg truck is hauling a 27 000-kg trailer along a level road, as Figure 4.32*a* illustrates. The acceleration is 0.78 m/s². Ignoring the retarding forces of friction and air resistance, determine (a) the magnitude of the tension in the horizontal drawbar between the trailer and the truck and (b) the force **D** that propels the truck forward.

REASONING AND SOLUTION

(a) Since the truck and the trailer accelerate along the horizontal direction and friction is being ignored, only forces that have components in the horizontal direction are of interest here. Therefore, Figure 4.32 omits the weight and the normal force, since they act vertically. Note, however, that these vertical forces balance, since there is no vertical component to the acceleration.

We begin with the trailer, whose free-body diagram is shown in Figure 4.32*b*. There is only one horizontal force acting on the trailer, the tension force **T** due to the drawbar. Therefore, it is straightforward to obtain the tension from $\Sigma F_x = ma_x$, since the mass of the trailer and the acceleration are known:

$$T = (27\ 000\ \text{kg})(0.78\ \text{m/s}^2) = \boxed{21\ 000\ \text{N}}$$

(b) Two horizontal forces act on the truck, as the free-body diagram in Figure 4.32*b* shows. One is the desired force **D**. The other is the force **T′**. According to Newton's third law, **T′** is the force with which the trailer pulls back on the truck, in reaction to the truck pulling forward. If the drawbar has negligible mass, the magnitude of **T′** is equal to the magnitude of **T**, namely, 21 000 N. Since the magnitude of **T′**, the mass of the truck, and the acceleration are known, $\Sigma F_x = ma_x$ can be used to determine the drive force:

$$D - (21\ 000\ \text{N}) = (8500\ \text{kg})(+0.78\ \text{m/s}^2)$$

$$\boxed{D = 28\ 000\ \text{N}}$$

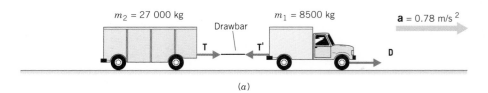

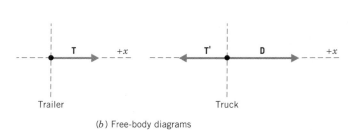

(a)

Figure 4.32 (a) The force **D** acts on the truck and propels it forward. The drawbar exerts the tension force **T′** on the truck and the tension force **T** on the trailer. (b) The free-body diagrams for the truck and the trailer, ignoring the vertical forces.

The force of gravity is often present among the forces that affect the acceleration of an object. Examples 17 and 18 deal with typical situations.

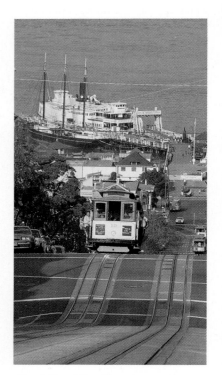

The hills of San Francisco; a number of forces, including the weight of the car, affect the acceleration of a car on these hills.

Example 17   Driving in San Francisco

San Francisco is famous for its hills. A car (mass = 1350 kg) is being propelled up one of these hills by a drive force that is parallel to the hill and has a magnitude of 7200 N. As Figure 4.33a illustrates, the hill makes a 25° angle with the horizontal. Assuming that the retarding forces of friction and air resistance are negligible, find (a) the acceleration of the car and (b) the normal force exerted on the car by the hill.

REASONING AND SOLUTION
(a) Figure 4.33b gives the free-body diagram for the car. The 7200-N drive force acts along the $+x$ axis, parallel to the hill. The normal force $\mathbf{F_N}$ acts along the $+y$ axis, perpendicular to the hill. The weight of the car [$W = mg = (1350 \text{ kg})(9.80 \text{ m/s}^2) = 13\,200$ N] has the x and y components shown in the diagram. Newton's second law can be used to determine the acceleration of the car from the sum of the x components of the forces:

$$\Sigma F_x = 7200 \text{ N} - W \sin 25° = ma_x$$
$$7200 \text{ N} - (13\,200 \text{ N}) \sin 25° = (1350 \text{ kg})a_x$$

$$\boxed{a_x = 1.2 \text{ m/s}^2}$$

(b) Since the acceleration of the car has no component perpendicular to the hill, the sum of the y components of the forces must be zero:

$$\Sigma F_y = F_N - (13\,200 \text{ N}) \cos 25° = 0$$

$$\boxed{F_N = 12\,000 \text{ N}}$$

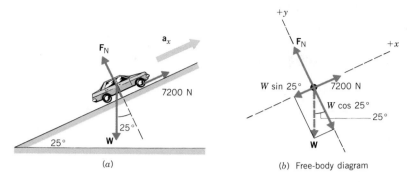

(a)   (b) Free-body diagram

Figure 4.33 (a) The car accelerates up the hill under the influence of the 7200 N drive force, the normal force $\mathbf{F_N}$, and the weight $\mathbf{W}$ of the car. (b) The free-body diagram for the car.

Example 18   Accelerating Blocks

A block (mass $m_1 = 8.00$ kg) is moving on a frictionless incline plane whose angle is 30.0°. This block is connected to a second block (mass $m_2 = 22.0$ kg) by a cord that passes over a small, frictionless pulley (see Figure 4.34a). Find the acceleration of each block and the tension in the cord.

REASONING AND SOLUTION   Since both blocks accelerate, there must be a net force acting on each one. The key to solving this problem is to realize that Newton's second law can be used separately for each block to relate the net force and the acceleration. It is also important to realize that both blocks have accelerations of the same magnitude a, since they move as a unit.

We assume that m_1 accelerates up the incline and choose this direction to be the $+x$ axis for m_1. If m_1 in reality accelerates down the incline, then the value obtained for the acceleration will be a negative number. There are three forces that act on m_1: (1) $\mathbf{W_1}$ is the weight [$W_1 = m_1 g = (8.00$ kg$)(9.80$ m/s$^2) = 78.4$ N], (2) $\mathbf{T}$ is the force applied because of the tension in the cord, and (3) $\mathbf{F_N}$ is the normal force that the incline exerts. Figure 4.34b shows the free-body diagram for m_1. The components of the weight are given in this diagram, and it is the only force that does not point along the x, y axes. Applying the second law to the motion of m_1 shows that

$$\Sigma F_x = -W_1 \sin 30.0° + T = m_1 a_x$$

$$-(78.4 \text{ N}) \sin 30.0° + T = (8.00 \text{ kg})a$$

where we have set $a_x = a$. This equation cannot be solved as it stands, since both T and a are unknown quantities. To complete the solution, we next consider block m_2.

There are two forces that act on m_2, as the free-body diagram in Figure 4.34b indicates: (1) $\mathbf{W_2}$ is the weight [$W_2 = m_2 g = (22.0$ kg$)(9.80$ m/s$^2) = 216$ N] and (2) $\mathbf{T'}$ is exerted as a result of m_1 pulling back on the connecting cord. If the masses of the cord and the frictionless pulley are negligible, the magnitudes of $\mathbf{T'}$ and $\mathbf{T}$ are the same: $T' = T$. Applying the second law to m_2 reveals that

$$\Sigma F_y = T' - W_2 = m_2 a_y$$

$$+T - 216 \text{ N} = (22.0 \text{ kg})(-a)$$

The acceleration a_y has been set equal to $-a$ since block m_2 moves downward along the $-y$ axis in the free-body diagram, consistent with the original assumption that block m_1 moves up the incline. Now there are two equations in two unknowns, and following the procedure discussed in Appendix C, they may be solved simultaneously to give the values of the tension T and the acceleration a:

$$T = 86.3 \text{ N} \quad \text{and} \quad a = 5.89 \text{ m/s}^2$$

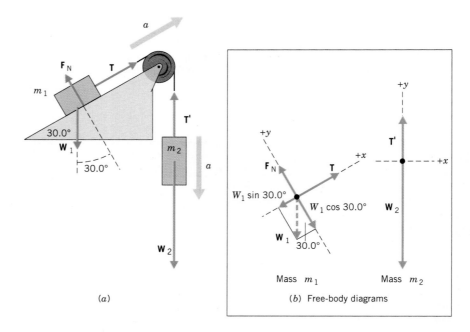

Figure 4.34 *(a)* Three forces act on m_1: its weight $\mathbf{W}_1$, the normal force $\mathbf{F_N}$, and the force $\mathbf{T}$ due to the tension in the cord. Two forces act on m_2: its weight $\mathbf{W}_2$ and the force $\mathbf{T}'$ due to the tension. The acceleration is labeled $\mathbf{a}$. *(b)* The free-body diagrams for the two masses.

INTEGRATION OF CONCEPTS

NEWTON'S SECOND LAW AND KINEMATICS

In the present chapter, we have seen that when a net force acts on an object, the object accelerates. Newton's second law specifies how to determine the acceleration from values of the net force and the mass of the object. On the other hand, in Chapters 2 and 3, we have seen how the equations of kinematics are used to describe the motion of an object under the condition of constant acceleration. In the equations of kinematics, the acceleration plays a central role, for it tells us how rapidly the velocity of a moving object changes as time passes. By themselves, however, the equations of kinematics offer no insight as to *why* the acceleration has a particular value. For that insight, we turn to Newton's second law, which reveals that the acceleration is determined by the net force $\Sigma \mathbf{F}$ that acts on the object and by the mass m of the object. Specifically, we can use the second law to determine the acceleration as $\mathbf{a} = \Sigma \mathbf{F}/m$ and then use that value in the equations of kinematics. We see, then, that Newton's second law and the equations of kinematics are not separate and unrelated parts of physics. Quite the opposite. They are closely connected by the idea of acceleration. Together, they provide a description and fundamental understanding of motion under the condition of constant acceleration. You will find that a number of the problems at the end of the present chapter (e.g., problems 6 and 28) combine Newton's second law with the equations of kinematics.

NEWTON'S LAW OF UNIVERSAL GRAVITATION, NEWTON'S SECOND LAW, AND THE KINEMATICS OF FREE-FALL MOTION

Gravity pulls objects downward toward the earth. This phenomenon is explained by Newton's law of universal gravitation, which we have studied in the present chapter. This law reveals that the magnitude of the gravitational force pulling downward on an object of mass m near the earth's surface is $F = GM_E m/R_E^2$, where G is the universal gravitational constant, M_E is the mass of the earth, and R_E is the radius of the earth. When a single force such as this acts on an object, we know from Newton's second law that an acceleration results that has a magnitude of $a = F/m$. The magnitude of the acceleration caused by the gravitational force, then, is $a = (GM_E m/R_E^2)/m = GM_E/R_E^2$. In Chapters 2 and 3, however, we have seen that the equations of kinematics can be used to describe free-fall motion near the earth's surface, if a value of $a = g = 9.80$ m/s^2 is used for the magnitude of the acceleration due to gravity. If the known values for G, M_E, and R_E are substituted into $a = GM_E/R_E^2$, we find that just this value of 9.80 m/s^2 results for the acceleration due to gravity. Thus, it is because of Newton's law of universal gravitation that the acceleration due to gravity has the value it has in the equations of kinematics for free-fall motion.

SUMMARY

Newton's first law of motion or **law of inertia** states that an object continues in a state of rest or in a state of motion at a constant speed along a straight line unless compelled to change that state by a net force. **Inertia** is the natural tendency of an object to remain at rest or in motion at a constant speed along a straight line. The **mass** of a body is a quantitative measure of inertia and is measured in an SI unit called the **kilogram** (kg). An **inertial reference frame** is one in which Newton's law of inertia is valid.

Newton's second law of motion states that the acceleration **a** of an object is directly proportional to the net force $\Sigma\mathbf{F}$ acting on the object and inversely proportional to the mass m of the object; $\mathbf{a} = \Sigma\mathbf{F}/m$ or $\Sigma\mathbf{F} = m\mathbf{a}$. The SI unit of force is the **newton** (N). When determining the net force, a **free-body diagram** is helpful. A free-body diagram is a diagram that represents the object and the forces acting on it.

Newton's third law of motion, often called the **action–reaction law,** states that whenever one object exerts a force on a second object, the second object exerts an oppositely directed force of equal magnitude on the first object.

Newton's law of universal gravitation states that every particle in the universe exerts an attractive force on every other particle. According to the law, the magnitude F of the force between two particles of masses m_1 and m_2 is directly proportional to the product of the masses and inversely proportional to the square of the distance r between them; $F = Gm_1m_2/r^2$. The force is directed along the line between the two particles. The constant G has a value of $G = 6.673 \times 10^{-11}$ N·m^2/kg^2 and is called the universal gravitational constant. The gravitational force is one of nature's four fundamental forces. The **weight** of an object on earth is the gravitational force that the earth exerts on the object.

The **apparent weight** is the force that an object exerts on the platform of a scale and may be larger or smaller than the true weight, if the object and the scale are accelerating.

A surface exerts a force on an object with which it is in contact. The component of the force perpendicular to the surface is called the **normal force.** The component parallel to the surface is called friction. The **force of static friction** between two surfaces opposes any impending relative motion of the surfaces. The magnitude of the static friction force depends on the magnitude of the applied force and can assume any value up to a maximum of $f_s^{MAX} = \mu_s F_N$, where μ_s is the **coefficient of static friction** and F_N is the magnitude of the normal force. The **force of kinetic friction** between two surfaces sliding against one another opposes the relative motion of the surfaces. This force has a magnitude given by $f_k = \mu_k F_N$, where μ_k is the **coefficient of kinetic friction.**

The word **tension** is commonly used to mean the tendency of a rope to be pulled apart due to forces that are applied at each end. Because of tension, a rope transmits a force from one end to the other. When a rope is accelerating, the force is transmitted undiminished only if the rope is massless. Neither the tension force, the normal force, nor the friction force is one of nature's fundamental forces.

An object is in **equilibrium** when the object moves at a constant velocity (which may be zero), or, in other words, when it is not accelerating. The sum of the forces that act on an object in equilibrium is zero. Under equilibrium conditions in two dimensions, the separate sums of the force components in the x direction and in the y direction must each be zero: $\Sigma F_x = 0$ and $\Sigma F_y = 0$. If an object is not in equilibrium, then Newton's second law must be used to account for the acceleration: $\Sigma F_x = ma_x$ and $\Sigma F_y = ma_y$. Sections 4.11 and 4.12 discuss five steps that facilitate the use of Newton's laws of motion.

SOLVED PROBLEMS

Solved Problem 1   Hoisting a Scaffold
Related Problems: *68   *69   **74

A window washer on a scaffold is hoisting the scaffold up the side of a skyscraper by pulling downward on a rope, as part *a* of the drawing illustrates. The magnitude of the pulling force is 540 N, and the combined mass of the person and the scaffold is 155 kg. (a) Find the upward acceleration of the unit. (b) Find the tension in the rope when the scaffold is hanging motionless.

REASONING In this problem, the person and the scaffold are considered together as a single unit or object, since the rope and pulley arrangement is attached to both of them. The rope exerts a force on the unit in three places. The left end of the rope exerts an upward force **T** on the person's hands. This force arises because the person pulls downward on the rope with a 540-N force, and the rope exerts an oppositely directed force of equal magnitude on the person, in accord with Newton's third law. Thus, the magnitude of the upward force **T** is 540 N, which is equal to the magnitude of the tension in the rope, as Section 4.10 discusses. If the masses of the rope and each pulley are negligible and if the pulleys are friction-free, the tension is transmitted undiminished along the rope. Then, a 540-N tension force **T** acts upward on the left side of the scaffold pulley (see part *a* of the drawing). The tension is also

transmitted to the point P, where the rope attaches to the roof. The roof pulls back on the rope in accord with the third law, and this pull leads to the force **T** = 540 N that acts on the right side of the scaffold pulley. In addition to the three upward forces, the weight of the entire unit must be taken into account [$W = mg = (155$ kg$)(9.80$ m/s$^2) = 1520$ N]. Part *b* of the drawing shows the free-body diagram and a set of x, y axes.

SOLUTION

(a) Newton's second law can be applied to calculate the acceleration a_y ($\Sigma F_y = ma_y$):

$$+T + T + T - 1520 \text{ N} = (155 \text{ kg})a_y$$
$$3(540 \text{ N}) - 1520 \text{ N} = (155 \text{ kg})a_y$$

$$\boxed{a_y = 0.65 \text{ m/s}^2}$$

(b) To find the tension in the rope when the unit is motionless, we need only set a_y equal to zero in the previous analysis:

$$3T - 1520 \text{ N} = 0 \qquad \boxed{T = 507 \text{ N}}$$

Thus, the tension is less when the scaffold is at rest than when it is accelerating upward ($T = 540$ N). When the scaffold is at rest, the tension serves only to support the weight of the unit, by virtue of the three upward forces **T**. But additional tension is needed to accelerate the unit.

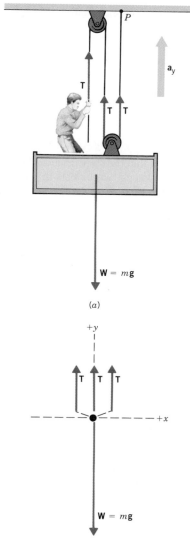

(a)

(b) Free-body diagram of the unit

SUMMARY OF IMPORTANT POINTS The essential point here is that the tension is the same everywhere in a massless rope, even when the rope is wrapped around pulleys (which are assumed to be massless and frictionless). Because of the tension, the rope on each side of a pulley exerts a force on the pulley. In this problem, for instance, a net upward force of $2T$ acts on the scaffold pulley.

> ***Solved Problem 2   Hauling a Crate***
> Related Problems: *70   *71   *72   **75

A flatbed truck is carrying a crate up a 10.0° hill, as part *a* of the drawing shows. The coefficient of static friction between the truck bed and the crate is $\mu_s = 0.350$. Find the maximum acceleration that the truck can attain before the crate begins to slip backward relative to the truck.

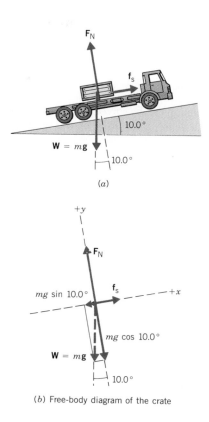

(a)

(b) Free-body diagram of the crate

REASONING As the truck accelerates up the hill, the crate will not slip as long as it has the same acceleration as the truck. Therefore, a net force must act on the crate to accelerate it, and the static frictional force $\mathbf{f}_s$ contributes in a major way to this net force. As the acceleration of the truck increases, $\mathbf{f}_s$ must also increase to produce a corresponding increase in the acceleration of the crate. However, the static frictional force can increase only until its maximum magnitude $f_s^{MAX} = \mu_s F_N$ is reached, at which point the crate and the truck have the maximum acceleration a^{MAX}. If the acceleration of the truck increases even more, the crate will not be able to "keep up" and slipping will occur.

To find a^{MAX}, we focus attention on the crate, and part *b* of the drawing shows the free-body diagram, along with a convenient set of x, y axes. The three forces acting on the crate at the instant slipping begins are (1) its weight $\mathbf{W} = m\mathbf{g}$, (2) the normal force $\mathbf{F}_N$ exerted by the truck bed, and (3) the maximum static frictional force $\mathbf{f}_s^{MAX}$.

SOLUTION The x and y components of the forces in the free-body diagram are listed as follows:

Force	x Component	y Component
W	$-mg \sin 10.0°$	$-mg \cos 10.0°$
F_N	0	$+F_N$
f_s	$+\mu_s F_N$	0

Using the sum of the x components of the forces and the fact that $a_x = a^{MAX}$ in Newton's second law ($\Sigma F_x = ma_x$), we find that

$$-mg \sin 10.0° + \mu_s F_N = ma^{MAX}$$

Before this equation can be solved for a^{MAX}, however, a value is needed for F_N, the magnitude of the normal force. This value can be obtained by considering the force components along the y axis. Since the crate does not accelerate along this axis, the sum of the y components of the forces must be zero according to Newton's second law ($\Sigma F_y = ma_y = 0$):

$$-mg \cos 10.0° + F_N = 0$$

It can be seen that $F_N = mg \cos 10.0°$. Substituting this value

into the previous equation reveals that

$$-g \sin 10.0° + \mu_s g \cos 10.0° = a^{MAX}$$

The mass m does not appear here, because it occurs in each term on either side of the equation and, thus, is eliminated algebraically. Letting $g = 9.80$ m/s² and $\mu_s = 0.350$, we see that $\boxed{a^{MAX} = 1.68 \text{ m/s}^2}$.

SUMMARY OF IMPORTANT POINTS   One of the main points in this problem is that the static frictional force (not the kinetic frictional force) keeps two objects that are in contact from slipping relative to each other. However, there is a limit beyond which slipping cannot be prevented. This limit occurs when the static frictional force attains its maximum magnitude f_s^{MAX}. Another important point is that f_s^{MAX} depends on the normal force. When finding the normal force, take into account all force components that are perpendicular to the surface where the friction is.

QUESTIONS

1. The instructions for mounting a phono cartridge on the tone arm of a stereo turntable say to "adjust the tracking force of the cartridge so it is less than three grams." From the point of view of correct physics, is there anything wrong with this statement? Explain.

2. Why do you lunge forward when your car suddenly comes to a halt? Why are you "thrown backward" when your car rapidly accelerates? In your explanation, refer to the most appropriate one of Newton's three laws of motion.

3. A bird feeder of large mass is hung from a tree limb, as the drawing shows. A cord attached to the bottom of the feeder has been left dangling free. Curiosity gets the best of a child, who pulls on the dangling cord in an attempt to see what's in the feeder. The dangling cord is cut from the same source as the cord attached to the limb. Is the cord between the feeder and the limb more likely to snap with a slow continuous pull or a sudden downward pull? Explain.

4. Is a net force being applied to an object when the object is moving downward (a) with a constant acceleration of 9.80 m/s² and (b) with a constant velocity of 9.80 m/s? Give your reasoning.

5. Newton's second law indicates that when a body is accelerating, a net force must be acting on the body. Does this mean that when two or more forces are applied to an object simultaneously, the object must always accelerate? Explain.

6. A father and his seven-year-old daughter are facing each other on ice skates. With their hands, they push off against one another. (a) Compare the magnitudes of the forces that they experience. (b) Which one, if either, experiences the larger acceleration? Account for your answers.

Question 3

7. A gymnast is bouncing on a trampoline. After a high bounce the gymnast comes down and hits the elastic surface of the trampoline. In so doing the gymnast applies a force to the trampoline. (a) Describe the effect this force has on the elastic surface. (b) The surface applies a reaction force to the gymnast. Describe the effect that this reaction force has on the gymnast.

8. According to Newton's third law, when you push on an object, the object pushes back on you with an oppositely directed force of equal magnitude. If the object is a massive crate resting on the floor, it will probably not move. Some people think that the reason the crate does not move is that the two oppositely directed pushing forces cancel. Explain why this logic is faulty and why the crate does not move.

9. Three point particles have identical masses. Each particle experiences only the gravitational forces due to the other two particles. How should the particles be arranged so each one experiences a net gravitational force that has the same magnitude? Give your reasoning.

10. When a body is moved from sea level to the top of a mountain, what changes—the body's mass, its weight, or both? Explain.

11. Suppose you wish to calculate the acceleration of gravity on the surface of a distant planet. What properties of the planet must you know? Justify your answers.

12. Does the acceleration of a freely falling object depend to any extent on the location, i.e., whether the object is on top of Mt. Everest or in Death Valley, California? Explain.

13. A 10-kg suitcase is placed on a scale that is in an elevator. Is the elevator accelerating up or down when the scale reads (a) 75 N and (b) 120 N? Justify your answers.

14. A stack of books whose true weight is 165 N is placed on a scale in an elevator. The scale reads 165 N. Can you tell from this information whether the elevator is moving with a constant velocity of 2 m/s upward or 2 m/s downward or whether the elevator is at rest? Explain.

15. Suppose you are in an elevator that is moving upward with a constant velocity. A scale inside the elevator shows your weight to be 600 N. (a) Does the scale register a value that is greater than, less than, or equal to 600 N during the time when the elevator slows down as it comes to a stop? (b) What is the reading when the elevator is stopped? (c) How does the value registered on the scale compare to 600 N during the time when the elevator picks up speed again on its way back down? Give your reasoning in each case.

16. A person has a choice of either pushing or pulling a sled at a constant velocity, as the drawing illustrates. If the angle θ is the same in both cases, does it require less force to push or to pull? Account for your answer.

17. A box is resting on a platform, and the coefficient of static friction between the two touching surfaces is μ_s. In an experiment, an astronaut pushes horizontally on the box so it just begins to move. He practices this maneuver on earth and performs it on the moon. In which case, if either, does he exert a greater pushing force? Justify your answer.

18. A rope is used in a tug-of-war between two teams of five people each. Both teams are equally strong, so neither team wins. An identical rope is tied to a tree and the same ten people pull just as hard on the loose end as they did in the contest. In both cases, the people pull steadily with no jerking. Which rope, if either, is more likely to break? Explain.

19. A stone is thrown from the top of a cliff. As the stone falls, is it in equilibrium? Explain, ignoring air resistance.

20. Can an object ever be in equilibrium if the object is acted on by (a) a single nonzero force, (b) two forces that point in mutually perpendicular directions, and (c) two forces that point in directions that are not perpendicular? Account for your answers.

21. During the final stages of descent, a sky diver with an open parachute approaches the ground with a constant velocity. The wind does not blow him from side to side. Is the sky diver in equilibrium and, if so, what forces are responsible for the equilibrium?

22. A weight hangs from a ring at the middle of a rope, as the drawing illustrates. Can the person who is pulling on the right end of the rope ever make the rope perfectly horizontal? Explain your answer in terms of the forces that act on the ring.

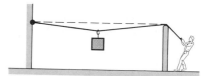

PROBLEMS

Section 4.3 Newton's Second Law of Motion

1. An empty airplane, whose mass is 30 400 kg has a take-off acceleration of 1.20 m/s². What is the net force accelerating the plane?

2. A bicycle has a mass of 13.1 kg and its rider has a mass of 81.7 kg. The rider is pumping hard, so that a horizontal net force of 9.78 N accelerates them. What is the acceleration?

3. A Porsche 911 accelerates from rest to 27 m/s (60 mph) due north in 5.8 s. The mass of the car is 1400 kg. What is the magnitude and direction of the average net force that acts on the Porsche?

4. A 1580-kg car is traveling with a speed of 15.0 m/s. What is the magnitude of the horizontal net force that is required to bring the car to a halt in a distance of 50.0 m?

5. During a circus performance, a 72-kg human cannonball is shot out of an 18-m-long cannon. If the human cannonball spends 0.95 s in the cannon, determine the average net force exerted on him in the barrel of the cannon.

6. A catapult on an aircraft carrier is capable of accelerating a plane from 0 to 56.0 m/s in a distance of 80.0 m. Find the average net force that the catapult exerts on a 13 300-kg jet.

*7. A net force acts on mass m_1 and produces an acceleration. A mass m_2 is added to mass m_1. The same net force acting on the two masses together produces one-third the acceleration. Determine the ratio m_2/m_1.

Section 4.4 The Vector Nature of Newton's Second Law of Motion, Section 4.5 Newton's Third Law of Motion

8. A force vector has a magnitude of 720 N and a direction of 38° north of east. Determine the magnitude and direction of the components of the force that point along the north–south line and along the east–west line.

9. Two skaters, an 82-kg man and a 48-kg woman, are standing on ice. Neglect any friction between the skate blades and the ice. The woman pushes on the man with a force of 45 N due east. Determine the accelerations (magnitude and direction) of the man and the woman.

10. Find the magnitude and direction (relative to the x axis) of the net force acting on each of the objects in the drawing.

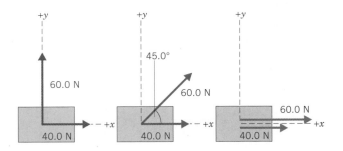

11. If the masses of the objects in problem 10 are each 5.00 kg, find their accelerations (both magnitude and direction).

12. A water-skier of mass 49 kg is being pulled due south by a horizontal towrope. The rope exerts a force of 228 N, due south. The water and air exert a combined frictional force of 165 N that is directed due north. What is the magnitude and direction of the skier's acceleration?

*13. A 36.5-kg sphere is acted upon by a 133-N force that is directed along the +x axis. What must be the magnitude and direction of a second force, such that the sphere experiences an acceleration of 3.96 m/s² directed 21.0° above the +x axis?

**14. A 4160-kg space probe is traveling with a speed of $v_0 = 2170$ m/s in the direction shown in the drawing. The probe has four engines, A, B, C, and D, as indicated. Each engine delivers a thrust of 68 300 N when turned on. Which engines should be fired and for how long, to change the velocity of the probe so the velocity has twice the original magnitude and points in a direction 90° clockwise relative to its original direction?

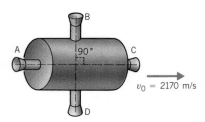

Section 4.6 The Gravitational Force, Section 4.7 Weight

15. Our solar system is in the Milky Way galaxy. The nearest galaxy is Andromeda, a distance of 2×10^{22} m away. The masses of the Milky Way and Andromeda galaxies are 7×10^{41} and $6 \times ^{41}$ kg, respectively. Treat the galaxies as particles and find the magnitude of the gravitational force exerted on the Milky Way by the Andromeda galaxy.

16. When a parachute opens, it develops a large drag force with the air. This upward force is initially greater than the weight of the sky diver and, thus, slows him down. Suppose the weight of the sky diver is 915 N and the drag force has a magnitude of 1027 N. What is the magnitude and direction of the acceleration?

17. Three objects are positioned along a straight line. From left to right their masses are 181, 70.0, and 405 kg. The 70.0-kg object is 0.310 m from the 181-kg object and 0.460 m from the 405-kg object. What is the net gravitational force acting on the 181-kg object?

18. The drawing (not to scale) shows one alignment of the sun, earth, and moon. The gravitational force $\mathbf{F}_{SM}$ that the sun exerts on the moon is perpendicular to the force $\mathbf{F}_{EM}$ that the earth exerts on the moon. The masses are: mass of sun = 1.99×10^{30} kg, mass of earth = 5.98×10^{24} kg, mass of moon = 7.35×10^{22} kg. The distances shown in the drawing are $r_{SM} = 1.50 \times 10^{11}$ m and $r_{EM} = 3.85 \times 10^8$ m. Determine the magnitude of the net gravitational force on the moon.

19. A space traveler whose mass is 115 kg leaves earth. What is his weight and mass (a) on earth and (b) in interplanetary space where there are no nearby planetary objects?

20. A person has a mass of 42.0 kg. What is the person's weight on the moon?

21. A can of coffee has a mass of 2.27 kg. What is its weight?

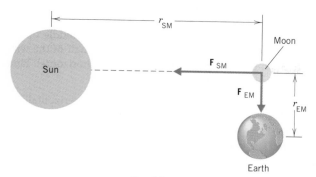

Problem 18

22. A space traveler weighs 580 N on earth. What will the traveler weigh on another planet whose radius is three times that of the earth and whose mass is twice that of the earth?

23. Mount Everest has an altitude of 8850 m above sea level. Approximately, what is the ratio between the acceleration due to gravity on the mountaintop and that at sea level?

24. A 95.0-kg person stands on a scale in an elevator. What is the apparent weight when the elevator is (a) accelerating upward with an acceleration of 1.80 m/s², (b) moving upward at a constant speed, and (c) accelerating downward with an acceleration of 1.30 m/s²?

25. A woman stands on a scale in a moving elevator. Her mass is 60.0 kg, and the combined mass of the elevator and scale is an additional 815 kg. Starting from rest, the elevator accelerates upward. During the acceleration, there is a tension of 9410 N in the hoisting cable. What is the reading on the scale during the acceleration?

26. A person carrying a briefcase (weight = 25 N) steps into an elevator. The elevator accelerates upward at 0.81 m/s². (a) What is the magnitude of the force that the person's hand exerts on the briefcase? (b) What is the apparent weight of the briefcase?

27. A rocket blasts off from rest and attains a speed of 45 m/s in 15 s. An astronaut has a mass of 57 kg. What is the astronaut's apparent weight during takeoff?

*28. Two identical space probes, each of mass 2.00×10^4 kg, are "parked" close to each other in outer space. If they are separated initially by 1.00 km, how long (in hours) will it take for the distance between them to decrease by 1.00 m, due to gravitational attraction? Assume that the probes behave as point particles and that the gravitational force remains constant as the probes move closer to each other.

*29. A spacecraft is on a journey to the moon. The masses of earth and the moon are, respectively, 5.98×10^{24} kg and 7.35×10^{22} kg. The distance between the centers of earth and the moon is 3.85×10^8 m. At what point, as measured from the center of earth, does the gravitational force exerted on the craft by earth balance the gravitational force exerted by the moon?

*30. Two identical masses m_1 are fixed to opposite corners of a square and exert a gravitational force on one another. Identical masses m_2 are added to each of the remaining corners. The net gravitational force acting on either mass m_1 is observed to be twice what it was originally. Determine the mass ratio m_1/m_2.

*31. A 55.0-kg person is riding in a hot-air balloon, and a scale shows this person's weight to be 549 N. The acceleration due to gravity at the location of the balloon is 9.79 m/s². Determine the magnitude and direction of the vertical component of the balloon's acceleration.

*32. A person has an apparent weight of 750 N when accelerating upward and 650 N when accelerating downward. In each case, the magnitude of the acceleration is the same. What is the person's true weight?

Section 4.8 The Normal Force, Section 4.9 Static and Kinetic Frictional Forces

33. A 60.0-kg crate rests on a level floor at a shipping dock. The coefficients of static and kinetic friction are 0.760 and 0.410, respectively. What horizontal pushing force is required to (a) just start the crate moving and (b) slide the crate across the dock at a constant speed?

34. A block whose weight is 45.0 N rests on a horizontal table. A horizontal force of 36.0 N is applied to the block. The coefficients of static and kinetic friction are 0.650 and 0.420, respectively. Will the block move under the influence of the force, and, if so, what will be the block's acceleration? Explain your reasoning.

35. A person is attempting to push a freezer across the room with a horizontal force of 267 N, but the freezer does not move. What is the static frictional force that the floor exerts on the freezer?

36. A 65.0-kg man is about to run across an icy pond. The coefficient of static friction between his shoes and the ice is 0.160. What is his maximum possible acceleration?

37. A 92-kg baseball player slides into second base. The coefficient of kinetic friction between the player and the ground is $\mu_k = 0.61$. (a) What is the magnitude of the frictional force? (b) If the player comes to rest after 1.2 s, what is his initial speed?

*38. A skater with an initial speed of 7.60 m/s is gliding across the ice. Air resistance is negligible. (a) The coefficient of kinetic friction between the ice and the skate blades is 0.100. Find the deceleration caused by kinetic friction. (b) How far will the skater travel before coming to rest?

*39. A block rests on a horizontal surface and weighs 425 N. A force is applied to the block and has a magnitude of 142 N. The force is directed upward at an angle θ relative to the horizontal. The block begins to move horizontally when

$\theta = 60.0°$. Determine the coefficient of static friction between the block and the surface.

**40. A 661-N force is being applied to a 121-kg block, as in the drawing. The coefficient of static friction between the block and the surface is $\mu_s = 0.410$. What is the minimal amount of additional mass that must be added on top of the block to prevent the block from moving?

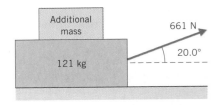

Section 4.11 Equilibrium Applications of Newton's Laws of Motion

41. A 12.0-kg lantern is suspended from the ceiling by two vertical wires. What is the tension in each wire?

42. A supertanker (mass $= 1.70 \times 10^8$ kg) is moving with a constant velocity. Its engines generate a forward thrust of 7.40×10^5 N. Determine (a) the magnitude of the resistive force exerted on the tanker by the water and (b) the magnitude of the upward buoyant force exerted on the tanker by the water.

43. The helicopter in the drawing is moving horizontally to the right at a constant velocity. The weight of the helicopter is $W = 53\ 800$ N. The lift force **L** generated by the rotating blade makes an angle of 21.0° with respect to the vertical. (a) What is the magnitude of the lift force? (b) Determine the magnitude of the air resistance **R** that opposes the motion.

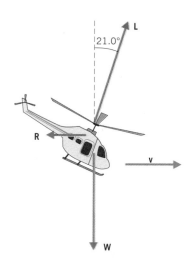

44. A wire is stretched between the tops of two identical buildings. When a tightrope walker is at the middle of the wire, the tension in the wire is 2220 N. Each half of the wire makes an angle of 8.00° with respect to the horizontal. Find the weight of the performer.

45. A 1.40-kg bottle of vintage wine is lying horizontally in the rack shown in the drawing. The two surfaces on which the bottle rests are 90.0° apart, and each exerts a force on the bottle that is perpendicular to the surface. What is the magnitude of each of these forces?

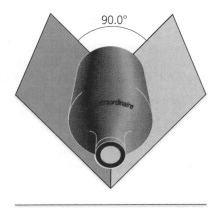

46. An 82.0-kg crate is resting on a frictionless surface that is inclined at an angle of 25.0° above the horizontal. The crate is held in place by a rope that is parallel to the incline. What is the tension in the rope?

47. A Mercedes-Benz 300SL ($m = 1700$ kg) is parked on a road that rises 15° above the horizontal. What is the magnitude of the static frictional force exerted on the tires by the road?

48. A 20.0-kg sled is being pulled across a horizontal surface at a constant velocity. The pulling force has a magnitude of 80.0 N and is directed at an angle of 30.0° above the horizontal. Determine the coefficient of kinetic friction.

49. A climber of weight 650 N is rappelling down a cliff. The rope makes an angle of 18.0° with the vertical, and the cliff exerts a force $\mathbf{F}_N$ on the feet of the climber; this force is directed 18.0° above the horizontal (see the drawing). Determine the tension in the rope.

50. A bicyclist coasts at a constant velocity along a road that slopes downward at an angle of 20.0° with respect to the horizontal. The combined mass of the bicycle and rider is 75.0 kg. Find the magnitude of the resistive force that opposes the motion.

*51. A 425-kg crate is hanging motionless from the end of a massless horizontal strut, as the drawing indicates. Find the tension in the cable that supports the strut. (*Hint: Consider the forces that act on the stationary ring, and asssume that the strut exerts a force on it that is directed horizontally to the left.*)

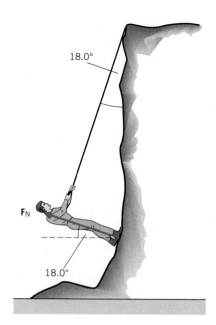

Problem 49

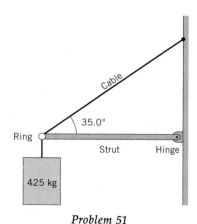

Problem 51

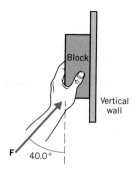

**54. A 0.600-kg kite is being flown at the end of a string. The string is straight and makes an angle of 55.0° above the horizontal. The kite is stationary and the tension in the string is 35.0 N. Determine the force (both magnitude and direction) that the wind exerts on the kite.

Section 4.12 Nonequilibrium Applications of Newton's Laws of Motion

55. A 350-kg sailboat has an acceleration of 0.62 m/s² at an angle of 64° north of east. Find the magnitude and direction of the net force that acts on the sailboat.

56. A car is towing a boat on a trailer. The driver starts from rest and accelerates to a speed of 11 m/s in a time of 28 s. The combined mass of the boat and trailer is 410 kg. What is the tension in the hitch that connects the trailer to the car?

57. A 1380-kg car is moving due east with an initial speed of 27.0 m/s. After 8.00 s the car has slowed down to 17.0 m/s. Find the magnitude and direction of the net force that produces the deceleration.

58. In the drawing, the weight of the block on the table is 111 N and that of the hanging block is 258 N. Ignoring all frictional effects and assuming the pulley to be massless, find (a) the acceleration of the two blocks and (b) the tension in the cord.

*52. A skier is pulled up a slope at a constant velocity by a tow bar. The slope is inclined at 25.0° with respect to the horizontal. The force applied to the skier by the tow bar is parallel to the slope. The skier's mass is 55.0 kg, and the coefficient of kinetic friction between the skis and the snow is 0.120. Find the magnitude of the force that the tow bar exerts on the skier.

53. The weight of the block in the drawing is 88.9 N. The coefficient of static friction between the block and the vertical wall is 0.560. What minimum force **F is required to (a) prevent the block from sliding down the wall *(Hint: the static frictional force is directed up the wall.)* and (b) start the block moving up the wall? *(Hint: the static frictional force is now directed down the wall.)*

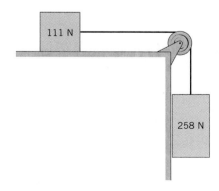

59. A rescue helicopter is lifting a man (weight = 822 N) from a capsized boat by means of a cable and harness.

(a) What is the tension in the cable when the man is given an initial upward acceleration of 1.10 m/s²? (b) What is the tension during the remainder of the rescue when he is pulled upward at a constant velocity?

60. A student is skateboarding down a ramp that is 6.0 m long and inclined at 18° with respect to the horizontal. The initial speed of the skateboarder at the top of the ramp is 2.6 m/s. Neglect friction and find the speed at the bottom of the ramp.

61. A lunar landing craft (mass = 11 400 kg) is about to touch down on the surface of the moon, where the acceleration due to gravity is 1.60 m/s². At an altitude of 165 m the craft's downward velocity is 18.0 m/s. To slow down the craft, a retrorocket is firing to provide an upward thrust. Assuming the descent is vertical, find the magnitude of the thrust needed to reduce the velocity to zero at the instant when the craft touches the lunar surface.

62. A cable is lifting a construction worker and a crate, as the drawing shows. The weights of the worker and crate are 965 and 1510 N, respectively. The acceleration of the cable is 0.620 m/s², upward. What is the tension in the cable (a) below the worker and (b) above the worker?

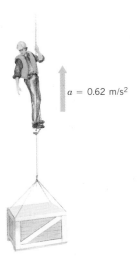

$a = 0.62 \text{ m/s}^2$

63. A passenger is pulling on the strap of a 15.0-kg suitcase with a force of 70.0 N. The strap makes an angle of 35.0° above the horizontal. A 37.8-N friction force opposes the motion (horizontal) of the suitcase. Determine the acceleration of the suitcase.

64. A locomotive is pulling two freight cars with an acceleration of 0.520 m/s². The mass of the first car is 51 300 kg, while that of the second car is 18 400 kg. Find the tension in the coupling mechanism between the engine and the first car and between the first car and the second car.

*65. Solve problem 58, assuming a coefficient of kinetic friction of 0.300 between the 111-N block and the table.

*66. A 205-kg log is pulled up a ramp by means of a rope that is parallel to the surface of the ramp. The ramp is inclined at 30.0° with respect to the horizontal. The coefficient of kinetic friction between the log and the ramp is 0.900, and the log has an acceleration of 0.800 m/s². Find the tension in the rope.

*67. A girl is sledding down a slope that is inclined at 30.0° with respect to the horizontal. A moderate wind is aiding the motion by providing a steady force of 105 N that is parallel to the motion of the sled. The combined mass of the girl and sled is 65.0 kg, and the coefficient of kinetic friction between the runners of the sled and the snow is 0.150. How much time is required for the sled to travel down a 175-m slope, starting from rest?

*68. As the drawing indicates, an electric motor is lowering a 452-kg crate with an acceleration of 1.60 m/s². Determine the tension in the cable. (*See Solved Problem 1 for a related problem.*)

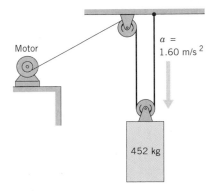

Motor

$a = 1.60 \text{ m/s}^2$

452 kg

*69. A 185-kg cart is pulled up a 25.0° incline by means of the cable and pulley arrangement shown in the drawing. The tension in the cable is 304 N. Assuming there is no friction, find the acceleration of the cart. (*See Solved Problem 1 for a related problem.*)

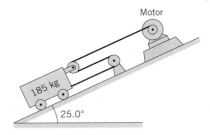

Motor

185 kg

25.0°

*70. A book is resting on a piece of paper. The paper is resting on a flat table. The coefficient of static friction between the book and the paper is 0.72. If a person pulls on the paper, what is the maximum acceleration that the book can have, before the book begins to slip relative to the paper? (*See Solved Problem 2 for a related problem.*)

*71. A crate is resting on the bed of a moving truck. The coefficient of static friction between the crate and the truck bed is 0.40. The driver hits the brakes. Assuming the truck is traveling on level ground, determine the maximum deceleration that the truck can have without the crate slipping forward relative to the truck. (*See Solved Problem 2 for a related problem.*)

*72. A truck is traveling at a speed of 25.0 m/s. A crate is resting on the bed of the truck, and the coefficient of static friction between the crate and the truck bed is 0.650. Determine the shortest distance in which the truck can come to a halt without causing the crate to slip forward relative to the truck. (*See Solved Problem 2 for a related problem.*)

**73. As part *a* of the drawing shows, two blocks are connected by a rope that passes over a set of pulleys. One block has a weight of 412 N, and the other has a weight of 908 N. The rope and the pulleys are massless and there is no friction. (a) What is the acceleration of the lighter block? (b) Suppose that the heavier block is removed, and a downward force of 908 N is provided by someone pulling on the rope, as part *b* of the drawing shows. Find the acceleration of the remaining block. (c) Explain why the answers in (a) and (b) are different.

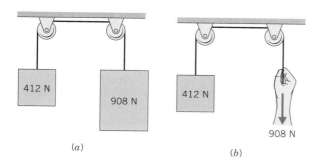

(a) (b)

**74. In the drawing, the rope and the pulleys are massless, and there is no friction. Find (a) the tension in the rope and (b) the acceleration of the 10.0-kg block. (*Hint: The larger mass moves twice as far as the smaller mass.*) (*See Solved Problem 1 for a related problem.*)

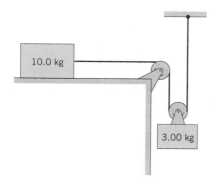

**75. A 5.00-kg block is placed on top of a 12.0-kg block that rests on a frictionless table. The coefficient of static friction between the two blocks is 0.600. What is the maximum hori-

zontal force that can be applied before the 5.00-kg block begins to slip relative to the 12.0-kg block, if the force is applied to (a) the more massive block and (b) the less massive block? (*See Solved Problem 2 for a related problem.*)

ADDITIONAL PROBLEMS

76. Saturn has an equatorial radius of 6.00×10^7 m and a mass of 5.67×10^{26} kg. (a) Compute the acceleration of gravity at the equator of Saturn. (b) How many times greater is a person's weight on Saturn compared to that on earth?

77. A net force of 525 N gives an object an acceleration of 4.20 m/s². (a) What net force is needed to give the object an acceleration of 13.7 m/s²? (b) What net force is required to keep the object moving at a constant velocity?

78. A 0.015-kg bullet is fired from a rifle. It takes 2.50×10^{-3} s for the bullet to travel the length of the barrel, and it exists the barrel with a speed of 715 m/s. Assuming that the acceleration of the bullet is constant, find the average net force exerted on the bullet.

79. A stuntman is being pulled along a rough road at a constant velocity, by a cable attached to a moving truck. The cable is parallel to the ground. The mass of the stuntman is 109 kg, and the coefficient of kinetic friction between the road and him is 0.870. Find the tension in the cable.

80. The mass of one small ball is 0.00150 kg, and the mass of another is 0.870 kg. If the center-to-center distance between these two balls is 0.100 m, find the magnitude of the gravitational force that each exerts on the other.

81. Three forces act on a moving object. One force has a magnitude of 80.0 N and is directed due north. Another has a magnitude of 60.0 N and is directed due west. What must be the magnitude and direction of the third force, such that the object continues to move with a constant velocity?

82. A fisherman is fishing from a bridge and is using a "45-N test line." In other words, the line will sustain a maximum force of 45 N without breaking. (a) What is the heaviest fish that can be pulled up vertically, when the line is reeled in at a constant speed? (b) Repeat part (a), assuming that the line is given an upward acceleration of 2.0 m/s².

83. A 55-kg person crouches on a scale and jumps straight up. As the person springs up, the reading on the scale suddenly rises to 622 N. What is the acceleration of the person at this instant?

84. At what altitude above the earth's surface would the value of *g* be one-half that at the surface?

85. A spacecraft has a mass of 3.50×10^4 kg and is drifting along a straight line through deep space. Its speed is 1820 m/s. An engine is suddenly turned on and provides a

thrust of 2240 N in the direction of the motion. (a) Find the acceleration of the spacecraft. (b) What is the distance (in km) traveled while the spacecraft increases its speed to 2310 m/s?

86. When a 0.058-kg tennis ball is served, it accelerates from rest to a speed of 45 m/s. The impact with the racket gives the ball a constant acceleration over a distance of 0.44 m. What is the magnitude of the force exerted on the ball by the racket?

*87. Two objects (45.0 and 21.0 kg) are connected by a massless string that passes over a massless, frictionless pulley. The pulley hangs from the ceiling. Find (a) the acceleration of the objects and (b) the tension in the string.

*88. Traveling at a speed of 16.1 m/s, the driver of an automobile suddenly locks the wheels by slamming on the brakes. The coefficient of kinetic friction between the tires and the road is 0.720. How far does the car skid before coming to a halt? Ignore the effects of air resistance.

*89. Several people are riding in a hot-air balloon. The combined mass of the people and balloon is 310 kg. The balloon is motionless in the air, because the downward-acting weight of the people and balloon is balanced by an upward-acting "buoyant" force. If the "buoyant" force remains constant, how much mass should be thrown overboard so the balloon acquires an upward acceleration of 0.15 m/s²?

*90. A toboggan slides down a hill and has a constant velocity. The angle of the hill is 11.3° with respect to the horizontal. What is the coefficient of kinetic friction between the surface of the hill and the toboggan?

*91. A 43.8-kg sign is suspended by two ropes, as the drawing shows. Find the tension in each rope.

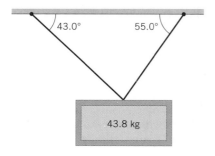

*92. A person whose weight is 5.20 × 10² N is being pulled up vertically by a rope from the bottom of a cave that is 35.1 m deep. The maximum tension that the rope can withstand

without breaking is 569 N. What is the shortest time, starting from rest, in which the person can be brought out of the cave?

*93. A 45.0-kg box is sliding up an incline that makes an angle of 15.0° with respect to the horizontal. The coefficient of kinetic friction between the box and the surface of the incline is 0.180. The initial speed of the box at the bottom of the incline is 1.50 m/s. How far does the box travel along the incline before coming to rest?

*94. A person is trying to judge if a picture (mass = 1.10 kg) is properly positioned by temporarily pressing it against a wall. The pressing force is perpendicular to the wall. The coefficient of static friction between the picture and the wall is 0.660. What is the minimum amount of pressing force that must be used?

**95. A small sphere is hung by a string from the ceiling of a van. When the van is stationary, the sphere hangs vertically. However, when the van accelerates, the sphere swings backward so that the string makes an angle of θ with respect to the vertical. (a) Derive an expression for the magnitude a of the acceleration of the van in terms of the angle θ and the magnitude g of the acceleration due to gravity. (b) Find the acceleration of the van when $\theta = 10.0°$. (c) What is the angle θ when the van moves with a constant velocity?

**96. The drawing shows three objects. They are connected by strings that pass over massless and friction-free pulleys. The objects move, and the coefficient of kinetic friction between the middle object and the surface of the table is 0.100. (a) What is the acceleration of the three masses? (b) Find the tension in each of the two strings.

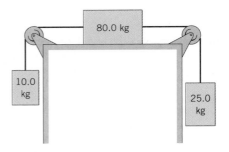

**97. A 225-kg crate rests on a surface that is inclined above the horizontal at an angle of 20.0°. A horizontal force (parallel to the ground, not the incline) whose magnitude is 535 N is required to start the crate moving down the incline. What is the coefficient of static friction between the crate and the surface of the incline?

CHAPTER 5

DYNAMICS OF UNIFORM CIRCULAR MOTION

At amusement parks, many rides whirl you around on circular paths. Motion on a circular path occurs in other circumstances too. A model airplane attached to a guideline often flies on a circle. In the "pairs" figure skating event, the beautiful "death spiral" is frequently part of the routine. In one version of this trick, the woman, held at arm's length by the man and bent nearly parallel to the ice, is whirled around in a circle on a single skate. A race car travels for a while on a circular arc when negotiating a banked turn. And many satellites, including the moon, orbit the earth on nearly circular paths. Our goal in this chapter is to understand the nature of the acceleration and forces that are present when an object moves on a circular path. In working toward this goal, we will use Newton's laws of motion, which apply just as well when the path of a moving object is a circle as when it is a straight line. We will begin by searching for any acceleration that exists and then use Newton's second law to tell us how much net force is required to maintain the acceleration.

5.1 UNIFORM CIRCULAR MOTION

There are many examples of motion on a circular path. Of the many possibilities, we want to single out those that satisfy the following definition:

Definition of Uniform Circular Motion
Uniform circular motion is the motion of an object traveling at a constant (uniform) speed on a circular path.

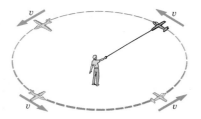

Figure 5.1 The motion of an airplane flying at a constant speed on a horizontal circular path is an example of uniform circular motion.

As an example of uniform circular motion, Figure 5.1 shows a model airplane on a guideline. The speed of the plane is the magnitude of the velocity vector **v**, and since the speed is constant, the vectors in the drawing have the same magnitude at various points on the circle.

Sometimes it is more convenient to describe uniform circular motion by specifying the period of the motion, rather than the speed. The *period* T is the time required for the object to travel once around the circle, that is, to make one complete revolution. There is a relationship between period and speed, since speed v is the distance traveled (circumference of the circle = $2\pi r$) divided by the time T:

$$v = \frac{2\pi r}{T} \tag{5.1}$$

If the radius is known, as in Example 1, the speed can be calculated from the period or vice versa.

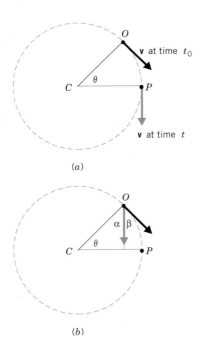

(a)

(b)

Figure 5.2 (a) For an object in uniform circular motion, the velocity vector **v** has different directions at different points on the circle.
(b) The velocity vector has been removed from point P, shifted parallel to itself, and redrawn with its tail at point O.

Example 1   A Tire-Balancing Machine

The wheel of a car has a radius of 0.29 m and is being rotated at 830 revolutions per minute (rpm) on a tire-balancing machine. Determine the speed at which the outer edge of the wheel is moving.

<u>*REASONING AND SOLUTION*</u>   The speed v can be obtained directly from $v = 2\pi r/T$, but first the period T is needed. Since the tire makes 830 revolutions in one minute, the number of minutes required for a single revolution is

$$\frac{1}{830 \text{ revolutions/min}} = 1.2 \times 10^{-3} \text{ min/revolution}$$

Therefore, the period is $T = 1.2 \times 10^{-3}$ min, which corresponds to 0.072 s. Equation 5.1 can now be used to find the speed:

$$v = \frac{2\pi r}{T} = \frac{2\pi(0.29 \text{ m})}{0.072 \text{ s}} = \boxed{25 \text{ m/s}}$$

The definition of uniform circular motion emphasizes that the magnitude of the velocity vector is constant. It is equally significant that the direction of the vector is *not constant*. In Figure 5.1, for instance, the velocity vector changes direction as the plane moves around the circle. Any change in the velocity vector, even if it is only a change in direction, means that an acceleration is occurring. This particular acceleration is called "centripetal acceleration," because it points toward the center of the circle, as explained in the next section.

5.2 CENTRIPETAL ACCELERATION

In this section we will determine how the magnitude a_c of the centripetal acceleration depends upon the speed v of the object and the radius r of the circular path. We will see that $a_c = v^2/r$.

In Figure 5.2a an object (symbolized by a dot •) is in uniform circular motion, and the velocity vector is drawn at two different times. At time t_0 the velocity is tangent to the circle at point O, while at a later time t the velocity is tangent at point P. As the object moves from O to P, the radius traces out the angle θ, and the velocity vector changes direction. To emphasize the change in direction, part b of the picture shows the velocity vector removed from point P, shifted parallel to itself, and redrawn with its tail at point O. The change in direction is indicated by the angle β between the two velocity vectors. Since the radii CO and CP are perpendicular to the tangents at points O and P, it follows that $\alpha + \beta = 90°$ and $\alpha + \theta = 90°$. Therefore, angle β and angle θ are equal.

As always, acceleration is the change $\Delta \mathbf{v}$ in velocity divided by the elapsed time Δt, or $\mathbf{a} = \Delta \mathbf{v}/\Delta t$. Figure 5.3$a$ shows the two velocity vectors oriented at the angle θ with respect to one another, together with the vector $\Delta \mathbf{v}$ that represents the change in velocity. The change $\Delta \mathbf{v}$ is the increment that must be added to the velocity at time t_0, so that the resultant velocity has the new direction after an elapsed time $\Delta t = t - t_0$. Figure 5.3b shows the sector of the circle COP. In the limit of a very small elapsed time Δt, the arc length OP can be approximated as a straight line whose length is the distance $v\,\Delta t$ traveled by the object. In this limit, COP is an isosceles triangle, as is the triangle in part a of the drawing. Since both triangles have equal apex angles θ, they are similar, so that

$$\frac{\Delta v}{v} = \frac{v\,\Delta t}{r}$$

This equation can be solved for $\Delta v/\Delta t$, to show that the magnitude a_c of the centripetal acceleration is given by $a_c = v^2/r$.

Centripetal acceleration is a vector quantity and has a direction as well as a magnitude. The direction is toward the center of the circle, a fact that can be understood with the aid of Figure 5.4. This drawing illustrates that an object in uniform circular motion would fly off on a tangent line if suddenly released from the circular path at point O. The object would move on a straight line (Newton's first law) to point A in the time it would have taken to travel on the circle to point P. It is as if the object drops through the distance AP in the process of remaining on the circle, and AP is directed toward the center of the circle in the limit that the angle θ is small. Thus, the object accelerates toward the center of the circle at every moment. The acceleration is called **centripetal acceleration,** because the word "centripetal" means "center-seeking."

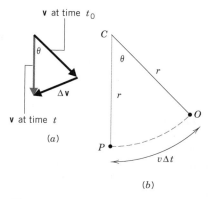

Figure 5.3   (*a*) The direction of the velocity vector at time *t* differs from the direction at time t_0 by an amount given by the angle θ. (*b*) When the object moves along the circumference of the circle from O to P, the radius r traces out the same angle θ.

Centripetal Acceleration

Magnitude: The centripetal acceleration of an object moving with a speed v on a circular path of radius r has a magnitude a_c given by

$$a_c = \frac{v^2}{r} \tag{5.2}$$

Direction: The centripetal acceleration vector always points toward the center of the circle and continually changes direction as the object moves.

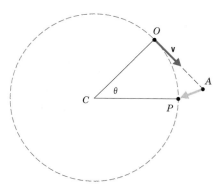

Figure 5.4   The object must accelerate in a direction that points toward the center of the circle, if the object is to remain on the circular path.

The following example illustrates the effect of the radius r on the centripetal acceleration.

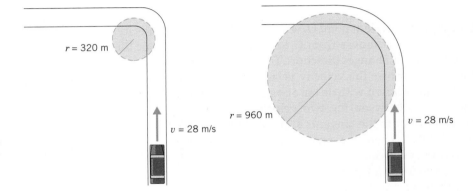

Figure 5.5 A car travels at the same speed around two curves with different radii. For the turn with the larger radius, the car has the smaller centripetal acceleration.

Example 2 **Tight Turns and Gentle Turns**

A car is driven around one curve whose radius is 320 m and then another whose radius is 960 m. Figure 5.5 shows the two curves. In both cases the speed is 28 m/s. Compare the centripetal accelerations for both turns.

REASONING AND SOLUTION In each case, the magnitude of the acceleration can be obtained from $a_c = v^2/r$:

[Radius = 320 m] $a_c = \dfrac{(28 \text{ m/s})^2}{320 \text{ m}} = \boxed{2.5 \text{ m/s}^2}$

[Radius = 960 m] $a_c = \dfrac{(28 \text{ m/s})^2}{960 \text{ m}} = \boxed{0.82 \text{ m/s}^2}$

When the radius is larger, the centripetal acceleration is smaller. In fact, when r becomes very large, the centripetal acceleration approaches zero. The motion along the arc of an infinitely large circle entails no acceleration, because it is just like motion at a constant speed along a straight line.

Because of the different centripetal accelerations, driving around each turn in Figure 5.5 would feel different, as most drivers know from their experiences with tight turns (smaller r) and gentle turns (larger r). The feeling is associated with the force that must be present in uniform circular motion, and we now turn to this topic.

5.3 CENTRIPETAL FORCE

Newton's second law indicates that whenever an object accelerates, there must be a net force to create the acceleration. Thus, in uniform circular motion there must be a net force to produce the centripetal acceleration. Moreover, the magnitude F_c of the net force can be calculated in the usual fashion, by multiplying the mass of the object by the magnitude of the acceleration: $F_c = ma_c = mv^2/r$. This net force is called the *centripetal force* and points in the same direction as the centripetal acceleration, that is, toward the center of the circle.

The "death spiral" in figure skating. The pull of the man's arm keeps the woman on a circular path.

Centripetal Force

Magnitude: The centripetal force is the name given to the net force required to keep an object of mass m, moving at a speed v, on a circular path of radius r and has a magnitude of

$$F_c = \frac{mv^2}{r} \qquad (5.3)$$

Direction: The centripetal force always points toward the center of the circle and continually changes direction as the object moves.

The phrase "centripetal force" does not denote a new and separate force created by nature. The phrase merely labels the net force pointing toward the center of the circular path, and this net force is the vector sum of all the force components that point along the radial direction. As we will see in Section 5.7, it is possible that more than one force component contributes to the centripetal force at the same time.

In some cases it is easy to identify the source of the centripetal force. For instance, when a model airplane flies on a guideline, the only force pulling the airplane inward is the tension force of the guideline, so this force is the centripetal force. (See Figure 5.6.) Examples 3 and 4 illustrate the fact that higher speeds and smaller circles require greater tension in the guideline.

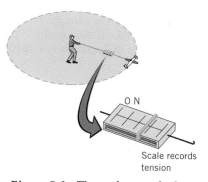

Figure 5.6 The scale records the tension in the guideline. See Examples 3 and 4.

Example 3   The Effect of Speed on Centripetal Force

The model airplane in Figure 5.6 has a mass of 0.90 kg and moves at a constant speed. Find the tension T in the guideline for speeds of 19 and 38 m/s with a 17-m line.

REASONING AND SOLUTION   Equation 5.3 gives the tension directly: $F_c = T = mv^2/r$.

[Speed = 19 m/s] $T = \dfrac{(0.90 \text{ kg})(19 \text{ m/s})^2}{17 \text{ m}} = \boxed{19 \text{ N}}$

[Speed = 38 m/s] $T = \dfrac{(0.90 \text{ kg})(38 \text{ m/s})^2}{17 \text{ m}} = \boxed{76 \text{ N}}$

Example 4   The Effect of Radius on Centripetal Force

The model airplane in Figure 5.6 has a constant speed of 28 m/s and a mass of 0.90 kg. Find the tension T in guidelines whose lengths are 11 and 22 m.

REASONING AND SOLUTION   According to Equation 5.3, $F_c = T = mv^2/r$.

[Radius = 11 m] $T = \dfrac{(0.90 \text{ kg})(28 \text{ m/s})^2}{11 \text{ m}} = \boxed{64 \text{ N}}$

[Radius = 22 m] $T = \dfrac{(0.90 \text{ kg})(28 \text{ m/s})^2}{22 \text{ m}} = \boxed{32 \text{ N}}$

When a car moves at a steady speed around an unbanked curve, the centripetal force keeping the car on the curve comes from the static friction between the road and the tires, as Figure 5.7a indicates. It is static, rather than kinetic friction, because the tires are not slipping with respect to the radial direction. If the static frictional force is insufficient, given the speed and the radius of the turn, the car will skid off the road. Example 5 shows how wet or icy conditions can limit safe driving.

Figure 5.7   (a) When the car moves without skidding around a curve, static friction between the road and the tires provides the centripetal force to keep the car on the road. (b) If the upholstery on the seat cannot provide enough static friction, the side of the car ultimately keeps the passenger on the circular path.

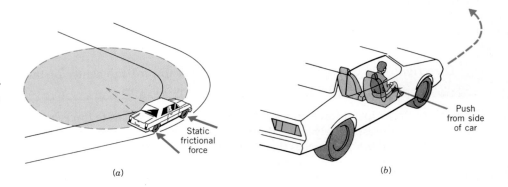

Static frictional force

Push from side of car

(a)   (b)

Example 5   Road Conditions and Safe Driving

Compare the maximum speeds at which a car can safely negotiate an unbanked turn (radius = 50.0 m) in dry weather (coefficient of static friction = 0.900) and icy weather (coefficient of static friction = 0.100).

REASONING   At the maximum speed, the maximum centripetal force acts on the tires, and static friction must provide it. The magnitude of the maximum force of static friction is specified by Equation 4.7 as $f_s^{\text{MAX}} = \mu_s F_N$, where μ_s is the coefficient of static friction and F_N is the magnitude of the normal force. Thus, we begin by finding the normal force.

SOLUTION   Since the car does not accelerate in the vertical direction, the weight mg of the car is balanced by the normal force, so $F_N = mg$. From Equation 5.3 it follows, then, that

$$F_c = \mu_s F_N = \mu_s mg = \frac{mv^2}{r}$$

Consequently, $\mu_s g = v^2/r$, and

$$v = \sqrt{\mu_s g r}$$

The mass m of the car has been eliminated algebraically from this result. All cars, heavy or light, have the same maximum speed. The maximum speeds can now be calculated:

Dry road ($\mu_s = 0.900$)	Icy road ($\mu_s = 0.100$)
$v = \sqrt{(0.900)(9.80 \text{ m/s}^2)(50.0 \text{ m})}$	$v = \sqrt{(0.100)(9.80 \text{ m/s}^2)(50.0 \text{ m})}$
$v = 21.0 \text{ m/s}$	$v = 7.00 \text{ m/s}$

As most people have learned by experience, the icy turn requires a slower speed for safe driving.

The passenger in Figure 5.7 must also experience a centripetal force to remain on the circular path. However, if the upholstery is very slippery, there may not be enough static friction to keep the passenger in place as the driver negotiates a tight turn at high speed. Then, when viewed from inside the car, the passenger appears to be thrown toward the outside of the turn. What really happens is that he slides off on a tangent to the circle, until he encounters a source of centripetal force to keep him in place while the car turns. In Figure 5.7 this occurs when the passenger bumps into the side of the car, which pushes on him with the necessary force. Clearly, the centripetal force can come from different sources. It can come from the static friction between the passenger and the upholstery, from a seat belt, or from the side of the car when the passenger bumps against it.

Sometimes the source of the centripetal force is not obvious. A pilot making a turn, for instance, banks or tilts the plane at an angle to create the centripetal force. As a plane flies, the air pushes upward on the wing surfaces with a net lifting force **L** that is perpendicular to the wing surfaces, as Figure 5.8a shows. Part b of the drawing illustrates that when the plane is banked at an angle θ, a component $L \sin \theta$ of the lifting force is directed toward the center of the turn. It is this component of the lift that provides the centripetal force. Greater speeds and/or tighter turns require greater centripetal forces. In such situations, the pilot must bank the plane at a larger angle, so that a larger component of the lift points toward the center of the turn. The technique of banking into a turn also has an application in the construction of high-speed roadways, where the road itself is banked to achieve a similar effect, as the next section discusses.

THE PHYSICS OF . . .

flying an airplane in a banked turn.

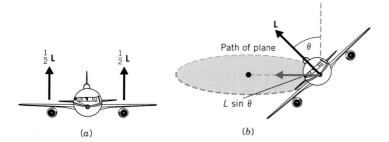

(a)   (b)

Figure 5.8   (a) The air exerts an upward lifting force $\frac{1}{2}$L on each wing. (b) When a plane executes a circular turn, the plane banks at an angle θ. The component $L \sin \theta$ of the lifting force is directed toward the center of the circle and provides the required centripetal force.

5.4 BANKED CURVES

When a car travels without skidding around an unbanked curve, the static frictional force between the tires and the road provides the centripetal force. From day to day, however, the friction may vary, depending on the condition of the pavement (ice, oil slicks, etc.). The reliance on friction can be eliminated completely for a given speed if the curve is banked at an angle relative to the horizontal, much in the same way that a plane is banked while making a turn.

Figure 5.9a shows a car going around a friction-free banked curve. The radius of the curve is r, where r is measured parallel to the horizontal and not to the slanted road surface. Part b shows the normal force $\mathbf{F_N}$ that the road applies to the car, the normal force being directed perpendicular to the road surface. Because the roadbed makes an angle θ with respect to the horizontal, the normal force has a component $F_N \sin \theta$ pointing toward the center C of the circle, and this component provides the centripetal force:

$$F_c = F_N \sin \theta = \frac{mv^2}{r}$$

Since the car does not accelerate in the vertical direction, the vertical component $F_N \cos \theta$ of the normal force must balance the weight mg of the car, so $F_N \cos \theta = mg$. Dividing this equation into the previous one shows that

$$\frac{F_N \sin \theta}{F_N \cos \theta} = \frac{mv^2/r}{mg}$$

$$\tan \theta = \frac{v^2}{rg} \qquad (5.4)$$

Equation 5.4 indicates that, for a given speed v, the centripetal force needed for a turn of radius r can be obtained from the normal force by banking the turn at an angle θ, independent of the mass of the vehicle. Greater speeds and smaller radii require more steeply banked curves, that is, larger values of θ. However, at a speed that is too small for a given θ, a car would slide down a frictionless banked curve; at a speed that is too large, a car would slide off the top. A typical banking condition is determined in the next example.

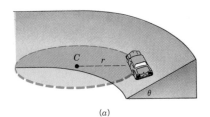

(a)

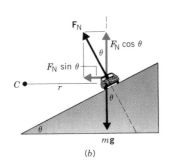

(b)

Figure 5.9 (a) A car travels in a circular path of radius r on a frictionless banked road. The banking angle is θ, and the center of the circle is at C. (b) The forces that act on the car are its weight mg and the normal force $\mathbf{F_N}$. The component of the normal force pointing toward the center of the circle ($F_N \sin \theta$) provides the centripetal force.

Example 6   A Banked Curve

Determine the angle at which a frictionless curve should be banked for a speed of 35 m/s and a turning radius of 550 m.

REASONING AND SOLUTION   Equation 5.4 gives the answer.

$$\tan \theta = \frac{v^2}{rg} = \frac{(35 \text{ m/s})^2}{(550 \text{ m})(9.80 \text{ m/s}^2)} = 0.23$$

$$\theta = \tan^{-1} 0.23 = \boxed{13°}$$

5.5 SATELLITES IN CIRCULAR ORBITS

THE RELATION BETWEEN ORBITAL RADIUS AND ORBITAL SPEED

Today there are many satellites in orbit about the earth. The ones in circular orbits are examples of uniform circular motion. Like a model airplane on a guideline, each satellite is kept on its circular path by a centripetal force. The gravitational pull of the earth provides the centripetal force and acts like an invisible guideline for the satellite.

There is only one speed that a satellite can have if the satellite is to remain in an orbit with a fixed radius. To see how this fundamental characteristic arises, consider the gravitational force acting on the satellite of mass m in Figure 5.10. In accord with Newton's law of gravitation (Equation 4.3), the gravitational force is directed toward the center of the earth and has a magnitude given by $F = GmM_E/r^2$. Here G is the universal gravitational constant, M_E is the mass of the earth, and r is the distance from the center of the earth to the satellite. Since the gravitational force is the only force acting on the satellite in the radial direction, it alone provides the centripetal force. Therefore,

$$F_c = G\frac{mM_E}{r^2} = \frac{mv^2}{r}$$

Solving for the speed v of the satellite gives

$$v = \sqrt{\frac{GM_E}{r}} \qquad (5.5)$$

If the satellite is to remain in an orbit of radius r, the speed must have precisely this value. The closer the satellite is to the earth, the smaller is the value for r and the greater the orbital speed must be. Once in orbit at the correct speed, the satellite continues in uniform circular motion forever, assuming that effects such as friction due to residual atmosphere do not reduce the speed.

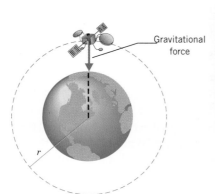

Figure 5.10 For a satellite in circular orbit around the earth, the gravitational force provides the required centripetal force.

An astronaut working on a satellite in orbit around the earth.

Equation 5.5 applies to man-made earth satellites or to the moon, which is a natural satellite of the earth. This equation also holds for satellites in circular orbits about the sun or another planet, provided M_E is replaced by the mass of the object about which the satellite moves.

The mass m of the satellite does not appear in Equation 5.5, having been eliminated algebraically. *Consequently, for a given orbit, a satellite with a large mass has exactly the same orbital speed as a satellite with a small mass.* However, more effort is certainly required to lift the larger-mass satellite into orbit. The orbital speed of an artificial satellite is determined in the following example.

Example 7   Orbital Speed of the Hubble Space Telescope

Determine the speed of the Hubble space telescope orbiting at a height of 596 km above the earth's surface.

REASONING   Before Equation 5.5 can be applied, the orbital radius r must be determined *relative to the center of the earth*. Since the radius of the earth is approximately 6.38×10^6 m, and the height of the telescope above the earth's surface is 0.596×10^6 m, the orbital radius is $r = 6.98 \times 10^6$ m.

SOLUTION   The orbital speed is

$$v = \sqrt{\frac{GM_E}{r}} = \sqrt{\frac{(6.67 \times 10^{-11} \text{ N} \cdot \text{m}^2/\text{kg}^2)(5.98 \times 10^{24} \text{ kg})}{6.98 \times 10^6 \text{ m}}}$$

$$\boxed{v = 7.56 \times 10^3 \text{ m/s } (16\,900 \text{ mi/h})}$$

THE PERIOD OF THE SATELLITE

The period T of a satellite is the time required for one orbital revolution. As in any uniform circular motion, the period is related to the speed of the motion by $v = 2\pi r/T$. Substituting v from Equation 5.5 shows that

$$\sqrt{\frac{GM_E}{r}} = \frac{2\pi r}{T}$$

Solving this expression for the period T gives

$$T = \frac{2\pi r^{3/2}}{\sqrt{GM_E}} \qquad (5.6)$$

Although derived for earth orbits, Equation 5.6 can also be used for calculating the periods of those planets that have nearly circular orbits about the sun, if M_E is replaced by the mass M_S of the sun and r is interpreted as the distance between the center of the planet and the center of the sun. The fact that the period is proportional to the three-halves power of the orbital radius is known as Kepler's third law, for it is one of the laws discovered by Johannes Kepler (1571–1630) during his studies of the motions of the planets. Kepler's third law also holds for elliptical orbits, which will be discussed in Chapter 9.

An important application of Equation 5.6 occurs in the field of communications, where "synchronous satellites" are put into a circular orbit that is in the

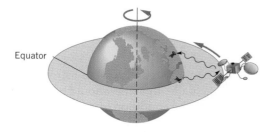

Figure 5.11   A synchronous satellite orbits the earth in a circular path that is in the plane of the equator. The period of the satellite is chosen to be one day, the same as the period of the earth's rotation. For clarity, the drawing is not to scale.

Equator

plane of the equator, as Figure 5.11 shows. The period of such a satellite is chosen to be one day, which is also the time it takes for the earth to turn once about its axis. Therefore, these satellites move around their orbits in a way that is synchronized with the rotation of the earth. For earth-based observers, synchronous satellites have the useful characteristic of appearing in fixed positions in the sky and can serve as "stationary" relay stations for communication signals sent up from the earth's surface. According to Equation 5.6, the orbit of a synchronous satellite can have only one radius, and this value is determined in the next example.

THE PHYSICS OF . . .

a synchronous satellite.

Example 8   The Orbital Radius for Synchronous Satellites

What is the height above the earth's surface at which all synchronous satellites (regardless of mass) must be placed in orbit?

REASONING   We can use Equation 5.6 to find the distance r of the satellite from the center of the earth. However, the problem asks for the height H of the satellite above the earth's surface. Therefore, to find H, we will have to take into account the fact that the earth itself has a radius of 6.38×10^6 m.

SOLUTION   The period T of a synchronous satellite is known to be one day* ($T = 8.64 \times 10^4$ s). In using this value it is convenient to rearrange the equation $T = 2\pi r^{3/2}/\sqrt{GM_E}$ as follows:

$$r^{3/2} = \frac{T\sqrt{GM_E}}{2\pi}$$

$$r^{3/2} = \frac{(8.64 \times 10^4 \text{ s}) \sqrt{(6.67 \times 10^{-11} \text{ N·m}^2/\text{kg}^2)(5.98 \times 10^{24} \text{ kg})}}{2\pi}$$

By squaring and then taking the cube root, we can solve this equation to show that $r = 4.23 \times 10^7$ m. Since the radius of the earth is approximately 6.38×10^6 m, the height of the satellite above the earth's surface is

$$H = 4.23 \times 10^7 \text{ m} - 0.64 \times 10^7 \text{ m} = \boxed{3.59 \times 10^7 \text{ m} \ (22\,300 \text{ mi})}$$

---

* Successive appearances of the sun define the solar day of 24 h or 8.64×10^4 s. The sun moves against the background of the stars, however, and the time required for the earth to turn once on its axis relative to the fixed stars is 23 h 56 min, which is called the sidereal day. The sidereal day should be used in Example 8, but the neglect of this effect introduces an error of less than 0.4 percent in the answer.

5.6 APPARENT WEIGHTLESSNESS AND ARTIFICIAL GRAVITY

The idea of life on board an orbiting satellite conjures up visions of astronauts floating around in a state of "weightlessness." Actually, this state should be called "apparent weightlessness," because it is similar to the condition of zero apparent weight that occurs in an elevator during free-fall. Figure 5.12a reviews apparent weight, the force that an object exerts on the platform of a scale. In free-fall, the apparent weight becomes zero, because both the scale and a person on it fall together and hence cannot push against one another. In contrast, the true weight is the gravitational force ($F = GmM_\text{E}/r^2$) that the earth exerts on an object, and this force is not zero in a freely falling elevator or aboard an orbiting satellite.

Within an orbiting satellite, apparent weightlessness has a meaning similar to that in the freely falling elevator. In uniform circular motion, an object continually accelerates or falls toward the center of the circle, in order to remain on the circular path. Therefore, the astronaut in Figure 5.12b and the scale beneath his feet both fall at the same rate toward the center of the orbit at all times and cannot push against one another. As a result, the apparent weight is zero. The only difference between the satellite and the elevator is that the satellite moves in a circle, so that its "falling" does not bring it closer to the earth.

The physiological effects of prolonged apparent weightlessness are only partially known. To minimize such effects, it is likely that artificial gravity will be provided in large space stations of the future. To help explain artificial gravity, Figure 5.13 shows a space station rotating about an axis. Because of the rotational motion, any object located at a point P on the interior surface of the station experiences a centripetal force directed toward the axis. The surface of the station provides this force by pushing on the feet of an astronaut. The centripetal force can be adjusted to match the astronaut's earth weight by properly selecting the rotational speed of the space station, as Example 9 illustrates.

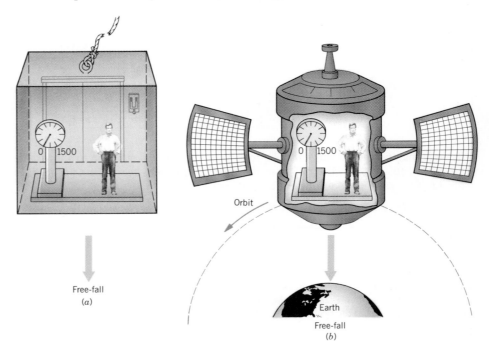

Free-fall
(a)

Orbit

Earth

Free-fall
(b)

Figure 5.12 (a) During free-fall, the elevator accelerates downward with the acceleration of gravity, and the apparent weight of the person is zero. (b) The orbiting space station is also in free-fall toward the center of the earth.

Example 9   Artificial Gravity

At what speed must the surface of the space station ($r = 1700$ m) move in Figure 5.13, so that the astronaut at point P experiences a push on his feet that equals his earth weight?

REASONING AND SOLUTION   The earth weight of the astronaut (mass = m) is mg, and with this substitution, Equation 5.3 can be used to determine the required speed: $F_c = mg = mv^2/r$. Solving this equation for the speed, we find that

$$v = \sqrt{rg} = \sqrt{(1700 \text{ m})(9.80 \text{ m/s}^2)} = \boxed{130 \text{ m/s}}$$

Figure 5.13   The surface of the rotating space station pushes on an object with which it is in contact and thereby provides the centripetal force needed to keep the object moving on a circular path.

*5.7 VERTICAL CIRCULAR MOTION

Motorcycle stunt drivers often perform a feat in which they drive their cycles around a vertical circular track, as in Figure 5.14a. Usually, the speed varies in this stunt, decreasing as the cycle moves upward and increasing as the cycle comes downward. When the speed of travel on a circular path changes from moment to moment, the motion is said to be "nonuniform." Using the concepts we have developed so far, it is possible to gain considerable insight into the nonuniform circular motion that typically occurs on a vertical circle.

As the speed changes in the motorcycle trick, the magnitude of the centripetal force also changes. There are four points on the circle where the centripetal force can be identified easily, as Figure 5.14b indicates. But remember, the centripetal force is not a new and separate force of nature. Instead, at each point the centripetal force is the net sum of all the force components oriented along the radial direction and points toward the center of the circle. The drawing shows only the weight of the cycle plus rider (magnitude = mg) and the normal force pushing on the cycle (magnitude = F_N). The propulsion and braking forces are omitted for simplicity and, in any event, do not act in the radial direction. The magnitude of

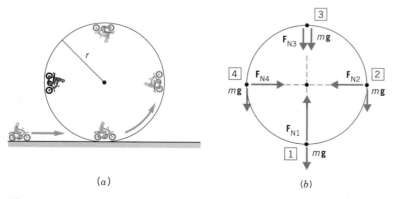

(a)   (b)

Figure 5.14   (a) A vertical loop-the-loop motorcycle stunt. (b) The normal force F_N and the weight mg of the cycle and the rider are shown here at four locations.

During part of his descent, this skydiver has traveled on a vertical circular arc.

the centripetal force at each of the four points is given below in terms of mg and F_N:

$$(1) \quad F_{c1} = \frac{mv_1{}^2}{r} = F_{N1} - mg$$

$$(2) \quad F_{c2} = \frac{mv_2{}^2}{r} = F_{N2}$$

$$(3) \quad F_{c3} = \frac{mv_3{}^2}{r} = F_{N3} + mg$$

$$(4) \quad F_{c4} = \frac{mv_4{}^2}{r} = F_{N4}$$

PROBLEM SOLVING INSIGHT

The centripetal force is the name given to the net force that points toward the center of a circular path. As shown here, there may be several forces that contribute to this net force.

Only the brave ride the "Corkscrew" at Knott's Berry Farm amusement part in Buena Park, California.

As the cycle goes around, the magnitude of the normal force changes. It changes because the speed changes and because the weight does not have the same effect at every point on the circle. At the bottom, the normal force and the weight oppose one another, giving a centripetal force of magnitude $F_{N1} - mg$. At the top, in contrast, the normal force and the weight reinforce each other to provide a centripetal force whose magnitude is $F_{N3} + mg$. At points 2 and 4 on either side, only F_{N2} and F_{N4} provide the centripetal force. The weight is tangent to the circle at points 2 and 4 and has no component pointing toward the center. If the speed at each of the four points is known, along with the mass and the radius, the normal force at each point can be determined from the equations above.

Riders who perform the loop-the-loop trick know that they must have at least a minimum speed at the top of the circle to remain on the track. This speed can be determined by considering the centripetal force at point 3. The speed v_3 in the equation $mv_3{}^2/r = F_{N3} + mg$ is a minimum when F_{N3} is zero. Then, the speed is given by $v_3 = \sqrt{rg}$. At this speed, the track does not exert a normal force to keep the cycle on the circle at point 3, because the weight mg provides all the centripetal force. For a radius of 9.0 m, for example, this expression predicts a minimum speed of 9.4 m/s (21 mi/h). Under these conditions, the rider experiences an apparent weightlessness like that discussed in Section 5.6, because for an instant the rider and the cycle are falling freely toward the center of the circle.

INTEGRATION OF CONCEPTS

UNIFORM CIRCULAR MOTION AND ACCELERATION

We first encountered the concept of acceleration in Chapter 2, where we learned that an acceleration occurs whenever there is a change in the velocity of an object. Since velocity is a vector quantity, it has both a magnitude and a direction. Thus, a velocity vector can change in one or both of two ways. The magnitude of the velocity can change, perhaps from 5 to 20 m/s, and in Chapter 2 we dealt exclusively with this kind of change. The second way in which a velocity can change is for its direction to change, even though the magnitude of the velocity remains the same. Here in Chapter 5 we find that such a change occurs in uniform circular motion, where the velocity has a constant magnitude, but continually changes direction from moment to moment on a circular path. True, we give the resulting acceleration the name of "centripetal acceleration" to denote that it is directed toward the center of the circle. However, the name should not obscure the fact that this particular acceleration arises just as does any other acceleration, namely, from a change in the velocity.

UNIFORM CIRCULAR MOTION AND NEWTON'S SECOND LAW OF MOTION

The centripetal acceleration of an object moving in uniform circular motion has a magnitude of $a_c = v^2/r$, where v is the speed of the object and r is the radius of the circle. In Chapter 4 we saw that whenever an object is accelerating, there must be a net force acting on it in the direction of the acceleration. The relation between the net force and acceleration is given by Newton's second law as $\Sigma F = ma$. It is important to recognize that uniform circular motion is just one particular kind of motion that obeys the second law, that is, $F_c = ma_c$, where $F_c = \Sigma F$ is the centripetal force, or the net force, pointing toward the center of the circle.

UNIFORM CIRCULAR MOTION AND THE VECTOR COMPONENTS OF NEWTON'S SECOND LAW

In Chapter 4 we also saw how to apply Newton's second law in two dimensions by using the x and y components of the net force and the acceleration: $\Sigma F_x = ma_x$ and $\Sigma F_y = ma_y$. Sometimes, there is an acceleration in the x direction, but not in the y direction, so the equations representing the second law in component form become $\Sigma F_x = ma_x$ and $\Sigma F_y = 0$. In other words, there is a net force in the x direction, but the force components balance (or cancel) in the y direction. Here in Chapter 5, there are also times when uniform circular motion requires that Newton's second law be used in a similar fashion. Homework problem 16, for instance, deals with one example of such a case. In this problem we choose the x direction as the radial direction of the circular motion and write $\Sigma F_x = ma_x$ as $F_c = m(v^2/r)$. We choose the y direction to be the one in which no acceleration occurs, namely, perpendicular to the plane of the circle. Then, the equation $\Sigma F_y = 0$ represents the balance of the force components in this direction. In the specific case of problem 16 (see the drawing), the vertical component of the tension in the support cable balances the weight of the chair and its occupant. Thus, the two equations, $F_c = m(v^2/r)$ and $\Sigma F_y = 0$, together express Newton's second law of motion for uniform circular motion.

SUMMARY

In **uniform circular motion,** an object of mass m travels at a constant speed v on a circular path of radius r. The **period** T of the motion is the time required to make one revolution. The speed, the period, and the radius are related according to $v = 2\pi r/T$. The velocity vector in such motion is always changing direction, and, therefore, an acceleration exists. This acceleration is called **centripetal acceleration,** and its magnitude is $a_c = v^2/r$, while its direction is toward the center of the circle. To create this acceleration, a net force pointing toward the center of the circle is needed. This net force is called the **centripetal force,** and its magnitude is $F_c = mv^2/r$.

There are many examples of uniform circular motion, including the **banking of curves** and the **orbiting of**

satellites. The angle θ at which a friction-free curve is banked depends on the radius r of the curve and the speed v at which the curve is to be negotiated, according to $\tan \theta = v^2/(rg)$. The speed and the period of a satellite in a circular orbit about the earth depend on the radius of the orbit, according to $v = \sqrt{GM_E/r}$ and $T = 2\pi r^{3/2}/\sqrt{GM_E}$, where G is the universal gravitational constant and M_E is the mass of the earth. **Apparent weightlessness** and **artificial gravity** in satellites can be explained in terms of uniform circular motion.

SOLVED PROBLEMS

Solved Problem 1   A Rotating Space Station

Related Problems: **10   **18   **30

A space station is rotating to create artificial gravity, as the drawing indicates. The rate of rotation is chosen so the outer ring ($r_A = 2150$ m) simulates the acceleration of gravity on the surface of Venus (8.62 m/s^2). (a) How long does it take the space station to turn once around its axis; in other words, what is its period T? (b) What should be the radius r_B, so the inner ring simulates the acceleration of gravity on the surface of Mercury (3.63 m/s^2)?

REASONING Since the various sections of the space station are rigidly connected, they rotate as a unit. As the outer ring turns through one revolution, so does the inner ring. Both rings, therefore, have the same period, and we will use this fact in solving this problem.

SOLUTION
(a) Equation 5.1 ($v_A = 2\pi r/T$) can be used with the given radius r_A to calculate the period T, once the speed v_A is known. This speed can be obtained by solving the equation for centripetal acceleration ($a_c = v_A^2/r_A$) and using the given acceleration and radius:

$$a_c = \frac{v_A^2}{r_A}$$

$$v_A = \sqrt{a_c r_A} = \sqrt{(8.62 \text{ m/s}^2)(2150 \text{ m})} = 136 \text{ m/s}$$

$$T = \frac{2\pi r_A}{v_A} = \frac{2\pi(2150 \text{ m})}{136 \text{ m/s}} = \boxed{99.3 \text{ s}}$$

(b) Since the inner ring has the same period as the outer ring, it follows that

$$T = \frac{2\pi r_A}{v_A} = \frac{2\pi r_B}{v_B}$$

$$r_B = \frac{r_A v_B}{v_A}$$

Considering that r_A and v_A are known, a value for v_B is needed to calculate the radius of the inner ring. The speed v_B is such that the centripetal acceleration of the inner ring is 3.63 m/s^2. Therefore,

$$3.63 \text{ m/s}^2 = \frac{v_B^2}{r_B}$$

Solving this expression for v_B in terms of r_B and substituting the result into $r_B = r_A v_B/v_A$ reveals that

$$r_B = \frac{r_A \sqrt{(3.63 \text{ m/s}^2)r_B}}{v_A}$$

Squaring both sides of this expression, we find

$$r_B = \frac{r_A^2(3.63 \text{ m/s}^2)}{v_A^2} = \frac{(2150 \text{ m})^2(3.63 \text{ m/s}^2)}{(136 \text{ m/s})^2} = \boxed{907 \text{ m}}$$

SUMMARY OF IMPORTANT POINTS There are two important ideas in this problem. One is that the speed in uniform circular motion can always be represented as the circumference of the circle divided by the period. The other is that when a rigid object rotates about an axis, all points on the object have the same period.

QUESTIONS

1. A car is moving with a constant speed along a circular path. According to the definition of equilibrium given in Section 4.11, is the car in equilibrium? If not, why not?

2. For uniform circular motion, complete the following table by answering "yes" or "no" in the appropriate spaces. Provide a reason for each of your answers.

	Velocity vector	Acceleration vector
Constant magnitude?		
Constant direction?		

3. The equations of kinematics (Equations 3.3–3.6) describe the motion of an object that has a constant acceleration. These equations cannot be applied to uniform circular motion. Why not?

4. Is it possible for an object to have an acceleration when the velocity of the object is constant? When the speed of the object is constant?

5. A car is traveling at a constant speed along the road *ABCDE* shown in the drawing. Sections *AB* and *DE* are straight. Rank the accelerations in each of the four sections according to magnitude, listing the smallest first.

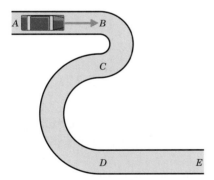

6. At an amusement park there is a rotating circular platform that resembles a large turntable. For the price of a ticket, anyone can try to remain in place without falling off, as the platform picks up speed. Initially each person stands at the same distance from the center. Explain why people with rubber soles on their shoes have an easier time on this ride than people whose shoes have hard leather soles.

7. What is the chance of a light car safely rounding an unbanked curve on an icy road as compared to that for a heavy car: worse, the same as, or better? Assume that both cars have the same speed and are equipped with identical tires. Explain.

8. In a circus act, a man hangs upside down from a trapeze, legs bent over the bar and arms downward. He is holding his partner, who is hanging beneath him. Is it easier to hold the partner while stationary or while swinging back and forth? Why?

9. A propeller, operating under test conditions, is being made to rotate at ever faster speeds. Explain what is likely to happen, and why, when the maximum rated speed of the propeller is exceeded.

10. A penny is placed on a rotating turntable. Where on the platter does the penny require the largest centripetal force to remain in place? Give your reasoning.

11. Explain why a real airplane must bank as it flies in a circle, while a model airplane on a guideline can fly in a circle without banking.

12. Would a change in the earth's mass affect (a) the banking of airplanes as they turn, (b) the banking of roadbeds, (c) the speeds with which satellites are put into circular orbits, and (d) the performance of the loop-the-loop motorcycle stunt? In each case, give your reasoning.

13. A space station is in circular orbit about the earth and has no artificial gravity. A book is on a table in this space station. Is any kinetic frictional force encountered when the book slides across the table? Explain.

14. Two cars are identical, except for the type of tread design on their tires. The cars are driven at the same speed and enter the same turn. Car A cannot negotiate the turn, but car B can. Which tread design yields a larger coefficient of static friction between the tires and the road? Give your reasoning in terms of centripetal force.

PROBLEMS

Section 5.1 Uniform Circular Motion,
Section 5.2 Centripetal Acceleration

1. The tips of the blades in a food blender are moving with a speed of 21 m/s in a circle that has a radius of 0.053 m. How much time does it take for the blades to make one revolution?

2. A horse races once around a circular track in a time of 118 s, with a speed of 17 m/s. What is the radius of the track?

3. Magnetic tape is being spooled from a supply reel to a takeup reel. The tape is moving with a speed of 0.191 m/s, and the radius of the takeup reel is 0.0762 m. What is the magnitude and direction of the centripetal acceleration of the tape as it is wound on the outer layer of the takeup reel?

4. A bicycle chain is wrapped around a rear sprocket ($r = 0.039$ m) and a front sprocket ($r = 0.10$ m). The chain moves with a speed of 1.4 m/s around the sprockets, while the bike moves at a constant velocity. Find the magnitude of the acceleration of a chain link that is in contact with (a) the rear sprocket, (b) neither sprocket, and (c) the front sprocket.

5. Two cars are going around curves, one car traveling at 26 m/s and the other traveling at 13 m/s. Each car experiences a centripetal acceleration of the same magnitude. What is the ratio of the radii of the two curves?

6. How long does it take a plane, traveling at a speed of 110 m/s, to fly once around a circle whose radius is 2850 m?

7. A race car travels at a constant speed around a circular track whose radius is 2.6 km. If the car goes once around the track in 360 s, what is the magnitude of the centripetal acceleration of the car?

*8. The earth rotates once per day about an axis passing through the north and south poles, an axis that is perpendicular to the plane of the equator. Assuming the earth is a sphere with a radius of 6.38×10^6 m, determine the speed and centripetal acceleration of a person situated (a) at the equator and (b) at a latitude of $45.0°$.

*9. A centrifuge is a device in which a small container of material is rotated at a high speed on a circular path. Such a device is used in medical laboratories, for instance, to cause the more dense red blood cells to settle through the less dense blood serum and collect at the bottom of the container. Suppose the centripetal acceleration of the sample is 6.25×10^3 times larger than the acceleration due to gravity. How many revolutions per minute is the sample making, if it is located at a radius of 5.00 cm from the axis of rotation?

**10. A boy is riding a merry-go-round at a distance of 7.00 m from its center. The boy experiences a centripetal acceleration of 7.50 m/s². What centripetal acceleration is experienced by another person who is riding at a distance of 3.00 m from the center? (See Solved Problem 1 for a related problem.)

Section 5.3 Centripetal Force

11. A 0.015-kg ball is shot from the plunger of a pinball machine. Because of a centripetal force of 0.028 N, the ball follows a circular arc whose radius is 0.25 m. What is the speed of the ball?

12. A car is safely negotiating an unbanked circular turn at a speed of 21 m/s. The maximum static frictional force acts on the tires. Suddenly a wet patch in the road reduces the maximum static frictional force by a factor of three. If the car is to continue safely around the curve, to what speed must the driver slow down the car?

13. A child is twirling a 0.0120-kg ball on a string in a horizontal circle whose radius is 0.100 m. The ball travels once around the circle in 0.500 s. (a) Determine the centripetal force acting on the ball. (b) If the speed is doubled, does the centripetal force double? If not, by what factor does the centripetal force increase?

14. A model airplane is being flown on a guideline that can sustain at most 180 N of tension. At a speed of 28 m/s, what is the radius of the smallest horizontal circle in which a 0.75 kg plane can be flown?

15. A car rounds an unbanked curve (radius = 92 m) without skidding at a speed of 26 m/s. What is the smallest possible coefficient of static friction between the tires and the road?

*16. A "swing" ride at a carnival consists of chairs that are swung in a circle by 12.0-m cables attached to a vertical rotating pole, as the drawing shows. Suppose the total mass of a chair and its occupant is 220 kg. (a) Determine the tension in the cable attached to the chair. (b) Find the speed of the chair.

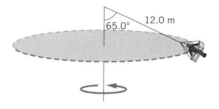

*17. What is the minimum coefficient of static friction necessary to allow a penny to rotate along with a $33\frac{1}{3}$-rpm record (diameter = 0.300 m), when the penny is placed anywhere on the record?

**18. A rigid massless rod is rotated about one end in a horizontal circle. There is a mass m_1 attached to the center of the rod and a mass m_2 attached to the outer end of the rod. The inner section of the rod sustains twice as much tension as the outer section. Find the ratio m_2/m_1. (See Solved Problem 1 for a related problem.)

Section 5.4 Banked Curves

19. At what angle should a curve of radius 150 m be banked, so cars can travel safely at 25 m/s without relying on friction?

20. A curve of radius 120 m is banked at an angle of $18°$. At what speed can it be negotiated under icy conditions where friction is negligible?

*21. There is a similarity between a plane banking into a

turn and a car going around a banked curve. The lifting force **L** in Figure 5.8 plays the same role as the normal force **F**$_N$ in Figure 5.9. (a) Derive an expression that relates the banking angle to the speed of the plane, the radius of the turn, and the acceleration due to gravity. (b) At what angle with respect to the horizontal should a plane be banked when traveling at 195 m/s around a turn whose radius is 8250 m?

**22. Refer to problem 21 before attempting to solve this problem. A jet ($m = 2.00 \times 10^5$ kg), flying at 123 m/s, banks to make a horizontal circular turn. The radius of the turn is 3810 m. Calculate the necessary lifting force.

Section 5.5 Satellites in Circular Orbits, Section 5.6 Apparent Weightlessness and Artificial Gravity

23. A satellite is placed in orbit 6.00×10^5 m above the surface of Jupiter. Jupiter has a mass of 1.90×10^{27} kg and a radius of 7.14×10^7 m. Find the orbital speed of the satellite.

24. Suppose the surface (radius $= r$) of the space station in Figure 5.13 is rotating at 35.8 m/s. What must be the value of r for the astronauts to weigh one-half of their earth weight?

25. Venus rotates slowly about its axis, the period being 243 days. The mass of Venus is 4.87×10^{24} kg. Determine the radius for a synchronous satellite in orbit about Venus.

26. The moon orbits the earth at a distance of 3.85×10^8 m. Assume that this distance is between the centers of the earth and the moon and that the mass of the earth is 5.98×10^{24} kg. Find the period for the moon's motion around the earth. Express the answer in days and compare it to the length of a month.

27. A rocket is used to place a synchronous satellite in orbit about the earth. What must be the speed of the rocket when it releases the satellite?

*28. If the earth had no atmosphere and were a perfectly smooth sphere of radius 6.38×10^6 m and mass 5.98×10^{24} kg, what would be the greatest orbital speed that a satellite could have in a circular orbit about the earth?

*29. The earth orbits the sun once a year at a distance of 1.50×10^{11} m. Venus orbits the sun at a distance of 1.08×10^{11} m. These distances are between the centers of the planets and the sun. How long does it take for Venus to make one orbit around the sun?

**30. To create artificial gravity, the space station shown in the drawing is rotating at a rate of 1.00 rpm. The radii of the cylindrically shaped chambers have the ratio $r_A/r_B = 4.00$. Each chamber A simulates an acceleration due to gravity of 10.0 m/s^2. Find values for (a) r_A, (b) r_B, and (c) the acceleration due to gravity that is simulated in chamber B. (*See Solved Problem 1 for a related problem.*)

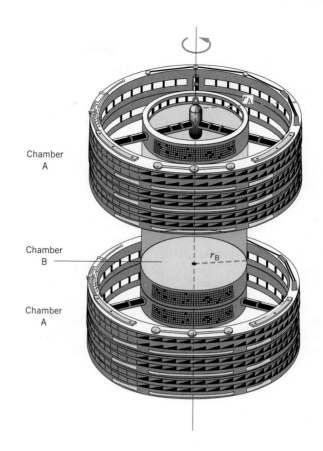

Chamber A

Chamber B

Chamber A

Section 5.7 Vertical Circular Motion

31. The condition of apparent weightlessness can be created for a brief instant when a plane flies over the top of a vertical circle. At a speed of 215 m/s, what is the radius of the vertical circle that the pilot must use?

32. A motorcycle has a constant speed of 25.0 m/s as it passes over the top of a hill whose radius of curvature is 126 m. The mass of the motorcycle and driver is 342 kg. Find the magnitude of (a) the centripetal force and (b) the normal force that acts on the cycle.

33. A roller coaster at an amusement park has a dip that bottoms out in a vertical circle of radius r. Passengers feel the seat of the car pushing on them with a force equal to twice their weight as they go through the dip. If $r = 20.0$ m, how fast is the roller coaster car traveling at the bottom of the dip?

34. A downhill skier, whose mass is 50.0 kg, attains a speed of 21.0 m/s just as she reaches the point where a jump is necessary (point A in the drawing). When she leaves the ground, her velocity is horizontal. In other words, point A is at the bottom of the circular arc AB (radius $= 27.0$ m). Determine the normal force acting on the skis at point A.

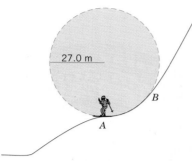

Problem 34

35. A 2100-kg demolition ball swings at the end of a 15-m cable on the arc of a vertical circle. At the lowest point of the swing, the ball is moving at a speed of 7.6 m/s. Determine the tension in the cable. *(Hint: The tension serves the same purpose as the normal force at point 1 in Figure 5.14.)*

*   **36.** A motorcycle is traveling up one side of a hill and down the other side. The crest is a circular arc with a radius of 45.0 m. Determine the maximum speed that the cycle can have while moving over the crest without losing contact with the road.

ADDITIONAL PROBLEMS

37. A 125-kg crate rests on the flatbed of a truck that moves at a speed of 15.0 m/s around a curve whose radius is 66.0 m. The crate does not slip relative to the truck. Obtain the magnitude of the static frictional force that the truck bed exerts on the crate.

38. An unbanked curve in the road has a radius of 75.0 m. The greatest speed that a motorcycle can have without skidding around this curve is 25.0 m/s. Another curve has an identical surface but has a radius of 125 m. What is the maximum speed that the motorcycle can have around the second curve without skidding?

39. The earth travels around the sun once a year in an approximately circular orbit whose radius is 1.50×10^{11} m. From these data determine (a) the orbital speed of the earth and (b) the mass of the sun.

40. Speedboat A negotiates a curve whose radius is 120 m. Speedboat B negotiates a curve whose radius is 240 m. Each boat experiences the same centripetal acceleration. What is the ratio v_A/v_B of the speeds of the boats?

41. A satellite circles the earth in an orbit whose radius is twice the earth's radius. The earth's mass is 5.98×10^{24} kg and its radius is 6.38×10^6 m. What is the period of the satellite?

42. A dust particle of mass 1.5×10^{-7} kg is located on the outer edge of a compact disc recording (radius = 0.060 m). The disc is rotating at 3.5 rev/s. Find the magnitude of the centripetal force that acts on the dust particle.

43. The coefficient of static friction between the tires of a motorcycle and the road is $\mu_s = 0.56$. At a speed of 19 m/s, what is the radius of the tightest turn that the driver can hope to handle?

44. The maximum tension that a 0.50-m string can tolerate is 14 N. A 0.25-kg ball attached to this string is being whirled in a vertical circle. What is the maximum speed that the ball can have (a) at the top of the circle and (b) at the bottom of the circle? *(Hint: The tension serves the same purpose as the normal force in Figure 5.14.)*

*   **45.** A racetrack has the shape of an inverted cone, as the drawing shows. On this surface the cars race in circles that are parallel to the ground. For a speed of 27.0 m/s, at what value of the distance d should a driver locate his car, if he wishes to stay on a circular path without depending on friction?

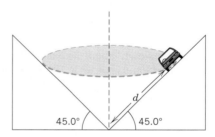

*   **46.** A helicopter rotor turns at a rate of 315 rpm. The tip of one of the blades moves on a circle whose radius is 7.50 m. (a) What is the speed of the tip of the blade? (b) What is the centripetal acceleration of the tip of the blade? Express the answer in terms of multiples of $g = 9.80$ m/s^2.

* * **47.** At amusement parks, there is a popular ride where the floor of a rotating cylindrical room falls away, leaving the riders "plastered" against the wall. Suppose the radius of the room is 3.30 m and the speed of the wall is 10.0 m/s when the floor falls away. (a) What is the source of the centripetal force acting on the riders? (b) How much centripetal force acts on a 55.0-kg rider? (c) What is the minimum coefficient of static friction that must exist between a rider's back and the wall, if the rider is to remain in place when the floor drops away?

* * **48.** A block is hung by a string from the inside roof of a van. When the van goes straight ahead at a speed of 28 m/s, the block hangs vertically down. But when the van maintains this same speed around a curve (radius = 150 m), the block swings toward the outside of the curve. Then the string makes an angle θ with the vertical. Find θ.

WORK AND ENERGY

World-class oarsmen, shown here in the "straight pair" competition, are incredibly well-conditioned athletes. To achieve a competitive speed, they work hard and expend an enormous amount of energy in overcoming the retarding effect of the water. To accomplish the same result, motorboats use engines instead of oarsmen and burn gasoline to obtain the energy for doing the work. In physics the concepts of work and energy are closely related, and we will find that there are a number of kinds of energy. A moving boat, for instance, has kinetic energy. But a stationary object can also have energy by virtue of its position relative to the earth. For example, a skier about to begin a downhill run, has potential energy due to gravity, because he or she is high up on a mountain slope. The relationship between work and energy will lead us to an important fundamental principle called the principle of conservation of energy. This principle will provide us with a powerful method for analyzing motion in the present chapter and for analyzing other situations in later chapters.

6.1 WORK

WORK DONE BY A CONSTANT FORCE THAT POINTS IN THE DIRECTION OF THE MOTION

Work is a familiar concept. For example, it takes work to push a heavy object such as a stalled car. In fact, the more pushing force that is used and the greater the displacement over which the car is moved, the greater the amount of work that is done. Force and displacement, then, are the two essential elements of work, as

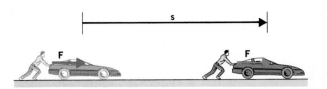

Figure 6.1 Work is done when a force **F** pushes a car through a displacement **s**.

Figure 6.1 illustrates. The drawing shows a situation in which a constant force **F** points in the same direction as the displacement **s** of the car. In such a case, the work W is defined as the magnitude F of the force times the magnitude s of the displacement: $W = Fs$. The work done to push a car is the same whether the car is moved north to south or east to west, provided that the amount of force used and the distance moved are the same. Work does not convey directional information and, therefore, is a scalar quantity.

The equation above indicates that the unit of work is force times distance, or the newton·meter in SI units. One newton·meter is referred to as a *joule* (J) (pronounced "jewel"), in honor of James Joule (1818–1889) and his research into the nature of work, energy, and heat. Table 6.1 summarizes the units for work in several systems of measurement.

The definition of work as $W = Fs$ does have one surprising feature. If the distance s is zero, the work is zero, even if a force is applied. Pushing on an immovable object, such as a brick wall, may tire your muscles, but there is no work done of the type we are discussing. In physics, the idea of work is intimately tied up with the idea of motion. If there is no movement of the object, the work done by the force acting on the object is zero.

Table 6.1 Units of Measurement for Work

System	Force	×	Distance	=	Work
SI	newton (N)		meter (m)		joule (J)
CGS	dyne (dyn)		centimeter (cm)		erg
BE	pound (lb)		foot (ft)		foot·pound (ft·lb)

WORK DONE BY A CONSTANT FORCE THAT POINTS AT AN ANGLE RELATIVE TO THE DIRECTION OF THE MOTION

Often, the force and displacement do not point in the same direction. For instance, Figure 6.2a shows a luggage carrier being pulled to the right by a force that is applied along the handle. The force is directed at an angle θ relative to the displacement. In such a case, only the component of the force along the displacement is used in defining work. As part b of the drawing illustrates, this component is $F \cos \theta$, and it appears in the general definition of work given below.

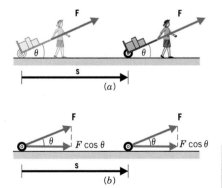

Figure 6.2 (a) Work can be done by a force **F** that points at an angle θ relative to the displacement **s**. (b) The force component that points along the displacement is $F \cos \theta$.

Definition of Work

The work done on an object by a constant force **F** is

$$W = (F \cos \theta)s \qquad (6.1)$$

where F is the magnitude of the force, s is the magnitude of the displacement, and θ is the angle between the force and the displacement.

SI Unit of Work: newton·meter = joule (J)

When the force points in the same direction as the displacement, then $\theta = 0°$, and Equation 6.1 reduces to $W = Fs$. The following example illustrates how Equation 6.1 is used to calculate work.

Example 1   Pulling a Luggage Carrier

Find the work done by a 45.0-N force in pulling the luggage carrier in Figure 6.2a at an angle $\theta = 50.0°$ for a distance $s = 75.0$ m.

REASONING AND SOLUTION   According to Equation 6.1, the work done on the luggage carrier by the 45.0-N force is

$$W = (F \cos \theta)s = [(45.0 \text{ N}) \cos 50.0°](75.0 \text{ m}) = \boxed{2170 \text{ J}}$$

The answer is expressed in newton·meters or joules (J).

The definition of work in Equation 6.1 takes into account only the component of the force in the direction of the displacement. The force component perpendicular to the displacement does no work. To do work, there must be a force _and_ a displacement, and since there is no displacement in the perpendicular direction, there is no work produced by the perpendicular component of the force.

POSITIVE AND NEGATIVE WORK

Work can be either positive or negative, depending on whether a component of the force points in the same direction as the displacement or in the opposite direction. Example 2 illustrates how positive and negative work arise.

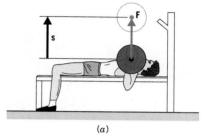

(a)

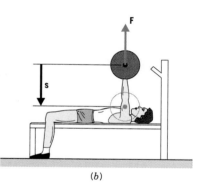

(b)

Figure 6.3   (a) During the lifting phase, the force **F** does positive work on the barbell. (b) During the lowering phase, the force does negative work.

Example 2   Bench-pressing

The weight lifter in Figure 6.3 is bench-pressing a barbell whose weight is 710 N. He raises the barbell a distance of 0.65 m above his chest and then lowers the barbell the same distance. The weight is raised and lowered at a constant velocity. Determine the work done on the barbell by the weight lifter during (a) the lifting phase and (b) the lowering phase.

REASONING   To calculate the work, it is necessary to know the force exerted by the weight lifter. The barbell is raised and lowered at a constant velocity and, therefore, is in equilibrium. Consequently, the force **F** exerted by the weight lifter must balance the weight of the barbell, so $F = 710$ N.

SOLUTION
(a) As part _a_ of the drawing shows, the force and the displacement are in the same direction. So the angle between them is $\theta = 0°$, and the work done by the force is

$$W = (F \cos \theta)s = [(710 \text{ N}) \cos 0°](0.65 \text{ m}) = \boxed{460 \text{ J}}$$

(b) When the barbell is lowered, the force and the displacement are in opposite directions, as part _b_ of the drawing indicates. The angle between the force and displacement is now $\theta = 180°$, and the work is

$$W = (F \cos \theta)s = [(710 \text{ N}) \cos 180°](0.65 \text{ m}) = \boxed{-460 \text{ J}}$$

THE PHYSICS OF . . .

positive and negative "reps" in weight lifting.

Bench-pressing; the work done on the barbell during the lifting phase is positive.

since cos 180° = −1. The work is negative, because the force is opposite to the displacement. Weight lifters call each complete up-and-down movement of the barbell a repetition, or "rep." The lifting of the weight is referred to as the positive part of the rep, and the lowering is known as the negative part of the rep.

As Example 2 illustrates, work can be positive or negative. If the force has a component in the *same* direction as the displacement of the object, the work done by the force is *positive*. On the other hand, if a force component points in the direction *opposite* to the displacement, the work is *negative*. If the force is perpendicular to the displacement, the force has no component in the direction of the displacement, and the work is zero ($\theta = 90°$ in Equation 6.1). The next example deals with the work done by a static frictional force when it accelerates an object.

Example 3   Accelerating a Crate

Figure 6.4 shows a 120-kg crate sitting on the flatbed of a truck that is moving with an acceleration of $a = +1.5$ m/s² along the positive x axis. The crate does not slip with respect to the truck. What is the work done on the crate when the truck moves a distance of $s = 65$ m?

REASONING   To determine the work done on the crate, it is first necessary to find the net force exerted on the crate in the direction of the displacement. Part b of the drawing shows the free-body diagram for the crate. The weight **W** of the crate and the normal force $\mathbf{F_N}$ are directed perpendicular to the displacement, so they do no work on the crate. Only the static frictional force $\mathbf{f_s}$ between the truck bed and the crate does work, since it is the only force acting on the crate in the x direction. We can determine the frictional force from the knowledge that the crate does not slip relative to the truck. The fact that no slippage occurs means that the crate has the same acceleration of $a = +1.5$ m/s² as does the truck. The force creating this acceleration is the static frictional force, which points in the same direction as the acceleration, the forward direction. After all, if there were no static friction between the crate and the truck, the truck could be driven out from under the crate. Knowing the mass of the crate and its acceleration, we can use Newton's second law to obtain the magnitude of the frictional force. Then, knowing the frictional force, we can determine the work done on the crate.

SOLUTION   From Newton's second law, we find that
$$f_s = ma = (120 \text{ kg})(1.5 \text{ m/s}^2) = 180 \text{ N}$$
The work done on the crate by the static frictional force is
$$W = (f_s \cos \theta)s = (180 \text{ N})(\cos 0°)(65 \text{ m}) = \boxed{1.2 \times 10^4 \text{ J}} \tag{6.1}$$
The work is positive, because the frictional force is in the same direction as the displacement ($\theta = 0°$).

6.2 THE WORK–ENERGY THEOREM AND KINETIC ENERGY

Most people expect that if you do work, you should get something as a result. In physics, when a net force performs work on an object, there is always a result

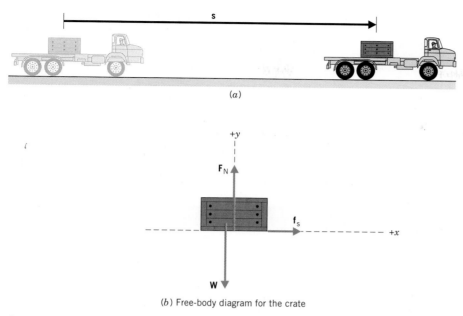

(a)

(b) Free-body diagram for the crate

Figure 6.4 (a) The truck and crate are accelerating to the right for a distance of $s = 65$ m. (b) The free-body diagram for the crate.

from the effort. The result is a change in the **kinetic energy** of the object. The important relationship that relates work to the change in kinetic energy is known as the **work–energy theorem.**

To gain some insight into the idea of kinetic energy and the work–energy theorem, look at Figure 6.5, where a constant net force **F** is acting on an airplane of mass m. The net force is the vector sum of all the forces acting on the plane, and for simplicity its direction is assumed to be parallel to the displacement **s**. According to Newton's second law, the net force produces an acceleration whose magnitude a is given by $a = F/m$. Consequently, the speed of the plane changes from an initial value of v_0 to a final value of v_f. Multiplying both sides of $F = ma$ by the distance s gives

$$Fs = mas$$

The term as on the right side can be related to v_0 and v_f by using Equation 2.9 ($v_f^2 = v_0^2 + 2as$), which is one of the equations of kinematics. Solving this equa-

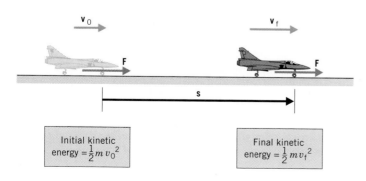

Initial kinetic
energy $= \frac{1}{2}m v_0^2$

Final kinetic
energy $= \frac{1}{2}m v_f^2$

Figure 6.5 When a constant net force **F** acts over a displacement **s**, the force does work on the plane. As a result of the work done, the kinetic energy of the plane increases.

tion to give $as = \frac{1}{2}(v_f^2 - v_0^2)$ and substituting this result into $Fs = mas$ shows that

$$\underbrace{Fs}_{\text{Work}} = \underbrace{\tfrac{1}{2}mv_f^2}_{\substack{\text{Final}\\\text{kinetic}\\\text{energy}}} - \underbrace{\tfrac{1}{2}mv_0^2}_{\substack{\text{Initial}\\\text{kinetic}\\\text{energy}}}$$

This expression is the work–energy theorem. Its left side is the work W done by the net force, while its right side involves the difference between two terms, each of which has the form $\frac{1}{2}(\text{mass})(\text{speed})^2$. The quantity $\frac{1}{2}(\text{mass})(\text{speed})^2$ is called kinetic energy and plays a significant role in physics, as we will soon see.

Definition of Kinetic Energy

The kinetic energy KE of an object with mass m and speed v is given by

$$KE = \tfrac{1}{2}mv^2 \tag{6.2}$$

SI Unit of Kinetic Energy: joule (J)

The SI unit of kinetic energy is the same as the unit for work, the joule. Kinetic energy, like work, is a scalar quantity. These observations are not surprising, for work and kinetic energy are closely related, as is clear from the following statement of the work–energy theorem.

The Work–Energy Theorem

When a net force does work W on an object, the kinetic energy of the object changes from its initial value of KE_0 to a final value of KE_f, the difference between the two values being equal to the work:

$$W = KE_f - KE_0 = \tfrac{1}{2}mv_f^2 - \tfrac{1}{2}mv_0^2 \tag{6.3}$$

The work–energy theorem may be derived for any direction of the force relative to the displacement, not just the situation in Figure 6.5. In fact, the force may even vary from point to point along a path that is curved rather than straight, and the theorem remains valid. According to the work–energy theorem, a moving object has kinetic energy, because work was done to accelerate the object from rest to a speed v. Conversely, an object with kinetic energy can perform work, if it is allowed to push or pull on another object.

Example 4 illustrates the work–energy theorem and considers a single force that does work to change the kinetic energy of a space probe.

Example 4   A Space Probe

A space probe of mass $m = 5.00 \times 10^4$ kg is traveling at a speed of $v_0 = 1.10 \times 10^4$ m/s through deep space. No forces act on the probe except that generated by its own engine. The engine exerts a constant force $\mathbf{F}$ of 4.00×10^5 N, directed parallel to the displacement (Figure 6.6). The engine fires continually while the probe moves in a straight line for a displacement $\mathbf{s}$ of 2.50×10^6 m. Determine the final speed of the probe.

REASONING AND SOLUTION   The final speed v_f of the probe can be obtained from the work–energy theorem, but first the work done on the probe by the engine must be calculated. The work is obtained from its definition (Equation 6.1):

$$W = (F \cos \theta)s$$

$$= [(4.00 \times 10^5 \text{ N}) \cos 0°](2.50 \times 10^6 \text{ m}) = 1.00 \times 10^{12} \text{ J}$$

The work is positive, because the force and displacement are in the same direction (see the drawing). Since $W = KE_f - KE_0$ according to the work–energy theorem, the final kinetic energy of the probe is

$$KE_f = W + KE_0$$

$$= (1.00 \times 10^{12} \text{ J}) + \tfrac{1}{2}(5.00 \times 10^4 \text{ kg})(1.10 \times 10^4 \text{ m/s})^2 = 4.03 \times 10^{12} \text{ J}$$

The final kinetic energy is $KE_f = \tfrac{1}{2}mv_f^2$, so the final speed is

$$v_f = \sqrt{\frac{2(KE_f)}{m}} = \sqrt{\frac{2(4.03 \times 10^{12} \text{ J})}{(5.00 \times 10^4 \text{ kg})}} = \boxed{1.27 \times 10^4 \text{ m/s}}$$

Since the force of the engine does positive work, the final speed of the probe is greater than its initial speed, in accord with the work–energy theorem.

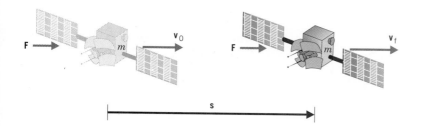

Figure 6.6   The engine of the space probe generates a force **F** that points in the same direction as the displacement **s**. The force performs positive work, causing the probe to gain kinetic energy.

In Example 4 only one force, that of the engine, does work on the space probe. If several forces act on an object, they must be added together vectorially to give the net force. The work done by the net force can then be related to the change in the object's kinetic energy by using the work–energy theorem, as in the next example.

Example 5   Downhill Skiing

A 58-kg skier is coasting down a 25° slope, as Figure 6.7a shows. A kinetic frictional force of magnitude $f_k = 70$ N opposes her motion. Near the top of the slope, the skier's speed is $v_0 = 3.6$ m/s. Ignoring air resistance, determine the speed v_f at a point that is displaced 57 m downhill.

REASONING   The final speed of the skier can be calculated from the work–energy theorem, provided the work done by the net force acting on the skier is known. To find the net force we use the free-body diagram of the skier.

SOLUTION   The free-body diagram in part b of the drawing shows all the forces acting on the skier. The normal force $\mathbf{F_N}$ and the component of the skier's weight perpendicular to the slope, $mg \cos 25°$, do no work, because they are perpendicular to the displacement **s**. Only force components along the displacement do work. The net

force along the displacement is

$$\mathbf{F} = mg \sin 25° - f_k = (58 \text{ kg})(9.80 \text{ m/s}^2) \sin 25° - 70 \text{ N} = +170 \text{ N}$$

The work done by the net force is

$$W = (F \cos \theta)s = [(170 \text{ N}) \cos 0°](57 \text{ m}) = 9700 \text{ J} \qquad (6.1)$$

The work is positive, because the net force and the displacement point in the same direction. From the work–energy theorem ($W = \text{KE}_f - \text{KE}_0$), it follows that the final kinetic energy of the skier is

$$\text{KE}_f = W + \text{KE}_0 = 9700 \text{ J} + \tfrac{1}{2}(58 \text{ kg})(3.6 \text{ m/s})^2 = 10\ 100 \text{ J}$$

Since the final kinetic energy is $\text{KE}_f = \tfrac{1}{2}mv_f^2$, the final speed of the skier is

$$v_f = \sqrt{\frac{2(\text{KE}_f)}{m}} = \sqrt{\frac{2(10\ 100 \text{ J})}{58 \text{ kg}}} = \boxed{19 \text{ m/s}}$$

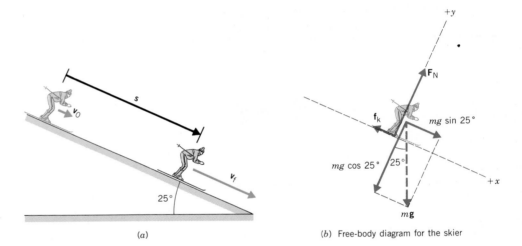

Figure 6.7 (*a*) A skier coasting downhill. (*b*) The free-body diagram for the skier.

(*a*)

(*b*) Free-body diagram for the skier

PROBLEM SOLVING INSIGHT

Example 5 emphasizes that *the work–energy theorem deals with the work done by the net force when a number of forces act on an object. The work–energy theorem does not apply to the work done by an individual force,* unless that force happens to be the only one present, in which case it is the net force. If the work done by the net force is *positive,* as in Example 5, the kinetic energy of the object *increases.* If the work done by the net force is *negative,* the kinetic energy *decreases.*

6.3 GRAVITATIONAL POTENTIAL ENERGY

WORK DONE BY THE FORCE OF GRAVITY

The gravitational force is a well-known force that can do positive or negative work, and Figure 6.8 helps to show how the work can be determined. This drawing depicts a basketball of mass m moving vertically downward, the force of gravity mg being the only force acting on the ball. The initial height of the ball is h_0, and the final height is h_f, both distances measured from the earth's surface. The displacement **s** is downward and has a magnitude of $s = h_0 - h_f$. To calculate the work W_{gravity} done on the ball by the force of gravity, we use $W = (F \cos \theta)s$ with $F = mg$ and $\theta = 0°$, since the force and displacement are in the same direction:

$$W_{\text{gravity}} = (mg \cos 0°)(h_0 - h_f) = mg(h_0 - h_f) \qquad (6.4)$$

Equation 6.4 is valid for *any path* taken between the initial and final heights, and not just for the straight-down path shown in Figure 6.8. For example, the same expression can be derived for the three paths shown in Figure 6.9. Thus, only the *change in vertical distance* $(h_0 - h_f)$ need be considered when calculating the work done by gravity. Since the change in the vertical distance is the same for each path in the drawing, the work done by gravity is the same in each case.

In Equation 6.4 only the difference between h_0 and h_f appears. Therefore, the vertical distances themselves need not be measured from the earth. For instance, they could be measured relative to a zero level that is one meter above the ground, and $h_0 - h_f$ would still have the same value. We assume here that the object remains close to the surface of the earth, so we can ignore the dependence of g on height and use the value of $g = 9.80 \text{ m/s}^2$. Example 6 illustrates how the work done by gravity is used in conjunction with the work–energy theorem.

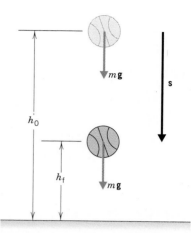

Figure 6.8   Gravity exerts a force $m\mathbf{g}$ on the basketball. Work is done by the gravitational force as the basketball falls from a height of h_0 to a height of h_f.

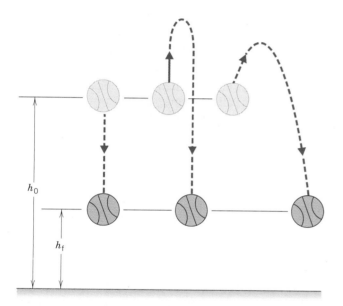

Figure 6.9   An object can be moved along any number of different paths in going from an initial height of h_0 to a final height of h_f. In each case, the work done by the gravitational force is the same $[W_{\text{gravity}} = mg(h_0 - h_f)]$, since the change in vertical distance $(h_0 - h_f)$ is the same.

Example 6   A Gymnast on a Trampoline

A gymnast ($m = 48.0$ kg) springs vertically upward from a trampoline. The gymnast leaves the trampoline at a height of 1.20 m and reaches a maximum height of 4.80 m before falling back down. All heights are measured with respect to the ground. Ignoring air resistance, determine (a) the initial speed v_0 with which the gymnast leaves the trampoline and (b) the speed of the gymnast after falling back to a height of 3.50 m.

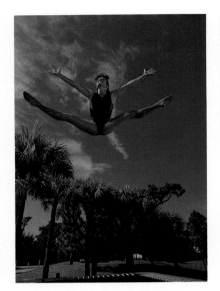

Bouncing off a trampoline; only the gravitational force acts on the gymnast in the air.

REASONING We can find the speed of the gymnast by using the work–energy theorem, provided the work done by the net force can be determined. Since only the gravitational force acts on the gymnast in the air, the gravitational force is the net force, and we can evaluate the work by using the relation $W_{gravity} = mg(h_0 - h_f)$.

SOLUTION

(a) Part _a_ of Figure 6.10 shows the gymnast moving upward. The initial and final heights are $h_0 = 1.20$ m and $h_f = 4.80$ m, respectively. The work done by the gravitational force is

$$W_{gravity} = mg(h_0 - h_f) \tag{6.4}$$
$$W_{gravity} = (48.0 \text{ kg})(9.80 \text{ m/s}^2)(1.20 \text{ m} - 4.80 \text{ m}) = -1690 \text{ J}$$

The work is negative, because the gravitational force (downward) is opposite to the displacement (upward) of the gymnast. The initial speed v_0 follows from the work–energy theorem, $W = KE_f - KE_0$. This theorem indicates that $W = -KE_0$, since the final speed and kinetic energy are zero at the highest point. Thus, $W = -\frac{1}{2}mv_0^2$ and

$$v_0 = \sqrt{\frac{-2W}{m}} = \sqrt{\frac{-2(-1690 \text{ J})}{48.0 \text{ kg}}} = \boxed{8.39 \text{ m/s}}$$

(b) Part _b_ of the drawing shows the gymnast on the way down. The calculation here proceeds in the same fashion as in part (a), except now the initial position is at the top of the flight path, so $h_0 = 4.80$ m and $h_f = 3.50$ m:

$$W_{gravity} = mg(h_0 - h_f) = (48.0 \text{ kg})(9.80 \text{ m/s}^2)(4.80 \text{ m} - 3.50 \text{ m}) = 612 \text{ J}$$

The work is now positive, because the force of gravity and the displacement are in the _same_ direction (downward) as the gymnast falls down. Since the initial speed and kinetic energy are zero at the top of the flight path, the work–energy theorem indicates that $W = KE_f - KE_0 = KE_f = \frac{1}{2}mv_f^2$. The final speed at a height of 3.50 m is

$$v_f = \sqrt{\frac{2W}{m}} = \sqrt{\frac{2(612 \text{ J})}{48.0 \text{ kg}}} = \boxed{5.05 \text{ m/s}}$$

Figure 6.10 (a) The gymnast bounces upward with an initial speed v_0 and reaches maximum height with a final speed of zero. (b) Starting at the top with an initial speed of zero, the gymnast falls back down. Notice in the two parts of the drawing that h_0 has different values, and so does h_f.

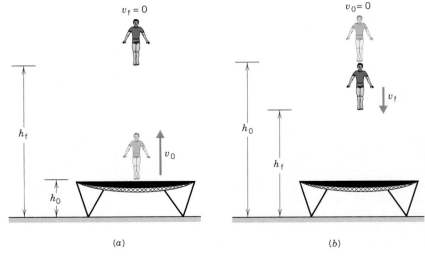

GRAVITATIONAL POTENTIAL ENERGY

We have seen that an object in motion has kinetic energy. Energy also occurs in other forms. For example, an object may possess energy by virtue of its position relative to the earth; such an object is said to have gravitational potential energy. A pile driver, for instance, is used by construction workers to pound "piles" or structural support beams into the ground. The pile driver contains a massive hammer that is raised to a height h above the ground and then dropped (see Figure 6.11). As a result, the hammer has the potential to do the work of driving the pile into the ground. The greater the height of the hammer, the greater is the potential for doing work, and the greater is the gravitational potential energy.

Now, let's obtain an expression for the gravitational potential energy of an object at a given height above the ground. Our starting point is Equation 6.4 for the work done by the gravitational force as an object moves from an initial height h_0 to a final height h_f: $W_{\text{gravity}} = mgh_0 - mgh_f$. This equation indicates that the work done by the gravitational force is equal to the amount by which the quantity mgh changes as the object falls. The value of mgh is larger when the height is larger and smaller when the height is smaller. We are led, then, to identify the quantity mgh as the *gravitational potential energy*.

Definition of Gravitational Potential Energy

The gravitational potential energy PE is the energy that an object of mass m has by virtue of position above the surface of the earth, that position being measured by the height h of the object relative to an arbitrary zero level:

$$PE = mgh \tag{6.5}$$

SI Unit of Gravitational Potential Energy: joule (J)

Gravitational potential energy, like work and kinetic energy, is a scalar quantity, and has the same SI unit as they do, the joule. The work done by the force of

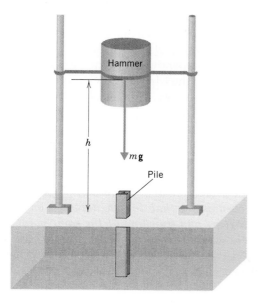

Figure 6.11 The gravitational potential energy of the hammer relative to the ground is PE $= mgh$.

gravity, as expressed in Equation 6.4, shows that only the *difference* between two potential energies is significant. Therefore, the zero level for the heights can be taken anywhere, as long as both h_0 and h_f are measured relative to the same zero level. The gravitational potential energy depends on both the object and the earth (m and g, respectively), as well as the height h. Therefore, the gravitational potential energy belongs to the object and the earth as a system, although one often speaks of the object alone as possessing the gravitational PE potential energy.

6.4 CONSERVATIVE FORCES AND NONCONSERVATIVE FORCES

CONSERVATIVE FORCES

The gravitational force has an interesting property that when an object is moved from one place to another, the work done by the gravitational force does not depend on the choice of path. In Figure 6.9, for instance, an object moves from an initial height h_0 to a final height h_f along several different paths. As Section 6.3 discusses, the work done by gravity depends only on the initial and final heights, and not on the path between these heights. With each height a gravitational potential energy can be associated, PE = mgh, and the work done by gravity is the difference between the initial and final potential energies:

$$W_{\text{gravity}} = mgh_0 - mgh_f = \text{PE}_0 - \text{PE}_f$$

If the work done by a force in moving an object between two positions is independent of the path of the motion, the force is called a "conservative force." The gravitational force is our first example of a conservative force. Later we will encounter other examples, such as the elastic force of a spring and the electrical force between electrically charged particles. As with the gravitational force, a potential energy can be associated with any of these conservative forces.

Figure 6.12 illustrates another way to describe a conservative force. The picture shows a roller coaster car racing through dips and double dips, ultimately returning to its starting point. This kind of path, which begins and ends at the same place, is called a *closed* path. Gravity provides the only force that does work on the car, assuming that there is no friction or air resistance. Of course, the track exerts a normal force, but this force is always directed perpendicular to the motion and, hence, does no work. On the downward parts of the trip, the gravitational force does positive work, increasing the car's kinetic energy. Conversely, on the upward parts of the motion, the gravitational force does negative work, decreasing the car's kinetic energy. Over the entire trip, the gravitational force does as much positive work as negative work, so the net work is zero, and the car returns to its starting point with the same kinetic energy it had at the start. In terms of the gravitational potential energy, $W_{\text{gravity}} = \text{PE}_0 - \text{PE}_f$. But PE_0 equals PE_f, since the initial and final positions of a closed path are the same. Therefore, $W_{\text{gravity}} = 0$ for a closed path. *A conservative force, then, is one that does no net work on an object moving around a closed path, starting and finishing at the same point.* We see, then, that there are two equivalent ways to describe a conservative force. One way is in terms of an object moving between two points. The other way is in terms of an object moving around a closed path.

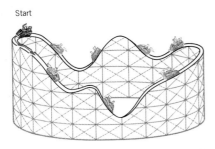

Start

Figure 6.12 A roller coaster track is an example of a closed path.

NONCONSERVATIVE FORCES

Not all forces are conservative forces. A force is nonconservative if the work it does on a moving object depends on the path of the motion. The kinetic frictional force is one example of a nonconservative force. When an object slides over a surface, the kinetic frictional force points opposite to the sliding motion and does negative work equal in magnitude to the kinetic frictional force multiplied by the length of the path. Between any two points, greater amounts of work are done over longer paths between the points. The work, thus, depends on the choice of path, and so the kinetic frictional force is nonconservative. Air resistance is another nonconservative force. The concept of potential energy is not defined for a nonconservative force.

For a closed path, the total work done by a nonconservative force is not zero as it is for a conservative force. In Figure 6.12, for instance, a frictional force would oppose the motion and slow down the car. Unlike gravity, friction would do negative work on the car throughout the entire trip, on *both* the up and down parts of the motion. Assuming that the car makes it back to the starting point, the car will have *less* kinetic energy than it had originally. Table 6.2 gives some examples of conservative and nonconservative forces.

THE WORK–ENERGY THEOREM

In most everyday situations, both conservative forces (such as gravity) and nonconservative forces (such as friction and air resistance) contribute to the net force acting on an object. Therefore, we may write the work W done by the net force as $W = W_c + W_{nc}$, where W_c is the work done by the conservative forces and W_{nc} is the work done by the nonconservative forces. According to the work–energy theorem, the work done by the net force is equal to the change in the object's kinetic energy, or $W_c + W_{nc} = \frac{1}{2}mv_f^2 - \frac{1}{2}mv_0^2$. If the only conservative force acting on the object is the gravitational force, then $W_c = W_{gravity} = mg(h_0 - h_f)$, and the work–energy theorem becomes

$$mg(h_0 - h_f) + W_{nc} = \tfrac{1}{2}mv_f^2 - \tfrac{1}{2}mv_0^2$$

The work done by the gravitational force can be moved to the right side of this equation, with the result that

$$W_{nc} = (\tfrac{1}{2}mv_f^2 - \tfrac{1}{2}mv_0^2) + (mgh_f - mgh_0) \qquad (6.6)$$

Table 6.2   Examples of Conservative and Nonconservative Forces

Conservative Forces	Nonconservative Forces
Gravitational force (Ch. 4)	Static and kinetic frictional forces
Elastic spring force (Ch. 10)	Air resistance
Electric force (Ch. 18, 19)	Tension
	Normal force
	Propulsion force of a rocket or a motor

In terms of kinetic and potential energies, we find that

$$W_{nc} = \underbrace{(KE_f - KE_0)}_{\substack{\text{Change in} \\ \text{kinetic energy}}} + \underbrace{(PE_f - PE_0)}_{\substack{\text{Change in} \\ \text{gravitational} \\ \text{potential energy}}} \tag{6.7a}$$

Equation 6.7a states that the work done by all the nonconservative forces equals the change in the object's kinetic energy plus the change in its gravitational potential energy. It is customary to use the delta symbol (Δ) to denote such changes; thus, $\Delta KE = (KE_f - KE_0)$ and $\Delta PE = (PE_f - PE_0)$. With the delta notation, the work–energy theorem takes the form

$$W_{nc} = \Delta KE + \Delta PE \tag{6.7b}$$

In the next two sections, we will show why the form of the work–energy theorem expressed by Equations 6.7a and 6.7b is useful.

6.5 *THE CONSERVATION OF MECHANICAL ENERGY*

The work–energy theorem has led us to consider kinetic and potential energy. The sum of these two kinds of energy is called the *total mechanical energy* E, so that $E = KE + PE$. The work–energy theorem of Equation 6.7a can be expressed in terms of the total mechanical energy as

$$W_{nc} = (KE_f - KE_0) + (PE_f - PE_0)$$
$$= \underbrace{(KE_f + PE_f)}_{E_f} - \underbrace{(KE_0 + PE_0)}_{E_0}$$

or

$$W_{nc} = E_f - E_0 \tag{6.8}$$

Equation 6.8 states that W_{nc}, the work done by nonconservative forces, changes the total mechanical energy from an initial value of E_0 to a final value of E_f.

The conciseness of the work–energy theorem in the form $W_{nc} = E_f - E_0$ allows an important basic principle of physics to stand out. To see how this principle arises, suppose that only the gravitational force does work on the object. Then, the net work done by the nonconservative forces is zero ($W_{nc} = 0$), and Equation 6.8 reduces to

$$E_f = E_0 \tag{6.9}$$

This result indicates that the final mechanical energy is equal to the initial mechanical energy. Consequently, the total mechanical energy *remains constant all along the path between the initial and final points*, never varying from the initial value of E_0. A quantity that stays constant throughout the motion is said to be "conserved." The fact that the total mechanical energy is conserved when $W_{nc} = 0$ is called the *principle of conservation of mechanical energy*.

Launching a four-man bobsled.

The Principle of Conservation of Mechanical Energy

The total mechanical energy ($E = $ KE $+$ PE) of an object remains constant as the object moves, provided that the net work done by nonconservative forces is zero.

The principle of conservation of mechanical energy offers keen insight into the way in which the physical universe operates. While the sum of the kinetic and potential energies at any point is conserved, the two forms may be interconverted or transformed into one another. Kinetic energy of motion is converted into potential energy of position, for instance, when a moving object coasts up a hill. Conversely, potential energy is converted into kinetic energy when an object above the earth's surface is allowed to fall. Figure 6.13 illustrates such transformations of energy for a bobsled run, assuming that nonconservative forces, such as friction and wind resistance, can be ignored. The normal force, being directed

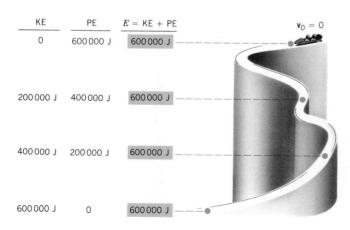

KE	PE	$E = $ KE $+$ PE
0	600 000 J	600 000 J
200 000 J	400 000 J	600 000 J
400 000 J	200 000 J	600 000 J
600 000 J	0	600 000 J

$\mathbf{v}_0 = 0$

Figure 6.13 If friction and wind resistance are ignored, a bobsled run illustrates how kinetic and potential energy can be interconverted, while the total mechanical energy remains constant at each point along the run. The total mechanical energy is 600 000 J, being all potential energy at the top and all kinetic energy at the bottom.

perpendicular to the path, does no work. Only the force of gravity does work, so the total mechanical energy E remains constant at all points along the run. The conservation principle is well known for the ease with which it can be applied, as in the next three examples.

Example 7   A Daredevil Motorcycle Rider

A motorcycle rider is trying to leap across the canyon shown in Figure 6.14 by driving horizontally off the cliff. When it leaves the cliff, the cycle has a speed of 38.0 m/s. Ignoring air resistance, find the speed with which the cycle strikes the ground on the other side.

REASONING   Once the cycle leaves the cliff, no forces other than gravity act on the cycle, since air resistance is being ignored. Thus, $W_{nc} = 0$, and the principle of conservation of mechanical energy can be used to find the final speed of the motorcyclist.

SOLUTION   The principle of conservation of mechanical energy is written as

$$\underbrace{\tfrac{1}{2}mv_f^2 + mgh_f}_{E_f} = \underbrace{\tfrac{1}{2}mv_0^2 + mgh_0}_{E_0} \qquad (6.9)$$

The mass m of the rider and cycle can be eliminated algebraically from this equation, since m appears as a factor in every term. Solving for v_f gives

$$v_f = \sqrt{v_0^2 + 2g(h_0 - h_f)}$$

$$v_f = \sqrt{(38.0 \text{ m/s})^2 + 2(9.80 \text{ m/s}^2)(70.0 \text{ m} - 35.0 \text{ m})} = \boxed{46.2 \text{ m/s}}$$

Examples 8 and 9 emphasize that the principle of conservation of mechanical energy can be applied even when forces act perpendicular to the path of a moving object.

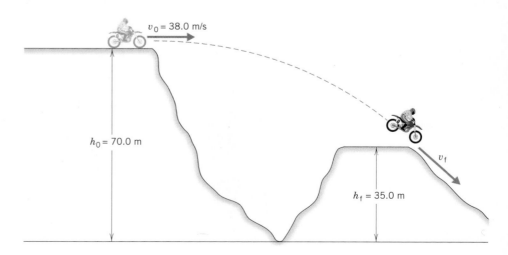

Figure 6.14   A daredevil jumping a canyon.

Example 8   *The Favorite Swimming Hole*

A 6.00-m rope is tied to a tree limb and used as a swing. A person starts from rest with the rope held in a horizontal orientation, as in Figure 6.15. Ignoring friction and air resistance, determine how fast the person is moving at the lowest point on the circular arc of the swing.

REASONING   First, we need to decide if the conservation of mechanical energy applies to this problem. After all, a nonconservative force does act on the person, namely, the tension **T** in the rope. The tension, however, does no work, because it points perpendicular to the circular path of the motion. Thus, $W_{nc} = 0$, and the conservation principle is applicable.

SOLUTION   From Equation 6.9 we have

$$\underbrace{\tfrac{1}{2}mv_f^2 + mgh_f}_{E_f} = \underbrace{\tfrac{1}{2}mv_0^2 + mgh_0}_{E_0} \qquad (6.9)$$

The mass m can be eliminated algebraically, and since $v_0 = 0$, it follows that $\tfrac{1}{2}v_f^2 + gh_f = gh_0$. Therefore,

$$v_f = \sqrt{2g(h_0 - h_f)} = \sqrt{2(9.80 \text{ m/s}^2)(6.00 \text{ m})} = \boxed{10.8 \text{ m/s}}$$

where we have used the fact that $h_0 - h_f = 6.00$ m (see the drawing).

PROBLEM SOLVING INSIGHT

When nonconservative forces are perpendicular to the motion, we can still use the principle of conservation of mechanical energy, because such "perpendicular" forces do no work.

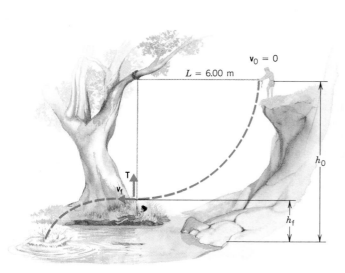

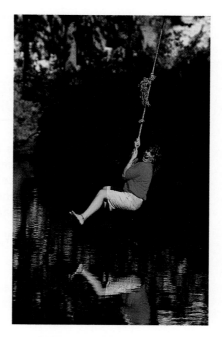

Figure 6.15   A fun way to get into the water. The tension **T** in the rope acts perpendicular to the circular arc and, hence, does no work on the person. Therefore, the principle of conservation of mechanical energy applies, and the speed v_f can be determined.

One way to cool off.

Figure 6.16 The Magnum XL-200 roller coaster; one of the fastest roller coasters in the world, includes a vertical drop of 59.3 m.

Example 9   *The Magnum XL-200*

One of the fastest roller coasters in the world is the Magnum XL-200 at Cedar Point Park in Sandusky, Ohio (Figure 6.16). This ride includes a vertical drop of 59.3 m. Assume that the roller coaster has a speed of nearly zero as it crests the top of the hill. Neglect friction and find the speed of the riders at the bottom of the hill.

REASONING   Since we are neglecting friction, we may set the work done by the frictional force equal to zero. A normal force acts on each rider, but this force is perpendicular to the motion, so it does not do any work. Thus, the work done by nonconservative forces is zero, and we may use the principle of conservation of mechanical energy to find the speed of the riders at the bottom of the hill.

SOLUTION   The principle of conservation of mechanical energy states that

$$\underbrace{\tfrac{1}{2}mv_f^2 + mgh_f}_{E_f} = \underbrace{\tfrac{1}{2}mv_0^2 + mgh_0}_{E_0} \qquad (6.9)$$

The mass m of the rider appears as a factor in every term in this equation and can be eliminated algebraically. Solving for the final speed gives

$$v_f = \sqrt{v_0^2 + 2g(h_0 - h_f)}$$

The initial speed of the roller coaster is assumed to be zero, $v_0 = 0$, and the vertical height of the hill is $h_0 - h_f = 59.3$ m:

$$v_f = \sqrt{2(9.80 \text{ m/s}^2)(59.3 \text{ m})} = \boxed{34.1 \text{ m/s (about 76 mph)}}$$

6.6 NONCONSERVATIVE FORCES AND THE WORK–ENERGY THEOREM

Most moving objects experience one or more nonconservative forces, such as friction, air resistance, and propulsive forces. The work W_{nc} done by the net nonconservative force is not zero, and, consequently, the total mechanical energy of the object is not conserved. In these situations, the difference between the final and initial total mechanical energies is equal to W_{nc}, as expressed by Equation 6.8:

$$W_{nc} = E_f - E_0$$
$$= (\tfrac{1}{2}mv_f^2 + mgh_f) - (\tfrac{1}{2}mv_0^2 + mgh_0)$$

The next two examples illustrate how Equation 6.8 is used when nonconservative forces are present.

Example 10   *The Magnum XL-200, Revisited*

In Example 9, we ignored friction. In reality, however, friction is present when the Magnum XL-200 roller coaster descends the hill. The actual speed of the riders at the bottom is 32.2 m/s, which is less than that determined in Example 9. How much work is done by the nonconservative frictional force on a 55.0-kg rider during the descent from a height h_0 to a height h_f, where $h_0 - h_f = 59.3$ m?

REASONING AND SOLUTION   The work–energy theorem, expressed by Equation 6.8, can be used to determine the work W_{nc} done by the nonconservative frictional force.

$$W_{nc} = (\tfrac{1}{2}mv_f^2 + mgh_f) - (\tfrac{1}{2}mv_0^2 + mgh_0) \qquad (6.8)$$

$$\underbrace{\phantom{(\tfrac{1}{2}mv_f^2 + mgh_f)}}_{E_f} \quad \underbrace{\phantom{(\tfrac{1}{2}mv_0^2 + mgh_0)}}_{E_0}$$

Since $v_0 = 0$ at the top of the hill, this equation can be written as

$$W_{nc} = \tfrac{1}{2}mv_f^2 - mg(h_0 - h_f)$$

$$W_{nc} = \tfrac{1}{2}(55.0\text{ kg})(32.2\text{ m/s})^2 - (55.0\text{ kg})(9.80\text{ m/s}^2)(59.3\text{ m}) = \boxed{-3450\text{ J}}$$

The work done by friction is negative, because the frictional force acts opposite to the motion of the riders.

Example 11   Fireworks

A Fourth-of-July rocket (0.20 kg) is launched from rest and follows an erratic flight path to reach the point P, as Figure 6.17 shows. Point P is 29 m above the starting point. In the process, 425 J of work is done on the rocket by the nonconservative force generated by the burning propellant. Ignoring air resistance and the mass lost due to the burning propellant, find the speed v_f of the rocket at the point P.

REASONING AND SOLUTION   The only nonconservative force acting on the rocket is the force generated by the burning propellant, and the work done by this force is $W_{nc} = 425$ J. We may use Equation 6.8 to find the final speed of the rocket.

$$W_{nc} = (\tfrac{1}{2}mv_f^2 + mgh_f) - (\tfrac{1}{2}mv_0^2 + mgh_0) \qquad (6.8)$$

Setting $v_0 = 0$ and solving for the final speed of the rocket, we get

$$v_f = \sqrt{\frac{2[W_{nc} - mg(h_f - h_0)]}{m}}$$

$$v_f = \sqrt{\frac{2[425\text{ J} - (0.20\text{ kg})(9.80\text{ m/s}^2)(29\text{ m})]}{0.20\text{ kg}}} = \boxed{61\text{ m/s}}$$

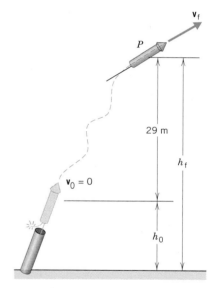

Figure 6.17   A Fourth-of-July rocket, moving along an erratic flight path, reaches a point P that is 29 m above the launch point.

6.7 POWER

In many situations, the time it takes to do work is just as important as the amount of work that is done. Consider two automobiles that are identical in all respects (e.g., same mass), except that one has a "souped-up" engine. The car with the "souped-up" engine can go from zero to sixty miles per hour in 4 seconds, while the other car requires 8 seconds to do the same thing. Each engine does work in accelerating its car, but one does it more quickly. Where cars are concerned, we have come to associate the quicker performance with an engine that has a larger horsepower rating. A large horsepower rating means that the engine can do a large amount of work in a short time. In physics, the horsepower rating is just one way to measure an engine's ability to generate power. The idea of *power* incorporates both the concepts of work and time, for power is work done per unit time.

Definition of Average Power

Average power $\overline{P}$ is the average rate at which work W is done, and it is obtained by dividing W by the time t required to perform the work:

$$\overline{P} = \frac{\text{Work}}{\text{Time}} = \frac{W}{t} \qquad (6.10)$$

SI Unit of Power: joule/s = watt (W)

Since both work and time are scalar quantities, power is also a scalar quantity. The unit in which power is expressed is that of work divided by time, or a joule per second in SI units. One joule per second is called a watt (W), in honor of James Watt (1736–1819), the developer of the steam engine. The unit of power in the BE system is the foot-pound per second (ft·lb/s), although the familiar horsepower (hp) unit is frequently used for specifying the power generated by electric motors and internal combustion engines:

1 horsepower = 550 foot·pounds/second = 746 watts

Table 6.3 summarizes the units for power in the various systems of measurement.

An alternative expression for power can be obtained from Equation 6.1. The work W done when a constant net force of magnitude F points in the same direction as the displacement, is $W = (F \cos 0°)s = Fs$. Dividing both sides of this equation by t, the time it takes the force to move the object through the distance s, gives

$$\frac{W}{t} = \frac{Fs}{t}$$

Recognizing W/t as the average power $\overline{P}$, and s/t as the average speed $\overline{v}$, we have

$$\overline{P} = F\overline{v} \qquad (6.11)$$

The next example illustrates the use of Equation 6.11.

Table 6.3 Units of Measurement for Power

System	Work	÷ Time	= Power
SI	joule (J)	second (s)	watt (W)
CGS	erg	second (s)	erg per second (erg/s)
BE	foot·pound (ft·lb)	second (s)	foot·pound per second (ft·lb/s)

Example 12   The Power to Accelerate a Car

A 1.10×10^3-kg car, starting from rest, accelerates for 5.00 s. The magnitude of the acceleration is $a = 4.60$ m/s^2. Determine the average power generated by the net force that accelerates the vehicle.

REASONING   We can find the average power by using the relation $\bar{P} = F\bar{v}$, provided the magnitude F of the net force and the average speed $\bar{v}$ of the car can be determined. The net force can be obtained from Newton's second law, and the average speed can be calculated from the equations of kinematics.

SOLUTION   According to Newton's second law, the magnitude of the net force that acts on the car is

$$F = ma = (1.10 \times 10^3 \text{ kg})(4.60 \text{ m/s}^2) = 5060 \text{ N}$$

Since the car starts from rest ($v_0 = 0$) and has a constant acceleration, the average speed $\bar{v}$ of the car is one-half of its final speed v:

$$\bar{v} = \tfrac{1}{2}(v_0 + v) = \tfrac{1}{2}v \qquad (2.6)$$

Because the initial speed of the car is zero, the final speed of the car after 5.00 s is the product of its acceleration and time:

$$v = v_0 + at = (4.60 \text{ m/s}^2)(5.00 \text{ s}) = 23.0 \text{ m/s} \qquad (2.4)$$

Thus, the average speed is $\bar{v} = 11.5$ m/s, and the average power is

$$\bar{P} = F\bar{v} = (5060 \text{ N})(11.5 \text{ m/s}) = \boxed{5.82 \times 10^4 \text{ W } (78.0 \text{ hp})}$$

You can verify that the average power can also be obtained from Equation 6.10, $\bar{P} = W/t$. In this case, the work W done by the net force must be determined and then divided by the time of 5.00 s.

6.8 OTHER FORMS OF ENERGY AND THE CONSERVATION OF ENERGY

Up to now, we have considered only two kinds of energy, kinetic energy and gravitational potential energy. There are many other types of energy, however. Electrical energy is used to run electrical appliances. Thermal energy (heat) is utilized in cooking food. Moreover, the work done by the kinetic frictional force often appears as thermal energy, as you can experience by rubbing your hands back and forth. Chemical energy is the energy stored in the molecules of fuels and food. When gasoline is burned, some of the stored chemical energy is released and does the work of moving cars, airplanes, and boats. The chemical energy stored in food provides the energy needed for metabolic processes.

One of the most controversial forms of energy is nuclear energy. The research of many scientists, most notably Albert Einstein, led to the discovery that mass itself is one manifestation of energy. Einstein's famous equation, $E_0 = mc^2$, describes how mass m and energy E_0 are related, where c is the speed of light and has a value of 3.00×10^8 m/s. Because the speed of light is so large, this equation implies that very small masses are equivalent to large amounts of energy. The relationship between mass and energy will be discussed further in Chapter 28.

We have seen that kinetic energy can be converted into gravitational potential energy and vice versa. In general, energy of all types can be converted from one form to another. Part of the chemical energy stored in food is transformed into the kinetic energy of walking and into the thermal energy needed to keep our bodies at a temperature near 98.6 °F. Similarly, in a moving car the chemical energy of gasoline is converted into kinetic energy, as well as electrical energy (to power the radio, headlights, and air conditioner), and thermal energy (to heat the car during the winter). Whenever energy is transformed from one form to another, it is found that no energy is gained or lost in the process; the sum total of all the energies before the process is equal to the sum total of the energies after the process. This observation leads to the following important principle:

The Principle of Conservation of Energy

Energy can neither be created nor destroyed, but can only be converted from one form to another.

Learning how to convert energy from one form to another more efficiently is one of the main goals of modern science and technology.

INTEGRATION OF CONCEPTS

ENERGY AND FORCE

Energy and force are two of the most fundamental concepts in science. We have studied kinetic and potential energy in the present chapter and have encountered a number of types of forces in Chapter 4. Both concepts enable us to obtain insights into a wide variety of natural phenomena. In particular, the insights obtained with the principle of conservation of energy are often difficult to obtain in any other way. And the concept of force, as described in Newton's three laws of motion, allows us to understand why the motion of an object remains the same or changes from moment to moment. It is important to remember that the concepts of energy and force are not separate and unrelated parts of physics. As you use them, keep in mind that they are united by the idea of work and the work–energy theorem. This theorem shows that the work done on an object by a net force causes the kinetic energy of the object to change. Often, we classify forces into conservative and nonconservative types. If only conservative forces do work on an object, the work–energy theorem reveals that the total mechanical energy of the object remains constant at all times. This means that the kinetic energy can be transformed into potential energy and vice versa, but the sum of the two energies always remains the same.

THE WORK–ENERGY THEOREM AND KINETIC FRICTION

As we have seen, the work–energy theorem takes into account the work done by both conservative and nonconservative forces. Kinetic friction is a nonconservative force that arises whenever two surfaces slide against one another. The kinetic frictional force opposes the relative sliding motion. In opposing the motion of a baseball player sliding into home plate, for example, the kinetic frictional force does negative work on the player. According to the work–energy theorem, this negative work acts to reduce the kinetic energy and, hence, the speed of the player. As Section 4.9 discusses, the magnitude of the kinetic frictional force is the product of the coefficient of kinetic friction and the magnitude of the normal force pressing the surfaces together. In a number of homework problems at the end of the chapter (e.g., problems 18 and 19), you will need to use this expression for the kinetic frictional force. You will use it in determining the work done by the frictional force, in order to account for this work in the work–energy theorem.

SUMMARY

The **work** W done by a constant force acting on an object is $W = (F \cos\theta)s$, where F is the magnitude of the force, s is the magnitude of the displacement, and θ is the angle between the force and the displacement. Work can be positive or negative, depending on whether the force component along the displacement points in the same direction as the displacement or in the opposite direction.

The **kinetic energy** KE of an object of mass m and speed v is $\text{KE} = \frac{1}{2}mv^2$. The **work–energy theorem** states that the work W done by the net force acting on an object equals the difference between the final kinetic energy KE_f and the initial kinetic energy KE_0 of the object: $W = \text{KE}_f - \text{KE}_0$. If the net force does positive work, the kinetic energy increases; if the net force does negative work, the kinetic energy decreases.

The **work done by the force of gravity** on an object of mass m is $W_{\text{gravity}} = mg(h_0 - h_f)$, where h_0 and h_f are the initial and final heights of the object, respectively.

Gravitational potential energy PE is the energy that an object has by virtue of its position. For an object near the surface of the earth, the gravitational potential energy is given by $\text{PE} = mgh$, where h is the height of the object relative to an arbitrary zero level.

A **conservative force** is one that, in moving an object between two points, does work that is independent of the path taken between the points. Alternatively, a force is conservative if the work it does in moving an object around any closed path is zero. A force is a **nonconservative force** if the work it does on a moving object depends on the path of the motion.

The **total mechanical energy** E is the sum of the kinetic energy and the gravitational potential energy: $E = \text{KE} + \text{PE}$. The work–energy theorem can be expressed in an alternative form as $W_{\text{nc}} = E_f - E_0$, where W_{nc} is the net work done by the nonconservative forces, and E_f and E_0 are the final and initial total mechanical energies, respectively.

The **principle of conservation of mechanical energy** states that the total mechanical energy remains constant along the path of an object, provided that the net work done by nonconservative forces is zero. While E is constant, however, KE and PE may be transformed into one another.

Average power $\bar{P}$ is the work done per unit time, $\bar{P} = \text{Work}/\text{Time}$, or the rate at which work is done.

The **principle of conservation of energy** states that energy can neither be created nor destroyed, but can only be transformed from one form to another.

SOLVED PROBLEMS

Solved Problem 1   A Diving Plane
Related Problems: **10   *60   *62

A 6.00×10^3-kg plane is diving at an angle of $10.0°$ for a distance of 1.70×10^3 m, as in the drawing. Four forces act on the plane: its weight $\mathbf{W} = m\mathbf{g}$, the lift force $\mathbf{L}$ acting perpendicular to the surfaces of the wings, the forward thrust $\mathbf{T}$ (magnitude $= 1.80 \times 10^4$ N) generated by the plane's engine, and the force of air resistance $\mathbf{R}$ opposing the plane's motion. The net work done by these four forces is $+2.90 \times 10^7$ J. (a) Find the work done by $\mathbf{R}$ alone and (b) the magnitude of $\mathbf{R}$.

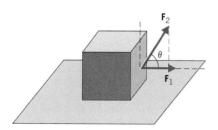

REASONING The net work of 2.90×10^7 J is the work done by the four forces that act on the plane. There are sufficient data to evaluate the work done by the three forces **W**, **L**, and **T**. By subtracting this work from the net work, we can determine the work W_R done by the resistive force **R**. Once W_R is found, it becomes possible to find the magnitude of **R** from the definition of work, since the displacement of the plane is known.

SOLUTION

(a) The work that each force contributes to the total is listed in the table below:

Force	Work $= (F \cos \theta)s$
W	$[(6.00 \times 10^3 \text{ kg})(9.80 \text{ m/s}^2) \cos 80.0°] \times (1.70 \times 10^3 \text{ m}) = +1.74 \times 10^7$ J
L	$(L \cos 90.0°)(1.70 \times 10^3 \text{ m}) = 0$
T	$[(1.80 \times 10^4 \text{ N}) \cos 0°](1.70 \times 10^3 \text{ m}) = +3.06 \times 10^7$ J
R	Work $= W_R$

The work done by gravity is positive, reflecting the fact that a component of the plane's weight ($mg \cos 80.0°$) is in the same direction as the displacement. The lift force **L** does no work, since it is perpendicular to the displacement. The individual contributions add to equal the net work of $+2.90 \times 10^7$ J:

$$1.74 \times 10^7 \text{ J} + 3.06 \times 10^7 \text{ J} + W_R = 2.90 \times 10^7 \text{ J}$$

Solving for W_R gives $\boxed{W_R = -1.90 \times 10^7 \text{ J}}$. The work done by the resistive force **R** is negative, since **R** points opposite to the displacement.

(b) Now that W_R is known, the magnitude of **R** can be found from Equation 6.1 [$W_R = (R \cos \theta)s$]:

$$R = \frac{W_R}{s \cos \theta} = \frac{-1.90 \times 10^7 \text{ J}}{(1.70 \times 10^3 \text{ m}) \cos 180°} = \boxed{1.12 \times 10^4 \text{ N}}$$

SUMMARY OF IMPORTANT POINTS Part (a) emphasizes that the net work W is the sum of the individual work contributions, one for each force that acts on the object. The work done by an individual force may be positive, negative, or zero. When all but one of these contributions are known, the unknown contribution can be determined. Part (b) shows that knowing the work done by a single force can lead to a value for the magnitude of the force, if the distance and the angle θ are known.

QUESTIONS

1. Can work be done on an object that remains at rest? Explain.

2. Two forces F_1 and F_2 are acting on the box shown in the drawing, causing the box to move across the floor. The two force vectors are drawn to scale. Which force does more work? Justify your answer.

3. A train, traveling at a constant speed, makes a 180° turn on a semicircular section of track and heads in a direction that is opposite to its initial direction. Even though a centripetal force acts on the train during the turn, this force does no work on the train. Why?

4. A box is being moved with a velocity **v** by a force **P** (parallel to **v**) along a level horizontal floor. The normal force is F_N, the kinetic frictional force is f_k, and the weight of the box is $m\mathbf{g}$. Decide which forces do positive, zero, or negative work. Provide a reason for each of your answers.

5. A sailboat is moving at a constant velocity. (a) Is work being done by a net force acting on the boat? Explain. (b) Recognizing that the wind propels the boat forward and the water resists the boat's motion, what does your answer in part (a) imply about the work done by the wind's force compared to the work done by the water's resistive force?

6. A ball has a speed of 15 m/s. Only one force acts on the ball. After this force acts, the speed of the ball is 7 m/s. Has the force done positive or negative work? Explain.

7. Three forces, F_A, F_B, and F_C, act on a bicycle and do positive, zero, and negative work, respectively. The bicycle moves on a horizontal surface. (a) Describe what each force would do to the kinetic energy of the bicycle, if that force alone acted. (b) With respect to the displacement of the bicycle, what can be said about the directions of these forces?

8. By measuring the speed of a boat at one moment and comparing it to the speed at another moment, one can tell whether a net force has done work on the boat. This statement is possible because of the work–energy theorem. Suppose the two speeds are measured to be exactly the same, and the boat moves horizontally. Is it safe to conclude that no force has acted on the boat? Explain, making sure to distinguish between the phrases "no force" and "no net force."

9. A motorcycle is being driven at a steady speed up a hill. (a) Is the term W_{nc} in Equation 6.7a positive, negative, or zero? Why? (b) The motorcycle is acted on by two nonconservative forces, a drive force that moves it forward and a frictional force that retards its motion. What does your answer to part (a) imply about the relative amounts of work done by each of these forces?

10. Suppose the total mechanical energy of an object is conserved. (a) If the kinetic energy decreases, what must be true about the gravitational potential energy? (b) If the potential energy decreases, what must be true about the kinetic energy? (c) If the kinetic energy does not change, what must be true about the potential energy?

11. Consider the following two situations in which the retarding effects of friction and air resistance are negligible. Car A approaches a hill. The driver turns off the engine at the bottom of the hill, and the car coasts up the hill. Car B, its engine running, is driven up the hill at a constant speed. Which situation is an example of the principle of conservation of mechanical energy? Provide a reason for your answer.

12. A trapeze artist, starting from rest, swings downward on the bar, lets go at the bottom of the swing, and falls freely to the net. An assistant, standing on the same platform as the trapeze artist, jumps from rest straight downward. Friction and air resistance are negligible. (a) On which person, if either, does gravity do the greatest amount of work? Explain. (b) Who, if either, strikes the net with a greater speed? Why?

13. The drawing shows an empty fuel tank being released by three different jet planes. At the moment of release, each plane has the same speed and each tank is at the same height above the ground. However, the directions of travel are different. In the absence of air resistance, do the tanks have different speeds when they hit the ground? If so, which tank has the largest speed and which has the smallest speed? Explain.

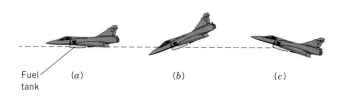

Fuel tank   (a)   (b)   (c)

14. Is it correct to conclude that one engine is doing twice the work of another just because it is generating twice the power? Explain, neglecting friction and taking into account the time of operation of the engines.

PROBLEMS

Section 6.1 Work

1. A person pulls a toboggan for a distance of 35.0 m along the snow with a rope directed 25.0° above the snow. The tension in the rope is 94.0 N. (a) How much work is done on the toboggan by the tension force? (b) How much work is done if the same tension is directed parallel to the snow?

2. The cable of a large crane applies a force of 2.2×10^4 N to a demolition ball as it lifts the ball vertically upward a distance of 7.6 m. (a) How much work does this force do on the ball? (b) Is the work positive or negative? Explain.

3. When spring arrives, a woman packs her winter clothes in a box and lifts it at a constant velocity to the top shelf of her closet, a distance of 1.8 m above the floor. The clothes weigh 150 N. How much work does she do in lifting the clothes?

4. Suppose in Figure 6.2 that $+1.10 \times 10^3$ J of work are done by the force F (magnitude = 30.0 N) in moving the luggage carrier a distance of 50.0 m. At what angle θ is the force oriented with respect to the ground?

5. You are moving into an apartment at the beginning of the semester. Your weight is 685 N and that of your belongings is 915 N. (a) How much work does the elevator do in lifting you and your belongings up five stories (15.2 m) at a constant velocity? (b) How much work does the elevator do on you (without belongings) on the downward trip, which is also made at a constant velocity? Be sure you include the correct sign for the work.

6. The drawing shows a boat being pulled by two locomotives through a canal of length 2.00 km. The tension in each cable is 5.00×10^3 N, and $\theta = 20.0°$. What is the net work done on the boat by the two locomotives?

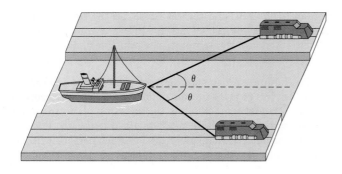

7. From a balcony, a piano is lowered through a distance of 4.00 m, by means of two ropes. Each rope makes an angle of 15° with respect to the vertical and sustains a tension of 1500 N. How much work is done on the piano by *each rope*?

*8. A 2.40×10^2-N force is pulling an 85.0-kg refrigerator across a horizontal surface. The force acts at an angle of 20.0° above the surface. The coefficient of kinetic friction is 0.200, and the refrigerator moves a distance of 8.00 m. Find (a) the work done by the pulling force, and (b) the work done by the kinetic frictional force.

*9. A 1.00×10^2-kg crate is being pulled across a horizontal floor by a force **P** that makes an angle of 30.0° above the horizontal. The coefficient of kinetic friction is 0.200. What should be the magnitude of **P**, so that the net work done by it and the kinetic frictional force is zero?

10. A 1200-kg car is being driven up a 5.0° hill, as the drawing illustrates. The frictional force is directed opposite to the motion of the car and has a magnitude of $f = 5.0 \times 10^2$ N. The force **F is applied to the car by the road and propels the car forward. In addition to these two forces, two other forces act on the car: its weight **W**, and the normal force $\mathbf{F_N}$ directed perpendicular to the road surface. The length of the road up the hill is 3.0×10^2 m. What should be the magnitude of **F**, so that the net work done by all the forces acting on the car is $+150\,000$ J? *(See Solved Problem 1 for a related problem.)*

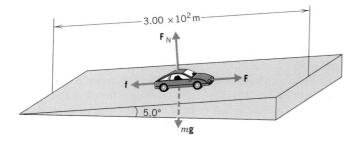

Section 6.2 The Work–Energy Theorem and Kinetic Energy

11. A 0.075-kg arrow is fired horizontally. The bow string exerts an average force of 65 N on the arrow over a distance of 0.90 m. With what speed does the arrow leave the bow?

12. A 65.0-kg jogger is running at a speed of 5.30 m/s. (a) What is the kinetic energy of the jogger? (b) How much work is done by the force that accelerates the jogger to 5.30 m/s from rest?

13. A water-skier whose mass is 70.3 kg has an initial speed of 6.10 m/s. Later, the speed of the skier is 11.3 m/s. Determine the work done by the net force acting on the skier.

14. A pitcher hurls a 0.25-kg softball at a speed of 25 m/s. How much work is done on the softball by the hurler's arm?

15. A 1.20×10^3-kg automobile coasts through a 50.0-m-long snowdrift that has been blown onto the road. The automobile has a speed of 20.0 m/s as it approaches the drift and emerges with a speed of 8.00 m/s. Find the average net force acting on the car in the drift. Relative to the motion of the car, what is the direction of this net force?

16. A 5.0×10^4-kg space probe is traveling at a speed of 11 000 m/s through deep space. Retrorockets are fired along the line of motion to reduce the probe's speed. The retrorockets generate a force of 4.0×10^5 N over a distance of 2500 km. What is the final speed of the probe?

17. When a 0.045-kg golf ball takes off after being hit, its speed is 41 m/s. (a) How much work is done on the ball by the club? (b) Assume that the force of the golf club acts parallel to the motion of the ball and that the club is in contact with the ball for a distance of 0.010 m. Ignore the weight of the ball and determine the average force applied to the ball by the club.

*18. The speed of a hockey puck decreases from 45.00 to 44.67 m/s in coasting 16 m across the ice. Find the coefficient of kinetic friction between the puck and the ice.

*19. In screeching to a halt, a car leaves skid marks that are 65 m long. The coefficient of kinetic friction between the tires and the road is $\mu_k = 0.71$. How fast was the car going before the driver applied the brakes?

*20. A wind-driven iceboat has a mass of 4.00×10^2 kg. The boat starts from rest and reaches a speed of 16.0 m/s after traveling a distance of 60.0 m. The coefficient of kinetic friction between the ice and the runners of the boat is 0.100. Determine the work done on the boat by the wind.

*21. The head of a sledge hammer weighs 22 N and is moving at a speed of 7.6 m/s when it strikes a stake. The stake moves 0.025 m into the ground in response. Assume that forty percent of the hammer's kinetic energy is converted into the initial kinetic energy of the stake. Apply the work–energy theorem to the stake, and obtain the average resistive force applied to the stake by the ground.

Section 6.3 Gravitational Potential Energy

22. A 0.15-kg ball is thrown 9.0 m straight up. (a) Find the work done by the gravitational force. Be sure to include the correct sign. (b) What is the change ($\Delta PE = PE_f - PE_0$) in the gravitational potential energy?

23. A shot-putter puts a shot (weight = 71.1 N) that leaves his hand at a distance of 1.52 m above the ground. (a) Find the work done by the gravitational force when the shot has risen to a height of 2.13 m. Include the correct sign for the work. (b) Determine the change ($\Delta PE = PE_f - PE_0$) in the gravitational potential energy of the shot.

24. Relative to the ground, what is the gravitational potential energy of a 55.0-kg person who is at the top of the Sears Tower, a height of 443 m above the ground?

25. A 75.0-kg skier rides a 2830-m-long lift to the top of a mountain. The lift makes an angle of 14.6° with the horizontal. What is the change in the skier's gravitational potential energy?

26. The longest escalator in the world is in Hong Kong and has a length of 227 m. A 52.0-kg person rides the escalator from bottom to top, and her gravitational potential energy changes by 5.86×10^4 J. What is the angle of the escalator above the horizontal?

Section 6.5 The Conservation of Mechanical Energy

27. A gymnast is swinging on a high bar. The distance between his waist and the bar is 1.1 m, as the drawing shows. At the top of the swing his speed is momentarily zero. Ignore friction and find the speed of his waist at the bottom of the swing.

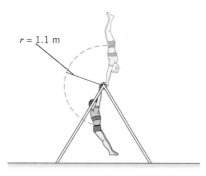

$r = 1.1$ m

28. A pole-vaulter approaches the takeoff point at a speed of 9.00 m/s. Assuming that only this speed determines the height to which he can rise, find the maximum height at which the vaulter can clear the bar.

29. A 2.00-kg rock is released from rest at a height of 20.0 m. Ignore air resistance and determine the kinetic energy, gravitational potential energy, and total mechanical energy at each of the following heights: 20.0, 15.0, 10.0, 5.00, and 0 m.

30. A person is sled-riding down a hill that is 4.6 m high. Starting at the top with a speed of 3.1 m/s, the sled reaches the bottom with a speed of 7.6 m/s. (a) Determine whether mechanical energy has been conserved. (b) Why might mechanical energy not be conserved?

31. A woman runs with a speed of 5.4 m/s off a platform that is 10.0 m above the water. Ignore air resistance. How fast is she moving when she hits the water?

32. A cyclist approaches the bottom of a gradual hill at a speed of 11 m/s. The hill is 5.0 m high, and the cyclist estimates that she is going fast enough to coast up and over it without peddling. Ignoring air resistance and friction, find the speed at which the cyclist crests the hill.

33. A water-skier lets go of the tow rope upon leaving the end of a jump ramp at a speed of 14.0 m/s. As the drawing indicates, the skier has a speed of 13.0 m/s at the highest point of the jump. Ignoring air resistance, determine the skier's height H above the *top of the ramp* at the highest point.

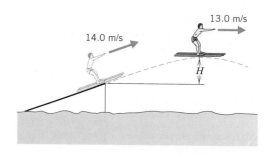

*34. A grappling hook, attached to a 1.5-m rope, is whirled in a circle that lies in the vertical plane. The hook is whirled at a constant rate of three revolutions per second. In the absence of air resistance, to what maximum height can the hook be cast?

*35. A very small metal ball, starting from point A in the drawing, is projected down the curved runway. Upon leaving the runway at point B, the ball is traveling straight upward and reaches a height of 4.00 m above the floor before falling back down. Ignoring friction and air resistance, find the speed of the ball at point A.

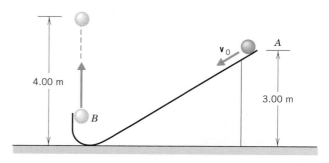

*36. A water slide is constructed so that swimmers, starting from rest at the top of the slide, leave the end of the slide traveling horizontally. As the drawing shows, one person is observed to hit the water 5.00 m from the end of the slide in

0.500 s after leaving the slide. Ignoring friction and air resistance, find the height H in the drawing.

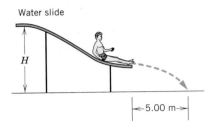

Water slide

H

←5.00 m→

**37. A swing is made from a rope that will tolerate a maximum tension of 8.00×10^2 N without breaking. Initially, the swing hangs vertically. The swing is then pulled back at an angle of 60.0° with respect to the vertical and released from rest. What is the mass of the heaviest person who can ride the swing?

Section 6.6 Nonconservative Forces and the Work–Energy Theorem

38. A basketball player makes a jump shot. The 0.60-kg ball is released at a height of 2.0 m above the floor with a speed of 7.2 m/s. The ball goes through the net 3.0 m above the floor at a speed of 4.2 m/s. How much work is done on the ball by air resistance, a nonconservative force?

39. A roller coaster (375 kg) moves from A (5.00 m above the ground) to B (20.0 m above the ground). Two nonconservative forces are present: friction does -2.00×10^4 J of work on the car, and a chain mechanism does $+3.00 \times 10^4$ J of work to help the car up a long climb. What is the change in the car's kinetic energy, $\Delta KE = KE_f - KE_0$, from A to B?

40. A 74.0-kg student, starting from rest, slides down a 11.8-m-high water slide. On the way down, friction (a nonconservative force) does -5.60×10^3 J of work on him. How fast is he going at the bottom of the slide?

41. A 55.0-kg skateboarder starts out with a speed of 1.80 m/s. He does $+80.0$ J of work on himself by pushing with his feet against the ground. In addition, friction does -265 J of work on him. In both cases, the forces doing the work are nonconservative. The final speed of the skateboarder is 6.00 m/s. (a) Calculate the change ($\Delta PE = PE_f - PE_0$) in the gravitational potential energy. (b) How much has the vertical height of the skater changed, and is the skater above or below the starting point?

42. A 5.00×10^2-kg hot-air balloon takes off from rest at the surface of the earth. The nonconservative wind and lift forces take the balloon up, doing $+9.70 \times 10^4$ J of work on the balloon in the process. At what height above the surface of the earth does the balloon have a speed of 8.00 m/s?

*43. A gymnast is bouncing on a trampoline. On the upward part of the motion, the mat of the trampoline pushes on the gymnast with a nonconservative force over a distance of 0.300 m. After leaving the mat, the 55.0-kg gymnast rises into the air for an additional 2.00 m before falling back down. What average force does the mat exert on the gymnast?

*44. A bowler releases a bowling ball with an initial speed of 6.7 m/s. The ball slides down the alley for a distance of 4.0 m before beginning to roll. Kinetic friction (a nonconservative force) acts on the ball, and the coefficient of kinetic friction between the ball the alley is $\mu_k = 0.20$. Find the speed of the ball just before it begins to roll.

*45. At a carnival, you can try to ring a bell by striking a target with a 9.00-kg hammer. In response, a 0.400-kg metal piece is sent upward toward the bell, which is 5.00 m above. Suppose that 25.0% of the hammer's kinetic energy is used to do the (nonconservative) work of sending the metal piece upward. How fast must the hammer be moving when it strikes the target, so that the bell just barely rings?

**46. A 3.00-kg model rocket is launched vertically straight up with sufficient initial speed to reach a maximum height of 1.00×10^2 m, even though air resistance (a nonconservative force) performs -8.00×10^2 J of work on the rocket. How high would the rocket have gone without air resistance?

Section 6.7 Power

47. One kilowatt·hour (kWh) is the amount of work or energy generated when one kilowatt of power is supplied for a time of one hour. A kilowatt·hour is the unit of energy used by power companies when figuring your electric bill. Determine the number of joules of energy in one kilowatt·hour.

48. A person is making homemade ice cream. She exerts a force of magnitude 22 N on the free end of the crank handle, and this end moves in a circular path of radius 0.28 m. The force is always applied parallel to the motion of the handle. If the handle is turned once every 1.3 s, what is the average power being expended?

49. The floors in a typical house are separated by a vertical distance of approximately 2.4 m. A teenager (weight = 440 N) climbs the stairs between floors. Find the average power necessary to accomplish this, if the stairs are climbed in (a) 10.0 s and (b) 2.0 s. Express the answers in units of horsepower.

50. A 3.00×10^2-kg piano is being lifted at a steady speed from ground level straight up to an apartment 10.0 m above the ground. The crane that is doing the lifting produces a steady power of 4.00×10^2 W. How much time does it take to lift the piano?

*51. A car accelerates uniformly from rest to 29 m/s in 12 s along a level stretch of road. Ignoring friction, determine the average power required to accelerate the car if (a) the weight of the car is 1.2×10^4 N, and (b) if the weight of the car is 1.6×10^4 N. Express your answers in units of horsepower.

*52. A 73-kg sprinter, starting from rest, reaches a speed of 7.0 m/s in 1.8 s, with a negligible effect due to air resistance. The sprinter then runs the remainder of the race at a steady speed of 7.0 m/s under the influence of a 35-N force due to air resistance. What is the average power needed (a) to accelerate the runner and (b) to sustain the steady speed at which most of the race is run?

**53. A motorcycle (mass of cycle plus rider = 2.50×10^2 kg) is traveling at a steady speed of 20.0 m/s over a 1.00-km stretch of road. The force of air resistance acting on the cycle and rider is 2.00×10^2 N. Find the power necessary to sustain this speed if (a) the road is level and (b) if the road is sloped upward at 37.0° with respect to the horizontal.

**54. A 1.20×10^3-kg car has a speed of 11.0 m/s at the bottom and a speed of 23.0 m/s at the top of a hill that makes an angle of 5.00° with respect to the horizontal. The length of the hill is 1.50 km, and the force of friction opposing the car's motion has a magnitude of 6.00×10^2 N. Determine the average power required to accelerate the car up the hill.

ADDITIONAL PROBLEMS

55. A slingshot fires a pebble from the top of a building at a speed of 10.0 m/s. The building is 20.0 m tall. Ignoring air resistance, find the speed with which the pebble strikes the ground when the pebble is fired (a) horizontally, (b) vertically straight up, and (c) vertically straight down.

56. A softball pitcher has a "windmill" windup in which a 0.25-kg ball moves on a vertical arc of radius $r = 0.51$ m. To make the ball accelerate, she exerts a 28-N force parallel to the ball's motion along the circular arc. The speed of the ball is 12 m/s at the top of the arc. With what speed is the ball released one-half a revolution later at the bottom of the arc?

57. Two cars, A and B, are traveling with the same speed of 40.0 m/s, each having started from rest. Car A has a mass of 1.20×10^3 kg, and car B has a mass of 2.00×10^3 kg. Compared to the work required to bring car A up to speed, how much *additional* work is required to bring car B up to speed?

58. The brakes of a truck cause it to slow down by applying a retarding force of 3.0×10^3 N to the truck over a distance of 850 m. How much work does this force perform on the truck? Is the work positive or negative? Why?

59. A pitcher throws a 0.140-kg baseball, and it approaches the bat at a speed of 40.0 m/s. The bat does $W_{nc} = 1.40 \times 10^2$ J of work on the ball in hitting it. Ignoring air resistance, determine the speed of the ball after the ball leaves the bat and is 25.0 m above the point of impact.

*60. The (nonconservative) force propelling a 1.50×10^3-kg car up a mountain road does 4.70×10^6 J of work on the car. The car starts from rest at sea level and has a speed of 27.0 m/s at an altitude of 2.00×10^2 m above sea level. Obtain the work done on the car by friction and air resistance, both of which are nonconservative forces. (*See Solved Problem 1 for a related problem.*)

*61. A 95-kg refrigerator is resting on a frictionless horizontal surface and is attached by a rope to a motor-driven winch. The winch is turned on and for 4.0 s supplies an average power of 110 W while the refrigerator is pulled across the surface. What is the speed of the refrigerator at the end of the time interval?

*62. A 55-kg box is being pushed a distance of 7.0 m across the floor by a force **P** whose magnitude is 150 N. The force **P** is parallel to the displacement of the box. The coefficient of kinetic friction is 0.25. Determine the work done on the box by each of the *four* forces that act on the box. Be sure to include the proper plus or minus sign for the work done by each force. (*See Solved Problem 1 for a related problem.*)

*63. A wrecking ball swings at the end of a 10.0-m cable on a vertical circular arc. The crane operator manages to give the ball a speed of 6.00 m/s as the ball passes through the lowest point of its swing and then gives the ball no further assistance. Friction and air resistance are negligible. What speed v_f does the ball have when the cable makes an angle of 30.0° with respect to the vertical?

*64. A 55.0-kg diver dives off a 10.0-m-high tower straight down into the water. Neglect air resistance during the descent. She comes to rest 3.00 m under the surface of the water. Determine the average force that the water exerts on the diver. This force is nonconservative.

**65. The drawing shows a version of the loop-the-loop trick for a small windup car. If the car is given an initial speed of 4.0 m/s, what is the largest value that the radius r can have if the car is to remain in contact with the circular track at all times?

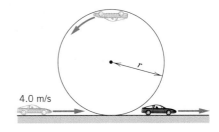

**66. A truck is traveling at 11.1 m/s down a hill when the brakes on all four wheels lock. The hill makes an angle of 15.0° with respect to the horizontal. The coefficient of kinetic friction between the tires and the road is 0.750. How far does the truck skid before coming to a stop?

CHAPTER 7

IMPULSE AND MOMENTUM

In principle, it is always possible to use Newton's second law to predict how an object will accelerate under the influence of a net force. In practice, however, it is only when the net force is constant or known to be changing in a predictable manner that the application of the second law is straightforward. In many situations, the net force does not behave in such a convenient manner. For example, during a fireworks display, each rocket explodes in midair, sending out a colorful shower of burning fragments. During the explosion each of the many fragments interacts with the others, and the force that any one experiences is neither constant nor does it change in a predictable way. Even when only two objects interact, there is often little information to indicate what happens from moment to moment. When a rifle is fired, the burning gunpowder drives the bullet forward, while giving the rifle a backward "kick." A moment-to-moment description of the force on the bullet, however, would be hard to obtain. Similarly, when a bat hits a baseball, scant information is available to indicate the exact details of how the force changes during the collision. In circumstances such as these, where one or more objects interact and detailed information about the forces is not available, it is not practical to use Newton's second law to determine values for the acceleration at each instant. As we will see in this chapter, however, there is a method by which information can be obtained about the motion of interacting objects. The method has its roots in Newton's second law, but also builds on Newton's third law, which deals with the action–reaction forces that occur when objects interact.

7.1 THE IMPULSE–MOMENTUM THEOREM

Figure 7.1 shows a baseball being hit by a bat. The ball has an initial velocity $\mathbf{v_0}$ just before contact is made and a final velocity $\mathbf{v_f}$ just after leaving the bat. In general, the final velocity does not equal the initial velocity, either in magnitude or direction. During the time interval $\Delta t = t_f - t_0$, the bat and the ball are in contact, and the force $\mathbf{F}$ exerted on the ball changes. As the graph in the drawing indicates, the magnitude of the force rises from zero at the instant the bat touches the ball, reaches a maximum value, and then returns to zero as the ball leaves the bat. The graph also shows the magnitude $\overline{F}$ of the average force, for the sake of comparison.

If a baseball is to be hit well, both the size of the force and the time of contact are important. When a sufficiently large average force acts on the ball for a long enough time, the ball is hit solidly. Therefore, we are motivated to bring together the average force and the time of contact, calling the product of the two the *impulse* of the force.

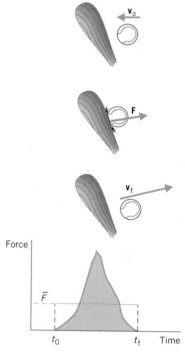

Figure 7.1 When a bat hits a ball, a force $\mathbf{F}$ is applied to the ball. As a result, the ball's velocity changes from an initial value of $\mathbf{v_0}$ to a final value of $\mathbf{v_f}$.

Definition of Impulse

The impulse of a force is the product of the average force $\overline{\mathbf{F}}$ and the time interval Δt during which the force acts:

$$\mathbf{Impulse} = \overline{\mathbf{F}}\,\Delta t \tag{7.1}$$

Impulse is a vector quantity and has the same direction as the average force.

  SI Unit of Impulse: newton·second (N·s)

When a ball is hit, it responds to the value of the impulse. A large impulse produces a large response, that is, a well-hit ball. Of course, the phrase "large response" means that the ball departs from the bat with a large velocity, although the more massive the ball, the less velocity it picks up in a given interaction with the bat. Therefore, both mass and velocity play a role in how an object responds to a given impulse. The effect of mass and velocity is included in the concept of *momentum,* which is defined as follows:

Definition of Linear Momentum

The linear momentum $\mathbf{p}$ of an object is the product of the object's mass m and velocity $\mathbf{v}$:

$$\mathbf{p} = m\mathbf{v} \tag{7.2}$$

Linear momentum is a vector quantity that points in the same direction as the velocity.

  SI Unit of Momentum: kilogram·meter/second (kg·m/s)

Newton's second law can be used to reveal a relationship between impulse and momentum. When the velocity of an object changes from $\mathbf{v}_0$ to $\mathbf{v}_f$ during a time interval Δt, the average acceleration $\overline{\mathbf{a}}$ is given by Equation 2.4 as

$$\overline{\mathbf{a}} = \frac{\Delta \mathbf{v}}{\Delta t} = \frac{\mathbf{v}_f - \mathbf{v}_0}{\Delta t}$$

According to Newton's second law, the cause of the acceleration is an average net force, $\overline{\mathbf{F}} = m\overline{\mathbf{a}}$. Thus,

$$\overline{\mathbf{F}} = m\left(\frac{\mathbf{v}_f - \mathbf{v}_0}{\Delta t}\right) = \frac{m\mathbf{v}_f - m\mathbf{v}_0}{\Delta t} \tag{7.3}$$

In this result, the numerator on the right is the final momentum minus the initial momentum, so the average net force is given by the change in momentum per unit of time.* Multiplying both sides of Equation 7.3 by Δt yields Equation 7.4, which is known as the *impulse–momentum theorem.*

Impulse–Momentum Theorem

When a net force $\mathbf{F}$ acts on an object, the impulse of the net force is equal to the change in momentum of the object:

$$\overline{\mathbf{F}}\,\Delta t = \underbrace{m\mathbf{v}_f}_{\substack{\text{Final} \\ \text{momentum}}} - \underbrace{m\mathbf{v}_0}_{\substack{\text{Initial} \\ \text{momentum}}} \tag{7.4}$$

Impulse = Change in momentum

Examples 1 and 2 illustrate how this important theorem is used.

Example 1   A Well-Hit Ball

A baseball ($m = 0.14$ kg) has an initial velocity of $\mathbf{v}_0 = -38$ m/s as it approaches the bat. We have chosen the direction of approach as the negative direction. The bat applies a force that is much larger than the weight of the ball, and the ball departs from the bat with a final velocity of $\mathbf{v}_f = +58$ m/s. The contact time between the bat and ball is $\Delta t = 4.0 \times 10^{-3}$ s. Find the average force exerted on the ball.

REASONING   In hitting the ball, the bat imparts an impulse to it. If the impulse can be determined, we can use Equation 7.1 (**Impulse** $= \overline{\mathbf{F}}\,\Delta t$) to find the average force exerted on the ball, since the contact time is given. To find the impulse, we turn to the impulse–momentum theorem, which indicates that the impulse of the net force is the final momentum minus the initial momentum. The force of the bat is the net force acting on the ball, assuming that the ball's weight is negligible in comparison. We begin, then, by finding the initial and final momenta of the ball.

---

* The equality between force and the time rate of change of momentum is the version of the second law of motion presented originally by Newton.

SOLUTION

$$\text{Initial momentum: } \mathbf{p_0} = m\mathbf{v_0} = (0.14 \text{ kg})(-38 \text{ m/s}) \tag{7.2}$$
$$= -5.3 \text{ kg·m/s}$$
$$\text{Final momentum: } \mathbf{p_f} = m\mathbf{v_f} = (0.14 \text{ kg})(+58 \text{ m/s}) \tag{7.2}$$
$$= +8.1 \text{ kg·m/s}$$
$$\textbf{Impulse} = m\mathbf{v_f} - m\mathbf{v_0} \tag{7.4}$$
$$= (8.1 \text{ kg·m/s}) - (-5.3 \text{ kg·m/s})$$
$$= +13.4 \text{ kg·m/s}$$

Now that the impulse is known, the contact time can be used in Equation 7.1 to find the average force:

$$\overline{\mathbf{F}} = \frac{\textbf{Impulse}}{\Delta t} = \frac{+13.4 \text{ kg·m/s}}{4.0 \times 10^{-3} \text{ s}} = \boxed{+3400 \text{ N}}$$

The force is positive, indicating that it points opposite to the velocity of the approaching ball. A force of 3400 N corresponds to 760 lb, such a large value being necessary to change the ball's momentum during the brief contact time.

Example 2   A Rain Storm

During a storm, rain comes straight down with a velocity of -15 m/s and hits the roof of a car perpendicularly. The mass of rain that strikes the car roof per second is 0.060 kg/s. Assuming that the rain comes to rest upon striking the car roof, find the average force exerted by the rain on the car roof.

REASONING   This example differs from Example 1 in an important way. Example 1 gives information about the ball and asks for the force applied to the ball. In contrast, the present example gives information about the rain, but doesn't ask for the force applied to the rain. Instead, it asks for the force acting on the car roof. However, the force exerted on the roof by the rain and the force exerted on the rain by the roof have equal magnitudes and opposite directions, according to Newton's law of action and reaction. So our approach here will be to find the force exerted on the rain and then apply the law of action and reaction to obtain the force on the roof.

SOLUTION   The average force needed to reduce the rain's momentum to zero is given by Equation 7.3 as

$$\overline{\mathbf{F}} = \frac{m\mathbf{v_f} - m\mathbf{v_0}}{\Delta t} = -\left(\frac{m}{\Delta t}\right)\mathbf{v_0}$$

since the rain comes to rest upon hitting the car ($\mathbf{v_f} = 0$). The term $m/\Delta t$ is the mass of rain that strikes the car roof per second. Thus, the average force acting on the rain is

$$\overline{\mathbf{F}} = -(0.060 \text{ kg/s})(-15 \text{ m/s}) = +0.90 \text{ N}$$

This force is in the positive or upward direction. According to the action–reaction law, the force exerted on the car roof also has a magnitude of 0.90 N but points downward:

Force on roof = $\boxed{-0.90 \text{ N}}$.

It is worth comparing the impulse–momentum theorem to the work–energy theorem discussed in Chapter 6. The impulse–momentum theorem states that the impulse produced by a net force is equal to the change in the object's momentum, while the work–energy theorem states that the work done by a net force is equal to the change in the object's kinetic energy. Since the work–energy theorem leads directly to the principle of conservation of mechanical energy, it is not surprising that the impulse–momentum theorem also leads to a useful principle called the *principle of conservation of linear momentum*, which the next section considers.

7.2 THE PRINCIPLE OF CONSERVATION OF LINEAR MOMENTUM

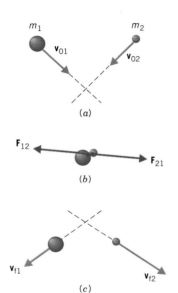

Figure 7.2 (*a*) The velocities of two objects are $\mathbf{v}_{01}$ and $\mathbf{v}_{02}$ before a collision. (*b*) During the collision, each object exerts a force on the other object. These forces are labeled $\mathbf{F}_{12}$ and $\mathbf{F}_{21}$. (*c*) The velocities are $\mathbf{v}_{f1}$ and $\mathbf{v}_{f2}$ after the collision.

To explain the conservation of linear momentum, we apply the impulse–momentum theorem to a midair collision between two objects. The two objects (masses m_1 and m_2) are approaching each other with initial velocities $\mathbf{v}_{01}$ and $\mathbf{v}_{02}$, as Figure 7.2*a* shows. The collection of objects being studied is referred to as the "system." The objects interact during the collision in part *b* of the drawing and then depart with the final velocities $\mathbf{v}_{f1}$ and $\mathbf{v}_{f2}$ shown in part *c*. Because of the collision, the initial and final velocities are not the same.

There are two types of forces acting on the system:

1. *Internal forces*—Forces that the objects within the system exert on each other.
2. *External forces*—Forces exerted on the objects by agents that are external to the system.

The forces $\mathbf{F}_{12}$ and $\mathbf{F}_{21}$ in Figure 7.2*b* are action–reaction forces that arise during the collision and are internal forces. The force $\mathbf{F}_{12}$ is exerted on object 1 by object 2, while the force $\mathbf{F}_{21}$ is exerted on object 2 by object 1. The force of gravity also acts on the objects, their weights being $\mathbf{W}_1$ and $\mathbf{W}_2$. These weights are external forces, because they are applied by the earth, which is outside the system. Friction and air resistance would also be considered external forces, although these forces are ignored here for the sake of simplicity. The impulse–momentum theorem, then, gives the following results:

[Object 1]
$$\Big(\underbrace{\mathbf{W}_1}_{\substack{\text{External} \\ \text{force}}} + \underbrace{\overline{\mathbf{F}}_{12}}_{\substack{\text{Internal} \\ \text{force}}} \Big) \Delta t = m_1 \mathbf{v}_{f1} - m_1 \mathbf{v}_{01}$$

[Object 2]
$$\Big(\underbrace{\mathbf{W}_2}_{\substack{\text{External} \\ \text{force}}} + \underbrace{\overline{\mathbf{F}}_{21}}_{\substack{\text{Internal} \\ \text{force}}} \Big) \Delta t = m_2 \mathbf{v}_{f2} - m_2 \mathbf{v}_{02}$$

Adding these equations produces a single result for the system as a whole:

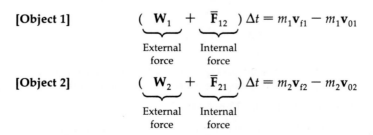

$$(\underbrace{\mathbf{W}_1 + \mathbf{W}_2}_{\substack{\text{External} \\ \text{forces}}} + \underbrace{\overline{\mathbf{F}}_{12} + \overline{\mathbf{F}}_{21}}_{\substack{\text{Internal} \\ \text{forces}}}) \Delta t = \underbrace{(m_1 \mathbf{v}_{f1} + m_2 \mathbf{v}_{f2})}_{\substack{\text{Total final} \\ \text{momentum } \mathbf{P}_f}} - \underbrace{(m_1 \mathbf{v}_{01} + m_2 \mathbf{v}_{02})}_{\substack{\text{Total initial} \\ \text{momentum } \mathbf{P}_0}}$$

On the right side of this equation, the quantity $m_1\mathbf{v}_{f1} + m_2\mathbf{v}_{f2}$ is the vector sum of the final momenta for each object, or the total final momentum $\mathbf{P}_f$ of the system. Likewise, $m_1\mathbf{v}_{01} + m_2\mathbf{v}_{02}$ is the total initial momentum $\mathbf{P}_0$. Therefore, the result above can be rewritten as

$$\left(\begin{array}{c}\text{Sum of average}\\ \text{external forces}\end{array} + \begin{array}{c}\text{Sum of average}\\ \text{internal forces}\end{array}\right)\Delta t = \mathbf{P}_f - \mathbf{P}_0 \qquad (7.5)$$

The advantage of the internal/external force classification is that the internal forces always add together to give zero, as a consequence of Newton's law of action–reaction; $\mathbf{F}_{12} = -\mathbf{F}_{21}$, so that $\mathbf{F}_{12} + \mathbf{F}_{21} = 0$. Cancellation of the internal forces occurs no matter how many parts there are to the system and allows us to ignore the internal forces, as Equation 7.6 indicates:

$$\text{(Sum of average external forces)}\ \Delta t = \mathbf{P}_f - \mathbf{P}_0 \qquad (7.6)$$

We developed this result with gravity as the only external force. But, in general, the sum of the external forces on the left includes *all* external forces.

With the aid of Equation 7.6, it is possible to see how the conservation of linear momentum arises. Suppose that the sum of the external forces is zero. A system for which this is true is called an *isolated system.* For example, in a system composed of two billiard balls colliding on a frictionless pool table, the weight of each ball and the normal forces provided by the table are the external forces. Since the weights and the normal forces balance, the sum of the external forces is zero, and the balls constitute an isolated system. In such a case, Equation 7.6 indicates that

$$0 = \mathbf{P}_f - \mathbf{P}_0 \qquad \text{or} \qquad \mathbf{P}_f = \mathbf{P}_0 \qquad (7.7)$$

In other words, the final total momentum of the isolated system after the balls collide is the same as the initial total momentum. This result is known as the *principle of conservation of linear momentum.*

Principle of Conservation of Linear Momentum

The total linear momentum of an isolated system remains constant (is conserved). An isolated system is one for which the vector sum of the external forces acting on the system is zero.

The total linear momentum of the arrow and the apple is the same just before and just after the collision (assumed to be very short in duration).

This principle applies to a system containing any number of objects, regardless of the internal forces, provided the system is isolated. Whether a force is considered to be internal depends on what objects are included as members of the system. In the case of two billiard balls, the collision forces are considered to be internal if both balls are included. However, if *only one ball* is included, the collision force exerted on it by the other ball is an external force, since the other ball is then outside the system. Clearly, the total linear momentum of a one-ball system is *not* conserved in the presence of this external collision force; the momentum (and, hence, velocity) of a single billiard ball always changes during a collision. Example 3 illustrates an application of momentum conservation.

Example 3   Assembling a Freight Train

A freight train is being assembled in a switching yard, and Figure 7.3 shows two boxcars in the process of being coupled together. Car 1 has a mass of $m_1 = 65 \times 10^3$ kg and moves at a velocity of $v_{01} = +0.80$ m/s. Car 2, with a mass of $m_2 = 92 \times 10^3$ kg and a velocity of $v_{02} = +1.2$ m/s, overtakes car 1 and couples to it. Neglecting friction, find the common velocity v_f of the two cars after they become coupled.

REASONING   The two boxcars constitute the system. The sum of the external forces acting on the system is zero, because the weight of each car is balanced by a corresponding normal force, and friction is being neglected. Thus, the system is isolated, and the principle of conservation of linear momentum applies. The coupling forces that each car exerts on the other are internal forces and do not affect the applicability of this principle.

SOLUTION   Momentum conservation indicates that

$$\underbrace{(m_1 + m_2)v_f}_{\substack{\text{Total momentum} \\ \text{after collision}}} = \underbrace{m_1 v_{01} + m_2 v_{02}}_{\substack{\text{Total momentum} \\ \text{before collision}}}$$

This equation can be solved for v_f, the common velocity of the two cars after the collision:

$$v_f = \frac{m_1 v_{01} + m_2 v_{02}}{m_1 + m_2}$$

$$= \frac{(65 \times 10^3 \text{ kg})(0.80 \text{ m/s}) + (92 \times 10^3 \text{ kg})(1.2 \text{ m/s})}{(65 \times 10^3 \text{ kg} + 92 \times 10^3 \text{ kg})}$$

$$= \boxed{+1.0 \text{ m/s}}$$

PROBLEM SOLVING INSIGHT

The conservation of linear momentum is applicable only when the net external force acting on the system is zero. Therefore, the first step in applying momentum conservation to problem solving is to be sure that the net external force is zero.

Figure 7.3   (a) One boxcar eventually catches up with the other and couples to it. (b) The coupled cars move together with a common velocity after the collision.

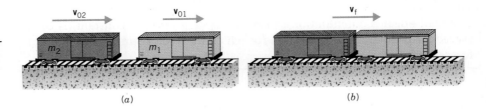

(a)   (b)

   In the previous example it can be seen that the velocity of car 1 increases, while the velocity of car 2 decreases as a result of the collision. The acceleration and deceleration arise at the moment the cars become coupled, because the cars exert internal forces on each other. These forces are equal in magnitude and opposite in direction, in accord with Newton's third law. The powerful feature of the momentum conservation principle is that it allows us to determine the changes in velocity without knowing what the internal forces are. Example 4 further illustrates this feature.

Example 4   Ice Skaters

Starting from rest, two skaters "push off" against each other on smooth level ice, where friction is negligible. One is a woman ($m_1 = 54$ kg), and one is a man ($m_2 = 88$ kg). As Figure 7.4b shows, the woman moves away with a velocity of $v_{f1} = +2.5$ m/s. Find the "recoil" velocity v_{f2} of the man.

REASONING   For a system consisting of the two skaters, the sum of the external forces is zero, because the weight of each skater is balanced by a corresponding normal force and the ice is assumed to be frictionless. The skaters, then, constitute an isolated system, and the principle of conservation of linear momentum applies.

SOLUTION   The total momentum of the skaters before they push on each other is zero, since they are at rest. Momentum conservation requires that the total momentum remains zero after the skaters have separated, as in part b of the drawing:

$$\underbrace{m_1 v_{f1} + m_2 v_{f2}}_{\substack{\text{Total momentum} \\ \text{after pushing}}} = \underbrace{0}_{\substack{\text{Total momentum} \\ \text{before pushing}}}$$

Solving for the recoil velocity of the man gives

$$v_{f2} = \frac{-m_1 v_{f1}}{m_2} = \frac{-(54 \text{ kg})(2.5 \text{ m/s})}{88 \text{ kg}}$$

$$= \boxed{-1.5 \text{ m/s}}$$

The minus sign indicates that the man moves to the left in the drawing. After the skaters separate, the total momentum of the system remains zero, because momentum is a vector quantity, and the momenta of the man and the woman have equal magnitudes but opposite directions.

(a)

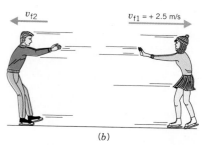

(b)

Figure 7.4   (a) In the absence of friction, two skaters pushing on each other constitute an isolated system. (b) As the skaters move away, the total linear momentum of the system remains zero, which is what it was initially.

   It is important to realize that the total linear momentum may be conserved even when the total kinetic energy is not constant. In Example 4, for instance, the initial kinetic energy is zero since the skaters are stationary. But after they push off, the skaters are moving, so each has kinetic energy. The kinetic energy changes, because work is done by the internal force that each skater exerts on the other. This work causes the kinetic energy to increase, as required by the work–energy theorem (see Section 6.2). However, internal forces cannot change the total linear momentum of a system, since the total linear momentum of an isolated system is conserved in the presence of such forces.

7.3   COLLISIONS IN ONE DIMENSION

As discussed in the last section, linear momentum is conserved when two objects collide, provided they constitute an isolated system. When the objects are atoms or subatomic particles, it is often found, in addition, that the total kinetic energy of the system is conserved. In other words, the total kinetic energy of the particles before the collision equals the total kinetic energy of the particles after the colli-

An inelastic collision at the ball park.

sion. In such a case, whatever kinetic energy is gained by one particle is lost by the other.

In general, however, when two macroscopic objects collide, such as two cars, the total kinetic energy after the collision is less than that before the collision. During a collision, kinetic energy is lost mainly in two ways. First, it can be converted into heat because of friction. Second, kinetic energy is lost whenever an object suffers permanent distortion and does not return to its original shape. In this case, energy is spent in creating the damage, as in an automobile collision. With very hard objects, such as a solid steel ball and a marble floor, the permanent distortion suffered upon collision is much smaller than with softer objects and, consequently, less kinetic energy is spent.

Collisions are often classified according to whether kinetic energy changes during the collision:

1. *Elastic collision*—One in which the total kinetic energy of the system after the collision is equal to the total kinetic energy before the collision.

2. *Inelastic collision*—One in which the total kinetic energy of the system is *not* the same before and after the collision; if the objects stick together after colliding, the collision is said to be completely inelastic.

The boxcars coupling together in Figure 7.3 is an example of a completely inelastic collision. When a collision is completely inelastic, the greatest amount of kinetic energy is lost. Example 5 shows how one particular elastic collision is described using the conservation of linear momentum and the fact that no kinetic energy is lost during the collision.

Example 5   A Collision in One Dimension

As Figure 7.5 illustrates, a ball of mass $m_1 = 0.250$ kg and velocity $v_{01} = +5.00$ m/s collides head-on with a ball of mass $m_2 = 0.800$ kg that is initially at rest. No external forces act on the balls. If the collision is elastic, what are the velocities of the balls after the collision?

REASONING AND SOLUTION   The total linear momentum of the two-ball system is conserved, because no external forces act on the system. Momentum conservation applies whether or not the collision is elastic:

$$\underbrace{m_1 v_{f1} + m_2 v_{f2}}_{\substack{\text{Total momentum} \\ \text{after collision}}} = \underbrace{m_1 v_{01} + 0}_{\substack{\text{Total momentum} \\ \text{before collision}}}$$

For an elastic collision, the total kinetic energy is the same before and after the collision:

$$\underbrace{\tfrac{1}{2}m_1 v_{f1}{}^2 + \tfrac{1}{2}m_2 v_{f2}{}^2}_{\substack{\text{Total kinetic energy} \\ \text{after collision}}} = \underbrace{\tfrac{1}{2}m_1 v_{01}{}^2 + 0}_{\substack{\text{Total kinetic energy} \\ \text{before collision}}}$$

There are now two equations containing the two unknown quantities v_{f1} and v_{f2}. These equations can be solved simultaneously to give

$$v_{f1} = \left(\frac{m_1 - m_2}{m_1 + m_2}\right) v_{01} \tag{7.8a}$$

Before

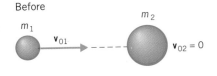

After

Figure 7.5   A 0.250-kg ball, traveling with an initial velocity of $v_{01} = +5.00$ m/s, undergoes an elastic collision with a 0.800-kg ball that is initially at rest.

$$v_{f2} = \left(\frac{2m_1}{m_1 + m_2}\right) v_{01} \qquad (7.8b)$$

With the given values for m_1, m_2, and v_{01}, Equations 7.8 yield the following values for v_{f1} and v_{f2}:

$$\boxed{v_{f1} = -2.62 \text{ m/s}} \quad \text{and} \quad \boxed{v_{f2} = +2.38 \text{ m/s}}$$

The negative value for v_{f1} indicates that m_1 rebounds to the left after the collision in Figure 7.5, while the positive value for v_{f2} indicates that m_2 moves to the right.

We can get a feel for an elastic collision by dropping a steel ball onto a hard surface, such as a marble slab resting on the ground. If the collision is elastic, the ball will rebound to its original height, as Figure 7.6a illustrates. This observation is consistent with the results of Example 5, the colliding objects being the steel ball (mass $= m_1$) and the earth ($m_2 = 6 \times 10^{24}$ kg). With v_{01} being the velocity of the ball just before impact, Equation 7.8a gives the velocity v_{f1} of the ball just after impact. Since m_1 is negligible compared with m_2, this equation shows that

$$v_{f1} = -\left(\frac{m_2}{m_2}\right) v_{01} = -v_{01}$$

The velocity v_{f1} is directed opposite to v_{01} but has the same magnitude. Thus, the kinetic energy of the steel ball is the same before and after the elastic collision, and the ball rebounds to its original height as kinetic energy is converted into gravitational potential energy. In contrast, a partially deflated basketball exhibits little rebound from a relatively soft asphalt surface, as in part b, indicating that a large fraction of the ball's kinetic energy is dissipated during the collision. The completely deflated basketball in part c has no bounce at all, and a maximum amount of kinetic energy is lost during the collision.

The next example illustrates a completely inelastic collision in a device called a "ballistic pendulum." This device can be used to measure the speed of a bullet.

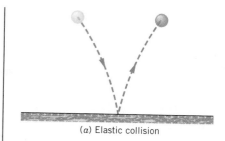

(a) Elastic collision

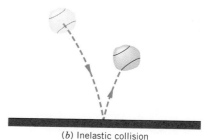

(b) Inelastic collision

(c) Completely inelastic collision

Figure 7.6 (a) A hard steel ball would rebound to its original height after striking a hard marble surface if the collision were elastic. (b) A partially deflated basketball has little bounce on a soft asphalt surface. (c) A deflated basketball has no bounce at all.

Example 6   A Ballistic Pendulum

A ballistic pendulum is sometimes used in laboratories to measure the speed of a projectile, such as a bullet. The ballistic pendulum shown in Figure 7.7a consists of a block of wood (mass $m_2 = 2.50$ kg) suspended by a wire of negligible mass. A bullet (mass $m_1 = 0.0100$ kg) is fired with a speed v_{01} into the block of wood. Just after the bullet collides with the block, the block (with the bullet in it) has a speed v_f and then swings to a maximum height of 0.650 m above the initial position (see part b of the drawing). Find the speed of the bullet.

REASONING   The physics of the ballistic pendulum can be divided into two parts. First, there is the completely inelastic collision between the bullet and the block. Second, there is the resulting motion of the block and bullet as they swing upward. The total momentum of the system (block plus bullet) is conserved during the collision, because the sum of the external forces acting on the system is nearly zero.* Further-

THE PHYSICS OF . . .

measuring the speed of a bullet.

---

* The sum of the external forces acting on the system before the collision is not exactly zero, because the force of gravity acts on the incoming bullet and is not balanced by another force. However, the force of gravity changes the momentum of the bullet by a negligibly small amount during the collision, since the collision occurs so quickly. Therefore, momentum conservation is a very good approximation.

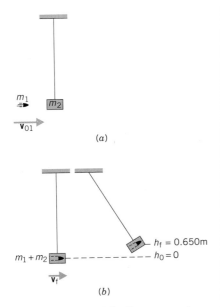

Figure 7.7 (*a*) A bullet approaches a ballistic pendulum. (*b*) The block and bullet swing upward after the collision.

more, as the system swings upward, the principle of conservation of mechanical energy applies, because nonconservative forces do no work. The tension force in the wire does no work because it acts perpendicular to the motion. And air resistance is negligible during the swing.

SOLUTION Applying the momentum conservation principle to the collision, we find that

$$\underbrace{(m_1 + m_2)v_f}_{\substack{\text{Total momentum} \\ \text{after collision}}} = \underbrace{m_1 v_{01}}_{\substack{\text{Total momentum} \\ \text{before collision}}}$$

This equation can be solved for the initial speed v_{01} of the bullet:

$$v_{01} = \frac{m_1 + m_2}{m_1} v_f$$

Before this result can be used to determine v_{01}, a value is needed for the speed v_f immediately after the collision. This value can be obtained from the maximum height to which the system swings, by using the principle of conservation of mechanical energy:

$$\underbrace{(m_1 + m_2)gh_f}_{\substack{\text{Total mechanical energy} \\ \text{at the top of the swing,} \\ \text{all potential}}} = \underbrace{\tfrac{1}{2}(m_1 + m_2)v_f^2}_{\substack{\text{Total mechanical energy} \\ \text{at the bottom of the} \\ \text{swing, all kinetic}}}$$

It follows that

$$v_f = \sqrt{2gh_f} = \sqrt{2(9.80 \text{ m/s}^2)(0.650 \text{ m})} = 3.57 \text{ m/s}$$

With this value for v_f, it is now possible to determine the speed of the bullet:

$$v_{01} = \frac{m_1 + m_2}{m_1} v_f = \frac{0.0100 \text{ kg} + 2.50 \text{ kg}}{0.0100 \text{ kg}} (3.57 \text{ m/s})$$

$$= \boxed{+896 \text{ m/s}}$$

7.4 COLLISIONS IN TWO DIMENSIONS

The "break," a series of two-dimensional collisions, starts a game of pool.

The collisions discussed so far have been "head-on" or one-dimensional collisions, in the sense that the velocities of the objects point along a single line before and after contact is made. Collisions often occur, however, in two or three dimensions. Figure 7.8 shows a two-dimensional case in which two balls collide on a horizontal frictionless table.

For the system consisting of the two balls, the external forces include the weights of the balls and the corresponding normal forces produced by the table. Since each weight is balanced by a normal force, the sum of the external forces is zero, and the total momentum of the system is conserved, as Equation 7.7 indicates ($\mathbf{P}_f = \mathbf{P}_0$). Momentum is a vector quantity, however, and in two dimensions

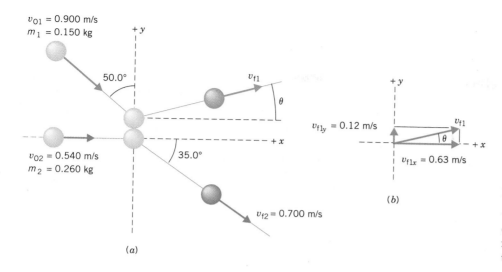

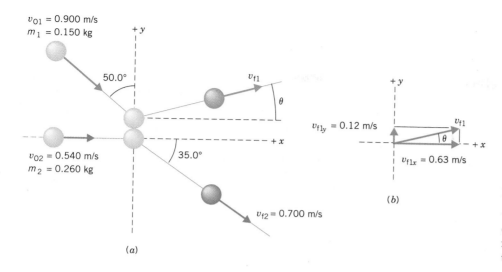

Figure 7.8 (*a*) Top view of two balls colliding on a horizontal frictionless table. (*b*) This part of the drawing shows the x and y components of the velocity of ball 1 after the collision.

the x and y components of the total momentum are conserved separately. In other words, Equation 7.7 is equivalent to the following two equations:

[x Component] $P_{fx} = P_{0x}$ (7.9a)

[y Component] $P_{fy} = P_{0y}$ (7.9b)

Example 7 shows how to deal with a two-dimensional collision when the total linear momentum of the system is conserved.

Example 7   A Collision in Two Dimensions

For the data given in Figure 7.8, use momentum conservation to determine the magnitude and direction of the velocity of ball 1 after the collision.

REASONING AND SOLUTION   To determine the velocity of ball 1 after the collision, we calculate the components of this velocity, which are v_{f1x} and v_{f1y}. Applying momentum conservation (Equation 7.9a) to the x component of the total momentum, we find that

[x Component]

$\underbrace{(0.150 \text{ kg})(v_{f1x})}_{\text{Ball 1, after}} + \underbrace{(0.260 \text{ kg})(0.700 \text{ m/s})(\cos 35.0°)}_{\text{Ball 2, after}}$

$= \underbrace{(0.150 \text{ kg})(0.900 \text{ m/s})(\sin 50.0°)}_{\text{Ball 1, before}} + \underbrace{(0.260 \text{ kg})(0.540 \text{ m/s})}_{\text{Ball 2, before}}$

This equation can be solved to show that $v_{f1x} = +0.63$ m/s. Applying momentum conservation (Equation 7.9b) to the y component of the total momentum, we find that

---

PROBLEM SOLVING INSIGHT

Momentum, being a vector quantity, has a magnitude and a direction. For motion in two dimensions, the most convenient way to take into account the direction of the momentum is to use vector components and to assign a plus or minus sign to each component, as illustrated in this example.

[y Component]

$$\underbrace{(0.150 \text{ kg})(v_{f1y})}_{\text{Ball 1, after}} + \underbrace{(0.260 \text{ kg})[-(0.700 \text{ m/s})(\sin 35.0°)]}_{\text{Ball 2, after}}$$

$$= \underbrace{(0.150 \text{ kg})[-(0.900 \text{ m/s})(\cos 50.0°)]}_{\text{Ball 1, before}} + \underbrace{0}_{\text{Ball 2, before}}$$

The solution to this equation reveals that $v_{f1y} = +0.12$ m/s.

Figure 7.8b shows the x and y components of the final velocity of ball 1. The magnitude of the velocity is

$$v_{f1} = \sqrt{(0.63 \text{ m/s})^2 + (0.12 \text{ m/s})^2} = \boxed{0.64 \text{ m/s}}$$

The direction of the velocity is given by the angle θ:

$$\tan \theta = \frac{0.12 \text{ m/s}}{0.63 \text{ m/s}} = 0.19 \qquad \theta = \tan^{-1} 0.19 = \boxed{11°}$$

THE PHYSICS OF . . .

rocket propulsion.

*7.5 ROCKET PROPULSION

Space exploration and the widespread use of artificial satellites have been made possible by advances in rocket propulsion. Rocket engines are based on the familiar notion of "recoil," the same idea that is involved when the two skaters in Example 4 push off against each other. Recoil is also involved when a bullet is fired from a rifle. The rifle and the bullet push off against each other with the aid of the burning gunpowder, and the gun recoils as the bullet is propelled forward. In all examples of recoil, internal forces act on the two objects that constitute the system, accelerating them in opposite directions.

In a rocket, fuel is burned to create fast moving hot gases that are forced out the rear of the rocket. According to Newton's third law, the hot gases, in turn, apply an oppositely directed force of equal magnitude on the rocket, making it recoil. In this sense, the rocket behaves like a large rifle aimed downward. When the bullet (the hot gases) is fired downward, the rifle (the rocket) recoils upward, as in Figure 7.9a. The internal force applied to the rocket by the ejected gases is called the "thrust," and the impulse–momentum theorem can be applied to identify the factors that determine the thrust.

In Figure 7.9b a rocket moves away from the earth at a velocity $\mathbf{v_0}$. The drawing focuses on the momentum of a small mass Δm of propellant before and after a burn. The gravitational force acting on Δm is negligible. The propellant is changed into hot gases in a time interval Δt. According to the impulse–momentum theorem, the impulse that causes the momentum of this propellant to change is given by

$$\overline{\mathbf{F}} \, \Delta t = (\Delta m)\mathbf{v_f} - (\Delta m)\mathbf{v_0} \tag{7.4}$$

$\overline{\mathbf{F}}$ is the average net force exerted on Δm during the time interval Δt, $\mathbf{v_f}$ is the final velocity of the propellant (in the form of the ejected gases), and $\mathbf{v_0}$ is the initial

"Rocket Man" at the '84 Olympic Games; his propulsion is based on recoil.

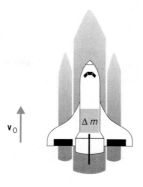

(b) Before a burn

(a)

(c) After a burn

Figure 7.9 (a) A launch of the space shuttle. (b) Before a burn, the mass Δm of propellant has an upward velocity of $\mathbf{v}_0$ relative to the earth. (c) As a result of a burn, hot gases of mass Δm are ejected with a velocity of $\mathbf{v}_f$ relative to the earth.

velocity of the propellant (and rocket) before the burn. Both velocities are measured with respect to the earth.

Since the gravitational force acting on Δm is negligible, the average net force $\overline{\mathbf{F}}$ is just the downward force $\mathbf{T}'$ exerted by the engine on the propellant. Consequently, the impulse–momentum theorem can be rewritten as

$$\mathbf{T}' = \left(\frac{\Delta m}{\Delta t}\right)(\mathbf{v}_f - \mathbf{v}_0)$$

If the thrust that the ejected gases exert *on* the rocket is labeled $\mathbf{T}$, then according to Newton's third law $\mathbf{T} = -\mathbf{T}'$. Therefore, multiplying both sides of the result above by -1, we find that

$$\mathbf{Thrust} = \mathbf{T} = -\left(\frac{\Delta m}{\Delta t}\right)(\mathbf{v}_f - \mathbf{v}_0) \qquad (7.10)$$

One factor that determines the thrust is $\Delta m/\Delta t$, which is the mass of propellant burned per second. A larger burn rate creates a larger thrust. Another factor is $\mathbf{v}_f - \mathbf{v}_0$, which is the velocity of the ejected gases *relative to the moving rocket*. Example 8 illustrates the use of Equation 7.10.

Example 8   Rocket Thrust

A rocket is moving away from the earth at a velocity of $v_0 = +4.0 \times 10^3$ m/s and is burning propellant at a rate of $\Delta m/\Delta t = 550$ kg/s. The hot gases are ejected at a velocity of $v_f = -8.0 \times 10^3$ m/s with respect to the earth. Find the thrust exerted on the rocket by the ejected gases.

REASONING AND SOLUTION   The thrust is given directly by Equation 7.10:

$$T = -\left(\frac{\Delta m}{\Delta t}\right)(v_f - v_0)$$

$$T = -(550 \text{ kg/s})(-8.0 \times 10^3 \text{ m/s} - 4.0 \times 10^3 \text{ m/s})$$

$$= \boxed{+6.6 \times 10^6 \text{ N}}$$

INTEGRATION OF CONCEPTS

CONSERVATION OF LINEAR MOMENTUM AND CONSERVATION OF ENERGY

In physics, conservation principles are among the most important of all principles. They deal with quantities that remain unchanged throughout the motion, the initial and final values of the quantity being the same. We have discussed two of these conserved quantities so far. Here in Chapter 7, we have studied the principle of conservation of linear momentum, and in Chapter 6, we learned about the principle of conservation of energy. While these two conservation principles focus on different aspects of the motion, they deal with them in similar ways. In both principles, it is the *total quantity* (linear momentum or energy) that is conserved. When one part of the total increases, the other part decreases by the same amount, so that the sum of the two parts remains unchanged. Suppose, for example, that a system consists of two objects, each having a linear momentum. Momentum conservation dictates that when the momentum of one object increases, the momentum of the other object decreases by the same amount. The total linear momentum of the two objects, however, remains the same throughout the motion. Similarly, the total mechanical energy of a system consists of two parts, kinetic energy and potential energy. Energy conservation requires that when one part of the total energy increases, the other part decreases by the same amount, the total mechanical energy remaining the same throughout the motion. It is important to note that the two conservation principles may both apply to a given situation. You will need to use both momentum conservation and energy conservation in a number of homework problems at the end of the present chapter (e.g., problems 19 and 23).

SUMMARY

The **impulse** of a force is the product of the average force $\overline{F}$ and the time interval Δt during which the force acts: **Impulse** $= \overline{F}\,\Delta t$. Impulse is a vector quantity.

The **linear momentum p** of an object is the product of the object's mass m and velocity **v**: $p = mv$. Linear momentum is a vector quantity. The total momentum of a system of objects is the vector sum of the momenta of the individual objects.

The **impulse–momentum theorem** states that an impulse produces a change in an object's momentum, according to $\overline{F}\,\Delta t = mv_f - mv_0$, where mv_f is the final momentum and mv_0 is the initial momentum.

The **principle of conservation of linear momentum** states that the total linear momentum of an isolated system remains constant. An isolated system is one for which the sum of the external forces acting on the system is zero.

An **elastic collision** is one in which the total kinetic energy is the same before and after the collision. An **inelastic collision** is one in which the total kinetic energy is not the same before and after the collision. If the objects stick together after colliding, the collision is said to be **completely inelastic.**

The **thrust T** developed by a rocket engine can be determined with the aid of the impulse–momentum theorem: $T = -(\Delta m/\Delta t)(v_f - v_0)$, where $\Delta m/\Delta t$ is the mass of propellant burned per second, and $v_f - v_0$ is the velocity of the ejected gases relative to the moving rocket. The terms v_f and v_0 are, respectively, the velocities of the ejected gases and the rocket relative to the earth.

QUESTIONS

1. Two identical automobiles have the same speed, one traveling east and one traveling west. Do these cars have the same momentum? Explain.

2. If two different objects have the same momentum, do they necessarily have the same kinetic energy? Give a reason for your answer.

3. Can a single object have kinetic energy but no momentum? Can a system of two or more objects have a total kinetic energy that is not zero but a total momentum that is zero? Account for your answers.

4. An airplane is flying horizontally with a constant momentum during a time interval Δt. (a) With the aid of Equation 7.4, decide whether a net impulse is acting on the plane during this time interval. (b) In the horizontal direction, the thrust and air resistance both act on the plane. What does the answer in part (a) imply about the impulse of the resistive force and the impulse of the thrust?

5. An object slides along the surface of the earth and slows down because of kinetic friction. If the object itself is considered as the system, the kinetic frictional force must be identified as an external force that, according to Equation 7.4, decreases the momentum of the system. If *both* the object and the earth are considered to be part of the system, is the force of kinetic friction still an external force? Can the friction force change the total linear momentum of the two-body system? Give your reasoning for both answers.

6. When driving a golf ball, a good "follow-through" helps to increase the distance of the drive. A good follow-through means that the club head is kept in contact with the ball as long as possible. Using the impulse–momentum theorem, explain why this technique allows you to hit the ball farther.

7. In movies, Superman hovers stationary in midair, grabs a villain by the neck, and throws him forward. Superman, however, remains stationary. Using the conservation of linear momentum, explain what is wrong with this scene.

8. A satellite explodes in outer space, far from any other body, sending thousands of pieces in all directions. How does the linear momentum of the satellite before the explosion compare with the total linear momentum of all the pieces after the explosion? Account for your answer.

9. A blank cartridge is one in which the lead bullet is replaced by a thin paper cap. When a gun fires a blank, is the recoil greater than, the same as, or less than when the gun fires a standard bullet? Give your reasoning in terms of the law of conservation of linear momentum.

10. A collision occurs between three moving billiard balls such that no net external force acts on the three-ball system. Is the momentum of *each* ball conserved during the collision? If so, explain why. If not, what quantity is conserved?

11. In an elastic collision, is the kinetic energy of *each* object the same before and after the collision? Explain.

12. The drawing shows a garden sprinkler that whirls around a vertical axis. Using the impulse–momentum theorem, explain how the water causes the whirling motion.

13. On a distant asteroid, a large catapult is used to "throw" large chunks of stone into space. Could such a device be used as a propulsion system to move the asteroid closer to the earth? Explain.

14. Review Example 5. Now, suppose both objects have the same mass, $m_1 = m_2$. Describe what happens to the velocities of both objects as a result of the collision, using Equations 7.8a and 7.8b to justify your answers.

PROBLEMS

Section 7.1 The Impulse–Momentum Theorem

1. What is the impulse of a $+35$-N force that acts on an object for 0.40 s?

2. (a) What is the momentum and the kinetic energy of a car (mass $= 2.00 \times 10^3$ kg) that is moving due north at a speed of 15.0 m/s? If the speed is tripled, by what factor does (b) the momentum increase and (c) the kinetic energy increase?

3. (a) The earth travels with an approximate speed of 29.9 km/s (66 900 mi/h) in its journey around the sun. The mass of the earth is 5.98×10^{24} kg. Find the magnitude of its linear momentum. (b) Is the direction of the earth's linear momentum constant? If not, describe how it changes and specify the force that causes it to change.

4. A freight train moves due north with a speed of 17 m/s. The mass of the train is 6.0×10^6 kg. How fast would a 1500-kg automobile have to be moving due north to have the same momentum as the train?

5. Two arrows are fired horizontally with the same speed of 30.0 m/s. Each arrow has a mass of 0.100 kg. One is fired due east and the other due south. Find the magnitude and direction of the total momentum of this two-arrow system.

6. Two men pushing a stalled car generate a net force of $+840$ N for 5.0 s. What is the final momentum of the car?

7. A woman, driving a golf ball off the tee, gives the ball a speed of 28 m/s. The mass of the ball is 0.045 kg, and the duration of the impact with the golf club is 6.0×10^{-3} s. (a) What is the change in momentum of the ball? (b) Determine the average force applied by the club.

*8. A 1220-kg car is moving due south at a speed of 20.0 m/s. A 1540-kg car is moving due east at a speed of 30.0 m/s. A third car has a mass of 935 kg and is heading 45.0° south of east at a speed of 15.0 m/s. (a) Find the total linear momentum (magnitude and direction) of the three-car system. (b) Find the total kinetic energy of the system.

*9. A 0.060-kg ball is dropped from a height of 4.0 m above the floor. Neglecting air resistance, find the momentum (magnitude and direction) of the ball just before impact.

*10. A stream of water strikes a stationary turbine blade, as the drawing illustrates. The incident water stream has a velocity of $+18.0$ m/s, while the exiting water stream has a velocity of -18.0 m/s. The mass of water per second that strikes the blade is 25.0 kg/s. Find the magnitude of the net force exerted on the water by the blade.

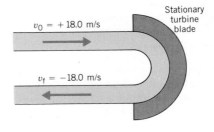

**11. A 1080-kg car moves with a speed of 28.0 m/s and is headed 30.0° north of east. A second car has a mass of 1630 kg and is headed due south. A third car has a mass of 1350 kg and is headed due west. What must be the speeds of the second and third cars, so that the total linear momentum of the three-car system is zero?

**12. A dump truck is being filled with sand. The sand falls straight downward from rest from a height of 2.00 m above the truck bed, and the mass of sand that hits the truck per second is 55.0 kg/s. The truck is parked on the platform of a weight scale. By how much does the scale reading exceed the weight of the truck and sand?

Section 7.2 The Principle of Conservation of Linear Momentum

13. For tests using a *ballistocardiograph,* a patient lies on a horizontal platform that is supported on jets of air. Because of the air jets, the friction impeding the horizontal motion of the platform is negligible. Each time the heart beats, blood is pushed out of the heart in a direction that is nearly parallel to the platform. Since momentum must be conserved, the body and the platform recoil, and this recoil can be detected to provide information about the heart. For each beat, suppose that 0.050 kg of blood is pushed out of the heart with a velocity of $+0.25$ m/s and that the mass of the patient and platform is 85 kg. Assuming that the patient does not slip with respect to the platform, determine the recoil velocity.

14. A 55-kg swimmer is standing on a stationary 210-kg floating raft. The swimmer then runs off the raft horizontally with a velocity of $+4.6$ m/s relative to the shore. Find the recoil velocity that the raft would have if there were no friction and resistance due to the water.

15. An astronaut is motionless in outer space. Upon command, the propulsion unit strapped to his back ejects some gas with a velocity of $+14$ m/s, and the astronaut recoils with a velocity of -0.50 m/s. After the gas is ejected, the mass of the astronaut is 160 kg. What is the mass of the ejected gas?

16. In a science fiction novel two enemies, Bonzo and Ender, are fighting in outer space. From stationary positions they push against each other. Bonzo flies off with a velocity of $+1.5$ m/s, while Ender recoils with a velocity of -2.5 m/s. (a) Without doing any calculations, decide which person has the greater mass. Give your reasoning. (b) Determine the ratio of the masses (m_{Bonzo}/m_{Ender}) of these two people.

17. A two-stage rocket moves in space at a constant velocity of 4900 m/s. The two stages are then separated by a small explosive charge placed between them. Immediately after the explosion the velocity of the 1200-kg upper stage is 5700 m/s in the same direction as before the explosion. What is the velocity of the 2400-kg lower stage after the explosion?

***18.** Show that the kinetic energy KE_2 of the recoiling man in Example 4 is related to the kinetic energy KE_1 of the recoiling woman according to $KE_2 = (m_1/m_2)KE_1$.

***19.** Adolf and Ed are wearing harnesses and are hanging from the ceiling by means of ropes attached to them. They are face to face and push off against one another. Adolf has a mass of 110 kg, and Ed has a mass of 73 kg. Following the push, Adolf swings upward to a height of 0.30 m above his starting point. To what height above his starting point does Ed rise?

****20.** A wagon is coasting at a speed v_A along a straight and level road. When ten percent of the wagon's mass is thrown

off the wagon, parallel to the ground and in the forward direction, the wagon is brought to a halt. If the direction in which this mass is thrown is exactly reversed, everything else remaining the same, the wagon accelerates to a new speed v_B. Calculate v_B/v_A.

Section 7.3 and Section 7.4 Collisions in One and Two Dimensions

21. A 31-kg swimmer runs with a horizontal velocity of $+4.0$ m/s off a boat dock into a stationary 8.0-kg rubber raft. Find the velocity that the swimmer and raft would have after the impact, if there were no friction and resistance due to the water.

22. In a football game, a receiver is standing still, having just caught a pass. Before he can move, a tackler, running at a velocity of $+4.5$ m/s, grabs him. The tackler holds onto the receiver, and the two move off together with a velocity of $+2.6$ m/s. The mass of the tackler is 115 kg. Find the mass of the receiver.

23. A 2.50×10^{-3}-kg bullet, traveling at a velocity of $+425$ m/s, strikes the wooden block of a ballistic pendulum. The block has a mass of 0.200 kg. (a) Find the velocity of the bullet/block combination immediately after the collision. (b) How high does the combination rise above its initial position?

24. A golf ball bounces down a flight of steel stairs, striking each stair once on the way down. The ball starts at the top step with a vertical velocity component of zero. If all the collisions with the stairs are elastic, and if the vertical height of the staircase is 3.00 m, determine the bounce height when the ball reaches the bottom of the stairs. Neglect air resistance.

25. Batman (mass = 91 kg) jumps straight down from a bridge into a boat (mass = 510 kg) in which a criminal is fleeing. The velocity of the boat is initially $+11$ m/s. What is the velocity of the boat after Batman lands in it?

26. A 0.150-kg projectile is fired with a velocity of $+715$ m/s at a 2.00-kg wooden block that rests on a frictionless table. The velocity of the block, immediately after the projectile passes through it, is $+40.0$ m/s. Find the velocity with which the projectile exits from the block.

27. A 5.00-kg ball, moving to the right at a velocity of $+2.00$ m/s on a frictionless table, collides head-on with a stationary 7.50-kg ball. Find the final velocities of the balls if the collision is (a) elastic and (b) completely inelastic.

***28.** A 60.0-kg person, running horizontally with a velocity of $+3.80$ m/s, jumps onto a 12.0-kg sled that is initially at rest. (a) Ignoring the effects of friction, find the velocity of the sled and person as they move away. (b) The sled and person coast 30.0 m on level snow before coming to rest. What is the coefficient of kinetic friction between the sled and the snow?

*29. By accident, a large plate is dropped and breaks into three pieces. The pieces fly apart parallel to the floor. As the plate falls, its momentum has only a vertical component, and no component parallel to the floor. After the collision, the component of the total momentum parallel to the floor must remain zero, since the external force acting on the plate has no component parallel to the floor. Using the data shown in the drawing, find the masses of pieces 1 and 2.

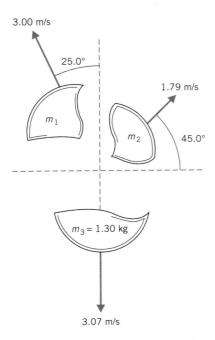

*30. A mine car, whose mass is 440 kg, rolls at a speed of 0.50 m/s on a horizontal track, as the drawing shows. A 150-kg chunk of coal has a speed of 0.80 m/s when it leaves the chute. Determine the velocity of the car/coal system after the coal has come to rest in the car.

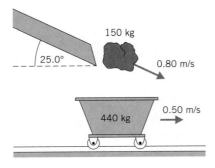

*31. A 50.0-kg skater is traveling due east at a speed of 3.00 m/s. A 70.0-kg skater is moving due south at a speed of 7.00 m/s. They collide and hold on to each other after the collision, managing to move off at an angle θ south of east, with a speed of v_f. Find (a) the angle θ and (b) the speed v_f, assuming that friction can be ignored.

**32. Two identical balls are traveling toward each other with speeds of 4.0 and 7.0 m/s, and they experience an elastic head-on collision. Obtain the velocities (magnitude and direction) of each ball after the collision.

**33. A ball is dropped from rest at the top of a 6.10-m-tall building, falls straight downward, collides inelastically with the ground, and bounces back. The ball loses 10.0% of its kinetic energy every time it collides with the ground. How many bounces can the ball make and still reach a window sill that is 2.44 m above the ground?

Section 7.5 Rocket Propulsion

34. A rocket burns propellant at a rate of $\Delta m/\Delta t = 1.5$ kg/s, ejecting gases with a speed of 7800 m/s relative to the rocket. Find the magnitude of the thrust.

35. At what rate $\Delta m/\Delta t$ must a rocket burn propellant to achieve a thrust whose magnitude is 14 000 N, if the speed of the ejected gases is 6400 m/s relative to the rocket?

36. During a launch, a rocket (mass $= 2.72 \times 10^4$ kg) lifts off vertically from rest. The engine burns propellant at a rate of $\Delta m/\Delta t = 584$ kg/s and ejects gases with a speed of 1680 m/s relative to the rocket. (a) Determine the magnitude of the thrust developed by the engine. (b) Find the initial upward acceleration of the rocket. (c) Noting that the mass of the rocket is smaller because fuel has been consumed, determine the acceleration of the rocket at the end of 10.0 s.

ADDITIONAL PROBLEMS

37. In the James Bond movie *Diamonds Are Forever*, the lead female character fires a machine gun while standing at the edge of an off-shore oil rig. As she fires the gun, she is driven back over the edge and into the sea. Suppose the mass of a bullet is 0.010 kg, and its velocity is $+720$ m/s. If her mass (including the gun) is 51 kg, what recoil velocity does she acquire in response to a single shot from a stationary position?

38. A volleyball is spiked so that its incoming velocity of $+4.0$ m/s is changed to an outgoing velocity of -21 m/s. The mass of the volleyball is 0.35 kg. What impulse does the player apply to the ball?

39. A 1550-kg car, traveling with a velocity of $+12.0$ m/s, plows into a 1220-kg stationary car. During the collision, the two cars lock bumpers and then move together as a unit. (a) What is their common velocity just after the impact? (b) What fraction of the initial kinetic energy remains after the collision?

40. In problem 39, suppose that both cars have special bumpers, so the collision is elastic. (a) What is the total linear momentum of the two-car system before and after the collision? (b) What is the total kinetic energy of the system before and after the collision? (c) What is the velocity of each car after the collision?

41. A 46-kg skater is standing still in front of a wall. By pushing against the wall she propels herself backward with a velocity of -1.2 m/s. Her hands are in contact with the wall for 0.80 s. Ignore friction and wind resistance. Find the average force she exerts on the wall (which has the same magnitude, but opposite direction, as the force that the wall applies to her).

42. Two joggers are running with the same velocity of 4.00 m/s due south. Their masses are 90.0 and 55.0 kg. (a) Find the magnitude of the total linear momentum and the total kinetic energy of the two-jogger system. (b) Repeat part (a), assuming that one of the joggers is running due north at 4.00 m/s.

43. With the engines off, a spaceship is coasting at a velocity of $+230$ m/s through outer space. It fires a rocket straight ahead at an enemy vessel. The mass of the rocket is 1300 kg, and the mass of the spaceship (not including the rocket) is 4.0×10^6 kg. The firing of the rocket brings the spaceship to a halt. What is the velocity of the rocket?

44. A 1200-kg car has an initial velocity of 13 m/s, eastward. The driver applies the brakes and brings the car to rest in 6.0 s. Determine the average net force (magnitude and direction) exerted on the car.

*45. A person stands in a stationary canoe and throws a 5.00-kg stone with a velocity of 8.00 m/s at an angle of 30.0° above the horizontal. The person and canoe have a combined mass of 105 kg. Ignoring air resistance and effects of the water, find the horizontal recoil velocity of the canoe.

*46. A 3.00-kg block of wood rests on the muzzle opening of a vertically oriented rifle, the stock of the rifle being firmly planted on the ground. When the rifle is fired, an 8.00-g bullet (velocity = 8.00×10^2 m/s, straight upward) is completely embedded in the block. (a) Using the conservation of linear momentum, find the velocity of the block/bullet system immediately after the collision. (b) How high does the block/bullet system rise above the muzzle opening of the rifle?

*47. A machine gun is mounted on a cart that is free to roll on a flat, frictionless surface. The gun fires parallel to the surface at a rate of ten bullets per second, each bullet having a mass of 0.0097 kg and a speed of 790 m/s relative to the cart. Determine the magnitude of the average thrust exerted on the machine gun by the bullets.

*48. A Fourth-of-July rocket is moving at a speed of 50.0 m/s. The rocket suddenly breaks into two pieces of equal mass, and they fly off with velocities $\mathbf{v}_1$ and $\mathbf{v}_2$, as shown in the drawing. What are the magnitudes of $\mathbf{v}_1$ and $\mathbf{v}_2$?

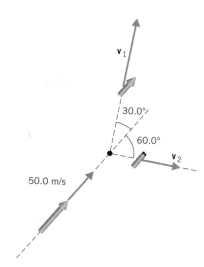

**49. Starting with an initial speed of 5.00 m/s at a height of 0.300 m, a 1.50-kg ball swings downward and strikes a 4.60-kg ball that is at rest, as the drawing shows. (a) Using the principle of conservation of mechanical energy, find the speed of the 1.50-kg ball just before impact. (b) Assuming that the collision is elastic, find the velocities (magnitude and direction) of both balls just after the collision. (c) How high does each ball swing after the collision, ignoring air resistance?

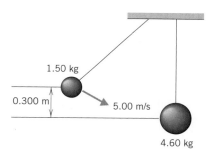

**50. A .22-caliber bullet is fired from a rifle that has a 0.61-m barrel. The bullet has a mass of 0.0026 kg and it exits the barrel with a velocity of 410 m/s, eastward. (a) Find the impulse of the force that acts on the bullet. (b) Assuming the acceleration of the bullet is constant, use the appropriate equation of kinematics from Chapter 2 and find the time that the bullet spends in the barrel. (c) What is the average net force (magnitude and direction) that acts on the bullet?

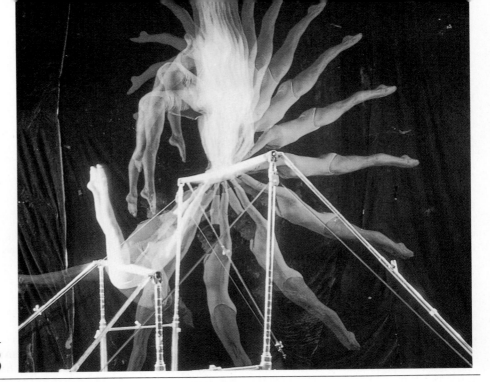

CHAPTER 8

ROTATIONAL KINEMATICS

It is easy to find examples of rotational motion. For example, the time-lapse photo shown above captures a gymnast rotating on the uneven parallel bars. Or the motion could be that of a rotating compact disc recording, the tires on a moving automobile, or the propeller on an airplane in flight. Our analysis of such motions begins with a discussion of kinematics, in which the motion of an object is described without reference to the forces that act on the object. We follow the same sequence as we did in Chapter 2, where linear motion was discussed. There, we introduced the concepts of displacement, velocity, and acceleration and then developed the equations of kinematics for constant acceleration. In the present chapter, we begin with the concepts of angular displacement, angular velocity, and angular acceleration and then turn to the equations of rotational kinematics for constant angular acceleration.

8.1 ROTATIONAL MOTION AND ANGULAR DISPLACEMENT

When a rigid object rotates, points on the object move on circular paths. In Figure 8.1, for example, we see the circular paths traversed by points *A*, *B*, and *C* on a spinning skater. The centers of all such circular paths define a line, called the *axis of rotation.*

The angle through which a rigid object rotates about a fixed axis is called the *angular displacement.* Figure 8.2 shows how the angular displacement is measured for the rotating take-up reel on a tape deck. Recognizing that the axis of rotation is through the spindle on which the reel is mounted, we draw a radial line on the reel. A radial line is one that intersects the axis of rotation perpendicularly. As the reel turns, we observe the angle through which this line marked r moves relative to a convenient reference line. The radial line moves from its initial orientation at angle θ_0 to a final orientation at angle θ (Greek letter theta). In the process, the line sweeps out the angle $\Delta\theta = \theta - \theta_0$. The angle $\Delta\theta$ is the angular displacement. A rotating object may rotate either counterclockwise or clockwise, and standard convention distinguishes between these alternatives by calling the angular displacement positive when it is counterclockwise and negative when it is clockwise.

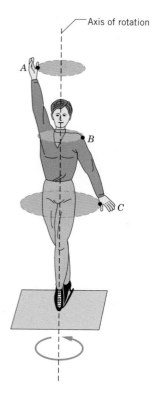

Definition of Angular Displacement

When a rigid body rotates about a fixed axis, the angular displacement is the angle $\Delta\theta$ swept out by a line passing through any point on the body and intersecting the axis of rotation perpendicularly. By convention, the angular displacement is positive if it is counterclockwise and negative if it is clockwise.

SI Unit of Angular Displacement: radian (rad)*

Angular displacement is often expressed in one of three units. The first is the familiar *degree,* and it is well known that there are 360 degrees in a circle. The second unit is the *revolution (rev),* one revolution representing one complete turn of 360°. The most useful unit from a scientific viewpoint is the SI unit called the *radian (rad).* Figure 8.3 shows how the radian is defined, again using the take-up reel of a tape deck as an example. The picture focuses attention on a point P on the reel. This point starts out on the horizontal axis, so that $\theta_0 = 0$, and the angular displacement is $\Delta\theta = \theta - \theta_0 = \theta$. As the reel rotates, the point traces out an arc of

Figure 8.1 When an object rotates, points on the object, such as A, B, or C, move on circular paths. The centers of the circles form a line that is the axis of rotation.

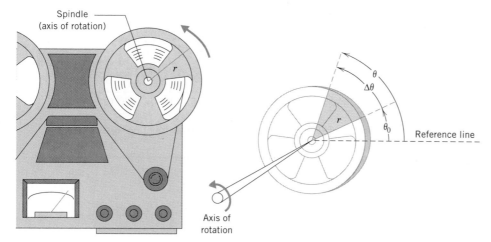

Figure 8.2 For the take-up reel of a tape deck, the angular displacement is the angle $\Delta\theta$ swept out by the radial line r as the reel turns about its axis of rotation.

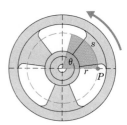

Figure 8.3 In radian measure, the angle θ is defined to be the arc length s divided by the radius r.

* The radian is neither a base nor a derived SI unit. It is regarded as a supplementary SI unit.

length s, which is measured along a circle of radius r. Equation 8.1 defines the angle θ in radians:

$$\theta \text{ (in radians)} = \frac{\text{Arc length}}{\text{Radius}} = \frac{s}{r} \qquad (8.1)$$

According to this definition, an angle in radians is the ratio of two lengths, for example, meters/meters. In calculations, therefore, the radian is treated as a unitless number and has no effect on other units that it multiplies or divides.

To convert between degrees and radians, it is only necessary to remember that the arc length of an entire circle of radius r is the circumference $2\pi r$. Therefore, according to Equation 8.1, *the number of radians that corresponds to 360° or one revolution is*

$$360° = \frac{2\pi r}{r} = 2\pi \text{ rad}$$

Hence, the number of degrees in one radian is

$$1 \text{ rad} = \frac{360°}{2\pi} = 57.3°$$

It is useful to express an angle θ in radians, because then the arc length s subtended at any radius r can be calculated by multiplying θ by r. Example 1 illustrates this point and shows how to convert between degrees and radians.

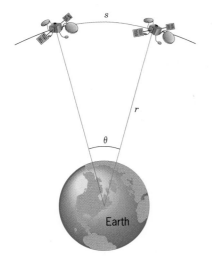

Figure 8.4 Two adjacent synchronous satellites have an angular separation of $\theta = 2.00°$. The distances and angles have been exaggerated for clarity.

Example 1   Adjacent Synchronous Satellites

Synchronous or "stationary" communications satellites are put into an orbit whose radius is $r = 4.23 \times 10^7$ m, as Figure 8.4 illustrates. The orbit is in the plane of the equator, and two adjacent satellites have an angular separation of $\theta = 2.00°$. Find the arc length s (see the drawing) that separates the satellites.

<u>REASONING AND SOLUTION</u>   Once the angle θ is converted into radians, the relation $\theta = s/r$ can be used to find the arc length between the satellites:

$$2.00° = (2.00 \text{ degrees})\left(\frac{2\pi \text{ radians}}{360 \text{ degrees}}\right) = 0.0349 \text{ radians}$$

From Equation 8.1, it follows that

$$s = r\theta = (4.23 \times 10^7 \text{ m})(0.0349 \text{ rad}) = \boxed{1.48 \times 10^6 \text{ m (920 miles)}}$$

The radian unit, being a unitless quantity, is dropped from the final result, leaving the answer expressed in meters.

8.2 ANGULAR VELOCITY AND ANGULAR ACCELERATION

ANGULAR VELOCITY

According to Equation 2.2 ($\overline{\mathbf{v}} = \Delta\mathbf{s}/\Delta t$), the average linear velocity is the linear displacement of the object divided by the time required for the displacement to occur. For rotational motion about a fixed axis, the *average angular velocity* $\overline{\omega}$ (Greek letter omega) is obtained in an analogous way, as the angular displacement divided by the elapsed time during which the displacement occurs.

Definition of Average Angular Velocity

$$\frac{\text{Average}}{\text{angular}} = \frac{\text{Angular displacement}}{\text{Elapsed time}}$$

$$\overline{\omega} = \frac{\theta - \theta_0}{t - t_0} = \frac{\Delta\theta}{\Delta t} \qquad (8.2)$$

SI Unit of Angular Velocity: radian per second (rad/s)

The SI unit for angular velocity is the radian per second (rad/s), although other units such as revolutions per minute (rev/min) are encountered. In agreement with the sign convention adopted for angular displacement, angular velocity is positive when the rotation is counterclockwise and negative when the rotation is clockwise. Example 2 shows how the concept of average angular velocity is applied to a gymnast.

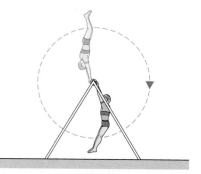

The angular velocity of a Frisbee is one reason why some people have trouble catching it.

Example 2 Gymnast on a High Bar

A gymnast on a high bar swings through two revolutions in a time of 1.90 s, as Figure 8.5 suggests. Find the average angular velocity (in rad/s) of the gymnast.

REASONING AND SOLUTION   The average angular velocity of the gymnast is the angular displacement divided by the elapsed time. The angular displacement (in radians) is

$$\Delta\theta = -2.00 \text{ revolutions}$$

$$\Delta\theta = -2.00 \text{ revolutions}\left(\frac{2\pi \text{ radians}}{1 \text{ revolution}}\right) = -12.6 \text{ radians}$$

where the minus sign denotes that the gymnast rotates clockwise. The average angular velocity is

$$\overline{\omega} = \frac{\Delta\theta}{\Delta t} = \frac{-12.6 \text{ rad}}{1.90 \text{ s}} = \boxed{-6.63 \text{ rad/s}} \qquad (8.2)$$

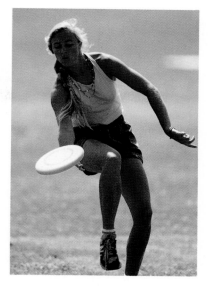

Figure 8.5   Swinging on a high bar.

The *instantaneous angular velocity* ω is the angular velocity that exists at any given instant. To measure it, we follow the same procedure used in Chapter 2 for the instantaneous linear velocity. In this procedure, a small angular displacement $\Delta\theta$ occurs during a small time interval Δt. The time interval is so small that it approaches zero ($\Delta t \rightarrow 0$), and in this limit, the measured average angular velocity, $\overline{\omega} = \Delta\theta/\Delta t$, becomes the instantaneous velocity ω:

$$\omega = \lim_{\Delta t \to 0} \overline{\omega} = \lim_{\Delta t \to 0} \frac{\Delta\theta}{\Delta t} \qquad (8.3)$$

The magnitude of the instantaneous angular velocity, without reference to whether it is a positive or negative quantity, is called the *instantaneous angular speed*. If a rotating object has a constant angular velocity, the instantaneous value and the average value are the same.

ANGULAR ACCELERATION

In linear motion, a changing velocity means that an acceleration is occurring. Such is also the case in rotational motion; a changing angular velocity means that an *angular acceleration* α (Greek letter alpha) is occurring. There are many examples of angular acceleration. For instance, as a compact disc recording is played, the disc turns with an angular velocity that is continually decreasing. And when the push buttons of an electric blender are changed from a lower setting to a higher setting, the angular velocity of the blades increases.

When the linear velocity of an object changes, Equation 2.4 ($\bar{\mathbf{a}} = \Delta \mathbf{v} / \Delta t$) defines the average linear acceleration as the change in velocity per unit time. When the angular velocity changes from an initial value of ω_0 at time t_0 to a final value of ω at time t, the average angular acceleration is defined similarly:

Definition of Average Angular Acceleration

$$\frac{\text{Average angular}}{\text{acceleration}} = \frac{\text{Change in angular velocity}}{\text{Elapsed time}}$$

$$\bar{\alpha} = \frac{\omega - \omega_0}{t - t_0} = \frac{\Delta \omega}{\Delta t} \qquad (8.4)$$

SI Unit of Average Angular Acceleration: radian per second squared (rad/s^2)

The SI unit for average angular acceleration is the unit for angular velocity divided by the unit for time, or $(\text{rad/s})/\text{s} = \text{rad/s}^2$. An angular acceleration of 5 rad/s^2, for example, means that the angular velocity of the rotating object changes by 5 radians per second during each second of acceleration.

The *instantaneous angular acceleration* is the angular acceleration at a given instant. Previously, in discussing linear motion, a condition of constant acceleration was assumed, so that the average and instantaneous accelerations were identical ($\bar{\mathbf{a}} = \mathbf{a}$). Similarly, we will assume here that the angular acceleration is constant. Consequently, the instantaneous angular acceleration α and the average angular acceleration $\bar{\alpha}$ are the same ($\bar{\alpha} = \alpha$). The next example illustrates the concept of angular acceleration.

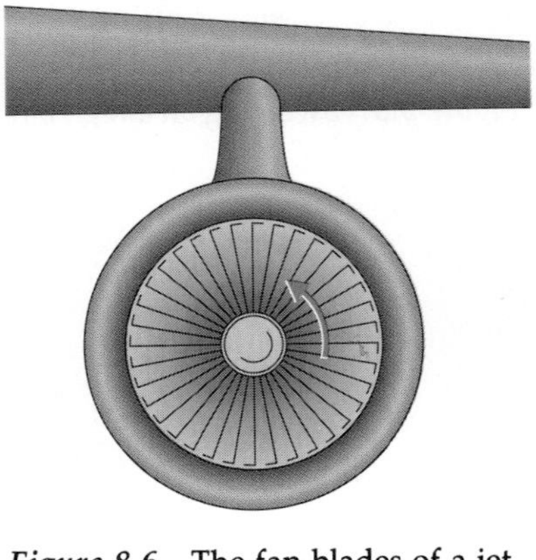

Figure 8.6 The fan blades of a jet engine are accelerating in a counter-clockwise direction.

Example 3   A Jet Revving Its Engines

A jet awaiting clearance for takeoff is momentarily stopped on the runway. As the engines idle, the fan blades are rotating with an angular velocity of $+110 \text{ rad/s}$, where the positive sign indicates a counterclockwise rotation (see Figure 8.6). As the plane takes off, the angular velocity of the blades reaches $+330 \text{ rad/s}$ in a time of 14 s. Find the angular acceleration, assuming it to be constant.

REASONING AND SOLUTION   The angular acceleration is constant, so it is equal to the average acceleration. Applying Equation 8.4, we find

$$\bar{\alpha} = \frac{\omega - \omega_0}{t - t_0} = \frac{(330 \text{ rad/s}) - (110 \text{ rad/s})}{14 \text{ s}} = \boxed{+16 \text{ rad/s}^2}$$

Thus, the angular velocity of the fan blades increases by 16 rad/s during each second that the blades are accelerating.

8.3 THE EQUATIONS OF ROTATIONAL KINEMATICS

A complete description of rotational motion requires values for the angular displacement $\Delta\theta$, the angular acceleration α, the final angular velocity ω, the initial angular velocity ω_0, and the elapsed time Δt. In Example 3, for instance, only the angular displacement of the fan blades during the 14-s interval is missing. Such missing information can be calculated, however. For convenience in the calculations, we assume that the orientation of the rotating object is given by $\theta_0 = 0$ at time $t_0 = 0$. Then, the angular displacement becomes $\Delta\theta = \theta - \theta_0 = \theta$, and the time interval becomes $\Delta t = t - t_0 = t$.

In Example 3, the angular velocity of the fan blades increases at a constant rate from an initial value of $\omega_0 = +110$ rad/s to a final value of $\omega = +330$ rad/s. Therefore, the average angular velocity is midway between the initial and final values:

$$\bar{\omega} = \tfrac{1}{2}[(110 \text{ rad/s}) + (330 \text{ rad/s})] = +220 \text{ rad/s}$$

In other words, when the angular acceleration is constant, the average angular velocity is given by

$$\bar{\omega} = \tfrac{1}{2}(\omega_0 + \omega) \qquad (8.5)$$

With a value for the average angular velocity, Equation 8.2 can be used to obtain the angular displacement of the fan blades:

$$\theta = \bar{\omega}t = (220 \text{ rad/s})(14 \text{ s}) = +3100 \text{ rad}$$

In general, when the angular acceleration is constant, the angular displacement can be obtained from

$$\theta = \bar{\omega}t = \tfrac{1}{2}(\omega_0 + \omega)t \qquad (8.6)$$

This equation and Equation 8.4 provide a complete description of rotational motion under the condition of constant angular acceleration. Equation 8.4 (with $t_0 = 0$) and Equation 8.6 are compared below with the analogous results in linear kinematics:

If the angular acceleration is constant, the motion of a roulette wheel can be described by using the concepts of kinematics.

Rotational motion (α = constant)		Linear motion (a = constant)	
$\omega = \omega_0 + \alpha t$	(8.4)	$v = v_0 + at$	(2.4)
$\theta = \tfrac{1}{2}(\omega_0 + \omega)t$	(8.6)	$s = \tfrac{1}{2}(v_0 + v)t$	(2.7)
$\theta = \omega_0 t + \tfrac{1}{2}\alpha t^2$	(8.7)	$s = v_0 t + \tfrac{1}{2}at^2$	(2.8)
$\omega^2 = \omega_0^2 + 2\alpha\theta$	(8.8)	$v^2 = v_0^2 + 2as$	(2.9)

The purpose of this comparison is to emphasize that the mathematical forms of Equations 8.4 and 2.4 are identical, as are the forms of Equations 8.6 and 2.7. Of

Table 8.1 Symbols Used in Rotational and Linear Kinematics

Quantity	Rotational Motion	Linear Motion
Displacement	θ	s
Initial velocity	ω_0	v_0
Final velocity	ω	v
Acceleration	α	a
Time	t	t

course, the symbols used for the rotational variables are different than those used for the linear variables, as Table 8.1 indicates.

In Chapter 2, Equations 2.4 and 2.7 are used to derive the remaining two equations of kinematics (Equations 2.8 and 2.9). These additional equations convey no new information but are convenient to have when solving problems. Similar derivations can be carried out here. The results are listed above as Equations 8.7 and 8.8 and can be inferred directly from their counterparts in linear motion by making the substitution of symbols indicated in Table 8.1. Equations 8.4, 8.6, 8.7, and 8.8 are called the *equations of rotational kinematics for constant angular acceleration.* The following example illustrates that they are used in the same fashion as the equations of linear kinematics.

Figure 8.7 The angular velocity of the blades in an electric blender changes each time a different push button is chosen.

Example 4 Blending with a Blender

The blades of an electric blender are whirling with an angular velocity of $+375$ rad/s while the "puree" button is pushed in, as Figure 8.7 shows. When the "blend" button is pressed, the blades accelerate and reach a greater angular velocity in $+44.0$ rad (seven revolutions). The angular acceleration has a constant value of $+1740$ rad/s². Find the final angular velocity of the blades.

REASONING   The three known variables are listed in the table below, along with a question mark indicating that a value for the final angular velocity ω is being sought.

θ	α	ω	ω_0	t
$+44.0$ rad	$+1740$ rad/s²	?	$+375$ rad/s	

We can use Equation 8.8, because it relates the angular variables θ, α, ω, and ω_0.

SOLUTION   From Equation 8.8 we find that

$$\omega^2 = \omega_0^2 + 2\alpha\theta$$
$$= (375 \text{ rad/s})^2 + 2(1740 \text{ rad/s}^2)(44.0 \text{ rad}) = 2.94 \times 10^5 \text{ rad}^2/\text{s}^2$$

$$\boxed{\omega = +542 \text{ rad/s}}$$

The negative root is disregarded, since the blades do not reverse their direction of rotation.

The equations of rotational kinematics can be used with any self-consistent set of units for θ, α, ω, ω_0, and t. Radians are used in Example 4 only because data are given in terms of radians. Had the data for θ, α, and ω_0 been provided in rev, rev/s^2, and rev/s, respectively, then Equation 8.8 could have been used to determine the answer for ω directly in rev/s.

8.4 ANGULAR VARIABLES AND TANGENTIAL VARIABLES

In the familiar ice-skating stunt known as "crack-the-whip," a number of skaters attempt to maintain a straight line as they skate around the one person (the pivot) who remains in place. Figure 8.8 shows each skater moving on a circular arc and includes the corresponding velocity vector at the instant portrayed in the picture. For every individual skater, the vector is drawn tangent to the appropriate circle and, therefore, is called the *tangential velocity* $\mathbf{v}_T$. The magnitude of the tangential velocity is referred to as the *tangential speed.*

Of all the skaters involved in the stunt, the one farthest from the pivot has the hardest job. Why? Because, in keeping the line straight, this skater covers more distance than anyone else. To accomplish this, he must skate faster than anyone else or, in other words, must have the largest tangential speed. In fact, the line remains straight only if each person skates with the correct tangential speed. Those skaters closer to the pivot must move with smaller tangential speeds than those farther out, as indicated by the magnitudes of the vectors drawn in Figure 8.8.

With the aid of Figure 8.9, it is possible to show that the tangential speed of any skater is directly proportional to his distance r from the pivot, assuming a given angular speed for the rotating line. When the line rotates as a rigid unit for a time t, it sweeps out the angle θ. The distance s through which a skater moves along a circular arc can be calculated from the relation $s = r\theta$, provided θ is measured in radians. Dividing both sides of this equation by t gives $s/t = r(\theta/t)$. The term s/t is the tangential speed v_T (e.g., in meters/second) of the skater, while the term θ/t is the angular speed ω (in radians/second) at which the line rotates:

$$v_T = r\omega \qquad (\omega \text{ in rad/s}) \qquad (8.9)$$

Thus, for a given angular speed ω, the tangential speed v_T is directly proportional to the radius r. In this expression, the terms v_T and ω refer to the magnitudes of the tangential and angular velocities, respectively, and are numbers without algebraic signs.

It is important to emphasize that the angular speed ω in $v_T = r\omega$ must be expressed in radian measure (e.g., in rad/s); no other units, such as revolutions per second, are acceptable. This restriction arises because the equation was derived by using the definition of radian measure, $s = r\theta$.

The real challenge for the "crack-the-whip" skaters is to keep the line straight, while making it pick up angular speed, that is, while giving it an angular acceleration. To make the angular speed of the line increase, each skater must increase his tangential speed, since the two quantities are related according to $v_T = r\omega$. Of course, the fact that a skater must skate faster and faster means that he must accelerate in the tangential direction, and his tangential acceleration a_T can be related to the angular acceleration α of the straight line. If time is measured relative to $t_0 = 0$, the definition of linear acceleration is given by Equation 2.4 as

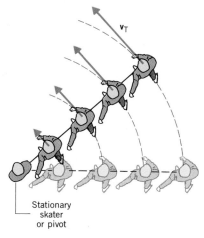

Figure 8.8 When doing a stunt known as "crack-the-whip," each skater along the radial line moves on a circular arc. The tangential velocity $\mathbf{v}_T$ of each skater is represented by an arrow that is tangent to each arc.

THE PHYSICS OF . . .

"crack-the-whip."

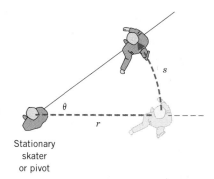

Figure 8.9 During a time t, the line of skaters sweeps through an angle θ. An individual skater, located at a distance r from the stationary skater, moves through a distance s on a circular arc.

$a_T = (v_T - v_{T0})/t$, where v_T and v_{T0} are the final and initial tangential speeds, respectively. Substituting $v_T = r\omega$ for the tangential speed shows that

$$a_T = \frac{v_T - v_{T0}}{t} = \frac{(r\omega) - (r\omega_0)}{t} = r\left(\frac{\omega - \omega_0}{t}\right)$$

Since $\alpha = (\omega - \omega_0)/t$ according to Equation 8.4, it follows that

$$a_T = r\alpha \qquad (\alpha \text{ in rad/s}^2) \qquad (8.10)$$

This result shows that for a given value of α the tangential acceleration a_T is proportional to the radius r, so the skater farthest from the pivot must generate the largest tangential acceleration. In this expression, the terms a_T, r, and α refer to the magnitudes of the numbers involved, without reference to any algebraic sign. Moreover, as is the case for ω in $v_T = r\omega$, only radian measure can be used for α in $a_T = r\alpha$.

There is an advantage to using the angular velocity ω and the angular acceleration α to describe the rotational motion of a rigid object. These angular quantities describe the motion of the *entire object*. In contrast, the tangential quantities v_T and a_T describe only the motion of a single point on the object, and Equations 8.9 and 8.10 indicate that different points located at different distances r have different tangential velocities and accelerations. Example 5 stresses this advantage.

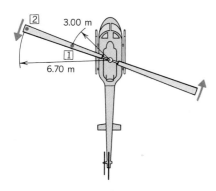

Figure 8.10 Points 1 and 2 on the rotating blade of the helicopter have the same angular speed and acceleration, but they have *different* tangential speeds and accelerations.

Example 5   A Whirlybird

A helicopter blade starts from rest and, with a constant angular acceleration, reaches an angular speed of 6.50 rev/s in 5.00 s. For points 1 and 2 on the blade in Figure 8.10, find the magnitudes of (a) the tangential speeds and (b) the tangential accelerations.

REASONING AND SOLUTION

(a) Using the radii shown in the drawing, we can compute the tangential speeds of each point with the aid of $v_T = r\omega$. However, since this equation can only be used with radian measure, the given angular speed must first be converted to rad/s from rev/s:

$$\omega = \left(6.50\ \frac{\text{rev}}{\text{s}}\right)\left(\frac{2\pi\ \text{rad}}{1\ \text{rev}}\right) = 40.8\ \frac{\text{rad}}{\text{s}}$$

The tangential speed of each point is

[Point 1] $v_T = r\omega = (3.00\ \text{m})(40.8\ \text{rad/s}) = \boxed{122\ \text{m/s (273 mph)}}$ (8.9)

[Point 2] $v_T = r\omega = (6.70\ \text{m})(40.8\ \text{rad/s}) = \boxed{273\ \text{m/s (611 mph)}}$ (8.9)

The rad unit, being dimensionless, does not appear in the final answers.

(b) The tangential accelerations of points 1 and 2 can be determined using $a_T = r\alpha$. First, however, it is necessary to determine the angular acceleration α, which is the same for both points since the blade is rigid. Since the blade starts from rest and attains an angular velocity of $+40.8$ rad/s in 5.00 s, α can be obtained directly from the definition of angular acceleration:

$$\alpha = \frac{\omega - \omega_0}{t} = \frac{40.8\ \text{rad/s} - 0}{5.00\ \text{s}} = +8.16\ \text{rad/s}^2 \qquad (8.4)$$

The tangential accelerations can now be determined:

[Point 1]	$a_T = r\alpha = (3.00 \text{ m})(8.16 \text{ rad/s}^2) = \boxed{24.5 \text{ m/s}^2}$	(8.10)
[Point 2]	$a_T = r\alpha = (6.70 \text{ m})(8.16 \text{ rad/s}^2) = \boxed{54.7 \text{ m/s}^2}$	(8.10)

8.5 CENTRIPETAL ACCELERATION AND TANGENTIAL ACCELERATION

The tangential acceleration discussed in the last section should not be confused with the centripetal acceleration discussed in Chapter 5. That chapter deals with *uniform circular motion,* in which a particle moves at a constant tangential speed on a circular path. The tangential speed v_T is the magnitude of the tangential velocity vector. Even though the magnitude of the tangential velocity is constant, an acceleration is present, since the direction of the velocity changes continually. Because the resulting acceleration points toward the center of the circle, it is called the centripetal acceleration. Figure 8.11*a* shows the centripetal acceleration $\mathbf{a}_c$ for a model airplane flying on a guide wire, the magnitude of $\mathbf{a}_c$ being

$$a_c = \frac{v_T^2}{r} \qquad (5.2)$$

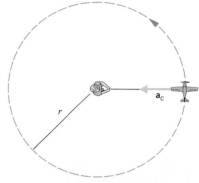

(a) Uniform circular motion

The subscript "T" has been included in this equation as a reminder that it is the tangential speed that appears in the numerator.

The centripetal acceleration can be expressed in terms of the angular speed ω by using the substitution $v_T = r\omega$:

$$a_c = \frac{v_T^2}{r} = \frac{(r\omega)^2}{r} = r\omega^2 \qquad (\omega \text{ in rad/s}) \qquad (8.11)$$

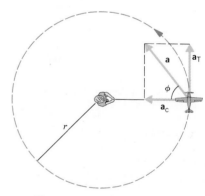

(b) Nonuniform circular motion

Only radian measure, such as rad/s, can be used for ω in this expression, since the derivation depends on the relation $v_T = r\omega$, which presumes radian measure.

While considering uniform circular motion in Chapter 5, we ignored the details of how the motion is established in the first place. In Figure 8.11*a*, for instance, the engine of the plane produced a thrust in the tangential direction, and this force led to a tangential acceleration. In response, the tangential speed of the plane increased from moment to moment, until the situation shown in the drawing resulted. While the tangential speed is changing, the motion is called *nonuniform circular motion.*

Figure 8.11*b* illustrates an important feature of nonuniform circular motion. Since both the direction and the magnitude of the tangential velocity are changing, the airplane experiences two acceleration components simultaneously. The changing direction means that there is a centripetal acceleration $\mathbf{a}_c$. The magnitude of $\mathbf{a}_c$ at any moment can be calculated using the value of the instantaneous angular speed and the radius: $a_c = r\omega^2$. The fact that the magnitude of the tangential velocity is changing means that there is also a tangential acceleration $\mathbf{a}_T$. The magnitude of $\mathbf{a}_T$ can be determined from the angular acceleration α according to $a_T = r\alpha$, as the previous section explains. Alternatively, a_T can be calculated using Newton's second law, $F_T = ma_T$, if the magnitude F_T of the net tangential force and the mass m are known. Figure 8.11*b* shows the two acceleration compo-

Figure 8.11   (*a*) A top view of a model airplane flying on a guide wire. If the plane has a constant tangential speed, it is an example of uniform circular motion. (*b*) Nonuniform circular motion occurs when the tangential speed changes, in which case there is a tangential acceleration $\mathbf{a}_T$ in addition to the centripetal acceleration $\mathbf{a}_c$.

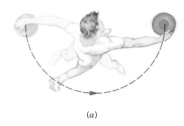

(a)

(b)

Figure 8.12 (a) A discus thrower. (b) The total acceleration **a** of the discus just before the discus is released. The total acceleration is the vector sum of the centripetal acceleration **a**$_c$ and the tangential acceleration **a**$_T$.

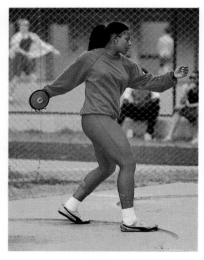

A discus thrower at a college track and field meet.

nents. The total acceleration is given by the vector sum of **a**$_c$ and **a**$_T$. Since **a**$_c$ and **a**$_T$ are perpendicular, the magnitude of the total acceleration **a** can be obtained from the Pythagorean theorem as $a = \sqrt{a_c^2 + a_T^2}$, while the angle ϕ can be determined from $\tan \phi = a_T/a_c$. The next example applies these concepts to a discus thrower.

Example 6   A Discus Thrower

Discus throwers often warm up by standing with both feet flat on the ground and throwing the discus with a twisting motion of their bodies. Figure 8.12a illustrates a top view of such a warm-up throw. Starting from rest, the thrower accelerates the discus to a final angular velocity of +15.0 rad/s in a time of 0.270 s before releasing it. During the acceleration, the discus moves on a circular arc of radius 0.810 m. Find (a) the magnitude a of the total acceleration just before the discus is released and (b) the angle ϕ that the total acceleration makes with the radius at this moment.

REASONING   The magnitude of the total acceleration is $a = \sqrt{a_c^2 + a_T^2}$, where a_c and a_T are the magnitudes of the centripetal and tangential accelerations. The angle ϕ in Figure 8.12b is given by $\tan \phi = a_T/a_c$. The centripetal acceleration can be evaluated from $a_c = r\omega^2$. The tangential acceleration follows from $a_T = r\alpha$, where α can be found from the definition of angular acceleration in Equation 8.4.

SOLUTION
(a) The magnitude of the centripetal acceleration is

$$a_c = r\omega^2 = (0.810 \text{ m})(15.0 \text{ rad/s})^2 = 182 \text{ m/s}^2 \qquad (8.11)$$

The magnitude of the tangential acceleration is $a_T = r\alpha$. But $\alpha = (\omega - \omega_0)/t$ according to Equation 8.4, so

$$a_T = r\alpha = r\left(\frac{\omega - \omega_0}{t}\right)$$

$$a_T = (0.810 \text{ m})\left(\frac{15.0 \text{ rad/s} - 0}{0.270 \text{ s}}\right) = 45.0 \text{ m/s}^2$$

Just before the moment of release, the total acceleration of the discus is

$$a = \sqrt{a_c^2 + a_T^2} = \sqrt{(182 \text{ m/s}^2)^2 + (45.0 \text{ m/s}^2)^2} = \boxed{187 \text{ m/s}^2}$$

(b) The angle ϕ in Figure 8.12b is given by

$$\phi = \tan^{-1}\left(\frac{a_T}{a_c}\right) = \tan^{-1}\left(\frac{45.0 \text{ m/s}^2}{182 \text{ m/s}^2}\right) = \boxed{13.9°}$$

8.6 ROLLING MOTION

Rotation plays an important role in many situations, and one familiar situation, that of rolling motion, deserves special attention. Figure 8.13 shows the rolling motion of an automobile tire. The essence of rolling motion is that there is *no slipping* at the point of contact where the tire touches the ground. To a good approximation, the tires on a normally moving automobile roll and do not slip. On the other hand, the squealing tires that accompany the start of a drag race are rotating, but they are not rolling as they rapidly spin and slip against the ground.

When the tires in Figure 8.13 roll, there is a relationship between the angular speed at which the tires rotate and the linear speed at which the car moves forward. With the help of part *b* of the drawing and the assumption that the car has a constant linear speed, it is possible to determine this relationship. As a tire rolls along the ground, the axle moves through the linear distance *d*. Provided that the tire does not slip, the distance *d* must be equal to the circular arc length *s*, measured along the outer edge of the tire: $d = s$. Dividing both sides of this equation by the elapsed time *t* shows that $d/t = s/t$. The term d/t is the speed at which the axle moves parallel to the ground, namely, the linear speed *v* of the car. The term s/t is the tangential speed v_T at which a point on the outer edge of the tire moves relative to the axle. In addition, v_T is related to the angular speed ω about the axle according to $v_T = r\omega$ (Equation 8.9). Therefore, it follows that

$$\underbrace{v}_{\substack{\text{Linear} \\ \text{speed}}} = \underbrace{r\omega}_{\substack{\text{Tangential} \\ \text{speed, } v_T}} \qquad (\omega \text{ in rad/s}) \qquad (8.12)$$

If the car in Figure 8.13 has a linear acceleration **a** parallel to the ground, a point on the tire's outer edge experiences a tangential acceleration $\mathbf{a}_T$ relative to the axle. The same kind of reasoning used in the last paragraph reveals that the magnitudes of these accelerations are the same and that they are related to the angular acceleration α of the wheel relative to the axle:

$$\underbrace{a}_{\substack{\text{Linear} \\ \text{acceleration}}} = \underbrace{r\alpha}_{\substack{\text{Tangential} \\ \text{acceleration, } a_T}} \qquad (\alpha \text{ in rad/s}^2) \qquad (8.13)$$

Equations 8.12 and 8.13 may be applied to any rolling motion, as long as the object does not slip against the surface on which it is rolling. Example 7 illustrates the basic features of rolling motion.

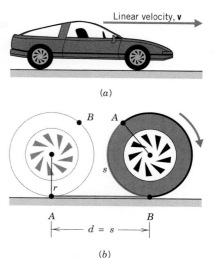

Figure 8.13 (*a*) As an automobile moves with a linear speed *v*, its tires roll along the ground. (*b*) If the tires roll and do not slip, the distance *d*, through which an axle moves, equals the circular arc length *s* measured along the outer edge of a tire.

Example 7 An Accelerating Car

An automobile starts from rest and for 20.0 s has a constant linear acceleration of 0.800 m/s² to the right. During this period, the tires do not slip. The radius of the tires is 0.330 m. At the end of the 20.0-s interval what is the angle through which each wheel has rotated?

REASONING As the car accelerates and goes faster and faster, the tires rotate faster and faster. Thus, each tire has an angular acceleration, which must be taken into account when we determine the angular displacement of each wheel. Since the tires roll without slipping, the magnitude α of the angular acceleration is related to the magnitude *a* of the linear acceleration of the car by $a = r\alpha$ (Equation 8.13). Therefore, we find that

$$\alpha = \frac{a}{r} = \frac{0.800 \text{ m/s}^2}{0.330 \text{ m}} = 2.42 \text{ rad/s}^2$$

The angular data, then, are as follows:

θ	α	ω	ω_0	t
?	-2.42 rad/s²		0	20.0 s

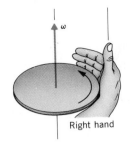

Right hand

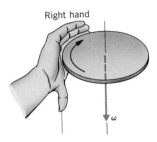

Right hand

Figure 8.14 The angular velocity vector ω of a rotating object points along the axis of rotation. The direction along the axis depends on the sense of the rotation and can be determined with the aid of a right-hand rule (see text).

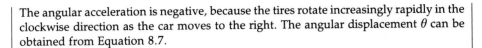

The angular acceleration is negative, because the tires rotate increasingly rapidly in the clockwise direction as the car moves to the right. The angular displacement θ can be obtained from Equation 8.7.

SOLUTION From Equation 8.7, we find that

$$\theta = \omega_0 t + \tfrac{1}{2}\alpha t^2 = \tfrac{1}{2}(-2.42 \text{ rad/s}^2)(20.0 \text{ s})^2 = \boxed{-484 \text{ rad}}$$

*8.7 THE VECTOR NATURE OF ANGULAR VARIABLES

We have presented angular velocity and angular acceleration by making the analogy between angular variables and linear variables. Like the linear velocity and the linear acceleration, the angular quantities are also vectors and have a direction as well as a magnitude. As yet, however, we have not discussed the directions of these vectors.

When a rigid object rotates about a fixed axis, it is the axis that identifies the motion, and the angular velocity vector points along this axis. Figure 8.14 shows how to determine the direction using a *right-hand rule*:

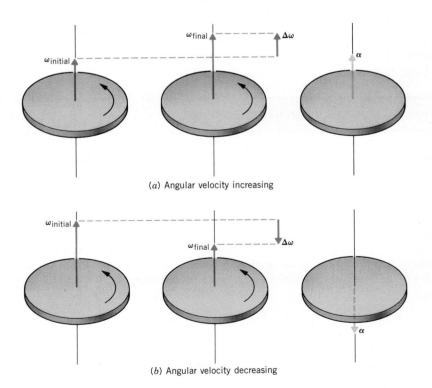

(a) Angular velocity increasing

(b) Angular velocity decreasing

Figure 8.15 The angular acceleration vector α of a rotating object points along the axis of rotation. The direction of the angular acceleration is the same as the direction of the change $\Delta\omega$ in the angular velocity, where $\Delta\omega = \omega_{\text{final}} - \omega_{\text{initial}}$.

Right-Hand Rule. Grasp the axis of rotation with your right hand, so that your fingers circle the axis in the same sense as the rotation. Your extended thumb points along the axis in the direction of the angular velocity vector.

Note that no part of the rotating object moves in the direction of the angular velocity vector.

Angular acceleration arises when the angular velocity changes, and the acceleration vector also points along the axis of rotation. The acceleration vector has the same direction as the *change* in the angular velocity, as Figure 8.15 illustrates.

INTEGRATION OF CONCEPTS

ROTATIONAL MOTION AND LINEAR MOTION

In Chapter 2, we began our study of motion by considering linear motion, or motion along a line. Here in Chapter 8, we have continued our study by considering rotational motion. Although these two kinds of motion are different, they also have much in common. In both, the position of an object changes, and to describe the change, the concept of displacement is used. In the linear case, we locate the position of the object by drawing a vector from a coordinate origin to the spot where the object happens to be. As the object moves, this vector changes, and the change is called the displacement. For rotational motion of a rigid object about a fixed axis, we imagine a line passing through any point on the object and intersecting the axis perpendicularly. As the object rotates, this line changes its position by sweeping out an angle. The change in position, as given by that angle, is called the angular displacement. In both the linear and the rotational cases, the fundamental idea is that of displacement as a change in position. The two kinds of motion also have another aspect in common. In each type, the changes in the position of an object can occur slowly or rapidly, and we use the concept of velocity to indicate this aspect of the motion. For linear motion, the velocity is the displacement per unit time, while for rotational motion, the angular velocity is the angular displacement per unit time. In both cases, the idea of velocity indicates the displacement per unit time that occurs as the object moves. Another feature that linear and rotational motion have in common is acceleration. When the velocity changes from one value to another, we use the concept of acceleration to describe how rapidly the change occurs. For linear motion, the acceleration is the change in velocity per unit time, while for rotational motion, the angular acceleration is the change in angular velocity per unit time. For both types, the idea of acceleration conveys the change in velocity per unit time that occurs as the object moves. We see, then, that the two kinds of motion do indeed have much in common.

THE EQUATIONS OF LINEAR KINEMATICS AND THE EQUATIONS OF ROTATIONAL KINEMATICS

In Chapters 2 and 3, we saw how to use the equations of linear kinematics to describe the linear motion of an object. These equations relate the displacement s, the acceleration a, the final velocity v, the initial velocity v_0, and the time t. Each equation contains 4 of these 5 variables and applies only under the condition of constant acceleration. In the present chapter, we have seen how to use the equations of rotational kinematics to describe the rotational motion of a rigid object about a fixed axis. The equations of rotational kinematics relate the angular displacement θ, the angular acceleration α, the final angular velocity ω, the initial angular velocity ω_0, and the time t. As in the linear case, each equation contains 4 of these 5 variables and applies only when the angular acceleration is constant. The two sets of equations of kinematics have the same mathematical form, with only the algebraic symbols being different. Therefore, all of the things you have learned about applying the equations of linear kinematics are useful here in Chapter 8.

COMBINED MOTIONS AND THE EQUATIONS OF KINEMATICS

In our study of motion, we have dealt primarily with situations in which the motion has been either linear or rotational. The one exception has been here in Chapter 8, when we discussed rolling motion, in which linear and rotational motions occur simultaneously. There are also other situations in which both kinds of motion occur together. For instance, a baton twirler can throw a spinning baton directly upward. The baton rotates all the while it moves upward and then downward along a line. When linear and rotational motions occur simultaneously, we can use the equations of linear kinematics from Chapters 2 and 3 to describe the linear motion and the equations of rotational kinematics to describe the rotational motion. At the end of the present chapter, you will find a number of homework problems that combine the two types of motion (e.g., problems 14, 47, and 54).

SUMMARY

When a rigid body rotates about a fixed axis, the **angular displacement** is the angle swept out by a line passing through any point on the body and intersecting the axis of rotation perpendicularly. The **radian (rad)** is the SI unit of angular displacement. In radians, the angle θ is defined as the circular arc length s traveled by a point on the rotating body divided by the radial distance r of the point from the axis: $\theta = s/r$.

The **average angular velocity** $\overline{\omega}$ is the angular displacement $\Delta\theta$ divided by the elapsed time Δt: $\overline{\omega} = \Delta\theta/\Delta t$. When Δt is infinitesimally short, the average angular velocity becomes equal to the **instantaneous angular velocity** ω. The magnitude of the instantaneous angular velocity is called the **instantaneous angular speed**.

The **average angular acceleration** $\overline{\alpha}$ is the change $\Delta\omega$ in the angular velocity divided by the elapsed time

Δt: $\overline{\alpha} = \Delta\omega/\Delta t$. When Δt is infinitesimally short, the average angular acceleration becomes equal to the **instantaneous angular acceleration** α.

When a rigid body rotates with constant angular acceleration about a fixed axis, the angular displacement θ, final angular velocity ω, initial angular velocity ω_0, angular acceleration α, and the elapsed time t are related as follows, assuming that $\theta_0 = 0$ at $t_0 = 0$:

$$\omega = \omega_0 + \alpha t \quad \text{and} \quad \theta = \tfrac{1}{2}(\omega_0 + \omega)t$$

These two equations can be combined algebraically to give two additional equations, Equations 8.7 and 8.8. Together, the four equations are known as the **equations of rotational kinematics for constant angular acceleration.** These equations may be used with any self-consistent set of units and are not restricted to radian measure.

When a rigid body rotates about a fixed axis, any single point on the body moves on a circular path of radius r. Such a point moves through an arc length s and has a tangential velocity $\mathbf{v}_T$ and, possibly, a tangential acceleration $\mathbf{a}_T$. The **angular and tangential variables are related** by the equations $s = r\theta$, $v_T = r\omega$, and $a_T = r\alpha$. These equations refer to the magnitudes of the variables involved, without reference to positive or negative signs, and only radian measure can be used when applying them.

A point on an object rotating with **nonuniform circular motion** experiences a total acceleration that is the vector sum of two perpendicular acceleration compo-

nents, the tangential acceleration $\mathbf{a}_T$ and the centripetal acceleration $\mathbf{a}_c$; $\mathbf{a} = \mathbf{a}_c + \mathbf{a}_T$.

The essence of **rolling motion** is that there is no slipping at the point where the object touches the surface upon which it is rolling. As a result, the tangential speed v_T (relative to the axis through the center of the object) of a point on the outer edge of a rolling object (radius r) is equal to the linear speed v with which the object moves parallel to the surface. The relation $v = v_T$ ($= r\omega$) expresses this conclusion. The magnitudes of the tangential acceleration a_T and the linear acceleration a of a rolling object are similarly related: $a = a_T$ ($= r\alpha$).

QUESTIONS

1. In the drawing, the flat triangular sheet ABC is lying in the plane of the paper. This sheet is going to rotate about an axis that also lies in the plane of the paper and passes through point A. Draw two such axes that are oriented so that points B and C will move on circular paths having the same radii.

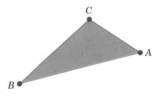

2. Are the equations of rotational kinematics valid when the angular displacement, angular velocity, and angular acceleration are expressed in terms of degrees, rather than radians? Explain why or why not.

3. Starting with Equations 8.4 and 8.6, derive Equations 8.7 and 8.8, assuming that $\theta_0 = 0$ at $t_0 = 0$.

4. A thin rod rotates at a constant angular speed. Consider the tangential speed of each point on the rod for the case when the axis of rotation is perpendicular to the rod (a) at its center and (b) at one end. Explain for each case whether there are any points on the rod that have the same tangential speeds.

5. A car is up on a hydraulic lift at a garage. The wheels are free to rotate and the drive wheels are rotating with a constant angular velocity. Does a point on the rim of a wheel have (a) a tangential acceleration and (b) a centripetal acceleration? In each case, give your reasoning.

6. Two points are located on a rigid wheel that is rotating with an increasing angular velocity about a fixed axis. The axis is perpendicular to the wheel at its center. Point 1 is located on the rim and point 2 is halfway between the rim and the axis.

(a) Which point (if either) turns through the greater angle in a given amount of time? At any given instant, which point (if either) has the greater (b) angular velocity, (c) angular acceleration, (d) tangential speed, (e) tangential acceleration, and (f) centripetal acceleration? Provide a reason for each of your answers.

7. Section 5.6 discusses how the uniform circular motion of a space station can be used to create "artificial" gravity for the astronauts. This can be done by adjusting the angular speed of the space station, so the centripetal acceleration at the astronaut's feet equals g, the acceleration due to gravity (see Figure 5.13). If such an adjustment is made, will the acceleration due to the "artificial" gravity also equal g at the astronaut's head? Account for your answer.

8. A hammer-thrower at a track meet whirls the hammer in a circular path, as the drawing shows. If the wire were to be cut at the instant shown in the drawing, would the hammer move toward point A, B, or C? Justify your choice.

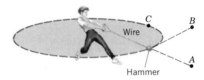

9. Explain why a given point on the rim of a tire has an acceleration when the tire is on a car that is moving at a constant linear velocity.

10. A bicycle is turned upside down, the front wheel is spinning (see the drawing), and there is an angular acceleration. At the instant shown, there are six points on the wheel that have arrows associated with them. Which of the following quantities could the arrows represent: (a) tangential velocity, (b) tangential acceleration, (c) centripetal acceleration? In each case, answer why the arrows do or do not represent the quantity.

11. Suppose that the speedometer of a truck is set to read the linear speed of the truck, but uses a device that actually measures the angular speed of the tires. If larger diameter tires are mounted on the truck, will the reading on the speedometer

be correct? If not, will the reading be greater than or less than the true linear speed of the truck? Why?

12. The automobile tire in the drawing rolls and does not slip. The axle of the wheel at point C moves with a constant linear velocity **v** parallel to the ground. For any point on the wheel, two velocity vectors contribute to the total velocity of that point with respect to the ground: (1) the velocity **v** of the forward (linear) motion of the wheel and (2) the tangential velocity caused by the rotation of the wheel. (a) Using vectors, explain why the *total* velocity of point A with respect to the ground has a magnitude of $2v$ at the instant shown. (b) Also using vectors, explain why the total velocity of point B with respect to the ground is zero.

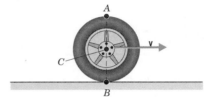

PROBLEMS

**Section 8.1 Rotational Motion and Angular Displacement,
Section 8.2 Angular Velocity and Angular Acceleration**

1. For each of the following angles, give its equivalent in radians: (a) $45°$, (b) $180°$, and (c) $360°$.

2. A gymnast on a trampoline does a double backward somersault. What is the magnitude of the angular displacement (in radians) of the gymnast?

3. The moon has a diameter of 3.48×10^6 m and is 3.85×10^8 m from the earth. The sun has a diameter of 1.39×10^9 m and is 1.50×10^{11} m from the earth. Show that the angle θ subtended by the moon, as measured by a person standing on the earth, is approximately equal to the angle subtended by the sun.

4. On a wristwatch, what is the angular velocity (magnitude and direction) of (a) the second hand and (b) the hour hand? Express your answers in rad/s.

5. A string trimmer is a grass and weed cutting tool that utilizes a 0.21-m length of nylon "string," rotating at 650 rad/s about an axis perpendicular to one end of the string. (a) What is the time needed for the string to sweep out an angle of 0.61 rad? (b) Assuming that the length of the string does not change, find the distance through which the tip of the string moves during this time.

6. A diver completes $3\frac{1}{2}$ somersaults in 1.7 s. What is the average angular speed (in rad/s) of the diver?

7. After being turned on, a turntable reaches its rated an-

gular speed of 4.71 rad/s (45 rpm) in a time of 4.10 s. What is the average angular acceleration?

8. An automatic drier spins wet clothes at an angular speed of 6.8 rad/s. Starting from rest, the drier reaches its operating speed with an average angular acceleration of 7.0 rad/s². How long does it take the drier to come up to speed?

9. An electric fan is running on HIGH. After the LOW button is pressed, the angular speed of the fan decreases to 83.8 rad/s in 1.75 s. The deceleration is 42.0 rad/s². Determine the initial angular speed of the fan.

*10. Two people start at the same place and walk around a circular lake in opposite directions. One has an angular speed of 1.7×10^{-3} rad/s, while the other has an angular speed of 3.4×10^{-3} rad/s. How long will it be before they meet?

*11. A space station consists of two donut-shaped living

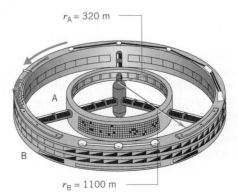

chambers, A and B, that have the radii shown in the drawing. As the station rotates to create artificial gravity, an astronaut in chamber A is moved 240 m along a circular arc. How far along a circular arc is an astronaut in chamber B moved during the same time?

*12. Suppose that wedge-shaped pieces are cut from a pie, so that the arc length along the outer crust of each piece exactly equals the radius of the piece. After all such pieces are eaten, what is the apex angle (in radians) of the remaining portion?

*13. A stroboscope is a light that flashes on and off at a constant rate. It can be used to illuminate a rotating object, and if the flashing rate is adjusted properly, the object can be made to appear stationary. What is the shortest time between flashes of light that will make a three-bladed propeller appear stationary when it is rotating with an angular speed of 16.7 rev/s?

**14. A baton twirler throws a spinning baton directly upward. As it goes up and returns to the twirler's hand, the baton turns through four revolutions. Ignoring air resistance and assuming that the average angular speed of the baton is 1.80 rev/s, determine the height to which the center of the baton travels above the point of release.

Section 8.3 The Equations of Rotational Kinematics

15. The drill bit of a variable-speed electric drill has a constant angular acceleration of 2.50 rad/s². The initial angular speed of the bit is 5.00 rad/s. After 4.00 s, (a) what angle has the bit turned through and (b) what is bit's angular speed?

16. A turntable is switched from 4.71 rad/s (45 rpm) to 3.49 rad/s (33⅓ rpm), and the platter rotates through an angle of 11.5 rad in reaching the new angular speed. What is the angular acceleration of the platter?

17. The angular speed of the rotor in a centrifuge increases from 420 to 1420 rad/s in a time of 5.00 s. (a) Obtain the angle through which the rotor turns during this time. (b) What is the angular acceleration?

18. A wheel accelerates so that its angular speed increases from 150 to 580 rad/s in 16 revolutions. What is the angular acceleration in rad/s²?

19. A flywheel has a constant angular deceleration of 2.0 rad/s². (a) Find the angle through which the flywheel turns as it comes to rest from an angular speed of 220 rad/s. (b) Find the time required for the flywheel to come to rest.

20. An airliner arrives at the terminal, and the pilot shuts off the engines. The initial angular velocity of the fan blades is 1800 rad/s, and it takes 120 s for them to come to rest. What is the angular displacement of the blades?

*21. After 10.0 s, a spinning roulette wheel has slowed down to an angular speed of 1.88 rad/s. During this time, the wheel rotates through an angle of 44.0 rad. Determine the angular acceleration of the wheel.

*22. A fan blade, whose angular acceleration is a constant 2.00 rad/s², rotates through an angle of 285 radians in 11.0 s. How long did it take the blade, starting from rest, to reach the *beginning* of the 11.0-s interval?

*23. The drive propeller of a ship starts from rest and accelerates at 3.50×10^{-3} rad/s² for 1.80×10^3 s. For the next 3.60×10^3 s, the propeller rotates at a constant angular speed. Then it decelerates at 2.00×10^{-3} rad/s², until it slows (without reversing direction) to an angular speed of 5.00 rad/s. Find the angular displacement of the propeller.

**24. A child, hunting for his favorite wooden horse, is running on the ground around the edge of a stationary merry-go-round. The angular speed of the child has a constant value of 0.250 rad/s. At the instant the child spots the horse, one-quarter of a turn away, the merry-go-round begins to move (in the direction the child is running) with a constant angular acceleration of 0.0100 rad/s². What is the shortest time it takes for the child to catch up with the horse?

Section 8.4 Angular Variables and Tangential Variables

25. The earth has a radius of 6.38×10^6 m and turns on its axis once every 23.9 h. What is the tangential speed (in m/s) of a person living in Ecuador, a country in South America that lies on the equator?

26. Two fans, whose blade diameters are 0.152 and 0.254 m, run at the same constant angular speed of 125 rad/s. (a) By how much does the tangential speed at the tip of the longer blade exceed that at the end of the shorter blade? (b) Is there any point on the longer blade with the same tangential speed as the tip of the shorter blade? If so, where?

27. A fisherman is hauling in a fish at a speed of 0.14 m/s. The fishing line is being spooled onto a reel whose radius is 0.030 m. What is the angular speed of the reel?

28. A small disk (radius = 2.00 mm) is attached to a high-speed drill at a dentist's office and is turning at 7.85×10^4 rad/s. Determine the tangential speed of a point on the outer edge of this disk.

29. A meter stick is rotating with a constant angular acceleration of 12.0 rad/s² about an axis that passes through one end of the stick and is perpendicular to it. What point on the stick has a tangential acceleration whose magnitude equals that of the acceleration due to gravity?

*30. A compact disc (CD) recording contains music on a spiral track. Music is put onto a CD with the assumption that, during playback, the music will be detected at a *constant tangential speed* at any point. Since $v_T = r\omega$, a CD rotates at a smaller angular speed for music near the outer edge and a larger angular speed for music near the inner part of the disc. A CD has a radius of about 0.060 m and rotates at 3.5 rev/s for music at the outer edge. Find (a) the constant tangential speed at which music is detected and (b) the angular speed (in rev/s) for music at a distance of 0.025 m from the center of a CD.

*31. A thin straight rod is rotating with an angular speed of 3.14 rad/s about a vertical axis, as the drawing shows. In one second, the end of the rod moves through a circular arc whose length equals the length of the rod. Find the angle θ.

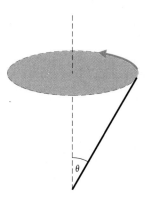

**32. One type of slingshot can be made from a length of rope and a leather pocket for holding the stone. The stone can be thrown by whirling it rapidly in a horizontal circle and releasing it at the right moment. Such a slingshot is used to throw a stone from the edge of a cliff, the point of release being 20.0 m above the base of the cliff. The stone lands on the ground below the cliff at a point X. The horizontal distance of point X from the base of the cliff (directly beneath the point of release) is thirty times the radius of the circle on which the stone is whirled. Determine the angular speed of the stone at the moment of release.

Section 8.5 Centripetal Acceleration and Tangential Acceleration

33. During a tennis serve, a racket is given an angular acceleration of magnitude 160 rad/s². At the top of the serve, the racket has an angular speed of 14 rad/s. If the distance between the top of the racket and the shoulder is 1.5 m, find the magnitude of the total acceleration of the top of the racket.

34. A circular disk of radius 0.100 m, is rotating at a constant angular speed about an axis that is perpendicular to the disk at its center. Determine the ratio of the centripetal acceleration at point A on the circumference to that at point B located 0.0700 m from the center.

35. A floppy disk for a personal computer rotates with a constant angular speed of 31.4 rad/s about an axis perpendicular to the disk at its center. (a) Find the tangential speed of a point that is 0.0508 m from the center of the disk. (b) What is the magnitude of the centripetal acceleration at this point?

36. A 220-kg speedboat is negotiating a circular turn (radius = 32 m) around a buoy. During the turn, the engine causes a net tangential force to be applied to the boat. The magnitude of the force is 550 N. The initial tangential speed of the boat going into the turn is 5.0 m/s. (a) Find the tangential acceleration. (b) After the boat is 2.0 s into the turn, find the centripetal acceleration.

*37. A disk has a constant angular acceleration of 4.00 rad/s² about an axis perpendicular to the disk at its center. Find the radius at a point on the disk where, 0.500 s after the disk begins to rotate, the magnitude of the total acceleration (centripetal plus tangential) equals that of the acceleration due to gravity.

*38. A thin rigid rod is rotating with a constant angular acceleration about an axis that passes perpendicularly through one of its ends. At one instant, the total acceleration vector (centripetal plus tangential) at the other end of the rod makes a 60.0° angle with respect to the rod and has a magnitude of 15.0 m/s². The rod has an angular speed of 2.00 rad/s at this instant. What is the rod's length?

**39. The blades of a windmill start from rest and rotate with an angular acceleration of 22.0 rad/s². At any point on a blade, how much time passes before the magnitude of the tangential acceleration equals the magnitude of the centripetal acceleration?

Section 8.6 Rolling Motion

Note: All problems in this section assume that there is no slipping of the surfaces in contact during the rolling motion.

40. Suppose you are riding a stationary exercise bicycle, and the electronic meter indicates that the wheel is rotating at 11 rad/s (105 rpm). The wheel has a radius of 0.45 m. If you ride the bike for 1810 s, how far would you have gone if the bike could move?

41. A bicycle travels a linear distance of 515 m. Through what angle (in radians) does each tire ($r = 0.356$ m) rotate?

42. An automobile tire has a radius of 0.330 m, and its center moves forward with a linear speed of $v = 15.0$ m/s. (a) Determine the angular speed of the wheel. (b) Relative to the axle, what is the tangential speed of a point located 0.175 m from the axle?

43. On an open-reel tape deck, the tape is being pulled past the playback head at a constant linear speed of 0.381 m/s. (a) Using the data in part *a* of the drawing, find the angular speed of the take-up reel. (b) After 2.40×10^3 s, the take-up reel is almost full, as part *b* of the drawing indicates. Find the average angular acceleration of the reel and specify whether the acceleration indicates an increasing or decreasing angular velocity.

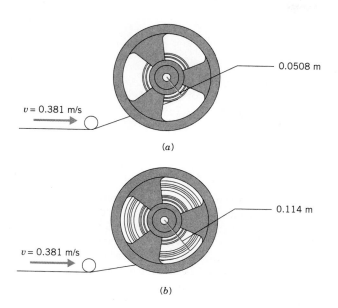

(a)

(b)

gears is negligible compared to the radii. Determine the angular velocity (magnitude and direction) of the smaller gear.

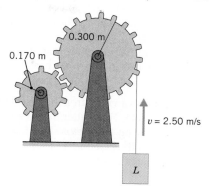

44. A car is traveling with a speed of 15.0 m/s along a straight horizontal road. The wheels of the car have a radius of 0.480 m. If the car speeds up with a linear acceleration of 2.00 m/s² for 5.00 s, find the angular displacement of each wheel during this period.

*45. A person lowers a bucket into a well by turning the hand crank, as the drawing illustrates. The crank handle moves with a constant tangential speed of 1.20 m/s on its circular path. Find the linear speed with which the bucket moves down the well.

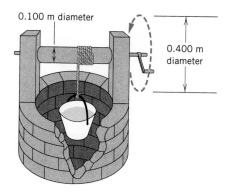

*46. The two-gear combination shown in the drawing is being used to hoist the load L with a constant upward speed of 2.50 m/s. The rope attached to the load is being wound onto a cylinder behind the big gear. The depth of the teeth of the

**47. A bicycle, moving along the ground, has a constant linear acceleration. What must be the angular acceleration of the bike tires, so that in 5.00 s the bike moves through a distance equal to ten times the radius of the tires and, in the process, doubles its forward linear speed?

ADDITIONAL PROBLEMS

48. A race car travels with a constant tangential speed of 75.0 m/s around a circular track of radius 625 m. Find (a) the magnitude of the car's total acceleration and (b) the direction of its total acceleration relative to the radial direction.

49. A circular disk rotates with a constant angular speed about an axis perpendicular to the disk at its center. Point A on the disk is located 0.100 m from the center and has a tangential speed of 1.50 m/s. (a) What is the angular speed of the disk? (b) Find the tangential speed of point B, which is located 0.170 m from the center.

50. The earth rotates once about its axis in 23.9 hours and orbits the sun once in 365 days. Find the average angular speed (in rad/s) of the earth's (a) rotational motion and (b) orbital motion.

51. A train is rounding a circular curve whose radius is 2.00×10^2 m. At one instant, the train has an angular acceleration of 1.50×10^{-3} rad/s² and an angular speed of 0.0500 rad/s. (a) Find the magnitude of the total acceleration (centripetal plus tangential) of the train. (b) Determine the angle of the total acceleration relative to the radial direction.

52. The shaft of a pump starts from rest and has an angular acceleration of 3.00 rad/s² for 18.0 s. At the end of this interval, what is (a) the shaft's angular speed and (b) the angle through which the shaft has turned?

53. An electric circular saw is designed to reach its final angular speed, starting from rest, in 1.50 s. Its average angular acceleration is 328 rad/s². Obtain its final angular speed.

*54. A quarterback throws a pass that is a perfect spiral. In other words, the football does not wobble, but spins smoothly about an axis passing through each end of the ball. Suppose the ball spins at 7.7 rev/s. In addition, the ball is thrown with a linear speed of 19 m/s at an angle of 45° with respect to the ground. If the ball is caught at the same height at which it left the quarterback's hand, how many times has the ball rotated while in the air?

*55. The drawing shows a view (from beneath) of the platter on a belt-drive turntable. The platter has an angular speed of 3.49 rad/s. The pulley on the motor shaft has a radius of 0.0127 m. Assuming that the belt does not slip, determine the angular speed of the motor shaft.

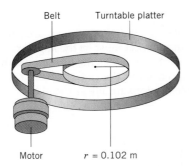

*56. A dentist causes the bit of a high-speed drill to accelerate from an angular speed of 1.05×10^4 rad/s to an angular speed of 3.14×10^4 rad/s. In the process, the bit turns through 1.88×10^4 rad. Assuming a constant angular acceleration, how long would it take the bit to reach its maximum speed of 7.85×10^4 rad/s, starting from rest?

*57. A rectangular plate is rotating with a constant angular acceleration about an axis that passes perpendicularly through one corner, as the drawing shows. The tangential acceleration measured at corner A has twice the magnitude of that measured at corner B. What is the ratio L_1/L_2 of the lengths of the sides of the rectangle?

*58. At the local swimming hole, a favorite trick is to run horizontally off a cliff that is 2.7 m above the water. One show-off claims to have done this trick in a new way, which involves tucking into a "ball" and (he claims) rotating through three revolutions before hitting the water. Ignoring air resistance and assuming the "ball" is already rotating at the instant it begins falling vertically, evaluate the likelihood of doing such a trick by calculating the constant angular speed (in rev/s) that would be necessary.

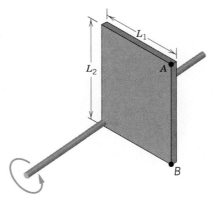

Problem 57

**59. Two disks start from rest and rotate as the drawing shows. The angular accelerations of these disks are constant but have different magnitudes. The solid (not dashed) colored reference lines on the disks were directly above each other at the start. The drawing shows the disks at a later time, when the lines are separated by 0.524 rad, the angular speed of disk A exceeds that of disk B by 1.50 rad/s, and each disk has made less than one revolution. The angular acceleration of disk A is 2.50 rad/s². Find the angle through which disk A has turned.

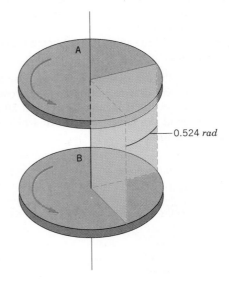

CHAPTER 9

ROTATIONAL DYNAMICS

Sailing has its moments, especially when you have to worry about being capsized by the wind. The sailors here are using their weights to counteract the effect of the wind by hanging out over the edge of the boat. But weight alone is not the only thing that matters, because the location of a weight can be equally important. For instance, the thin sailor hanging the farthest out to the right may well have more effect in keeping the boat upright than the heavier one with his feet on the side of the boat. In hanging out over the edge, these sailors are involved with a branch of physics known as rotational dynamics, because rotational dynamics deals with what causes rotational motion or what changes it once it has begun. As we discuss rotational dynamics in this chapter, we will encounter a number of concepts that will be familiar from earlier chapters on linear dynamics. These concepts, however, will be in a form suitable for rotational motion.

9.1 THE EFFECTS OF FORCES AND TORQUES ON THE MOTION OF RIGID OBJECTS

Most rigid objects, like a propeller or a wheel, have their mass spread out and not concentrated at a single point. Such objects can move in a number of ways. Figure 9.1a illustrates one possibility called translational motion, in which all points on the body travel on parallel paths (not necessarily straight lines). In pure translation there is no rotation of any line in the body. Because translational motion can occur along a curved line, it is often called curvilinear motion or, more simply, linear motion. Another possibility is rotational motion, which may occur in combination with translational motion, as is the case for the somersaulting gymnast in Figure 9.1b.

We have seen many examples of how a net force affects the linear motion of an object by causing it to accelerate. When there is no net force, there is no accelera-

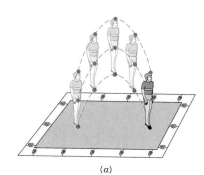

(a)

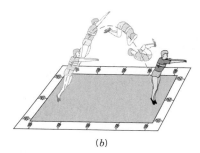

(b)

Figure 9.1 An example of (a) translational motion and (b) combined translational and rotational motion.

215

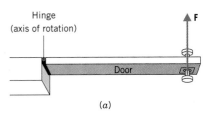
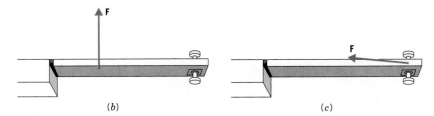

(a)   (b)   (c)

Figure 9.2 It is easier to open the door with a force of a given magnitude by (a) pushing at the door's outer edge than by (b) pushing closer to the axis of rotation (the hinge). (c) Pushing nearly into the hinge makes it very difficult to open the door.

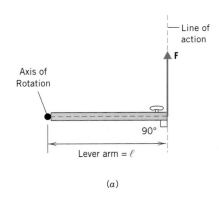

(a)

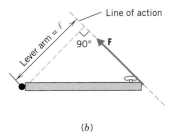

(b)

Figure 9.3 In this top view, the hinges of a door (axis of rotation) appear as a dot. The line of action and lever arm are illustrated for a force applied (a) perpendicular to the door and (b) at an angle with respect to the door.

tion and the object is in equilibrium. Since rigid objects can also rotate, however, we need to expand our concept of equilibrium and take into account the possibility of rotation. A net force causes linear motion to change, but what causes rotational motion to change? For example, something causes the propeller on a ship to change its rotational speed while the ship is speeding up. Is it simply a net force? Or is it something else? As it turns out, it is a net torque, not a net force, that causes rotational speed to change. Just as greater net forces cause greater linear accelerations, greater net torques cause greater rotational or angular accelerations. To represent torque, we use the symbol τ (Greek letter *tau*). To explain the idea of torque, we turn to Figure 9.2.

When you push on a door with a force **F**, as in Figure 9.2a, the door opens more quickly when the force is larger, because a larger force generates a larger torque, other things being equal. However, the door does not open as quickly if you apply the same force at a point closer to the hinge, as in part *b* of the drawing, because the force now produces less torque. Furthermore, if your push is directed nearly at the hinge, as in part *c*, you will have a hard time opening the door, because the torque is nearly zero. In summary, then, the torque depends on the magnitude of the force, on the point where the force is applied relative to the axis of rotation, and on the direction of the force.

For simplicity, we deal with situations in which the force lies in a plane that is perpendicular to the axis of rotation. In Figure 9.3, for instance, the axis is perpendicular to the page and the force lies in the plane of the paper. The drawing shows the line of action and the lever arm of the force, two concepts that are important in the definition of torque. The *line of action* is an extended line drawn colinear with the force. The *lever arm* is the distance ℓ between the line of action and the axis of rotation, measured on a line that is perpendicular to both. The magnitude of the torque is defined as the magnitude of the force times the lever arm:

Definition of Torque

Magnitude:

Torque = (Magnitude of the force) × (Lever arm)

$$\tau = F\ell \tag{9.1}$$

Direction: The torque is a positive quantity if the force tends to produce a counterclockwise rotation* about the axis, and negative if the force tends to produce a clockwise rotation.

SI Unit of Torque: newton·meter (N·m)

---

* We shall see in Section 9.4 that a net torque produces an angular acceleration, which is a change in the angular velocity per unit time. Thus, a net torque is not needed to sustain the angular velocity, but is needed to change it.

Equation 9.1 indicates that forces of the same magnitude can produce *different* torques, depending on the value of the lever arm, and Example 1 illustrates this feature.

Example 1 *Different Lever Arms, Different Torques*

In Figure 9.4 a force whose magnitude is 55 N is applied to a door. However, the lever arms are different in the four parts of the drawing: (a) $\ell = 0.80$ m, (b) $\ell = 0.20$ m, (c) $\ell = 0.60$ m, and (d) $\ell = 0$. Find the magnitude of the torque in each case.

REASONING AND SOLUTION   Because the lever arm is different in each case, Equation 9.1 gives different values for the torque, even though the magnitude of the force is the same.

(a) $\tau = F\ell = (55 \text{ N})(0.80 \text{ m}) = \boxed{44 \text{ N} \cdot \text{m}}$

(b) $\tau = F\ell = (55 \text{ N})(0.20 \text{ m}) = \boxed{11 \text{ N} \cdot \text{m}}$

(c) $\tau = F\ell = (55 \text{ N})(0.60 \text{ m}) = \boxed{33 \text{ N} \cdot \text{m}}$

(d) $\tau = F\ell = (55 \text{ N})(0) = \boxed{0}$

In part *d* the line of action of *F* passes through the axis of rotation (the hinge). Hence, the lever arm is zero, and the torque is zero.

Muscles and tendons in our bodies produce torques that cause rotation about various joints. Example 2 illustrates how the Achilles tendon produces a torque about the ankle joint.

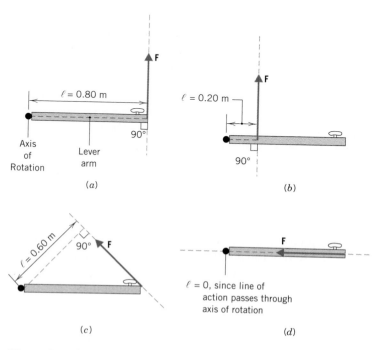

Figure 9.4 In each case shown here, the force has the same magnitude, but produces different torques, because the lever arms are different.

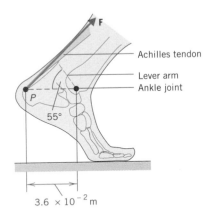

Achilles tendon

Lever arm
Ankle joint

P

55°

3.6×10^{-2} m

Figure 9.5   The force **F** generated by the Achilles tendon produces a clockwise (negative) torque about the ankle joint.

Example 2   *The Achilles Tendon*

Figure 9.5 shows the Achilles tendon exerting a force of magnitude $F = 720$ N on the heel at the point *P*. Determine the torque (magnitude and direction) of this force about the ankle joint.

REASONING AND SOLUTION   To calculate the magnitude of the torque, it is necessary to have a value for the lever arm ℓ. From the drawing, it can be seen that the lever arm is $\ell = (3.6 \times 10^{-2} \text{ m}) \cos 55° = 2.1 \times 10^{-2}$ m. The magnitude of the torque is

$$\tau = F\ell = (720 \text{ N})(2.1 \times 10^{-2} \text{ m}) = 15 \text{ N·m} \qquad (9.1)$$

The force **F** tends to produce a clockwise rotation about the ankle joint, so the torque is negative: $\boxed{\tau = -15 \text{ N·m}}$.

Now that we have introduced the idea of torque, the next step is to incorporate it into the concept of equilibrium. The following section shows that torques, as well as forces, must be balanced in order for a rigid object to be in equilibrium.

9.2 RIGID OBJECTS IN EQUILIBRIUM

EQUILIBRIUM

If a rigid body is in equilibrium, its motion does not change. By "motion" we mean both linear and rotational motion. An object whose motion is not changing has no acceleration of any kind. Therefore, the net force $\Sigma \mathbf{F}$ applied to the object must be zero, since $\Sigma \mathbf{F} = m\mathbf{a}$ and $\mathbf{a} = 0$. For two-dimensional motion, the condition $\Sigma \mathbf{F} = 0$ means that the x and y components of the net force are separately zero: $\Sigma F_x = 0$ and $\Sigma F_y = 0$ (Equations 4.9a and 4.9b). When calculating the net force, we include only *external forces*, that is, those forces applied to the object by external agents. The internal forces that occur between the internal parts of an object need not be included, since they always occur in action–reaction pairs. As far as the motion of the entire object is concerned, the action–reaction pairs of internal forces do not have any effect, since each pair always consists of oppositely directed forces of equal magnitude, and the effect of one force cancels the effect of the other.

A net torque causes rotational motion to change. But there is no change in the motion of a rigid body in equilibrium, so there can be no net torque acting under equilibrium conditions. The sum of the positive torques must be balanced by the sum of the negative torques. Using the symbol $\Sigma \tau$ to represent the net torque (the sum of all positive and negative torques), we write this condition as

$$\Sigma \tau = 0$$

The conditions that must be met if a rigid body is to be in equilibrium are summarized below.

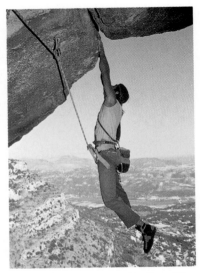

A breathtaking moment of equilibrium.

Equilibrium of a Rigid Body

A rigid body is in equilibrium if it has zero translational acceleration and zero angular acceleration. In equilibrium, the sum of the externally applied forces is zero, and the sum of the externally applied torques is zero:

$$\Sigma F_x = 0 \qquad \Sigma F_y = 0 \qquad \text{(4.9a and 4.9b)}$$

$$\Sigma \tau = 0 \qquad\qquad\qquad \text{(9.2)}$$

The procedure used to analyze the forces and torques acting on a body in equilibrium is similar to the procedure outlined in Section 4.11. For use here, the first four steps of that procedure are summarized below. Two additional steps account for any torques that may be present:

Step 1. Select the object to which the conditions for equilibrium are to be applied.

Step 2. Draw a free-body diagram that shows all the external forces acting on the object, each force with its proper direction.

Step 3. Choose a convenient set of x, y axes and resolve all forces into components that lie along these axes.

Step 4. Apply the conditions that specify the balance of forces at equilibrium: $\Sigma F_x = 0$ and $\Sigma F_y = 0$.

Step 5. Select a convenient axis of rotation. Identify the point where each force acts on the object, and calculate the torque produced by each force about the axis of rotation. Set the sum of the torques about this axis equal to zero: $\Sigma \tau = 0$.

Step 6. Solve the equations in Steps 4 and 5 for the desired unknown quantities.

Example 3 illustrates how the conditions for equilibrium are applied to a diving board.

Example 3   A Diving Board

A woman whose weight is 531 N is poised at the right end of a diving board, whose length is 4.50 m. The board has negligible weight and is bolted down at the left end, while being supported 1.60 m away by a fulcrum, as Figure 9.6a shows. Find the forces $\mathbf{F}_1$ and $\mathbf{F}_2$ that the bolt and the fulcrum, respectively, exert on the board.

REASONING   Part b of the figure shows the free-body diagram of the diving board. Three forces act on the board: $\mathbf{F}_1$, $\mathbf{F}_2$, and the force due to the diver's weight $\mathbf{W}$. In choosing the directions of $\mathbf{F}_1$ and $\mathbf{F}_2$ we have used our intuition: $\mathbf{F}_1$ points downward, because the bolt must pull in that direction to counteract the tendency of the board to rotate clockwise about the fulcrum; $\mathbf{F}_2$ points upward because the board pushes downward against the fulcrum and, in reaction, the fulcrum pushes upward on the board.

SOLUTION   Since the board is in equilibrium, the sum of the vertical forces must be zero:

$$\Sigma F_y = -F_1 + F_2 - 531 \text{ N} = 0 \qquad \text{(4.9b)}$$

Similarly, the sum of the torques must be zero, $\Sigma \tau = 0$. For calculating torques, we select an axis that passes through the left end of the board and is perpendicular to the page. (We will see shortly that this choice is arbitrary.) The force $\mathbf{F}_1$ produces no torque

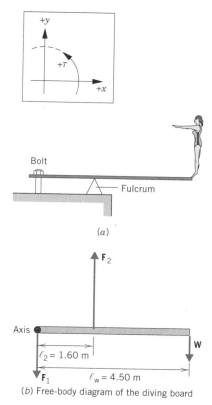

Figure 9.6   (a) A diver stands at the end of a diving board that is in equilibrium. (b) The free-body diagram shows the three forces that act on the board. The box at the upper left shows the positive x and y directions for the forces, as well as the positive (counterclockwise) direction for the torques.

since it passes through the axis and has a zero lever arm, while $\mathbf{F}_2$ creates a counterclockwise (positive) torque, and $\mathbf{W}$ produces a clockwise (negative) torque. The free-body diagram shows the lever arms for the torques:

$$\Sigma\tau = +F_2\ell_2 - W\ell_w = +F_2(1.60 \text{ m}) - (531 \text{ N})(4.50 \text{ m}) = 0 \qquad (9.2)$$

$$\boxed{F_2 = 1490 \text{ N}}$$

This value for F_2 can be substituted into Equation 4.9b above to show that $\boxed{F_1 = 960 \text{ N}}$.

THE AXIS USED FOR CALCULATING TORQUES IS ARBITRARY

In Example 3 the sum of the external torques is calculated using an axis that passes through the left end of the diving board. *However, the location of the axis is completely arbitrary, for an object in equilibrium is in equilibrium with respect to any axis whatsoever.* Thus, the sum of the external torques is zero, no matter where the axis is placed. Example 4 illustrates this important point.

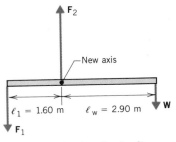

Figure 9.7 The free-body diagram for the diving board shown in Figure 9.6*a*. The rotational axis is now at the fulcrum.

Example 4   The Diving Board, Revisited

Repeat Example 3, choosing another rotational axis for computing the torques.

REASONING AND SOLUTION   A new axis is selected so that it passes through the fulcrum and is perpendicular to the page, as Figure 9.7 indicates. The equation representing the balance of the vertical forces is the same as it is in Example 3:

$$\Sigma F_y = -F_1 + F_2 - 531 \text{ N} = 0$$

The equation representing the balance of torques with respect to the new axis is

$$\Sigma\tau = +F_1\ell_1 - W\ell_w = +F_1(1.60 \text{ m}) - (531 \text{ N})(2.90 \text{ m}) = 0$$

Solving for F_1 yields $\boxed{F_1 = 960 \text{ N}}$. This value for F_1 can be substituted into the equation for the balance of the vertical forces to show that $\boxed{F_2 = 1490 \text{ N}}$. These answers are identical to those in Example 3 and illustrate the fact that the axis for computing torques may be chosen arbitrarily.

PROBLEM SOLVING INSIGHT

While Example 4 shows that the location of the rotational axis is arbitrary, *in practice one usually chooses its location so the lines of action of one or more of the unknown forces pass through the axis.* Such a choice simplifies the torque equation, because the torques produced by these forces are zero. For instance, in Example 4 the torque due to the force $\mathbf{F}_2$ does not appear in the equation representing the balance of torques.

DETERMINATION OF THE LEVER ARMS

In a calculation of torque, the lever arm of the force must be determined relative to the axis of rotation. In Examples 3 and 4 the lever arms are obvious, but sometimes this is not the case. Example 5 deals with a situation in which a little care is needed in determining lever arms.

Example 5 Fighting a Fire

In Figure 9.8a an 8.00-m ladder of weight $W_L = 355$ N leans against a smooth vertical wall. The term "smooth" means that the wall can exert only a normal force directed perpendicular to the surface and cannot exert a friction force directed parallel to the surface. A firefighter, whose weight is $W_F = 875$ N, stands 6.30 m from the bottom of the ladder. The weight of the ladder may be assumed to act at the ladder's center, and the hose may be neglected. Find the forces that the wall and the ground exert on the ladder.

REASONING Part b of the figure shows the free-body diagram of the ladder. The following forces are exerted on the ladder:

1. Its weight $\mathbf{W_L}$
2. The weight $\mathbf{W_F}$ of the firefighter
3. The force $\mathbf{P}$ applied to the top of the ladder by the wall and directed perpendicular to the wall
4. The forces $\mathbf{G_x}$ and $\mathbf{G_y}$, which are the horizontal and vertical components of the force exerted by the ground on the bottom of the ladder.

The force $\mathbf{G_x}$ is produced by static friction and prevents the ladder from slipping. The force $\mathbf{G_y}$ is the normal force applied to the ladder by the ground.

SOLUTION Since the ladder is in equilibrium, the net force acting on it is zero:

$$\Sigma F_x = G_x - P = 0 \tag{4.9a}$$
$$\Sigma F_y = G_y - W_L - W_F \tag{4.9b}$$
$$= G_y - 355\text{ N} - 875\text{ N} = 0$$

$$\boxed{G_y = 1230\text{ N}}$$

Equation 4.9a cannot be solved as it stands, because it contains two unknown variables. However, another equation can be obtained from the fact that the net torque acting on an object in equilibrium is zero. In calculating torques, it is convenient to use an axis at the left end of the ladder, directed perpendicular to the page, as Figure 9.8c indicates. This axis is convenient, because $\mathbf{G_x}$ and $\mathbf{G_y}$ produce no torques about it, their lever arms being zero. Consequently, these forces will not appear in the equation representing the balance of torques. The lever arms for the remaining forces are shown in part c as red, dashed lines. The following list summarizes these forces, the lever arms, and the torques:

Force	Lever arm	Torque
$W_L = 355$ N	$\ell_L = (4.00\text{ m})\cos 50.0°$	$-W_L\ell_L$
$W_F = 875$ N	$\ell_F = (6.30\text{ m})\cos 50.0°$	$-W_F\ell_F$
P	$\ell_P = (8.00\text{ m})\sin 50.0°$	$+P\ell_P$

Setting the sum of the torques equal to zero gives

$$\Sigma\tau = -W_L\ell_L - W_F\ell_F + P\ell_P \tag{9.2}$$
$$= -(355\text{ N})(4.00\cos 50.0°\text{ m}) - (875\text{ N})(6.30\cos 50.0°\text{ m})$$
$$+ P(8.00\sin 50.0°\text{ m}) = 0$$

This equation can be solved to give $\boxed{P = 727\text{ N}}$. Equation 4.9a indicates that $G_x = P$, so $\boxed{G_x = 727\text{ N}}$.

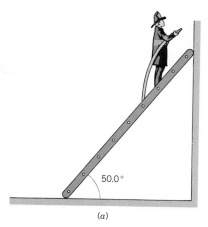

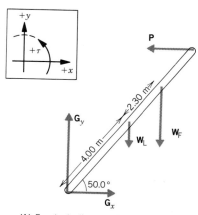

(b) Free-body diagram of the ladder

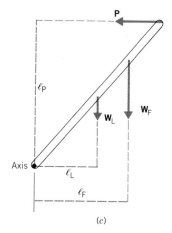

Figure 9.8 (a) A ladder leaning against a smooth (frictionless) wall. (b) The free-body diagram for the ladder. (c) The forces and their lever arms. The axis of rotation is at the lower end of the ladder and is perpendicular to the page.

SELECTING THE DIRECTIONS OF THE FORCES IN THE FREE-BODY DIAGRAM

To a large extent the directions of the forces acting on an object in equilibrium can be deduced using intuition. Sometimes, however, the direction of an unknown force is not obvious, and it is inadvertently drawn reversed in the free-body diagram. This kind of mistake causes no difficulty. *Choosing the direction of an unknown force backward in the free-body diagram simply means that the value determined for the force will be a negative number,* as the next example illustrates.

Example 6   Bodybuilding

A bodybuilder, strengthening his shoulder muscles, holds a dumbbell of weight W_d as in Figure 9.9a. His arm is extended horizontally and weighs $W_a = 31.0$ N. The weights W_d and W_a act on the arm at the points identified in part b of the drawing. (See part b for distances.) The deltoid muscle is assumed to be the only muscle acting and is attached to the arm as shown. The maximum force M that the deltoid muscle can supply to keep the arm horizontal has a magnitude of 1840 N. What is the weight of the largest dumbbell that can be held, and what are the horizontal and vertical force components, S_x and S_y, that the shoulder joint applies to the arm?

REASONING   Figure 9.9b shows the free-body diagram of the arm. Note that S_x is directed to the right, because the deltoid muscle pulls the arm in toward the shoulder joint, and the joint pushes back in accordance with Newton's third law. The direction of the force S_y, however, is less obvious, and we are alert for the possibility that the direction chosen in the free-body diagram might be backward. If so, the value obtained for S_y will turn out negative.

SOLUTION   The arm is in equilibrium, so the forces must balance:

$$\Sigma F_x = S_x - (1840 \text{ N}) \cos 13.0° = 0 \qquad (4.9a)$$

$$\boxed{S_x = 1790 \text{ N}}$$

$$\Sigma F_y = S_y + (1840 \text{ N}) \sin 13.0° - 31.0 \text{ N} - W_d = 0 \qquad (4.9b)$$

Equation 4.9b cannot be solved at this point, because it contains two unknowns. However, since the arm is in equilibrium, the torques acting on the arm must balance, and this fact provides another equation. To calculate torques, we choose an axis through the left end of the arm and perpendicular to the page. With this axis, the torques due to S_x and S_y are zero. The list below summarizes the forces, their lever arms, and the torques.

Force	Lever arm	Torque
$W_a = 31.0$ N	$\ell_a = 0.280$ m	$-W_a\ell_a$
W_d	$\ell_d = 0.620$ m	$-W_d\ell_d$
$M = 1840$ N	$\ell_t = (0.150 \text{ m}) \sin 13.0°$	$+M\ell_t$

The condition specifying a zero net torque is

$$\Sigma \tau = -W_a\ell_a - W_d\ell_d + M\ell_t \qquad (9.2)$$
$$= -(31.0 \text{ N})(0.280 \text{ m}) - W_d(0.620 \text{ m})$$
$$+ (1840 \text{ N})(0.150 \text{ m}) \sin 13.0° = 0$$

Equation 9.2 can be solved to show that the maximum dumbbell weight is $\boxed{W_d = 86.1 \text{ N}}$. This value for W_d can be substituted in Equation 4.9b to show that the value for S_y is $\boxed{S_y = -297 \text{ N}}$. The minus sign indicates that the choice of direction for S_y in the free-body diagram is wrong. In reality, S_y has a magnitude of 297 N but is directed downward, not upward.

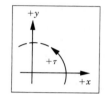

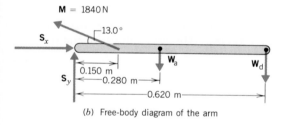

(a)

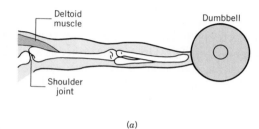

(b) Free-body diagram of the arm

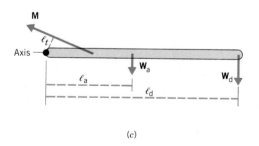

(c)

Figure 9.9 (a) The fully extended, horizontal arm of a bodybuilder supports a dumbbell. (b) The free-body diagram for the arm. (c) The forces that act on the arm and their lever arms. The axis of rotation is at the left end of the arm and is perpendicular to the page.

9.3 CENTER OF GRAVITY

Often, it is important to know the torque produced by the weight of an *extended* body. In Examples 5 and 6, for instance, it is necessary to determine the torques caused by the weight of the ladder and arm, respectively. In both cases the weight of the object is considered to act at a definite point for the purpose of calculating the torque. This point is called the *center of gravity* (abbreviated "cg").

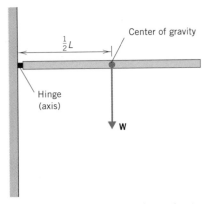

Figure 9.10 A thin, uniform, horizontal rod of length L is attached to a vertical wall by a hinge. The center of gravity of the rod is at its geometrical center.

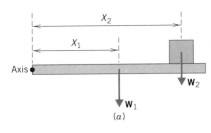

(a)

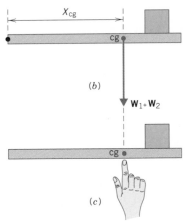

(b)

(c)

Figure 9.11 (a) A box rests near the right end of a horizontal board. (b) The total weight acts at the center of gravity of the group. (c) The group can be balanced by applying an external force at the center of gravity.

Definition of Center of Gravity

The center of gravity of a rigid body is the point at which its weight can be considered to act when calculating the torque due to the weight.

When an object has a symmetric shape and its weight is distributed uniformly, the center of gravity lies at its geometrical center. For instance, consider the thin, uniform, horizontal rod of length L attached to a vertical wall by a hinge in Figure 9.10. The center of gravity of the rod is located at the geometrical center. The lever arm for the weight **W** is $L/2$, and the magnitude of the torque is $\tau = W(L/2)$. In a similar fashion, the center of gravity of any symmetrically shaped and uniform object, such as a sphere, disk, cube, cylinder, etc., is located at its geometrical center. The center of gravity need not necessarily lie within the object itself. The center of gravity of a stereo record, for instance, lies at the center of the spindle hole and, hence, is "outside" the record.

Suppose we have a group of objects, with known weights and centers of gravity, and it is necessary to know the center of gravity for the group as a whole. As an example, Figure 9.11a shows a group composed of two parts: a horizontal uniform board (weight W_1) and a uniform box (weight W_2) near the right end of the board. The center of gravity of the group can be determined by calculating the net torque created by the board and box about an axis that is arbitrarily picked to lie at the left end of the board. Part a of the figure shows the weights W_1 and W_2 and their corresponding lever arms X_1 and X_2. The net torque is $\Sigma\tau = W_1 X_1 + W_2 X_2$. It is also possible to calculate the net torque by treating the total weight $W_1 + W_2$ as if it were located at the center of gravity and had the lever arm X_{cg}, as part b of the drawing indicates: $\Sigma\tau = (W_1 + W_2)X_{cg}$. The two values for the net torque must be the same, so that

$$W_1 X_1 + W_2 X_2 = (W_1 + W_2)X_{cg}$$

This expression can be solved for X_{cg}, which locates the center of gravity relative to the axis:

$$\begin{bmatrix} \text{Center} \\ \text{of} \\ \text{gravity} \end{bmatrix} \qquad X_{cg} = \frac{W_1 X_1 + W_2 X_2 + \cdots}{W_1 + W_2 + \cdots} \qquad (9.3)$$

The notation "$+\cdots$" indicates that the reasoning above can be extended to account for any number of weights distributed along a horizontal line. Figure 9.11c illustrates that the group can be balanced by a single external force, if the line of action of the force passes through the center of gravity, and if the force is equal in magnitude, but opposite in direction, to the weight of the group. Example 7 demonstrates how to calculate the center of gravity for the human arm.

Example 7 The Center of Gravity of an Arm

The horizontal arm in Figure 9.12 is composed of three parts: the upper arm (weight $W_1 = 17$ N), the lower arm ($W_2 = 11$ N), and the hand ($W_3 = 4.2$ N). The drawing shows the center of gravity of each part, measured with respect to the shoulder joint. Find the center of gravity of the entire arm, relative to the shoulder joint.

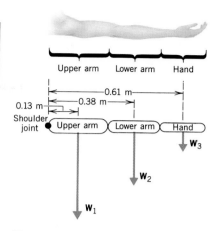

Figure 9.12 The three parts of the human arm, and the weight and center of gravity for each.

REASONING AND SOLUTION The coordinate X_{cg} of the center of gravity is given by

$$X_{cg} = \frac{W_1 X_1 + W_2 X_2 + W_3 X_3}{W_1 + W_2 + W_3} \qquad (9.3)$$

$$= \frac{(17 \text{ N})(0.13 \text{ m}) + (11 \text{ N})(0.38 \text{ m}) + (4.2 \text{ N})(0.61 \text{ m})}{17 \text{ N} + 11 \text{ N} + 4.2 \text{ N}}$$

$$= \boxed{0.28 \text{ m}}$$

The center of gravity of an object with an irregular shape and a nonuniform weight distribution can be found by suspending the object from two different points P_1 and P_2, one at a time. Figure 9.13a shows the object at the moment of release, when its weight **W**, acting at the center of gravity, has a nonzero lever arm ℓ. At this instant the weight produces a torque about the axis. The tension force **T**, applied to the object by the suspension cord produces no torque, because its line of action passes through the axis. Hence, in part *a* there is a net torque applied to the object, and the object begins to rotate. Friction eventually brings the object to rest as in part *b*, where the center of gravity lies directly below the point of suspension. In such an orientation, the line of action of the weight passes through the axis, so there is no longer any net torque. In the absence of a net torque the object remains at rest.

By suspending the object from a second point (see Figure 9.13c), a second line through the object can be established, along which the center of gravity must also lie. The center of gravity, then, must be at the intersection of the two lines.

The center of gravity is closely related to another concept known as the *center of mass*. If the weight $W = mg$ of each object in Equation 9.3 is replaced by its mass m, the resulting equation gives the x coordinate of the center of mass. The two points are identical if the acceleration g due to gravity does not vary over the physical extent of the objects. Otherwise, the center of gravity and the center of mass are different. For ordinary-sized objects the two centers coincide.

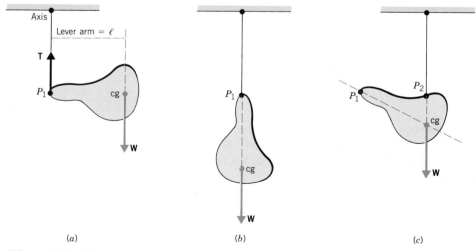

(a)   (b)   (c)

Figure 9.13 The center of gravity (cg) of an object can be located experimentally by suspending the object from two different points, one at a time.

9.4 NEWTON'S SECOND LAW FOR ROTATIONAL MOTION ABOUT A FIXED AXIS

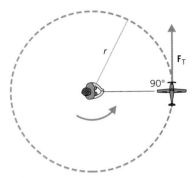

Figure 9.14 A model airplane of mass m is flying on a guideline of length r. A net tangential force F_T acts on the plane.

The goal of this section is to put Newton's second law into a form that is suitable for describing the rotational motion of a rigid object about a fixed axis. We begin by considering a particle moving on a circular path. Figure 9.14 presents a good approximation of this situation by using a small model plane on a guideline of negligible mass. The plane's engine produces a net tangential force F_T that gives the plane a tangential acceleration a_T, in accord with Newton's second law, $F_T = ma_T$. The torque τ produced by the net tangential force is $\tau = F_T r$, where the radius r of the circular path is also the lever arm. As a result, the torque is $\tau = ma_T r$. But the tangential acceleration is related to the angular acceleration according to $a_T = r\alpha$ (Equation 8.10), where α must be expressed in rad/s^2. With this substitution for a_T, the torque becomes

$$\tau = \underbrace{(mr^2)}_{\substack{\text{Moment} \\ \text{of inertia } I}}\alpha \tag{9.4}$$

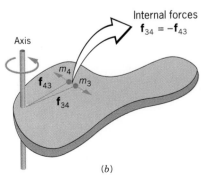

(a)

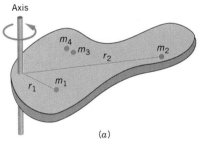

Internal forces
$\mathbf{f}_{34} = -\mathbf{f}_{43}$

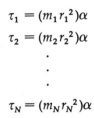

(b)

Figure 9.15 (a) A rigid body consists of a large number of particles, four of which are shown here. (b) The internal forces that particles 3 and 4 exert on each other obey Newton's law of action and reaction.

Equation 9.4 is the form of Newton's second law we have been seeking. It indicates that the net torque τ is directly proportional to the angular acceleration α. The constant of proportionality is $I = mr^2$, which is called the **moment of inertia of the particle**. The SI unit for moment of inertia is kg·m^2.

If all objects were particles, it would be just as convenient to use the second law in the form $F_T = ma_T$, as in the form $\tau = I\alpha$. The advantage in using $\tau = I\alpha$ is that it can be applied to any rigid body rotating about a fixed axis, and not just to a particle. To illustrate how this advantage arises, Figure 9.15a shows a flat sheet of material that rotates about an axis perpendicular to the sheet. The sheet is composed of a number of mass particles, $m_1, m_2, \ldots, m_N$, where N is very large. Only four particles are shown for the sake of clarity. Each particle behaves in the same way as the model airplane in Figure 9.14 and obeys the relation $\tau = I\alpha$:

$$\tau_1 = (m_1 r_1{}^2)\alpha$$
$$\tau_2 = (m_2 r_2{}^2)\alpha$$
$$\vdots$$
$$\tau_N = (m_N r_N{}^2)\alpha$$

In these equations each particle has the same angular acceleration α, since the rotating object is assumed to be rigid. Adding together the N equations and factoring out the common value of α, we find that

$$\underbrace{\Sigma\tau}_{\substack{\text{Net} \\ \text{torque}}} = \underbrace{(\Sigma mr^2)}_{\substack{\text{Moment} \\ \text{of inertia}}}\alpha \tag{9.5}$$

where $\Sigma\tau = \tau_1 + \tau_2 + \cdots + \tau_N$ is the sum of the torques, and $\Sigma mr^2 = m_1 r_1{}^2 + m_2 r_2{}^2 + \cdots + m_N r_N{}^2$ represents the sum of the individual moments of inertia.

The latter quantity is the *moment of inertia of the body* and is represented by the symbol I:

$$\begin{bmatrix} \text{Moment of} \\ \text{inertia of} \\ \text{a body} \end{bmatrix} \qquad\qquad I = \Sigma m r^2 \qquad\qquad (9.6)$$

In Equation 9.6, r is the perpendicular radial distance of each particle from the axis of rotation. Combining Equation 9.6 with Equation 9.5 gives the following result:

Rotational Analog of Newton's Second Law for Rigid Bodies Rotating About a Fixed Axis

$$\text{Net torque} = \begin{pmatrix} \text{Moment of} \\ \text{inertia} \end{pmatrix} \times \begin{pmatrix} \text{Angular} \\ \text{acceleration} \end{pmatrix}$$

$$\Sigma \tau = I\alpha \qquad\qquad (9.7)$$

Requirement: α must be expressed in rad/s^2.

The form of the second law for rotational motion, $\Sigma \tau = I\alpha$, is similar to that for translational (linear) motion, $\Sigma F = ma$, and is valid only in an inertial frame. The moment of inertia I plays the same role for rotational motion that the mass m does for translational motion. Thus, I is a measure of the rotational inertia of a body. When using Equation 9.7, α must be expressed in rad/s^2, because the relation $a_T = r\alpha$ (which requires radian measure) was used in the derivation.

When calculating the sum of torques in Equation 9.7, it is necessary to include only the *external torques*, those applied by agents outside the body. The torques produced by internal forces need not be considered, because they always combine to produce a net torque of zero. Internal forces are those that one particle within the body exerts on another particle and always occur in pairs of oppositely directed forces of equal magnitude, in accord with Newton's third law (see m_3 and m_4 in Figure 9.15b). The forces in such a pair have the same line of action, so the forces have identical lever arms and produce torques of equal magnitudes. One torque is counterclockwise, while the other is clockwise, the net torque from the pair being zero.

It can be seen from Equation 9.6 that the moment of inertia depends on both the mass of each particle and its distance from the axis of rotation. The farther a particle is from the axis, the greater is its contribution to the moment of inertia. Therefore, although a rigid object possesses a unique total mass, it does not have a unique moment of inertia, for *the moment of inertia depends on the location and orientation of the axis relative to the particles that make up the object.* Example 8 shows how the moment of inertia can change when the axis of rotation changes.

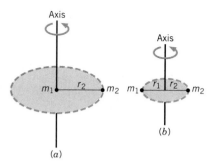

Figure 9.16 Two particles, masses m_1 and m_2, are attached to the ends of a massless rigid rod. The moment of inertia of this object is different, depending on whether the rod rotates about an axis passing through (a) the end or (b) the center of the rod.

Example 8   The Moment of Inertia Depends on Where the Axis Is

Two particles each have a mass m and are fixed to the ends of a thin rigid rod, whose mass can be ignored. The length of the rod is L. Find the moment of inertia when this object rotates relative to an axis that is perpendicular to the rod at (a) one end and (b) the center. (See Figure 9.16.)

REASONING AND SOLUTION

(a) Particle 1 lies on the axis, as part *a* of the drawing shows, and has a zero radial distance: $r_1 = 0$. In contrast, particle 2 moves on a circle whose radius is $r_2 = L$. The moment of inertia is

$$I = \Sigma mr^2 = m_1 r_1^2 + m_2 r_2^2 = \boxed{mL^2} \tag{9.6}$$

(b) Part *b* of the drawing shows that particle 1 no longer lies on the axis but now moves on a circle of radius $r_1 = L/2$. Particle 2 moves on a circle with the same radius, $r_2 = L/2$. Therefore,

$$I = \Sigma mr^2 = m_1 r_1^2 + m_2 r_2^2 = m(L/2)^2 + m(L/2)^2 = \boxed{\tfrac{1}{2}mL^2}$$

This value differs from that in part (a), because the axis of rotation is different.

The procedure illustrated in Example 8 can be extended using integral calculus to evaluate the moment of inertia of a rigid object with a continuous mass distribution, and Table 9.1 gives some typical results. These results depend on the total mass of the object, its shape, and the location and orientation of the axis.

In Example 9, the relation $\Sigma\tau = I\alpha$ is applied to the platter of a stereo turntable.

Example 9 The Torque of a Turntable Motor

Most turntables can bring a record from rest up to the rated angular speed of $33\tfrac{1}{3}$ rev/min in one-half a revolution. The platter of one turntable has a moment of inertia of 0.0500 kg·m² (including the effect of the record). Neglecting frictional effects, what torque (assumed constant) must the turntable motor apply to the platter to achieve this performance?

REASONING Newton's second law for rotational motion can be used to find the torque, once the angular acceleration is determined. The angular acceleration can be calculated from the data in the table below and the appropriate equation of rotational kinematics.

θ	α	ω	ω_0	t
$-\pi$ rad ($-\tfrac{1}{2}$ rev)	?	-3.49 rad/s ($-33\tfrac{1}{3}$ rev/min)	0	

The rotational variables θ and ω are negative, because the rotation of a turntable platter is clockwise when viewed from above. The data for θ and ω have been converted to radian measure, because $\Sigma\tau = I\alpha$ requires that α be expressed in radian measure.

SOLUTION From $\omega^2 = \omega_0^2 + 2\alpha\theta$ (Equation 8.8) it follows that

$$\alpha = \frac{\omega^2 - \omega_0^2}{2\theta} = \frac{(-3.49 \text{ rad/s})^2}{2(-\pi \text{ rad})} = -1.94 \text{ rad/s}^2$$

The second law for rotational motion can now be used to obtain the torque:

$$\Sigma\tau = I\alpha = (0.0500 \text{ kg·m}^2)(-1.94 \text{ rad/s}^2) = \boxed{-0.0970 \text{ N·m}} \tag{9.7}$$

Sometimes, rotational motion and translational motion occur together. The next example deals with an interesting situation in which both angular acceleration and translational acceleration must be considered.

Table 9.1 Moments of Inertia for Various Rigid Objects of Mass *M*

Thin-walled hollow cylinder or hoop		$I = MR^2$
Solid cylinder or disk		$I = \frac{1}{2}MR^2$
Thin rod, axis perpendicular to rod and passing through center		$I = \frac{1}{12}ML^2$
Thin rod, axis perpendicular to rod and passing through one end		$I = \frac{1}{3}ML^2$
Solid sphere, axis through center		$I = \frac{2}{5}MR^2$
Solid sphere, axis tangent to surface		$I = \frac{7}{5}MR^2$
Thin-walled spherical shell, axis through center		$I = \frac{2}{3}MR^2$
Thin rectangular sheet, axis parallel to one edge and passing through center of other edge		$I = \frac{1}{12}ML^2$
Thin rectangular sheet, axis along one edge		$I = \frac{1}{3}ML^2$

Example 10   Hoisting a Crate

A 451-kg crate is being lifted by the hoisting mechanism in Figure 9.17*a*. The two cables are wrapped around their respective pulleys, which have radii of 0.600 and 0.200 m. The pulleys are fastened together to form a "dual" pulley and turn as a single unit about the center axle, relative to which the combined moment of inertia is $I = 50.0$ kg·m². If a tension of magnitude $T_1 = 2150$ N is maintained in the cable attached to the motor, find the angular acceleration of the "dual" pulley and the tension in the cable connected to the crate.

REASONING   Three external forces act on the dual pulley, as its free-body diagram shows (Figure 9.17b). These forces are (1) the tension $\mathbf{T_1}$ in the cable connected to the motor, (2) the tension $\mathbf{T_2}$ in the cable attached to the crate, and (3) the reaction force $\mathbf{P}$ exerted on the dual pulley by the axle. The force $\mathbf{P}$ arises because the two cables pull the pulley down and to the left into the axle, and the axle pushes back, thus keeping the pulley in place. Notice that $\mathbf{P}$ has a zero lever arm with respect to the axle, since the line of action of $\mathbf{P}$ passes directly through the axle.

SOLUTION   Using the lever arms shown in part b of the figure, we can apply the second law to the rotational motion of the pulley:

$$\Sigma \tau = I\alpha \tag{9.7}$$

$$(2150 \text{ N})(0.600 \text{ m}) - T_2(0.200 \text{ m}) = (50.0 \text{ kg} \cdot \text{m}^2)\alpha$$

This equation contains two unknown quantities, so a second equation is needed and may be obtained by considering the crate. The crate accelerates upward under the action of its weight [$W = (451 \text{ kg})(9.80 \text{ m/s}^2) = 4420 \text{ N}$] and the cable tension $\mathbf{T_2}$, as the free-body diagram in part c of the drawing indicates. Applying Newton's second law to the translational motion of the crate gives

$$\Sigma F_y = ma_y \tag{4.2b}$$

$$T_2 - (4420 \text{ N}) = (451 \text{ kg})(a_y)$$

Because the cable attached to the crate rolls on the pulley without slipping, the linear acceleration a_y of the crate is related to the angular acceleration of the pulley via Equation 8.13: $a_y = r\alpha = (0.200 \text{ m})\alpha$. With this substitution for a_y, Equation 4.2b becomes

$$T_2 - (4420 \text{ N}) = (451 \text{ kg})(0.200 \text{ m})\alpha$$

This result and Equation 9.7 can be solved simultaneously to yield

$$\boxed{T_2 = 4960 \text{ N}} \quad \text{and} \quad \boxed{\alpha = 6.0 \text{ rad/s}^2}$$

Figure 9.17   (a) The crate is lifted upward by the motor and pulley arrangement. The free-body diagram for (b) the dual pulley and (c) the crate.

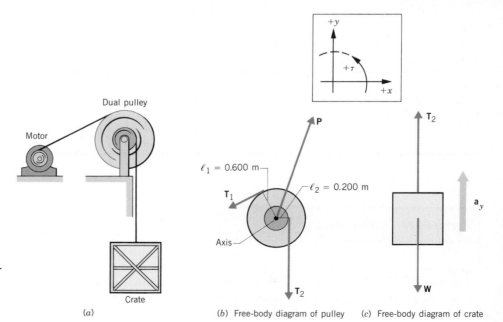

(a)

(b) Free-body diagram of pulley

(c) Free-body diagram of crate

Table 9.2   Analogies between Rotational and Translational Concepts

Physical Concept	Rotational	Translational
Displacement	θ	s
Velocity	ω	v
Acceleration	α	a
The cause of acceleration	Torque τ	Force F
Inertia	Moment of inertia I	Mass m
Newton's second law	$\Sigma\tau = I\alpha$	$\Sigma F = ma$
Work	$\tau\theta$	Fs
Kinetic energy	$\frac{1}{2}I\omega^2$	$\frac{1}{2}mv^2$
Momentum	$L = I\omega$	$p = mv$

We have seen that Newton's second law for rotational motion, $\Sigma\tau = I\alpha$, has the same form as that for translational motion, $\Sigma F = ma$, so each rotational variable has a translational analog: torque τ and force F are analogous quantities, as are moment of inertia I and mass m, and angular acceleration α and linear acceleration a. The other physical concepts developed for studying translational motion, such as kinetic energy and momentum, also have rotational analogs. For future reference, Table 9.2 itemizes these concepts and their rotational analogs.

9.5   ROTATIONAL WORK AND ENERGY

ROTATIONAL WORK

Work and energy are among the most fundamental and useful concepts in physics. Chapter 6 discusses their application to translational motion. These concepts are equally useful for rotational motion, provided they are expressed in terms of angular variables.

The work W done by a constant force that points in the same direction as the displacement is $W = Fs$ (Equation 6.1), where F and s are the magnitudes of the force and displacement, respectively. In Figure 9.18 a rope is wrapped around a wheel and is under a constant tension F. If the rope is pulled out a distance s, the wheel rotates through an angle $\theta = s/r$ (Equation 8.1), where r is the radius of the wheel and θ is in radians. The work done by the tension force in turning the wheel is $W = Fs = Fr\theta$. But Fr is the torque τ applied to the wheel by the tension, so the rotational work can be written in angular variables as follows:

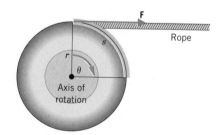

Figure 9.18   The force **F** does work in rotating the wheel through the angle θ.

Definition of Rotational Work

The rotational work $W_\mathbf{R}$ done by a constant torque τ in turning an object through an angle θ is

$$W_\mathbf{R} = \tau\theta \qquad (9.8)$$

Requirement: θ must be expressed in radians.

SI Unit of Rotational Work: joule (J)

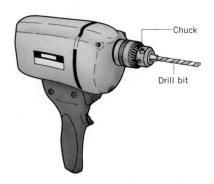

Figure 9.19 An electric drill.

Example 11 considers the rotational work done by a common power tool.

Example 11   *An Electric Drill*

An electric drill is turned on (Figure 9.19), and the chuck exerts a constant torque of 4.0×10^{-4} N·m to turn the drill bit through 9.0 revolutions. Find the rotational work done by the chuck.

REASONING AND SOLUTION   Before we can use $W_R = \tau\theta$ for calculating work, it is necessary to express the angular displacement θ in radians: $\theta = (9.0 \text{ rev}) \times [2\pi \text{ rad}/(1 \text{ rev})] = 57$ rad. The rotational work done by the chuck is

$$W_R = \tau\theta = (4.0 \times 10^{-4} \text{ N·m})(57 \text{ rad}) = \boxed{2.3 \times 10^{-2} \text{ J}}$$

ROTATIONAL KINETIC ENERGY

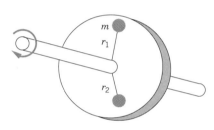

Figure 9.20 The rotating wheel is composed of many particles, two of which are shown.

A rotating body possesses kinetic energy, because its constituent particles are in motion. If the body is rotating with an angular speed ω, the tangential speed v_T of a particle at a distance r from the axis is $v_T = r\omega$ (Equation 8.9). Figure 9.20 shows two such particles. If a particle's mass is m, its kinetic energy is $\frac{1}{2}mv_T^2 = \frac{1}{2}mr^2\omega^2$. The kinetic energy of the entire rotating body, then, is the sum of the kinetic energies of the particles:

$$\text{Rotational KE} = \Sigma\tfrac{1}{2}mr^2\omega^2 = \tfrac{1}{2}(\Sigma mr^2)\omega^2$$

In this result, the angular velocity ω is the same for all particles in a rigid body and, therefore, has been factored outside the summation. The term in parentheses is the moment of inertia, $I = \Sigma mr^2$, so the rotational kinetic energy takes the form given below:

Definition of Rotational Kinetic Energy

The rotational kinetic energy of a rigid object rotating with an angular speed ω about a fixed axis and having a moment of inertia I is

$$\text{Rotational KE} = \tfrac{1}{2}I\omega^2 \qquad\qquad (9.9)$$

Requirement: ω must be expressed in rad/s.

SI Unit of Rotational Kinetic Energy: joule (J)

When a bicycle or a car is in motion, its tires are both translating and rotating. The total kinetic energy KE of an object that is simultaneously translating and rotating is the sum of its translational and rotational kinetic energies:

$$\text{Total KE} = \tfrac{1}{2}mv^2 + \tfrac{1}{2}I\omega^2$$

where v is the translational speed of the object's center of mass, m is the total mass, I is the moment of inertia about an axis through the center of mass, and ω is the angular speed. The next example deals with combined translational and rotational motion.

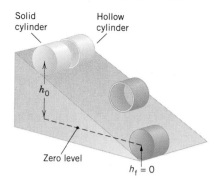

Example 12 Rolling Cylinders

A hollow cylinder (mass $= m_h$, radius $= r_h$) and a solid cylinder (mass $= m_s$, radius $= r_s$) start from rest at the top of an inclined plane (Figure 9.21). Both cylinders start at the same vertical height h_0. All heights are measured relative to an arbitrarily chosen zero level that passes through the center of mass of a cylinder when it is at the bottom of the incline (see the drawing). Ignoring energy losses due to retarding forces, determine which cylinder has the greatest translational speed upon reaching the bottom of the incline.

REASONING Only the conservative force of gravity does work on the cylinders, so the total mechanical energy is conserved as they roll down the incline. However, in using the principle of conservation of mechanical energy, we must include the rotational kinetic energy. The total mechanical energy E at any height h above the zero level is the sum of the translational kinetic energy ($\frac{1}{2}mv^2$), the rotational kinetic energy ($\frac{1}{2}I\omega^2$), and the gravitational potential energy (mgh):

$$E = \tfrac{1}{2}mv^2 + \tfrac{1}{2}I\omega^2 + mgh$$

As the cylinders roll down, potential energy is converted into kinetic energy. But the kinetic energy is shared between the translational form ($\frac{1}{2}mv^2$) and the rotational form ($\frac{1}{2}I\omega^2$). The object with more of its kinetic energy in the translational form will have the greater translational speed at the bottom of the incline. As we shall see, the solid cylinder has the greater translational speed.

SOLUTION The total mechanical energy E_f at the bottom ($h_f = 0$) is the same as the total mechanical energy E_0 at the top ($h = h_0$, $v_0 = 0$, $\omega_0 = 0$):

$$\tfrac{1}{2}mv_f^2 + \tfrac{1}{2}I\omega_f^2 + mgh_f = \tfrac{1}{2}mv_0^2 + \tfrac{1}{2}I\omega_0^2 + mgh_0$$
$$\tfrac{1}{2}mv_f^2 + \tfrac{1}{2}I\omega_f^2 = mgh_0$$

Since each cylinder rolls without slipping, the final rotational speed ω_f and the final translational speed v_f of its center of mass are related according to Equation 8.12, $\omega_f = v_f/r$, where r is the radius of the cylinder. Substituting this expression for ω_f into the equation above and solving for v_f yields

$$v_f = \sqrt{\frac{2mgh_0}{m + I/r^2}}$$

Setting $m = m_h$, $r = r_h$ and $I = m r_h^2$ for the hollow cylinder and then setting $m = m_s$, $r = r_s$ and $I = \frac{1}{2}m r_s^2$ for the solid cylinder (see Table 9.1), we find that the two cylinders have the following translational speeds at the bottom of the incline:

[Hollow cylinder] $\qquad v_f = \sqrt{gh_0}$

[Solid cylinder] $\qquad v_f = \sqrt{\dfrac{4gh_0}{3}} = 1.15\sqrt{gh_0}$

The solid cylinder, having the greatest translational speed, arrives at the bottom first, a result that is independent of the masses and radii of the cylinders.

Figure 9.21 A hollow cylinder and a solid cylinder each have the same mass and start together from rest at the top of an incline. The conservation of mechanical energy can be used to show that the solid cylinder, having the greatest translational speed, reaches the bottom first.

9.6 ANGULAR MOMENTUM

In Chapter 7 the linear momentum p of an object is defined as the product of its mass m and linear velocity v, $p = mv$. For rotational motion the analogous concept is called the **angular momentum** L. The mathematical form of angular momentum

This satellite image shows a severe storm over the British Isles. The swirling air mass has a large angular momentum.

is identical to that of linear momentum, with the mass m and the linear velocity v being replaced with their rotational counterparts, the moment of inertia I and the angular velocity ω.

Definition of Angular Momentum

The angular momentum L of a body rotating about a fixed axis is the product of the body's moment of inertia I and its angular velocity ω:

$$L = I\omega \qquad (9.10)$$

Requirement: ω must be expressed in rad/s.

SI Unit of Angular Momentum: kg·m²/s

Linear momentum is an important concept in physics, because the total momentum P of a system is conserved when the net external force acting on the system is zero. Then, the final linear momentum P_f and the initial linear momentum P_0 of the system are the same: $P_f = P_0$. Similarly, when no net external torque acts on a system, the final and the initial angular momenta are the same: $L_f = L_0$. This is the **principle of conservation of angular momentum.**

Principle of Conservation of Angular Momentum

The total angular momentum of a system remains constant (is conserved) if the net external torque acting on the system is zero.

Since the angular momentum is the product of the moment of inertia I and the angular velocity ω, one interesting consequence of the conservation principle is that any change in one of these variables must be accompanied by a corresponding change in the other to keep the product constant. For example, the skater in Figure 9.22a is spinning with both arms and a leg outstretched. If the external torques produced by air resistance and friction are so small that they can be ignored, the skater would spin forever at the same angular velocity, because her angular momentum would be conserved. It is also because of the conservation of angular momentum that the skater can spin at a greater angular velocity by pulling in her arms and leg, as part b of the drawing shows. When the mass of each arm and the leg is moved closer to the rotational axis, the skater's moment of inertia decreases. The angular velocity, therefore, must increase to keep the angular momentum L constant ($L = I\omega$ = constant), as in Example 13.

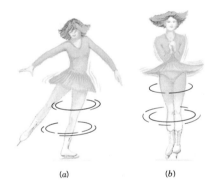

Figure 9.22 (a) A skater spins slowly on one skate, with both arms and one leg outstretched. (b) As she pulls her arms and leg in toward the rotational axis, her moment of inertia I decreases, and the angular speed ω increases.

THE PHYSICS OF . . .

a spinning ice skater.

Example 13   A Spinning Skater

A skater is spinning at an angular velocity of $\omega_0 = 9.0$ rad/s. By pulling her arms and legs inward, she reduces her moment of inertia to 41% of its initial value. What is the final angular velocity of her spin?

REASONING AND SOLUTION   We assume that any external torques due to friction and wind resistance are negligible. Then, the principle of conservation of angular momentum applies, and the skater's final and initial angular momenta are the same: $I_f\omega_f = I_0\omega_0$. Since she has reduced her moment of inertia to 41% of its initial value,

$I_f = 0.41\ I_0$, and we find that

$$I_f \omega_f = (0.41\ I_0)\omega_f = I_0 \omega_0$$

As a result, $\omega_f = \omega_0/0.41 = (9.0\ \text{rad/s})/0.41 = \boxed{22\ \text{rad/s}}$.

The last example in this chapter involves a satellite and illustrates another application of the principle of conservation of angular momentum.

Example 14   A Satellite in an Elliptical Orbit

An artificial satellite is placed into an elliptical orbit about the earth, as in Figure 9.23. Telemetry data indicate that its point of closest approach (called the *perigee*) is $r_P = 8.37 \times 10^6$ m from the center of the earth, while its point of greatest distance (called the *apogee*) is $r_A = 25.1 \times 10^6$ m from the center of the earth. The speed of the satellite at the perigee is $v_P = 8450$ m/s. Find its speed v_A at the apogee.

REASONING   The only force of any significance that acts on the satellite is the gravitational force of the earth. However, at any instant, this force is directed toward the center of the earth and passes through the axis about which the satellite instantaneously rotates. Therefore, the gravitational force exerts *no torque* on the satellite (the lever arm is zero). Consequently, the angular momentum of the satellite remains constant at all times.

SOLUTION   Since the angular momentum is the same at the apogee (A) and the perigee (P), it follows that $I_A \omega_A = I_P \omega_P$. Furthermore, the orbiting satellite can be considered a point mass, so its moment of inertia is $I = mr^2$. In addition, the angular speed ω of the satellite is related to its tangential speed v_T by $\omega = v_T/r$. If these relations are used at the apogee and perigee, the conservation of angular momentum gives the following result:

$$I_A \omega_A = I_P \omega_P$$

$$(mr_A{}^2)\left(\frac{v_A}{r_A}\right) = (mr_P{}^2)\left(\frac{v_P}{r_P}\right)$$

$$v_A = \frac{r_P v_P}{r_A} = \frac{(8.37 \times 10^6\ \text{m})(8450\ \text{m/s})}{25.1 \times 10^6\ \text{m}} = \boxed{2820\ \text{m/s}}$$

The answer is independent of the mass of the satellite. The satellite behaves just like the skater in Figure 9.22, because its speed is greater at the perigee, where the moment of inertia is smaller.

THE PHYSICS OF . . .

a satellite in orbit about the earth.

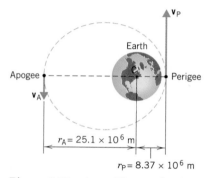

Figure 9.23   A satellite moving in an elliptical orbit about the earth. The gravitational force exerts no torque on the satellite, so the angular momentum of the satellite is conserved.

The result in Example 14 indicates that a satellite does not have a constant speed in an elliptical orbit. The speed changes from a maximum at the perigee to a minimum at the apogee; the closer the satellite comes to the earth, the faster it travels. Planets moving around the sun in elliptical orbits exhibit the same kind of behavior, and Johannes Kepler (1571–1630) formulated his famous second law based on observations of such characteristics of planetary motion. Kepler's second law states that, in a given amount of time, a line joining any planet to the sun sweeps out the same amount of area no matter where the planet is on its elliptical orbit, as Figure 9.24 illustrates. The conservation of angular momentum can be used to show why the law is valid, by means of a calculation similar to that in Example 14.

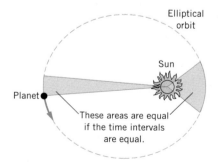

Figure 9.24   Kepler's second law of planetary motion states that a line joining a planet to the sun sweeps out equal areas in equal time intervals.

INTEGRATION OF CONCEPTS

NEWTON'S SECOND LAW FOR ROTATIONAL MOTION AND THE EQUATIONS OF ROTATIONAL KINEMATICS

In Chapter 4 we discussed Newton's second law of motion, $\Sigma\mathbf{F} = m\mathbf{a}$. This law shows how the net force $\Sigma\mathbf{F}$ acting on an object of mass m determines the acceleration $\mathbf{a}$ that appears in the equations of linear kinematics. Now, for a rigid body rotating about a fixed axis, we have seen that the rotational analog of Newton's law takes the form $\Sigma\tau = I\alpha$. A net torque $\Sigma\tau$ acting on an object that has a moment of inertia I causes an angular acceleration α. As is the case for linear motion, the angular acceleration provides the link between Newton's second law and the equations of rotational kinematics. If the net torque and the moment of inertia are known, the angular acceleration can be determined from the second law and then used in the equations of rotational kinematics. Conversely, if the angular acceleration can be found from the equations of rotational kinematics, it can then be used in Newton's second law to provide information about the net torque or the moment of inertia. The linkage between the rotational analog of Newton's second law and the equations of rotational kinematics is the focus of a number of homework problems for this chapter (e.g., problems 27, 30, and 64).

ROTATIONAL MOTION AND THE CONSERVATION OF MECHANICAL ENERGY

One of the important concepts that we encountered in Chapter 6 is the principle of conservation of mechanical energy. This principle states that the total mechanical energy E of an object remains constant when the net work done by nonconservative forces is zero. In Chapter 6, the total mechanical energy E is the sum of the translational kinetic energy $\frac{1}{2}mv^2$, and the gravitational potential energy mgh. When rotational motion also occurs and there is no net work done by nonconservative forces, the total mechanical energy is still conserved. Now, however, the rotational kinetic energy, $\frac{1}{2}I\omega^2$, must be included in the total mechanical energy: $E = \frac{1}{2}I\omega^2 + \frac{1}{2}mv^2 + mgh$. While E remains constant during the motion, the two types of kinetic energies and the potential energy may be transformed into one another. A number of homework problems for this chapter utilize the conservation of mechanical energy with combined rotational and translational motions (e.g., problems 42, 43, and 44).

ROTATIONAL MOTION AND THE CONSERVATION OF ANGULAR MOMENTUM

We discussed the principle of conservation of linear momentum in Chapter 7. This principle states that the total linear momentum of an isolated system remains constant, an isolated system being one in which the sum of the external forces acting on the system is zero. A rotating system, such as a spinning ice skater, also has momentum, which is called angular momentum and is the product of the moment of inertia and angular velocity. In a manner similar to that for the conservation of linear momentum, the angular momentum remains constant if the net external torque acting on the system is zero. Therefore, if the moment of inertia of an isolated system should increase, for example, the magnitude of the angular velocity must decrease in order that the angular momentum remain unchanged during the rotational motion. Homework problems that deal with the conservation of angular momentum include problems 46, 48, and 50, for instance.

SUMMARY

The **torque** τ of a force is the magnitude F of the force times the lever arm ℓ: $\tau = F\ell$. The lever arm is the perpendicular distance between the line of action of the force and the axis of rotation.

A rigid body is in **equilibrium** if it has zero translational acceleration and zero angular acceleration, in which case the net external force and the net external torque acting on the body are zero. For forces acting only in the x, y plane, the conditions for equilibrium are $\Sigma F_x = 0$, $\Sigma F_y = 0$, and $\Sigma \tau = 0$.

The **center of gravity** of a rigid object is the point where its entire weight can be considered to act when calculating the torque due to the weight of the object. For a symmetrical body with uniformly distributed weight, the center of gravity is located at the geometrical center of the body. When a number of objects whose weights W_1, W_2, . . . are distributed along the x axis at locations X_1, X_2, . . . , the center of gravity is located at $X_{cg} = (W_1 X_1 + W_2 X_2 + \cdots)/(W_1 + W_2 + \cdots)$.

Newton's second law for rotational motion is $\Sigma \tau = I\alpha$, where $\Sigma \tau$ is the net external torque applied to a body,

I is the moment of inertia of the body, and α is the angular acceleration (in rad/s²). The **moment of inertia** I of an object composed of N particles is $I = (m_1 r_1^2 + m_2 r_2^2 + \cdots + m_N r_N^2)$, where m_1, m_2, . . . , m_N are the masses of the particles and r_1, r_2, . . . , r_N are the perpendicular distances of the particles from the axis of rotation.

The **rotational work** W_R done by a constant torque τ in turning a rigid body through an angular displacement θ (in radians) is expressed by $W_R = \tau\theta$. The **rotational kinetic energy** of an object with angular speed ω (in rad/s) and moment of inertia I is $KE = \frac{1}{2}I\omega^2$. The **total mechanical energy** E of a rigid object is the sum of its translational kinetic energy, its rotational kinetic energy, and its gravitational potential energy.

The **angular momentum** L of a body rotating with angular velocity ω (in rad/s) about a fixed axis and having a moment of inertia I is $L = I\omega$. The **principle of conservation of angular momentum** states that the total angular momentum of a system remains constant if the net external torque acting on the system is zero.

SOLVED PROBLEMS

Solved Problem 1   The Iron Cross
Related Problems: **21   **22   **63

In gymnastics a number of beautiful routines are performed on the still rings, one of them being the "iron cross." A gymnast who performs this feat weighs 655 N, and each arm is 0.600 m long, as measured from the shoulder joint. Holding his arms parallel to the ground, he uses his latissimus dorsi muscles and pushes with each hand against the rings to support himself, as in part a of the drawing. The latissimus dorsi muscle attaches to the arm at a distance of 0.0700 m from the shoulder joint and, as an approximation, is assumed to be the only muscle acting. It pulls the arm with a force **M** downward and toward the body, as in part b of the drawing. Ignoring the

THE PHYSICS OF . . .

the "iron cross" gymnastics routine.

weight of each arm, find (a) the tension in each supporting rope and (b) the forces that are applied to each arm by the shoulder joint and the latissimus dorsi muscle.

REASONING AND SOLUTION
(a) Part a of the figure shows the free-body diagram for the gymnast and indicates that the pertinent forces are the tensions $\mathbf{T}_1$ and $\mathbf{T}_2$ in each rope and the 655-N weight acting at

Gymnast Chainley Umphrey.

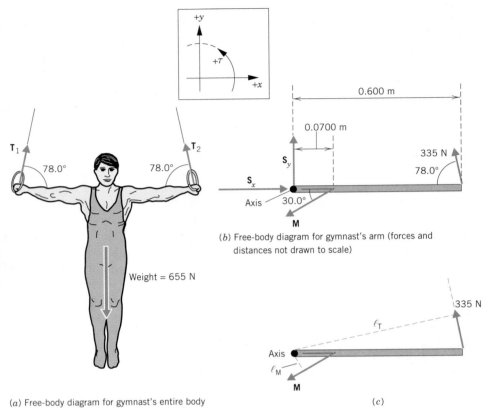

(a) Free-body diagram for gymnast's entire body

(b) Free-body diagram for gymnast's arm (forces and distances not drawn to scale)

(c)

the gymnast's center of gravity. The tension in each rope has the same magnitude ($T_1 = T_2 = T$), assuming that both sides of the gymnast's body are identical, so that his center of gravity is located midway between the two rings. Since the gymnast is in equilibrium, the magnitude T of the tension can be obtained by setting the sum of the vertical force components equal to zero:

$$\Sigma F_y = T \sin 78.0° + T \sin 78.0° - 655 \text{ N} = 0 \quad (4.9b)$$

$$T = \frac{655 \text{ N}}{2 \sin 78.0°} = \boxed{335 \text{ N}}$$

(b) For this part of the problem we select one arm as the object of our analysis, and part b of the figure shows its free-body diagram. The forces acting on the arm are as follows:

1. The 335-N force applied to the arm by the tension in the rope [see part (a)]
2. The muscle force **M**
3. The horizontal and vertical force components S_x and S_y that are exerted on the arm by the shoulder joint.

The directions of S_x and S_y are drawn using our intuition, and if they should be wrong, the values for these forces will turn out to be negative numbers. Since the arm is in equilibrium, the net horizontal force, the net vertical force, and the net

torque are zero. When calculating torques, we select an axis parallel to the ground and perpendicular to the arm at the shoulder joint. For this axis, only the muscle force **M** and the 335-N tension force contribute torques. The corresponding lever arms are shown in part c of the drawing (red, dashed lines): $\ell_M = (0.0700 \text{ m}) \sin 30.0°$ and $\ell_T = (0.600 \text{ m}) \sin 78.0°$.

$$\Sigma F_x = S_x - M \cos 30.0° - (335 \text{ N}) \cos 78.0° = 0 \quad (4.9a)$$
$$\Sigma F_y = S_y - M \sin 30.0° + (335 \text{ N}) \sin 78.0° = 0 \quad (4.9b)$$
$$\Sigma \tau = -M(0.0700 \text{ m}) \sin 30.0°$$
$$+ (335 \text{ N})(0.600 \text{ m}) \sin 78.0° = 0 \quad (9.2)$$

Equation 9.2 can be solved directly to show that $\boxed{M = 5620 \text{ N}}$. With this value for M, Equations 4.9a and 4.9b can be used to show that

$$\boxed{S_x = 4940 \text{ N}} \quad \text{and} \quad \boxed{S_y = 2480 \text{ N}}$$

SUMMARY OF IMPORTANT POINTS Two points are emphasized in this problem. One is that the selection of the object to be analyzed is a very important step in problem solving. Sometimes, as is the case here, it is necessary to choose different objects for analysis at different stages of the same problem. A second important point is that large forces may be necessary to keep a rigid object in equilibrium, if the lever arms of the forces are small. The large force (more than eight times the body weight) exerted by the latissimus dorsi muscle is needed because the muscle attaches to the upper arm so close to the shoulder joint.

QUESTIONS

1. Explain (a) how it is possible for a large force to produce only a small, or even zero, torque, and (b) how it is possible for a small force to produce a large torque.

2. A magnetic tape is being played on a cassette deck. The tension in the tape applies a torque to the supply reel. Assuming the tension remains constant during playback, discuss how this torque varies as the reel becomes empty.

3. A flat rectangular sheet of plywood is fixed so that it can rotate about an axis perpendicular to the sheet through one corner. How should a force (acting in the plane of the sheet) be applied to the plywood so as to create the largest possible torque? Give your reasoning.

4. A torque is the product of a force and a distance (lever arm). Work is also the product of a force and a distance. Yet, torque and work *are different*. What is it about the distances that makes torque and work different?

5. Starting in the spring, fruit begins to grow on the outer end of a branch on a pear tree. Explain how the center of gravity of the pear-growing branch shifts during the course of the summer.

6. The free-body diagram shown in the drawing has been made by a student to show the forces that act on a thin rod in equilibrium. According to the student, the three forces are drawn to scale and lie in the plane of the paper. Are these forces sufficient to keep the rod in equilibrium, or are additional forces necessary? Explain.

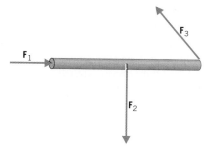

7. An A-shaped step ladder is standing on frictionless ground. The ladder consists of two sections joined at the top and kept from spreading apart by a horizontal crossbar. Draw a free-body diagram showing the forces that keep *one* section of the ladder in equilibrium.

8. For each of the two examples shown in the drawing, which rotating system, (a) or (b), has the *larger* moment of inertia? Give the reason for your answers.

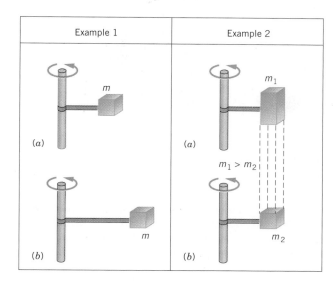

9. Two wheels have the same shape and radii, and they rotate about axes through their centers. The wheels are made from different substances and have different masses. In each case the material is distributed uniformly throughout the wheel. Which wheel, if any, has the larger moment of inertia? Account for your answer.

10. A flat triangular sheet of uniform material is shown in the drawing. There are three possible axes of rotation, each perpendicular to the sheet and passing through one corner, A, B, or C. For which axis is the greatest torque required to bring the triangle up to an angular speed of 100 rev/min in 10.0 s, starting from rest? Explain, assuming that the torque is kept constant while it is being applied.

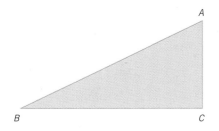

11. An object has an angular velocity. It also has an angular acceleration due to torques that are present. Therefore, the angular velocity is changing. What happens to the angular velocity if (a) additional torques are applied so as to make the net torque suddenly equal to zero and (b) all the torques are suddenly removed?

12. The satellite shown in the drawing is initially moving with a constant translational velocity and zero angular velocity through outer space. (a) When the two engines are fired, each generating a thrust of magnitude T, will the translational velocity increase, decrease, or remain the same? Why? (b) Explain what will happen to the angular velocity.

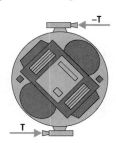

13. Can the mass of a rigid body be considered as concentrated at its center of mass for purposes of computing (a) the body's translational kinetic energy, and (b) the body's moment of inertia? If not, why not?

14. A contest is held between identical twins. Each uses an identical automobile tire. The goal is to roll down a hill in the least amount of time, starting from rest at the top. One twin fashions a crude unicycle, using a negligible mass of material

to build the seat and axle system. The other twin simply curls up inside the tire. Ignoring friction and air resistance, who wins? Give your reasoning.

15. A woman is sitting on the spinning seat of a piano stool with her arms folded. What happens to her angular velocity and her angular momentum when she extends her arms outward? Justify your answers.

16. Suppose the ice cap at the South Pole melted and the water was distributed uniformly over the earth's oceans. Would the earth's angular velocity increase, decrease, or remain the same? Explain.

17. Many rivers, like the Mississippi River, flow from north to south toward the equator. These rivers often carry a large amount of sediment that they deposit when entering the ocean. What effect does this redistribution of the earth's soil have on the angular velocity of the earth? Why?

18. A person is sitting in a chair and swinging his leg back and forth. Is the angular momentum of the leg itself conserved? Explain.

19. A person is hanging motionless from a vertical rope over a swimming pool. He lets go of the rope and drops straight down. After letting go, is it possible for him to curl into a ball and start spinning? Justify your answer.

20. A hoop, a solid cylinder, a spherical shell, and a solid sphere are placed at rest at the top of an inclined plane. All the objects have the same mass and radius. They are then released at the same time. What is the order in which they reach the bottom? Justify your answer.

PROBLEMS

Section 9.1 The Effects of Forces and Torques on the Motion of Rigid Objects

1. In San Francisco a very simple technique is used to turn around a cable car when it reaches the end of its route. The car rolls onto a turntable, which can rotate about a vertical axis through its center. Then, two people push perpendicularly on the car, one at each end, as in the drawing. The turntable is

rotated one-half of a revolution to turn the car around. If the length of the car is 9.20 m and each person pushes with a 185-N force, what is the net torque applied to the car?

2. A force of 110 N is applied perpendicularly to the left edge of the rectangle shown in the drawing. (a) Find the torque (magnitude and direction) produced by this force with respect to an axis perpendicular to the plane of the rectangle at corner A and (b) with respect to a similar axis at corner B.

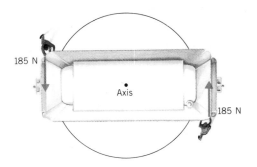

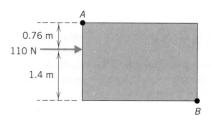

3. A thin, square sheet of plywood is 0.60 m on a side. This sheet can rotate about an axis perpendicular to the square at its center. What is the magnitude of the torque due to a 5.0-N force applied at one corner and directed along a side of the square?

4. Find the net torque (magnitude and direction) produced by the forces F_1 and F_2 about the rotational axis shown in the drawing. The forces are acting on a thin rigid rod, and the axis is perpendicular to the page.

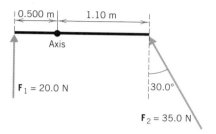

*5. One end of a meter stick is pinned to a table, so the stick can rotate freely in a plane parallel to the tabletop. Two forces, both parallel to the tabletop, are applied to the stick in such a way that the net torque is zero. One force has a magnitude of 2.00 N and is applied perpendicular to the length of the stick at the free end. The other force has a magnitude of 6.00 N and acts at a 30.0° angle with respect to the length of the stick. Where along the stick should the 6.00-N force be applied? Express this distance with respect to the end that is pinned.

*6. A pair of forces with equal magnitudes, opposite directions, and different lines of action is called a "couple." When a couple acts on a rigid object, it produces a torque that does *not* depend on the location of the axis. The drawing shows an example of a couple acting on a rigid rod. The axis is located at the left end of the rod and is perpendicular to the plane containing the two force vectors. Determine an expression for the torque produced by the couple, thereby showing that the torque depends only on F and d.

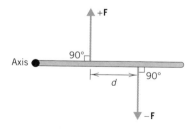

**7. A rotational axis is directed perpendicular to the plane of a square and is located somewhere within the square. Two forces, F_1 and F_2, are applied to diagonally opposite corners, and act along the sides of the square, first as shown in part *a* and then as shown in part *b* of the drawing. In each case the net torque produced by the forces is zero. The square is one meter on a side, and the magnitude of F_2 is three times that of F_1. Locate the point where the axis intersects the plane of the square; measure this point relative to the lower right-hand corner of the square.

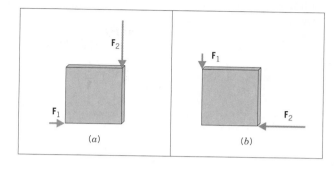

(a) (b)

Section 9.2 Rigid Objects in Equilibrium, Section 9.3 Center of Gravity

8. A 71.0-kg boulder is placed on a 2.00-m-long board, at a point that is 1.40 m from one end. Cliff and Will support the board at each end so that it is horizontal. Cliff is nearest the boulder. If the weight of the board is negligible, what force does each apply to the board?

9. A uniform ladder is 2.4 m long. It is used at a playground, balanced at its center on vertical posts, so that it is horizontal. The ladder tilts when a child who weighs 210 N hangs on one end of it. Where on the ladder should a second child hang, so that the ladder can again be horizontal, if the second child weighs 280 N? Express your answer relative to the center of the ladder.

10. The tonearm on a stereo turntable is positioned horizontally with the stylus resting in the record groove and a fulcrum providing support near the other end. The tracking force is defined to be the magnitude of the normal force F_N that acts at the point where the stylus touches the record. Usually F_N is adjusted by positioning a counterweight on the other end of the tonearm, as the drawing shows. Suppose the tonearm has a mass of 0.0600 kg (excluding the counterweight), with a center of gravity as indicated in the drawing. Find the location X for a 0.100-kg counterweight, so that the tracking force has a value of 9.80×10^{-3} N.

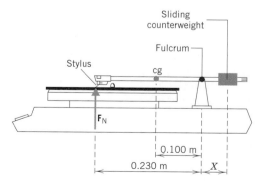

11. In an isometric exercise a person places a hand on a scale and pushes vertically downward, keeping the forearm horizontal. This is possible because the triceps muscle applies an upward force **M** perpendicular to the arm, as the drawing indicates. The forearm weighs 22.0 N and has a center of gravity as indicated. The scale registers 111 N. Determine the magnitude of **M**.

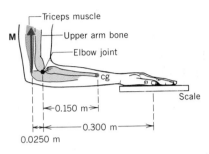

Triceps muscle
M
Upper arm bone
Elbow joint
cg
Scale
0.150 m
0.300 m
0.0250 m

12. The wheels, axle, and handles of a wheelbarrow weigh 60.0 N. The load chamber and its contents weigh 525 N. It is

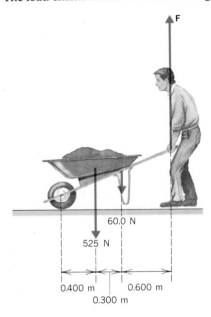

F
60.0 N
525 N
0.400 m
0.600 m
0.300 m

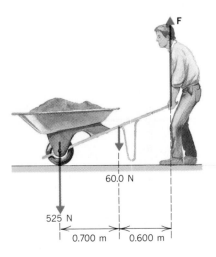

F
60.0 N
525 N
0.700 m
0.600 m

well known that the wheelbarrow is much easier to use if the center of gravity of the load is placed directly over the axle, as in the lower wheelbarrow in the drawing. Verify this fact by calculating the vertical lifting force **F** required to support each of the two wheelbarrows shown.

13. A lunch tray is being held in one hand, as the drawing illustrates. The mass of the tray itself is 0.200 kg, and its center of gravity is located at its geometrical center. On the tray is a 1.00-kg plate of food and a 0.250-kg cup of coffee. Obtain the force **T** exerted by the thumb and the force **F** exerted by the four fingers. Both forces act perpendicular to the tray, which is being held parallel to the ground.

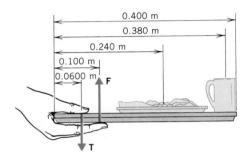

0.400 m
0.380 m
0.240 m
0.100 m
0.0600 m
F
T

14. Three objects are situated on the *x* axis. Their masses and positions are as follows: (a) 4.00 kg at $x = +1.00$ m, (b) 2.00 kg at $x = -0.500$ m, and (c) 2.50 kg at $x = -1.50$ m. Where on the *x* axis is the center of gravity of these objects?

15. A jet transport has a weight of 1.00×10^6 N and is at rest on the runway. The two rear wheels are 15.0 m behind the front wheel, and the plane's center of gravity is 12.6 m behind the front wheel. Determine the normal force exerted on the front wheel and on each of the two rear wheels.

*16. A man holds a 178-N ball in his hand, with the forearm horizontal (see the drawing). He can support the ball in this position because of the flexor muscle force **M**, which is applied perpendicular to the forearm. The forearm weighs 22.0 N and has a center of gravity as indicated. Find (a) the magnitude of **M** and (b) the magnitude and direction of the force applied by the upper arm bone to the forearm at the elbow joint.

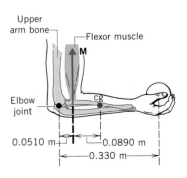

Upper arm bone
Flexor muscle
M
Elbow joint
cg
0.0510 m
0.0890 m
0.330 m

*17. A woman who weighs 5.00×10^2 N is leaning against a smooth vertical wall, as the drawing shows. Find the force $\mathbf{F_N}$ (directed perpendicular to the wall) exerted on her shoulder by the wall and the horizontal and vertical components of the force exerted on her shoes by the ground.

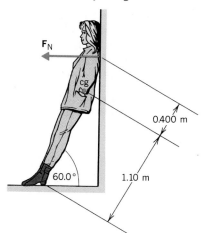

*18. A person is sitting with one leg outstretched, so that it makes an angle of 30.0° with the horizontal, as the drawing indicates. The weight of the leg below the knee is 44.5 N with the center of gravity located below the knee joint. The leg is being held in this position because of the force $\mathbf{M}$ applied by the quadriceps muscle, which is attached 0.100 m below the knee joint (see the drawing). Obtain the magnitude of $\mathbf{M}$.

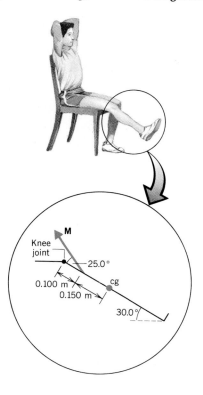

*19. A 125-kg uniform beam is attached to a vertical wall and is supported by a wire. The beam is 3.00 m long, and a 2.00×10^2-kg crate hangs from it. Using the data shown in the drawing, find (a) the tension T in the wire and (b) the horizontal and vertical components of the force that the wall exerts on the left end of the beam.

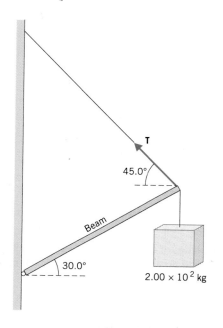

*20. A massless, rigid board is placed across two bathroom scales that are separated by a distance of 2.00 m. A person lies on the board. The scale under his head reads 425 N and the scale under his feet reads 315 N. (a) Find the weight of the person. (b) Locate the center of gravity of the person relative to the scale beneath his head.

**21. An inverted "vee" is made of uniform boards and weighs 356 N. Each side has the same length and makes a 30.0° angle with the vertical, as the drawing shows. Find the force of friction that acts on the lower end of each leg of the "vee." (See Solved Problem 1 for a related problem.)

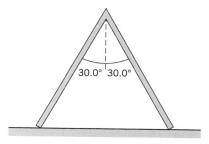

**22. The drawing shows an A-shaped ladder. Both sides of the ladder are equal in length. This ladder is standing on a frictionless horizontal surface and only the crossbar (which has a negligible mass) of the "A" keeps the ladder from collapsing. The ladder is uniform and has a mass of 20.0 kg. Determine the tension in the crossbar of the ladder. *(See Solved Problem 1 for a related problem.)*

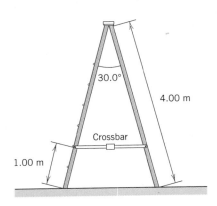

Section 9.4 Newton's Second Law for Rotational Motion about a Fixed Axis

23. The blades of a ceiling fan have a moment of inertia of $0.16 \text{ kg} \cdot \text{m}^2$ and an angular acceleration of 7.0 rad/s^2. What net torque is being applied to the blades?

24. A phonograph record has a diameter of 0.305 m and a mass of 0.122 kg. Assuming the record to be a uniform solid disk, determine the moment of inertia relative to the axis about which the record rotates.

25. A uniform solid disk with a mass of 30.0 kg and a radius of 0.200 m is free to rotate about a frictionless axle. Forces of 90.0 and 125 N are applied to the disk, as the drawing illustrates. What is (a) the net torque produced by the two forces and (b) the angular acceleration of the disk?

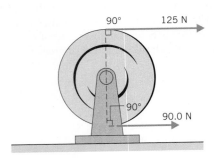

26. A bicycle wheel has a radius of 0.330 m and a rim whose mass is 1.20 kg. The wheel has 50 spokes, each with a mass of 0.010 kg. (a) Calculate the moment of inertia of the rim about the axle. (b) Determine the moment of inertia of *any one spoke*, assuming it to be a long thin rod that can rotate about one end. (c) Find the *total* moment of inertia of the wheel, including the rim and all 50 spokes.

27. The circular blade on a radial arm saw is turning at 262 rad/s at the instant the motor is turned off. In 18.0 s the speed of the blade is reduced to 85 rad/s. Assume the blade to be a uniform solid disk of radius 0.130 m and mass 0.400 kg. Find the net torque applied to the blade.

28. A turntable platter has a radius of 0.150 m and is rotating at 3.49 rad/s. When the power is shut off, the platter slows down and comes to rest in 15.0 s, due to a net retarding torque of $6.20 \times 10^{-3} \text{ N} \cdot \text{m}$. Assume the platter to be a uniform solid disk. Determine the mass of the platter.

29. A clay vase on a potter's wheel experiences an angular acceleration of 8.00 rad/s^2 due to the application of a 10.0-N·m net torque. Find the total moment of inertia of the vase and potter's wheel.

*30. Refer to the drawing for question 12 (not problem 12) and assume that the radius of the satellite is 1.2 m. Suppose that the magnitude of the thrust generated by each of the two engines is $T = 720$ N and that the moment of inertia of the satellite is $650 \text{ kg} \cdot \text{m}^2$. If the satellite is rotating at an angular velocity of 1.1 rad/s at the time the engines begin to fire, how many revolutions will the satellite make during a 3.0-s burst from the engines?

*31. The drawing shows a model for the motion of the human forearm in throwing a dart. Because of the force **M** applied by the triceps muscle, the forearm can rotate about an axis at the elbow joint. Assume that the forearm has the dimensions shown in the drawing and a moment of inertia of $0.065 \text{ kg} \cdot \text{m}^2$ (including the effect of the dart) relative to the axis at the elbow. Assume also that the force **M** acts perpendicular to the forearm. Ignoring the effect of gravity and any frictional forces, determine the magnitude of the force **M** needed to give the dart a tangential speed of 5.0 m/s in 0.10 s, starting from rest.

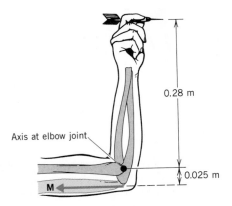

*32. A thin, rigid, uniform rod has a mass of 2.00 kg and a length of 2.00 m. (a) Find the moment of inertia of the rod relative to an axis that is perpendicular to the rod at one end. (b) Suppose all the mass of the rod were located at a single point. Determine the perpendicular distance of this point from the axis in part (a), such that this point particle has the same moment of inertia as the rod. This distance is called the *radius of gyration* of the rod.

*33. One part of a fireworks display uses a 0.10-m-square platform, on which tubes of gun powder are mounted along each outer edge (see the drawing). As the powder burns, the square spins about an axis perpendicular to its center. The unit reaches an angular speed of 31 rad/s in one-half second, starting from rest, because each of three tubes generates a force of 0.30 N and the fourth tube generates an unknown force. The moment of inertia of the entire unit has a constant value of 1.0×10^{-3} kg·m², and the forces are constant. What is the magnitude of the force generated by the fourth tube?

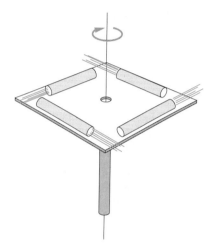

*34. The *parallel axis theorem* provides a useful way to calculate the moment of inertia I about an arbitrary axis. The theorem states that $I = I_{cm} + Mh^2$, where I_{cm} is the moment of inertia of the object relative to an axis that passes through the center of mass and is parallel to the axis of interest, M is the total mass of the object, and h is the perpendicular distance between the two axes. Use this theorem and information to determine an expression for the moment of inertia of a solid cylinder of radius R relative to an axis that lies on the surface of the cylinder and is perpendicular to the circular ends.

**35. A thin uniform rod has a length of 3.00 m and is cut into two pieces. The moment of inertia of each piece is measured relative to an axis that is perpendicular to one end of the piece. These moments of inertia are found to have a ratio of 2:1. Find the lengths of the two pieces.

**36. By means of a rope whose mass is negligible, two blocks are suspended over a pulley, as the drawing shows. The pulley is a uniform solid cylindrical disk. The downward acceleration of the 44.0-kg mass is observed to be exactly one-half the acceleration due to gravity. Noting that the tension in

the rope is not the same on each side of the pulley, determine the mass of the pulley.

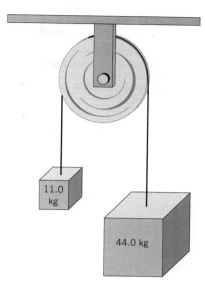

Section 9.5 Rotational Work and Energy

37. The mass of the earth is 6.0×10^{24} kg, its radius is 6.4×10^6 m, and it has an angular speed of 1 rev/day. Assuming the earth to be a uniform sphere, what is its rotational kinetic energy (in joules)?

38. A four-bladed ceiling fan rotates with an angular speed of 30.0 rad/s. The length of each blade is 0.600 m, and the mass of each blade is 2.00 kg. Each blade can be approximated as a uniform thin rod that rotates about one end. Determine the total rotational kinetic energy of the four blades.

39. A car is moving with a speed of 20.0 m/s. Each wheel has a radius of 0.300 m and a moment of inertia of 0.700 kg·m². The car has a total mass (including the wheels) of 1.90×10^3 kg. Find (a) the translational kinetic energy of the entire car, (b) the total rotational kinetic energy of the four wheels, and (c) the total kinetic energy of the car.

40. A solid disk has a mass of 162 kg and a radius of 1.30 m. This disk rotates about an axis through its center, like a record, and has an angular speed of 18.0 rad/s. If all the kinetic energy of the disk were used to lift a 3.00-kg block, how high could the block be lifted?

*41. A thin-walled spherical shell is rolling on a surface. What fraction of its total kinetic energy is in the form of rotational kinetic energy about the center of mass?

*42. A marble and a cube have the same mass. Starting from rest at the same height, the marble rolls and the cube slides (no kinetic friction) down a ramp. Determine the ratio of the center of mass speed of the cube to the center of mass speed of the marble at the bottom of the ramp. (*See Example 12.*)

*43. A solid cylinder and a thin-walled hollow cylinder (see Table 9.1) have the same mass and radius. They are rolling horizontally toward the bottom of an inclined plane. The center of mass of each has the same translational speed. The cylinders roll up the incline and reach their highest points. Calculate the ratio of the distances (s_{solid}/s_{hollow}) along the incline through which each center of mass moves.

*44. A bowling ball encounters a 0.760 m vertical rise on the way back to the ball rack, as the drawing illustrates. Neglect frictional losses and assume the mass of the ball is distributed uniformly. If the translational speed of the ball is 3.50 m/s at the bottom of the rise, find the translational speed at the top.

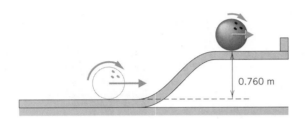

0.760 m

**45. A yo-yo consists of two uniform solid cylinders, each of which has a radius of 3.50×10^{-2} m. The two cylinders are joined at their centers by a short rod (negligible mass) that is perpendicular to each cylinder. The radius of this rod is 5.00×10^{-3} m. As usual, a string (negligible thickness) is wound around the center rod, and the yo-yo is allowed to roll vertically down the string starting from rest. (a) Assuming that the yo-yo does not slip on the string, how fast is the center of mass of the yo-yo moving after it has descended through a vertical distance of 1.00 m? (b) What is the angular speed of the yo-yo at this point?

Section 9.6 Angular Momentum

46. For a certain satellite with an apogee distance of $r_A = 1.30 \times 10^7$ m, the ratio of the orbital speed at perigee to the orbital speed at apogee is 1.20. Find the perigee distance r_P.

47. A woman stands at the center of a platform. The woman and the platform rotate with an angular speed of 5.00 rad/s. Friction is negligible. Her arms are outstretched, and she is holding a dumbbell in each hand. In this position the total moment of inertia of the rotating system (platform, woman, and dumbbells) is 5.40 kg·m². By pulling in her arms, the moment of inertia is reduced to 3.80 kg·m². Find her new angular speed.

48. A baggage carousel at an airport is rotating with an angular speed of 0.20 rad/s when the baggage begins to be loaded onto it. The moment of inertia of the carousel is 1500 kg·m². Ten pieces of baggage with an average mass of 15 kg each are dropped vertically onto the carousel and come to rest at a perpendicular distance of 2.0 m from the axis of rotation.

(a) Assuming that no net external torque acts on the system of carousel and baggage, find the final angular speed. (b) In reality, the angular speed of a baggage carousel does not change. Therefore, speaking qualitatively, what can you say about the external torque acting on this kind of system?

49. A thin rod has a length of 0.20 m and rotates in a circle on a frictionless tabletop. The axis is fixed to the table and is through one end of the rod. The rod has an angular velocity of 0.40 rad/s and a moment of inertia of 1.2×10^{-3} kg·m². A bug standing on the axis decides to crawl out to the other end of the rod. When the bug (mass = 5.0×10^{-3} kg) gets where he's going, what is the angular velocity of the rod?

*50. A cylindrically shaped space station is rotating about the axis of the cylinder to create artificial gravity. The radius of the cylinder is 82.5 m. The moment of inertia of the station without people is 3.00×10^9 kg·m². Suppose 500 people, with an average mass of 70.0 kg each, live on this station. As they move radially from the outer surface of the cylinder toward the axis, the angular speed of the station changes. What is the maximum possible percentage change in the station's angular speed due to the radial movement of the people?

**51. A small 0.500-kg object moves on a frictionless horizontal table in a circular path of radius 1.00 m. The angular speed is 6.28 rad/s. The object is attached to a string of negligible mass that passes through a small hole in the table at the center of the circle. Someone under the table begins to pull the string downward to make the circle smaller. If the string will tolerate a tension of no more than 105 N, what is the radius of the smallest possible circle on which the object can move?

ADDITIONAL PROBLEMS

52. A cylinder is rotating about an axis that passes through the center of each circular end piece. The cylinder has a radius of 0.0750 m, an angular speed of 88.0 rad/s, and a moment of inertia of 0.850 kg·m². A brake shoe presses against the surface of the cylinder and applies a tangential frictional force to it. The frictional force reduces the angular speed of the cylinder by a factor of two during a time of 5.00 s. (a) Find the angular deceleration of the cylinder. (b) Find the force of friction applied by the brake shoe.

53. A post is driven perpendicularly into the ground and serves as the axis about which a gate rotates. A force of 12 N is applied perpendicular to the gate and acts parallel to the ground. How far from the post should the force be applied to produce a torque of 3.0 N·m?

54. The mass of an automobile is 1160 kg and the horizontal distance between its front and rear axles is 2.54 m. The center of gravity of the car is between the front and rear tires, and the horizontal distance between the center of gravity and the front axle is 1.02 m. Determine the normal force that the

ground applies to *each* of the two front wheels and to *each* of the two rear wheels.

55. A playground carousel is free to rotate about its center on frictionless bearings. The carousel has an angular speed of 3.14 rad/s, a moment of inertia of 125 kg·m², and a radius of 1.50 m. A 40.0-kg person, standing still next to the carousel, jumps onto it very close to the outer edge. Use the conservation of angular momentum to find the resulting angular speed of the carousel and person.

56. Three point masses are located on the x axis as follows: (1) 4.00 kg at $x = 0.200$ m, (2) 10.0 kg at $x = 0.500$ m, and (3) 1.50 kg at $x = 0.900$ m. (a) Calculate the moment of inertia of *each* mass with respect to the y axis. (b) Find the total moment of inertia. (c) Based on the results above, decide whether it is true that the smallest mass necessarily contributes the smallest amount to the total moment of inertia. Explain.

57. One side of a rotating rectangle is three times the length of other side. A force $\mathbf{F_A}$ is applied to one corner of the rectangle and is parallel to the short side. A second force $\mathbf{F_B}$ is applied to the diagonally opposite corner and is parallel to the long side. Find the ratio of the magnitudes of the forces F_B/F_A so that the rectangle rotates at a constant angular velocity about an axis perpendicular to the rectangle at its center.

58. Three objects lie in the x, y plane. Each rotates about the z axis with an angular speed of 6.00 rad/s. The mass m of each object and its perpendicular distance r from the z axis are as follows: (1) $m_1 = 6.00$ kg and $r_1 = 2.00$ m, (2) $m_2 = 4.00$ kg and $r_2 = 1.50$ m, (3) $m_3 = 3.00$ kg and $r_3 = 3.00$ m. (a) Find the tangential speed of each object. (b) Determine the total kinetic energy of this system using the expression KE = $\frac{1}{2}m_1v_1^2 + \frac{1}{2}m_2v_2^2 + \frac{1}{2}m_3v_3^2$. (c) Obtain the total moment of inertia of the system. (d) Find the rotational kinetic energy of the system using the relation $\frac{1}{2}I\omega^2$ to verify that the answer is the same as that in (b).

***59.** A uniform steel beam of length 5.00 m has a weight of 4.50×10^3 N. One end of the beam is bolted to a vertical wall. The beam is held in a horizontal position by a cable attached between the other end of the beam and a point on the wall. The cable makes an angle of 25.0° above the horizontal. A load whose weight is 12.0×10^3 N is hung from the beam at a point that is 3.50 m from the wall. Find (a) the magnitude of the tension in the supporting cable and (b) the magnitude of the force exerted on the end of the beam by the bolt that attaches the beam to the wall.

***60.** A thin uniform stick is initially positioned in the vertical direction, with its lower end attached to a frictionless axis that is mounted on the floor. The stick has a length of 2.00 m and is allowed to fall, starting from rest. Find the linear speed

of the free end of the stick, just before the stick hits the floor after rotating through 90°.

***61.** A flat uniform circular disk (radius = 2.00 m, mass = 1.00×10^2 kg) is initially stationary and fixed so that it can rotate in the horizontal plane about a frictionless axis perpendicular to the center of the disk. A 40.0-kg person, standing 1.25 m from the axis, begins to run on the disk in a circular path and has a tangential speed of 2.00 m/s relative to the ground. (a) Find the resulting angular speed of the disk (in rad/s) and describe the direction of the rotation. (b) Determine the time it takes for a spot marking the starting point to pass again beneath the runner's feet.

****62.** The drawing shows the top view of two doors. The doors are uniform and have identical widths. Door A rotates about an axis through its left edge, while door B rotates about an axis through the center. The same force **F** is applied perpendicular to each door at its right edge, and the force remains perpendicular to each door as the door turns. Starting from rest, door A rotates through a certain angle in 3.00 s. How long does it take door B to rotate through the same angle?

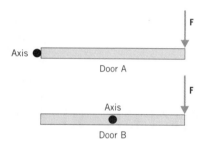

****63.** The drawing shows an inverted "A" that is suspended from the ceiling by two vertical ropes. Each leg of the "A" has a length of 2L and a weight of 120 N. The horizontal crossbar has a negligible weight. Find the force that the crossbar applies to each leg. (*See Solved Problem 1 for a related problem.*)

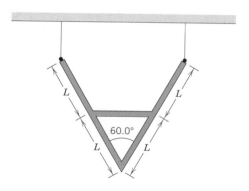

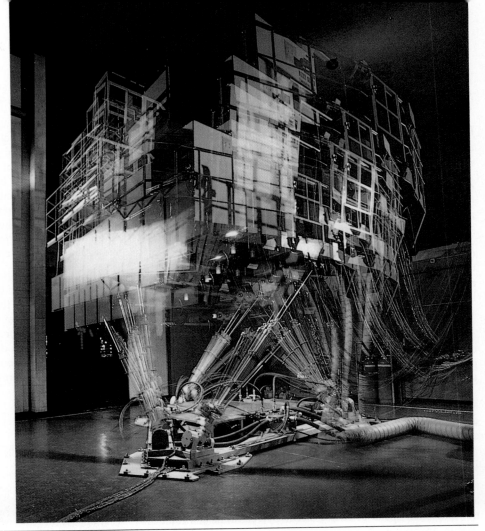

CHAPTER 10

ELASTICITY AND SIMPLE HARMONIC MOTION

A flight simulator is a chamber containing an artificial cockpit, in which pilots are trained to cope with all types of flying conditions. For instance, the chamber can be shaken vigorously in multiple directions, as in the photograph above, to simulate the intense and complicated vibrations that accompany adverse weather conditions. In this chapter, we will study one important kind of vibrational motion, called simple harmonic motion. Simple harmonic motion is an idealized form of the vibratory motion that results when you push or pull on an object attached to a spring and then release it. The spring has a natural tendency to return to its original shape after the force distorting it is removed, and this tendency leads to the vibratory motion. We will see that simple harmonic motion also plays a role in the way a number of familiar devices operate, including loudspeakers, turntables, and clocks.

Many materials besides the metals from which springs are made have the ability to return to their original shapes. Such materials are said to be "elastic." From an atomic viewpoint, elastic behavior has its origin in the forces between the atoms that comprise the material. Figure 10.1 symbolizes these forces with the aid of springs that are drawn between the atoms. It is well known that a spring exerts a force when its coils are pulled apart or pushed together. Thus, because of atomic-level "springs," a material tends to return to its initial shape once the forces that cause a deformation are removed. The exact details of a material's elastic behavior depend on the interatomic forces that are present. In this chapter, our study of elasticity begins with a description of the kinds of deformations that elastic materials can experience. Then, we will turn our attention to the simple harmonic motion of an object attached to a spring.

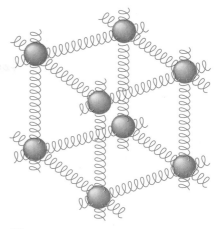

Figure 10.1   The forces between atoms act like springs. The atoms are represented by black spheres, and the springs between some atoms have been omitted for clarity.

10.1 ELASTIC DEFORMATION

STRETCHING, COMPRESSION, AND YOUNG'S MODULUS

The forces that hold the atoms of a solid together are particularly strong, so considerable force must be applied to stretch a solid object. The force needed depends on several factors, as Figure 10.2 illustrates. For one thing, it depends on the extent of the stretching. For two identical rods, part *a* of the drawing indicates that more force is required to produce a greater amount of stretch than a smaller amount. Part *b* indicates another fact: To stretch two equally long rods by the same amount, more force is required for the rod with the larger cross-sectional area, assuming the rods are made from the same material. Finally, part *c* shows that, for a given amount of stretch, more force is required for a shorter rod than for a longer rod, provided the rods are made from the same material and have the same cross-sectional area. Simply take a long rubber band, for example, and

The tennis ball and the strings of the racquet are made from elastic materials.

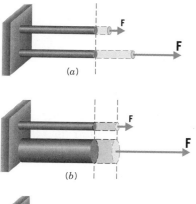

Figure 10.2   The amount of force *F* needed to stretch a solid rod depends on (*a*) the amount by which the rod is stretched, (*b*) the cross-sectional area of the rod, and (*c*) the unstretched length of the rod.

stretch it an arbitrary amount. Then stretch a much shorter rubber band with the same cross-sectional area by the same amount. You will be easily convinced that more force is required to stretch the shorter rubber band.

Experiments have shown that the elastic behavior described above can be expressed by the following relation, provided that the amount of stretching is small compared to the original length of the object:

$$F = Y \left(\frac{\Delta L}{L_0} \right) A \qquad (10.1)$$

As Figure 10.3 shows, F denotes the magnitude of the stretching force applied perpendicular to the cross-sectional area A, ΔL is the increase in length, and L_0 is the original length. The term Y is a proportionality constant called *Young's modulus*, after Thomas Young (1773–1829), and its value depends on the nature of the material, as Table 10.1 reveals. Solving Equation 10.1 for Y shows that Young's modulus has units of force per unit area (N/m²).

PROBLEM SOLVING INSIGHT

It should be noted that the magnitude of the force in Equation 10.1 is proportional to the fractional increase in length $\Delta L / L_0$, rather than the absolute increase ΔL. The magnitude of the force is also proportional to the cross-sectional area, which need not be circular, but can have any shape (e.g., rectangular).

Forces that are applied as in Figure 10.3 and cause stretching are called "tensile" forces because they create a tension in the material, much like the tension in a rope. Equation 10.1 also applies when the force compresses the material along its length. In this situation, the force is applied in a direction opposite to that shown in Figure 10.3, and ΔL stands for the amount by which the original length L_0 decreases.

Most solids have Young's moduli that are rather large, reflecting the fact that a large force is needed to change the length of a solid object by even a small amount, as Example 1 illustrates.

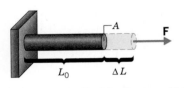

Figure 10.3 In this diagram, **F** denotes the stretching force, A the cross-sectional area, L_0 the original length of the rod, and ΔL the amount of stretch.

Table 10.1 Values for the Young's Modulus of Solid Materials

Material	Young's Modulus Y (N/m²)
Aluminum	6.9×10^{10}
Bone (compression)	9.4×10^{9}
Bone (tension)	1.6×10^{10}
Brass	9.0×10^{10}
Brick	1.4×10^{10}
Copper	1.1×10^{11}
Mohair	2.9×10^{9}
Nylon	3.7×10^{9}
Pyrex glass	6.2×10^{10}
Steel	2.0×10^{11}
Teflon	3.7×10^{8}
Tungsten	3.6×10^{11}

Example 1   Bone Compression

In a circus act, a performer supports the combined weight (1640 N) of a number of colleagues (see Figure 10.4). Each thighbone (femur) of this performer has a length of 0.55 m and an effective cross-sectional area of 7.7×10^{-4} m^2. Determine the amount by which each thighbone compresses under the extra weight.

REASONING AND SOLUTION   The additional weight supported by each bone is 820 N, and Table 10.1 indicates that Young's modulus for compression is 9.4×10^9 N/m^2. The amount of compression can be obtained from Equation 10.1:

$$\Delta L = \frac{FL_0}{YA} = \frac{(820 \text{ N})(0.55 \text{ m})}{(9.4 \times 10^9 \text{ N/m}^2)(7.7 \times 10^{-4} \text{ m}^2)} = \boxed{6.2 \times 10^{-5} \text{ m}}$$

This is a very small change in length, the fractional decrease being $\Delta L/L_0 = 0.00011$.

Example 2 illustrates how the structural design of a bridge must take into account the large forces that can come into play when the length of a steel beam changes by even a small amount.

Example 2   An Expanding Steel Beam

A steel beam, 9.6 m in length and 0.10 m^2 in cross-sectional area, is used in the roadbed of a bridge. The beam is mounted between two concrete supports, as Figure 10.5 shows, and is fitted perfectly into place, with no room for expansion. In response to a rise in temperature of 19 Celsius degrees, the beam would expand by 2.2×10^{-3} m if it were free to do so. To prevent this small expansion, what force must be supplied by the concrete supports?

REASONING   Imagine that the beam is expanded to begin with. To fit it between the concrete supports, it would have to be compressed by 2.2×10^{-3} m. To accomplish the compression, the concrete supports apply a compressive force to each end of the beam. We can use Equation 10.1 to find the force.

SOLUTION   Table 10.1 indicates that Young's modulus for steel is 2.0×10^{11} N/m^2. Using this value in Equation 10.1, we find that the compressive force is

$$F = Y\left(\frac{\Delta L}{L_0}\right)A = (2.0 \times 10^{11} \text{ N/m}^2)\left(\frac{2.2 \times 10^{-3} \text{ m}}{9.6 \text{ m}}\right)(0.10 \text{ m}^2) = \boxed{4.6 \times 10^6 \text{ N}}$$

If the concrete supports cannot supply this large a force (about one million pounds), they will move or crack to provide the needed expansion space. Notice how large the force is, even though 2.2×10^{-3} m is only 0.023% of the original length of the beam.

SHEAR DEFORMATION AND THE SHEAR MODULUS

It is possible to deform a solid object in a way other than a simple stretch or compression. For instance, place a book on a rough table and push on the book as in Figure 10.6a. The resulting deformation is called *shear deformation.* Part *b* of the drawing indicates that the deformation occurs because of the combined effect of the force **F** applied to the top of the book and the force $-$**F** applied to the bottom of the book by the table. These two forces have equal magnitudes, but

Figure 10.4   A balancing act at the Cirque du Soleil (Circus of the Sun).

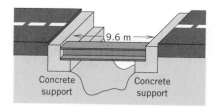

Figure 10.5   The steel beam is fitted between concrete supports with no room provided for expansion of the beam.

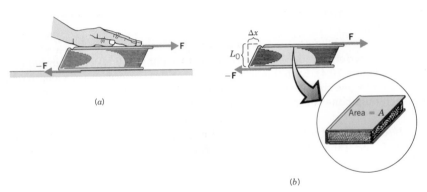

(a)

(b)

Figure 10.6   (a) An example of a shear deformation. (b) The shearing forces **F** and −**F**, applied parallel to the cross-sectional area A and perpendicular to the thickness L_0, cause a solid object to change shape. The shear deformation is ΔX.

opposite directions, and ensure that the book remains in equilibrium. Equation 10.2 gives the magnitude F of the force needed to produce an amount of shear ΔX for an object with cross-sectional area A and thickness L_0:

$$F = S\left(\frac{\Delta X}{L_0}\right) A \qquad (10.2)$$

This equation is very similar to Equation 10.1. The constant of proportionality S is called the **shear modulus** and, like Young's modulus, has units of force per unit area (N/m^2). The value of S depends on the nature of the material, and Table 10.2 gives some representative values. Example 3 illustrates how to determine the shear modulus of a favorite dessert.

Table 10.2   Values for the Shear Modulus of Solid Materials

Material	Shear Modulus S (N/m²)
Aluminum	2.4×10^{10}
Bone	8.0×10^{10}
Brass	3.5×10^{10}
Copper	4.2×10^{10}
Lead	5.4×10^{9}
Nickel	7.3×10^{10}
Steel	8.1×10^{10}
Tungsten	1.5×10^{11}

Example 3   J-E-L-L-O

A block of Jell-O is resting on a plate. Figure 10.7a gives the dimensions of the block. You are bored, impatiently waiting for dinner, and push tangentially across the top surface with a force of $F = 0.45$ N. The top surface moves a distance $\Delta X = 6.0 \times 10^{-3}$ m relative to the bottom surface (part b of the drawing). Use this idle gesture to measure the shear modulus of Jell-O.

REASONING AND SOLUTION   The shear modulus is given by Equation 10.2 as $S = FL_0/(A\,\Delta X)$, where $A = (0.070\text{ m})(0.070\text{ m})$ is the area of the top surface, and $L_0 = 0.030$ m is the thickness of the block:

$$S = \frac{FL_0}{A\,\Delta X} = \frac{(0.45\text{ N})(0.030\text{ m})}{(0.070\text{ m})(0.070\text{ m})(6.0 \times 10^{-3}\text{ m})} = \boxed{460\text{ N/m}^2}$$

Jell-O can be deformed easily, so its shear modulus is significantly less than that of a more rigid material like steel (see Table 10.2).

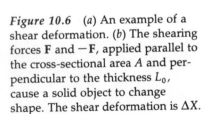

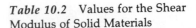

(a)

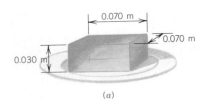

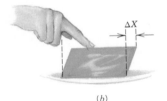

(b)

Figure 10.7   (a) A block of Jell-O and (b) a shearing force applied to it.

Although Equations 10.1 and 10.2 are similar, they refer to different kinds of deformations. The shearing force in Figure 10.6 is parallel to the area A, whereas the tensile force in Figure 10.3 is perpendicular to the area A. Furthermore, the ratio $\Delta X/L_0$ in Equation 10.2 is different than the ratio $\Delta L/L_0$ in Equation 10.1: The distances ΔX and L_0 are perpendicular, whereas ΔL and L_0 are parallel. Young's modulus refers to a _change in length_ of one dimension of a solid object as a result of tensile or compressive forces. In contrast, the shear modulus refers to a _change in shape_ of a solid object as a result of shearing forces.

VOLUME DEFORMATION AND THE BULK MODULUS

When a compressive force is applied along one dimension of a solid, the length of that dimension decreases. It is also possible to apply compressive forces so that the size of every dimension (length, width, and depth) decreases, leading to a decrease in volume, as Figure 10.8 illustrates. This kind of overall compression occurs, for example, when an object is submerged in a liquid, and the liquid presses inward everywhere on the object. The forces acting in such situations are applied perpendicular to every surface, and it is more convenient to speak of the perpendicular force per unit area, rather than the amount of any one force in particular. The perpendicular force per unit area is called the *pressure P.*

The result of increasing the pressure on an object by an amount ΔP is that the volume of the object decreases by an amount ΔV (see Figure 10.8). Such a pressure increase occurs, for example, when a swimmer dives deeper into the water. Experiment reveals that the amount of pressure increase needed to decrease the volume is directly proportional to the fractional volume change $\Delta V/V_0$, where V_0 is the initial volume:

$$\Delta P = -B\left(\frac{\Delta V}{V_0}\right) \qquad (10.3)$$

This relation is analogous to Equations 10.1 and 10.2, except that the area A in those equations does not appear here explicitly; the area is already taken into account by the concept of pressure (force per unit area). The proportionality constant B is known as the **bulk modulus.** The minus sign occurs because an increase in pressure (ΔP positive) always creates a decrease in volume (ΔV negative), and B is given as a positive quantity. Like Young's modulus and the shear modulus, the bulk modulus has units of force per unit area (N/m^2), and its value depends on the nature of the material. Table 10.3 gives representative values of the bulk modulus and includes liquids as well as solids.

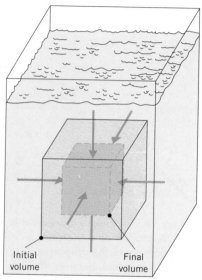

Figure 10.8 The arrows denote the forces that push perpendicularly on every surface of an object immersed in a liquid. The force per unit area is the pressure. When the pressure increases, the volume of the object decreases from a value of V_0 to a value of $V_0 - \Delta V$.

10.2 STRESS, STRAIN, AND HOOKE'S LAW

Equations 10.1–10.3 specify the amount of force needed for a given amount of elastic deformation, and they are repeated below to emphasize their common features:

$$\frac{F}{A} = Y\left(\frac{\Delta L}{L_0}\right) \qquad (10.1)$$

$$\frac{F}{A} = S\left(\frac{\Delta X}{L_0}\right) \qquad (10.2)$$

$$\Delta P = -B\left(\frac{\Delta V}{V_0}\right) \qquad (10.3)$$

Stress | is proportional to | Strain

Table 10.3 Values for the Bulk Modulus of Solid and Liquid Materials

Material	Bulk Modulus B (N/m^2)
Solids	
Aluminum	7.1×10^{10}
Brass	6.7×10^{10}
Copper	1.3×10^{11}
Lead	4.2×10^{10}
Nylon	6.1×10^{9}
Pyrex glass	2.6×10^{10}
Steel	1.4×10^{11}
Liquids	
Ethanol	8.9×10^{8}
Oil	1.7×10^{9}
Water	2.2×10^{9}

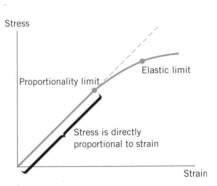

Figure 10.9 Hooke's law (stress is directly proportional to strain) is valid only up to the proportionality limit of a material. Between the proportionality limit and the elastic limit, stress is no longer proportional to strain. Beyond the elastic limit, the material remains deformed even when the stress is removed.

The left side of each equation is the magnitude of the force per unit area required to cause an elastic deformation. In general, the ratio of the force to the area is called the *stress*. The right side of each equation involves the change in a quantity (ΔL, ΔX, or ΔV) divided by a quantity (L_0 or V_0) relative to which the change is compared. The terms $\Delta L/L_0$, $\Delta X/L_0$, and $\Delta V/V_0$ are unitless ratios, and each is referred to as the *strain* that results from the applied stress. In the case of stretch and compression, the strain is the fractional change in length, whereas in volume deformation it is the fractional change in volume. In shear deformation the strain refers to a change in shape of the object. Experiments show that these three equations, with constant values for Young's modulus, the shear modulus, and the bulk modulus, apply to a wide range of materials. Therefore, stress and strain are directly proportional to one another, a relationship first discovered by Robert Hooke (1635–1703) and now referred to as *Hooke's law.*

Hooke's Law for Stress and Strain

Stress is directly proportional to strain.

SI Unit of Stress: newton per square meter = pascal (Pa)

SI Unit of Strain: Strain is a unitless quantity.

The SI unit of stress is the *pascal (Pa),* named for the French scientist Blaise Pascal (1623–1662); $1 \text{ Pa} = 1 \text{ N/m}^2$.

In reality, materials obey Hooke's law only up to a certain limit, as Figure 10.9 shows. As long as stress remains proportional to strain, a plot of stress versus strain is a straight line. The point on the graph where the material deviates from straight line behavior is called the "proportionality limit." Beyond the proportionality limit stress and strain are no longer directly proportional. However, if the stress does not exceed the "elastic limit" of the material, the object will return to its original size and shape once the stress is removed. The "elastic limit" is the point beyond which the object no longer returns to its original size and shape when the stress is removed; the object remains permanently deformed.

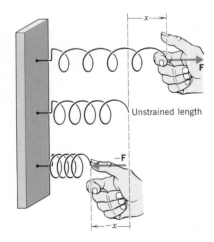

Figure 10.10 An ideal spring obeys the equation, $F = kx$, where **F** is the force applied to the spring, x is the amount of stretch or compression, and k is the spring constant.

10.3 THE IDEAL SPRING AND SIMPLE HARMONIC MOTION

Springs are familiar objects that exhibit elastic behavior. They have a great variety of applications, from pogo sticks to automobile suspension systems. Provided the applied force is not large enough to cause permanent deformation, the spring in Figure 10.10 will return to its original length after being stretched or compressed. For relatively small deformations, the force F required to stretch or compress a spring obeys the following equation:

$$F = kx \qquad (10.4)$$

In this expression x denotes the amount by which the spring is stretched or compressed from its unstrained length. The term k is a proportionality constant called the *spring constant* and has dimensions of force per unit length (N/m). Sometimes k is referred to as the *stiffness* of the spring, because a large value for k means the spring is "stiff," in the sense that a large force is required to stretch or compress it.

Equation 10.4 is a Hooke's law type of relationship, as can be seen by comparing it to Equation 10.1 for the stretching or compressing of a solid rod:

$$F = \frac{YA}{L_0} \Delta L$$
$$F = \underbrace{k}\ \ \underbrace{x}$$

This comparison shows that x is analogous to ΔL, whereas k is analogous to the term YA/L_0, which is a constant for a given object. A spring that behaves according to the Hooke's law relationship $F = kx$ is said to be an *ideal spring.* Example 4 illustrates one application of a spring.

Example 4   A Tire Pressure Gauge

In a tire pressure gauge, the air in the tire pushes against a spring when the gauge is attached to the tire valve, as in Figure 10.11. Suppose the spring constant of the spring is $k = 320$ N/m and the bar indicator of the gauge extends 2.0 cm when the gauge is pressed against the air valve. What force does the air in the tire apply to the spring?

REASONING AND SOLUTION   Since the spring constant is known, the force applied to the spring can be obtained from Equation 10.4:

$$F = kx = (320 \text{ N/m})(0.020 \text{ m}) = \boxed{6.4 \text{ N}}$$

Thus, the exposed length of the bar indicator gives a measure of the force that the air pressure in the tire exerts on the spring. Since pressure is force per unit area and the area of the plunger surface (see drawing) is fixed, the bar indicator can be marked in units of pressure.

THE PHYSICS OF . . .

a tire pressure gauge.

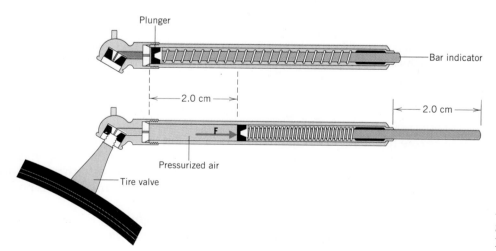

Figure 10.11   A tire pressure gauge is one of many practical applications of springs.

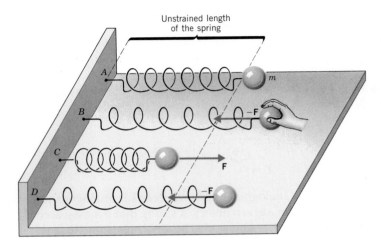

Figure 10.12 The restoring force (see blue arrows) produced by an ideal spring always points opposite to the direction in which the spring is deformed and leads to a back-and-forth motion of the object.

To stretch or compress a spring, a force must be applied to it. In accord with Newton's third law, the spring exerts an oppositely directed force of equal magnitude. This reaction force is applied by the spring to the agent that does the pulling or pushing. In other words, the reaction force is applied to the object attached to the spring. The reaction force is also called a "restoring force," for a reason that will be clarified shortly. The restoring force of an ideal spring is obtained from the relation $F = kx$ by including the minus sign required by Newton's action–reaction law, as indicated below.

Hooke's Law Restoring Force of an Ideal Spring

The restoring force of an ideal spring is

$$F = -kx \qquad (10.5)$$

The minus sign indicates that the restoring force always points in a direction opposite to that in which the spring is deformed.

Figure 10.12 helps to explain why the phrase "restoring force" is used. In the picture, an object of mass m is attached to a spring on a frictionless table. In part A, the object is at rest. The spring is undeformed and, hence, applies no force to the object, which is in equilibrium. In part B, the spring has been stretched to the right, so it exerts the leftward-pointing force $-\mathbf{F}$. When the object is released, this force pulls it to the left, restoring it toward its equilibrium position. However, consistent with Newton's first law, the moving object has inertia and coasts beyond the equilibrium position, compressing the spring as in part C. The force exerted by the spring now points to the right and, after bringing the object to a momentary halt, acts to restore the object to its equilibrium position. But the object's inertia again carries it beyond the equilibrium position, this time stretching the spring and leading to the restoring force $-\mathbf{F}$ shown in part D. The back-and-forth motion illustrated in the drawing then repeats itself, continuing forever, since there is no friction in this example.

When the restoring force has the mathematical form given by $F = -kx$, the type of friction-free motion illustrated in Figure 10.12 is designated as "simple harmonic motion." By attaching a pen to the object and moving a strip of paper

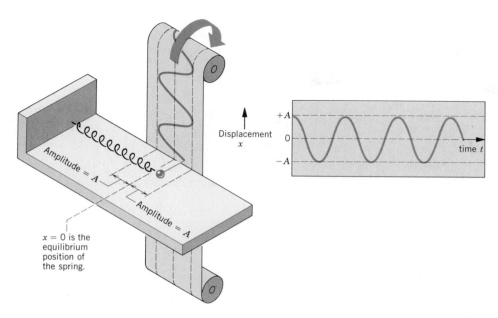

Figure 10.13 When an object moves in simple harmonic motion, a graph of its position as a function of time has a sinusoidal shape with an amplitude A.

past it at a steady rate, we can record the position of the vibrating object as time passes. Figure 10.13 illustrates the resulting graphical record of simple harmonic motion. The maximum excursion from equilibrium is the *amplitude A* of the motion. The shape of this graph is characteristic of simple harmonic motion and is called "sinusoidal," because it has the shape of a trigonometric sine or cosine function.

When an object attached to a horizontal spring is moved from its equilibrium position and released, the restoring force $F = -kx$ leads to simple harmonic motion. The restoring force also leads to simple harmonic motion when the object is attached to a vertical spring. When the spring is vertical, however, the weight of the object stretches the spring, and the motion occurs with respect to the equilibrium position on the stretched spring, as Figure 10.14 indicates. The amount of initial stretching d_0 caused by the weight of the object can be calculated by equating the weight to the magnitude of the restoring force that supports it; thus, $mg = kd_0$, which gives $d_0 = mg/k$.

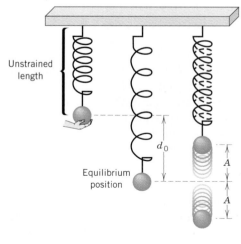

Figure 10.14 The weight of an object on a vertical spring stretches the spring by an amount d_0. Simple harmonic motion of amplitude A occurs with respect to the equilibrium position of the stretched spring.

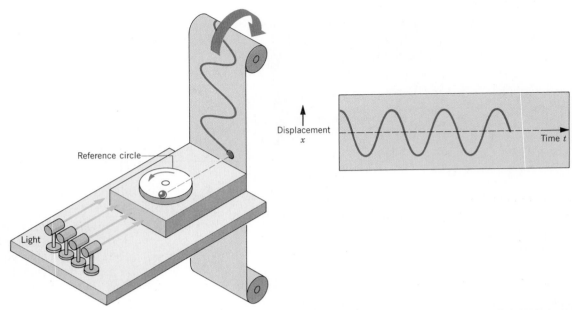

Figure 10.15 The ball mounted on the turntable moves in uniform circular motion, and its shadow, projected on a moving strip of film, executes simple harmonic motion.

10.4 SIMPLE HARMONIC MOTION AND THE REFERENCE CIRCLE

Simple harmonic motion, like any motion, can be described in terms of displacement, velocity, and acceleration, and the model in Figure 10.15 is helpful in explaining these characteristics. This model consists of a small ball attached to the top of a rotating turntable. The ball is moving in uniform circular motion (see Chapter 5) on a circle known as the *reference circle.* As the ball moves, its shadow falls on a strip of film, which is moving upward at a steady rate and records the position of the shadow as time passes. A comparison of the film with the paper in Figure 10.13 reveals identical patterns, indicating that the shadow of the ball is a good model for simple harmonic motion.

DISPLACEMENT

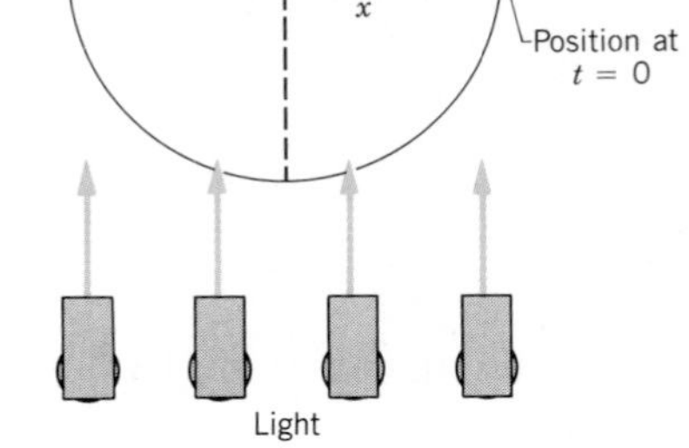

Figure 10.16 The ball's shadow on the film has a displacement x that depends on the angle θ through which the ball has moved on the reference circle.

Figure 10.16 takes a closer look at the reference circle (radius $= A$) and indicates how to determine the displacement of the shadow on the film. The ball starts on the x axis and moves through the angle θ in a time t. Since the circular motion is uniform, the ball moves with a constant angular speed ω (in rad/s). Therefore, the angle has a value (in rad) of $\theta = \omega t$. The position of the shadow on the film is just x, the projection of the radius A onto the x axis:

$$x = A \cos \theta = A \cos \omega t \qquad (10.6)$$

Figure 10.17 shows a graph of this equation. As time passes, the shadow of the ball oscillates between the values of $x = +A$ and $x = -A$, corresponding to the limiting values of $+1$ and -1 for the cosine of an angle. The radius A of the reference circle, then, is the amplitude of the simple harmonic motion.

As the ball moves one revolution or cycle around the reference circle, its shadow executes one cycle of back-and-forth motion. For any object in simple harmonic motion, the time required to complete one cycle is the *period T*, as Figure 10.17 indicates. The value of T depends on the angular speed ω of the ball. Since $\omega = \theta/t$ and 2π radians of angular displacement correspond to the one cycle that occurs in T seconds, we find that

$$\omega = \frac{2\pi}{T} \qquad (10.7)$$

Often, instead of the period, it is more convenient to speak of the *frequency f* of the motion, the frequency being just the number of cycles of the motion per second. For example, if an object on a spring completes 10 cycles in one second, the frequency is $f = 10$ cycles/s. The period T, or the time for one cycle, would be $\frac{1}{10}$ s. Thus, frequency and period are related according to

$$f = \frac{1}{T} \qquad (10.8)$$

Usually one cycle per second is referred to as one hertz (Hz), the unit being named after Heinrich Hertz (1857–1894). One thousand cycles per second is denoted as one kilohertz (kHz), with the result that five thousand cycles per second can be written as 5 kHz, for instance.

Using the relationships $\omega = 2\pi/T$ and $f = 1/T$, we can relate the angular speed ω (in rad/s) to the frequency f (in cycles/s or Hz):

$$\omega = \frac{2\pi}{T} = 2\pi f \qquad (10.9)$$

Because ω is directly related to the frequency f, it is often called the *angular frequency* of the simple harmonic motion.

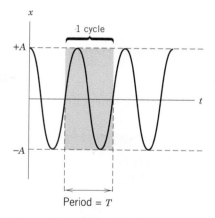

Figure 10.17 For simple harmonic motion, the graph of displacement x versus time t is a sinusoidal curve. The period T is the time required for one complete cycle of the motion.

VELOCITY

The reference circle model can also be used to determine the velocity of an object in simple harmonic motion. Figure 10.18 shows the tangential velocity $\mathbf{v_T}$ of the ball on the reference circle. The drawing indicates that the velocity $\mathbf{v}$ of the shadow is just the x component of the vector $\mathbf{v_T}$, that is, $v = -v_T \sin \theta$, where $\theta = \omega t$. The minus sign is necessary, since $\mathbf{v}$ points to the left, in the direction of the negative x axis. Since the tangential speed v_T is related to the angular speed ω by $v_T = r\omega$ (Equation 8.9) and since $r = A$, it follows that $v_T = A\omega$. Therefore, the velocity in simple harmonic motion is given by

$$v = -A\omega \sin \theta = -A\omega \sin \omega t \qquad (10.10)$$

This velocity is *not* constant, but varies between maximum and minimum values as time passes. When the shadow changes direction at either end of the oscillatory motion, the velocity is zero. When the shadow passes through the $x = 0$ position, the velocity has a maximum magnitude of $A\omega$, since the sine of an angle is between $+1$ and -1:

$$v_{max} = A\omega \qquad (\omega \text{ in rad/s}) \qquad (10.11)$$

Both the amplitude A and the angular frequency ω determine the maximum velocity, as Example 5 emphasizes.

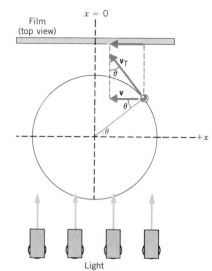

Figure 10.18 The velocity $\mathbf{v}$ of the ball's shadow is the x component of the tangential velocity $\mathbf{v_T}$ of the ball on the reference circle.

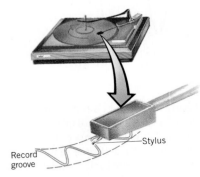

Record
groove
                                      Stylus

Figure 10.19 A phono stylus (monaural) generates electrical signals by vibrating in simple harmonic motion within a record groove.

Example 5   *The Maximum Speed of a Phono Stylus*

In response to a record groove, a phono stylus (monaural) generates electrical signals by vibrating back and forth in simple harmonic motion. (See Figure 10.19.) In a standard test, it is arranged that the frequency of the motion is $f = 1.0$ kHz and the amplitude is $A = 8.0 \times 10^{-6}$ m. What is the maximum speed of the stylus?

REASONING AND SOLUTION   The maximum speed of the vibrating stylus can be obtained from Equation 10.11 and the fact that $\omega = 2\pi f$:

$$v_{max} = A\omega = (8.0 \times 10^{-6} \text{ m})[2\pi(1.0 \times 10^3 \text{ Hz})] = \boxed{0.050 \text{ m/s}}$$

On test records that are used to check out the performance of your stereo system, you will find this value for v_{max} written as 5 cm/s.

ACCELERATION

In simple harmonic motion, the velocity is not constant; consequently, there must be an acceleration. This acceleration can also be determined with the aid of the reference circle model. As Figure 10.20 shows, the ball on the reference circle moves in uniform circular motion, and, therefore, has a centripetal acceleration $\mathbf{a}_c$ that points toward the center of the circle. The acceleration $\mathbf{a}$ of the shadow is the x component of the centripetal acceleration; $a = -a_c \cos \theta$. The minus sign is needed because the acceleration of the shadow points to the left. Recalling that the centripetal acceleration is related to the angular speed ω by $a_c = r\omega^2$ (Equation 8.11) and using $r = A$, we find that $a_c = A\omega^2$. With this substitution, the acceleration in simple harmonic motion becomes

$$a = -A\omega^2 \cos \theta = -A\omega^2 \cos \omega t \qquad (10.12)$$

The acceleration, like the velocity, does *not* have a constant value as time passes. The maximum magnitude of the acceleration is

$$a_{max} = A\omega^2 \qquad (\omega \text{ in rad/s}) \qquad (10.13)$$

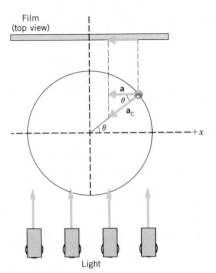

Figure 10.20 The acceleration **a** of the ball's shadow is the x component of the centripetal acceleration $\mathbf{a}_c$ of the ball on the reference circle.

Although both the amplitude A and the angular frequency ω determine the maximum value, the frequency has a particularly strong effect, for it is squared. Example 6 shows that the acceleration can be remarkably large in a practical situation.

Example 6   The Maximum Acceleration of a Loudspeaker Diaphragm

The diaphragm of a loudspeaker moves back and forth in simple harmonic motion to create sound, as in Figure 10.21. The frequency of the motion is $f = 1.0$ kHz, and the amplitude is $A = 2.0 \times 10^{-4}$ m. Find the maximum acceleration of the diaphragm.

REASONING AND SOLUTION   Using Equation 10.13 and the fact that $\omega = 2\pi f$, we find that

$$a_{max} = A\omega^2 = (2.0 \times 10^{-4} \text{ m})[2\pi(1.0 \times 10^3 \text{ Hz})]^2 = \boxed{7.9 \times 10^3 \text{ m/s}^2}$$

This acceleration is more than 800 times that due to gravity and is much larger than the acceleration experienced by astronauts during a rocket launch. The construction of the diaphragm must be sturdy enough to withstand such large accelerations.

THE PHYSICS OF . . .

a loudspeaker diaphragm.

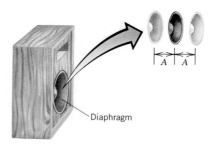

Diaphragm

Figure 10.21 The diaphragm of a loudspeaker generates a 1.0-kHz sound by moving back and forth in simple harmonic motion.

FREQUENCY OF VIBRATION

With the aid of Newton's second law ($F = ma$), it is possible to determine the frequency at which an object of mass m vibrates on a spring. The mass of the spring itself is assumed to be negligible. The force that the spring applies to the object is the Hooke's law restoring force $F = -kx$, so that Newton's second law becomes $-kx = ma$. Using Equation 10.6 for the displacement x and Equation 10.12 for the acceleration a, we find that

$$-k(A \cos \omega t) = m(-A\omega^2 \cos \omega t)$$

which yields

$$\omega = \sqrt{\frac{k}{m}} \qquad (\omega \text{ in rad/s}) \tag{10.14}$$

In this expression, the angular frequency ω must be in radians per second. Larger spring constants k and smaller masses m result in larger frequencies. Example 7 illustrates the effect of the mass on the frequency.

Example 7   A Vibrating Test Platform

Electronic equipment for high-performance aircraft like the Space Shuttle must be able to withstand large vibrations. Figure 10.22 shows a platform that can be used to test such equipment. The platform has a mass of 3.00 kg and is mounted on four springs, each of which has a spring constant of $k = 1150$ N/m. If electronic equipment is placed on the platform, find the frequency f at which the system vibrates when the electronic equipment has a mass of (a) 4.00 kg and (b) 7.00 kg.

REASONING   The vibrational frequency f is related to the angular frequency ω by $f = \omega/2\pi$ (Equation 10.9). The angular frequency of each spring is given by Equation 10.14 as $\omega = \sqrt{k/m}$, where m is the mass supported by each spring. Since there are four springs, each supports one-fourth of the total mass.

SOLUTION

(a) The mass supported by each spring is $m = \tfrac{1}{4}(3.00 \text{ kg} + 4.00 \text{ kg}) = 1.75 \text{ kg}$. The frequency of vibration is

$$f = \frac{1}{2\pi}\sqrt{\frac{k}{m}} = \frac{1}{2\pi}\sqrt{\frac{1150 \text{ N/m}}{1.75 \text{ kg}}} = \boxed{4.08 \text{ Hz}}$$

(b) The mass supported by each spring is now $m = \tfrac{1}{4}(3.00 \text{ kg} + 7.00 \text{ kg}) = 2.50 \text{ kg}$, and the frequency is

$$f = \frac{1}{2\pi}\sqrt{\frac{k}{m}} = \frac{1}{2\pi}\sqrt{\frac{1150 \text{ N/m}}{2.50 \text{ kg}}} = \boxed{3.41 \text{ Hz}}$$

The frequency is lower here because of the greater inertia of the larger mass.

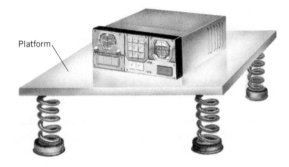

Platform

Figure 10.22   A piece of electronic equipment on a vibrating test platform.

THE PHYSICS OF . . .

detecting and measuring small amounts of chemicals.

Example 7 indicates that the mass of the vibrating object influences the frequency of simple harmonic motion. Electronic sensors are being developed that take advantage of this effect in detecting and measuring small amounts of chemicals. These sensors utilize tiny quartz crystals that vibrate when an electric current passes through them. If the crystal is coated with a substance that absorbs a particular chemical, then its mass increases as the chemical is absorbed and, according to Equation 10.14, the frequency of the simple harmonic motion decreases. The change in frequency is detected electronically, and the sensor is calibrated to give the mass of the absorbed chemical as a function of the change in frequency.

10.5 ENERGY AND SIMPLE HARMONIC MOTION

ELASTIC POTENTIAL ENERGY

We saw in Chapter 6 that an object above the surface of the earth has gravitational potential energy. Therefore, when the object is allowed to fall, like the hammer of the pile driver in Figure 6.11, it can do work. A spring also has potential energy when the spring is stretched or compressed, and we refer to it as *elastic potential energy*. Because of elastic potential energy, a stretched or compressed spring can do work on an object that is attached to the spring. For instance, Figure 10.23 shows a door-closing unit that is often found on screen doors. When the door is opened, a spring inside the unit is compressed and has elastic potential energy. When the door is released, the compressed spring expands and does the work of closing the door.

To find an expression for the elastic potential energy, we will determine the work done by the spring force on an object. Figure 10.24 shows a stretched spring contracting from an initial position x_0 to a final position x_f after the object is released. The work done by the spring force can be obtained from the definition of work as $W = (F \cos \theta)s$ (Equation 6.1). In this definition, s is the magnitude of the displacement, which is $x_0 - x_f$. The term F is the magnitude of the spring force. Equation 10.5 gives the spring force as $F = -kx$, with the magnitude being just kx. As the spring contracts from x_0 to x_f, the magnitude of the spring force changes from kx_0 to kx_f. To account for the changing magnitude, we can use the average of kx_0 and kx_f, because the dependence on x is linear: $\bar{F} = \frac{1}{2}(kx_0 + kx_f)$. The work W_{elastic} done by the average spring force is, then,

$$W_{\text{elastic}} = (\bar{F} \cos \theta)s = [\tfrac{1}{2}(kx_0 + kx_f)] \cos 0°(x_0 - x_f)$$
$$W_{\text{elastic}} = \tfrac{1}{2}kx_0^2 - \tfrac{1}{2}kx_f^2 \tag{10.15}$$

In the calculation above, θ is $0°$, since the spring force points in the same direction as the displacement. Equation 10.15 indicates that the work done by the spring force is equal to the amount by which the quantity $\frac{1}{2}kx^2$ changes as the spring contracts. The value of $\frac{1}{2}kx^2$ is larger when the magnitude of x is larger and smaller when the magnitude of x is smaller. The quantity $\frac{1}{2}kx^2$ is analogous to the quantity mgh, which we identified in Chapter 6 as the gravitational potential energy. Similarly, we identify the quantity $\frac{1}{2}kx^2$ as the elastic potential energy.

Definition of Elastic Potential Energy

The elastic potential energy PE_{elastic} is the energy that a spring has by virtue of being stretched or compressed. For an ideal spring that has a spring constant k and is stretched or compressed by an amount x relative to its unstrained length, the elastic potential energy is

$$PE_{\text{elastic}} = \tfrac{1}{2}kx^2 \tag{10.16}$$

SI Unit of Elastic Potential Energy: joule (J)

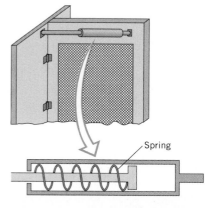

Figure 10.23   A door-closing unit.

Elastic potential energy stored in the drawn bow will propel the arrow.

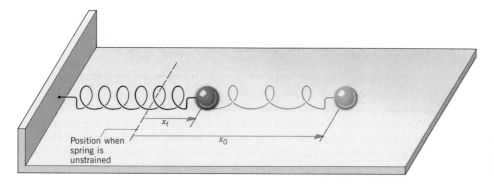

Figure 10.24   When the object is released, the extent to which the spring is stretched changes from an initial value of x_0 to a final value of x_f.

Equation 10.16 indicates that the elastic potential energy is a maximum for a fully stretched or compressed spring and zero for a spring that is neither stretched nor compressed ($x = 0$).

THE CONSERVATION OF MECHANICAL ENERGY

In Chapter 6 we discussed the work–energy theorem, which relates the work W done by the net force acting on an object to the change in the object's kinetic energy:

$$W = \tfrac{1}{2}mv_f^2 - \tfrac{1}{2}mv_0^2 \qquad (6.3)$$

Remember that forces can be categorized into two groups: (1) conservative forces, such as the gravitational force and the elastic spring force, and (2) nonconservative forces, such as friction and air resistance. Consequently, the work is the sum of two contributions, $W = W_c + W_{nc}$, where W_c is the work done by the conservative forces and W_{nc} is the work done by the nonconservative forces. If the only conservative forces acting on the object are the elastic spring force and the gravitational force, the work–energy theorem takes the following form:

$$\underbrace{W_{elastic} + W_{gravity}}_{W_c} + W_{nc} = \tfrac{1}{2}mv_f^2 - \tfrac{1}{2}mv_0^2$$

By substituting $W_{elastic} = \tfrac{1}{2}kx_0^2 - \tfrac{1}{2}kx_f^2$ (Equation 10.15) and $W_{gravity} = mgh_0 - mgh_f$ (Equation 6.4) and rearranging terms, we obtain

$$W_{nc} = (\tfrac{1}{2}mv_f^2 - \tfrac{1}{2}mv_0^2) + (\tfrac{1}{2}kx_f^2 - \tfrac{1}{2}kx_0^2) + (mgh_f - mgh_0) \qquad (10.17)$$

This relation is the same as Equation 6.6, except that terms involving the elastic potential energy ($\tfrac{1}{2}kx^2$) have now been taken into account.

In Chapter 6, we introduced the total mechanical energy E as the sum of the kinetic energy and the gravitational potential energy: $E = \text{KE} + \text{PE}_{gravity}$. We now expand this definition to include the elastic potential energy: $E = \text{KE} + \text{PE}_{gravity} + \text{PE}_{elastic}$. The total mechanical energy, then, becomes

$$E = \tfrac{1}{2}mv^2 + mgh + \tfrac{1}{2}kx^2 \qquad (10.18)$$

With this definition, Equation 10.17 becomes

$$W_{nc} = E_f - E_0$$

which shows that the work done by nonconservative forces changes the total mechanical energy from an initial value of E_0 to a final value of E_f.

If the nonconservative forces do no work, $W_{nc} = 0$, the equation above yields $E_f = E_0$, and the total mechanical energy of an object does not change as the object moves. The principle of conservation of total mechanical energy is the subject of the next example.

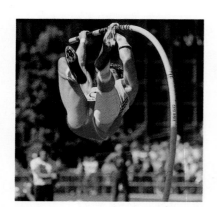

The bent pole has elastic potential energy that is converted to gravitational potential energy and kinetic energy as the vaulter goes up, clears the bar, and comes down.

Example 8 An Object on a Horizontal Spring

In Figure 10.13 an object of mass m is vibrating horizontally on a frictionless surface. The simple harmonic motion has an amplitude A, because the spring was stretched initially to $x = A$ and then released from rest. Determine the kinetic energy at $x = A$, $x = A/2$, and $x = 0$ and compare these results to the potential energy at the same points.

REASONING   The conservation principle indicates that, in the absence of friction (a nonconservative force), the initial and final total mechanical energies are the same:

$$E_f = E_0$$

$$KE_f + \tfrac{1}{2}kx_f^2 + mgh_f = KE_0 + \tfrac{1}{2}kx_0^2 + mgh_0$$

Since the spring is horizontal, gravitational potential energy plays no role. Algebraically, $h_f = h_0$, and the above equation becomes

$$KE_f + \tfrac{1}{2}kx_f^2 = KE_0 + \tfrac{1}{2}kx_0^2$$

Initially at $x_0 = A$, the object is stationary, so $KE_0 = 0$. Consequently, the conservation principle indicates that

$$KE_f + \tfrac{1}{2}kx_f^2 = \tfrac{1}{2}kA^2$$

The initial total mechanical energy consists only of elastic potential energy $\tfrac{1}{2}kA^2$. At any other position, the object may have both kinetic and elastic potential energies, but the total of both types must always add up to equal the initial energy of $\tfrac{1}{2}kA^2$.

SOLUTION   The expression above can be used to calculate the final kinetic energy KE_f at $x_f = A$, $x_f = A/2$, and $x_f = 0$, and the results are compared to the elastic potential energy in Table 10.4. All of the energy is in the form of elastic potential energy at $x_f = A$ and in the form of kinetic energy at $x_f = 0$. At intermediate points the energy is part kinetic and part potential. In the absence of friction, the simple harmonic motion merely converts the energy between one form and the other, the total always being equal to the amount $\tfrac{1}{2}kA^2$ present initially.

Table 10.4   Energy Distribution for an Object in Simple Harmonic Motion on a Horizontal Spring

Spring	KE	$PE_{elastic}$	$E = KE + PE_{elastic}$
A	0	$\tfrac{1}{2}kA^2$	$\tfrac{1}{2}kA^2$
$A/2$	$\tfrac{3}{8}kA^2$	$\tfrac{1}{8}kA^2$	$\tfrac{1}{2}kA^2$
$x_f = 0$	$\tfrac{1}{2}kA^2$	0	$\tfrac{1}{2}kA^2$

In the previous example, gravitational potential energy plays no role because the spring is horizontal. The next example illustrates that gravitational potential energy must be taken into account when a spring is oriented vertically.

Example 9   A Falling Ball on a Vertical Spring

A 0.20-kg ball is attached to a vertical spring, as in Figure 10.25. The spring constant of the spring is 28 N/m. The ball, supported initially so that the spring is neither stretched nor compressed, is released from rest. In the absence of air resistance, how far does the ball fall before being brought to a momentary stop by the spring?

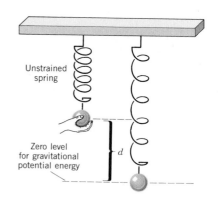

Figure 10.25   The ball is supported initially so that the spring is unstrained. After being released from rest, the ball falls through the distance d before being momentarily stopped by the spring.

REASONING AND SOLUTION   Since air resistance is absent, only the conservative forces of gravity and the spring act on the ball. Therefore, the principle of conservation of mechanical energy applies:

$$E_f = E_0$$

$$\tfrac{1}{2}mv_f^2 + \tfrac{1}{2}kx_f^2 + mgh_f = \tfrac{1}{2}mv_0^2 + \tfrac{1}{2}kx_0^2 + mgh_0$$

As Figure 10.25 indicates, the spring is unstrained to begin with and has no elastic potential energy; $\tfrac{1}{2}kx_0^2 = 0$. As the ball falls through the distance d, the spring is stretched, leading to an elastic potential energy of $\tfrac{1}{2}kx_f^2 = \tfrac{1}{2}kd^2$ at the bottom. Initially, the ball is located a distance d above its final position, so the initial and final gravitational potential energies are $mgh_0 = mgd$ and $mgh_f = 0$. Because the ball is released from rest and comes to a halt at the bottom, the initial and final kinetic energies are zero. The conservation of energy principle now becomes

$$\tfrac{1}{2}kd^2 = mgd$$

reflecting the fact that gravitational potential energy (mgd) has been completely converted into elastic potential energy ($\tfrac{1}{2}kd^2$) when the ball comes to a halt. Solving this equation for d gives

$$d = \frac{2mg}{k} = \frac{2(0.20 \text{ kg})(9.80 \text{ m/s}^2)}{(28 \text{ N/m})} = \boxed{0.14 \text{ m}}$$

The distance d calculated in Example 9 is not the same as the distance d_0 at which the ball would hang stationary on the spring. Determine d_0 to see whether it is greater or less than d and explain why.

10.6 THE PENDULUM

As Figure 10.26 shows, a *simple pendulum* consists of a particle of mass m, attached to a frictionless pivot P by a support of length L, whose mass is negligible. When the particle is pulled away from its equilibrium position by an angle θ and released, the particle swings back and forth. By attaching a pen to the bottom of the swinging particle and moving a strip of paper beneath it at a steady rate, we can record the position of the particle as time passes. The graphical record reveals a pattern that is similar (not identical) to the sinusoidal pattern for simple harmonic motion.

The force of gravity is responsible for the back-and-forth rotation about the axis at P. The rotation speeds up as the particle approaches the lowest point on the arc and slows down on the upward part of the swing. Eventually the rotational speed is reduced to zero, and the particle swings back. As Section 9.4 discusses, a net torque is required to change the rotational speed. The gravitational force $m\mathbf{g}$ produces this torque. (The tension $\mathbf{T}$ in the support creates no torque, because it points directly at the pivot and, therefore, has a zero lever arm.) Since Torque = Force $\times$ Lever arm, the torque τ is the gravitational force multiplied by the perpendicular distance d to the axis: $\tau = -(mg)d$. The minus sign is included since the torque is a restoring torque; that is, it acts to reduce the angle θ. The perpendicular distance d is very nearly equal to the arc length on the circular path if only small values ($10°$ or less) of θ are considered. Furthermore, if θ is expressed in radians, the arc length and the radius L of circular path are related, according to Arc

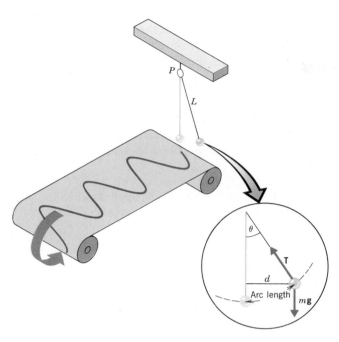

Figure 10.26 A simple pendulum swinging back and forth about the pivot *P*. If the angle θ is small, the swinging is approximately simple harmonic motion.

length $= L\theta$. Under these conditions, it follows that $d \approx L\theta$, and the torque created by gravity is

$$\tau \approx -\underbrace{mgL}_{k'}\,\theta \qquad (10.19)$$

In Equation 10.19, the term mgL has a constant value k', independent of θ. *For small angles,* then, the torque that restores the pendulum to its vertical equilibrium position is proportional to the angular displacement θ. The expression $\tau = -k'\theta$ has the same form as the Hooke's law restoring force for a particle on a spring, $F = -kx$. Therefore, we expect the frequency of the back-and-forth movement of the pendulum to be given by an equation analogous to Equation 10.14 ($\omega = 2\pi f = \sqrt{k/m}$). In place of the spring constant k, the constant $k' = mgL$ will appear, and, as usual in rotational motion, in place of the mass m, the moment of inertia I will appear:

$$\omega = 2\pi f = \sqrt{\frac{mgL}{I}} \qquad \text{(small angles only)} \qquad (10.20)$$

The moment of inertia of a particle of mass m, rotating at a radius $R = L$ about an axis, is given by $I = mL^2$ (Equation 9.6). Substituting this expression for I into Equation 10.20 reveals for a simple pendulum that

$$\omega = 2\pi f = \sqrt{\frac{g}{L}} \qquad \text{(small angles only)} \qquad (10.21)$$

The mass of the particle has been eliminated algebraically from this expression, so only the length L and the acceleration g due to gravity determine the frequency of a simple pendulum. Equation 10.21 does not apply if the angle of oscillation is large, for then the pendulum does not exhibit simple harmonic motion. For small-angle motion, Equation 10.21 provides the basis for using a pendulum to keep time, as Example 10 demonstrates.

Example 10   Keeping Time

Determine the length of a simple pendulum that will swing back and forth in simple harmonic motion with a period of 1.00 s.

REASONING AND SOLUTION   Equation 10.21 can be applied to find the period, if we remember that the relationship between frequency f and period T is $f = 1/T$:

$$2\pi f = \frac{2\pi}{T} = \sqrt{\frac{g}{L}}$$

$$L = \frac{T^2 g}{4\pi^2} = \frac{(1.00 \text{ s})^2 (9.80 \text{ m/s}^2)}{4\pi^2} = \boxed{0.248 \text{ m}}$$

Figure 10.27 shows an elegant grandfather clock that uses a pendulum to keep time.

Figure 10.27   A grandfather clock.

It is not necessary that the object in Figure 10.26 be a simple particle. It may be an extended object, in which case the pendulum is called a *physical pendulum*. For small oscillations, Equation 10.20 still applies, but the moment of inertia I is no longer mL^2. The proper value for the rigid object must be used. (See Section 9.4 for a discussion of moment of inertia.) In addition, the length L for a physical pendulum is the distance between the axis at P and the center of gravity of the object.

10.7 DAMPED HARMONIC MOTION

In simple harmonic motion, an object oscillates with a constant amplitude, because there is no mechanism for dissipating energy. In reality, however, friction or some other energy-dissipating mechanism is always present. In the presence of energy dissipation, the amplitude of oscillation decreases as time passes, and the motion is no longer simple harmonic motion. Instead, it is referred to as *damped harmonic motion,* the decrease in amplitude being called "damping."

THE PHYSICS OF . . .

a shock absorber.

One widely used application of damped harmonic motion is in the suspension system of an automobile. Figure 10.28*a* shows a shock absorber attached to a main suspension spring of a car. A shock absorber is designed to introduce damping forces. As part *b* of the drawing shows, a shock absorber consists of a piston in a reservoir of oil. When the piston moves in response to a bump in the road, holes in the piston head permit the piston to pass through the oil. Viscous forces that arise during this movement cause the damping.

Figure 10.29 illustrates the different degrees of damping that can exist. As applied to the example of a car's suspension system, these graphs show the vertical position of the chassis after it has been pulled upward by the amount A_0 at time $t_0 = 0$ and then released. Part *a* of the figure compares undamped or simple

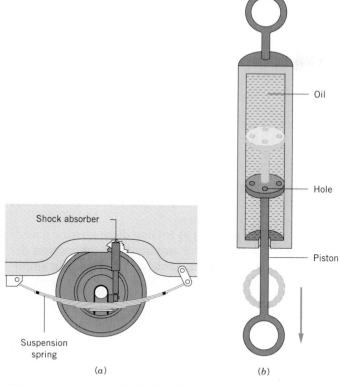

Figure 10.28 (a) A shock absorber mounted in the suspension system of an automobile and (b) a simplified, cutaway view of the shock absorber.

harmonic motion in curve 1 to slightly damped motion in curve 2. In damped harmonic motion, the chassis oscillates with decreasing amplitude until it eventually comes to rest. As the degree of damping is increased from curve 2 to curve 3, the car makes fewer oscillations before coming to a halt. Part *b* of the drawing shows that as the degree of damping is increased still further, there comes a point when the car does not oscillate at all after it is released but, rather, settles directly

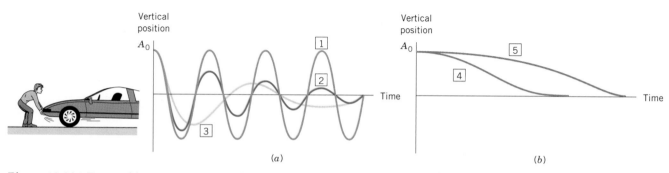

Figure 10.29 Damped harmonic motion. The degree of damping increases from curve 1 to curve 5. Curve 1 represents undamped or simple harmonic motion, while curve 4 represents critically damped harmonic motion. Curves 2 and 3 show underdamped motion, while curve 5 illustrates overdamped motion.

back to its equilibrium position, as in curve 4. The smallest degree of damping that completely eliminates the oscillations is termed "critical damping," and the motion is said to be *critically damped*.

Figure 10.29*b* also shows that the car takes the longest time to return to its equilibrium position in curve 5, where the degree of damping is increased above the critical value for curve 4. When the damping exceeds the critical value, the motion is said to be *overdamped*. In contrast, when the damping is less than the critical level, the motion is said to be *underdamped* (curves 2 and 3). Typical automobile shock absorbers are designed to produce underdamped motion somewhat like that in curve 3.

10.8 DRIVEN HARMONIC MOTION AND RESONANCE

In damped harmonic motion, a mechanism such as friction dissipates the energy of an oscillating system, with the result that the amplitude of the motion decreases. This section discusses the opposite effect, namely, the increase in amplitude that results when energy is continually added to an oscillating system.

To set an object on an ideal spring into simple harmonic motion, some agent must apply a force that stretches or compresses the spring initially. Suppose that this force is applied at all times, not just for a brief initial moment. The force could be provided, for example, by a person who simply pushes and pulls the object back and forth. The resulting motion is known as **driven harmonic motion**, because the additional force drives or controls the behavior of the object to a large extent. The additional force is identified as the **driving force**.

Figure 10.30 illustrates one particularly important example of driven harmonic motion. Here, the driving force has the same frequency as the spring system and always points in the direction of the object's velocity. The frequency of the spring system is $f = (1/2\pi)\sqrt{k/m}$ and is called a natural frequency, because it is the frequency at which the spring system naturally oscillates. Since the driving force and the velocity always have the same direction, positive work is done on the object at all times, and the total mechanical energy of the system increases. As a result, the amplitude of the vibration becomes larger and will increase without limit, if there is no damping force to dissipate the energy being added by the driving force. The situation in Figure 10.30 is known as **resonance**.

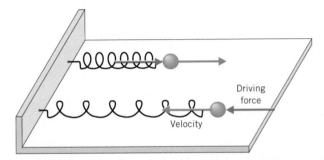

Figure 10.30 Resonance occurs when the frequency of the driving force matches a frequency at which the object naturally vibrates. The blue arrows represent the driving force, and the red arrows represent the velocity of the object.

Resonance

Resonance is the condition under which an oscillating force can transmit large amounts of energy to an oscillating object, leading to a large amplitude motion. In the absence of damping, resonance occurs when the frequency of the force matches a natural frequency at which the object will oscillate.

The role played by the frequency of a driving force is a critical one. The matching of this frequency with a natural frequency of vibration allows even a relatively weak force to produce a large amplitude vibration, because the effect of each push–pull cycle is cumulative.

Resonance can occur with any object that can oscillate, and springs need not be involved. For example, the diaphragm of a loudspeaker has a natural frequency of vibration. When designing loudspeakers, audio engineers must take this natural frequency into account, or else resonance will cause the diaphragm to vibrate with a much greater amplitude at one sound frequency than at others, leading to severe distortion in the reproduction of music.

INTEGRATION OF CONCEPTS

ELASTIC POTENTIAL ENERGY AND GRAVITATIONAL POTENTIAL ENERGY

We first talked about potential energy in Chapter 6, when we discussed gravitational potential energy ($PE_{gravity} = mgh$). Now in Chapter 10, we have studied elastic potential energy ($PE_{elastic} = \frac{1}{2}kx^2$). These two types of potential energy have some common features that characterize all types of potential energy. Both are associated with conservative forces, the gravitational force on one hand, and the elastic force of a spring on the other. The concept of potential energy has no meaning for nonconservative forces. Another common feature is that potential energy is energy of position. Gravitational potential energy is energy that an object has by virtue of the object's position relative to the earth. Elastic potential energy is energy that a spring has, because the spring is stretched or compressed relative to its unstrained position. A third feature that characterizes both types of potential energy is that each is used in the same way in the principle of conservation of mechanical energy. Mechanical energy is the sum of kinetic and potential energies, and the conservation principle indicates that this sum remains constant throughout the motion of a system, if nonconservative forces do no work. Under such a condition, kinetic energy and potential energies may be interconverted, but the sum of the two always remains the same. The conservation principle applies in the same way when only gravitational potential energy is present, when only elastic potential energy is present, or when both types are present simultaneously, as they are in a number of homework problems at the end of this chapter (e.g., problems 42, 48, and 49). The fact that all types of potential energy are treated in the same way in the conservation principle is one reason why the principle is so useful.

SIMPLE HARMONIC MOTION AND NEWTON'S SECOND LAW

Newton's second law lies at the heart of mechanics, which is the branch of physics that deals with the motion of objects and the forces that change it. In earlier chapters, we have applied the second law ($\Sigma\mathbf{F} = m\mathbf{a}$) to situations in which a constant net force $\Sigma\mathbf{F}$ acts on an object. With a constant value for $\Sigma\mathbf{F}$, the second law predicts a constant or uniform value for the acceleration $\mathbf{a}$ of the object. A constant acceleration means that the equations of kinematics can be applied to the motion. In our present study of simple harmonic motion, Newton's second law plays a similar role. A net force also acts on an object in simple harmonic motion, and we can substitute the net force into the second law to find the acceleration. The net force acting in simple harmonic motion is the Hooke's law restoring force, $F = -kx$. This force, however, varies from place to place and is not constant. Correspondingly, the acceleration that Newton's second law predicts is not constant. Since the acceleration is not constant, the equations of kinematics cannot be used to analyze simple harmonic motion. Nonetheless, the role of the second law is the same in simple harmonic motion as it is in uniformly accelerated motion. In both types of motion, a net force causes the velocity to change, and the change in velocity per unit time is the acceleration, which can be obtained from Newton's second law.

SUMMARY

The possible kinds of **elastic deformation** include stretch and compression, shear deformation, and volume deformation. The forces required to create them are given by Equations 10.1–10.3 and are characterized with the aid of proportionality constants called, respectively, **Young's modulus,** the **shear modulus,** and the **bulk modulus.**

Stress is the force per unit area applied to an object and causes **strain.** For stretch/compression and volume deformation, strain is the resulting fractional change in length or volume. For shear deformation, strain reflects the change in shape of an object. **Hooke's law** states that stress is directly proportional to strain, up to a limit called the **proportionality limit.**

The force F that must be applied to stretch or compress an **ideal spring** is $F = kx$, where k is the spring constant and x is the distance by which the spring is stretched or compressed from its unstrained length. A spring exerts a **restoring force** on an object attached to the spring. The restoring force produced by an ideal spring is $F = -kx$, where the minus sign indicates that the restoring force points opposite to the direction of the stretch or compression.

Simple harmonic motion is the oscillatory motion that occurs when a restoring force of the form $F = -kx$ acts on an object. A graphical record of position versus time for an object in simple harmonic motion is sinusoidal. The **amplitude** A of the motion is the maximum distance that the object moves away from its equilibrium position. The **period** T is the time required to complete one cycle of the motion, while the **frequency** f is the number of cycles per second that occur. Frequency and period are related according to $f = 1/T$. The frequency f (in Hz) is related to the angular frequency ω (in rad/s) according to $\omega = 2\pi f$. For an object of mass m on a spring with spring constant k, the frequency is determined by $2\pi f = \sqrt{k/m}$. The velocity and the acceleration in simple harmonic motion are continually changing with time; the maximum speed is $v_{max} = A\omega$, and the maximum acceleration is $a_{max} = A\omega^2$.

The **elastic potential energy** of an object attached to an ideal spring is $PE_{elastic} = \frac{1}{2}kx^2$. The total mechanical energy E of an object is the sum of its kinetic energy, gravitational potential energy, and elastic potential energy; $E = \frac{1}{2}mv^2 + mgh + \frac{1}{2}kx^2$. If nonconservative forces like friction do no work, the total mechanical energy of an object is conserved, $E_f = E_0$.

A **simple pendulum** is a particle of mass m attached to a frictionless pivot by a support whose length is L and whose mass is negligible. The small-angle ($\leq 10°$) back-and-forth swinging of a simple pendulum is simple harmonic motion, while large-angle motion is not. The frequency of small-angle motion is given by $2\pi f = \sqrt{g/L}$. A **physical pendulum** consists of a rigid object, with moment of inertia I and mass m, suspended from a frictionless pivot. For small-angle displacements, the frequency of simple harmonic motion for a physical pendulum is determined by $2\pi f = \sqrt{mgL/I}$, where L is the distance between the axis of rotation and the center of gravity of the rigid object.

Damped harmonic motion is motion in which the amplitude of oscillation decreases as time passes. **Critical damping** is the minimum degree of damping that eliminates any oscillations in the motion as the object returns to its equilibrium position.

Driven harmonic motion occurs when an additional driving force is applied to an object along with the restoring force. **Resonance** is the condition under which a driving force can transmit large amounts of energy to an oscillating object, leading to large amplitude motion. In the absence of damping, resonance occurs when the frequency of the driving force matches a natural frequency at which the object oscillates.

QUESTIONS

1. Three rods, made of the same material, have equal lengths and have square, circular, and triangular cross sections, as the drawing shows. Rank the rods according to the amount of force (smallest first) required to stretch each rod by the same amount. Give your reasoning.

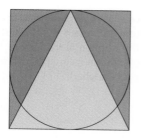

2. A trash compactor crushes empty aluminum cans, thereby reducing the total volume of the cans by 75%. Can the value given in Table 10.3 for the bulk modulus of aluminum be used to calculate the pressure generated in the trash compactor? Explain.

3. Both sides of the relation $F = S(\Delta X/L_0)A$ (Equation 10.2) can be divided by the area A to give F/A on the left side. Why can this force per unit area *not* be called a pressure, such as the pressure that appears in $\Delta P = -B(\Delta V/V_0)$ (Equation 10.3)?

4. The block in the drawing rests on the ground. Which face, A, B, or C, experiences the largest and which experiences the smallest stress when the block is resting on it? Explain.

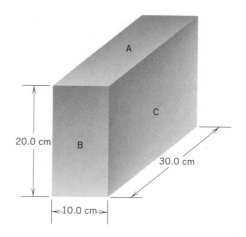

20.0 cm

30.0 cm

10.0 cm

5. Where on the path followed by an object in simple harmonic motion is (a) the velocity equal to zero and (b) the acceleration equal to zero? Refer to Figures 10.18 and 10.20.

6. Ignoring the damping introduced by the shock absorbers, explain why the number of passengers in a car affects the vibration frequency of the car's suspension system.

7. Explain how an astronaut in an orbiting satellite can use a spring with a known spring constant to measure his mass. Normally he could hang motionless from the spring to stretch it and, thus, determine his weight mg from the spring constant and the amount of stretch. By dividing his weight by g, he could determine his mass. This method will not work since his apparent weight is zero.

8. A block is attached to a horizontal spring and slides back and forth in simple harmonic motion on a frictionless horizontal surface. A second identical block is suddenly attached to the first block. In one case, the attachment is accomplished by joining the blocks at one extreme end of the oscillation cycle. In another case, the attachment is accomplished by joining the blocks at the point when the spring is unstrained. In either case, the velocities of the blocks are exactly matched at the instant of joining. For each method of attachment, explain how the amplitude, the frequency, and the maximum speed of the oscillation change.

9. Over a nightclub entrance is mounted a horizontal straight strip of equally spaced light bulbs, each one of which turns on in sequence for one-half second. Thus, the lighted bulb appears to move from left to right to left to right, etc. Is the apparent motion of the lighted bulb simple harmonic motion? Give your reasoning.

10. An electric saber saw consists of a blade that is driven back and forth by a pin mounted on the circumference of a rotating circular disk. As the disk rotates at a constant angular speed, the pin engages a slot and forces the blade back and forth, as the drawing illustrates. Is the motion of the blade simple harmonic motion? Explain, remembering that simple harmonic motion results only when the force applied to the moving object is given by $F = -kx$, which predicts a maximum force at the extreme ends of the motion.

11. Is more elastic potential energy stored in a spring when the spring is compressed by one centimeter than when it is stretched by the same amount? Explain.

12. In principle, the motion of a simple pendulum and an object on an ideal spring can both be used to provide the basic time interval or period used in a clock. Which of the two kinds of clocks is likely to become more inaccurate when carried to the top of a high mountain? Justify your answer.

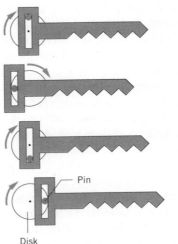

Question 10

Disk, Pin

13. Suppose two people are swinging on identical playground swings that are next to one another. They are talking to each other as the swings move back and forth in synchronism. Why do the swings remain synchronized, even though the two people have different masses?

14. Suppose you were kidnapped and held prisoner by space invaders in a completely isolated room, with nothing but a watch and a pair of shoes (including two shoelaces of known length). Explain how you might determine whether this room is on earth or on the moon.

15. A car travels over a road that contains a series of equally spaced bumps. Explain why a particularly "jarring" ride can result if the horizontal velocity of the car, the bump spacing, and the oscillation frequency of the car's suspension system are properly "matched."

PROBLEMS

Note: Unless otherwise indicated, the values for Young's modulus Y, the shear modulus S, and the bulk modulus B are given, respectively, in Table 10.1, Table 10.2, and Table 10.3.

Section 10.1 Elastic Deformation,
Section 10.2 Stress, Strain, and Hooke's Law

1. An 82-kg mountain climber hangs freely on a 4.0-mm radius nylon rope. If the rope stretches by 0.10 m, what is the unstretched length of the rope?

2. A rectangular solid block has the dimensions shown in the drawing for question 4 and a shear modulus of 7.00×10^9 N/m². A force of 4.2×10^4 N is applied as in Figure 10.6, with one face of the block fixed to an immovable horizontal surface. Determine the resulting shear deformation ΔX when the face fixed to the horizontal surface is (a) face A and (b) face C.

3. A CD player is mounted on four cylindrical rubber blocks. Each cylinder has a height of 0.030 m and a cross-sectional area of 1.2×10^{-3} m², and the shear modulus for rubber is 2.6×10^6 N/m². If a horizontal force of magnitude 32 N is applied to the CD player, how far will the unit move sideways? Assume that each block receives one-fourth of the force.

4. How much change in pressure is required to produce a change in volume of a block of steel that is 0.050% of the original volume.

5. A piece of aluminum is surrounded by air at a pressure of 1.01×10^5 Pa. The aluminum is placed in a vacuum chamber where the pressure is reduced to zero. Determine the fractional change $\Delta V/V_0$ in the volume of the aluminum.

6. An 1800-kg car is being lifted at a steady speed by a crane and hangs at the end of a cable whose radius is 6.0×10^{-3} m. The cable is 15 m in length and stretches by 8.0×10^{-3} m because of the weight of the car. Determine (a) the stress, (b) the strain, and (c) Young's modulus for the cable.

7. At the elastic limit for stretching a copper wire, the stress is 1.5×10^8 Pa. (a) Use this value in Hooke's law to calculate the corresponding strain. (b) Strictly speaking, Hooke's law does not apply at the elastic limit. Nevertheless, if strain values are kept below the value calculated in part (a), a copper wire will not be permanently deformed by stretching. Referring to Figure 10.9, explain why.

8. Two metal beams are joined together by four steel rivets, as the drawing indicates. Each rivet has a radius of 5.0×10^{-3} m and is to be exposed to a shearing stress of no more than 5.0×10^8 Pa. What is the maximum tension **T** that can be applied to each beam, assuming that each rivet carries one-fourth of the total load?

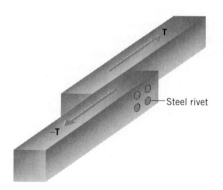

9. The shovel of a backhoe is controlled by hydraulic cylinders that are moved by oil under pressure. Determine the volume strain $\Delta V/V_0$ experienced by the oil, when the pressure increases from 1.0×10^5 Pa to 7.2×10^5 Pa while the shovel is digging a trench.

10. A copper cube, 0.30 m on a side, is subjected to a shearing force of $F = 6.0 \times 10^6$ N, as the drawing shows. Find the angle θ, which is one measure of how the shape of the block has been altered by the resulting shear deformation.

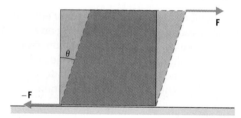

***11.** A helicopter is lifting a 2100-kg jeep. The steel suspension cable is 48 m long and has a radius of 5.0×10^{-3} m. (a) Find the amount that the cable is stretched when the jeep is suspended motionless in the air. (b) What is the amount of cable stretch when the jeep is hoisted upward with an acceleration of 1.5 m/s²?

***12.** A gymnast does a one-arm handstand. The humerus, which is the upper arm bone between the elbow and the shoulder joint, may be approximated as a 0.30-m-long cylinder with an outer radius of 1.00×10^{-2} m and a hollow inner core with a radius of 4.0×10^{-3} m. Excluding the arm, the mass of the gymnast is 63 kg. (a) What is the compressional strain of the humerus? (b) By how much is the humerus compressed?

***13.** An 8.0-kg stone at the end of a steel wire is being whirled at a constant speed of 12 m/s in a circle. The stone is moving on the surface of a frictionless horizontal table. The wire is 4.0 m long and has a radius of 1.0×10^{-3} m. Find the strain in the wire.

***14.** A 1.0×10^{-3}-kg spider is hanging vertically by a thread that has a Young's modulus of 4.5×10^9 N/m² and a radius of 13×10^{-6} m. Suppose that a 95-kg person is hanging vertically on an aluminum wire. What is the radius of the wire that would exhibit the same strain as the spider's thread, when the thread is stressed by the full weight of the spider?

***15.** A square plate is 1.0×10^{-2} m thick, measures 3.0×10^{-2} m on a side, and has a mass of 7.2×10^{-2} kg. The shear modulus of the material is 2.0×10^{10} N/m². One of the square faces rests on a flat horizontal surface, and the coefficient of static friction between the plate and the surface is 0.90. A force is applied as in Figure 10.6. Determine (a) the maximum possible amount of shear stress, (b) the maximum possible amount of shear strain, and (c) the maximum possible amount of shear deformation ΔX (see Figure 10.6) that can be created by the applied force just before the plate begins to move.

***16.** Suppose that F newtons of force are required to stretch a wire (circular cross section) by an amount ΔL. This wire is melted down, and a new wire (circular cross section) is formed that has half the length of the first. How much force is needed to stretch this second wire by the same amount ΔL?

****17.** A thin rod is made of two sections joined together end to end. The sections are identical, except that one is steel and the other is tungsten. One end of this composite rod is attached to an immovable wall, while the other is pulled until the total length changes by one millimeter. By how much does the length of each section increase?

****18.** A solid brass sphere is subjected to a pressure of 1.0×10^5 N/m² due to the earth's atmosphere. On Venus the pressure due to the atmosphere is 9.0×10^6 N/m². By what fraction $\Delta r/r_0$ does the radius of the sphere decrease when it is exposed to the Venusian atmosphere? Assume that the change in radius is very small relative to the initial radius.

Section 10.3 The Ideal Spring and Simple Harmonic Motion

19. A spring has a spring constant of 248 N/m. Find the magnitude of the force needed (a) to stretch the spring by 3.00×10^{-2} m from its unstrained length and (b) to compress the spring by the same amount.

20. A 0.61-kg object is supported by a spring whose spring constant is 1200 N/m. The system is set into oscillation. What is its angular frequency ω?

21. A spring stretches by 0.030 m when an 8.0-kg object is hung from it. By how much will the spring stretch when it is used to suspend a 4.0-kg object?

22. In a room that is 2.44 m high, a spring (unstrained length = 0.305 m) hangs from the ceiling. From this spring a board hangs, so that its 1.98-m length is perpendicular to the floor, the lower end just reaching to, but not touching, the floor. The board weighs 102 N. What is the spring constant of the spring?

23. A hand exerciser utilizes a coiled spring. A force of 89.0 N is required to compress the spring 0.0191 m. Determine the force needed to compress the spring by 0.0508 m.

24. A car is hauling a 92-kg trailer, to which it is connected by a spring. The spring constant is 2300 N/m. The car accelerates with an acceleration of 0.30 m/s². By how much does the spring stretch?

*25. To measure the static friction coefficient between a 2.3-kg block and a vertical wall, the setup shown in the drawing is used. A spring (spring constant = 480 N/m) is attached to the block. Someone pushes on the end of the spring in a direction perpendicular to the wall until the block does not slip downward. If the spring in such a setup is compressed by 0.051 m, what is the coefficient of static friction?

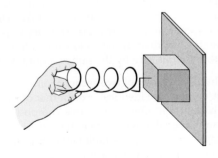

*26. An 11.5-kg uniform board is wedged into a corner and held by a spring at a 45.0° angle, as the drawing shows. The spring has a spring constant of 152 N/m and is parallel to the floor. Find the amount by which the spring is stretched from its unstrained length.

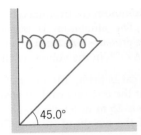

**27. A 15.0-kg block rests on a horizontal table and is attached to one end of a massless, horizontal spring. By pulling horizontally on the other end of the spring, someone causes the block to accelerate uniformly and reach a speed of 5.00 m/s in 0.500 s. In the process, the spring is stretched by 0.200 m. The block is then pulled at a *constant speed* of 5.00 m/s, during which time the spring is stretched by only 0.0500 m. Find (a) the spring constant of the spring and (b) the coefficient of kinetic friction between the block and the table.

28. A 30.0-kg block is resting on a flat horizontal table. On top of this block is resting a 15.0-kg block, to which a horizontal spring is attached, as the drawing illustrates. The spring constant of the spring is 325 N/m. The coefficient of kinetic friction between the lower block and the table is 0.600, while the coefficient of static friction between the two blocks is 0.900. A horizontal force **F is applied to the lower block as shown. This force is increasing in such a way as to keep the blocks moving at a *constant speed*. At the point where the upper block begins to slip on the lower block determine (a) the amount by which the spring is compressed and (b) the magnitude of the force **F**.

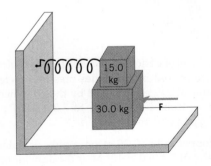

Section 10.4 Simple Harmonic Motion and the Reference Circle

29. A small ball is attached to the outer edge of a $33\frac{1}{3}$-rpm (3.49 rad/s) stereo record that is revolving on a turntable. The record has a radius of 0.152 m. In a fashion similar to that in Figure 10.15, the shadow of the ball is projected onto a screen. For the simple harmonic motion of the shadow, what is (a) the amplitude and (b) the maximum acceleration?

30. The shock absorbers in the suspension system of a car are in such bad shape that they have no effect on the behavior of the springs attached to the axles. Each of the identical springs attached to the front axle supports 320 kg. A person pushes down on the middle of the front end of the car and notices that it vibrates through five cycles in 3.0 s. Find the spring constant of either spring.

31. The diaphragm of a loudspeaker (see Figure 10.21) is moving back and forth in simple harmonic motion with a frequency of 128 Hz and an amplitude of 2.54×10^{-3} m. Determine the maximum speed of the diaphragm.

32. A satellite circles the earth once every 5300 s in an orbit that passes over the north and south poles. The radius of the orbit is 6.5×10^6 m. The sun is shining and the satellite casts a shadow on the earth, much like the ball on the turntable in Figure 10.15 casts its shadow on the screen. What is the speed of the shadow as the satellite passes over the equator, where the sun's rays strike the earth perpendicularly?

33. A computer to be used in a satellite must be able to withstand accelerations of up to 25g. In a test to see if it meets this specification, the computer is bolted to a frame that is vibrated back and forth in simple harmonic motion at a frequency of 9.5 Hz. What is the minimum amplitude of vibration that must be used in this test?

*34. In Figure 10.16, the radius of the reference circle is 0.500 m. Suppose the frequency of the simple harmonic motion of the shadow is 2.00 Hz. At time $t = 0.0500$ s calculate (a) the position x, (b) the magnitude of the velocity, and (c) the magnitude of the acceleration of the shadow.

*35. Suppose that an object on a vertical spring oscillates up and down at a frequency of 5.00 Hz. By how much would this object, hanging at rest, stretch the spring?

*36. When an object of mass m_1 is hung on a vertical spring and set into vertical simple harmonic motion, its frequency is 10.0 Hz. When another object of mass m_2 is hung on the spring along with m_1, the frequency of the motion is 5.00 Hz. Find the ratio m_2/m_1 of the masses.

*37. A spring stretches by 0.018 m when a 2.8-kg object is suspended from its end. How much mass should be attached to this spring so that its frequency of vibration is $f = 3.0$ Hz?

**38. A spring (spring constant = 80.0 N/m) is mounted on the floor and is oriented vertically. A 0.600-kg block is placed on top of this spring and pushed down to start it oscillating in simple harmonic motion. The block is not attached to the spring. (a) Obtain the frequency (in Hz) of the motion. (b) Determine the amplitude at which the block will lose contact with the spring.

Section 10.5 Energy and Simple Harmonic Motion

39. An archer pulls the bow string back for a distance of 0.470 m before releasing the arrow. The bow and string act like a spring whose spring constant is 425 N/m. (a) What is the elastic potential energy of the drawn bow and arrow? (b) The arrow has a mass of 0.0300 kg. How fast is it traveling when it leaves the bow?

40. A 1.00×10^{-2}-kg block is resting on a horizontal frictionless surface and is attached to a horizontal spring whose spring constant is 124 N/m. The block is shoved parallel to the spring axis and is given an initial speed of 8.00 m/s, while the spring is initially unstrained. What is the amplitude of the resulting simple harmonic motion?

41. The spring constant for a spring in a dart gun is 1400 N/m. When the gun is cocked, the spring is compressed 0.075 m. What is the speed of a 2.4×10^{-2}-kg dart when it leaves the gun horizontally?

42. In preparation for shooting a ball in a pinball machine, a spring ($k = 675$ N/m) is compressed by 0.0650 m relative to its unstrained length. The ball ($m = 0.0585$ kg) is at rest against the spring at point A. When the spring is released, the ball slides (without rolling) to point B, which is 0.300 m higher than point A. How fast is the ball moving at B?

43. A 2.00-kg object is hanging from the end of a vertical spring. The spring constant is 50.0 N/m. The object is pulled 0.200 m downward and released from rest. Complete the table below by calculating the kinetic energy, the gravitational potential energy, the elastic potential energy, and the total mechanical energy E for each of the vertical positions indicated. The vertical positions h indicate distances above the point of release, where $h = 0$.

h (meters)	KE	PE (gravity)	PE (elastic)	E
0				
0.100				
0.200				
0.300				
0.400				

44. A 0.22-kg ball is oscillating on a horizontal spring. The period and amplitude of the oscillation are 0.50 s and 0.15 m, respectively. How much work was done to start the simple harmonic motion?

45. A heavy-duty stapling gun uses a 0.150-kg metal rod that rams against the staple to eject it. The rod is pushed by a stiff spring called a "ram spring" ($k = 34\,000$ N/m). The mass of this spring may be ignored. Squeezing the handle of the gun first compresses the ram spring by 3.5×10^{-2} m from its unstrained length and then releases it. Assuming that the ram spring is oriented vertically and is still compressed by 1.0×10^{-2} m when the downward-moving ram hits the staple, find the speed of the ram at the instant of contact.

*46. A spring is compressed by 0.0800 m and is used to launch an object horizontally with a speed of 2.40 m/s. At what angular frequency (in rad/s) would the object oscillate on the spring?

*47. A 14.6-kg block and a 29.2-kg block are resting on a horizontal frictionless surface. Between the two is squeezed a spring (spring constant = 1170 N/m). The spring is compressed by 0.152 m from its unstrained length. With what speed does each block move away when the mechanism keeping the spring squeezed is released?

*48. A 1.1-kg object is suspended from a vertical spring whose spring constant is 120 N/m. (a) Find the amount by which the spring is stretched from its unstrained length. (b) The object is pulled straight down by an additional distance of 0.20 m and released from rest. Find the speed with which the object passes through its original position on the way up.

**49. A 70.0-kg circus performer is fired from a cannon that is elevated at an angle of 45.0° above the horizontal. The cannon uses strong elastic bands to propel the performer, much in the same way that a slingshot fires a stone. Setting up for this stunt involves stretching the bands by 3.00 m from their unstrained length. At the point where the performer flies free of the bands, his height above the floor is the same as that of the net into which he is shot. He takes 4.00 s to travel the horizontal distance of 50.0 m between this point and the net. Ignore friction and air resistance and determine the effective spring constant of the firing mechanism.

**50. A spring is mounted vertically on the floor. The mass of the spring is negligible. A certain object is placed on the spring to compress it. When the object is pushed down further by just a bit and then released, one up/down oscillation cycle occurs in 0.250 s. However, when the object is pushed down by 5.00×10^{-2} m to point P and then released, the object flies entirely off the spring. To what height above point P does the object rise in the absence of air resistance?

**51. A 1.00×10^{-2}-kg bullet is fired horizontally into a 2.50-kg wooden block attached to one end of a massless, horizontal spring ($k = 845$ N/m). The other end of the spring is fixed in place, and the spring is unstrained initially. The block rests on a horizontal, frictionless surface. The bullet strikes the block perpendicularly and quickly comes to a halt within it. As a result of this completely inelastic collision, the spring is compressed along its axis and causes the block/bullet to oscillate with an amplitude of 0.200 m. What is the speed of the bullet?

Section 10.6 The Pendulum

52. If the period of a simple pendulum is to be 2.0 s, what should be its length?

53. A grandfather clock can be approximated as a simple pendulum of length 1.00 m and keeps accurate time at a location where $g = 9.83$ m/s². In a location where $g = 9.78$ m/s²,

what must be the new length of the pendulum, such that the clock continues to keep accurate time?

54. Astronauts on a distant planet set up a simple pendulum of length 1.2 m. The pendulum executes simple harmonic motion and makes 100 complete vibrations in 280 s. What is the acceleration due to gravity?

55. A wrecking ball is hanging at the end of a long cable on a crane. A bright student wants to estimate the length of the cable and, therefore, improvises by using a simple pendulum made from a 0.500-m length of string and a stone. The student observes that, in swinging back and forth over a small amplitude, the wrecking ball makes one complete oscillation cycle in the time it takes the stone to complete five cycles. What is the length of the cable?

*56. The period of a simple pendulum is 0.200% longer at location A than it is at location B. Find the ratio g_A/g_B of the acceleration due to gravity at these two locations.

*57. Pendulum A is a physical pendulum made from a thin, rigid, and uniform rod whose length is 1.00 m. One end of this rod is attached to the ceiling by a frictionless hinge, so the rod is free to swing back and forth. Pendulum B is a simple pendulum whose length is also 1.00 m. Obtain the ratio T_A/T_B of their periods for small-angle oscillations.

**58. A point on the surface of a solid sphere (radius = R) is attached directly to a pivot on the ceiling. The sphere swings back and forth as a physical pendulum with a small amplitude. What is the length of a simple pendulum that has the same period as this physical pendulum? Give your answer in terms of R.

ADDITIONAL PROBLEMS

59. The drawing shows how a piston in an automobile engine is attached to the crankshaft, which is rotating with an angular speed of $\omega = 126$ rad/s. If the shadow of point P could be projected onto a screen, the shadow would move in simple harmonic motion. Find (a) the amplitude, (b) the period, and (c) the maximum speed of the motion.

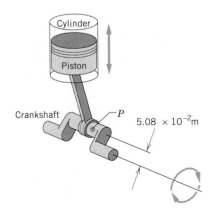

60. A die is designed to punch holes of radii 1.00×10^{-2} m in a metal sheet that is 3.0×10^{-3} m thick, as the drawing illustrates. To punch through the sheet, the die must exert a shearing stress of 3.5×10^8 Pa. What force **F** must be applied to the die? (*Hint: Consider carefully which area is used in your calculation.*)

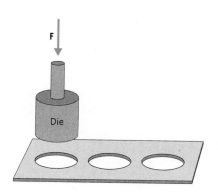

61. A rifle fires a 1.00×10^{-2}-kg pellet straight upward, because the pellet rests on a compressed spring that is released when the trigger is pulled. The spring has a negligible mass and is compressed by 7.50×10^{-2} m from its unstrained length. The pellet rises to a height of 5.00 m above its position on the compressed spring. Ignoring air resistance, determine the spring constant of the spring.

62. A 2.5-m length of aluminum cable is to be used in an attempt to pull a car out of a ditch. The maximum tension that the cable will have to sustain is 15 000 N, and, to be on the safe side, only 1.0×10^{-3} m of cable stretch will be tolerated. What is the radius of the thinnest cable that should be used?

63. A 9.0-kg object hangs on a copper rod (radius $= 3.0 \times 10^{-3}$ m) from a support that is located 2.0 m directly above. The rod acts as a "spring," and the object oscillates vertically with a small amplitude. Find the frequency f of the simple harmonic motion. (*Hint: Section 10.3 shows that the spring constant for the rod is $k = YA/L_0$.*)

64. Atoms in a solid are not stationary, but vibrate about their equilibrium positions. Typically, the frequency of vibration is about 2.0×10^{12} Hz and the amplitude is about 1.1×10^{-11} m. For a typical atom, what is its (a) maximum speed and (b) maximum acceleration?

65. When used in an exercise apparatus, a spring is stretched 0.24 m when a bodybuilder exerts a force of 410 N. When used vertically to support a 12-kg object, by how much does this spring compress?

66. The pressure increases by 9800 N/m² for every meter of depth beneath the surface of the ocean. At what depth does the volume of a Pyrex glass cube, 1.0×10^{-2} m on an edge at the ocean's surface, decrease by 1.0×10^{-10} m³?

*67. A block is attached to a horizontal spring and oscillates

back and forth on a frictionless horizontal surface at a frequency of 3.00 Hz. The amplitude of the motion is 5.08×10^{-2} m. At the point where the block has its maximum speed, it suddenly splits into two identical parts, only one part remaining attached to the spring. (a) What is the amplitude and the frequency of the simple harmonic motion that exists after the block splits? (b) Repeat part (a), assuming that the block splits when it is at one of its extreme positions.

*68. A piece of mohair from an Angora goat has a radius of 31×10^{-6} m. What is the least number of identical pieces of mohair that should be used to suspend a 75-kg person, so the strain $\Delta L/L_0$ experienced by each piece is less than 0.010? Assume that the tension is the same in all the pieces.

*69. In 0.750 s, a 7.00-kg block is pulled through a distance of 4.00 m on a frictionless horizontal surface, starting from rest. The block has a constant acceleration and is pulled by means of a horizontal spring that is attached to the block. The spring constant of the spring is 415 N/m. By how much does the spring stretch?

*70. Two solid cubes of the same size, one tungsten and one lead, are subjected to the same shearing force, as in the drawing that accompanies problem 10. Determine the ratio $\theta_{lead}/\theta_{tungsten}$ for the angles that characterize the resulting shear deformation, assuming that the angles are small.

**71. A 0.200-m uniform bar has a mass of 0.750 kg and is released from rest in the vertical position, as the drawing indicates. The spring is initially unstrained and has a spring constant of $k = 25.0$ N/m. Find the speed with which end A strikes the horizontal surface.

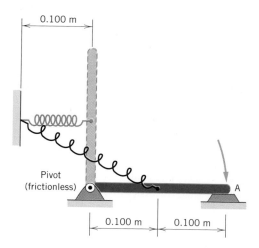

**72. A tray is moved horizontally back and forth in simple harmonic motion at a frequency of 2.00 Hz. On this tray is an empty cup. Obtain the coefficient of static friction between the tray and the cup, given that the cup is observed to begin slipping when the amplitude of the motion is 5.00×10^{-2} m.

CHAPTER 11

FLUIDS

Fluids are materials that can flow, and they include both gases and liquids. Air, for instance, is primarily a mixture of nitrogen and oxygen gases, and we call it the wind when it flows from place to place. Skydivers can take advantage of the air flowing over their parafoils (a kind of parachute) to adjust their descent and "fly" in formation on the way down. The most familiar fluid in the liquid state is water, and flowing water has many uses, ranging from the fun of white-water rafting to the generation of hydroelectric power.

Our discussion of fluids has two parts, statics and dynamics. Fluid statics concentrates on the properties of fluids at rest, while fluid dynamics focuses on fluids in motion. Fluids at rest display remarkable characteristics. For example, static fluids in the hydraulic systems of cars, trucks, and heavy construction equipment transmit the large forces that such equipment uses so cleverly for pushing, pulling, lifting, and digging. And the property of static fluids that enables water in the ocean to keep a supertanker afloat also enables the air to keep a hot-air balloon floating above the ground. Moving fluids have equally remarkable properties. They provide the lift force that enables an airplane to fly, the force that allows a baseball pitcher to throw a curve ball, and even the force that can tear the roof from a house in a storm. The first six sections in this chapter deal with fluid statics, while the remaining sections treat fluid dynamics.

11.1 MASS DENSITY

The **mass density** of a liquid or gas is one factor that determines its behavior as a fluid. As indicated below, the mass density is the mass per unit volume and is denoted by the Greek letter rho (ρ).

Table 11.1 Mass Densities[a] of Common Substances

Substance	Mass Density ρ (kg/m³)
Solids	
Aluminum	2 700
Brass	8 470
Concrete	2 200
Copper	8 890
Diamond	3 520
Gold	19 300
Ice	917
Iron (steel)	7 860
Lead	11 300
Quartz	2 660
Silver	10 500
Wood (yellow pine)	550
Liquids	
Blood (whole, 37 °C)	1 060
Ethyl alcohol	806
Mercury	13 600
Oil (hydraulic)	800
Water (4 °C)	1.000×10^3
Gases	
Air	1.29
Carbon dioxide	1.98
Helium	0.179
Hydrogen	0.0899
Nitrogen	1.25
Oxygen	1.43

[a] Unless otherwise noted, densities are given at 0 °C and 1 atm pressure.

Definition of Mass Density

The mass density ρ is the mass m of a substance divided by its volume V:

$$\rho = \frac{m}{V} \tag{11.1}$$

SI Unit of Mass Density: kg/m³

Equal volumes of different substances generally have different masses, so the density depends on the nature of the material. Table 11.1 lists the densities of some common liquids and gases and also includes some solids for comparison. Gases have the smallest densities, because gas molecules are relatively far apart and a given volume of gas contains a large fraction of empty space. In contrast, the molecules are much more tightly packed in liquids and solids, and the tighter packing leads to larger densities.

The density of a substance also depends on temperature and pressure. However, for the range of temperatures and pressures encountered in this text, the densities of liquids and solids do not differ much from the values in Table 11.1. On the other hand, the densities of gases are particularly sensitive to changes in temperature and pressure.

It is the mass of a substance, not its weight, that enters into the definition of density. In situations where weight is needed, it can be calculated from the mass density, the volume, and the acceleration of gravity, as Example 1 illustrates.

Example 1 Blood as a Fraction of Body Weight

The body of a man whose weight is about 690 N typically contains about 5.2×10^{-3} m³ (5.5 qt) of blood. (a) Calculate the weight of the blood and (b) express it as a percentage of the body weight.

REASONING AND SOLUTION
(a) The weight W of the blood can be obtained with the aid of the acceleration due to gravity g ($W = mg$), provided the mass of the blood is known. The mass can be found by using the density of blood from Table 11.1:

$$m = \rho V = (1060 \text{ kg/m}^3)(5.2 \times 10^{-3} \text{ m}^3) = 5.5 \text{ kg} \tag{11.1}$$

$$W = mg = (5.5 \text{ kg})(9.80 \text{ m/s}^2) = \boxed{54 \text{ N}}$$

(b) The percentage of body weight contributed by the blood is

$$\text{Percentage} = \frac{54 \text{ N}}{690 \text{ N}} \times 100 = \boxed{7.8\%}$$

A convenient way to compare densities is to use the concept of *specific gravity*. The specific gravity of a substance is its density divided by the density of a standard reference material, usually chosen to be water at 4 °C:

$$\text{Specific gravity} = \frac{\text{Density of substance}}{\text{Density of water at 4 °C}} = \frac{\text{Density of substance}}{1.000 \times 10^3 \text{ kg/m}^3} \quad (11.2)$$

Being the ratio of two densities, specific gravity has no units. For example, Table 11.1 reveals that diamond has a specific gravity of 3.52, since the density of diamond is 3.52 times greater than the density of water at 4 °C.

The next two sections deal with the important concept of pressure. We will see that the density of a fluid is one factor determining the pressure that a fluid exerts.

11.2 PRESSURE

Most people who have fixed a flat tire know something about pressure. The final step in fixing a flat is to reinflate the tire to the proper pressure. The underinflated tire is soft, because it contains an insufficient number of molecules of air to push outward against the rubber and give the tire that solid feel. When air is added from a pump, the number of molecules inside the tire and the collective force they exert are increased. When the tire is inflated to the proper pressure, the air pushes outward with enough force to give the tire the shape it needs to roll properly.

The air molecules within a tire are free to wander throughout its entire volume, and in the course of their wandering they collide with one another and with the inner walls of the tire. The collisions lead to the zigzag paths shown in Figure 11.1a. And the collisions with the walls, in particular, allow the air to exert a force against every part of the wall area, as part b of the figure shows. The idea of *pressure* takes into account the force, as well as the area over which the force acts.

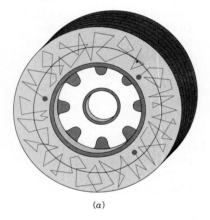

(a)

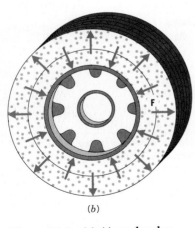

(b)

Figure 11.1 (a) Air molecules within a tire follow zigzag paths because of collisions. (b) In colliding with the inner walls of the tire, the air molecules exert a force on every part of the wall surface.

Definition of Pressure
The pressure P is the magnitude F of the force acting perpendicular to a surface divided by the area A over which the force acts: $$P = \frac{F}{A} \quad (11.3)$$ **SI Unit of Pressure:** N/m² = pascal (Pa)

Equation 11.3 indicates that the unit for pressure is the unit of force divided by the unit of area. The SI unit for pressure, then, is newton/meter² (N/m²), a combination that is referred to as the *pascal (Pa)*. The pascal is also the unit for measuring stress (see Section 10.2). A pressure of 1 Pa is a small amount. Many common situations involve pressures of approximately 10^5 Pa, an amount re-

ferred to as one *bar* of pressure. Alternatively, force can be measured in pounds and area in square inches, so another unit for pressure is pounds per square inch (psi = lb/inch²).

Because of its pressure, the air in a tire applies a force to any surface with which it is in contact. Suppose, for instance, that a small cube is inserted inside the tire. As Figure 11.2 shows, the air pressure causes a force to act on each face of the cube, the force in each case being perpendicular to the face. In a similar fashion, a liquid also exerts pressure. Water pressure, for example, causes a force to be applied perpendicularly to the side of a swimming pool or the body of a swimmer. A swimmer feels the water pushing inward everywhere on her body, as Figure 11.3 illustrates. In general, a static fluid cannot produce a force parallel to a surface, for if it did, the surface would apply a reaction force to the fluid, consistent with Newton's action–reaction law. In response, the fluid would flow and would not then be static.

While fluid pressure can generate a force, the pressure itself is not a vector quantity, as is the force. In the definition of pressure, $P = F/A$, the symbol F refers only to the magnitude of the force, so that pressure has no directional characteristic. The force generated by the pressure of a static fluid is always perpendicular to the surface on which the fluid acts, as Example 2 illustrates.

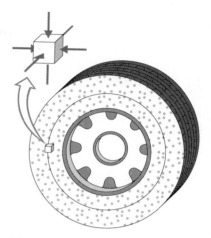

Figure 11.2   A small cube inserted inside a tire experiences forces acting perpendicular to each of its six faces. The air pressure in the tire generates these forces.

Example 2   The Force on a Swimmer

Suppose the pressure acting on the back of a swimmer's hand is 1.2×10^5 Pa, a realistic value near the bottom of the diving end of a pool. The surface area of the back of the hand is 8.4×10^{-3} m². (a) Determine the magnitude of the force that acts on it. (b) Discuss the direction of the force.

REASONING AND SOLUTION
(a) A pressure of 1.2×10^5 Pa is 1.2×10^5 N/m², and from the definition of pressure in Equation 11.3, we can see that the force is pressure times area:

$$F = PA = (1.2 \times 10^5 \text{ N/m}^2)(8.4 \times 10^{-3} \text{ m}^2) = \boxed{1.0 \times 10^3 \text{ N}}$$

This is a rather large force, about 230 lb.

(b) In Figure 11.3, the hand (palm downward) is oriented parallel to the bottom of the pool. Since the water pushes perpendicularly against the back of the hand, the force **F** is directed downward in the drawing. This downward-acting force is balanced by an upward-acting force on the palm, thus keeping the hand in equilibrium. If the hand were rotated by 90°, the direction of these forces would also be rotated by 90°, always being perpendicular to the hand.

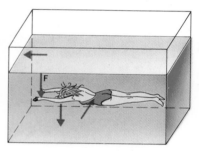

Figure 11.3   Water applies a force perpendicular to each surface within the water, including the walls and bottom of the swimming pool, and all parts of the swimmer's body.

A person need not be under water to experience the effects of pressure. Walking about on land, we are at the bottom of the earth's atmosphere, which, being a fluid, pushes inward on our bodies just like the water in a swimming pool. As Figure 11.4 indicates, there is enough air above the surface of the earth to create the following pressure at sea level:

$$\begin{bmatrix} \textbf{Atmospheric} \\ \textbf{pressure at} \\ \textbf{sea level} \end{bmatrix} \qquad 1.013 \times 10^5 \text{ Pa} = 1 \text{ atmosphere}$$

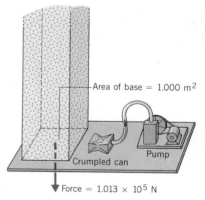

Figure 11.4   Atmospheric pressure at sea level is 1.013×10^5 Pa, which is sufficient to crumple a can if the inside air is pumped out.

Figure 11.5 Workmen at a limestone quarry topple a huge slice of stone with the aid of air bags fitted into slots and then inflated.

Figure 11.6 Lynx have large paws that act as natural snowshoes.

This amount of pressure corresponds to 14.70 lb/in.² and is referred to as one *atmosphere (atm)*. Just how significant one atmosphere of pressure is can be appreciated by looking in Figure 11.4 at the results of pumping out all the air from inside a gasoline can. Without the air inside to push outward, the external air pushes hard enough on the can to crumple it.

Figure 11.5 shows one way that the force generated by fluid pressure is used. At limestone quarries workmen cut slots into the stone and then topple huge slices of it with the aid of air bags fitted into the slots. Pumps inflate the air bags, and the pressure of the air generates the force to push over the cut stone.

In contrast to the situation in Figure 11.5, it is sometimes useful to reduce pressure. Lynx, for example, are especially well suited for hunting on snow because of their oversize paws (see Figure 11.6). The large paws function as snowshoes that distribute the weight over a large area. Thus, they reduce the weight per unit area or pressure that the cat applies to the surface and prevent the cat from sinking into the snow.

THE PHYSICS OF . . .

cutting limestone.

THE PHYSICS OF . . .

lynx paws.

11.3 THE RELATION BETWEEN PRESSURE AND DEPTH IN A STATIC FLUID

The pressure that an underwater swimmer experiences depends on how far beneath the surface he is. The deeper he goes, the stronger the water pushes on his body. To help determine the exact relation between pressure and depth, Figure 11.7 shows a container of fluid and focuses attention on one particular column of fluid. The fluid, being at rest, is in equilibrium, so the net force acting on

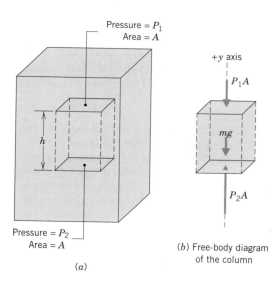

Pressure = P_1
Area = A

h

Pressure = P_2
Area = A

(a)

+y axis

P_1A

mg

P_2A

(b) Free-body diagram
of the column

Figure 11.7 (a) A container of fluid in which one particular column of the fluid is outlined. The fluid is at rest. (b) The free-body diagram, showing the vertical forces acting on the column.

the column must be zero. The free-body diagram in the figure shows all the vertical forces acting on the column.

On the top face (area $= A$), the fluid pressure P_1 generates a downward force whose magnitude is P_1A. Similarly, on the bottom face, the pressure P_2 generates an upward force of magnitude P_2A. The pressure P_2 is greater than the pressure P_1, because the bottom face supports the weight of more fluid than the upper one does. In fact, the excess weight supported by the bottom face is exactly the weight of the fluid within the column. As the free-body diagram indicates, this excess weight is mg, where m is the mass of the fluid and g is the acceleration due to gravity. Setting the sum of the vertical forces equal to zero, we find that

$$\Sigma F_y = P_2A - P_1A - mg = 0 \quad \text{or} \quad P_2A = P_1A + mg$$

The mass m is related to the density ρ and the volume V of the column by $m = \rho V$. Since the volume is the cross-sectional area A times the vertical dimension h, we have $m = \rho A h$. With this substitution, the condition for equilibrium becomes $P_2A = P_1A + \rho A h g$. The area A can be eliminated algebraically from this expression, with the result that

$$P_2 = P_1 + \rho g h \qquad (11.4)$$

This equation indicates that if the pressure P_1 is known at a higher level, the larger pressure P_2 at a deeper level can be calculated by adding the increment $\rho g h$. In determining the pressure increment $\rho g h$, we assumed that the density ρ is the same at any vertical distance h; that is, the fluid is incompressible. This assumption is reasonable for liquids, since the bottom layers of a liquid can support the upper layers with little compression. In a gas, however, the lower layers are compressed markedly by the weight of the upper layers, with the result that the density varies with vertical distance. For example, the density of our atmosphere is larger near the earth's surface than it is at higher altitudes. When applied to gases, the relation $P_2 = P_1 + \rho g h$ can be used only when h is small enough that any variation in ρ can be neglected. Example 3 illustrates how to apply this relation between pressure and depth.

These divers are talking inside a pressure "bubble." The air in the "bubble" has a pressure that is greater than atmospheric pressure, because the bubble is under water.

Example 3 The Swimming Hole

Figure 11.8 shows the cross section of a swimming hole. Points A and B are both located at a distance of $h = 5.50$ m below the surface of the water. Find the pressure at each of these two points.

REASONING   The pressure at point B is the same as that at point A, since both are located at the *same vertical distance* beneath the surface. The fact that point B is displaced horizontally to the right is of no concern, because only the vertical distance h affects the pressure increment $\rho g h$ in Equation 11.4. To understand this important feature more clearly, consider the path $AA'B'B$ in Figure 11.8. The pressure decreases on the way up along the vertical segment AA' and increases by the same amount on the way back down along segment $B'B$. Since no change in pressure occurs along the horizontal segment $A'B'$, the pressure is the same at A and B.

SOLUTION   The pressure acting on the surface of the water is the atmospheric pressure of 1.01×10^5 Pa. Using this value as P_1 in Equation 11.4, we can determine a value for the pressure P_2 at either point A or B, both of which are located 5.50 m under the water. Table 11.1 gives the density of water as 1.000×10^3 kg/m³.

$$P_2 = P_1 + \rho g h$$
$$P_2 = 1.01 \times 10^5 \text{ Pa} + (1.000 \times 10^3 \text{ kg/m}^3)(9.80 \text{ m/s}^2)(5.50 \text{ m}) = \boxed{1.55 \times 10^5 \text{ Pa}}$$

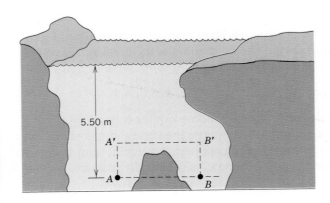

Figure 11.8 The pressures at points A and B are the same, since both points are located at the same vertical distance of 5.50 m beneath the surface of the water.

Figure 11.9 shows an irregularly shaped container of liquid. Reasoning similar to that used in Example 3 leads to the conclusion that the pressure is the same at points A, B, C, and D since each is at the same vertical distance h beneath the surface. In effect, the arteries in our bodies constitute an irregularly shaped "container" for the blood. The next example examines the blood pressure at different places in this "container."

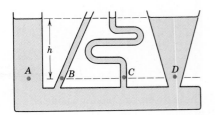

Figure 11.9 Since points A, B, C, and D are at the same distance h beneath the liquid surface, the pressure at each of them is the same.

Example 4 Blood Pressure

Blood in the arteries is flowing, but as a first approximation, the effects of this flow can be ignored and the blood treated as a static fluid. Estimate the amount by which the blood pressure P_2 in the anterior tibial artery at the foot exceeds the blood pressure P_1 in the aorta at the heart when the body is (a) reclining horizontally as in Figure 11.10a and (b) standing as in Figure 11.10b.

REASONING AND SOLUTION

(a) When the body is horizontal, there is little or no vertical separation between the feet and the heart. Since $h = 0$,

$$P_2 - P_1 = \rho g h = \boxed{0} \qquad (11.4)$$

(b) When an adult is erect, the vertical separation between the feet and the heart is about 1.35 m, as Figure 11.10b indicates. Table 11.1 gives the density of blood as 1060 kg/m³, so that

$$P_2 - P_1 = \rho g h = (1060 \text{ kg/m}^3)(9.80 \text{ m/s}^2)(1.35 \text{ m}) = \boxed{1.40 \times 10^4 \text{ Pa}}$$

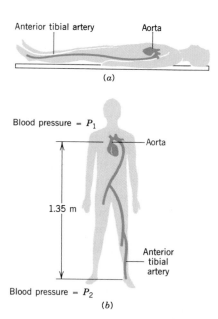

Figure 11.10 The blood pressure in the feet can exceed the blood pressure in the heart, depending on whether the body is (a) reclining horizontally or (b) standing erect.

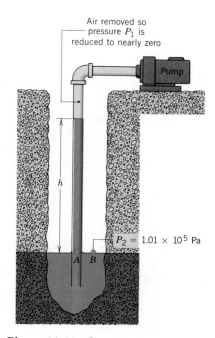

Figure 11.11 One way to pump water out of a well.

Sometimes, fluid pressure places limits on how a job can be done. Figure 11.11, for instance, shows one method of pumping water out of a well. Example 5 shows that this method can only be used when the well is not too deep.

Example 5   Pumping Water

Determine the maximum height to which water can be pumped from a well with the arrangement in Figure 11.11.

THE PHYSICS OF . . .

pumping water from a well.

REASONING The job of the pump is to draw the air out of the pipe. With the pipe empty of air, the atmospheric pressure in the well shaft pushes the water upward. A similar process takes place when you drink from a straw. The best the pump can do is to remove most of the air, in which case the pressure P_1 at the top of the water in the pipe is nearly zero. The pressure P_2 at the bottom of the pipe at point A is the same as that at point B, namely, the atmospheric pressure of 1.01×10^5 Pa, because the two points are at the same elevation. Equation 11.4 can be applied to obtain the maximum pumping height h.

SOLUTION

$$P_2 = P_1 + \rho g h = \rho g h$$

$$h = \frac{P_2}{\rho g} = \frac{1.01 \times 10^5 \text{ Pa}}{(1.000 \times 10^3 \text{ kg/m}^3)(9.80 \text{ m/s}^2)} = \boxed{10.3 \text{ m}}$$

If water is located more than 10.3 m underground, schemes other than that shown in the drawing must be used to obtain it.

11.4 PRESSURE GAUGES

One of the simplest pressure gauges is the mercury barometer used for measuring atmospheric pressure. As Figure 11.12 shows, this device is a tube sealed at one end, filled completely with mercury, and then inverted, so that the open end is under the surface of a pool of mercury. Except for a negligible amount of mercury vapor, the space above the mercury in the tube is empty, and the pressure P_1 is zero there. The pressure P_2 at point A at the bottom of the mercury column is the same as that at point B, namely, atmospheric pressure, for these two points are at the same level. With $P_1 = 0$ and $P_2 = P_{atm}$, it follows from Equation 11.4 that $P_{atm} = 0 + \rho g h$. Thus, the atmospheric pressure can be determined from the height h of the mercury in the tube, the density ρ of mercury, and the acceleration due to gravity. Usually weather forecasters report the pressure in terms of the height h, expressing it in millimeters or inches of mercury. For instance, using $P_{atm} = 1.013 \times 10^5$ Pa and $\rho = 13.6 \times 10^3$ kg/m³ for the density of mercury, we find that $h = P_{atm}/(\rho g) = 760$ mm (29.9 inches).* Slight variations from this value occur, depending on weather conditions and altitude.

Could water or some other liquid be used in a barometer instead of mercury? In principle, yes. However, the density of water is 13.6 times smaller than that of mercury, so the vertical column of water needed to achieve the pressure increment $\rho g h$ would have to be 13.6 times longer. Such a length (about 34 feet for atmospheric pressure) is inconvenient, and water barometers are rarely used.

Figure 11.13 shows another kind of pressure gauge, the open-tube manometer. The phrase "open-tube" refers to the fact that one side of the U-tube is open to atmospheric pressure. The tube contains a liquid, often mercury, and its other side is connected to the container whose pressure P_2 is to be measured. When the pressure in the container is equal to the atmospheric pressure, the liquid levels in both sides of the U-tube are the same, as Figure 11.13a indicates. When the pressure in the container is greater than atmospheric pressure, as in Figure 11.13b,

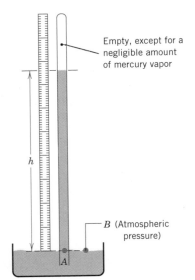

Empty, except for a negligible amount of mercury vapor

h

B (Atmospheric pressure)

A

Figure 11.12 A mercury barometer.

* A pressure of one millimeter of mercury is sometimes referred to as one *torr*, to honor the inventor of the barometer, Evangelista Torricelli (1608–1647). Thus, one atmosphere of pressure is 760 torr.

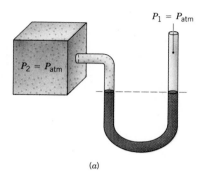
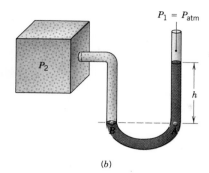

Figure 11.13 An open-tube manometer is used to measure the pressure P_2 in a container. In (a) the pressure P_2 equals P_{atm}, while in (b) the pressure P_2 exceeds P_{atm}.

the liquid in the tube is pushed downward on the left side and rises on the right side. The relation $P_2 = P_1 + \rho g h$ can be used to determine the container pressure. Atmospheric pressure exists at the top of the right column, so that $P_1 = P_{atm}$. The pressure P_2 is the same at points A and B, so we find that $P_2 = P_{atm} + \rho g h$ or

$$P_2 - P_{atm} = \rho g h$$

The height h is proportional to $P_2 - P_{atm}$, which is called the *gauge pressure*. The gauge pressure is the amount by which the container pressure exceeds atmospheric pressure. The actual value for P_2 is called the *absolute pressure*.

The sphygmomanometer is a familiar device for measuring blood pressure. As Figure 11.14 illustrates, a squeeze bulb can be used to inflate the cuff with air, so that the pressure applied to the arm cuts off the flow of blood through the artery below the cuff. When the release valve is opened, the cuff pressure starts to drop. Blood begins to flow again when the pressure created by the heart at the peak of its beating cycle exceeds the cuff pressure. Using a stethoscope to listen for the initial flow, the operator can measure the corresponding cuff gauge pressure with an open-tube manometer. This cuff gauge pressure is called the *systolic* pressure. As the amount of air in the cuff continues to drop, there comes a point when even the pressure created by the heart at the low point of its beating cycle is sufficient to cause blood to flow. Identifying this point with the stethoscope, the operator can again measure the corresponding cuff gauge pressure, which is referred to as the *diastolic* pressure. The systolic and diastolic pressures are reported in millimeters of mercury, and values of 120 and 80, respectively, are typical of a young healthy heart.

11.5 PASCAL'S PRINCIPLE

As we have seen, the pressure in a fluid increases with depth, due to the weight of the fluid above the point of interest. In addition, a confined fluid may be subjected to an additional pressure by the application of an external force. Figure 11.15 illustrates this important aspect of fluid behavior.

Part *a* of the drawing shows two interconnected cylindrical chambers.The chambers have different diameters and, together with the connecting tube, are completely filled with a liquid. The larger chamber is sealed at the top with a cap,

PROBLEM SOLVING INSIGHT

When solving problems involving pressure, be sure to note the distinction between gauge pressure and absolute pressure.

THE PHYSICS OF . . .

measuring blood pressure.

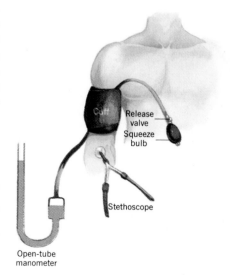

Figure 11.14 A sphygmomanometer is used to measure blood pressure.

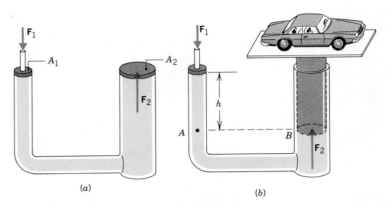

Figure 11.15   (a) An external force $\mathbf{F_1}$ is applied to the piston on the left. As a result, a force $\mathbf{F_2}$ is exerted on the cap on the chamber on the right. (b) The familiar hydraulic car lift.

while the smaller one is fitted with a movable piston. What determines the pressure P_1 at a point immediately beneath the piston? According to the definition of pressure, it is the magnitude F_1 of the external force divided by the area A_1 of the piston: $P_1 = F_1/A_1$. If it is necessary to know the pressure P_2 *at any other place in the liquid*, we just add to the value of P_1 the increment ρgh, which takes into account the depth below the piston: $P_2 = P_1 + \rho gh$. The important feature here is this: The pressure P_1 adds to the pressure ρgh due to the depth of the liquid at any point, whether that point is in the smaller chamber, the connecting tube, or the larger chamber. Therefore, if the applied pressure P_1 is increased or decreased, the pressure at any other point within the confined liquid changes correspondingly. This behavior is in accord with *Pascal's principle.*

Pascal's Principle

Any change in the pressure applied to a completely enclosed fluid is transmitted undiminished to all parts of the fluid and the enclosing walls.

The real usefulness of the arrangement in Figure 11.15a becomes apparent when the force F_2 applied by the liquid to the cap on the right side is calculated. The area of the cap is A_2 and the pressure there is P_2. As long as the tops of the left and right chambers are at the same level, the pressure increment ρgh is zero, so that $P_2 = P_1 + \rho gh$ becomes $P_2 = P_1$. Consequently, $F_2/A_2 = F_1/A_1$, and

$$F_2 = F_1\left(\frac{A_2}{A_1}\right) \tag{11.5}$$

THE PHYSICS OF . . .

a hydraulic car lift.

If area A_2 is larger than area A_1, a large force $\mathbf{F_2}$ can be applied to the cap on the right chamber, starting with a smaller force $\mathbf{F_1}$ on the left. Depending on the ratio of the areas A_2/A_1, the force $\mathbf{F_2}$ can be large indeed, as in the familiar hydraulic car lift shown in part b. In this device the force $\mathbf{F_2}$ is not applied to a cap that seals the larger chamber, but, rather, to a movable plunger that lifts a car. Example 6 deals with a hydraulic car lift.

Example 6   A Car Lift

In a hydraulic car lift, the input piston has a radius of $r_1 = 0.0120$ m and a negligible weight. The output plunger has a radius of $r_2 = 0.150$ m. The combined weight of the car and the output plunger is $F_2 = 20\ 500$ N. The lift uses hydraulic oil that has a density of 8.00×10^2 kg/m³. What input force F_1 is needed to support the car and the output plunger when the bottom surfaces of the piston and plunger are at (a) the same level and (b) the levels shown in Figure 11.15b with $h = 1.10$ m?

REASONING   When the bottom surfaces of the piston and plunger are at the same levels, as in part (a), Equation 11.5 applies. However, this equation does not apply in part (b), where the bottom surface of the output plunger is $h = 1.10$ m below the input piston. Therefore, our solution in part (b) will have to take into account the pressure increment $\rho g h$. In any event, we will see that the input force is less than the combined weight of the plunger and car.

SOLUTION
(a) Using $A = \pi r^2$ for the circular areas of the piston and plunger, we find from Equation 11.5 that

$$F_1 = F_2 \left(\frac{A_1}{A_2}\right) = F_2 \left(\frac{\pi r_1^2}{\pi r_2^2}\right)$$

$$= (20\ 500\ \text{N}) \frac{(0.0120\ \text{m})^2}{(0.150\ \text{m})^2} = \boxed{131\ \text{N}}$$

(b) In Figure 11.15b, the bottom surface of the plunger at point B on the right is at the same level as point A on the left, which is at a depth h beneath the input piston. Therefore, we can apply $P_2 = P_1 + \rho g h$, with $P_2 = F_2/(\pi r_2^2)$ and $P_1 = F_1/(\pi r_1^2)$:

$$\frac{F_2}{(\pi r_2^2)} = \frac{F_1}{(\pi r_1^2)} + \rho g h$$

Solving for F_1 gives

$$F_1 = F_2 \left(\frac{r_1^2}{r_2^2}\right) - \rho g h (\pi r_1^2)$$

$$= (20\ 500\ \text{N}) \frac{(0.0120\ \text{m})^2}{(0.150\ \text{m})^2} - (8.00 \times 10^2\ \text{kg/m}^3)$$

$$\times (9.80\ \text{m/s}^2)(1.10\ \text{m})\, \pi\, (0.0120\ \text{m})^2 = \boxed{127\ \text{N}}$$

The answer here is less than in part (a), because the weight of the 1.10-m column of hydraulic oil provides some of the input force to support the car.

PROBLEM SOLVING INSIGHT

Note that the relation $F_1/A_1 = F_2/A_2$, which results from Pascal's principle, applies only when the points 1 and 2 lie at the same depth ($h = 0$) in the fluid.

In a device such as a hydraulic car lift, the same amount of work is done by both the input and output forces in the absence of friction. The larger output force $\mathbf{F_2}$ moves through a smaller distance, while the smaller input force $\mathbf{F_1}$ moves through a larger distance. The work, being the product of the magnitude of the force and the distance, is the same in either case, as it must be, since mechanical energy is conserved.

An enormous variety of clever devices use hydraulic fluids to create large forces, starting with small ones. In addition to the hydraulic car lift, such devices include the landing gear on an airplane and a backhoe used for digging. Figure 11.16 shows these applications.

Figure 11.16 Three examples of devices that employ a hydraulic fluid to generate a large output force, starting with a small input force.

11.6 ARCHIMEDES' PRINCIPLE

Anyone who has tried to push a beach ball under the water has felt how the water pushes back with a strong upward force. This upward force is called the *buoyant force,* and all fluids apply such a force to objects that are immersed in them. The buoyant force exists because fluid pressure is larger at greater depths.

In Figure 11.17*a* a cylinder of height h is being held under the surface of a liquid. The pressure P_1, acting on the top face, generates the downward force P_1A, where A is the area of the face. Similarly, the pressure P_2, acting on the bottom face, generates the upward force P_2A. Since the pressure is greater at greater depths, the upward force exceeds the downward force. Consequently, the liquid applies to the cylinder a net upward force, or buoyant force, whose magnitude F_B is

$$F_B = P_2A - P_1A = (P_2 - P_1)A$$

Substituting $P_2 - P_1 = \rho gh$ from Equation 11.4, we find that the buoyant force equals ρghA. In this result hA is the volume of liquid that the cylinder moves aside or displaces in being submerged, and ρ denotes the density of the liquid, not the density of the material from which the cylinder is made. Therefore, ρhA gives the

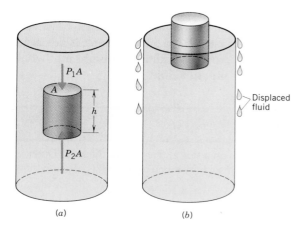

Figure 11.17 (*a*) A cylinder beneath the surface of a liquid (fluid). The fluid applies an upward force P_2A to the bottom face of the cylinder and a downward force P_1A to the top face. (*b*) The weight of the displaced fluid is the weight of the fluid that would spill out if the container were filled to the brim before the cylinder is inserted into the liquid.

mass m of the displaced fluid, so that the buoyant force equals mg, the weight of the displaced fluid. Part *b* of the drawing helps to clarify the meaning of the phrase "weight of the displaced fluid." This phrase refers to the weight of the fluid that would spill out, if the container were filled to the brim before the cylinder is inserted into the liquid. The buoyant force is not a new type of force. It is just the name given to the net upward force exerted by the fluid on the object.

The cylindrical shape of the object in Figure 11.17 is not important. No matter what the shape, the buoyant force arises in a similar fashion, in accord with *Archimedes' principle*. It was an impressive accomplishment that the Greek scientist Archimedes (ca. 287–212 B.C.) discovered the essence of this principle so long ago.

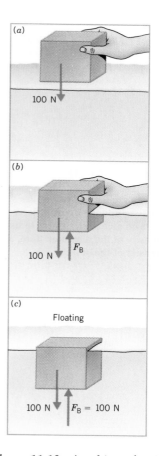

Figure 11.18 An object of weight 100 N is being inserted into a liquid. The deeper the object is, the more liquid it displaces, and the stronger the buoyant force is. In part *c*, the buoyant force matches the 100-N weight of the object, so the object floats.

Archimedes' Principle

Any fluid applies a buoyant force to an object that is partially or completely immersed in it; the magnitude of the buoyant force equals the weight of the fluid that the object displaces:

$$\underbrace{F_B}_{\substack{\text{Magnitude of} \\ \text{the buoyant} \\ \text{force}}} = \underbrace{W_{\text{fluid}}}_{\substack{\text{Weight of} \\ \text{the displaced} \\ \text{fluid}}} \qquad (11.6)$$

The effect that the buoyant force has depends on how strong it is compared with the other forces that are acting. For example, if the buoyant force is strong enough to balance the force of gravity, an object will float in a fluid. Figure 11.18 explores this possibility. In part *a*, a block that weighs 100 N is held above a liquid and displaces none of it, so the liquid exerts no buoyant force. In part *b*, the block displaces some liquid, and the liquid applies a buoyant force F_B to the block, according to Archimedes' principle. Nevertheless, if the block were released, it would sink, because the buoyant force is not sufficiently strong to balance the weight. Finally, in part *c*, enough of the block is submerged to provide a buoyant force strong enough to balance the 100-N weight, so the block is in equilibrium and does not sink when released. The block floats. If the buoyant force were not sufficiently large to balance the weight, even with the block completely submerged, the block would sink when released. Even if an object sinks, however, there is still a buoyant force acting on it; it's just that the buoyant force is not large enough to balance the weight. Example 7 provides additional insight into what determines whether an object will float or sink in a fluid.

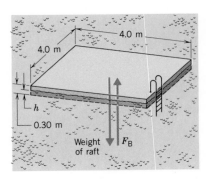

Figure 11.19 If the raft floats, it will do so with a distance h beneath the surface of the water.

PROBLEM SOLVING INSIGHT

When you are using Archimedes' principle to find the buoyant force F_B that acts on an object, be sure to use the density of the displaced fluid, not the density of the object.

Example 7   A Swimming Raft

A solid, square, pinewood raft measures 4.0 m on a side and is 0.30 m thick. (a) Determine whether the raft floats in water, and (b) if so, how much of the raft is beneath the surface (see the distance h in Figure 11.19).

REASONING   To determine whether the raft floats, we will compare the weight of the raft to the maximum possible buoyant force and see if there could conceivably be enough buoyant force to balance the weight. If so, then the value of the distance h in Figure 11.19 can be obtained by utilizing the fact that the floating raft is in equilibrium, with the magnitude of the buoyant force equaling the raft's weight.

SOLUTION
(a) The weight of the raft can be calculated from the density $\rho_{pine} = 550$ kg/m³ (Table 11.1), the volume of the wood, and the acceleration due to gravity. The volume of the wood is $V_{pine} = 4.0$ m $\times$ 4.0 m $\times$ 0.30 m $= 4.8$ m³, so that

$$\text{Weight of raft} = (\rho_{pine} V_{pine})g = (550 \text{ kg/m}^3)(4.8 \text{ m}^3)(9.80 \text{ m/s}^2) = 26\,000 \text{ N}$$

The maximum possible buoyant force occurs when the entire raft is under the surface, displacing 4.8 m³ of water. The weight of this volume of water is the maximum buoyant force and can be obtained using the density of water:

$$F_B^{max} = \rho_{water}(4.8 \text{ m}^3)g$$
$$F_B^{max} = (1.000 \times 10^3 \text{ kg/m}^3)(4.8 \text{ m}^3)(9.80 \text{ m/s}^2) = 47\,000 \text{ N}$$

Since the maximum possible buoyant force exceeds the weight of the raft (26 000 N), the raft will float only partially submerged, with the distance h beneath the water.

(b) We now find the value of h. The buoyant force balances the weight of the floating raft, so the magnitude of the buoyant force must be $F_B = 26\,000$ N. But according to Equation 11.6, the magnitude of the buoyant force is also the weight of the displaced fluid, $F_B = W_{fluid}$. Thus, we find that $W_{fluid} = 26\,000$ N. Using the density of water, we can also express the weight of the displaced water as $W_{fluid} = \rho_{water} V_{water} g$, where the volume is $V_{water} = 4.0$ m $\times$ 4.0 m $\times$ h. As a result,

$$W_{fluid} = \rho_{water}(4.0 \text{ m} \times 4.0 \text{ m} \times h)g = 26\,000 \text{ N}$$

$$h = \frac{26\,000 \text{ N}}{\rho_{water}(4.0 \text{ m} \times 4.0 \text{ m})g}$$

$$h = \frac{26\,000 \text{ N}}{(1.000 \times 10^3 \text{ kg/m}^3)(4.0 \text{ m} \times 4.0 \text{ m})(9.80 \text{ m/s}^2)} = \boxed{0.17 \text{ m}}$$

In deciding whether the raft floats in part (a) of Example 7, we compared the raft's weight $[(\rho_{pine} V_{pine})g]$ to the maximum possible buoyant force $[(\rho_{water} V_{pine})g]$. The comparison depends only on the densities ρ_{pine} and ρ_{water}. The take-home message is this: Any object that is *solid throughout* will float in a liquid

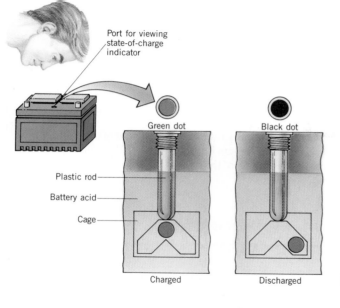

Port for viewing state-of-charge indicator

Green dot

Black dot

Plastic rod

Battery acid

Cage

Charged

Discharged

Figure 11.20 A state-of-charge indicator for a car battery.

if the density of the object is less than or equal to the density of the liquid. For instance, at 0 °C ice has a density of 917 kg/m³, while water has a density of 1000 kg/m³. Therefore, ice floats in water.

Although a solid piece of a high-density material like steel will sink in water, such materials can, nonetheless, be used to make floating objects. A supertanker, for example, floats because it is *not* solid metal. Such a ship contains enormous amounts of empty space and, because of its shape, displaces enough water to balance its own large weight.

A useful application of Archimedes' principle can be found in car batteries. To alert the owner that recharging is necessary, some batteries include a state-of-charge indicator, such as the one illustrated in Figure 11.20. The battery includes a viewing port that looks down through a plastic rod, which extends into the battery acid. Attached to the end of this rod is a "cage," containing a green ball that floats or sinks in the acid. When the battery is charged, the density of the acid is large enough so that its buoyant force makes the ball rise to the top of the cage, to just beneath the plastic rod. The viewing port shows a green dot, indicating that the battery is sufficiently charged. As the battery discharges, the density of the acid decreases. Since the buoyant force is the weight of the acid displaced by the ball, the buoyant force also decreases. As a result, the ball sinks into one of the chambers oriented at an angle. With the ball no longer visible, the viewing port shows a dark or black dot, warning the owner that the battery charge is low.

Archimedes' principle has allowed us to determine how an object can float in a liquid. This principle also applies to gases, as the next example illustrates.

THE PHYSICS OF . . .

a state-of-charge battery indicator.

THE PHYSICS OF . . .

a Goodyear airship.

Example 8   A Goodyear Airship

Normally, a Goodyear airship, such as that in Figure 11.21, contains about 5.40×10^3 m³ of helium whose density is 0.179 kg/m³. Find the weight of the load W_L that the airship can carry in equilibrium at an altitude where the density of air is 1.20 kg/m³.

REASONING   The airship and its load are in equilibrium. Thus, the buoyant force F_B applied to the airship by the surrounding air balances the weight W_{helium} of the helium and the weight W_L of the load, including the solid parts of the airship. The free-body diagram in Figure 11.21b shows these forces.

<u>SOLUTION</u> Because the forces in the free-body diagram balance, we have $W_{helium} + W_L = F_B$, so that

$$W_L = F_B - W_{helium}$$

According to Archimedes' principle, the buoyant force is the weight of the displaced air: $F_B = W_{air} = \rho_{air}V_{ship}g$. The weight of the helium is $W_{helium} = \rho_{helium}V_{ship}g$. The volume of the airship is nearly the same as the volume of the helium, $V_{ship} = 5.40 \times 10^3$ m³, so we find that

$$W_L = (\rho_{air} - \rho_{helium})V_{ship}g$$

$$W_L = (1.20 \text{ kg/m}^3 - 0.179 \text{ kg/m}^3)(5.40 \times 10^3 \text{ m}^3)(9.80 \text{ m/s}^2) = \boxed{5.40 \times 10^4 \text{ N}}$$

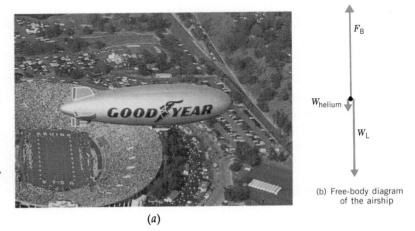

(a)

(b) Free-body diagram
of the airship

Figure 11.21 (*a*) A helium-filled Goodyear airship. (*b*) The free-body diagram of the airship, including the load weight W_L.

11.7 FLUIDS IN MOTION AND STREAMLINES

There are many examples of fluid motion or fluid flow. Water may flow smoothly and slowly in a quiet stream or violently over a waterfall. The air may form a gentle breeze or a raging tornado. To deal with such diversity, it helps to identify some of the basic types of fluid flow.

Fluid flow can be steady or unsteady. In steady flow the velocity of the fluid particles at any point is constant as time passes. For instance, in Figure 11.22 a fluid particle flows with a velocity of $v_1 = +2$ m/s past point 1. In steady flow every particle passing through this point has this same velocity. At another location the velocity may be different, as in a river, which usually flows fastest near its center and slowest near its banks. Thus, at point 2, the fluid velocity is $v_2 = +0.5$ m/s, and if the flow is steady, all particles passing through this point have a velocity of $+0.5$ m/s.

Unsteady flow exists whenever the velocity at a point in the fluid changes as time passes. *Turbulent flow* is an extreme kind of unsteady flow and occurs when

Figure 11.22 Two fluid particles in a stream. At different locations in the stream the particle velocities may be different, as indicated by v_1 and v_2.

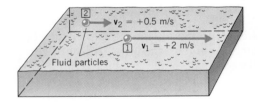

there are sharp obstacles or bends in the path of a fast-moving fluid, as in the rapids in Figure 11.23. In turbulent flow, the velocity at any particular point changes erratically from moment to moment, both in magnitude and direction.

Fluid flow can be compressible or incompressible. Most liquids are nearly incompressible; that is, the density of a liquid remains almost constant as the pressure changes. To a good approximation, then, liquids flow in an incompressible manner. In contrast, gases are highly compressible. However, there are situations in which the density of a flowing gas remains constant enough that the flow can be considered incompressible.

Fluid flow can be viscous or nonviscous. A viscous fluid, such as honey, does not flow readily and is said to have a large viscosity. In contrast, water is less viscous and flows more readily; water has a smaller viscosity than honey. The flow of a viscous fluid is an energy-dissipating process and is analogous to the energy-dissipating motion that occurs in the presence of kinetic friction. The viscosity hinders neighboring layers of fluid from sliding freely past one another. A fluid with zero viscosity flows in an unhindered manner with no dissipation of energy. Although no real fluid has zero viscosity at normal temperatures, many fluids have negligibly small viscosities. An incompressible, nonviscous fluid is called an *ideal fluid.*

Fluid flow can be rotational or irrotational. The flow is rotational when a part of the fluid has rotational as well as translational motion. To test for rotational flow, a small paddle wheel can be immersed in the fluid. If the wheel rotates, as in Figure 11.24a, the flow is rotational. If the paddle wheel does not rotate, as in part *b* of the figure, the flow is irrotational, and the fluid exhibits only translational motion.

When the flow is steady, *streamlines* are often used to represent the trajectories of the fluid particles. A streamline is a line drawn in the fluid such that a tangent to the streamline at any point is parallel to the fluid velocity at that point. Figure 11.25 shows the velocity vectors at three points along a streamline. The fluid velocity can vary (in both magnitude and direction) from point to point

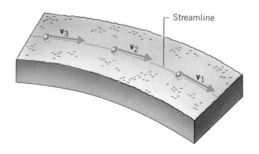

Figure 11.23 The flow of water in a rapids is an example of turbulent flow.

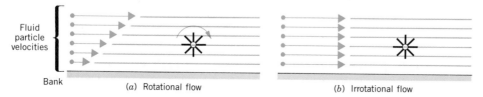

Figure 11.24 (*a*) In a river, water near the bank usually has a smaller velocity than water near the center. This top view shows that a small paddle wheel placed near the bank would rotate, indicating rotational flow. (*b*) In an ideal fluid in which all portions of the fluid have the same velocity, a paddle wheel would not rotate, indicating irrotational flow.

Figure 11.25 At any point along a streamline, the velocity vector of the fluid particle at that point is tangent to the streamline.

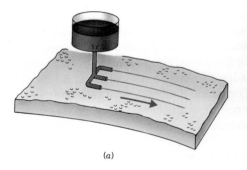

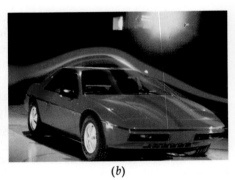

Figure 11.26 (*a*) In the steady flow of a liquid, a colored dye reveals the streamlines. (*b*) A smoke streamer reveals a streamline pattern for the air flowing around a car.

(*a*)

(*b*)

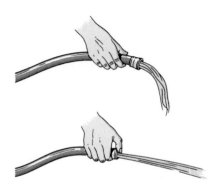

Figure 11.27 A tubular region of a moving fluid, whose sidewalls consist of streamlines, is known as a tube of flow.

along a streamline, but at any given point, the velocity is constant in time, as required by the condition of steady flow. In fact, steady flow is often called *streamline flow.*

Figure 11.26*a* illustrates a method for making streamlines visible by using small tubes to release a colored dye. The dye does not immediately mix with the liquid and is carried along a streamline. In the case of a flowing gas, such as that in a wind tunnel, streamlines are often revealed by smoke streamers, as part *b* of the figure shows.

In steady flow, the pattern of streamlines is steady in time, and, as Figure 11.26*a* indicates, no two streamlines cross one another. If they did, every particle arriving at the crossing point could go one way or another. This would mean that the velocity at the crossing point would change from moment to moment, a condition that does not exist in steady flow.

Streamlines provide a useful way of looking at fluid flow. Imagine, for instance, a tubular region of fluid whose sidewalls consist of streamlines, as in Figure 11.27. This region is known as a *tube of flow.* In accord with the meaning of streamlines, a tube of flow has the property that no fluid can flow through its sidewalls, for the fluid velocity is parallel to the sidewalls. In this sense, a tube of flow is analogous to a pipe of the same shape.

11.8 THE EQUATION OF CONTINUITY

Have you ever used your thumb to control the water flowing from the end of a hose, as in Figure 11.28? If so, you probably have seen that the water velocity increases when your thumb reduces the cross-sectional area of the hose opening. This kind of fluid behavior is described by the *equation of continuity*, which expresses the simple idea that the mass of fluid entering one end of a pipe must leave at the other end. For example, if fluid enters a pipe at a *mass flow rate* of 5 kilograms per second, then fluid must also leave at the same rate, assuming that there are no places between the entry and exit points to add or remove fluid. Thus, when you reduce the opening in the hose with your finger, the velocity with which the water leaves the hose must increase if the mass flow rate remains constant.

Figure 11.29 shows a small mass of fluid or fluid element moving along a tube. Upstream at position 2, where the tube has a cross-sectional area A_2, the fluid has a speed v_2 and a density ρ_2. Downstream at location 1, the corresponding quanti-

Figure 11.28 When the end of a hose is partially closed off, thus reducing its cross-sectional area, the fluid velocity increases.

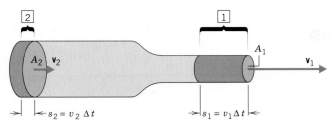

Figure 11.29   A fluid flowing in a tube that has different cross-sectional areas at positions 1 and 2.

ties are v_1, ρ_1, and A_1. During a small time interval Δt, the fluid at point 2 moves a distance of $s_2 = v_2 \Delta t$. The volume of fluid that has flowed past this point is $A_2 s_2 = A_2 v_2 \Delta t$. The mass Δm of this fluid element is the product of the density and volume: $\Delta m = \rho_2 A_2 v_2 \Delta t$. Dividing Δm by Δt gives the mass flow rate (the mass per second):

$$\text{Mass flow rate at position 2} = \frac{\Delta m}{\Delta t} = \rho_2 A_2 v_2 \qquad (11.7a)$$

Similar reasoning leads to the mass flow rate at position 1:

$$\text{Mass flow rate at position 1} = \rho_1 A_1 v_1 \qquad (11.7b)$$

Since no fluid can cross the sidewalls of the tube, the mass flow rates at positions 1 and 2 must be equal. But these positions were selected arbitrarily, so the mass flow rate has the same value everywhere in the tube, an important result known as the equation of continuity.

Equation of Continuity

The mass flow rate ($\rho A v$) has the same value at every position along a tube that has a single entry and a single exit point for fluid flow. For two positions along such a tube

$$\rho_1 A_1 v_1 = \rho_2 A_2 v_2 \qquad (11.8)$$

where ρ = fluid density (kg/m³)
$\quad\quad A$ = cross-sectional area of tube (m²)
$\quad\quad v$ = fluid speed (m/s)

SI Unit of Mass Flow Rate: kg/s

The density of an incompressible fluid does not change during flow, so that $\rho_1 = \rho_2$, and the equation of continuity reduces to

$$A_1 v_1 = A_2 v_2 \qquad (11.9)$$

The quantity Av represents the volume of fluid per second that passes through the tube and is referred to as the *volume flow rate Q*:

$$Q = \text{volume flow rate} = Av \qquad (11.10)$$

Closing down the end of a hose with your thumb gives you a distinct advantage in a water fight.

Equation 11.9 shows that where the tube area is large, the fluid speed is small, and, conversely, where the tube area is small, the speed is large. Example 9 explores this behavior in more detail for the hose in Figure 11.28.

Example 9   A Garden Hose

The water from a garden hose fills a bucket in 30.0 s. The volume of the bucket is 8.00×10^{-3} m³ (about two gallons). Find the speed of the water that leaves the hose through (a) an unobstructed opening with a cross-sectional area of 2.85×10^{-4} m² and (b) an obstructed opening that has only half as much area.

REASONING AND SOLUTION

(a) Once the volume flow rate Q is known, the speed of the water can be obtained from $Q = Av$:

$$Q = \frac{8.00 \times 10^{-3} \text{ m}^3}{30.0 \text{ s}} = 2.67 \times 10^{-4} \text{ m}^3/\text{s}$$

The speed is

$$v = \frac{Q}{A} = \frac{2.67 \times 10^{-4} \text{ m}^3/\text{s}}{2.85 \times 10^{-4} \text{ m}^2} = \boxed{0.937 \text{ m/s}}$$

(b) Water can be considered incompressible, so the equation of continuity can be applied in the form $A_1 v_1 = A_2 v_2$. Since $A_2 = \frac{1}{2} A_1$, we find that

$$v_2 = \left(\frac{A_1}{A_2}\right) v_1 = \left(\frac{A_1}{\frac{1}{2} A_1}\right)(0.937 \text{ m/s}) = \boxed{1.87 \text{ m/s}} \qquad (11.9)$$

The next example applies the equation of continuity to the flow of blood.

THE PHYSICS OF . . .

a clogged artery.

PROBLEM SOLVING INSIGHT

The equation of continuity in the form $A_1 v_1 = A_2 v_2$ applies only when the density of the fluid is constant. If the density is not constant, the equation of continuity is $\rho_1 A_1 v_1 = \rho_2 A_2 v_2$.

Example 10   A Clogged Artery

In the condition known as atherosclerosis, a deposit or atheroma forms on the arterial wall and reduces the opening through which blood can flow. In the carotid artery in the neck, blood flows three times faster through a partially blocked region than it does through an unobstructed region. Determine the ratio of the effective radii of the artery at the two places.

REASONING AND SOLUTION   Blood, like most liquids, is incompressible, and the equation of continuity in the form of Equation 11.9 can be applied. Since the area of a circle is πr^2, it follows that

$$\underbrace{(\pi r_U^2) v_U}_{\text{Unobstructed}} = \underbrace{(\pi r_A^2) v_A}_{\substack{\text{Obstructed by} \\ \text{atheroma}}}$$

The ratio of the radii is

$$\frac{r_U}{r_A} = \sqrt{\frac{v_A}{v_U}} = \sqrt{3} = \boxed{1.7}$$

The unobstructed artery has an effective radius that is 70% larger than the afflicted region.

11.9 BERNOULLI'S EQUATION

For *steady, irrotational* flow, the speed, pressure, and elevation of an *incompressible, nonviscous* fluid are related by an equation discovered by Daniel Bernoulli (1700–1782). To set the stage for our derivation of **Bernoulli's equation,** let us make two observations about a moving fluid. First, whenever a fluid flowing in a horizontal pipe encounters a region of reduced cross-sectional area, the pressure of the fluid drops, as Figure 11.30a indicates. When moving from region 2 downstream to region 1 in the picture, the fluid speeds up (accelerates), as required by the conservation of mass and expressed by the equation of continuity. According to Newton's second law, the accelerating fluid must be subjected to an unbalanced force. Such an unbalanced force can exist only if the pressure in region 2 is greater than the pressure in region 1. The difference between the pressures is given by Bernoulli's equation. Second, if the fluid experiences a rise in elevation, as in part b of the figure, the pressure at the bottom is greater than the pressure at the top. The basis for this statement is our previous study of static fluids, and Bernoulli's equation confirms it, provided that the cross-sectional area of the pipe does not change.

To derive Bernoulli's equation, consider Figure 11.31a. This drawing shows a small portion of fluid (called a fluid element) of mass m, upstream in region 2 of a pipe. Both the cross-sectional area and the elevation are different at different places along this pipe. The speed, pressure, and elevation in this region are v_2, P_2, and y_2, respectively. Downstream in region 1 these variables have the values v_1, P_1, and y_1. As Chapter 6 discusses, an object moving under the influence of gravity possesses a mechanical energy E that is the sum of the kinetic energy KE and the gravitational potential energy PE: $E = KE + PE = \frac{1}{2}mv^2 + mgy$. If work W_{nc} is done on the fluid element by nonconservative forces, the mechanical energy changes. According to the work–energy theorem (see Section 6.4), the work done equals the change in mechanical energy:

$$W_{nc} = E_1 - E_2 = \underbrace{(\tfrac{1}{2}mv_1^2 + mgy_1)}_{\substack{\text{Mechanical}\\ \text{energy in region 1}}} - \underbrace{(\tfrac{1}{2}mv_2^2 + mgy_2)}_{\substack{\text{Mechanical}\\ \text{energy in region 2}}} \qquad (6.8)$$

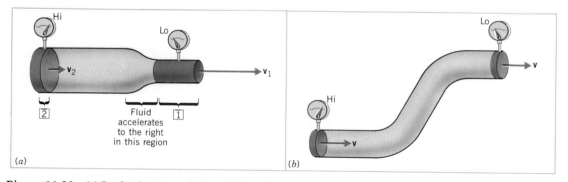

(a)

(b)

Figure 11.30 (a) In this horizontal pipe, the pressure in region 2 is greater than that in region 1. The difference in pressure accelerates the fluid to the right. (b) When the fluid changes elevation, the pressure at the bottom is greater than that at the top, assuming the cross-sectional area of the pipe remains constant.

Figure 11.31 (*a*) A fluid element moving through a pipe whose cross-sectional area and elevation both change. (*b*) The fluid element experiences a force −**F** on its top surface due to the fluid above it, and a force **F** + Δ**F** on its bottom surface due to the fluid below it.

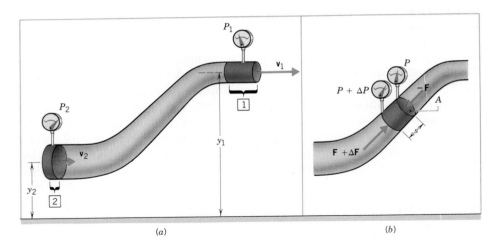

(*a*)   (*b*)

Now we need to examine the term W_{nc} and see what nonconservative forces do work on the fluid. Since the flow is assumed to be nonviscous, there are no viscous forces. However, there are forces that come from the fluid surrounding the element of mass m. To see why, refer to Figure 11.31*b*. On the top surface of the fluid element, the surrounding fluid exerts a pressure P. This pressure gives rise to a force of magnitude $F = PA$, where A is the cross-sectional area. On the bottom surface, the surrounding fluid exerts a slightly greater pressure, $P + \Delta P$, where ΔP is the pressure difference between the ends of the element. As a result, the force on the bottom surface is $F + \Delta F = (P + \Delta P)A$. The *net* force pushing the fluid element up the tube is $\Delta F = (\Delta P)A$. When the fluid element moves through its own length s, the work done is the product of the net force and the distance, according to Equation 6.1: Work $= (\Delta F)s = (\Delta P)As$. The quantity As is the volume V of the element, so the work is $(\Delta P)V$. The total work done on the fluid element in moving it from region 2 to region 1 is the sum of the small increments of work $(\Delta P)V$ done as the element moves along the tube. This sum amounts to $W_{nc} = (P_2 - P_1)V$, where $P_2 - P_1$ is the pressure difference between the two regions. With this expression for W_{nc}, the work–energy theorem becomes

$$W_{nc} = (P_2 - P_1)V = (\tfrac{1}{2}mv_1{}^2 + mgy_1) - (\tfrac{1}{2}mv_2{}^2 + mgy_2)$$

By dividing both sides of this result by the volume V, recognizing that m/V is the density ρ of the fluid, and rearranging terms, we obtain Bernoulli's equation.

Bernoulli's Equation

For any two points (1 and 2) in the steady, irrotational flow of a nonviscous, incompressible fluid, the pressure P, the fluid speed v, and the elevation y are related by

$$P_1 + \tfrac{1}{2}\rho v_1{}^2 + \rho g y_1 = P_2 + \tfrac{1}{2}\rho v_2{}^2 + \rho g y_2 \qquad (11.11)$$

Bernoulli's equation is a direct consequence of the work–energy theorem and applies only if the flow is nonviscous, so that viscous losses are absent. Since the points 1 and 2 were selected arbitrarily, the term $P + \tfrac{1}{2}\rho v^2 + \rho g y$ has a constant value at all positions in the pipe. For this reason, Bernoulli's equation is sometimes expressed as $P + \tfrac{1}{2}\rho v^2 + \rho g y = $ constant.

Equation 11.11 can be regarded as an extension of the earlier result that specifies how the pressure varies with depth in a static fluid ($P_2 = P_1 + \rho gh$), the terms $\frac{1}{2}\rho v_1^2$ and $\frac{1}{2}\rho v_2^2$ accounting for the effects of fluid speed. Bernoulli's equation reduces to the result for static fluids when the speed of the fluid is the same everywhere ($v_1 = v_2$), as it is when the cross-sectional area remains constant. Under such conditions, Bernoulli's equation is $P_1 + \rho gy_1 = P_2 + \rho gy_2$. After rearrangement, this result becomes

$$P_2 = P_1 + \rho g(y_1 - y_2) = P_1 + \rho gh$$

which is the result (Equation 11.4) for static fluids.

11.10 APPLICATIONS OF BERNOULLI'S EQUATION

When a moving fluid is contained in a horizontal pipe, all parts of it have the same elevation ($y_1 = y_2$), and Bernoulli's equation simplifies to

$$P_1 + \frac{1}{2}\rho v_1^2 = P_2 + \frac{1}{2}\rho v_2^2 \qquad (11.12)$$

Thus, the quantity $P + \frac{1}{2}\rho v^2$ remains constant throughout a horizontal pipe; if v decreases, P increases and vice versa. This is exactly the result that we deduced qualitatively from Newton's second law at the beginning of Section 11.9. Example 11 illustrates the use of Equation 11.12.

Example 11   An Enlarged Blood Vessel

THE PHYSICS OF . . .

an aneurysm.

An aneurysm is an abnormal enlargement of a blood vessel such as the aorta. Suppose that, because of an aneurysm, the cross-sectional area A_1 of the aorta increases to a value $A_2 = 1.7A_1$. The speed of the blood ($\rho = 1060$ kg/m^3) through a normal portion of the aorta is $v_1 = 0.40$ m/s. Assuming the aorta is horizontal (the person is lying down), determine the amount by which the pressure in the enlarged region exceeds that in the normal region.

REASONING   According to the equation of continuity, the speed of the blood in the enlarged section of the artery is smaller than the speed in a healthy section. In turn, Bernoulli's equation for horizontal flow (Equation 11.12) indicates that the smaller speed leads to a greater pressure.

SOLUTION   From the equation of continuity we find the speed v_2 of the blood in the aneurysm to be

$$v_2 = \left(\frac{A_1}{A_2}\right) v_1 = \left(\frac{A_1}{1.7A_1}\right)(0.40 \text{ m/s}) = 0.24 \text{ m/s} \qquad (11.9)$$

From Bernoulli's equation for horizontal flow it follows that $P_1 + \frac{1}{2}\rho v_1^2 = P_2 + \frac{1}{2}\rho v_2^2$. Solving for $P_2 - P_1$, we find

$$P_2 - P_1 = \frac{1}{2}\rho(v_1^2 - v_2^2)$$

$$= \frac{1}{2}(1060 \text{ kg/m}^3)[(0.40 \text{ m/s})^2 - (0.24 \text{ m/s})^2] = \boxed{54 \text{ Pa}}$$

The excess pressure puts added stress on the already weakened tissue of the arterial wall at the aneurysm.

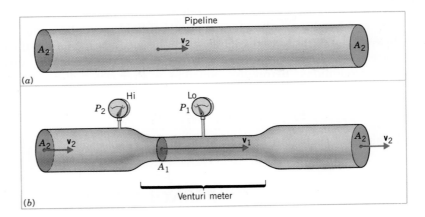

Figure 11.32 A Venturi meter can be used to measure the fluid speed v_2 in a gas pipeline.

THE PHYSICS OF . . .

a Venturi meter.

A Venturi meter is a device for measuring the speed of a fluid within a pipe. The speed can be obtained by substituting a Venturi meter for a section of the pipe, as Figure 11.32 illustrates. As the fluid moves into the narrow part of the Venturi meter, the speed of the fluid increases from v_2 to v_1, and the pressure of the fluid decreases from P_2 to P_1. From a measurement of these pressures, the fluid speed (and volume flow rate) can be determined, as the next example shows.

Example 12   **Measuring the Speed of a Flowing Gas**

In Figure 11.32a gas is flowing through a horizontal pipe whose cross-sectional area is $A_2 = 0.0700 \text{ m}^2$. The gas has a density of $\rho = 1.30 \text{ kg/m}^3$. In part b, a Venturi meter is substituted for a section of the pipe. The pressure difference measured with the Venturi meter (cross-sectional area $A_1 = 0.0500 \text{ m}^2$) is $P_2 - P_1 = 120 \text{ Pa}$. Find (a) the speed v_2 of the gas in the pipe and (b) the volume flow rate Q of the gas.

REASONING   With the measured pressure difference, Bernoulli's equation for horizontal flow can be used to find the speed v_2 of the gas in the pipe. However, Bernoulli's equation also contains the speed v_1 of the gas in the Venturi meter. Therefore, we begin our solution by expressing v_1 in terms of v_2 with the aid of the equation of continuity, assuming that the gas can be treated as an incompressible fluid.

SOLUTION
(a) From the equation of continuity (Equation 11.9) it follows that $v_1 = A_2 v_2 / A_1$, so that $v_1 = (0.0700 \text{ m}^2)v_2/(0.0500 \text{ m}^2) = 1.40 v_2$. Substituting this relation into Bernoulli's equation, we find that

$$P_1 + \tfrac{1}{2}\rho(1.40\ v_2)^2 = P_2 + \tfrac{1}{2}\rho v_2^2 \tag{11.12}$$

$$\tfrac{1}{2}\rho(1.40^2 - 1)v_2^2 = P_2 - P_1$$

$$v_2 = \sqrt{\frac{2(P_2 - P_1)}{\rho(1.40^2 - 1)}} = \sqrt{\frac{2(120 \text{ Pa})}{(1.30 \text{ kg/m}^3)(1.40^2 - 1)}} = \boxed{14 \text{ m/s}}$$

(b) The volume flow rate is

$$Q = A_2 v_2 = (0.0700 \text{ m}^2)(14 \text{ m/s}) = \boxed{0.98 \text{ m}^3/\text{s}} \tag{11.10}$$

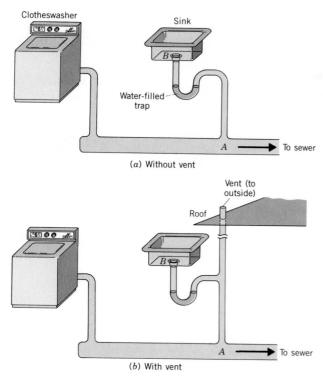

Figure 11.33 In a household plumbing system, a vent is necessary to equalize the pressures at points *A* and *B*, thus preventing the trap from being emptied. An empty trap allows sewer gas to enter the house.

The impact of fluid flow on pressure is amazingly widespread. Figure 11.33, for instance, illustrates how household plumbing takes into account the implications of Bernoulli's equation. The U-shaped section of pipe beneath the sink is called a "trap," because it traps water, which serves as a barrier to prevent sewer gas from leaking into the house. Part *a* of the drawing shows poor plumbing. When water from the clotheswasher rushes through the sewer pipe, the high speed flow causes the pressure at point *A* to drop. The pressure at point *B* in the sink, however, remains at the higher atmospheric pressure. As a result, the water is pushed out of the trap and into the sewer line, leaving no protection against sewer gas. A correctly designed system should be vented to the outside of the house, as in Figure 11.33*b*. The purpose of the vent is to ensure that the pressure at *A* remains the same as that at *B* (atmospheric pressure), even when water from the clotheswasher is rushing through the pipe. Thus, the purpose of the vent is to prevent the trap from being emptied, not to provide an escape route for sewer gas.

One of the most spectacular examples of the impact of fluid flow on pressure is the dynamic lift on airplane wings. Figure 11.34*a* shows a wing (in cross section) moving to the right, with the air flowing past the wing to the left. Because of the shape of the wing, the air travels faster over the curved upper surface than it does over the flatter lower surface. According to Bernoulli's equation, the pressure above the wing is lower (faster moving air), while the pressure below the wing is higher (slower moving air). Thus, the wing is lifted upward. Part *b* of the figure shows the wing of an airplane.

THE PHYSICS OF . . .

household plumbing.

THE PHYSICS OF . . .

airplane wings.

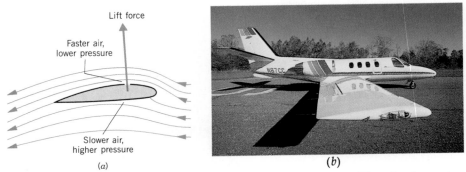

Faster air,
lower pressure

Lift force

Slower air,
higher pressure

(a)

(b)

Figure 11.34 (*a*) Streamlines of air flow around an airplane wing. The wing is moving to the right. (*b*) An airplane.

Figure 11.35 A ski jumper curves his body to take advantage of the dynamic lift force due to the flow of air around him.

THE PHYSICS OF . . .

ski jumping.

THE PHYSICS OF . . .

a curve ball.

Ski jumpers use dynamic lift, just like airplane wings do, to help themselves stay in the air longer. Figure 11.35 shows a jumper who has curved his body to mimic the cross-sectional shape of an airplane wing.

The curve ball, one of the main weapons in the arsenal of a baseball pitcher, is another illustration of the effects of fluid flow. Figure 11.36*a* shows a baseball moving to the right with no spin. The view is from above, looking down toward the ground. In this situation, air flows with the same speed around both sides of the ball, and the pressure is the same on both sides. No net force exists to make the ball curve. However, when the ball is given a spin, the air close to its surface is dragged around with it; the air on one half of the ball is speeded up (lower pressure), while that on the other half is slowed down (higher pressure). Part *b* of the picture illustrates the effects of a counterclockwise spin. The baseball experiences a net deflection force and curves on its way from the pitcher's mound to the plate, as part *c* shows.

As a final application of Bernoulli's equation, Figure 11.37*a* shows a large tank from which water is emerging through a small pipe near the bottom. Bernoulli's equation can be used to determine the speed (called the efflux speed) at which the water leaves the pipe, as the next example shows.

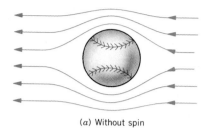

(a) Without spin

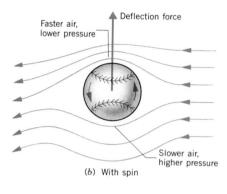

Deflection force

Faster air, lower pressure

Slower air, higher pressure

(b) With spin

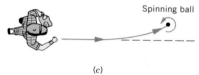

Spinning ball

(c)

Figure 11.36 These views of a baseball are from above, looking down toward the ground, with the ball moving to the right. (*a*) Without spin, the baseball does not curve to either side. (*b*) A spinning baseball curves in the direction of the deflection force. (*c*) The spin in part *b* causes the ball to curve as shown here.

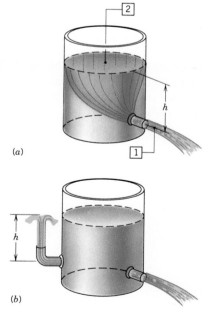

(a)

(b)

Figure 11.37 (*a*) Bernoulli's equation can be used to determine the speed of the water leaving the small pipe. (*b*) An ideal fluid (no viscosity) will rise to the fluid level in the tank after leaving a vertical outlet nozzle.

Example 13   Efflux Speed

Find an expression for the speed of the water leaving the pipe in Figure 11.37*a*.

REASONING   In the drawing, point 2 is at one end of a tube of flow that begins at the top surface of the water. Point 1 is just outside the efflux pipe, at the other end of the tube of flow. We will apply Bernoulli's equation to the water in this tube of flow.

SOLUTION   At both points 1 and 2 the pressure is atmospheric pressure, so that $P_1 = P_2$, and Bernoulli's equation becomes $\frac{1}{2}\rho v_1^2 + \rho g y_1 = \frac{1}{2}\rho v_2^2 + \rho g y_2$. The density ρ can be eliminated algebraically from this result, which can then be solved for the square of the efflux speed v_1:

$$v_1^2 = v_2^2 + 2g(y_2 - y_1) = v_2^2 + 2gh$$

where we have substituted $h = y_2 - y_1$ for the height of the liquid above the efflux tube. If the tank is very large, the water level changes only slowly and the speed at point 2 can be set equal to zero, so that $\boxed{v_1 = \sqrt{2gh}}$.

The speed of the liquid coming out of the tank is the same as if the liquid had freely fallen through a height h (see Equation 2.9 with $s = h$ and $a = g$), a result known as *Torricelli's theorem.* If the outlet pipe were pointed directly upward, as in part *b* of the drawing, the liquid would rise to a height h equal to the fluid level above the pipe. However, if the liquid is not an ideal fluid, its viscosity cannot be neglected. Then, the efflux speed would be less than that given by Bernoulli's equation, and the liquid would rise to a height less than *h*.

*11.11 VISCOUS FLOW

VISCOSITY

In an ideal fluid there is no viscosity to hinder the fluid layers as they slide past one another. Within a pipe of uniform cross section, every layer of an ideal fluid moves with the same velocity, even the layer next to the wall, as Figure 11.38a shows. When viscosity is present, the fluid layers do not all have the same velocity, as part *b* of the drawing illustrates. The fluid closest to the wall does not move at all while the fluid at the center of the pipe has the greatest velocity. The fluid layer next to the wall surface does not move, because it is held tightly by intermolecular forces. So strong are these forces that if a solid surface moves, the adjacent fluid layer moves along with it and remains at rest *relative* to the moving surface. That is why a fine layer of dust remains on a car even at high driving speeds. The layer of air in contact with the car has no velocity relative to the car, and, thus, does not blow off the dust.

To help introduce viscosity in a quantitative fashion, Figure 11.39a shows a viscous fluid between two parallel plates. The top plate is free to move while the bottom one is stationary. If the top plate is to move with a velocity **v** relative to the bottom plate, a force **F** is required. For a highly viscous fluid, like thick honey, a large force is needed; for a less viscous fluid, like water, a smaller force is necessary. As part *b* of the drawing suggests, we may imagine the fluid to be composed of many thin horizontal layers. When the top plate moves, the intermediate fluid layers slide over each other. The velocity of each layer is different, changing uniformly from **v** at the top plate to zero at the bottom plate. The resulting flow is called *laminar flow*, since a thin layer is often referred to as a lamina. As each layer moves, it is subjected to viscous forces from its neighbors, and the purpose of the force **F** is to compensate for the effect of these forces, so that any layer can move with a constant velocity.

The amount of force required in Figure 11.39a depends on several factors. Larger areas *A*, being in contact with more fluid, require larger forces, so that the force is proportional to the contact area ($F \propto A$). For a given area, greater speeds require larger forces, with the result that the force is proportional to the speed ($F \propto v$). The force is also inversely proportional to the perpendicular distance *y* between the top and bottom plates ($F \propto 1/y$). The larger the distance *y*, the smaller is the force required to achieve a given speed with a given contact area.

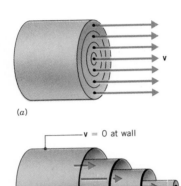

(a)

v = 0 at wall

v is a maximum at the center

(b)

Figure 11.38 (a) In ideal (nonviscous) fluid flow, all fluid particles across the pipe have the same velocity. (b) In viscous flow, the speed of the fluid is zero at the surface of the pipe and increases to a maximum along the center axis.

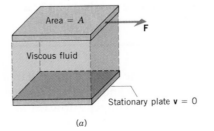

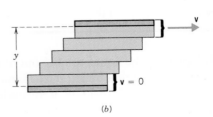

Figure 11.39 (a) A force **F** is applied to the top plate, which is in contact with a viscous fluid. (b) Because of the force **F**, the top plate and the adjacent layer of fluid move with a constant velocity **v**.

These three proportionalities can be expressed together in the following manner: $F \propto Av/y$. Equation 11.13 expresses this relationship with the aid of a proportionality constant η (Greek letter *eta*), which is called the **coefficient of viscosity** or simply the *viscosity*.

Force Needed to Move a Layer of Viscous Fluid with a Constant Velocity

The tangential force **F** required to move a fluid layer at a constant speed v, when the layer has an area A and is located a perpendicular distance y from an immobile surface, is given by

$$F = \frac{\eta A v}{y} \qquad\qquad (11.13)$$

where η is the coefficient of viscosity.

  SI Unit of Viscosity: Pa·s

  Common Unit of Viscosity: poise (P)

By solving this equation for the viscosity, $\eta = Fy/(vA)$, it can be seen that the SI unit for viscosity is $N \cdot m/[(m/s) \cdot m^2] = Pa \cdot s$. Another common unit for viscosity is the *poise* (P), which is used in the cgs system of units and is named after the French physician Jean Poiseuille (1797–1869, pronounced, approximately, as Pwah-zoy'). The following relation exists between the two units:

$$1 \text{ poise (P)} = 0.1 \text{ Pa} \cdot \text{s}$$

Values of viscosity depend on the nature of the fluid, and Table 11.2 lists data for some liquids and gases. Under ordinary conditions, the viscosities of gases are significantly *smaller* than those of liquids. Moreover, the viscosities of either liquids or gases depend markedly on temperature. Usually, the viscosities of liquids decrease as the temperature is increased. Anyone who has heated honey or oil, for example, knows that these fluids flow much more freely at an elevated

Table 11.2   Coefficients of Viscosity of Common Fluids

Fluid	Temperature (°C)	Viscosity η (Pa·s)
Gases		
Air	0	0.0171×10^{-3}
	20	0.0182×10^{-3}
	40	0.0193×10^{-3}
Carbon dioxide	20	0.0147×10^{-3}
Helium	20	0.0196×10^{-3}
Liquids		
Whole blood	37	4×10^{-3}
Glycerine	20	1500×10^{-3}
Methanol	20	0.584×10^{-3}
Water	0	1.78×10^{-3}
	20	1.00×10^{-3}
	40	0.651×10^{-3}

temperature. In contrast, the viscosities of gases increase as the temperature is raised. In general, fluids having smaller values of η are more nearly ideal fluids, because they flow more readily with only relatively weak viscous forces impeding their movement; an ideal fluid has $\eta = 0$.

POISEUILLE'S LAW

Viscous flow occurs in a wide variety of situations, such as oil moving through a pipeline, a liquid being forced through the needle of a hypodermic syringe, or blood moving in the human circulatory system. Figure 11.40 identifies the factors that determine the volume flow rate Q (m³/s) of the fluid in cases like these.

First, a difference in pressure $P_2 - P_1$ must be maintained between any two locations along the pipe in order for the fluid to flow. In fact, Q is proportional to $P_2 - P_1$, a greater pressure difference leading to a larger flow rate. Second, a long pipe offers greater resistance to the flow than a short pipe does, and Q is inversely proportional to the length L. Because of this fact, long pipelines, such as the Alaskan pipeline, have pumping stations at various places along the line to compensate for a drop in pressure. Third, high-viscosity fluids flow less readily than low viscosity fluids do, and Q is inversely proportional to the viscosity η. Finally, the volume flow rate is larger in a pipe of larger radius, other things being equal. The dependence on the radius R is a surprising one, Q being proportional to R^4. If, for instance, the pipe radius is reduced to one-half of its original value, the volume flow rate is reduced to one-sixteenth of its original value, assuming the other variables remain constant. The mathematical relation for Q in terms of these parameters was discovered by Poiseuille and is known as *Poiseuille's law:*

$$Q = \frac{\pi R^4 (P_2 - P_1)}{8\eta L} \tag{11.14}$$

Example 14 illustrates the use of Poiseuille's law.

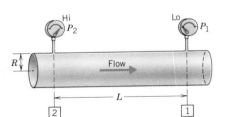

Figure 11.40 For viscous flow, the difference in pressure $P_2 - P_1$, the radius R and length L of the tube, and the viscosity η of the fluid influence the volume flow rate.

THE PHYSICS OF . . .

a hypodermic syringe.

Example 14   Giving an Injection

A hypodermic syringe is filled with a solution whose viscosity is 1.5×10^{-3} Pa·s. As Figure 11.41 shows, the plunger area of the syringe is 8.0×10^{-5} m², and the length of the needle is 0.025 m. The internal radius of the needle is 4.0×10^{-4} m. The gauge pressure in a vein is 1900 Pa (14 mm of mercury). What force must be applied to the plunger, so that 1.0×10^{-6} m³ of solution can be injected in 3.0 s?

REASONING   Because the needle has a much smaller radius than the syringe barrel, the time needed to empty the syringe is determined primarily by the volume flow rate through the needle. It is to the needle, then, that we apply Poiseuille's equation to determine the pressure difference $P_2 - P_1$ (see drawing). Once this pressure difference is known, we will be able to obtain the applied pressure and, hence, the applied force.

SOLUTION   The volume flow rate is $Q = (1.0 \times 10^{-6} \text{ m}^3)/(3.0 \text{ s}) = 3.3 \times 10^{-7}$ m³/s. According to Poiseuille's law, the required pressure difference is

$$P_2 - P_1 = \frac{8\eta L Q}{\pi R^4}$$

$$= \frac{8(1.5 \times 10^{-3} \text{ Pa·s})(0.025 \text{ m})(3.3 \times 10^{-7} \text{ m}^3/\text{s})}{\pi (4.0 \times 10^{-4} \text{ m})^4} = 1200 \text{ Pa} \quad (11.14)$$

Area = 8.0×10^{-5} m²

0.025 m

F

P_2 P_1

Figure 11.41   The difference in pressure $P_2 - P_1$ required to sustain the fluid flow through a hypodermic needle can be found with the aid of Poiseuille's law.

Since $P_1 = 1900$ Pa, the pressure P_2 must be $P_2 = 1200$ Pa $+ 1900$ Pa $= 3100$ Pa. This pressure is nearly equal to the pressure at the plunger, because the barrel of the syringe is so large that little pressure difference is required to sustain the flow up to point 2, where the fluid encounters the narrow needle. Since pressure is force per unit area, the force that must be applied to the plunger is the pressure times the plunger area:

$$F = (3100 \text{ Pa})(8.0 \times 10^{-5} \text{ m}^2) = \boxed{0.25 \text{ N}}$$

INTEGRATION OF CONCEPTS

PROPERTIES OF A STATIC FLUID AND NEWTON'S SECOND LAW

Fluids have many interesting and useful properties. A static fluid is one that is at rest and, therefore, is not accelerating. Such a fluid is in equilibrium, and Newton's second law is the cornerstone for analyzing its properties. According to the second law, there is no net force acting on any object in equilibrium. Using this condition for equilibrium, we are able to determine how the pressure in an incompressible fluid varies with depth, $P_2 = P_1 + \rho g h$. From this relation, we are able to formulate Pascal's principle and Archimedes' principle, which tell us how hydraulic lifts work and how objects float. However, both of these principles are rooted in Newton's second law.

THE CONSERVATION OF MASS AND THE EQUATION OF CONTINUITY

In Chapters 6, 7, and 9 we introduced three important conservation principles: the conservation of mechanical energy, the conservation of linear momentum, and the conservation of angular momentum. We saw that, under certain conditions, the total mechanical energy, the total linear momentum, and the total angular momentum remain constant throughout the motion of a system. Here in this chapter, we encounter another type of conservation law for a moving fluid, the conservation of mass. If a fluid is moving in a tube that has a single entry point and a single exit point, the mass of fluid per second passing any point along the tube is a constant. The mass of fluid per second is $\rho A v$, where ρ is the density of the fluid, A is the cross-sectional area of the tube, and v is the speed of the fluid. Thus, as a fluid flows, its density, cross-sectional area, and speed may change from place to place in the tube. But the product of these three variables remains constant if the mass of fluid per second flowing in the tube is conserved.

BERNOULLI'S EQUATION AND THE WORK–ENERGY THEOREM

The speed of a flowing fluid may change from place to place. For an ideal fluid, Bernoulli's equation relates the speed to the pressure and elevation at each point along a streamline. However, Bernoulli's equation is not a new and separate law of physics, but is an outgrowth of the work–energy theorem that we considered in Chapter 6. This theorem states that the kinetic energy of an object changes when work is performed on the object by a net force. Similarly, when a net force does work on a fluid, the kinetic energy and, hence, speed of the fluid change. For an ideal fluid, there are two kinds of forces that do work on it, the conservative gravitational force and the nonconservative force due to pressure differences within the fluid. When the work done by these forces is incorporated into the work–energy theorem, the result is Bernoulli's equation.

SUMMARY

Fluids are materials that can flow. The **mass density** ρ of any substance is its mass m divided by its volume V: $\rho = m/V$.

Pressure P is the magnitude F of the force acting perpendicular to a surface divided by the area A over which the force acts: $P = F/A$. The SI unit of pressure is the pascal (Pa); $1\,\text{Pa} = 1\,\text{N/m}^2$. In the presence of gravity, the upper layers of a fluid push downward on the layers beneath, with the result that **fluid pressure is related to depth.** In an incompressible static fluid whose density is ρ, the relation is $P_2 = P_1 + \rho g h$, where P_1 is the pressure at one level and P_2 is the pressure at a level that is h meters deeper. The **gauge pressure** is the amount by which a pressure P exceeds atmospheric pressure. The **absolute pressure** is the actual value for P.

According to **Pascal's principle,** any change in the pressure applied to a completely enclosed fluid is transmitted undiminished to all parts of the fluid and the enclosing walls.

The **buoyant force** is the net upward force that a fluid applies to any object that is immersed partially or completely in it. **Archimedes' principle** states that the magnitude of the buoyant force equals the weight of the fluid that the immersed object displaces.

The **mass flow rate** (kg/s) of a fluid with a density ρ, flowing with a speed v in a pipe of cross-sectional area A, is the mass per second flowing past a point and is given by $\rho A v$. The **equation of continuity** expresses the fact that mass is conserved; what flows into one end of a pipe flows out the other end, assuming there are no

additional entry or exit points in between. In terms of the mass flow rate, the equation of continuity is $\rho_1 A_1 v_1 = \rho_2 A_2 v_2$, where the subscripts 1 and 2 denote two points along the pipe. If the fluid is incompressible, $\rho_1 = \rho_2$. The equation of continuity then becomes $A_1 v_1 = A_2 v_2$, where the product of the cross-sectional area and speed is called the **volume flow rate** Q; $Q = Av$.

Bernoulli's equation is a direct consequence of the work–energy theorem and describes the steady irrotational flow of an ideal fluid whose density is ρ. For any two points (1 and 2) in the fluid, this equation relates the pressure P, the speed v, and the elevation y: $P_1 + \frac{1}{2}\rho v_1^2 + \rho g y_1 = P_2 + \frac{1}{2}\rho v_2^2 + \rho g y_2$. When the flow is horizontal, Bernoulli's equation indicates that higher fluid speeds are associated with lower fluid pressures.

The **coefficient of viscosity** η is the proportionality constant that determines how much tangential force **F** is required to move a fluid layer at a constant speed v, when the layer has an area A and is located a perpendicular distance y from an immobile surface. The magnitude of the force is $F = \eta A v/y$. The SI unit of viscosity is Pa·s. To make a viscous fluid flow from location 2 to location 1 along a pipe of radius R and length L, the pressure P_2 at location 2 must exceed the pressure P_1 at location 1. **Poiseuille's law** gives the volume flow rate Q that results from such a pressure difference: $Q = \pi R^4(P_2 - P_1)/(8\eta L)$. Bernoulli's equation does not apply to viscous flow.

SOLVED PROBLEMS

Solved Problem 1   A Hollow Rock

Related Problems: *39   *40   **41

To verify his suspicion that a rock specimen is hollow, a geologist weighs the specimen in air and in water. He finds that the specimen weighs twice as much in air as it does in water. The solid part of the specimen has a density of 5.0×10^3 kg/m³. What is the fraction of the specimen's apparent volume that is solid?

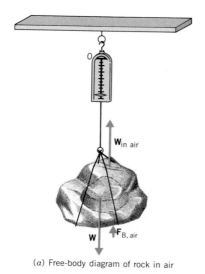

(a) Free-body diagram of rock in air

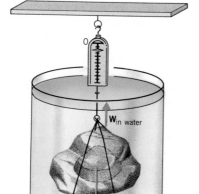

(b) Free-body diagram of rock in water

REASONING Parts a and b of the drawing show the free-body diagrams that correspond to the rock being weighed in air and in water. These two situations can be analyzed in exactly the same fashion. In each, the measured weight is the true weight W of the specimen diminished by the buoyant force of the surrounding fluid, air in one case and water in the other. The buoyant force $F_{B, \text{air}}$ of the air can be ignored (the weight of the displaced air is small compared to the weight of

the specimen), but the buoyant force $F_{B, \text{water}}$ of the water must be taken into account (the weight of the displaced water is not small).

SOLUTION According to Archimedes' principle, the buoyant force of the water is the weight of the water displaced by the total volume of the rock ($V_R + V_H$), where V_R denotes the volume of the solid part of the rock and V_H the hollow part:

$$F_{B, \text{water}} = \rho_{\text{water}}(V_R + V_H)g$$

As usual, we have used Equation 11.1 to express the mass as density times volume and then multiplied the mass by the acceleration due to gravity to obtain the weight. The weight of the rock in air $W_{\text{in air}}$ and the weight in water $W_{\text{in water}}$ are

[In air] $\quad W_{\text{in air}} = W - F_{B, \text{air}} = W$

[In water] $\quad W_{\text{in water}} = W - F_{B, \text{water}}$

$$= W - \rho_{\text{water}}(V_R + V_H)g$$

The weight in air is twice the weight in water, so that

$$W = 2[W - \rho_{\text{water}}(V_R + V_H)g]$$
$$W = 2\rho_{\text{water}}(V_R + V_H)g$$

The next step is to write the true weight W of the specimen as the sum of the weights of its parts, the solid rock (density $= \rho_R$) and the gas in the hollow space (density $= \rho_H$): $W = \rho_R V_R g + \rho_H V_H g$. Since the density of any gas in the hollow space is negligibly small, the true weight of the rock is simply $W = \rho_R V_R g$. With this value for W, the previous result can be used to find the fraction $V_R/(V_R + V_H)$ of the specimen that is solid:

$$\rho_R V_R g = 2\rho_{\text{water}}(V_R + V_H)g$$
$$\frac{V_R}{V_R + V_H} = \frac{2\rho_{\text{water}}}{\rho_R} = \frac{2(1.000 \times 10^3 \text{ kg/m}^3)}{5.0 \times 10^3 \text{ kg/m}^3} = \boxed{0.40}$$

SUMMARY OF IMPORTANT POINTS When an object is composed of two parts that have different densities, information about the individual parts can be obtained from the apparent weight of the object in a fluid. The apparent weight will be less than the true weight because of the buoyant force, which can be taken into account with the aid of Archimedes' principle. An object that floats in a fluid has a zero apparent weight, since the buoyant force balances the true weight. An expression for the apparent weight can often be used along with other known information to obtain the solution of a problem that deals with the individual parts of the object.

QUESTIONS

1. A pile of empty aluminum cans has a volume of 1.0 m³. The density of aluminum is 2700 kg/m³. Explain why the mass of the pile of cans is *not* equal to $\rho_{Al} V = (2700$ kg/m³)(1.0 m³) = 2700 kg.

2. The part of a magnetic phonograph cartridge that rests on a rotating stereo record is called the stylus and is usually a small polished diamond. The tip (radius $\approx 1.8 \times 10^{-5}$ m) of this diamond touches the record. Explain why the pressure that the stylus applies to the record is large, even though the force the stylus applies to the record is quite small.

3. As you climb a mountain, your ears "pop" because of the changes in atmospheric pressure. In which direction does your eardrum move (a) as you climb up and (b) as you climb down? Give your reasoning.

4. A bottle of fruit juice is sealed under partial vacuum, with a lid on which a red dot or "button" is painted. Around the button the following phrase is printed: "Button pops up when seal is broken." Explain why the button remains pushed in when the seal is intact.

5. A sealed container of water is full, except for a tall thin tube that is attached to it, as the drawing shows. Water is poured into this tube, and before the tube is full the container bursts. Explain why.

6. The drawing shows two tanks at a marine exhibit, each with identical observation windows. Each tank is filled to the top with seawater and exposed to the air, but one contains a much greater volume of water than the other, as shown. Compare the total force that each window must sustain. Account for your answer.

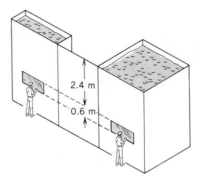

7. Why does the cork fly out with a loud "pop" when a bottle of champagne is opened? To answer this question on an exam a student says, "Because the gas pressure in the bottle is about 0.3×10^5 Pa." Explain whether the student is referring to absolute or gauge pressure.

8. A scuba diver is below the surface of the water when a storm approaches, dropping the air pressure above the water. Would a sufficiently sensitive pressure gauge attached to his wrist register this drop in air pressure? Give your reasoning.

9. Is either one (or both) of the following statements true? (1) Any object that floats in mercury also floats in water. (2) Any object that floats in water also floats in mercury. Justify your answers.

10. A glass beaker, filled to the brim with water, is resting on a scale. A block of wood is carefully placed in the water and floats. The water that spills over the beaker is wiped away, and the beaker is still filled to the brim. How do the initial and final readings on the scale compare? Explain.

11. A glass of water has an ice cube floating in it. The glass is filled to the brim. When the ice cube melts, will the water level drop, remain the same, or rise, causing water to spill out? Give your reasoning.

12. A solid 10.0-kg sphere and a solid 10.0-kg cube are completely immersed in the same liquid. Which object (if either) experiences the larger buoyant force (a) when each object is made from lead and (b) the sphere is made from lead and the cube is made from aluminum? Justify your answers.

13. Could there be any truth to the statement that it is easier to float (and, hence, swim) in salty ocean water than in fresh water? Explain.

14. In steady flow, the velocity **v** of a fluid particle at any point is constant in time. On the other hand, a fluid accelerates when it moves into a region of smaller cross-sectional area. (a) Explain what causes the acceleration. (b) Explain why the condition of steady flow does not rule out such acceleration.

15. The cross-sectional area of a stream of water becomes smaller as the water falls from a faucet. Account for this phenomenon in terms of the equation of continuity. What would you expect to happen to the cross-sectional area when the water is shot upward, as it is in a fountain?

16. Suppose you are driving your car along side a moving truck. Using your knowledge of the relation between air speed and pressure, determine whether the truck and car will tend to pull together or push apart. Give your reasoning.

17. Can Bernoulli's equation describe the flow of water that is cascading down a rock-strewn spillway? Explain.

18. Hold two sheets of paper loosely together, one on top of the other, and blow air between them. Instead of being blown further apart, the two sheets will come together. Discuss this phenomenon in terms of Bernoulli's equation.

19. Which way would you have to spin a baseball, so that it curves upward on its way to the plate? In describing the spin, state how you are viewing the ball. Justify your answer.

20. A passenger is smoking a cigarette in the backseat of a moving car. To remove the smoke, the driver opens a window just a bit. Explain why the smoke is drawn to and out of the driver's window.

21. The airport in Phoenix, Arizona, has occasionally been closed to large planes because of weather conditions that do not entail storms or low visibility. Instead, the conditions combine to create an unusually low air density. Using Bernoulli's equation, explain why such an air density would make it difficult for a large, heavy plane to take off, especially if the runway were not exceptionally long.

22. When a tarpaulin-covered semitrailer speeds down the highway, the canvas over the cargo area bulges outward. Account for the outward bulge.

23. To change the oil in a car, you remove a plug beneath the engine and let the old oil run out. Your car has been sitting in the garage on a cold day. Before changing the oil, it is advisable to run the engine for a while. Why?

PROBLEMS

Section 11.1 Mass Density

1. A water bed has dimensions of $1.83 \text{ m} \times 2.13 \text{ m} \times 0.229 \text{ m}$. The floor of the bedroom will tolerate an additional load of no more than 6660 N (1500 lb). Find the weight of the bed and determine whether it should be purchased.

2. Numerous jewelry items of solid silver are melted down and cast into a solid circular disk that is 0.0200 m thick. The total mass of the jewelry is 10.0 kg. Find the radius of the disk.

3. A pirate in a movie is carrying a chest ($0.30 \text{ m} \times 0.30 \text{ m} \times 0.20 \text{ m}$) that is supposed to be filled with gold. To see how ridiculous this is, determine the weight of the gold that this fellow is carrying. To judge how large this weight is, remember that 1 N = 0.225 lb.

4. The *karat* is a dimensionless unit that is used to indicate the proportion of gold in a gold-containing alloy. An alloy that is one karat gold contains a weight of pure gold that is one part in twenty-four. What is the volume of gold in a 14.0-karat gold necklace whose weight is 1.00 N?

***5.** An irregularly shaped chunk of concrete has a hollow spherical cavity inside. The mass of the chunk is 33 kg, and the volume enclosed by the outside surface of the chunk is 0.025 m³. What is the radius of the spherical cavity?

***6.** Planners of an experiment are evaluating the design of a helium-filled (0 °C, 1 atm pressure) sphere of radius R. Ultra-thin silver foil of thickness T is used to make the sphere, and the designers claim that the mass of helium in the sphere equals the mass of silver used. Assuming that T is much less than R, calculate the ratio T/R for such a sphere.

***7.** An antifreeze solution is made by mixing ethylene glycol ($\rho = 1116 \text{ kg/m}^3$) with water. Suppose the specific gravity of such a solution is 1.0730. Assuming that the total volume of the solution is the sum of its parts, determine the volume percentage of ethylene glycol in the solution.

Section 11.2 Pressure

8. A rectangular table measures 0.80 m by 1.2 m. What is the magnitude and direction of the force that atmospheric pressure applies to (a) the top surface and (b) the bottom surface of the table?

9. The circular top of a can of soda has a radius of 0.0320 m. The pull-tab has an area of $3.80 \times 10^{-4} \text{ m}^2$. The absolute pressure of the carbon dioxide in the can is 1.40×10^5 Pa. Find the force that this gas generates (a) on the top of the can (including the pull-tab) and (b) on the pull-tab itself.

10. High-heeled shoes can cause tremendous pressure to be applied to a floor. Suppose the radius of a heel is 6.00×10^{-3} m. At times during a normal walking motion, nearly the entire body weight acts perpendicular to the surface of such a heel. Find the pressure that is applied to the floor under the heel because of the weight of a 50.0-kg woman.

11. A glass bottle of soda is sealed with a screw cap. The absolute pressure of the carbon dioxide in the bottle is

1.80×10^5 Pa. Assuming that the top and bottom surfaces of the cap each have an area of 5.20×10^{-4} m^2, obtain the force that the screw threads must exert to keep the cap on the bottle.

12. A brick weighs 17.8 N and is resting on the ground. The dimensions of the brick are 0.203 m $\times$ 0.0890 m $\times$ 0.0570 m. A number of the bricks are then stacked on top of this one. What is the smallest number of whole bricks (including the one on the ground) that could be used, so that their weight creates a pressure of at least one atmosphere on the ground beneath the first brick? (*Hint: First decide which face of the brick is in contact with the ground.*)

*13. A 75.0-kg person is sitting on a beach ball of negligible weight, as the drawing shows. Suppose the absolute pressure in the ball is 1.20×10^5 Pa. Determine the radius of the circular area where the beach ball is in contact with the ground.

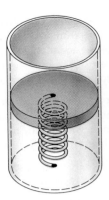

Problem 14

body diagram of the piston as the first step in your solution, and then find the pressure of the gas.

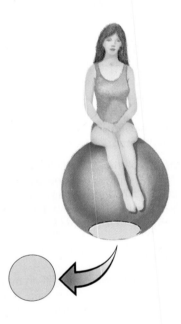

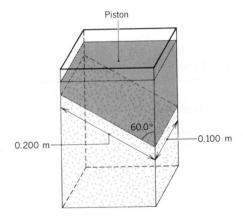

Piston

0.200 m 60.0° 0.100 m

Section 11.3 The Relation between Pressure and Depth in a Static Fluid, Section 11.4 Pressure Gauges

16. The deep end of a swimming pool has a depth of 2.00 m. The atmospheric pressure above the pool is 1.01×10^5 Pa. What is the pressure at the bottom of the pool?

17. Express a pressure of 450 mm of mercury in Pa.

18. The Mariana trench is located in the Pacific Ocean and has a depth of approximately 11 000 m. The density of seawater is 1025 kg/m^3. (a) If a diving chamber were to explore such depths, what force would the water exert on the chamber's observation window (radius = 0.10 m)? (b) For comparison, determine the weight of a jetliner whose mass is 1.2×10^5 kg.

19. A meat baster consists of a squeeze bulb attached to a plastic tube. When the bulb is squeezed and released, with the open end of the tube under the surface of the basting sauce, the sauce rises in the tube and can then be squirted over the

*14. A cylinder is fitted with a piston, beneath which is a spring, as in the drawing. Friction is absent. The spring constant of the spring is 2900 N/m. The piston has a negligible mass and a radius of 0.030 m. (a) When air beneath the piston is completely pumped out, how much does the atmospheric pressure cause the spring to compress? (b) How much work does the atmospheric pressure do in compressing the spring?

*15. The piston chamber shown in the drawing is oriented vertically. The piston itself has a mass of 10.0 kg, fits into the chamber without friction, and is in equilibrium in the position shown. The piston surface in contact with the gas in the chamber is a rectangle whose dimensions are given in the drawing. There is no air above the piston. Construct a free-meat. Suppose water rises 0.15 m in the tube when this device is tested (see drawing). (a) Find the absolute pressure in the bulb, assuming that atmospheric pressure has the value of 1.013×10^5 Pa. (b) Would you expect this device to work better or worse on top of a mountain? Explain.

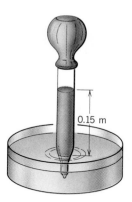

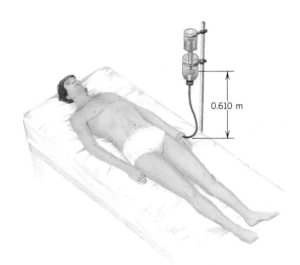

Problem 21

20. A water tower is a familiar sight in many towns. The purpose of such a tower is to provide storage capacity and to provide sufficient pressure in the pipes that deliver the water to customers. The drawing shows a spherical reservoir that contains 5.25×10^5 kg of water when full. The reservoir is vented to the atmosphere at the top. For a full reservoir, find the gauge pressure that the water has at the faucet in (a) house A and (b) house B. Ignore the diameter of the delivery pipes.

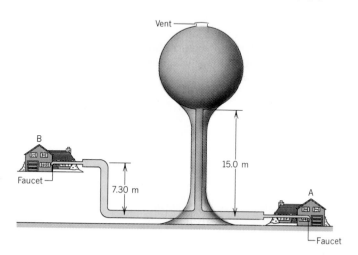

21. The drawing shows a setup for intravenous feeding. With the distance shown, nutrient solution ($\rho = 1030$ kg/m³) can just barely enter the blood in the vein. What is the gauge pressure of the venous blood? Express your answer in millimeters of mercury.

*22. A mercury barometer reads 750.0 mm on the roof of a building and 760.0 mm on the ground. Assuming a constant value of 1.29 kg/m³ for the density of air, determine the height of the building.

*23. A 1.00-m-tall container is filled to the brim, part way with mercury and the rest of the way with water. The container is open to the atmosphere. What must be the depth of each layer, so the absolute pressure on the bottom of the container is twice the atmospheric pressure?

**24. Two identical containers are open at the top and are connected at the bottom via a tube of negligible volume and a valve which is closed. Both containers are filled initially to the same height of 1.00 m, one with water, the other with mercury, as the drawing indicates. The valve is then opened. Water and mercury are immiscible. Determine the fluid level in each container when equilibrium is reestablished.

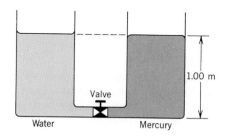

**25. As the drawing illustrates, a pond has the shape of an inverted cone with the tip sliced off and has a depth of 5.00 m. The atmospheric pressure above the pond is 1.01×10^5 Pa. The circular top surface (radius = R_2) and circular bottom surface (radius = R_1) of the pond are both parallel to the ground. The magnitude of the force acting on the top surface is the same as the magnitude of the force acting on the bottom surface. Obtain R_2 and R_1.

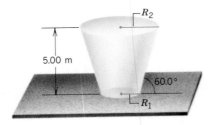

Section 11.5 Pascal's Principle

26. A hydraulic system operates the brakes on a car. The driver applies a force of $F_1 = 5.0$ N to the input piston by pressing the brake pedal. As a result, the output plunger applies a force $F_2 = 340$ N to each brake shoe. Ignore any difference in heights between the piston and plunger. What is the ratio A_2/A_1 of the plunger and piston areas?

27. The drawing shows a hydraulic press used in a trash compactor, where the radii of the input piston and the output plunger are 6.4×10^{-3} m and 5.1×10^{-2} m, respectively. If the height difference between the input piston and the output plunger can be neglected, what force is applied to the trash when the input force is 330 N?

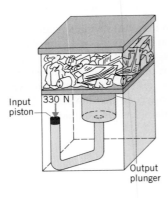

28. The atmospheric pressure above a swimming pool changes from 755 to 765 mm of mercury. The bottom of the pool is a 12 m $\times$ 24 m rectangle. By how much does the force on the bottom of the pool increase?

29. The drawing shows two blocks (masses m_1 and m_2) resting on hydraulic cylinders. The bottom surfaces of the cylinders are at the same level. The radius of the cylinder on the right is twice that of the cylinder on the left. Find the ratio m_2/m_1 of the masses.

*30. A dump truck uses a hydraulic cylinder, as the drawing illustrates. When activated by the operator, a pump injects hydraulic oil into the cylinder at an absolute pressure of 3.54×10^6 Pa and drives the output plunger, which has a

Problem 29

radius of 0.150 m. Assuming the plunger remains perpendicular to the floor of the load bed, find the torque that the plunger creates about the axis identified in the drawing.

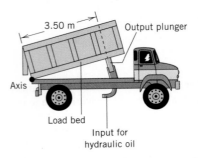

*31. The hydraulic oil in a car lift has a density of 8.00×10^2 kg/m³. The weight of the input piston is negligible. The radii of the input piston and output plunger are 8.00×10^{-3} m and 0.140 m, respectively. What input force **F** is needed to support the 22 300-N combined weight of a car and the output plunger, when the bottom surfaces of the piston and plunger are at (a) the same level and (b) the levels shown in the drawing?

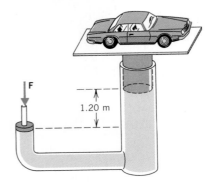

Section 11.6 Archimedes' Principle

32. A 0.10 m $\times$ 0.20 m $\times$ 0.30 m block is suspended from a wire and is completely under water. What buoyant force acts on the block?

33. A buoyant force of 26 N acts on a piece of quartz that is completely immersed in ethyl alcohol. What is the volume of the quartz?

34. Only a small part of an iceberg protrudes above the water, while the bulk lies below the surface. The density of ice is 917 kg/m³ and that of seawater is 1025 kg/m³. Find the percentage of the iceberg's volume that lies below the surface.

35. What is the total mass of swimmers that the raft in Example 7 can carry and float with its top surface at water level?

36. A hydrometer is a device used to measure the density of a liquid. The hydrometer is a cylindrical tube that is weighted at one end, so that it floats with the heavier end downward. As the drawing illustrates, the hydrometer is often contained inside a large "medicine dropper," into which the liquid is drawn using a squeeze bulb. The weighted hydrometer tube has a mass of 6.00×10^{-3} kg and a radius of 5.00×10^{-3} m. How far from the bottom of the tube should the mark be put that denotes (a) battery acid whose density is 1280 kg/m³ and (b) antifreeze solution whose density is 1073 kg/m³?

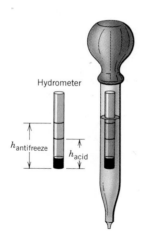

Hydrometer

$h_{antifreeze}$

h_{acid}

37. A person can change the volume of his body by taking air into his lungs. The amount of change can be determined by weighing the person under water. Suppose that under water a person weighs 20.0 N with partially full lungs and 40.0 N with empty lungs. Find the change in body volume.

*38. A spring has a spring constant of 578 N/m. When used to suspend an object in air, the spring stretches by 0.0640 m. When used to suspend the same object in water, the spring stretches by 0.0520 m. (a) What buoyant force acts on the object? (b) What is the volume of the part of the object that is covered by water?

*39. A 1967 Kennedy half-dollar has a mass of 1.150×10^{-2} kg. The coin is a mixture of silver and copper, and in water the coin weighs 0.1011 N. Determine the mass of silver in the coin. (See Solved Problem 1 for a related problem.)

*40. A paperweight is made of glass ($\rho = 2.60 \times 10^3$ kg/m³) and has a hollow cavity within it. In air, the paperweight weighs 12.0 N. In ethyl alcohol, the paperweight weighs 4.00 N more than it does in water. What percentage of the total volume of the paperweight is empty? (See Solved Problem 1 for a related problem.)

**41. One kilogram of glass ($\rho = 2.60 \times 10^3$ kg/m³) is shaped into a hollow spherical shell that just barely floats in water. What are the inner and outer radii of the shell? Do not assume the shell is thin. (See Solved Problem 1 for a related problem.)

**42. A cylinder (radius = 0.150 m, height = 0.120 m) has a mass of 7.00 kg. This cylinder is floating in water. Then oil ($\rho = 725$ kg/m³) is poured on top of the water until the situation shown in the drawing results. How much of the height of the cylinder is in the oil?

Oil

Water

Section 11.8 The Equation of Continuity

43. A patient recovering from surgery is being given fluid intravenously. The fluid has a density of 1030 kg/m³, and 9.5×10^{-4} m³ of it flow into the patient every six hours. Find the mass flow rate in kg/s.

44. Suppose that blood flows through the aorta with a speed of 0.35 m/s. The cross-sectional area of the aorta is 2.0×10^{-4} m². (a) Find the volume flow rate of the blood. (b) The aorta branches into tens of thousands of capillaries whose total cross-sectional area is about 0.28 m². What is the average blood speed through them?

45. Oil is flowing with a speed of 1.22 m/s through a pipeline with a radius of 0.305 m. How many gallons of oil (1 gal = 3.79×10^{-3} m³) flow in one day?

46. Water flows with a volume flow rate of 1.50 m³/s in a pipe. Find the speed of the water at a point where the pipe radius is 0.500 m.

*47. Three fire hoses are connected to a fire hydrant. Each hose has a radius of 0.020 m. Water enters the hydrant through an underground pipe of radius 0.080 m. In this pipe the water has a speed of 3.0 m/s. (a) How many kilograms of water are poured onto a fire in one hour? (b) Find the water speed in each hose.

Section 11.9 Bernoulli's Equation,
Section 11.10 Applications of Bernoulli's Equation

48. Prairie dogs are burrowing rodents. They do not suffocate in their burrows, because the effect of air speed on pressure creates sufficient air circulation. The animals maintain a difference in the shapes of the two entrances to the burrow, and because of this difference, the air ($\rho = 1.29$ kg/m^3) blows past the openings at different speeds, as the drawing indicates. Assuming that the openings are at the same vertical level, find the difference in air pressure between the openings and indicate which way the air circulates.

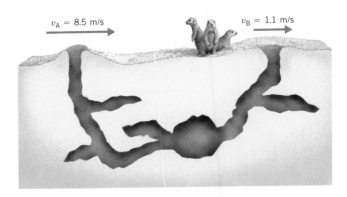

$v_A = 8.5$ m/s $v_B = 1.1$ m/s

49. Suppose that a 15-m/s wind is blowing across the roof of your house. The density of air is 1.29 kg/m^3. (a) Determine the reduction in pressure (below atmospheric pressure of stationary air) that accompanies this wind. (b) Explain why some roofs are "blown outward" when the wind speed is high.

50. A small crack occurs at the base of a 15.0-m-high dam. The effective crack area through which water leaves is 1.00×10^{-3} m^2. (a) Ignoring viscous losses, what is the speed of the water flowing through the crack? (b) How many cubic meters of water per second leave the dam?

51. Water flows at a speed of 0.500 m/s through a hose with a radius of 0.0200 m. The hose is horizontal. (a) At what speed does the water pass through a nozzle with an effective radius of 3.00×10^{-3} m? (b) What must be the absolute pressure of the water entering the hose if the pressure at the nozzle is atmospheric pressure?

52. A fountain sends a stream of water 5.00 m into the air. (a) Neglecting air resistance and any viscous effects, what must be the speed of the water at the point where the water leaves the pipe feeding the fountain? At that point, the pressure is atmospheric pressure. (b) The effective cross-sectional area of the pipe is 5.00×10^{-4} m^2. How many gallons per minute are being used by the fountain? (1 gal = 3.79×10^{-3} m^3)

53. Water flows downward within a pipe of constant cross-sectional area. At a point somewhere down the pipe, the speed of the water is 7.00 m/s and the absolute pressure is

1.50×10^5 Pa. At a height of 3.00 m above this point, find (a) the speed of the water and (b) the absolute pressure.

*54. An airplane wing is designed so that the speed of the air across the top of the wing is 248 m/s when the speed of the air below the wing is 225 m/s. The density of the air is 1.29 kg/m^3. What is the lifting force on a wing of area 20.0 m^2?

*55. Water is running out of a faucet, falling straight down, with an initial speed of 0.50 m/s. At what distance below the faucet is the radius of the stream reduced to one-half its value at the faucet?

*56. A water rocket is partially filled with water, as the drawing illustrates. The region above the water is filled with compressed air from a hand pump. The gauge pressure of the compressed air is 3.0×10^5 Pa. Neglecting the variation in pressure with water height and assuming that the area A_2 is much larger than A_1, find the speed v_1 at which the water leaves the nozzle.

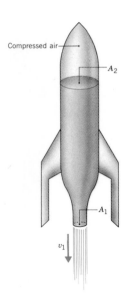

Compressed air

A_2

A_1

v_1

*57. In a closed tank, the absolute pressure of the air above the water is 6.01×10^5 Pa. The water leaves the bottom of the tank through a nozzle that is directed straight upward. The opening of the nozzle is 4.00 m below the surface of the water. (a) Find the speed at which the water leaves the nozzle. (b) Ignoring air resistance and viscous effects, determine the height to which the water rises.

*58. A liquid is flowing through a horizontal pipe whose radius is 0.0200 m. The pipe bends straight upward through a height of 10.0 m and joins another horizontal pipe whose radius is 0.0400 m. What volume flow rate will keep the pressures in the two horizontal pipes the same?

**59. Two circular holes, one larger than the other, are cut in the side of a large water tank whose top is open to the atmosphere. The center of one of these holes is located twice as far beneath the surface of the water as the other. The volume flow rate of the water coming out of the holes is the same. (a) Decide which hole is located where. (b) Calculate the ratio of the radius of the larger hole to the radius of the smaller hole.

**60. A siphon tube is useful for removing liquid from a tank. The siphon tube is first filled with liquid, and then one end is inserted into the tank. Liquid then drains out the other end, as the drawing illustrates. (a) Using reasoning similar to that employed in obtaining Torricelli's theorem, derive an expression for the speed v of the fluid emerging from the tube. This expression should give v in terms of the vertical height y and the acceleration due to gravity g. (Note that this speed does not depend on the depth d of the tube below the surface of the liquid.) (b) At what value of the vertical distance y will the siphon stop working? (c) Derive an expression for the absolute pressure at the highest point in the siphon (point A) in terms of the atmospheric pressure P_0, the fluid density ρ, g, and the heights h and y. (Note that the fluid speed at point A is the same as the speed of the fluid emerging from the tube, because the cross-sectional area of the tube is the same everywhere.)

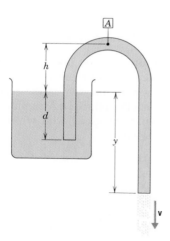

Section 11.11 Viscous Flow

61. A blood vessel is 0.10 m in length and has a radius of 1.5×10^{-3} m. Blood flows at a rate of 1.0×10^{-7} m³/s through this vessel. Determine the difference in pressure that must be maintained between the two ends of the vessel.

62. A pressure difference of 1.1×10^3 Pa is needed to drive water ($\eta = 1.0 \times 10^{-3}$ Pa·s) through a pipe whose radius is 6.4×10^{-3} m. The volume flow rate of the water is 3.2×10^{-4} m³/s. What is the length of the pipe?

63. Poiseuille's law remains valid so long as the fluid flow is laminar. For sufficiently high speed, however, the flow becomes turbulent, even if the fluid is moving through a smooth pipe with no restrictions. It is found experimentally that the flow is laminar as long as the Reynolds number Re is less than about 2000: Re $= 2\bar{v}\rho R/\eta$. Here $\bar{v}$, ρ, and η are, respectively, the average speed, density, and viscosity of the fluid, and R is the radius of the pipe. Calculate the highest average speed that blood ($\rho = 1060$ kg/m³) could have and still remain in laminar flow when it flows through the aorta ($R = 8.0 \times 10^{-3}$ m).

64. A certain volume of water ($\eta = 1.00 \times 10^{-3}$ Pa·s) is observed to flow through a tube in 115 s. The same volume of another liquid flows through the same tube, under the same conditions, in 175 s. What is the viscosity of the liquid?

*65. When an object moves through a fluid, as when a ball falls through air or a glass sphere falls through water, the fluid exerts a viscous force **F** on the object that tends to slow it down. For a small sphere of radius R, moving slowly with a speed v, the magnitude of the viscous force is given by Stoke's law, $F = 6\pi\eta R v$, where η is the viscosity of the fluid. (a) What is the viscous force on a glass sphere of radius $R = 1.0 \times 10^{-3}$ m falling through water ($\eta = 1.00 \times 10^{-3}$ Pa·s) when the sphere has a speed of 3.0 m/s? (b) The speed of the falling sphere increases until the viscous force balances the weight of the sphere. Thereafter, no net force acts on the sphere, and it falls with a constant speed called the terminal speed. If the sphere has a mass of 1.0×10^{-5} kg, what is its terminal speed?

*66. Two hoses are connected to the same outlet by means of a Y-connector. One hose has a radius that is 1.50 times larger than the other, but each has the same length. Find the ratio of the average speed of the water in the larger hose to that in the smaller hose.

ADDITIONAL PROBLEMS

67. At what depth beneath the surface of a lake is the absolute pressure three times the atmospheric pressure of 1.10×10^5 Pa that acts on the lake's surface?

68. The water tower in the drawing is drained by a pipe that extends to the ground. Assume that the flow is nonviscous. (a) What is the absolute pressure at point 1 if the valve is *closed*, assuming that the top surface of the water at point 2 is at atmospheric pressure? (b) What is the absolute pressure at point 1 when the valve is opened and the water is flowing? (c) Assuming the effective cross-sectional area of the valve opening is 2.00×10^{-2} m², find the volume flow rate at point 1.

69. What is the radius of a hydrogen-filled balloon that would carry a load of 5750 N when the density of air is 1.29 kg/m³?

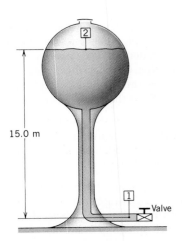

Problem 68

70. To remove water from a deep well, the pumping arrangement discussed in Example 5 cannot be used. Instead a submersible pump is put under the water at the bottom of the well and is used to push the water up through a pipe. What minimum output gauge pressure must the pump generate to make the water reach the nozzle at ground level, 71 m above the pump?

71. In ten minutes, 0.17 m³ of water pours into a tub through a pipe whose radius is 6.5×10^{-3} m. What is the speed of the water in the pipe?

72. The ice on a lake is 0.010 m thick. The lake is circular, with a radius of 480 m. Find the mass of the ice.

73. Accomplished silver workers in India can pound silver into incredibly thin sheets, as thin as 3.00×10^{-7} m (about one hundredth of the thickness of this sheet of paper). Find the number of square meters of such a sheet that can be formed from 1.00 kg of silver.

74. During a heavy rain, a 3.0 m × 4.6 m family room is flooded to a depth of 0.15 m. To remove the water ($\eta = 1.00 \times 10^{-3}$ Pa·s), a pump is used that does the job in two hours. The water flows through a horizontal pipe of radius 6.4×10^{-3} m and length 6.7 m. What gauge pressure does the pump produce?

75. Determine the buoyant force that the air applies (a) to a solid 16-kg piece of aluminum and (b) to a hollow cube, 0.80 m on a side, made from this piece of aluminum.

76. (a) A mercury barometer is used at a location where the atmospheric pressure is 1.013×10^5 Pa and the acceleration due to gravity is 9.78 m/s². How high (in mm) does the mercury rise in the barometer? (b) At another location the atmo-

spheric pressure is also 1.013×10^5 Pa, but the acceleration due to gravity is 9.83 m/s². How high (in mm) does the mercury rise here?

*77. A water line with an internal radius of 6.5×10^{-3} m is connected to a shower head that has 12 holes. The speed of the water in the line is 1.2 m/s. (a) What is the volume flow rate in the line? (b) At what speed does the water leave one of the holes (effective hole radius = 4.6×10^{-4} m) in the head?

*78. A cylinder (with circular ends) and a hemisphere are solid throughout and made from the same material. They are resting on the ground, the cylinder on one of its ends and the hemisphere on its flat side. The weight of each causes the same pressure to act on the ground. The cylinder is 0.500 m high. What is the radius of the hemisphere?

*79. What is the smallest integral number of logs ($\rho = 725$ kg/m³, radius = 0.0800 m, length = 3.00 m) that can be used to build a raft that will carry four people, each of whom has a mass of 80.0 kg?

*80. The vertical face of a reservoir dam is 120 m wide and 11 m high. Find the total force that the water in a completely full reservoir exerts on this vertical face. (*Hint: The pressure varies linearly with depth, so you must use an average pressure.*)

*81. A geologist finds a solid rock that is composed solely of quartz and gold. The rock has a mass of 12.0 kg and a volume of 4.00×10^{-3} m³. What mass of gold is contained in the rock?

*82. A pump draws water into a horizontal pipe located 12 m beneath the surface of a reservoir. The speed of the water in the pipe causes the pressure in the pipe to decrease, in accord with Bernoulli's principle. Assuming nonviscous flow, what is the maximum speed with which water can flow through the intake pipe?

**83. A spring is attached permanently to the bottom of an empty swimming pool, with the axis of the spring oriented vertically. An 8.00-kg block of wood ($\rho = 840$ kg/m³) is fixed to the top of the spring and compresses it. Then the pool is filled with water, completely covering the block. The spring is now observed to be stretched twice as much as it had been compressed. Determine the percentage of the block's total volume that is hollow. Ignore any air in the hollow space.

**84. A house has a roof (colored pink) with the dimensions shown in the drawing. Determine the magnitude and direction of the net force that the atmosphere applies to the roof, when the outside pressure rises suddenly by 10.0 mm of mercury before the pressure in the attic can adjust.

**85. The air speed of a plane can be measured with a Pitot-static tube, an example of which is illustrated in the drawing. The Pitot-static tube consists of two concentric tubes: The inner one is the static tube, and the outer one with holes in it is the Pitot tube. The difference in air pressure between the two

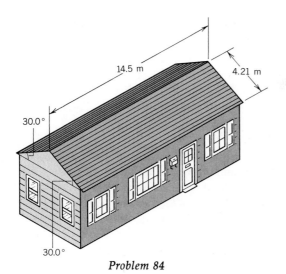

14.5 m

4.21 m

30.0°

30.0°

Problem 84

the air rushing past the holes in the Pitot tube has a high speed. (a) By applying Bernoulli's equation, show that the speed of the air is $v = \sqrt{2(P_2 - P_1)/\rho}$, where $P_2 - P_1$ is the difference in pressure and ρ is the air density at the altitude of the plane. (b) By expressing $P_2 - P_1$ in terms of the height h and density ρ_0 of the fluid in the U-tube, show that the air speed can be expressed as $v = \sqrt{2gh\rho_0/\rho}$.

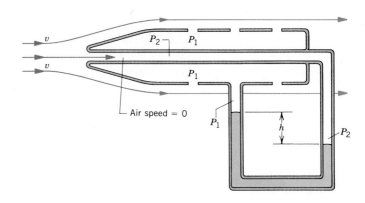

v

P_2 P_1

v

P_1

Air speed = 0

P_1

h

P_2

is measured by the U-tube manometer in the drawing. The air speed inside the static tube is zero, because the closed tube presents an immovable obstacle to the flow of air. In contrast,

Physics & Law Enforcement

Thousands of years ago, when people started living together in large communities, they found that they needed protection not only from outside groups of people but also from members of their own populations. As a result, police forces came into existence. Poverty, drug abuse, and other social factors conspire to make police an essential part of everyday life. Now, at the end of the twentieth century, with higher population densities than ever before and growing profits in crime, the need for effective protection has never been greater. Fortunately, applications of physics to criminology have greatly strengthened the hand of law enforcers.

Imagine that you are a police officer in a major city. It's a hot, muggy evening, and you have just received a call for a bank break-in that is in progress. The perpetrators were detected by a burglar alarm, sensitive to the heat given off by their bodies. The alarm senses electromagnetic waves in the infrared region of the spectrum. You type the bank's coordinates into the car's computer, which immediately displays the fastest route to the scene. You call the dispatcher on a scrambled, FM radio, which also uses electromagnetic waves. She arranges to have all the lights on the route set green. You arrive at the bank and check that your bulletproof, body armor is securely fastened. You put on infrared night vi-

sion equipment, and through it the streets light up as if it were noontime. Through a loudspeaker you call to the bank robbers. They do not respond, but the thermal and sound sensors convince you they are there. You fire in tear gas, they emerge, coughing and choking, and you arrest them without firing a shot. The infrared and sound sensors, FM radio, loudspeaker, and high-impact body armor are all based on physics principles.

Now you are on routine traffic patrol in an unmarked car, traveling westbound on the six-lane highway cutting through your city. Your radar unit is keeping track of the speed of the traffic. The radar uses the Doppler shift in the frequency of the emitted radar signal to measure the speed of vehicles as they approach or recede from you. By knowing your car's speed, the radar can indicate the true speed of the vehicles. Suddenly, you are passed by a red sports car traveling at 110 mph. While in pursuit, you call in the registration number and within minutes the dispatcher advises you that the car was stolen. By the time you catch up to the car, it is abandoned on a side street and the thief has fled. You call in a forensics team, whose members check it carefully for fingerprints and other clues to the thief's identity. The fingerprints are enlarged and FAXed via electromagnetic waves to headquarters, and the other material, including hair and mud samples, is taken for analysis. They will be scanned by a microscope and a chemical analyzer.

Now you are called to the scene of an automobile accident. One of the drivers is drunk and is being tested for alcohol in his blood. No one was seriously injured, but two cars were "totaled." The amount of damage indicates that at least one of the vehicles was traveling above the speed limit. To determine the speeds of the cars at the time of the accident, you measure the skid marks. When a car is moving normally, the static friction between its tires and the road prevents the tires from sliding. But when the breaks are locked at speed, the kinetic friction both slows the car and wears down the tire. With the aid of the equations of kinematics, it will be possible to use the length of the skid marks to estimate the speed of the car when the breaks were applied. The driver you suspect of being intoxicated had been going 30 mph over the limit.

The last call of the night is to a home in which a shot has been fired through a second story window. The homeowner suspects his neighbor of shooting at him. The bullet hole in the wall is below the hole in the window, and they do line up with the neighbor's house.

However, the difference in heights between where the bullet entered the room and where it embedded itself in the wall suggests that the shot came from farther away. Knowing the difference in heights and the type of bullet (i. e., its initial velocity), and assuming it was

Night vision equipment is becoming increasingly important in the field of law enforcement. One technique for enhancing night vision takes advantage of the infrared radiation emitted by objects. Another utilizes methods for detecting small amounts of visible light emitted by dimly lit objects.

shot from ground level, it will be possible to calculate the distance from which it was shot by using the equations of two-dimensional kinematics and the acceleration due to gravity. You get a ballistics expert and within the hour have identified a probable point of discharge. It happens to coincide with a known crack house.

Much of the basic physics in these essays has been applied to everyday life by engineers. The crucial point to remember is that our "high tech" society, with all its complex electrical, electronic, and mechanical devices, relies on an enormous foundation of basic physics to function and to grow. We invite you to think about how applied physics directly affects your life, from your kitchen, to your car, to your career.

FOR INFORMATION ABOUT CAREERS IN LAW ENFORCEMENT

Career Information Center, *Vol. 11, 4th edition, 1990, Glencoe/Macmillan, publishers, has descriptions of careers that include those of police officers, detectives, crime laboratory technicians, criminologists, FBI special agents, and other law enforcement personnel.*

Encyclopedia of Careers and Vocational Guidance, *Vol. 3, 8th edition, William E. Hopke, ed., 1990, J. G. Ferguson, publisher, has several in-depth descriptions of the working conditions that police, FBI, and other law enforcement officers encounter.*

You can also write to the following, among many others listed in the books above:

■ *Federal Bureau of Investigation, U. S. Department of Justice, Washington, DC 20535*

■ *American Sociological Association, 1722 N St., NW Washington, DC 20036.*

■ *International Association of Chiefs of Police, 13 Firstfield Rd., Gaithersburg, MD 20878*

Copyright © 1991 by Neil F. Comins.

PART TWO

THERMAL PHYSICS

We have now completed our discussion of mechanics, and the next four chapters deal with the study of thermal physics. In these chapters, the concepts of temperature and heat play important roles. We will see that temperature can be understood by using the ideas introduced in mechanics, except that now they will be applied at the microscopic level of atoms and molecules. And we will see that heat is a form of energy. In fact, thermal physics is actually the study of energy and energy transfer.

Heat plays a vital role in our lives. First and foremost, the human body must be maintained at a temperature near 98.6 °F. The purpose of clothing, as well as heating and air conditioning systems, is to help achieve the necessary temperature control by regulating the transfer of heat into and out of our bodies. Heat is important in the food we eat, for some foods must be cooked, while others must be refrigerated to keep them fresh. Heat is also involved in many other aspects of our lives, including, for example, the generation of electrical energy at power plants and the operation of automobile engines.

We will discuss the interesting things that can happen to an object when it is heated or cooled. It can, for instance, change its size, change from a solid into a liquid (as when an ice cube melts), or glow white-hot (as does the sun or an incandescent light bulb). We will also discuss how heat is transferred from one place to another by the processes of convection, conduction, and radiation.

Many devices, including heat engines, refrigerators, air conditioners, and heat pumps, operate by using heat and work simultaneously. A major portion of thermal physics is concerned with the interrelationship between heat and work and is called thermodynamics. The underlying basis for thermodynamics is found in several fundamental laws. Of these, one will be familiar from our study of mechanics. This particular one is known as the first law of thermodynamics and is a restatement of the principle of conservation of energy, as applied to heat and work. The other laws deal with temperature and heat and will be explained for the first time in this part of the text. As an outgrowth of the law known as the second law of thermodynamics, we will encounter the curious concept of entropy, which has fascinated scientists and nonscientists alike.

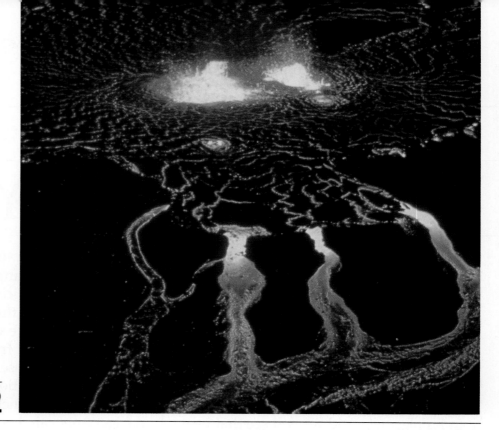

CHAPTER 12

TEMPERATURE AND HEAT

An erupting volcano is one of nature's most spectacular displays, and the flowing molten lava can be stunningly beautiful when viewed from afar. Up close, however, molten lava just presents the overwhelming sensation of being hot. To most people, the experience of being hot means that the temperature is uncomfortably high. When things are too hot, we think about cooling down by removing excess heat in some way. In fact, in common usage, the ideas of temperature and heat are closely interrelated and sometimes even used interchangeably. In this chapter, we will see that this common usage has arisen because the temperature of an object can often be made to change by the addition or removal of heat. However, we will also see that temperature and heat are different concepts and that it is not correct to use the terms interchangeably. Additional effects of temperature and heat that will be explored in this chapter include thermal expansion and the changes that can occur between the solid, liquid, and gaseous states of matter.

12.1 COMMON TEMPERATURE SCALES

To measure temperature we use a thermometer. Many thermometers make use of the fact that materials usually expand when their temperatures increase. Figure 12.1 shows the common mercury-in-glass thermometer, which utilizes the expansion of liquid mercury to indicate the temperature. The thermometer consists of a mercury-filled glass bulb connected to a capillary tube. When the mercury is heated, it expands into the capillary tube, the amount of expansion being proportional to the change in temperature. The outside of the glass is marked with an appropriate scale for reading the temperature.

A number of different temperature scales have been devised, two popular choices being the *Celsius* (formerly, centigrade) and *Fahrenheit scales.* Figure 12.1 illustrates these scales. Historically,* both scales were defined by assigning two temperature points on the scale and then dividing the distance between them into a number of equally spaced intervals. One point was chosen to be that at which ice melts under one atmosphere of pressure (the "ice point"), and the other was that at which water boils under one atmosphere of pressure (the "steam point"). On the Celsius scale, an ice point of 0 °C (0 degrees Celsius) and a steam point of 100 °C were selected. On the Fahrenheit scale, an ice point of 32 °F (32 degrees Fahrenheit) and a steam point of 212 °F were chosen. The Celsius scale is used worldwide, while the Fahrenheit scale is used mostly in the United States, often in home medical thermometers.

There is a subtle difference in the way the temperature of an object is reported, as compared to a *change* in its temperature. For example, the temperature of the human body is about 37 °C, where the symbol °C stands for "degrees Celsius." However, the *change* between two temperatures is specified in "Celsius degrees" (C°)—not in "degrees Celsius." Thus, if the body temperature rises to 39 °C, the change in temperature is 2 Celsius degrees or 2 C°, not 2 °C.

As Figure 12.1 indicates, the separation between the ice and steam points on the Celsius scale is divided into 100 Celsius degrees, while on the Fahrenheit scale the separation is divided into 180 Fahrenheit degrees. Therefore, the size of the Celsius degree is larger than that of the Fahrenheit degree by a factor of $\frac{180}{100}$, or $\frac{9}{5}$. Examples 1 and 2 illustrate how to convert between the Celsius and Fahrenheit scales using this factor.

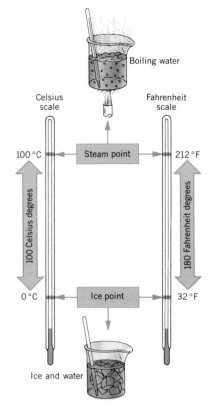

Figure 12.1 The Celsius and Fahrenheit temperature scales.

Example 1   Converting from a Fahrenheit to a Celsius Temperature

A healthy person has an oral temperature of 98.6 °F. What would this reading be on the Celsius scale?

REASONING AND SOLUTION   A temperature of 98.6 °F is 66.6 Fahrenheit degrees above the ice point of 32.0 °F. Since 1 C° = $\frac{9}{5}$ F°, the difference of 66.6 F° is equivalent to

$$(66.6 \text{ F}°) \frac{(1 \text{ C}°)}{(\frac{9}{5} \text{ F}°)} = 37.0 \text{ C}°$$

Adding 37.0 Celsius degrees to the ice point of 0 °C on the Celsius scale gives a Celsius temperature of $\boxed{37.0 \text{ °C}}$.

---

* Today, the Celsius and Fahrenheit scales are defined in terms of the Kelvin temperature scale; Section 12.2 discusses the Kelvin scale.

Example 2   Converting from a Celsius to a Fahrenheit Temperature

A time and temperature sign on a bank indicates the outdoor temperature is −20.0 °C. Find the corresponding temperature on the Fahrenheit scale.

REASONING AND SOLUTION   The temperature of −20.0 °C is 20.0 Celsius degrees *below* the ice point of 0 °C. This number of Celsius degrees corresponds to

$$(20.0 \text{ C}°) \frac{(\frac{9}{5} \text{ F}°)}{(1 \text{ C}°)} = 36.0 \text{ F}°$$

Subtracting 36.0 Fahrenheit degrees from the ice point of 32.0 °F on the Fahrenheit scale gives a Fahrenheit temperature of $\boxed{-4.0 \text{ °F}}$.

The procedure used in Examples 1 and 2 for converting between the Celsius and Fahrenheit scales can be summarized as follows:

1. Determine the difference between the stated temperature and the ice point on the given scale.
2. Convert this temperature difference from one scale to the other scale by using the fact that one Celsius degree is $\frac{9}{5}$ larger than one Fahrenheit degree.
3. Add or subtract the temperature difference on the new scale to or from the ice point on the new scale.

12.2 THE KELVIN TEMPERATURE SCALE

THE SCALE ITSELF

Although the Celsius and Fahrenheit scales are widely used, the *Kelvin temperature scale* has greater scientific significance. It was introduced by the Scottish physicist William Thompson (Lord Kelvin, 1824–1907), and in his honor each degree on the scale is called a kelvin (K). By international agreement, the symbol K is not written with a degree sign (°), nor is the word "degrees" used when quoting temperatures. For example, a temperature of 300 K (not 300 °K) is read as "three hundred kelvins," not "three hundred degrees kelvin." Because temperature cannot be expressed in terms of the three SI base units for mass (kilogram), length (meter), and time (second), it is necessary to define a fourth base unit for temperature measurement. The kelvin is the SI base unit for temperature.

The size of one kelvin is identical to that of one Celsius degree, for there are one hundred divisions between the ice and steam points on both scales. As we will discuss shortly, experiments have shown that there exists a lowest possible temperature, below which no substance can be cooled. This lowest temperature is defined to be the zero point on the Kelvin scale, and is referred to as absolute zero. Moreover, the ice point (0 °C) occurs at 273.15 K on the Kelvin scale. Thus, the Kelvin temperature T and the Celsius temperature T_c are related by

$$T = T_c + 273.15 \tag{12.1}$$

Figure 12.2 compares the Kelvin and Celsius scales.

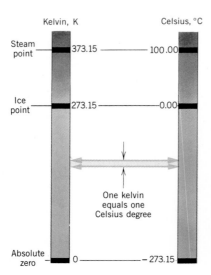

Figure 12.2   A comparison of the Kelvin and Celsius temperature scales.

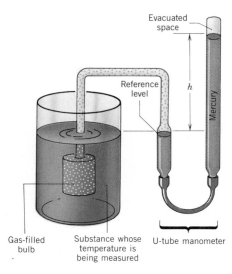

Figure 12.3 A constant-volume gas thermometer.

THE CONSTANT-VOLUME GAS THERMOMETER

When a gas is heated, its pressure increases, and when a gas is cooled, its pressure decreases, assuming the gas is confined to a fixed volume. For example, the air pressure in automobile tires can rise by as much as 20% after the car has been driven a few miles and the tires become warm. The change in gas pressure with temperature is the basis for the *constant-volume gas thermometer.*

A constant-volume gas thermometer consists of a gas-filled bulb to which a pressure gauge is attached, as in Figure 12.3. The gas is often hydrogen or helium at a low density, and the pressure gauge can be a U-tube manometer filled with mercury. The bulb is placed in thermal contact with the substance whose temperature is being measured. The volume of the gas is held constant by raising or lowering the *right* column of the U-tube manometer in order to keep the mercury level in the *left* column at the same reference level. The pressure of the gas is proportional to the height h of the mercury on the right. As the temperature changes, the pressure changes and can be used to indicate the temperature, once the constant-volume gas thermometer has been calibrated.

ABSOLUTE ZERO

Suppose the pressure of the gas in Figure 12.3 is measured at different temperatures. If the results are plotted on a pressure versus temperature graph, a straight line is obtained, as in Figure 12.4. If the straight line is extended or extrapolated to lower and lower temperatures, the line crosses the temperature axis at −273.15 °C. In reality, no gas can be cooled to this temperature, because all gases liquify before reaching it. However, helium and hydrogen liquify at such low temperatures that they are often used in the thermometer. This kind of graph can be obtained for different amounts of low-density gas and for different types of gas. In all cases, it is found that the straight line extrapolates back to −273.15 °C on the temperature axis, which suggests that the value of −273.15 °C has fundamental significance. The significance of this number is that it is the *absolute zero point* for temperature measurement. The phrase "absolute zero" means that temperatures lower than −273.15 °C cannot be reached by continually cooling a

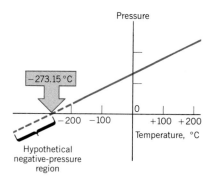

Figure 12.4 A plot of pressure versus temperature for a low-density gas at constant volume. The graph is a straight line and, when extrapolated (dashed line), crosses the temperature axis at −273.15 °C.

Table 12.1 Temperatures of Various Phenomena

Temperature (K)	Phenomenon
4.2	Helium liquifies
20	Hydrogen liquifies
77	Nitrogen liquifies
273	Water freezes
310	Human body temperature
373	Water boils
600	Lead melts
1336	Gold melts
6000	Surface temperature of the sun
16 000	Core temperature of the earth
10^7	Core temperature of the sun
10^9	Core temperature of hottest stars

gas or any other substance. If lower temperatures could be reached, then further extrapolation of the straight line in Figure 12.4 would suggest that negative absolute gas pressures could exist. Such a situation would be impossible, because a negative absolute gas pressure has no meaning. Thus, the Kelvin scale is chosen so that its zero temperature point is the lowest temperature attainable. Table 12.1 gives the Kelvin temperatures associated with various phenomena.

12.3 THERMOMETERS

All thermometers make use of the change in some physical property with temperature. A property that changes with temperature is called a ***thermometric property***. For example, the thermometric property of the mercury thermometer is the length of the mercury column, while in the constant-volume gas thermometer the thermometric property is the pressure of the gas. Several other thermometers and their thermometric properties are discussed below.

The *thermocouple* is a thermometer used extensively in scientific laboratories. The thermocouple consists of thin wires of different metals, welded together at the ends to form two junctions, as Figure 12.5 illustrates. Often the metals are copper and constantan (a copper–nickel alloy). One of the junctions, called the "hot" junction, is placed in thermal contact with the object whose temperature is being measured. The other junction, termed the "reference" junction, is kept at a known constant temperature (usually an ice–water mixture at 0 °C). The thermocouple generates a "voltage" that depends on the *difference in temperature* between the two junctions. This voltage is the thermometric property and is measured by a voltmeter, as the drawing indicates. With the aid of calibration tables, the temperature of the hot junction can be obtained from the voltage. Thermocouples are used to measure temperatures as high as 2300 °C or as low as −270 °C.

Most substances offer resistance to the flow of electricity. Because this electrical resistance changes with temperature, electrical resistance is another thermometric property. *Electrical resistance thermometers* are often made from platinum

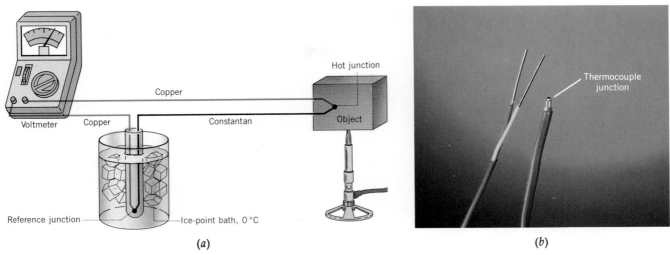

Figure 12.5 (*a*) A thermocouple is made from two different types of wires, copper and constantan in this case. (*b*) A thermocouple junction between two different wires.

wire, because platinum has excellent mechanical and electrical properties in the temperature range from -270 °C to $+700$ °C. Since the electrical resistance of platinum wire is known as a function of temperature, the temperature of a substance can be determined by placing the resistance thermometer in thermal contact with the substance and measuring the resistance of the platinum wire.

Radiation emitted by an object can also be used to indicate the temperature of the object. At low to moderate temperatures, the predominant type of radiation emitted is infrared radiation. As the temperature is raised, the intensity of the radiation increases substantially. In one interesting application, an infrared camera registers the intensity of the infrared radiation produced at different locations on the human body. The camera is connected to a color monitor that displays the different infrared intensities as different colors. This "thermal painting" of the body is called a *thermograph*. Thermography is an important diagnostic tool in medicine. For example, breast cancer may be indicated in a thermograph by the elevated temperatures often generated by malignant tissue. Figure 12.6 shows typical thermographs used to diagnose breast cancer. Figure 12.7 shows a ther-

THE PHYSICS OF . . .

thermography.

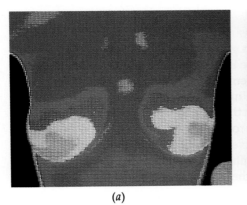 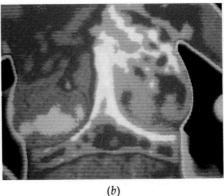

(*a*)   (*b*)

Figure 12.6 In these thermographs, blue represents the coolest and yellow/white the hottest temperatures. (*a*) Healthy breasts register a predominantly bluish color, indicating relatively cool temperatures. (*b*) The breast on the right in this thermograph has an invasive carcinoma (cancer) and registers colors from red to yellow/white, indicating markedly elevated temperatures.

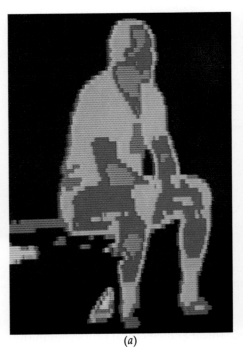 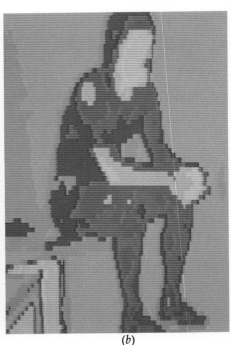

(a)   (b)

Figure 12.7 (*a*) A thermogram of a man immediately before playing a game of squash, indicating normal body temperatures. The color code runs from white/yellow at the hottest extreme, through red, blue, green, purple, to black, the coolest color. (*b*) A thermogram of a squash player cooling down just after a game. The player's face and arms are still hot (yellow) compared with the rest of his body.

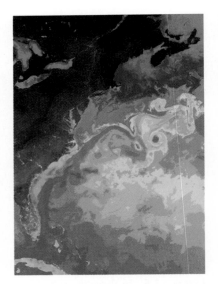

Figure 12.8 A thermogram of the eastern United States, Canada, and the Atlantic Ocean. The warmest water is red (about 80 °F), cooling to yellow, green, blue, and purple (about 40 °F) in the northern region.

mogram of a person before and after exercising, revealing regions of elevated temperatures on the body.

Oceanographers and meteorologists use thermographs extensively to map the temperature distribution on the surface of the earth. For example, Figure 12.8 shows a satellite image of the United States, Canada, and a portion of the Atlantic Ocean. Cuba and Florida are at the lower left of the picture and the Great Lakes are at the upper left. The image vividly shows the temperature variations in the ocean, from a warm 80 °F in the south to a cool 40 °F in the extreme north. The Gulf Stream is also evident in the picture, showing up as a deep red color. It moves up the eastern edge of Florida, Georgia, South Carolina, and North Carolina, before heading out into the Atlantic near the middle of the picture.

12.4 *LINEAR THERMAL EXPANSION*

NORMAL SOLIDS

Have you ever found the metal lid on a glass jar too tight to open? One way to loosen it is to run hot water over it. The lid loosens, because the metal expands more than the glass does. To varying extents, most materials expand when heated and contract when cooled. The increase in any one dimension of a solid is called *linear expansion*, linear in the sense that the expansion occurs along a line. Figure 12.9 illustrates the linear expansion of a rod whose length is L_0 when the temperature is T_0. When the temperature increases to $T_0 + \Delta T$, the length becomes $L_0 + \Delta L$, where ΔT and ΔL are the magnitudes of the changes in temperature and length, respectively. Conversely, when the temperature decreases to $T_0 - \Delta T$, the length decreases to $L_0 - \Delta L$.

The linear expansion of a solid depends on the amount of the temperature change. The greater the change in temperature, the greater the change in length. For modest temperature changes, experiments show that the change in length is directly proportional to the change in temperature ($\Delta L \propto \Delta T$). In addition, the change in length is proportional to the initial length of the rod, a fact that can be understood by considering Figure 12.10. Part *a* of the drawing shows two identical rods. Each rod has a length L_0 and expands by ΔL when the temperature increases by ΔT. Part *b* shows the two heated rods combined into a single rod, for which the total expansion is the sum of the expansions of each part, namely, $\Delta L + \Delta L = 2\,\Delta L$. Clearly, the amount of expansion doubles if the rod is twice as long to begin with. In other words, the change in length ΔL is directly proportional to the original length L_0 ($\Delta L \propto L_0$). Equation 12.2 expresses the fact that ΔL is proportional to both L_0 and ΔT ($\Delta L \propto L_0 \Delta T$) by using a proportionality constant α, which is called the *coefficient of linear expansion.*

Linear Thermal Expansion of a Solid

The length L_0 of an object changes by an amount ΔL when its temperature changes by an amount ΔT:

$$\Delta L = \alpha L_0\, \Delta T \qquad (12.2)$$

where α is the coefficient of linear expansion.

Common Unit for the Coefficient of Linear Expansion: $\dfrac{1}{C^\circ} = (C^\circ)^{-1}$

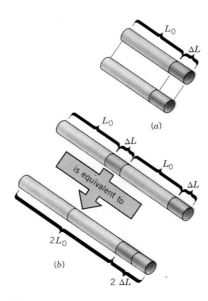

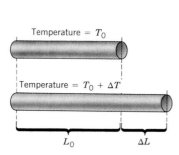

Figure 12.9 When the temperature of a rod is raised, the length of the rod increases.

Figure 12.10 (*a*) Each of the two rods expands by an amount ΔL when heated. (*b*) When the two rods are combined into a single rod of length $2L_0$, the "combined" rod expands by $2\,\Delta L$.

Table 12.2   Coefficients of Thermal Expansion for Solids and Liquids[a]

Substance	Coefficient of Thermal Expansion, $(C°)^{-1}$	
	Linear (α)	Volumetric (β)
Solids		
Aluminum	23×10^{-6}	69×10^{-6}
Brass	19×10^{-6}	57×10^{-6}
Concrete	12×10^{-6}	36×10^{-6}
Copper	17×10^{-6}	51×10^{-6}
Glass (common)	8.5×10^{-6}	26×10^{-6}
Glass (Pyrex)	3.3×10^{-6}	9.9×10^{-6}
Gold	14×10^{-6}	42×10^{-6}
Iron or steel	12×10^{-6}	36×10^{-6}
Lead	29×10^{-6}	87×10^{-6}
Nickel	13×10^{-6}	39×10^{-6}
Quartz (fused)	0.50×10^{-6}	1.5×10^{-6}
Silver	19×10^{-6}	57×10^{-6}
Liquids[b]		
Benzene	—	1240×10^{-6}
Carbon tetrachloride	—	1240×10^{-6}
Ethyl alcohol	—	1120×10^{-6}
Gasoline	—	950×10^{-6}
Mercury	—	182×10^{-6}
Methyl alcohol	—	1200×10^{-6}
Water	—	207×10^{-6}

[a] The values for α and β pertain to a temperature near 20 °C.

[b] Since liquids do not have fixed shapes, the coefficient of linear expansion is not defined for them.

Solving Equation 12.2 for α shows that $\alpha = \Delta L/(L_0 \, \Delta T)$. Since the length units of ΔL and L_0 cancel, the coefficient of linear expansion α has the unit of $(C°)^{-1}$ when the temperature difference ΔT is expressed in Celsius degrees (C°). Different materials with the same initial length expand and contract by different amounts as the temperature changes, so the value of α depends on the nature of the material. Table 12.2 shows some typical values. A bar of silver expands 38 times more than a similar bar of fused quartz. Thus, silver has a value for α that is 38 times greater than that for fused quartz. Coefficients of linear expansion vary somewhat depending on the range of temperatures involved, but the values in Table 12.2 are adequate approximations. Example 3 deals with a situation where a dramatic effect due to thermal expansion can be observed, even though the change in temperature is small.

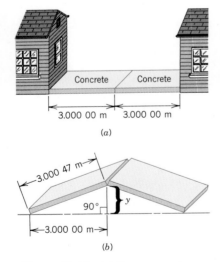

Figure 12.11   (a) Two concrete slabs completely fill the space between the buildings. (b) When the temperature increases, each slab expands in length, causing the sidewalk to buckle.

Example 3   Buckling of a Sidewalk

A concrete sidewalk is constructed between two buildings on a day when the temperature is 25 °C. The sidewalk consists of two slabs, each three meters in length and of negligible thickness (Figure 12.11a). As the temperature rises to 38 °C, the slabs expand, but no space is provided for thermal expansion. The buildings do not move, so the slabs buckle upward. Determine the vertical distance y in part b of the drawing.

REASONING AND SOLUTION   The change in length of each slab can be determined by using the coefficient of linear expansion for concrete given in Table 12.2 and noting that the change in temperature is 13 C°:

$$\Delta L = \alpha L_0\, \Delta T = [12 \times 10^{-6}\ (\text{C}°)^{-1}](3.0\ \text{m})(13\ \text{C}°) = 0.000\ 47\ \text{m} \qquad (12.2)$$

The expanded length of each slab is 3.000 47 m. The vertical distance *y* can be obtained by applying the Pythagorean theorem to the right triangle in part *b* of the drawing:

$$y = \sqrt{(3.000\ 47\ \text{m})^2 - (3.000\ 00\ \text{m})^2} = \boxed{0.053\ \text{m}}$$

The buckling of a sidewalk is one consequence of not providing sufficient room for thermal expansion, and Figure 12.12*a* shows another. It is common for builders to incorporate expansion joints or spaces at intervals along railroad tracks and bridge roadbeds to alleviate such problems. Part *b* of the figure shows such an expansion joint in a bridge.

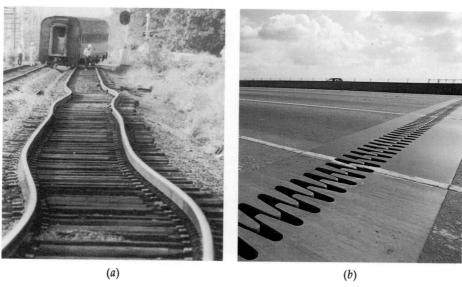

(a)   (b)

Figure 12.12   (*a*) The rails buckled because inadequate allowance was made for thermal expansion. (*b*) An expansion joint in a bridge.

While Example 3 shows how thermal expansion can cause problems, there are also times when it can be useful. For instance, each year thousands of children are taken to emergency rooms suffering from burns caused by scalding tap water. Such accidents can be reduced with the aid of thermal expansion, as illustrated in the antiscalding device shown in Figure 12.13. This device screws onto the end of a faucet and quickly shuts off the flow of water when it becomes too hot. As the water temperature rises, the actuator spring expands and pushes the plunger forward, shutting off the flow. When the water cools, the spring retracts and the water flow resumes.

THERMAL STRESS

If the concrete slabs in Figure 12.11 had not buckled upward, they would have been subjected to immense forces from the buildings. The forces needed to keep a solid object from expanding must be strong enough to counteract any change in

THE PHYSICS OF . . .

an antiscalding device.

Movable plunger   Activator spring

Water flow

Figure 12.13   An antiscalding device.

length that would occur due to a change in temperature. Although the change in temperature may be small, the forces—and hence the stresses—can be enormous, as Example 4 illustrates.

THE PHYSICS OF . . .

thermal stress.

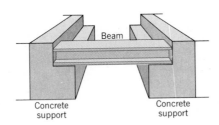

Figure 12.14 A steel beam is mounted between concrete supports with no room provided for thermal expansion of the beam.

Example 4   The Stress on a Steel Beam

A steel beam is used in the roadbed of a bridge. The beam is mounted between two concrete supports when the temperature is 23 °C, with no room provided for thermal expansion (Figure 12.14). What compressional stress must the concrete supports apply to each end of the beam, if they are to keep the beam from expanding when the temperature rises to 42 °C?

REASONING   Recall from Section 10.2 that the stress (force per unit cross-sectional area) required to change the length L_0 of an object by an amount ΔL is

$$\text{Stress} = Y \frac{\Delta L}{L_0} \qquad (10.1)$$

where Y is Young's modulus. If the steel beam were free to expand because of the change in temperature, the length would change by $\Delta L = \alpha L_0 \, \Delta T$. Because the concrete supports do not permit any expansion, they must supply a stress to compress the beam. Thus,

$$\text{Stress} = Y \frac{\Delta L}{L_0} = Y \frac{\alpha L_0 \, \Delta T}{L_0} = Y \alpha \, \Delta T$$

SOLUTION   Young's modulus and the coefficient of linear expansion for steel are $Y = 2.0 \times 10^{11}$ N/m² (Table 10.1) and $\alpha = 12 \times 10^{-6}$ (C°)⁻¹, respectively. The change in temperature from 23 to 42 °C is $\Delta T = 19$ C°. The thermal stress is

$$\text{Stress} = Y\alpha \, \Delta T = (2.0 \times 10^{11} \text{ N/m}^2)[12 \times 10^{-6} \text{ (C°)}^{-1}](19 \text{ C°}) = \boxed{4.6 \times 10^7 \text{ N/m}^2}$$

This stress is enormous. If the beam has a cross-sectional area of $A = 0.10$ m², the force applied to each end by a concrete support is $F = (\text{Stress})A = 4.6 \times 10^6$ N (over one million pounds).

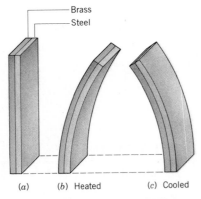

Figure 12.15   (a) A bimetallic strip and its behavior when (b) heated and (c) cooled.

(a)   (b) Heated   (c) Cooled

THE BIMETALLIC STRIP

A *bimetallic strip* is made from two thin strips of metal that have *different* coefficients of linear expansion, as Figure 12.15a shows. Often brass [$\alpha = 19 \times 10^{-6}$ (C°)⁻¹] and steel [$\alpha = 12 \times 10^{-6}$ (C°)⁻¹] are selected. The two pieces are welded or riveted together. When the bimetallic strip is heated, the brass, having the larger value of α, expands more than the steel. Since the two metals are bonded together, the bimetallic strip bends into an arc as in part b, with the longer brass piece having a larger radius than the steel piece. When the strip is cooled, the bimetallic strip bends in the opposite direction, as in part c.

Bimetallic strips are frequently used as adjustable automatic switches in electrical appliances. Figure 12.16 shows an automatic coffee maker that turns off when the coffee is brewed to the selected strength. In part a, while the brewing cycle is on, electricity passes through the heating coil that warms the water. The electricity can flow, because the contact mounted on the bimetallic strip touches the contact mounted on the "strength" adjustment knob, thus providing a continuous path for the electricity. When the bimetallic strip gets hot enough to bend away, as in part b of the drawing, the contacts separate. The electricity stops,

THE PHYSICS OF . . .

an automatic coffee maker.

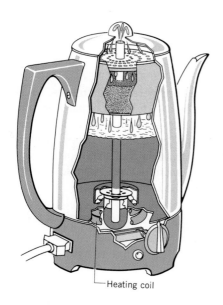

Heating coil

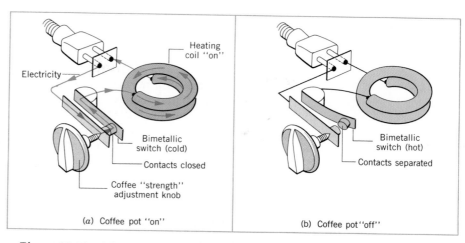

(a) Coffee pot "on"

(b) Coffee pot "off"

Figure 12.16 A bimetallic strip controls the brewing time on this automatic coffee maker. (a) When the bimetallic strip is cold, electricity flows through the heater. (b) When sufficiently hot, the bimetallic strip bends away from the control knob contact, thus turning off the electricity.

because it no longer has a continuous path along which to flow, and the brewing cycle is shut off. Turning the "strength" knob adjusts the brewing time by adjusting the distance through which the bimetallic strip must bend for the contact points to separate.

THE EXPANSION OF HOLES

An interesting example of linear expansion occurs when there is a hole in a piece of solid material. For example, when a circular ring is heated, does the hole itself expand or contract? Surprisingly, the hole expands. One way to visualize how the expansion arises is to imagine that the unheated ring is cut and straightened out to form a linear strip (Figure 12.17a). The strip is then heated and expands. When the expanded strip is rolled up again, it forms a larger ring with a larger hole (part b of the drawing). Thus, *a hole in a piece of solid material expands when heated and contracts when cooled, just as if it were filled with the material that surrounds it.* The thermal expansion of the ring and its hole is analogous to a photographic enlargement; in both situations everything is enlarged, including holes. If the hole is circular, the equation $\Delta L = \alpha L_0 \, \Delta T$ can be used to find the change in any linear dimension of the hole, such as its radius or diameter. Example 5 illustrates this type of linear expansion.

PROBLEM SOLVING INSIGHT

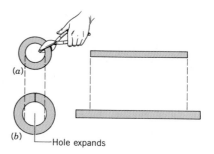

Figure 12.17 (a) A circular ring is "unrolled" hypothetically to form a strip. (b) When heated, the strip expands. The heated strip is "rolled" back into a ring that has a larger hole.

Example 5   A Heated Engagement Ring

A gold engagement ring has a diameter of 1.5×10^{-2} m and a temperature of 27 °C. The ring falls into a sink of hot water whose temperature is 49 °C. What is the change in the diameter of the hole in the ring?

REASONING AND SOLUTION   The hole expands as if it were filled with gold, so the change in the diameter is given by Equation 12.2, the coefficient of linear expansion for gold being $\alpha = 14 \times 10^{-6}$ (C°)$^{-1}$:

$$\Delta L = \alpha L_0 \, \Delta T = [14 \times 10^{-6} \, (C°)^{-1}](1.5 \times 10^{-2} \text{ m})(49 \text{ °C} - 27 \text{ °C}) = \boxed{4.6 \times 10^{-6} \text{ m}}$$

12.5 VOLUME THERMAL EXPANSION

NORMAL MATERIALS

The volume of a normal material increases as the temperature increases. Most solids and liquids behave in this fashion. By analogy with linear expansion, the change in volume ΔV over modest temperature changes is proportional to the change in temperature ΔT and to the initial volume V_0. These two proportionalities can be converted into Equation 12.3 below with the aid of a proportionality constant β, known as the *coefficient of volume expansion*. The algebraic form of this equation is similar to that for linear expansion, $\Delta L = \alpha L_0 \Delta T$.

Volume Thermal Expansion

The volume V_0 of an object changes by an amount ΔV when its temperature changes by an amount ΔT:

$$\Delta V = \beta V_0 \Delta T \qquad (12.3)$$

where β is the coefficient of volume expansion.

Common Unit for the Coefficient of Volume Expansion: $(C°)^{-1}$

The unit for β, like that for α, is $(C°)^{-1}$. Values for β depend on the nature of the material, and Table 12.2 lists some examples measured near 20 °C. The values of β for liquids are substantially larger than those for solids, because liquids typically expand more than solids, given the same initial volumes and temperature changes. Table 12.2 also shows that, for most solids, the coefficient of volume expansion is three times greater than the coefficient of linear expansion: $\beta = 3\alpha$. (See problem 32.)

If a cavity exists within a solid object, the volume of the cavity increases when the object expands, just as if the cavity were filled with the surrounding material. The expansion of the cavity is analogous to the expansion of a hole in a circular ring. Accordingly, the change in volume of a cavity can be found using the relation $\Delta V = \beta V_0 \Delta T$, where β is the coefficient of volume expansion of the material that surrounds the cavity. Example 6 illustrates this important point.

THE PHYSICS OF . . .

the overflow of an automobile radiator.

PROBLEM SOLVING INSIGHT

The way in which the level of a liquid in a container changes with temperature depends on both the change in volume of the liquid and the change in volume of the container.

Example 6   An Automobile Radiator

A small plastic container, called the coolant reservoir, catches the radiator fluid that overflows when an automobile engine becomes hot (see Figure 12.18). The radiator is made of copper, and the coolant has a coefficient of volume expansion of $\beta = 410 \times 10^{-6}\ (C°)^{-1}$. If the radiator is filled to its 15-quart capacity when the engine is "cold" (6.0 °C), how much overflow from the radiator will spill into the reservoir when the coolant reaches its operating temperature of 92 °C?

REASONING   When the temperature increases, both the coolant and radiator expand. If they were to expand by the same amount, there would be no overflow, because the increase in the radiator's volume would hold the increase in the coolant's volume. However, the liquid coolant expands more than the radiator, and the amount of overflow is the amount of coolant expansion *minus* the expansion of the radiator cavity.

SOLUTION When the temperature increases by 86 C°, the coolant expands by an amount

$$\Delta V = \beta V_0\, \Delta T \tag{12.3}$$

$$\Delta V = [410 \times 10^{-6}\ (\text{C}°)^{-1}](15\ \text{quarts})(86\ \text{C}°) = 0.53\ \text{quarts}$$

The volume of the radiator cavity expands as if it were filled with the surrounding material (copper). The expansion of the radiator cavity is

$$\Delta V = \beta V_0\, \Delta T = [51 \times 10^{-6}\ (\text{C}°)^{-1}](15\ \text{quarts})(86\ \text{C}°) = 0.066\ \text{quarts}$$

The amount of coolant overflow is 0.53 quarts − 0.066 quarts = $\boxed{0.46\ \text{quarts}}$.

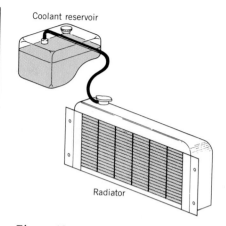

Figure 12.18 An automobile radiator and a coolant reservoir for catching the overflow from the radiator.

THE ANOMALOUS BEHAVIOR OF WATER NEAR 4 °C

While most substances expand when heated, a few do not. For instance, if water at 0 °C is heated, its volume *decreases* until the temperature reaches 4 °C. Above 4 °C water behaves normally, and its volume increases as the temperature increases. Because a given mass of water has a minimum volume at 4 °C, the density (mass per unit volume) of water is greatest at 4 °C, as Figure 12.19 shows.

The fact that water has its greatest density at 4 °C, rather than at 0 °C, has important consequences for the way in which a lake freezes. When the air temperature drops, the surface layer of water is chilled. As the temperature of the surface layer drops toward 4 °C, this layer becomes more dense than the warmer water below. The denser water sinks and pushes up the deeper and warmer water, which in turn is chilled at the surface. This process continues until the temperature of the entire lake reaches 4 °C. Further cooling of the surface water below 4 °C makes it *less dense* than the deeper layers; consequently, the surface layer does not sink, but stays on top. Continued cooling of the top layer to 0 °C leads to the formation of ice that floats on the water, because ice has a smaller density than water at any temperature. Below the ice, however, the water temperature remains above 0 °C. The sheet of ice acts as an insulator that reduces the loss of heat from the lake, especially if the ice is covered with a blanket of snow, which is also an insulator. Furthermore, heat transferred from the ground beneath the lake helps to keep the water under the ice sheet from freezing. As a result, lakes usually do not freeze solid, even during prolonged cold spells, so fish and other aquatic life can survive during the winter.

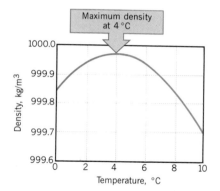

Figure 12.19 The density of water in the temperature range from 0 to 10 °C. Water has a maximum density of 999.973 kg/m³ at 4 °C. (This value for the density is equivalent to the often-quoted density of 1.000 00 grams per milliliter.)

12.6 HEAT AND INTERNAL ENERGY

An object with a high temperature is said to be hot, and the word "hot" brings to mind the word "heat." *Heat* flows from a hotter object to a cooler object when the two are placed in contact. It is for this reason that a pot of boiling water feels hot to the touch, while a glass of ice water feels cold. The temperature of boiling water is higher than the normal body temperature of 37 °C, while the temperature of ice water is lower than 37 °C. When the person in Figure 12.20a touches the pot, heat flows from the hotter pot into the cooler hand. When the person touches the glass in part *b* of the drawing, heat again flows from hot to cold, in this case from the warmer hand into the colder glass. The response of the nerves in the hand to the

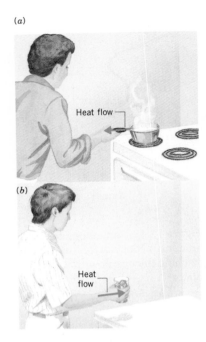

(a)

(b)

Figure 12.20 Heat is energy in transit from hot to cold. (*a*) Heat flows from the hotter pot to the colder hand. (*b*) Heat flows from the warmer hand to the colder glass of ice water.

arrival or departure of heat prompts the brain to identify the pot as being hot and the glass as being cold.

But just what is heat? As the definition below indicates, heat is a form of energy, energy in transit from hot to cold.

Definition of Heat

Heat is energy that flows from a higher-temperature object to a lower-temperature object because of the difference in temperatures.

SI Unit of Heat: joule (J)

Being a kind of energy, heat is measured in the same units used for work, kinetic energy, and potential energy. Thus, the SI unit for heat is the joule.

The heat that flows from hot to cold in Figure 12.20 originates in the **internal energy** of the hot substance. The internal energy of a substance is the sum of the molecular kinetic energy (due to the random motion of the molecules), the molecular potential energy (due to forces that act between the atoms of a molecule and between molecules), and other kinds of molecular energy. When heat flows in circumstances where the work done is negligible, the internal energy of the hot substance decreases and the internal energy of the cold substance increases. While heat may originate in the internal energy supply of a substance, *it is not correct to say that a substance contains heat.* The substance contains internal energy, not heat. The word "heat" is used only when referring to the energy actually in transit from hot to cold.

12.7 SPECIFIC HEAT CAPACITY

SOLIDS AND LIQUIDS

Greater amounts of heat are needed to raise the temperature of solids or liquids to higher values. A greater amount of heat is also required to raise the temperature of a greater mass of a material. Similar comments apply when the temperature is lowered, except that heat must be removed. For limited temperature ranges, experiment shows that the amount of heat Q is directly proportional to the change in temperature ΔT and to the mass m. These two proportionalities are expressed below in Equation 12.4, with the help of a proportionality constant c that is referred to as the *specific heat capacity* of the material.

Heat Supplied or Removed in Changing the Temperature of a Substance

The heat Q that must be supplied or removed to change the temperature of a substance of mass m by an amount ΔT is

$$Q = cm\,\Delta T \qquad (12.4)$$

where c is the specific heat capacity of the substance.

Common Unit for Specific Heat Capacity: $J/(kg \cdot C°)$

Table 12.3 Specific Heat Capacities[a] of Some Solids and Liquids

Substance	Specific Heat Capacity, c	
	J/(kg·C°)	kcal/(kg·C°)[b]
Solids		
Aluminum	9.00×10^2	0.215
Copper	387	0.0924
Glass	840	0.20
Human body (37 °C, average)	3500	0.83
Ice (−15 °C)	2.00×10^3	0.478
Iron or steel	452	0.108
Lead	128	0.0305
Silver	235	0.0562
Liquids		
Benzene	1740	0.415
Ethyl alcohol	2450	0.586
Glycerin	2410	0.576
Mercury	139	0.0333
Water (15 °C)	4186	1.000

[a] Except as noted, the values are for 25 °C and 1 atm of pressure.

[b] The values given are the same in units of cal/(g·C°).

Solving Equation 12.4 for the specific heat capacity shows that $c = Q/(m \, \Delta T)$, so the unit for specific heat capacity is J/(kg·C°). Table 12.3 reveals that the value of the specific heat capacity depends on the nature of the material. Example 7 demonstrates the use of Equation 12.4.

Example 7   A Hot Jogger

In a half hour, a 65-kg jogger can generate 8.0×10^5 J of heat. This heat is removed from the jogger's body by a variety of means, including the body's own temperature regulating mechanisms. If the heat were not removed, how much would the body temperature increase?

REASONING AND SOLUTION   Table 12.3 gives the average specific heat capacity of the human body as 3500 J/(kg·C°). With this value, Equation 12.4 shows that the temperature increase would be

$$\Delta T = \frac{Q}{cm} = \frac{8.0 \times 10^5 \text{ J}}{[3500 \text{ J/(kg·C°)}](65 \text{ kg})} = \boxed{3.5 \text{ C°}}$$

An increase in body temperature of 3.5 °C would be life threatening.

Tennis player Gabriela Sabatini. One way her body gets rid of excess heat is by perspiration.

HEAT UNITS OTHER THAN THE JOULE

There are three heat units other than the joule in common use. One kilocalorie (1 kcal) was defined historically as the amount of heat needed to raise the temperature of one kilogram of water by one Celsius degree.* With $Q = 1.00$ kcal,

* From 14.5 to 15.5 °C.

$m = 1.00$ kg, and $\Delta T = 1.00$ C°, the equation $Q = cm \, \Delta T$ shows that such a definition is equivalent to a specific heat capacity for water of $c = 1.00$ kcal/(kg·C°). Similarly, one calorie (1 cal) was defined as the amount of heat needed to raise the temperature of one gram of water by one Celsius degree, which yields a value of $c = 1.00$ cal/(g·C°). (Nutritionists use the word "Calorie," with a capital C, to specify the energy content of foods; this use is unfortunate, since 1 Calorie = 1000 calories = 1 kcal.) The British thermal unit (Btu) is the other commonly used heat unit and was defined historically as the amount of heat needed to raise the temperature of one pound of water by one Fahrenheit degree.

It was not until the time of James Joule (1818–1889) that the relationship between energy in the form of work (in units of joules) and energy in the form of heat (in units of kilocalories) was firmly established. Joule's experiments revealed that the performance of mechanical work can make the temperature of a substance rise, just as the absorption of heat can. His experiments and those of later workers have shown that

$$1 \text{ kcal} = 4186 \text{ joules} \quad \text{or} \quad 1 \text{ cal} = 4.186 \text{ joules}$$

Because of its historical significance, this conversion factor is known as the *mechanical equivalent of heat*. Example 8 illustrates the use of the joule and the kilocalorie as heat units.

Example 8 Taking a Hot Shower

Cold water at a temperature of 15 °C enters a heater, and the resulting hot water has a temperature of 61 °C. A person uses 120 kg of hot water in taking a shower. Find the number of (a) joules and (b) kilocalories needed to heat the water. (c) Assuming that the power utility charges $0.10 per kilowatt·hour for electrical energy, determine the cost of heating the water.

REASONING AND SOLUTION

(a) The number of joules of heat can be determined from Equation 12.4, since the specific heat capacity of water is known:

$$Q = cm \, \Delta T = [4186 \text{ J/(kg·C°)}](120 \text{ kg})(61 \text{ °C} - 15 \text{ °C}) = \boxed{2.3 \times 10^7 \text{ J}}$$

(b) The corresponding number of kilocalories can be obtained by using the mechanical equivalent of heat:

$$Q = (2.3 \times 10^7 \text{ J})\left(\frac{1 \text{ kcal}}{4186 \text{ J}}\right) = \boxed{5.5 \times 10^3 \text{ kcal}}$$

Alternatively, the same answer can be obtained directly from $Q = cm \, \Delta T$, if $c = 1.000$ kcal/(kg·C°) is used for the specific heat capacity of water.

(c) The kilowatt·hour (kWh) is the unit of energy that utility companies use in your electric bill. To calculate the cost, we need to determine the number of joules in one kilowatt·hour. Recalling that Energy = Power × Time (Equation 6.10), and that 1 watt = 1 joule/second, we find that 1 kWh = (1000 J/s)(3600 s) = 3.60×10^6 J. The number of kilowatt·hours of energy used to heat the water is

$$(2.3 \times 10^7 \text{ J})\left(\frac{1 \text{ kWh}}{3.60 \times 10^6 \text{ J}}\right) = 6.4 \text{ kWh}$$

At a cost of $0.10 per kWh, the bill for the heat is $\boxed{\$0.64}$.

Table 12.4 Specific Heat Capacities[a] of Gases

Gas	Specific Heat Capacity	
	Constant Pressure, c_P [J/(kg·C°)]	Constant Volume, c_V [J/(kg·C°)]
Ammonia	2190	1670
Carbon dioxide	833	638
Nitrogen	1040	739
Oxygen	912	651
Water vapor (100 °C)	2020	1520

[a] Except as noted, the values are for 15 °C and 1 atm of pressure.

GASES

As we will see in Chapter 15, the value of the specific heat capacity depends on whether the pressure or volume is held constant while energy in the form of heat is added to or removed from a substance. The distinction between constant pressure and constant volume is usually not important for solids and liquids but is significant for gases. Different values are obtained when the specific heat capacity for a gas is measured under conditions of *constant pressure* as compared to conditions of *constant volume*. Table 12.4 illustrates the difference for several gases and indicates that the value c_P at constant pressure is greater than the value c_V at constant volume.

CALORIMETRY

The specific heat capacity of a substance can be determined using the technique of *calorimetry.* Figure 12.21 shows one kind of experimental apparatus, called a calorimeter. Essentially, a calorimeter is an insulated container like a thermos. A perfect thermos would prevent any heat from leaking out or in. However, energy in the form of heat can flow *between* materials inside the thermos to the extent that they have different temperatures, for example, between ice cubes and warm tea. Such heat flow satisfies the conservation of energy, for the colder materials gain the energy that the hotter materials lose. The colder materials warm up and the hotter materials cool down, until a common temperature is reached at thermal equilibrium. The next example shows how the specific heat capacity of a substance can be determined from the temperature changes that occur inside a calorimeter as thermal equilibrium is established.

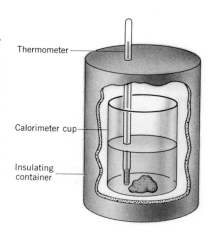

Thermometer

Calorimeter cup

Insulating container

Figure 12.21 A calorimeter.

Example 9 *Measuring the Specific Heat Capacity*

The calorimeter cup in Figure 12.21 is made from 0.15 kg of aluminum and contains 0.20 kg of water. Initially, the water and the cup have a common temperature of 18.0 °C. An unknown material ($m = 0.040$ kg) is heated to a temperature of 97.0 °C and then added to the water. The temperature of the water, the cup, and the unknown material is 22.0 °C after thermal equilibrium is reestablished. Ignoring the small amount of heat gained by the thermometer, find the specific heat capacity of the unknown material.

REASONING   Since energy is conserved and there is negligible heat flow between the calorimeter and the outside surroundings, the heat gained by the cold water and the aluminum cup as they warm up is equal to the heat lost by the unknown material as it cools down. Each quantity of heat can be calculated using $Q = cm\,\Delta T$.

SOLUTION

Heat gained   Heat lost by
by aluminum = unknown
and water   material

$$(cm\,\Delta T)_{\text{aluminum}} + (cm\,\Delta T)_{\text{water}} = (cm\,\Delta T)_{\text{unknown}}$$

$$[9.00 \times 10^2\ \text{J/(kg·C°)}](0.15\ \text{kg})(22.0\ °\text{C} - 18.0\ °\text{C})$$
$$+ [4186\ \text{J/(kg·C°)}](0.20\ \text{kg})(22.0\ °\text{C} - 18.0\ °\text{C})$$
$$= c_{\text{unknown}}(0.040\ \text{kg})(97.0\ °\text{C} - 22.0\ °\text{C})$$

$$\boxed{c_{\text{unknown}} = 1300\ \text{J/(kg·C°)}}$$

An important feature of the calculation in Example 9 is the way in which the temperature changes ΔT are written. Each is written as a positive number, that is, the higher temperature minus the lower temperature, so that the heat contribution is a positive number. In this fashion, we ensure that both sides of the equation, Heat gained = Heat lost, have the same algebraic sign. This important feature is also present in the next example.

Example 10   A Beach Party

An ice chest at a beach party contains 24 cans of soda at 4.0 °C. Each can of soda has a mass of 0.35 kg and a specific heat capacity of 3800 J/(kg·C°). Someone adds a 5.0-kg watermelon at 29 °C to the chest. The specific heat capacity of watermelon is nearly the same as that of water. Ignore the specific heat capacity of the chest and determine the final temperature T of the soda and watermelon.

REASONING   We assume that no heat is lost through the chest to the outside. Then, energy conservation dictates that the heat gained by the soda is equal to the heat lost by the watermelon. Each quantity of heat is given by $Q = cm\,\Delta T$, where we write the change in temperature ΔT as the higher temperature minus the lower temperature.

PROBLEM SOLVING INSIGHT

In the equation "Heat gained = Heat lost," both sides must have the same algebraic sign. Therefore, when calculating heat contributions, always write any corresponding temperature changes as the higher minus the lower temperature.

SOLUTION

Heat gained by soda = Heat lost by watermelon

$$(cm\,\Delta T)_{\text{soda}} = (cm\,\Delta T)_{\text{watermelon}}$$

$$[3800\ \text{J/(kg·C°)}](24 \times 0.35\ \text{kg})(T - 4.0\ °\text{C}) = [4186\ \text{J/(kg·C°)}](5.0\ \text{kg})(29\ °\text{C} - T)$$

Solving for the final temperature yields $\boxed{T = 14\ °\text{C}}$.

12.8   THE LATENT HEAT OF PHASE CHANGE

Surprisingly, there are situations in which the addition or removal of heat does not cause a temperature change. Consider a well-stirred glass of iced tea that has come to thermal equilibrium. Even though heat enters the glass from the warmer

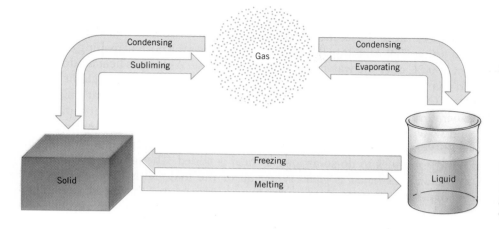

Figure 12.22 Three familiar phases of matter—solid, liquid, and gas—and the phase changes that can occur between any two of them.

room, the temperature of the tea does not rise above 0 °C as long as ice cubes are present. Apparently the heat is being used for some purpose other than raising the temperature. In fact, the heat is being used to melt the ice, and only when all of it is melted will the temperature of the liquid begin to rise.

One important point illustrated by the example above is that there is more than one type or phase of matter. For instance, some of the water in the glass is in the solid phase (ice) and some in the liquid phase. The gas or vapor phase is the third familiar phase of matter. In the gas phase, water is referred to as water vapor or steam.

A second important point in the iced tea example is that matter can change from one phase to another, and heat plays a role in the change. Figure 12.22 summarizes the various possibilities for phase changes between solids, liquids, and gases. A solid can *melt* or *fuse* into a liquid if heat is added, while the liquid can *freeze* into a solid if heat is removed. Similarly, a liquid can *evaporate* into a gas if heat is supplied, while the gas can *condense* into a liquid if heat is taken away. Rapid evaporation, with the formation of vapor bubbles within the liquid, is called boiling. Finally, a solid can sometimes change directly into a gas if heat is provided. We say that the solid *sublimes* into a gas. Examples of sublimation are (1) solid carbon dioxide, CO_2 (dry ice), turning into gaseous CO_2 and (2) solid naphthalene (moth balls) turning into naphthalene fumes. Conversely, if heat is removed under the right conditions, the gas will condense directly into a solid.

Figure 12.23 displays a graph that indicates what typically happens when heat is added to a material that changes phases. The graph records temperature versus

Ice climbing is one of the more breath-taking sports that uses the solid phase of water.

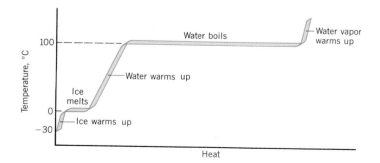

Figure 12.23 The graph shows the way the temperature of water changes as heat is added, starting with ice at −30 °C. The pressure is atmospheric pressure.

heat and refers to water at the normal atmospheric pressure of 1.01×10^5 Pa. The water starts off as ice at the subfreezing temperature of -30 °C. As heat is added, the temperature of the ice increases, in accord with the specific heat capacity of ice [2000 J/(kg·C°)]. Not until the temperature reaches the normal melting/freezing point of 0 °C does the water begin to change phase. As long as heat is added, the solid changes into the liquid, the temperature staying at 0 °C until *all the ice has melted*. Once all the material is in the liquid phase, additional heat causes the temperature to increase again, now in accord with the specific heat capacity of liquid water [4186 J/(kg·C°)]. When the temperature reaches the normal boiling/condensing point of 100 °C, the water begins to change from the liquid to the gas and continues to do so as long as heat is added. The temperature remains 100 °C *until all liquid is gone*. When all of the material is in the gas phase, additional heat once again causes the temperature to rise, this time according to the specific heat capacity of water vapor at constant atmospheric pressure [2020 J/(kg·C°)].

When a substance changes from one phase to another, the amount of heat that must be added or removed depends on the type of material and the nature of the phase change. The heat per kilogram associated with a phase change is referred to as *latent heat*:

Definition of Latent Heat

The latent heat L is the heat per kilogram that must be added or removed when a substance changes from one phase to another.

SI Unit of Latent Heat: J/kg

The *latent heat of fusion* L_f refers to the change between solid and liquid phases, the *latent heat of vaporization* L_v applies to the change between liquid and gas phases, and the *latent heat of sublimation* L_s refers to the change between solid and gas phases.

Table 12.5 Latent Heats[a] of Fusion and Vaporization

Substance	Melting Point (°C)	Latent Heat of Fusion, L_f (J/kg)	Boiling Point (°C)	Latent Heat of Vaporization, L_v (J/kg)
Ammonia	-77.8	33.2×10^4	-33.4	13.7×10^5
Benzene	5.5	12.6×10^4	80.1	3.94×10^5
Copper	1083	20.7×10^4	2566	47.3×10^5
Ethyl alcohol	-114.4	10.8×10^4	78.3	8.55×10^5
Gold	1063	6.28×10^4	2808	17.2×10^5
Lead	327.3	2.32×10^4	1750	8.59×10^5
Mercury	-38.9	1.14×10^4	356.6	2.96×10^5
Nitrogen	-210.0	2.57×10^4	-195.8	2.00×10^5
Oxygen	-218.8	1.39×10^4	-183.0	2.13×10^5
Water	0.0	33.5×10^4	100.0	22.6×10^5

[a] The values pertain to 1 atm pressure.

Table 12.5 gives some typical values of latent heats of fusion and vaporization. For instance, the latent heat of fusion for water is 3.35×10^5 J/kg. Thus, 3.35×10^5 J of heat must be supplied to melt one kilogram of ice at 0 °C into liquid water at 0 °C; conversely, this amount of heat must be removed from one kilogram of liquid water at 0 °C to freeze the liquid into ice at 0 °C. In comparison, the latent heat of vaporization for water has the much larger value of 22.6×10^5 J/kg. When water boils at 100 °C, 22.6×10^5 J of heat must be supplied for each kilogram of liquid turned into vapor. And when water vapor condenses at 100 °C, this amount of heat is released for each kilogram of vapor that changes back into liquid. Liquid water at 100 °C is hot enough by itself to cause a bad burn, and the additional effect of the large latent heat can cause severe tissue damage if condensation occurs on the skin. Examples 11 and 12 illustrate how to take into account the effect of latent heat when using the conservation of energy principle.

THE PHYSICS OF . . .

steam burns.

Example 11   *Ice-cold Lemonade*

Ice at 0 °C is placed in a Styrofoam cup containing 0.32 kg of lemonade at 27 °C. The specific heat capacity of lemonade is virtually the same as that of water. After the ice and lemonade reach an equilibrium temperature, some ice still remains. Neglect the specific heat capacity of the cup and any heat lost to the surroundings. Determine the mass of ice that has melted.

REASONING   According to the principle of energy conservation, the heat gained by the melting ice equals the heat lost by the cooling lemonade. The heat gained by the melting ice is $Q = mL_f$, since the latent heat L_f is defined as the heat per kilogram that must be added when a solid changes to a liquid. The heat lost by the lemonade is given by $Q = cm\,\Delta T$, where ΔT is the higher temperature of 27 °C minus the lower equilibrium temperature. The equilibrium temperature is 0 °C, because there is some ice remaining, and ice is in equilibrium with liquid water when the temperature is 0 °C.

SOLUTION

$$\begin{matrix} \text{Heat gained} \\ \text{by ice} \end{matrix} = \begin{matrix} \text{Heat lost by} \\ \text{lemonade} \end{matrix}$$

$$(mL_f)_{ice} = (cm\,\Delta T)_{lemonade}$$

$$m_{ice} = \frac{(cm\,\Delta T)_{lemonade}}{L_f}$$

$$m_{ice} = \frac{[4186\ \text{J/(kg} \cdot \text{C}°)](0.32\ \text{kg})(27\ °\text{C} - 0\ °\text{C})}{33.5 \times 10^4\ \text{J/kg}} = \boxed{0.11\ \text{kg}}$$

Example 12   *Getting Ready for a Party*

A 7.00-kg glass bowl [$c = 840$ J/(kg·C°)] contains 16.0 kg of punch at 25.0 °C. Two-and-a-half kilograms of ice [$c = 2.00 \times 10^3$ J/(kg·C°)] are added to the punch. The ice has an initial temperature of -20.0 °C, having been kept in a very cold freezer. The punch may be treated as if it were water [$c = 4186$ J/(kg·C°)], and it may be assumed that there is no heat flow between the punch bowl and the external environment. What is the temperature of the punch, ice, and bowl when they reach thermal equilibrium?

REASONING   First, it is necessary to check whether any ice is left at equilibrium. If so, the final temperature will be 0 °C, the melting/freezing point of water. We will see, however, that all the ice does melt and the final temperature of the mixture is above 0 °C. The final temperature can be determined by using the conservation of energy to equate the total heat gained to the total heat lost.

SOLUTION   To check whether any ice is left at equilibrium, we apply the equation $Q = cm\,\Delta T$ to calculate the heat that would be available if the punch and the bowl were cooled from 25.0 °C to 0.0 °C:

$$Q = \underbrace{[4186\ \text{J}/(\text{kg}\cdot\text{C}°)](16.0\ \text{kg})(25.0\ °\text{C} - 0.0\ °\text{C})}_{\text{Punch}}$$

$$+ \underbrace{[840\ \text{J}/(\text{kg}\cdot\text{C}°)](7.00\ \text{kg})(25.0\ °\text{C} - 0.0\ °\text{C})}_{\text{Glass bowl}} = 1.82 \times 10^6\ \text{J}$$

Is this enough heat to melt all the ice? The heat needed to melt the ice is

$$\text{Heat needed} = \underbrace{[2.00 \times 10^3\ \text{J}/(\text{kg}\cdot\text{C}°)](2.50\ \text{kg})[0.0\ °\text{C} - (-20.0\ °\text{C})]}_{\text{Heat to warm the ice from } -20.0\ °\text{C to } 0.0\ °\text{C}}$$

$$+ \underbrace{(2.50\ \text{kg})(3.35 \times 10^5\ \text{J}/\text{kg})}_{\text{Heat to melt ice at } 0.0\ °\text{C}} = 0.938 \times 10^6\ \text{J}$$

where the heat to melt the ice at 0.0 °C is the mass of the ice (2.50 kg) times the latent heat of fusion (3.35 × 10⁵ J/kg). Clearly, there is enough heat to melt all the ice and raise the final temperature above 0.0 °C. The final temperature T can be determined by equating the total heat gained to the total heat lost:

(a) Heat gained when $= [2.00 \times 10^3\ \text{J}/(\text{kg}\cdot\text{C}°)](2.50\ \text{kg})[0.0\ °\text{C} - (-20.0\ °\text{C})]$
    ice warms to 0.0 °C

(b) Heat gained when $= (2.50\ \text{kg})(3.35 \times 10^5\ \text{J}/\text{kg})$
    ice melts at 0.0 °C

(c) Heat gained
    when melted ice (liquid) $= [4186\ \text{J}/(\text{kg}\cdot\text{C}°)](2.50\ \text{kg})(T - 0.0\ °\text{C})$
    warms to temperature T

(d) Heat lost
    when punch cools to $= [4186\ \text{J}/(\text{kg}\cdot\text{C}°)](16.0\ \text{kg})(25.0\ °\text{C} - T)$
    temperature T

(e) Heat lost
    when bowl cools to $= [840\ \text{J}/(\text{kg}\cdot\text{C}°)](7.00\ \text{kg})(25.0\ °\text{C} - T)$
    temperature T

$$\underbrace{(a) + (b) + (c)}_{\text{Heat gained}} = \underbrace{(d) + (e)}_{\text{Heat lost}}$$

This equation can be solved to show that $\boxed{T = 11\ °\text{C}}$.

*12.9 EQUILIBRIUM BETWEEN PHASES OF MATTER

Under specific conditions of temperature and pressure, a substance can exist in equilibrium in more than one phase at the same time. Consider Figure 12.24, which shows a container kept at a constant temperature by a large reservoir of heated sand. Initially the container is evacuated, and part *a* shows it just after it has been partially filled with a liquid. A few fast-moving molecules escape the liquid and form a vapor phase, as part *b* suggests. These molecules pick up the required energy (the latent heat of vaporization) during collisions with neighboring molecules in the liquid. However, the reservoir of heated sand replenishes the energy carried away, thus maintaining the constant temperature. At first, the movement of molecules is predominantly from liquid to vapor, although some molecules in the vapor phase do reenter the liquid. As the molecules accumulate in the vapor, the number reentering the liquid eventually equals the number entering the vapor, and equilibrium becomes established, as in part *c*. From this point on, the concentration of molecules in the vapor phase does not change, and the vapor pressure remains constant. The pressure of the vapor that coexists in equilibrium with the liquid is called the ***equilibrium vapor pressure*** of the liquid.

The equilibrium vapor pressure does not depend on the volume of space above the liquid. If more space were provided, more liquid would vaporize, until equilibrium was reestablished at the same vapor pressure, assuming the same temperature is maintained. In fact, the equilibrium vapor pressure depends only on the temperature of the liquid; a higher temperature causes a higher pressure, as the graph in Figure 12.25 indicates for the specific case of water. Only when the temperature and vapor pressure correspond to a point on the curved line, which is called the ***vapor pressure curve*** or the ***vaporization curve***, can liquid and vapor phases coexist in equilibrium.

To illustrate the use of a vaporization curve, let's see what happens when water boils in a pot that is *open to the air*. Assume the air pressure acting on the water is 1.01×10^5 Pa (one atmosphere). When boiling occurs, bubbles of water vapor form throughout the liquid, rise to the surface, and break. For these bubbles to form and rise, the pressure of the vapor inside them must at least equal the air

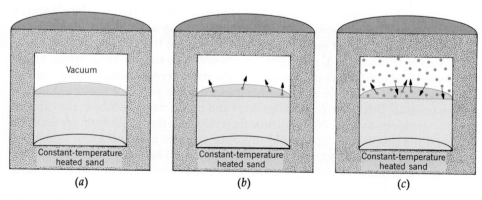

Figure 12.24 (*a*) Initially, only a liquid is in the evacuated container. (*b*) Some of the molecules begin entering the vapor phase. (*c*) Equilibrium is reached when the number of molecules entering the vapor phase equals the number returning to the liquid.

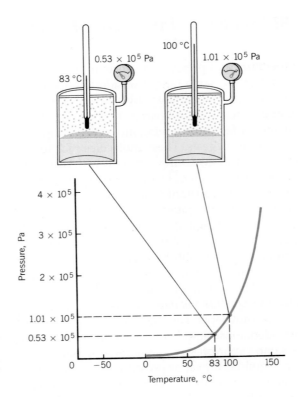

Figure 12.25 A plot of the equilibrium vapor pressure versus temperature is called the vapor pressure curve or the vaporization curve, the example shown being that for liquid water.

pressure acting on the surface of the water. According to Figure 12.25, a value of 1.01×10^5 Pa corresponds to a temperature of 100 °C. Consequently, water boils at 100 °C at 1 atmosphere of pressure. In general, a *liquid boils at the temperature for which its vapor pressure equals the external pressure.* Water will not boil, then, at sea level if the temperature is only 83 °C, because at this temperature the vapor pressure of water is only 0.53×10^5 Pa (see Figure 12.25), a value less than the external pressure of 1.01×10^5 Pa. However, water does boil at 83 °C on a mountain at an altitude of just under five kilometers, because the atmospheric pressure there is 0.53×10^5 Pa.

THE PHYSICS OF . . .

spray cans.

The operation of spray cans is based on the equilibrium between a liquid and its vapor. Figure 12.26*a* shows that a spray can contains a liquid propellant that is mixed with the product (such as hair spray). Inside the can, propellant vapor forms over the liquid. A propellant is chosen that has an equilibrium vapor pressure that is greater than atmospheric pressure at room temperature. Consequently, when the nozzle of the can is pressed, as in part *b* of the drawing, the vapor pressure forces the liquid propellant and product up the tube in the can and out the nozzle as a spray. When the nozzle is released, the coiled spring reseals the can and the propellant vapor builds up once again to its equilibrium value.

As is the case for liquid/vapor equilibrium, a solid can only be in equilibrium with its liquid phase at specific conditions of temperature and pressure. For each temperature, there is a single pressure at which the two phases can coexist in equilibrium. A plot of the equilibrium pressure versus equilibrium temperature is referred to as the *fusion curve*, and Figure 12.27*a* shows a typical curve for a normal substance. A normal substance expands upon melting (e.g., carbon dioxide and sulfur). Since higher pressures make it more difficult for such materials to

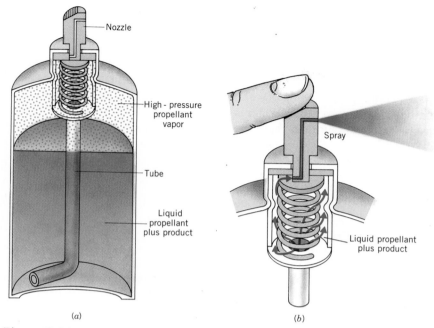

Figure 12.26 (a) A closed spray can containing liquid and vapor in equilibrium. (b) An open spray can.

expand, a higher melting temperature is needed for a higher pressure, and the fusion curve slopes upward to the right. Part *b* of the picture illustrates the fusion curve for water, one of the few substances that contract when they melt. Higher pressures make it easier for such substances to melt. Consequently, a lower melting temperature is associated with a higher pressure, and the fusion curve slopes downward to the right.

It should be noted that just because two phases can coexist in equilibrium does not necessarily mean that they will. Other factors may prevent it. For example, water in an *open* bowl may never come into equilibrium with water vapor if air currents are present. What happens is that the liquid, perhaps at a temperature of 25 °C, attempts to establish the corresponding equilibrium vapor pressure of

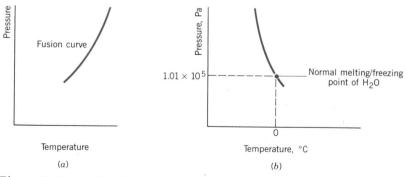

Figure 12.27 (a) The fusion curve for a normal substance that expands on melting. (b) The fusion curve for water, one of the few substances that contract on melting.

3.2 × 10³ Pa. If air currents continually blow the water vapor away, however, equilibrium will never be established, and eventually the water will evaporate completely. Each kilogram of water that goes into the vapor phase takes along the latent heat of vaporization. Because of this heat loss, the remaining liquid would become cooler, except for the fact that the surroundings replenish the loss and prevent the cooling from taking place. In the case of the human body, however, evaporative cooling does occur. Water is exuded by the sweat glands and evaporates from a much larger area than the surface of a typical bowl of water. The removal of heat along with the water vapor is one mechanism that the body uses to maintain its constant temperature.

THE PHYSICS OF . . .

evaporative cooling of the human body.

*12.10 HUMIDITY

Air is a mixture of several types of gases, such as nitrogen, oxygen, and water vapor. The total pressure of a gas mixture is the sum of the partial pressures of the component gases. The partial pressure of a gas is the pressure it would exert if it alone occupied the entire volume at the same temperature as the mixture. The partial pressure of water vapor in air depends on weather conditions. It can be as low as zero (although this rarely happens), or it can be as high as the equilibrium vapor pressure of water at the given temperature.

To provide an indication of how much water vapor is in the air, weather forecasters usually give the *relative humidity*. If the relative humidity is too low, the air contains such a small amount of water vapor that our skin and mucus membranes tend to dry out. If the relative humidity is too high, especially on a hot day, we become very uncomfortable and our skin feels "sticky." Under such conditions, the air holds so much water vapor that the water exuded by our sweat glands cannot evaporate efficiently.

THE PHYSICS OF . . .

relative humidity.

The relative humidity can be defined as the ratio (expressed as a percentage) of the partial pressure of water vapor in the air to the equilibrium vapor pressure at a given temperature.

$$\text{Percent relative humidity} = \frac{\text{Partial pressure of water vapor}}{\text{Equilibrium vapor pressure of water at the existing temperature}} \times 100 \quad (12.5)$$

The term in the denominator on the right of Equation 12.5 is given by the vaporization curve of water and is the pressure of the water vapor in equilibrium with the liquid. At a given temperature, the pressure of the water vapor in the air cannot exceed this value. If it did, the vapor would not be in equilibrium with the liquid and would condense as dew or rain to reestablish equilibrium.

When the partial pressure of the water vapor equals the equilibrium vapor pressure of water at a given temperature, the relative humidity is 100%. In such a situation, the vapor is said to be *saturated*, because it is present in the maximum amount, as it would be above a pool of liquid at equilibrium in a closed container. If the relative humidity is less than 100%, the water vapor is said to be *unsaturated*. Example 13 demonstrates how to find the relative humidity.

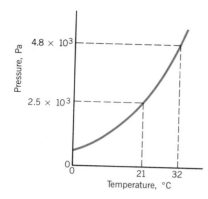

Figure 12.28 The vaporization curve of water.

Example 13 Relative Humidities

One day, the partial pressure of water vapor in the air is 2.0×10^3 Pa. Using the vaporization curve for water in Figure 12.28, determine the relative humidity if the temperature is (a) 32 °C and (b) 21 °C.

REASONING AND SOLUTION

(a) According to Figure 12.28, the equilibrium vapor pressure of water at 32 °C is 4.8×10^3 Pa. Equation 12.5 reveals that the relative humidity is

$$\text{Relative humidity at 32 °C} = \frac{2.0 \times 10^3 \text{ Pa}}{4.8 \times 10^3 \text{ Pa}} \times 100 = \boxed{42\%}$$

(b) A similar calculation shows that

$$\text{Relative humidity at 21 °C} = \frac{2.0 \times 10^3 \text{ Pa}}{2.5 \times 10^3 \text{ Pa}} \times 100 = \boxed{80\%}$$

Whatever the partial pressure of water vapor in the air, it is possible to locate this value on the vertical axis of the vapor pressure curve and identify a corresponding temperature. This temperature is known as the *dew point*. Figure 12.29 shows that if the partial pressure of water vapor is 3.2×10^3 Pa, the dew point is

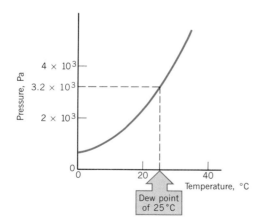

Figure 12.29 On the vaporization curve of water, the dew point is the temperature that corresponds to the actual partial pressure of water vapor in the air.

Figure 12.30 For fog to form, the air temperature must drop below the dew point.

25 °C. The partial pressure would correspond to a relative humidity of 100%, if the ambient temperature were equal to the temperature at the dew point. Hence, the dew point is the temperature below which water vapor in the air condenses in the form of liquid drops (dew or fog). The closer the actual temperature is to the dew point, the closer the relative humidity is to 100%. Thus, for fog to form above the Hudson River in Figure 12.30, the air temperature must drop below the dew point. Similarly, water condenses on the outside of a cold glass when the temperature of the air next to the glass falls below the dew point. And the cold coils in a home dehumidifier (see Figure 12.31) function very much in the same way that the cold glass does. The coils are kept cold by a circulating refrigerant. When the air blown across them by the fan cools below the dew point, water vapor condenses in the form of droplets, which collect in a receptacle.

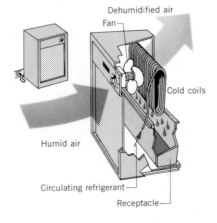

Figure 12.31 The cold coils of a dehumidifier cool the air blowing across them below the dew point, and water vapor condenses out of the air.

INTEGRATION OF CONCEPTS

HEAT AND THE CONSERVATION OF ENERGY

The fact that energy can be neither created nor destroyed, but can only be converted from one form to another is a remarkable insight into how our physical universe operates. We call this fact the principle of conservation of energy and first encountered it in Chapter 6. There, we dealt only with kinetic and potential energies. Now in Chapter 12, we have expanded our concept of energy to include heat, which is energy that flows from a higher-temperature object to a lower-temperature object because of the difference in temperature. Even though heat is neither kinetic nor potential energy, it shares with them a common behavior according to the conservation principle. No matter what its form, energy can be neither created nor destroyed. This fact governs the way objects at different temperatures come to an equilibrium temperature when they are placed in contact. Hotter objects cool down and cooler ones warm up. If there is no heat lost to the external surroundings, conservation of energy requires that the heat lost by the hotter objects equals the heat gained by the cooler ones. If the objects undergo phase changes as they lose or gain heat, the latent heat of the phase changes must also be taken into account according to the conservation principle. To put this behavior into perspective, remember that we saw our first example of the conservation of energy in Chapter 6, when we talked about mechanical energy. Mechanical energy is related to work, which, in turn, is built upon the concepts of force and displacement. Heat is far removed from such origins. And yet, because it is so fundamental, the conservation principle applies. From it, we gain an insight into how objects at different temperatures come to an equilibrium temperature when they are placed in contact. This wide applicability is a hallmark of a truly fundamental principle.

SUMMARY

On the **Celsius temperature scale,** there are 100 equal divisions between the ice point (0 °C) and the steam point (100 °C). On the **Fahrenheit temperature scale,** there are 180 equal divisions between the ice point (32 °F) and the steam point (212 °F). For most scientific work, the **Kelvin temperature scale** is the scale of choice. One kelvin is equal in size to one Celsius degree; however, the temperature T on the Kelvin scale differs from the temperature T_c on the Celsius scale by an additive constant of 273.15: $T = T_c + 273.15$. The lower limit of temperature is called **absolute zero** and is designated as 0 K on the Kelvin scale. The operation of any thermometer is based on the change in some physical property with temperature; this physical property is called a **thermometric property.**

Most substances expand when heated. For **linear expansion,** an object of length L_0 experiences a change in length ΔL when the temperature changes by ΔT: $\Delta L = \alpha L_0 \Delta T$, where α is the **coefficient of linear expansion.** For **volume expansion,** the change in volume ΔV of an object of volume V_0 is given by $\Delta V = \beta V_0 \Delta T$, where β is the **coefficient of volume expansion.** When the temperature changes, a hole in a plate or a cavity in a piece of solid material expands or contracts as if the hole or cavity were filled with the surrounding material.

For an object held rigidly in place, a **thermal stress** can occur when the object attempts to expand or contract. The thermal stress can be extremely large, even when the temperature change is small.

In the temperature range from 0 to 4 °C water is

unlike most substances, because its volume decreases as the temperature increases. Above 4 °C, water behaves normally, because its volume increases as the temperature increases. Water has a minimum volume and a maximum density at 4 °C.

The **internal energy** of a substance is the sum of the internal kinetic, potential, and other kinds of energy that the molecules of the substance have. **Heat** is energy that flows from a higher-temperature substance to a lower-temperature substance because of the difference in temperatures. The SI unit for heat is the joule.

The **specific heat capacity** c of a substance of mass m determines how much heat Q must be supplied to or removed from the substance to change its temperature by an amount ΔT: $Q = cm \, \Delta T$. When materials are placed in thermal contact within a perfectly insulated container, the **principle of energy conservation** requires that heat lost by the warmer materials equals the heat gained by the cooler materials.

Heat must be supplied or removed to make a material change from one phase to another. The amount of heat per kilogram of material is called the latent heat of the phase change. The **latent heats of fusion, vaporization,** and **sublimation** refer, respectively, to the solid/liquid, liquid/vapor, and solid/vapor phase changes.

The **equilibrium vapor pressure** of a substance is the pressure of the vapor phase that is in equilibrium with the liquid phase. Vapor pressure depends only on temperature. For a liquid, a plot of the equilibrium vapor pressure versus temperature is called the **vapor pressure curve.** The vapor pressure curve gives those combinations of temperature and pressure at which the corresponding two phases can coexist in equilibrium. The **fusion curve** gives the combinations of temperature and pressure for equilibrium between solid and liquid phases.

The **relative humidity** is the ratio (expressed as a percentage) of the partial pressure of water vapor in the air to the equilibrium vapor pressure at a given temperature. The **dew point** is the temperature below which the water vapor in the air condenses. On the vaporization curve of water, the dew point is the temperature that corresponds to the actual pressure of water vapor in the air.

QUESTIONS

1. For the highest accuracy, would you choose an aluminum or a steel tape rule for year-round outdoor use? Why?

2. The first international standard of length was a metal bar kept at the International Bureau of Weights and Measures. One meter of length was defined to be the distance between two fine lines engraved near the ends of the bar. Why was it important that the bar be kept at a constant temperature?

3. A circular hole is cut through a flat aluminum plate. A spherical brass ball has a diameter that is slightly *smaller* than the diameter of the hole. The plate and the ball have the same temperature at all times. Should the plate and ball both be heated or both be cooled so the ball *cannot* fall through the hole? Give your reasoning.

4. For added strength, many highways and buildings are constructed with reinforced concrete (concrete that is reinforced with embedded steel rods). Table 12.2 shows that the coefficient of linear expansion for concrete is nearly the same as that for steel. Why is it important that these two coefficients be nearly the same?

5. At a certain temperature, an aluminum rod is hung from an aluminum frame, as the drawing shows. A small gap exists between the rod and the floor. The frame and rod are heated uniformly. (a) Is it possible that the rod will ever touch the floor? Explain. (b) Repeat part (a), assuming the frame is aluminum and the rod is lead.

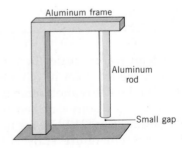

Aluminum frame

Aluminum rod

Small gap

6. A simple pendulum is made using a long thin metal wire. When the temperature drops, does the period of the pendulum increase, decrease, or remain the same? Account for your answer.

7. The drawing shows a cross-sectional view of three cylinders A, B, and C. Each is made from a different material; one is steel, one is brass, and one is lead. All three have the same temperature. Cylinder A just barely fits into the hole in cylinder B, and cylinder B just fits into the hole in cylinder C. As the cylinders are heated to the same, but higher, temperature, cylinder C falls off, while cylinder A becomes tightly wedged to cylinder B. Which cylinder is made from which material? Give your reasoning. There are two possible answers.

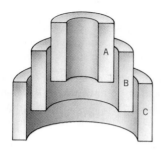

8. The metal sheet shown in part *a* of the drawing has three holes cut through it. Each hole has a different shape. Someone claims that the sheet is heated uniformly to produce the sheet shown in part *b*. Does part *b* of the drawing correctly represent the heated sheet? Justify your answer.

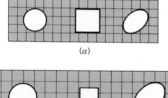

(*a*)

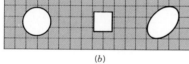

(*b*)

9. A hot steel ring fits snugly over a cold brass cylinder. The temperatures of the ring and cylinder are, respectively, above and below room temperature. Account for the fact that it is nearly impossible to pull the ring off the cylinder once the assembly has equilibrated at room temperature.

10. For glass baking dishes, Pyrex glass is used instead of common glass. A cold Pyrex dish taken from the refrigerator can be put directly into a hot oven without cracking from thermal stress. A dish made from common glass would crack. With the aid of Table 12.2, explain why Pyrex is better in this respect than common glass.

11. Suppose liquid mercury and glass both had the same coefficient of volume expansion. Explain why such a mercury-in-glass thermometer would not work.

12. When the bulb of a mercury-in-glass thermometer is inserted into boiling water, the mercury column first drops slightly before it begins to rise. Account for this phenomenon. (*Hint: Consider what happens initially to the glass envelope.*)

13. Two different objects are supplied with equal amounts of heat. Give the reason(s) why their temperature changes would not necessarily be the same.

14. Two objects are made from the same material. The more massive object has a higher temperature than the less massive object. If the two are placed in contact, which object will experience the greater temperature change? Explain.

15. Near a large body of water, the fluctuations in air temperature are usually less extreme than they are far away from the water. Explain why.

16. To help lower the high temperature of a sick patient, an alcohol rub is sometimes used. Isopropyl alcohol is rubbed over the patient's back, arms, legs, etc., and allowed to evaporate. Why does the procedure work?

17. Supposed the latent heat of vaporization of H_2O were one-tenth its actual value. (a) Other things being equal, would it take the same time, a shorter time, or a longer time for a pot of water on a stove to boil away? (b) Would the evaporative cooling mechanism of the human body be as effective? Account for both answers.

18. Fruit blossoms are permanently damaged when the temperature drops below about -4 °C (a "hard freeze"). Orchard owners sometimes spray a film of water over the blossoms to protect them when a hard freeze is expected in the spring. From the point of phase changes, give a reason for the protection.

19. If a bowl of water is placed in a closed container and water vapor is pumped away rapidly enough, the remaining liquid will turn to ice. Explain why the ice appears.

20. Water is boiling in a flask. The flask is removed from the source of heat, and a cork is placed in the neck of the flask. The boiling stops. However, if cold water is poured over the top of the flask, boiling restarts. Explain why this happens.

21. A bowl of water is covered tightly and allowed to sit at a constant temperature of 23 °C for a long time. What is the relative humidity in the space between the surface of the water and the cover? Justify your answer.

22. Is it possible for dew to form on Tuesday night and not on Monday night, even though Monday night is the cooler night? Incorporate the idea of the dew point into your answer.

23. Two rooms in a house have the same temperature. One of the rooms contains an indoor swimming pool. On a cold day the windows of the pool room are "steamed up," while those of the other room are clear. Explain.

24. A jar is half filled with boiling water. The lid is then screwed on the jar. After the jar has cooled to room temperature, the lid is difficult to remove. Why?

PROBLEMS

Note: For problems in this set, use the values of α and β given in Table 12.2, and the values of c, L_f, and L_v given in Tables 12.3 and 12.5, unless stated otherwise.

Section 12.1 Common Temperature Scales, Section 12.2 The Kelvin Temperature Scale

1. A personal computer is designed to operate over the temperature range from 10.0 to 105 °F. To what do these temperatures correspond (a) on the Celsius scale and (b) on the Kelvin scale?

2. Liquid nitrogen boils at a temperature of 77 K. What is this temperature on the Celsius scale?

3. A comfortable temperature for most people is around 24 °C. What is this temperature (a) on the Fahrenheit scale and (b) on the Kelvin scale?

4. A hamburger at 25 °C is heated to 175 °C. What is the *change* in its temperature in (a) F ° and (b) kelvins?

5. A temperature of absolute zero occurs at −273.15 °C. What is this temperature on the Fahrenheit scale?

*6. At what temperature will the reading on the Fahrenheit scale be numerically equal to that on the Celsius scale?

**7. Space invaders land on earth. On the invaders' temperature scale, the ice point is at 15 °I (I = invader), and the steam point is at 165 °I. The invaders' thermometer shows the temperature on earth to be 42 °I. Using logic similar to that in Example 1 in the text, what would this temperature be on the Celsius scale?

Section 12.4 Linear Thermal Expansion

8. The Concorde is 62 m long when its temperature is 23 °C. In flight, the outer skin of this supersonic aircraft can reach 105 °C due to air friction. The coefficient of linear expansion of the skin is 2.0×10^{-5} (C °)$^{-1}$. Find the amount by which the Concorde expands.

9. A steel aircraft carrier is 370 m long when moving through the icy North Atlantic at a temperature of 2.0 °C. By how much does the carrier lengthen when it is traveling in the warm Mediterranean Sea at a temperature of 21 °C?

10. Find the approximate length of the Golden Gate bridge if it is known that the steel in the roadbed expands by 0.53 m when the temperature changes from +2 to +32 °C.

11. An aluminum baseball bat has a length of 0.86 m at a temperature of 25 °C. When the temperature of the bat is raised, the bat expands by 0.000 16 m. Determine the final temperature of the bat.

12. A steel beam is used in the construction of a skyscraper.

By what fraction $\Delta L/L_0$ does the length of the beam increase when the temperature changes from that on a cold winter day (−15 °F) to that on a summer day (+105 °F)?

13. A 45-m sidewalk is made from a special type of concrete. The sidewalk is a single piece and is observed to expand by 0.014 m when the temperature changes from 1.7 to 35 °C. What is the coefficient of linear expansion of this concrete?

14. A rod made from a particular alloy is heated from 25.0 °C to the boiling point of water. Its length increases by 8.47×10^{-4} m. The rod is then cooled from 25.0 °C to the freezing point of water. By how much does the rod shrink?

15. A commonly used method of fastening one part to another part is called "shrink fitting." A steel rod has a diameter of 2.0026 cm, and a flat plate contains a hole whose diameter is 2.0000 cm. The rod is cooled so that it just fits into the hole. When the rod warms up, the enormous thermal stress exerted by the plate holds the rod securely to the plate. By how many Celsius degrees should the rod be cooled?

*16. The brass bar and the aluminum bar in the drawing are each attached to an immovable wall. At 22 °C the air gap between the rods is 1.0×10^{-3} m. At what temperature will the gap be closed?

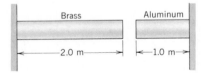

*17. A lead sphere has a diameter that is 0.050% larger than the inner diameter of a steel ring when each has a temperature of 70.0 °C. Thus, the ring will not slip over the sphere. At what common temperature will the ring just slip over the sphere?

*18. A simple pendulum consists of a relatively massive ball connected to one end of a thin aluminum wire. The period of the pendulum is 1.000 00 s. The temperature rises by 120 C°, and the length of the wire increases. Determine the period of the heated pendulum.

*19. A steel ruler is accurate when the temperature is 25 °C. When the temperature drops to −15 °C, the ruler no longer reads correctly, but it can be made to read correctly if a stress is applied to each end of the ruler. (a) Should the stress be a compression or a tension? Why? (b) What is the magnitude of the necessary stress?

**20. An aluminum wire of radius 3.0×10^{-4} m is stretched between the ends of a concrete block, as the drawing illustrates. When the system (wire and concrete) is at 35 °C, the tension in the wire is 50.0 N. What is the tension in the wire when the system is heated to 185 °C?

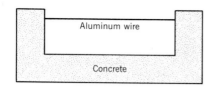

**21. A thin uniform aluminum rod is rotating freely at a constant angular speed about an axis perpendicular to its center. The temperature of the rod increases by 195 °C. (a) Does the angular speed increase or decrease? Why? (b) By what percentage does the angular speed change?

Section 12.5 Volume Thermal Expansion

22. A swimming pool contains 110 m³ (about 30 000 gal) of water. The sun heats the water from 17 to 27 °C. What is the change in the volume of the water?

23. A container completely filled with carbon tetrachloride is heated from 22 to 72 °C. The change in the volume of the carbon tetrachloride is 6.8×10^{-4} m³. What is the volume of the container, assuming it does not expand?

24. Suppose that the gas tank in your car is completely filled when the temperature is 17 °C. How many gallons will spill out of the twenty-gallon steel tank when the temperature rises to 35 °C?

25. Many hot-water heating systems have a reservoir tank connected directly to the pipeline, so as to allow for expansion when the water becomes hot. The heating system of a house has 76 m of copper pipe whose inside radius is 9.5×10^{-3} m. When the water and pipe are heated from 24 to 78 °C, what must be the minimum volume of the reservoir tank to hold the overflow of water?

26. Suppose you were selling apple cider for two dollars a gallon when the temperature is 4.0 °C. The coefficient of volume expansion of the cider is 280×10^{-6} (C°)$^{-1}$. If the expansion of the container is neglected, how much more money (in pennies) would you make per gallon by refilling the container on a day when the temperature is 26 °C?

27. During an all-night cram session, a student heats up a one-liter (1.0×10^{-3} m³) Pyrex beaker of cold coffee. Initially, the temperature is 24 °C, and the beaker is filled to the brim. A short time later when the student returns, the temperature has risen to 85 °C. The coefficient of volume expansion of the coffee is that of water. How much coffee has spilled out of the beaker? Be sure to take the expansion of the beaker into account.

28. A thin spherical shell of silver has an inner radius of 1.5×10^{-2} m when the temperature is 25 °C. The shell is heated to 135 °C. Find the change in the interior volume of the shell.

*29. A mercury-in-glass thermometer, designed to measure body temperature orally, contains 1.0×10^{-3} kg of mercury.

The density of mercury is 13 600 kg/m³. The capillary tube attached to the bulb reservoir has a radius of 2.3×10^{-5} m. Neglect the expansion of the glass, and calculate the distance between the 98 and the 99 °F marks.

*30. The bulk modulus of water is $B = 2.2 \times 10^{9}$ N/m². How many atmospheres of pressure are required to keep water from expanding when it is heated from 15 to 25 °C?

*31. A solid aluminum sphere has a radius of 0.50 m and a temperature of 75 °C. The sphere is then completely immersed in a pool of water whose temperature is 25 °C. The sphere cools, while the water temperature remains nearly at 25 °C, because the pool is very large. The sphere is weighed in the water immediately after being submerged and then again after cooling to 25 °C. (a) Which weight is larger? Why? (b) Use Archimedes' principle to find the magnitude of the *difference* between the weights.

*32. Each side of a cube has a length L_0. When the temperature of the cube is increased by ΔT, the enlarged volume of the cube becomes $(L_0 + \Delta L)^3$. Expand this expression for the enlarged volume and show that the change in volume ΔV is given approximately by $\Delta V = 3\alpha V_0 \Delta T$, where $V_0 = L_0^3$ is the initial volume and α is the coefficient of linear expansion. (*Hint: The terms involving* $(\alpha \Delta T)^2$ *and* $(\alpha \Delta T)^3$ *are much smaller than the term involving* $\alpha \Delta T$ *and, therefore, can be neglected; try substituting any reasonable numbers for* α *and* ΔT *to convince yourself of this fact.*) Comparing this expression with $\Delta V = \beta V_0 \Delta T$, we see that $\beta = 3\alpha$ for solids.

**33. Two identical thermometers are made of Pyrex glass and they contain, respectively, identical volumes of mercury and methyl alcohol. If the expansion of the glass is taken into account, how many times greater is the distance between the degree marks on the methyl alcohol thermometer than that on the mercury thermometer?

**34. The column of mercury in a barometer has a height of 0.760 m when the pressure is one atmosphere and the temperature is 0.0 °C. What will be the height of the mercury column for the same one atmosphere of pressure when the temperature rises to 38.0 °C on a hot day? (*Hint: The pressure in the barometer is given by* Pressure $= \rho g h$, *and the density* ρ *of the mercury changes when the temperature changes.*)

Section 12.6 Heat and Internal Energy, Section 12.7 Specific Heat Capacity

35. Passive solar homes often have a concrete wall that collects and stores solar energy. One such wall has a mass of 1.5×10^4 kg, and the concrete has a specific heat capacity of 880 J/(kg·C°). If the temperature of the wall increases by 12 C° during the day, how much energy has the wall absorbed?

36. Blood can carry excess energy from the interior to the surface of the body, where the energy is dispersed in a number of ways. While a person is exercising, 0.6 kg of blood flows to the surface of the body and releases 2000 J of energy. The blood arriving at the surface has the temperature of the body interior, 37.0 °C. Assuming that blood has the same specific heat capacity as water, determine the temperature of the blood that leaves the surface and returns to the interior.

37. A gallon (1 gal $= 3.785 \times 10^{-3}$ m³) of water at 25 °C is placed in a refrigerator to make "ice water." How much heat must the refrigerator remove from the water to cool it to 5 °C?

38. Into a 0.200-kg copper cup (20.0 °C) is put 0.100 kg of aluminum at 50.0 °C and 0.250 kg of water at 85.0 °C. Assuming there is no heat flow between the cup and its outside environment, find the final equilibrium temperature. The final temperature is greater than 50.0 °C.

39. The air inside a car has a mass of 2.6 kg and a specific heat capacity of 720 J/(kg·C°). The driver loses heat at the rate of 120 joules per second. If all the heat goes into increasing the air temperature, how long will it take for the temperature to change from 21 to 37 °C?

40. When you take a bath, how many kilograms of hot water (60.0 °C) and cold water (25.0 °C) must you mix, so that the temperature of the bath is 40.0 °C? The total mass of water (hot plus cold) is 185 kg. Ignore any heat flow between the water and its external surroundings.

41. If the price of electrical energy is $0.10 per kilowatt·hour, what is the cost of using electrical energy to heat the water in a swimming pool (12.0 m × 9.00 m × 1.5 m) from 15 to 27 °C?

***42.** Lead shot (0.600 kg, 90.0 °C) and steel shot (0.100 kg, 60.0 °C) are put into a can. How many kilograms of water at 74.0 °C must be added, so that in reaching thermal equilibrium the lead and the steel experience a temperature change of the *same magnitude*? Ignore the specific heat capacity of the can and any heat exchanged with the environment.

***43.** The box of a well-known breakfast cereal states that one ounce of the cereal contains 110 Calories (1 food Callorie = 4186 J). If all this energy could be converted by a weight lifter's body into work done in lifting a barbell, what is the heaviest barbell that could be lifted a distance of 2.0 m?

***44.** An electric hot water heater takes in cold water at 25.0 °C and delivers hot water. The hot water has a constant temperature of 30.0 °C, when the "hot" faucet is left wide open all the time and the volume flow rate is 1.5×10^{-4} m³/s. What is the minimum power rating of the hot water heater?

****45.** A steel rod ($\rho = 7860$ kg/m³) has a length of 2.0 m. It is bolted at both ends between immobile supports. Initially there is no tension in the rod, because the rod just fits between the supports. Find the tension that develops when the rod loses 3300 J of heat.

Section 12.8 The Latent Heat of Phase Change

46. An ice cube tray holds 0.39 kg of water at 0 °C. How much heat must a freezer remove to make ice cubes at 0 °C?

47. The heat needed to vaporize an amount of liquid mercury at 357 °C is 3.9×10^4 J. What is the mass of the mercury?

48. A 10.0-kg block of ice has a temperature of −10.0 °C. The pressure is one atmosphere. The block absorbs 4.11×10^6 J of heat. What is the final temperature of the liquid water?

49. Assume the pressure is one atmosphere and determine the heat required to produce 2.00 kg of water vapor at 100.0 °C, starting with (a) 2.00 kg of water at 100.0 °C and (b) 2.00 kg of water at 0.0 °C.

50. Suppose the amount of heat removed when 3.0 kg of water freezes at 0 °C were removed from ethyl alcohol at its freezing/melting point of −114 °C. How many kilograms of ethyl alcohol would freeze?

51. A woman finds the front windshield of her car covered with ice at −10.0 °C. The ice has a thickness of 5.00×10^{-4} m, and the windshield has an area of 1.10 m². The density of ice is 917 kg/m³. How much heat is required to melt the ice?

52. The latent heat of vaporization of H_2O at body temperature (37.0 °C) is 2.42×10^6 J/kg. To cool the body of a 75-kg jogger [average specific heat capacity = 3500 J/(kg·C°)] by 1.0 C°, how many kilograms of water in the form of sweat have to be evaporated?

53. When it rains, water vapor in the air condenses into liquid water, and energy is released. How much energy is released when 0.0254 m (one inch) of rain falls over an area of 2.59×10^6 m² (one square mile)? The energy released is enough to heat at least one thousand homes for a year in the United States.

54. Ice at −10.0 °C and steam at 130 °C are brought together at atmospheric pressure in a perfectly insulated container. After thermal equilibrium is reached, the liquid phase at 50.0 °C is present. Ignoring the container and the equilibrium vapor pressure of the liquid at 50.0 °C, find the ratio of the mass of steam to the mass of ice.

55. In solar-assisted greenhouses some provision is usually made for storing solar energy in order to minimize heating costs. One way of storing energy is to use large containers of water. Another way is to use hydrated sodium sulfate (Glauber salt), which has a normal melting/freezing point of 32.4 °C, a latent heat of fusion of 239 000 J/kg, and a specific heat capacity of 2850 J/(kg·C°) in the liquid phase and 1900 J/(kg·C°) in the solid phase. On an equal volume basis, the sodium sulfate may have an advantage over water. To show this advantage, calculate the heat released when (a) 1.0 m³ of water (1.0×10^3 kg) and (b) 1.0 m³ of sodium sulfate (1.6×10^3 kg) cool from 35.0 to 18.0 °C. (c) Repeat parts (a) and (b) for a temperature interval of 30.0 to 18.0 °C and comment on how the two storage media compare under these conditions.

***56.** It is claimed that if a lead bullet goes fast enough, it can melt completely when it comes to a halt suddenly, and all its kinetic energy is converted into heat via friction. Find the minimum speed of a lead bullet (30.0 °C) for such an event to happen.

*57. An unknown material has a normal melting/freezing point of −25.0 °C, and the liquid phase has a specific heat capacity of 160 J/(kg·C°). One-tenth of kilogram of the solid at −25.0 °C is put into a 0.150-kg aluminum calorimeter cup that contains 0.100 kg of glycerin. The temperature of the cup and the glycerin is initially 27.0 °C. All the unknown material melts, and the final temperature at equilibrium is 20.0 °C. The calorimeter loses no energy to the external environment. What is the latent heat of fusion of the unknown material?

*58. In the British engineering system, the latent heat of fusion of H_2O is 144 Btu/lb. Suppose an air conditioner can remove heat at a rate of 12 000 Btu/h. Determine the number of tons (1 ton = 2000 lb) of water at 0 °C that such an air conditioner could freeze into ice at 0 °C in 24 h. This is the number of "tons of air-conditioning" that the unit can provide, a phrase that is sometimes used to rate the cooling capacity of air conditioners.

**59. A locomotive wheel is 1.00 m in diameter. A 25.0-kg steel band has a temperature of 20.0 °C and a diameter that is 6.00×10^{-4} m less than that of the wheel. What is the smallest mass of water vapor at 100 °C that can be condensed on the steel band to heat it, so that it will fit onto the wheel? Do not ignore the water that results from the condensation.

Section 12.9 Equilibrium Between Phases of Matter

60. The pressure is 5.0×10^6 Pa. The vapor pressure curve of carbon dioxide that accompanies this problem is to scale. Use the diagram to determine the temperature at which liquid carbon dioxide will exist in equilibrium with its vapor phase.

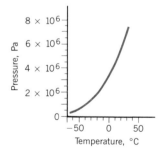

61. The equilibrium vapor pressure of liquid sulfur dioxide is 5.07×10^5 Pa at 32.1 °C. What phase is present at equilibrium at a temperature of 32.1 °C when the pressure is one atmosphere? Explain.

*62. A container is fitted with a movable piston of negligible mass and radius $r = 0.050$ m. Inside the container is liquid water in equilibrium with its vapor, as the drawing shows. The piston remains stationary with a 120-kg block on top of it. The air pressure acting on the top of the piston is one atmosphere. By using the vaporization curve for water in Figure 12.25, find the temperature of the water.

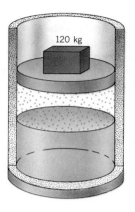

Section 12.10 Humidity

63. Using the vapor pressure curve for water that accompanies this problem, find the partial pressure of water vapor on a day when the weather forecast gives the relative humidity as 70.0% and the temperature as 38.0 °C.

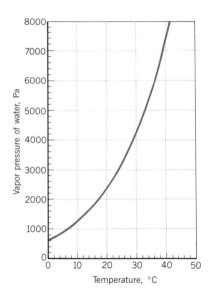

64. What is the relative humidity on a day when the temperature is 20 °C and the dew point is 5 °C? Use the vapor pressure curve that accompanies problem 63.

65. The relative humidity is 35% when the temperature is 27 °C. Using the vapor pressure curve for water that accompanies problem 63, determine the dew point.

*66. A woman has been outdoors where the temperature is 10 °C. She walks into a 25 °C house, and her glasses "steam up." Using the vapor pressure curve for water that accompanies problem 63, find the smallest possible value for the relative humidity of the room.

**67. At a picnic, a glass contains 0.300 kg of tea at 30.0 °C, which is the air temperature. To make iced tea, someone adds 0.0670 kg of ice at 0.0 °C and stirs the mixture continually.

When all the ice melts and the final temperature is reached, the glass begins to fog up, because water vapor condenses on the outer glass surface. Using the vapor pressure curve for water that accompanies problem 63, ignoring the specific heat capacity of the glass, and treating the tea as if it were water, estimate the relative humidity.

ADDITIONAL PROBLEMS

68. When the temperature of a coin is raised by 75 °C, the coin's diameter increases by 2.3×10^{-5} m. If the original diameter is 1.8×10^{-2} m, find the coefficient of linear expansion.

69. How much heat must be added to 0.45 kg of aluminum to change it from a solid at 130 °C to a liquid at 660 °C (its melting point)? The latent heat of fusion for aluminum is 4.0×10^5 J/kg.

70. A hole is drilled through a copper plate whose temperature is 11 °C. (a) When the temperature of the plate is increased, will the radius of the hole be larger or smaller than the radius at 11 °C? Why? (b) When the plate is heated to 110 °C, by what fraction $\Delta r / r_0$ will the radius of the hole change?

71. At a fabrication plant, a hot metal forging has a mass of 75 kg and a specific heat capacity of 430 J/(kg·C°). To harden it, the forging is quenched by immersion in 710 kg of oil that has a temperature of 32 °C and a specific heat capacity of 2700 J/(kg·C°). The final temperature of the oil and forging at thermal equilibrium is 47 °C. Assuming that heat flows only between the forging and the oil, determine the initial temperature of the forging.

72. A 1.0×10^{-3}-m³ container made of common glass is filled with mercury at a temperature of 0.0 °C. The density of mercury is 13 600 kg/m³. How many kilograms of mercury overflow when the temperature is raised to 25 °C?

73. What is the length of a fused quartz rod whose length changes by the same amount as that of a 0.10-m lead rod when both experience the same temperature change?

74. The outdoor temperature is 15 °C, and the relative humidity is 45%. The partial pressure of water vapor inside a house is the same as outdoors, but the temperature in the house is 25 °C. Using the vapor pressure curve for water that accompanies problem 63, determine the relative humidity in the house.

75. A precious-stone dealer wishes to find the specific heat capacity of a 0.030-kg gemstone. The specimen is heated to 95.0 °C and then placed in a 0.15-kg copper vessel that contains 0.080 kg of water at equilibrium at 25.0 °C. The loss of heat to the external environment is negligible. When equilibrium is established, the water temperature is 28.5 °C. What is the specific heat capacity of the specimen?

76. On the Rankine temperature scale, which is sometimes used in engineering applications, the ice point is at 491.67 °R and the steam point is at 671.67 °R. Determine a relationship (analogous to Equation 12.1) between the Rankine and Fahrenheit temperature scales.

*77. An aluminum can is filled to the brim with 3.5×10^{-4} m³ of soda at 5 °C. When the can and soda are heated to 78 °C, 3.6×10^{-6} m³ of soda spills over. What is the coefficient of volume expansion of the soda? Be sure to take into account the expansion of the can itself.

*78. To help keep his barn warm on cold days, a farmer stores 840 kg of solar-heated water in barrels. For how many hours would a 2.0-kW electric space heater have to operate to provide the same amount of heat as the water does, when it cools from 10.0 to 0.0 °C and completely freezes?

*79. A can is filled with a liquid to 97.0% of its capacity. The temperature of the can and the liquid is 0.0 °C. The material from which the can is made has a coefficient of volume expansion of 85×10^{-6} (C°)$^{-1}$. At a temperature of 100.0 °C, the can is observed to be filled to the brim. Determine the coefficient of volume expansion of the liquid.

*80. A 0.25-kg coffee mug is made from a material that has a specific heat capacity of 950 J/(kg·C°) and contains 0.30 kg of water. The cup and the water are at 25 °C. To make a cup of coffee, a small electric heater is immersed in the water and brings it to a boil in two minutes. Assume that the cup and the water always have the same temperature and determine the minimum power rating of this heater.

*81. A bar of aluminum and a bar of copper each have the same length when the temperature is 25 °C. The aluminum bar is heated to 65 °C. To what temperature must the copper bar be heated so it has the same length as the heated aluminum bar?

**82. Water is moving with a speed of 5.00 m/s just before it passes over the top of a waterfall. At the bottom, 5.00 m below, the water flows away with a speed of 3.00 m/s. What is the largest amount by which the temperature of the water at the bottom could exceed the temperature of the water at the top?

**83. A steel ruler is calibrated to read true at 20.0 °C. A draftsman uses the ruler at 40.0 °C to draw a line on a 40.0 °C copper plate. As indicated on the warm ruler, the length of the line is 0.50 m. To what temperature should the plate be cooled, such that the length of the line truly becomes 0.50 m?

**84. A 1.0-kg steel sphere will not fit through a circular hole in a 1.0-kg aluminum plate, because the radius of the sphere is 0.10% larger than the radius of the hole. If both the sphere and the plate are always kept at the same temperature, how much heat must be put into the two so the ball just passes through the hole?

CHAPTER 13

THE TRANSFER OF HEAT

These male walruses on Round Island, just off the southwest coast of Alaska, pack themselves tightly together even when there is plenty of space available. It has been suggested that the packing behavior has evolved as a technique for conserving body heat in arctic surroundings. With their bodies so close together, some of the heat lost by one animal is gained by another, so that less heat on average is transferred out of each walrus. The transfer of heat from one place to another is also important in our own lives. For instance, most of our energy (except for small amounts of nuclear energy) originates in the sun and is transferred to us over a distance of 150 million kilometers through the void of space. The sunlight of today provides the energy to drive photosynthesis in plants that provide food and, hence, metabolic energy, while the sunlight of eons ago nurtured the organic matter that eventually produced the important fossil fuels of oil, natural gas, and coal. Combustion engines extract energy from burning fuel to run cars, trucks, and other vehicles and could not operate without the transfer of heat. Within the home, energy transfer occurs routinely. A heating unit transfers heat throughout a house on a cold day, while an air conditioner removes heat on a hot day. And our bodies constantly transfer heat in one direction or another, to prevent the dangerous effects of excessive heating or cooling. There are three fundamental processes by which heat is transferred: convection, conduction, and radiation. This chapter considers each of them.

Figure 13.1 The smoke from a campfire rises because of convection.

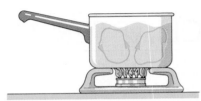

Figure 13.2 Convection currents are set up when a pan of water is heated.

13.1 CONVECTION

When part of a fluid is warmed, such as the air above a fire, the volume of the fluid expands, and the density decreases. According to Archimedes' principle (see Section 11.6), the surrounding cooler and denser fluid exerts a buoyant force on the warmer fluid and pushes it upward. As warmer fluid is pushed upward, the surrounding cooler fluid replaces it. This cooler fluid, in turn, is warmed and pushed upward. Thus, a continuous flow is established. The fluid flow carries along heat and is called a *convection current.* Whenever heat is transferred by the bulk movement of a gas or a liquid, the heat is said to be transferred by *convection.*

> **Convection**
>
> Convection is the process in which heat is carried from place to place by the bulk movement of a fluid.

The smoke rising above the campfire in Figure 13.1 is one visible result of convection. Figure 13.2 shows an example of convection currents in a pan of water being heated on a gas burner. The currents distribute the heat from the burning gas to all parts of the water.

THE PHYSICS OF . . .

heating and cooling by convection.

Certain kinds of home heating systems take advantage of convection to distribute heat throughout a room. Figure 13.3*a* illustrates the air convection current originating from a baseboard heating unit near the floor. Had the heating unit been located near the ceiling, the warm air would remain there, and very little convection current would be generated to distribute the heat. Part *b* of the figure shows an analogous situation in a refrigerator, where the convection current is set up by the cooling coils. The coils are near the *top* of the refrigerator, in direct contrast to the placement of the heating unit in part *a*. As the temperature of the air in contact with the coils decreases, the volume decreases, and the density increases. This cooler and denser air sinks to the bottom and forces warmer, less dense air upward to the cooling coils. The resulting convection current keeps all parts of the refrigerator uniformly cool. Placing the coils at the bottom of the

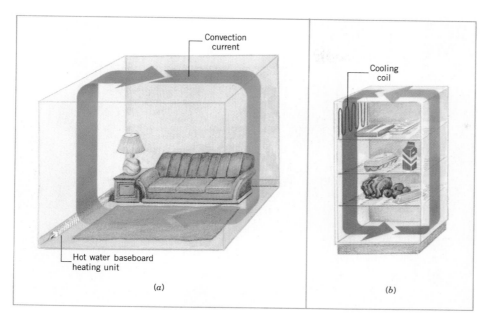

(a)   (b)

Figure 13.3   (*a*) Air warmed by the baseboard heating unit is pushed to the top of the room by the cooler and denser air. (*b*) Air cooled by the cooling coils sinks to the bottom of the refrigerator. In both (*a*) and (*b*) a convection current is established.

refrigerator would produce stagnant, cool air at the bottom and lead to inefficient cooling at the top.

Another example of convection occurs when the ground, heated by the sun's rays, warms the neighboring air. Surrounding cooler and denser air pushes the heated air upward. The resulting updraft or "thermal" can be quite strong, depending on the amount of heat that the ground can supply. As Figure 13.4 illustrates, these thermals can be used by glider pilots to gain considerable altitude. Birds such as eagles utilize thermals in a similar fashion.

Strong thermals are found where the temperature near ground level is markedly warmer than it is at higher altitudes. Although strong thermals are not found everywhere, it is usual for air temperature to decrease with increasing altitude, and the resulting upward convection currents are important for dispersing pollutants from industrial sources and automobile exhaust systems. Sometimes, however, meteorological conditions cause a layer to form in the atmosphere where the temperature increases with increasing altitude. Such a layer is called an *inversion*

THE PHYSICS OF . . .

"thermals."

Bald eagles often soar for considerable distances with the aid of thermals.

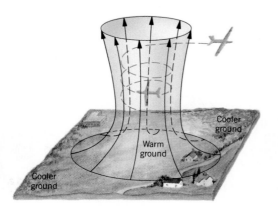

Figure 13.4   Updrafts, or thermals, are produced by the convective movement of air that has been warmed by the ground.

Figure 13.5 In the absence of convection currents in the air, pollutants often accumulate and form a smog layer in Los Angeles, California.

layer, because its temperature profile is inverted compared to the usual situation. An inversion layer arrests the normal upward convection currents, causing a stagnant-air condition in which the concentration of pollutants increases, as in Figure 13.5.

We have been discussing *natural convection,* in which a temperature difference causes the density at one place in a fluid to be different than at another. Sometimes, natural convection is inadequate to transfer sufficient amounts of heat. In such cases *forced convection* is often used, and an external device such as a fan mixes the warmer and cooler portions of the fluid. Figure 13.6 shows two examples of forced convection. In one, a fan mounted on a computer creates the forced convection that removes heat produced by the electrical components. In the other, a pump circulates radiator fluid through an automobile engine to remove the excessive heat from the combustion process.

THE PHYSICS OF . . .

cooling by forced convection.

Figure 13.6 The forced convection created by a fan removes heat from a computer. The forced convection generated by a pump circulates radiator fluid through an automobile engine to remove excessive heat.

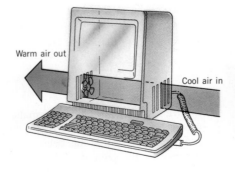

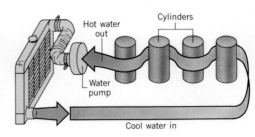

13.2 CONDUCTION

Anyone who has fried a hamburger in an all-metal skillet knows that the metal handle becomes hot. Somehow, heat is transferred from the burner to the handle. Clearly, heat is not being transferred by the bulk movement of the metal or the surrounding air, so convection can be ruled out. Instead, heat is transferred directly through the metal by a process called *conduction.*

Conduction
Conduction is the process whereby heat is transferred directly through a material, any bulk motion of the material playing no role in the transfer.

One mechanism for conduction occurs when the atoms or molecules in a hotter part of the material vibrate or move with greater energy than those in a cooler part. By means of collisions, the more energetic molecules pass on some of their energy to their less energetic neighbors. Figure 13.7 illustrates this conduction mechanism in a gas. Molecules that strike the hotter wall absorb energy from it and rebound with a greater kinetic energy than when they arrived. As these more energetic molecules collide with their less energetic neighbors, they transfer some of their energy to the neighbors. Through such molecular collisions, heat is conducted from the hotter to the cooler wall.

A similar mechanism for the conduction of heat occurs in metals. Metals are different from most substances in having a pool of electrons that are more or less free to wander throughout the metal. These free electrons can transport energy and allow metals to transfer heat very well. The free electrons are also responsible for the excellent electrical conductivity that metals have.

Materials that conduct heat well are called *thermal conductors.* Most metals, such as aluminum, copper, gold, and silver, are excellent thermal conductors. Substances that conduct heat poorly are called *thermal insulators.* Some common thermal insulators are wood, glass, and most plastics. Thermal insulators have many important applications. Virtually all new housing construction incorporates thermal insulation in attics and walls to reduce heating and cooling costs. And the wooden or plastic handles on many pots and pans reduce the flow of heat to the cook's hand.

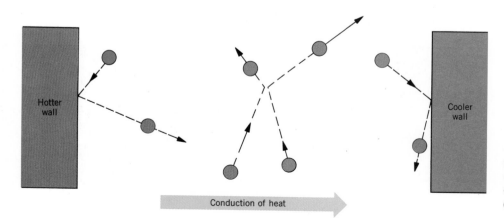

Conduction of heat

Figure 13.7   Heat conduction in a gas occurs when energetic molecules (red) transfer some of their energy to less energetic molecules (blue) through collisions.

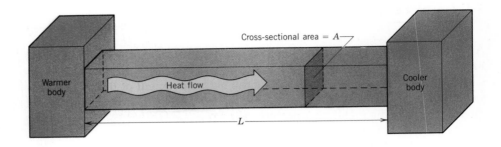

Figure 13.8 Heat is conducted through the bar when the ends of the bar are maintained at different temperatures. The heat flows from the warmer to the cooler end.

Table 13.1 Thermal Conductivities[a] of Selected Materials

Substance	Thermal Conductivity, k [J/(s·m·C°)]
Metals	
Aluminum	240
Brass	110
Copper	390
Iron	79
Lead	35
Silver	420
Steel (stainless)	14
Gases	
Air	0.0256
Hydrogen (H_2)	0.180
Nitrogen (N_2)	0.0258
Oxygen (O_2)	0.0265
Other Materials	
Asbestos	0.090
Body fat	0.20
Concrete	1.1
Diamond	2450
Glass	0.80
Goose down	0.025
Ice (0 °C)	2.2
Styrofoam	0.010
Water	0.60
Wood (oak)	0.15
Wool	0.040

[a] Except as noted, the values pertain to temperatures near 20 °C.

To illustrate the factors that influence the conduction of heat, Figure 13.8 displays a rectangular bar. The ends of the bar are in thermal contact with two bodies, one of which is kept at a constant higher temperature, while the other is kept at a constant lower temperature. Although not shown for the sake of clarity, the sides of the bar are insulated, so the heat lost through them is negligible. The amount of heat Q conducted through the bar from the warmer end to the cooler end depends on a number of factors:

1. Q is proportional to the length of time t during which conduction has been taking place ($Q \propto t$). More heat flows in longer time periods.

2. Q is proportional to the temperature difference ΔT between the two ends of the bar ($Q \propto \Delta T$). A larger temperature difference causes more heat to flow. No heat flows when both ends have the same temperature, so that $\Delta T = 0$.

3. Q is proportional to the cross-sectional area A of the bar ($Q \propto A$). Figure 13.9 helps to explain this fact by showing two identical bars (insulated sides not shown) placed between the warmer and cooler bodies. Clearly, twice as much heat flows through two bars as through one. Since two bars are equivalent to one bar with twice the cross-sectional area, doubling the area doubles the heat flow. In other words, Q is proportional to A.

4. Q is inversely proportional to the length L of the bar ($Q \propto 1/L$). Greater lengths of material conduct less heat. To experience this effect, put two insulated mittens (the kind that cooks keep around the stove) on the *same hand*. Then, touch a hot pot and notice that it feels cooler than when you wear only one mitten, signifying that less heat passes through the greater thickness ("length") of material.

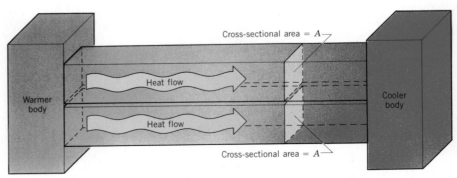

Figure 13.9 Twice as much heat flows through two identical bars as through one bar.

The proportionalities above can be stated together as $Q \propto A\, \Delta Tt/L$. Equation 13.1 expresses this result with the aid of a proportionality constant k, which is called the *thermal conductivity*.

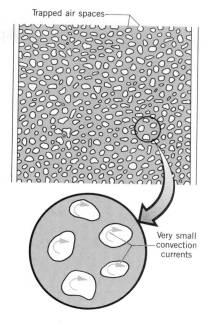

Trapped air spaces

Very small convection currents

Conduction of Heat Through a Material

The heat Q conducted during a time t through a bar of length L and cross-sectional area A is

$$Q = \frac{kA\,\Delta Tt}{L} \qquad (13.1)$$

where ΔT is the temperature difference between the ends of the bar and k is the thermal conductivity of the material.

SI Unit of Thermal Conductivity: $J/(s \cdot m \cdot C°)$

Since $k = QL/(tA\,\Delta T)$, the SI unit for thermal conductivity is $J \cdot m/(s \cdot m^2 \cdot C°)$ or $J/(s \cdot m \cdot C°)$. The SI unit of power is the joule/second (J/s) or watt (W), so the thermal conductivity is also given in units of $J/(s \cdot m \cdot C°) = W/(m \cdot C°)$.

Different materials have different thermal conductivities, and Table 13.1 gives some representative values. Because metals are such good thermal conductors, they have the largest thermal conductivities. In comparison, liquids and gases generally have small thermal conductivities. In fact, in most fluids the heat transferred by conduction is negligible compared to that transferred by convection when there are strong convection currents. Air, for instance, with its small thermal conductivity, is an excellent thermal insulator when confined to small spaces where no appreciable convection currents can be established. Goose down, Styrofoam, and wool derive their fine insulating properties in part from the small dead-air spaces within them, as Figure 13.10 illustrates. We also take advantage of dead-air spaces when we dress "in layers" during very cold weather and put on several layers of relatively thin clothing, rather than one thick layer. The air trapped between the layers acts as an excellent insulator.

Example 1 deals with the role that conduction through body fat plays in regulating body temperature.

Figure 13.10 Styrofoam contains many small, dead-air spaces. These small spaces inhibit heat transfer by convection currents, and since air has a low thermal conductivity, Styrofoam is an excellent thermal insulator.

THE PHYSICS OF . . .

dressing warm.

Example 1   Heat Transfer in the Human Body

When excessive heat is produced within the body, it must be transferred to the skin and dispersed if the temperature at the body interior is to be maintained at the normal value of 37.0 °C. One possible mechanism for transfer is conduction through body fat. Suppose that heat travels through 0.030 m of fat in reaching the skin, which has a total surface area of 1.7 m² and a temperature of 34.0 °C. Find the amount of heat that reaches the skin in half an hour (1800 s).

<u>REASONING AND SOLUTION</u>   In Table 13.1 the thermal conductivity of body fat is given as $k = 0.20\ J/(s \cdot m \cdot C°)$. According to Equation 13.1,

$$Q = \frac{kA\,\Delta Tt}{L}$$

$$Q = \frac{[0.20\ J/(s \cdot m \cdot C°)](1.7\ m^2)(37.0\ °C - 34.0\ °C)(1800\ s)}{0.030\ m} = \boxed{6.1 \times 10^4\ J}$$

Goose down is a first-rate thermal insulator.

THE PHYSICS OF . . .

heat transfer in the human body.

For comparison, a jogger can generate over ten times this amount of heat in a half hour. Thus, conduction through body fat is not a particularly effective way of removing excess heat. Heat transfer via blood flow to the skin is more effective and has the added advantage that the body can vary the blood flow as needed.

Example 2 examines the heat conduction through two common thermal insulators, Styrofoam and wood.

Example 2   An Ice Chest

A portable ice chest has walls 0.020 m thick. The area of the walls is 0.66 m². For a picnic the chest is loaded with 3.0 kg of ice at 0 °C. The temperature at the outside surface of the chest is 35 °C. Find the time required to melt the ice when the chest is made from (a) Styrofoam and (b) wood.

REASONING AND SOLUTION
(a) Recall from Section 12.8 that the latent heat of fusion L_f is the heat per kilogram needed to melt a solid. Table 12.5 gives $L_f = 3.35 \times 10^5$ J/kg for ice. Thus, the heat necessary to melt 3.0 kg of ice is

$$Q = mL_f = (3.0 \text{ kg})(3.35 \times 10^5 \text{ J/kg}) = 1.0 \times 10^6 \text{ J}$$

PROBLEM SOLVING INSIGHT

When a substance changes phase, remember to take the latent heat of the phase change into account when determining the amount of energy gained or lost by the substance.

The time it takes for this amount of heat to penetrate the chest can be determined by using Equation 13.1 with the value of $k = 0.010$ J/(s·m·C°) for the thermal conductivity of Styrofoam (see Table 13.1):

$$t = \frac{QL}{kA \, \Delta T}$$

$$t = \frac{(1.0 \times 10^6 \text{ J})(0.020 \text{ m})}{[0.010 \text{ J/(s·m·C°)}](0.66 \text{ m}^2)(35 \text{ °C} - 0 \text{ °C})} = \boxed{87\,000 \text{ s or 24 h}}$$

(b) Table 13.1 indicates that the thermal conductivity of wood is 15 times greater than that of Styrofoam. Therefore, other factors being the same, heat will penetrate wood 15 times more quickly than it penetrates Styrofoam: $t = (24 \text{ h})/15 = \boxed{1.6 \text{ h}}$. Styrofoam, with its smaller thermal conductivity, is clearly the better thermal insulator.

13.3 RADIATION

Energy from the sun is brought to earth by large amounts of visible light waves, as well as by substantial amounts of infrared and ultraviolet waves. These waves belong to a class of waves known as electromagnetic waves, a class that also includes the microwaves used for cooking and the radio waves used for AM and FM broadcasts. The sunbathers in Figure 13.11 feel hot, because their bodies absorb energy from the sun's electromagnetic waves. And anyone who has stood by a roaring fire or put a hand near an incandescent light bulb has experienced a similar effect. Thus, fires and light bulbs also emit electromagnetic waves, and when the energy of such waves is absorbed, it can have the same effect as heat. The process of transferring energy via electromagnetic waves is called *radia-*

Figure 13.11 Suntans are produced by ultraviolet rays.

tion and, unlike convection or conduction, it does not require a material medium. Electromagnetic waves from the sun, for example, travel through the void of space during their journey to earth.

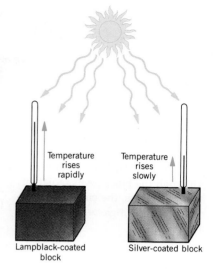

Radiation

Radiation is the process in which energy is transferred by means of electromagnetic waves.

All bodies continuously radiate energy in the form of electromagnetic waves. Even an ice cube radiates energy, although so little of it is in the form of visible light that an ice cube cannot be seen in the dark. Likewise, the human body emits insufficient visible light to be seen in the dark. However, as we saw in Figure 12.7, the infrared waves radiating from the body can be detected in the dark by electronic cameras. Generally, an object does not emit much visible light until the temperature of the object exceeds about 1000 K. Then a characteristic red glow appears, like that of a heating coil on an electric stove. When its temperature reaches about 1700 K, an object begins to glow white-hot, like the tungsten filament in an incandescent light bulb.

In the transfer of energy by radiation, the absorption of electromagnetic waves is just as important as the emission. The surface of an object plays a significant role in determining how much radiant energy the object will absorb or emit. The two blocks in Figure 13.12, for example, are identical, except that one has a rough surface coated with lampblack (a fine black soot), while the other has a highly polished silver surface. A thermometer is inserted into each block, and the blocks are placed in direct sunlight. It is found that the temperature of the black block rises at a much faster rate than that of the silvery block. The rapid temperature rise occurs because lampblack absorbs about 97% of the incident radiant energy, while the silvery surface absorbs only about 10%. As Figure 13.13 indicates, the remaining part of the incident energy is reflected in each case. We see the lampblack as black in color because it reflects so little of the light falling on it, while the silvery surface looks like a mirror because it reflects so much light. Since the color black is associated with nearly complete absorption of visible light, the term

Figure 13.12 The temperature of the block coated with lampblack rises faster than the temperature of the block coated with silver, because the black surface absorbs radiant energy from the sun at a greater rate than does the silver surface.

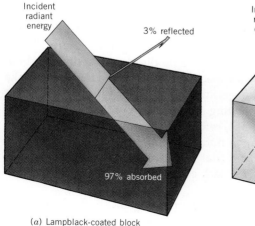

(*a*) Lampblack-coated block

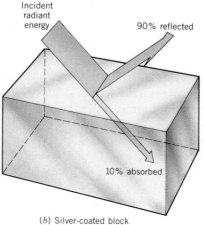

(*b*) Silver-coated block

Figure 13.13 (*a*) The lampblack surface absorbs about 97% of the incident radiant energy, while reflecting 3%. (*b*) The polished silver surface absorbs about 10% of the incident radiant energy and reflects 90%.

Figure 13.14 When a block and its surroundings have the same constant temperature, the block emits the same amount of radiant energy that it absorbs in a given time interval. This is true for a block coated with (a) lampblack and (b) silver. Although the block emits radiation in all directions, the emission is represented by a single arrow in these drawings.

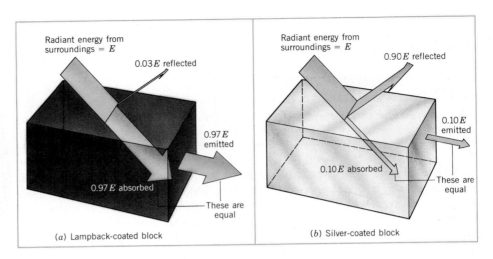

perfect blackbody or, simply, *blackbody* is used when referring to an object that absorbs *all* the electromagnetic waves falling on it.

All objects emit and absorb electromagnetic waves simultaneously. When a body has the same constant temperature as its surroundings, the amount of radiant energy being absorbed must balance the amount being emitted in a given interval of time. As Figure 13.14 illustrates, the block coated with lampblack absorbs and emits the same amount of radiant energy, and the silvery block does too. In either case, if absorption were greater than emission, the block would experience a net gain in energy. As a result, the temperature of the block would rise and not be constant. Similarly, if emission were greater than absorption, the temperature would fall. Since absorption and emission are balanced, *a material that is a good absorber, like lampblack, is also a good emitter, and a material that is a poor absorber, like polished silver, is also a poor emitter.* A perfect blackbody, being a perfect absorber, is also a perfect emitter.

The fact that a black surface is both a good absorber and a good emitter is the reason people are uncomfortable wearing dark clothes during the summer. Dark clothes absorb a large fraction of the sun's radiation and then reemit it in all directions. About one-half of the emitted radiation is directed inward toward the body and creates the sensation of warmth. Light-colored clothes, in contrast, are cooler to wear, since they absorb relatively little of the incident radiation.

The use of light colors for comfort also occurs in nature. Most lemurs, for instance, are nocturnal and have dark fur like the one shown in Figure 13.15a. Since they are active at night, the dark fur poses no disadvantage in absorbing excessive sunlight. Figure 13.15b shows a species of lemur called the white sifaka that lives in semiarid regions where there is little shade. The white color of the fur may help in thermoregulation, by reflecting sunlight during the hot part of the day. However, during the cool mornings, reflection of sunlight would be a hindrance in warming up. It is interesting to note that these lemurs have black skin and only sparse fur on their bellies, and that to warm up in the morning, they show their dark bellies to the sun. The dark color enhances the absorption of sunlight.

The amount of radiant energy Q emitted by a perfect blackbody depends on several factors. To begin with, Q is proportional to the radiation time interval t

THE PHYSICS OF . . .

summer clothing.

THE PHYSICS OF . . .

a white sifaka lemur warming up.

($Q \propto t$). The longer the time, the greater the amount of energy radiated. Experiment shows that Q is also proportional to the surface area A ($Q \propto A$). An object with a large surface area radiates more energy than one with a small surface area, other things being equal. Finally, experiment reveals that Q is proportional to the *fourth power of the Kelvin temperature T* ($Q \propto T^4$), so the emitted energy increases markedly with increasing temperature. If, for example, the Kelvin temperature of an object doubles, the object emits 2^4 or 16 times more energy. Combining these factors into a single proportionality, we see that $Q \propto T^4At$. This proportionality is converted into an equation by inserting a proportionality constant σ, known as the *Stefan–Boltzmann constant*. It has been found experimentally that $\sigma = 5.67 \times 10^{-8}$ J/(s·m²·K⁴):

$$Q = \sigma T^4At$$

The relationship above holds only for a perfect emitter. Most objects are not perfect emitters, however. For instance, a dark-colored human body radiates only about 80% of the visible light energy that a perfect emitter would radiate, so Q (for dark skin) $= (0.80)\sigma T^4At$. The factor such as the 0.80 in this equation is called the *emissivity e* and is a dimensionless number between zero and one. The emissivity is the ratio of the energy an object actually radiates to the energy the object would radiate if it were a perfect emitter. For visible light, the value of e for the human body varies between about 0.65 and 0.80, the smaller values pertaining to lighter skin colors. For infrared radiation, e is nearly one for all skin colors. For a perfect blackbody emitter, $e = 1$. Including the factor e on the right side of the expression $Q = \sigma T^4At$ leads to the *Stefan–Boltzmann law of radiation.*

(a)

(b)

Figure 13.15 (a) Most lemurs, like this one, are nocturnal and have dark fur. (b) The species of lemur called the white sifaka, however, is active during the day and has white fur.

The Stefan–Boltzmann Law of Radiation

The radiant energy Q, emitted in a time t by an object that has a Kelvin temperature T, a surface area A, and an emissivity e, is given by

$$Q = e\sigma T^4At \qquad (13.2)$$

where σ has a value of 5.67×10^{-8} J/(s·m²·K⁴).

In Equation 13.2, the Stefan–Boltzmann constant σ is a universal constant in the sense that its value is the same for all bodies, regardless of the nature of their surfaces. The emissivity e, however, depends on the condition of the surface. Example 3 shows how the Stefan–Boltzmann law can be used to determine the size of a star.

Example 3   A Supergiant Star

The supergiant star Betelgeuse has a surface temperature of about 2900 K and emits a radiant power (in joules per second or watts) of approximately 4×10^{30} W. The temperature is about half and the power about 10 000 times that of our sun. Assuming that Betelgeuse is a perfect emitter (emissivity $e = 1$) and spherical, find its radius.

REASONING According to the Stefan–Boltzmann law, the power emitted is $Q/t = e\sigma T^4A$. A star with a relatively small temperature T can have a relatively large radiant power Q/t only if the area A is large. As we will see, Betelgeuse indeed has a very large area, so its radius is tremendous.

SOLUTION   Solving the Stefan–Boltzmann law for the area, we find

$$A = \frac{Q/t}{e\sigma T^4}$$

But the surface area of a sphere is $A = 4\pi r^2$, so $r = \sqrt{A/4\pi}$. Therefore, we have

$$r = \sqrt{\frac{Q/t}{4\pi e\sigma T^4}}$$

$$r = \sqrt{\frac{4 \times 10^{30}\ \text{W}}{4\pi(1)[5.67 \times 10^{-8}\ \text{J}/(\text{s}\cdot\text{m}^2\cdot\text{K}^4)](2900\ \text{K})^4}} = \boxed{3 \times 10^{11}\ \text{m}}$$

For comparison, Mars orbits the sun at a distance of 2.28×10^{11} m. Betelgeuse is certainly a "supergiant."

The next example explains how to apply the Stefan–Boltzmann law when an object, such as a wood stove, simultaneously emits and absorbs radiant energy.

Example 4   A Wood-Burning Stove

A wood-burning stove stands unused in a room where the temperature is 18 °C (291 K). A fire is started inside the stove. Eventually, the temperature of the stove surface reaches a constant 198 °C (471 K), and the room warms to a constant 29 °C (302 K). The stove has an emissivity of 0.900 and a surface area of 3.50 m². Determine the *net* radiant power generated by the stove when the stove (a) is unheated and has a temperature equal to room temperature and (b) has a temperature of 198 °C.

REASONING   The stove emits more radiant power when heated than when unheated. In both cases, however, the Stefan–Boltzmann law can be used to determine the amount of power emitted. Power is energy per unit time or Q/t. But in this problem we need to find the *net* power produced by the stove. The net power is the power the stove emits minus the power the stove absorbs. The power the stove absorbs comes from the walls, ceiling, and floor of the room, all of which emit radiation.

SOLUTION
(a) Remembering that temperature must be expressed in kelvins when using the Stefan–Boltzmann law, we find that

$$\text{Power emitted by unheated stove at 18 °C} = \frac{Q}{t} = e\sigma T^4 A \tag{13.2}$$

$$= (0.900)[5.67 \times 10^{-8}\ \text{J}/(\text{s}\cdot\text{m}^2\cdot\text{K}^4)](291\ \text{K})^4(3.50\ \text{m}^2)$$

$$= 1280\ \text{W}$$

The fact that the unheated stove emits 1280 W of power and yet maintains a constant temperature means that the stove also absorbs 1280 W of radiant power. Thus, the *net* power generated by the unheated stove is zero:

$$\text{Net power generated by stove at 18 °C} = \underbrace{1280\ \text{W}}_{\substack{\text{Power emitted}\\\text{by stove at}\\\text{18 °C}}} - \underbrace{1280\ \text{W}}_{\substack{\text{Power emitted by}\\\text{room at 18 °C and}\\\text{absorbed by stove}}} = \boxed{0}$$

(b) The hot stove (198 °C or 471 K) emits more radiant power than it absorbs from the cooler room. The radiant power the stove emits is

Power emitted by stove at 198 °C $= \dfrac{Q}{t} = e\sigma T^4 A$

$$= (0.900)[5.67 \times 10^{-8}\ \text{J}/(\text{s}\cdot\text{m}^2\cdot\text{K}^4)](471\ \text{K})^4(3.50\ \text{m}^2) = 8790\ \text{W}$$

The radiant power the stove absorbs from the room is identical to the power that the stove would emit at the constant room temperature of 29 °C (302 K). The reasoning here is exactly like that in part (a):

Power emitted by room at 29 °C and absorbed by stove $= \dfrac{Q}{t} = e\sigma T^4 A$

$$= (0.900)[5.67 \times 10^{-8}\ \text{J}/(\text{s}\cdot\text{m}^2\cdot\text{K}^4)](302\ \text{K})^4(3.50\ \text{m}^2) = 1490\ \text{W}$$

The *net* radiant power the stove produces is

Net power generated by stove at 198 °C $=$ 8790 W $-$ 1490 W $=$ $\boxed{7300\ \text{W}}$

Power emitted by stove at 198 °C   Power emitted by room at 29 °C and absorbed by stove

Example 4 illustrates that when an object has a higher temperature than its surroundings, the object emits a net radiant power $P_{net} = (Q/t)_{net}$. The net power is the power the object emits minus the power the object absorbs. Applying the Stefan–Boltzmann law as in Example 4 leads to the following expression for P_{net} when the temperature of the object is T and the temperature of the environment is T_0:

$$P_{net} = e\sigma A(T^4 - T_0{}^4) \tag{13.3}$$

13.4 APPLICATIONS

To keep heating and air-conditioning bills to a minimum, it pays to use good thermal insulation in your home. Well-insulated walls and attics minimize heat transfer by conduction. The logic behind home insulation comes directly from Equation 13.1. According to this equation, the heat per unit time Q/t flowing through a thickness of material is $Q/t = kA\,\Delta T/L$. Keeping the value for Q/t to a minimum means using materials that have small thermal conductivities k and large thicknesses L. Construction engineers, however, prefer to use Equation 13.1 in the slightly different form shown below:

$$\frac{Q}{t} = \frac{A\,\Delta T}{L/k}$$

The term L/k in the denominator is called the R value of the insulation. It is convenient to talk about R values for building materials, because the R value expresses in a single number the combined effects of thermal conductivity and thickness. Larger R values reduce the heat per unit time flowing through the

THE PHYSICS OF . . .

rating thermal insulation by R values.

Figure 13.16 The highly reflecting metal foil that covers the Hubble space telescope minimizes the temperature changes that would otherwise occur during each orbit around the earth.

THE PHYSICS OF . . .

regulating the temperature of an orbiting satellite.

THE PHYSICS OF . . .

a solar collector.

material and, therefore, mean better insulation. It is also convenient to use *R* values to describe layered slabs formed by sandwiching together a number of materials with different thermal conductivities and different thicknesses. The *R* values for the individual layers can be added to give a single *R* value for the entire slab. It should be noted, however, that *R* values are expressed using units of feet, hours, F°, and BTU for thickness, time, temperature, and heat, respectively.

When it is in the earth's shadow, an orbiting satellite is shielded from the intense electromagnetic waves emitted by the sun. But when a satellite moves out of the earth's shadow, the satellite experiences the full effect of these waves. As a result, the temperature within a satellite would increase and decrease sharply during an orbital period, to the detriment of sensitive electronic circuitry, unless precautions are taken. To minimize temperature fluctuations, satellites are often covered with a highly reflecting and, hence, poorly absorbing metal foil, as Figure 13.16 shows. By reflecting much of the sunlight, the foil minimizes temperature rises. Being a poor absorber, the foil is also a poor emitter and reduces radiant energy losses. Reducing these losses keeps the temperature from falling excessively when the satellite is in the earth's shadow.

The design of solar collectors takes into account all three methods of energy transfer to capture radiant energy from the sun. As Figure 13.17 illustrates, cool water is pumped into the collector, heated by solar energy, and then sent into the living quarters. The entire inside of the collector, including the water pipes, is coated with a highly absorptive black paint, to capture as much radiant energy as possible. The copper from which the pipes are made has a large thermal conductivity, and, therefore, readily conducts the absorbed energy to the water. The glass cover minimizes the loss of heat due to air convection.

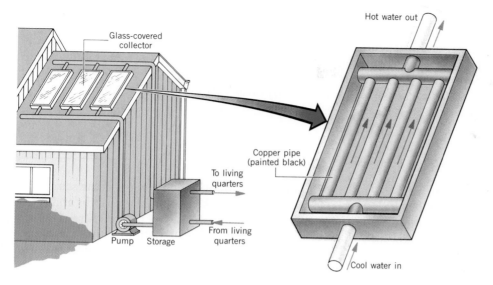

Figure 13.17 The design of a hot water solar collector takes into account energy transfer via convection, conduction, and radiation.

A thermos bottle, sometimes referred to as a Dewar flask, reduces the rate at which hot liquids cool down or cold liquids warm up. A thermos accomplishes its job by minimizing heat transfer via convection, conduction, and radiation. A thermos usually consists of a double-walled glass vessel with silvered inner walls (see Figure 13.18). The space between the walls is evacuated to minimize energy losses due to conduction and convection. The silvered surfaces reflect most of the radiant energy that would otherwise enter or leave the liquid in the thermos. Finally, little heat is lost through the glass or the rubberlike gaskets and stopper, since these materials have relatively small thermal conductivities.

While metals conduct heat especially well, diamond conducts it even better. The thermal conductivity of diamond is much greater than that of silver (see Table 13.1). The high thermal conductivity of diamond is one reason why computer

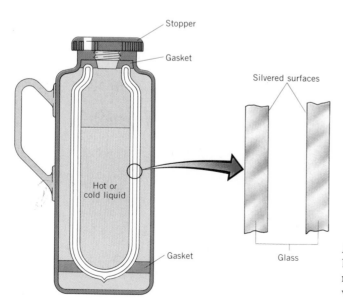

Figure 13.18 A thermos bottle or Dewar flask. A thermos bottle minimizes energy transfer due to convection, conduction, and radiation.

chip manufacturers are excited about the recent discovery of how to grow diamond coatings on silicon. They are hopeful that the discovery will lead to "faster" chips, ones that can perform more calculations per second than are currently possible. Most computer chips are made from silicon and the more rapidly a chip performs, the more heat it generates. A diamond coating, with its high thermal conductivity, would help significantly in dissipating the greater heat generated by faster chips.

SUMMARY

Convection is the process in which heat is carried by the bulk movement of a fluid. During natural convection, the warmer, less dense part of a fluid is pushed upward by the buoyant force provided by the surrounding cooler and denser part. Forced convection occurs when an external device, such as a fan or a pump, causes the fluid to move.

Conduction is the process whereby heat is transferred directly through a material, any bulk motion of the material playing no role in the transfer. The heat Q conducted during a time t through a bar of length L and cross-sectional area A is expressed as $Q = kA \, \Delta T t / L$, where ΔT is the difference in temperature between the ends of the bar and k is the **thermal conductivity** of the material. Materials that have large values of k, such as most metals, are known as thermal conductors. Materials that have small values, such as Styrofoam and wood, are referred to as thermal insulators.

Radiation is the process in which energy is transferred by electromagnetic waves. All objects, regardless of their temperature, simultaneously absorb and emit electromagnetic waves. A body that is at the same constant temperature as its surroundings absorbs and emits equal amounts of radiant energy per unit time. **Objects that are good absorbers of radiant energy are also good emitters, and objects that are poor absorbers are also poor emitters.** An object that absorbs all the radiation incident upon it is called a **perfect blackbody.** A perfect blackbody, being a perfect absorber, is also a perfect emitter.

The amount of radiant energy Q emitted during a time t by an object whose surface area is A and whose Kelvin temperature is T is given by the **Stefan–Boltzmann law,** $Q = e\sigma T^4 A t$. In this equation, σ is the Stefan–Boltzmann constant $[\sigma = 5.67 \times 10^{-8} \text{ J/} (\text{s} \cdot \text{m}^2 \cdot \text{K}^4)]$ and e is the emissivity, a dimensionless number characterizing the surface of the object. The emissivity lies between 0 and 1, being zero for a nonemitting surface and one for a perfect blackbody. The net radiant power emitted by an object of temperature T located in an environment of temperature T_0 is $P_{net} = e\sigma A(T^4 - T_0^4)$.

SOLVED PROBLEMS

Solved Problem 1   Layered Insulation

Related Problems: *11   *12   *32

One wall of a house being remodeled consists of 0.019-m-thick plywood backed by 0.076-m-thick insulation, as the drawing shows. The temperature at the inside surface is 25.0 °C, while the temperature at the outside surface is 4.0 °C, both being constant. The thermal conductivities of the insulation and plywood are, respectively, 0.030 and 0.080 J/ (s·m·C°), and the area of the wall is 35 m². Find the heat conducted through the wall in one hour.

<u>REASONING</u> The temperature at the insulation–plywood interface (see drawing) must be determined before the heat

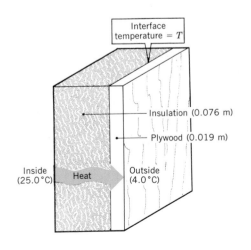

conducted through the wall can be obtained. In calculating this temperature T, we observe that no heat is accumulating in the wall, for the inner and outer temperatures are constant. Therefore, the heat conducted through the insulation must equal the heat conducted through the plywood during the same time. Moreover, the heat conducted through either material can be expressed by the relation $Q = kA\,\Delta Tt/L$.

SOLUTION Since the heat conducted through either material is the same, we have $Q_{insulation} = Q_{plywood}$. It follows from Equation 13.1 that

$$\left(\frac{kA\,\Delta Tt}{L}\right)_{insulation} = \left(\frac{kA\,\Delta Tt}{L}\right)_{plywood}$$

$$\frac{[0.030\ J/(s\cdot m\cdot C°)]A(25.0\ °C - T)t}{0.076\ m}$$

$$= \frac{[0.080\ J/(s\cdot m\cdot C°)]A(T - 4.0\ °C)t}{0.019\ m}$$

Eliminating the area A and time t algebraically and solving this equation for T reveals that the temperature at the insulation–plywood interface is $T = 5.8\ °C$.

The heat conducted through the wall can now be found by using $T = 5.8\ °C$ in the expression for either $Q_{insulation}$

or $Q_{plywood}$, since the two quantities are equal. Choosing $Q_{insulation}$, we find that

$$\text{Heat conducted through wall} = Q_{insulation}$$

$$= \frac{[0.030\ J/(s\cdot m\cdot C°)](35\ m^2)(25.0\ °C - 5.8\ °C)(3600\ s)}{0.076\ m}$$

$$= \boxed{9.5 \times 10^5\ J}$$

It is straightforward to calculate that the amount of heat flowing through the plywood in one hour would be $Q_{plywood} = 110 \times 10^5\ J$, if the insulation were absent. Clearly, the insulation helps to reduce substantially the loss of heat through the wall.

SUMMARY OF IMPORTANT POINTS The same amount of heat is conducted through each layer in a multiple-layer material when the temperatures at the layer interfaces are constant. Using this fact and applying the conduction equation ($Q = kA\,\Delta Tt/L$) to each layer makes it possible to determine the temperature at the interface between adjacent layers. With a knowledge of the interface temperature, the heat conducted through any one layer can be obtained.

QUESTIONS

1. One often hears about heat transfer by convection in gases and liquids, but not in solids. Why?

2. A heavy drape, hung close to a cold window, reduces heat loss considerably, by interfering primarily with one of the three processes of heat transfer. Explain which one.

3. The *windchill factor* is a term used by weather forecasters. Roughly speaking, it refers to the fact that you feel colder when the wind is blowing than when it is not, even though the air temperature is the same in either case. Which of the three processes for heat transfer plays the principal role in the windchill factor? Explain your reasoning.

4. Often the following warning sign is seen on bridges: "Caution—Bridge surface freezes before road surface." Account for the warning, singling out the most appropriate of the three heat transfer processes and remembering that, unlike the road, a bridge has both its top and bottom surfaces exposed to the air.

5. One way that heat is transferred from place to place inside the human body is by the flow of blood. Which one of the three heat transfer processes best describes this action of the blood? Justify your answer.

6. A poker used in a fireplace is held at one end, while the other end is in the fire. Why are pokers made of iron rather

than copper? Ignore the fact that iron may be cheaper and stronger.

7. For heat conducted along a bar, the term ΔT in the equation $Q = kA\,\Delta Tt/L$ represents the temperature difference between the two ends of the bar. Does the term ΔT in the equation for the linear expansion of a bar, $\Delta L = \alpha L_0\,\Delta T$, also represent the temperature difference between the two ends of the bar? If not, what does ΔT represent?

8. Your head feels colder under an air-conditioning vent when your hair is wet than when it is dry. Why?

9. Grandma says that it is quicker to bake a potato if you put a nail into it. In fact, she is right. Justify her baking technique in terms of one of the three processes of heat transfer.

10. Several days after a snowstorm, the roof on a house is uniformly covered with snow. On a neighboring house, however, the snow on the roof has completely melted. Which house is probably better insulated? Give your reasoning.

11. The metal cooling coils in a freezer have become coated with ice. Explain whether the ice helps or hinders the transfer of heat from warm food to the coolant in the coils.

12. Many high-quality pots have copper bases and polished stainless steel sides. A high-quality pot is designed so

heat can enter readily and be distributed evenly, while the rate of energy loss from the pot is kept to a minimum. Based on conduction and radiation principles, explain why this design is better than all-copper or all-steel units.

13. Two objects have the same shape. Object A has an emissivity of 0.3, and object B has an emissivity of 0.6. Each radiates the same power. Is the Kelvin temperature of A twice that of B? Give your reasoning.

14. A concave mirror can be used to start a fire by directing sunlight onto a small spot on a piece of paper. Explain why the mirror does not get as hot as the paper.

15. Two strips of material, A and B, are identical, except they have emissivities of 0.4 and 0.7, respectively. The strips are heated to the same temperature and have a red glow. A brighter glow signifies that more energy per second is being radiated. Which strip has the brighter glow? Explain.

16. You have to leave a hot cup of coffee sitting for five minutes. To have the warmest coffee to drink, should you add cold cream at the beginning or the end of the five-minute period? Defend your choice.

17. What is the principal means by which heat is transferred from a hot body to a cold body when they are separated by (a) a vacuum and (b) a silver bar? Account for each answer.

18. Two identical hot cups of cocoa are sitting on a kitchen table. One has a metal spoon in it and one does not. After five minutes, which cup is cooler? Explain which of the three heat transfer processes play roles here.

19. (a) Would a hot solid cube cool more rapidly if it were left intact or cut in half? Explain your answer in terms of one or more of the three heat transfer processes. (b) Using reasoning similar to that used in answering part (a), decide which cools faster, one pound of wide and flat lasagna noodles or one pound of spaghetti noodles. Assume that both kinds of noodles are made from the same pasta and start out with the same temperature.

20. If you were stranded in the mountains in bitter cold weather, it would help to minimize energy losses from your body by curling up into the tightest ball possible. Which of the factors in Equation 13.2 are you using to the best advantage by curling into a ball? Explain.

PROBLEMS

Note: For problems in this set, use the values for thermal conductivities given in Table 13.1 unless stated otherwise.

Section 13.2 Conduction

1. One end of an iron poker is placed in a fire where the temperature is 502 °C, and the other end is kept at a temperature of 26 °C. The poker is 1.2 m long and has a radius of 5.0×10^{-3} m. Ignoring the heat lost along the length of the poker, find the amount of heat conducted from one end of the poker to the other in 5.0 s.

2. A glass windowpane is 1.5 m high, 1.2 m wide, and 4.0 mm thick. In one hour, 5.5×10^6 J of heat is conducted through the glass. The temperature at the inside surface is 21 °C. What is the temperature at the outside surface? There are two answers, depending on the direction of the heat flow.

3. At a picnic, a Styrofoam cup contains lemonade and ice at 0 °C. The thickness of the cup is 2.0×10^{-3} m, and the area is 0.016 m^2. The temperature at the outside surface of the cup is 35 °C. (a) How much heat is conducted through the Styrofoam in one hour? (b) The latent heat of fusion for ice is 3.35×10^5 J/kg. What mass of ice would the heat in part (a) melt?

4. The temperature in an electric oven is 160 °C. The temperature at the outer surface in the kitchen is 50 °C. The oven (surface area = 1.6 m^2) is insulated with material that has a thickness of 0.020 m and a thermal conductivity of 0.045 J/ (s·m·C°). (a) How much energy is used to operate the oven for six hours? (b) At a price of $0.10 per kilowatt·hour for electrical energy, what is the cost of operating the oven?

5. A skier wears a jacket filled with goose down that is 15 mm thick. Another skier wears a wool sweater that is 5.0 mm thick. Both have the same surface area. Assuming the temperature difference between the inner and outer surfaces of each garment is the same, calculate the ratio (wool/goose down) of the heats lost due to conduction during same time interval.

6. Two materials have the same insulating value if the same amount of heat per second per square meter flows through each due to the same temperature difference. Ignoring air convection, what thickness of body fat is required to give the same insulating value as a 0.010-m thickness of air?

7. Heat is conducted by two bars, one made from asbestos and the other from copper. One end of each bar is at 125 °C, and the other end is at 25 °C. The bars have the same cross-sectional area of 4.00×10^{-4} m^2 and the same length of 0.200 m. Ignore any heat lost through the sides of the bars. (a) Find the *total* power or energy per second conducted by the two bars. (b) Identify the bar through which the greater percentage of the total power is transmitted. For this bar, which single parameter (length, cross-sectional area, or thermal conductivity) is responsible for the larger heat conduction?

8. A refrigerator has a surface area of 5.0 m². It is lined with 0.080-m-thick insulation whose thermal conductivity is 0.040 J/(s·m·C°). The interior temperature is kept at 5 °C, while the temperature at the outside surface is 25 °C. At what rate (in J/s or watts) is heat being removed from the unit?

***9.** In the conduction equation $Q = kA\,\Delta Tt/L$, the combination of factors kA/L is called the *conductance*. The human body has the ability to vary the conductance of the tissue beneath the skin by means of vasoconstriction and vasodilation, in which the flow of blood to the veins and capillaries underlying the skin is decreased and increased, respectively. The conductance can be adjusted over a range such that the tissue beneath the skin is equivalent to a thickness of 0.080 mm of Styrofoam or 3.5 mm of air. By what factor can the body adjust the conductance?

***10.** One end of a brass bar is maintained at 295 °C, while the other end is kept at a constant but lower temperature. The cross-sectional area of the bar is 4.0×10^{-4} m². Because of insulation, there is negligible heat loss through the sides of the bar. Heat flows through the bar, however, at the rate of 2.7 J/s. What is the temperature of the bar at a point 0.20 m from the hot end?

***11.** Two rods, one of aluminum and the other of copper, are joined end to end. The cross-sectional area of each is 4.0×10^{-4} m², and the length of each is 0.040 m. The free end of the aluminum rod is kept at 302 °C, while the free end of the copper rod is kept at 25 °C. The loss of heat through the sides of the rods may be ignored. (a) What is the temperature at the aluminum–copper interface? (b) How much heat is conducted through the unit in 2.0 s? (c) What is the temperature in the aluminum rod at a distance of 0.015 m from the hot end? *(See Solved Problem 1 for a related problem.)*

***12.** In an aluminum pot, 0.20 kg of water at 100 °C boils away in five minutes. The bottom of the pot is 2.5×10^{-3} m thick and has a surface area of 0.020 m². To prevent the water from boiling too rapidly, a stainless steel plate has been placed between the pot and the heating element. The plate is 1.2×10^{-3} m thick, and its area matches that of the pot. Assuming that heat is conducted into the water only through the bottom of the pot, find the temperature at (a) the aluminum–steel interface and (b) the steel surface in contact with the heating element. *(See Solved Problem 1 for a related problem.)*

****13.** A 0.30-m-thick sheet of ice covers a lake. The air temperature at the ice surface is -15 °C. In five minutes, the ice thickens by a small amount. Assume that no heat flows from the ground below into the water and that the added ice is very thin compared to 0.30 m. Find the number of millimeters by which the ice thickens.

****14.** The drawing shows a solid cylindrical rod made from a center cylinder of lead and an outer concentric jacket of copper. Except for its ends, the rod is insulated (not shown), so that the loss of heat from the curved surface is negligible.

When a temperature difference is maintained between its ends, this rod conducts one-half the amount of heat that it would conduct if it were solid copper. Determine the ratio of the radii r_1/r_2.

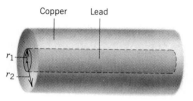

Copper   Lead

r_1

r_2

Section 13.3 Radiation

15. Assume that the earth has an average surface temperature of 22 °C and is a perfect emitter of radiation (emissivity $e = 1$). Find the energy per second per square meter that the earth radiates into space.

16. An object emits 30 W of radiant power. If it were a perfect blackbody, other things being equal, it would emit 90 W of radiant power. What is the emissivity of the object?

17. Two light bulbs are identical, except that one has a filament that operates at 2500 K, while the other operates at 2200 K. Find the ratio of the energy radiated by the hotter bulb to the energy radiated by the cooler bulb.

18. The amount of radiant power produced by the sun is approximately 3.9×10^{26} W. Assuming the sun to be a perfect blackbody sphere with a radius of 6.96×10^{8} m, find its surface temperature.

19. A perfect blackbody has a temperature of 605 °C. An identically shaped object whose emissivity is 0.400 emits the same radiant power as the blackbody. What is the Celsius temperature of this second object?

20. By what factor should the Kelvin temperature of an object be increased to double the radiant energy per second emitted by the object?

21. How many days does it take a perfect blackbody cube (0.0100 m on a side, 30.0 °C) to radiate the amount of energy that a one-hundred-watt light bulb uses in one hour?

22. Suppose the skin temperature of a naked person is 34 °C when the person is standing inside a room whose temperature is 25 °C. The skin area of the individual is 1.5 m². (a) Assuming the emissivity is 0.80, find the net loss of radiant power from the body. (b) Determine the number of food Calories of energy (1 food Calorie = 4186 J) that is lost in one hour due to the net loss rate obtained in part (a). Metabolic conversion of food into energy replaces this loss.

***23.** Our sun has a radius of 6.96×10^{8} m. Sirius B is a white dwarf star that emits only 0.04 times the radiant power of the sun, even though it has a surface temperature that is four times that of the sun. (a) What is the radius of Sirius B? (b) The mass of Sirius B is 2×10^{30} kg, nearly the same as the sun's. What is the mass density of this white dwarf star? For comparison, remember that the density of water is 1000 kg/m³.

**24. A solid aluminum sphere is coated with lampblack (emissivity = 0.97) and hung inside an evacuated container. The sphere has a radius of 0.020 m and is initially at 20.0 °C. The container is maintained at a temperature of 70.0 °C. (a) Assuming that the temperature of the sphere does not change very much, what is the *net energy* gained by the sphere in 10.0 s? (b) Estimate the change in temperature of the sphere.

**25. A small sphere (emissivity = 0.90, radius = r_1) is located at the center of a spherical asbestos shell (thickness = 1.0 cm, outer radius = r_2). The thickness of the shell is small compared to the inner and outer radii of the shell. The temperature of the small sphere is 800.0 °C, while the temperature of the inner surface of the shell is 600.0 °C, both temperatures remaining constant. Assuming that $r_2/r_1 = 10.0$ and ignoring any air inside the shell, find the temperature of the outer surface of the shell.

ADDITIONAL PROBLEMS

26. A person's body is covered with 1.6 m² of wool clothing. The thickness of the wool is 2.0×10^{-3} m. The temperature at the outside surface of the wool is 11 °C, and the skin temperature is 36 °C. How much heat per second does the person lose due to conduction?

27. The filament of a light bulb has a temperature of 3.0×10^3 °C and radiates sixty watts of power. The emissivity of the filament is 0.36. Find the surface area of the filament.

28. A car parked in the sun absorbs energy at a rate of 560 watts per square meter of surface area. The car reaches a temperature at which it radiates energy at this same rate. Treating the car as a perfect radiator ($e = 1$), find the temperature.

29. A rod, made from an unknown metal, is 1.2 m long and has a cross-sectional area of 3.0×10^{-4} m². Except for its ends, the rod is insulated, so that no heat escapes through the sides of the rod. The rod conducts heat at a rate equal to or greater than 3.7 J/s, when a temperature difference of 75 C° is maintained between its ends. From which one or more of the metals listed in Table 13.1 could the rod be made?

30. In an electrically heated home, the temperature of the ground in contact with a concrete basement wall is 12.8 °C. The temperature at the inside surface of the wall is 20.0 °C. The wall is 0.10 m thick and has an area of 9.0 m². Assume one kilowatt·hour of electrical energy costs $0.10. How many hours are required for one dollar's worth of energy to be conducted through the wall?

31. One day, the temperature of a radiator has to be 66 °C to keep the surrounding walls of a room at 27 °C. The next day is warmer outside, so the temperature of the radiator needs to be only 52 °C to keep the walls at 27 °C. Assuming the room is heated only by radiation, determine the ratio of the *net* power radiated by the unit on the colder day to that radiated on the warmer day.

*32. Three building materials, plasterboard [$k = 0.30$ J/(s·m·C°)], brick [$k = 0.60$ J/(s·m·C°)], and wood [$k = 0.10$ J/(s·m·C°)], are sandwiched together as the drawing illustrates. The temperatures at the inside and outside surfaces are 27 °C and 0 °C, respectively. Each material has the same thickness and cross-sectional area. Find the temperature at the plasterboard–brick interface and at the brick–wood interface. (*See Solved Problem 1 for a related problem.*)

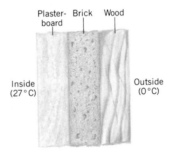

*33. A solid sphere has a temperature of 773 K. The sphere is melted down and recast into a cube that has the same emissivity and emits the same radiant power as the sphere. What is the cube's temperature?

**34. Two cylindrical rods have the same mass. One is made of silver (density = 10 500 kg/m³), and one is made of iron (density = 7860 kg/m³). Both rods conduct the same amount of heat per second when the same temperature difference is maintained across their ends. What is the ratio (silver-to-iron) of (a) the lengths and (b) the radii of these rods?

CHAPTER 14

THE IDEAL GAS LAW AND KINETIC THEORY

*Scuba divers rely on a supply of compressed air to remain under water for extended periods. The relatively large volume of air that a diver would normally breathe on land can be compressed into a small scuba tank because air is a gas. There is a lot of empty space between the molecules in a gas, and under the influence of a high pressure, this space can be reduced, so that the gas can be compressed into a smaller volume. The amount of air that can be put into a scuba tank depends on the volume of the tank, the pressure used to compress the gas, and the temperature. The fact that temperature has an effect is not surprising, for we have studied in previous chapters how changes in temperature can cause a solid, liquid, or a gaseous material to change size. We now study gases in greater detail by using a model that is referred to as an **ideal gas.** With the aid of this model, we will explore the relationship between pressure, volume, and temperature. By applying Newton's second and third laws to the motion of the particles of an ideal gas, we will obtain the great insight that the Kelvin temperature of the gas is proportional to the average kinetic energy of one of its particles. Because this approach focuses on the motion of the particles it is known as the **kinetic theory of gases,** where the word "kinetic" has the same connotation as in the phrase "kinetic energy." Since normal volumes of gas contain extremely large numbers of particles, we begin with some preliminary concepts that facilitate dealing with such large numbers.*

14.1 MOLECULAR MASS, THE MOLE, AND AVOGADRO'S NUMBER

ATOMIC AND MOLECULAR MASSES

Often it is necessary to know the masses of the atoms or molecules from which a material is made. The relative masses of the atoms of different elements can be expressed in terms of their *atomic masses,** which indicate how massive one atom is compared to another. The atomic masses of all the elements are listed in the periodic table, part of which is shown in Figure 14.1. The complete periodic table is given on the inside of the back cover. In general, the masses listed are average values and take into account the various types or isotopes of an element that exist naturally. The periodic table indicates, for example, that the helium atom (He) has an atomic mass of 4.00260, while the corresponding value for the fluorine atom (F) is 18.9984; thus, atomic fluorine is more massive than atomic helium by a factor of $18.9984/4.00260 = 4.74651$.

To set up the atomic mass scale, a value must be chosen for one of the elements. The reference element has been chosen to be the most abundant isotope of carbon, called carbon-12, and its atomic mass is defined to be exactly twelve. The units on this scale are called *atomic mass units* (u). Thus, one atomic mass unit is exactly one-twelfth the mass of a carbon-12 atom. In Figure 14.1 the atomic mass of carbon (C) is given as 12.011 u, rather than exactly 12 u, because a small amount (about 1%) of the naturally occurring material is an isotope called carbon-13. The value 12.011 u is an average that reflects the small contribution of carbon-13.

The *molecular mass* of a molecule is the sum of the atomic masses of its atoms. For instance, the molecular mass of an H_2O molecule is the sum of the atomic masses of the two hydrogen atoms and the oxygen atom; molecular mass of water $= 2(1.00794 \text{ u}) + 15.9994 \text{ u} = 18.0153 \text{ u}$.

THE MOLE AND AVOGADRO'S NUMBER

Macroscopic amounts of materials contain large numbers of atoms or molecules. Even in a small volume of gas, 1 cm³, for example, the number is enormous. It is

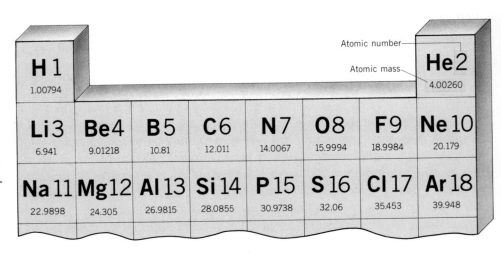

Figure 14.1 A portion of the periodic table showing the atomic number and atomic mass of each element. In the periodic table it is customary to omit the symbol "u" denoting atomic mass units.

* In chemistry the expression "atomic weight" is frequently used in place of "atomic mass."

convenient to express such large numbers in terms of a single unit, the *gram-mole*, or simply the *mole* (symbol: *mol*). *One gram-mole of a substance contains as many particles (atoms or molecules) as there are atoms in 12 grams of the isotope carbon-12.* Experiment shows that 12 grams of carbon-12 contain 6.022×10^{23} atoms. This number is called *Avogadro's number* N_A, after the Italian scientist Amedeo Avogadro (1776–1856). Although defined in terms of carbon atoms, the concept of a mole can be applied to any collection of objects by noting that one mole is Avogadro's number of objects. Thus, one mole of atomic sulfur contains 6.022×10^{23} sulfur atoms, one mole of water contains 6.022×10^{23} H_2O molecules, and one mole of golf balls contains 6.022×10^{23} golf balls. Just as the meter is the SI base unit for length, the mole is the SI base unit for expressing "the amount of substance."

Although one mole of any substance contains Avogadro's number of particles, it is *not true* in general that a mole of one substance has the same mass as a mole of another substance. For example, one mole of carbon-12 atoms is defined to have a mass of 12 grams. However, one mole of aluminum atoms has a mass of 26.9815 grams for the following reason. An aluminum atom is more massive than a carbon-12 atom by the ratio of their atomic masses, (26.9815 u)/(12 u). Since Avogadro's number (1 mole) of carbon-12 atoms has a mass of 12 grams, then Avogadro's number (1 mole) of aluminum atoms must have a mass of

$$\left(\frac{26.9815 \text{ u}}{12 \text{ u}} \right) (12 \text{ grams}) = 26.9815 \text{ grams}$$

Thus, one mole of aluminum has a mass that is equal to its atomic mass, expressed in grams. This reasoning can be extended to any other atom or molecule, with the result that *one mole of a substance has a mass in grams that is equal to the atomic or molecular mass of the substance.*

Example 1 illustrates how to determine the number of atoms and molecules present in two famous gemstones.

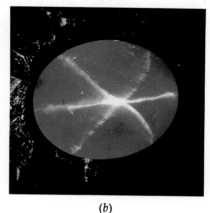

(a)

(b)

Figure 14.2 (a) The oval-shaped Hope diamond surrounded by 16 smaller diamonds. (b) The Rosser Reeves ruby. Both gems are on display at the Smithsonian Institution in Washington, D.C.

Example 1   *The Hope Diamond and the Rosser Reeves Ruby*

Figure 14.2a shows the Hope diamond (44.5 carats), which is almost pure carbon. Figure 14.2b shows the Rosser Reeves ruby (138 carats), which is primarily aluminum oxide (Al_2O_3). One carat is equivalent to a mass of 0.200 g. Determine (a) the number of carbon atoms in the diamond and (b) the number of Al_2O_3 molecules in the ruby.

REASONING   The number of atoms (or molecules) in a known mass of material is the number of moles of material times the number of atoms per mole (Avogadro's number). We can find the number of moles by dividing the known mass by the atomic mass of the substance.

SOLUTION
(a) The mass of the diamond is (44.5 carats)[(0.200 g)/(1 carat)] = 8.90 g. Since the average atomic mass of naturally occurring carbon is 12.011 u, one mole of it has a mass of 12.011 g. The number of moles of carbon in the diamond is

$$(8.90 \text{ g}) \left(\frac{1 \text{ mol}}{12.011 \text{ g}} \right) = 0.741 \text{ mol}$$

THE PHYSICS OF . . .

gemstones.

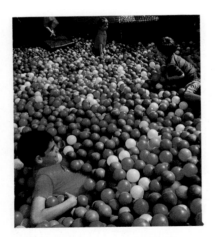

Although there are a lot of balls here, the number is extremely small compared with Avogadro's number.

Because one mole contains Avogadro's number N_A of carbon atoms, the number of atoms in the diamond is $(0.741)N_A$, or

$$(0.741 \text{ mol}) \left(\frac{6.022 \times 10^{23} \text{ atoms}}{1 \text{ mol}} \right) = \boxed{4.46 \times 10^{23} \text{ atoms}}$$

(b) The mass of the ruby is $(138 \text{ carats})[(0.200 \text{ g})/(1 \text{ carat})] = 27.6 \text{ g}$. The molecular mass of aluminum oxide (Al_2O_3) is the sum of the atomic masses of its atoms:

$$\text{Molecular mass} = 2(26.9815 \text{ u}) + 3(15.9994 \text{ u}) = 101.9612 \text{ u}$$

Thus, one mole of Al_2O_3 has a mass of 101.9612 g. Calculations like those in part (a) reveal that the ruby contains 0.271 mol or $\boxed{1.63 \times 10^{23} \text{ molecules of } Al_2O_3}$.

14.2 THE IDEAL GAS LAW AND THE BEHAVIOR OF GASES

THE IDEAL GAS LAW

The ideal gas law expresses the relationship between the pressure, the Kelvin temperature, the volume, and the number of moles of an ideal gas. An ideal gas is an idealized model for real gases. Real gases behave according to this model if their densities are sufficiently low. The condition of low density means that the molecules of the gas are so far apart that they do not interact (except during collisions that are effectively elastic).

In discussing the constant volume gas thermometer, Section 12.2 has already explained the relationship between the absolute pressure and Kelvin temperature of a low-density gas. This thermometer utilizes a small amount of gas (e.g., hydrogen or helium) placed inside a bulb of constant volume. Since the density is kept low, the gas behaves as an ideal gas. Experiment reveals that a plot of gas pressure versus temperature is a straight line, as in Figure 12.4. This plot is redrawn in Figure 14.3, with the change that the temperature axis is now labeled in kelvins rather than in degrees Celsius. The graph indicates that the absolute pressure P is directly proportional to the Kelvin temperature T ($P \propto T$), for a fixed volume and a fixed number of molecules.

The relation between absolute pressure and the number of molecules of an ideal gas is simple. Experience indicates that it is possible to increase the pressure

Figure 14.3 The pressure inside a constant-volume gas thermometer is directly proportional to the Kelvin temperature, because the gas behaves as an ideal gas.

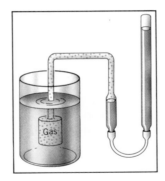

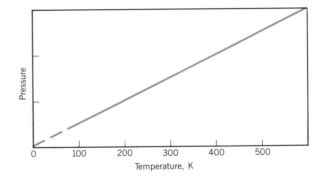

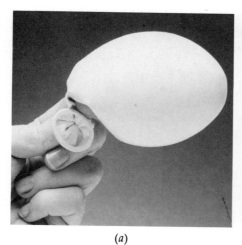
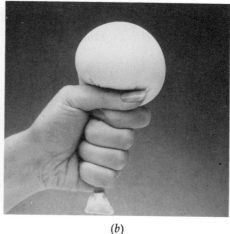

(a)   (b)

Figure 14.4 (*a*) The air pressure in the partially filled balloon can be increased by decreasing the volume of the balloon, as illustrated in (*b*).

of a gas by adding more molecules; this is exactly what happens when a tire is pumped up. When the volume and temperature of a low-density gas are kept constant, doubling the number of molecules doubles the pressure, so the absolute pressure of an ideal gas is proportional to the number of molecules or the number of moles n of the gas ($P \propto n$).

To see how the absolute pressure of a gas depends on the volume of the gas, look at the partially filled balloon in Figure 14.4*a*. This balloon is "soft," because the pressure of the air is too low to expand the balloon to its fullest. However, if all the air in the balloon is squeezed into a small "bubble," as in part *b* of the figure, the "bubble" has a very tight feel, indicating that the pressure in the smaller volume is high enough to stretch the rubber to the limit. Thus, it is possible to increase the pressure of a gas by reducing its volume, and if the number of molecules and the temperature are kept constant, the pressure of an ideal gas is inversely proportional to its volume V ($P \propto 1/V$).

The three relations discussed above for the absolute pressure of an ideal gas can be expressed as a single proportionality, $P \propto nT/V$. This proportionality can be written as an equation by inserting a proportionality constant R, called the ***universal gas constant.*** The value of R has been determined experimentally to be 8.31 J/(mol·K) for any real gas whose density is sufficiently low to ensure ideal gas behavior. The resulting equation is known as the ***ideal gas law.***

Ideal Gas Law

The absolute pressure P of an ideal gas is directly proportional to the Kelvin temperature T and the number of moles n of the gas and is inversely proportional to the volume V of the gas: $P = R(nT/V)$. In other words,

$$PV = nRT \qquad (14.1)$$

where $R = 8.31$ J/(mol·K) is the universal gas constant.

Sometimes, it is convenient to express the ideal gas law in terms of the total number of particles N, instead of the number of moles n. To obtain such an

The size or volume of a hot-air balloon depends on the pressure, temperature, and number of moles of air inside the balloon.

expression, we multiply and divide the right side of the ideal gas law by Avoga-dro's number $N_A = 6.022 \times 10^{23}$ particles/mol* and recognize that the product nN_A is equal to the total number N of particles:

$$PV = nRT = nN_A \left(\frac{R}{N_A}\right) T = N \left(\frac{R}{N_A}\right) T$$

The constant term R/N_A is referred to as *Boltzmann's constant*, in honor of the Austrian physicist Ludwig Boltzmann (1844–1906), and is represented by the symbol k:

$$k = \frac{R}{N_A} = \frac{8.31 \text{ J/(mol·K)}}{6.022 \times 10^{23} \text{ mol}^{-1}} = 1.38 \times 10^{-23} \text{ J/K}$$

With this substitution, the ideal gas law becomes

$$PV = NkT \qquad (14.2)$$

Example 2 presents an application of the ideal gas law.

THE PHYSICS OF . . .

oxygen in the lungs.

Example 2   Oxygen in the Lungs

In the lungs, the respiratory membrane separates tiny sacs of air (absolute pressure = 1.00×10^5 Pa) from the blood in the capillaries. These sacs are called alveoli, and it is from them that oxygen enters the blood. The average radius of the alveoli is 0.125 mm, and the air inside contains 14% oxygen, which is a somewhat smaller amount than in fresh air. Assuming that the air behaves as an ideal gas at body temperature (310 K), find the number of oxygen molecules in one of the sacs.

REASONING AND SOLUTION   The volume of a sac is $V = \frac{4}{3}\pi r^3 = 8.18 \times 10^{-12}$ m³. The form of the ideal gas law given in Equation 14.2 is convenient to use here, because it explicitly contains the total number N of molecules:

$$N = \frac{PV}{kT} = \frac{(1.00 \times 10^5 \text{ Pa})(8.18 \times 10^{-12} \text{ m}^3)}{(1.38 \times 10^{-23} \text{ J/K})(310 \text{ K})} = 1.9 \times 10^{14}$$

The number of oxygen molecules is 14% of this value or $0.14N = \boxed{2.7 \times 10^{13}}$.

PROBLEM SOLVING INSIGHT

In the ideal gas law, the temperature T must be expressed on the Kelvin scale. The Celsius and Fahrenheit scales cannot be used.

The next example shows that one mole of an ideal gas occupies a volume of 22.4 liters at a temperature of 0 °C and a pressure of one atmosphere (1.013×10^5 Pa). These conditions of temperature and pressure are known as *standard temperature and pressure (STP)*.

Example 3   Standard Temperature and Pressure Conditions

Find the volume occupied by one mole of an ideal gas at STP conditions.

REASONING AND SOLUTION   Before the ideal gas law can be used to deter-mine the volume, the standard temperature of 0 °C must be converted to kelvins:

---

* Since "particles" is not an SI unit, it is often omitted. Then, particles/mol = 1/mol = mol⁻¹.

$T = 0 + 273 = 273$ K. At this Kelvin temperature and at a pressure of 1.013×10^5 Pa, the volume of one mole of gas is

$$V = \frac{nRT}{P} = \frac{(1.00 \text{ mol})[8.31 \text{ J/(mol} \cdot \text{K)}](273 \text{ K})}{1.013 \times 10^5 \text{ Pa}} = 22.4 \times 10^{-3} \text{ m}^3 \qquad (14.1)$$

Since 1 liter = 1000 cm³ = 10^{-3} m³, the volume occupied by one mole of an ideal gas at STP is 22.4 liters .

The ideal gas law serves as a guide for the designers of the safety airbags with which many cars are now equipped. During a collision, these bags inflate (Figure 14.5) and cushion the impact of the driver with the steering column. To expand the bags, nitrogen is forced into them. But how much nitrogen is needed? Once designers have determined the proper pressure and volume for the expanded bag at a temperature appropriate for the coldest of driving conditions, the number n of moles of nitrogen can be estimated from the ideal gas law; $n = PV/RT$.

BOYLE'S LAW AND CHARLES' LAW

Historically the work of several investigators led to the formulation of the ideal gas law. The Irish scientist Robert Boyle (1627–1691) discovered that the absolute pressure of a fixed mass (fixed number of moles) of a low-density gas at constant temperature is inversely proportional to its volume ($P \propto 1/V$). This fact is often called Boyle's law and can be derived from the ideal gas law by noting that $P = nRT/V = \text{constant}/V$ when n and T are constants. Alternatively, if an ideal gas changes from an initial pressure and volume (P_i, V_i) to a final pressure and volume (P_f, V_f), it is possible to write $P_i V_i = nRT$ and $P_f V_f = nRT$, or

Figure 14.5 A driver's side airbag.

$$\begin{bmatrix} \text{Constant } T, \\ \text{constant } n \end{bmatrix} \qquad P_i V_i = P_f V_f \qquad (14.3)$$

This equation is a concise way of expressing Boyle's law.

Figure 14.6 illustrates how pressure and volume change according to Boyle's law for a fixed number of moles of an ideal gas at a constant temperature of 100 K. The gas begins with an initial pressure and volume of P_i and V_i and is compressed. The pressure increases as the volume decreases, according to $P = nRT/V$, until the final pressure and volume of P_f and V_f are reached. The curve that passes through the initial and final points is called an *isotherm*, meaning "same temper-

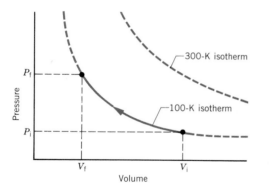

Figure 14.6 A pressure-versus-volume plot for a compression of an ideal gas at a constant temperature. Each isotherm is a plot of the equation $P = nRT/V = \text{constant}/V$, one with $T = 100$ K and the other with $T = 300$ K.

ature." If the temperature had been 300 K, rather than 100 K, the compression would have occurred along the 300-K isotherm. Different isotherms do not intersect. Example 4 deals with an application of Boyle's law to scuba diving.

THE PHYSICS OF . . .

scuba diving.

Example 4   Scuba Diving

In scuba* diving, a greater water pressure acts on a diver at greater depths. The air pressure inside the body cavities (e.g., lungs, sinuses) must be maintained at the same pressure as that of the surrounding water, otherwise they might collapse. A special valve automatically adjusts the pressure of the air breathed from a scuba tank to ensure that the air pressure equals the water pressure at all times. The scuba gear in Figure 14.7a consists of a 0.0150-m^3 tank filled with compressed air at an absolute pressure of 2.02×10^7 Pa. Assuming that air is consumed at a rate of 0.0300 m^3 per minute and that the temperature is the same at all depths, determine how long the diver can stay under at a depth of (a) 10.0 m and (b) 30.0 m.

REASONING   The time (in minutes) that a scuba diver can remain under water is equal to the volume of air that is available divided by the volume per minute consumed by the diver. The available volume is the volume of air at the pressure P_2 breathed by the diver. This pressure is determined by the depth beneath the surface, according to $P_2 = P_1 + \rho gh$, where P_1 is the atmospheric pressure at the surface (see Figure 14.7b). Since we know the pressure and volume of air in the scuba tank, and since the temperature is constant, we can use Boyle's law to find the volume of air available at the pressure P_2.

SOLUTION
(a) Using $\rho = 1025$ kg/m^3 for the density of seawater, we find that the pressure P_2 at the depth of $h = 10.0$ m is

$$P_2 = P_1 + \rho gh \tag{11.4}$$

$$P_2 = 1.01 \times 10^5 \text{ Pa} + (1025 \text{ kg/}m^3)(9.80 \text{ m/}s^2)(10.0 \text{ m}) = 2.01 \times 10^5 \text{ Pa}$$

The pressure and volume of the air in the tank are $P_i = 2.02 \times 10^7$ Pa and $V_i = 0.0150$ m^3, respectively. According to Boyle's law, the volume of air V_f available at a pressure of $P_f = 2.01 \times 10^5$ Pa is

PROBLEM SOLVING INSIGHT

When using the ideal gas law, either directly or in the form of Boyle's law, remember that the pressure P must be the absolute pressure, not the gauge pressure.

$$V_f = \frac{P_i V_i}{P_f} = \frac{(2.02 \times 10^7 \text{ Pa})(0.0150 \text{ }m^3)}{2.01 \times 10^5 \text{ Pa}} = 1.51 \text{ }m^3 \tag{14.3}$$

Of this volume, only 1.51 m^3 − 0.0150 m^3 = 1.50 m^3 is available for breathing, because 0.0150 m^3 of air always remains in the tank. At a consumption rate of 0.0300 m^3/min, the compressed air will last for

$$t = \frac{1.50 \text{ }m^3}{0.0300 \text{ }m^3/\text{min}} = \boxed{50.0 \text{ min}}$$

(b) The calculation here is like that in part (a). Equation 11.4 indicates that at a depth of 30.0 m, the water pressure is 4.02×10^5 Pa. Because this pressure is twice that at the 10.0-m depth, Boyle's law reveals that the volume of air provided by the tank is now only $V_f = 0.754$ m^3. The air available for use is 0.754 m^3 − 0.0150 m^3 = 0.739 m^3. At a consumption rate of 0.0300 m^3/min, the air will last for $\boxed{t = 24.6 \text{ min}}$, so the deeper dive must have a shorter duration.

* The word is an acronym for self-contained underwater breathing apparatus.

Another investigator whose work contributed to the formulation of the ideal gas law was the Frenchman, Jacques Charles (1746–1823). He discovered that the volume of a fixed mass (fixed number of moles) of a low-density gas at constant pressure is directly proportional to the Kelvin temperature ($V \propto T$). This relationship is known as Charles' law and can be obtained from the ideal gas law by noting that $V = nRT/P = (\text{constant})T$, if n and P are constant. Equivalently, when an ideal gas changes from an initial volume and temperature (V_i, T_i) to a final volume and temperature (V_f, T_f), it is possible to write $V_i/T_i = nR/P$ and $V_f/T_f = nR/P$, or

$$\begin{bmatrix} \text{Constant } P, \\ \text{constant } n \end{bmatrix} \qquad \frac{V_i}{T_i} = \frac{V_f}{T_f} \qquad\qquad (14.4)$$

Equation 14.4 is one way of stating Charles' law.

14.3 KINETIC THEORY OF GASES

As useful as it is, the ideal gas law provides little insight as to how pressure and temperature are related to properties of the molecules themselves, such as their speeds. To show how such microscopic properties are related to the pressure and temperature of an ideal gas, this section examines the dynamics of molecular motion with the aid of Newton's second and third laws.

THE DISTRIBUTION OF MOLECULAR SPEEDS

A macroscopic container filled with a gas at standard temperature and pressure contains a large number of particles (atoms or molecules). These particles are in constant, random motion, colliding with each other and with the walls of the container. In the course of one second, a particle undergoes many collisions, and each one changes the particle's speed and direction of motion. As a result, the atoms or molecules, in general, have different speeds. It is possible, however, to speak about an average particle speed. At any given instant, some particles have speeds less than the average, some near the average, and some greater than the average. For conditions of low gas density, the distribution of speeds within a large collection of molecules at a constant temperature was calculated by the Scottish physicist James Clerk Maxwell (1831–1879). Figure 14.8 displays the Maxwellian speed distribution curves for O_2 gas at two different temperatures.

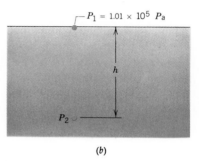

(a)

(b)

Figure 14.7   (a) The air pressure inside the body cavities of scuba divers must be maintained at the same pressure as that of the surrounding water. (b) The pressure P_2 at a depth h below the surface of the ocean is given by $P_2 = P_1 + \rho gh$, where ρ is the density of seawater.

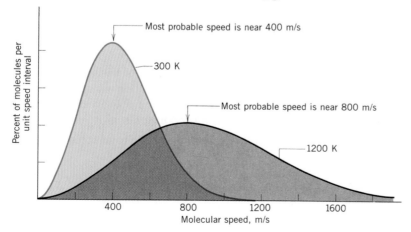

Figure 14.8   The Maxwellian distribution curves for particle speeds in oxygen gas at temperatures of 300 and 1200 K.

When the temperature is 300 K, the maximum in the curve indicates that the most probable speed is about 400 m/s. At a temperature of 1200 K, the distribution curve is shifted to the right, and the most probable speed increases to about 800 m/s.

KINETIC THEORY

If a ball is thrown against a wall, it exerts a force on the wall. As Figure 14.9 illustrates, gas particles do the same thing, except that their masses are smaller and their speeds are greater. The number of particles is so great and they strike the wall so often that the effect of their individual impacts appears as a continuous force. Dividing the magnitude of this force by the area of the wall gives the pressure exerted by the gas.

To calculate the force, consider an ideal gas composed of N identical particles in a cubical container whose sides have length L. Except for elastic* collisions, these particles do not interact. Figure 14.10 focuses attention on one particle of mass m as it strikes the right wall perpendicularly and rebounds elastically. While approaching the wall, the particle has a velocity $+v$ and linear momentum $+mv$ (see Section 7.1 for a review of linear momentum). The particle rebounds with a velocity $-v$ and momentum $-mv$, travels to the left wall, rebounds again, and heads back toward the right. The time t between collisions with the right wall is the round-trip distance $2L$ divided by the speed of the particle, that is, $t = 2L/v$. According to Newton's second law of motion, in the form of the impulse-momentum theorem, the average force exerted on the particle by the wall is given by the change in the particle's momentum per unit time:

$$\text{Average force} = \frac{\text{Final momentum} - \text{Initial momentum}}{\text{Time between successive collisions}} \tag{7.4}$$

$$= \frac{(-mv) - (+mv)}{2L/v} = \frac{-mv^2}{L}$$

According to Newton's action–reaction law, the force applied to the wall by the particle is equal in magnitude to this value, but oppositely directed (i.e., $+mv^2/L$). The magnitude F of the *total* force exerted on the right wall is equal to the number of particles that collide with the wall during the time t multiplied by the average force exerted by each particle. Since the N particles move randomly in three dimensions, one-third of them on the average strike the right wall during the time t. Therefore, the total force is

$$F = \left(\frac{N}{3}\right)\left(\frac{m\overline{v^2}}{L}\right)$$

In this result v^2 has been replaced by $\overline{v^2}$, the *average* value of the squared speed. The collection of particles possesses a Maxwell distribution of speeds, so an average value for v^2 must be used, rather than a value for any individual particle. The square root of the quantity $\overline{v^2}$ is called the **root-mean-square speed**, or, for short, the *rms speed*; $v_{\text{rms}} = \sqrt{\overline{v^2}}$. With this substitution, the total force becomes

$$F = \left(\frac{N}{3}\right)\left(\frac{mv_{\text{rms}}^2}{L}\right)$$

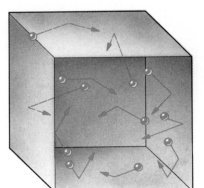

Figure 14.9 The pressure that a gas exerts is caused by the impact of its molecules on the walls of the container.

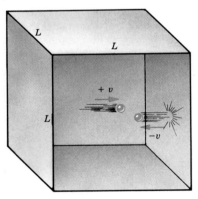

Figure 14.10 A gas particle is shown colliding elastically with the right wall of the container and rebounding from it.

* The term "elastic" is used here to mean that *on the average*, in a large number of particles, there is no gain or loss of translational kinetic energy because of collisions.

Pressure is force per unit area. Therefore, the pressure P acting on a wall of area L^2 is

$$P = \frac{F}{L^2} = \left(\frac{1}{L^2}\right)\left(\frac{N}{3}\right)\left(\frac{mv_{rms}^2}{L}\right)$$

Since the volume of the box is $V = L^3$, this equation can be written as

$$PV = \tfrac{2}{3}N(\tfrac{1}{2}mv_{rms}^2) \tag{14.5}$$

Equation 14.5 relates the macroscopic properties of the gas, its pressure and volume, to the microscopic properties of the constituent particles, their mass and speed. Since the term $\tfrac{1}{2}mv_{rms}^2$ is the average translational kinetic energy $\overline{KE}$ of an individual particle, it follows that

$$PV = \tfrac{2}{3}N(\overline{KE})$$

This result is similar to the ideal gas law, $PV = NkT$. Both equations have identical terms on the left, so the terms on the right must be equal: $\tfrac{2}{3}N(\overline{KE}) = NkT$. Therefore,

$$\overline{KE} = \tfrac{1}{2}mv_{rms}^2 = \tfrac{3}{2}kT \tag{14.6}$$

Equation 14.6 is significant, for it allows us to interpret temperature in terms of the motion of gas particles. Equation 14.6 indicates that the Kelvin temperature is directly proportional to the average translational kinetic energy of an individual particle in an ideal gas, no matter what the pressure and volume are. On the average, the particles in an ideal gas have greater kinetic energies when the gas is hotter than when it is cooler.

If two ideal gases have the same temperature, the relation $\tfrac{1}{2}mv_{rms}^2 = \tfrac{3}{2}kT$ indicates that the average kinetic energy of each kind of gas particle is the same. In general, however, the rms speeds of the different particles are not the same, for the masses may be different. The next two examples illustrate these facts and show how rapidly gas particles move at normal temperatures.

Example 5   The Kinetic Energy of Molecules in Air

Air is primarily a mixture of nitrogen N_2 and oxygen O_2. Assume that each behaves as an ideal gas. At a temperature of 293 K, find the average translational kinetic energy for each type of molecule.

REASONING AND SOLUTION   According to kinetic theory, the particles in each gas have the *same* average translational kinetic energy, since each gas has the same temperature:

$$\overline{KE} = \tfrac{3}{2}kT = \tfrac{3}{2}(1.38 \times 10^{-23} \text{ J/K})(293 \text{ K}) = \boxed{6.07 \times 10^{-21} \text{ J}} \tag{14.6}$$

Example 6   The Speed of Molecules in Air

In Example 5 we found that the average translational kinetic energy of each molecule in air is 6.07×10^{-21} J at 293 K. Determine the rms speed of the nitrogen (molecular mass = 28.0 u) and oxygen (molecular mass = 32.0 u) molecules in the air at this temperature.

REASONING The average translational speed can be obtained from Equation 14.6 $v_{rms} = \sqrt{2(\overline{KE})/m}$, once the mass of each type of gas particle is known. Since Avogadro's number of particles (1 mole) has a mass in grams equal to the molecular mass, we can determine the mass of one particle.

SOLUTION Using Avogadro's number and the molecular mass of nitrogen, we find a particle mass of

$$m = \frac{28.0 \text{ g/mol}}{6.022 \times 10^{23} \text{ mol}^{-1}} = 4.65 \times 10^{-23} \text{ g} = 4.65 \times 10^{-26} \text{ kg}$$

Similarly, we find the particle mass for oxygen to be 5.31×10^{-26} kg. The calculation of the rms speed for nitrogen is shown below, the calculation for oxygen being analogous:

[Nitrogen] $\qquad v_{rms} = \sqrt{\dfrac{2(\overline{KE})}{m}} = \sqrt{\dfrac{2(6.07 \times 10^{-21} \text{ J})}{4.65 \times 10^{-26} \text{ kg}}} = \boxed{511 \text{ m/s}}$

[Oxygen] $\qquad v_{rms} = \boxed{478 \text{ m/s}}$

For comparison, the speed of sound at a temperature of 293 K is 343 m/s (767 mi/h).

PROBLEM SOLVING INSIGHT

The average translational kinetic energy is the same for all ideal-gas molecules at the same temperature, regardless of the molecular mass. The rms translational speed of the molecules is not the same, however, but depends on the mass.

The equation $\overline{KE} = \frac{3}{2}kT$ has also been applied to particles that are much larger than atoms or molecules. The English botanist, Robert Brown (1773–1858) observed through a microscope that pollen grains suspended in water move on very irregular, zigzag paths. This Brownian motion can also be observed with other particle suspensions, such as fine smoke particles in air. In 1905, Albert Einstein (1879–1955) showed that Brownian motion could be explained as a response of the large suspended particles to impacts from the moving molecules of the fluid medium (e.g., water or air). As a result of the impacts, the suspended particles have the same average translational kinetic energy as the fluid molecules, namely, $\overline{KE} = \frac{3}{2}kT$. But unlike the molecules, the particles are large enough to be seen through a microscope and have a comparatively small average velocity because of their relatively large mass.

THE INTERNAL ENERGY OF A MONATOMIC IDEAL GAS

Chapter 15 deals with the science of thermodynamics, in which the concept of internal energy plays an important role. Using the results just developed for the average translational kinetic energy, we conclude this section by expressing the internal energy of a monatomic ideal gas in a form that is suitable for use later on.

The internal energy of a substance is the sum of the various kinds of energy that the atoms or molecules of the substance possess. A monatomic ideal gas is composed of single atoms. These atoms are assumed to be so small that the mass is concentrated at a point, with the result that the moment of inertia I about the center of mass is negligible. Thus, the rotational kinetic energy $\frac{1}{2}I\omega^2$ is also negligible. Vibrational kinetic and potential energies are absent, because the atoms are not connected by chemical bonds and, except for elastic collisions, do not interact. As a result, the internal energy U is the total translational kinetic energy of the N atoms that constitute the gas: $U = N(\frac{1}{2}mv_{rms}^2)$. Since $\frac{1}{2}mv_{rms}^2 = \frac{3}{2}kT$ according to Equation 14.6, the internal energy can be written in terms of the Kelvin temperature as

$$U = N(\tfrac{1}{2}mv_{rms}^2) = N(\tfrac{3}{2}kT)$$

Usually, U is expressed in terms of the number of moles n, rather than the number of atoms N. Using the fact that Boltzmann's constant is $k = R/N_A$, where R is the universal gas constant and N_A is Avogadro's number, and realizing that $N/N_A = n$, we find that

$$\begin{bmatrix} \text{Monatomic} \\ \text{ideal gas} \end{bmatrix} \qquad U = \tfrac{3}{2}nRT \qquad\qquad (14.7)$$

The internal energy of the helium in a Goodyear blimp is determined in the next example.

Example 7 A Goodyear Blimp

The helium in a Goodyear blimp typically has a volume of 5400 m³. At a temperature of 283 K and a pressure of 1.1×10^5 Pa, what is the internal energy of this helium?

REASONING The internal energy U of an ideal monatomic gas, like helium, is the sum of the average translational kinetic energies of all the atoms. This sum is given by Equation 14.7 as $U = \tfrac{3}{2}nRT$, where n is the number of moles of helium in the blimp. The number of moles can be determined with the aid of the ideal gas law as $n = PV/RT$.

SOLUTION The number n of moles of helium is

$$n = \frac{PV}{RT} = \frac{(1.1 \times 10^5 \text{ Pa})(5400 \text{ m}^3)}{[8.31 \text{ J/(mol·K)}](283 \text{ K})} = 2.5 \times 10^5 \text{ mol}$$

The internal energy of the helium is

$$U = \tfrac{3}{2}nRT = \tfrac{3}{2}(2.5 \times 10^5 \text{ mol})[8.31 \text{ J/(mol·K)}](283 \text{ K}) = \boxed{8.8 \times 10^8 \text{ J}}$$

---

*14.4 DIFFUSION

You can smell the fragrance of a perfume at some distance from an open bottle, because perfume molecules evaporate from the liquid, where they are relatively concentrated, and spread out into the air, where they are less concentrated. During their journey, they collide with other molecules, so their paths resemble the zigzag paths characteristic of Brownian motion. The process in which molecules move from a region of higher concentration to one of lower concentration is called *diffusion*. Diffusion also occurs in liquids (see Figure 14.11) and solids. However, compared to the rate of diffusion in gases, the rate is generally smaller in liquids and even smaller in solids. The host medium, such as the air or water in the examples above, is referred to as the *solvent*, while the diffusing substance, like the perfume molecules or the ink in Figure 14.11, is known as the *solute*.

The diffusion process can be described in terms of the arrangement in Figure 14.12a. A hollow channel of length L and cross-sectional area A is filled with a fluid. The left end of the channel is connected to a container in which the solute concentration C_2 is relatively high, while the right end is connected to a container in which the solute concentration C_1 is lower. These concentrations are defined as the total mass of the solute molecules divided by the volume of the solution (e.g., 0.1 kg/m³). Because of the difference in concentration between the ends of the

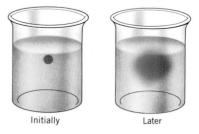

Initially   Later

Figure 14.11 A drop of ink placed in water eventually becomes completely dispersed because of diffusion.

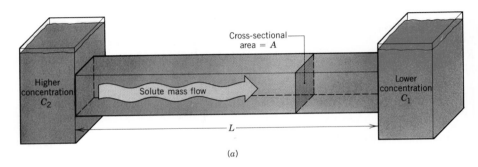

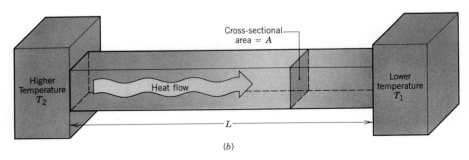

Figure 14.12 (a) Solute mass diffuses through the channel from the region of higher concentration to the region of lower concentration. (b) Heat is conducted along a bar whose ends are maintained at different temperatures.

channel, $\Delta C = C_2 - C_1$, there is a net diffusion of the solute from the left end to the right end.

Figure 14.12a is similar to Figure 13.8 for the conduction of heat along a bar, which, for convenience, is reproduced in Figure 14.12b. When the ends of the bar are maintained at different temperatures, T_2 and T_1, the heat Q conducted along the bar in a time t is

$$Q = \frac{kA\,\Delta Tt}{L} \tag{13.1}$$

where $\Delta T = T_2 - T_1$, and k is the thermal conductivity. Whereas conduction is the flow of heat from a region of higher temperature to a region of lower temperature, diffusion is the mass flow of solute from a region of higher concentration to a region of lower concentration. By analogy with Equation 13.1, it is possible to write an equation for diffusion: (1) replace Q by the mass m of solute that is diffusing through the channel, (2) replace $\Delta T = T_2 - T_1$ by the difference in concentrations $\Delta C = C_2 - C_1$, and (3) replace k by a constant known as the diffusion constant D. The resulting equation, first formulated by the German physiologist Adolf Fick (1829–1901), is referred to as *Fick's law of diffusion.*

Fick's Law of Diffusion

The mass m of solute that diffuses in a time t through a solvent contained in a channel of length L and cross-sectional area A is

$$m = \frac{DA\,\Delta Ct}{L} \tag{14.8}$$

where ΔC is the concentration difference between the ends of the channel and D is the diffusion constant.

SI Unit for the Diffusion Constant: m^2/s

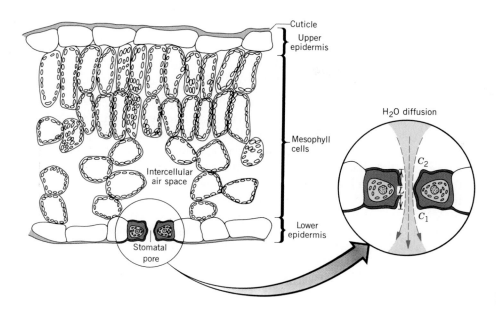

Figure 14.13 A cross-sectional view of a leaf. Water vapor diffuses out of the leaf through a stomatal pore.

It can be verified from Equation 14.8 that the diffusion constant has units of m²/s, the exact value depending on the nature of the solute and the solvent. For example, the diffusion constant for ink in water is different than that for ink in benzene. Example 8 illustrates an important application of Fick's law.

Example 8   Water Given Off by Plant Leaves

Large amounts of water can be given off by plants. It has been estimated, for instance, that a single sunflower plant can lose up to a pint of water a day during the growing season. Figure 14.13 shows a cross-sectional view of a leaf. Inside the leaf, water passes from the liquid phase to the vapor phase at the walls of the mesophyll cells. The water vapor then diffuses through the intercellular air spaces and eventually exits the leaf through small openings, called stomatal pores. The diffusion constant for water vapor in air is $D = 2.4 \times 10^{-5}$ m²/s. A stomatal pore has a cross-sectional area of about $A = 8.0 \times 10^{-11}$ m² and a length of about $L = 2.5 \times 10^{-5}$ m. The concentration of water vapor on the interior side of a pore is roughly $C_2 = 0.022$ kg/m³, while that on the outside is approximately $C_1 = 0.011$ kg/m³. Determine the mass of water vapor that passes through a stomatal pore in one hour.

REASONING AND SOLUTION   Fick's law of diffusion shows that

$$m = \frac{DA\,\Delta Ct}{L} \tag{14.8}$$

$$m = \frac{(2.4 \times 10^{-5}\ \text{m}^2/\text{s})(8.0 \times 10^{-11}\ \text{m}^2)(0.022\ \text{kg/m}^3 - 0.011\ \text{kg/m}^3)(3600\ \text{s})}{2.5 \times 10^{-5}\ \text{m}}$$

$$= \boxed{3.0 \times 10^{-9}\ \text{kg}}$$

This amount of water may not seem significant. However, a single leaf may have a million or so stomatal pores, so the water lost by an entire plant can be substantial.

THE PHYSICS OF . . .

water loss from plant leaves.

INTEGRATION OF CONCEPTS

KINETIC THEORY OF GASES AND NEWTON'S LAWS OF MOTION

The kinetic theory of gases provides us with an understanding of temperature on a microscopic basis. It reveals that the Kelvin temperature is proportional to the average translational kinetic energy of an individual particle of an ideal gas. To put this result in perspective, remember that temperature is a property that characterizes, for example, the gas in a weather balloon. In the sense that the balloon contains a large number of gas particles, it is a macroscopic object, and temperature is a macroscopic property. Thus, kinetic theory bridges the gap between the macroscopic world, where large numbers of particles are important, and the world of individual atoms and molecules. And Newton's laws of motion are the pillars upon which the bridge is built. These laws apply to the collision of an individual gas particle with the walls of a container. During the collision, the wall applies a force to the particle, which causes the particle to rebound. Newton's second law, in the form of the impulse–momentum theorem, leads to an expression for the force exerted by the wall on the particle. The third law indicates that the particle exerts an oppositely directed force of equal magnitude on the wall. When a large number of gas particles strike a surface, the collective force per unit area that they exert is the pressure of the gas. The pressure, in turn, is related to the Kelvin temperature by the ideal gas law. The basis of kinetic theory, then, is to be found in Newton's laws of motion.

SUMMARY

Each element in the periodic table is assigned an **atomic mass.** One **atomic mass unit** (u) is exactly one-twelfth the mass of an atom of carbon-12. The **molecular mass** of a molecule is the sum of the atomic masses of its atoms. One **mole** of a substance contains **Avogadro's number** N_A of particles, where $N_A = 6.022 \times 10^{23}$ particles per mole. The mass in grams of one mole of a substance is equal to the atomic or molecular mass of its particles.

The **ideal gas law** relates the absolute pressure P, the volume V, the number of moles n, and the Kelvin temperature T of an ideal gas according to $PV = nRT$, where $R = 8.31$ J/(mol·K) is the universal gas constant. An alternative form of the ideal gas law is $PV = NkT$, where N is the number of particles and $k = R/N_A$ is Boltzmann's constant. A real gas behaves as an ideal gas when the density of the real gas is low enough that its particles do not interact, except via elastic collisions.

The distribution of particle speeds in an ideal gas at constant temperature is the **Maxwell speed distribution.** According to the **kinetic theory of gases,** an ideal gas consists of a large number of particles (atoms or molecules) that are in constant random motion. The particles are far apart compared to their dimensions, so they do not interact except when elastic collisions occur. The pressure on the walls of a container is produced by the impact of the particles with the walls. According to the kinetic theory of gases, the Kelvin temperature T of an ideal gas is a measure of the average translational kinetic energy $\overline{KE}$ per particle through the relation $\overline{KE} = \frac{3}{2}kT$, where k is Boltzmann's constant. The **internal energy** U of n moles of a monatomic ideal gas is $U = \frac{3}{2}nRT$.

Diffusion is the process whereby solute molecules move through a solvent from a region of higher concentration to a region of lower concentration. **Fick's law of diffusion** states that the mass m of solute that diffuses in a time t through a solvent contained in a channel of length L and cross-sectional area A is given by $m = (DA\, \Delta Ct)/L$, where ΔC is the concentration difference between the ends of the channel and D is the diffusion constant.

QUESTIONS

1. (a) Which, if either, contains a greater number of molecules, a mole of hydrogen (H_2) or a mole of oxygen (O_2)? (b) Which one has more mass? Give reasons for your answers.

2. Suppose two different substances, A and B, have the same mass densities. (a) In general, does one mole of substance A have the same mass as one mole of substance B? (b) Does 1 m^3 of substance A have the same mass as 1 m^3 of substance B? Justify each answer.

3. The entrance to a restaurant consists of two doors, arranged as in the drawing. When door A is opened *suddenly* to the outside, door B also swings briefly toward the outside, even though no one touches it. Explain this behavior in terms of the air pressure in the restaurant and in the space between the doors.

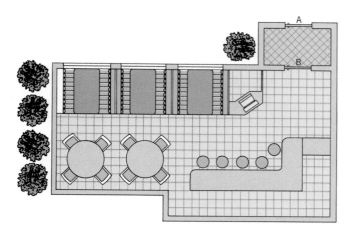

4. Assuming that air behaves like an ideal gas, explain what happens to the pressure in a tightly sealed house when the electric furnace turns on for a while.

5. Airplanes have emergency oxygen bags that fall from the ceiling when the air pressure in the cabin suddenly drops. This could happen, for example, if part of the fuselage rips open. At high altitudes, where the air pressure outside the plane is relatively low, the oxygen bags expand to their fullest. However, at low altitudes, the bags expand only partially. Why don't the bags expand to their fullest at low altitudes?

6. A slippery cork is being pressed into a very full (but not 100% full) bottle of wine. When released, the cork slowly slides back out. However, if some wine is removed from the bottle before the cork is inserted, the cork does not slide out. Account for these observations in terms of the ideal gas law.

7. The bubbles emitted by the breathing apparatus of a scuba diver expand as they rise to the surface of the water. Why? In giving your explanation, state any assumptions you are making.

8. A commonly used packing material consists of "bubbles" of air trapped between bonded layers of plastic, as the photograph shows. Using the ideal gas law, explain why this packing material offers less protection on cold days than on warm days.

9. If the translational speed of each molecule in an ideal gas were tripled, would the Kelvin temperature also triple? If not, by what factor would the Kelvin temperature increase? Account for your answer.

10. Suppose that the atoms in a container of helium (He) have the same translational rms speed as the molecules in a container of argon (Ar). Treating each gas as an ideal gas, explain which, if either, has the greater temperature.

11. Example 6 in the text shows that, near room temperature, a gas molecule has a translational rms speed on the order of hundreds of meters per second. At such a speed, a molecule could travel across an ordinary room in just a fraction of a second. Yet, it often takes several seconds, and sometimes minutes, for the smell of a perfume to travel across a room. Why does it take so long?

12. In the lungs, oxygen in very small sacs called alveoli diffuses into the blood. The diffusion occurs directly through the walls of the sacs. The walls are very thin, so the oxygen diffuses over a distance L that is quite small. Because there are so many alveoli, the effective area A across which diffusion occurs is very large. Use this information, together with Fick's law of diffusion, and explain why the mass of oxygen per second that diffuses into the blood is large.

PROBLEMS

Note: The pressures referred to in these problems are absolute pressures, unless indicated otherwise.

Section 14.1 Molecular Mass, the Mole, and Avogadro's Number

1. Methane (CH_4) is a gas found on some planets and in interstellar space. What is (a) the molecular mass of methane and (b) the mass (in kg) of a methane molecule?

2. Hemoglobin has a molecular mass of 64 500 u. Find the mass (in kg) of one molecule of hemoglobin.

3. Glucose is a sugar whose chemical formula is $C_6H_{12}O_6$. (a) Determine the molecular mass of glucose. (b) How many molecules are contained in 350 g of glucose?

4. A can of soda has a mass of 354 g. Assume that the soda is mainly water (H_2O), and find the number of water molecules in the can.

5. The hydrogen in a certain amount of hydrochloric acid (HCl) is equivalent to four grams of hydrogen gas (H_2). To how many grams of chlorine gas (Cl_2) is the chlorine in the acid equivalent?

*6. A mass of 135 g of an element is known to contain 30.1×10^{23} atoms. What is the element?

*7. Ethyl alcohol (C_2H_5OH) has a density of 806 kg/m³, and a volume of 2.00×10^{-3} m³. (a) Determine the mass (in kg) of a molecule of ethyl alcohol, and (b) find the number of molecules in the liquid.

**8. Estimate the spacing between the centers of neighboring atoms in a piece of solid aluminum, based on a knowledge of the density (2700 kg/m³) and atomic mass (26.9815 u) of aluminum. (*Hint: Assume the volume of the solid is filled with many small cubes, with one atom at the center of each.*)

Section 14.2 The Ideal Gas Law and the Behavior of Gases

9. In a portable oxygen system, the oxygen (O_2) is contained in a cylinder whose volume is 0.0028 m³. A full cylinder has an absolute pressure of 1.5×10^7 Pa when the temperature is 296 K. Find the mass of oxygen in the cylinder.

10. At the start of a trip, a driver adjusts the pressure in her tires to be 2.81×10^5 Pa when the outdoor temperature is 284 K. At the end of the trip she measures the pressure to be 3.01×10^5 Pa. Neglecting the expansion of the tires, find the air temperature inside the tires at the end of the trip.

11. In a diesel engine, the piston compresses air at 305 K to a volume that is one-sixteenth of the original volume and a

pressure that is 55.0 times the original pressure. What is the temperature of the air after the compression?

12. A 0.010-m³ container is initially evacuated, and then 2.0 g of water is placed in it. After a time, all the water evaporates, and the temperature is 348 K. Find the pressure.

13. Oxygen for hospital patients is kept in special tanks, where the oxygen has a pressure of 65.0 atmospheres and a temperature of 288 K. The tanks are stored in a separate room, and the oxygen is pumped to the patient's room, where it is administered at a pressure of 1.00 atmosphere and a temperature of 297 K. What volume does 1.00 m³ of oxygen in the tanks occupy at the conditions in the patient's room?

14. The air inside a refrigerator has a temperature of 280 K, a volume of 0.44 m³, and a pressure of 1.01×10^5 Pa. The molecular mass of air is 29 u. What is the mass of the air?

15. In certain regions of outer space the temperature is about 3 K, and there are approximately 5×10^6 molecules per cubic meter. (a) How many moles are there per cubic meter? (b) What is the pressure exerted by this gas?

16. A young male adult takes in about 5.0×10^{-4} m³ of fresh air during a normal breath. Fresh air contains approximately 21% oxygen. Assuming that the pressure in the lungs is 1.0×10^5 Pa and air is an ideal gas at a temperature of 310 K, find the number of oxygen molecules in a normal breath.

17. A frictionless gas-filled cylinder is fitted with a movable piston, as the drawing shows. The block resting on the top of the piston determines the constant pressure that the gas has. The height h is 12.0 cm when the temperature is 273 K and increases as the temperature increases. What is the value of h when the temperature reaches 318 K?

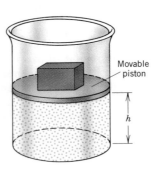

Movable piston

h

18. The relative humidity is 67% on a day when the temperature is 34 °C. Using the graph that accompanies problem 63 in Chapter 12, determine the number of moles of water vapor per cubic meter of air.

*19. An air bubble is located at a depth h beneath the surface of a pond. The bubble rises and reaches the surface, where the volume of the bubble has expanded to twice its initial value. Assuming the temperature of the bubble remains constant and atmospheric pressure is 1.01×10^5 Pa, find the depth h.

*20. An ideal gas exerts a pressure of 5.00×10^5 Pa when its temperature is 4.00×10^2 K. What is the mass density if the gas is helium (He)?

*21. A primitive diving bell consists of a cylindrical tank with one end open and one end closed. The tank is lowered into a freshwater lake, open end downward. Water rises into the tank, compressing the trapped air, whose temperature remains constant during the descent. The tank is brought to a halt when the distance between the surface of the water in the tank and the surface of the lake is 40.0 m. Atmospheric pressure at the surface of the lake is 1.01×10^5 Pa. Find the fraction of the tank's volume that is filled with air.

*22. One assumption of the ideal gas law is that the atoms or molecules themselves occupy a negligible volume. Verify that this assumption is reasonable by considering gaseous argon (Ar). Argon has an atomic radius of 0.70×10^{-10} m. For STP conditions, calculate the percentage of the total volume occupied by the atoms.

**23. A spherical balloon is made from a material whose mass is 2.00 kg. The thickness of the material is negligible compared to the 1.20-m radius of the balloon. The balloon is filled with helium gas at a temperature of 296 K and just floats in air, neither rising nor falling. The density of the surrounding air is 1.19 kg/m³. Find the absolute pressure of the helium gas.

**24. A gas fills the right portion of a horizontal cylinder whose radius is 5.00 cm. The initial pressure of the gas is 1.01×10^5 Pa. A frictionless movable piston separates the gas from the left portion of the cylinder that is evacuated and contains an ideal spring, as the drawing shows. The piston is initially held in place by a pin. The spring is initially unstrained, and the length of the gas-filled chamber is 20.0 cm. When the pin is removed and the gas is allowed to expand at constant temperature, the length of the gas-filled chamber doubles. Determine the spring constant of the spring.

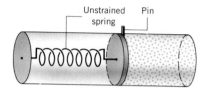

Unstrained   Pin
spring

Section 14.3 Kinetic Theory of Gases

25. The surface of the sun has a temperature of about 6.0×10^3 K. This hot gas contains hydrogen atoms ($m = 1.67 \times 10^{-27}$ kg). Find the rms speed of these atoms.

26. Very fine smoke particles are suspended in air. The translational rms speed of a smoke particle is 4.9×10^{-3} m/s, and the temperature is 295 K. Find the mass of a particle.

27. The temperature of a gas is raised from 303 K to 373 K. (a) What is the ratio of the translational kinetic energy of a gas molecule at the higher temperature to that at the lower temperature? (b) What is the ratio of the molecule's rms speeds at the two temperatures?

28. The average value of the squared speed $\overline{v^2}$ does not equal the square of the average speed $(\overline{v})^2$. To verify this fact, consider three particles with the following speeds: $v_1 = 2.0$ m/s, $v_2 = 5.0$ m/s, and $v_3 = 8.0$ m/s. Calculate (a) $\overline{v^2} = \frac{1}{3}(v_1^2 + v_2^2 + v_3^2)$ and (b) $(\overline{v})^2 = [\frac{1}{3}(v_1 + v_2 + v_3)]^2$.

29. The *escape velocity* is the minimum velocity that an object must have so it can escape from the earth without being pulled back by the gravitational force. The escape velocity is 1.12×10^4 m/s. At what temperature is the rms speed large enough for hydrogen molecules (H_2) to have enough translational velocity to leave the earth?

30. At what temperature would the translational rms speed of hydrogen molecules (H_2) be equal to that of oxygen molecules (O_2) at 3.0×10^2 K?

31. Neon (Ne) is a monatomic gas. At a temperature of 323 K, what is the internal energy of two grams of neon?

*32. Helium (He), a monatomic gas, fills a 0.010-m³ container. The pressure of the gas is 6.2×10^5 Pa. How long would a 0.25-hp engine have to run (1 hp = 746 W) to produce an amount of energy equal to the internal energy of this gas?

*33. In 10.0 s, 200 bullets strike and embed themselves in a wall. The bullets strike the wall perpendicularly. Each bullet has a mass of 5.0×10^{-3} kg and a speed of 1200 m/s. (a) What is the average change in momentum per second for the bullets? (b) Determine the average force exerted on the wall. (c) Assuming the bullets are spread out over an area of 3.0×10^{-4} m², obtain the average pressure they exert on this region of the wall.

*34. The pressure of oxygen (O_2) in a volume of 50.0 m³ is 2.12×10^4 Pa. The volume contains 421 moles of oxygen. Find the translational rms speed of the oxygen molecules.

Section 14.4 Diffusion

35. Ammonia, which has a strong smell, is diffusing through air contained in a tube of length L and cross-sectional area of 4.0×10^{-4} m². The diffusion constant for ammonia in air is 4.2×10^{-5} m²/s. In a certain time, 8.4×10^{-8} kg of ammonia diffuses through the air when the difference in ammonia concentration between the ends of the tube is 3.5×10^{-2} kg/m³. Find the speed at which ammonia diffuses through the air, that is, the length of the tube divided by the time to travel this length.

36. The diffusion constant of ethanol in water is 12.4×10^{-10} m^2/s. A cylinder has a cross-sectional area of 4.00 cm^2 and a length of 2.00 cm. A difference in ethanol concentration of 1.50 kg/m^3 is maintained between the ends of the cylinder. In one hour, what mass of ethanol diffuses through the cylinder?

37. It is found that the amino acid glycine diffuses through water at a mass rate of $m/t = 6.00 \times 10^{-14}$ kg/s. The diffusion constant is 10.6×10^{-10} m^2/s. A tube of water has a radius of 1.40 cm. What difference in concentration per unit length of the tube $\Delta C/L$ must be maintained to give this flow?

38. Carbon tetrachloride (CCl_4) is diffusing through benzene (C_6H_6), as the drawing illustrates. The concentration of CCl_4 at the left end of the tube is maintained at 1.00×10^{-2} kg/m^3, and the diffusion constant is 20.0×10^{-10} m^2/s. The CCl_4 enters the tube at a mass rate of 5.00×10^{-13} kg/s. Using this data and that shown in the drawing, find (a) the mass of CCl_4 per second that passes point A and (b) the concentration of CCl_4 at point A.

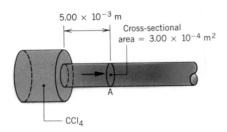

5.00 × 10^{-3} m

Cross-sectional area = 3.00 × 10^{-4} m^2

A

CCl_4

*39. It is possible to convert Fick's law into a form that is useful in situations where the concentration is essentially zero at one end of the diffusion channel ($C_1 = 0$ in Figure 14.12a). In such circumstances, Fick's law becomes $m = DAC_2t/L$. (a) Noting that AL is the volume V of the channel and that m/V is the average concentration of solute in the channel, show that Fick's law becomes $t = L^2/(2D)$. This form of Fick's law can be used to estimate the time required for the first solute molecules to traverse the channel. (b) A bottle of perfume is opened in a room where convection currents are absent. Assuming the diffusion constant for perfume in air is 1.0×10^{-5} m^2/s, estimate the minimum time required for the perfume to be smelled 2.5 cm away.

ADDITIONAL PROBLEMS

40. An ultrahigh vacuum pump can reduce the absolute pressure to 1.2×10^{-7} Pa. In a volume of 2.0 m^3 and at a temperature of 3.0×10^2 K, how many molecules of gas are present at this pressure?

41. A bicycle tire whose volume is 4.1×10^{-4} m^3 has a temperature of 296 K and a pressure of 4.8×10^5 Pa. A cyclist brings the pressure up to 6.2×10^5 Pa without changing the temperature. How many moles of air must be pumped into the tire?

42. The chlorophyll-a molecule, $C_{55}H_{72}MgN_4O_5$, is important in photosynthesis. (a) Determine its molecular mass (in atomic mass units). (b) What is the mass (in grams) of 3.00 moles of chlorophyll-a molecules?

43. A tube has a length of 0.015 m and a cross-sectional area of 7.0×10^{-4} m^2. The tube is filled with a solution of sucrose in water. The diffusion constant of sucrose in water is 5.0×10^{-10} m^2/s. A difference in concentration of 3.0×10^{-3} kg/m^3 is maintained between the ends of the tube. How much time is required for 8.0×10^{-13} kg of sucrose to be transported through the tube?

44. If the translational rms speed of the water vapor molecules (H_2O) in air is 648 m/s, what is the translational rms speed of the carbon dioxide molecules (CO_2)? Both gases are at the same temperature.

*45. A closed cylindrical tank has a height of 0.80 m. The tank is initially filled with air at a pressure of 2.0 atm. Water is pumped into the tank until the air pressure reaches 6.0 atm. Assuming the temperature of the air remains constant, determine the height of the compressed air.

*46. A drop of water has a radius of 1.20 mm. How many water molecules are in the drop?

*47. In a TV, electrons with a speed of 8.4×10^7 m/s strike the screen from behind, causing it to glow. The electrons come to a halt after striking the screen. Each electron has a mass of 9.11×10^{-31} kg, and there are 6.2×10^{16} electrons per second hitting the screen over an area of 1.2×10^{-7} m^2. What is the pressure that the electrons exert on the screen?

*48. Assume the pressure in a room remains constant at 1.01×10^5 Pa and the air is composed only of nitrogen (N_2). The volume of the room is 60.0 m^3. When the temperature increases from 289 to 302 K, what mass of air escapes from the room?

**49. At the normal boiling point of a material, the liquid phase has a density of 958 kg/m^3, and the vapor phase has a density of 0.598 kg/m^3. Determine the ratio of the distance between neighboring molecules in the gas phase to that in the liquid phase. (Hint: Assume the volume of each phase is filled with many small cubes, with one molecule at the center of each cube.)

**50. A cylindrical glass beaker of height 1.520 m rests on a table. The bottom half of the beaker is filled with a gas, and the top half is filled with liquid mercury that is exposed to the atmosphere. The gas and mercury do not mix, because they are separated by a frictionless, movable piston of negligible mass and thickness. The initial temperature is 273 K. The temperature is increased until a value is reached when one-half of the mercury has spilled out. Ignore the thermal expansion of the glass and mercury, and find this temperature.

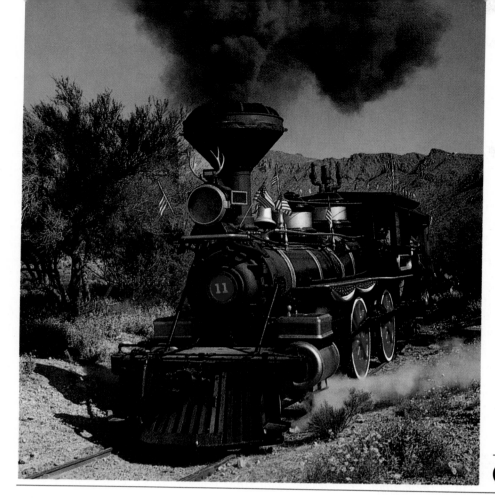

CHAPTER 15

THERMODYNAMICS

The need for energy in one form or another has always existed and always will. Early steam-driven locomotives, for example, obtained their energy by burning coal or wood, whereas today's trains use electricity or diesel fuel. Vast amounts of money are spent to meet the energy demands of transportation industries like the railroads, as well as those of other industrial and domestic customers. The search for new sources of energy and for more efficient ways of using current supplies is never-ending. The use of energy, however, must be consistent with natural laws. One law that we have already talked about is the conservation of energy. But other laws also apply to energy in the form of heat and work. **Thermodynamics** *is the branch of physics that is built upon the fundamental laws of nature that heat and work obey. We will see that thermodynamics has some surprising things to say about how energy can be used, beyond the statement that it can be neither created nor destroyed.*

In thermodynamics, the collection of objects upon which attention is being focused is called the **system,** *while everything else in the environment is called the* **surroundings.** *Usually, the system and its surroundings are separated by walls of some kind. For example, in a car engine the system may be the burning gas contained within the engine block, while the radiator and the outside air are parts of the surroundings.*

Figure 15.1 The hot air in a balloon is one example of a thermodynamic system.

Walls such as those of the engine block permit heat to flow through them and are called **diathermal walls.** *In contrast, perfectly insulating walls that do not permit heat to flow between the system and its surroundings are called* **adiabatic walls.**

To understand what the laws of thermodynamics have to say about the use of energy, we need to describe the physical condition or **state of a system.** *We might be interested, for instance, in the hot air within one of the balloons in Figure 15.1. The hot air itself would be the system, and the skin of the balloon provides the walls that separate this system from the surrounding cooler air. The state of the system would be specified by giving values for the pressure, volume, and temperature of the hot air, assuming that its mass is known.*

As we shall see in this chapter, there are four laws of thermodynamics. We begin with the one known as the zeroth law and then consider the remaining three.

15.1 THE ZEROTH LAW OF THERMODYNAMICS

The zeroth law of thermodynamics deals with the concept of *thermal equilibrium.* Two systems are said to be in thermal equilibrium if there is no net flow of heat between them when they are brought into thermal contact. For instance, if you dive into Lake Michigan in January, you find out quickly that your body is *not* in thermal equilibrium with the water. Your body, being quite warm compared to the water, loses heat fast. To help explain the central idea of the zeroth law of thermodynamics, Figure 15.2*a* shows two systems labeled X and Z. Each is within a container whose adiabatic walls are made from slabs of insulation that prevent the flow of heat, and each has the same temperature, as indicated by the thermometers labeled Y. In part *b*, one wall of each container is replaced by a thin silver sheet, and the two silver sheets are touched together. Silver has a large thermal conductivity, so heat flows through it readily and the silver sheets behave as diathermal walls. Even though the diathermal walls would permit it, no net

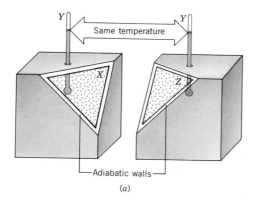

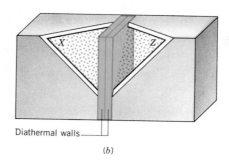

Figure 15.2 (*a*) Systems X and Z are surrounded by adiabatic walls and register the same temperature on the thermometers labeled Y. (*b*) When X is put into thermal contact with Z through diathermal walls, no net flow of heat occurs between the systems.

flow of heat occurs in part *b*, indicating that system X and system Z are in thermal equilibrium. There is no net flow of heat, because the two systems have the same temperature. Thus, *temperature is the indicator of thermal equilibrium in the sense that there is no net flow of heat between two systems in thermal contact that have the same temperature.*

In Figure 15.2 the role of the thermometer is an important one, and in recognition of it, we call the thermometer system Y. When the thermometer is in thermal equilibrium with system X, it registers the same temperature as it does when it is in equilibrium with system Z. In effect, systems X and Z are both in thermal equilibrium with the thermometer, system Y. In each case, the thermometer is in the same state, revealing that X and Z are equally hot. Consequently, X and Z are found to be in thermal equilibrium with each other. This finding is an example of the *zeroth law of thermodynamics.*

The Zeroth Law of Thermodynamics

Two systems individually in thermal equilibrium with a third system are in thermal equilibrium with each other.

The zeroth law establishes temperature as the indicator of thermal equilibrium and implies that all parts of a system must be in thermal equilibrium if the system is to have a definable single temperature. In other words, there can be no net flow of heat within a system in thermal equilibrium.

15.2 THE FIRST LAW OF THERMODYNAMICS

Heat flow is one way that a system can gain energy from or lose energy to its surroundings. A system also *gains* energy if the surroundings do work on the system, and *loses* energy by doing work on the surroundings. Suppose that a system gains heat and does no work. Energy conservation dictates that the internal energy of the system increases from U_i to U_f, the change being $\Delta U = U_f - U_i = Q$. In writing this equation, we use the convention that *heat Q is positive when the system absorbs heat and negative when the system gives off heat.* If a system does work W on its surroundings, and there is no heat flow, energy conservation requires that the internal energy decreases from U_i to U_f. Now the change in internal energy is $\Delta U = U_f - U_i = -W$. The minus sign is included because we follow the convention that *work is positive when it is done by the system and negative when it is done on the system.* If a system gains energy Q in the form of heat and simultaneously loses energy W by doing work, the change in internal energy due to both factors is given below by Equation 15.1. Thus, the **first law of thermodynamics** is just the conservation of energy principle applied to heat, work, and internal energy.

The First Law of Thermodynamics

As dictated by the principle of conservation of energy, the internal energy of a system changes from an initial value U_i to a final value of U_f when the system absorbs an amount of heat Q and performs an amount of work W:

$$\Delta U = U_f - U_i = Q - W \qquad (15.1)$$

Examples 1 and 2 illustrate the use of Equation 15.1 and the sign conventions for Q and W.

Example 1 Positive Work

A system absorbs 1500 J of heat from its surroundings and performs 2200 J of work on its surroundings. Determine the change in the internal energy of the system.

REASONING Figure 15.3 illustrates this situation. Since the system loses more energy in doing work than it gains in the form of heat, the internal energy of the system decreases. In other words, ΔU is negative.

SOLUTION The heat is $Q = +1500$ J, since it is absorbed by the system. The work is $W = +2200$ J, since it is done *by* the system. According to the first law of thermodynamics

$$\Delta U = Q - W = (+1500 \text{ J}) - (+2200 \text{ J}) = \boxed{-700 \text{ J}} \qquad (15.1)$$

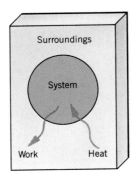

Figure 15.3 A system gains energy in the form of heat and loses energy in doing work.

Example 2 Negative Work

A system absorbs 1500 J of heat from its surroundings. Simultaneously, the surroundings perform 2200 J of work on the system. Find the change in the internal energy of the system.

REASONING Figure 15.4 shows the system and its surroundings. Since the system gains energy in the form of heat and in the form of work, the internal energy of the system increases, and ΔU is positive.

SOLUTION The heat is $Q = +1500$ J, since it is absorbed by the system. But the work is $W = -2200$ J, since it is done *on* the system. From the first law of thermodynamics, we find that

$$\Delta U = Q - W = (+1500 \text{ J}) - (-2200 \text{ J}) = \boxed{+3700 \text{ J}} \qquad (15.1)$$

PROBLEM SOLVING INSIGHT

When using the first law of thermodynamics, as expressed by Equation 15.1, be careful to adhere to the proper sign conventions for the heat Q and the work W.

In the first law of thermodynamics, the internal energy U, heat Q, and work W are energy quantities, and each is expressed in energy units such as joules. However, there is a fundamental difference between U, on the one hand, and Q and W on the other. The next example sets the stage for explaining this difference.

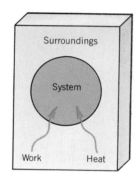

Figure 15.4 A system gains energy in the form of heat and in the form of work performed by the surroundings.

Example 3   An Ideal Gas

The temperature of three moles of a monatomic ideal gas is reduced from $T_i = 540$ K to $T_f = 350$ K by two different methods. In the first method 5500 J of heat flows into the gas, while in the second method, 1500 J of heat flows into it. In each case find (a) the change in the internal energy of the gas and (b) the work done by the gas.

REASONING   Since the internal energy of the ideal gas is $U = \frac{3}{2}nRT$ and since the number of moles n is fixed, only a change in temperature T can alter the internal energy. In both methods, the change in temperature is the same, so the change in internal energy is the same. As the temperature decreases, the internal energy decreases. Once the change in internal energy is known, the work can be obtained from the first law of thermodynamics, using the given values for the heat.

SOLUTION
(a) Using Equation 14.7 for the internal energy of a monatomic ideal gas, we find for each method of adding heat that

$$\Delta U = \tfrac{3}{2}nR(T_f - T_i) = \tfrac{3}{2}(3.0 \text{ mol})[8.31 \text{ J/(mol} \cdot \text{K)}](350 \text{ K} - 540 \text{ K}) = \boxed{-7100 \text{ J}}$$

(b) Since ΔU is now known and the heat is given in each method, Equation 15.1 can be used to determine the work:

[1st method] $W = Q - \Delta U = 5500 \text{ J} - (-7100 \text{ J}) = \boxed{12\ 600 \text{ J}}$

[2nd method] $W = Q - \Delta U = 1500 \text{ J} - (-7100 \text{ J}) = \boxed{8600 \text{ J}}$

In each method the gas does work, but it does more work in the first method.

To understand the difference between U, on the one hand, and Q and W on the other, consider the value for ΔU in Example 3. In both methods ΔU is the same. The value of ΔU is determined once the initial and final temperatures are specified, because the internal energy of an ideal gas depends only on the Kelvin temperature. Temperature is one of the variables (along with pressure and volume) that define the state of a system. The internal energy depends only on the state of a system, not on the method by which the system arrives at a given state. In recognition of this characteristic, internal energy is referred to as a _function of state_.* In contrast, heat and work are not functions of state, because they have different values for each different method used to make the system change from one state to another, as in Example 3.

* The fact that an ideal gas is used in Example 3 does not restrict our conclusion here. Had a real (nonideal) gas or other material been used, the only difference would have been that the expression for the internal energy would have been more complicated. It might have involved the volume V, as well as the temperature T, for instance.

15.3 THERMAL PROCESSES INVOLVING PRESSURE, VOLUME, AND TEMPERATURE

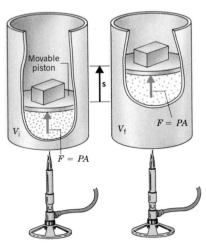

Figure 15.5 The substance in the chamber is expanding isobarically, because the pressure is held constant by the external atmosphere and the weight of the movable piston and the block.

A system can interact with its surroundings in many ways, and the heat and work that come into play always obey the first law of thermodynamics. This section introduces four common thermal processes. In each case, it is assumed that the process is *quasi-static;* that is, the process occurs slowly enough that a uniform pressure and temperature exist throughout all regions of the system at all times.

An isobaric process is one that occurs at constant pressure. For instance, Figure 15.5 shows a substance (solid, liquid, or gas) contained in a chamber fitted with a movable frictionless piston. The substance itself is the system, while everything else (the container, piston, block, and burner) is regarded as the surroundings. The pressure P experienced by the system is always the same and is determined by the external atmosphere and by the weight of the piston and the block on the piston. Heating the substance makes it expand and do work W in lifting the piston and block through the displacement s. The work can be calculated from $W = Fs$ (Equation 6.1), where F is the magnitude of the force and s is the magnitude of the displacement. The force is generated by the pressure P acting on the cross-sectional area A of the piston, according to $F = PA$. With this substitution for F, the work becomes $W = (PA)s$. But the product $A \cdot s$ is the change in volume of the material, $\Delta V = V_f - V_i$, where V_f and V_i are the final and initial volumes, respectively. Thus, the expression for the work is

$$\begin{bmatrix} \text{Isobaric} \\ \text{process} \end{bmatrix} \qquad W = P\,\Delta V = P(V_f - V_i) \qquad (15.2)$$

Consistent with our sign convention for work, this result predicts a positive value for the work done *by a system* when the system expands isobarically (V_f exceeds V_i). Equation 15.2 also applies to an isobaric compression (V_f less than V_i). Then, the work is negative, again consistent with the sign convention, since work must be done *on the system* to compress it. Example 4 emphasizes that $W = P\,\Delta V$ applies to any system, solid, liquid, or gas, as long as the pressure remains constant while the volume changes.

Example 4 Isobaric Expansion of Water

One gram of water is placed in the cylinder in Figure 15.5, and the pressure is maintained at 2.0×10^5 Pa. The temperature of the water is raised by 31 C°. In one case, the water is in the liquid phase and expands by the small amount of 1.0×10^{-8} m³. In another case, the water is in the gas phase and expands by the much greater amount of 7.1×10^{-5} m³. For the water in each case, find (a) the work done and (b) the change in the internal energy.

REASONING In both cases the material expands, so work is done by the system (water) on the surroundings and is, therefore, positive. The work can be found from $W = P\,\Delta V$. Liquids (and solids), however, change volume much less than gases do under comparable conditions. Hence, the work of expansion or compression for liquids is negligible compared to that for gases. Once the work is known, the first law of thermodynamics $\Delta U = Q - W$ can be used to find the change in internal energy, provided a value for the heat Q can be found. The heat needed to raise the temperature of each material can be obtained from $Q = cm\,\Delta T$ (Equation 12.4), where the specific heat capacity of liquid water is $c = 4186 \text{ J/(kg} \cdot \text{C}°)$ and the specific heat capacity of water vapor at constant pressure is $c_P = 2020 \text{ J/(kg} \cdot \text{C}°)$.

SOLUTION

(a) In both cases, the process is isobaric, so the work done is given by Equation 15.2:

$$W_{\text{liquid}} = P\,\Delta V = (2.0 \times 10^5 \text{ Pa})(1.0 \times 10^{-8} \text{ m}^3) = \boxed{0.0020 \text{ J}}$$

$$W_{\text{gas}} = P\,\Delta V = (2.0 \times 10^5 \text{ Pa})(7.1 \times 10^{-5} \text{ m}^3) = \boxed{14 \text{ J}}$$

(b) Using $\Delta U = Q - W = cm\,\Delta T - W$, we find

$$\Delta U_{\text{liquid}} = [4186 \text{ J/(kg} \cdot \text{C}°)](0.0010 \text{ kg})(31 \text{ C}°) - 0.0020 \text{ J}$$

$$= 130 \text{ J} - 0.0020 \text{ J} = \boxed{130 \text{ J}}$$

$$\Delta U_{\text{gas}} = [2020 \text{ J/(kg} \cdot \text{C}°)](0.0010 \text{ kg})(31 \text{ C}°) - 14 \text{ J}$$

$$= 63 \text{ J} - 14 \text{ J} = \boxed{49 \text{ J}}$$

Virtually all the 130 J of heat added to the liquid serves to change the internal energy, since the volume change and the corresponding work of expansion are so small. In contrast, a significant fraction of the 63 J of heat added to the vapor causes work of expansion to be done, so that noticeably less than 63 J is left for the internal energy change.

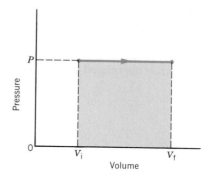

Figure 15.6 For an isobaric process, a pressure-versus-volume plot is a horizontal straight line, and the work done $[W = P(V_f - V_i)]$ is the colored rectangular area under the graph.

It is often convenient to display thermal processes graphically. For instance, Figure 15.6 shows a plot of pressure versus volume for an isobaric expansion. Since the pressure is constant, the graph is a horizontal straight line, beginning at the initial volume V_i and ending at the final volume V_f. In terms of such a plot, the work $W = P(V_f - V_i)$ is the area under the graph, which is the shaded rectangle of height P and width $V_f - V_i$.

Another common thermal process is an *isochoric process, one that occurs at constant volume.* Figure 15.7a illustrates an isochoric process in which a substance (solid, liquid, or gas) is heated. The substance would expand if it could, but the rigid container keeps the volume constant. The expansion of the container itself is negligible. Because the volume is constant, the pressure inside rises, and the substance exerts more and more force on the walls. While enormous forces can be generated in the closed container, no work is done, since the walls do not move. The pressure–volume plot shown in part b of the drawing is a vertical straight line, for the volume is constant. The work is the area under the graph, as it is for an isobaric process, and the area under the vertical line is zero, because the work is zero. Since no work is done, the first law of thermodynamics indicates that the heat in an isochoric process serves only to change the internal energy of the system: $\Delta U = Q - W = Q$.

A third important thermal process is an *isothermal process, one that takes place at constant temperature.* The next section illustrates the details of an isothermal process when the system is an ideal gas.

Last, there is the *adiabatic process, one that occurs without the transfer of heat.* Since there is no heat transfer, Q equals zero, and the first law indicates that $\Delta U = Q - W = -W$. Thus, when a system does work adiabatically, the internal energy of the system decreases by exactly the amount of the work done. When

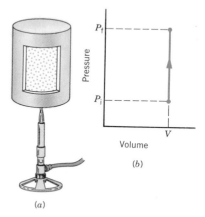

Figure 15.7 (a) The substance in the chamber is being heated isochorically, because the rigid chamber keeps the volume constant. (b) The pressure–volume plot for an isochoric process is a vertical straight line. The area under the graph is zero, indicating that no work is done.

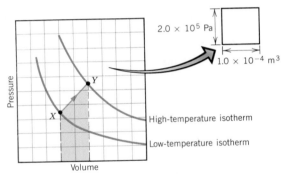

Figure 15.8 The thermal process that occurs from X to Y is not one of the four discussed in the text. However, the work done by the gas is given by the colored area.

work is done on a system adiabatically, the internal energy increases correspondingly. The next section discusses an adiabatic process in more detail.

A process may be complex enough that no part of it is recognizable as one of the four discussed above. For instance, Figure 15.8 shows a process for a gas in which the pressure, volume, and temperature are changed along the straight line from X to Y. With the aid of integral calculus, it can be shown that *the area under a pressure–volume graph is the work for any kind of process,* so the area representing the work has been colored in the drawing. The volume increases, and the gas does work. This work is positive by convention, as is the area. In contrast, if a process reduces the volume, work is done on the gas, and this work is negative by convention. Correspondingly, the area under the pressure–volume graph is assigned a negative value. In Example 5, we determine the work for the case shown in Figure 15.8.

Example 5   *The Area under a Pressure–Volume Graph*

Determine the work for the process in which the pressure, volume, and temperature of a gas are changed along the straight line from X to Y in Figure 15.8.

REASONING   The work is given by the area (in color) under the straight line between X and Y. Since the volume increases, the gas does work against the surroundings, so work is positive. The area can be found by counting squares in Figure 15.8 and multiplying by the area per square.

SOLUTION   We estimate that there are 9.0 colored squares in the drawing. The area of one square is $(2.0 \times 10^5 \text{ Pa})(1.0 \times 10^{-4} \text{ m}^3) = 2.0 \times 10^1 \text{ J}$, so the work is

$$W = +(9.0 \text{ squares})(2.0 \times 10^1 \text{ J/square}) = \boxed{+180 \text{ J}}$$

15.4 THERMAL PROCESSES THAT UTILIZE AN IDEAL GAS

ISOTHERMAL EXPANSION OR COMPRESSION

When a system performs work isothermally, the temperature remains constant. In Figure 15.9a, for instance, a metal cylinder contains n moles of an ideal gas, and the large mass of hot sand maintains the cylinder and gas at a constant Kelvin

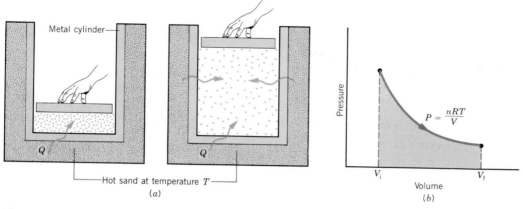

Figure 15.9 (*a*) The ideal gas in the cylinder is expanding isothermally at temperature *T*. The force holding the piston in place is reduced slowly, so the expansion occurs quasi-statically. (*b*) The work done by the gas is given by the colored area.

temperature *T*. The piston is held in place initially so the volume of the gas is V_i. As the external force applied to the piston is reduced quasi-statically, the gas expands to the final volume V_f. Figure 15.9*b* gives a plot of pressure ($P = nRT/V$) versus volume for the process. The work *W* done by the gas is *not* given by $W = P \, \Delta V = P(V_f - V_i)$, because the pressure is not constant. Nevertheless, the work is equal to the area under the graph. The techniques of integral calculus lead to the following result* for *W*:

$$
\begin{bmatrix}
\text{Isothermal} \\
\text{expansion or} \\
\text{compression of} \\
\text{an ideal gas}
\end{bmatrix}
\qquad
W = nRT \ln \left(\frac{V_f}{V_i} \right)
\qquad (15.3)
$$

Where does the energy for this work originate? Since the internal energy of an ideal gas is proportional to the Kelvin temperature ($U = \tfrac{3}{2} nRT$ for a monatomic ideal gas), the internal energy remains constant throughout an isothermal process, and the change in internal energy is zero. The first law of thermodynamics becomes $\Delta U = 0 = Q - W$. In other words, $Q = W$, and the energy for the work originates in the hot sand. Heat flows into the gas from the sand, as Figure 15.9*a* illustrates. If the gas is compressed isothermally, Equation 15.3 still applies, and heat flows out of the gas into the sand. The following example deals with the isothermal expansion of an ideal gas.

Example 6   Isothermal Expansion of an Ideal Gas

Two moles of argon gas at 298 K expand isothermally from an initial volume of $V_i = 0.025 \text{ m}^3$ to a final volume of $V_f = 0.050 \text{ m}^3$. Assuming that argon is an ideal gas, find (a) the work done by the gas, (b) the change in the internal energy of the gas, and (c) the heat supplied to the gas.

---

* In this result, "ln" denotes the natural logarithm to the base $e = 2.71828$. The natural logarithm is related to the common logarithm to the base ten by $\ln(V_f/V_i) = 2.303 \log(V_f/V_i)$.

REASONING AND SOLUTION

(a) The work done by the gas can be found from Equation 15.3:

$$W = nRT \ln \left(\frac{V_f}{V_i} \right)$$

$$W = (2.0 \text{ mol})[8.31 \text{ J/(mol} \cdot \text{K)}](298 \text{ K}) \ln \left(\frac{0.050 \text{ m}^3}{0.025 \text{ m}^3} \right) = \boxed{+3400 \text{ J}}$$

(b) There is no change in the internal energy of an ideal gas during an isothermal process: $\boxed{\Delta U = 0}$.

(c) The heat Q added to the gas can be determined from the first law of thermodynamics:

$$Q = \Delta U + W = 0 + 3400 \text{ J} = \boxed{+3400 \text{ J}} \qquad (15.1)$$

All the heat supplied is used for doing work, since the internal energy does not change.

ADIABATIC EXPANSION OR COMPRESSION

When a system performs work adiabatically, no heat flows into or out of the system. Figure 15.10a shows an arrangement in which n moles of an ideal gas do work under adiabatic conditions, expanding quasi-statically from an initial volume V_i to a final volume V_f. The arrangement is similar to that in Figure 15.9 for isothermal expansion. However, a different amount of work is done here, because the cylinder is now surrounded by insulating material that prevents the flow of heat, so $Q = 0$. According to the first law of thermodynamics, the change in internal energy is $\Delta U = Q - W = -W$. Since the internal energy of an ideal monatomic gas is $U = \frac{3}{2}nRT$, it follows that $\Delta U = U_f - U_i = \frac{3}{2}nR(T_f - T_i)$, where T_i and T_f are the initial and final Kelvin temperatures. With this substitution, the relation $\Delta U = -W$ becomes

$$\left[\begin{array}{l} \textbf{Adiabatic} \\ \textbf{expansion or} \\ \textbf{compression of} \\ \textbf{a monatomic} \\ \textbf{ideal gas} \end{array} \right] \qquad W = \tfrac{3}{2}nR(T_i - T_f) \qquad\qquad (15.4)$$

Equation 15.4 indicates that if the ideal gas expands adiabatically and does work (W is positive), the final temperature of the gas is less than the initial temperature. The internal energy of the gas is reduced to provide the necessary energy, and because the internal energy is proportional to the Kelvin temperature, the temperature decreases. Figure 15.10b shows a plot of pressure versus volume for this process. The adiabatic curve (red) intersects the isotherms (blue) at the higher initial temperature [$T_i = P_iV_i/(nR)$] and the lower final temperature [$T_f = P_fV_f/(nR)$]. The colored area under the adiabatic curve represents the work done.

The reverse of an adiabatic expansion is an adiabatic compression (W is negative), and Equation 15.4 indicates that the final temperature exceeds the initial temperature. The energy provided by the agent doing the work increases the internal energy of the gas. As a result, the gas becomes hotter.

The equation that gives the adiabatic curve (red) between the initial pressure

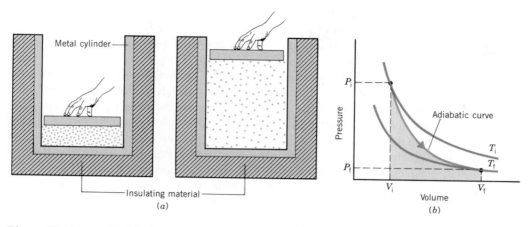

Figure 15.10 (*a*) The ideal gas in the cylinder is expanding adiabatically. The force holding the piston in place is reduced slowly, so the expansion occurs quasi-statically. (*b*) A plot of pressure versus volume yields the adiabatic curve shown in red, which intersects the isotherms (blue) at the initial temperature T_i and the final temperature T_f. The work done by the gas is given by the colored area.

and volume (P_i, V_i) and the final pressure and volume (P_f, V_f) in Figure 15.10*b* can be derived using integral calculus. The result is

$$
\begin{bmatrix}
\textbf{Adiabatic} \\
\textbf{expansion or} \\
\textbf{compression of} \\
\textbf{an ideal gas}
\end{bmatrix}
\qquad
P_i V_i^{\gamma} = P_f V_f^{\gamma}
\qquad\qquad (15.5)
$$

where the exponent γ is the ratio of the specific heat capacities at constant pressure and constant volume, $\gamma = c_P/c_V$. Equation 15.5 applies in conjunction with the ideal gas law, for *each point* on the adiabatic curve satisfies the relation $PV = nRT$.

15.5 SPECIFIC HEAT CAPACITIES AND THE FIRST LAW OF THERMODYNAMICS

In this section the first law of thermodynamics is used to gain some understanding of the factors that determine the specific heat capacity of a material. Remember, when the temperature of a substance changes as a result of heat flow, the change in temperature ΔT and the amount of heat Q are related according to $Q = cm\,\Delta T$. In this expression c denotes the specific heat capacity in units of J/(kg·C°), and m is the mass in kilograms. It is more convenient now to express the amount of material as the number of moles n, rather than the number of kilograms. Therefore, we replace the expression $Q = cm\,\Delta T$ with the following analogous expression:

$$
Q = Cn\,\Delta T \qquad\qquad (15.6)
$$

where the capital letter C (as opposed to the lowercase c) refers to the *molar specific heat capacity* in units of J/(mol·K). In addition, the unit for measuring the temperature change ΔT is the Kelvin (K) rather than the Celsius degree (C°).

For gases it is necessary to distinguish between the molar specific heat capacity C_P for conditions of constant pressure and C_V for conditions of constant volume. With the help of the first law and an ideal gas as an example, it is possible to see why C_P and C_V differ.

According to the first law, $Q = \Delta U + W$, so the heat needed to raise the temperature of an ideal gas from an initial temperature T_i to a final temperature T_f can be calculated by evaluating ΔU and W. The internal energy of an ideal gas depends only on temperature, and for a monatomic ideal gas $U = \frac{3}{2}nRT$ (Equation 14.7). As a result, $\Delta U = U_f - U_i = \frac{3}{2}nR(T_f - T_i)$. When the heating process occurs at constant pressure, the work done is given by Equation 15.2: $W = P\,\Delta V = P(V_f - V_i)$. For an ideal gas $PV = nRT$, so the work becomes $W = nR(T_f - T_i)$. On the other hand, when the volume is constant, $\Delta V = 0$, and the work done is zero. The calculation of the heat is summarized below:

$$Q = \Delta U + W$$

$$Q_{\text{constant pressure}} = \tfrac{3}{2}nR(T_f - T_i) + nR(T_f - T_i)$$

$$Q_{\text{constant volume}} = \tfrac{3}{2}nR(T_f - T_i) + 0$$

The molar specific heat capacities can now be determined, since Equation 15.6 indicates that $C = Q/[n(T_f - T_i)]$:

$$
\begin{bmatrix}
\textbf{Constant pressure} \\
\textbf{for a monatomic} \\
\textbf{ideal gas}
\end{bmatrix}
\qquad
C_P = \tfrac{3}{2}R + R = \tfrac{5}{2}R
\qquad (15.7)
$$

$$
\begin{bmatrix}
\textbf{Constant volume} \\
\textbf{for a monatomic} \\
\textbf{ideal gas}
\end{bmatrix}
\qquad
C_V = \tfrac{3}{2}R
\qquad (15.8)
$$

The ratio γ of the specific heats is

$$
\begin{bmatrix}
\textbf{Monatomic} \\
\textbf{ideal gas}
\end{bmatrix}
\qquad
\gamma = \frac{C_P}{C_V} = \frac{\tfrac{5}{2}R}{\tfrac{3}{2}R} = \frac{5}{3}
\qquad (15.9)
$$

For real monatomic gases near room temperature, experimental values of C_P and C_V give ratios very close to the theoretical value of $\frac{5}{3}$.

The difference between C_P and C_V arises because work is done when the gas expands in response to the addition of heat under conditions of constant pressure, whereas no work is done under conditions of constant volume. For a monatomic ideal gas, C_P exceeds C_V by an amount equal to R, the ideal gas constant:

$$C_P - C_V = R \qquad (15.10)$$

In fact, it can be shown that Equation 15.10 applies to any kind of ideal gas—monatomic, diatomic, etc.

15.6 THE SECOND LAW OF THERMODYNAMICS

Ice cream melts when left out on a warm day. A cold can of soda warms up on a hot day at a picnic. This is always the case. Neither the ice cream nor the soda ever becomes cooler when left somewhere hot, because heat does not flow spontaneously from a cooler object into a warmer environment. Rather, heat flows spontaneously the other way, from hot to cold. Examples such as these relate to one of the most profound laws in all of science, the *second law of thermodynamics*.

The Second Law of Thermodynamics: The Heat Flow Statement

Heat flows spontaneously from a substance at a higher temperature to a substance at a lower temperature and does not flow spontaneously in the reverse direction.

The second law of thermodynamics refers to the natural tendency of heat to flow from hot to cold. Of course, heat can be forced to flow in the reverse direction, if work is done to make it do so. For instance, an air conditioner makes heat flow from the cool interior of a house to the hot outdoors with the aid of the work done by electrical energy. Refrigerators and heat pumps are also devices that use work to make heat flow from cold to hot, against its natural tendency.

The second law of thermodynamics has important implications for the manner in which many devices operate, and we now turn our attention to one of them, the heat engine.

15.7 HEAT ENGINES

ESSENTIAL FEATURES

A *heat engine* is any device that uses heat to perform work. To illustrate features common to all heat engines, Figure 15.11 shows a simplified steam engine. The boiler receives heat from a high-temperature source, such as an oil or gas burner. In the boiler, liquid water is converted into steam at a high temperature and pressure. The steam passes through the intake valve into the cylinder and ex-

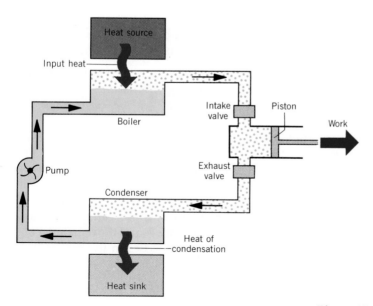

Figure 15.11   A steam engine.

pands against the piston, forcing it to move. The piston may be coupled to a wheel in order to turn it, for example. The work done on the moving piston comes from the internal energy of the steam, so the temperature of the steam in the cylinder drops. As the piston returns to its original position, the cooler steam is forced out the exhaust valve and into the condenser, where the steam is changed back into a liquid. The heat of condensation is given up to a heat sink, which may be the atmosphere. The water is then pumped back to the boiler, and the cycle is repeated. The combination of steam and water is known as the *working substance* of the engine, for it is the agent that does the work. Heat engines, such as the steam engine, operate in repeating cycles. In other words, the conclusion of one cycle is the start of another, so the working substance is in the same state at the end of one cycle and beginning of the next.

The essential features of a steam engine are listed below, because they are shared by all types of heat engines:

1. Heat is supplied to the engine at a relatively high temperature.
2. Part of the input heat is used to perform work.
3. The remainder of the input heat is rejected at a temperature lower than the input temperature.

Figure 15.12 emphasizes these essential features in a schematic fashion. In this drawing the symbol Q_H refers to the magnitude of the input heat. The subscript H stands for "hot," and the place from which the input heat comes is called the "hot reservoir." The symbol Q_C denotes the magnitude of the rejected heat. The subscript C means "cold," and the place in the environment where the rejected heat goes is known as the "cold reservoir." The symbol W denotes the magnitude of the work done by the engine. The symbols Q_H, Q_C, and W refer to magnitudes only, without reference to algebraic signs, so negative values are not used for these quantities in the equations in which they appear.

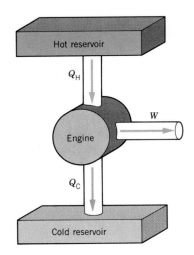

Figure 15.12 This schematic representation of a heat engine shows the input heat (magnitude = Q_H) that originates from the hot reservoir, the work (magnitude = W) done by the engine, and the heat (magnitude = Q_C) that the engine rejects to the cold reservoir.

EFFICIENCY

From a practical point of view an important characteristic of an engine is its *efficiency.* An engine that converts most of the input heat into work is efficient. The efficiency of a heat engine is defined as the ratio of the work W done by the engine to the input heat Q_H:

$$\text{Efficiency} = \frac{\text{Work done}}{\text{Input heat}} = \frac{W}{Q_H} \tag{15.11}$$

If the input heat were converted entirely into work, the engine would have an efficiency of 1.00, since $W = Q_H$; such an engine would be 100% efficient. *Efficiencies are often quoted as percentages obtained by multiplying Equation 15.11 by a factor of 100.* Thus, an efficiency of 68% would mean that a value of 0.68 is used for the value of the efficiency in the equation.

An engine, like any device, must obey the principle of conservation of energy. Some of an engine's input heat Q_H is converted into work W and the remainder Q_C is rejected to the cold reservoir. If there are no other losses in the engine, the principle of energy conservation requires that

$$Q_H = W + Q_C \tag{15.12}$$

Solving this equation for W and substituting the result into Equation 15.11 leads to the following alternative expression for the efficiency of a heat engine:

$$\text{Efficiency} = \frac{Q_H - Q_C}{Q_H} = 1 - \frac{Q_C}{Q_H} \qquad (15.13)$$

Example 7 illustrates how the concepts of efficiency and energy conservation are applied to a heat engine.

Example 7   An Automobile Engine

An automobile engine has an efficiency of 22.0% and produces 2510 J of work. How much heat is rejected by the engine?

REASONING AND SOLUTION   The amount Q_C of heat rejected to the cold reservoir can be calculated by using $W = 2510$ J in the expression $Q_C = Q_H - W$, provided the input heat Q_H is known. According to Equation 15.11,

$$Q_H = \frac{W}{\text{Efficiency}} = \frac{2510 \text{ J}}{0.220} = 11\ 400 \text{ J}$$

The amount of rejected heat is

$$Q_C = Q_H - W = 11\ 400 \text{ J} - 2510 \text{ J} = \boxed{8900 \text{ J}} \qquad (15.12)$$

In Example 7, less than one-quarter of the input heat is converted into work, because the efficiency of the automobile engine is only 22.0%. If the engine were 100% efficient, all the input heat would be converted into the work of moving the car. Unfortunately, nature does not permit 100% efficient heat engines to exist, as the next two sections discuss.

15.8 CARNOT'S PRINCIPLE AND THE CARNOT ENGINE

CARNOT'S PRINCIPLE

What is it that makes a heat engine operate with maximum efficiency? The answer to this fundamental question came from the French engineer Sadi Carnot (1796–1832). Carnot proposed that a heat engine has a maximum efficiency when the processes within the engine are reversible. Figure 15.13 illustrates the essence of a *reversible process* by using a gas-filled cylinder and a frictionless piston. In part *a*, the gas supports the piston and has a pressure, volume, and temperature of P_1, V_1, and T_1. When 100 J of heat are added to the gas, as in part *b*, the gas expands, pushes the piston upward and thereby does work. When the gas stops expanding, its pressure, volume, and temperature are P_2, V_2, and T_2. In part *c*, the piston falls back to its original position and does work in compressing the gas as heat is removed. If the process is reversible, 100 J of heat flows out of the gas and back into the environment as the gas returns to its original state (P_1, V_1, and T_1). Thus, *a reversible process is one in which both the system (the gas) and its environment*

Although today's automobile engines do not approach the ideal efficiency of a reversible heat engine, they perform much better than the models of the past. However, pollutants still remain a problem. In an attempt to reduce unwanted emmission, automobiles of the future, such as this General Motor's HX3 concept car, may use electric motors.

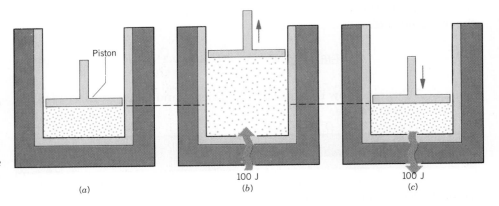

Figure 15.13 (*a*) The frictionless piston is supported by the gas. (*b*) When 100 J of heat is added, the gas expands and does work in lifting the piston. (*c*) If the process is reversible, the system (the gas) and the environment (the piston and the rest of the universe) can both be returned to their initial states.

(*a*) (*b*) (*c*)

(the piston and the rest of the universe) can be returned to exactly the states they were in before the process occurred.

In a reversible process, *both* the system and its environment can be returned to their initial states. Therefore, a process that involves an energy-dissipating mechanism, such as friction, cannot be reversible, because the energy wasted due to friction would alter the system or the environment or both. For example, suppose there were friction between the piston and the wall of the cylinder in Figure 15.13. When the piston returns to its initial position (part *c*), the pressure and temperature would not be P_1 and T_1, so the gas would not be in the initial state it had in part *a*. There are also reasons other than friction why a real process may not be reversible. For instance, the spontaneous flow of heat from a hot substance to a cold substance is irreversible, even though friction is not present. For heat to flow in the reverse direction, work must be done. The agent doing such work must be located in the environment of the hot and cold substances, and, therefore, the environment must change while the heat is moved back from cold to hot. Since the system and the environment cannot *both* be returned to their initial states, the process of spontaneous heat flow is irreversible. In fact, all spontaneous processes are irreversible, such as the explosion of an unstable chemical or the bursting of a bubble. When the word "reversible" is used in connection with engines, it does not mean just a gear that allows the engine to operate a device in reverse. All cars have a reverse gear, for instance, but no automobile engine is thermodynamically reversible, since friction exists no matter which way the car moves.

Today, the idea that the efficiency of a heat engine is a maximum when the engine operates reversibly is referred to as *Carnot's principle.*

> ### Carnot's Principle: An Alternative Statement of the Second Law of Thermodynamics
>
> No irreversible engine operating between two reservoirs at constant temperatures can have a greater efficiency than a reversible engine operating between the same temperatures. Furthermore, all reversible engines operating between the same temperatures have the same efficiency.

Carnot's principle is quite remarkable, for no mention is made of the working substance of the engine. It does not matter if the working substance is a gas, a liquid, or a solid. As long as the process is reversible, the efficiency of the engine is a maximum. Carnot's principle does *not* state, or even imply, that a reversible engine has an efficiency of 100%.

It can be shown that if Carnot's principle were not valid, it would be possible for heat to flow spontaneously from a cold substance to a hot substance, in violation of the second law of thermodynamics. In effect, then, Carnot's principle is another way of expressing the second law of thermodynamics.

THE CARNOT ENGINE

No real engine operates reversibly. Nonetheless, the idea of a reversible engine provides a useful standard for judging the performance of real engines. Figure 15.14 shows a reversible engine, called a **Carnot engine,** that is particularly useful as an idealized model. An important feature of the Carnot engine is that all input heat Q_H originates from a hot reservoir *at a single temperature* T_H and all rejected heat Q_C goes into a cold reservoir *at a single temperature* T_C.

Carnot's principle implies that the efficiency of a reversible engine is independent of the working substance of the engine, and therefore can depend only on the temperatures of the hot and cold reservoirs. Since Efficiency $= 1 - Q_C/Q_H$ according to Equation 15.13, the ratio Q_C/Q_H can depend only on the reservoir temperatures. This observation led Kelvin to define a **thermodynamic temperature scale.** He proposed that the thermodynamic temperatures of the cold and hot reservoirs be defined such that the ratio of these temperatures is equal to Q_C/Q_H. Thus, the thermodynamic temperature scale is related to the heats absorbed and rejected by a Carnot engine, and is independent of the working substance. If a reference point on the thermodynamic temperature scale is properly chosen, it can be shown that the scale is identical to the Kelvin temperature scale introduced in Section 12.2 and used in the ideal gas law. As a result, the ratio of the rejected heat Q_C to the input heat Q_H is

$$\frac{Q_C}{Q_H} = \frac{T_C}{T_H} \qquad (15.14)$$

where the temperatures T_C and T_H *must be expressed in kelvins.*

The efficiency of a Carnot engine can be written in a particularly useful way by substituting Equation (15.14) into the relation Efficiency $= 1 - Q_C/Q_H$:

$$\text{Efficiency of a Carnot engine} = 1 - \frac{T_C}{T_H} \qquad (15.15)$$

This relation gives the maximum possible efficiency for a heat engine operating between two Kelvin temperatures T_C and T_H, and the next example illustrates its application.

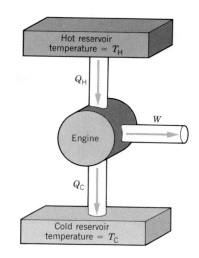

Figure 15.14 A Carnot engine is a reversible engine in which all input heat Q_H originates from a hot reservoir at a single temperature T_H, and all rejected heat Q_C goes into a cold reservoir at a single temperature T_C. The work done by the engine is W.

Example 8   A Tropical Ocean as a Heat Engine

Water near the surface of a tropical ocean has a temperature of 298.2 K (25.0 °C), while water 700 m beneath the surface has a temperature of 280.2 K (7.0 °C). It has been proposed that the warm water be used as the hot reservoir and the cool water as the cold reservoir of a heat engine. (a) Find the maximum possible efficiency for such an engine. (b) Determine the minimum input heat Q_H that would be needed if a number of these engines were to produce an amount of work equal to the 8.1×10^{19} J of energy that the United States consumed in 1987.

THE PHYSICS OF . . .

extracting work from a warm ocean.

REASONING The maximum possible efficiency is the efficiency that a Carnot engine would have (Equation 15.15) operating between higher and lower temperatures of $T_H = 298.2$ K and $T_C = 280.2$ K. Once the maximum efficiency is known, we can use Equation 15.11 to obtain the minimum input heat Q_H needed for the engines to produce $W = 8.1 \times 10^{19}$ J of work.

SOLUTION

(a) Using $T_H = 298.2$ K and $T_C = 280.2$ K in Equation 15.5, we find that

$$\text{Efficiency of a Carnot Engine} = 1 - \frac{T_C}{T_H} = 1 - \frac{280.2 \text{ K}}{298.2 \text{ K}} = \boxed{0.060 \ (6.0\%)}$$

(b) According to Equation 15.11, the input heat Q_H needed for the engines to produce 8.1×10^{19} J of work is

$$Q_H = \frac{W}{\text{Efficiency}} = \frac{8.1 \times 10^{19} \text{ J}}{0.060} = \boxed{1.4 \times 10^{21} \text{ J}}$$

Real engines, being less efficient than Carnot engines, would require an input energy that is greater than 1.4×10^{21} J to produce the same amount of work.

In Example 8 the maximum possible efficiency is only 6.0%. The small efficiency arises because the Kelvin temperatures of the hot and cold reservoirs are nearly the same. A greater efficiency is possible only when there is a greater difference between the reservoir temperatures. In any event, Equation 15.15 indicates that a *perfect heat engine has an efficiency that is less than 1.0 or 100%*. We note in this regard that when T_C approaches absolute zero (0 K), the maximum possible efficiency approaches 1.0. However, experiments have shown that it is not possible to cool a substance to absolute zero, so nature does not permit the existence of a 100% efficient heat engine. As a result, there will always be heat rejected to a cold reservoir whenever a heat engine is used to do work, even if friction and other irreversible processes are eliminated completely.

15.9 REFRIGERATORS, AIR CONDITIONERS, AND HEAT PUMPS

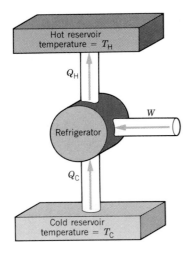

Figure 15.15 In a refrigeration process, work W is used to remove heat Q_C from the cold reservoir and deposit heat Q_H into the hot reservoir.

THE PHYSICS OF . . .

a refrigerator.

Refrigerators, air conditioners, and heat pumps are familiar devices that make heat flow from cold to hot, and each operates in a similar fashion. As Figure 15.15 illustrates, these devices use work W to "reach into" a cold reservoir, "grab onto" an amount of heat Q_C, and deposit an amount of heat Q_H into a hot reservoir. Generally speaking, such a process is called a *refrigeration process*. Often the work is done by a motor. A comparison of this drawing with Figure 15.14 shows that the directions of the arrows symbolizing heat and work in a refrigeration process are opposite to those in an engine process. Energy is conserved during a refrigeration process, just as it is in an engine process, so $Q_H = W + Q_C$. Moreover, if the process occurs reversibly, we have ideal devices that are called Carnot refrigerators, Carnot air conditioners, and Carnot heat pumps. For these ideal devices, the relation $Q_C/Q_H = T_C/T_H$ applies, just as it does for the Carnot engine.

In a *refrigerator,* the interior of the unit is the cold reservoir, while the warmer exterior is the hot reservoir. As Figure 15.16 illustrates, the refrigerator takes heat from the food inside and deposits it into the kitchen, along with the energy needed to do the work of making the heat flow from cold to hot. For this reason,

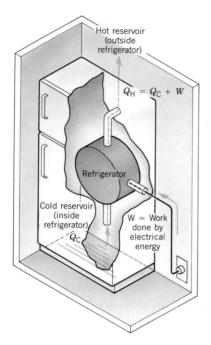

Figure 15.16 In a refrigerator the cold reservoir is the interior region where the food is kept, and the hot reservoir is the kitchen itself.

the outside surfaces (usually the sides and back) of most refrigerators are warm to the touch while they are operating. Thus, a refrigerator warms the kitchen.

An *air conditioner* is like a refrigerator, except that the room itself is the cold reservoir and the outdoors is the hot reservoir. Figure 15.17 illustrates a window air conditioner, which cools a room by removing heat from it and depositing that heat outdoors. The work used to make the heat flow from cold to hot is also deposited outdoors. Clearly, if the air conditioner in the drawing were removed

THE PHYSICS OF . . .

an air conditioner.

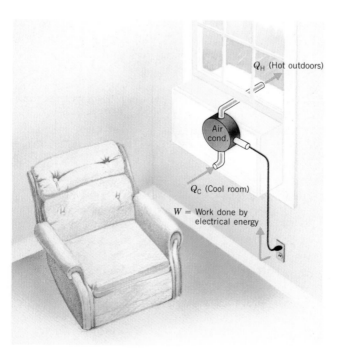

Figure 15.17 A window air conditioner removes heat from a room, the cold reservoir, and deposits heat outdoors, which is the hot reservoir.

from the window and put on the floor of the room, it would not cool the living space. Quite the opposite! It would warm the room. Why? Because the heat pumped out the back of the air conditioner is $Q_H = Q_C + W$ and is greater than the heat Q_C pulled into the front of the unit.

In a sense, refrigerators and air conditioners operate like pumps. They pump heat "uphill" from a lower temperature to a higher temperature, just as a water pump forces water uphill from a lower elevation to a higher elevation. It would be appropriate to call them heat pumps. However, the name "heat pump" is reserved for the device illustrated in Figure 15.18, which is a home heating appliance. The heat pump uses work W to make heat Q_C from the wintry outdoors (the cold reservoir) flow up the temperature "hill" into a warm house (the hot reservoir). According to the conservation of energy, the heat pump deposits inside the house an amount of heat equal to $Q_H = Q_C + W$. The air conditioner and the heat pump do closely related jobs. The air conditioner refrigerates the inside of the house and heats up the outdoors, while the heat pump refrigerates the outdoors and heats up the inside. These jobs are so closely related that most heat pump systems serve in a dual capacity, being equipped with a switch that converts them from heaters in the winter into air conditioners in the summer.

Heat pumps are popular for home heating in today's energy-conscious world, and it is easy to understand why. Suppose 1000 J of energy is available to use for home heating. Figure 15.19a shows that a conventional electric heating system uses this 1000 J to heat a coil of wire, just as in a toaster. A fan blows air across the hot coil, and forced convection carries the 1000 J of heat into the living room. Part b of the drawing shows that a heat pump does not use the 1000 J directly as heat. Instead, the heat pump uses the 1000 J to do the work of pumping heat Q_C from the cooler outdoors into the warmer room. The pump delivers to the inside an amount of energy $Q_H = Q_C + 1000$ J. Obviously, the heat pump puts more than

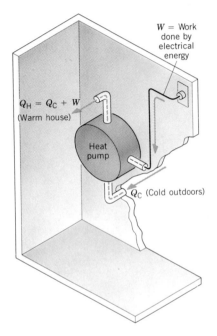

W = Work done by electrical energy

$Q_H = Q_C + W$
(Warm house)

Heat pump

Q_C (Cold outdoors)

Figure 15.18 In a heat pump the cold reservoir is the wintry outdoors, and the hot reservoir is the inside of the house.

Figure 15.19 (a) This conventional electric heating system is delivering 1000 J of heat to the living room. (b) In a heat pump, 1000 J of energy is used as work to obtain heat Q_C from the outdoors, the amount of energy delivered to the room being $Q_H = Q_C + 1000$ J.

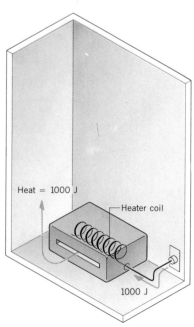

Heater coil

Heat = 1000 J

1000 J

(a) Conventional electric heating

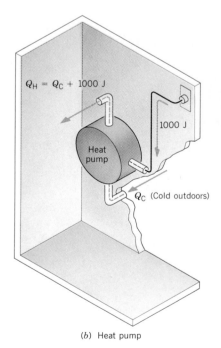

$Q_H = Q_C + 1000$ J

1000 J

Heat pump

Q_C (Cold outdoors)

(b) Heat pump

1000 J of heat into the room, whereas the conventional electric heating system provides only 1000 J. The next example shows how the basic relations $Q_H = W + Q_C$ and $Q_C/Q_H = T_C/T_H$ are used.

Example 9   A Heat Pump

An ideal or Carnot heat pump is used to heat a house to a temperature of $T_H = 294$ K (21 °C). How much work must be done by the pump to deliver $Q_H = 3350$ J of heat into the house when the outdoor temperature T_C is (a) 273 K (0 °C) and (b) 252 K (-21 °C)?

REASONING   The conservation of energy ($Q_H = W + Q_C$) certainly applies to the heat pump. Thus, the work can be determined from $W = Q_H - Q_C$, provided we can obtain a value for Q_C, the heat taken by the pump from the outside. To determine Q_C, we use the fact that the pump is a Carnot heat pump and operates reversibly. Therefore, $Q_C/Q_H = T_C/T_H$ applies, and $Q_C = Q_H(T_C/T_H)$. The work, then, is $W = Q_H - Q_H(T_C/T_H) = Q_H[1 - (T_C/T_H)]$.

SOLUTION

(a) At an indoor temperature of $T_H = 294$ K and an outdoor temperature of $T_C = 273$ K, the work needed is

$$W = Q_H\left(1 - \frac{T_C}{T_H}\right) = (3350 \text{ J})\left(1 - \frac{273 \text{ K}}{294 \text{ K}}\right) = \boxed{239 \text{ J}}$$

(b) This solution is identical to that in part (a), except that it is now cooler outside, so $T_C = 252$ K. Consequently, more work must be done by the heat pump: $W = \boxed{479 \text{ J}}$.

More work must be done because the heat is pumped up a greater temperature "hill" when the outside is colder than when it is warmer.

A measure of the performance of a heat pump can be obtained by specifying the ratio of the heat Q_H delivered into the house to the work W required to deliver it. This ratio is known as the *coefficient of performance*:

[Heat pump]
$$\text{Coefficient of performance} = \frac{Q_H}{W} \qquad (15.16)$$

The coefficient of performance depends on the indoor and outdoor temperatures, as the next example illustrates.

Example 10   How Well Does a Heat Pump Perform?

A Carnot heat pump maintains the temperature in a house at $T_H = 293$ K (20 °C). Find the coefficient of performance of the pump when the outdoor temperature is $T_C = 281$ K (8 °C).

REASONING   Equation 15.16 defines the coefficient of performance as Q_H/W and applies whether or not the heat pump is a Carnot heat pump. In addition, however, Carnot devices obey the relation $Q_C/Q_H = T_C/T_H$. Since temperature data is given here, we wish to take advantage of this additional relation. Therefore, we begin by using the conservation of energy to express the work as $W = Q_H - Q_C$. With this substitution, the coefficient of performance becomes $Q_H/W = Q_H/(Q_H - Q_C)$. Dividing the numerator and denominator on the right by Q_H gives

$$\frac{Q_H}{W} = \frac{1}{1 - Q_C/Q_H} = \frac{1}{1 - T_C/T_H}$$

This result applies only to Carnot heat pumps, because they are reversible and obey the relation $Q_C/Q_H = T_C/T_H$.

SOLUTION When $T_C = 281$ K outside and $T_H = 293$ K inside,

$$\text{Coefficient of performance} = \frac{1}{1 - T_C/T_H} = \frac{1}{1 - (281\text{ K})/(293\text{ K})} = \boxed{24}$$

The large coefficient of performance found in Example 10 occurs because the heat pump is an ideal or Carnot pump. Commercially available units have coefficients of about 3–4 under favorable conditions.

It is also possible to specify a coefficient of performance for refrigerators and air conditioners. However, unlike a heat pump, the job of these two devices is to cool, not to heat. As a result, the coefficient of performance of a refrigerator or an air conditioner is the ratio of the heat Q_C removed from the cold reservoir to the work W needed to remove it:

$$\begin{bmatrix}\textbf{Refrigerator}\\\textbf{or}\\\textbf{air conditioner}\end{bmatrix} \qquad \begin{matrix}\text{Coefficient}\\\text{of}\\\text{performance}\end{matrix} = \frac{Q_C}{W} \qquad (15.17)$$

Commercially available refrigerators and air conditioners have coefficients of performance in the range 2–6, depending on the temperatures involved, values that are less than those for ideal or Carnot devices.

15.10 ENTROPY AND THE SECOND LAW OF THERMODYNAMICS

ENTROPY

A Carnot engine has the maximum possible efficiency for its operating conditions, because the processes occurring within it are reversible. Irreversible processes, such as friction, always cause real engines to operate at less than maximum efficiency, for they reduce our ability to use heat for performing work. As an extreme example, imagine that a hot object is placed in thermal contact with a cold object, so heat flows spontaneously, and hence irreversibly, from hot to cold. Eventually both objects reach the same temperature, and $T_C = T_H$. A Carnot engine using these two objects as heat reservoirs is unable to do work, because the efficiency of the engine is zero [Efficiency $= 1 - (T_C/T_H) = 0$]. In general, irreversible processes cause us to lose some, but not necessarily all, of the ability to perform work. This partial loss can be expressed in terms of a concept called _entropy_.

To introduce the idea of entropy we recall the relation $Q_C/Q_H = T_C/T_H$ that applies to a Carnot engine. This equation can be rearranged as $Q_C/T_C = Q_H/T_H$, which focuses attention on the heat Q divided by the Kelvin temperature T. The quantity Q/T is called the change in the entropy ΔS:

$$\Delta S = \left(\frac{Q}{T}\right)_R \qquad (15.18)$$

In this expression the temperature T must be in kelvins, and the subscript R refers to the word "reversible." It can be shown that Equation 15.18 applies to any process in which heat Q enters or leaves a system reversibly at a constant temperature. Such is the case for the heat that flows into and out of the reservoirs of a Carnot engine. Equation 15.18 indicates that the SI unit for entropy is a joule per kelvin (J/K).

Entropy, like internal energy, is a function of the state or condition of the system. Only the state of a system determines the entropy S that a system has. Therefore, the change in entropy ΔS is equal to the entropy of the final state of the system minus the entropy of the initial state.

We can now describe what happens to the entropy of a Carnot engine. As the engine operates, the entropy of the hot reservoir decreases, since heat Q_H departs at a Kelvin temperature T_H. The change in the entropy of the hot reservoir is $\Delta S_H = -Q_H/T_H$, where the minus sign is needed to indicate a decrease in entropy, since the symbol Q_H denotes only the magnitude of the heat. In contrast, the entropy of the cold reservoir increases by an amount $\Delta S_C = +Q_C/T_C$, for the rejected heat enters the cold reservoir at a Kelvin temperature T_C. The total change in entropy is

$$\Delta S_C + \Delta S_H = \frac{Q_C}{T_C} - \frac{Q_H}{T_H} = 0$$

because $Q_C/T_C = Q_H/T_H$.

The fact that the total change in entropy is zero for a Carnot engine is one specific illustration of a general result. It can be proved that when *any* reversible process occurs, the change in the entropy of the universe is zero; $\Delta S_{universe} = 0$ for a reversible process. The word "universe" means that $\Delta S_{universe}$ takes into account the entropy changes of all parts of the system and all parts of the environment. *Reversible processes, then, do not alter the total entropy of the universe.* To be sure, the entropy of one part of the universe may change because of a reversible process, but if so, the entropy of another part must change in the opposite way by the same amount.

To understand what happens to the entropy of the universe when an *irreversible* process occurs is more difficult, for the expression $\Delta S = (Q/T)_R$ does not apply directly to such a process. However, if a system changes irreversibly from an initial state to a final state, this expression can be used to calculate ΔS indirectly, as Figure 15.20 indicates. We imagine a hypothetical reversible process that causes the system to change between the *the same initial and final states* and then find ΔS for this reversible process. The value obtained for ΔS also applies to the irreversible process that actually occurs, since only the natures of the initial and final states, and not the path between them, determines ΔS. Example 11 illustrates this indirect method and shows that spontaneous (irreversible) processes cause the entropy of the universe to increase.

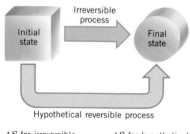

Figure 15.20 Although the relation $\Delta S = (Q/T)_R$ applies to reversible processes, it can be used as part of an indirect procedure to find the entropy change for an irreversible process. This drawing illustrates the procedure discussed in the text.

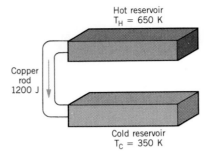

Figure 15.21 Heat flows sponta-neously from a hot reservoir to a cold reservoir.

Example 11   The Entropy of the Universe Increases

Figure 15.21 shows that 1200 J of heat flows spontaneously from a hot reservoir at 650 K to a cold reservoir at 350 K. Determine the amount by which this irreversible process changes the entropy of the universe, assuming that no other changes occur.

REASONING   The hot-to-cold heat flow is irreversible, so the relation $\Delta S = (Q/T)_R$ is applied to a hypothetical process whereby the 1200 J of heat is taken reversibly from the hot reservoir and added reversibly to the cold reservoir.

SOLUTION   The total entropy change of the universe is the algebraic sum of the entropy changes for each reservoir:

$$\Delta S_{universe} = -\underbrace{\frac{1200\ J}{650\ K}}_{\substack{\text{Entropy lost} \\ \text{by the hot} \\ \text{reservoir}}} + \underbrace{\frac{1200\ J}{350\ K}}_{\substack{\text{Entropy gained} \\ \text{by the cold} \\ \text{reservoir}}} = \boxed{+1.6\ J/K}$$

The irreversible process causes the entropy of the universe to increase by 1.6 J/K.

Example 11 is a specific illustration of a general result: *any irreversible process increases the entropy of the universe.* In other words, $\Delta S_{universe} > 0$ for an irreversible process. Reversible processes do not alter the entropy of the universe, whereas irreversible processes cause the entropy to increase. Therefore, the entropy of the universe continually increases, like time itself, and entropy is sometimes even called "time's arrow." It can be shown that the behavior of the entropy of the universe constitutes a completely general statement of the second law of thermodynamics, which applies not only to heat flow but also to all kinds of other processes.

The Second Law of Thermodynamics Stated in Terms of Entropy

The total entropy of the universe does not change when a reversible process occurs ($\Delta S_{universe} = 0$) and increases when an irreversible process occurs ($\Delta S_{universe} > 0$).

ENERGY THAT IS UNAVAILABLE FOR DOING WORK

When an irreversible process occurs and the entropy of the universe increases, the energy available for doing work decreases, as the next example illustrates.

Example 12   Energy Unavailable for Doing Work

Suppose that 1200 J of heat is used as input for an engine under two different conditions. In Figure 15.22a the heat is supplied by a hot reservoir whose temperature is 650 K. In part b of the drawing, the heat flows irreversibly through a copper rod into a second reservoir whose temperature is 350 K and then enters the engine. In either case, a 150-K reservoir is used as the cold reservoir. For each case, determine the maximum amount of work that can be obtained from the 1200 J of heat.

REASONING The work obtained from the engine is the product of its efficiency and the input heat: $W = (\text{Efficiency})Q_H = (\text{Efficiency}) \times (1200 \text{ J})$. The maximum amount of work is obtained when the efficiency is a maximum, that is, when the engine is a Carnot engine. In each case, the efficiency can be determined from Equation 15.15 and the Kelvin temperatures of the hot and cold reservoirs.

SOLUTION

$$\left[\begin{array}{l}\text{Before}\\\text{irreversible}\\\text{heat flow}\end{array}\right] \quad \text{Efficiency} = 1 - \frac{T_C}{T_H} = 1 - \frac{150 \text{ K}}{650 \text{ K}} = 0.77$$

$$W = (\text{Efficiency})(1200 \text{ J}) = (0.77)(1200 \text{ J}) = \boxed{920 \text{ J}}$$

$$\left[\begin{array}{l}\text{After}\\\text{irreversible}\\\text{heat flow}\end{array}\right] \quad \text{Efficiency} = 1 - \frac{T_C}{T_H} = 1 - \frac{150 \text{ K}}{350 \text{ K}} = 0.57$$

$$W = (\text{Efficiency})(1200 \text{ J}) = (0.57)(1200 \text{ J}) = \boxed{680 \text{ J}}$$

When the 1200 J of input heat is taken from the 350-K reservoir instead of the 650-K reservoir, the efficiency of the Carnot engine is smaller. As a result, less work (680 J versus 920 J) can be extracted from the input heat. In this sense, the input heat is of lower "quality" when it is taken from a 350-K reservoir compared to a 650-K reservoir.

Example 12 shows that 240 J less work (920 J − 680 J) can be performed when the input heat is obtained from the reservoir with the smaller temperature. In other words, the irreversible process of heat flow through the copper rod causes 240 J of energy to become unavailable for doing work. The irreversible process also causes the entropy of the universe to increase, and the change in entropy

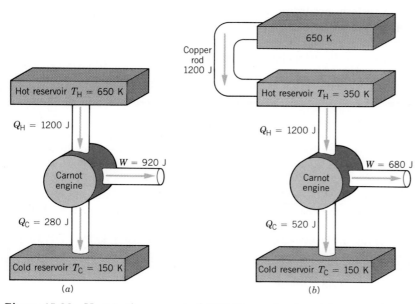

Figure 15.22 Heat in the amount of 1200 J is used as input for an engine under two different conditions in parts *a* and *b*.

$\Delta S_{\text{universe}}$ is related to the energy that is unavailable for doing work $W_{\text{unavailable}}$. Example 11 deals with the same irreversible heat flow that is present in Example 12. If the change in entropy calculated in Example 11 ($\Delta S_{\text{universe}} = +1.6$ J/K) is multiplied by the lowest Kelvin temperature in Example 12 (150 K), an answer is obtained that is identical to the 240 J of "unavailable" energy just discussed. This is a specific illustration of the following general result:

$$W_{\text{unavailable}} = T_0 \Delta S_{\text{universe}} \qquad (15.19)$$

where T_0 is the Kelvin temperature of the coldest heat reservoir. Since irreversible processes cause the entropy of the universe to increase, they cause energy to be degraded, for part of the energy becomes unavailable for the performance of work. In contrast, there is no penalty when reversible processes occur, because for them $\Delta S_{\text{universe}} = 0$, and there is no loss of work, $W_{\text{unavailable}} = T_0 \Delta S_{\text{universe}} = 0$. However, the entropy of the universe is always increasing, so the amount of energy that is unavailable for work is also increasing. Therefore, at some time in the future, it is conceivable that all energy will be degraded to the point where none can be used to do work, leading to the so-called "heat death" of the universe. "Heat death" will arrive when irreversible heat flow from the hot to the cold parts of the universe has occurred to such an extent that all regions have a common temperature. Under such a condition, the hot and cold heat reservoirs that engines must have to produce work would not exist.

ORDER AND DISORDER

Entropy can also be interpreted in terms of order and disorder. As an example, consider a block of ice (Figure 15.23) with each of its H_2O molecules fixed rigidly in place in a highly structured and ordered arrangement. In comparison, the puddle of water into which the ice melts is disordered and unorganized, for the molecules in a liquid are free to move from place to place. Heat is required to melt the ice and produce the disorder. Moreover, heat flow into a system increases the entropy of the system, according to $\Delta S = (Q/T)_R$. We associate an increase in entropy, then, with an increase in disorder. Conversely, we associate a decrease in entropy with a decrease in disorder or a greater degree of order. Example 13 illustrates an order-to-disorder change and the increase of entropy that accompanies it.

Demolition experts are causing this building to go from an ordered state (lower entropy) to a disordered state (higher entropy).

Example 13   Order to Disorder

Find the change in entropy that results when a 2.3-kg block of ice melts slowly (reversibly) at 273 K (0 °C).

REASONING AND SOLUTION   Since the phase change occurs reversibly at a constant temperature, the change in entropy can be found by using $\Delta S = (Q/T)_R$, where Q is the heat absorbed by the melting ice. This heat can be determined by using the latent heat of fusion of water L_f: $Q = mL_f = (2.3 \text{ kg})(3.35 \times 10^5 \text{ J/kg}) = 7.7 \times 10^5$ J. The change in entropy is

$$\Delta S = \frac{7.7 \times 10^5 \text{ J}}{273 \text{ K}} = \boxed{+2.8 \times 10^3 \text{ J/K}}$$

a result that is positive, since the ice absorbs heat as it melts.

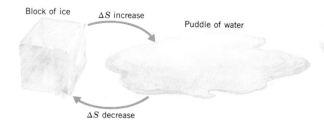

Figure 15.23   A block of ice is an example of an ordered system relative to a puddle of water.

15.11 THE THIRD LAW OF THERMODYNAMICS

To the zeroth, first, and second laws of thermodynamics we add the third (and last) law. The *third law of thermodynamics* indicates that it is impossible to reach a temperature of absolute zero.

The Third Law of Thermodynamics
It is not possible to lower the temperature of any system to absolute zero in a finite number of steps.

This law, like the second law, can be expressed in a number of ways, but a discussion of them is beyond the scope of this text. There is nothing in the second law of thermodynamics that prohibits the temperature of a system from being lowered to absolute zero. The third law is needed to explain a number of experimental observations that cannot be explained by the other laws of thermodynamics.

INTEGRATION OF CONCEPTS

THE FIRST LAW OF THERMODYNAMICS AND THE CONSERVATION OF ENERGY

The atoms and molecules of which materials are composed exhibit different kinds of motions and exert forces on one another. As a result, they have kinetic and potential energy. These and other kinds of molecular energy constitute the internal energy of a substance. When a substance participates in a process involving energy in the form of heat and work, the internal energy of the substance can change. The change in the internal energy equals the algebraic sum of the heat and work, in accordance with the conservation of energy principle. In thermodynamics, this principle—as applied to internal energy, heat, and work—is known as the first law of thermodynamics.

THE SECOND LAW OF THERMODYNAMICS AND THE TRANSFER OF HEAT

In Section 13.2 we discuss the transfer of heat through a material by the mechanism of conduction. The conduction of heat through a rod, for instance, occurs when one end of the rod is maintained at a higher temperature than the other end. Thus, the idea of heat flowing spontaneously from a higher temperature to a lower temperature, and not the other way, is inherent in our earlier discussion. In the present chapter, this same idea about the natural tendency of heat to flow from a higher to a lower temperature is called the second law of thermodynamics. Now, however, we expand our view of heat flow to include the possibility that work may be done as the heat flows. Because of this expanded view, the second law of thermodynamics is a powerful and broadly applicable law. For instance, it applies to chemical reactions, as well as to heat engines. As applied to heat engines, the second law of thermodynamics places a limit on the maximum efficiency that a heat engine can have. In addition, the second law of thermodynamics reveals that heat can be made to flow from a lower to a higher temperature if work is provided, perhaps by a motor. This type of reverse heat flow forms the basis of a number of useful devices, such as air conditioners and heat pumps.

SUMMARY

Thermodynamics is the branch of physics built upon the laws obeyed by energy in the form of work and heat. A thermodynamic **system** is the collection of objects on which attention is being focused, and the **surroundings** are everything else. The **state of the system** is the physical condition of the system, as described by values for physical parameters, usually pressure, volume, and temperature.

Two systems are in **thermal equilibrium** if there is no net flow of heat between them when they are brought into thermal contact. The **zeroth law of thermodynamics** states that two systems individually in thermal equilibrium with a third system are in thermal equilibrium with each other. **Temperature** is the indicator of thermal equilibrium in the sense that there is no net flow of heat between two systems in thermal contact that have the same temperature.

The **first law of thermodynamics** states that when a system absorbs an amount of heat Q and performs an amount of work W, the internal energy of the system changes from its initial value of U_i to a final value of U_f, according to $\Delta U = U_f - U_i = Q - W$. The first law is the conservation of energy principle applied to heat, work, and internal energy. The internal energy is called a **function of state**, because it depends only on the state of the system and not on the method by which the system

came to be in a given state. Heat and work are not functions of state, because they depend on how the system is changed from one state to another.

Thermal processes are **quasi-static** when they occur slowly enough that a uniform pressure and temperature exist throughout the system. An **isobaric process** is one that occurs at constant pressure. The work W done when a system changes from an initial volume V_i to a final volume V_f at a constant pressure P is $W = P(V_f - V_i)$. An **isochoric process** is one that takes place at constant volume, and no work is done in such a process. An **isothermal process** is one that occurs at constant temperature. An **adiabatic process** is one that takes place without the transfer of heat. The work done in a quasi-static thermal process is given by the area under the pressure-versus-volume graph for the process.

When n moles of an ideal gas change quasi-statically from an initial volume V_i to a final volume V_f at a constant Kelvin temperature T, the work done is $W = nRT \ln(V_f/V_i)$. The internal energy of an ideal gas is proportional to the Kelvin temperature and does not change during an isothermal process. When n moles of a monatomic ideal gas change quasi-statically and adiabatically from an initial temperature T_i to a final temperature T_f, the work done is $W = \frac{3}{2}nR(T_i - T_f)$. Along with the ideal gas law, an ideal gas also obeys the relation

$P_iV_i^\gamma = P_fV_f^\gamma$ in an adiabatic process, where $\gamma = C_P/C_V$ is the ratio of the specific heat capacities at constant pressure and constant volume.

The **molar specific heat capacity** C of a substance determines how much heat Q is added or removed when the temperature of n moles of the substance changes by an amount ΔT: $Q = Cn\,\Delta T$. For a monatomic ideal gas, the molar specific heat capacities at constant pressure and constant volume are, respectively, $C_P = \frac{5}{2}R$ and $C_V = \frac{3}{2}R$, where R is the ideal gas constant. For any type of ideal gas $C_P - C_V = R$.

A **reversible process** is one in which both the system and its environment can be returned to exactly the initial states they were in before the process occurred. All spontaneous processes, such as the conduction of heat and any process involving friction, are irreversible.

The **second law of thermodynamics** can be stated in a number of equivalent forms. In terms of heat flow, the second law declares that heat flows spontaneously from a substance at a higher temperature to a substance at a lower temperature and does not flow spontaneously in the reverse direction. In the form known as **Carnot's principle,** the second law states that no irreversible engine operating between two reservoirs at constant temperatures can have a greater efficiency than a reversible engine operating between the same temperatures. Furthermore, all reversible engines operating between the same temperatures have the same efficiency. In terms of **entropy,** the second law states that the total entropy of the universe does not change when a reversible process occurs and increases when an irreversible process occurs.

A **heat engine** operates in cycles and produces work W from input heat Q_H that is extracted from a heat reservoir at a relatively high temperature. The engine rejects heat Q_C into a reservoir at a relatively low temperature. The **efficiency** of a heat engine is defined as Efficiency = W/Q_H. In addition, the principle of the conservation of energy requires that $Q_H = W + Q_C$.

A **Carnot engine** is a reversible engine in which all input heat Q_H originates from a hot reservoir at a single Kelvin temperature T_H and all rejected heat Q_C goes into a cold reservoir at a single Kelvin temperature T_C. For a Carnot engine, $Q_C/Q_H = T_C/T_H$. The **efficiency of a Carnot engine** is the maximum efficiency that an engine operating between two fixed temperatures can have: Efficiency of a Carnot engine = $1 - T_C/T_H$. The maximum efficiency of a Carnot engine is less than one, so even an ideal engine has an efficiency less than 100%.

Refrigerators, air conditioners, and **heat pumps** are devices that utilize work W to make heat Q_C flow from a lower Kelvin temperature T_C to a higher Kelvin temperature T_H. In the process (the refrigeration process) they deposit an amount of heat Q_H at the higher temperature. The principle of the conservation of energy requires that $Q_H = W + Q_C$. If the refrigeration process is perfect, in the sense that it occurs reversibly, the devices are called Carnot devices and the relation $Q_C/Q_H = T_C/T_H$ holds. The **coefficient of performance of a heat pump** is Q_H/W, while the **coefficient of performance of a refrigerator or an air conditioner** is Q_C/W.

The **change in entropy** ΔS for a process in which heat Q enters or leaves a system reversibly at a constant Kelvin temperature T is $\Delta S = (Q/T)_R$, where R stands for "reversible." Irreversible processes cause energy to be degraded in the sense that part of the energy becomes unavailable for the performance of work. The **energy that is unavailable for doing work** because of an irreversible process is $W_{unavailable} = T_0\Delta S_{universe}$, where $\Delta S_{universe}$ is the total entropy change of the universe and T_0 is the Kelvin temperature of the coldest reservoir into which heat is rejected. Increased entropy is associated with a greater degree of disorder and decreased entropy with a lesser degree of disorder (more order).

The **third law of thermodynamics** states that it is not possible to lower the temperature of any system to absolute zero in a finite number of steps.

SOLVED PROBLEMS

| **Solved Problem 1   A Subsidence Inversion** |
| Related Problems: *25   *26   **27   **28 |

THE PHYSICS OF . . .

a temperature inversion in meteorology.

The temperature of the atmosphere usually decreases with increasing altitude. Sometimes, however, meteorologists speak about a *temperature inversion.* In a temperature inversion, the temperature increases, rather than decreases, with increasing altitude. The drawing illustrates how a so-called subsidence inversion comes about. Cool, low-pressure air at a higher altitude subsides or drops to a lower altitude where the pressure is greater. In falling into the region of greater pressure, the air is compressed adiabatically. The compression is adiabatic, because the drop is rapid and there is little time for heat to flow into or out of the falling air. As a result of the adiabatic compression, the air becomes warmer, and its temperature can become greater than that of the air at a lower altitude. In the drawing, the air at the lower altitude of 0.5 km, for example, has a temperature of 285 K. Assume the air is a diatomic ideal gas for which $\gamma = 7/5$. Suppose that air from an altitude of two kilometers ($T_i = 280$ K, $P_i = 79\ 500$ Pa) drops to an altitude of one kilometer ($P_f = 89\ 900$ Pa). Determine the temperature T_f that results because of the drop.

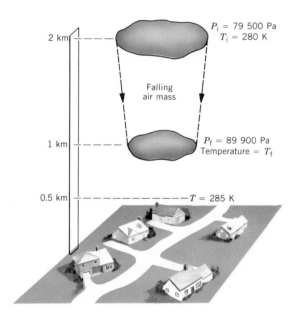

REASONING There are two main steps in the solution to this problem. Since the air it compressed adiabatically, we first determine the ratio of the final and initial volumes using Equation 15.5 ($P_iV_i^{\gamma} = P_fV_f^{\gamma}$). Then, it is necessary to convert this volume ratio into a temperature ratio. This conversion is possible because the ideal gas law [$T = PV/(nR)$] can be used for both the final and the initial conditions.

SOLUTION From $P_iV_i^{\gamma} = P_fV_f^{\gamma}$ we find that

$$\left(\frac{V_f}{V_i}\right)^{7/5} = \frac{P_i}{P_f} = \frac{79\ 500\ \text{Pa}}{89\ 900\ \text{Pa}} = 0.884$$

$$\frac{V_f}{V_i} = (0.884)^{5/7} = 0.916$$

Using the ideal gas law, we find the ratio of final and initial temperatures and, then, the final temperature:

$$\frac{T_f}{T_i} = \frac{P_fV_f/nR}{P_iV_i/nR} = \left(\frac{P_f}{P_i}\right)\left(\frac{V_f}{V_i}\right) = \left(\frac{89\ 900\ \text{Pa}}{79\ 500\ \text{Pa}}\right)(0.916) = 1.04$$

$$T_f = (1.04)T_i = (1.04)(280\ \text{K}) = \boxed{290\ \text{K}}$$

The temperature of 290 K at an altitude of 1 km is greater than that of 285 K at an altitude of 0.5 km (see the drawing). Thus, as altitude increase from 0.5 to 1 km, the temperature increases. This increase in temperature is the temperature inversion.

SUMMARY OF IMPORTANT POINTS The main intent here is to illustrate how the initial pressure, volume, and temperature are related to the final pressure, volume, and temperature in an adiabatic process involving an ideal gas. Since the relation $P_iV_i^{\gamma} = P_fV_f^{\gamma}$ does not include temperature explicitly, problems that deal with the initial and final temperatures can be solved by bringing the ideal gas law into play.

QUESTIONS

1. Ignore friction and assume that air behaves as an ideal gas. The plunger of a bicycle tire pump is pushed down rapidly with the end of the pump sealed so no air escapes. Explain why the cylinder of the pump becomes warm to the touch.

2. One hundred joules of heat is added to a gas, and the gas expands at constant pressure. Is it possible that the internal energy increases by 200 J? Account for your answer with the aid of the first law of thermodynamics.

3. In an isobaric expansion of an ideal gas, is it possible for heat to flow out of the gas? Explain, using the first law of thermodynamics.

4. listed below are five values of heat and work that result when a system interacts with its environment. In each case,

state whether the internal energy of the system increases, decreases, or remains the same, and justify your choices using the first law of thermodynamics:

(a) $W = -500$ J and $Q = 0$
(b) $W = 0$ and $Q = -200$ J
(c) $W = +100$ J and $Q = +100$ J
(d) $W = -100$ J and $Q = -100$ J
(e) $W = +300$ J and $Q = +500$ J

5. (a) Is it possible for the temperature of a substance to rise without heat flowing into the substance? (b) Does the temperature of a substance necessarily have to change because heat flows into or out of it? In each case, give your reasoning and use the example of an ideal gas.

6. The drawing shows an arrangement for an adiabatic free expansion or "throttling" process. The process is adiabatic because the entire arrangement is contained within perfectly insulating walls. The gas in chamber A rushes suddenly into chamber B through a hole in the partition. Chamber B is suddenly evacuated, so the gas expands there under zero external pressure and the work $W = P \Delta V$ is zero. Assume the gas is an ideal gas and explain how the final temperature of the gas after expansion compares to its initial temperature.

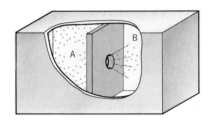

7. Suppose a material contracts when it is heated. Follow the same line of reasoning used in the text to reach Equations 15.7 and 15.8 and deduce which specific heat capacity for the material is larger, C_P or C_V.

8. When a solid melts at constant pressure, the volume of the resulting liquid does not differ much from the volume of the solid. Using what you know about the first law of thermodynamics and about the latent heat of fusion, describe how the internal energy of the liquid compares to the internal energy of the solid.

9. Two *irreversible* engines operate between the same hot and cold reservoirs. Does Carnot's principle require that these two engines have the same efficiency? Justify your answer.

10. Consider a hypothetical engine that takes 10 000 J of heat from a hot reservoir and 5000 J of heat from a cold reservoir and produces 15 000 J of work. (a) Does this engine violate the first law of thermodynamics? Explain. (b) Does the engine violate the second law of thermodynamics? Explain.

11. The second law of thermodynamics, in the form of Carnot's principle, indicates that the most efficient heat engine operating between two temperatures is a reversible one. Does this mean that a reversible engine operating between the temperatures of 600 and 400 K must be more efficient than an *irreversible* engine operating between 700 and 300 K? Provide a reason for your answer.

12. Three reversible engines, A, B, and C, use the same cold reservoir for their exhaust heats. However, they use different hot reservoirs that have the following temperatures: (A) 1000 K, (B) 1100 K, and (C) 900 K. Rank these engines in order of increasing efficiency (smallest efficiency first). Account for your answer.

13. Suppose you wish to improve the efficiency of a Carnot engine. Compare the improvement to be realized via each of the following alternatives: (a) lower the Kelvin temperature of the cold reservoir by a factor of four, (b) raise the Kelvin temperature of the hot reservoir by a factor of four, (c) cut the Kelvin temperature of the cold reservoir in half and double the Kelvin temperature of the hot reservoir. Give your reasoning.

14. Explain why you cannot (a) cool your kitchen by leaving the refrigerator door open and (b) cool your bedroom by putting a window air conditioner on the floor.

15. Is it possible for a Carnot heat pump to have a coefficient of performance that is less than one? Justify your answer.

16. Explain why the coefficient of performance of a heat pump cannot be determined from a statement such as "the pump delivers 15 000 J of heat per cycle."

17. It has been said that heat pumps can't possibly deliver more energy into your house than they consume in operating. Can they or can't they? Explain.

18. An event happens somewhere in the universe and, as a result, the entropy of an object changes by −5 J/K. According to the second law of thermodynamics, which one (or more) of the following is a possible value for the entropy change for the rest of the universe: −5 J/K, 0 J/K, +5 J/K, +10 J/K? Account for your choice(s).

19. When water freezes from a less-ordered liquid to a more-ordered solid, its entropy decreases. Why doesn't this decrease in entropy violate the entropy version of the second law of thermodynamics?

20. In each of the following cases, which has the greater entropy: (a) a handful of popcorn kernels or the popcorn that results from them, (b) a salad before or after it has been tossed, (c) a messy apartment with clothes strewn all over or a neat apartment? Why?

21. A glass of water contains a teaspoon of dissolved sugar. After a while, the water evaporates, leaving behind sugar crystals. The entropy of the sugar crystals is less than the entropy of the dissolved sugar, because the sugar crystals are in a more ordered state. Why doesn't this process violate the entropy version of the second law of thermodynamics?

22. A builder uses lumber to construct a building, which is unfortunately destroyed in a fire. Thus, the lumber existed at one time or another in three different states: (1) as unused building material, (2) as a building, and (3) as a burned-out shell of a building. Rank these three states in order of decreasing entropy (largest first). Provide a reason for the ranking.

23. A refrigerator is advertised as being easier to "live with" during the summer, because it puts into your kitchen only the heat that it removes from the food. Does this advertising claim violate the second law of thermodynamics? Account for your answer.

PROBLEMS

Section 15.2 The First Law of Thermodynamics

1. A jogger does 2.5×10^5 J of work during an exercise routine. Her internal energy decreases by 6.8×10^5 J. How much heat did she absorb or give off? (You specify which.)

2. When one gallon of gasoline is burned in a car engine, 1.19×10^8 J of internal energy is released. Suppose that 1.00×10^8 J of this energy flows directly into the surroundings (engine block and exhaust system) in the form of heat. If 6.0×10^5 J of work is required to make the car go one mile, how many miles can the car travel on one gallon of gas?

3. One-half mole of a monatomic ideal gas absorbs 1200 J of heat while performing 2500 J of work. By how much does the temperature of the gas change? Is the change an increase or a decrease?

4. The drawing shows two objects, A and B. The arrows in the drawing symbolize the flow of heat into or out of an object, as well as the work done on or by an object. Find the change in the internal energy of the system when it is (a) object A only, (b) object B only, and (c) objects A and B together. In each case, specify whether the change is a decrease or an increase.

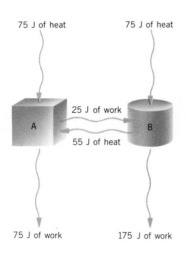

*5. In exercising, a weight lifter loses 0.100 kg of water through evaporation, the heat required to evaporate the water coming from the weight lifter's body. The work done in lifting weights is 1.00×10^5 J. (a) Assuming the latent heat of vaporization of perspiration is 2.42×10^6 J/kg, find the amount by which the internal energy of the weight lifter decreases. (b) Determine the minimum number of nutritional calories of food (1 nutritional calorie = 4186 J) that must be consumed to replace the loss of internal energy.

Section 15.3 Thermal Processes Involving Pressure, Volume, and Temperature

6. The internal energy of a system increases by 185 J during an adiabatic process. Determine whether work is done on or by the system, and find the magnitude of the work.

7. Under isobaric conditions a gas expands and does 470 J of work. The change in volume of the gas is 3.6×10^{-3} m³. What is the pressure?

8. The internal energy of a system increases by 1350 J when the system absorbs 1150 J of heat at a constant pressure of 1.01×10^5 Pa. By how much does the volume of the system change? Does the volume increase or decrease?

9. (a) Using the data presented in the accompanying pressure-versus-volume graph, estimate the work done when the system changes from A to B to C along the path shown. (b) Determine whether the work is done by the system or on the system and, hence, whether the work is positive or negative.

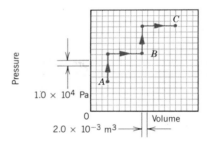

10. A gas is contained in a chamber such as that in Figure 15.5. Suppose the region outside the chamber is evacuated and the total mass of the block and the movable piston is 120 kg. When 1750 J of heat flows into the gas, the internal energy of the gas increases by 1550 J. What is the distance s through which the piston rises?

11. The specific heat capacity of a solid material is 1100 J/(kg·C°). The temperature of 2.0 kg of this material is raised by 6.0 C°. Ignoring the work that corresponds to the small change in the volume of the material, determine the change in the internal energy of the material.

12. When a .22-caliber rifle is fired, the expanding gas from the burning gunpowder creates a pressure behind the bullet. This pressure causes the force that pushes the bullet through the barrel. The barrel has a length of 0.61 m and an opening whose radius is 2.8×10^{-3} m. A bullet (mass = 2.6×10^{-3} kg) has a speed of 370 m/s after passing through this barrel. Ignore kinetic friction and determine the average pressure of the expanding gas.

13. The pressure and volume of a gas are changed along the straight line from A to B in the accompanying graph. From this graph determine an expression for the work done by the gas in terms of the parameters P_A, P_B, V_A, and V_B. (Hint: The *work is represented by the colored area.)*

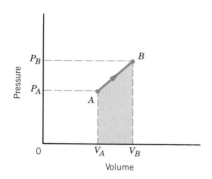

*14. When a monatomic ideal gas expands at a constant pressure of 2.0×10^5 Pa, the volume of the gas increases by 5.0×10^{-3} m^3. Determine how much heat flows into or out of the gas. Specify the direction of the flow.

*15. A piece of aluminum has a volume of 1.4×10^{-3} m^3. The coefficient of volume expansion for aluminum is $\beta = 69 \times 10^{-6}$ (C°)$^{-1}$. The temperature of this object is raised from 20 to 320 °C. How much work does the expanding aluminum do on the surrounding air, if the air pressure is 1.01×10^5 Pa?

*16. Refer to the drawing that accompanies problem 9. When a system changes from A to B along the path shown on the pressure-versus-volume graph, it absorbs 2700 J of heat. What is the change in the internal energy of the system?

*17. The pressure and volume of an ideal monatomic gas change from A to B to C, as the drawing shows. Determine the total heat for the process and state whether the flow of heat is into or out of the gas.

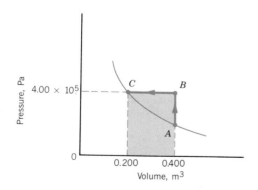

**18. The drawing refers to one mole of monatomic ideal gas and shows a process that has four steps, two isobaric (A to B, C to D) and two isochoric (B to C, D to A). Complete the following table by calculating ΔU, W, and Q (including the algebraic signs) for each of the four steps. Note that the gas has returned to its initial state at the end of the process, so the value for the total ΔU can be predicted in advance without any calculations.

	ΔU	W	Q
A to B			
B to C			
C to D			
D to A			
Total			

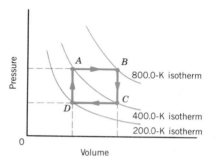

Section 15.4 Thermal Processes That Utilize an Ideal Gas

19. The temperature of three moles of an ideal gas is 373 K. How much work does the gas do in expanding isothermally to four times its initial volume?

20. A monatomic ideal gas does work isothermally. In the process, 4700 J of heat flows into the gas. How much work does the gas do?

21. Two grams of helium (molecular mass = 4.0 u) expand isothermally at 350 K and do 1600 J of work. Assuming helium is an ideal gas, determine the ratio of the final volume of the gas to the initial volume.

22. When 2.00×10^3 J of work is performed adiabatically to compress one-half mole of a monatomic ideal gas, the Kelvin temperature of the gas doubles. Determine the initial temperature of the gas.

23. A monatomic ideal gas ($\gamma = 5/3$) is compressed adiabatically and its volume is reduced by a factor of two. Determine the factor by which its pressure increases.

*24. A bubble from the tank of a scuba diver in a lake contains 3.5×10^{-4} mol of gas. The bubble expands as it rises to the surface from a freshwater depth of 10.3 m. Assuming the gas is an ideal gas and the temperature remains constant at 291 K, find the amount of heat that flows into the bubble.

*25. A diesel engine does not use spark plugs to ignite the fuel and air in the cylinders. Instead, the temperature required to ignite the fuel occurs because the pistons compress the air in the cylinders. Suppose air at an initial temperature of 27 °C is compressed adiabatically to a temperature of 681 °C. Assume the air to be an ideal gas for which $\gamma = 7/5$. Find the compression ratio, which is the ratio of the initial volume to the final volume. (*See Solved Problem 1 for a related problem.*)

*26. A monatomic ideal gas ($\gamma = 5/3$) is contained within a perfectly insulated cylinder that is fitted with a movable piston. The initial pressure of the gas is 1.50×10^5 Pa. The piston is pushed so as to compress the gas, with the result that the Kelvin temperature doubles. What is the final pressure of the gas? (*See Solved Problem 1 for a related problem.*)

**27. The work done by one mole of a monatomic ideal gas ($\gamma = 5/3$) in expanding adiabatically is 825 J. The initial temperature and volume of the gas are 393 K and 0.100 m³. Obtain the final temperature and volume of the gas. (*See Solved Problem 1 for a related problem.*)

**28. The drawing shows an adiabatically isolated cylinder that is divided initially into two identical parts by an adiabatic partition. Both sides contain one mole of a monatomic ideal gas ($\gamma = 5/3$), with the initial temperature being 525 K on the left and 275 K on the right. The partition is then allowed to move slowly (i.e., quasi-statically) to the right, until the pressures on each side of the partition are the same. Find the final temperatures on the left and right. (*See Solved Problem 1 for a related problem.*)

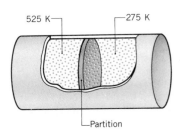

525 K — 275 K

— Partition

Section 15.5 Specific Heat Capacities and the First Law of Thermodynamics

29. How much heat is required to change the temperature of 1.5 mol of a monatomic ideal gas by 77 K, if the pressure is held constant?

30. The temperature of 2.5 mol of helium (a monatomic gas) is lowered by 35 K under conditions of constant volume. Assuming that helium behaves as an ideal gas, how much heat is removed from the gas?

31. Argon is a monatomic gas whose molecular mass is 39.9 u. The temperature of eight grams of argon is raised by 75 K under conditions of constant pressure. Assuming that argon is an ideal gas, how much heat is required?

32. The volume of a monatomic ideal gas is held constant while heat is added. The temperature of the gas increases by 45 K. If the same amount of heat were added to the same sample of gas under conditions of constant pressure, by how much would the temperature increase?

33. Heat Q is added to a monatomic ideal gas at constant pressure. As a result, the gas does work W. Find the ratio Q/W.

*34. The temperature of two moles of an ideal gas (not monatomic) increases, because 7.50×10^2 J of heat is absorbed under conditions of constant volume. Under conditions of constant pressure, the temperature increases by the same amount when 1.00×10^3 J of heat is absorbed. By how many kelvins does the temperature change? (*Hint: The gas is not monatomic, but the relation $C_P - C_V = R$ still applies.*)

*35. Even at rest, the human body generates heat. The heat arises because of the body's metabolism, that is, the chemical reactions that are always occurring in the body to generate energy. In rooms designed for use by large groups, adequate ventilation or air conditioning must be provided to remove this heat. Consider a classroom containing 200 students. Assume that the metabolic rate of generating heat is 130 W for each student and that the heat accumulates during a fifty-minute lecture. In addition, assume that the air has a molar specific heat of $C_V = \frac{5}{2}R$ and that the room (volume = 1200 m³, initial pressure = 1.01×10^5 Pa, and initial temperature = 21 °C) is sealed shut. If all the heat generated by the students were absorbed by the air, by how much would the air temperature rise during a lecture?

*36. A monatomic ideal gas expands at constant pressure. (a) What percentage of the heat being supplied to the gas is used to increase the internal energy of the gas? (b) What percentage is used for doing the work of expansion?

**37. One mole of neon, a monatomic gas, starts out at conditions of standard temperature and pressure. The gas is heated at constant volume until its pressure is tripled, then further heated at constant pressure until its volume is doubled. Assume that neon behaves as an ideal gas. For the entire process, find the heat added to the gas.

Section 15.7 Heat Engines

38. The input heat for an engine is 8700 J. The work done by the engine is 5300 J. What is the efficiency?

39. The input heat for an engine is 2.41×10^4 J, and the rejected heat is 5.86×10^3 J. Find the work done by the engine.

40. An engine has an efficiency of 71% and produces 4800 J of work. Determine (a) the input heat and (b) the rejected heat.

41. In doing 16 600 J of work, an engine rejects 9700 J of heat. What is the efficiency of the engine?

*42. Engine A discards 72% of its input heat into a cold reservoir. Engine B has twice the efficiency as engine A. What percentage of its input heat does engine B discard?

Section 15.8 Carnot's Principle and the Carnot Heat Engine

43. An engine has a hot reservoir temperature of 970 K and a cold reservoir temperature of 570 K. The engine operates at three-fourths maximum efficiency. What is the efficiency of the engine?

44. Five thousand joules of heat is put into a Carnot engine whose hot and cold reservoirs have temperatures of 500 and 200 K, respectively. How much heat is converted into work?

45. Electric motors convert electrical energy into work with efficiencies approaching 95%. For a Carnot engine to have such an efficiency, what must be the ratio of the Kelvin temperatures of the cold and hot reservoirs?

46. The ratio of the input heat to the discarded heat of a Carnot engine is 1.5. (a) What is the efficiency of the engine? (b) What is the ratio of the Kelvin temperature of the hot reservoir to the Kelvin temperature of the cold reservoir?

47. A Carnot engine operates between temperatures of 650 and 350 K. To improve the efficiency of the engine, it is decided either to raise the temperature of the hot reservoir by 40 K or to lower the temperature of the cold reservoir by 40 K. Which change gives the greatest improvement? Justify your answer by calculating the efficiency in each case.

48. An engine does 20 900 J of work and rejects 7330 J of heat into a cold reservoir at 298 K. What is the smallest possible temperature of the hot reservoir?

*49. A power plant taps steam superheated by geothermal energy to 505 K (the temperature of the hot reservoir) and uses the steam to do work in turning the turbine of an electric generator. The steam is then converted back into water in a condenser at 323 K (the temperature of the cold reservoir), after which the water is pumped back down into the earth where it is heated again. The output power of the plant is 84 000 kilowatts. Determine (a) the maximum efficiency at which this plant can operate and (b) the minimum amount of rejected heat that must be removed from the condenser every twenty-four hours.

**50. The hot and cold reservoirs of a Carnot engine have temperatures of 905 and 405 K, respectively. The engine does the work of lifting a 10.0-kg block straight up from rest, so that at a height of 4.00 m the block has a speed of 8.00 m/s. How much heat must be put into the engine?

**51. A nuclear-fueled electric power plant utilizes a so-called "boiling water reactor." In this type of reactor, nuclear energy causes water under pressure to boil at 285 °C (the temperature of the hot reservoir). After the steam does the work of turning the turbine of an electric generator, the steam is converted back into water in a condenser at 40 °C (the temperature of the cold reservoir). To keep the condenser at 40 °C, the rejected heat must be carried away by some means, for example, by water from a river. The plant operates at three-fourths of its Carnot efficiency, and the electrical output power of the plant is 1.2×10^9 watts. A river with a water flow rate of 1.0×10^5 kg/s is available to remove the rejected heat from the plant. Find the number of Celsius degrees by which the temperature of the river rises.

Section 15.9 Refrigerators, Air Conditioners, and Heat Pumps

52. What is the coefficient of performance of an air conditioner that uses 3500 J of electrical energy to remove 11 200 J of heat from a room?

53. The coefficient of performance of a refrigerator is 4.6. How much electrical energy is used in removing 4100 J of heat from the food inside?

54. The water in a deep underground well is used as the cold reservoir of a Carnot heat pump that maintains the temperature of a house at 298 K. To deposit 12 600 J of heat in the house, the heat pump requires 806 J of work. Determine the temperature of the well water.

55. The temperatures indoors and outdoors are 299 and 312 K, respectively. A Carnot air conditioner deposits 6.12×10^5 J of heat outdoors. How much heat is removed from the house?

56. A Carnot refrigerator maintains the food inside it at 276 K, while the temperature of the kitchen is 298 K. The refrigerator removes 3.00×10^4 J of heat from the food. How much heat is delivered to the kitchen?

57. A Carnot engine has an efficiency of 0.70. If this engine were run backward as a heat pump, what would be the coefficient of performance?

58. A heat pump removes 2090 J of heat from the outdoors and delivers 3140 J of heat to the inside of a house. (a) How much work does the heat pump need? (b) What is the coefficient of performance of the heat pump?

*59. Two kilograms of liquid water at 0 °C is put into the freezer compartment of a Carnot refrigerator. The temperature of the compartment is −15 °C and the temperature of the kitchen is 27 °C. If the cost of electrical energy is ten cents per kilowatt·hour, how much does it cost to make two kilograms of ice at 0 °C?

*60. How long would a 3.00-kW space heater have to run to put into a kitchen the same amount of heat as a refrigerator (coefficient of performance = 3.00) does when it freezes 1.50 kg of water at 20.0 °C into ice at 0.0 °C?

Section 15.10 Entropy and the Second Law of Thermodynamics

61. Suppose the entropy of a system decreases by 25 J/K because of some process. (a) Based on the second law of thermodynamics, what can you conclude about the entropy change of the environment that surrounds this system? (b) Interpret your answer to part (a) in terms of order and disorder of the environment.

62. Four kilograms of carbon dioxide sublimes from solid "dry ice" to a gas at a pressure of 1.00 atm and a temperature of 194.7 K. The latent heat of sublimation is 5.77×10^5 J/kg. Find the change in entropy of the carbon dioxide.

63. A process occurs in which the entropy of a system increases by 125 J/K. During the process, the energy that becomes unavailable for doing work is zero. (a) Is this process reversible or irreversible? Give your reasoning. (b) Determine the change in the entropy of the environment of the system.

*64. (a) Find the equilibrium temperature that results when one kilogram of liquid water at 373 K is added to two kilograms of liquid water at 283 K in a perfectly insulated container. (b) When heat is added to or removed from a solid or liquid of mass m and specific heat capacity c, the change in entropy can be shown to be $\Delta S = mc \ln(T_f/T_i)$, where T_i and T_f are the initial and final Kelvin temperatures. Use this equation to calculate the entropy change for each amount of water. Then combine the two entropy changes algebraically to obtain the total entropy change. Note that the process is irreversible, so the total entropy change of the universe is greater than zero. (c) Assuming the coldest reservoir at hand has a temperature of 273 K, determine the amount of energy that becomes unavailable for doing work because of the irreversible process.

*65. (a) Five kilograms of water at 80.0 °C is mixed in a perfect thermos with 2.00 kg of ice at 0.0 °C, and the mixture is allowed to reach equilibrium. Using the expression $\Delta S = mc \ln(T_f/T_i)$ [see problem 64] and the change in entropy for melting, find the change in entropy that occurs. (b) Should the entropy of the universe increase or decrease as a result of the mixing process? Give your reasoning and state whether your answer in part (a) is consistent with your answer here.

ADDITIONAL PROBLEMS

66. Five moles of oxygen expands isothermally from 0.100 to 0.400 m³. To maintain the constant temperature, 2.50×10^4 J of heat is added to the system. Assuming oxygen to be an ideal gas, determine the temperature.

67. The work done to compress one mole of a monatomic ideal gas is 6200 J. The temperature of the gas changes from 350 to 550 K. How much heat flows between the gas and its surroundings? Determine whether the heat flows into or out of the gas.

68. An engine rejects three times more heat than it converts into work. What is the efficiency of the engine?

69. A system gains 1500 J of heat, while the internal energy of the system increases by 4500 J and the volume decreases by 0.010 m³. Assume the pressure is constant and find its value.

70. What is the maximum coefficient of performance of a refrigerator that operates between temperatures of 277 and 302 K?

71. Find the change in entropy of the H_2O molecules when (a) three kilograms of ice melts into water at 273 K and (b) three kilograms of water changes into steam at 373 K. (c) On the basis of the answers to parts (a) and (b), discuss which change creates more disorder in the collection of H_2O molecules.

72. Suppose 550 J of heat is removed from two moles of a monatomic ideal gas. What drop in temperature occurs when the energy is removed under conditions of (a) constant volume and (b) constant pressure?

73. A Carnot heat pump operates between an outdoor temperature of 265 K and an indoor temperature of 298 K. Find its coefficient of performance.

74. A Carnot engine has an efficiency of 0.700, and the temperature of its cold reservoir is 378 K. (a) Determine the temperature of its hot reservoir. (b) If 5230 J of heat are rejected to the cold reservoir, what amount of heat is put into the engine?

75. The volume of a gas is changed along the curved line between A and B in the drawing. Find the work for the process, and determine whether the work is positive or negative.

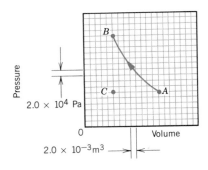

*76. Refer to the drawing in problem 75, where the curve between A and B is an isotherm. An ideal gas begins at A and is changed along the horizontal line from A to C and then along the vertical line from C to B. Find the heat for the process ACB and determine whether it flows into or out of the gas. (*Hint: The area under the graph is the work.*)

*77. An engine is run in reverse as a heat pump. An identical engine (with the same values of Q_H, Q_C, and W) is run in reverse as a refrigerator. The coefficient of performance of the heat pump is three times greater than the coefficient of performance of the refrigerator. Obtain (a) the coefficient of performance of the refrigerator, (b) the coefficient of performance of the heat pump, and (c) the efficiency of the engine.

*78. Using the relationship for an adiabatic expansion or compression of an ideal gas ($P_i V_i^\gamma = P_f V_f^\gamma$) together with the ideal gas law, derive an expression similar to the one above, but involving only volume, temperature, and γ.

*79. Engine A receives three times more input heat, produces five times more work, and rejects two times more heat than engine B. Find the efficiency of each engine.

*80. A ten-watt heater is used to heat a monatomic ideal gas at a constant pressure of 2.50×10^5 Pa. During the process, the 1.00×10^{-3}-m³ volume of the gas increases by 20.0%. How long was the heater on?

*81. Suppose the gasoline in a car engine burns at 631 °C, while the exhaust temperature is 139 °C and the outdoor temperature is 27 °C. Assume the engine can be treated as a Carnot engine (a gross oversimplification). In an attempt to increase mileage performance, an inventor builds a second engine that functions between the exhaust and outdoor temperatures and uses the exhaust heat to produce additional work. Assume the inventor's engine can also be treated as a Carnot engine. By what percentage does the inventor's device increase the work obtained from the burning gasoline?

**82. Suppose a monatomic ideal gas is contained within a vertical cylinder that is fitted with a movable piston. The piston is frictionless and has a negligible mass. The area of the piston is 3.14×10^{-2} m², and the pressure outside the cylinder is 1.01×10^5 Pa. Heat (2093 J) is removed from the gas. Through what distance does the piston drop?

Physics & Geology

Geologists have made many profound discoveries over the past few decades about how the earth works, including the existence of tectonic plates, the dynamics of volcanoes and earthquakes, the general structure of the earth's deep interior (mantle and core), and details of its surface layer (the crust). What impact have principles of physics had on geology? Imagine that it is 5:00 P.M., October 17, 1989, and you are a geophysicist who has just returned to San Francisco from Santa Barbara, after three exhausting days searching for clues to new oil and natural gas deposits.

Oil and natural gas are trapped in certain porous rocks that have relatively low densities compared with neighboring regions. Knowing this fact enables geologists and geophysicists to search for such oil-bearing strata. They explore the interior of the earth's surface layer or crust in a variety of ways.

One method, called seismic reflection profiling, is done by setting off a series of explosions on the surface of the earth, or on ocean bottoms, and then detecting the vibrations of the earth caused by the explosions. This is what you have been doing. The energy of the blasts is transferred into the earth as longitudinal waves, like sound, and transverse waves, like waves traveling down a string. These waves travel through the various layers of the earth, reflecting and refracting as they encounter rocks of different density. Some of the reflected energy is then picked up by sensitive microphones called geophones or hydrophones that have been placed around the region where the explosions occurred. Timing the return of waves from different explosion sites

Earthquakes, like the one that caused this damage in San Francisco in 1989, occur when seismic waves are generated by motions of the earth's crust that occur along geological faults.

around a local area, knowing the speed of the waves in different rock, and using sophisticated computer programs, enable you to determine the interior distribution of rock in the earth's crust, which then helps you to determine where to look for the oil and gas. You are especially interested in an upward fold, or pocket, of porous rock.

Another method of locating lower density rock is to measure the local gravitational pull from the rocks. Because the oil-bearing rock has a lower density (lower mass per volume), it exerts less gravitational pull at the earth's surface than does an equal volume of higher density rock. Oil-seeking geologists use instruments called gravimeters to measure the gravitational pull of the earth at different locations. By combining the gravimetric data and the underground maps made by seismic reflection profiling, geophysicists can predict the locations of many oil deposits. Natural science is not so simple that all oil deposits can be determined just by using basic physics like this, but many of the bigger deposits have been found in this way.

You are in your office, studying the preliminary data on your computer terminal when the room starts to shake. You are thrown violently onto the floor and barely fend off the computer as it slides off the table. You are in the midst of a major earthquake. Here, too, physics explains geological events.

The earth has a surprisingly thin, solid surface called the crust, which is no more than 60 km thick. Under it is thousands of kilometers of molten (liquid) rock that is heated from

radioactivity and by compression due to the gravity of the earth pushing down on it. This heating causes the molten rock, or magma, to rise and sink as convection currents, like boiling water. The rising magma pushes against the bottom of the crust, causing the crust to move in different directions. As a result, the earth's surface is broken into numerous pieces called tectonic plates. Like ice floating on water, the earth's crust is less dense than the magma below it. The plates are either moving apart, as along the center of the Atlantic Ocean; moving together, as is the Indian-Australian plate crunching into the Eurasian plates; or rubbing against each other, as the North American Plate does against the Pacific Plate. This last effect is most famous as the San Andreas Fault, which moves 3 cm per year, on average.

The San Andreas Fault, from which you are only a few miles, can be idealized in terms of friction between the rock of the two plates. If the two plates slid with no friction, then they would be continually sliding with respect to each other and jarring earthquakes would not occur there. Everyone would keep clear of the sliding zone and that would be that. But since the plates are pressing against each other with great force, there is tremendous static friction and the plates cannot move continuously; strain exists at their boundary. Earthquakes, such as the one you are experiencing, essentially occur when the sideways, or lateral, force on the plates overcomes the static friction between them, causing sudden movements of the land.

Where the crust is thinnest, where it has structural defects, or where the

magma is hottest, volcanoes often occur. Here the rising magma exerts enough pressure (force per area) to break through the surface. Sometimes, such as the recent eruption of Mount Saint Helens in Washington State, the pressure can build up until it is strong enough to blow off the solid rock on the top of a volcano, much like a cannon. When this happened to Mount Saint Helens, three cubic kilometers of volcanic matter were blown over 20 km into the sky, coming down over the next few years under the influence of gravity.

The hard disk containing your data was not hurt by the fall it took. You clean up your work area as the quake subsides and drive home, with an unforgettable reminder of the fragile hold humans have on the earth's surface.

FOR INFORMATION ABOUT CAREERS IN GEOLOGY

Job descriptions of geology and geology-related fields, including geography, geophysics, oceanography, and petrology can be found in the **Encyclopedia of Careers and Vocational Guidance,** *Vol. 2, 8th edition, William E. Hopke, ed., 1990, J. G. Ferguson, publisher.*

Many references, with addresses and phone numbers, for career information in geology and geology-related fields are provided in the **Professional Careers Sourcebook,** *K. M. Savage and C. A. Dorgan, eds., 1990, Gale Research, Inc., publisher.*

Other places to write for career information are the following:

- *American Geological Institute, 4220 King Street, Alexandria, VA 22302*
- *American Physical Society, 335 E. 45th St., New York, NY 10017*
- *Association of Engineering Geologists, 323 Boston Post Rd., Sudbury, MA 01776*
- *Geological Society of America, P.O. Box 9140, Boulder, CO 80301*

Copyright © 1991 by Neil F. Comins.

PART THREE

WAVE MOTION

Waves. The word conjures up images of summers at the beach, where the surf can be fun. But water waves are not the only kinds of waves that can and do affect us. Waves and wave-related effects occur frequently. Two human senses, for example, utilize waves; we hear because of sound waves and see because of light waves. Musical instruments operate by using waves of various types, and information is sent from place to place with the aid of radio waves. Ultrasonic waves are used in medicine to obtain images of the developing fetus. Sonar locates objects under water with sound waves. Waves even play a role in the kitchen, when food is cooked in a microwave oven.

There is a great variety of waves, and Part Three of this text discusses some of the basic types. We will find that a large and important group called periodic waves can be described in terms of a few characteristics. In the discussion, a number of concepts from earlier parts of the book will recur, including velocity, frequency, temperature, pressure, and power.

One of the most intriguing abilities of waves is that one wave can pass directly through another. Thus, two or more waves can exist at the same place at the same time. This ability leads to a number of surprising, unique, and useful phenomena. At the heart of each of these lies the fact that one wave can either augment or diminish the effect of another wave at a given place, depending on the circumstances. The phenomena that can arise when two or more waves come together are important in a variety of areas, including acoustics, the design of noise-free environments, radio and television technology, stereophonic sound reproduction, and photography. To explain how multiple waves combine, we will use a simple but fundamental principle called the principle of linear superposition.

CHAPTER 16

WAVES AND SOUND

Fennec foxes are the smallest of all foxes, with an average length of only fourteen inches. Yet, as you can see in this photograph, they have relatively enormous ears, about four inches long. One possible purpose for such large ears is to funnel more sound energy into the animal's hearing mechanism. In this chapter, we will see that sound is a kind of wave and, as such, it can carry energy from place to place. There are also other kinds of waves besides sound waves, and many of them share a number of basic features. We will begin by discussing these common features and the relationships among them. Then, we consider sound waves and related phenomena in greater detail, including the familiar concepts of the speed of sound, pitch, loudness, and decibels.

16.1 THE NATURE OF WAVES

Water waves have two features common to all waves. First, *a wave is a traveling disturbance.* In Figure 16.1 the wave created by the motorboat travels across the lake and disturbs the fisherman. It should be noted that *there is no bulk flow of water* outward from the motorboat. The wave is not a bulk movement of water such as a river, but, rather, a disturbance traveling on the surface of the lake. Second, *a wave carries energy from place to place.* Part of the wave's energy in Figure 16.1 is transferred to the fisherman and his boat.

We will consider two basic types of waves, transverse and longitudinal. Figure 16.2 illustrates how a transverse wave can be generated using a Slinky, a remarkable toy that is a long, loosely coiled spring. If one end of the Slinky is jerked up

Figure 16.1 The wave created by the motorboat travels across the lake and disturbs the fisherman.

and down, as in part *a*, an upward pulse is sent traveling toward the right. If the end is then jerked down and up, as in part *b*, a downward pulse is generated and also moves to the right. If the end is continually moved up and down in simple harmonic motion, an entire wave is produced. As part *c* illustrates, the wave consists of a series of alternating upward and downward sections that propagate to the right, disturbing the vertical position of the Slinky in the process. To focus attention on the disturbance, a colored dot is attached to the Slinky in part *c* of the drawing. As the wave advances to the right, the dot is displaced up and down in simple harmonic motion. The motion of the dot occurs perpendicular, or transverse, to the direction in which the wave travels. This example shows that *a transverse wave is one in which the disturbance is perpendicular to the direction of travel of the wave.* Radio waves, light waves, and microwaves are transverse waves. Transverse waves also travel on the strings of instruments such as guitars and banjos.

A longitudinal wave can also be generated with a Slinky, and Figure 16.3 demonstrates how. When one end of the Slinky is pushed forward along its length (i.e., longitudinally) and then returned to its starting point, as in part *a*, a compressed region where the coils are squeezed together is sent traveling to the

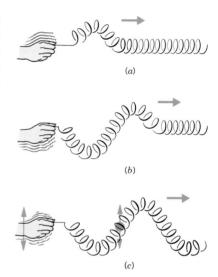

Figure 16.2 (*a*) An upward pulse moves to the right, followed by (*b*) a downward pulse. (*c*) When the end of the Slinky is moved up and down continuously, a transverse wave is produced.

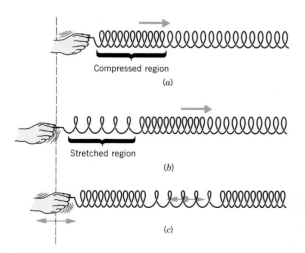

Figure 16.3 (*a*) A compressed region moves to the right, followed by (*b*) a stretched region. (*c*) When the end of the Slinky is moved back and forth continuously, a longitudinal wave is produced.

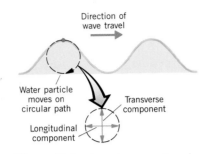

Figure 16.4 A water wave is neither transverse nor longitudinal, since water particles at the surface move clockwise on circular paths as the wave moves from left to right.

right. If the end is pulled backward and then returned to its starting point, as in part *b*, a stretched region where the coils are pulled apart is formed and also moves to the right. If the end is continually moved back and forth in simple harmonic motion, an entire wave is created. As part *c* shows, the wave consists of a series of alternating compressed and stretched regions that travel to the right and disturb the separation between adjacent coils. A colored dot is once again attached to the Slinky to emphasize the vibratory nature of the disturbance. In response to the wave, the dot is displaced back and forth in simple harmonic motion along the line of travel of the wave. Thus, *a longitudinal wave is one in which the disturbance is parallel to the line of travel of the wave.* A sound wave is a longitudinal wave.

Some waves are neither transverse nor longitudinal. For instance, in a water wave the motion of the water particles is not strictly perpendicular or strictly parallel to the line along which the wave travels. Instead, the motion includes both transverse and longitudinal components, since the water particles at the surface move on circular paths, as Figure 16.4 indicates.

16.2 PERIODIC WAVES

The transverse and longitudinal waves that we have been discussing consist of patterns that are produced over and over again by the source. These kinds of waves are called *periodic waves.* In Figures 16.2 and 16.3 the repetitive patterns occur as a result of the simple harmonic motion of the left end of the Slinky. Therefore, some of the terminology (cycle, amplitude, period, and frequency) used to describe periodic waves is the same as that used in connection with simple harmonic motion.

To introduce this terminology, Figure 16.5 presents a graphical representation of a transverse wave on a Slinky. One *cycle* of the wave is shaded in color in both parts of the drawing. In part *a* the vertical position of the Slinky is plotted on the ordinate, while the corresponding distance along the length of the Slinky is

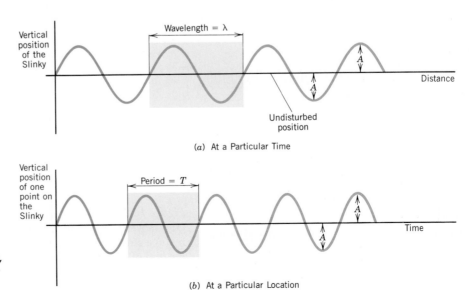

Figure 16.5 In parts *a* and *b*, one cycle of the wave is shaded in color, and the amplitude of the wave is denoted as *A*.

A surfer catches a wave for a ride. Sound and light are also waves, and "catching" these waves gives us our senses of hearing and sight.

plotted on the abscissa. Such a graph is equivalent to a photograph of the wave taken at a particular instant in time and shows the disturbance that exists at each point along the Slinky's length. As marked on this graph, the *amplitude* A is the maximum excursion of a particle of the medium from the particle's undisturbed position. The amplitude is the distance between a crest, or highest point on the wave pattern, and the undisturbed position; the amplitude is also the distance between a trough, or lowest point on the wave pattern, and the undisturbed position. The *wavelength* λ is the horizontal length of one cycle of the wave, as shown in color in Figure 16.5*a*. The wavelength is also the horizontal distance between two successive crests, two successive troughs, or any two successive equivalent points on the wave. A wave is a series of many cycles.

Part *b* of Figure 16.5 shows a graph in which time, rather than distance, is plotted on the abscissa. This graph is obtained by observing a single point on the Slinky. As the wave passes, the point moves up and down as time passes. As indicated on the graph, the *period* T is the time required for one cycle of the wave to move past an observer. Since the point under observation moves up and down in simple harmonic motion, the period is also the time required for one complete up/down cycle, just as it is for an object on a spring. Equivalently, the period is the time required for the wave to travel a distance of one wavelength. The period T is related to the *frequency* f, just as it is for any example of simple harmonic motion:

$$f = \frac{1}{T} \tag{10.8}$$

The period is commonly measured in seconds, and frequency is measured in cycles per second or hertz (Hz). If, for instance, one cycle of a wave takes one-tenth of a second to pass an observer, then ten cycles pass during each second, as Equation 10.8 indicates ($f = 1/(0.1 \text{ s}) = 10$ cycles/s $= 10$ Hz).

A simple relation exists between the period, the wavelength, and the speed of a wave, a relation that Figure 16.6 helps to introduce. Imagine waiting at a railroad crossing, while a freight train moves by at a constant speed. The train consists of a long line of identical boxcars, each of which has a length λ and requires a time T to pass. You can determine the speed v of the train by dividing the length by the time; that is, $v = \lambda/T$. This same equation applies for a wave and relates the speed

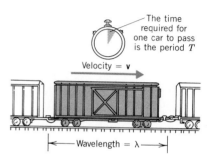

Figure 16.6 A train moving at a constant speed serves as an analogy for a traveling wave.

of the wave to the wavelength λ and the period T. Since the frequency of a wave is $f = 1/T$, the expression for the speed is often written as

$$v = f\lambda \qquad (16.1)$$

The terminology discussed above and the fundamental relations $f = 1/T$ and $v = f\lambda$ apply to longitudinal as well as to transverse waves. Example 1 illustrates how the wavelength of a wave is determined by the wave speed and the frequency established by the source.

Example 1   The Wavelength of Radio Waves

AM and FM radio waves are transverse waves that consist of electric and magnetic disturbances. These waves travel at a speed of 3.00×10^8 m/s. A station broadcasts an AM radio wave whose frequency is 1230×10^3 Hz (1230 kHz on the dial) and an FM radio wave whose frequency is 91.9×10^6 Hz (91.9 MHz on the dial). Find the distance between adjacent crests in each wave.

REASONING   The distance between adjacent crests is the wavelength λ. Since the speed of each wave is $v = 3.00 \times 10^8$ m/s and the frequencies stated are established at the broadcasting station, the relation $v = f\lambda$ can be used to determine the wavelengths:

SOLUTION

[AM] $\lambda = \dfrac{v}{f} = \dfrac{3.00 \times 10^8 \text{ m/s}}{1230 \times 10^3 \text{ Hz}} = \boxed{244 \text{ m}}$

[FM] $\lambda = \dfrac{v}{f} = \dfrac{3.00 \times 10^8 \text{ m/s}}{91.9 \times 10^6 \text{ Hz}} = \boxed{3.26 \text{ m}}$

Notice that the wavelength of an AM radio wave is longer than two and one-half football fields!

PROBLEM SOLVING INSIGHT

The equation $v = f\lambda$ applies to all kinds of periodic waves.

16.3 THE SPEED OF A WAVE ON A STRING

THE DEPENDENCE OF WAVE SPEED ON PROPERTIES OF THE STRING

The properties of the material* or medium through which a wave travels determine the speed of the wave. For example, Figure 16.7 shows a transverse wave on a string and draws attention to four string particles that have been drawn as colored dots. As the wave moves to the right, each particle is displaced, one after the other, from its undisturbed position. In the drawing, particles 1 and 2 have already been displaced upward, while particles 3 and 4 are not yet affected by the wave. Particle 3 will be the next one to move upward because the section of string immediately to its left (i.e., particle 2) will pull it upward.

Figure 16.7 leads us to conclude that the speed with which the wave moves to the right depends on how quickly one particle of the string is accelerated upward

---

* Electromagnetic waves can move through a vacuum, as well as through materials such as glass and water.

Figure 16.7 As a transverse wave moves to the right with speed v, each string particle is displaced, one after the other, from its undisturbed position.

in response to the net pulling force exerted by its adjacent neighbors. In accord with Newton's second law, a stronger net force results in a greater acceleration, and, thus, a faster-moving wave. The ability of one particle to pull on its neighbors depends on how tightly the string is stretched, that is, on the tension (see Section 4.10 for a review of tension). The greater the tension, the greater the pulling force the particles exert on each other, and the faster the wave travels, other things being equal.

In addition to the tension, there is another factor that influences the wave speed. According to Newton's second law, the inertia or mass of particle 3 in Figure 16.7 also affects how quickly it responds to the upward pull of particle 2. For a given net pulling force, a smaller mass has a greater acceleration than a larger mass. Therefore, other things being equal, a wave travels faster on a string whose particles have a small mass, or, as it turns out, on a string that has a small mass per unit length. The mass per unit length is called the *linear density* of the string.

The effects of the tension F and the mass per unit length m/L are evident in the following expression for the speed v of a small-amplitude wave on a string:

$$v = \sqrt{\frac{F}{m/L}} \qquad (16.2)$$

Clearly, a larger tension F and a smaller linear density m/L lead to a larger value for the speed. We can verify that Equation 16.2 gives the correct unit for speed. The SI unit for the force F is that of mass times acceleration, or $\text{kg}(\text{m/s}^2)$. The unit for the linear density m/L is kg/m. Thus, the unit of $\sqrt{F/(m/L)}$ is $\sqrt{(\text{kg}\cdot\text{m/s}^2)/(\text{kg/m})} = \sqrt{\text{m}^2/\text{s}^2} = \text{m/s}$, which is the unit for speed.

The motion of transverse waves along a string is important in the operation of musical instruments, such as the guitar, the violin, and the piano. In these instruments, the strings are either plucked, bowed, or struck to produce transverse waves. Example 2 discusses the speed of the waves on the strings of a guitar.

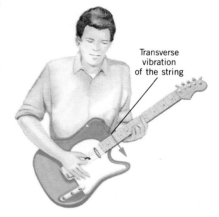

Figure 16.8 Transverse waves are generated on a guitar string by plucking it.

THE PHYSICS OF . . .

waves on guitar strings.

Example 2 Waves Traveling on Guitar Strings

Transverse waves travel on the strings of an electric guitar after the strings are plucked. (See Figure 16.8.) The length of each string between its two fixed ends is 0.628 m, and the mass is 0.208 g for the highest pitched E string and 3.32 g for the lowest pitched E string. Each string is under a tension of 226 N. Find the speeds of the waves on the two strings.

REASONING AND SOLUTION   The tension F, the mass m, and the length L are known for each string, and the speeds of the waves are given by Equation 16.2:

$$\begin{bmatrix} \text{High-} \\ \text{pitched} \\ \text{E} \end{bmatrix} \qquad v = \sqrt{\frac{F}{m/L}} = \sqrt{\frac{226 \text{ N}}{(0.208 \times 10^{-3} \text{ kg})/(0.628 \text{ m})}} = \boxed{826 \text{ m/s}}$$

$$\begin{bmatrix} \text{Low-} \\ \text{pitched} \\ \text{E} \end{bmatrix} \qquad v = \sqrt{\frac{F}{m/L}} = \sqrt{\frac{226 \text{ N}}{(3.32 \times 10^{-3} \text{ kg})/(0.628 \text{ m})}} = \boxed{207 \text{ m/s}}$$

Notice how fast the waves move; the speeds correspond to 1850 mi/h and 463 mi/h, respectively.

*16.4 THE MATHEMATICAL DESCRIPTION OF A WAVE

When a wave travels through a medium, it disturbs the particles of the medium by displacing them from their undisturbed positions. Suppose a particle is located a distance x from a coordinate origin. We would like to know the displacement y of this particle from its undisturbed position at any time t as the wave passes. For periodic waves that result from simple harmonic motion of the source, the expression for the displacement involves a sine or cosine, a fact that is not surprising. After all, in Chapter 10 simple harmonic motion is described using sinusoidal equations, and the graphs for a wave in Figure 16.5 look like a plot of displacement versus time for an object oscillating on a spring (see Figure 10.13).

Our tack will be to present the expression for the displacement and then show graphically that it gives a correct description. Equation 16.3 represents the displacement of a wave that travels in the $+x$ direction (to the right) and has an amplitude A, frequency f, and wavelength λ. Equation 16.4 applies to a wave moving in the $-x$ direction (to the left).

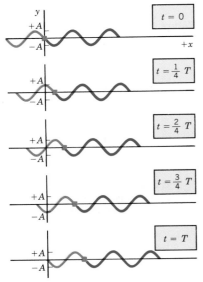

$$\begin{bmatrix} \textbf{Wave motion} \\ \textbf{toward } + x \end{bmatrix} \qquad y = A \sin\left(2\pi ft - \frac{2\pi x}{\lambda}\right) \qquad (16.3)$$

$$\begin{bmatrix} \textbf{Wave motion} \\ \textbf{toward } - x \end{bmatrix} \qquad y = A \sin\left(2\pi ft + \frac{2\pi x}{\lambda}\right) \qquad (16.4)$$

These equations apply to transverse or longitudinal waves and assume that $y = 0$ when $x = 0$ and $t = 0$.

Consider a transverse wave moving in the $+x$ direction along a string. The term $(2\pi ft - 2\pi x/\lambda)$ in Equation 16.3 is called the *phase angle* of the wave. A string particle located at the origin ($x = 0$) exhibits simple harmonic motion with a phase angle of $2\pi ft$, that is, its displacement as a function of time is $y = A \sin(2\pi ft)$. A particle located at a distance x also exhibits simple harmonic motion, but its phase angle is $2\pi f[t - x/(f\lambda)] = 2\pi f(t - x/v)$, where the quantity x/v is the time needed for the wave to travel the distance x. In other words, the simple harmonic motion that occurs at x is delayed by the time interval x/v compared to that occurring at the origin.

Figure 16.9 shows the displacement y plotted as a function of position x along the string at a series of time intervals separated by one-fourth of the period T ($t = 0, \frac{1}{4}T, \frac{2}{4}T, \frac{3}{4}T, T$). These graphs are constructed by substituting the corresponding value for t in Equation 16.3, remembering that $f = 1/T$, and then calculating y at a series of values for x. The graphs are like photographs taken at various times as the wave moves to the right. For reference, the colored square on each graph marks the place on the wave that is located at $x = 0$ when $t = 0$. As time passes, the colored square moves to the right, along with the wave. It should be noted that the phase angle ($2\pi ft - 2\pi x/\lambda$) is measured in *radians*, not degrees.

Figure 16.9 Equation 16.3 is plotted here at a series of times separated by one-fourth of the period T. The colored square in each of the graphs marks the place on the wave that is located at $x = 0$ when $t = 0$. As time passes, the wave moves to the right.

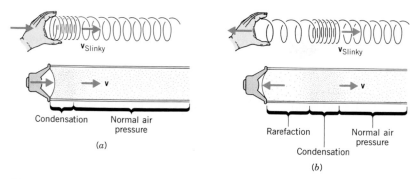

Figure 16.10 (a) When the speaker diaphragm moves outward, it creates a condensation. (b) When the diaphragm moves inward, it creates a rarefaction. The condensation and rarefaction on the Slinky are included for comparison. In reality, the velocity of the wave on the Slinky v_{Slinky} is much smaller than the velocity of sound in air **v**. For simplicity, the two waves are shown here to have the same velocity.

Therefore, when using a calculator to evaluate the function $\sin(2\pi ft - 2\pi x/\lambda)$, the calculator must be set to its radian mode. In a similar manner, it can be shown that Equation 16.4 represents a wave moving in the $-x$ direction.

16.5 THE NATURE OF SOUND

LONGITUDINAL SOUND WAVES

Sound is a longitudinal wave that is created by a vibrating object, such as a guitar string, the human vocal cords, or the diaphragm of a loudspeaker. Moreover, sound can be created or transmitted only in a medium, such as a gas, liquid, or solid. As we will see, the particles of the medium must be present for the disturbance of the wave to move from place to place. Sound cannot exist in a vacuum.

To see how sound waves are produced and why they are longitudinal, consider the vibrating diaphragm of a loudspeaker. When the diaphragm moves outward, it compresses the air directly in front of it, as in Figure 16.10a. This compression causes the air pressure to rise slightly. The region of increased pressure, called a *condensation*, travels away from the speaker at the speed of sound. After producing a condensation, the diaphragm reverses its motion and moves inward, as in part *b* of the drawing. The inward motion produces a region known as a *rarefaction*, where the air pressure is slightly less than normal. Following immediately behind the condensation, the rarefaction also travels away from the speaker at the speed of sound.

For comparison, Figure 16.10 includes the condensation and rarefaction of a longitudinal wave on a Slinky. Figure 16.11 emphasizes that the sound wave, like the Slinky wave, is longitudinal. As the wave passes, the colored dots attached both to the Slinky and to an air molecule execute simple harmonic motion about their undisturbed positions. The colored arrows on either side of the dots indicate that the simple harmonic motion occurs parallel to the line of travel. The drawing also illustrates that the wavelength λ is the distance between the centers of two successive condensations; λ is also the distance between the centers of two successive rarefactions.

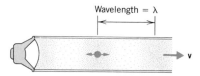

Figure 16.11 Both the wave on the Slinky and the sound wave are longitudinal waves. The colored dots attached to the Slinky and to an air molecule vibrate back and forth parallel to the line of travel of the wave.

THE PHYSICS OF . . .

how a loudspeaker diaphragm produces sound.

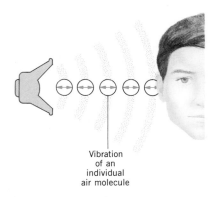

Figure 16.12 Although the condensations and rarefactions travel from the speaker to the listener, the individual air molecules do not move with the wave. A given molecule vibrates back and forth about a fixed location.

Figure 16.12 illustrates a sound wave spreading out in space as it leaves a loudspeaker and travels toward a listener. When the condensations and rarefactions arrive at the ear, they force the eardrum to vibrate at the same frequency as the speaker diaphragm. The vibratory motion of the eardrum is interpreted by the brain as sound.

It should be emphasized that sound is not a mass movement of air, such as occurs on a windy day. In Figure 16.12, for example, the air molecules on which the speaker diaphragm pushes are not the same ones that push on the eardrum. As the condensations and rarefactions of the sound wave travel outward from the vibrating diaphragm, the individual air molecules are not carried along with the wave. Rather, each molecule executes simple harmonic motion about a fixed location. In doing so, one molecule collides with its neighbor and passes the condensations and rarefactions on. The neighbor, in turn, repeats the process.

THE FREQUENCY OF A SOUND WAVE

Each cycle of a sound wave includes one condensation and one rarefaction, and the *frequency* is the number of cycles per second that passes by a given location. For example, if the diaphragm of a speaker vibrates back and forth in simple harmonic motion at a frequency of 1000 Hz, then 1000 condensations, each followed by a rarefaction, are generated every second, thus forming a sound wave whose frequency is also 1000 Hz. A sound with a single frequency is called a *pure tone*. Experiments have shown that a healthy young person hears all sound frequencies in the range from approximately 20 to 20 000 Hz (20 kHz). The ability to hear the high frequencies decreases with age, however, and a normal middle-aged adult hears frequencies only up to 12–14 kHz.

Pure tones are used in push-button telephones, such as that shown in Figure 16.13. These phones contain electric circuits that simultaneously produce two pure tones when each button is pressed, a different pair of tones for each different button. The tones are transmitted electronically to the central telephone office, where they activate switching circuits that complete the call. The drawing also shows the two pure tones that are generated when the various buttons are

THE PHYSICS OF . . .

push-button telephones.

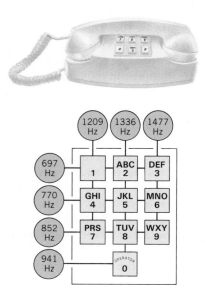

Figure 16.13 A push-button telephone and a schematic showing the two pure tones produced when each button is pressed.

pressed. For example, pressing the '5' button simultaneously produces pure tones of 770 and 1336 Hz, while the '9' button generates tones of 852 and 1477 Hz.

Sound can be generated whose frequency lies below 20 Hz and above 20 kHz, although humans normally do not hear it. In contrast, some dogs hear frequencies as high as 30 kHz, and, therefore, can respond to dog whistles that humans cannot hear. Bats can hear even higher frequencies and depend on high-frequency sound (up to a frequency of 100 kHz) for locating their prey and navigating (Figure 16.14). Sound waves whose frequencies lie above 20 kHz are called *ultrasonic waves,* while those that lie below 20 Hz are referred to as *infrasonic waves.*

Frequency is an objective property of a sound wave, because frequency can be measured with an electronic frequency counter. A listener's perception of frequency, however, is subjective. The brain interprets the frequency detected by the ear primarily in terms of the subjective quality called *pitch.* A pure tone with a large (high) frequency is interpreted as a high-pitched sound, while a pure tone with a small (low) frequency is interpreted as a low-pitched sound. A piccolo produces high-pitched sounds, and a tuba produces low-pitched sounds.

Figure 16.14 Bats use ultrasonic sound waves for navigating and locating prey.

THE PRESSURE AMPLITUDE OF A SOUND WAVE

Figure 16.15 illustrates a pure-tone sound wave traveling in a tube. Attached to the tube is a series of pressure gauges that indicate the pressure variations along the wave. The graph shows that the air pressure varies sinusoidally along the length of the tube. Although this graph has the appearance of a transverse wave, remember that the sound wave itself is a longitudinal wave. The graph also shows the *pressure amplitude* of the wave, which is the maximum change in pressure, measured relative to the undisturbed or atmospheric pressure. The pressure fluctuations in a sound wave are normally very small. For instance, in a normal conversation between two people the pressure amplitude is about 3×10^{-2} Pa, certainly a small amount compared with the atmospheric pressure of $1.01 \times 10^{+5}$ Pa. The ear is remarkable in being able to detect such small changes.

Loudness is an important attribute of sound that depends primarily on the amplitude of the wave: the larger the amplitude the louder the sound. The pressure amplitude is an objective property of a sound wave, since it can be measured with electronic equipment. Loudness, on the other hand, is subjective. Each individual determines what is loud, depending on the acuteness of his or her hearing.

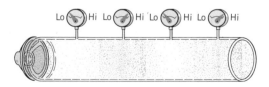

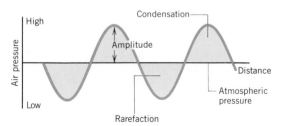

Figure 16.15 A sound wave is a series of alternating condensations and rarefactions. The graph shows that the condensations in a sound wave are regions of higher-than-normal air pressure, and the rarefactions are regions of lower-than-normal pressure.

16.6 THE SPEED OF SOUND

GASES

Dolphins are highly intelligent mammals that communicate with each other by transmitting audible "squeaks" and "clicks" through water.

Sound travels through gases, liquids, and solids at considerably different speeds, as Table 16.1 reveals. Near room temperature, the speed of sound in air is 343 m/s (767 mi/h) and is markedly greater in liquids and solids. For example, sound travels more than four times faster in water and more than seventeen times faster in steel than it does in air. In general, sound travels slowest in gases, faster in liquids, and fastest in solids.

Like the speed of a wave on a guitar string, the speed of sound depends on the properties of the medium. In a gas, it is only when molecules collide that the condensations and rarefactions of a sound wave can move from place to place. It is reasonable, then, to expect the speed of sound in a gas to have the same order of magnitude as the average molecular speed between collisions. For an ideal gas this average speed is the translational rms-speed given by Equation 14.6: $v_{rms} = \sqrt{3kT/m}$, where T is the Kelvin temperature, m is the mass of a molecule, and k is Boltzmann's constant. Although the expression for v_{rms} overestimates the speed of sound, it does give the correct dependence on Kelvin temperature and particle mass. Careful analysis shows that the speed of sound in an ideal gas is given by

[Ideal gas]
$$v = \sqrt{\frac{\gamma kT}{m}}$$
(16.5)

where $\gamma = c_P/c_V$ is the ratio of the specific heat capacity at constant pressure c_P to the specific heat capacity at constant volume c_V.

The factor γ is introduced in Section 15.4, where the adiabatic compression and expansion of an ideal gas is discussed. The factor γ enters into Equation 16.5 because the condensations and rarefactions of a sound wave are formed by adiabatic compressions and expansions of the gas. When a sound wave travels through a gas, the regions that are compressed (the condensations) become

Table 16.1   Speed of Sound in Gases, Liquids, and Solids

Substance	Temperature (°C)	Speed (m/s)
Gases		
Air	0	331
Air	20	343
Carbon dioxide	0	259
Oxygen	0	316
Helium	0	965
Liquids		
Chloroform	20	1004
Ethanol	20	1162
Mercury	20	1450
Fresh water	20	1482
Solids (bulk)		
Copper	—	5010
Glass (Pyrex)	—	5640
Lead	—	1960
Steel	—	5960

slightly warmed, and the regions that are expanded (the rarefactions) become slightly cooled. However, no appreciable heat flows from a condensation to an adjacent rarefaction, because the distance between the two (half a wavelength) is relatively large for most audible sound waves and a gas is a poor thermal conductor. Thus, the compression and expansion process is adiabatic. Example 3 uses Equation 16.5 to illustrate how camera technology takes advantage of the speed of sound.

Example 3   An Autofocusing Camera

Some cameras have a mechanism that automatically focuses the camera with the aid of sound waves. Figure 16.16 illustrates the central idea of this feature. To initiate the focusing process, the camera generates a pulse of ultrasonic sound that travels to the subject being photographed. Like an echo, the pulse reflects off the subject and returns to the camera. By measuring the time it takes the sound to make the round-trip and using a preset value for the speed of sound, the camera can determine the distance to the subject and set the lens to its proper focus. Suppose the time required for the sound pulse to make the round-trip is 20.0×10^{-3} s on a day when the air temperature is 296 K (23 °C). Assuming air behaves as an ideal gas for which $\gamma = 1.40$ and the average molecular mass of an air molecule is 28.9 u, find the distance s between the camera and the subject.

REASONING   The distance between the camera and the subject is $s = vt$, where v is the speed of sound and t is the time for the sound pulse to reach the subject. The time t is just one-half the round-trip time, so $t = 10.0 \times 10^{-3}$ s. The speed of sound in air can be obtained directly from Equation 16.5, once the mass of an air molecule is known. The mass of an air molecule is the average molecular mass of air (expressed in kilograms) divided by Avogadro's number N_A (see Section 14.1).

SOLUTION   The mass of an air molecule is

$$m = \frac{28.9 \times 10^{-3} \text{ kg/mol}}{N_A} = \frac{28.9 \times 10^{-3} \text{ kg/mol}}{6.022 \times 10^{23} \text{ mol}^{-1}} = 4.80 \times 10^{-26} \text{ kg}$$

For the speed of sound, we find

$$v = \sqrt{\frac{\gamma kT}{m}} = \sqrt{\frac{(1.40)(1.38 \times 10^{-23} \text{ J/K})(296 \text{ K})}{4.80 \times 10^{-26} \text{ kg}}} = 345 \text{ m/s} \qquad (16.5)$$

The distance from the camera to the subject is

$$s = vt = (345 \text{ m/s})(10.0 \times 10^{-3} \text{ s}) = \boxed{3.45 \text{ m}}$$

THE PHYSICS OF . . .

an autofocusing camera.

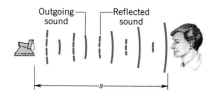

Figure 16.16   To set its focus automatically, the camera determines the distance s with the aid of ultrasonic sound waves. The camera measures the time for the sound to travel to the subject and back and computes s from a knowledge of the speed of sound.

PROBLEM SOLVING INSIGHT

When using the equation $v = \sqrt{\gamma kT/m}$ to calculate the speed of sound in an ideal gas, be sure to express the temperature T in kelvins and not in degrees Celsius or Fahrenheit.

LIQUIDS

In a liquid, the speed of sound depends on the density ρ and the *adiabatic* bulk modulus B_{ad} of the liquid:

[Liquid] $$v = \sqrt{\frac{B_{ad}}{\rho}} \qquad (16.6)$$

The bulk modulus is introduced in Section 10.1 in a discussion of the volume deformation of liquids and solids. There it is tacitly assumed that the temperature remains constant while the volume of the material changes; i.e., the compression or expansion is isothermal. The values for the bulk moduli in Table 10.3 reflect this experimental condition and, consequently, are known as isothermal bulk moduli. However, the condensations and rarefactions in a sound wave occur under *adiabatic* rather than isothermal conditions. Thus, the adiabatic bulk modulus B_{ad} must be used when calculating the speed of sound in liquids. Values of B_{ad} will be provided as needed in this text. The next example emphasizes that sound travels much faster in a liquid than in a gas.

THE PHYSICS OF . . .

an ultrasonic ruler.

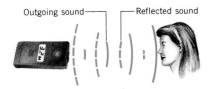

Outgoing sound ———— ———— Reflected sound

Figure 16.17 An ultrasonic ruler.

Example 4   An Ultrasonic Ruler

Figure 16.17 shows an ultrasonic ruler that measures the distance between two points by sending out an ultrasonic pulse of sound and detecting the reflected pulse. The ultrasonic ruler works like the camera autofocusing mechanism in Example 3, except the distance is displayed on a digital readout. The ruler is calibrated to measure distances in air. Will it correctly measure distances under the ocean? If not, determine the factor by which it is in error.

REASONING   Since the ultrasonic sound travels much faster in seawater than it does in air, the reflected pulse returns in a much shorter time. This quicker return time fools the ultrasonic ruler into believing the object is much closer than it actually is. Therefore, the ruler will not measure underwater distances correctly. The error depends on the speed of sound in seawater.

SOLUTION   The speed of sound in seawater can be determined from Equation 16.6:

$$v = \sqrt{\frac{B_{ad}}{\rho}} = \sqrt{\frac{2.31 \times 10^9 \text{ Pa}}{1025 \text{ kg/m}^3}} = 1500 \text{ m/s}$$

In comparison, the speed of sound in air is about 343 m/s. Thus, the ultrasonic ruler calculates, erroneously, that the object is $1500/343 = 4.4$ times closer under water than it actually is.

SOLID BARS

When sound travels through a long slender solid bar, the speed of the sound depends on the properties of the medium according to

$$\begin{bmatrix} \textbf{Long slender} \\ \textbf{solid bar} \end{bmatrix} \qquad\qquad v = \sqrt{\frac{Y}{\rho}} \qquad\qquad (16.7)$$

where Y is Young's modulus (defined in Section 10.1) and ρ is the density. Example 5 illustrates how much faster sound travels in a steel bar than in air.

Example 5   Listening for a Train

Have you ever listened for an approaching train by kneeling next to the track and putting your "ear to the rail"? On a day when the temperature is 20 °C, how many times greater is the speed of sound in the rail than in the air?

REASONING AND SOLUTION   A rail can be approximated as a long slender bar, so the speed of sound in the rail is given by Equation 16.7. Table 10.1 lists

Young's modulus for steel as $Y = 2.0 \times 10^{11}$ N/m², and Table 11.1 gives the density of steel as $\rho = 7860$ kg/m³. The speed of sound in the rail is

$$v = \sqrt{\frac{Y}{\rho}} = \sqrt{\frac{2.0 \times 10^{11} \text{ N/m}^2}{7860 \text{ kg/m}^3}} = 5.0 \times 10^3 \text{ m/s} \qquad (16.7)$$

This value for v is considerably less than the value of 5960 m/s given for bulk steel in Table 16.1. The speed of sound in air at 20 °C is 343 m/s. Therefore, the speed of sound in the rail is $(5.0 \times 10^3 \text{ m/s})/(343 \text{ m/s}) = \boxed{15 \text{ times greater}}$ than in air.

16.7 SOUND INTENSITY

Sound waves carry energy that can be used to do work, like forcing the eardrum to vibrate. Or, in an extreme case such as a sonic boom, the energy can be sufficient to cause damage to windows and buildings. The amount of energy transported per second by a sound wave is called the *power* of the wave and is measured in SI units of joules per second (J/s) or watts (W).

When a sound wave leaves a source, such as the loudspeaker in Figure 16.18, the power spreads out and passes through increasingly larger areas. The *sound intensity I* is defined as the sound power P that passes perpendicularly through a surface divided by the area A of the surface:

$$I = \frac{P}{A} \qquad (16.8)$$

The sound intensity is greatest for music lovers who have "front-row seats."

The unit of sound intensity is power per unit area, or W/m². The next example illustrates how the sound intensity changes as the distance from a loudspeaker changes.

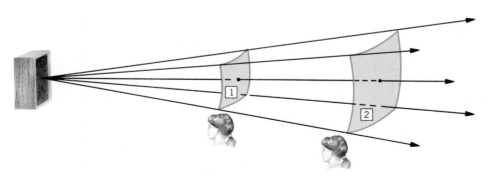

Figure 16.18 The intensity of a sound wave is the sound power that passes perpendicularly through a surface, such as surface 1 or surface 2, divided by the area of the surface.

Example 6   Sound Intensities

In Figure 16.18, 12×10^{-5} W of sound power passes perpendicularly through the surfaces labeled 1 and 2. These surfaces have areas of $A_1 = 4.0$ m² and $A_2 = 12$ m². Determine the sound intensity at each surface and discuss why listener 2 hears a quieter sound than listener 1.

REASONING AND SOLUTION   The sound intensity at each surface can be computed from Equation 16.8:

[Surface 1] $I_1 = \dfrac{P}{A_1} = \dfrac{12 \times 10^{-5} \text{ W}}{4.0 \text{ m}^2} = \boxed{3.0 \times 10^{-5} \text{ W/m}^2}$

[Surface 2] $I_2 = \dfrac{P}{A_2} = \dfrac{12 \times 10^{-5} \text{ W}}{12 \text{ m}^2} = \boxed{1.0 \times 10^{-5} \text{ W/m}^2}$

The sound intensity is less at the more-distant surface, where the same power passes through a threefold greater area. The ear of a listener, with its fixed area, intercepts less power where the intensity, or power per unit area, is smaller. Thus, listener 2 intercepts less of the sound power than listener 1. With less power striking the ear, the sound is quieter.

For a 1000-Hz tone, the smallest sound intensity that the human ear can detect is about 1×10^{-12} W/m²; this intensity is called the **threshold of hearing.** On the other extreme, continuous exposure to intensities greater than 1 W/m² can be painful and result in permanent hearing damage. The human ear is remarkable for the wide range of intensities to which it is sensitive.

If a source emits sound *uniformly in all directions,* the sound intensity depends on distance in a simple way. Figure 16.19 shows such a source at the center of an imaginary sphere of radius r. Since all the radiated sound power P passes through the spherical surface of area $A = 4\pi r^2$, the intensity at a distance r is

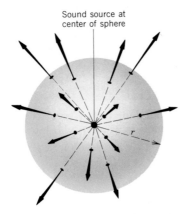

Figure 16.19   A source that emits sound uniformly in all directions.

$$\begin{bmatrix} \text{Spherically} \\ \text{uniform} \\ \text{radiation} \end{bmatrix} \qquad\qquad I = \dfrac{P}{4\pi r^2} \qquad\qquad (16.9)$$

From this we see that the intensity of a source that radiates sound uniformly in all directions varies as $1/r^2$. For example, if the distance increases by a factor of two, the sound intensity decreases by a factor of $2^2 = 4$. Equation 16.9 is valid only when no walls, ceilings, floors, etc., are present to reflect the sound and cause it to pass through the same surface more than once. Example 7 illustrates the effect of the $1/r^2$ dependence of intensity on distance.

Example 7   Fireworks

During a fireworks display, a rocket explodes high in the air, as Figure 16.20 illustrates. Assume the sound spreads out uniformly in all directions and reflections from the ground can be ignored. When the sound reaches listener 2, who is $r_2 = 640$ m away from the explosion, the sound has an intensity of $I_2 = 0.10$ W/m². What is the sound intensity detected by listener 1, who is $r_1 = 160$ m away from the explosion?

REASONING AND SOLUTION   Listener 1 is four times closer to the explosion than listener 2. Therefore, the sound intensity detected by listener 1 is $4^2 = 16$ times greater than that detected by listener 2, as the calculation below indicates:

$$\frac{I_1}{I_2} = \frac{\dfrac{P}{4\pi r_1^2}}{\dfrac{P}{4\pi r_2^2}} = \frac{r_2^2}{r_1^2} = \frac{(640 \text{ m})^2}{(160 \text{ m})^2} = 16$$

As a result, $I_1 = (16)I_2 = (16)(0.10 \text{ W/m}^2) = \boxed{1.6 \text{ W/m}^2}$.

Figure 16.20 If an explosion in a fireworks display radiates sound uniformly in all directions, the intensity at any distance r is $I = P/(4\pi r^2)$, where P is the sound power of the explosion.

16.8 DECIBELS

COMPARING SOUND INTENSITIES USING DECIBELS

The *decibel* (*dB*) is a measurement unit encountered frequently in connection with stereo systems. For example, the decibel is used extensively on specification sheets to describe the performance characteristics of receivers, CD players, speakers, and cassette decks. The main application of the decibel concept is for comparing two sound intensities. The simplest method of comparison would be to compute the ratio of the intensities. For instance, we could compare $I = 8 \times 10^{-12}$ W/m^2 to $I_0 = 1 \times 10^{-12}$ W/m^2 by computing $I/I_0 = 8$ and stating that I is eight times greater than I_0. However, because of the way in which the human hearing mechanism responds to intensity, it is more appropriate to use a logarithmic scale for the comparison. For this purpose, the *intensity level* β (expressed in decibels) is defined as follows:

$$\beta \text{ (in decibels)} = 10 \log\left(\frac{I}{I_0}\right) \qquad (16.10)$$

where "log" denotes the logarithm to the base ten. I_0 is the intensity of the reference level to which I is being compared and is often the threshold of hearing, $I_0 = 1.00 \times 10^{-12}$ W/m^2. With the aid of a calculator, the intensity level can be evaluated for the values of I and I_0 given above:

$$\beta = 10 \log\left(\frac{8 \times 10^{-12} \text{ W/m}^2}{1 \times 10^{-12} \text{ W/m}^2}\right) = 10 \log 8 = 10(0.9) = 9 \text{ dB}$$

This result indicates that I is 9 decibels greater than I_0.

Although β is called the "intensity level," it is *not* an intensity and does *not* have intensity units of W/m^2. In fact, the decibel unit, like the radian, is unitless, since it is the product of the number 10 and a logarithm, both of which are pure numbers without units.

Notice that when I is at the threshold of hearing, i.e., when $I = I_0$, the intensity level is 0 dB according to Equation 16.10:

$$\beta = 10 \log\left(\frac{I_0}{I_0}\right) = 10 \log 1 = 0$$

Table 16.2 Typical Sound Intensities and Intensity
Levels Relative to the Threshold of Hearing

	Intensity I (W/m^2)	Intensity level β (dB)
Threshold of hearing	1.0×10^{-12}	0
Rustling leaves	1.0×10^{-11}	10
Whisper	1.0×10^{-10}	20
Normal conversation (1 meter)	3.2×10^{-6}	65
Inside car in city traffic	1.0×10^{-4}	80
Car without muffler	1.0×10^{-2}	100
Live rock concert	1.0	120
Threshold of pain	10	130

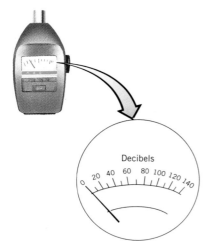

Figure 16.21 A sound level meter and a close-up view of its decibel scale.

since log 1 = 0. Thus, *an intensity level of zero decibels does not mean that the sound intensity I is zero; it means that I = I₀.*

Table 16.2 lists the intensities I and the associated intensity levels β for some common sounds, using the threshold of hearing as the reference level. Intensity levels can be measured with a sound level meter, such as the one in Figure 16.21. The intensity level β is displayed on its scale, assuming that the threshold of hearing is 0 dB.

INTENSITY LEVEL CHANGES AND LOUDNESS CHANGES

When a sound wave reaches a listener's ear, the sound is interpreted by the brain as loud or soft, depending on the intensity of the wave. Greater intensities give rise to louder sounds. However, the relation between intensity and loudness is not a simple proportionality, for doubling the intensity does *not* double the loudness. The correlation between intensity and loudness is a fascinating one, as we will now see.

Suppose you are sitting in front of a stereo system that is producing an intensity level of 90 dB. If the volume control on the amplifier is turned up slightly to produce a 91 dB level, you would just barely notice the change in loudness. *Hearing tests have revealed that a one decibel (1 dB) change in the intensity level is approximately the smallest change in loudness that an average listener can detect.* Since 1 dB is the smallest perceivable increment in loudness, a change of 3 dB, say, from 90 to 93 dB, is still a rather small change in loudness. However, a 3-dB increase in loudness, while small, corresponds to a doubling of the sound intensity, as Example 8 illustrates.

Example 8   Comparing Sound Intensities

Figure 16.22 shows two audio systems. System 1 produces an intensity level of $\beta_1 = 90.0$ dB, while system 2 produces an intensity level of $\beta_2 = 93.0$ dB. The corresponding intensities (in W/m^2) are I_1 and I_2, respectively. Determine the ratio I_2/I_1.

REASONING   Intensity levels are related to logarithms. Therefore, in solving this problem, we take advantage of the following property of logarithms: log A − log B = log (A/B). Subtracting the two intensity levels and using this property, we find that

$$\beta_2 - \beta_1 = 10 \log \left(\frac{I_2}{I_0}\right) - 10 \log \left(\frac{I_1}{I_0}\right) = 10 \log \left(\frac{I_2/I_0}{I_1/I_0}\right) = 10 \log \left(\frac{I_2}{I_1}\right)$$

SOLUTION   Using the result above for the difference in the intensity levels, we obtain 93.0 dB − 90.0 dB = 10 log (I_2/I_1) or 0.30 = log (I_2/I_1). As a result,

$$\frac{I_2}{I_1} = 10^{0.30} = \boxed{2.0}$$

We see that increasing the loudness by only a small amount (3 dB) corresponds to doubling the intensity. Conversely, doubling the intensity changes the loudness only slightly and does _not_ double it.

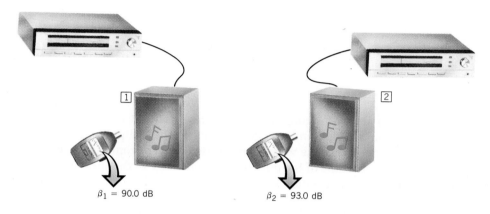

β_1 = 90.0 dB β_2 = 93.0 dB

Figure 16.22   System 2 sounds slightly louder than system 1, since β_2 is only 3.0 dB greater than β_1.

To double the loudness of a sound, the intensity must be increased by more than a factor of two. _**Experiment shows that if the intensity level increases by 10 dB, the new sound seems approximately twice as loud as the original sound.**_ For instance, a 70-dB intensity level sounds about twice as loud as a 60-dB level, and an 80-dB intensity level sounds about twice as loud as a 70-dB level. The factor by which the sound intensity must be increased to double the loudness can be determined by the method used in Example 8:

$$\beta_2 - \beta_1 = 10.0 \text{ dB} = 10 \left[\log \left(\frac{I_2}{I_0}\right) - \log \left(\frac{I_1}{I_0}\right) \right]$$

Solving this equation reveals that $I_2/I_1 = 10.0$. Thus, increasing the sound intensity by a factor of ten will only double the perceived loudness. Consequently, the 200-watt loudspeaker system in Figure 16.23 will sound only about twice as loud as the much cheaper 20-watt system.

Figure 16.23   In spite of its tenfold greater power, the 200-watt audio system has only about double the loudness of the 20-watt system.

16.9 APPLICATIONS OF SOUND

SONAR

Sonar (**so**und **na**vigation **r**anging) is a technique for determining water depth and locating underwater objects, such as reefs, submarines, and schools of fish. The core of a sonar unit consists of an ultrasonic transmitter and receiver mounted on

THE PHYSICS OF . . .

sonar.

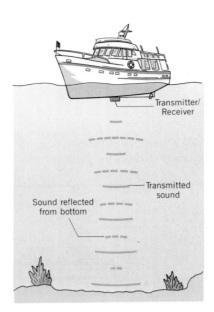

Figure 16.24 Sonar uses ultra-sonic sound to measure water depth.

the bottom of a ship, as Figure 16.24 illustrates. The transmitter emits a short pulse of ultrasonic sound, and at a later time the reflected pulse returns and is detected by the receiver. The water depth is determined from the electronically measured round-trip time of the pulse and a knowledge of the speed of sound in water; the depth registers automatically on an appropriate meter. Such a depth measurement is similar to the distance measurement discussed for the autofocusing camera in Example 3 and the ultrasonic ruler in Example 4.

THE PHYSICS OF . . .

the use of ultrasound in medicine.

ULTRASOUND IN MEDICINE

When ultrasonic waves are used in medicine for diagnostic purposes, high-frequency sound pulses are produced by a transmitter and directed into the body. As in sonar, reflections occur. They occur each time a pulse encounters a boundary between two tissues that have different densities or a boundary between a tissue and the adjacent fluid. By scanning ultrasonic waves across the body and detecting the echoes generated from various locations within the body, it is possible to obtain a "picture" or sonogram of the inner anatomy. Ultrasonic waves are employed extensively in obstetrics to examine the developing fetus (Figure 16.25). The fetus, surrounded by the amniotic sac, can be distinguished from other anatomical features so that fetal size, position, and possible abnormalities can be detected.

Ultrasound is also used in other medically related areas. For instance, malignancies in the liver, kidney, brain, and pancreas can be detected with ultrasound. In cases where internal hemorrhaging occurs, it is possible to identify the bleeding area and even obtain a gross estimate of blood loss using ultrasonic techniques. Yet another application involves monitoring the real-time movement of pulsating structures, such as heart valves ("echocardiography") and large blood vessels.

When ultrasound is used to locate internal anatomical features or foreign objects in the body, the wavelength of the sound wave must be about the same size, or smaller, than the object to be located. Therefore, high frequencies in the range from 1 to 15 MHz (1 MHz = 1 megahertz = 1×10^6 Hz) are the norm, so that the wavelengths are small. For instance, the wavelength of 5 MHz ultra-

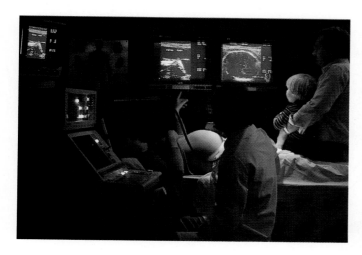

Figure 16.25 An ultrasonic scanner produces an image of a 17-week-old fetus.

sound can be calculated from $\lambda = v/f$ to be 0.3 mm if a value of 1540 m/s is used for the speed of sound through tissue. A sound wave with a frequency higher than 5 MHz and a correspondingly shorter wavelength is required for locating objects smaller than 0.3 mm.

Ultrasound also has applications other than locating objects in the body. Neurosurgeons use a device called a **c**avitron **u**ltrasonic **s**urgical **a**spirator (CUSA) to remove brain tumors once thought to be inoperable. Ultrasonic sound waves cause the slender tip of the CUSA probe (see Figure 16.26) to vibrate at approximately 23 kHz. The probe shatters any section of the tumor that it touches, and the fragments are flushed out of the brain with a saline solution. Because the tip of the probe is small, the surgeon can selectively remove small bits of malignant tissue without damaging the surrounding healthy tissue.

THE PHYSICS OF . . .

the cavitron ultrasonic surgical aspirator.

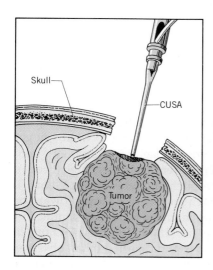
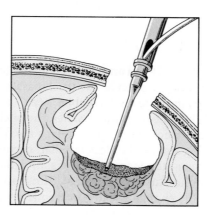

Figure 16.26 Neurosurgeons use a cavitron ultrasonic surgical aspirator (CUSA) to "cut out" brain tumors without adversely affecting the surrounding healthy tissue.

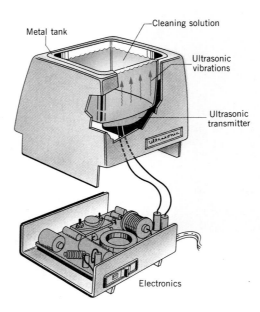

Figure 16.27 An ultrasonic cleaner.

ULTRASONIC CLEANERS

Ultrasonic cleaners are especially popular with jewelers and scientists for cleaning delicate instruments and objects with hard-to-get-at places. Figure 16.27 illustrates the main features of an ultrasonic cleaner, which includes a metal tank filled with a cleaning fluid. Attached to the bottom of the tank is an ultrasonic transmitter that typically produces 40-kHz sound waves. The high-frequency sound causes a phenomenon called cavitation, in which small, partially evacuated spaces form in the cleaning fluid. The partial vacuum in these spaces causes dirt and grease particles to be pulled from even the tiniest cracks and crevices.

16.10 THE DOPPLER EFFECT

INTRODUCTION

Have you ever heard an approaching fire truck and noticed the distinct change in the sound of the siren as the truck passes? The effect is similar to what you get when you put the two syllables "eee" and "yow" together to produce "eee-yow." While the truck approaches, the pitch of the siren is relatively high ("eee"), but as the truck passes and moves away, the pitch suddenly drops ("yow"). Something similar, but less familiar, occurs when an observer moves toward or away from a stationary source of sound. Such phenomena were first identified in 1842 by the Austrian physicist Christian Doppler (1803–1853). The *Doppler effect* is the change in pitch or frequency of the sound detected by an observer because the sound source and the observer have different velocities with respect to the medium of sound propagation.

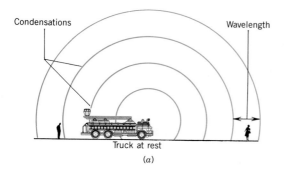

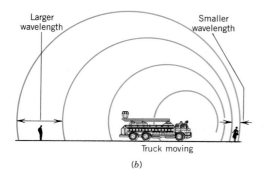

Figure 16.28   (*a*) When the truck is stationary, the wavelength of the sound is the same in front of and behind the truck. (*b*) When the truck is moving, the wavelength in front of the truck becomes smaller, while the wavelength behind the truck becomes larger.

MOVING SOURCE

To see how the Doppler effect arises, consider the sound emitted by a siren on a fire truck. In Figure 16.28*a*, the sound spreads out in a spherical pattern after leaving the siren of a stationary fire truck. Like the truck, the air is assumed to be stationary with respect to the earth. Each circular line in the drawing represents a condensation of the sound wave. Since the sound pattern is symmetrical, listeners standing in front of or behind the truck detect the same number of condensations per second and, consequently, hear the same frequency. Once the truck begins to move, however, the situation changes, as part *b* of the picture illustrates. Ahead of the truck the condensations are closer together, resulting in a decrease in the wavelength of the sound. This "bunching-up" effect occurs because the truck is moving with respect to the air. Thus, the truck "gains ground" on a previously emitted condensation before emitting the next one. Since the condensations are closer together, the observer standing in front of the truck senses more of them arriving per second than he does when the truck is stationary. The increased rate of arrival corresponds to a greater sound frequency, which the observer hears as a higher pitch.

Behind the moving truck, the condensations are farther apart than they are when the truck is stationary. This increase in the wavelength occurs because the truck pulls away from the condensations emitted toward the rear. Consequently, fewer condensations per second arrive at the ear of an observer behind the truck, corresponding to a smaller sound frequency or lower pitch.

We can determine how much the frequency changes because of the Doppler effect. If the stationary siren emits one condensation at the time $t = 0$, it will emit the next one at time T, where T is the period of the wave. The distance between

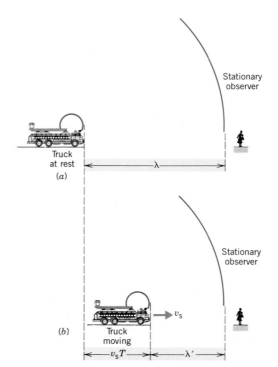

Figure 16.29 (a) When the fire truck is stationary, the distance between successive condensations is one wavelength λ. (b) When the truck moves with a speed v_s, the wavelength of the sound in front of the truck is shortened to λ'.

these two condensations is the wavelength λ of the sound produced by the stationary source, as Figure 16.29a indicates. When the truck is moving with a speed v_s (the subscript s stands for the "source" of sound) toward a stationary observer, the siren also emits two successive condensations, one at $t = 0$ and one at time T. However, prior to emitting the second condensation, the truck moves closer to the observer by a distance $v_s T$, as Figure 16.29b shows. As a result, the distance between successive condensations is no longer the wavelength λ created by the stationary siren, but, rather, a wavelength λ' that is shortened by the amount $v_s T$:

$$\lambda' = \lambda - v_s T$$

The frequency f' of the sound wave, as perceived by the stationary observer, is the speed of sound v divided by the shortened wavelength λ', according to Equation 16.1:

$$f' = \frac{v}{\lambda'} = \frac{v}{\lambda - v_s T}$$

But it is also true for the stationary siren that $\lambda = v/f$ and $T = 1/f$, where f is the frequency at which the source emits the sound (not the frequency f' perceived by the listener). With the aid of these substitutions for λ and T, the expression for f' can be arranged to give the following result:

$$\begin{bmatrix} \textbf{Sound source} \\ \textbf{moving toward} \\ \textbf{stationary} \\ \textbf{listener} \end{bmatrix} \qquad f' = f\left(\frac{1}{1 - \dfrac{v_s}{v}} \right) \qquad\qquad (16.11)$$

The denominator in Equation 16.11 is less than one, so the listener hears a frequency f' that is *greater* than the frequency f emitted by the source. The difference between these two frequencies, $f' - f$, is called the *Doppler shift*, and its magnitude depends on the ratio of the speed of the source v_s to the speed of sound v.

When the siren moves away from the listener, rather than toward the listener, the wavelength λ' becomes *greater* than λ according to

$$\lambda' = \lambda + v_s T$$

Notice the presence of the "+" sign in this equation, in contrast to the "−" sign that appeared earlier. The same reasoning that led to Equation 16.11 can be used to obtain an expression for the observed frequency f':

$$\left[\begin{array}{c} \text{Sound source} \\ \text{moving away from} \\ \text{stationary} \\ \text{listener} \end{array}\right] \qquad f' = f\left(\frac{1}{1 + \dfrac{v_s}{v}}\right) \qquad (16.12)$$

The denominator in Equation 16.12 is greater than one, so the listener hears a frequency f' that is *less* than the frequency f emitted by the source. The next example illustrates the Doppler effect.

Example 9   The Sound of a Passing Train

A high-speed train is traveling at a speed of 44.7 m/s (100 mi/h) when the engineer sounds the 415-Hz warning horn. The speed of sound is 343 m/s. What are the frequency and wavelength of the sound, as perceived by a person standing at a crossing, when the train is (a) approaching and (b) leaving the crossing?

REASONING   When the train approaches, the person at the crossing hears a sound whose frequency is greater than 415 Hz, because of the Doppler effect. As the train moves away, the person hears a frequency that is less than 415 Hz. We may use Equations 16.11 and 16.12, respectively, to determine these frequencies. In either case, the observed wavelength can be obtained according to Equation 16.1 as the speed of sound divided by the observed frequency.

SOLUTION
(a) When the train approaches, the observed frequency is

$$f' = f\left(\frac{1}{1 - \dfrac{v_s}{v}}\right) = (415 \text{ Hz})\left[\frac{1}{1 - \dfrac{44.7 \text{ m/s}}{343 \text{ m/s}}}\right] = \boxed{477 \text{ Hz}} \qquad (16.11)$$

The observed wavelength is

$$\lambda' = \frac{v}{f'} = \frac{343 \text{ m/s}}{477 \text{ Hz}} = \boxed{0.719 \text{ m}} \qquad (16.1)$$

(b) When the train leaves the crossing, the observed frequency is

$$f' = f\left(\frac{1}{1 + \dfrac{v_s}{v}}\right) = (415 \text{ Hz})\left[\frac{1}{1 + \dfrac{44.7 \text{ m/s}}{343 \text{ m/s}}}\right] = \boxed{367 \text{ Hz}} \qquad (16.12)$$

In this case, the observed wavelength is

$$\lambda' = \frac{v}{f'} = \frac{343 \text{ m/s}}{367 \text{ Hz}} = \boxed{0.935 \text{ m}}$$

MOVING OBSERVER

Figure 16.30 shows how the Doppler effect arises when the sound source is stationary and the observer moves, again assuming the air is stationary. In part *a* of the drawing, both the source of sound and the observer are stationary, and the observer hears the frequency *f* emitted by the source. In part *b* the observer moves with a speed v_o ("o" stands for observer) toward the stationary source and covers a distance $v_o t$ in a time *t*. During this time, the moving observer encounters all the condensations detected by the stationary observer, *plus an additional number.* The additional number of condensations encountered is the distance $v_o t$ divided by the distance λ between successive condensations, or $v_o t / \lambda$. Thus, the additional number of condensations encountered per second is v_o / λ, and the moving observer hears a higher frequency f' given by

$$f' = f + \frac{v_o}{\lambda} = f\left(1 + \frac{v_o}{f\lambda}\right)$$

Using the fact that $v = f\lambda$, we find that

$$\begin{bmatrix} \textbf{Observer moving} \\ \textbf{toward stationary} \\ \textbf{sound source} \end{bmatrix} \qquad f' = f\left(1 + \frac{v_o}{v}\right) \qquad (16.13)$$

An observer moving *away from* a stationary source moves in the same direction as the sound wave, and, as a result, intercepts *fewer* condensations per second than a stationary observer does. In this case, the moving observer hears a smaller frequency f' that is given by

$$\begin{bmatrix} \textbf{Observer moving} \\ \textbf{away from stationary} \\ \textbf{sound source} \end{bmatrix} \qquad f' = f\left(1 - \frac{v_o}{v}\right) \qquad (16.14)$$

The physical mechanism producing the Doppler effect in the case of the moving observer is different from that in the case of the moving source. When the source moves and the observer is stationary, the wavelength of the sound changes, giving rise to the frequency f' heard by the observer. On the other hand,

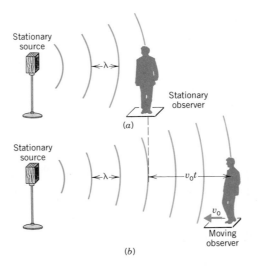

Figure 16.30 An observer moving with a speed v_o toward a stationary source of sound, as in part *b*, intercepts more wave condensations per unit of time than does the stationary observer in part *a*.

when the observer moves and the source is stationary, *the wavelength of the sound does not change.* Instead, a moving observer intercepts a different number of wave condensations per second than does a stationary observer and, therefore, detects a different frequency f'.

GENERAL CASE

It is possible for both the sound source and the observer to move with respect to the medium of sound propagation. If the medium is stationary, Equations 16.11 – 16.14 may be combined to give the observed frequency f' as

$$\begin{bmatrix} \textbf{Sound source} \\ \textbf{moving and} \\ \textbf{observer moving} \end{bmatrix} \qquad f' = f\left(\frac{1 \pm \dfrac{v_o}{v}}{1 \mp \dfrac{v_s}{v}}\right) \qquad (16.15)$$

In the numerator, the plus sign applies when the observer moves toward the source, and the minus sign applies when the observer moves away from the source. In the denominator, the minus sign is used when the source moves toward the observer, and the plus sign is used when the source moves away from the observer. The symbols v_o, v_s, and v denote numbers without an algebraic sign, because the direction of travel has been taken into account by the plus and minus signs that appear directly in this equation.

DOPPLER FLOW METER

The Doppler flow meter is an interesting medical application of the Doppler effect. This device measures the speed of blood flow, using transmitting and receiving elements that are placed directly on the skin, as in Figure 16.31. The transmitter emits a continuous sound whose frequency is typically about 5 MHz. When the sound is reflected from the red blood cells, its frequency is changed in a kind of Doppler effect, because the cells are moving. The receiving element detects the reflected sound, and an electronic counter measures its frequency, which is Doppler-shifted relative to the transmitter frequency. From the change in frequency the speed of the blood flow can be determined. Typically, the change in frequency is around 6000 Hz for flow speeds of about 0.1 m/s. The Doppler flow meter can be used to locate regions where blood vessels have narrowed, since greater flow speeds occur in the narrowed regions, according to the equation of continuity (see Section 11.8). In addition, the Doppler flow meter can be used to detect the motion of a fetal heart as early as 8 – 10 weeks after conception.

The Doppler effect is also employed in radar devices to measure the speed of moving vehicles. However, electromagnetic waves, rather than sound waves, are used for such purposes.

*16.11 THE SENSITIVITY OF THE HUMAN EAR

Although the ear is capable of detecting sound intensities as small as 1×10^{-12} W/m², it is *not* equally sensitive to all frequencies, as Figure 16.32 shows. This figure displays a series of graphs that are known as the *Fletcher – Munson curves,*

THE PHYSICS OF . . .

the Doppler flow meter.

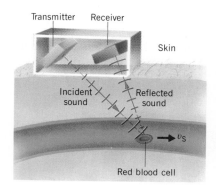

Figure 16.31 A Doppler flow meter measures the speed of red blood cells.

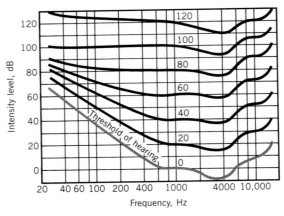

Figure 16.32 Each curve represents the intensity levels at which sounds of various frequencies have the same loudness. The curves are labeled by their intensity levels at 1000 Hz. These curves are known as the Fletcher–Munson curves.

Figure 16.33 The loudness switch on a stereo receiver.

THE PHYSICS OF . . .

the loudness switch on a
stereo receiver.

after H. Fletcher and M. Munson, who first determined them in 1933. In these graphs the audible sound frequencies are plotted on the horizontal axis, while the sound intensity levels (in decibels) are plotted on the vertical axis. Each curve is a *constant loudness* curve, in the sense that it shows the sound intensity level needed at each frequency to make the sound appear to have the same loudness. For example, the lowest (colored) curve represents the threshold of hearing. It shows the intensity levels at which sounds of different frequencies just become audible. The graph indicates that the intensity level of a 100-Hz sound must be about 37 dB greater than the intensity level of a 1000-Hz sound to be at the threshold of hearing. Therefore, the ear is *less sensitive* to a 100-Hz sound than it is to a 1000-Hz sound. In general, Figure 16.32 reveals that the ear is most sensitive in the range from about 1 to 5 kHz, and becomes progressively less sensitive at higher and lower frequencies.

Each curve in Figure 16.32 represents a different loudness, and each is labeled according to its intensity level at 1000 Hz. For instance, the curve labeled ''60'' represents all sounds that have the same loudness as that of a 1000-Hz sound whose intensity level is 60 dB. These constant-loudness curves become ''flatter'' as the loudness increases, the relative flatness indicating that the ear is nearly equally sensitive to all frequencies when the sound is loud. Thus, when you play your stereo at very loud levels, you hear the low frequencies, the middle frequencies, and the high frequencies equally well. However, when you turn down the volume control on the receiver so the sound becomes quiet, the high and low frequencies seem to ''disappear,'' for the ear is relatively insensitive to these frequencies under such conditions.

Many manufacturers of audio equipment include a loudness switch on their receivers (see Figure 16.33) to compensate for the loss of hearing sensitivity at quiet listening levels. When this switch is turned on, the intensities of quiet high- and low-frequency sounds are increased automatically. For loud sounds, the loudness function is designed to have little effect, consistent with the fact that our ears hear all frequencies more or less uniformly under such conditions.

INTEGRATION OF CONCEPTS

SINUSOIDAL WAVES AND SIMPLE HARMONIC MOTION

As we have seen, a wave is a traveling disturbance. In particular, a periodic wave consists of patterns or cycles that are produced over and over again by the source of the disturbance. The source may be a hand vibrating the end of a string or a loudspeaker diaphragm vibrating against the air. As the source moves through one cycle of its repetitive motion, one cycle of the wave is produced. In Chapter 10 we studied simple harmonic motion, which is one particular kind of repetitive motion. When a source of a periodic wave exhibits simple harmonic motion, each wave cycle has the shape of a sine or cosine function, and the wave is said to be sinusoidal. The common ground between sinusoidal waves and simple harmonic motion is evident in the fact that the terms "cycle," "amplitude," "period," and "frequency" are used for both. Furthermore, when a sinusoidal wave travels along a string or a sinusoidal sound wave passes through the air, each particle of the medium is, in turn, set into simple harmonic motion. Therefore, we can describe the displacement, velocity, and acceleration of each particle using the equations that specify these quantities for any object in simple harmonic motion (see Section 10.4). The maximum speed and the maximum acceleration of each particle of the medium are given by Equations 10.11 and 10.13, respectively. This fact comes into play in homework problems such as problems 18 and 82 at the end of the current chapter. We see, then, that periodic sinusoidal waves are another illustration of simple harmonic motion.

SOUND AND PRESSURE

Sound is a longitudinal wave that consists of alternating condensations and rarefactions and can exist only in a medium (e.g., air). A condensation is a region where the pressure in the medium is higher than normal, while a rarefaction is a region where the pressure is less than normal. As a result, when a sound wave travels through a spot, the pressure there fluctuates above and below the norm as the condensations and rarefactions pass through. We have studied the concept of pressure before, and it is worthwhile to compare what we learned then to what we now know about sound waves. In Chapter 11 we saw that the pressure at a given depth in a fluid is determined by the weight of the fluid above a unit area located at that depth. In the case of air, this weight of fluid is what causes the normal air pressure to be 1.01×10^5 N/m² at sea level. In Chapter 11 we also saw that the pressure is constant at a given place, assuming that the density of the fluid does not fluctuate in time. When a sound wave passes through, however, the fluid density does fluctuate in time. As the source of the wave vibrates against the fluid, some fluid particles are pushed together and others are made to move apart. Where the particles are crowded together, the density and pressure increase above the norm. Where the particles are moved apart, the density and pressure decrease below the norm. These changes in pressure, superimposed on the normal static pressure of the fluid, are what constitute the sound wave.

SUMMARY

A **wave** is a traveling disturbance and carries energy from place to place. In a **transverse wave,** the disturbance is perpendicular to the direction of travel of the wave. In a **longitudinal wave,** the disturbance is parallel to the line along which the wave travels.

In a **periodic wave,** the pattern of the disturbance is produced over and over again by the source of the wave. The **amplitude** of the wave is the maximum excursion of a particle of the medium from the particle's undisturbed position. The **wavelength** λ is the distance along the length of the wave between two successive equivalent points, such as two crests or two troughs. The **period** T is the time required for the wave to travel a distance of one wavelength. The **frequency** f (in hertz) is the number of wave cycles per second that pass an observer and is the reciprocal of the period (in seconds): $f = 1/T$. The **speed** v of a wave is related to the wavelength and the frequency according to $v = f\lambda$.

The **speed of a wave** depends on the properties of the medium in which the wave travels. For a transverse wave on a string that has a tension F and a mass per unit length m/L, the wave speed is $v = \sqrt{F/(m/L)}$.

Sound is a longitudinal wave that consists of alternating regions of greater-than-normal pressure (condensations) and less-than-normal pressure (rarefactions). Each cycle of a sound wave includes one condensation and one rarefaction. A sound wave with a large frequency is interpreted by the brain as a high-pitched sound, while one with a small frequency is interpreted as a low-pitched sound. The **pressure amplitude** of a sound wave is the maximum change in pressure, measured relative to the undisturbed pressure. The larger the pressure amplitude, the louder the sound.

The **speed of sound** v depends on the properties of the medium. In an ideal gas, the speed of sound is $v = \sqrt{\gamma kT/m}$, where $\gamma = c_P/c_V$ is the ratio of the specific heat capacities at constant pressure and constant volume, k is

Boltzmann's constant, T is the Kelvin temperature, and m is the mass of a molecule of the gas. In a liquid, the speed of sound is $v = \sqrt{B_{ad}/\rho}$, where B_{ad} is the adiabatic bulk modulus and ρ is the mass density. For a solid in the shape of a long slender bar, the expression for the speed of sound is $v = \sqrt{Y/\rho}$, where Y is Young's modulus.

The **intensity** I of a sound wave is the power P that passes perpendicularly through a surface divided by the area A of the surface: $I = P/A$. The SI unit of intensity is watts per square meter (W/m²). The smallest sound intensity that humans can detect is known as the **threshold of hearing** and is about 1×10^{-12} W/m² for a 1-kHz sound. When a source emits sound uniformly in all directions, the intensity of the sound is inversely proportional to the square of the distance from the source.

The **intensity level** β (in decibels) is used to compare a sound intensity I to the sound intensity I_0 of a reference level: $\beta = 10 \log (I/I_0)$.

The **Doppler effect** is the change in frequency detected by an observer because the sound source and the observer have different velocities with respect to the medium of sound propagation. If the observer and source move with speeds v_o and v_s, respectively, and if the medium is stationary, the frequency f' detected by the observer is

$$f' = f\left(\frac{1 \pm v_o/v}{1 \mp v_s/v}\right)$$

where f is the frequency of the sound emitted by the source, and v is the speed of sound. In the numerator, the plus sign applies when the observer moves toward the source, and the minus sign applies when the observer moves away from the source. In the denominator, the minus sign is used when the source moves toward the observer, and the plus sign is used when the source moves away from the observer.

QUESTIONS

1. Considering the nature of a water wave (see Figure 16.4), describe the motion of a fishing float on the surface of a lake when a wave passes beneath the float. Is it really correct to say that the float bobs straight "up and down"? Explain.

2. "Domino Toppling" is one entry in the *Guiness Book of World Records.* The event consists of lining up an incredible number of dominoes and then letting them topple, one after another. Is the disturbance that propagates along the line of dominoes transverse, longitudinal, or partly both. Explain.

3. Suppose that a longitudinal wave moves along a Slinky at a speed of 5 m/s. Does one coil of the Slinky move through a distance of 5 m in one second? Justify your answer.

4. A transverse wave on a horizontal wire travels with a speed of 1000 m/s. The particles of the wire move up and down in simple harmonic motion. (a) Explain whether the particles themselves are moving at a speed of 1000 m/s, using Equation 10.11 to justify your answer. (b) If you say that the particles are *not* traveling at 1000 m/s, then explain what factors determine the speed of the particles.

5. A wire is strung tightly between two immovable posts. Discuss how an increase in temperature affects the speed of a transverse wave on this wire. Give your reasoning.

6. One end of each of two identical strings is attached to a wall. Each string is being pulled tight by someone at the other end. A transverse pulse is sent traveling along one of the strings. A bit later an identical pulse is sent traveling along the other string. What, if anything, can be done to make the second pulse catch up with and pass the first pulse? Account for your answer.

7. In Section 4.10 the concept of a "massless" rope is discussed. How long would it take for a transverse wave to travel the length of a massless rope? Justify your answer.

8. There is a rule of thumb for estimating how far away a storm is. Following a lightning flash, count off the seconds until the thunder is heard. Divide the number of seconds by five. The result gives the approximate distance *in miles*. Explain why this rule works, noting that light waves from the lightning flash travel at a speed that is much greater than the speed of sound.

9. Do you expect an echo to return to you more quickly or less quickly on a hot day as compared to a cold day, other things being equal? Account for your answer.

10. A loudspeaker produces a sound wave. Does the wavelength of the sound increase, decrease, or remain the same, when the wave travels from air into water? Justify your answer. (*Hint: The frequency does not change as the sound enters the water.*)

11. Some animals rely on an acute sense of hearing for survival, and the visible part of the ear on such animals is often relatively large. Explain how this anatomical feature helps to increase the sensitivity of the animal's hearing for low-intensity sounds.

12. The sound intensity I produced by a loudspeaker in your living room is *not* given by $I = P/4\pi r^2$, where P is the sound power emitted by the speaker and r is the distance from the speaker. Give a reason why this relation does not apply.

13. If the sound intensity level is measured relative to the threshold of hearing, what does it mean for the intensity level to be negative, e.g., -20 dB?

14. If two people talk simultaneously and each creates an intensity level of 65 dB at a certain point, does the total intensity level at this point equal 130 dB? Account for your answer.

15. Suppose you are swinging on a swing. Somewhere in front of you a stationary whistle is blowing. Specify where in your swinging motion you would hear the highest pitch, where you would hear the lowest pitch, and where you would hear the same pitch as that heard by a stationary observer. Note that there may be more than one place in the motion where each of the above is heard. Give your reasoning.

16. Two cars, one behind the other, are traveling in the same direction at the same speed. Does either driver hear the other's horn at a frequency that is different than that heard when both cars are at rest? Justify your answer.

PROBLEMS

Section 16.2 Periodic Waves

1. A person standing in the ocean notices that after a wave crest passes by, ten more crests pass in a time of 120 s. What is the frequency of the wave?

2. Sound travels at a speed of 343 m/s in air at 20 °C. The wavelength of a sound wave is 1.31 m. Find the period of the wave.

3. A light wave travels through air at a speed of 3.0×10^8 m/s. Red light has a wavelength of about 6.6×10^{-7} m. What is the frequency of red light?

4. A longitudinal wave with a frequency of 3.0 Hz takes 1.7 s to travel the length of a 2.5-m Slinky (see Figure 16.3). Determine the wavelength of the wave.

5. The right-most key on a piano produces a sound wave that has a frequency of 4185.6 Hz. Assuming the speed of sound in air is 343 m/s, find the corresponding wavelength.

6. A wave has a frequency of 45 Hz and a speed of 22 m/s. Determine, if possible, (a) its period, (b) its wavelength, and (c) its amplitude. If it is not possible to determine any of these quantities, then so state.

7. A person fishing from a pier observes that four wave crests pass by in 7.0 s and estimates the distance between two successive crests as 4.0 m. The timing starts with the first crest and ends with the fourth. What is the speed of the wave?

8. Suppose the amplitude and frequency of the transverse wave in Figure 16.2c are, respectively, 1.3 cm and 5.0 Hz. Find the *total vertical distance* through which the colored dot moves in 3.0 s.

*9. The amplitude of a transverse wave on a string is 0.010 m. A particle of the string moves a total distance of 0.120 m in 0.10 s. The wavelength of the wave is 1.5 m. What is the speed of the wave?

*10. In Figure 16.3c the colored "dot" exhibits simple harmonic motion as the longitudinal wave passes. The wave has an amplitude of 5.4×10^{-3} m and a frequency of 4.0 Hz. Find the maximum acceleration of the dot.

**11. A water-skier is moving at a speed of 12.0 m/s. When she skis in the same direction as a traveling wave, she springs upward every 0.500 s because of the wave crests. When she skis in the direction opposite to that in which the wave moves, she springs upward every 0.400 s in response to the crests. Determine (a) the speed and (b) the wavelength of the wave.

Section 16.3 The Speed of a Wave on a String

12. A transverse wave is traveling with a speed of 300 m/s on a horizontal string. If the tension in the string is increased by a factor of four, what is the speed of the wave?

13. The linear density of the A string on a violin is 7.8×10^{-4} kg/m. A wave on the string has a frequency of 440 Hz and a wavelength of 65 cm. What is the tension in the string?

14. A 0.200-kg wire is stretched between two posts 20.0 m apart and has a tension of 90.0 N. The wire is struck at one end, and a transverse pulse travels toward the other end. How long does the pulse take to travel the length of the wire?

15. A 0.50-m string is stretched so the tension is 1.7 N. A transverse wave of frequency 120 Hz and wavelength 0.31 m travels on the string. What is the mass of the string?

16. The middle C string on a piano is under a tension of 944 N. The period and wavelength of a wave on this string are 3.82×10^{-3} s and 1.26 m, respectively. Find the linear density of the string.

*17. To measure the acceleration due to gravity on a distant planet, an astronaut hangs a 0.085-kg ball from the end of a wire. The wire has a length of 1.5 m and a linear density of 3.1×10^{-4} kg/m. Using electronic equipment, the astronaut measures the time for a transverse pulse to travel the length of the wire and obtains a value of 0.083 s. The mass of the wire is negligible compared to the mass of the ball. Determine the acceleration due to gravity.

*18. A horizontal wire is under a tension of 315 N and has a mass per unit length of 6.50×10^{-3} kg/m. A transverse wave with an amplitude of 2.50 mm and a frequency of 585 Hz is traveling on this wire. As the wave passes, a particle of the wire moves up and down in simple harmonic motion. Obtain (a) the speed of the wave and (b) the maximum speed with which the particle moves up and down.

*19. A wire has a cross-sectional area of 4.2×10^{-8} m² and is made from a material whose density is 7900 kg/m³. Determine the wire's *linear density (m/L)*.

**20. A copper wire, whose cross-sectional area is 1.1×10^{-6} m², has a linear density of 7.0×10^{-3} kg/m and is strung between two walls. At the ambient temperature, a transverse wave travels with a speed of 46 m/s on this wire. The coefficient of linear expansion for copper is 17×10^{-6} (C°)⁻¹, and Young's modulus for copper is 1.1×10^{11} N/m². What will be the speed of the wave when the temperature is lowered by 14 C°?

Section 16.4 The Mathematical Description of a Wave

21. The displacement (in meters) of a wave is $y = (0.26) \sin (\pi t - 3.7\pi x)$, where t is in seconds and x is in meters. (a) Is the wave traveling in the $+x$ or $-x$ direction? (b) What is the displacement y when $t = 38$ s and $x = 13$ m?

22. A wave has the following properties: amplitude = 0.37 m, period = 0.77 s, wave speed = 12 m/s. The wave is traveling in the $-x$ direction. What is the mathematical expression (similar to Equation 16.3 or 16.4) for the wave?

23. A wave has a displacement (in meters) of $y = (0.45) \sin (8.0\pi t + \pi x)$, where t and x are expressed in seconds and meters, respectively. (a) Find the amplitude, the frequency, the wavelength, and the speed of the wave. (b) Is this wave traveling in the $+x$ or $-x$ direction?

24. The drawing shows two graphs that represent a transverse wave on a string. The wave is moving in the $+x$ direction. Using the information contained in these graphs, write the mathematical expression (similar to Equation 16.3 or 16.4) for the wave.

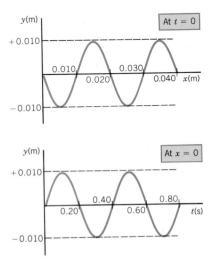

25. A transverse wave on a string is traveling in the $+x$ direction and has an amplitude of 2.00 mm, a wavelength of 0.300 m, and a speed of 15.0 m/s. Write the equation that gives the displacement y as a function of time for a point located on the string at $x = +0.400$ m.

*26. Refer to the graphs that accompany problem 24. From the data in these graphs, determine the speed of the wave.

Section 16.6 The Speed of Sound

27. Suppose someone said: "Sound whose wavelength is larger than the size of your ear cannot be heard." (a) Assume the speed of sound is 343 m/s and compute the wavelength of sound at the limits of human hearing, 20 Hz and 20 kHz. (b) Compare these values with the (estimated) width of your ear. Based on this comparison, is the statement above correct?

28. The magnetic tape of a cassette deck moves with a speed of 0.048 m/s ($1\frac{7}{8}$ inches per second). The recording head records a 15 000-Hz tone on the tape. What is the wavelength λ of the magnetized regions?

29. Argon (molecular mass = 39.9 u) is a monatomic gas. Assuming that it behaves like an ideal gas at 298 K, find (a) the rms-speed of argon atoms and (b) the speed of sound in argon.

30. The speed of sound in a container of hydrogen at 201 K is 1220 m/s. What would be the speed of sound if the temperature were raised to 405 K? Assume that hydrogen behaves like an ideal gas.

31. Aluminum has a Young's modulus of 6.9×10^{10} N/m² and a density of 2700 kg/m³. How much time does it take for sound to travel the 1.4-m length of an aluminum rod?

32. The wavelength of a sound wave in air is 2.74 m at 20 °C. What is the wavelength of this sound wave in fresh water at 20 °C? (Hint: The frequency of the sound is the same in both media.)

33. At 20 °C the densities of fresh water and ethanol are, respectively, 998 and 789 kg/m³. Find the ratio of the adiabatic bulk modulus of fresh water to the adiabatic bulk modulus of ethanol at 20 °C.

34. The speed v of longitudinal waves in a liquid is given by $v = \sqrt{B_{ad}/\rho}$, where B_{ad} is the adiabatic bulk modulus and ρ is the density. Show that $\sqrt{B_{ad}/\rho}$ has the units of speed.

35. As the drawing illustrates, a siren can be made by blowing a jet of air through 20 equally spaced holes in a rotating disk. If the siren is to produce a 2200-Hz tone, what must be the angular speed of the disk?

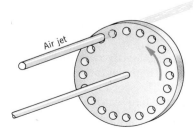

36. An explosion occurs at the end of a pier. The sound reaches the other end of the pier by traveling through three media: air, fresh water, and a slender handrail of solid steel. The speeds of sound in air, water, and the handrail are 343, 1482, and 5040 m/s, respectively. The sound travels a distance of 125 m in each medium. (a) Through which medium does the sound arrive first, second, and third? (b) After the first sound arrives, how much later do the second and third sounds arrive?

*37. A long slender bar is made from an unknown material. The length of the bar is 0.83 m, its cross-sectional area is 1.3×10^{-4} m², and its mass is 2.1 kg. A sound wave travels from one end of the bar to the other end in 1.9×10^{-4} s. From which one of the materials listed in Table 10.1 is the bar most likely to be made?

*38. A sonar unit on a boat is capable of detecting the return of an ultrasonic pulse during times up to 1.500 s after the pulse is transmitted. (a) Assuming the freshwater temperature is 20 °C, what is the maximum water depth that can be measured? (b) If the sonar unit measures time with an error of ±0.004 s, what is the error (in meters) in the depth measurement?

*39. At a height of ten meters above the surface of a lake, a sound pulse is generated. The echo from the bottom of the lake returns to the point of origin 0.140 s later. The air and water temperatures are 20 °C. How deep is the lake?

*40. Both krypton (Kr) and neon (Ne) can be approximated as monatomic ideal gases. The atomic mass of krypton is 83.8 u, while that of neon is 20.2 u. A loudspeaker produces a sound whose wavelength in krypton is 1.25 m. If the loudspeaker were used in neon at the same temperature, what would be the wavelength? (Hint: The sound frequency is the same in both gases.)

**41. As a prank, someone drops a water-filled balloon out of a window. The balloon is released from rest at a height of 10.0 m above the ears of a man who is the target. Because of a guilty conscience, however, the prankster shouts a warning after the balloon is released. The warning will do no good, however, if shouted after the balloon reaches a certain point, even if the man could react infinitely quickly. Assuming the air temperature is 20 °C and ignoring the effect of air resistance on the balloon, determine how far above the man's ears this point is.

**42. In a mixture of argon (atomic mass = 39.9 u) and neon (atomic mass = 20.2 u), the speed of sound is 363 m/s at 3.00×10^2 K. Assume that both monatomic gases behave as ideal gases. Find the percentage of the atoms that are argon and the percentage that are neon.

Section 16.7 Sound Intensity

43. A typical adult ear has a surface area of 2.1×10^{-3} m². The sound intensity during a normal conversation is about

3.2×10^{-6} W/m² at the listener's ear. Assume the sound strikes the surface of the ear perpendicularly. How much power is intercepted by the ear?

44. The average sound intensity inside a busy restaurant is 3.2×10^{-5} W/m². How much energy goes into each ear (area $= 2.1 \times 10^{-3}$ m²) during a one-hour meal?

45. Suppose that sound is emitted uniformly in all directions by a public address system. The intensity at a location 22 m away from the sound source is 3.0×10^{-4} W/m². What is the intensity at a spot that is 78 m away?

46. At a distance of 3.8 m from a siren, the sound intensity is 3.6×10^{-2} W/m². Assuming the siren radiates sound uniformly in all directions, find the total power radiated.

47. A loudspeaker has a circular opening with a radius of 0.0950 m. The electrical power needed to operate the speaker is 25.0 W. The average sound intensity at the opening is 17.5 W/m². What percentage of the electrical power is converted by the speaker into sound power?

***48.** When a helicopter is hovering 1100 m directly overhead, an observer on the ground measures a sound intensity I. Assume that sound is radiated uniformly from the helicopter and that ground reflections are negligible. How far must the helicopter fly in a straight line parallel to the ground before the observer measures a sound intensity of $\frac{1}{3}I$?

***49.** A source radiates 1.2×10^{-5} W of sound power uniformly in all directions. A cube and a sphere (radius $= 0.56$ m) are centered on this source. The sphere just fits within the cube. What is the sound intensity at a corner of the cube?

****50.** A rocket, starting from rest, travels straight up with an acceleration of 58.0 m/s². When the rocket is at a height of 562 m, it produces sound that eventually reaches a ground-based monitoring station directly below. The sound is emitted uniformly in all directions. The monitoring station measures a sound intensity I. Later on, the station measures an intensity $\frac{1}{3}I$. Assuming the speed of sound is 343 m/s, find the time that has elapsed between the two measurements.

Section 16.8 Decibels

51. The sound intensity level of a jet engine is 138 dB above the threshold of hearing. What is the sound intensity?

52. An amplified guitar has a sound intensity level that is 14 dB greater than the same unamplified sound. What is the ratio of the amplified intensity to the unamplified intensity?

53. One of the important specifications of a cassette tape deck is its signal-to-noise rating. This specification indicates how much of the sound intensity created when playing a tape is due to the musical tones (the signal) and how much is due to the hissing sound produced by the moving tape (the noise). Suppose the sound intensity due to the musical tones

is 1.2×10^{-4} W/m², while that due to the tape hiss is 3.2×10^{-11} W/m². What is the signal-to-noise rating, which is the number of *decibels* by which the signal exceeds the noise?

54. The intensity level of the sound produced at a rock concert often reaches 120 dB. The intensity level of a quiet flute is about 67 dB. What is the ratio of the sound intensity of a rock concert to the sound intensity of a quiet flute?

55. The equation $\beta = 10 \log (I/I_0)$, which defines the decibel, is sometimes written in terms of power P (in watts) rather than intensity I (in watts/meter²). The form $\beta = 10 \log (P/P_0)$ can be used to compare two power levels in terms of decibels. Suppose that stereo amplifier A is rated at $P = 250$ watts per channel, while amplifier B has a rating of $P_0 = 45$ watts per channel. (a) Expressed in decibels, how much more powerful is A compared to B? (b) Will A sound more than twice as loud as B? Justify your answer.

56. For information, read problem 55 before working this problem. Stereo manufacturers express the power output of an audio amplifier using the decibel, abbreviated as dBW, where the "W" indicates that a reference power level of $P_0 = 1.00$ W has been used. If an amplifier has a power rated at 17.5 dBW, how many watts of power can this amplifier deliver?

***57.** The intensity level of sound A is 5.0 dB greater than that of sound B and 3.0 dB less than that of sound C. Determine the ratio (I_C/I_B) of the intensity of sound C to the intensity of sound B.

***58.** The sound intensity level of a person speaking normally is about 65 dB above the threshold of hearing. What is the minimum number of people speaking simultaneously, each with this intensity level, that is necessary to produce a sound intensity level at least 78 dB above the threshold of hearing?

****59.** A source emits sound uniformly in all directions. A radial line is drawn from this source. On this line, determine the positions of two points, 1.00 m apart, such that the intensity level at one point is 2.00 dB greater than that at the other.

Section 16.10 The Doppler Effect

60. At a football game, a stationary spectator is watching the halftime show. A trumpet player in the band is playing a 784-Hz tone while marching directly toward the spectator at a speed of 0.83 m/s. On a day when the speed of sound is 343 m/s, what frequency does the spectator hear?

61. A hawk is flying directly away from a bird watcher at a speed of 11.0 m/s. The hawk produces a shrill cry whose frequency is 865 Hz. The speed of sound is 343 m/s. What is the frequency that the bird watcher hears?

62. A train is blowing its whistle while traveling at a speed of 33.0 m/s. The speed of sound is 343 m/s. Observer A is directly in front of the train, while observer B is directly behind it. Find the ratio of the whistle frequency heard by A to that heard by B.

63. Suppose you are stopped for a traffic light, and an ambulance approaches you from behind with a speed of 18 m/s. The siren on the ambulance produces sound with a frequency of 955 Hz. The speed of sound in air is 343 m/s. What is the wavelength of the sound reaching your ears?

64. An aircraft carrier has a speed of 13.0 m/s relative to the water. A jet is catapulted from the deck and has a speed of 67.0 m/s relative to the water. The engines produce a 1550-Hz whine, and the speed of sound is 343 m/s. What is the frequency of the sound heard by the crew on the ship?

***65.** Two trucks travel at the same speed. They are far apart on adjacent lanes and approach each other essentially head-on. One driver hears the horn of the other truck at a frequency that is 1.20 times the frequency he hears when the trucks are stationary. The speed of sound is 343 m/s. At what speed is each truck moving?

***66.** A motorcycle starts from rest and accelerates along a straight line at 2.81 m/s², on a day when the speed of sound is 343 m/s. A siren is located at the starting point and remains stationary. How far has the motorcycle gone when the driver hears the frequency of the siren at 90.0% of the value it has when the motorcycle is stationary?

****67.** A microphone is attached to a spring that is suspended from the ceiling, as the drawing illustrates. Directly below on the floor is a stationary 440-Hz source of sound. The microphone vibrates up and down in simple harmonic motion with a period of 2.0 s. The difference between the maximum and minimum sound frequencies detected by the microphone is 2.1 Hz. Ignoring any reflections of sound in the room and using 343 m/s for the speed of sound, determine the amplitude of the simple harmonic motion.

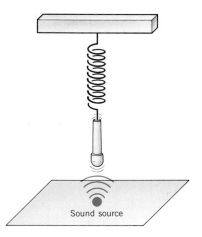
Sound source

ADDITIONAL PROBLEMS

68. The volume control on a stereo amplifier is adjusted so the sound intensity level increases from 23 to 61 dB. What is the ratio of the final sound intensity to the original sound intensity?

69. A bat emits sound whose frequency is 91 kHz. The air temperature is 35 °C, so the speed of sound is not 343 m/s. Find the wavelength of the sound.

70. A vibrator moves one end of a rope up and down to generate a wave. The tension in the rope is 58 N. The frequency is then doubled. To what value must the tension be adjusted, so the new wave has the same wavelength as the old one?

71. An observer is moving away from a stationary source of sound. In fresh water at 20 °C, the observer hears the frequency at a value that is 99.60% of the vibration frequency of the source. How fast is the observer moving?

72. A person lying on an air mattress in the ocean rises and falls through one complete cycle every five seconds. The crests of the wave causing the motion are 20.0 m apart. Determine (a) the frequency and (b) the speed of the wave.

73. A source radiates 0.38 W of sound power uniformly in all directions and is at the center of a sphere. How much power passes through a 0.22-m² patch of surface on the sphere if the radius of the sphere is 1.3 m?

74. The distance between a loudspeaker and the left ear of a listener is 2.70 m. (a) Calculate the time required for sound to travel this distance if the air temperature is 20 °C. (b) Assuming the sound frequency is 523 Hz, how many wavelengths of sound are contained in this distance?

75. A wave is moving in the + x direction. Assuming the wave has the following properties, write the equation of the wave (similar to Equation 16.3 or 16.4): speed = 7.1 m/s, amplitude = 0.15 m, wavelength = 0.28 m.

76. A speeder looks in his rearview mirror. He notices that a police car has pulled behind him and is matching his speed of 38 m/s. The siren on the police car has a frequency of 860 Hz when the police car and the listener are stationary. The speed of sound is 343 m/s. What frequency does the speeder hear when the siren is turned on in the moving police car?

77. In Figure 16.2 the hand moves the end of the Slinky up and down through two complete cycles in one second. The wave moves along the Slinky at a speed of 0.50 m/s. Find the distance between two adjacent crests on the wave.

78. A listener doubles his distance from a source that emits sound uniformly in all directions. By how many decibels does the sound intensity level decrease?

***79.** The drawing shows a frictionless incline and pulley.

The two blocks are connected by a wire (mass per unit length = 0.0250 kg/m) and remain stationary. A transverse wave on the wire has a speed of 75.0 m/s. Neglecting the weight of the wire relative to the tension in the wire, find the masses m_1 and m_2 of the blocks.

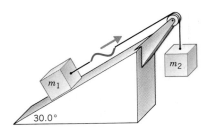

*80. A 3.49 rad/s ($33\frac{1}{3}$ rpm) record has a 5.00 kHz tone cut in the groove. If the groove is located 0.100 m from the center of the record (see drawing), what is the "wavelength" in the groove?

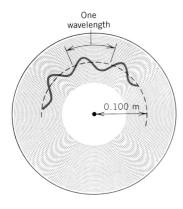

*81. A steel cable, of cross-sectional area 2.83×10^{-3} m^2, is kept under a tension of 1.00×10^4 N. The density of steel is 7860 kg/m^3. At what speed does a transverse wave move along the cable?

*82. The amplitude and frequency of a transverse wave on a string are, respectively, 7.8×10^{-3} m and 6.0 Hz. Remembering that a particle of the string exhibits simple harmonic motion as the wave passes, determine the maximum speed of a particle. (Hint: This speed is not the speed of the wave.)

*83. A loudspeaker is generating sound in a room. At a certain point, the sound waves coming directly from the speaker (without reflecting from the walls) create an intensity level of 75.0 dB. The waves reflected from the walls create, by themselves, an intensity level of 72.0 dB at the same point. These levels are relative to the threshold of hearing. Relative to the threshold of hearing, what is the total intensity level? (Hint: The answer is not 147.0 dB.)

*84. Civil engineers use a transit theodolite when surveying. A modern version of this device determines distance by measuring the time required for an ultrasonic pulse to reach a target, reflect from it, and return. Effectively, such a theodolite is calibrated properly when it is programmed with the speed of sound appropriate for the ambient air temperature. (a) Suppose the round-trip time for the pulse is 0.580 s on a day when the air temperature is 293 K, the temperature for which the instrument is calibrated. How far is the target from the theodolite? (b) Assume that air behaves as an ideal gas. If the air temperature were 298 K, rather than the calibration temperature of 293 K, what percentage error would there be in the distance measured by the theodolite?

**85. At an outdoor party, Arnold is talking at a distance of 1.5 m from the beer keg. A number of other people at a distance of 3.5 m from the keg are also talking. Each individual, including Arnold, is producing the same sound power, which is assumed to spread out uniformly in all directions. The sound intensity at the beer keg due to all the others is at least as large as that produced by Arnold. What is the minimum number of other people talking?

CHAPTER 17

THE PRINCIPLE OF LINEAR SUPERPOSITION AND INTERFERENCE PHENOMENA

As we have seen, a wave is a traveling disturbance. On a lake, a motorboat generates a wave that travels outward and rocks a sunbather floating on a raft. A bird sings a morning song to a hiker, who hears it because the sound wave travels through the air and pushes on her eardrums ever so gently. Such examples, which involve only a single wave, form the basis of the discussion in the last chapter. Often, however, there is more than one wave present at the same time, as in the performance by Pink Floyd captured in the photograph above. The guitars, drums, and saxophone each generate sound waves, and these waves combine to form a total sound wave that reaches the listener. In this chapter we will explore some of the interesting phenomena—such as diffraction, beats, and standing waves—that can arise when a number of waves combine at the same place at the same time.

17.1 THE PRINCIPLE OF LINEAR SUPERPOSITION

There is a straightforward way to deal with situations in which two or more waves pass through the same place simultaneously. To see how, examine Figures 17.1 and 17.2, which show two transverse pulses of equal heights moving toward each other along a Slinky. In Figure 17.1 both pulses are "up," while in Figure 17.2 one

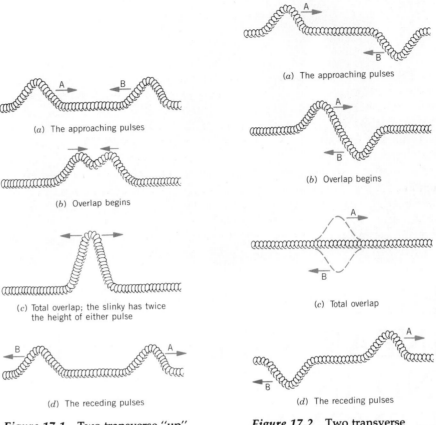

(a) The approaching pulses

(b) Overlap begins

(c) Total overlap; the slinky has twice
the height of either pulse

(d) The receding pulses

Figure 17.1 Two transverse "up"
pulses passing through each other.

(a) The approaching pulses

(b) Overlap begins

(c) Total overlap

(d) The receding pulses

Figure 17.2 Two transverse
pulses, one "up" and one "down,"
passing through each other.

is "up" and the other is "down." Part *b* of each drawing shows the two pulses just beginning to overlap, as the dashed lines indicate. The pulses merge, and the Slinky assumes a shape that is *the sum of the shapes of the individual pulses.* Thus, when the two "up" pulses overlap completely, as in Figure 17.1*c*, the Slinky has a pulse height that is twice the height of an individual pulse. Likewise, when the "up" pulse and the "down" pulse overlap exactly, as in Figure 17.2*c*, they momentarily cancel, and the Slinky becomes straight. In either case, the two pulses move apart after overlapping, and the Slinky once again conforms to the shapes of the individual pulses.

The adding together of individual pulses to form a resultant pulse is an example of a more general concept called the *principle of linear superposition.*

The Principle of Linear Superposition

When two or more waves are present simultaneously at the same place, the resultant wave is the sum of the individual waves.

This principle can be applied to all types of waves, including sound waves, water waves, and electromagnetic waves such as light. It embodies one of the most important concepts in all of physics, and the remainder of this chapter deals with examples related to the superposition of waves.

17.2 CONSTRUCTIVE AND DESTRUCTIVE INTERFERENCE OF SOUND WAVES

Suppose that the sounds from two speakers overlap in the middle of a listening area, as in Figure 17.3, and that each speaker produces a sound wave of the same amplitude and the same frequency. For convenience, the wavelength of the sound is chosen to be $\lambda = 1$ m. In addition, assume the diaphragms of the speakers vibrate synchronously; that is, they move outward together and inward together. If the distance of each speaker from the overlap point is the same (3 m in the drawing), the condensations of one wave always meet the condensations of the other when the waves come together; similarly, rarefactions always meet rarefactions. Figure 17.4 shows the pressure patterns of the individual sounds, as well as the combined pressure pattern at the overlap point. According to the principle of linear superposition, the combined pattern is the sum of the individual patterns. As a result, the pressure fluctuations at the overlap point have twice the amplitude that the individual waves have, and a listener at this spot hears a louder sound than that coming from either speaker alone. When two waves always meet condensation-to-condensation and rarefaction-to-rarefaction (or crest-to-crest and trough-to-trough), they are said to be *exactly in phase* and to exhibit *constructive interference.*

Now consider what happens if one of the speakers is moved. The result is surprising. In Figure 17.5, the left speaker is moved away* from the overlap point by a distance equal to one-half of the wavelength, or 0.5 m. Therefore, at the

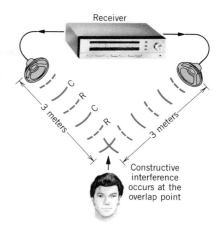

Figure 17.3 As a result of constructive interference between the two sound waves, a loud sound is heard at an overlap point located equally distant from the two speakers.

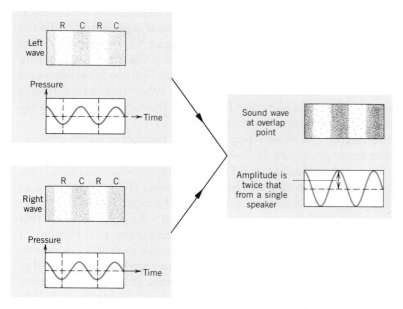

Figure 17.4 When sound waves from the left and right speakers in Figure 17.3 arrive at the overlap point, condensations always meet condensations and rarefactions always meet rarefactions, leading to constructive interference and a loud sound. The drawings in red are obtained by adding the two drawings in blue.

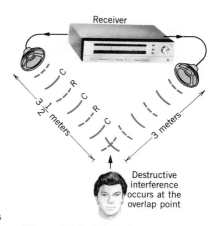

Figure 17.5 The left speaker is one-half of a wavelength ($\frac{1}{2}$ m) farther from the overlap point than the right speaker. Because of destructive interference, no sound is heard at the overlap point.

* When the left speaker is moved back, its sound intensity and, hence, its pressure amplitude decrease at the overlap point. In this chapter assume that the power delivered to the left speaker by the receiver is increased slightly to keep the amplitudes equal at the overlap point.

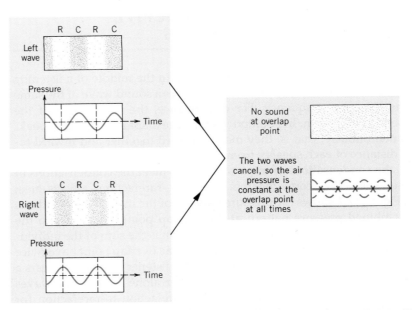

Figure 17.6 The sound waves from the left and right speakers in Figure 17.5 arrive at the overlap point in such a way that the condensations from one speaker always meet the rarefactions from the other, leading to destructive interference and an absence of sound. The drawings in red are obtained by adding the drawings in blue.

THE PHYSICS OF . . .

noise-canceling head-phones.

overlap point, a condensation arriving from the left meets a rarefaction arriving from the right. Likewise, a rarefaction arriving from the left meets a condensation arriving from the right. According to the principle of linear superposition, the net effect is a mutual cancellation of the two waves, as Figure 17.6 emphasizes. The condensations from one wave offset the rarefactions from the other, leaving only a *constant air pressure.* A constant air pressure, devoid of condensations and rarefactions, means that a listener detects no sound. When two waves always meet condensation-to-rarefaction (or crest-to-trough), they are said to be *exactly out of phase* and to exhibit *destructive interference.*

Destructive interference can be very beneficial, as a number of recent applications illustrate. For instance, Figure 17.7 shows a pair of noise-canceling headphones. A small microphone is mounted inside the headphones and detects noise such as the engine noise that an airplane pilot would hear. The headphones also contain circuitry to process the electronic signal from the microphone and reproduce the noise in a form that is exactly out of phase with the original noise. This out-of-phase version is played back through the headphones and, because of destructive interference, combines with the original noise to produce a quieter background. Similar kinds of noise-canceling techniques are being developed to "quiet" the ride in automobiles and passenger sections of airplanes. In these applications, the vehicle itself is envisioned as a "giant" headphone, within which the passenger sits.

It should be apparent that if the left speaker in Figure 17.5 were moved away from the overlap point by *another* one-half wavelength ($3\frac{1}{2}$ m + $\frac{1}{2}$ m = 4 m), the two waves would again be in phase, and the listener would again hear a loud sound. In such a case, constructive interference results because the left sound wave travels one whole wavelength (1 m) farther than the right wave and, at the overlap point, condensation meets condensation and rarefaction meets rarefac-

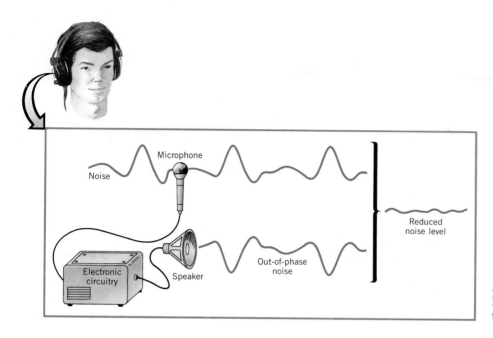

Figure 17.7 Noise-canceling headphones utilize destructive interference.

tion. In general, the important issue is the *difference* in the distances traveled by each wave in reaching the overlap point. *A difference that is an integer number (1, 2, 3, . . .) of wavelengths leads to constructive interference; a difference that is a half-integer number ($\frac{1}{2}$, $1\frac{1}{2}$, $2\frac{1}{2}$, . . .) of wavelengths leads to destructive interference.*

Interference effects can also be detected if the two speakers are fixed in position and the listener moves about the room. Consider Figure 17.8, where the sound waves spread outward from each speaker, as indicated by the concentric circular arcs. Each solid arc represents the center of a condensation, while each dashed arc represents the center of a rarefaction. Where the two waves overlap, there are places of constructive interference and places of destructive interference. Constructive interference occurs at any spot where two condensations or two rarefactions intersect, and the drawing shows four such places as solid dots. A listener stationed at any one of these locations hears a loud sound. On the other hand, destructive interference occurs at any place where a condensation and a rarefaction intersect, such as the two open dots in the picture. A listener situated at a point of destructive interference hears no sound. At locations where neither constructive nor destructive interference occurs, the two waves partially reinforce or partially cancel, depending on the position relative to the speakers. Listeners at such places hear a sound whose loudness is between that heard at the points of constructive and destructive interference. Example 1 illustrates how to decide what a listener hears.

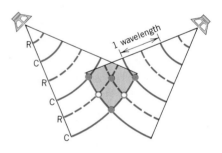

Figure 17.8 The overlapping of two sound waves produces interference effects in the shaded region. The solid lines denote the centers of condensations (C), while the dashed lines denote the centers of rarefactions (R). Constructive interference occurs at each solid dot, and destructive interference occurs at each open dot.

Example 1 What Does a Listener Hear?

In Figure 17.9 two loudspeakers, *A* and *B*, are separated by 3.20 m. A listener is stationed at point *C*, which is 2.40 m directly in front of speaker *B*. The triangle *ABC* is a right triangle. Both speakers are playing identical 214-Hz tones, and the speed of sound is 343 m/s. Does the listener hear a loud sound or no sound?

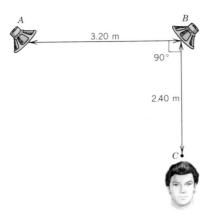

Figure 17.9 Example 1 discusses whether this setup leads to constructive or destructive interference at C for 214-Hz sound waves.

REASONING The listener will hear either a loud sound or no sound, depending upon whether the interference occurring at point C is constructive or destructive. To determine which it is, we need to find the difference in the distances traveled by the two sound waves that reach point C and determine whether the difference is an integer or half-integer number of wavelengths. In either event, the wavelength can be found from Equation 16.1 ($\lambda = v/f$).

SOLUTION Since the triangle ABC is a right triangle, the distance AC is given by the Pythagorean theorem as $\sqrt{(3.20 \text{ m})^2 + (2.40 \text{ m})^2} = 4.00$ m. The distance BC is given as 2.40 m. Thus, we find that the difference in the travel distances for the waves is 4.00 m − 2.40 m = 1.60 m. The wavelength of the sound, however, is

$$\lambda = \frac{v}{f} = \frac{343 \text{ m/s}}{214 \text{ Hz}} = 1.60 \text{ m} \qquad (16.1)$$

Since the difference in the distances is one wavelength, constructive interference occurs at point C, and the listener hears a loud sound.

Since there are many places of constructive and destructive interference within the overlap region in Figure 17.8, it is possible for a listener to walk about the area and hear marked variations in loudness. Interference redistributes the sound intensity by taking the intensity from places of destructive interference and giving it to places of constructive interference. The phenomenon is exhibited by all types of waves, not just sound waves.

17.3 DIFFRACTION

When a wave encounters an obstacle or the edges of an opening, it bends around them. For instance, a sound wave produced by a stereo system bends around the edges of an open doorway, as Figure 17.10a illustrates. If such bending did not occur, sound could be heard outside the room only at locations directly in front of the doorway, as part b of the drawing suggests. (It is assumed that no sound is transmitted directly through the walls.) The bending of a wave around an obstacle or the edges of an opening is called *diffraction*. All kinds of waves exhibit diffraction.

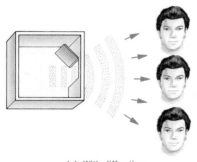

(a) With diffraction

(b) Without diffraction

Figure 17.10 (a) The bending of a sound wave around the edges of the doorway is an example of diffraction. The source of the sound within the room is not shown. (b) If diffraction did not occur, the sound wave would not bend as it passes through the doorway.

To demonstrate how the bending of waves arises, Figure 17.11 shows an expanded view of Figure 17.10. When the sound wave reaches the doorway, the air in the doorway is set into longitudinal vibration. In effect, each air molecule in the doorway is a source of a sound wave in its own right, and, for purposes of illustration, the drawing shows two of the molecules. Each molecule produces a sound wave that expands spherically outward, much like the water wave generated when a stone is dropped into a pond. The sound waves generated by all the molecules in the doorway must be added together to obtain the total sound intensity at any location outside the room, as prescribed by the principle of linear superposition. However, even considering only the waves from the two molecules in the picture, it is clear that the expanding wave patterns reach locations off to either side of the doorway. The net effect is a "bending" or diffraction of the sound around the edges of the opening. Further insight into the origin of diffraction can be obtained with the aid of Huygens' principle (see Section 27.5).

When the sound waves generated by every molecule in the doorway are added together, it is found that there are places where the intensity is a maximum and

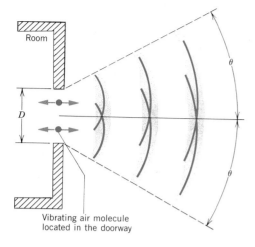

Room

D

Vibrating air molecule
located in the doorway

Figure 17.11 Each vibrating air molecule in the doorway generates a sound wave that expands outward and bends, or diffracts, around the edges of the doorway. Because of interference effects among the sound waves produced by all the molecules in the doorway, the sound intensity is mostly confined to the region on either side of the doorway defined by the angle θ.

places where it is zero, in a fashion similar to that discussed in the previous section. Analysis shows that at a great distance from the doorway the intensity is a maximum directly opposite the center of the opening. As the distance to either side of the center increases, the intensity decreases and reaches zero, then rises again to a maximum, falls again to zero, rises back to a maximum, and so on. Only the maximum at the center is a strong one. The other maxima are weak and become progressively weaker at greater distances from the center. In Figure 17.11 the angle θ defines the location of the first minimum intensity point on either side of the center. Equation 17.1 gives θ in terms of the wavelength λ and the width D of the doorway and assumes that the doorway can be treated like a slit whose height is very large compared to its width:

$$\begin{bmatrix} \textbf{Single slit—} \\ \textbf{first minimum} \end{bmatrix} \qquad \sin \theta = \frac{\lambda}{D} \qquad\qquad (17.1)$$

Waves also bend around the edges of openings other than single slits. Particularly important is the diffraction of sound by a circular opening. In this case, the angle θ is related to the wavelength λ and the diameter D of the opening by

$$\begin{bmatrix} \textbf{Circular opening} \\ \textbf{—first minimum} \end{bmatrix} \qquad \sin \theta = 1.22 \, \frac{\lambda}{D} \qquad\qquad (17.2)$$

An important point to remember about Equations 17.1 and 17.2 is that the extent of the diffraction depends on the ratio of the wavelength to the size of the opening. If the ratio λ/D is small, then θ is small. Little diffraction occurs, and the waves are beamed in the forward direction as they leave an opening, much like the light from a flashlight. Such sound waves are said to have "narrow dispersion." Since high-frequency sound has a relatively small wavelength, it tends to have a narrow dispersion. On the other hand, for larger values of the ratio λ/D, the angle θ is larger. The waves spread out over a larger region, and are said to have a "wide dispersion." Low-frequency sound, with its relatively large wavelength, typically has a wide dispersion.

Loudspeaker designers, for instance, utilize a large value of λ/D in the type of speaker known as a diffraction horn. They choose the width D of the horn (see Figure 17.12) to be much smaller than the wavelengths of the sounds that are to

THE PHYSICS OF . . .

a diffraction horn loudspeaker.

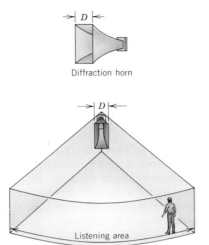

Figure 17.12 The type of loud-speaker known as a diffraction horn is mounted as in part *a* to achieve a wide dispersion of sound into the listening area.

be reproduced. Figure 17.12*a* shows a diffraction horn mounted so that diffraction spreads the sound into the widest possible listening area. If diffraction is to be an effective spreading mechanism, the width *D* must be parallel to the floor, as in part *a* of the drawing, and not perpendicular to the floor as in part *b*. The orientation in part *b* assumes that the wide flaring of the horn spreads out the sound, which is not the case in a diffraction horn.

Diffraction horns are only one type of loudspeaker. Other types allow the sound to emerge through circular openings, and Example 2 illustrates one way in which wide dispersion is achieved in such speakers.

Example 2   Designing a Loudspeaker for Wide Dispersion

(a) A 1500-Hz sound and a 8500-Hz sound each come from a loudspeaker whose diameter is 0.30 m. Assuming the speed of sound in air is 343 m/s, find the diffraction angle θ for each sound. (b) If the 8500-Hz sound is to be produced by a second speaker, how small should this speaker be, so the sound has the same wide dispersion as does the low-pitched sound coming from the 0.30-m speaker?

<u>REASONING</u>   The diffraction angle θ for each sound wave is given by $\sin \theta = 1.22(\lambda/D)$. However, to use this equation, we must first calculate the wavelengths from $\lambda = v/f$.

SOLUTION

(a) The wavelengths of the two sounds are

$$\lambda_{1500} = \frac{343 \text{ m/s}}{1500 \text{ Hz}} = 0.23 \text{ m} \quad \text{and} \quad \lambda_{8500} = \frac{343 \text{ m/s}}{8500 \text{ Hz}} = 0.040 \text{ m}$$

The diffraction angles can now be determined:

$\begin{bmatrix} \textbf{1500-Hz} \\ \textbf{sound} \end{bmatrix}$ $\sin \theta = 1.22 \dfrac{\lambda_{1500}}{D} = 1.22 \left(\dfrac{0.23 \text{ m}}{0.30 \text{ m}} \right) = 0.94$ (17.2)

$\theta = \sin^{-1} 0.94 = \boxed{70°}$

$\begin{bmatrix} \textbf{8500-Hz} \\ \textbf{sound} \end{bmatrix}$ $\sin \theta = 1.22 \dfrac{\lambda_{8500}}{D} = 1.22 \left(\dfrac{0.040 \text{ m}}{0.30 \text{ m}} \right) = 0.16$ (17.2)

$\theta = \sin^{-1} 0.16 = \boxed{9.2°}$

Figure 17.13 illustrates these results.

(b) To find the speaker diameter that will give a $\theta = 70°$ dispersion to the 8500-Hz sound, we again use the relation $\sin \theta = 1.22(\lambda/D)$:

$$D = \frac{1.22 \, \lambda_{8500}}{\sin \theta} = \frac{1.22 \, (0.040 \text{ m})}{\sin (70°)} = \boxed{0.052 \text{ m}}$$

This result shows that high-frequency sound can have a wide dispersion, *provided the diameter of the speaker is small enough.* Accordingly, loudspeaker designers use a small diameter speaker called a tweeter to generate high-frequency sound, as Figure 17.14 indicates.

THE PHYSICS OF . . .

tweeter loudspeakers.

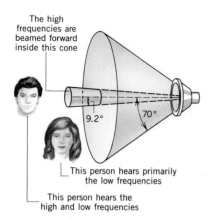

Figure 17.13 Because the dispersion of high frequencies is less than that of low frequencies, you should be directly in front of the speaker to hear both the high and low frequencies equally well.

Figure 17.14 Small-diameter speakers, called tweeters, are used to produce high-frequency sound. The small diameter helps to promote a wider dispersion of the sound.

17.4 BEATS

Constructive or destructive interference can occur when two sound waves with *the same frequency* overlap. Here we consider what happens when two waves with *slightly different frequencies* overlap. Once again, the principle of linear superposition will be our guide. In Figure 17.15 the waves come from two tuning forks placed side by side. A tuning fork has the property of producing a single-frequency sound wave when struck with a sharp blow. The two tuning forks in the drawing are identical, and each is designed to produce a 440-Hz tone. However, a small piece of putty has been attached to one fork, whose frequency is lowered to 438 Hz because of the added mass. When the forks are sounded simultaneously, the loudness of the resulting sound rises and falls periodically — faint, then loud, then faint, then loud, and so on. The periodic variations in loudness are called *beats* and result from the interference between two sound waves with slightly different frequencies.

For clarity, Figure 17.15 shows the condensations and rarefactions of the sound waves separately. In reality, however, the waves spread out and overlap. In accord with the principle of linear superposition, the ear detects the combined total of the two. Notice that there are places where the waves interfere constructively and places where they interfere destructively. When a region of constructive interference reaches the ear, a loud sound is heard. When a region of destructive interference arrives, the sound intensity drops to zero (assuming each of the waves has the same amplitude). The number of times per second that the loudness rises and falls is the *beat frequency* and is the *difference* between the two sound frequencies. Thus, in the situation illustrated in Figure 17.15, an observer hears the sound loudness rise and fall at the rate of 2 times per second (440 Hz − 438 Hz).

Figure 17.16 helps to explain why the beat frequency is the difference between the two frequencies. The drawing displays graphical representations of the pressure patterns of a 10-Hz wave and a 12-Hz wave, along with the pressure pattern that results when the two overlap. These frequencies have been chosen for convenience, even though they lie below the audio range and are inaudible. Audible sound waves behave in exactly the same way. The top two drawings show the pressure variations in a 1-second interval of each wave. The third drawing shows the result of adding together the first two patterns according to the

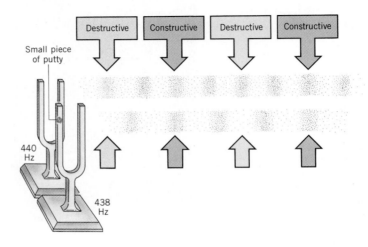

Figure 17.15 Two tuning forks have slightly different frequencies of 440 and 438 Hz. The phenomenon of beats occurs when the forks are sounded simultaneously. The sound waves are not drawn to scale.

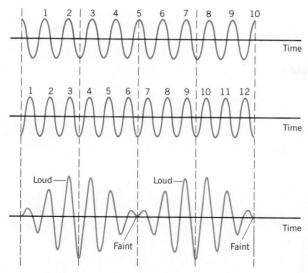

Figure 17.16 A 10-Hz sound wave and a 12-Hz sound wave, when added together, produce a wave with a beat frequency of 2 Hz. The drawings show the pressure patterns of the individual waves and the pressure pattern (in red) that results when the two overlap. The time interval shown is one second.

principle of linear superposition. Notice that the amplitude in the third drawing is not constant, as it is in the individual waves. Instead, the amplitude changes from a minimum to a maximum, back to a minimum, etc. When such pressure variations reach the ear and occur in the audible frequency range, they produce a loud sound when the amplitude is a maximum, and a faint sound when the amplitude is a minimum. Two loud-faint cycles, or beats, occur in the 1-second interval shown in the drawing, corresponding to a beat frequency of 2 Hz. This is consistent with our earlier statement that the beat frequency is the difference between the frequencies of the individual waves, that is, 12 Hz − 10 Hz = 2 Hz.

Musicians often tune their instruments by listening to a beat frequency. For instance, a guitar player sounds an out-of-tune string along with a tone from a source known to have the correct frequency. The guitarist adjusts the tension in the string until the beats vanish, ensuring that the string is vibrating at the correct frequency.

THE PHYSICS OF . . .

tuning a musical instrument.

17.5 TRANSVERSE STANDING WAVES

A standing wave is another effect that can occur when two waves travel through the same place at the same time. Figure 17.17 shows some of the essential features of transverse standing waves. In this figure one end of a string is attached to a wall, and the other end is vibrated back and forth. Regions of the string move so fast that they appear only as a "blur" in the photographs. Each of the patterns shown is called a ***standing wave pattern.*** Notice that the patterns include special places called nodes and antinodes. The ***nodes*** are places that do not vibrate at all, and the ***antinodes*** are places where maximum vibration occurs. Figure 17.18 shows a series of superimposed drawings for each of the patterns in Figure 17.17. These drawings freeze the shape of the string at various times and emphasize the maximum vibration that occurs at an antinode with the aid of a red "dot" attached to the string.

Each standing wave pattern is produced at a unique frequency of vibration. These unique frequencies form a series, the smallest frequency f_1 corresponding to

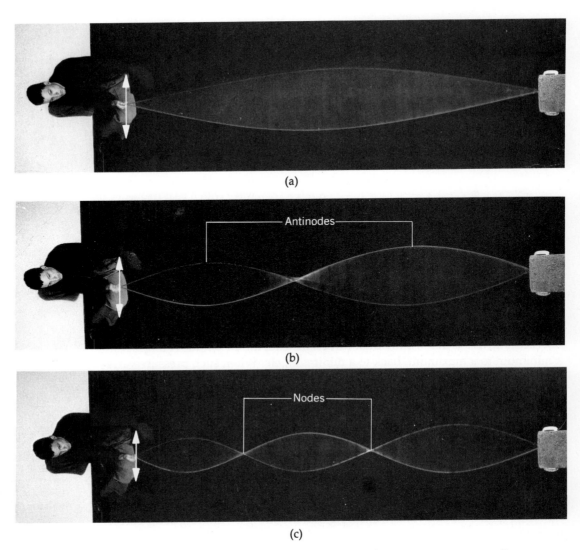

Figure 17.17 Vibrating a string at certain unique frequencies sets up standing wave patterns, such as the three shown here. Patterns (*b*) and (*c*) result, respectively, by vibrating the string at two and three times the frequency used in (*a*).

Figure 17.18 This illustration shows a series of superimposed drawings for each of the standing wave patterns in Figure 17.17. The red dots attached to the string at an antinode focus attention on the maximum vibration that occurs there.

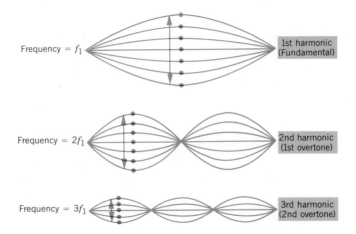

the one-loop pattern and the larger frequencies being integer multiples of f_1, as Figure 17.18 indicates. Thus, if f_1 is 10 Hz, the frequency needed to establish the 2-loop pattern is $2f_1$ or 20 Hz, while that needed to create the 3-loop pattern is $3f_1$ or 30 Hz, and so on. The frequencies in this series (f_1, $2f_1$, $3f_1$, etc.) are called *harmonics.* The lowest frequency f_1 is called the first harmonic, and the higher frequencies are labeled as the second harmonic ($2f_1$), the third harmonic ($3f_1$), and so forth. The harmonic number (1st, 2nd, 3rd, etc.) corresponds to the number of loops in the standing wave pattern. The frequencies in this series are also referred to as the fundamental frequency, the first overtone, the second overtone, and so on. Thus, frequencies above the fundamental are *overtones* (see Figure 17.18).

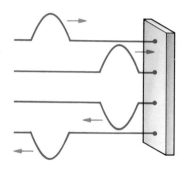

Figure 17.19 In reflecting from the wall, a forward-traveling half-cycle becomes a backward-traveling half-cycle that is inverted.

The standing wave patterns in Figure 17.17 arise because *identical* waves travel on the string in *opposite directions* and combine in accord with the principle of linear superposition. A standing wave is said to be "standing," since it does not travel in one direction or the other, as do the individual waves that produce it. Figure 17.19 shows why there are waves traveling in both directions on the string. At the top of the picture, one-half of a wave cycle (the remainder of the wave is omitted for clarity) is moving toward the wall on the right. When the half-cycle reaches the wall, it causes the string to pull upward on the wall. Consistent with Newton's action–reaction law, the wall pulls downward on the string, and a downward-pointing half-cycle is sent back toward the left. Thus, the wave reflects from the wall. Upon arriving back at the point of origin, the wave reflects again, this time from the hand vibrating the string. The hand is essentially fixed (the vibration amplitude of the hand is assumed to be small) and behaves as the wall does in causing reflections. Repeated reflections at both ends of the string create a multitude of wave cycles traveling in both directions.

As each new cycle is formed by the vibrating hand, previous cycles that have reflected from the wall arrive and reflect again from the hand. Unless the timing is right, however, the new cycles and the reflected cycles tend to offset one another, and the formation of a standing wave is inhibited. Think about pushing someone on a swing and timing your pushes so that the effect of one push reinforces that of another. Such reinforcement in the case of wave cycles on the string leads to a large amplitude standing wave. Suppose the string has a length L and its left end is being vibrated at a frequency f_1. The time required to create a new wave cycle is the period T of the wave, where $T = 1/f_1$. On the other hand, the time needed for a cycle to travel from the hand to the wall and back, a distance of $2L$, is $2L/v$, where v is the wave speed. Reinforcement between new and reflected cycles occurs if these two times are equal; that is, if $1/f_1 = 2L/v$. Thus, a standing wave is established when the string is vibrated with a frequency of $f_1 = v/(2L)$.

Repeated reinforcement between newly created and reflected cycles causes a large amplitude standing wave to develop on the string, *even though the hand itself vibrates with only a small amplitude.* Thus, the motion of the string is a resonance effect, analogous to that discussed in Section 10.8 for an object attached to a spring. The frequency f_1 at which resonance occurs is sometimes called a *natural frequency* of the string, similar to the frequency at which an object oscillates on a spring.

There is a difference between the resonance of the string and the resonance of a spring system, however. An object on a spring has only a single natural frequency, whereas the string has a *series* of natural frequencies. The series arises because a reflected wave cycle need not return to its point of origin in time to reinforce *every* newly created cycle. Reinforcement can occur, for instance, on *every other* new cycle, as it does if the string is vibrated at twice the frequency f_1, or $f_2 = 2f_1$. Likewise, if the vibration frequency is $f_3 = 3f_1$, reinforcement occurs on *every third* new cycle. Similar arguments apply for any frequency $f_n = nf_1$, where n is an integer. As a result, the series of natural frequencies that lead to standing waves on a string fixed at both ends is given by

$$\begin{bmatrix} \text{String} \\ \text{fixed at} \\ \text{both ends} \end{bmatrix} \qquad f_n = n\left(\frac{v}{2L}\right) \qquad n = 1, 2, 3, 4, \ldots \qquad (17.3)$$

It is also possible to obtain Equation 17.3 in another way. In Figure 17.18, one-half of a wave cycle is outlined in red for each of the harmonics, to show that each "loop" in a standing wave pattern corresponds to one-half a wavelength. Since the two fixed ends of the string are nodes, the length L of the string must contain an integer number n of half-wavelengths: $L = n(\frac{1}{2}\lambda_n)$ or $\lambda_n = 2L/n$. Using this result for the wavelength in the relation $f_n\lambda_n = v$ shows that $f_n(2L/n) = v$, which can be rearranged to give Equation 17.3.

Standing waves on a string play an important role in the way many musical instruments produce sound. For instance, a guitar string is stretched between two supports and, when plucked, vibrates according to the series of natural frequencies given by Equation 17.3. The next example illustrates how this series of frequencies governs the design and playing of a guitar.

THE PHYSICS OF . . .

playing a guitar.

Example 3 Playing a Guitar

The heaviest string on an electric guitar has a linear density of $m/L = 5.28 \times 10^{-3}$ kg/m and is stretched with a tension $F = 226$ N. This string produces the musical note E when vibrating along its entire length in a standing wave at the fundamental frequency of 164.8 Hz. (a) Find the length L of the string between its two fixed ends (see Figure 17.20*a*). (b) A guitar player wants the string to vibrate at a fundamental frequency of 2×164.8 Hz = 329.6 Hz, as it must if the musical note E is to be sounded one octave higher in pitch. To accomplish this, he presses the string against the proper fret and then plucks the string (see part *b* of the drawing). Find the distance L between the fret and the bridge of the guitar.

REASONING   The fundamental frequency f_1 is given by Equation 17.3 with $n = 1$: $f_1 = v/(2L)$. Since f_1 is known in both parts (a) and (b), the length L in each case can be calculated directly from this expression, once the speed v is known. The speed, in turn, is related to the tension F and the linear density m/L according to Equation 16.2.

SOLUTION
(a) The speed is

$$v = \sqrt{\frac{F}{m/L}} = \sqrt{\frac{226 \text{ N}}{5.28 \times 10^{-3} \text{ kg/m}}} = 207 \text{ m/s} \qquad (16.2)$$

According to $f_1 = v/(2L)$, the length of the string is

$$L = \frac{v}{2f_1} = \frac{207 \text{ m/s}}{2(164.8 \text{ Hz})} = \boxed{0.628 \text{ m}}$$

(b) The distance L that locates the fret can be determined exactly as in part (a) by using the wave speed $v = 207$ m/s and noting that the frequency is now $f_1 = 329.6$ Hz: $\boxed{L = 0.314 \text{ m}}$. This length is exactly half that determined in part (a), because the frequencies have a ratio of 2:1. Similarly, all the frets on the neck of a guitar have precise locations, according to the notes associated with them (see problem 50).

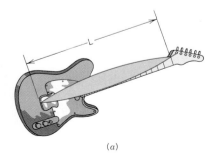

(a)

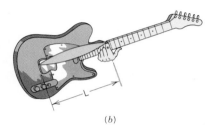

(b)

Figure 17.20   These drawings show the standing waves that exist on a guitar string under different playing conditions.

17.6 LONGITUDINAL STANDING WAVES

Standing wave patterns can also be formed from longitudinal waves traveling in opposite directions. For example, when sound reflects from a wall, the forward- and backward-going waves can produce a standing wave. Figure 17.21 illustrates the vibrational motion in a longitudinal standing wave on a Slinky. As in a transverse standing wave, there are nodes and antinodes. At the nodes the coils of the Slinky do not vibrate at all, while at the antinodes the coils vibrate with maximum amplitude, as indicated by the red dots in the picture. The vibration occurs along the line of travel of the individual waves, as is to be expected for longitudinal waves. In a standing wave of sound, the molecules behave as the red dots do.

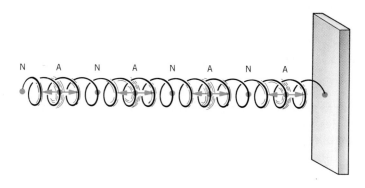

Figure 17.21 A longitudinal standing wave on a Slinky showing the displacement nodes (N) and antinodes (A).

Musical instruments in the wind family depend on longitudinal standing waves in producing sound. Since wind instruments (trumpet, flute, clarinet, pipe organ, etc.) are modified tubes or columns of air, it is useful to examine the standing waves that can be set up in such tubes. Figure 17.22 shows two cylindrical columns of air that are open at both ends. Sound waves, originating from a tuning fork, travel up and down within each tube, since they reflect from the ends of the tubes, even though the ends are open. If the frequency f of the tuning fork matches one of the natural frequencies of the air column, the downward- and upward-traveling waves combine to form a standing wave, and the sound of the

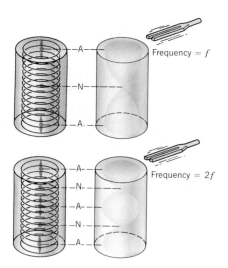

Figure 17.22 A pictorial representation of longitudinal standing waves on a Slinky (left side) and in a tube (right side) open at both ends (A, antinode; N, node).

tuning fork becomes markedly louder. To emphasize the longitudinal nature of the standing wave patterns, the left side of the drawing replaces the air in the tubes with Slinkies, on which the nodes and antinodes are indicated with red dots. As an additional aid in visualizing the standing waves, the right side of the drawing shows blurred blue patterns within each tube. These patterns symbolize the amplitude of the vibrating air molecules at various locations. Wherever the pattern is widest, the amplitude of vibration is greatest (a displacement antinode), and wherever the pattern is narrowest there is no vibration (a displacement node).

To determine the natural frequencies of the air columns in Figure 17.22, notice that there is a displacement antinode at each end of the tube, because the air molecules there are free to move.* As in a transverse standing wave, the distance between two successive antinodes is one-half of a wavelength, so the length L of the tube must be an integer number n of half-wavelengths: $L = n(\frac{1}{2}\lambda_n)$ or $\lambda_n = 2L/n$. Using this wavelength in the relation $f_n = v/\lambda_n$ shows that the natural frequencies f_n of the tube are

$$\begin{bmatrix} \text{Tube open} \\ \text{at both} \\ \text{ends} \end{bmatrix} \qquad f_n = n\left(\frac{v}{2L}\right) \qquad n = 1, 2, 3, 4, \ldots \qquad (17.4)$$

At these frequencies, large amplitude standing waves develop within the tube, due to resonance. Examples 4 and 5 illustrate how Equation 17.4 is involved when a flute and a piccolo are played.

* In reality, the antinode does not occur exactly at the open end. However, if the diameter of the tube is small compared to the length of the tube, little error is made in assuming that the antinode is located right at the end.

THE PHYSICS OF . . .

a flute.

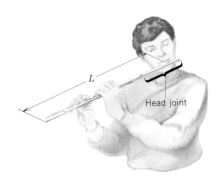

Figure 17.23 When all the holes are closed on a flute, the fundamental frequency of the lowest note is determined by the length L of the tube between the mouthpiece and the end, as well as the speed of sound in air. Geometrical factors within the head joint also have an effect, but they can be ignored as a first approximation.

Example 4   Playing a Flute

When all the holes are closed on a standard flute, the lowest note it can sound is a middle C, whose fundamental frequency is 261.6 Hz. (a) The air temperature is 293 K, and the speed of sound is 343 m/s. Assuming the flute is a cylindrical tube open at both ends, determine the distance L in Figure 17.23, that is, the distance from the mouthpiece to the end of the tube. (This distance is only approximate, since the antinode does not occur exactly at the mouthpiece.) (b) A flautist can alter the length of the flute by adjusting the extent to which the head joint is inserted into the main stem of the instrument. If the air temperature rises to 305 K, to what length must a flute be adjusted to play a middle C?

REASONING   The fundamental frequency f_1 is given by Equation 17.4 with $n = 1$: $f_1 = v/(2L)$. This expression can be used to calculate the length as $L = v/(2f_1)$. When the speed of sound v changes, as it does when the temperature changes, the length of the flute must be changed. The effect of temperature on the speed of sound in air is given by $v = \sqrt{\gamma kT/m}$ (Equation 16.5), assuming air behaves as an ideal gas. Thus, the speed is proportional to the square root of the Kelvin temperature ($v \propto \sqrt{T}$), a fact that we can use to find the speed at the higher temperature.

SOLUTION
(a) At a temperature of 293 K, when the speed of sound is $v = 343$ m/s, the length of the flute is

$$L = \frac{v}{2f_1} = \frac{343 \text{ m/s}}{2(261.6 \text{ Hz})} = \boxed{0.656 \text{ m}}$$

(b) Since $v \propto \sqrt{T}$, it follows that

$$\frac{v_{305\,K}}{v_{293\,K}} = \frac{\sqrt{305\ K}}{\sqrt{293\ K}} = 1.02$$

As a result, $v_{305\,K} = 1.02\,(v_{293\,K})$, so $v_{305\,K} = 1.02\,(343\ \text{m/s}) = 3.50 \times 10^2$ m/s. The adjusted flute length is

$$L = \frac{v}{2f_1} = \frac{3.50 \times 10^2 \text{ m/s}}{2(261.6 \text{ Hz})} = \boxed{0.669 \text{ m}}$$

Thus, to play in tune at the higher temperature, a flautist must lengthen the flute by 0.013 m.

Example 5 A Piccolo

THE PHYSICS OF . . .

a piccolo.

Assume that both a piccolo and a flute are cylindrical tubes with both ends open. The lowest fundamental frequency produced by a piccolo is 587.3 Hz, while that produced by a flute is 261.6 Hz. How long is a piccolo compared to a flute?

REASONING AND SOLUTION   The length of the piccolo can be compared to the length of the flute by noting that L is inversely proportional to the fundamental frequency, according to Equation 17.4: $L \propto 1/f_1$. Therefore,

$$\frac{L_{\text{piccolo}}}{L_{\text{flute}}} = \frac{(1/f_1)_{\text{piccolo}}}{(1/f_1)_{\text{flute}}} = \frac{1/(587.3 \text{ Hz})}{1/(261.6 \text{ Hz})} = \boxed{0.445}$$

The piccolo is slightly less than one-half the length of the flute.

Standing waves can also exist in a tube with only one end open, as the patterns in Figure 17.24 indicate. Note the difference between these patterns and those in Figure 17.22. Here the standing waves have a displacement antinode at the open end and a displacement node at the closed end, where the air molecules are not free to move. Since the distance between a node and an adjacent antinode is one-fourth of a wavelength, the length L of the tube must be an odd number of quarter-wavelengths: $L = 1(\frac{1}{4}\lambda)$ and $L = 3(\frac{1}{4}\lambda)$ for the two standing wave patterns in Figure 17.24. In general, then, $L = n(\frac{1}{4}\lambda_n)$, where n is any odd integer

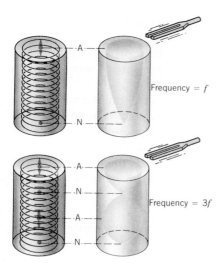

Frequency $= f$

Frequency $= 3f$

Figure 17.24 A pictorial representation of the longitudinal standing waves on a Slinky (left side) and in a tube of air (right side) open only at one end (A, antinode; N, node).

$(n = 1, 3, 5, \ldots)$. From this result it follows that $\lambda_n = 4L/n$, and the natural frequencies f_n can be obtained from the relation $f_n = v/\lambda_n$:

$$\begin{bmatrix} \text{Tube open} \\ \text{at only} \\ \text{one end} \end{bmatrix} \qquad f_n = n\left(\frac{v}{4L}\right) \qquad n = 1, 3, 5, \ldots \qquad (17.5)$$

A tube open at only one end can develop standing waves only at the odd harmonic frequencies f_1, f_3, f_5, etc. In contrast, a tube open at both ends can develop standing waves at all harmonic frequencies f_1, f_2, f_3, etc. Moreover, the fundamental frequency f_1 of a tube is half as high when the tube has only one end open compared to when it has both ends open. In other words, a tube open at both ends must be *twice* as long as a tube open at only one end in order to produce the *same* fundamental frequency.

*17.7 COMPLEX SOUND WAVES

Musical instruments, such as those in the Stanford University band shown here, produce complex sound waves that consist of a fundamental frequency and overtone frequencies.

Musical instruments produce sound in a way that depends on standing waves. Example 3 illustrates the role of transverse standing waves on the string of an electric guitar, while Examples 4 and 5 stress the role of longitudinal standing waves in the air column within a flute and a piccolo. In each example sound is produced at the fundamental frequency of the instrument.

In general, however, a musical instrument does not produce just the fundamental frequency when it plays a note, but generates a number of harmonics as well. Different instruments, such as a flute and a trumpet, generate harmonics to different extents, and the harmonics give the instruments their characteristic sound qualities or timbres. Suppose, for instance, that a flute player and a trumpet player both sound concert A, a note whose fundamental frequency is 440 Hz. Even though both instruments are playing the same note, most people can distinguish the sound of the flute from that of the trumpet. The instruments sound different because the relative amplitudes of the harmonics (880 Hz, 1320 Hz, etc.) that the instruments create are different.

The sound wave corresponding to a note produced by a musical instrument or a singer is called a **complex sound wave**, because it consists of a mixture of the fundamental and harmonic frequencies. The pattern of pressure fluctuations in a complex wave can be obtained by using the principle of linear superposition, as Figure 17.25 indicates. This drawing shows a bar graph in which the heights of the bars give the relative amplitudes of the harmonics contained in a note such as a singer might produce. When the individual pressure patterns for each of the six harmonics are added together, they yield the complex pressure pattern shown on the right side of the picture.*

In practice, a bar graph such as that in Figure 17.25 is determined with the aid of an electronic instrument known as a spectrum analyzer. When the note is produced, the complex sound wave is detected by a microphone that converts the wave into an electrical signal. The electrical signal, in turn, is fed into the spectrum analyzer, as Figure 17.26 illustrates. The spectrum analyzer then determines the amplitude and frequency of each harmonic present in the complex wave and displays the results on its screen.

---

* In carrying out the addition, we assume that each individual pattern begins at zero at the origin when the time equals zero.

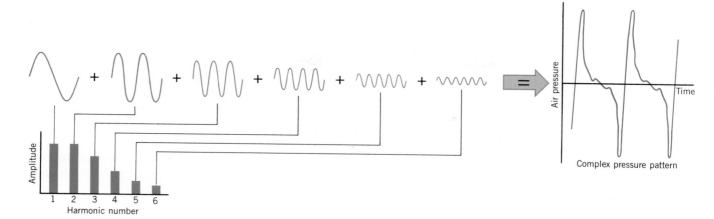

Figure 17.25   The graph on the right shows the pattern of pressure fluctuations such as a singer might produce in singing a note. The pattern is the sum of the first six harmonics. The relative amplitudes of the harmonics correspond to the heights of the vertical bars in the bar graph.

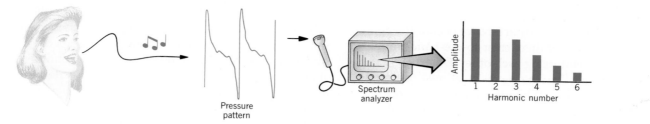

Figure 17.26   A microphone detects a complex sound wave, and a spectrum analyzer determines the amplitude and frequency of each harmonic present in the wave.

INTEGRATION OF CONCEPTS

INTERFERENCE OF WAVES AND THE CONSERVATION OF ENERGY

In Section 17.2 we have seen how sound waves from two loudspeakers interfere with each other to produce a resultant sound wave that is the sum of the individual waves. The individual sound waves from the speakers carry energy as they move, and the resultant wave contains the same energy as the sum of the energies of the individual waves. This fact is consistent with the principle of conservation of energy, which we first encountered in Section 6.8. This principle states that energy can neither be created nor destroyed, but can only be converted from one form to another. One of the interesting consequences of interference is that the energy is redistributed, so there are places within the listening area where the sound is loud and other places where there is no sound at all. Interference, so to speak, "robs Peter to pay Paul," but it always conserves energy in the process. Energy is also conserved when a standing wave is produced, either on a string or in a tube of air. The energy of the standing wave is the sum of the energies of the individual waves that comprise the standing wave. Once again, interference redistributes the energy of the individual waves to create locations of greatest energy (antinodes) and locations of no energy (nodes). The redistribution of energy also occurs when a wave is diffracted around an obstacle or the edges of an opening. As we have seen, diffraction is also an interference effect, one in which part of the energy of the wave is directed into regions that would not be accessible had interference not occurred. Interference and the redistribution of energy occurs for all types of waves, and we will see another example with light waves in Chapter 27. In all cases, the redistribution of energy that occurs is in accord with the principle of conservation of energy.

SUMMARY

The **principle of linear superposition** states that when two or more waves are present simultaneously at the same place, the resultant wave is the sum of the individual waves. **Constructive interference** occurs at a point when two waves meet there crest-to-crest and trough-to-trough, thus reinforcing each other. **Destructive interference** occurs when the waves meet crest-to-trough and cancel each other. When the waves meet crest-to-crest and trough-to-trough, they are **exactly in phase.** When they meet crest-to-trough, they are **exactly out of phase.**

Diffraction is the bending of a wave around an obstacle or the edges of an opening. The angle through which the wave bends depends on the ratio of the wavelength λ of the wave to the width D of the opening; the greater the ratio λ/D, the greater the angle.

Beats are the periodic variations in amplitude that arise from the linear superposition of two waves that have slightly different frequencies. When the waves are sound waves, the variations in amplitude cause the loudness to vary at the **beat frequency,** which is the difference between the frequencies of the waves.

A transverse or longitudinal **standing wave** is the pattern of disturbance that results when oppositely traveling waves of the same frequency and amplitude pass through each other. A standing wave has places of minimum and maximum vibration called, respectively, **nodes** and **antinodes.** Under resonance conditions, standing waves can be established only at certain frequencies f_n, known as the **natural frequencies.** For a string that is fixed at both ends and has a length L, the natural frequencies are $f_n = n(v/2L)$, where v is the speed of the wave on the string and n is any positive integer, $n = 1, 2, 3, \ldots$. For a gas in a cylindrical tube open at both ends, the natural frequencies are given by the same expression, where v is the speed of sound in the gas and $n = 1, 2, 3, \ldots$. However, if the cylindrical tube is open only at one end, the natural frequencies are $f_n = n(v/4L)$, where $n = 1, 3, 5, \ldots$.

A **complex sound wave** consists of a mixture of a fundamental frequency and overtone frequencies.

QUESTIONS

1. Does the principle of linear superposition imply that two sound waves, passing through the same place at the same time, always create a louder sound than either wave alone? Explain.

2. Suppose you are sitting at the overlap point between the two speakers in Figure 17.5. Because of destructive interference, you hear no sound, even though both speakers are emitting identical sound waves. One of the two speakers is suddenly shut off. Describe what you would hear.

3. In Figures 17.3 and 17.5, it has been assumed that the speaker diaphragms vibrate synchronously, that is, they move outward together and inward together. This is achieved by connecting the wires between the amplifier and each speaker in exactly the same way. Then, the speakers are said to be "phased correctly." It is possible, however, to connect the wires so that one diaphragm moves outward when the other moves inward. Explain what the listeners in Figure 17.3 and 17.5 would then hear.

4. Refer to Example 1 in Section 16.2. Which type of radio wave, AM or FM, diffracts more readily around a given obstacle? Give your reasoning.

5. There are three tuning forks, each with a different frequency. Two of the three are sounded together and produce beats. Is it possible for the beat frequency to be the same, no matter which pair of tuning forks are chosen? Explain.

6. Take a rubber band and fasten one end to an immovable object. Slowly stretch the rubber band, all the while plucking it to produce a sound. Explain why the frequency of the sound changes.

7. A string is attached to a wall and vibrated back and forth, as in Figure 17.17. The vibration frequency and length of the string are fixed. The tension in the string is changed, and it is observed that at certain values of the tension a standing wave pattern develops. Account for the fact that no standing waves are observed once the tension is increased beyond a certain value.

8. By blowing across the top of an empty soda bottle, a relatively loud sound can be created, because a standing wave develops in the bottle. Explain why the pitch of the sound is higher when the bottle is partially filled with water.

9. In terms of length, why is it an advantage to use a pipe closed at one end, rather than a pipe open at both ends, to generate the lowest audible frequency on a pipe organ?

10. Standing waves can ruin the acoustics of a concert hall if there is excessive reflection of the sound waves that the performers generate. For example, suppose a performer generates a 2093-Hz tone. If a large amplitude standing wave is present, it is possible for a listener to move a distance of only 4.1 cm and hear the loudness of the tone change from loud to faint. Account for this observation in terms of standing waves, pointing out why the distance is 4.1 cm.

11. The tones produced by a typical orchestra are complex sound waves, and most have fundamental frequencies less than 5000 Hz. However, a high-quality stereo system must be able to reproduce frequencies up to 20 000 Hz accurately. Explain why.

12. The sound of a normal voice is a complex wave, consisting of a fundamental frequency plus various overtones that create a characteristic voice quality. When the lungs are filled with helium instead of air, the voice has a high-pitched quality, somewhat like Donald Duck's voice. Assume the sound of the voice is generated by the vocal cords vibrating above a gas-filled tube closed at one end. Using data from Table 16.1, explain why helium has the effect of raising the pitch.

PROBLEMS

Section 17.2 Constructive and Destructive Interference of Sound Waves

1. The drawing shows a string on which two rectangular pulses are traveling at a constant speed of 1 cm/s at time $t = 0$. Using the principle of linear superposition, draw the shape of the string at $t = 1$ s, 2 s, 3 s, 4 s, and 5 s.

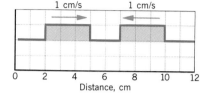

2. Repeat problem 1, assuming the pulse on the right is pointing downward, rather than upward.

3. Repeat problem 1, assuming the pulses have the shape (half up and half down) shown in the drawing.

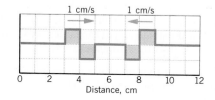

4. In Figure 17.9, the two speakers are separated by 3.20 m, and both are reproducing identical 214-Hz tones. The speed of sound is 343 m/s. Suppose point C is 6.00 m directly in front of speaker B, instead of the 2.40 m shown in the drawing. Does constructive or destructive interference occur at point C?

5. Two loudspeakers are set up as in Figure 17.9, and point C is located as shown there. The speed of sound is 343 m/s. The speakers play the same tone. What is the smallest frequency that will produce destructive interference at point C?

*6. Refer to the drawing for problem 1. What is the value of the time t when the centers of the pulses coincide?

*7. Suppose the two speakers in Figure 17.8 are separated by a distance of 3.000 m. Consider a point P that is in front of the speakers and whose perpendicular distance from the line joining the speakers is 2.200 m. The point P is between the speakers and is 2.500 m from the speaker on the right. The speed of sound is 343 m/s. Does constructive or destructive interference occur at P when the speakers produce identical sound waves whose frequency is (a) 1466 Hz and (b) 977 Hz?

Section 17.3 Diffraction

8. The width D of a diffraction horn loudspeaker is 0.050 m. The speed of sound is 343 m/s. At what frequency is the diffraction angle θ equal to 45°?

9. A speaker has a diameter of 0.30 m. (a) Assuming the speed of sound is 343 m/s, find the diffraction angle θ for a 2.0-kHz tone and for a 6.0-kHz tone. (b) What speaker diameter D should be used to generate a 6.0-kHz tone whose diffraction angle is as wide as that for the 2.0-kHz tone in part (a)?

10. A circular speaker produces a 1250-Hz tone, and the diffraction angle θ is 65°. The speed of sound is 343 m/s. Determine the diameter of the speaker.

*11. A 4.00-kHz tone is being produced by a speaker with a diameter of 0.150 m. The air temperature changes from 0 to 35 °C. Assuming air to be an ideal gas, find the *change* in the diffraction angle θ.

Section 17.4 Beats

12. Two ultrasonic sound waves combine and form a beat frequency that is in the range of human hearing. The frequency of one of the ultrasonic waves is 70 kHz. What is (a) the smallest possible and (b) the largest possible value for the frequency of the other ultrasonic wave?

13. A tuning fork vibrates at a frequency of 521 Hz. An out-of-tune piano string vibrates at 519 Hz. How much time separates successive beats?

14. Two guitars are slightly out of tune. When they play the same note simultaneously, the sounds they produce have wavelengths of 0.776 and 0.769 m. On a day when the speed of sound is 343 m/s, what beat frequency is heard?

15. Two pure tones are sounded together. The drawing shows the pressure variations of the two sound waves, measured with respect to atmospheric pressure. What is the beat frequency?

16. When a guitar string is sounded along with a 440-Hz tuning fork, a beat frequency of 5 Hz is heard. When the same string is sounded along with a 436-Hz tuning fork, the beat frequency is 9 Hz. What is the frequency of the string?

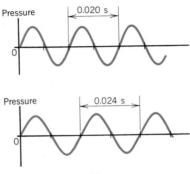

Problem 15

*17. Each of three tuning forks A, B, and C has a slightly different frequency. When A and B are sounded together, they produce a beat frequency of 2 Hz. When A and C are sounded together they produce a beat frequency of 5 Hz. What is the beat frequency that occurs when B and C are sounded together? There are two possible answers.

**18. Two loudspeakers are mounted on a merry-go-round whose radius is 9.00 m. The speakers both play a tone whose true frequency is 100.0 Hz. As the drawing illustrates, they are situated at opposite ends of a diameter. The speed of sound is 343 m/s, and the merry-go-round revolves once every 20.0 s. What is the beat frequency that is detected by the listener when the merry-go-round is near the position shown?

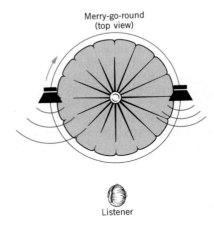

Merry-go-round
(top view)

Listener

Section 17.5 Transverse Standing Waves

19. A stretched rubber band has a length of 0.10 m and a fundamental frequency of 440 Hz. What is the speed at which waves travel on the rubber band?

20. If the end of the string in Figure 17.17 is vibrated at a frequency of 3.0 Hz and the distance between two successive nodes is 0.23 m, what is the speed of the waves on the string?

21. On a cello, the string with the largest linear density $(1.56 \times 10^{-2} \text{ kg/m})$ is the C string. This string produces a fundamental frequency of 65.4 Hz and has a length of 0.800 m between the two fixed ends. Find the tension in the string.

22. For the cello C string in problem 21, find the time required for a wave to travel the length of the string.

23. The lowest note on a piano has a fundamental frequency of 27.5 Hz and is produced by a wire that has a length of 1.18 m. The speed of sound in air is 343 m/s. Determine the ratio of the wavelength of the sound wave to the wavelength of the waves that travel on the wire.

24. The G string on a guitar has a fundamental frequency of 196 Hz and a length of 0.62 m. This string is pressed against the proper fret to produce the note C, whose fundamental frequency is 262 Hz. What is the distance L between the fret and the end of the string at the bridge of the guitar (see Figure 17.20b)?

25. Ideally, the strings on a violin are stretched with the same tension. Each has the same length between its two fixed ends. The musical notes and corresponding fundamental frequencies of two of these strings are G (196.0 Hz) and E (659.3 Hz). The linear density of the E string is $3.47 \times 10^{-4} \text{ kg/m}$. What is the linear density of the G string?

*26. A length of steel wire for a bass guitar has the following specifications: cross-sectional area $= 5.0 \times 10^{-7} \text{ m}^2$, density $= 7860 \text{ kg/m}^3$. (This is *not* the linear density; it is the mass per unit volume.) The wire is stretched between two fixed supports that are 0.90 m apart, and the second harmonic is 196 Hz. What is the tension in the wire?

*27. A copper block is suspended in air from a wire. As the drawing shows, a container of mercury is then raised up around the block, until the fundamental frequency of the wire is reduced by a factor of two. Determine the ratio h/h_0 that gives the fraction of the block immersed in the mercury. *(Hint: See Table 11.1 for density values and take advantage of Archimedes' principle.)*

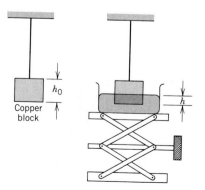

**28. The note that is three octaves above middle C is supposed to have a fundamental frequency of 2093 Hz. On a certain piano the steel wire that produces this note has a cross-sectional area of $7.85 \times 10^{-7} \text{ m}^2$. The wire is stretched between two pegs. When the piano is tuned properly to produce the correct frequency at 25.0 °C, the wire is under a tension of 818.0 N. Suppose the temperature drops to 20.0 °C. In addition, as an approximation, assume the wire is kept from contracting as the temperature drops. Consequently, the tension in the wire changes. What beat frequency is produced when this piano and another instrument (properly tuned) sound the note simultaneously?

Section 17.6 Longitudinal Standing Waves

29. A cylindrical tube sustains standing waves at the following frequencies: 500, 700, and 900 Hz. There are no standing waves at frequencies of 600 and 800 Hz. (a) What is the fundamental frequency? (b) Is the tube open at both ends or just at one end?

30. The fundamental frequency of a vibrating system is 400 Hz. For each of the following systems, give the three lowest frequencies (excluding the fundamental) at which standing waves can occur: (a) a string fixed at both ends, (b) a cylindrical pipe with both ends open, and (c) a cylindrical pipe with only one end open.

31. A tube of air is open at only one end and has a length of 1.5 m. This tube sustains a standing wave at its third harmonic. What is the distance between one node and the adjacent antinode?

32. An organ pipe, open at both ends, produces the middle C note (262 Hz) when sustaining a standing wave at its third harmonic. The speed of sound is 343 m/s. What is the length of the pipe?

33. The fundamental frequencies of two air columns are the same. Column A is open at both ends, while column B is open at only one end. The length of column A is 0.60 m. What is the length of column B?

34. One method for measuring the speed of sound uses standing waves. A cylindrical tube is open at both ends, and one end is placed against a loudspeaker. A movable plunger is inserted into the other end. The distance between the loudspeaker and the plunger is L. When the loudspeaker generates a 485-Hz tone, the smallest value of L for which a standing wave is formed is 0.264 m. What is the speed of sound in the gas in the tube? *(Hint: The plunger closes one end of the tube.)*

35. A tube is open only at one end, has a length of 2.50 m, and is at a temperature of 310 K. The speed of sound at 293 K is 343 m/s. Assuming air is an ideal gas, find the fundamental frequency of the tube.

36. A tube, open at both ends, contains an unknown ideal gas for which $\gamma = 1.40$. At 293 K, the shortest tube in which a standing wave can be set up with a 294-Hz tuning fork has a length of 0.248 m. Find the mass of a molecule.

* **37.** A cylindrical pipe is *closed at both ends*. Derive an expression for the frequencies of the allowed standing waves, similar in form to Equations 17.4 and 17.5, in terms of the speed of sound v, the length of the pipe L, and the harmonic number n. State which integer values of n are allowed.

* **38.** A tunnel leading straight through a hill makes tones at 135 and 165 Hz especially loud because of standing waves. Assuming the speed of sound is 343 m/s, find the smallest length the tunnel could have. (*Hint: The ratio of the frequencies is 9/11.*)

* **39.** Two loudspeakers facing each other are producing identical 440-Hz tones. A listener walks from one speaker toward the other at a constant speed and hears the loudness change at a frequency of 3.0 Hz. The speed of sound is 343 m/s. What is the walking speed?

** **40.** A tube, open at only one end, is cut into two shorter (nonequal) lengths. The piece open at both ends has a fundamental frequency of 425 Hz, while the piece open only at one end has a fundamental frequency of 675 Hz. What is the fundamental frequency of the original tube?

ADDITIONAL PROBLEMS

41. The A string on a string bass is tuned to vibrate at a fundamental frequency of 55 Hz. If the tension in the string were changed so the fundamental frequency became 110 Hz, what would be the ratio of the new tension to the original tension?

42. The range of human hearing is roughly from twenty hertz to twenty kilohertz. Based on these limits and a value of 343 m/s for the speed of sound, what are the lengths of the longest and shortest pipes (open at both ends and producing sound at their fundamental frequencies) that you expect to find in a pipe organ?

43. Review Example 1 in the text. Speaker *A* is moved further to the left, while *ABC* remains a right triangle. What is the separation between the speakers when constructive interference occurs again at point *C*?

44. When a tuning fork is sounded together with a 492-Hz tone, a beat frequency of 2 Hz is heard. Then a small piece of putty is stuck to the tuning fork, and the tuning fork is again sounded along with the 492-Hz tone. The beat frequency decreases. What is the frequency of the tuning fork?

45. The sound produced by the loudspeaker in the drawing has a frequency of 5700 Hz and arrives at the microphone

via two different paths. The sound travels through the left tube *LXM*, which has a fixed length. Simultaneously, the sound travels through the right tube *LYM*, the length of which can be changed by moving the sliding section. As the length of the path *LYM* is changed, the sound loudness detected by the microphone changes. When the sliding section is pulled out by 0.025 m, the loudness changes from a maximum to a minimum. Find the speed at which sound travels through the gas in the tube.

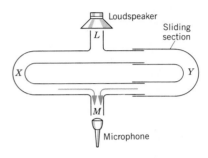

46. Both neon (Ne) and helium (He) are monatomic gases and can be assumed to be ideal gases. The fundamental frequency of a tube of neon is 268 Hz. What is the fundamental frequency of the tube if the tube is filled with helium, all other factors remaining the same?

47. Sometimes, when a wind blows across a long wire, a low-frequency "moaning" sound is produced. This sound arises because a standing wave is set up on the wire, like a standing wave on a guitar string. Assume that a wire (linear density $= 0.029$ kg/m) sustains a tension of 22 N, because the wire is stretched between two poles that are 45 m apart. The lowest frequency that a human ear can detect is about 20.0 Hz. What is the lowest harmonic number that could be responsible for the "moaning" sound?

* **48.** Two tuning forks X and Y have different frequencies and produce an 8-Hz beat frequency when sounded together. When X is sounded along with a 392-Hz tone, a 3-Hz beat frequency is detected. When Y is sounded along with the 392-Hz tone, a 5-Hz beat frequency is heard. What are the frequencies f_X and f_Y when (a) f_X is greater than f_Y and (b) f_X is less than f_Y?

* **49.** A vertical tube is closed at one end and open to air at the other end. The air pressure is 1.01×10^5 Pa. The tube has a length of 0.75 m. Mercury (mass density $= 13\ 600$ kg/m^3) is poured into it to shorten the effective length for standing waves. What is the absolute pressure at the bottom of the mercury column, when the fundamental frequency of the shortened, air-filled tube is equal to the third harmonic of the original tube?

**50. As the drawing shows, the length of a guitar string is 0.628 m. Note that the frets are numbered for convenience. A performer can play a musical scale on a single string, because the spacing *between the frets* is designed according to the following rule: When the string is pushed against any fret j, the fundamental frequency of the shortened string is larger by a factor of the twelfth root of two ($\sqrt[12]{2}$) than it is when the string is pushed against the fret $j - 1$. Assuming the tension in the string is the same for any note, find the spacing (a) between fret 1 and fret 0 and (b) between fret 7 and fret 6.

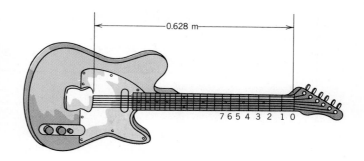

Physics & Music

For centuries, long before its physical principles were known, music brought pleasure to many people. Today, with the physics of sound firmly understood, the art of making musical sounds has become a science, resulting in a phenomenal growth in the variety and capabilities of musical instruments.

Imagine that you are a keyboard musician in a rock band. Your training probably began on the piano or guitar, which use vibrating strings to create standing waves of different wavelengths. From studies of acoustics (the science of sound) we know that each string on a piano or any stringed instrument creates overtones as well as its fundamental frequency. Overtones are multiples of the fundamental frequency, and combinations of overtones with different intensities give each

stringed instrument its distinctive sound.

You weren't satisfied with the normal range of sound on your instrument, so you began experimenting by striking its strings with different implements and at different places. They all created notes with the same fundamental frequencies but different timbres. But even that wasn't enough; climbing over a piano was cumbersome, and you wanted a greater range of sounds. In college you learned about synthesized sounds and electronic keyboards. By understanding how sound is generated, and how different instruments combine different frequencies of sound waves to create their characteristic tones, physicists and engineers have learned to duplicate the sounds electronically — not just by hit and

Compact disc (CD) recordings use digital techniques for reproducing music. Physics principles play a fundamental role in the way these techniques are used to transform the information on a CD into sound.

miss, but by frequency analysis of the actual sounds. The result is the electronic keyboard technology available today. By combining digital electronic re-creation of sound with high-quality stereo amplifiers and speakers, the sound of virtually any instrument can be produced. That, then, is your medium: computer-controlled electronic keyboards.

Your band works on the cutting edge of musical creativity and your lead guitarist, who writes most of the music, wants you to create sounds that no instrument has ever made. You sit down at your computer and program it to generate them. By understanding the physical principles on which music is based, you can extend the type of sounds beyond what has ever been heard before.

Consider, too, the recording of your band's music. The ability to reproduce sound accurately has come a long way in the past one hundred years. Up until a few years ago, you could be pretty sure that your music on a record would soon be accompanied by annoying clicks and scratches. The more a record was used, the more the needle damaged it, and the more noise you heard. Not any more. Your music from a compact disc (CD) sounds as good the thousandth time you play it as it did the first time, and on the first time the music sounds fantastic.

Consider some of the applications of physics to the recording of your work: Thomas Edison's first record player in 1877 was a strictly mechanical device, converting sound waves directly into mechanical vibrations, which were than stored on metal or glass cylinders. Playing back required the needle to press firmly into the vibration groove on the cylinder, so that the sound horn could mechanically amplify the needle's vibrations and could re-create sound (longitudinal waves). However, the recording process was noisy, and the record wear caused by the needle was rapid and extreme. But it was a start.

Electronic amplification of sound reduced the force that the needle had to exert on the groove, thereby dramatically decreasing record wear. The needle, attached to a ''cartridge,'' rests gently on the record. The needle is coupled to tiny magnets that are surrounded by coils of wire. In one design, the magnets move with the vibrations of the record. The changing magnetic field creates a tiny electric current in the wire, which is then electronically amplified. Magnets in the speakers, driven by currents from the amplifier, convert the electrical signal back into sound. Yet, inevitably, the needle, dropped or bumped, or running over pieces of dust on the record, causes records to wear. And besides record wear, the sound is often imperfect, because the recording process itself added noise that was then included on the record.

Finally, with the advent of CDs, the needle was removed completely and replaced by a laser beam. Recording devices and CDs were both digitized (the sound encoded in numbers), thereby adding minimal noise to any recorded material. Reconstructing sound waves from numbers on the CD is now a fine art, and even with scratches and dust on the disc, the electronic circuitry of the CD players is sophisticated enough to convert the numbers on the disc into excellent reproductions of your music. The result is music reproduced with a clarity and lack of noise unprecedented in sound reproduction history.

What other basic physics can be used to improve recording technology? The last major hurdle is to overcome the need to rotate a disc, for the equipment required is relatively cumbersome, heavy, expensive, and unreliable, especially when the players are portable units. With advances in solid state physics —the physics of material properties —solid state storage devices, from which the data can be taken without any physical movement, have now been developed. When these devices are commercially available, it will be possible to record and to play back with no moving parts.

FOR INFORMATION ABOUT CAREERS IN SOUND TECHNOLOGY

Careers applying sound to technology include engineering physics, electrical and electronics engineering, many technician's jobs, those of performing artists, and many jobs in the recording industry. Job Descriptions can be found in the **Encyclopedia of Careers and Vocational Guidance,** *Vol. 2, 8th Edition, William E. Hopke, ed., 1990, J. G. Ferguson, publisher.*

Many references, with addresses and phone numbers, for career information in engineering and performing arts are provided in the **Professional Careers Sourcebook.** *K. M. Savage and C. A. Dorgan, eds., 1990, Gale Research, Inc., publisher.*

Other places to write for career information are the following:

- *American Physical Society, 335 E. 45th St., New York, NY 10017*
- *Audio Engineering Society, 60 East 42nd St., Room 2520, New York, NY 10065*
- *National Academy of Recording Arts and Sciences, 303 N. Glenoaks Blvd., Burbank, CA 91502*

Copyright © 1991 by Neil F. Comins.

PART FOUR

ELECTRICITY AND MAGNETISM

In a very real sense, electricity and magnetism have made modern civilization what it is today. Take them away and you take away much of what fills our houses and our lives. For instance, electricity gives us virtually all of our artificial lighting, much of our heating, and many time-saving household appliances. Magnetism is an integral part of the motors found in an enormous number of useful devices, including fans, hair driers, vacuum cleaners, pumps, and food processors. It is also an integral part of the transformers that play a vital role in the distribution of electric power throughout the country.

The widespread impact of electricity and magnetism on society has come about because they are closely related. One of the most exciting discoveries in physics was that electricity and magnetism are both manifestations of the same fundamental force, which is called the electromagnetic force. The close interplay between electricity and magnetism has led to a far greater number of practical applications than would otherwise have been possible. The entire magnetic tape recording industry is based on the relationship between them. Medical diagnostic techniques, such as magnetic resonance imaging, depend on that relationship. The commercial generation of electric power depends on it too. And so does radio, television, and communications technology, as well as the information storage technology that is the foundation of the computer industry.

In this part of the text, we will begin by discussing the electrically charged particles that are found in nature. We will see that these particles, whether moving or at rest, exert forces on one another. These forces are the electric aspect of the electromagnetic force. In addition, the magnetic aspect appears when the charged particles are moving. Along with familiar concepts such as force and energy, we will encounter some new and useful concepts as we learn about the fundamental laws that govern electricity and magnetism.

Many of the applications of electricity and magnetism occur in the form of electric circuits, such as those in automobiles, airplanes, television sets, stereo systems, and computers. We will discuss circuits in two places. The first will occur after we study basic electricity. The second will follow after we study magnetism and are in a position to understand some of the circuit components that depend on magnetism.

CHAPTER 18

ELECTRIC FORCES AND ELECTRIC FIELDS

Lightning is nature's most spectacular display of electricity. The discovery of electricity can be traced to the Greek philosopher Thales (640? – 546 B.C.). He found that after amber (a petrified tree resin) is rubbed with wool or fur, the amber attracts small bits of leaves or straw. The origin of this attractive force is now known to be the electrical nature of matter itself. The word "electric" is, in fact, derived from the Greek word for amber (ēlektron). Despite the ancient roots of electricity, a correct understanding of its laws and the development of practical electrical devices is a scant 200 years old. In this chapter we will see that there are two kinds of electric charge, positive and negative, and they exert a special kind of force, called an electric force, on each other. It is this force that causes electric charges to flow between clouds and the earth and thereby causes lightning. It is also the electric force that holds together the positive and negative charges that make up atoms and molecules. Since our bodies are composed entirely of atoms and molecules, we owe our very existence to this force. In addition to exploring the properties of the electric force, we will show how it leads to the operation of many useful devices, such as photocopiers, laser printers, and electrostatic air cleaners.

18.1 THE ORIGIN OF ELECTRICITY

The electrical nature of matter is inherent in the atoms of all substances. An atom consists of a small, relatively massive nucleus that contains particles called protons and neutrons. Surrounding the nucleus is a diffuse cloud of orbiting particles known as electrons, as Figure 18.1 indicates. A proton has a mass of 1.67×10^{-27} kg, while an electron has a mass of 9.11×10^{-31} kg. Like mass, *electric charge* is an intrinsic property of protons and electrons, and only two types of charge have been discovered, positive and negative. A proton has a positive charge, and an electron has a negative charge.

Experiment reveals that the magnitude of the charge on the proton *exactly equals* the magnitude of the charge on the electron; the proton carries a charge $+e$, and the electron carries a charge $-e$. The SI unit for measuring the magnitude of an electric charge is the *coulomb* (C)*, and e has been determined experimentally to have the value

$$e = 1.60 \times 10^{-19} \text{ C}$$

In nature, atoms are normally found with equal numbers of protons and electrons. Usually, then, an atom carries no net charge, because the algebraic sum of the positive charge of the nucleus and the negative charge of the electrons is zero. When an atom, or any object, carries no net charge, the object is said to be *electrically neutral*. The neutrons in the nucleus are electrically neutral particles.

The charge on an electron or a proton is the *smallest* amount of free charge that has been discovered. Charges of larger magnitude are built up on an object by adding or removing electrons. Thus, any charge of magnitude q is an integer multiple of e; that is, $q = Ne$, where N is an integer. Because any electric charge q occurs in integer multiples of elementary, indivisible charges e, electric charge is said to be *quantized*. Example 1 emphasizes the quantized nature of electrical charge.

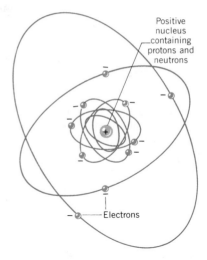

Figure 18.1 An atom contains a small, positively charged nucleus, about which the negatively charged electrons move. The closed-loop paths shown here are symbolic only. In reality, the electrons do not follow discreet paths, as Section 30.5 discusses.

Example 1   *A Lot of Electrons*

How many electrons are there in one coulomb of negative charge?

REASONING AND SOLUTION   A negative charge that has a magnitude of $q = 1.00$ C is composed of N electrons, according to $q = Ne$. Therefore,

$$N = \frac{q}{e} = \frac{1.00 \text{ C}}{1.60 \times 10^{-19} \text{ C}} = \boxed{6.25 \times 10^{18}}$$

Clearly, a negative charge of one coulomb contains an enormous number of electrons.

18.2 CHARGED OBJECTS AND THE ELECTRIC FORCES THAT THEY EXERT

THE SEPARATION OF CHARGES

Electricity has many useful applications, and they are related to the fact that it is possible to transfer electric charge from one object to another. Usually electrons are transferred, and the body that gains electrons acquires an excess of negative charge. The body that loses electrons has an excess of positive charge. Such

---

* At this time we omit a precise definition of the coulomb, since such a definition depends on electric currents and magnetic fields, concepts discussed in later chapters.

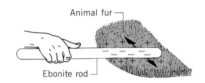

Figure 18.2 When an ebonite rod is rubbed against animal fur, electrons from the fur are transferred to the rod. This transfer gives the rod a negative charge (−) and leaves a positive charge (+) on the fur.

separation of charge occurs often when two unlike materials are rubbed together. For example, when an ebonite (hard, black rubber) rod is rubbed against animal fur, some of the electrons from the fur are transferred to the rod. The ebonite becomes negatively charged, and the fur becomes positively charged, as Figure 18.2 indicates. Similarly, if a glass rod is rubbed with a silk cloth, some of the electrons are removed from the glass and deposited on the silk, leaving the silk negatively charged and the glass positively charged. There are many familiar examples of charge separation, as when you walk across a nylon rug, vigorously run a comb through dry hair, or remove a pullover sweater. In each case, objects become "electrified" as surfaces rub against one another and a transfer of electrons occurs.

In the operation of electrical equipment, charge separation plays a fundamental role. For instance, batteries, microphones, alternators in automobile electrical systems, and electric power generators all depend on the separation of electric charges.

THE CONSERVATION OF CHARGE

When an ebonite rod is rubbed with animal fur, the rubbing process serves only to separate electrons and protons already present in the materials. No electrons or protons are created or destroyed. Whenever an electron is transferred to the rod, a proton is left behind on the fur. Since the charges on the electron and proton have identical magnitudes but opposite signs, the algebraic sum of the two charges is zero, and the transfer does not change the net charge of the fur/rod system. If each material contains an equal number of protons and electrons to begin with, the net charge of the system is zero initially and remains zero at all times during the rubbing process.

Electric charges play a role in many situations other than rubbing two surfaces together. They are involved, for instance, in chemical reactions, electric circuits, and radioactive decay. A great number of experiments have verified that in any situation, the *law of conservation of electric charge* is obeyed.

Law of Conservation of Electric Charge

During any process, the net electric charge of an isolated system remains constant (is conserved).

THE ELECTRIC FORCE THAT CHARGES EXERT ON EACH OTHER

It is easy to demonstrate that two electrically charged objects exert a force on one another. Consider Figure 18.3*a*, which shows two small balls that have been *oppositely charged* and are light and free to move. The balls are attracted toward each other. On the other hand, balls with the *same* type of charge, either both positive or both negative, repel each other, as part *b* of the drawing illustrates. It is a fundamental characteristic of electric charges that *like charges repel and unlike charges attract each other*.

The electrostatic air cleaner is one device that puts the electric force to good use, and Figure 18.4 shows how. This air cleaner is a great aid to people who suffer

THE PHYSICS OF . . .

an electrostatic air cleaner.

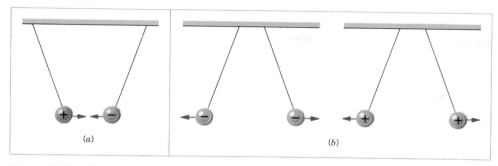

Figure 18.3   (a) A positive charge (+) and a negative charge (−) attract each other. (b) Two negative charges or two positive charges repel each other.

from respiratory problems, for it can remove up to 95% of all airborne particles, such as dust and pollen. A fan draws the contaminated air into the cleaner. The air first passes through an ordinary mesh filter that removes the larger particles, and then through a positively charged wire grid, known as the *charging electrode*. This electrode gives the contaminant particles a positive charge. The positively charged particles continue upward to the negatively charged grid, where they stick because of the electric force of attraction. Sometimes, after leaving the negative grid, the clean air passes through an activated charcoal filter before being discharged into the room. Activated charcoal is a good odor absorber and leaves the air smelling fresh.

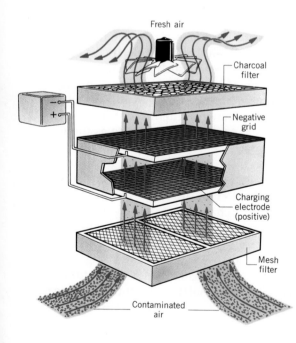

Figure 18.4   An electrostatic air cleaner. Unwanted airborne particles are given a positive charge as they pass the charging electrode. The positively charged particles are removed from the air stream, because they stick to the negative grid, which must be washed periodically.

THE PHYSICS OF . . .

decontaminating soils.

   The electric force that charges exert is also part of a technique for decontaminating soils deep underground. The technique, still in the experimental stage, also incorporates the use of sound waves. Soil particles typically have a negative charge, while contaminant particles such as petroleum sludge, often have a positive charge. A sound generator driven into the ground (see Figure 18.5) produces a

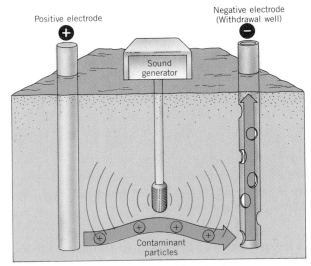

Figure 18.5 An experimental system for decontaminating underground soils.

low-frequency sound wave that is thought to help loosen the positive contaminants from the negative soil. Positive and negative electrodes are also driven into the ground to provide an electric force that drives the positive contaminants toward the negative electrode. The negative electrode also serves as a withdrawal well for removing the contaminants.

18.3 CONDUCTORS AND INSULATORS

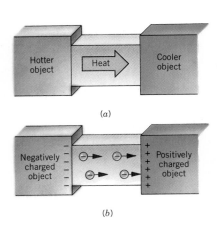

Figure 18.6 (a) Heat is conducted from the hotter end of the metal bar to the cooler end. (b) Electrons are conducted from the negatively charged end of the metal bar to the positively charged end.

Not only can electric charge exist *on an object*, but it can also move *through an object*. However, materials differ vastly in their ability to allow electric charge to move or be conducted through them. To help illustrate such differences in conductivity, Figure 18.6a recalls the conduction of heat through a bar of material whose ends are maintained at different temperatures. As Section 13.2 discusses, metals conduct heat readily and, therefore, are known as thermal conductors. On the other hand, substances that conduct heat poorly are referred to as thermal insulators.

A situation analogous to the conduction of heat arises when a metal bar is placed between two charged objects, as in Figure 18.6b. Electrons are conducted through the bar from the negatively charged object toward the positively charged object. Substances that readily conduct electric charge are called *electrical conductors.* Although there are exceptions, good thermal conductors are also good electrical conductors. Metals such as copper, aluminum, silver, and gold are excellent electrical conductors and, therefore, are used in electrical wiring. Materials that conduct electric charge *poorly* are known as *electrical insulators.* In many cases, thermal insulators are also electrical insulators. Common electrical insulators are rubber, many plastics, and wood. Insulators, such as the rubber or plastic that coats electrical wiring, prevent electric charge from going where it is not wanted.

The difference between electrical conductors and insulators is related to atomic structure. As electrons orbit the nucleus, those in the outer orbits experience a weaker force of attraction to the nucleus than do those in the inner orbits. Consequently, the outermost or valence electrons can be dislodged more easily than the inner ones. In a good conductor, some valence electrons actually become detached from a parent atom and wander more or less freely throughout the material, belonging to no one atom in particular. The exact number of electrons detached from each atom depends on the nature of the material, but is usually between one and three. When one end of a conducting bar is placed in contact with a negatively charged object and the other end in contact with a positively charged object, as in Figure 18.6b, the "free" electrons are able to move readily away from the negative end and toward the positive end. The ready movement of electrons is the hallmark of a good conductor. In an insulator the situation is different, for there are very few electrons free to move throughout the material. Virtually every electron remains bound to its parent atom. Without the "free" electrons, there is very little flow of charge when the material is placed between two oppositely charged bodies, so the material is an electrical insulator.

18.4 CHARGING BY CONTACT AND BY INDUCTION

CHARGING BY CONTACT

When a negatively charged ebonite rod is rubbed on a metal object, such as the sphere in Figure 18.7a, some of the excess electrons from the rod are transferred to the object. Once the electrons are on the metal sphere, where they can move readily, their mutual repulsive forces cause them to spread out over the surface of the sphere. The insulated stand prevents the electrons from flowing to the earth, where they could spread out even more. When the rod is removed, as in part b of the picture, the sphere is left with a uniformly distributed negative charge. In a similar manner, the sphere would be left with a positive charge after being in contact with a positively charged rod. In this case, electrons from the sphere would be transferred to the rod. The process of giving one object a net electric charge by placing it in contact with a charged object is known as *charging by contact.*

CHARGING BY INDUCTION

It is possible to charge a conductor in a way that does not involve contact. In Figure 18.8, a negatively charged rod is brought close to, *but does not touch,* a metal sphere. In the sphere, the free electrons closest to the rod move to the other side, as part a of the drawing indicates. As a result, the part of the sphere nearest the rod

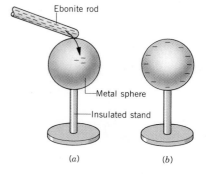

Figure 18.7 (a) Electrons are transferred by rubbing the negatively charged rod on the metal sphere. (b) When the rod is removed, the electrons distribute themselves uniformly over the surface of the sphere.

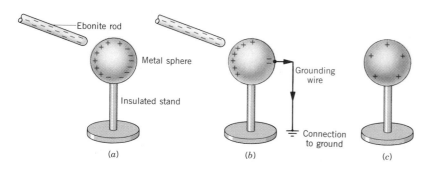

Figure 18.8 (a) When a charged rod is brought near the metal sphere without touching it, some of the positive and negative charges in the sphere are separated. (b) Some of the electrons leave the sphere through the ground wire, with the result (c) that the sphere acquires a positive net charge.

In this demonstration, the wand is electrically charged and causes induced charges to appear on the girl's hair. The hair, in turn, is attracted to the wand because of the attraction between unlike charges.

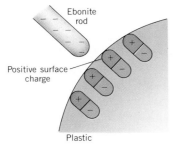

Figure 18.9 The negatively charged rod induces a slight positive surface charge on the plastic.

becomes positively charged and the part farthest away becomes negatively charged. These positively and negatively charged regions have been "induced" or "persuaded" to form because of the repulsive force between the negative rod and the free electrons in the sphere. If the rod were removed, the free electrons would return to their original places, and the charged regions on either side of the sphere would disappear.

Under most conditions the earth is a good electrical conductor. So when a metal wire is attached between the sphere and the ground, as in Figure 18.8*b*, some of the free electrons leave the sphere and distribute themselves over the much larger earth. If the grounding wire is then removed, followed by the ebonite rod, the sphere is left with a positive net charge, as part *c* of the picture shows. The process of giving one object a net electric charge *without* touching the object to a second charged object is called *charging by induction.* The process could also be used to give the sphere a negative net charge, if a positively charged rod were used. Then, electrons would be drawn up from the ground through the grounding wire and onto the sphere.

If the sphere in Figure 18.8 were made from an insulating material like plastic, instead of metal, the method of producing a net charge by induction would not work, because very little charge would flow through the insulating material and down the grounding wire. However, the electric force of the charged rod would have some effect, as Figure 18.9 illustrates. The electric force would cause the positive and negative charges in the molecules of the insulating material to separate slightly, with the positive charges being "pulled" toward the rod. Although no net charge is created, the surface of the plastic does acquire a slight induced positive charge and is attracted to the negative rod. For a similar reason, small dust particles are pulled out of the air and "stick" to the surface of a phonograph record that has been given a charge in the process of being wiped clean.

THE ELECTROSCOPE

The electroscope is a device for detecting small amounts of charge. The electroscope in Figure 18.10 consists of two thin strips, or leaves, of gold foil mounted at the bottom end of a metal rod. A metal knob caps the top of the rod, and glass windows enclose the leaves to prevent any effects due to air currents. An insulating rubber plug separates the metal rod from the metal case, so any charge on the rod does not leak away.

The electroscope can be used to determine if an insulator is charged and, if so, whether the charge is positive or negative. First, the electroscope is given a charge

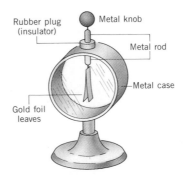

Figure 18.10 An electroscope.

of known polarity, by touching the metal knob with a negatively charged ebonite rod, for example. As can be seen in Figure 18.11a, the negative charge spreads out over the leaves, and the repulsive force between the like charges causes the leaves to spread apart. Then, the unknown charge is brought near the electroscope *without touching it*. If the unknown charge is positive, as in part b of the drawing, some of the electrons are drawn off the leaves and onto the metal knob. The loss of negative charge causes a reduction in the repulsive force between the leaves and, as a result, they partially collapse. Conversely, if the unknown charge is negative, as in part c, it forces free electrons to leave the knob and increase the negative charge on the leaves. Consequently, the leaves spread apart even further. In either case, the unknown charge is not brought close enough to the electroscope to cause the charge on the leaves to change polarity (see question 6).

18.5 COULOMB'S LAW

THE FORCE BETWEEN TWO POINT CHARGES

The electric force that stationary charged objects exert on each other is called the electrostatic force. This force depends on the amount of charge on and the distance between the objects. Experiments reveal that the greater the charges and the closer together they are, the greater is the force. To set the stage for explaining these features in more detail, Figure 18.12 shows two charged bodies. These objects are so small, compared to the separation distance r between them, that they can be regarded as mathematical points. The "point charges" have magnitudes q_1 and q_2. If the charges have *unlike* signs, as in part a of the picture, each charge is *attracted* to the other by a force that is directed along the line between them; $+\mathbf{F}$ is the electric force exerted on charge 1 by charge 2 and $-\mathbf{F}$ is the electric force exerted on charge 2 by charge 1. If the charges have the *same* sign (both positive or both negative), as in part b, each is repelled from the other. The repulsive forces, like the attractive forces, act along the line between the charges. Whether attractive or repulsive, the two forces are equal in magnitude but opposite in direction. These forces always exist as a pair, each one acting on a different object, in accord with Newton's action–reaction law.

The French physicist Charles-Augustin Coulomb (1736–1806) carried out a number of experiments to determine how the electric force between two point charges depends on the amount of each charge and the separation between them. His result, now known as **Coulomb's law**, is stated below.

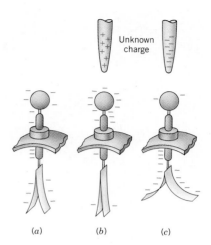

Figure 18.11 (a) A negative charge has been placed on the knob and leaves of an electroscope. (b) An unknown charge that is positive causes the leaves to collapse. (c) An unknown charge that is negative causes the leaves to diverge further.

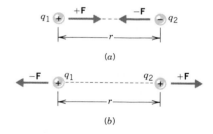

Figure 18.12 Each point charge exerts a force on the other. Regardless of whether the forces are (a) attractive or (b) repulsive, they are directed along the line between the charges and have equal magnitudes.

Coulomb's Law

The magnitude F of the electrostatic force exerted by one point charge on another point charge is directly proportional to the magnitudes q_1 and q_2 of the charges and inversely proportional to the square of the distance r between the charges:

$$F = k\frac{q_1 q_2}{r^2} \qquad (18.1)$$

where k is a proportionality constant whose value in SI units is $k = 8.99 \times 10^9$ N·m²/C².

Electropainting is standard technology in automobile manufacturing. The body of the car and the paint are given opposite charges. The resulting electrical attraction draws the paint onto the body, covering it thoroughly.

It is common practice to express k in terms of another constant ϵ_0, by writing $k = 1/(4\pi\epsilon_0)$; ϵ_0 is called the *permittivity of free space* and has a value of $\epsilon_0 = 1/(4\pi k) = 8.85 \times 10^{-12}$ C^2/(N·m^2). Equation 18.1 gives the magnitude of the electrostatic force that each point charge exerts on the other. When using this equation, then, it is important to remember to substitute only the charge magnitudes (without algebraic signs) for q_1 and q_2, as Example 2 illustrates.

Example 2 A Large Attractive Force

Two very small objects, whose charges are $+1.0$ and -1.0 C, are separated by 1.5 m. Find the magnitude of the attractive force that either charge exerts on the other.

REASONING AND SOLUTION Coulomb's law may be used to find the magnitude of the force, provided that only the *magnitudes of the charges* are used in the calculation:

$$F = k\frac{q_1 q_2}{r^2} = \frac{(8.99 \times 10^9 \text{ N·m}^2/\text{C}^2)(1.0 \text{ C})(1.0 \text{ C})}{(1.5 \text{ m})^2} = \boxed{4.0 \times 10^9 \text{ N}}$$

The force calculated in Example 2 corresponds to 900 million pounds and is enormous. However, charges as large as one coulomb are usually encountered only in the most severe conditions, as in a lightning bolt, where as much as 25 coulombs can be transferred between the cloud and the ground. The typical charges produced in the laboratory are much smaller and are measured conveniently in microcoulombs (1 microcoulomb = 1 μC = 10^{-6} C).

Coulomb's law has a form remarkably similar to Newton's law of gravitation ($F = Gm_1 m_2/r^2$). The force in both laws depends on the inverse square ($1/r^2$) of the distance between the two objects and is directed along the line between them. In addition, the force is proportional to the product of an intrinsic property of each of the objects, the charges q_1 and q_2 in Coulomb's law and the masses m_1 and m_2 in the gravitation law. But there is a major difference between the two laws. The electrostatic force can be either repulsive or attractive, depending on whether the charges have the same sign or not; in contrast, the gravitational force is always an attractive force.

Section 5.5 discusses how the gravitational attraction between the earth and a satellite provides the centripetal force that keeps the satellite in orbit. Example 3 illustrates that the electrostatic force of attraction plays a similar role in a famous model of the atom created by the Danish physicist Niels Bohr (1885–1962).

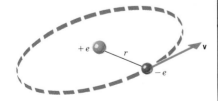

Figure 18.13 In the Bohr model of the hydrogen atom, the electron ($-e$) orbits the proton ($+e$) at a distance of $r = 5.29 \times 10^{-11}$ m. The velocity of the electron is **v**.

Example 3 A Model of the Hydrogen Atom

In the Bohr model of the hydrogen atom, the electron ($-e$) is in orbit about the nuclear proton ($+e$) at a radius of $r = 5.29 \times 10^{-11}$ m, as Figure 18.13 shows. Determine the speed of the electron, assuming the orbit to be circular.

REASONING Recall from Section 5.2 that any object moving with speed v on a circular path of radius r has a centripetal acceleration of $a_c = v^2/r$. This acceleration is directed toward the center of the circle. Newton's second law, then, specifies that the net force needed to create this acceleration is $\Sigma F = ma_c = mv^2/r$, where m is the mass of the object. This equation can be solved for the speed: $v = \sqrt{(\Sigma F)r/m}$. Since the mass of the electron is $m = 9.11 \times 10^{-31}$ kg and the radius is given, we can calculate the speed, once a value for the net force is available. For the electron in the hydrogen atom, the net force is provided almost exclusively by the electrostatic force, as given by Coulomb's law. The electron is also pulled toward the proton by a gravitational force. However, the gravitational force is negligible in comparison to the electrostatic force.

SOLUTION   The electron experiences an electrostatic force of attraction because of the proton, and magnitude of this force is

$$F = k\frac{q_1 q_2}{r^2} = \frac{(8.99 \times 10^9 \text{ N} \cdot \text{m}^2/\text{C}^2)(1.60 \times 10^{-19} \text{ C})(1.60 \times 10^{-19} \text{ C})}{(5.29 \times 10^{-11} \text{ m})^2}$$

$$= 8.22 \times 10^{-8} \text{ N}$$

Using this value for the net force, we find

$$v = \sqrt{\frac{(\Sigma F)r}{m}} = \sqrt{\frac{(8.22 \times 10^{-8} \text{ N})(5.29 \times 10^{-11} \text{ m})}{9.11 \times 10^{-31} \text{ kg}}} = \boxed{2.18 \times 10^6 \text{ m/s}}$$

This orbital speed is almost five million miles per hour.

THE FORCE ON A POINT CHARGE DUE TO TWO OR MORE OTHER POINT CHARGES

Up to now, we have been discussing the electrostatic force on a point charge q_1 due to a point charge q_2. Suppose that another point charge q_3 were also present. What would be the net force on q_1 due to both q_2 and q_3? It is convenient to deal with such a problem in parts. First, find the magnitude and direction of the force exerted on q_1 by q_2 (ignoring q_3). Then, determine the force exerted on q_1 by q_3 (ignoring q_2). The *net force* on q_1 is the *vector sum* of these two forces. Examples 4 and 5 illustrate this approach when the charges lie along a straight line and when they lie in a plane, respectively.

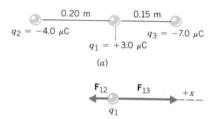

(a)

(b) Free-body diagram for q_1

Figure 18.14 (a) Three charges lying along the x axis. (b) The force exerted on q_1 by q_2 is $\mathbf{F}_{12}$, while the force exerted on q_1 by q_3 is $\mathbf{F}_{13}$.

Example 4   Three Charges on a Line

Figure 18.14a shows three point charges that lie along the x axis. Determine the magnitude and direction of the net electrostatic force on q_1.

REASONING   Part b of the drawing shows a free-body diagram of the forces that act on q_1. Since q_1 and q_2 have opposite signs, they attract one another. Thus, the force exerted on q_1 by q_2 is $\mathbf{F}_{12}$, and it points to the left. Similarly, the force exerted on q_1 by q_3 is $\mathbf{F}_{13}$ and is an attractive force. It points to the right in Figure 18.14b. The magnitudes of these forces can be obtained from Coulomb's law. The net force is the vector sum of $\mathbf{F}_{12}$ and $\mathbf{F}_{13}$.

SOLUTION   The magnitudes of the forces are

$$F_{12} = k\frac{q_1 q_2}{r_{12}^2} = \frac{(8.99 \times 10^9 \text{ N} \cdot \text{m}^2/\text{C}^2)(3.0 \times 10^{-6} \text{ C})(4.0 \times 10^{-6} \text{ C})}{(0.20 \text{ m})^2} = 2.7 \text{ N}$$

$$F_{13} = k\frac{q_1 q_3}{r_{13}^2} = \frac{(8.99 \times 10^9 \text{ N} \cdot \text{m}^2/\text{C}^2)(3.0 \times 10^{-6} \text{ C})(7.0 \times 10^{-6} \text{ C})}{(0.15 \text{ m})^2} = 8.4 \text{ N}$$

Since $\mathbf{F}_{12}$ points in the negative x direction, and $\mathbf{F}_{13}$ points in the positive x direction, the net force $\mathbf{F}$ is

$$\mathbf{F} = \mathbf{F}_{12} + \mathbf{F}_{13} = (-2.7 \text{ N}) + (8.4 \text{ N}) = \boxed{+5.7 \text{ N}}$$

Example 5 Three Charges in a Plane

Find the magnitude and direction of the net electrostatic force on q_1 in Figure 18.15a.

REASONING The force exerted on q_1 by q_2 is $\mathbf{F}_{12}$ and is an attractive force, because the two charges have opposite signs. It points along the line between the charges. The force exerted on q_1 by q_3 is $\mathbf{F}_{13}$ and is also an attractive force. It points along the line between q_1 and q_3. Coulomb's law gives the magnitudes of these forces. Since the forces do not point in the same direction (see Figure 18.15b) we will use vector components to find the net force.

SOLUTION The magnitudes of the forces are

$$F_{12} = k\frac{q_1q_2}{r_{12}^2} = \frac{(8.99 \times 10^9 \text{ N} \cdot \text{m}^2/\text{C}^2)(4.0 \times 10^{-6} \text{ C})(6.0 \times 10^{-6} \text{ C})}{(0.15 \text{ m})^2} = 9.6 \text{ N}$$

$$F_{13} = k\frac{q_1q_3}{r_{13}^2} = \frac{(8.99 \times 10^9 \text{ N} \cdot \text{m}^2/\text{C}^2)(4.0 \times 10^{-6} \text{ C})(5.0 \times 10^{-6} \text{ C})}{(0.10 \text{ m})^2} = 18 \text{ N}$$

The net force $\mathbf{F}$ is the vector sum of $\mathbf{F}_{12}$ and $\mathbf{F}_{13}$, as part *b* of the drawing shows. The components of $\mathbf{F}$ that lie in the x and y directions are $\mathbf{F}_x$ and $\mathbf{F}_y$, respectively. Our approach to finding $\mathbf{F}$ is the same as that used in Chapters 1 and 4. The forces $\mathbf{F}_{12}$ and $\mathbf{F}_{13}$ are resolved into x and y components. Then, the x components are combined to give $\mathbf{F}_x$, and the y components are combined to give $\mathbf{F}_y$. Once $\mathbf{F}_x$ and $\mathbf{F}_y$ are known, the magnitude and direction of $\mathbf{F}$ can be determined:

Force	x component	y component
$\mathbf{F}_{12}$	$+(9.6 \text{ N}) \cos 73° = +2.8 \text{ N}$	$+(9.6 \text{ N}) \sin 73° = +9.2 \text{ N}$
$\mathbf{F}_{13}$	$+18 \text{ N}$	0
$\mathbf{F}$	$F_x = +21 \text{ N}$	$F_y = +9.2 \text{ N}$

The magnitude F and the angle θ of the net force are

$$F = \sqrt{F_x^2 + F_y^2} = \sqrt{(21 \text{ N})^2 + (9.2 \text{ N})^2} = \boxed{23 \text{ N}}$$

$$\theta = \tan^{-1}\left(\frac{F_y}{F_x}\right) = \tan^{-1}\left(\frac{9.2 \text{ N}}{21 \text{ N}}\right) = \boxed{24°}$$

Figure 18.15 (a) Three charges lying in a plane. (b) The net force acting on q_1 is $\mathbf{F} = \mathbf{F}_{12} + \mathbf{F}_{13}$. The angle that $\mathbf{F}$ makes with the x axis is θ.

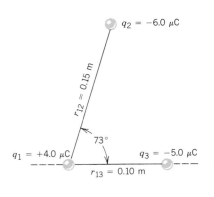

(a)

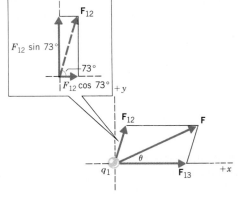

(b) Free-body diagram for q_1

18.6 THE ELECTRIC FIELD

DEFINITION OF THE ELECTRIC FIELD

A charge can experience an electrostatic force due to the presence of other charges. For instance, the positive charge q_0 in Figure 18.16, experiences a force **F**, which is the vector sum of the forces exerted by the charges on the rod and the two spheres. It is useful to think of q_0 as a **test charge** for determining the extent to which the surrounding charges generate a force. However, in using a test charge, we must be careful to select one with a very small magnitude, so that it does not alter the locations of the other charges. The next example illustrates how the concept of a test charge is applied.

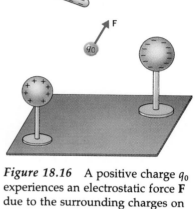

Figure 18.16 A positive charge q_0 experiences an electrostatic force **F** due to the surrounding charges on the ebonite rod and the two spheres.

Example 6 A Test Charge

The positive test charge in Figure 18.16 is $q_0 = +3.0 \times 10^{-8}$ C and experiences a force $F = 6.0 \times 10^{-8}$ N in the direction shown in the drawing. (a) Find the *force per coulomb* that the test charge experiences. (b) Using the result of part (a), predict the force that a charge of $+12 \times 10^{-8}$ C would experience if it replaced q_0.

REASONING AND SOLUTION
(a) The force per coulomb of charge is

$$\frac{F}{q_0} = \frac{6.0 \times 10^{-8} \text{ N}}{3.0 \times 10^{-8} \text{ C}} = \boxed{2.0 \text{ N/C}}$$

(b) The result from part (a) indicates that the surrounding charges can exert 2.0 newtons of force per coulomb of charge located where q_0 is. Thus, a charge of $+12 \times 10^{-8}$ C would experience a force whose magnitude is

$$F = (2.0 \text{ N/C})(12 \times 10^{-8} \text{ C}) = \boxed{24 \times 10^{-8} \text{ N}}$$

The direction of this force would be the same as that experienced by the test charge, since both have the same positive sign.

Sharks have electrically sensitive cells that allow them to detect weak electric fields, such as those produced by charges in the bodies of their prey.

The force per coulomb, $\mathbf{F}/q_0$, calculated in Example 6(a) is one illustration of a quantity that is very important in the study of electricity. This quantity is called the *electric field.*

Definition of the Electric Field

The electric field $\mathbf{E}$ that exists at a point is the electrostatic force $\mathbf{F}$ experienced by a small test charge q_0 placed at that point divided by the charge itself:

$$\mathbf{E} = \frac{\mathbf{F}}{q_0} \tag{18.2}$$

The electric field is a vector, and its direction is the same as the direction of the force $\mathbf{F}$ on a positive test charge.

SI Unit of the Electric Field: newton per coulomb (N/C)

Equation 18.2 indicates that the unit for the electric field is that of force divided by charge, which is a newton/coulomb (N/C) in SI units. The definition also emphasizes that the electric field is a vector with the same direction as the force on a *positive* test charge.

It is the surrounding charges that create an electric field at a given point. Any positive or negative charge placed at the point interacts with the field and, as a result, experiences a force, as the next example indicates.

Example 7   A Field Leads to a Force

In Figure 18.17a the charges on the two metal spheres and the ebonite rod create an electric field at the spot indicated. This field has a magnitude of 2.0 N/C and is directed as in the drawing. Determine the force on a charge placed at that spot, if the charge has a value of (a) $q_0 = +18 \times 10^{-8}$ C and (b) $q_0 = -24 \times 10^{-8}$ C.

REASONING   The electric field at a given spot can exert a variety of forces, depending on the magnitude and sign of the charge placed there. The charge is assumed to be small enough not to alter the locations of the surrounding charges that create the field.

SOLUTION
(a) The magnitude of the force is the product of the magnitudes of q_0 and $\mathbf{E}$:

$$F = q_0 E = (18 \times 10^{-8} \text{ C})(2.0 \text{ N/C}) = \boxed{36 \times 10^{-8} \text{ N}} \tag{18.2}$$

Since q_0 is positive, the force points in the same direction as the electric field, as part *b* of the drawing indicates.

(b) In this case, the magnitude of the force is

$$F = q_0 E = (24 \times 10^{-8} \text{ C})(2.0 \text{ N/C}) = \boxed{48 \times 10^{-8} \text{ N}} \tag{18.2}$$

The force on the negative charge points in the direction *opposite* to the force on the positive charge, that is, opposite to the electric field (see part *c* of the drawing).

At a particular point in space, each of the surrounding charges contributes to the net electric field that exists there. To determine the net field, it is necessary to

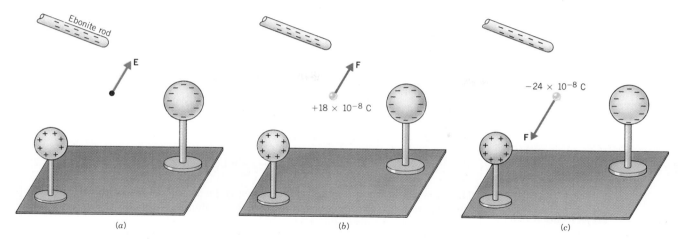

Figure 18.17   (a) The electric field **E** (2.0 N/C) that exists at a spot can exert a variety of forces at that spot, depending on the magnitude and sign of the charge placed there. (b) The force on a positive charge points in the same direction as **E**, while (c) the force on a negative charge points opposite to **E**.

obtain the various contributions separately and then find the vector sum of them all. Such an approach is an illustration of the principle of linear superposition, as applied to electric fields. (This principle is introduced in Section 17.1, in connection with waves.) Example 8 emphasizes the vector nature of the electric field.

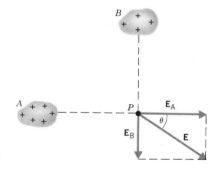

Figure 18.18   The electric field contributions **E**$_A$ and **E**$_B$, which come from the two charge distributions, are added vectorially to obtain the net field **E** at point *P*.

Example 8   Electric Fields Add as Vectors Do

Figure 18.18 shows two charged objects, *A* and *B*. Each contributes as follows to the net electric field at point *P*: **E**$_A$ = 3.00 N/C directed to the right, and **E**$_B$ = 2.00 N/C directed downward. Thus, **E**$_A$ and **E**$_B$ are perpendicular. What is the net field at *P*?

<u>REASONING AND SOLUTION</u>   The net field **E** is the vector sum of **E**$_A$ and **E**$_B$: **E** = **E**$_A$ + **E**$_B$. Since **E**$_A$ and **E**$_B$ are perpendicular, the magnitude of **E** is given by the Pythagorean theorem:

$$E = \sqrt{E_A^2 + E_B^2} = \sqrt{(3.00 \text{ N/C})^2 + (2.00 \text{ N/C})^2} = \boxed{3.61 \text{ N/C}}$$

The direction of **E** is given by the angle θ in the drawing:

$$\theta = \tan^{-1}\left(\frac{E_B}{E_A}\right) = \tan^{-1}\left(\frac{2.00 \text{ N/C}}{3.00 \text{ N/C}}\right) = \boxed{33.7°}$$

ELECTRIC FIELDS PRODUCED BY POINT CHARGES

A more complete understanding of the electric field concept can be gained by considering the field created by a point charge, as in the following example.

PROBLEM SOLVING INSIGHT

The electric field is a vector and has a direction as well as a magnitude. When adding electric fields together, take into account the directions of all fields, using vector components as needed.

Example 9   The Electric Field of a Point Charge

There is an isolated point charge of $q = +15\ \mu$C at the left in Figure 18.19*a*. Using a test charge of $q_0 = +0.80\ \mu$C, determine the electric field at point *P*, which is 0.20 m away.

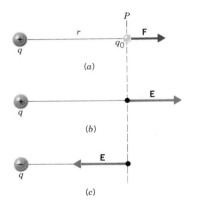

Figure 18.19 (a) At location P, a positive test charge q_0 experiences a repulsive force **F** due to the positive point charge q. (b) At P the electric field **E** is directed to the right. (c) If the charge q were negative rather than positive, the electric field would have the same magnitude as in (b) but would be directed to the left.

REASONING Following the definition of the electric field, we place the test charge q_0 at point P, determine the force acting on the test charge, and then divide the force by the test charge.

SOLUTION The magnitude of the force can be found from Coulomb's law:

$$F = \frac{kq_0 q}{r^2} = \frac{(8.99 \times 10^9 \text{ N} \cdot \text{m}^2/\text{C}^2)(0.80 \times 10^{-6} \text{ C})(15 \times 10^{-6} \text{ C})}{(0.20 \text{ m})^2} = 2.7 \text{ N} \quad (18.1)$$

The magnitude of the electric field is

$$E = \frac{F}{q_0} = \frac{2.7 \text{ N}}{0.80 \times 10^{-6} \text{ C}} = \boxed{3.4 \times 10^6 \text{ N/C}} \quad (18.2)$$

The electric field **E** points in the *same direction* as the force **F** on the positive test charge. Since the test charge experiences a force of repulsion directed to the right, the electric field vector also points to the right, as Figure 18.19b shows.

The electric field produced by a point charge q can be obtained in general terms from Coulomb's law. First, note that the magnitude of the force exerted by the charge q on a test charge q_0 is $F = kqq_0/r^2$. Then, divide this value by q_0 to obtain the magnitude of the field. Since q_0 is eliminated algebraically from the result, *the electric field does not depend on the test charge:*

[Point charge q] $$E = \frac{kq}{r^2} \quad (18.3)$$

As in Coulomb's law, only the magnitude of q is used in Equation 18.3, without regard to whether q is positive or negative. If q is positive, then **E** is directed away from q, as in Figure 18.19b. On the other hand, if q is negative, then **E** is directed toward q, since a negative charge attracts a positive test charge. For instance, Figure 18.19c shows the electric field that would exist at P if there were a charge of $q = -15 \ \mu\text{C}$ instead of $q = +15 \ \mu\text{C}$ at the left of the drawing. Example 10 reemphasizes the fact that all the surrounding charges make a contribution to the electric field that exists at a given place.

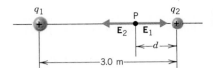

Figure 18.20 The two point charges q_1 and q_2 create electric fields E_1 and E_2 that cancel at a location P on the line between the charges.

Example 10   The Electric Fields from Separate Charges May Cancel

Two positive point charges, $q_1 = +16 \ \mu\text{C}$ and $q_2 = +4.0 \ \mu\text{C}$, are separated by a distance of 3.0 m, as Figure 18.20 illustrates. Find the spot on the line between the charges where the net electric field is zero.

REASONING Between the charges the two field contributions have opposite directions, and the electric field is zero at the place where the magnitude of E_1 equals that of E_2. However, since q_2 is smaller than q_1, this location must be *closer* to q_2, in order that the field of the smaller charge can balance the field of the larger charge. In the drawing, the cancellation spot is labeled P, and its distance from q_2 is d.

SOLUTION At P, $E_1 = E_2$, and using the expression $E = kq/r^2$, we have

$$\frac{k(16 \times 10^{-6} \text{ C})}{(3.0 \text{ m} - d)^2} = \frac{k(4.0 \times 10^{-6} \text{ C})}{d^2}$$

Rearranging this expression shows that $4.0d^2 = (3.0 \text{ m} - d)^2$, and taking the square root reveals that

$$\pm 2.0d = 3.0 \text{ m} - d$$

The plus and minus signs on the left occur because either the positive or negative root can be taken. Therefore, there are two possible values for d: $+1.0$ and -3.0 m. The negative value corresponds to a location off to the right of both charges, where the magnitudes of $\mathbf{E}_1$ and $\mathbf{E}_2$ are equal, but where the directions are the same. Thus, $\mathbf{E}_1$ and $\mathbf{E}_2$ do not cancel at this spot. The positive value corresponds to the location shown in the drawing and is the zero-field location: $\boxed{d = +1.0 \text{ m}}$.

THE ELECTRIC FIELD PRODUCED BY A PARALLEL PLATE CAPACITOR

Equation 18.3, which gives the electric field of a point charge, is a very useful result. With the aid of integral calculus, this equation can be applied in a variety of situations where point charges are distributed over a surface. One such example that has considerable practical importance is the *parallel plate capacitor.* As Figure 18.21 shows, this device consists of two parallel metal plates, each with area A. A charge $+q$ is spread uniformly over one plate, while a charge $-q$ is spread uniformly over the other plate. In the region between the plates and away from the edges, the electric field points from the positive plate toward the negative plate and is perpendicular to both. Using integral calculus to apply Equation 18.3 to every tiny part of the charge on each plate, it can be shown that the field has a magnitude of

$$\begin{bmatrix} \textbf{Parallel plate} \\ \textbf{capacitor} \end{bmatrix} \qquad E = \frac{q}{\epsilon_0 A} = \frac{\sigma}{\epsilon_0} \qquad (18.4)$$

In this expression the Greek symbol sigma (σ) denotes the charge per unit area ($\sigma = q/A$) and is sometimes called the charge density. Except in the region near the edges, the field has the same value at all places between the plates. The field does *not* depend on the distance from the charges, in distinct contrast to the field created by an isolated point charge.

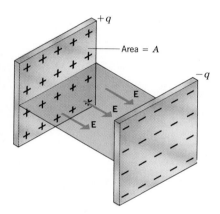

Figure 18.21   A parallel plate capacitor.

18.7 ELECTRIC FIELD LINES

As we have seen, electric charges create an electric field in the space surrounding them. It is useful to have a kind of "map" that gives the direction and indicates the strength of the field at various places. The great English physicist Michael Faraday (1791–1867) proposed an idea that provides such a "map," the idea of *electric field lines.* Since the electric field is the electric force per unit charge, the electric field lines also provide information about electric forces and are sometimes called *lines of force.*

To introduce the electric-field-line concept, Figure 18.22a shows a positive point charge $+q$. At the locations numbered 1–8, a positive test charge would experience a repulsive force, as the arrows in the drawing indicate. Therefore, the electric field created by the charge $+q$ is directed radially outward. The electric field lines are lines drawn to show this direction, as part b of the drawing illus-

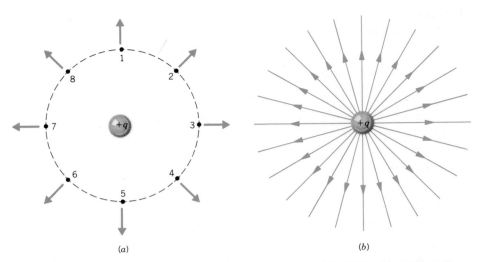

Figure 18.22 (*a*) At any of the eight marked spots around a positive point charge $+q$, a positive test charge would experience a repulsive force directed radially outward. (*b*) The electric field lines are directed radially outward from a positive point charge $+q$.

trates. The lines begin on the charge $+q$ and point radially *outward*. Figure 18.23 shows the electric field lines in the vicinity of a negative charge $-q$. In this case the lines are directed radially *inward*, because the force on a positive test charge is one of attraction, indicating that the electric field points inward. In general, *electric field lines are always directed away from positive charges and toward negative charges.*

The electric field lines in Figures 18.22 and 18.23 are drawn in only two dimensions, as a matter of convenience. Electric field lines radiate from the charges in three dimensions, and an infinite number of lines could be drawn. However, for clarity only a small number is ever included in pictures. The number is chosen to be proportional to the magnitude of the charge; thus, five times as many lines would emerge from a $+5q$ charge as from a $+q$ charge.

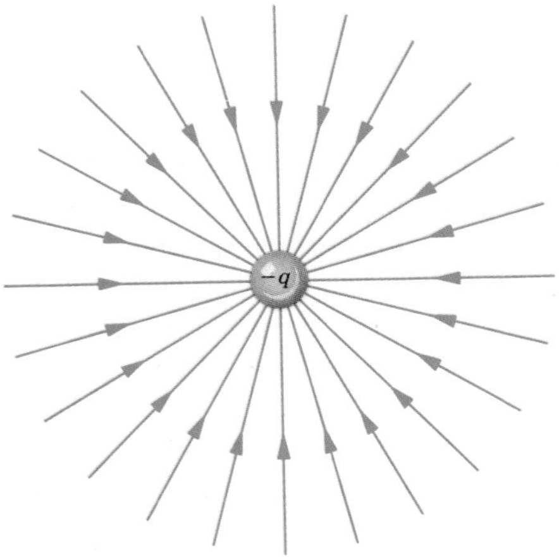

Figure 18.23 The electric field lines are directed radially inward toward a negative point charge $-q$.

The pattern of electric field lines also provides information about the strength of the field. Notice in Figures 18.22 and 18.23 that near the charges, where the electric field is strongest, the field lines are close together. Conversely, at distances far from the charges, where the electric field is weaker, the electric field lines are more spread out. It is true in general that the electric field is strongest in regions where the field lines are closest together. In fact, no matter how many charges are present, the number of lines per unit area passing perpendicularly through a surface is proportional to the magnitude of the electric field.

In regions where the electric field lines are equally spaced, there is the same number of lines per unit area everywhere, and the electric field has the same strength at all points. For example, Figure 18.24 shows the field lines between the plates of a parallel plate capacitor. The lines are parallel and equally spaced, except near the edges where they bulge outward. The equally spaced, parallel lines indicate that the electric field has the same magnitude and direction at all points in the central region of the capacitor.

Electric field lines are not always straight. More often they are curved, as in the case of an *electric dipole.* An electric dipole consists of two separated point charges that have the same magnitude but opposite signs. The electric field of a dipole is proportional to the product of the magnitude of one of the charges and the separation between the charges. This product is called the *dipole moment.* Many molecules, such as H_2O and HCl, have dipole moments. Figure 18.25 depicts the curved electric field lines in the vicinity of a dipole. For a curved field line, the electric field vector at a point is *tangent* to the line at that point (see points 1, 2, and 3 in the drawing). The pattern of the lines for the dipole indicates that the electric field is greatest in the region between and immediately surrounding the two charges, since the lines are closest together there.

Notice in Figure 18.25 that a given field line starts on the positive charge and ends on the negative charge. In general, *electric field lines always begin on a positive charge and end on a negative charge and do not start or stop in midspace. Furthermore, the number of field lines leaving a positive charge or enter-*

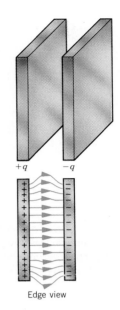

Figure 18.24 In the central region of a parallel plate capacitor the electric field lines are parallel and evenly spaced, indicating that the electric field there has the same magnitude and direction at all points.

PROBLEM SOLVING INSIGHT

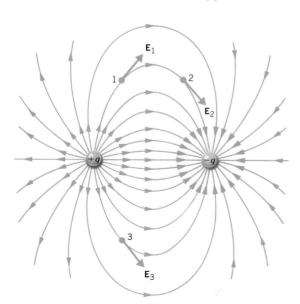

Figure 18.25 The electric field lines in the vicinity of an electric dipole are curved and extend from the positive charge to the negative charge. At any point, such as 1, 2, or 3, the electric field created by the dipole is tangent to the line that passes through the point.

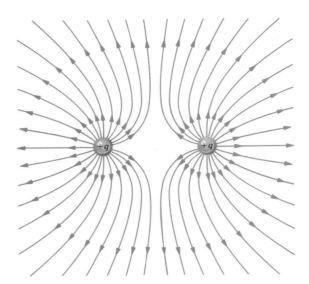

Figure 18.26 The electric field lines for two identical positive point charges. If the charges were both negative, the directions of the lines would be reversed.

ing a negative charge is proportional to the magnitude of the charge. This means, for example, that if 100 lines are drawn leaving a $+4\,\mu$C charge, then 75 lines would have to end on a $-3\,\mu$C charge and 25 lines on a $-1\,\mu$C charge. Thus, 100 lines leave the charge of $+4\,\mu$C and end on a *total charge* of $-4\,\mu$C, so the lines begin and end on equal amounts of total charge.

The electric field lines are also curved in the vicinity of two identical charges. Figure 18.26 shows the pattern associated with two positive point charges and reveals that there is an absence of lines in the region between the charges. The absence of lines indicates that the electric field is relatively weak between the charges.

18.8 THE ELECTRIC FIELD INSIDE A CONDUCTOR: SHIELDING

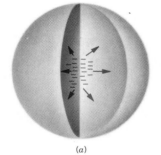

In conducting materials such as copper, electric charges move readily in response to the forces that electric fields exert. This characteristic property of conducting materials has a major effect on the electric field that can exist within and around them. Suppose that a piece of copper carries a number of excess electrons somewhere within it, as in Figure 18.27a. Each electron would experience a force of repulsion because of the electric field of its neighbors. And, since copper is a conductor, the excess electrons move readily in response to that force. In fact, as a consequence of the $1/r^2$ dependence on distance in Coulomb's law, they rush to the surface of the copper. Once static equilibrium is established with all of the excess charge on the surface, no further movement of charge occurs, as part b of the drawing indicates. Similarly, excess positive charge also moves to the surface of a conductor. In general, *at equilibrium under electrostatic conditions, any excess charge resides on the surface of a conductor.*

Now consider the interior of the copper in Figure 18.27b. The interior is electrically neutral, although there are still free electrons that can move under the

Figure 18.27 (a) Under electrostatic conditions, any excess charge within a conducting material quickly moves (b) to the surface where it resides at equilibrium.

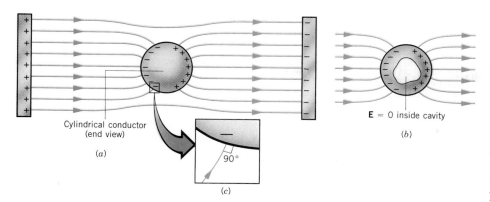

Cylindrical conductor
(end view)

(a)

90°

(c)

E = 0 inside cavity

(b)

Figure 18.28 (a) A cylindrical conductor (shown in cross section) is placed between the oppositely charged plates of a capacitor. The electric field lines do not penetrate the conductor. (b) The electric field is zero in a cavity within the conductor. (c) Just outside the conductor, the electric field lines are perpendicular to its surface.

influence of an electric field. The absence of a net movement of these free electrons indicates that there is no net electric field present within the conductor. In fact, the excess charges arrange themselves on the conductor surface precisely in the manner needed to make the total field zero within the material. Thus, *at equilibrium under electrostatic conditions, the electric field at any point within a conducting material is zero.* This fact has some fascinating implications.

Figure 18.28a shows an uncharged, solid, cylindrical conductor at equilibrium in the central region of a parallel plate capacitor. Induced charges on the surface of the cylinder alter the electric field lines of the capacitor. Since an electric field cannot exist within the conductor under these conditions, the electric field lines do not penetrate the cylinder and end or begin on the induced charges. Consequently, a test charge placed *inside* the conductor would feel no force due to the presence of the charges on the capacitor. In other words, *the conductor shields any charge within it from electric fields created outside the conductor.* The shielding results from the induced charges on the conductor surface.

Since the electric field is zero inside the conductor, nothing is disturbed if a cavity is cut from the interior of the material, as in part b of the drawing. Thus, the interior of the cavity is also shielded from external electric fields, a fact that has important applications, particularly for shielding electronic circuits. "Stray" electric fields are produced by various electrical appliances (e.g., hair driers, blenders, and vacuum cleaners) and these fields can interfere with the operation of sensitive electronic circuits, such as those in stereo amplifiers, televisions, and computers. To eliminate such interference, circuits are often enclosed within metal boxes that provide shielding from external fields.

Figure 18.28c shows another aspect of how conductors alter the electric field lines created by external charges. The lines are altered because *the electric field just outside the surface of a conductor is perpendicular to the surface at equilibrium under electrostatic conditions.* If the field were not perpendicular, there would be a component of the field parallel to the surface. Since the free electrons in the conductor can move in this direction, they would move under the force exerted by the parallel component. But, in reality, no electron flow occurs at equilibrium. Therefore, there can be no parallel component, and the electric field is perpendicular to the surface.

The preceding discussion deals with aspects of the electric field within and around a conductor at equilibrium under electrostatic conditions. These features are related to the fact that conductors contain electrons that can move freely. Since insulators do not contain free electrons, these features *do not apply to*

THE PHYSICS OF . . .

shielding electronic
circuits.

insulators. This section concludes with an example illustrating the behavior of a conducting material in the presence of an electric field.

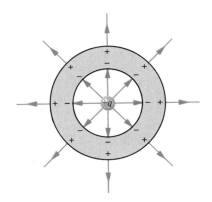

Figure 18.29 A positive charge $+q$ is suspended at the center of a hollow spherical conductor that is electrically neutral. Induced charges appear on the inner and outer surfaces of the conductor. The electric field within the conductor itself is zero.

Example 11   *A Conductor in a Field*

A charge $+q$ is suspended at the center of a hollow, electrically neutral, spherical conductor, as Figure 18.29 illustrates. Show that this charge induces (a) a charge of $-q$ on the interior surface and (b) a charge of $+q$ on the exterior surface of the conductor.

REASONING AND SOLUTION
(a) Electric field lines emanate from the positive charge $+q$. Since the electric field inside the metal conductor must be zero at equilibrium under electrostatic conditions, each field line ends when it reaches the conductor, as the picture shows. Consequently, there is an induced *negative* charge on the interior surface of the conductor, since field lines terminate only on negative charges. Furthermore, the lines begin and end on equal amounts of charge, so the magnitude of the total induced charge is the same as the magnitude of the charge at the center. Thus, the induced charge on the interior surface is $-q$.

(b) Before the charge $+q$ is introduced, the conductor is electrically neutral. Therefore, it carries no net charge, and since an induced charge of $-q$ appears on the interior surface, a charge $+q$ must be induced on the outer surface. As we have seen, there can be no excess charge within the metal. The positive charge on the outer surface generates electric field lines that radiate outward (see drawing) as if they originated from the central charge and the conductor were absent. The conductor does not shield the outside from the electric field produced by the charge on the inside.

18.9 COPIERS AND COMPUTER PRINTERS

THE PHYSICS OF . . .

xerography.

XEROGRAPHY

The electrostatic force that charged particles exert on one another plays the central role in an office copier. The copying process is called *xerography*, from the Greek *xeros* and *graphos*, meaning "dry writing." The heart of a copier is the xerographic drum, an aluminum cylinder coated with a layer of selenium (see Figure 18.30a). Aluminum is an excellent electrical conductor. Selenium, on the other hand, is a photoconductor; it is an insulator in the dark but becomes a conductor when exposed to light. Consequently, a positive charge deposited on the selenium surface will remain there, provided the selenium is kept in the dark. When the drum is exposed to light, however, electrons from the aluminum pass through the conducting selenium, and neutralize the positive charge.

The photoconductive property of selenium is critical to the xerographic process, as illustrated in Figure 18.30b. First, an electrode called a *corotron* gives the entire selenium surface a positive charge in the dark. Second, a series of lenses and mirrors focuses an image of a document onto the revolving drum. The dark and light areas of the document produce corresponding areas on the drum. The dark areas retain their positive charge, but the light areas become conducting and lose their positive charge, ending up neutralized. Thus, a positive-charge image of the document remains on the selenium surface. In the third step, a special dry, black powder, called the *toner*, is given a negative charge and then spread onto the

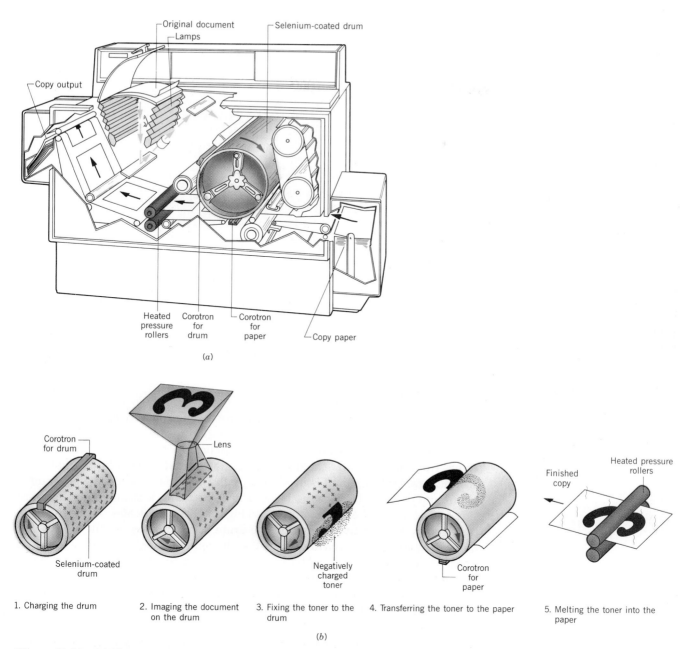

Figure 18.30 (*a*) This cutaway view shows the essential elements of a copying machine. (*b*) The five steps in the xerographic process.

drum, where it adheres selectively to the positively charged areas. The fourth step involves transferring the toner onto a blank piece of paper. However, the attraction of the positive-charge image holds the toner to the drum. To transfer the toner, the paper is given a *greater positive charge* than that of the image, with the aid of another corotron. Lastly, the paper and adhering toner pass through heated pressure rollers. As a result of the heat, the toner melts into the fibers of the paper and produces the finished copy.

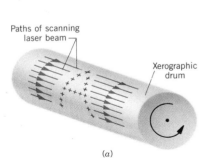

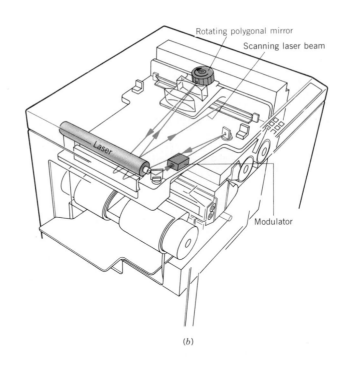

Figure 18.31 (*a*) As the laser beam scans back and forth across the surface of the xerographic drum, a positive-charge image of the letter "A" is created. (*b*) A laser printer showing the laser, modulator, and rotating mirror.

A LASER PRINTER

A laser printer is used with computers to provide high-quality copies of text and graphics. The laser printer is similar in operation to the xerographic machine, except the information to be reproduced is not on paper. Instead, the information comes from the computer's memory and is transferred to the printer through an electrical cable. Laser light is used to copy the information onto the selenium–aluminum drum. A laser beam, focused to a fine point, is scanned rapidly from side to side across the rotating drum, as Figure 18.31*a* indicates. While the light remains on, the positive charge on the drum is neutralized. As the laser beam moves, the computer turns the beam off at the right moments during each scan to produce the desired positive-charge image, which is the letter "A" in the picture.

Figure 18.31*b* shows the mechanism that turns the laser beam off and on and scans it across the xerographic drum. The light from the laser is sent through a device called a "modulator." The modulator also receives the information to be printed from the computer and, accordingly, allows the light to pass or blocks it. Thus, the laser output beam from the modulator is turned off and on, at rates often exceeding one million times per second. The output beam is then directed by a series of mirrors and lenses to a rotating polygonal mirror that causes the reflected beam to sweep from side to side. When the beam reflected from the rotating mirror is directed onto the xerographic drum, the beam scans across the drum and produces the positive-charge image.

AN INKJET PRINTER

An inkjet printer is another type of printer that uses electric charges in its operation. While shuttling back and forth across the paper, the inkjet printhead ejects a thin stream of ink. Figure 18.32 illustrates the elements of one type of printhead.

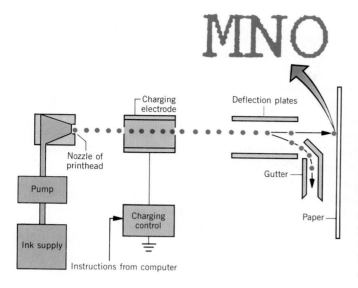

Figure 18.32 An inkjet printhead ejects a steady flow of ink droplets. The charging electrode is used to charge the droplets that are not needed on the paper. Charged droplets are deflected into a gutter by the deflection plates, while uncharged droplets fly straight onto the paper. The letters formed by an inkjet printer look normal, except when greatly enlarged, as they are here. Then the patterns from the drops become apparent.

The ink is forced out of a small nozzle and breaks up into extremely small droplets, with diameters less than 1×10^{-4} m (0.004 inch). About 150 000 droplets leave the nozzle each second and travel with a speed of approximately 64 km/h toward the paper. During their flight, the droplets pass through two electrical components, a *charging electrode* and the *deflection plates* (a parallel plate capacitor). When the printhead moves over regions of the paper that are not to be inked, the charging electrode is turned on and gives the ink droplets a net charge. The deflection plates divert the charged droplets into a gutter and thus prevent them from reaching the paper. Whenever ink is to be placed on the paper, the charging control, responding to instructions from the computer, turns off the charging electrode. The uncharged droplets fly straight through the deflection plates and strike the paper.

Inkjet printers are popular, because they can produce color copies. Full-color printheads use at least three nozzles for different colors (usually cyan or blue, magenta, and yellow). Often they also use a fourth nozzle for black ink. By adjusting the amount of ink that each nozzle produces, the printhead can deposit any color on the paper.

INTEGRATION OF CONCEPTS

THE ELECTROSTATIC FORCE AND NEWTON'S LAWS OF MOTION

In this chapter we have seen that electrically charged particles exert a force on each other. This electrostatic force is part of one of nature's four fundamental forces, the one known as the electromagnetic force. However, for all of its importance, the electrostatic force remains a force, and, like any force, can be used in Newton's laws of motion. These laws form the foundation of mechanics, the branch of physics that deals with the motion of objects and the forces that change it. The electrostatic force can cause an object to change its state of motion, that is, to accelerate, according to Newton's first and second laws. It can produce a change in the momentum of an object, in accordance with Newton's second law in the form known as the impulse–momentum theorem. It can do work and cause the kinetic energy of an object to change, according to the work–energy theorem. And the electrostatic force obeys Newton's third law, since the forces that two charged objects exert on each other are examples of action and reaction. The beauty of Newton's laws is that they apply to *all* types of forces, independent of the details that are specific to an individual kind of force, and independent of whether the force is one of nature's fundamental forces. Newton's laws relate to basic characteristics that forces of all types share, including the electrostatic force. It would be a complicated world indeed, if every type of force required a different type of Newton's laws to account for the effect of the force on the motion of an object. Fortunately, such is not the case.

SUMMARY

There are two kinds of **electric charge,** positive and negative; the SI unit of electric charge is the **coulomb** (C). The magnitude of the charge on an electron or a proton is $e = 1.60 \times 10^{-19}$ C. The electron carries a charge of $-e$, while the proton carries a charge of $+e$. The charge on any object, whether positive or negative, is **quantized,** in the sense that the charge consists of an integral number of protons or electrons. **The law of conservation of electric charge** states that the net electric charge of an isolated system remains constant during any process.

An **electrical conductor** is a material, such as copper, that conducts electric charge readily. An **electrical insulator** is a material, such as rubber, that conducts electric charge poorly. **Charging by contact** is the process of giving one object a net electric charge by placing it in contact with an object that is already charged. **Charging by induction** is the process of giving an object a net electric charge without touching it to a charged object.

One charge exerts an electric force on another charge. **For like charges the force is a repulsion, while for unlike charges the force is an attraction. Coulomb's law** gives the magnitude F of the electric force between two point charges as $F = kq_1q_2/r^2$, where q_1 and q_2 are the magnitudes of the charges, r is the distance between the charges, and $k = 8.99 \times 10^9$ N·m²/C². The force acts along the line between the two point charges. The permittivity of free space ϵ_0 is defined by the relation $k = 1/(4\pi\epsilon_0)$.

The **electric field E** at a given spot is a vector and is the electrostatic force **F** experienced by a small test charge q_0 placed at that spot divided by the charge itself: **E** = **F**/q_0. The direction of the electric field is the same as the direction of the force on a positive test charge. The SI unit for the electric field is the newton per coulomb (N/C). The source of the electric field at any spot is the charged objects that surround it. The magnitude of the electric field created by a point charge q is $E = kq/r^2$, where r is the distance from the charge. The magnitude of the electric field between the plates of a parallel plate capacitor is $E = \sigma/\epsilon_0$, where σ is the charge per unit area on either plate.

Electric field lines are lines that can be thought of as a "map" providing information about the direction and strength of the electric field. Electric field lines are directed away from positive charges and toward negative charges. The direction of the lines gives the direction of the electric field, since the electric field vector at a point is tangent to the line at that point. The electric field is strongest in regions where the number of lines per unit area passing perpendicularly through a surface is the greatest, that is, where the lines are closest together.

Excess negative or positive charge resides on the surface of a conductor at equilibrium under electrostatic conditions. In such a situation, the electric field at any point within the conducting material is zero, and the electric field just outside the surface of the conductor is perpendicular to the surface.

QUESTIONS

1. In Figure 18.8 the ground wire is removed first, followed by the rod, and the sphere is left with a positive charge. If the rod were removed first, followed by the ground wire, would the sphere be left with a charge? Account for your answer.

2. A rod made from insulating material carries a net charge, while a copper sphere is neutral. The rod and the sphere do not touch. Is it possible for the rod and the sphere to (a) attract one another and (b) repel one another? Explain.

3. An electroscope is charged so that its leaves are spread apart. The spread between the leaves decreases slightly when an electrically neutral copper object is brought near the metal knob of the electroscope without touching it. Explain.

4. The leaves of an electroscope are given a positive charge. As an insulator is brought from far away and moved toward the knob of the electroscope, the leaves are observed to spread apart even further. Is the insulator charged positively or negatively? Give your reasoning.

5. Blow up a balloon and rub it against your shirt a number of times. In so doing you give the balloon a net electric charge. Now touch the balloon to the ceiling. Upon being released, the balloon will remain stuck to the ceiling. Why?

6. A negatively charged insulator is brought near, but does not touch, the metal knob of a charged electroscope. As the insulator approaches, the leaves collapse. When the insulator is brought even closer, the leaves are observed to diverge. Deduce the sign of the original charge on the electroscope and account for the observed behavior.

7. Suppose an electroscope is electrically neutral, rather than being negatively charged as in Figure 18.11a. Can this neutral electroscope be used to determine (a) whether an object carries a net charge and (b) whether the net charge on the object is negative or positive? Justify your answers.

8. A particle is attached to a spring and is pushed so that the spring is compressed more and more. As a result, the spring exerts a greater and greater force on the particle. Similarly, a charged particle experiences a greater and greater force when pushed closer and closer to another particle that is fixed in position and has a charge of the same polarity. In spite of the similarity, the charged particle will *not* exhibit simple harmonic motion upon being released, as will the particle on the spring. Explain why not.

9. On a thin, nonconducting rod, positive charges are spread evenly, so that there is the same amount of charge per unit length at every point. On another identical rod, positive charges are spread evenly over only the left half, and the same amount of negative charges are spread evenly over the right half. For each rod, deduce the *direction* of the electric field at a point that is located directly above the midpoint of the rod. Give your reasoning.

10. There is an electric field at point *P*. A very small charge is placed at this point and experiences a force. Another very small charge is then placed at this point and experiences a force that differs in both magnitude and direction from that experienced by the first charge. How can these two different forces result from the single electric field that exists at point *P*?

11. The three drawings show four charges fixed to the corners of a rectangle in various ways. Consider the net electric field at the center *C* of the rectangle in each case. Rank the field magnitudes in decreasing order (largest first). Justify your ranking.

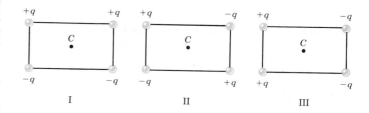

12. Drawings I and II show two examples of electric field lines. Decide which of the following statements are true and which are false, defending your choice in each case. (a) In both I and II the electric field is the same everywhere. (b) As you move from left to right in each case, the electric field becomes stronger. (c) The electric field in I is the same everywhere but becomes stronger in II as you move from left to right. (d) The electric fields in both I and II could be created by negative charges located somewhere on the left and positive charges somewhere on the right. (e) Both I and II arise from a single positive point charge located somewhere on the left.

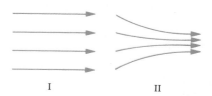

13. A positively charged particle is moving horizontally when it enters the region between the plates of a capacitor, as the drawing illustrates. (a) Draw the trajectory that the particle follows in moving through the capacitor. (b) When the particle is within the capacitor, which of the following four vectors, if any, are *parallel* to the electric field lines of the capacitor: the particle's displacement, its velocity, its linear momentum, its acceleration? For each vector explain why the vector is, or is not, parallel to the electric field.

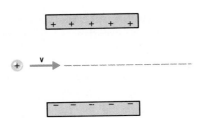

14. In all the pictures of electric field lines in the text there are no examples of two field lines that intersect. In fact, two field lines never intersect. Suppose that it were possible for two field lines to intersect. Discuss what would happen to a test charge placed at the crossover point.

PROBLEMS

Note: All charges are point charges, unless specified otherwise.

Section 18.1 The Origin of Electricity, Section 18.2 Charged Objects and the Electric Forces between Them, Section 18.3 Conductors and Insulators, Section 18.4 Charging by Contact and by Induction

1. How many electrons must be removed from an electrically neutral silver dollar to give it a charge of $+3.8 \ \mu C$?

2. A metal sphere has a charge of $+8.0 \ \mu C$. What is the net charge after 6.0×10^{13} electrons have been placed on it?

3. A rod has a charge of $-2.0 \ \mu C$. How many electrons must be removed so that the charge becomes $+3.0 \ \mu C$?

4. There are three identical metal spheres, A, B, and C. Sphere A carries a charge of $+5q$. Sphere B carries a charge of $-q$. Sphere C carries no net charge. Spheres A and B are touched together and then separated. Sphere C is then touched to sphere A and separated from it. Lastly, sphere C is touched to sphere B and separated from it. How much charge ends up on sphere C?

5. (a) In problem 4, what is the net charge on the three spheres before they are allowed to touch each other? (b) What is the net charge on the three spheres after they have touched?

Section 18.5 Coulomb's Law

6. The nucleus of the helium atom contains two protons that are separated by about 3.0×10^{-15} m. Find the magnitude of the electrostatic force that each proton exerts on the other. (The protons remain together in the nucleus because the repulsive electrostatic force is balanced by an attractive force called the strong nuclear force.)

7. Two very small spheres are initially neutral and separated by a distance of 0.50 m. Suppose that 3.0×10^{13} electrons are removed from one sphere and placed on the other. (a) What is the magnitude of the electrostatic force that acts on each sphere? (b) Is the force attractive or repulsive? Why?

8. At what separation distance do two point charges of $+1.0$ and $-1.0 \ \mu C$ exert a force of attraction on each other that is 440 N?

9. The force of repulsion between two like charges is 3.5 N. What will the force be if the distance between the charges is increased to five times its original value?

10. An object has a mass of 215 kg and is located at the surface of the earth (mass $= 5.98 \times 10^{24}$ kg, radius $= 6.38 \times 10^6$ m). Suppose this object and the earth each have an identical positive charge q. Assuming that the earth's charge is located at the center of the earth, determine q such that the electrostatic force exactly cancels the gravitational force.

11. Three charges are located on the $+x$ axis as follows: $q_1 = +25 \ \mu C$ at $x = 0$, $q_2 = +11 \ \mu C$ at $x = +2.0$ m, and $q_3 = +45 \ \mu C$ at $x = +3.5$ m. (a) Find the electrostatic force (magnitude and direction) acting on q_2. (b) Suppose q_2 were $-11 \ \mu C$, rather than $+11 \ \mu C$. Without performing any detailed calculations, specify the magnitude and direction of the force exerted on q_2. Give your reasoning.

12. A charge of $-3.00 \ \mu C$ is fixed at the center of a compass. Two additional charges are fixed on the circle of the compass (radius $= 0.100$ m). The charges on the circle are $-4.00 \ \mu C$ at the position due north and $+5.00 \ \mu C$ at the position due east. What is the magnitude and direction of the net electrostatic force acting on the charge at the center?

13. An equilateral triangle has sides of 0.15 m. Charges of -9.0, $+8.0$, and $+2.0 \ \mu C$ are located at the corners of the triangle. Find the magnitude and direction of the net electrostatic force exerted on the 2.0-μC charge.

14. Two particles, with identical positive charges and a separation of 2.60×10^{-2} m, are released from rest. Immediately after the release, particle 1 has an acceleration $\mathbf{a}_1$ whose magnitude is 4.60×10^3 m/s^2, while particle 2 has an acceleration $\mathbf{a}_2$ whose magnitude is 8.50×10^3 m/s^2. Particle 1 has a mass of 6.00×10^{-6} kg. Find (a) the charge on each particle and (b) the mass of particle 2.

*15. A charge $+q$ is fixed to each of three corners of a square. On the empty corner a charge is placed, such that there is no net electrostatic force acting on the diagonally opposite charge. What charge (magnitude and sign) is placed on the empty corner? Express your answer in terms of q.

*16. Two small objects A and B are fixed in place and separated by 2.00 cm. Object A has a charge of $+1.00\ \mu C$, and object B has a charge of $-1.00\ \mu C$. How many electrons must be removed from A and put onto B to make the electrostatic force that acts on each object an attractive force whose magnitude is 45.0 N?

*17. Two spheres are mounted on identical horizontal springs and rest on a frictionless table, as in the drawing. When the spheres are uncharged, the spacing between them is 5.0 cm, and the springs are unstressed. When each sphere has a charge of $+1.60\ \mu C$, the spacing doubles. Assuming the spheres have a negligible diameter, determine the spring constant of the springs.

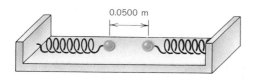

0.0500 m

*18. Two positive charges, when combined, give a total charge of $+9.00\ \mu C$. When the charges are separated by 3.00 m, the force exerted by one charge on the other has a magnitude of 8.00×10^{-3} N. Find the amount of each charge.

**19. There are four charges, each with a magnitude of $2.0\ \mu C$. Two are positive and two are negative. The charges are fixed to the corners of a 0.30-m square, one to a corner, in such a way that the force on any charge is directed toward the center of the square. Find the magnitude of the net electrostatic force experienced by any charge.

**20. Two identical, small insulating balls are suspended by separate 0.25-m threads that are attached to a common point on the ceiling. Each ball has a mass of 8.0×10^{-4} kg. Initially the balls are uncharged and hang straight down. They are then given identical positive charges and, as a result, spread apart with an angle of 36° between the threads. Determine (a) the charge on each ball and (b) the tension in the threads.

**21. Two objects are identical and small enough that their sizes can be ignored relative to the distance between them, which is 0.200 m. Each object carries a different charge, and they attract each other with a force of 1.20 N. The objects are brought into contact, so the net charge is shared equally, and then they are returned to their initial positions. Now it is found that the objects repel one another with a force whose magnitude is equal to that of the initial attractive force. What is the initial charge on each object? Note that there are two answers.

Section 18.6 The Electric Field, Section 18.7 Electric Field Lines, Section 18.8 Shielding

22. An electric field of 280 000 N/C points due south at a certain spot. What are the magnitude and direction of the force that acts on a charge of $-4.0\ \mu C$ at this spot?

23. A charge of $+3.0 \times 10^{-5}$ C is located at a place where there is an electric field that points due east and has a magnitude of 15 000 N/C. What are the magnitude and direction of the force acting on the charge?

24. An electric field with a magnitude of 160 N/C exists at a spot that is 0.15 m away from a charge. At a place that is 0.45 m away from this charge, what is the electric field strength?

25. The electric field at a distance of 0.50 m from a charge is 9.0×10^5 N/C, directed toward the charge. Find the magnitude and polarity of the charge.

26. Example 11 in the text deals with the hollow spherical conductor in Figure 18.29. The conductor is initially electrically neutral, and then a charge $+q$ is suspended at the center of the hollow space. Suppose the conductor initially has a net charge of $+2q$ instead of being neutral. What charges are then induced on the interior and exterior surfaces when the $+q$ charge is suspended at the center?

27. The helium atom contains two protons in its nucleus. When helium is singly ionized, one of its two electrons is completely removed. In the Bohr model of the singly ionized helium atom, the remaining electron orbits the nucleus at a radius of 2.65×10^{-11} m. (a) What is the electric field (magnitude and direction) created at the location of the electron by the positively charged nucleus? (b) What is the electrostatic force (magnitude and direction) acting on the electron? (c) Find the speed of the electron.

28. The magnitude of the electric field between the plates of a parallel plate capacitor is 2.4×10^5 N/C. Each plate carries a charge whose magnitude is $0.15\ \mu C$. What is the area of each plate?

29. A charge of $+3.5\ \mu C$ is fixed on the x axis at $x = +0.55$ m, while a charge of $-15\ \mu C$ is fixed at the origin. (a) Determine the net electric field (magnitude and direction) on the x axis at $x = +0.80$ m. (b) What force (magnitude and direction) would act on a charge of $-8.0\ \mu C$ placed on the x axis at $x = +0.80$ m?

30. Two charges each have a magnitude of $0.36\ \mu C$, one being positive and the other negative. These charges are fixed to the corners of an equilateral triangle, 0.75 m on a side. Determine the net electric field (magnitude and direction) that exists at the empty corner of the triangle.

*31. A small object has a mass of 2.0×10^{-3} kg and a charge of $-25\ \mu C$. It is placed at a certain spot where there is an electric field. When released, the object experiences an acceleration of 3.5×10^3 m/s^2 in the direction of the $+x$ axis. Determine the magnitude and direction of the electric field.

*32. A rectangle has a length of $2d$ and a height of d. Each of the following three charges is located at a corner of the rectangle: $+q_1$ (upper-left corner), $+q_2$ (lower-right corner), and $-q$ (lower-left corner). The net electric field at the (empty) upper-right corner is zero. Find the magnitude of q_1 and q_2. Express your answers in terms of q.

*33. The drawing shows two positive charges q_1 and q_2 fixed to a circle. At the center of the circle they produce a net electric field that is directed upward along the vertical axis. Determine the ratio q_2/q_1.

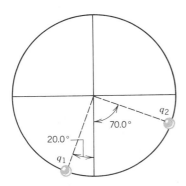

**34. The magnitude of the electric field between the plates of a parallel plate capacitor is 480 N/C. A silver dollar is placed between the plates and oriented parallel to the plates. (a) Ignoring the edges of the coin, find the induced charge density σ on each face of the coin. (b) Assuming the coin has a radius of 1.9 cm, find the total charge on each face of the coin.

**35. A rectangle has a length of $3L$ and a height of L. A charge of magnitude q_1 is placed at the upper-left corner, and another charge of magnitude q_2 is placed at the lower-right corner. The electric field at the (empty) lower-left corner is directed along the diagonal line that runs through the two empty corners. (a) Decide whether the charges have the same or different polarities. (b) Find the ratio q_2/q_1.

**36. A small plastic ball of mass 6.50×10^{-3} kg and charge $+0.150$ μC is suspended from an insulating thread and hangs between the plates of a capacitor (see the drawing). The ball is in equilibrium, with the thread making an angle of 30.0° with respect to the vertical. The area of each plate is 0.0150 m^2. What is the magnitude of the charge on each plate?

ADDITIONAL PROBLEMS

37. Two charges attract each other with a force of 1.5 N. What will be the force if the distance between them is reduced to one-ninth of its original value?

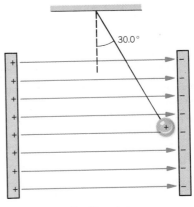

Problem 36

38. What electric field strength is needed to exert a 0.22-N force on a charge of 5.5×10^{-7} C?

39. Three positive charges are fixed along a line. From left to right they are q_1, q_2, and q_3. The charge q_2 is situated one-fourth of the way between q_1 and q_3 and experiences no net electrostatic force. Find the ratio q_3/q_1.

40. A tiny ball (mass = 0.012 kg) carries a charge of -18 μC. What electric field (magnitude and direction) is needed to cause the ball to float above the ground?

41. Three charges are fixed to an xy coordinate system. A charge of $+18$ μC is on the y axis at $y = +3.0$ m. A charge of -12 μC is at the origin. Lastly, a charge of $+45$ μC is on the x axis at $x = +3.0$ m. Determine the magnitude and direction of the net electrostatic force on the charge at $x = +3.0$ m.

42. Two charges, -16 and $+4.0$ μC, are fixed in place and separated by 3.0 m. (a) At what spot along the line between the charges is the net electric field zero? Locate this spot relative to the positive charge. (Hint: The spot does not necessarily lie between the two charges.) (b) What would be the force on a charge of $+14$ μC placed at this spot?

43. Two tiny spheres have the same mass and carry charges of the same magnitude. The mass of each sphere is 2.0×10^{-6} kg. The gravitational force that each sphere exerts on the other is balanced by the electric force. (a) What polarities can the charges have? (b) Determine the charge magnitude.

*44. A particle of mass 3.8×10^{-5} kg and charge $+12$ μC is released from rest in a region where there is a constant electric field of 470 N/C. (a) What is the acceleration of the particle due to the electric field? (b) How fast is the particle moving after traveling 0.020 m?

*45. Two small charged objects are attached to a horizontal spring, one at each end. The magnitudes of the charges are equal, and the spring constant is 180 N/m. The spring is observed to be stretched by 0.020 m relative to its unstrained length of 0.40 m. Determine (a) the possible polarities and (b) the magnitude of the charges.

*46. In the rectangle in the drawing, a charge is to be placed at the empty corner to make the net force on the charge at corner A point along the vertical direction. What charge (magnitude and sign) must be placed at the empty corner?

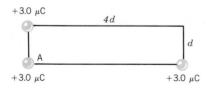

**47. A small spherical insulator of mass 8.00×10^{-2} kg and charge $+0.600 \ \mu C$ is hung by a thin wire of negligible mass. A charge of $-0.900 \ \mu C$ is held 0.150 m away from the sphere and directly to the right of it, so the wire makes an angle θ with the vertical (see drawing). Find (a) the angle θ and (b) the tension in the wire.

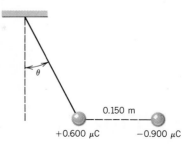

CHAPTER 19

ELECTRIC POTENTIAL ENERGY AND THE ELECTRIC POTENTIAL

High-voltage transmission lines, such as those shown above, are a common sight and help to make electricity one of today's technological marvels. Yet electricity is so common that we hardly think about it. It gives us our lights, heating and air conditioning, TV and radio, and keeps the ice cream in the refrigerator from melting. What's so remarkable about electricity is that, with the flick of a switch, it can instantly provide the power we need, be it a meager 60 watts for a light bulb or a hefty 20 000 watts for an air conditioner. Although electrical energy is typically generated many miles away, high-voltage transmission lines distribute it efficiently to consumers. In this chapter we will become acquainted with the concept of voltage and explore its relation to electrical energy and charge distributions. In addition, we will see that electrical energy and charge can be stored in an important device known as a capacitor.

19.1 POTENTIAL ENERGY

In the last chapter we discussed the electrostatic force between two point charges, $F = kq_1q_2/r^2$. The form of this equation is similar to that for the gravitational force between two point masses, which is $F = Gm_1m_2/r^2$, according to Newton's law of universal gravitation. Both of these forces are conservative forces, and as Section 6.4 explains, a potential energy can be associated with a conservative force. Thus, by analogy to the gravitational potential energy, we expect that an electric potential energy can be defined. To set the stage for a discussion of the electric potential energy, let's review some of the important aspects of the gravitational counterpart.

Figure 19.1 shows a ball of mass m located at point A, which is at a height h_A above the surface of the earth. Relative to the surface, the gravitational potential energy of the ball is given by Equation 6.5* as $GPE_A = mgh_A$, where g is the acceleration due to gravity. This expression assumes the height is small compared to the radius of the earth, so g is essentially constant at all heights. If the ball is to be lifted to a higher point B, an external force must be applied to compensate for the gravitational force. The hand in the drawing provides the external force. When the ball is lifted upward, it starts from rest and ends at rest. Therefore, there is no change in the ball's kinetic energy KE between the beginning and ending points, and the work W_{AB} done on the ball by the external force goes entirely into changing the gravitational potential energy:

$$W_{AB} = \underbrace{(KE_B - KE_A)}_{= 0} + (GPE_B - GPE_A) = mgh_B - mgh_A \qquad (6.7a)$$

If the ball is released from rest at B, it is accelerated toward the earth by the gravitational force. As the ball falls, its potential energy decreases and its kinetic energy increases. In the absence of dissipative forces such as friction, the total mechanical energy is conserved; that is, the sum of the gravitational potential energy and the kinetic energy remains constant at all times during the fall.

Figure 19.2 helps to clarify the analogy between electric potential energy and gravitational potential energy. In this drawing a positive test charge $+q_0$ is situated between two oppositely charged plates. Because of the charges on the plates, the test charge experiences an electric force that is directed toward the lower plate (the gravitational force is being neglected here). The hand in the drawing provides the external force needed to compensate for the electric force and move the test charge from A to B. The charge starts from rest and ends at rest, so that its kinetic energy does not change. Therefore, the work W_{AB} done on q_0 by the external force goes into changing the electric potential energy. The work equals the difference between the electric potential energy EPE at B and that at A:

$$W_{AB} = EPE_B - EPE_A \qquad (19.1)$$

Because the electric force is a conservative force, the path along which the charge is moved from A to B is of no consequence, for the work W_{AB} is the same for all paths.

19.2 THE ELECTRIC POTENTIAL DIFFERENCE

The work done to move the charge from A to B in Figure 19.2 depends on the magnitude of the charge, because the strength of the electric force opposing the motion depends on the magnitude of the charge. It is useful to express the work on a per-unit-charge basis by dividing both sides of Equation 19.1 by the charge q_0:

$$\frac{W_{AB}}{q_0} = \frac{EPE_B - EPE_A}{q_0} \qquad (19.2)$$

* The gravitational potential energy is denoted by GPE to distinguish it from the electric potential energy EPE.

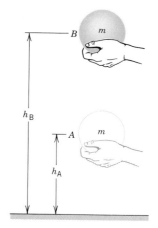

Figure 19.1 The gravitational potential energy of the ball at point A is $GPE_A = mgh_A$. Work W_{AB} (done by the hand) is required to raise the ball from A to B, where the gravitational potential energy is $GPE_B = mgh_B$.

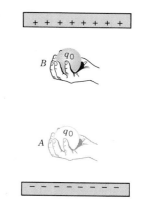

Figure 19.2 At locations A and B, the test charge $+q_0$ has electric potential energies of EPE_A and EPE_B. The work done to move the test charge from A to B at a constant speed is $W_{AB} = EPE_B - EPE_A$.

The *electric potential energy per unit charge* is an important concept in electricity and is known as the *electric potential*, or, simply, the *potential.*

Definition of Electric Potential

The electric potential V at a given point is the electric potential energy EPE of a small test charge q_0 situated at that point divided by the charge itself:

$$V = \frac{\text{EPE}}{q_0} \tag{19.3}$$

SI Unit of Electric Potential: joule/coulomb = volt (V)

The SI unit of electric potential is a joule per coulomb, a quantity known as a *volt.* The name honors Alessandro Volta (1745–1827), who invented the voltaic pile, the forerunner of the battery. Note that, in spite of the similarity in names, the electric potential energy EPE and the electric potential V are *not* the same. The electric potential energy, as its name implies, is an *energy* and, therefore, is measured in joules. In contrast, the electric potential is an *energy per unit charge,* and is measured in joules per coulomb, or volts.

According to Equations 19.2 and 19.3, the electric potential difference between two points A and B is related to the work per unit charge in the following way:

$$V_B - V_A = \frac{\text{EPE}_B - \text{EPE}_A}{q_0} = \frac{W_{AB}}{q_0} \tag{19.4}$$

Often, the "delta" notation is used to express the difference in potentials and the difference in potential energies: $\Delta V = V_B - V_A$ and $\Delta(\text{EPE}) = \text{EPE}_B - \text{EPE}_A$. In terms of this notation, Equation 19.4 takes the form

$$\Delta V = \frac{\Delta(\text{EPE})}{q_0} = \frac{W_{AB}}{q_0} \tag{19.4}$$

These night skiers at the Quebec Winter Carnival in Canada use battery-operated light torches to produce this serpentine display. The ability of a battery to provide energy for a light is indicated by the electric potential difference of the battery (expressed in volts).

Neither the potential V nor the potential energy EPE can be determined in an absolute sense, for only the *differences* ΔV and $\Delta(\text{EPE})$ are measurable in terms of the work W_{AB}. The gravitational potential energy has this same characteristic, since only the value at one height relative to that at some reference height has any significance. Example 1 emphasizes the relative nature of the electric potential.

Example 1   *Work, Electric Potential Energy, and Electric Potential*

In Figure 19.2, the work done to move the test charge ($q_0 = +2.0 \times 10^{-6}$ C) at a steady speed from A to B is $+5.0 \times 10^{-5}$ J. (a) Find the difference in the electric potential energies of the charge between the two points. (b) Determine the potential difference between the two points.

REASONING   Since work is required to move the test charge from A to B, this situation is like lifting an object uphill. At the top of the "hill" at B, the electric potential energy and the electric potential are greater than they are at the bottom at A.

SOLUTION

(a) The difference in potential energy between points A and B is equal to the work done in moving the charge from A to B. Therefore, $\text{EPE}_B - \text{EPE}_A = W_{AB} = \boxed{+5.0 \times 10^{-5} \text{ J}}$.

(b) The potential difference between A and B is the difference in potential energy divided by the charge

$$V_B - V_A = \frac{\text{EPE}_B - \text{EPE}_A}{q_0} = \frac{+5.0 \times 10^{-5} \text{ J}}{+2.0 \times 10^{-6} \text{ C}} = \boxed{+25 \text{ V}} \qquad (19.4)$$

The electric potential at B exceeds that at A by 25 V. It is not possible to determine separate values for V_B and V_A.

In Example 1, energy in the form of work is used to move the positive charge from the lower potential at A to the higher potential at B. According to the principle of the conservation of energy, this energy does not disappear. If the positive charge is released at B, it accelerates toward A because of the repulsion from the upper plate and the attraction to the lower plate. The speed of the charge increases as electric potential energy is converted into kinetic energy. Thus, *a positive charge accelerates from a region of higher potential toward a region of lower potential.* A negative charge behaves in the opposite fashion, since the electric force acting on it is directed opposite to that on a positive charge. *A negative charge accelerates from a region of lower potential toward a region of higher potential.* The next example illustrates the way positive and negative charges behave.

Example 2   *The Conservation of Energy*

In Figure 19.2, point B has an electric potential that is 25 V greater than that at point A, so $V_B - V_A = 25$ V. A particle has a mass of 1.8×10^{-5} kg and a charge whose magnitude is 3.0×10^{-5} C. The effects of gravity and friction are negligible. (a) If the particle has a positive charge and is released from rest at B, what speed v_A does the particle have when it arrives at A? (b) If the particle has a negative charge and is released from rest at A, what speed v_B does the particle have at B?

REASONING The only force acting on the moving charge is the conservative electric force. Therefore, the total energy of the charge (the sum of the kinetic energy KE and the electric potential energy EPE) is the same at points A and B:

$$KE_A + EPE_A = KE_B + EPE_B$$

Since $KE_A = \frac{1}{2}mv_A^2$ and the particle is at rest at point B, it follows that $\frac{1}{2}mv_A^2 = EPE_B - EPE_A$. The difference in potential energies is related to the difference in potentials by Equation 19.4, $EPE_B - EPE_A = q_0(V_B - V_A)$. As a result, we find that $\frac{1}{2}mv_A^2 = q_0(V_B - V_A)$, which allows us to determine the speed.

SOLUTION

(a) The speed is

$$v_A = \sqrt{\frac{2q_0(V_B - V_A)}{m}} = \sqrt{\frac{2(3.0 \times 10^{-5}\ C)(25\ V)}{1.8 \times 10^{-5}\ kg}} = \boxed{9.1\ m/s}$$

(b) A negative charge accelerates from a region of lower potential toward a region of higher potential. Therefore, when the negatively charged particle is released from rest at A, it accelerates toward B. The magnitude of the electric force causing this acceleration is the same as it is in part (a), but here the direction of the force is reversed. Thus, the speed v_B of the particle at B is the same as that calculated in part (a), $\boxed{v_B = 9.1\ m/s}$.

PROBLEM SOLVING INSIGHT

A positive charge accelerates when going from a higher potential to a lower potential. In contrast, a negative charge accelerates when going from a lower potential to a higher potential.

As a familiar application of electric potential energy and electric potential, Figure 19.3 shows a 12-V automobile battery with a headlight connected between the battery terminals. The positive terminal has a potential that is 12 V higher than the potential at the negative terminal. Positive charges are repelled from the positive terminal and travel through the wires and headlight toward the negative terminal.* As the charges pass through the headlight, virtually all their potential energy is converted into heat, which causes the filament to glow "white hot" and emit light. When the charges reach the negative terminal, they no longer have any potential energy. The battery then gives the charges an additional "shot" of potential energy by moving them to the higher-potential positive terminal, and the cycle is repeated. In raising the potential energy of the charges, the battery does work W_{AB} on them, and draws from its reserve of chemical energy to do so. Example 3 illustrates the concept of the electric potential difference as applied to a battery.

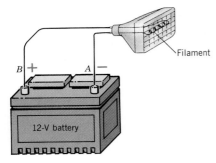

Figure 19.3 A headlight connected to a 12-V battery.

* Historically, it was believed that positive charges flow in the wires of an electric circuit. Today, it is known that negative charges flow in wires from the negative toward the positive terminal. However, it is customary to describe the flow of negative charges by specifying the opposite but equivalent flow of positive charges. This hypothetical flow of positive charges is called the conventional electric current, as we will see in Section 20.1.

Example 3   *Operating a Headlight*

Determine the number of particles, each carrying a charge of 1.60×10^{-19} C (the magnitude of the charge on an electron), that passes between the terminals of a 12-V car battery when a 60.0-W headlight burns for one hour.

REASONING To obtain the number of charged particles, we determine the total charge needed to provide the energy consumed by the headlight in one hour. Dividing the total charge by the charge on each particle gives the number of particles.

SOLUTION   Using energy at a rate of 60.0 joules per second (60.0 watts) for one hour, the headlight consumes a total energy of

$$\text{Energy} = \text{Power} \times \text{Time} = (60.0 \text{ W})(3600 \text{ s}) = 2.2 \times 10^5 \text{ J} \qquad (6.10)$$

Equation 19.4 gives the total amount of charge that delivers this much energy upon passing through a 12-V potential difference:

$$q_0 = \frac{\Delta(\text{EPE})}{\Delta V} = \frac{2.2 \times 10^5 \text{ J}}{12 \text{ V}} = 1.8 \times 10^4 \text{ C}$$

The number of particles whose individual charges combine to provide this total charge is $(1.8 \times 10^4 \text{ C})/(1.60 \times 10^{-19} \text{ C}) = \boxed{1.1 \times 10^{23}}$.

As used in connection with batteries, the volt is a familiar unit for measuring electric potential difference. The word "volt" also appears in another context, as part of a unit that is used to measure energy, particularly the energy of an atomic particle, such as an electron or a proton. This energy unit is called the _electron volt_ (eV). **One electron volt is the change in potential energy of an electron ($q_0 = 1.60 \times 10^{-19}$ C) when the electron moves through a potential difference of one volt.** Since the change in potential energy equals $q_0 \Delta V$, one electron volt is equal to $(1.60 \times 10^{-19} \text{ C}) \times (1.00 \text{ V}) = 1.60 \times 10^{-19}$ J; thus

$$1 \text{ eV} = 1.60 \times 10^{-19} \text{ J}$$

One million or 10^{+6} electron volts of energy is referred to as one MeV, and one billion or 10^{+9} electron volts of energy is one GeV, where the "G" stands for the prefix "giga" (pronounced "jig'a").

19.3 THE ELECTRIC POTENTIAL DIFFERENCE CREATED BY POINT CHARGES

A positive point charge $+q$ creates an electric potential in a fashion that Figure 19.4 helps to explain. This picture shows two locations A and B, at distances r_A and r_B from the charge. At any position between A and B an electrostatic force of repulsion acts on a positive test charge $+q_0$. The magnitude of the force is given by Coulomb's law as $F = kq_0q/r^2$. Thus, to move the test charge from A to B at a constant speed, an external agent must apply a force $F = kq_0q/r^2$ to balance the electric force. Since r varies between r_A and r_B, F also varies, and therefore the work done on q_0 is not simply the product of F and the distance between the points. However, the work can be found by using integral calculus. According to Equation 19.4, dividing the work W_{AB} by q_0 gives the potential difference between B and A. The result is

$$V_B - V_A = \frac{W_{AB}}{q_0} = \frac{kq}{r_B} - \frac{kq}{r_A} \qquad (19.5)$$

As point A is located farther and farther away from the charge q, r_A becomes larger and larger. In the limit that r_A becomes infinitely large, the term kq/r_A becomes zero, and it is customary to write $V_B = kq/r_B$. In this limit, then, we omit the subscripts and write the potential in the following form:

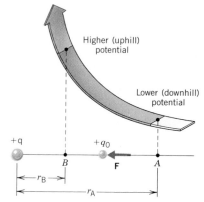

Figure 19.4 The positive test charge $+q_0$ experiences a repulsive force due to the positive point charge $+q$. As a result, work must be done by a force **F** to move the test charge from A to B. Consequently, the electric potential is higher (uphill) at B and lower (downhill) at A.

$$\begin{bmatrix} \text{Potential of} \\ \text{a point} \\ \text{charge} \end{bmatrix} \qquad V = \frac{kq}{r} \qquad (19.6)$$

The symbol V in this equation does not refer to the potential in any absolute sense. Rather, $V = kq/r$ stands for the amount by which the potential at a distance r from a point charge differs from the potential at an infinite distance away. In other words, V refers to a potential difference with the arbitrary assumption that the potential at infinity is zero.

With the aid of Equation 19.6, we can describe the effect that a point charge q has on the surrounding space. When q is positive, the value of $V = kq/r$ is also positive, indicating the positive charge has everywhere raised the potential above the zero reference value. Conversely, when q is negative, the potential V is also negative, indicating the negative charge has everywhere decreased the potential below the zero reference value. The next example deals with these effects quantitatively.

Example 4 The Potential of a Point Charge

Using a zero reference potential at infinity, determine the amount by which a point charge of 4.0×10^{-8} C alters the electric potential at a spot 1.2 m away when the charge is (a) positive and (b) negative.

REASONING AND SOLUTION
(a) Figure 19.5a shows the potential when the charge is positive:

$$V = \frac{kq}{r} = \frac{(8.99 \times 10^9 \text{ N} \cdot \text{m}^2/\text{C}^2)(+4.0 \times 10^{-8} \text{ C})}{1.2 \text{ m}} = \boxed{+300 \text{ V}} \qquad (19.6)$$

(b) Part b of the drawing illustrates the results when the charge is negative. A calculation similar to that in part (a) shows the potential is now negative: $\boxed{-300 \text{ V}}$.

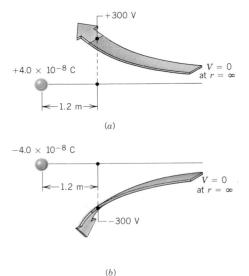

Figure 19.5 A point charge of 4.0×10^{-8} C alters the potential at a spot 1.2 m away. The potential is (a) increased by 300 V when the charge is positive and (b) decreased by 300 V when the charge is negative, relative to a zero reference potential at infinity.

The electric potential created by a point charge plays an important role in the structure of atoms, as Example 5 illustrates for the case of the hydrogen atom.

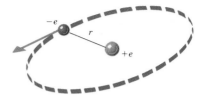

Figure 19.6 In the Bohr model of the hydrogen atom, the electron $(-e)$ orbits the proton $(+e)$ at a distance of $r = 5.29 \times 10^{-11}$ m.

Example 5   Ionization Energy

In the Bohr model of the hydrogen atom, the electron $(-e)$ is in an orbit around the nuclear proton $(+e)$ at a distance of $r = 5.29 \times 10^{-11}$ m, as Figure 19.6 shows. Find (a) the electric potential that the proton creates at this distance, (b) the total energy of the atom, and (c) the ionization energy for the atom. The ionization energy is the energy that must be put into the atom to remove the electron and place it at rest infinitely far from the proton. Express the answers to parts (b) and (c) in electron volts.

REASONING   The total energy of the atom is the sum of the electric potential energy and the kinetic energy of the electron. The potential energy (relative to a zero reference value at infinity) can be obtained from Equation 19.4 as EPE $= qV$, where $q = -e$ is the charge on the electron and V is the electric potential created by the proton. The electric potential is the point charge potential given by Equation 19.6. The kinetic energy is KE $= \frac{1}{2}mv^2$, where m is the mass of the electron and v is its speed.

SOLUTION
(a) The electric potential created by the proton charge of $+e = +1.60 \times 10^{-19}$ C is

$$V = \frac{kq}{r} = \frac{(8.99 \times 10^9 \text{ N} \cdot \text{m}^2/\text{C}^2)(+1.60 \times 10^{-19} \text{ C})}{5.29 \times 10^{-11} \text{ m}} = \boxed{+27.2 \text{ V}} \quad (19.6)$$

(b) The total energy E of the atom is $E =$ EPE $+$ KE. Using the electric potential of $+27.2$ V obtained in part (a), we find that the electric potential energy is

$$\text{EPE} = qV = (-1.60 \times 10^{-19} \text{ C})(+27.2 \text{ V}) = -4.35 \times 10^{-18} \text{ J}$$

From Example 3 in Section 18.5, we know that the speed of the electron is $v = 2.18 \times 10^6$ m/s. Since the mass of the electron is 9.11×10^{-31} kg, its kinetic energy is KE $= \frac{1}{2}mv^2 = 2.17 \times 10^{-18}$ J. The total energy, then, is

$$E = \text{EPE} + \text{KE} = -4.35 \times 10^{-18} \text{ J} + 2.17 \times 10^{-18} \text{ J} = -2.18 \times 10^{-18} \text{ J}$$

The total energy in electron volts (1 eV $= 1.60 \times 10^{-19}$ J) is $\boxed{E = -13.6 \text{ eV}}$.

(c) Removing the electron from its orbit and placing the electron at rest at infinity requires that $+13.6$ eV of energy be put into the atom. Then the total energy will be zero, the assumed reference value for the potential energy. Thus, the ionization energy of the hydrogen atom is $\boxed{+13.6 \text{ eV}}$, which agrees with the experimental value.

PROBLEM SOLVING INSIGHT

Be careful to distinguish between the concepts of potential V and electric potential energy EPE. Potential is electric potential energy per unit charge: $V =$ EPE/q.

A single point charge raises or lowers the potential at a given location, depending on whether the charge is positive or negative. *When two or more charges are present, the potential due to all the charges is obtained by adding together the individual potentials,* as Example 6 shows.

PROBLEM SOLVING INSIGHT

Example 6   The Total Electric Potential

At the locations A and B in Figure 19.7, find the total electric potential due to the two point charges.

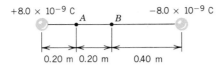

Figure 19.7 Both the positive and negative charges affect the electric potential at locations *A* and *B*.

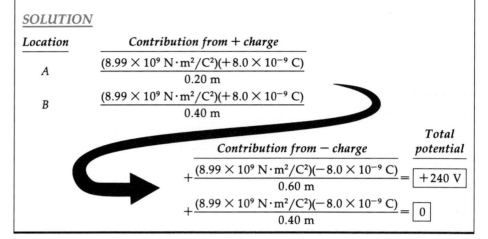

REASONING At each location, each charge contributes to the total electric potential. We obtain the individual contributions by using $V = kq/r$ and find the total potential by adding the individual contributions algebraically. The two charges have the same magnitude. Thus, at *A* the total potential is positive, because this spot is closer to the positive charge, whose effect dominates that of the more distant negative charge. At *B*, midway between the charges, the total potential is zero, since the potential of one charge exactly offsets that of the other.

SOLUTION

Location	Contribution from + charge
A	$\dfrac{(8.99 \times 10^9 \text{ N} \cdot \text{m}^2/\text{C}^2)(+8.0 \times 10^{-9} \text{ C})}{0.20 \text{ m}}$
B	$\dfrac{(8.99 \times 10^9 \text{ N} \cdot \text{m}^2/\text{C}^2)(+8.0 \times 10^{-9} \text{ C})}{0.40 \text{ m}}$

Contribution from − charge	Total potential
$+\dfrac{(8.99 \times 10^9 \text{ N} \cdot \text{m}^2/\text{C}^2)(-8.0 \times 10^{-9} \text{ C})}{0.60 \text{ m}}$	$= \boxed{+240 \text{ V}}$
$+\dfrac{(8.99 \times 10^9 \text{ N} \cdot \text{m}^2/\text{C}^2)(-8.0 \times 10^{-9} \text{ C})}{0.40 \text{ m}}$	$= \boxed{0}$

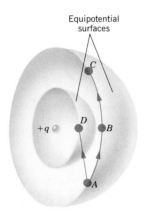

Figure 19.8 The equipotential surfaces that surround the point charge +*q* are spherical. No work is required to move a charge at a constant speed on a path that lies on an equipotential surface, such as the path *ABC*. However, work is required to move a charge between two equipotential surfaces, as along the path *AD*.

PROBLEM SOLVING INSIGHT

19.4 EQUIPOTENTIAL SURFACES AND THEIR RELATION TO THE ELECTRIC FIELD

EQUIPOTENTIAL SURFACES

An *equipotential surface* is a surface on which the electric potential is the same everywhere. The easiest equipotential surfaces to visualize are those that surround an isolated point charge. According to Equation 19.6, the potential at a distance *r* from a point charge *q* is $V = kq/r$. Thus, wherever *r* is the same, the potential is the same, and the equipotential surfaces are spherical surfaces centered on the charge. There are an infinite number of such surfaces, one for every value of *r*, and Figure 19.8 illustrates two of them. The larger the distance *r*, the smaller is the potential of the equipotential surface.

No work is required to move a charge at constant speed on an equipotential surface. This important characteristic arises because when work is done on an object, either the potential energy or the kinetic energy of the object changes. Since both types of energy remain unchanged when a charge is moved at constant speed on an equipotential surface, no work is needed to accomplish the motion. In Figure 19.8, for instance, no work is done in moving a test charge at constant speed along the circular arc *ABC*. The only force that must be applied to the test charge is that needed to counteract the electric force produced by *q*, and this applied force, being perpendicular to the path *ABC*, does no work. (See Section 6.1 for a review of work.) In contrast, work must be done to move a charge at a constant speed *between* equipotential surfaces, as from *A* to *D* in the picture. The work is the product of the charge and the difference between the potentials of the surfaces, in accord with Equation 19.4.

The spherical equipotential surfaces that surround an isolated point charge illustrate another characteristic of such surfaces. Figure 19.9 shows two equipotential surfaces around a positive point charge, along with some electric field lines. The electric field lines give the direction of the electric field, and for a positive point charge, the electric field is directed radially outward. Therefore, at each location on an equipotential sphere the electric field is perpendicular to the surface and points outward in the direction of decreasing potential, as the drawing emphasizes. This perpendicular relation is valid whether or not the shape of the equipotential surface is spherical; *the electric field created by any group of charges is everywhere perpendicular to the associated equipotential surfaces and points in the direction of decreasing potential.* For example, Figure 19.10 shows the electric field lines around an electric dipole, along with some equipotential surfaces. Since the field lines are not simply radial, the equipotential surfaces are no longer spherical, but, instead, have the necessary shape so as to be everywhere perpendicular to the field lines.

To see why an equipotential surface must be perpendicular to the electric field, consider Figure 19.11, which shows a hypothetical situation in which the per-

PROBLEM SOLVING INSIGHT

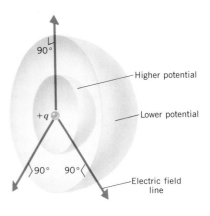

Figure 19.9 The radially directed electric field of a point charge is perpendicular to the spherical equipotential surfaces that surround the charge. The electric field points in the direction of *decreasing* potential.

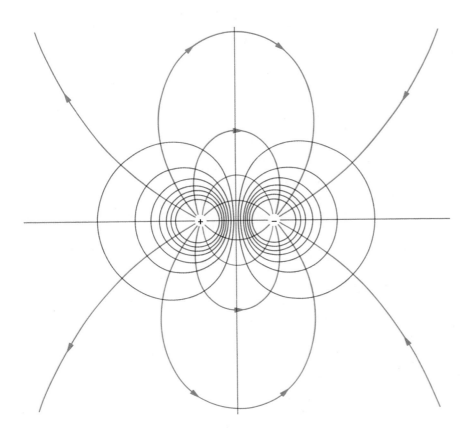

Figure 19.10 A cross-sectional view of the equipotential surfaces (in blue) of an electric dipole. The surfaces are drawn so that at every point they are perpendicular to the electric field lines (in red) of the dipole.

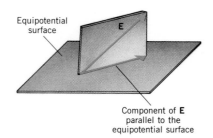

Equipotential surface

E

Component of **E** parallel to the equipotential surface

Figure 19.11 In this hypothetical situation, the electric field **E** is not perpendicular to the equipotential surface. As a result, there is a component of **E** parallel to the surface.

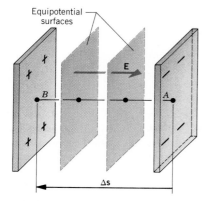

Equipotential surfaces

E

B

A

Δs

Figure 19.12 The metal plates of a parallel plate capacitor are equipotential surfaces. Two additional equipotential surfaces are shown between the plates. These two equipotential surfaces are parallel to the plates and are perpendicular to the electric field **E** between the plates.

pendicular relation does *not* hold. If **E** were not perpendicular to the equipotential surface, there would be a component of **E** parallel to the surface. This field component would exert an electric force on a test charge placed on the surface. To move the test charge at a constant speed along the surface, work would have to be done to counteract the effect of this force. Thus, the surface could not be an equipotential surface as assumed. The only way out of the dilemma is for the electric field to be perpendicular to the surface, so there is no component of the field parallel to the surface.

We have already encountered one equipotential surface in the last chapter. In Section 18.8, we found that the direction of the electric field just outside an electrical conductor is perpendicular to the conductor's surface, when the conductor is at equilibrium under electrostatic conditions. Thus, the surface of any conductor is an equipotential surface under such conditions. In fact, since the electric field is zero everywhere inside a conductor whose charges are in equilibrium, the entire conductor can be regarded as an equipotential volume.

THE RELATION BETWEEN THE ELECTRIC FIELD AND THE ELECTRIC POTENTIAL

There is a quantitative relation between the electric field and the equipotential surfaces. One example that illustrates this relation is the parallel plate capacitor in Figure 19.12. As Section 18.6 discusses, the electric field **E** between the metal plates is perpendicular to the plates and is the same everywhere, ignoring fringe fields at the edges. To be perpendicular to the electric field, the equipotential surfaces must be planes that are parallel to the plates, which themselves are equipotential surfaces. The potential difference between the plates is given by Equation 19.4 as $V_B - V_A = \Delta V = W_{AB}/q_0$, where A is a point on the negative plate and B is a point on the positive plate. The work required to move a positive test charge q_0 from A to B is $W_{AB} = F \Delta s$, where F is the magnitude of the applied force, directed to the left in opposition to the electric force, and Δs is the magnitude of the displacement along a line perpendicular to the plates. If the test charge is moved at a constant speed, the magnitude of the applied force equals the magnitude of the electric force, so $F = q_0 E$, and the work becomes $W_{AB} = F \Delta s = q_0 E \Delta s$. The potential difference between the capacitor plates is the work per unit charge, so

$$\Delta V = \frac{W_{AB}}{q_0} = \frac{q_0 E \Delta s}{q_0} \quad \text{or} \quad E = \frac{\Delta V}{\Delta s}$$

As the test charge is moved from the negative plate to the positive plate, the potential difference increases with distance ($\Delta V/\Delta s$ is positive). However, the electric field is in the opposite direction, for it points from the positive plate toward the negative plate. To account for the fact that the electric field is directed opposite to the direction in which the potential increases, Equation 19.7 includes a minus sign:

$$E = -\frac{\Delta V}{\Delta s} \tag{19.7}$$

The quantity $\Delta V/\Delta s$ is referred to as the *potential gradient* and has units of volts per meter. The next example deals further with the equipotential surfaces between the plates of a capacitor.

Example 7 *The Electric Field and Potential Are Related*

The plates of the capacitor in Figure 19.12 are separated by a distance of 0.032 m, and the potential difference between them is 64 V. Between the two equipotential surfaces shown in color there is a potential difference of 3.0 V. Find the spacing between the two colored surfaces.

REASONING   The magnitude of the electric field is $E = \Delta V / \Delta s$ (Equation 19.7 without the minus sign). To find the spacing between the two colored equipotential surfaces, we solve this equation for Δs, with $\Delta V = 3.0$ V and E equal to the magnitude of the electric field between the plates of the capacitor. A value for E can be obtained by using the values given for the distance and potential difference between the plates.

SOLUTION   The magnitude of the electric field between the capacitor plates is

$$E = \frac{\Delta V}{\Delta s} = \frac{64 \text{ V}}{0.032 \text{ m}} = 2.0 \times 10^3 \text{ V/m}$$

The spacing between the colored equipotential surfaces can now be determined:

$$\Delta s = \frac{\Delta V}{E} = \frac{3.0 \text{ V}}{2.0 \times 10^3 \text{ V/m}} = \boxed{1.5 \times 10^{-3} \text{ m}}$$

19.5 CAPACITORS AND DIELECTRICS

THE CAPACITANCE OF A CAPACITOR

In Section 18.6 we saw that a parallel plate capacitor consists of two parallel metal plates placed near one another, but not touching. This type of capacitor is only one among many. In general, a *capacitor* consists of two conductors of any shape placed near one another without touching. For a reason that will become clear later on, it is common practice to fill the region between the conductors or plates with an electrically insulating material called a *dielectric*, as Figure 19.13 illustrates.

A capacitor stores electric charge. Each capacitor plate carries a charge of the *same magnitude*, one being positive, while the other is negative. Because of the charges, the electric potential of the positive plate exceeds that of the negative plate by an amount V, as Figure 19.13 indicates. Experiment shows that when the magnitude q of the charge on each plate is doubled, the electric potential difference V is also doubled, so q is proportional to V: $q \propto V$. Equation 19.8 expresses this proportionality with the aid of a proportionality constant C, which is the *capacitance* of the capacitor.

The Relation between Charge and Potential Difference for a Capacitor

The magnitude q of the charge on each plate of a capacitor is directly proportional to the magnitude V of the potential difference between the plates:

$$q = CV \qquad (19.8)$$

where C is the capacitance.

   SI Unit of Capacitance: coulomb/volt = farad (F)

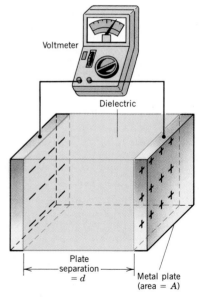

Figure 19.13 A parallel plate capacitor consists of two metal plates, one carrying a charge $+q$ and the other a charge $-q$. The potential of the positive plate exceeds that of the negative plate by an amount V. The region between the plates is filled with a dielectric.

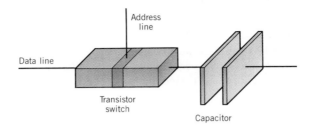

Figure 19.14 A transistor–capacitor combination is part of a RAM chip used in computer memories.

THE PHYSICS OF . . .

random-access memory (RAM) chips.

Equation 19.8 shows that the SI unit of capacitance is the coulomb per volt (C/V). This unit is called the *farad* (F), named after the English scientist Michael Faraday (1791–1867). One farad is an enormous amount of capacitance. Usually smaller amounts, such as a microfarad (1 μF = 10^{-6} F) or a picofarad (1 pF = 10^{-12} F), are used in electric circuits. The capacitance reflects the ability of the capacitor to store charge, in the sense that a larger capacitance C allows more charge q to be put onto the plates for a given value of the potential difference V.

The ability of a capacitor to store charge lies at the heart of the random-access memory (RAM) chips used in computers, where information is stored in the form of the "ones" and "zeros" that comprise binary numbers. Figure 19.14 illustrates the role of a capacitor in a RAM chip. The capacitor is connected to a transistor switch, to which two lines are connected, an address line and a data line. A single RAM chip often contains hundreds of thousands of such transistor–capacitor units. The address line is used by the computer to locate a particular transistor-capacitor combination, and the data line carries the data to be stored. A pulse on the address line turns on the transistor switch. With the switch turned on, a pulse coming in on the data line can cause the capacitor to charge. A charged capacitor means that a "one" has been stored, while an uncharged capacitor means that a "zero" has been stored.

THE DIELECTRIC CONSTANT

If a dielectric is inserted between the plates of a capacitor, the capacitance can increase markedly, because of the way in which the dielectric alters the electric field between the plates. Figure 19.15 shows how this effect comes about. In part *a*, the region between the charged plates is empty. The electric field lines point from the positive toward the negative plate. In part *b*, a dielectric has been inserted between the plates. Since the capacitor is not connected to anything, the charge on the plates remains constant as the dielectric is inserted. In many materials (e.g., water) the molecules possess permanent dipole moments, even though the molecules are electrically neutral. The dipole moment exists, because one end of a molecule has a slight excess of negative charge while the other end has a slight excess of positive charge. When such molecules are placed between the charged plates of the capacitor, the negative ends are attracted to the positive plate and the positive ends are attracted to the negative plate. As a result, the dipolar molecules tend to orient themselves end-to-end, as in part *b*. Whether or not a molecule has a permanent dipole moment, the electric field can cause the electrons to shift position within a molecule, making one end slightly negative and the opposite end slightly positive. Once again, the result is similar to that in part *b*. Because of the end-to-end orientation, the extreme left surface of the dielectric becomes positively charged, and the extreme right surface becomes negatively charged. The surface charges are shown in red in the picture.

Because of the surface charges on the dielectric, not all the electric field lines

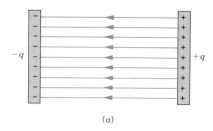

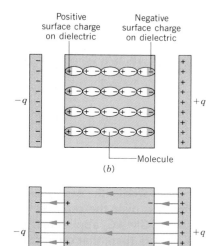

Figure 19.15 (a) The electric field
lines inside an empty capacitor.
(b) The electric field produced by
the charges on the plates creates an
end-to-end alignment of the molecular dipoles within the dielectric.
(c) The resulting positive and negative surface charges on the dielectric
cause a reduction in the electric
field within the dielectric. The space
between the dielectric and the
plates is added only for clarity. In
reality, the dielectric completely fills
the region between the plates.

generated by the charges on the plates pass through the dielectric. As Figure
19.15c shows, some of the field lines end on the negative surface charges and
begin again on the positive surface charges. Thus, the electric field inside the
dielectric is less than the electric field inside an empty capacitor, assuming the
charge on the plates remains constant. This reduction in the electric field is
described by the *dielectric constant* κ, which is the ratio of the field magnitude E_0
without the dielectric to the field magnitude E inside the dielectric:

$$\kappa = \frac{E_0}{E} \qquad (19.9)$$

Being a ratio of two field strengths, the dielectric constant is a number without
units. Moreover, since the field $\mathbf{E_0}$ without the dielectric is greater than the field $\mathbf{E}$
inside the dielectric, the dielectric constant is greater than unity. The value of κ
depends on the nature of the dielectric material, as Table 19.1 indicates.

Table 19.1 Dielectric Constants
of Some Common Substances[a]

Substance	Dielectric Constant, κ
Vacuum	1
Air	1.00054
Teflon	2.1
Benzene	2.28
Paper (royal gray)	3.3
Ruby mica	5.4
Neoprene rubber	6.7
Methyl alcohol	33.6
Water	80.4

[a] Near room temperature.

THE CAPACITANCE OF A PARALLEL PLATE CAPACITOR

The capacitance of a capacitor is affected by the geometry of the plates and the
dielectric constant of the material between them. For example, Figure 19.13
shows a parallel plate capacitor in which the area of each plate is A and the
separation between the plates is d. The magnitude of the electric field inside the
dielectric is given by Equation 19.7 (without the minus sign) as $E = V/d$, where V
is the potential difference between the plates. If the charge on each plate is kept

fixed, the electric field inside the dielectric is related to the electric field in the absence of the dielectric via Equation 19.9. Therefore,

$$E = \frac{E_0}{\kappa} = \frac{V}{d}$$

Since the electric field within an empty capacitor is $E_0 = q/(\epsilon_0 A)$ (see Equation 18.4), it follows that $q/(\epsilon_0 A\kappa) = V/d$, which can be solved for q to give

$$q = \left(\frac{\kappa\epsilon_0 A}{d}\right)V$$

A comparison of this expression with $q = CV$ (Equation 19.8) reveals that the capacitance C is

$$\begin{bmatrix} \textbf{Parallel plate} \\ \textbf{capacitor filled} \\ \textbf{with a dielectric} \end{bmatrix} \qquad C = \frac{\kappa\epsilon_0 A}{d} \qquad\qquad (19.10)$$

Notice that only the geometry of the plates (A and d) and the dielectric constant κ affect the capacitance. With C_0 representing the capacitance of the empty capacitor ($\kappa = 1$), Equation 19.10 shows that $C = \kappa C_0$. In other words, the capacitance with the dielectric present is increased by a factor of κ over the capacitance without the dielectric. It can be shown that the relation $C = \kappa C_0$ applies to any capacitor, not just to a parallel plate capacitor. One reason, then, that capacitors are filled with dielectric materials is to increase the capacitance. Example 8 illustrates the effect that increasing the capacitance can have on the amount of charge stored by a capacitor.

Example 8   Storing Electric Charge

The capacitance of an empty capacitor is 1.2 μF. The capacitor is connected to a 12-V battery and charged up. With the capacitor connected to the battery, a slab of dielectric material is inserted between the plates. As a result, 2.6×10^{-5} C of *additional* charge flows from one plate, through the battery, and on to the other plate. What is the dielectric constant κ of the material?

REASONING   The charge stored by a capacitor is $q = CV$, according to Equation 19.8. The battery maintains a constant potential difference of $V = 12$ volts between the plates of the capacitor, since the capacitor remains connected to the battery while the dielectric is inserted. Inserting the dielectric causes the capacitance C to increase, so that with V held constant, the charge q must increase. Thus, additional charge flows out of the battery and onto the capacitor plates. To find the dielectric constant, we apply Equation 19.8 to the empty capacitor and then to the capacitor filled with the dielectric material.

SOLUTION   The empty capacitor has a capacitance $C_0 = 1.2\ \mu$F and stores an amount of charge $q_0 = C_0 V$. With the dielectric material in place, the capacitor has a capacitance $C = \kappa C_0$ and stores an amount of charge $q = (\kappa C_0)V$. The additional charge that the battery supplies is $q - q_0 = (\kappa C_0)V - C_0 V = 2.6 \times 10^{-5}$ C. Solving for the dielectric constant, we find that

$$\kappa = \frac{2.6 \times 10^{-5}\ \text{C}}{C_0 V} + 1 = \frac{2.6 \times 10^{-5}\ \text{C}}{(1.2 \times 10^{-6}\ \text{F})(12\ \text{V})} + 1 = \boxed{2.8}$$

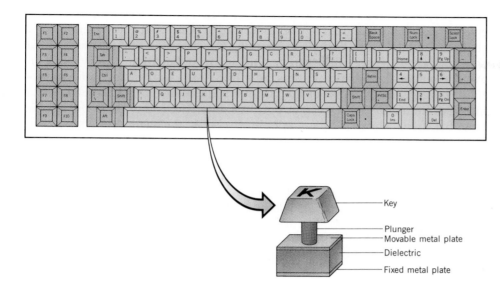

Figure 19.16 In one kind of computer keyboard, each key, when pressed, changes the separation between the plates of a capacitor.

Key
Plunger
Movable metal plate
Dielectric
Fixed metal plate

Capacitors are used often in electronic devices, and Example 9 deals with one familiar application.

THE PHYSICS OF . . .

a computer keyboard.

Example 9   A Computer Keyboard

One common kind of computer keyboard is based on the idea of capacitance. Each key is mounted on one end of a plunger, the other end being attached to a movable metal plate (see Figure 19.16). The two plates of the key form a capacitor. When the key is pressed, the movable plate is pushed closer to the fixed plate, and the capacitance increases. Electronic circuitry enables the computer to detect the *change* in capacitance and, thus, recognize which key has been pressed. The separation of the plates is normally 5.00×10^{-3} m, but decreases to 0.150×10^{-3} m when a key is pressed. The plate area is 9.50×10^{-5} m^2 and the capacitor is filled with a material whose dielectric constant is 3.50. Determine the change in capacitance that is detected by the computer.

REASONING AND SOLUTION   When the key is pressed, the capacitance is

$$C = \frac{\kappa \epsilon_0 A}{d} = \frac{(3.50)[8.85 \times 10^{-12}\ \text{C}^2/(\text{N} \cdot \text{m}^2)](9.50 \times 10^{-5}\ \text{m}^2)}{0.150 \times 10^{-3}\ \text{m}}$$

$$= 19.6 \times 10^{-12}\ \text{F}\quad (19.6\ \text{pF})$$

A similar calculation shows that when the key is *not* pressed, the capacitance is 0.589×10^{-12} F (0.589 pF). The *change* in capacitance is $\boxed{19.0 \times 10^{-12}\ \text{F}\ (19.0\ \text{pF})}$.

The presence of the dielectric increases the *change* in the capacitance. The greater the change in capacitance, the easier it is for the circuitry within the computer to detect it.

ENERGY STORAGE IN A CAPACITOR

A capacitor is a device for storing charge. Alternatively, it is possible to view the capacitor as a device for storing energy. After all, the charge on the plates possesses electric potential energy, which arises because work was done to deposit the charge on the plates. In fact, as each small increment of charge is deposited

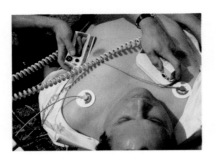

A defibrillator is a device used in emergency situations to revive a person who has suffered a heart attack. The device uses the energy stored in a capacitor to deliver a controlled electric shock that can restore normal heart rhythm.

during the charging process, the potential difference between the plates increases, and a larger amount of work is needed to bring up the next increment of charge. The total work W done in charging the capacitor, and hence the total electric potential energy EPE, can be obtained from Equation 19.4 by using an average potential difference $\overline{V}$; thus, $W = \text{EPE} = q\overline{V}$. As the capacitor becomes fully charged, the potential of the positive plate relative to the negative plate increases from 0 to V. The average potential difference is $\overline{V} = \frac{1}{2}V$, so the electric potential energy stored is $\text{EPE} = \frac{1}{2}qV$. Since $q = CV$, the energy stored becomes

$$\text{Energy} = \tfrac{1}{2}(CV)V = \tfrac{1}{2}CV^2 \qquad (19.11)$$

It is also possible to regard the energy as being stored in the electric field between the plates, rather than in the potential energy of the charge on the plates. The relation between energy and field strength can be obtained for a parallel plate capacitor by substituting $V = Ed$ and $C = \kappa\epsilon_0 A/d$ into Equation 19.11:

$$\text{Energy} = \frac{1}{2}\left(\frac{\kappa\epsilon_0 A}{d}\right)(Ed)^2$$

Since the area A times the separation d is the volume between the plates, the energy per unit volume or **energy density** is

$$\text{Energy density} = \frac{\text{Energy}}{\text{Volume}} = \tfrac{1}{2}\kappa\epsilon_0 E^2 \qquad (19.12)$$

It can be shown that this expression is valid for any electric field strength, not just that between the plates of a capacitor.

THE PHYSICS OF . . .

an electronic flash attachment for a camera.

The energy-storing capability of a capacitor is often put to good use in electronic circuits. For example, in an electronic flash attachment for a camera, energy from the battery pack is stored in capacitors. The capacitors are then discharged between the electrodes of the flash tube, which converts the energy into light. Flash duration times range from 1/200 to 1/1 000 000 second or less, with the shortest flashes being used in high-speed photography. Some flash attachments automatically control the flash duration by monitoring the light reflected from the photographic subject and quickly stopping or quenching the capacitor discharge when the reflected light reaches a predetermined level.

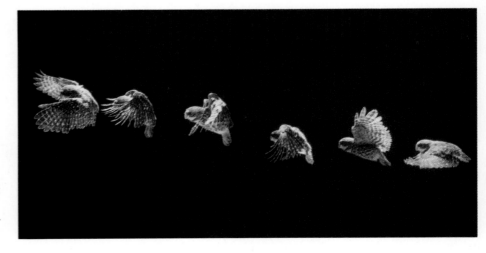

This time-lapse photo of an owl about to land was obtained by using an electronic flash attachment with the camera. The energy for each "flash" comes from the electrical energy stored in a capacitor.

*19.6 MEDICAL APPLICATIONS OF ELECTRIC POTENTIAL DIFFERENCES

Several important medical diagnostic techniques depend on the fact that the surface of the human body is *not* an equipotential surface. Between various points on the body there are small potential differences (approximately 30–500 μV), which provide the basis for electrocardiography, electroencephalography, and electroretinography. The potential differences can be traced to the electrical characteristics of muscle cells and nerve cells. In carrying out their biological functions, these cells utilize positively charged sodium and potassium ions and negatively charged chlorine ions that exist within the cells and in the intercellular fluid outside the cells. As a result of such charged particles, electric fields are generated that extend to the surface of the body and lead to small potential differences.

Figure 19.17 shows some locations on the body where electrodes are placed to measure potential differences in electrocardiography. The potential difference between two points changes as the heart beats and forms a repetitive pattern. The recorded pattern of potential difference versus time is called an electrocardiogram (ECG or EKG), and its shape depends on which pair of points in the picture (A and B, B and C, etc.) is used to locate the electrodes. The figure also shows some EKGs and indicates the regions (P, Q, R, S, and T) that can be associated with specific parts of the heart's beating cycle. The distinct differences between the EKGs of healthy and damaged hearts provide physicians with a valuable diagnostic tool.

In electroencephalography the electrodes are placed at specific locations on the head, as Figure 19.18 indicates, and they record the potential differences that characterize brain behavior. The graph of potential difference versus time is known as an electroencephalogram (EEG). The various parts of the patterns in an EEG are often referred to as "waves" or "rhythms." The drawing shows an

THE PHYSICS OF . . .

electrocardiography.

THE PHYSICS OF . . .

electroencephalography.

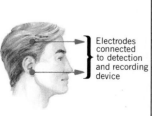

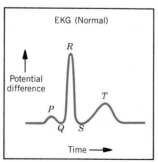

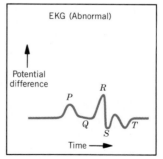

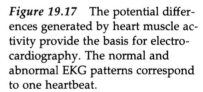

Figure 19.17 The potential differences generated by heart muscle activity provide the basis for electrocardiography. The normal and abnormal EKG patterns correspond to one heartbeat.

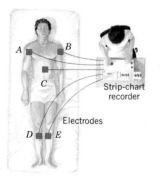

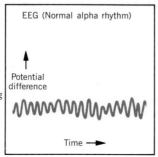

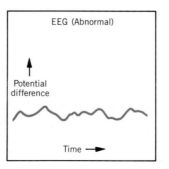

Figure 19.18 In electroencephalography the potential differences created by the electrical activity of the brain are used for diagnosing abnormal behavior.

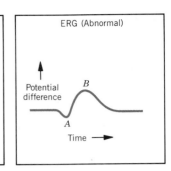

Figure 19.19 The electrical activity of the retina of the eye generates the potential differences used in electroretinography.

example of the main resting rhythm of the brain, the so-called alpha rhythm, and also illustrates the distinct differences that are found between the EEGs generated by healthy and diseased (abnormal) tissue.

The electrical characteristics of the retina of the eye lead to the potential differences measured in electroretinography. Figure 19.19 shows a typical electrode placement used to record the pattern of potential difference versus time that occurs when the eye is stimulated by a flash of light. One electrode is mounted on a contact lens, while the other is often placed on the forehead. The recorded pattern is called an electroretinogram (ERG), parts of the pattern being referred to as the "*A* wave" and the "*B* wave." As the graphs show, the ERGs of normal and diseased eyes can differ markedly.

INTEGRATION OF CONCEPTS

ELECTRIC POTENTIAL ENERGY AND THE CONSERVATION OF ENERGY

A number of times in this text we have returned to the principle of conservation of energy. The concept of energy includes kinetic and potential energies, and the conservation principle states that energy can neither be created nor destroyed, but can only be converted from one form to another. There are different types of potential energy, and in previous chapters we have encountered the gravitational and elastic forms. To this list we now add the electric potential energy. According to the conservation principle, energy may be interconverted between each of these potential energies and the kinetic energy as well. For example, homework problems 5, 6, 7, and 9 at the end of this chapter deal with the interconversion of electric potential energy and kinetic energy. Each type of potential energy is associated with a conservative force. It is the conservative nature of the electrostatic force, just as it is for the gravitational and the elastic forces, that allows us to introduce the concept of electric potential energy. Once the potential energies are known, the principle of conservation of energy provides a common framework that describes how the effects of conservative forces change the motion of a system. Certainly, the individual mathematical expressions for the various kinds of potential energy are different, but each is a form of energy. And the great value of the conservation principle is that it treats all kinds of potential energy, as well as kinetic energy, in the same way.

SUMMARY

When work W_{AB} is done to move a positive test charge $+q_0$ at constant speed from point A to point B, the work equals the difference between the **electric potential energy** EPE at B and that at A: $W_{AB} = \text{EPE}_B - \text{EPE}_A$. The **electric potential** V is the electric potential energy per unit charge, so the electric potential difference between two points is $V_B - V_A = (\text{EPE}_B - \text{EPE}_A)/q_0 = W_{AB}/q_0$. A positive charge accelerates from a region of higher potential toward a region of lower potential. Conversely, negative charge accelerates from a region of lower potential toward a region of higher potential. The electric potential at a distance r from a **point charge** q is $V = kq/r$. This expression for V assumes the potential is zero at an infinite distance away from the charge.

The **electron volt** (eV) is a unit of energy. One electron volt corresponds to 1.60×10^{-19} J and is the change in potential energy of an electron when the electron moves through a potential difference of one volt.

An **equipotential surface** is a surface on which the electric potential is the same everywhere. No work is needed to move a charge at a constant speed on an equipotential surface. The electric field created by any group of charges is always perpendicular to the associated equipotential surfaces and points in the direction of decreasing potential. The electric field is given by $E = -\Delta V/\Delta s$, where ΔV is the potential difference and Δs is the magnitude of the displacement perpendicular to the potential surfaces in the direction of increasing potential.

A **capacitor** is a device that can store charge and consists of two conductors that are near one another, but not touching. The magnitude q of the charge on each plate is given by $q = CV$, where V is the magnitude of the potential difference between the plates and C is the **capacitance**. The insulating material included between the plates is called a **dielectric**. The dielectric constant κ of the material is $\kappa = E_0/E$, where E_0 and E are, respectively, the magnitudes of the electric fields between the plates without and with a dielectric, assuming the charge on the plates is kept fixed. The capacitance of a parallel plate capacitor is $C = \kappa\epsilon_0 A/d$, where A is the area of each plate, and d is the distance between the plates. The **electric potential energy** stored in a capacitor is $\frac{1}{2}CV^2$. The **energy density** or energy stored per unit volume is $\frac{1}{2}\kappa\epsilon_0 E^2$.

SOLVED PROBLEMS

Solved Problem 1 The Potential Energy of a Group of Charges

Related Problems: *19 *20 **23

The drawing shows three point charges, initially very far apart, that are brought together and placed at the corners of an equilateral triangle. Each side of the triangle has a length of

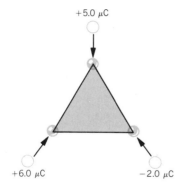

+5.0 μC

+6.0 μC   −2.0 μC

0.50 m. Determine the electric potential energy of the group. In other words, determine the amount by which the electric potential energy of the triangular group differs from that of the three charges in their initial, widely separated locations.

REASONING This problem is done by adding the charges to the triangle one at a time and calculating the energy needed to bring up each charge. The total potential energy of the triangular group is the sum of the energies needed to bring together the three charges.

SOLUTION The order in which the charges are put on the triangle does not matter; we begin with the charge of $+5.0$ μC. No energy is required to place this charge at a corner of the triangle, since the two other charges are initially very far away. Once the charge is in place, the potential it creates at either empty corner ($r = 0.50$ m) is

$$V = \frac{kq}{r} = \frac{(8.99 \times 10^9 \text{ N}\cdot\text{m}^2/\text{C}^2)(+5.0 \times 10^{-6} \text{ C})}{0.50 \text{ m}} \tag{19.6}$$

$$= +9.0 \times 10^4 \text{ V}$$

The energy required to bring up the charge of $+6.0\ \mu C$ and place it at one of the empty corners is

$$\text{EPE} = qV = (+6.0 \times 10^{-6}\ \text{C})(+9.0 \times 10^{4}\ \text{V}) \qquad (19.4)$$
$$= +0.54\ \text{J}$$

The electric potential at the remaining empty corner is the sum of the potentials due to the charges at the other two corners:

$$V = \frac{(8.99 \times 10^{9}\ \text{N}\cdot\text{m}^{2}/\text{C}^{2})(+5.0 \times 10^{-6}\ \text{C})}{0.50\ \text{m}}$$
$$+ \frac{(8.99 \times 10^{9}\ \text{N}\cdot\text{m}^{2}/\text{C}^{2})(+6.0 \times 10^{-6}\ \text{C})}{0.50\ \text{m}} = +2.0 \times 10^{5}\ \text{V}$$

The energy needed to bring up the charge of $-2.0\ \mu C$ and place it at the third corner is

$$\text{EPE} = qV = (-2.0 \times 10^{-6}\ \text{C})(+2.0 \times 10^{5}\ \text{V}) \qquad (19.4)$$
$$= -0.40\ \text{J}$$

The total potential energy of the triangular group differs from that of the widely separated charges by an amount that is the sum of the potential energies calculated above:

$$\text{Total potential energy} = 0.54\ \text{J} - 0.40\ \text{J} = \boxed{+0.14\ \text{J}}$$

This potential energy originates in the work that is done to bring the charges together.

SUMMARY OF IMPORTANT POINTS This problem shows how to find the electric potential energy of a group of point charges. Initially, the charges are widely separated. Then, each charge is brought up, one at a time, to form the group. The energy needed to add each charge to the group is determined. The total electric potential energy of the group is the sum of the individual energies. The value of the total energy does not depend on the order in which the individual charges are added to the group.

QUESTIONS

1. Three points, X, Y, and Z, are located from left to right on a line. A positive test charge is released from rest at X and moves toward and reaches Y. Subsequently, a positive test charge is released from rest at Y and moves toward and reaches Z. Assuming that only motion along the line is possible, what will a negative test charge do when it is released from rest at Y? Explain.

2. What charges, all having the same magnitude, would you place at the corners of a rectangle (one charge per corner), so that both the electric field and the electric potential (assuming a zero reference value at infinity) are zero at the center of the rectangle? Account for the fact that the charge distribution gives rise to both a zero field and a zero potential.

3. The electric field at a single point is zero. Does this fact necessarily mean that the electric potential at the same point is zero? Use a spot on the line between two identical point charges as an example to support your reasoning.

4. To measure the potential at the midpoint between the positive and negative charges of a dipole, a positive test charge is brought in from infinity at a constant speed. When the path followed is along the perpendicular bisector of the dipole, no work is required along any portion of the path. Why?

5. There are two points, A and B, on an equipotential surface. A positive test charge is moved from A to B along two different paths. One path lies entirely on the surface. The other path includes the point P, which does not lie on the surface. Compare the work done in moving the test charge from A to B along the two paths. Justify your answer.

6. The potential is constant throughout a given region of space. Is the electric field zero or nonzero in this region? Justify your answer.

7. In a region of space where the electric field is constant everywhere, as it is inside a parallel-plate capacitor, is the potential constant everywhere? Account for your answer.

8. A positive test charge is placed in an electric field. In what direction should the charge be moved relative to the field, such that the charge experiences a constant electric potential? Account for your answer.

9. The location marked P in the drawing lies midway between the point charges $+q$ and $-q$. The blue lines labeled A, B, and C are edge-on views of three planes. Which one of these planes could be an equipotential surface? Why?

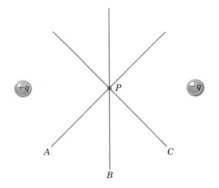

10. Imagine that you are moving a positive test charge along the line between two identical positive point charges. Is

the midpoint on the line analogous to the top of a mountain peak or to the bottom of a deep valley? Explain.

11. The electric field can be expressed in either of two units: newtons/coulomb or volts/meter. Show that these units for the electric field are equivalent.

12. A proton and an electron are released from rest at the midpoint between the plates of a charged parallel plate capacitor. Except for these particles, nothing else is between the plates. Which particle strikes a capacitor plate first? Why?

PROBLEMS

Note: All charges are assumed to be point charges unless specified otherwise.

Section 19.1 Potential Energy, Section 19.2 The Electric Potential Difference

1. The electric potential inside a living cell is lower than the electric potential outside the cell. Suppose the electric potential difference between the inner and outer cell wall is 0.095 V, a typical value. To maintain the internal electrical balance, the cell "pumps" out sodium ions. How much work must be done to remove a single sodium ion (charge $+e$)?

2. The anode (positive terminal) of an X-ray tube is at a potential of $+125\ 000$ V with respect to the cathode (negative terminal). (a) How much work (in joules) is done on an electron that is accelerated from the cathode to the anode? (b) If the electron is initially at rest, what kinetic energy does the electron have when it arrives at the anode?

3. An agent moves a charge of $+1.80 \times 10^{-4}$ C at a constant speed from point A to point B and performs 5.80×10^{-3} J of work on the charge. (a) What is the difference between the electric potential energies of the charge at the two points? (b) Determine the potential difference between the two points. (c) State which point is at the higher potential.

4. The potential of point A relative to point B is $V_A - V_B = +95$ V, while the potential of point C relative to point B is $V_C - V_B = +23$ V. How much work is needed to move a charge of $+45\ \mu$C from C to A?

5. In a television picture tube, electrons strike the screen after being accelerated from rest through a potential difference of 25 000 V. The speeds of the electrons are quite large, and in accurate calculations of the speeds, the effects of special relativity must be taken into account. Ignoring such effects and assuming an electron starts from rest, find the electron speed just before the electron strikes the screen.

6. A particle with a charge of $-1.5\ \mu$C and a mass of 2.5×10^{-6} kg is released from rest at point X and accelerates toward point Y, arriving there with a speed of 42 m/s. (a) What is the potential difference between X and Y? (b) Which point is at the higher potential? Give your reasoning.

7. A proton, released from rest, accelerates from one point to another and gains kinetic energy, because there is a poten-

tial difference of 1.5 μV between the two points. To acquire the same amount of kinetic energy, through what height would the proton have to fall freely, after being released from rest in the presence of the earth's gravity?

***8.** The energy in a lightning bolt is enormous. Consider, for example, a lightning bolt in which 25 C of charge moves through a potential difference of 1.2×10^8 V. With the amount of energy in this bolt, how many kilograms of water at 100 °C could be boiled into steam at 100 °C?

***9.** An electron and a proton, starting from rest, are accelerated through an electric potential difference of the same magnitude. In the process, the electron acquires a speed v_e, while the proton acquires a speed v_p. Find the ratio v_e/v_p.

Section 19.3 The Electric Potential Difference Created by Point Charges

10. To see how large a charge of one coulomb is, calculate the electric potential at a location that is 1.0 km away from a charge of $+1.0$ C.

11. There is an electric potential of $+130$ V at a spot that is 0.25 m away from a charge. Find the magnitude and sign of the charge.

12. Two charges of $+2.60 \times 10^{-8}$ and -5.50×10^{-8} C are separated by 1.40 m. What is the electric potential midway between them?

13. At a distance of 0.20 m from a charge, the electric potential is 164 V. What is the potential at a distance of 0.80 m?

14. Location A is 2.00 m from a charge of -3.00×10^{-8} C, while location B is 3.00 m from the charge. Find the potential difference $V_B - V_A$ between the two points, and state which point is at the higher potential.

15. Three charges are located at the corners of a square whose sides are 2.0 m in length. The charges are $+2.0$, $+14$, and $+5.0\ \mu$C. The empty corner of the square is opposite the 14-μC charge. How much work is required to bring up a fourth charge of $+8.0\ \mu$C and place it at the empty corner?

16. Two charges are fixed in place with a separation d. One charge is positive and has twice the magnitude of the other charge, which is negative. The positive charge lies to the left of the negative charge. Relative to the negative charge, locate the two spots on the line through the charges where the total potential is zero. One spot is (a) between the charges, and the other spot is (b) to the right of the negative charge.

17. Charges of $+2q$ and $-q$ are fixed in place and separated by a distance of 2.0 m. A dashed line is drawn through the negative charge, perpendicular to the line between the charges. Relative to the negative charge, where on the dashed line is the total potential equal to zero? There are two places.

*18. A charge of -3.00 μC is fixed in place. From a distance of 0.0450 m, a particle of mass 7.20×10^{-3} kg and charge -8.00 μC is fired with an initial speed of 65.0 m/s directly toward the fixed charge. How far does the particle travel before its speed is zero?

*19. A square is 0.50 m on a side. How much work is done to bring in four identical charges (5.0 μC each) from infinity and place them on the square, one to a corner? (See Solved Problem 1 for a related problem.)

*20. Determine the electric potential energy for the array of three charges shown in the drawing, relative to its value when the charges are infinitely far away. (See Solved Problem 1 for a related problem.)

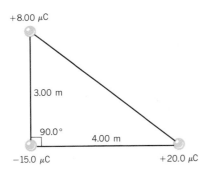

*21. According to the Bohr model, a singly ionized helium atom contains one electron in orbit at a distance of 2.65×10^{-11} m from a nucleus that contains two protons and two neutrons. (The neutrons are electrically neutral and play no role in this problem.) (a) Calculate the ionization energy (in joules) for the atom. (b) Express the answer in electron volts.

*22. A positive charge of $+q_1$ is located 3.00 m to the left of a negative charge $-q_2$. The charges have different magnitudes. On the line through the charges, the net *electric field* is zero at a spot 1.00 m to the right of the negative charge. On this line there are also two spots where the potential is zero. Locate these two spots relative to the negative charge.

**23. Charges q_1 and q_2 are fixed in place, q_2 being located at a distance d to the right of q_1. A third charge q_3 is then fixed to the line joining q_1 and q_2 at a distance d to the right of q_2. The third charge is chosen so the potential energy of the group is

zero; that is, the potential energy has the same value as that of the three charges when they are widely separated. Determine q_3, assuming that (a) $q_1 = q_2 = q$ and (b) $q_1 = q$ and $q_2 = -q$. (See Solved Problem 1 for a related problem.)

**24. A positive charge $+q_1$ is located to the left of a negative charge $-q_2$. On a line passing through the two charges, there are two places where the total potential is zero. The first place is between the charges and is 4.00 cm to the left of the negative charge. The second place is 7.00 cm to the right of the negative charge. (a) What is the distance between the charges? (b) Find q_1/q_2, the ratio of the magnitudes of the charges.

Section 19.4 Equipotential Surfaces and Their Relation to the Electric Field

25. What is the radius of the $+12$ V equipotential surface that surrounds a charge of $+2.0 \times 10^{-10}$ C?

26. Consider the equipotential surfaces that surround a point charge of $+1.50 \times 10^{-8}$ C. How far from the 190-V surface is the 75-V surface?

27. The inner and outer surfaces of a cell membrane carry a negative and positive charge, respectively. Because of these charges, a potential difference of about 0.095 V exists across the membrane. The thickness of the membrane is 7.5×10^{-9} m. What is the magnitude of the electric field in the membrane?

28. Two points, A and B, are separated by 0.016 m. The potential at A is $+28$ V, and that at B is $+95$ V. Determine the magnitude and direction of the electric field between the two points.

29. When you walk across a rug on a dry day, your body can become electrified, and its electric potential can change. When the magnitude of the potential becomes large enough, a spark can jump between your hand and a metal surface. A spark occurs when the electric field strength created by the charges on your body reaches the dielectric strength of the air. The dielectric strength of the air is 3.0×10^6 N/C and is the electric field strength at which the air suffers electrical breakdown. Suppose a spark 3.0 mm long jumps between your hand and a metal doorknob. Assuming that the electric field is uniform, find the potential difference between your hand and the doorknob.

*30. The drawing shows the potential at five points on a set of axes. Each of the four outer points is 6.0×10^{-3} m from the point at the origin. From the data shown, find the magnitude and direction of the electric field in the vicinity of the origin.

*31. Two equipotential surfaces surround a point charge of $+1.00 \times 10^{-9}$ C. The surfaces are separated by a distance of 1.00 m and have a potential difference of 1.00 V between them. (a) What are the radii of the surfaces? (b) What is the potential of each surface?

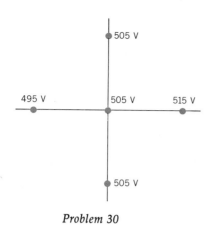

Problem 30

Section 19.5 Capacitors and Dielectrics

32. What voltage is required to store 7.2×10^{-5} C of charge on the plates of a 6.0-μF capacitor?

33. A capacitor has a capacitance of 2.5×10^{-8} F. In the charging process, electrons are removed from one plate and placed on the other plate. When the potential difference between the plates is 450 V, how many electrons have been transferred?

34. One farad is a large amount of capacitance. To see just how large, determine the area of each plate of an empty, one-farad parallel plate capacitor whose plate separation is one meter. Express your answer in square miles (1 square mile = 2.59×10^6 m^2).

35. An axon is the relatively long tail-like part of a neuron, or nerve cell. The outer surface of the axon membrane (dielectric constant = 5, thickness = 1×10^{-8} m) is charged positively, and the inner portion is charged negatively. Thus, the membrane is a kind of capacitor. Assuming an axon can be treated like a parallel plate capacitor with a plate area of 5×10^{-6} m^2, what is its capacitance?

36. A defibrillator is a device used in emergency situations to stimulate the heart muscle and start the heart beating again. Two "paddles" are placed on the body near the heart, and the energy stored in a capacitor is discharged through them. The energy discharged is 510 J and the capacitance is 120 μF. What is the potential difference across the capacitor plates?

37. The membrane that surrounds a certain type of living cell has a surface area of about 5×10^{-9} m^2 and a thickness of about 1×10^{-8} m. Assume that the membrane behaves like a parallel plate capacitor and has a dielectric constant of 5. (a) If the potential on the outer surface of the membrane is +60 mV greater than that on the inside surface, how much charge resides on the outer surface? (b) If the charge in part (a) is due to K$^+$ ions (charge + e), how many such ions are present on the outer surface?

38. A parallel plate capacitor is filled with ruby mica, and the area of each plate is 3.8 m^2. The capacitor stores 2.7 μC of charge when a 1.5-V flashlight battery provides the potential difference between the plates. What is the plate separation?

39. Each plate of a parallel plate capacitor has an area of 2.2×10^{-4} m^2 and stores a charge whose magnitude is 4.8×10^{-9} C. Determine the magnitude of the electric field between the plates when the capacitor is (a) empty and (b) filled with Teflon.

40. The electronic flash attachment for a camera contains a capacitor for storing the energy used to produce the flash. In one such unit, the potential difference between the plates of a 750-μF capacitor is 330 V. (a) Determine the energy that is used to produce the flash in this unit. (b) Assuming the flash lasts for 5.0×10^{-3} s, find the effective "wattage" of the flash.

*41. What is the potential difference between the plates of a 3.3-F capacitor that stores sufficient energy to operate a 75-W light bulb for one minute?

*42. The equipotential surfaces within the plates of an empty parallel plate capacitor are such that two surfaces (not the metal plates), 2.00 mm apart, have a potential difference of 1.20×10^{-3} V. The area of each plate is 7.50×10^{-4} m^2. How much charge is on each plate?

*43. The dielectric strength of an insulating material is the maximum electric field strength to which the material can be subjected without electrical breakdown occurring. Suppose a parallel plate capacitor is filled with a material whose dielectric constant is 3.5 and whose dielectric strength is 1.4×10^7 N/C. If this capacitor is to store 1.7×10^{-7} C of charge on each plate without suffering breakdown, what must be the radius of its circular plates?

*44. An empty capacitor is connected to a 12.0-V battery and charged up. The capacitor is then disconnected from the battery. A slab of dielectric material ($\kappa = 2.8$) is then inserted between the plates. Find the amount by which the potential difference across the plates changes. Specify whether the change is an increase or a decrease.

**45. The plate separation of a charged capacitor is 0.0800 m. A proton and an electron are released from rest at the midpoint between the plates. How far has the proton traveled by the time the electron strikes the positive plate?

ADDITIONAL PROBLEMS

46. A charge of +125 μC is fixed at the center of a square that is 0.64 m on a side. How much work is required to move a charge of +7.0 μC from one corner of the square to any other empty corner? Explain.

47. A parallel plate capacitor has a capacitance of 7.0 μF when filled with a dielectric. The area of each plate is 1.5 m^2

and the separation between the plates is 1.0×10^{-5} m. What is the dielectric constant of the dielectric?

48. Point A is at a potential of $+250$ V, and point B is at a potential of -150 V. An α-particle is a helium nucleus that contains two protons and two neutrons; the neutrons are electrically neutral. An α-particle starts from rest at A and accelerates toward B. When the α-particle arrives at B, what kinetic energy (in electron volts) does it have?

49. A capacitor stores 5.3×10^{-5} C of charge when connected to a 6.0-V battery. How much charge does the capacitor store when connected to a 9.0-V battery?

50. A charge of $+9q$ is fixed to one corner of a square, while a charge of $-8q$ is fixed to the opposite corner. Expressed in terms of q, what charge should be fixed to the center of the square, so the potential is zero at each of the two empty corners?

*51. An empty capacitor has a capacitance of 2.7 μF and is connected to a 12-V battery. A dielectric material ($\kappa = 4.0$) is inserted between the plates of this capacitor. What is the magnitude of the surface charge on the dielectric that is adjacent to either plate of the capacitor? *(Hint: The surface charge is equal to the difference in the charge on the plates with and without the dielectric.)*

*52. The electric field has a constant value of 3.0×10^3 V/m and is directed downward. The field is the same everywhere. The potential at a point P within this region is 135 V. Find the potential at the following points: (a) 8.0×10^{-3} m directly above P, (b) 3.3×10^{-3} m directly below P, (c) 5.0×10^{-3} m directly to the right of P.

**53. The potential difference between the plates of a capacitor is 175 V. Midway between the plates, a proton and an electron are released. The electron is released from rest. The proton is projected perpendicularly toward the negative plate with an initial speed. The proton strikes the negative plate at the same instant the electron strikes the positive plate. Find the initial speed of the proton.

**54. One particle has a mass of 3.00×10^{-3} kg and a charge of $+8.00$ μC. A second particle has a mass of 6.00×10^{-3} kg and the same charge. The two particles are initially held in place and then released. The particles fly apart, and when the separation between them is 0.100 m, the speed of the 3.00×10^{-3}-kg particle is 125 m/s. Find the initial separation between the particles. *(Hint: Both the energy and momentum of the two-particle system are conserved.)*

CHAPTER 20

ELECTRIC CIRCUITS

This interesting photograph of the skyline in Seattle, Washington, illustrates on a large scale the topic of this chapter, electric circuits. It's evident that electrical energy is performing a variety of services in these buildings. The energy is used for lighting, heating and cooling offices, operating elevators, pumping water to drinking fountains, and cooking food at restaurants. At the core of all such services are circuits, which are electrical "highways" that rout electrical energy to where it is needed, much like the freeways that rout people into and out of the city. In this chapter we will examine the concepts and principles that apply to electric circuits. Some of these ideas will be familiar from earlier chapters, such as voltage, power, and energy conservation. Others, like current and resistance, for example, will be introduced for the first time. We will also discuss the ways in which devices are connected into circuits.

20.1 ELECTROMOTIVE FORCE AND CURRENT

Look around you. Chances are that there is an electrical device nearby — a radio, a hair drier, a computer — something that needs electrical energy to operate. Such devices are common because electrical energy is convenient to use and because a

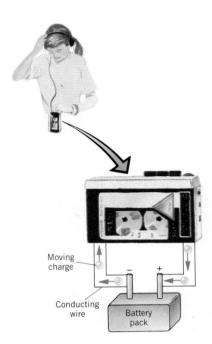

Figure 20.1 In an electric circuit, energy is transferred from a source (the battery pack) to a device (the cassette player) by moving charges.

Figure 20.2 Typical batteries and the symbol $\left(\dfrac{+}{|}\!\!\Big|\!\!\dfrac{-}{}\right)$ used to represent them in electric circuits.

variety of electrical energy sources are available. The battery is one of the more familiar sources of electrical energy. The energy needed to run a portable cassette player, for instance, comes from batteries, as Figure 20.1 illustrates. The transfer of energy takes place via an electric circuit, in which the source (the battery) and the energy-consuming device (the cassette player) are connected by conducting wires, through which electric charges move.

Within a battery, a chemical reaction occurs that transfers electrons from one terminal to the other, leaving one terminal negatively charged and the other positively charged. Figure 20.2 shows the two terminals of a car battery and of a flashlight battery. The drawing also illustrates the symbol $\left(\dfrac{+}{|}\!\!\Big|\!\!\dfrac{-}{}\right)$ used to represent a battery in circuit drawings. Because of the positive and negative charges on the battery terminals, an electric potential difference exists between them. The maximum potential difference is called the ***electromotive force* (emf)*** of the battery, for which the symbol $\mathscr{E}$ is used. In a typical car battery, the chemical reaction maintains the potential of the positive terminal at a maximum of 12 volts (12 joules/coulomb) higher than the potential of the negative terminal, so the emf is $\mathscr{E} = 12$ V. Thus, one coulomb of charge emerging from the battery and entering a circuit has at most 12 joules of energy. In a typical flashlight battery the emf is 1.5 V. In reality, the potential difference between the terminals of a battery is somewhat less than the maximum value indicated by the emf, for reasons that Section 20.9 discusses.

In a circuit such as that in Figure 20.1, the battery creates an electric field within† and parallel to the wire, directed from the positive toward the negative

---

* The word "force" appears in this context for historical reasons, even though it is incorrect; electric potential is energy per unit charge, which is not force.
† Here, an electric field exists inside a conductor, in contrast to the situation in electrostatics. The field exists here because the battery keeps the charges moving and prevents them from coming to equilibrium on the outer surface of the conductor, where they would cause the net electric field on the interior to be zero.

terminal. The field exerts a force on the free electrons in the wire, and they respond by moving. The resulting flow of charge is known as an *electric current.* Figure 20.3 shows charges moving inside a wire and crossing an imaginary surface that is perpendicular to their motion. The current is defined as the amount of charge per unit time that crosses this surface, in much the same sense that a river current is the amount of water flowing per unit of time. Suppose q is the amount of charge that passes through the surface in a time t. If the rate is constant, the current I is equal to the charge divided by the time:

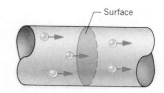

Figure 20.3   The electric current is the amount of charge per unit time that passes through a surface that is perpendicular to the motion of the charges.

$$I = \frac{q}{t} \qquad (20.1)$$

If charge does not flow at a constant rate, then Equation 20.1 gives the average current. Since the units for charge and time are the coulomb (C) and the second (s), the SI unit for current is a coulomb per second (C/s). One coulomb per second is referred to as an *ampere* (A), after the French mathematician André-Marie Ampère (1775–1836).

If the charges move around a circuit in the same direction at all times, the current is said to be *direct current (dc)*. Batteries, for example, produce direct current. In contrast, the current is said to be *alternating current (ac)* when the charges move first one way and then the opposite way, changing direction from moment to moment. Many energy sources produce alternating current, e.g., generators at power companies and microphones. Example 1 deals with the direct current produced by the battery in a pocket calculator.

Example 1   A Pocket Calculator

The current from the 3.0-V battery of a pocket calculator is 0.17 mA (1 mA = 10^{-3} A). In one hour of operation, (a) how much charge flows in the circuit and (b) how much energy does the battery deliver to the calculator circuit?

REASONING   Since current is defined as charge per unit time, the total charge that flows in one hour is the product of the current and the time (3600 s). The charge that leaves the 3.0-V battery has 3.0 joules of energy per coulomb of charge. Thus, the total energy delivered to the calculator circuit from the battery is the total charge in coulombs times the energy per coulomb.

SOLUTION

(a) The charge that flows in one hour can be determined from Equation 20.1:

$$q = It = (0.17 \times 10^{-3} \text{ A})(3600 \text{ s}) = \boxed{0.61 \text{ C}}$$

(b) The energy delivered to the calculator circuit is

$$\text{Energy} = \text{Charge} \times \underbrace{\frac{\text{Energy}}{\text{Charge}}}_{\text{Battery emf}} = (0.61 \text{ C})(3.0 \text{ V}) = \boxed{1.8 \text{ J}}$$

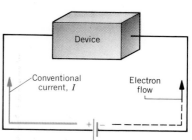

Figure 20.4   In a circuit, electrons actually flow through the metal wires. However, it is customary to use a conventional current I to describe the flow of charges.

Today, it is known that electrons flow in metal wires. Figure 20.4 shows the negative electrons emerging from the negative terminal of the battery and moving around the circuit toward the positive terminal. It is customary, however, *not* to

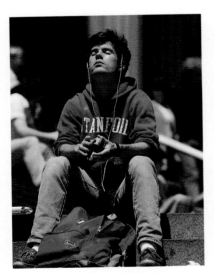

Thanks to miniaturized electric circuits, a variety of personal stereo systems can be used wherever you are.

use the flow of electrons when discussing circuits. Instead, a so-called *conventional current* is used, for reasons that date back to the time when it was believed that positive charges moved through metal wires. Conventional current is the hypothetical flow of positive charges that would have the same effect in the circuit as the movement of negative charges that actually does occur. In Figure 20.4, negative electrons *arrive* at the positive terminal of the battery. The same effect would have been achieved if an equivalent amount of positive charge had *left* the positive terminal. Therefore, the drawing shows the conventional current (in purple) originating from the positive terminal. A conventional current of hypothetical positive charges is consistent with our earlier use of a positive test charge for defining electric fields and potentials. The direction of conventional current is always from a point of higher potential toward a point of lower potential, that is, from the positive terminal toward the negative terminal. In this text, the symbol I stands for conventional current.

20.2 OHM'S LAW

The current that a battery can push through a wire is analogous to the water flow that a pump can push through a pipe. Greater pump pressures lead to larger water flow rates, and, similarly, greater battery voltages* lead to larger electric currents. In the simplest case, the current I is directly proportional to the voltage V; that is, $I \propto V$. Thus, a 12-V battery leads to twice as much current as a 6-V battery, when each is connected to the same circuit.

In a water pipe, the flow rate is not only determined by the pump pressure but is also affected by the length and diameter of the pipe. Longer and narrower pipes offer higher resistance to the moving water and lead to smaller flow rates for a given pump pressure. A similar situation exists in electrical circuits. In the electrical case, the *resistance* (R) is defined as the ratio of the voltage V applied across a piece of material to the current I through the material, or $R = V/I$. When only a small current results from a large voltage, there is a high resistance to the moving charge. For many materials (e.g., metals), the ratio V/I is the same for a given piece of material over a wide range of voltages and currents. In such a case, the resistance is a constant, and the relation $R = V/I$ is referred to as *Ohm's law*, after the German physicist Georg Simon Ohm (1789–1854), who discovered it.

Ohm's Law

The ratio V/I is a constant, where V is the voltage applied across a piece of material (such as a wire) and I is the current through the material:

$$\frac{V}{I} = R = \text{constant} \quad \text{or} \quad V = IR \qquad (20.2)$$

R is the resistance of the piece of material.

   SI Unit of Resistance: volt/ampere (V/A) = ohm (Ω)

---

* The potential difference between two points, such as the terminals of a battery, is commonly called the voltage between the points.

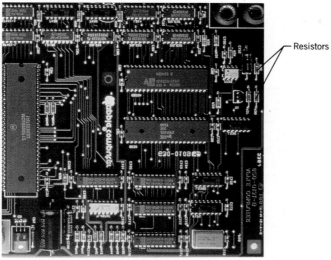

Resistors

Figure 20.5 Resistors are found on circuit boards of computers.

The SI unit for resistance is a volt per ampere, which is called an *ohm* and is represented by the Greek letter omega (Ω). Ohm's law is not a fundamental law of nature like Newton's laws of motion. It is only a statement of the way certain materials behave in electric circuits.

To the extent that a wire or an electrical device offers resistance to the flow of charges, it is called a ***resistor***. The resistance can have a wide range of values. The copper wires in a television set, for instance, have a negligibly small resistance. On the other hand, commercial resistors can have resistances up to many kiloohms (1 kΩ = 10^3 Ω) or megaohms (1 MΩ = 10^6 Ω). Such resistors play an important role in electric circuits, where they are used to limit the amount of current (see Figure 20.5).

In drawing electric circuits we follow the usual conventions: (1) a zigzag line (—$\wedge\wedge\wedge$—) represents a resistor and (2) a straight line (————) represents an ideal conducting wire, or one with a negligible resistance. Example 2 illustrates an application of Ohm's law to the circuit in a flashlight.

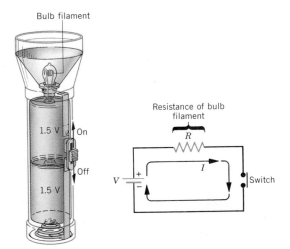

Figure 20.6 The circuit in this flashlight consists of a resistor (the filament of the light bulb) connected to a 3.0-V battery.

Example 2 A Flashlight

The filament in a light bulb is a resistor in the form of a thin piece of wire. The wire becomes hot enough to emit light because of the current in it. Figure 20.6 shows a flashlight that uses two 1.5-V batteries (effectively a single 3.0-V battery) to provide a current of 0.40 A in the filament. Determine the resistance of the glowing filament.

REASONING   The filament resistance is assumed to be the only resistance in the circuit. The potential difference applied across the filament is that of the 3.0-V battery. The resistance, given by Equation 20.2, is equal to this potential difference divided by the current.

SOLUTION   The resistance of the filament is

$$R = \frac{V}{I} = \frac{3.0 \text{ V}}{0.40 \text{ A}} = \boxed{7.5 \ \Omega} \tag{20.2}$$

20.3 RESISTANCE AND RESISTIVITY

In a water pipe, the length and cross-sectional area of the pipe determine the resistance the pipe offers to the flow of water. Longer pipes with smaller cross-sectional areas offer greater resistance. Analogous effects are found in the electrical case. For a wide range of materials, the resistance of a piece of material of length L and cross-sectional area A is

$$R = \rho \frac{L}{A} \tag{20.3}$$

where ρ is a proportionality constant known as the *resistivity* of the material. It can be seen from Equation 20.3 that the unit for resistivity is the ohm·meter ($\Omega \cdot$m), and Table 20.1 lists values for various materials. All the conductors in

Table 20.1 Resistivities[a] of Various Materials

Material	Resistivity ρ ($\Omega \cdot$m)	Material	Resistivity ρ ($\Omega \cdot$m)
Conductors		**Semiconductors**	
Aluminum	2.82×10^{-8}	Carbon	3.5×10^{-5}
Copper	1.72×10^{-8}	Germanium	0.5^b
Gold	2.44×10^{-8}	Silicon	$20-2300^b$
Iron	9.7×10^{-8}	**Insulators**	
Mercury	95.8×10^{-8}	Mica	$10^{11}-10^{15}$
Nichrome (alloy)	100×10^{-8}	Rubber (hard)	$10^{13}-10^{16}$
Silver	1.59×10^{-8}	Teflon	10^{16}
Tungsten	5.6×10^{-8}	Wood (maple)	3×10^{10}

[a] The values pertain to temperatures near 20 °C.
[b] Depending on purity.

Table 20.1 are metals and have small resistivities. Insulators such as rubber have large resistivities. Materials like germanium and silicon have intermediate resistivity values and are, accordingly, called *semiconductors*.

Resistivity is an inherent property of a material, inherent in the same sense that the density of a material is an inherent property. Resistance, on the other hand, depends on both the resistivity and the geometry of the material. Thus, two wires can be made from copper, which has a resistivity of 1.72×10^{-8} $\Omega \cdot$m, but Equation 20.3 indicates that a short wire with a large cross-sectional area offers a smaller resistance to current than a long, thin wire. Wires that carry large currents, such as main power cables, are thick rather than thin so that the resistance of the wires is kept as small as possible. Similarly, electric tools that are to be used far away from wall sockets require thicker extension cords, as Example 3 illustrates.

Example 3 Longer Extension Cords

THE PHYSICS OF . . .

electrical extension cords.

The instructions for an electric lawn mower suggest that a 20-gauge extension cord can be used for distances up to 35 m, but a thicker 16-gauge cord should be used for longer distances, to keep the resistance of the wire as small as possible. The cross-sectional area of 20-gauge wire is 5.2×10^{-7} m², while that of 16-gauge wire is 13×10^{-7} m². Determine the resistance of (a) 35 m of 20-gauge copper wire and (b) 75 m of 16-gauge copper wire.

REASONING AND SOLUTION We can use Equation 20.3, along with the resistivity of copper from Table 20.1, to find the resistance of the wires:

[20-gauge wire] $R = \dfrac{\rho L}{A} = \dfrac{(1.72 \times 10^{-8} \ \Omega \cdot m)(35 \ m)}{5.2 \times 10^{-7} \ m^2} = \boxed{1.2 \ \Omega}$

[16-gauge wire] $R = \dfrac{\rho L}{A} = \dfrac{(1.72 \times 10^{-8} \ \Omega \cdot m)(75 \ m)}{13 \times 10^{-7} \ m^2} = \boxed{0.99 \ \Omega}$

Even though it is more than twice as long, the thicker 16-gauge wire has less resistance than the thinner 20-gauge wire.

The resistivity of a material depends on temperature. In metals, the resistivity increases with increasing temperature, whereas in semiconductors the reverse is true. Certain materials have the property that their resistivity drops suddenly to

Table 20.2 Temperature Coefficients[a]
of Resistivity for Various Materials

Material	Temperature Coefficient of Resistivity α [(C°)$^{-1}$]
Aluminum	0.0039
Carbon	−0.0005
Copper	0.00393
Germanium	−0.05
Gold	0.0034
Iron	0.0050
Mercury	0.00089
Nichrome (alloy)	0.0004
Silicon	−0.07
Silver	0.0038
Tungsten	0.0045

[a] The values pertain to a temperature of 20 °C.

zero at very low temperatures. Such materials are called *superconductors,* because, with zero resistivity, they offer no resistance to electric current.

For many materials and limited temperature ranges it is possible to express the temperature dependence of the resistivity as follows:

$$\rho = \rho_0[1 + \alpha(T - T_0)] \tag{20.4}$$

In this expression ρ and ρ_0 are the resistivities at temperatures T and T_0, respectively. The term α has the unit of reciprocal temperature and is the *temperature coefficient of resistivity.* Table 20.2 gives values of α for various materials. When the resistivity increases with increasing temperature, α is positive, as it is for metals. When the resistivity decreases with increasing temperature, α is negative, as it is for carbon, germanium, and silicon. Since resistance is given by $R = \rho L/A$, both sides of Equation 20.4 can be multiplied by L/A to show that resistance depends on temperature according to

$$R = R_0[1 + \alpha(T - T_0)] \tag{20.5}$$

The next example illustrates the role of the resistivity and its temperature coefficient in determining the electrical resistance of a piece of material.

THE PHYSICS OF . . .

a heating element on an electric stove.

Example 4   The Heating Element of an Electric Stove

Figure 20.7 shows a heating element from an electric stove. The element contains a wire (length = 1.1 m, cross-sectional area = 3.1×10^{-6} m^2) through which electric charge flows. This wire is imbedded within an electrically insulating material that is contained within a metal casing. The wire becomes hot in response to the flowing charge and heats the casing. The material of the wire has a resistivity of $\rho_0 = 6.8 \times 10^{-5}$ $\Omega \cdot$m at $T_0 = 320$ °C and a temperature coefficient of resistivity of $\alpha = 2.0 \times 10^{-3}$ (C°)$^{-1}$. Determine the resistance of the heater wire at an operating temperature of 420 °C.

REASONING   We may use the relation $R = \rho L/A$ to find the resistance of the wire at 420 °C, provided the resistivity ρ can be determined at this temperature. Since the resistivity at 320 °C is given, Equation 20.4 can be employed to find the resistivity at 420 °C.

SOLUTION   At the operating temperature of 420 °C, the material of the wire has a resistivity of

$$\rho = \rho_0[1 + \alpha(T - T_0)] \tag{20.4}$$

$$\rho = (6.8 \times 10^{-5}\ \Omega\cdot\text{m})[1 + (2.0 \times 10^{-3}\ (\text{C}°)^{-1})(420\ °\text{C} - 320\ °\text{C})]$$

$$= 8.2 \times 10^{-5}\ \Omega\cdot\text{m}$$

This value of the resistivity can be used along with the length and cross-sectional area to find the resistance of the heater wire:

$$R = \frac{\rho L}{A} = \frac{(8.2 \times 10^{-5}\ \Omega\cdot\text{m})(1.1\ \text{m})}{3.1 \times 10^{-6}\ \text{m}^2} = \boxed{29\ \Omega} \tag{20.3}$$

The cherry-red glow of a heating element on an electric stove indicates that the element becomes very hot as a result of the electric current.

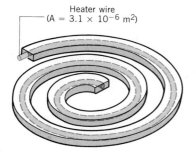

Heater wire
(A = 3.1 × 10⁻⁶ m²)

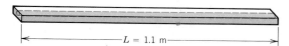

L = 1.1 m

Figure 20.7 A heating element from an electric stove.

20.4 ELECTRIC POWER

If the potential difference between the terminals of a battery is V, then a charge q emerging from the battery has an energy of qV, according to the definition of potential given in Equation 19.3. The product qV has units of energy (joules) since q is measured in coulombs and V is measured in joules/coulomb or volts. Since the amount of energy per second is the power, dividing qV by the time t gives the electric power qV/t provided to the circuit. The charge flowing per second q/t is the current I, so the electric power is IV, the product of current and voltage.

Electric Power

When there is a current I in a circuit as a result of a voltage V, the electric power P delivered to the circuit is

$$P = IV \tag{20.6}$$

SI Unit of Power: watt (W)

Power is measured in watts, and Equation 20.6 indicates, therefore, that the product of an ampere and a volt is equal to a watt.

Many electrical devices are essentially resistors that become hot when provided with sufficient electric power: toasters, irons, space heaters, heating elements on electric stoves, and incandescent light bulbs, to name just a few. In such cases, it is possible to obtain two additional, but equivalent, expressions for the power. These two expressions follow directly upon substituting $V = IR$, or equivalently $I = V/R$, into the relation $P = IV$:

$$P = IV \tag{20.6a}$$

$$P = I(IR) = I^2R \tag{20.6b}$$

$$P = \left(\frac{V}{R}\right)V = \frac{V^2}{R} \tag{20.6c}$$

Example 5 deals with the electric power utilized by the bulb of a flashlight.

Example 5   The Power and Energy Used in a Flashlight

In the flashlight in Figure 20.6, the current is 0.40 A, and the voltage is 3.0 V. Find (a) the power delivered to the bulb and (b) the energy dissipated in the bulb in 5.5 minutes of operation.

REASONING   The electrical power delivered to the bulb is the product of the current and voltage. Since power is energy per unit time, the energy delivered to the bulb is the product of the power and time.

SOLUTION
(a) The power is

$$P = IV = (0.40 \text{ A})(3.0 \text{ V}) = \boxed{1.2 \text{ W}} \tag{20.6a}$$

The "wattage" rating of this bulb would therefore be 1.2 W.

(b) The energy consumed in 5.5 minutes (330 s) follows from the definition of power as energy per unit time:

$$\text{Energy} = Pt = (1.2 \text{ W})(330 \text{ s}) = \boxed{4.0 \times 10^2 \text{ J}}$$

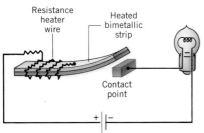

Figure 20.8   A bimetallic strip flasher.

Monthly electric bills specify the cost for the energy consumed during the month. Energy is the product of power and time, and electric companies compute your energy consumption by expressing power in kilowatts and time in hours. Therefore, a commonly used unit for energy is the *kilowatt-hour* (kWh). For instance, if you used an average power of 1440 watts (1.44 kW) for 30 days (720 h), your energy consumption would be (1.44 kW)(720 h) = 1040 kWh. At a cost of $0.10 per kWh, your monthly bill would be $104. One kilowatt-hour equals (1000 J/s)(3600 s) = 3.60×10^6 J of energy.

Figure 20.8 shows an interesting device that uses the heat generated when electric charge flows through a resistor. The device is called a bimetallic strip flasher. As Section 12.4 discusses, a bimetallic strip consists of two pieces of *dissimilar* metals fastened together. The metals expand by different amounts when heated, causing the strip to bend. The drawing shows a bimetallic strip with a resistance heater wire wrapped around it. While the strip is cool, its end touches the contact point. Charges from the battery pass directly through the strip and cause the light bulb to glow. However, as charges continue to flow, the resistance heater causes the bimetallic strip to become hot and bend away from the contact point, shutting off the current in the circuit. The light goes out and the heater shuts off. As the bimetallic strip cools, it bends back and touches the contact point again, turning the light back on. The on–off cycle repeats itself every second or so, and the result is a flashing light that is used as a warning device.

20.5   ALTERNATING CURRENT

Many electric circuits use batteries and involve direct current (dc). However, there are considerably more circuits that operate with alternating current (ac), in which the direction of charge flow reverses periodically. The common generators that create ac electricity depend on magnetic forces for their operation and are discussed in Chapter 22. In an ac circuit, these generators serve the same purpose as a battery serves in a dc circuit, that is, they give energy to the moving electric charges. This section deals with ac circuits that contain only resistance.

Since the electrical outlets in a house provide alternating current, all of us use ac circuits routinely. Figure 20.9 shows the ac circuit that is formed when a toaster is plugged into a wall socket. The heating element of a toaster is essentially a thin wire of resistance R and becomes red hot when energy is dissipated in it. The circuit schematic in the picture introduces the symbol ⊖ that is used to represent the generator. In this case, the generator is located at the electric power company.

Figure 20.10 shows a graph that records the voltage produced between the terminals of the most common kind of ac generator at each moment of time. The

Electric power companies send ac electricity to consumers by means of power cables that are suspended from high towers. Here, workers secure a cable into a pulley, after a helicopter had placed it in the groove 600 feet above the ground.

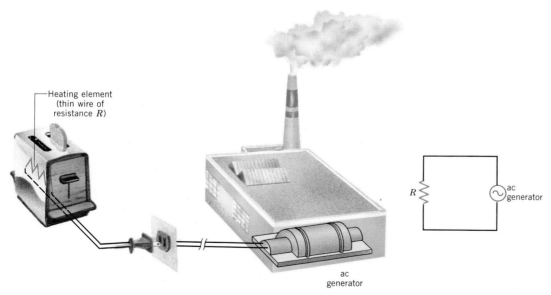

Figure 20.9   This circuit consists of a toaster (resistance = R) and an ac generator at the electric power company.

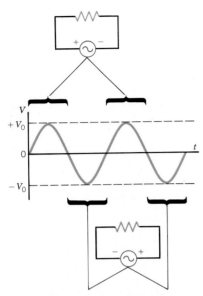

Figure 20.10 The voltage V produced between the terminals of an ac generator fluctuates sinusoidally in time, in the most common case. The circuits indicate the relative polarity of the generator terminals during the positive and negative parts of the sinusoidal graph.

voltage V fluctuates sinusoidally between positive and negative values as a function of time t;

$$V = V_0 \sin 2\pi f t \qquad (20.7)$$

where V_0 is the maximum or peak value of the voltage, and f is the frequency at which the voltage oscillates. In the United States, the voltage at most home wall outlets has a *peak value* of approximately $V_0 = 170$ volts and oscillates with a frequency of $f = 60$ Hz. Thus, the period of each cycle is $\frac{1}{60}$ s, and the polarity of the generator terminals reverses twice during each cycle, as Figure 20.10 indicates.

The current in an ac circuit also oscillates. In circuits that contain only resistance, the current reverses direction each time the polarity of the generator terminals reverses. Thus, the current in a circuit like that in Figure 20.10 would have a frequency of 60 Hz and changes direction twice during each cycle. Substituting $V = V_0 \sin 2\pi f t$ into $V = IR$ shows that the current can be represented as

$$I = \left(\frac{V_0}{R}\right) \sin 2\pi f t = I_0 \sin 2\pi f t \qquad (20.8)$$

The peak current is given by $I_0 = V_0/R$, so it can be determined if the peak voltage and the resistance are known.

The power delivered to an ac circuit by the generator is given by $P = IV$, just as it is in a dc circuit. However, since both I and V depend on time, the power fluctuates as time passes. Substituting Equations 20.7 and 20.8 for V and I into $P = IV$ gives

$$P = I_0 V_0 \sin^2 2\pi f t \qquad (20.9)$$

This expression is plotted in Figure 20.11.

Since the power fluctuates in an ac circuit, it is customary to consider the average power $\overline{P}$, which is one-half the peak power, as Figure 20.11 indicates:

$$\overline{P} = \tfrac{1}{2} I_0 V_0 \qquad (20.10)$$

On the basis of this expression, a kind of average current and average voltage can be introduced that are very useful when discussing ac circuits. A slight rearrangement of Equation 20.10 reveals that

$$\overline{P} = \left(\frac{I_0}{\sqrt{2}}\right)\left(\frac{V_0}{\sqrt{2}}\right) = I_{\text{rms}} V_{\text{rms}} \qquad (20.11)$$

I_{rms} and V_{rms} are called the *root mean square* (*rms*) current and voltage, respectively, and may be calculated from the peak values by dividing them by $\sqrt{2}$*:

$$I_{\text{rms}} = \frac{I_0}{\sqrt{2}} \qquad (20.12)$$

$$V_{\text{rms}} = \frac{V_0}{\sqrt{2}} \qquad (20.13)$$

For instance, in the United States the maximum ac voltage at a home wall socket is typically $V_0 = 170$ volts, and the corresponding rms voltage is $V_{\text{rms}} = 170$ volts/$\sqrt{2} = 120$ volts. When the instructions for an electric appliance specify 120 V, it is

---

* Equations 20.12 and 20.13 apply only for sinusoidal voltage and current.

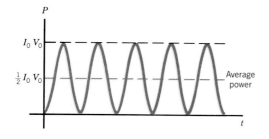

Figure 20.11 The power P in an ac circuit oscillates between zero and a peak value of I_0V_0, where I_0 and V_0 are the peak current and the peak voltage, respectively.

an rms voltage that is indicated. Similarly, when we specify an ac voltage or current in this text, it is an rms value, unless indicated otherwise. Likewise, when we specify ac power, it is an average power, unless stated otherwise.

Except for dealing with average quantities, the relation $\overline{P} = I_{rms}V_{rms}$ has the same form as Equation 20.6a $(P = IV)$. Moreover, Ohm's law can be written conveniently in terms of rms quantities:

$$V_{rms} = I_{rms}R \qquad (20.14)$$

Substituting Equation 20.14 into $\overline{P} = I_{rms}V_{rms}$ shows that the average power can be expressed in the following ways:

$$\overline{P} = I_{rms}V_{rms} \qquad (20.15a)$$

$$\overline{P} = I_{rms}^2 R \qquad (20.15b)$$

$$\overline{P} = \frac{V_{rms}^2}{R} \qquad (20.15c)$$

These expressions are completely analogous to $P = IV = I^2R = V^2/R$ for dc circuits. Example 6 deals with the average power in one familiar ac circuit.

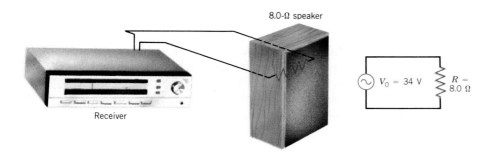

Figure 20.12 A receiver applies an ac voltage (peak value = 34 V) to an 8.0-Ω speaker.

Example 6   *Electrical Power Sent to a Loudspeaker*

A stereo receiver applies a peak ac voltage of 34 V to a speaker. The speaker is an 8.0-Ω speaker, in the sense that it behaves approximately* as an 8.0-Ω resistance. Figure 20.12 shows the circuit. Determine (a) the rms voltage, (b) the rms current, and (c) the average power for this circuit.

* Other factors besides resistance can affect the current and voltage in ac circuits; they are discussed in Chapter 23.

REASONING AND SOLUTION

(a) The peak value of the voltage is $V_0 = 34$ V, so the corresponding rms value is

$$V_{rms} = \frac{V_0}{\sqrt{2}} = \frac{34\text{ V}}{\sqrt{2}} = \boxed{24\text{ V}} \tag{20.13}$$

(b) The rms current can be obtained from Ohm's law:

$$I_{rms} = \frac{V_{rms}}{R} = \frac{24\text{ V}}{8.0\ \Omega} = \boxed{3.0\text{ A}} \tag{20.14}$$

(c) The average power is

$$\overline{P} = I_{rms}V_{rms} = (3.0\text{ A})(24\text{ V}) = \boxed{72\text{ W}} \tag{20.15a}$$

20.6 SERIES WIRING

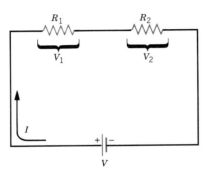

Figure 20.13 When two resistors are connected in series, the same current I is in both of them.

Thus far, we have dealt with circuits that include only a single device, such as a light bulb or a loudspeaker. There are, however, many circuits in which more than one device is connected to a voltage source. This section introduces one method by which such connections may be made, namely, series wiring. *Series wiring means that the devices are connected in such a way that there is the same electric current through each device.* Figure 20.13 shows a circuit in which two different devices, represented by resistors R_1 and R_2, are connected in series with a battery. Note that if the current in one resistor is interrupted, the current in the other is also interrupted. Such an interruption could occur, for example, if two light bulbs were connected in series, and the filament of one bulb broke. Because of the series wiring, the voltage V supplied by the battery is divided between the two resistors. The drawing indicates that the portion of the voltage across R_1 is V_1, while the portion across R_2 is V_2, so $V = V_1 + V_2$. Applying the definition of resistance to each resistor individually shows that

$$V = IR_1 + IR_2 = I(R_1 + R_2) = IR_S$$

where R_S is called the *equivalent resistance* of the series circuit. Thus, two resistors in series are equivalent to a single resistor whose resistance is $R_S = R_1 + R_2$, in the sense that there is the same current through R_S as there is through the series combination of R_1 and R_2. This line of reasoning can be extended to any number of resistors in series, with the result that

$$\begin{bmatrix}\text{Series}\\\text{resistors}\end{bmatrix} \qquad R_S = R_1 + R_2 + R_3 + \cdots \tag{20.16}$$

Example 7 illustrates the concept of the equivalent resistance in a series circuit.

Example 7   Resistors in a Series Circuit

A 6.00-Ω resistor and a 3.00-Ω resistor are connected in series with a 12.0-V battery, as Figure 20.14 indicates. Assuming the battery contributes no resistance to the circuit, find (a) the current, (b) the power dissipated in each resistor, and (c) the total power delivered to the resistors by the battery.

REASONING   The current I can be determined from Ohm's law as $I = V/R_S$, where

$R_S = R_1 + R_2$ is the equivalent resistance of the two resistors in series. The power delivered to each resistor is given by Equation 20.6b as $P = I^2R$, where R is the resistance of the resistor being considered and I is the current through it.

SOLUTION
(a) The equivalent resistance is

$$R_S = 6.00\ \Omega + 3.00\ \Omega = 9.00\ \Omega \qquad (20.16)$$

Applying Ohm's law yields the current as

$$I = \frac{V}{R_S} = \frac{12.0\ \text{V}}{9.00\ \Omega} = \boxed{1.33\ \text{A}} \qquad (20.2)$$

(b) Now that the current is known, the power dissipated in each resistor can be obtained from $P = I^2R$ (Equation 20.6b):

$$\begin{bmatrix} \textbf{6.00-}\Omega \\ \textbf{resistor} \end{bmatrix} \qquad P = I^2R = (1.33\ \text{A})^2(6.00\ \Omega) = \boxed{10.6\ \text{W}}$$

$$\begin{bmatrix} \textbf{3.00-}\Omega \\ \textbf{resistor} \end{bmatrix} \qquad P = I^2R = (1.33\ \text{A})^2(3.00\ \Omega) = \boxed{5.31\ \text{W}}$$

(c) The total power delivered by the battery to the two resistors is the sum of the contributions in part (b): $P = 10.6\ \text{W} + 5.31\ \text{W} = 15.9\ \text{W}$. Alternatively, the total power can be obtained directly by using the equivalent resistance $R_S = 9.00\ \Omega$ and the current from part (a):

$$P = I^2R_S = (1.33\ \text{A})^2(9.00\ \Omega) = \boxed{15.9\ \text{W}}$$

In general, the total power delivered to any number of resistors in series is equal to the power delivered to the equivalent resistor.

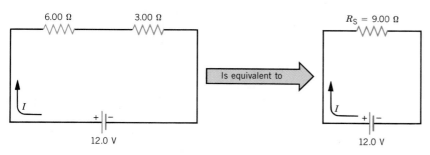

Figure 20.14 A 6.00-Ω and a 3.00-Ω resistor connected in series are equivalent to a single 9.00-Ω resistor.

20.7 PARALLEL WIRING

Parallel wiring is another method of connecting electrical devices together. *Parallel wiring means that the devices are connected in such a way that the same voltage is applied across each device.* Figure 20.15 shows two resistors connected in parallel between the terminals of a battery. Part *a* of the picture is drawn so as to emphasize that the entire voltage of the battery is applied across each resistor. Actually, parallel connections are rarely drawn in this manner; instead they are

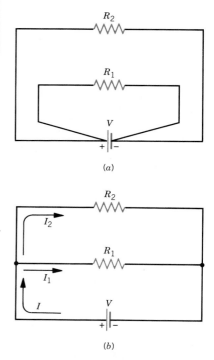

Figure 20.15 (a) When two resistors are connected in parallel, the same voltage V is applied across each resistor. (b) This circuit drawing is equivalent to that in part *a*. I_1 and I_2 are, respectively, the currents in R_1 and R_2.

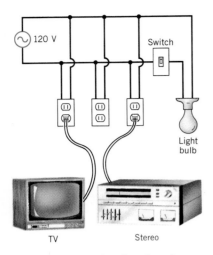

Figure 20.16 This drawing shows some of the parallel connections found in a typical home. Each wall socket provides 120 V to the appliance connected to it. In addition, 120 V is applied to the light bulb when the switch is turned on.

drawn as in part *b*, where the dots indicate the points where the wires for the two branches are joined together. Parts *a* and *b* are equivalent representations of the same circuit.

Parallel wiring is quite common. For example, when an electrical appliance is plugged into a wall socket, the appliance is connected in parallel with other appliances, as in Figure 20.16, where the entire voltage of 120 V is applied across the television, the stereo, and the light bulb. The presence of the unused socket or other devices that are turned off does not affect the operation of those devices that are turned on. Moreover, if the current in one device is interrupted (perhaps by an opened switch or a broken wire), the current in the other devices is not interrupted. In contrast, if household appliances were connected in series, there would be no current through any appliance if the current at any point in the circuit were halted.

When two resistors R_1 and R_2 are connected as in Figure 20.15, each receives current from the battery as if the other were not present. Therefore, R_1 and R_2 together draw more current from the battery than does either resistor alone. According to the definition of resistance, a larger current arises from a smaller resistance. Thus, the two parallel resistors behave as a single equivalent resistance that is *smaller* than either R_1 or R_2. Figure 20.17 returns to the water flow analogy to provide additional insight into this important feature of parallel wiring. In part *a*, two sections of pipe that have the same length are connected in parallel with a pump. In part *b* these two sections have been replaced with a single pipe of the same length, whose cross-sectional area equals the combined cross-sectional areas of section 1 and section 2. The pump can push more water per second through the wider pipe in part *b* than it can through *either* of the narrower pipes in part *a*. In effect, the wider pipe (sections 1 and 2 acting together) offers less resistance to the flow than either of the narrower pipes offers individually.

As in a series circuit, it is possible to replace a parallel combination of resistors with an equivalent resistor that results in the same total current and power for a given voltage as the original combination. The equivalent resistance of two resistors connected in parallel can be determined in the following fashion. First, notice in Figure 20.15 that the total current I from the battery is the sum of I_1 and I_2, where I_1 is the current in resistor R_1 and I_2 is the current in resistor R_2: $I = I_1 + I_2$.

Figure 20.17 (a) Two equally long pipe sections, with cross-sectional areas A_1 and A_2, are connected in parallel to a water pump. (b) The two pipe sections in parallel are equivalent to a single pipe of the same length whose cross-sectional area is $A_1 + A_2$.

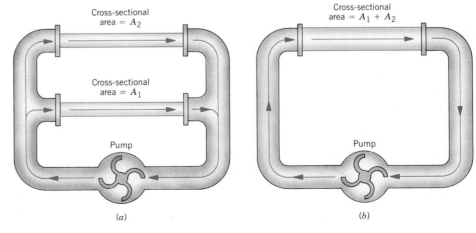

Since the same voltage V is applied across each resistor, the definition of resistance indicates that $I_1 = V/R_1$ and $I_2 = V/R_2$. Therefore,

$$I = I_1 + I_2 = \frac{V}{R_1} + \frac{V}{R_2} = V\left(\frac{1}{R_1} + \frac{1}{R_2}\right) = V\left(\frac{1}{R_P}\right)$$

where R_P is the equivalent resistance. Hence, when two resistors are connected in parallel, they are equivalent to a single resistor whose resistance R_P is given by $1/R_P = 1/R_1 + 1/R_2$. For any number of resistors in parallel, a similar line of reasoning shows that

$$\begin{bmatrix}\text{Parallel} \\ \text{resistors}\end{bmatrix} \qquad \frac{1}{R_P} = \frac{1}{R_1} + \frac{1}{R_2} + \frac{1}{R_3} + \cdots \qquad (20.17)$$

The next example deals with a parallel combination of resistors that occurs in a stereo system.

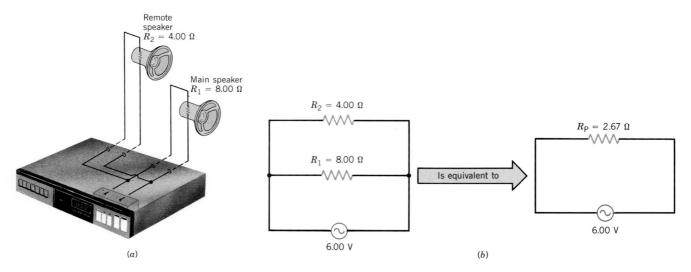

Figure 20.18   (*a*) The main and remote speakers in a stereo system are connected in parallel to the receiver. (*b*) The circuit schematic shows the situation when the ac voltage across the speakers is 6.00 V.

Example 8   Main and Remote Stereo Speakers

Most receivers allow the user to connect a pair of "remote" speakers (to play music in another room, for instance) in addition to the main speakers. Figure 20.18 shows that the remote speaker and the main speaker for the right stereo channel are connected to the receiver in parallel (for clarity, the speakers for the left channel are not shown). At the instant shown in the picture, the ac voltage across the speakers is 6.00 V. The main speaker has a resistance of 8.00 Ω, and the remote speaker has a resistance of 4.00 Ω.* Determine (a) the equivalent resistance of the two speakers, (b) the total current supplied by the receiver, (c) the current in each speaker, (d) the power dissipated in each speaker, and (e) the total power delivered by the receiver.

THE PHYSICS OF . . .

main and remote stereo speakers.

---

* In reality, frequency-dependent characteristics of the speaker (see Chapter 23) play a role in the operation of a loudspeaker. We assume here, however, that the frequency of the sound is low enough that the speakers behave as pure resistances.

REASONING   The total current supplied to the two speakers by the receiver can be obtained as $I_{rms} = V_{rms}/R_P$, where R_P is the equivalent resistance of the two speakers in parallel and is given by $1/R_P = 1/R_1 + 1/R_2$. The average power delivered to each speaker is the product of the current and the voltage. In the parallel connection the voltage applied to each speaker is the same, but the current in each speaker is different, a fact that we will have to take into account.

SOLUTION
(a) Since the speakers are in parallel, the equivalent resistance is

$$\frac{1}{R_P} = \frac{1}{8.00\ \Omega} + \frac{1}{4.00\ \Omega} = \frac{3}{8.00\ \Omega} \qquad (20.17)$$

$$R_P = \frac{8.00\ \Omega}{3} = \boxed{2.67\ \Omega}$$

This result is illustrated in part *b* of the drawing.

(b) The total current is

$$I_{rms} = \frac{V_{rms}}{R_P} = \frac{6.00\ V}{2.67\ \Omega} = \boxed{2.25\ A} \qquad (20.14)$$

(c) The current in each speaker can be determined from Ohm's law and the resistance of each speaker.

$$\begin{bmatrix} \textbf{8.00-}\Omega \\ \textbf{speaker} \end{bmatrix} \qquad I_{rms} = \frac{V_{rms}}{R} = \frac{6.00\ V}{8.00\ \Omega} = \boxed{0.750\ A}$$

$$\begin{bmatrix} \textbf{4.00-}\Omega \\ \textbf{speaker} \end{bmatrix} \qquad I_{rms} = \frac{V_{rms}}{R} = \frac{6.00\ V}{4.00\ \Omega} = \boxed{1.50\ A}$$

The sum of these currents is equal to the total current from the receiver, as determined in part (b).

(d) The average power dissipated in each speaker can be calculated using $\overline{P} = I_{rms}V_{rms}$ with the individual currents obtained in part (c):

$$\begin{bmatrix} \textbf{8.00-}\Omega \\ \textbf{speaker} \end{bmatrix} \qquad \overline{P} = (0.750\ A)(6.00\ V) = \boxed{4.50\ W} \qquad (20.11)$$

$$\begin{bmatrix} \textbf{4.00-}\Omega \\ \textbf{speaker} \end{bmatrix} \qquad \overline{P} = (1.50\ A)(6.00\ V) = \boxed{9.00\ W} \qquad (20.11)$$

(e) The total power delivered by the receiver is the sum of the individual values found in part (d), $\overline{P} = 4.50\ W + 9.00\ W = 13.5\ W$. Alternatively, the total power can be obtained using the total current supplied by the receiver, as determined in part (b):

$$\overline{P} = I_{rms}V_{rms} = (2.25\ A)(6.00\ V) = \boxed{13.5\ W} \qquad (20.11)$$

When a number of resistors are connected in parallel, the equivalent resistance is *less than* any of the individual resistances [see part (a) of Example 8]. In fact, it is the *smallest* resistance that has the largest impact in determining the equivalent resistance. If one resistance approaches zero, then according to Equation 20.17, the equivalent resistance also approaches zero. In such a case, the near-zero resistance is said to *short out* the other resistances, by providing a near-zero resistance path for the current to follow as a shortcut around the other resistances.

20.8 CIRCUITS WIRED PARTIALLY IN SERIES AND PARTIALLY IN PARALLEL

Often an electric circuit is wired partially in series and partially in parallel. The key to determining current, voltage, and power in such a case is to deal with the circuit in parts, with the resistances in each part being either in series or in parallel with each other. Example 9 shows how this analysis is carried out.

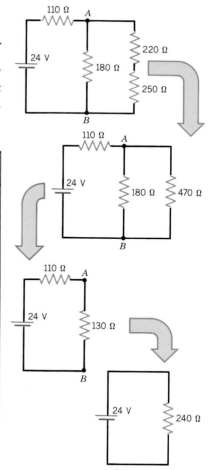

Figure 20.19 The four circuits shown in this picture are equivalent.

Example 9   A Four-Resistor Circuit

Figure 20.19 shows a circuit composed of a 24-V battery and four resistors, whose resistances are 110, 180, 220, and 250 Ω. Find (a) the total current supplied by the battery and (b) the voltage between points A and B in the circuit.

REASONING   The total current supplied by the battery can be obtained from Ohm's law, $I = V/R$, where R is the equivalent resistance of the four resistors. The equivalent resistance can be calculated by dealing with the circuit in parts. The voltage V_{AB} between the points A and B is also given by Ohm's law, $V_{AB} = IR$, where I is the current and R is the equivalent resistance between the two points.

SOLUTION
(a) The 220-Ω resistor and the 250-Ω resistor are in series, so they are equivalent to a single resistor whose resistance is 220 Ω + 250 Ω = 470 Ω (see Figure 20.19). The 470-Ω resistor is in parallel with the 180-Ω resistor. Their equivalent resistance is given by Equation 20.17:

$$\frac{1}{R_P} = \frac{1}{470 \ \Omega} + \frac{1}{180 \ \Omega} = 0.0077 \ \Omega^{-1}$$

$$R_P = \frac{1}{0.0077 \ \Omega^{-1}} = 130 \ \Omega$$

The circuit is now equivalent to a circuit containing a 110-Ω resistor in series with a 130-Ω resistor (see the drawing). This combination acts like a single resistor whose resistance is $R = 110 \ \Omega + 130 \ \Omega = 240 \ \Omega$. The total current from the battery is, then,

$$I = \frac{V}{R} = \frac{24 \ V}{240 \ \Omega} = \boxed{0.10 \ A}$$

(b) Ohm's law indicates that the voltage across the 130-Ω resistor between points A and B is

$$V_{AB} = IR = (0.10 \ A)(130 \ \Omega) = \boxed{13 \ V}$$

20.9 INTERNAL RESISTANCE

So far, the circuits we have considered include batteries or generators that add only their emfs to a circuit. In reality, however, such devices also add some resistance. This resistance is called the *internal resistance* of the battery or generator, because it is located inside the device. In a battery, the internal resistance is the resistance encountered by the current due to the materials from which the

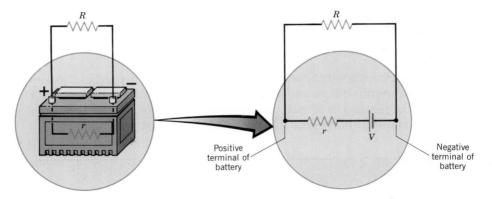

Figure 20.20   When an external resistance R is connected between the terminals of a battery, the resistance is connected in series with the internal resistance r of the battery.

battery is made. In a generator, the internal resistance is the resistance of wires and other components in the generator.

Figure 20.20 shows a schematic representation of the internal resistance r of a battery. The drawing emphasizes that when an external resistance R is connected to the battery, the resistance is connected *in series* with the internal resistance. The internal resistance of a functioning battery is typically small (several thousandths of an ohm for a new car battery). Nevertheless, the effect of the internal resistance may not be negligible. Example 10 illustrates that when current is drawn from a battery, the internal resistance causes the voltage between the terminals to drop below the maximum value specified by the battery's emf. The actual voltage between the terminals of a battery is known as the ***terminal voltage.***

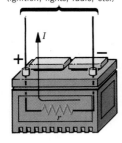

To car's electrical system
(ignition, lights, radio, etc.)

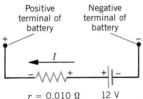

Positive terminal of battery   Negative terminal of battery

$r = 0.010\ \Omega$ 12 V

Figure 20.21   A car battery whose emf is 12 V and whose internal resistance is r.

Example 10   The Terminal Voltage of a Battery

Figure 20.21 shows a car battery whose emf is 12.0 V and whose internal resistance is $0.010\ \Omega$. This resistance is relatively large because the battery is old and the terminals are corroded. What is the terminal voltage when the current I drawn from the battery is (a) 10.0 A and (b) 100.0 A?

REASONING   The voltage between the terminals is not the entire 12.0-V emf, because part of the emf is needed to make the current go through the internal resistance. The amount of voltage needed can be determined from Ohm's law as the current I through the battery times the internal resistance r. For larger currents, a larger amount of voltage is needed, leaving less of the emf available between the terminals of the battery.

SOLUTION
(a) The amount of voltage needed to make a current of $I = 10.0$ A go through an internal resistance of $r = 0.010\ \Omega$ is $V = Ir = (10.0\ \text{A})(0.010\ \Omega) = 0.10$ V. To find the terminal voltage, remember that the direction of conventional current is always from high potential toward low potential. To emphasize this fact in the drawing, plus and minus signs have been included at the right and left ends, respectively, of the resistance r. The terminal voltage can be calculated by starting at the negative terminal of the battery and keeping track of how the voltage increases and decreases as we move

toward the positive terminal. The voltage rises by 12.0 V due to the battery's emf. However, the voltage drops by 0.10 V because of the potential difference across the internal resistance. Therefore, the terminal voltage is 12.0 V − 0.10 V = $\boxed{11.9 \text{ V}}$.

(b) When the current through the battery is 100.0 A, the amount of voltage needed to make the current go through the internal resistance is $V = (100.0 \text{ A})(0.010 \ \Omega) = 1.0$ V. The terminal voltage of the battery now decreases to 12.0 V − 1.0 V = $\boxed{11.0 \text{ V}}$.

Example 10 indicates that the terminal voltage of a battery is smaller when the current drawn from the battery is larger, an effect that any car owner can demonstrate. Turn the headlights on before starting your car, so the current through the battery is about 10 A, as in part (a) of Example 10. Then start the car. The starter motor draws a large amount of additional current from the battery, momentarily increasing the total current by an appreciable amount. Consequently, the terminal voltage of the battery decreases, causing the headlights to dim.

20.10 KIRCHHOFF'S RULES

Electric circuits that contain a number of resistors can often be analyzed by combining individual groups of resistors in series and parallel, as Section 20.8 discusses. However, there are many circuits in which no two resistors are in series or in parallel. To deal with such circuits it is necessary to employ methods other than the series–parallel method. One alternative is to take advantage of Kirchhoff's rules, named after their developer Gustav Kirchhoff (1824–1887). There are two rules, one dealing with the currents in a circuit and the other dealing with the voltages.

Figure 20.22 illustrates the basic idea behind Kirchhoff's first rule, or *junction rule*, as it is called. The picture shows a junction where several wires are connected together. As Section 18.2 discusses, electric charge is conserved. Therefore, since there is no continual accumulation of charges at the junction itself, the total charge per second flowing into the junction must equal the total charge per second flowing out of the junction. In other words, the junction rule states that *the total current directed into a junction must equal the total current directed out of the junction,* or 7 A = 5 A + 2 A for the specific case shown in the picture.

Figure 20.23 illustrates the basic idea behind Kirchhoff's second rule, or *loop rule,* as it is known. The drawing shows a circuit in which a 12-V battery is connected to a series combination of a 5-Ω and a 1-Ω resistor. The plus and minus signs associated with each resistor remind us that conventional current is directed from a higher potential toward a lower potential. Thus, from left to right, there is a potential drop of 10 V across the first resistor and another drop of 2 V across the second resistor. Keeping in mind that potential is the electric potential energy per unit charge, let us follow a positive test charge clockwise* around the circuit. Starting at the negative terminal of the battery, we see that the test charge gains potential energy because of the 12-V rise in potential due to the battery. The test charge then loses potential energy because of the 10-V drop in potential across the first resistor and the 2-V drop across the second resistor, ultimately arriving back

* The choice of the clockwise direction is arbitrary.

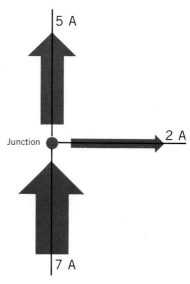

Figure 20.22 A junction is a point in a circuit where a number of wires are connected together. If 7 A of current is directed into the junction, then a total of 7 A of current must be directed out of the junction.

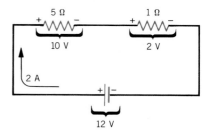

Figure 20.23 Following a positive test charge clockwise around the circuit, we see that the total voltage drop of 10 V + 2 V across the two resistors equals the voltage rise of 12 V provided by the battery. The plus and minus signs marking the ends of the resistors emphasize that the conventional current of 2 A is directed from a higher potential (+) toward a lower potential (−).

at the negative terminal. In traversing the closed circuit loop, the test charge is like a skier gaining gravitational potential energy in going up a hill on a chair lift and then losing it to friction in coming down and stopping. When the skier returns to the starting point, the gain equals the loss, so there is no net change in potential energy. Similarly, when the test charge arrives back at its starting point, there is no net change in electric potential energy, the gains matching the losses. This behavior of the test charge is an example of energy conservation, which the loop rule expresses in terms of the electric potential energy per unit charge. *The loop rule states that for a closed circuit loop, the total of all the potential rises* (12 V) *is the same as the total of all the potential drops* (10 V + 2 V).

Kirchhoff's rules can be applied to any circuit, even when the resistors are not in series or in parallel. The two rules are summarized below, and Examples 11 and 12 illustrate how to use them.

Kirchhoff's Rules

Junction Rule. The sum of the magnitudes of the currents directed into a junction equals the sum of the magnitudes of the currents directed out of the junction.

Loop Rule. Around any closed circuit loop, the sum of the potential drops equals the sum of the potential rises.

Example 11   Using Kirchhoff's Loop Rule

Figure 20.24 illustrates a circuit that contains two batteries and two resistors. Determine the current I in the circuit.

REASONING   The first step is to draw the direction of the current, which we have chosen to be clockwise around the circuit. This direction is *arbitrary*, and if it is incorrect, we will find that the value for I is a negative number.

The second step in the solution is to mark the two resistors with plus and minus signs, which serve as an aid in identifying the potential drops and rises for Kirchhoff's loop rule. *In marking the resistors, we remember that conventional current is always directed from a higher potential* (+) **toward a lower potential** (−). Thus, the plus and minus signs chosen for the two resistors *must* be those indicated in Figure 20.24, since the current is clockwise around the loop.

We may now apply Kirchhoff's loop rule to the circuit, starting at point A, proceeding clockwise around the loop, and identifying the potential drops and rises as we go. The potential across each resistor is given by Ohm's law as IR. The clockwise direction is arbitrary, and the same result is obtained with a counterclockwise path.

SOLUTION   Starting at point A, and moving clockwise around the loop, there is:

1. A potential drop of $IR = I(12\ \Omega)$ across the 12-Ω resistor as we go from the + side to the − side of the resistor
2. A potential drop of 6.0 V as we proceed from the positive terminal to the negative terminal of the 6.0-V battery
3. A potential drop of $IR = I(8.0\ \Omega)$ across the 8.0-Ω resistor
4. A potential rise of 24 V across the 24-V battery, as we proceed from the negative to the positive terminal.

PROBLEM SOLVING INSIGHT

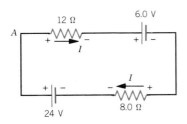

Figure 20.24   A single-loop circuit that contains two batteries and two resistors.

Setting the sum of the potential drops equal to the sum of the potential rises, as required by Kirchhoff's loop rule, gives

$$\underbrace{I(12\ \Omega) + 6.0\ \text{V} + I(8.0\ \Omega)}_{\text{Potential drops}} = \underbrace{24\ \text{V}}_{\text{Potential rises}}$$

Solving this equation for the current yields $\boxed{I = 0.90\ \text{A}}$. The current is a positive number, indicating that our initial choice for the (clockwise) direction of the current was a correct one.

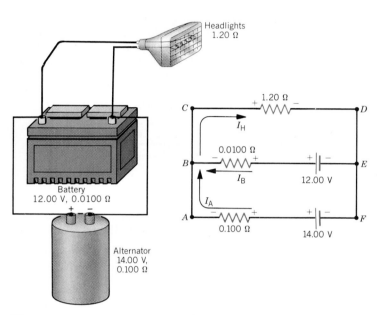

Figure 20.25 A circuit showing the headlight(s), battery, and alternator of a car.

Example 12   The Electrical System of a Car

THE PHYSICS OF . . .

an automobile electrical system.

In a car, the headlights are connected to the battery and would "run down" the battery if it were not for the alternator, which is run by the engine. Figure 20.25 indicates how the headlights and the alternator are connected to the battery. The picture also gives a circuit schematic in which the alternator is approximated as an additional 14.00-V battery for the sake of simplicity. The circuit includes an internal resistance of 0.0100 Ω for the battery and its leads, an internal resistance of 0.100 Ω for the alternator, and a resistance of 1.20 Ω for the headlights. Determine the current through the headlights, the battery, and the alternator.

<u>REASONING</u>   We begin by labeling the currents in the headlights (I_H), the battery (I_B), and the alternator (I_A). The drawing shows the directions chosen for these currents. The directions are arbitrary, and if any of them is incorrect, then the analysis will show that the corresponding value for the current is negative.

   Next, we mark the resistors with the plus and minus signs that serve as an aid in identifying the potential drops and rises for the loop rule, recalling that conventional current is always directed from a higher potential (+) toward a lower potential (−).

Thus, given the directions selected for I_H, I_B, and I_A, the plus and minus signs *must* be those indicated in Figure 20.25.

Kirchhoff's junction and loop rules can now be used.

SOLUTION The junction rule can be applied to junction B or junction E. In either case, the same equation results:

$$\begin{bmatrix} \text{Junction rule} \\ \text{applied} \\ \text{at } B \end{bmatrix} \qquad \underbrace{I_A + I_B}_{\substack{\text{Into} \\ \text{junction}}} = \underbrace{I_H}_{\substack{\text{Out of} \\ \text{junction}}}$$

In applying the loop rule to the lower loop *BEFA*, we start at point B, move clockwise around the loop, and identify the potential drops and rises. The clockwise direction is arbitrary, and the same result is obtained with a counterclockwise path. There is a potential rise of $I_B(0.0100\ \Omega)$ across the 0.0100-Ω resistor. This rise is followed by a drop of 12.00 V as we proceed from the positive to the negative terminal of the battery. Continuing around the loop, we find a 14.00-V rise across the alternator, followed by a drop of $I_A(0.100\ \Omega)$ across the 0.100-Ω resistor. Setting the sum of the potential drops equal to the sum of the potential rises gives the following result:

$$\begin{bmatrix} \text{Loop rule} \\ \text{applied} \\ \text{clockwise} \\ \text{around } BEFA \end{bmatrix} \qquad \underbrace{I_A(0.100\ \Omega) + 12.00\ \text{V}}_{\text{Potential drops}} = \underbrace{I_B(0.0100\ \Omega) + 14.00\ \text{V}}_{\text{Potential rises}}$$

Since there are three unknown variables in this problem, I_A, I_B, and I_H, a third equation is needed for a solution. To obtain the third equation, we apply the loop rule to the upper loop *CDEB*, choosing a clockwise path for convenience. The result is

$$\begin{bmatrix} \text{Loop rule} \\ \text{applied} \\ \text{clockwise} \\ \text{around } CDEB \end{bmatrix} \qquad \underbrace{I_B(0.0100\ \Omega) + I_H(1.20\ \Omega)}_{\substack{\text{Potential} \\ \text{drops}}} = \underbrace{12.00\ \text{V}}_{\substack{\text{Potential} \\ \text{rises}}}$$

These three equations can be solved simultaneously to show that

$$\boxed{I_H = 10.1\ \text{A},\ I_B = -9.0\ \text{A},\ I_A = 19.1\ \text{A}}$$

The negative answer for I_B indicates that the current through the battery is not directed from right to left, as drawn in Figure 20.25. Instead, the 9.0-A current is directed from left to right, opposite to the way current would be directed if the alternator were not connected. It is the left-to-right current created by the alternator that keeps the battery charged.

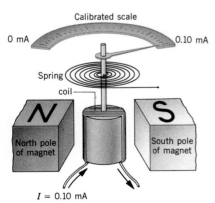

Figure 20.26 The essential parts of a dc galvanometer. The coil of wire and pointer rotate when there is a current in the coil.

20.11 THE MEASUREMENT OF CURRENT, VOLTAGE, AND RESISTANCE

THE GALVANOMETER

Current and voltage can be measured with devices known, respectively, as ammeters and voltmeters. There are two types of such devices, those that use digital electronics and those that do not. The essential feature of nondigital devices is the dc *galvanometer*. As Figure 20.26 illustrates, a galvanometer consists of a magnet,

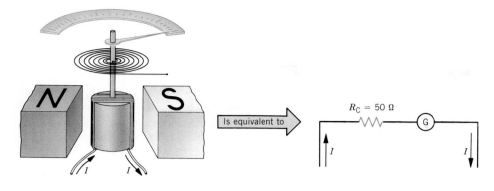

Figure 20.27 This picture shows how a galvanometer with a coil resistance of $R_C = 50\ \Omega$ is represented in a circuit diagram.

a coil of wire, a spring, a pointer, and a calibrated scale. The coil is mounted in such a way that it can rotate, which causes the pointer to move in relation to the scale. As Section 21.7 will discuss, the coil rotates in response to the torque applied by the magnet when there is a current in the coil. The coil stops rotating when this torque is balanced by the torque of the spring.

A galvanometer has two important characteristics that must be considered when it is used as part of a measurement device. First, the amount of dc current that causes a full-scale deflection of the pointer indicates the sensitivity of the galvanometer. For instance, Figure 20.26 shows an instrument that deflects full scale when the current in the coil is 0.10 mA. The second important characteristic is the resistance R_C of the wire in the coil. Figure 20.27 shows how a galvanometer with a coil resistance of $R_C = 50\ \Omega$ is represented in a circuit diagram. Both the full-scale current and the coil resistance depend on the details of the galvanometer construction.

THE AMMETER

An *ammeter* (*am*pere *meter*) is a device for measuring current and must be inserted into the circuit so the current passes directly through the ammeter, as Figure 20.28 shows. An ammeter includes a galvanometer and one or more *shunt resistors*. The purpose of a shunt resistor is to allow excess current to bypass the

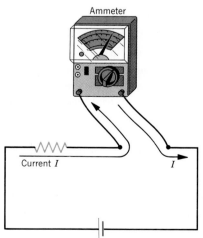

Figure 20.28 An ammeter measures the current I in a circuit. The ammeter is inserted into the circuit so that the current passes directly through the ammeter.

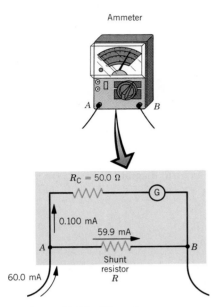

Figure 20.29 If a galvanometer with a full-scale deflection of 0.100 mA is to be used to measure a current of 60.0 mA, a shunt resistance R must be used, so the excess current of 59.9 mA can skirt around the galvanometer coil.

galvanometer coil, so the ammeter can be used to measure a current that exceeds the full-scale amount of the galvanometer. A shunt resistor is connected in parallel with the galvanometer coil, and the next example illustrates how the value of the shunt resistance is selected.

Example 13   An Ammeter

A galvanometer has a full-scale current of 0.100 mA and a coil resistance of R_C = 50.0 Ω. As Figure 20.29 shows, this galvanometer is used with a shunt resistor to form an ammeter that will register a full-scale deflection for a current of 60.0 mA. Determine the shunt resistance R.

REASONING   Since only 0.100 mA out of the available 60.0 mA is needed to cause a full-scale deflection of the galvanometer, the shunt resistor must allow the excess current of 59.9 mA to detour around the meter coil, as Figure 20.29 indicates. The value for the shunt resistance can be obtained by recognizing that the 50.0-Ω coil resistance and the shunt resistance are in parallel, both being connected between points A and B in the drawing. Thus, the voltage across each resistance is the same.

SOLUTION   Expressing voltage as the product of current and resistance, we find that

$$\underbrace{(59.9 \times 10^{-3} \text{ A})(R)}_{\substack{\text{Voltage across} \\ \text{shunt resistance}}} = \underbrace{(0.100 \times 10^{-3} \text{ A})(50.0 \text{ Ω})}_{\substack{\text{Voltage across} \\ \text{coil resistance}}}$$

$$R = \frac{(0.100 \times 10^{-3} \text{ A})(50.0 \text{ Ω})}{(59.9 \times 10^{-3} \text{ A})} = \boxed{0.0835 \text{ Ω}}$$

Typically, an ammeter includes a number of shunt resistors that provide several selectable current ranges.

When an ammeter is inserted into a circuit, the equivalent resistance of the coil and the shunt resistor adds to the circuit resistance. Any increase in circuit resistance causes a reduction in current, and this is a problem, for an ammeter should only measure the current, not change it. Therefore, an *ideal* ammeter would have zero resistance. In practice, a good ammeter is designed with a sufficiently small equivalent resistance, so there is only a negligible reduction of the current in the circuit when the ammeter is inserted.

THE PHYSICS OF . . .

a voltmeter.

THE VOLTMETER

A *voltmeter* is an instrument that measures the voltage between two points, A and B, in a circuit. Figure 20.30 shows that the voltmeter must be connected between the points and is *not* inserted into the circuit as an ammeter is. A voltmeter includes a galvanometer whose scale is calibrated in volts. Suppose, for instance, that the galvanometer in Figure 20.31 has a full-scale current of 0.1 mA and a coil resistance of 50 Ω. Under full-scale conditions, the voltage across the coil would be $V = IR_C = (0.1 \times 10^{-3} \text{ A})(50 \text{ Ω}) = 0.005$ V. Thus, this galvanometer could be used to register voltages in the range 0–0.005 V. A voltmeter, then, is a galvanometer used in this fashion, along with some provision for adjusting the range of voltages to be measured. Example 14 illustrates how the range of volt-

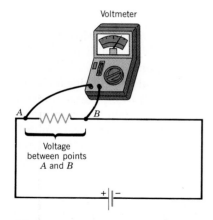

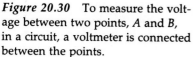

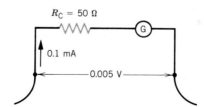

Figure 20.30 To measure the voltage between two points, A and B, in a circuit, a voltmeter is connected between the points.

Figure 20.31 The galvanometer shown has a full-scale deflection of 0.1 mA and a coil resistance of 50 Ω.

ages can be extended by the simple expedient of connecting a resistor in series with the coil.

Example 14   A Voltmeter

A galvanometer has a full-scale current of 0.100 mA and a coil resistance of $R_C = 50$ Ω. Determine the resistance R that must be connected in series with the coil to produce a voltmeter that will register a full-scale voltage of 0.500 V.

REASONING   Figure 20.32 shows the galvanometer connected in series with the resistance R. The resistance R is chosen so that when a voltage of 0.500 V is applied between points A and B in the drawing, the full-scale current of 0.100 mA will be in the galvanometer coil. The equivalent resistance of the series combination is $R + 50$ Ω. According to Ohm's law, the voltage across this combination is $V = I(R + 50$ $\Omega)$, from which R can be determined.

SOLUTION   Solving $V = I(R + 50$ $\Omega)$ for R yields

$$R = \frac{V}{I} - 50 \ \Omega = \frac{0.500 \ \text{V}}{0.100 \times 10^{-3} \ \text{A}} - 50 \ \Omega = \boxed{4950 \ \Omega}$$

Usually a voltmeter includes a number of additional series resistors that provide a variety of selectable voltage ranges.

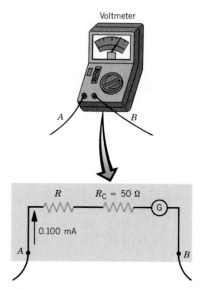

Figure 20.32 A voltmeter consists of a resistor R that is connected in series with the coil resistance R_C of the galvanometer.

Ideally, the voltage registered by a voltmeter should be the same as the voltage that exists when the voltmeter is not connected. However, a voltmeter takes some current from a circuit and, thus, alters the circuit voltage to some extent. An _ideal_ voltmeter would have infinite resistance and draw away only an infinitesimal amount of current. In reality, a good voltmeter is designed with a resistance that is large enough so the unit does not appreciably alter the voltage in the circuit to which it is connected.

THE WHEATSTONE BRIDGE

One way to measure resistance is to use a circuit called a *Wheatstone bridge,* after Charles Wheatstone (1802–1875), the English physicist who established its usefulness. The circuit illustrates a method of measurement known as the *null method.* In addition to the unknown resistance R, a Wheatstone bridge includes three other resistances R_1, R_2, and R_v, as Figure 20.33 shows. A galvanometer records any current between points A and B. To measure the unknown resistance, the variable resistance R_v is adjusted until the galvanometer registers zero or null current, in which case the Wheatstone bridge is said to be "balanced." Example 15 shows how the unknown resistance can be obtained from the value of R_v and the ratio R_1/R_2 in a balanced Wheatstone bridge.

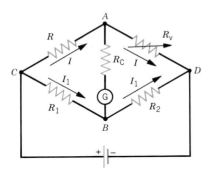

Figure 20.33 A Wheatstone bridge circuit. The resistor R_v marked with the arrow through it is the variable resistor.

Example 15   Measuring a Resistance with a Wheatstone Bridge

When the Wheatstone bridge in Figure 20.33 is balanced, the variable resistance is $R_v = 173\ \Omega$, and the ratio R_1/R_2 is 0.100. Determine the value of the unknown resistance R.

REASONING AND SOLUTION   The key to solving this problem lies in the information that the bridge is balanced. In other words, there is no current through the galvanometer. The direction of conventional current is from high potential toward low potential. But since there is no current through the galvanometer, points A and B in the circuit *must be at the same potential.* Furthermore, R_1 and R are connected at point C so the voltages across R_1 and R must be the same. Since voltage is current times resistance and since I_1 and I represent the currents in R_1 and R, it follows that

$$I_1 R_1 = IR$$

In a similar fashion, the voltage across R_2 must be the same as that across R_v:

$$I_1 R_2 = IR_v$$

Dividing these two equations shows that $R_1/R_2 = R/R_v$ or

$$R = \frac{R_1}{R_2} R_v = (0.100)(173\ \Omega) = \boxed{17.3\ \Omega}$$

20.12 CAPACITORS IN SERIES AND PARALLEL

Capacitors, like resistors, can be connected in series and in parallel. Parallel capacitors are simpler to understand and will be considered first. Figure 20.34 shows two capacitors connected in parallel to a battery. Since the capacitors are in parallel, they have the same voltage V across their plates. However, the capacitors *contain different amounts of charge.* The charge stored by a capacitor is $q = CV$ (Equation 19.8), so $q_1 = C_1 V$ and $q_2 = C_2 V$.

As with resistors, it is always possible to replace a parallel combination of capacitors with an *equivalent capacitor* that stores the same charge and energy for a given voltage as the combination does. To determine the equivalent capacitance C_P, note that the total charge q stored by the two capacitors is $q = q_1 + q_2$. Consequently,

$$q = q_1 + q_2 = C_1 V + C_2 V = (C_1 + C_2)V = C_P V$$

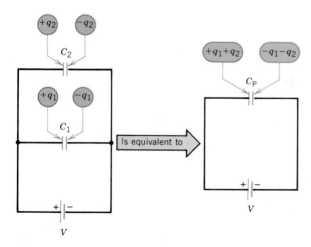

Figure 20.34 In a parallel combination of capacitances C_1 and C_2, the voltage V across each capacitor is the same, but the charges q_1 and q_2 on each capacitor are different.

This result indicates that two capacitors in parallel can be replaced by an equivalent capacitor whose capacitance is $C_P = C_1 + C_2$. For any number of capacitors in parallel, the equivalent capacitance is

$$\begin{bmatrix}\textbf{Parallel}\\\textbf{capacitors}\end{bmatrix} \qquad C_P = C_1 + C_2 + C_3 + \cdots \qquad (20.18)$$

Capacitances in parallel simply add together to give an equivalent capacitance. This behavior contrasts with that of resistors in parallel, which combine as reciprocals, according to Equation 20.17.

The equivalent capacitor not only stores the same amount of charge as the parallel combination of capacitors, but also stores the same amount of energy. For instance, the energy stored in a single capacitor is $\frac{1}{2}CV^2$ (Equation 19.11), so the total energy U stored by two capacitors in parallel is

$$U = \tfrac{1}{2}C_1V^2 + \tfrac{1}{2}C_2V^2 = \tfrac{1}{2}(C_1 + C_2)V^2 = \tfrac{1}{2}C_PV^2$$

which is equal to the energy stored in the equivalent capacitor C_P.

When capacitors are connected in series, the equivalent capacitance is different than when they are in parallel. As an example, Figure 20.35 shows two capacitors in series and reveals the following important fact: *All capacitors in series, regardless of their capacitances, contain charges of the same magnitude, $+q$ and $-q$, on their plates.* The battery places a charge of $+q$ on plate a of capacitor C_1,

PROBLEM SOLVING INSIGHT

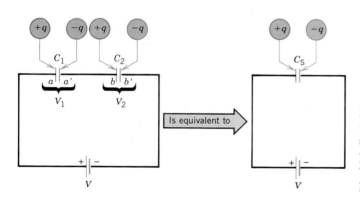

Figure 20.35 In a series combination of capacitances C_1 and C_2, the same amount of charge q is on the plates of each capacitor, but the voltages V_1 and V_2 across each capacitor are different.

and this charge induces a charge of $+q$ to depart from the opposite plate a', leaving behind a charge $-q$. The $+q$ charge that leaves plate a' is deposited on plate b of capacitor C_2 (since these two plates are connected by a wire), where it induces a $+q$ charge to move away from the opposite plate b', leaving behind a charge of $-q$. Thus, all capacitors in series contain charges of the same magnitude on their plates.

Note the difference between charging capacitors in parallel and in series. *In charging parallel capacitors, the battery moves a charge q that is the sum of the charges moved for each of the capacitors: $q = q_1 + q_2 + q_3 + \cdots$.* In contrast, in a series combination of n capacitors, the battery only moves a charge q, not nq, because the charge q passes by induction from one capacitor directly to the next one in line.

The equivalent capacitance C_S for the series connection in Figure 20.35 can be determined by observing that the battery voltage V is shared by the two capacitors, as it is in a series combination of resistors. The drawing indicates that the voltages across C_1 and C_2 are V_1 and V_2 and that $V = V_1 + V_2$. The voltage across each capacitor is related to the magnitude of the charge on each capacitor according to $V_1 = q/C_1$ and $V_2 = q/C_2$, the charge q being the same for each. Therefore,

$$V = \frac{q}{C_1} + \frac{q}{C_2} = q\left(\frac{1}{C_1} + \frac{1}{C_2}\right) = q\left(\frac{1}{C_S}\right)$$

where C_S is the equivalent capacitance. Thus, two capacitors in series can be replaced by a single capacitor whose capacitance C_S is given by $1/C_S = 1/C_1 + 1/C_2$. For any number of capacitors connected in series the result is

$$\begin{bmatrix} \text{Series} \\ \text{capacitors} \end{bmatrix} \qquad \frac{1}{C_S} = \frac{1}{C_1} + \frac{1}{C_2} + \frac{1}{C_3} + \cdots \qquad (20.19)$$

Equation 20.19 indicates that capacitances in series combine as reciprocals and do not simply add together as resistors in series do. It is left as an exercise (problem 84) to show that the equivalent capacitance stores the same electrostatic energy as the sum of the energies of the individual capacitors in the series combination.

It is possible to simplify circuits containing a number of capacitors in the same general fashion as that outlined for resistors in Example 9 and Figure 20.19. The capacitors in a parallel grouping can be combined according to Equation 20.18, and those in a series grouping can be combined according to Equation 20.19.

20.13 RC CIRCUITS

Many electric circuits contain both resistors and capacitors. Figure 20.36 illustrates an example of a resistor–capacitor or *RC* circuit. Part *a* of the drawing shows the circuit at a time t after the switch has been closed and the battery has begun to charge up the capacitor plates. The charge on the plates builds up gradually to its equilibrium value of $q_0 = CV_0$, where V_0 is the voltage of the battery. Assuming that the capacitor is uncharged at time $t = 0$ when the switch is closed, it can be shown that the magnitude q of the charge on the plates at time t is

$$\begin{bmatrix} \text{Capacitor} \\ \text{charging} \end{bmatrix} \qquad q = q_0[1 - e^{-t/(RC)}] \qquad (20.20)$$

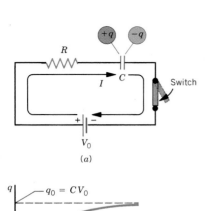

Figure 20.36 Charging a capacitor.

where the exponential *e* has the value of 2.718. . . . Part *b* of the drawing shows a graph of this expression, which indicates that the charge is $q = 0$ when $t = 0$ and increases gradually toward the equilibrium value of $q_0 = CV_0$. The voltage *V* across the capacitor at any time can be obtained from Equation 20.20 by dividing the charges *q* and q_0 by the capacitance *C*, since $V = q/C$ and $V_0 = q_0/C$.

The term *RC* in the exponent in Equation 20.20 is called the ***time constant*** τ of the circuit:

$$\tau = RC \qquad (20.21)$$

The time constant is measured in seconds; verification of the fact that an ohm times a farad is equivalent to a second is left as an exercise (see question 15). The time constant is the amount of time required for the capacitor to accumulate 63.2% of its equilibrium charge, as can be seen by substituting $t = \tau = RC$ in Equation 20.20; $q_0(1 - e^{-1}) = q_0(0.632)$. The charge approaches its equilibrium value rapidly when the time constant is small and slowly when the time constant is large.

Figure 20.37 shows a circuit at a time *t* after the switch is closed to allow a charged capacitor to begin discharging. There is no battery in this circuit, so the charge $+q$ on the left plate of the capacitor can flow counterclockwise through the resistor and neutralize the charge $-q$ on the right plate. Assuming the capacitor has a charge q_0 at time $t = 0$ when the switch is closed, it can be shown that

$$\begin{bmatrix} \textbf{Capacitor} \\ \textbf{discharging} \end{bmatrix} \qquad q = q_0 e^{-t/(RC)} \qquad (20.22)$$

where *q* is the amount of charge remaining on either plate at time *t*. The graph of this expression in part *b* of the drawing shows that the charge begins at q_0 when $t = 0$ and decreases gradually toward zero. Smaller values of the time constant *RC* lead to a more rapid discharge. Equation 20.22 indicates that when $t = \tau = RC$, the magnitude of the charge remaining on each plate is $q = q_0 e^{-1} = q_0(0.368)$. Therefore, the time constant is also the amount of time required for a charged capacitor to *lose* 63.2% of its charge.

The charging/discharging of a capacitor has many applications. For example, some automobiles come equipped with a feature that allows the windshield wipers to be used intermittently during a light drizzle. In this mode of operation the wipers remain off for a while and then turn on briefly. The timing of the on–off cycle is determined by the time constant of a resistor–capacitor combination.

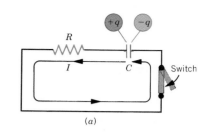

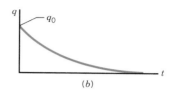

Figure 20.37   Discharging a capacitor.

THE PHYSICS OF . . .

windshield wipers.

*20.14 SAFETY AND THE PHYSIOLOGICAL EFFECTS OF CURRENT

THE PHYSICS OF . . .

electrical grounding.

Electric circuits, while very useful, can also be hazardous. To reduce the danger inherent in using circuits, proper ***electrical grounding*** is necessary. The next two figures help to illustrate what electrical grounding means and how it is achieved.

Figure 20.38*a* shows a clothes drier connected to a wall socket via an ordinary two-prong plug. The drier is operating normally, that is, the wires inside are insulated from the metal casing of the drier, so no charge flows through the casing itself. Notice that one terminal of the ac generator is customarily connected to ground (⏚) by the electric power company. Part *b* of the drawing shows the hazardous result that occurs if a wire comes loose and accidentally contacts the

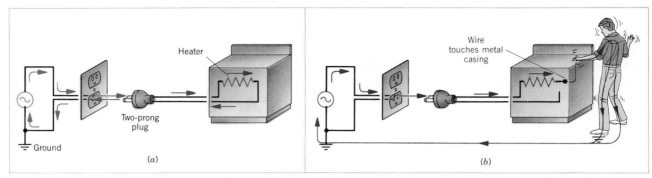

Figure 20.38 (*a*) A normally operating clothes drier that is connected to a wall socket via a two-prong plug. (*b*) The drier malfunctions, because an internal wire accidentally touches the metal casing. A person who touches the casing can receive an electrical shock.

metal casing of the drier. A person touching the casing receives a shock, since electric charge flows from the generator, through the casing, the person's body, and the ground on the way back to the generator, as the picture illustrates.

Figure 20.39 shows the same clothes drier connected to a wall socket via a three-prong plug that provides safe electrical grounding. The third prong on the plug connects the metal casing of the drier directly to a copper rod driven into the ground or to a copper water pipe that is in the ground. This arrangement protects against electrical shock in the event that a broken wire touches the metal casing. The charge flows from the generator, through the casing, through the third prong of the plug, and into the ground, returning eventually to the generator. No charge flows into the person's body, since the copper rod connected to the third prong of the plug provides much less electrical resistance than does the body.

Serious and sometimes fatal injuries can result from electric shock. The severity of the injury depends on the magnitude of the current and the parts of the body that the moving charges pass through. The amount of current that the body senses as a mild tingling sensation is about 0.001 A. Currents on the order of 0.01 – 0.02 A can cause muscle spasms, in which a person "can't let go" of the object causing the shock. Currents of approximately 0.2 A are potentially fatal, because they can make the heart fibrillate, or beat in an uncontrolled manner. Substantially larger currents stop the heart completely. However, since the heart often begins beating normally again after the current ceases, the larger currents can be less dangerous than the smaller currents that cause fibrillation.

Figure 20.39 When the drier malfunctions, a person touching it receives no shock, since electric charge flows through the third prong and into the ground, rather than through the person's body.

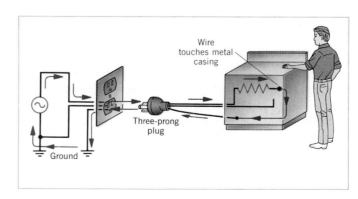

INTEGRATION OF CONCEPTS

KIRCHHOFF'S RULES AND CONSERVATION PRINCIPLES

Circuits are used to transfer electrical energy from sources to devices that use the energy. The currents and voltages in a circuit can be determined with the aid of Kirchhoff's two rules, the junction rule and the loop rule. Both of these rules are rooted in conservation principles.

The junction rule is based on the conservation of electric charge, which states that the net electric charge of an isolated system is conserved. When charge flows into a junction where two or more wires are connected, and if there is no buildup of charge at the junction itself, the net charge per second entering a junction equals the net charge per second leaving a junction. Since the charge flowing per second is the current, the junction rule expresses the conservation of charge by stating that the net current directed into a junction equals the net current directed out of a junction. This idea is very similar to that expressed in Section 11.8 by the equation of continuity for a flowing fluid. The equation of continuity states that the mass per unit time that enters one end of a pipe must equal that leaving at the other end.

Kirchhoff's loop rule is based on the conservation of energy, which states that energy can neither be created nor destroyed, but can only be converted from one form to another. When electric charge moves in a circuit, it gains energy from sources such as batteries and generators. The electrical energy is converted into heat as the charge moves through resistors, the production of heat being accompanied by an equal loss in electrical energy. When the charge moves around a complete circuit, arriving back at its starting point, there is no overall change in energy, the gains having been matched by the losses, as energy is converted from one form to another. Kirchhoff's loop rule expresses the conservation of energy by stating that, around any closed loop, the sum of the potential drops equals the sum of the potential gains.

SUMMARY

There must be at least one source or generator of electrical energy in an electric circuit. The **electromotive force (emf)** of a generator, such as a battery, is the maximum potential difference (in volts) that exists between the terminals of the generator.

The rate of flow of charge is called the **electric current.** If the rate is constant, the current I is given by $I = q/t$, where q is the magnitude of the charge crossing a surface in a time t, the surface being perpendicular to the motion of the charge. The SI unit for current is the coulomb per second (C/s), which is referred to as an ampere (A). When the charges flow only in one direction around a circuit, the current is called **direct current (dc).**

When the direction of charge flow changes from moment to moment, the current is known as **alternating current (ac).**

The definition of **electrical resistance** is $R = V/I$, where V is the voltage applied across a piece of material and I is the current through the material. If the ratio V/I is constant for all values of V and I, the relation $R = V/I$ or $V = IR$ is referred to as **Ohm's law.** Resistance is measured in volts per ampere, a unit called an ohm (Ω).

The resistance of a piece of material of length L and cross-sectional area A is $R = \rho L/A$, where ρ is the **resistivity** of the material. The resistivity of a material depends on the temperature. For many materials and limited temperature ranges, the temperature dependence is given by $\rho = \rho_0[1 + \alpha(T - T_0)]$, where ρ and ρ_0 are the resistivities at temperatures T and T_0, respectively, and α is the **temperature coefficient of resistivity.**

In a circuit in which a current I results from a voltage V, the **electric power** delivered to the circuit is $P = IV$. Since $V = IR$, the power dissipated in a resistance R is also given by $P = I^2R$ or $P = V^2/R$.

The **alternating voltage between the terminals of an ac generator** can be represented by $V = V_0 \sin 2\pi ft$, where V_0 is the peak value of the voltage, t is the time, and f is the frequency at which the voltage oscillates. Correspondingly, in a circuit containing only resistance, the **ac current** is $I = I_0 \sin 2\pi ft$, where I_0 is the peak value of the current and is related to the peak voltage via $I_0 = V_0/R$. The **root mean square (rms)** voltage and current are related to the peak values according to $V_{rms} = V_0/\sqrt{2}$ and $I_{rms} = I_0/\sqrt{2}$. The power in an ac circuit is the product of the current and the voltage and oscillates in time. The **average power** is $\overline{P} = I_{rms}V_{rms}$. Since $V_{rms} = I_{rms}R$, the average power can also be written as $\overline{P} = I_{rms}^2 R$ or $\overline{P} = V_{rms}^2/R$.

When devices are connected **in series,** there is the same electric current through each device. The **equivalent resistance R_S of a series combination of resistances** $(R_1, R_2, R_3,$ etc.) is $R_S = R_1 + R_2 + R_3 + \cdots$. The equivalent resistance dissipates the same total power as the series combination.

Connecting devices **in parallel** means that they are connected in such a way that the same voltage is applied across each device. In general, devices wired in parallel carry different currents. The **equivalent resistance R_P of a parallel combination of resistances** is $1/R_P = 1/R_1 + 1/R_2 + 1/R_3 + \cdots$. The equivalent resistance dissipates the same total power as the parallel combination.

The **internal resistance** of a battery or generator is the electrical resistance encountered in the battery or generator by the current. The **terminal voltage** is the voltage between the terminals of a battery or generator and is equal to the emf only when there is no current through the device. Because of the internal resistance r, the terminal voltage is less than the emf when there is current I, by an amount Ir.

Kirchhoff's rules may be used to analyze the currents and potential differences in electric circuits. The **junction rule** states that the sum of the magnitudes of the currents directed into a junction equals the sum of the magnitudes of the currents directed out of the junction. The **loop rule** states that, around any closed circuit loop, the sum of the potential drops equals the sum of the potential rises.

A **galvanometer** is a device that responds to electric current and is used in nondigital ammeters and voltmeters. An **ammeter** is an instrument that measures current and must be inserted into a circuit so the current passes directly through the ammeter. A **voltmeter** is an instrument for measuring the voltage between two points in a circuit. For this measurement a voltmeter must be connected between the two points and is not inserted into a circuit as an ammeter is. A **Wheatstone bridge** is a device that uses the **null method** for measuring resistance.

The **equivalent capacitance C_P for a parallel combination of capacitances** $(C_1, C_2, C_3,$ etc.) is $C_P = C_1 + C_2 + C_3 + \cdots$. In general, each capacitor in a parallel combination carries a different amount of charge. The equivalent capacitor carries the same total charge and stores the same total energy as the parallel combination.

The **equivalent capacitance C_S for a series combination of capacitances** is given by $1/C_S = 1/C_1 + 1/C_2 + 1/C_3 + \cdots$. Each capacitor in the combination carries the same amount of charge. The equivalent capacitor carries the same amount of charge as *any one* of the capacitors in the combination and stores the same total energy as the entire combination.

The **charging or discharging of a capacitor** in a dc series circuit (resistance R, capacitance C) does not occur instantaneously. Charge builds up gradually according to the relation $q = q_0[1 - e^{-t/(RC)}]$, where q is the charge on the capacitor at time t, q_0 is the charge on the capacitor at time $t = 0$, and RC is the **time constant** of the circuit. The discharging of a capacitor through a resistor is described by $q = q_0 e^{-t/(RC)}$.

QUESTIONS

1. The drawing shows a circuit in which a light bulb is connected to the household ac voltage via two switches S_1 and S_2. This is the kind of wiring, for example, that allows you to turn a carport light on and off from either inside the house or out in the carport. Explain which position A or B of S_2 turns the light on when S_1 is set to (a) position A and (b) position B.

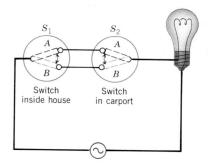

2. The resistance of a light bulb is not the same when the bulb is off as it is when the bulb is on. Why?

3. Two materials have different resistivities. Two wires of the same length are made, one from each of the materials. Is it possible for each wire to have the same resistance? Explain.

4. Does the resistance of a copper wire increase or decrease when both the length and the diameter of the wire are doubled? Justify your answer.

5. One electrical appliance operates with a voltage of 120 V, while another operates with 240 V. Based on this information alone, is it correct to say that the second appliance uses more power than the first? Give your reasoning.

6. Often, the instructions for an electrical appliance do not state how many watts of power the appliance uses. Instead, a statement such as "10 A, 120 V" is given. Explain why this statement is equivalent to telling you the power consumption.

7. A long extension cord is used to connect a light bulb to an electrical outlet. The current in the bulb is slightly less than that calculated using Ohm's law with the resistance of the light bulb and the voltage at the socket. Why?

8. The power rating of a 1000-W heater specifies the power consumed when the heater is connected to an ac voltage of 120 V. Explain why the power consumed by two of these heaters connected in series to a voltage of 120 V is not 2000 W.

9. A number of light bulbs are to be connected to a single electrical outlet. Will the bulbs provide more brightness if they are connected in series or in parallel? Why?

10. A car has two headlights. The filament of one burns out, so charges can no longer flow out of the battery and through the headlight. However, the other headlight stays on. Draw a circuit diagram that shows how the headlights are connected to the battery.

11. A normal light bulb has a single filament and a power rating that corresponds to the power dissipated in the filament resistance when the bulb is connected to an ac voltage of 120 V. Some bulbs, however, contain two separate filaments. These are the familiar three-way bulbs that, in the proper socket, can be switched to provide three different wattages. Observe that in the following three-way bulb the highest wattage is the sum of the other two choices: 30 W/70 W/ 100 W. With this observation in mind, explain how the two filaments can be connected to the voltage source to give three wattage ratings.

12. One of the circuits in the drawing contains resistors that are neither in series nor in parallel. Which is it?

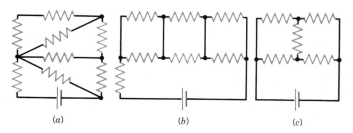

(a)   (b)   (c)

13. Compare the resistance of an ideal ammeter with the resistance of an ideal voltmeter and explain why the resistances are so different.

14. Describe what would happen to the current in a circuit if a voltmeter, inadvertently mistaken for an ammeter, were inserted into the circuit.

15. The time constant of a series RC circuit is $\tau = RC$. Verify that an ohm times a farad is equivalent to a second.

PROBLEMS

Note: For problems that involve ac conditions, the current and voltage are rms values and the power is an average value, unless indicated otherwise.

Section 20.1 Electromotive Force and Current, Section 20.2 Ohm's Law

1. A portable compact disc player is designed to play for 2.0 h on a fully charged battery pack. If the battery pack pro-

vides a total of 180 C of charge, how much current does the player use in operating?

2. Most of the wiring in a typical house can safely handle about 15 A of current. At this current level, how much charge flows through a wire in one hour?

3. A battery charger is connected to a dead battery and delivers a current of 8.0 A for 3.0 hours, keeping the voltage

across the battery terminals at 12 V in the process. How much energy is delivered to the battery?

4. A toaster has a resistance of 14 Ω and is plugged into a 120-V outlet. What is the current in the toaster?

5. The heating element of a clothes drier has a resistance of 11 Ω and is connected across a 240-V electrical outlet. What is the current in the heating element?

6. In the arctic, electric socks are useful. A pair of socks uses a 9.0-V battery pack for each sock. A current of 0.11 A is drawn from each battery pack by wire woven into the socks. Find the resistance of the wire in one sock.

*7. A car battery has a rating of 220 ampere·hours (A·h). This rating is one indication of the *total charge* that the battery can provide to a circuit before failing. (a) What is the total charge (in coulombs) that this battery can provide? (b) Determine the maximum current that the battery can provide for 38 minutes.

*8. The filament of a light bulb has a resistance of 192 Ω, and the bulb is operating from a 120-V outlet. How much energy is delivered to the bulb in 45 minutes?

Section 20.3 Resistance and Resistivity

9. High-voltage power lines are a familiar sight throughout the country. The aluminum wire used for some of these lines has a cross-sectional area of 4.9×10^{-4} m². What is the resistance of ten kilometers of this wire?

10. A coil of wire has a resistance of 38.0 Ω at 25 °C and 43.7 Ω at 55 °C. What is the temperature coefficient of resistivity?

11. Two wires have the same length and the same resistance. One is made from aluminum and the other from copper. Obtain the ratio of the cross-sectional area of the aluminum wire to that of the copper wire.

12. A copper wire has a cross-sectional area of 7.9×10^{-7} m². Find the resistance *per unit length* for this wire.

13. A copper cable carries a current of 1200 A. There is a potential difference of 1.6×10^{-2} V between two points on the cable that are 0.24 m apart. What is the radius of the cable?

14. A wire of unknown composition has a resistance of $R_0 = 35.0$ Ω when immersed in water at 20.0 °C. When the wire is placed in boiling water, its resistance rises to 47.6 Ω. What is the temperature of a hot summer day when the wire has a resistance of 37.8 Ω?

*15. A wire has a resistance of 21.0 Ω. It is melted down, and from the same volume of metal a new wire is made that is three times longer than the original wire. What is the resistance of the new wire?

*16. Liquid mercury in one rectangular container is poured into another rectangular container that has one-half the cross-sectional area. The resistance between the top and bottom surfaces of the mercury is measured in each case. Find the ratio of the resistance of the mercury in the new container to that of the mercury in the original container.

*17. An iron wire has a resistance of 5.90 Ω at 20.0 °C and a gold wire has a resistance of 6.70 Ω at the same temperature. At what temperature do the wires have the same resistance?

**18. A digital thermometer uses a thermistor as the temperature sensing element. A thermistor is a kind of semiconductor and has a large negative temperature coefficient of resistivity α. Suppose $\alpha = -0.060$ (C°)$^{-1}$ for the thermistor in a digital thermometer used to measure the temperature of a patient. The resistance of the thermistor decreases to 85% of its value at the normal body temperature of 37.0 °C. What is the patient's temperature?

Section 20.4 Electric Power

19. An automobile battery is being charged at a voltage of 12.0 V and a current of 19.0 A. How much power is being produced by the charger?

20. An electric alarm clock uses a 5.0-W motor and runs all day, every day. If electricity costs $0.10 per kWh, determine the yearly cost of running the clock.

21. The heating element in a toaster has a resistance of 14 Ω. The toaster is plugged into a 120-V outlet. What is the power dissipated by the toaster?

22. A 240-V clothes drier draws 16 A of current for a period of 45 min. How much energy, in kilowatt-hours, does the drier consume?

23. A cigarette lighter in a car is a resistor that, when activated, is connected across the 12-V battery. Suppose a lighter dissipates 33 W of power. Find (a) the resistance of the lighter and (b) the current that the lighter draws from the battery.

24. A commercial resistor can safely dissipate power only up to a certain rated value. Beyond this value, the resistor becomes excessively hot and often cracks apart. What is the largest voltage that can be applied across a 680-Ω resistor, when the resistor is rated at (a) 0.25 W and (b) 2.0 W?

*25. A stove is connected to a 240-V outlet and receives power P_{240}. When a restaurant owner uses the same stove with a 208-V outlet, the stove receives power P_{208}. Ignoring any change of resistance with temperature, find the ratio P_{240}/P_{208}.

**26. An iron wire has a resistance of 12 Ω at 20.0 °C and a mass of 1.3×10^{-3} kg. A current of 0.10 A is sent through the wire for one minute and causes the wire to become hot. Assuming all the electrical energy is dissipated in the wire and remains there, find the final temperature of the wire. (*Hint:*

Use the average resistance of the wire during the heating process, and see Table 12.3 for the specific heat capacity of iron.)

Section 20.5 Alternating Current

27. The current in a circuit is ac and has a peak value of 2.50 A. Determine the rms current.

28. In the wire connecting an electric clock to a wall socket, how many times during a day does the current reverse its direction?

29. An ac voltage of $V = (65 \text{ V}) \sin 2\pi (60 \text{ Hz})t$ is applied across a 25-Ω resistor. What is the rms current in the resistor?

30. A blow-drier and a vacuum cleaner each operate with an ac voltage of 120 V. The current rating of the blow-drier is 11 A, while that of the vacuum cleaner is 4.0 A. Determine the power consumed by (a) the blow-drier and (b) the vacuum cleaner. (c) Determine the ratio of the energy used by the blow-drier in 15 minutes to the energy used by the vacuum cleaner in one-half an hour.

31. The heating element in an iron has a resistance of 16 Ω and is connected to a 120-V wall socket. (a) What is the average power consumed by the iron, and (b) the peak power?

*32. The *recovery time* of a hot water heater is the time required to heat all the water in the unit to the desired temperature. Suppose that a 42-gal (1.00 gal = $3.79 \times 10^{-3} \text{ m}^3$) unit starts with cold water at 11 °C and delivers hot water at 55 °C. The unit is electric and utilizes a resistance heater (120 V ac, 3.2 Ω) to heat the water. Assuming no heat is lost to the environment, determine the recovery time (in hours) of the unit.

*33. On its highest setting, a heating element on an electric stove (see Figure 20.7) is connected to an ac voltage of 240 V. This element has a resistance of 29 Ω. (a) Find the power dissipated in the element. (b) Assuming that three-fourths of the heat produced by the element is used to heat a pot of water (the rest being wasted), find the time required to bring 1.9 kg of water (half a gallon) at 15 °C to a boil.

Section 20.6 Series Wiring

34. A 28-Ω resistor and a 62-Ω resistor are connected in series across a 48-V battery. What is the current in the circuit?

35. Three resistors, 25, 45, and 75 Ω, are connected in series, and a 0.51-A current passes through them. What is (a) the equivalent resistance and (b) the potential difference across the three resistors?

36. Three resistors (9.0, 5.0, and 1.0 Ω) are connected in series across a 24-V battery. Find (a) the current in, (b) the voltage across, and (c) the power dissipated in each resistor.

37. A battery dissipates 2.50 W of power in each of two

47.0-Ω resistors connected in series. What is the voltage of the battery?

38. A 16.0-Ω resistor and an 8.0-Ω resistor are connected in series across a 12.0-V battery. What is the voltage across each resistor?

39. The current in a series circuit is 15.0 A. When an additional 8.00-Ω resistor is inserted in series, the current drops to 12.0 A. What is the resistance in the original circuit?

*40. A 47-Ω resistor can dissipate up to 0.25 W of power without burning up. What is the smallest number of such resistors that can be connected in series across a 9.0-V battery without any one of them burning up?

*41. Three resistors are connected in series across a battery. The value of each resistance and its maximum power rating are as follows: 5.0 Ω and 20.0 W, 30.0 Ω and 10.0 W, and 15.0 Ω and 10.0 W. (a) What is the greatest voltage that the battery can have without one of the resistors burning up? (b) How much power does the battery deliver to the circuit in (a)?

Section 20.7 Parallel Wiring

42. A 16-Ω loudspeaker and an 8.0-Ω loudspeaker are connected in parallel across the terminals of an amplifier. Assuming the speakers behave as resistors, determine the equivalent resistance of the two speakers.

43. What resistance must be placed in parallel with a 155-Ω resistor to make the equivalent resistance 115 Ω?

44. Two resistors, 12.0 and 15.0 Ω, are connected in parallel. The current through the 15.0-Ω resistor is 4.0 A. (a) Determine the current in the other resistor. (b) What is the total power consumed by the two resistors?

45. How many 4.0-Ω resistors must be connected in parallel to create an equivalent resistance of one-sixteenth of an ohm?

46. A wire whose resistance is R is cut into three equally long pieces, which are then connected in parallel. In terms of R, what is the resistance of the parallel combination?

47. A 75-W lamp and a 15-W radio are connected in parallel to the same 120-V electrical outlet. (a) What is the total current coming from the outlet? (b) What is the equivalent resistance of these two devices?

*48. The total current delivered to a number of devices connected in parallel is the sum of the individual currents in each device. Circuit breakers are resettable automatic switches that protect against a dangerously large total current by "opening" to stop the current at a specified safe value. A 1650-W toaster, a 1090-W iron, and a 1250-W microwave oven are turned on in a kitchen. As the drawing shows, they are all connected through a 20-A circuit breaker to an ac voltage of 120 V. (a) Find the equivalent resistance of the three devices. (b) Obtain the total current delivered by the source and determine whether the breaker will "open" to prevent an accident.

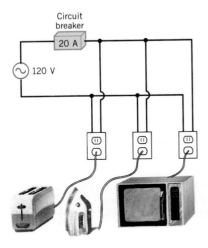

*49. A resistor (resistance = R) is connected first in parallel and then in series with a 2.00-Ω resistor. A battery delivers five times more current to the parallel combination than it does to the series combination. Determine the two possible values for R.

**50. The rear window defogger of a car consists of thirteen thin wires (resistivity = 88.0×10^{-8} Ω·m) embedded in the glass. The wires are connected in parallel to the 12.0-V battery, and each has a length of 1.30 m. The defogger can melt 2.10×10^{-2} kg of ice at 0 °C into water at 0 °C in two minutes. Assume that all the power dissipated in the wires is used immediately to melt the ice. Find the cross-sectional area of each wire.

Section 20.8 Circuits Wired Partially in Series and Partially in Parallel

51. For the combination of resistors shown in the drawing, determine the equivalent resistance between points A and B.

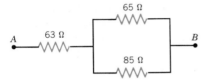

52. Circuit A has three resistors connected in series ($R_1 = 30$ Ω, $R_2 = 70$ Ω, and $R_3 = 210$ Ω). Circuit B has three resistors (different from any of those in circuit A) connected in parallel. In circuit B each resistor has the same resistance. What

is the resistance of each resistor in circuit B, such that the total resistance in B equals the total resistance in A?

53. Find the equivalent resistance between points A and B in the drawing.

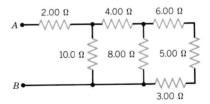

54. Two 25.0-Ω resistors are connected in series. This combination is connected between the terminals of a 75.0-V battery. A 50.0-Ω resistor is also connected across the battery, in parallel with the series combination. (a) How much current is supplied by the battery? (b) How much power is dissipated in one of the 25.0-Ω resistors?

55. Determine the equivalent resistance between the points A and B for the group of resistors in the drawing.

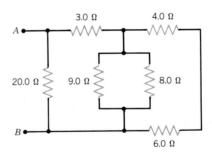

*56. Determine the power dissipated in the 2.0-Ω resistor in the circuit shown in the drawing.

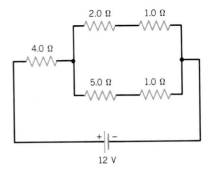

*57. Three identical resistors are connected in parallel. The equivalent resistance increases by 700 Ω when one resistor is removed and connected in series with the remaining two, which are still in parallel. Find the resistance of each resistor.

**58. The current in the 8.00-Ω resistor in the drawing is

0.500 A. Find the current in the 20.0-Ω resistor and in the 9.00-Ω resistor.

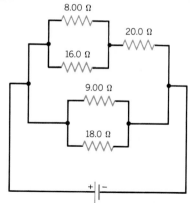

Section 20.9 Internal Resistance

59. A new "D" battery has an emf of 1.5 V. When a wire of negligible resistance is connected between the terminals of the battery, a current of 28 A is produced. Find the internal resistance of the battery.

60. A battery has an emf of 12.0 V and an internal resistance of 0.15 Ω. What is the terminal voltage when the battery is connected to a 1.50-Ω resistor?

61. A 2.00-Ω resistor is connected across a 6.00-V battery. The voltage between the terminals of the battery is observed to be only 4.90 V. Find the internal resistance of the battery.

62. A battery has an emf of 6.4 V and an internal resistance of 0.0048 Ω. An aluminum wire (length = 0.50 m, cross-sectional area = 2.0×10^{-6} m²) is connected between the terminals of the battery. What is the current in the wire?

63. A battery has an internal resistance of 0.50 Ω. A number of identical light bulbs, each with a resistance of 15 Ω, are connected in parallel across the battery terminals. The terminal voltage of the battery is observed to be one-half the emf of the battery. How many bulbs are connected?

*64. A 75.0-Ω and a 45.0-Ω resistor are connected in parallel. When this combination is connected across a battery, the current delivered by the battery is 0.294 A. When the 45.0-Ω resistor is disconnected, the current from the battery drops to 0.116 A. Determine (a) the emf and (b) the internal resistance of the battery.

Section 20.10 Kirchhoff's Rules

65. A current of 2.0 A exists in the partial circuit shown in the drawing. What is the magnitude of the potential difference between the points (a) A and B, and (b) A and C?

66. The drawing shows resistors that are partly in series and partly in parallel. (a) Find the current in the 4.0-Ω resistor without using Kirchhoff's rules. (b) Redetermine the current in the 4.0-Ω resistor, this time using Kirchhoff's rules. Verify that the answer obtained is the same as that in part (a).

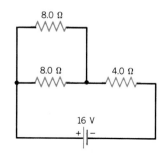

67. Two batteries, each with an internal resistance of 0.015 Ω, are connected as in the drawing. In effect, the 9.0-V battery is being used to charge the 8.0-V battery. What is the current in the circuit?

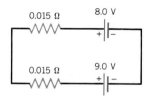

68. Find the magnitude and direction of the current in the 2.0-Ω resistor in the drawing.

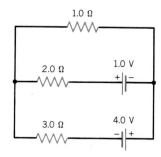

*69. Determine the voltage across the 5.0-Ω resistor in the drawing. Which end of the resistor is at the higher potential?

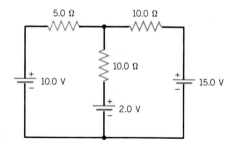

*70. For the circuit in the drawing, find the current in the 10.0-Ω resistor. Specify the direction of the current.

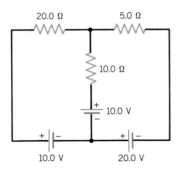

**71. Suppose the resistors in Figure 20.33 have the following resistances: $R = 10.0 \ \Omega$, $R_1 = 20.0 \ \Omega$, $R_2 = 30.0 \ \Omega$, $R_v = 40.0 \ \Omega$, and $R_C = 50.0 \ \Omega$. If the battery is a 10.0-V battery, what is the voltage between points A and B in the circuit? State which point is at the higher potential. *(Note: Do not assume that the Wheatstone bridge is balanced.)*

Section 20.11 The Measurement of Current, Voltage, and Resistance

72. A galvanometer has a coil resistance of 250 Ω and requires a current of 1.5 mA for full-scale deflection. This device is to be used in an ammeter that has a full-scale current of 25.0 mA. What is the value of the shunt resistance?

73. A galvanometer with a coil resistance of 16.0 Ω and a full-scale current of 0.250 mA is used with a shunt resistor to make an ammeter. The ammeter registers a maximum current of 6.00 mA. Find the equivalent resistance of the ammeter.

74. The equivalent resistance of a voltmeter is 140 000 Ω. The voltmeter uses a galvanometer that has a full-scale deflection of 180 μA. What is the maximum voltage that can be measured by the voltmeter?

75. A galvanometer has a coil resistance of 36 Ω. To make an ammeter, a 3.0-Ω shunt resistor is connected in parallel with the galvanometer. What percentage of the current entering the ammeter passes through the galvanometer?

*76. Two scales on a voltmeter measure voltages up to 20.0 and 30.0 V, respectively. The resistance connected in series with the galvanometer is 1680 Ω for the 20.0-V scale and 2930 Ω for the 30.0-V scale. Determine the coil resistance and the full-scale current of the galvanometer that is used in the voltmeter.

*77. In measuring a voltage, a voltmeter uses some current from the circuit. Consequently, the voltage measured is only an approximation to the voltage present when the voltmeter is not connected. Consider a circuit consisting of two 1550-Ω resistors connected in series across a 60.0-V battery. (a) Find the voltage across one of the resistors. (b) A voltmeter has a

full-scale voltage of 60.0 V and uses a galvanometer with a full-scale deflection of 5.00 mA. Determine the voltage that this voltmeter registers when it is connected across the resistor used in part (a).

Section 20.12 Capacitors in Series and Parallel

78. Three capacitors (3.0, 7.0, and 9.0 μF) are connected in series. What is their equivalent capacitance?

79. Determine the equivalent capacitance between A and B for the group of capacitors in the drawing.

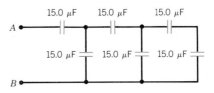

80. A 2.00-μF and a 4.00-μF capacitor are connected to a 60.0-V battery. How much charge is supplied by the battery in charging the capacitors when the wiring is (a) in parallel and (b) in series?

81. Three capacitors (4.0, 6.0, and 12.0 μF) are connected in series across a 50.0-V battery. Find the voltage across the 4.0-μF capacitor.

82. A 4.0-μF and an 8.0-μF capacitor are connected in parallel across a 25-V battery. Find (a) the equivalent capacitance and (b) the total charge stored on the two capacitors.

83. Three capacitors have identical geometries. One is filled with a material whose dielectric constant is 3.00. Another is filled with a material whose dielectric constant is 5.00. The third capacitor is filled with a material whose dielectric constant κ is such that this single capacitor has the same capacitance as the series combination of the two other capacitors. Determine κ.

84. Suppose two capacitors (C_1 and C_2) are connected in series. Show that the sum of the energies stored in these capacitors is equal to the energy stored in the equivalent capacitor. *(Hint: The energy stored in a capacitor can be expressed as $q^2/(2C)$.)*

*85. A 16.0-μF and a 4.0-μF capacitor are connected in parallel and charged by a 22-V battery. What voltage is required to charge a series combination of the two capacitors with the same total energy?

*86. A 3.00-μF and a 5.00-μF capacitor are connected in series across a 30.0-V battery. A 7.00-μF capacitor is then connected in parallel across the 3.00-μF capacitor. Determine the voltage across the 7.00-μF capacitor.

**87. The drawing shows two fully charged capacitors ($C_1 = 2.00 \ \mu$F, $q_1 = 6.00 \ \mu$C; $C_2 = 8.00 \ \mu$F, $q_2 = 12.0 \ \mu$C). The switch is closed, and charge flows until equilibrium is reestablished (i.e., until both capacitors have the same voltage across their plates). Find the resulting voltage across either capacitor.

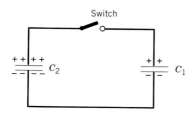

Section 20.13 *RC* Circuits

88. An electronic flash attachment for a camera produces a flash by using the energy stored in a 750-μF capacitor. Between flashes, the capacitor recharges through a resistor whose resistance is chosen so the capacitor recharges with a time constant of 3.0 s. Determine the value of the resistance.

89. The 150-μF capacitor in the drawing is fully charged. When the switch is opened, the capacitor begins to discharge. What is the time constant for the discharge?

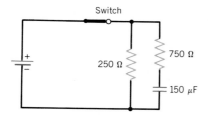

90. A charged capacitor is connected across a 9600-Ω resistor and allowed to discharge. The capacitor loses 63.2% of its original charge in a time of 8.3 s. What is the capacitance of the capacitor?

*91. Three identical capacitors are connected with a resistor in two different ways. When they are connected as in part *a* of the drawing, the time constant to charge up this circuit is 0.020 s. What is the time constant when they are connected with the same resistor as in part *b*?

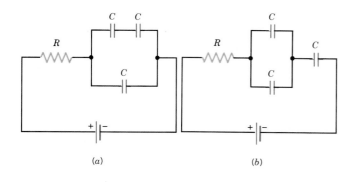

(a)   (b)

**92. How many time constants must elapse before a capacitor in a series *RC* circuit is charged to within 0.10% of its equilibrium charge?

ADDITIONAL PROBLEMS

93. A 3.0-μF capacitor and a 4.0-μF capacitor are connected in series across a 40.0-V battery. A 10.0-μF capacitor is also connected directly across the battery terminals. Find the total charge that the battery delivers to the capacitors.

94. An electric blanket is connected to a 120-V outlet and consumes 140 W of power. What is the current in the wire in the blanket?

95. A lightning bolt delivers a charge of 35 C to the ground in a time of 1.0×10^{-3} s. What is the current?

96. For the circuit shown in the drawing, find the current in the 3.00-Ω resistor. Be sure to specify the direction of the current.

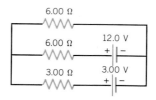

97. The filament of a light bulb has a resistance of 580 Ω. A voltage of 120 V is connected across the filament. How much current is in the filament?

98. A voltmeter utilizes a galvanometer that has a 180-Ω coil resistance and a full-scale current of 8.30 mA. The voltmeter measures voltages up to 30.0 V. Determine the resistance that is connected in series with the galvanometer.

99. The two headlights of a car consume a total power of 120 W. A driver parks the car but leaves the lights on. The 12-V battery is rated at 95 A·h. (See problem 7 for an explanation of this rating.) How long does it take for the battery to lose its charge?

100. A cylindrical aluminum pipe of length 1.50 m has an inner radius of 2.00×10^{-3} m and an outer radius of 3.00×10^{-3} m. The interior of the pipe is completely filled with copper. What is the resistance of this unit? (*Hint: Imagine that the pipe is connected between the terminals of a battery and decide whether the aluminum and copper parts of the pipe are in series or in parallel.*)

101. Eight different values of resistance can be obtained by connecting together three resistors (1.00, 2.00, and 3.00 Ω) in all possible ways. What are they?

102. An electric heater consumes 480 W of power when connected to a 120-V outlet. Two such heaters are connected in series, and the series combination is connected to a 120-V outlet. How much power does each heater now consume?

*103. An electric furnace runs nine hours a day to heat a house during January (31 days). The heating element has a resistance of 5.3 Ω and carries a current of 25 A. The cost of electricity is $0.10 per kWh. Find the monthly cost of running the furnace.

*104. An extension cord is used with an electric weed trimmer that has a resistance of 15.0 Ω. The extension cord is made of copper wire that has a cross-sectional area of 1.3×10^{-6} m^2. The combined length of the two wires in the extension cord is 92 m. (a) Determine the resistance of the extension cord. (b) The extension cord is plugged into a 120-V socket. What voltage is applied to the trimmer itself?

*105. A resistor has a resistance R, and a battery has an internal resistance r. When the resistor is connected across the battery, ten percent less power is dissipated in R than there would be if the battery had no internal resistance. Find the ratio r/R.

*106. A piece of nichrome wire has a radius of 6.5×10^{-4} m. It is used in a laboratory to make a heater that dissi-

pates 4.00×10^2 W of power when connected to a voltage source of 120 V. Ignoring the effect of temperature on resistance, estimate the necessary length of wire.

*107. Three resistors are connected in series to a battery. From left to right, the resistances are R_1, $R_2 = 5.0 \Omega$, and R_3. The voltage across R_1 and R_2 together is 8.0 V, while the voltage across R_2 and R_3 together is 4.0 V. The equivalent resistance of the three resistors is 22.0 Ω. Determine R_1, R_3, and the battery voltage.

**108. A sheet of gold foil (negligible thickness) is placed between the plates of a capacitor and has the same area as each of the plates. The foil is parallel to the plates, at a position one-third of the way from one to the other. Before the foil is inserted, the capacitance is C_0. What is the capacitance after the foil is in place?

**109. Two wires have the same cross-sectional area and are joined end to end to form a single wire. One is tungsten and the other carbon. The total resistance of the composite wire is the sum of the resistances of the pieces. The total resistance of the composite does not change with temperature. What is the ratio of the lengths of the tungsten and carbon sections?

CHAPTER 21

MAGNETIC FORCES AND MAGNETIC FIELDS

This photograph shows a scene from the film "The Hunt for Red October." The story of the film revolves about a fictional submarine, the Red October, that is powered by a revolutionary new drive system. Although Red October is pure fiction, there is a revolutionary new drive system based on magnetohydrodynamic propulsion that may soon power submarines and surface ships. This type of drive system depends on a magnetic force for propulsion, and a prototype ship using it has already been constructed. As we will see, magnetohydrodynamic propulsion is a good illustration of what this chapter is all about. It is about magnetic forces and how they are produced by magnetic fields. We will find that magnetic fields can be created by naturally occurring materials and also by electric currents. And we will see how a magnetic field can apply a magnetic force to moving electric charges. Magnetism is widely used today for many purposes, and a number of its applications in addition to magnetohydrodynamic propulsion will be discussed.

21.1 MAGNETIC FIELDS

PERMANENT MAGNETS

Permanent magnets have long been used in navigational compasses. As Figure 21.1 illustrates, the compass needle is a permanent magnet supported so it can rotate freely in a plane. When the compass is placed on a horizontal surface, the needle rotates until one end points approximately to the north. The end of the

Figure 21.1 The needle of a compass is a permanent magnet that has a north magnetic pole (N) at one end and a south magnetic pole (S) at the other end.

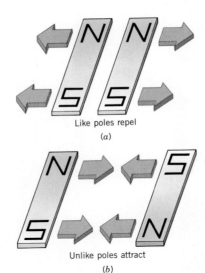

Figure 21.2 Bar magnets have a north magnetic pole at one end and a south magnetic pole at the other end. (*a*) Like poles repel each other, and (*b*) unlike poles attract each other.

needle that points north is labeled the **north magnetic pole;** the opposite end is the **south magnetic pole.**

Magnets can exert forces on each other. Figure 21.2 shows that the magnetic forces between north and south poles have the property that *like poles repel each other, and unlike poles attract each other.* This behavior is similar to that of like and unlike electric charges. However, there is a significant difference between magnetic poles and electric charges. It is possible to separate positive from negative electric charges and produce isolated charges of either kind. In contrast, no one has found a magnetic monopole (an isolated north or south pole). Any attempt to separate north and south poles by cutting a bar magnet in half fails, because each piece becomes a smaller magnet with its own north and south poles. Repeated cutting only produces more bar magnets, without yielding isolated magnetic poles.

Surrounding a magnet, there is a *magnetic field.* The magnetic field is analogous to the electric field that exists in the space around electric charges. Like the electric field, the magnetic field has both a magnitude and a direction. We postpone a discussion of the magnitude until Section 21.2, concentrating our attention here only on the direction of the field. *The direction of the magnetic field at any point in space is the direction indicated by the north pole of a small compass needle placed at that point.* Figure 21.3 shows how compasses can be used to map out the magnetic field in the space surrounding a bar magnet. Since like poles repel and unlike poles attract, the needle of each compass becomes aligned relative to the bar magnet in the manner shown in the picture. The compass needles provide a visual picture of the magnetic field that the bar magnet creates in the surrounding space.

As an aid in visualizing the electric field, we introduced the notion of electric field lines in Section 18.7. In a similar fashion, it is possible to draw magnetic field lines in the vicinity of a magnet. Figure 21.4*a* illustrates some magnetic field lines around a bar magnet. The lines appear to originate from the north pole and to end on the south pole; the lines do not start or stop in midspace. A visual image of the magnetic field lines in a plane can be created by sprinkling finely ground iron filings on a piece of paper that covers the magnet. Iron filings in a magnetic field

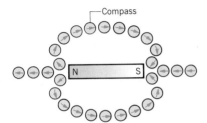

Figure 21.3 At any location in the vicinity of a magnet, the north pole of a small compass needle points in the direction of the magnetic field at that location.

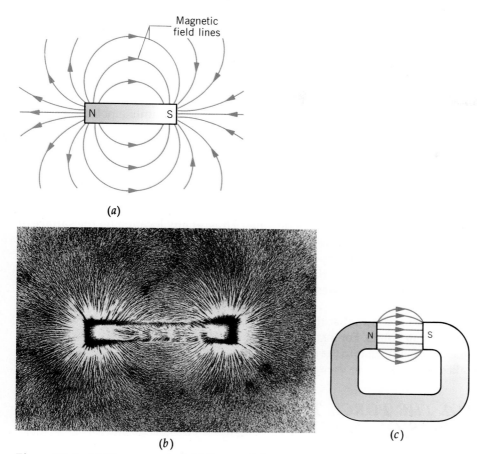

(a)

(b)

(c)

Figure 21.4   (a) The magnetic field lines and (b) the pattern of iron filings in the vicinity of a bar magnet. (c) The magnetic field lines in the gap of a horseshoe magnet.

behave like tiny compasses and align themselves along the magnetic field lines, as part *b* of the drawing shows.

As is the case with electric field lines, the magnetic field at any point is tangent to the magnetic field line at that point. Furthermore, the strength of the magnetic field is proportional to the number of lines per unit area that passes through a surface oriented perpendicular to the lines. Thus, the magnetic field is stronger in regions where the field lines are relatively close together and weaker where they are relatively far apart. For instance, in Figure 21.4*a* the lines are closest together near the north and south poles, so the strength of the magnetic field is greatest in these regions. Away from the poles, the magnetic field becomes weaker. Notice in part *c* of the drawing that the magnetic field lines in the gap between the poles of the horseshoe magnet are nearly parallel and equally spaced, indicating that the magnetic field there is approximately constant.

GEOMAGNETISM

Although the north pole of a compass needle points northward, it does not point exactly at the north geographic pole of the earth. The north geographic pole is that point where the earth's axis of rotation crosses the surface in the northern hemisphere (see Figure 21.5). Measurements of the magnetic field surrounding the earth show that the earth behaves magnetically almost as if it were a bar magnet.

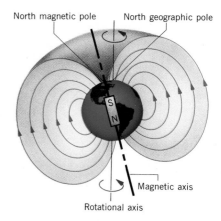

Figure 21.5   The earth behaves magnetically almost as if a bar magnet were located near its center. The axis of this fictitious bar magnet does not coincide with the earth's rotational axis; the two axes are currently about 11.5° apart.

The northern lights (aurora borealis) appear in the night sky at northern latitudes. They occur because the earth's magnetic field concentrates charged particles from the sun in these regions. The charged particles interact with the atoms and molecules in the upper atmosphere and produce the spectacular light displays.

As the drawing illustrates, the orientation of this fictitious bar magnet defines a magnetic axis for the earth. The location where the magnetic axis crosses the surface in the northern hemisphere is known as the north magnetic pole. The north magnetic pole of the earth is so named because it is the location toward which the north end of a compass needle points. Since unlike poles attract, the south pole of the earth's fictitious bar magnet lies beneath the north magnetic pole, as Figure 21.5 indicates.

The north magnetic pole does not coincide with the north geographic pole but, instead, lies in Hudson Bay, Canada, some 1300 km to the south. It is interesting to note that the position of the north magnetic pole is not fixed, but moves over the years. For example, the current location of the north magnetic pole is about 770 km northwest of its position in 1904. Thus, at any location on the surface of the earth, a compass needle points toward the north magnetic pole and deviates from the north geographic pole. The angle that a compass needle deviates is called the *angle of declination* for that location. For New York City, the present angle of declination is about 12° west, meaning that a compass needle points 12° west of geographic north.

Figure 21.5 shows that the earth's magnetic field lines are not parallel to the surface at all points. For instance, near the north magnetic pole the field lines are almost perpendicular to the surface of the earth. The angle that the magnetic field makes with respect to the surface at any point is known as the *angle of dip.*

21.2 THE FORCE THAT A MAGNETIC FIELD EXERTS ON A MOVING CHARGE

THE NATURE OF THE MAGNETIC FORCE

When a charge is placed in an electric field, the charge experiences an electric force. It is natural to ask, therefore, whether a charge placed in a magnetic field experiences a *magnetic force.* The answer is yes, provided two conditions are met:

1. The charge must be moving, for no magnetic force acts on a stationary charge.
2. The velocity of the moving charge must have a component that is perpendicular to the direction of the magnetic field.

To examine the second condition more closely, consider Figure 21.6, which shows a positive test charge $+q_0$ moving with a velocity **v** through a magnetic field labeled by the symbol **B**. The magnetic field is produced by an arrangement of magnets not shown in the drawing and is assumed to be constant in both magnitude and direction. If the charge moves *parallel or antiparallel* to the field, as in part *a* of the drawing, the charge experiences *no magnetic force.* If, on the other hand, the charge moves *perpendicular* to the field, as in part *b*, the charge experiences the *maximum possible force* **F**. In general, if a charge moves at an angle θ* with respect to the field (see part *c* of the drawing), only the velocity component $v \sin \theta$, which is perpendicular to the field, gives rise to a magnetic force. This force is smaller than the maximum possible force. The component of the velocity that is parallel to the magnetic field yields no force.

* The angle θ between the velocity of the charge and the magnetic field is chosen so that it lies in the range $0 \leq \theta \leq 180°$.

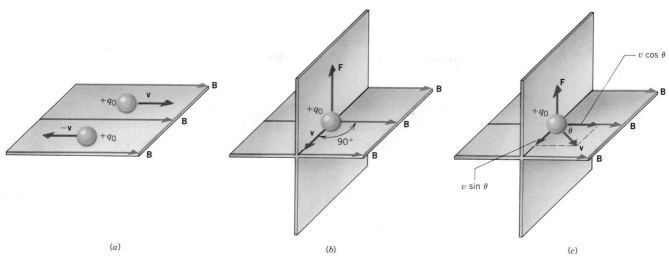

(a)   (b)   (c)

Figure 21.6 (a) No magnetic force acts on a charge moving with a velocity **v** that is parallel or antiparallel to a magnetic field **B**. (b) The charge experiences a maximum force **F** when the charge moves perpendicular to the field. (c) If the charge travels at an angle θ with respect to **B**, only the velocity component perpendicular to the field gives rise to a magnetic force. This component is $v \sin \theta$.

Figure 21.6 shows that the direction of the magnetic force **F** is perpendicular to both the velocity **v** and the magnetic field **B**; in other words, **F** is perpendicular to the plane defined by **v** and **B**. As an aid in remembering the direction of the force, it is convenient to use *Right-Hand Rule No. 1 (RHR-1)*, as Figure 21.7 illustrates:

> *Right-Hand Rule No. 1.* Extend the right hand so the fingers point along the direction of the magnetic field **B** and the thumb points along the velocity **v** of the charge. The palm of the hand then faces in the direction of the magnetic force **F** that acts on a positive charge.

In this rule, it is as if the open palm of the right hand is pushing on the positive charge in the direction of the magnetic force. If the moving charge is *negative* instead of positive, the direction of the magnetic force is *opposite* to that predicted by RHR-1. Thus, there is an easy method for finding the force on a moving negative charge. First, assume the charge is positive and use RHR-1 to find the direction of the force. Then, reverse this direction to find the direction of the force acting on the negative charge.

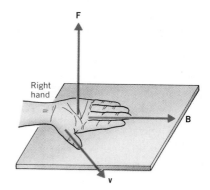

Figure 21.7 Right-Hand Rule No. 1 is illustrated. When the right hand (R.H.) is oriented so the fingers point along the magnetic field **B** and the thumb points along the velocity **v** of a positively charged particle, the palm faces in the direction of the magnetic force **F** applied to the particle.

DEFINITION OF THE MAGNETIC FIELD

It is observed experimentally that when a charge moves through a magnetic field, the charge experiences a magnetic force whose magnitude is directly proportional to (1) the magnitude of the charge and (2) the component of its velocity that is perpendicular to the magnetic field. Because of these facts, it is possible to define the magnitude of the magnetic field in a manner that is similar to that used for the electric field, although the details differ.

Recall that the electric field at any point in space is the force per unit charge that acts on a test charge q_0 placed at that point. In other words, to determine the electric field **E**, we divide the electric force **F** by the charge q_0: $\mathbf{E} = \mathbf{F}/q_0$. However, the magnetic force depends not only on the charge magnitude q_0, but also on the velocity component $v \sin \theta$ that is perpendicular to the magnetic field. Therefore, to determine the magnitude of the magnetic field, we divide the magnitude of the magnetic force not only by q_0, but also by $v \sin \theta$, according to the following definition:

Definition of the Magnetic Field

The magnitude B of the magnetic field at any point in space is defined as

$$B = \frac{F}{q_0(v \sin \theta)} \qquad (21.1)$$

where F is the magnitude of the magnetic force on a positive test charge q_0 whose velocity $\mathbf{v}$ makes an angle θ ($0 \leq \theta \leq 180°$) with the field. The magnetic field $\mathbf{B}$ is a vector, and its direction can be determined by using a small compass needle.

SI Unit of the Magnetic Field: $\dfrac{\text{newton} \cdot \text{second}}{\text{coulomb} \cdot \text{meter}} = 1$ tesla (T)

The unit of magnetic field strength that follows from Equation 21.1 is the $\text{N} \cdot \text{s}/(\text{C} \cdot \text{m})$. This unit is called the *tesla* (T), in tribute to the Croatian-born, American engineer Nikola Tesla (1856–1943). Thus, one tesla is the strength of the magnetic field in which a unit test charge, traveling perpendicular to the magnetic field with a speed of one meter per second, experiences a force of one newton. Because a coulomb per second is an ampere (1 C/s = 1 A), the tesla is often written as $1 \text{ T} = 1 \text{ N}/(\text{A} \cdot \text{m})$.

In many situations the magnetic field has a value that is considerably less than one tesla. For example, the strength of the magnetic field near the earth's surface is approximately 10^{-4} T. In such circumstances, a magnetic field unit called the *gauss* (G) is sometimes used. Although not an SI unit, the gauss is a convenient size for many applications involving magnetic fields. The relation between the gauss and the tesla is

$$1 \text{ gauss} = 10^{-4} \text{ tesla}$$

Example 1 deals with the magnetic force exerted on a moving proton and on a moving electron.

Example 1   Magnetic Forces on Charged Particles

A proton in a particle accelerator has a speed of 5.0×10^6 m/s. The proton encounters a magnetic field whose magnitude is 0.40 T and whose direction makes an angle of $\theta = 30.0°$ with respect to the proton's velocity (see Figure 21.6c). Find (a) the magnitude and direction of the magnetic force on the proton and (b) the acceleration of the proton. (c) What would be the force and acceleration if the particle were an electron instead of a proton?

REASONING AND SOLUTION

(a) Since the positive charge on a proton is 1.60×10^{-19} C, the magnitude of the magnetic force is

$$F = q_0 vB \sin \theta \qquad (21.1)$$

$$F = (1.60 \times 10^{-19}\ \text{C})(5.0 \times 10^6\ \text{m/s})(0.40\ \text{T})(\sin 30.0°) = \boxed{1.6 \times 10^{-13}\ \text{N}}$$

The direction of the magnetic force is given by RHR-1 and is directed upward in Figure 21.6c, with the magnetic field pointing to the right.

(b) The acceleration of the proton follows directly from Newton's second law as the magnetic force divided by the mass m_p of the proton:

$$a = \frac{F}{m_p} = \frac{1.6 \times 10^{-13}\ \text{N}}{1.67 \times 10^{-27}\ \text{kg}} = \boxed{9.6 \times 10^{13}\ \text{m/s}^2} \qquad (4.1)$$

(c) The magnitude of the magnetic force on the electron is the same as that on the proton, since both have the same speed and charge magnitude. However, the direction of the force on the electron is opposite to that on the proton, since the charge on the electron is negative. Furthermore, the electron has a smaller mass m_e and, therefore, experiences a significantly greater acceleration:

$$a = \frac{F}{m_e} = \frac{1.6 \times 10^{-13}\ \text{N}}{9.11 \times 10^{-31}\ \text{kg}} = \boxed{1.8 \times 10^{17}\ \text{m/s}^2}$$

> **PROBLEM SOLVING INSIGHT**
>
> The direction of the magnetic force exerted on a negative charge is opposite to that exerted on a positive charge, assuming both charges are moving in the same direction.

21.3 THE MOTION OF A CHARGED PARTICLE IN A MAGNETIC FIELD

COMPARING PARTICLE MOTION IN ELECTRIC AND MAGNETIC FIELDS

The motion of a charged particle in an electric field is noticeably different than the motion in a magnetic field. For example, Figure 21.8a shows a positive charge moving between the plates of a parallel plate capacitor. Initially, the charge is moving perpendicular to the direction of the electric field. Since the direction of the electric force on a positive charge is in the same direction as the electric field, the particle is deflected sideways in the drawing. Part b of the drawing shows the same particle traveling initially at right angles to a magnetic field. An application

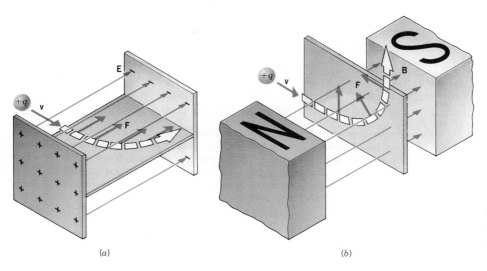

(a)   (b)

Figure 21.8   (a) The electric force **F** that acts on a positive charge is parallel to the electric field **E** and causes the particle's trajectory to bend sideways. (b) The magnetic force **F** is perpendicular to both the magnetic field **B** and the velocity **v** and causes the particle's trajectory to bend in a vertical plane.

of RHR-1 shows that when the charge enters the field, the charge is deflected upward (not sideways) by the magnetic force. As the charge moves upward, the direction of the magnetic force changes, always remaining perpendicular to both the magnetic field and the velocity. When a charged particle travels in a magnetic field, then, the charge never experiences a force that is parallel to the field, as it does in the electric case. Because of the difference in the way that electric and magnetic fields exert forces on charges, the work done on the particle by each field is different, as we will now see.

THE WORK DONE ON A CHARGED PARTICLE MOVING THROUGH ELECTRIC AND MAGNETIC FIELDS

In Figure 21.8*a* an electric field applies a force to a positively charged particle, and, consequently, the path of the particle bends in the direction of the force. Because there is a component of the particle's displacement in the direction of the electric force, the force does work on the particle. This work increases the kinetic energy, and hence the speed, of the particle, as specified by the work–energy theorem presented in Section 6.2.

In contrast to an electric field, *a constant magnetic field does no work on the moving charged particle* in Figure 21.8*b*. This fact arises because the magnetic force always acts in a direction that is perpendicular to the motion of the charge. Consequently, the displacement of the moving charge never has a component in the direction of the magnetic force. As a result, the magnetic force cannot do work and change the kinetic energy of the charge, although the force can alter the direction of motion.

THE CIRCULAR TRAJECTORY

To describe the motion of a charged particle in a magnetic field more completely, and to emphasize that the field does no work, we now discuss the special case in which the velocity of the particle is perpendicular to a uniform magnetic field.* As Figure 21.9 illustrates, the magnetic force serves to move the particle in a circular path. To understand why the path is circular, consider two points on the circumference labeled 1 and 2. When the positively charged particle is at point 1, the magnetic force **F** is perpendicular to the velocity **v** and points directly upward in the drawing. This force causes the trajectory to bend upward. When the particle reaches point 2, the magnetic force still remains perpendicular to the velocity, but is now directed to the left in the drawing. *The magnetic force always remains perpendicular to the velocity and is directed toward the center of the circular path.*

To find the radius of the circular path in Figure 21.9, we recall the concept of centripetal force from Section 5.3. The centripetal force is the net force, directed toward the center of the circle, that is needed to keep a particle moving along a

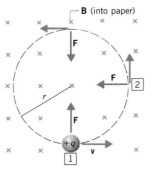

Figure 21.9 A positively charged particle is moving perpendicular to a constant magnetic field. The magnetic force **F** causes the particle to move on a circular path.

* In many instances it is convenient to orient the magnetic field **B** so its direction is perpendicular to the page. In these cases it is customary to use a dot to symbolize the magnetic field pointing out of the page (toward the reader); this dot symbolizes the tip of the arrow representing the **B** vector. A region where a constant magnetic field is directed *into the page* is drawn as a series of crosses that indicate the tail feathers of the arrows representing the **B** vectors. Therefore, regions where a magnetic field is directed out of the page or into the page are drawn as shown below:

Out of page   Into page

circular path. The magnitude F_c of the centripetal force depends on the speed v and mass m of the particle, as well as the radius r of the circle:

$$F_c = \frac{mv^2}{r} \qquad (5.3)$$

In the present situation, the magnetic force furnishes the centripetal force needed to keep the charge $+q$ on the circular path. In keeping the charge on the circular path, the magnetic force does no work, since it is perpendicular to the motion. According to Equation 21.1, the magnetic force is $qvB \sin 90°$, so $qvB = mv^2/r$ or

$$r = \frac{mv}{qB} \qquad (21.2)$$

Equation 21.2 shows that the radius of the circle is inversely proportional to the magnitude of the magnetic field, with stronger fields producing "tighter" circular paths.

Example 2 illustrates that an electric field can change the kinetic energy of a charged particle, but a magnetic field cannot, although a magnetic field can cause the particle to move along a circular path.

Example 2   The Motion of a Proton in Electric and Magnetic Fields

A proton starts from rest at the positive plate of a parallel plate capacitor and is accelerated toward the negative plate by the electric force. The potential difference between the plates is $V = 2100$ volts. The high-speed proton leaves the capacitor through a small hole in the negative plate. Once outside the capacitor, the proton travels at a constant velocity until it enters a region of constant magnetic field of magnitude 0.10 T. The velocity and magnetic field are perpendicular, as in Figure 21.9. Find (a) the speed of the proton when it leaves the capacitor, (b) the change in the proton's kinetic energy due to the magnetic field, and (c) the radius of the circular path on which the proton moves in the magnetic field.

REASONING   Initially, when the proton (charge $= +e$) is at the positive plate, the electric potential energy relative to the negative plate is eV. As the proton approaches the negative plate, all the potential energy is converted into kinetic energy, so $eV = \frac{1}{2}mv^2$ and the speed of the proton is $v = \sqrt{2eV/m}$. The proton enters the magnetic field with this speed. The magnetic field does no work on the proton, because the displacement of the proton is always perpendicular to the magnetic force. The magnetic field does, however, cause the proton to travel in a circular path.

SOLUTION
(a) The speed of the proton is

$$v = \sqrt{\frac{2eV}{m}} = \sqrt{\frac{2(1.60 \times 10^{-19}\ \text{C})(2100\ \text{V})}{1.67 \times 10^{-27}\ \text{kg}}} = \boxed{6.3 \times 10^5\ \text{m/s}}$$

(b) Since the magnetic field does no work on the moving proton, the kinetic energy of the proton does not change, according to the work–energy theorem.

(c) Since the kinetic energy remains constant, the speed of the proton does not change, and the radius of the circle can be found from Equation 21.2:

$$r = \frac{mv}{qB} = \frac{(1.67 \times 10^{-27}\ \text{kg})(6.3 \times 10^5\ \text{m/s})}{(1.60 \times 10^{-19}\ \text{C})(0.10\ \text{T})} = \boxed{6.6 \times 10^{-2}\ \text{m}}$$

21.4 THE MASS SPECTROMETER

THE PHYSICS OF . . .

a mass spectrometer.

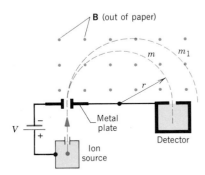

Figure 21.10   The basic features of a mass spectrometer. The dashed lines are the paths traveled by ions of different masses. Ions with mass m follow the path of radius r and enter the detector. Ions with the larger mass m_1 follow the outer path and miss the detector.

Physicists use mass spectrometers for determining the relative masses and abundances of isotopes.* Chemists use these instruments to help identify unknown molecules produced in chemical reactions. Mass spectrometers are also used during surgery, where they give the anesthesiologist information on the gases, including the anesthetic, in the patient's lungs.

In the type of mass spectrometer illustrated in Figure 21.10, the atoms or molecules are first vaporized and then ionized by the ion source. The ionization process removes one electron from the particle, leaving it with a net positive charge of $+e$. The positive ions are then accelerated through the potential difference V that is applied between the ion source and the metal plate. With a speed v, the ions pass through a hole in the plate and enter a region of constant magnetic field **B**, where they are deflected in semicircular paths. Only those ions following a path with the proper radius r strike the detector, which records the number of ions arriving per second.

The mass m of the detected ions can be expressed in terms of r, B, and v by recalling that the radius of the path followed by a particle of charge $+e$ is $r = mv/eB$ (Equation 21.2). In addition, the results of Example 2 show that the ion speed v can be expressed in terms of the accelerating potential V as $v = \sqrt{2eV/m}$. Algebraically eliminating v from these two equations and solving for the mass gives

$$m = \left(\frac{er^2}{2V}\right) B^2$$

This result shows that the mass of each ion reaching the detector is proportional to B^2. By experimentally changing the value of B, and keeping the term in the parentheses constant, ions of different masses are allowed to enter the detector. A plot of the detector output as a function of B^2 then gives an indication of what masses are present and the abundance of each mass.

Figure 21.11 shows a record obtained by a mass spectrometer for naturally occurring neon gas. The results show that the element neon has three isotopes whose atomic mass numbers are 20, 21, and 22. These isotopes occur because neon atoms exist with different numbers of neutrons in the nucleus. Notice that the isotopes have different abundances, with neon-20 being the most abundant.

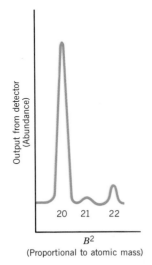

Figure 21.11   The mass spectrum of naturally occurring neon, showing three isotopes whose atomic mass numbers are 20, 21, and 22. The larger the peak, the more abundant the isotope.

*21.5 THE HALL EFFECT

The current in a metal conductor is due to the motion of electrons, and electrons carry a negative charge. However, there are important materials in which the electric current is not necessarily caused by the motion of negative charge carriers. For example, semiconductors—most notably silicon and germanium—are important in the technology of integrated circuits. In contrast to the situation in metals, the charge carriers in semiconductors can be either negative or positive, depending on how the semiconductors are fabricated. (See Section 23.6.) When

* Isotopes are atoms that have the same atomic number, but different atomic masses due to the presence of different numbers of neutrons in the nucleus.

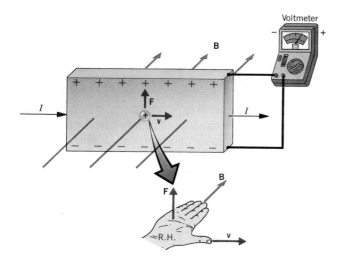

Figure 21.12 Positive charges moving to the right are deflected upward by the magnetic force **F**, giving the top surface of the slab a positive charge. The resulting potential difference between the top and bottom surfaces is called the Hall emf, and is registered by the voltmeter.

new types of semiconductors are developed, it is important to identify whether the charge carriers are negative or positive.

An experimental method for unambiguously determining the type of carrier was devised by Edwin H. Hall in 1879. Figure 21.12 illustrates Hall's method, which is widely used today. A thin, flat, conducting slab is placed in a constant magnetic field, such that the field is oriented perpendicular to the wide face of the slab. Suppose the current I in the drawing consists of moving positive charges. According to RHR-1, the charges are deflected upward by the magnetic force **F**. Thus, positive charges accumulate at the top edge of the slab, while corresponding negative charges accumulate at the bottom edge. Because of the buildup of positive and negative charges, an emf, called the *Hall emf* (or *Hall voltage*), appears across the slab, with the top of the slab being at a higher potential relative to the bottom. The Hall emf can be measured with a voltmeter, such as the one shown in the drawing. The emf builds up until the electric field produced by the separated positive and negative charges exerts an electric force on the current I that is equal and opposite to the magnetic force. Therefore, a current of positively charged carriers produces a situation in which the top of the slab becomes positively charged and the bottom becomes negatively charged.

On the other hand, the *same* current I could also have been caused by negative charge carriers moving to the *left* in Figure 21.12. An application of RHR-1 (with a reversal of the direction of the predicted force, since the moving charges are negative) shows that the top of the slab now becomes negatively charged and the bottom becomes positively charged. Therefore, negative charges moving to the left generate a Hall emf of opposite polarity to that produced by positive charges moving to the right. Thus, the polarity of the Hall emf reveals whether the charge carriers are positive or negative.

Another use of the Hall effect is found in an instrument known as a Hall probe, which measures the strength of a magnetic field. It has been determined experimentally that the Hall emf is directly proportional to the strength of the magnetic field into which the conducting slab is placed. A Hall probe is a convenient, hand-held instrument that has been calibrated to register the strength of the magnetic field, rather than the Hall emf.

THE PHYSICS OF . . .

a Hall probe.

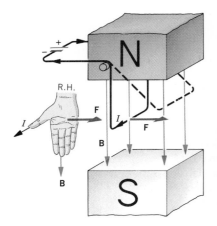

Figure 21.13 The wire carries a current I and the bottom segment of the wire is oriented perpendicular to a magnetic field **B**. A magnetic force **F** deflects the wire to the right.

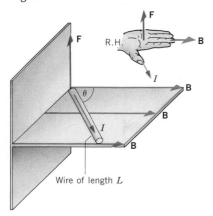

Figure 21.14 The current I in the wire, oriented at an angle θ with respect to a magnetic field **B**, is acted upon by a magnetic force **F**.

THE PHYSICS OF . . .

how a loudspeaker produces sound.

21.6 THE FORCE ON A CURRENT IN A MAGNETIC FIELD

As we have seen, a charge moving through a magnetic field can experience a magnetic force. Since an electric current is a collection of moving charges, a current in the presence of a magnetic field can also experience a magnetic force. In Figure 21.13, for instance, a current-carrying wire is placed between the poles of a magnet. When the direction of the current I is as shown, the moving charges experience a magnetic force that pushes the wire to the right in the drawing. The direction of the force is determined in the usual manner by using RHR-1, with the minor modification that the velocity of a positive charge is replaced by the direction of the conventional current I. If the direction of the current in the drawing were reversed by switching the leads to the battery, the direction of the force would be reversed, and the wire would be pushed to the left.

When a charge moves through a magnetic field, the magnitude of the force that acts on the charge is $F = qvB \sin \theta$. With the aid of Figure 21.14, this expression can be put into a form that is more suitable for use with an electric current. The drawing shows a wire of length L that carries a current I. The wire is oriented at an angle θ with respect to a magnetic field **B**. This picture is similar to Figure 21.6c, except that now the charges move in a wire. The magnitude F of the magnetic force exerted on this length of wire is the net force acting on the total charge q moving in the wire. Multiplying and dividing the right side of $F = qvB \sin \theta$ by t, the time needed for the total charge to travel the length of the wire, gives

$$F = \left(\frac{q}{t}\right)(vt)B \sin \theta$$

The term q/t is the current I in the wire, and the term vt is the length L of the wire. With these two substitutions, we arrive at the following expression for the magnetic force exerted on a current-carrying wire:

$$\begin{bmatrix} \textbf{Magnetic force on} \\ \textbf{a current-carrying} \\ \textbf{wire of length } L \end{bmatrix} \qquad F = ILB \sin \theta \qquad (21.3)$$

As in the case of a single charge traveling in a magnetic field, the magnetic force on a current-carrying wire is a maximum when the wire is oriented perpendicular to the field ($\theta = 90°$) and vanishes when the current is parallel or antiparallel to the field ($\theta = 0°$ or $180°$). The direction of the magnetic force is given by RHR-1.

Most loudspeakers operate on the principle that a magnetic field exerts a force on a current-carrying wire. Figure 21.15a shows a speaker design that consists of three parts: a cone, a voice coil, and a permanent magnet. The cone is usually made from specially treated, stiff paper and is mounted so it can vibrate back and forth. When vibrating, the cone pushes and pulls on the air in front of it, thereby creating sound waves. Attached to the apex of the cone is a hollow cardboard cylinder, around which many turns of wire are wound. This cylinder and its coils of wire are collectively called the "voice coil"; the voice coil is slipped over one pole of the permanent magnet, which is the north pole in the drawing. The permanent magnet itself does not move, but the voice coil is designed to move freely over the north pole. The two ends of the voice-coil wire are connected to the speaker terminals located at the rear panel of a receiver.

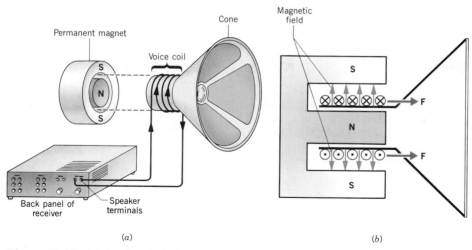

Figure 21.15 (*a*) An "exploded" view of one type of speaker design, which shows a cone, a voice coil, and a permanent magnet. (*b*) Because of the current in the voice coil (shown as ⊗ and ⊙), the magnetic field causes a force **F** to be exerted on the voice coil and cone.

The receiver acts as an ac generator, sending an alternating current to the voice coil. The alternating current interacts with the magnetic field to generate an alternating force that pushes and pulls on the voice coil and the attached cone. To see how the magnetic force arises, consider Figure 21.15*b*, which is a cross-sectional view of the voice coil and the magnet. In the cross-sectional view, the current is directed into the page in the upper half of the voice coil (⊗⊗⊗) and out of the page in the lower half (⊙⊙⊙). In both cases the magnetic field is perpendicular to the current, so the maximum possible force is exerted on the wire. An application of RHR-1 to both the upper and lower halves of the voice coil shows that the magnetic force **F** in the drawing is directed to the right, causing the cone to accelerate in that direction. One-half of a cycle later when the current is reversed, the direction of the magnetic force is also reversed, and the cone accelerates to the left. If, for example, the alternating current from the receiver has a frequency of 1000 Hz, the alternating magnetic force causes the cone to vibrate back and forth at the same frequency, and a 1000-Hz sound wave is produced. Thus, it is the magnetic force on a current-carrying wire that is responsible for converting an electrical signal into a sound wave. In Example 3 a typical force and acceleration in a loudspeaker are determined.

PROBLEM SOLVING INSIGHT

Whenever the current in a wire reverses direction, the force exerted on the wire by a given magnetic field also reverses direction.

Example 3   The Force and Acceleration in a Loudspeaker

The voice coil of a speaker has a diameter of $d = 0.025$ m, contains 55 turns of wire, and is placed in a 0.10-T magnetic field. The current in the voice coil is 2.0 A. (a) Determine the magnetic force that acts on the coil and cone. (b) If the voice coil and cone have a combined mass of 0.020 kg, find their acceleration.

REASONING   The magnetic force that acts on the current-carrying voice coil is given by Equation 21.3 as $F = ILB \sin \theta$. The effective length L of the wire in the voice coil is the number of turns N times the circumference (πd) of one turn: $L = N\pi d$. The acceleration of the voice coil and cone is given by Newton's second law as the magnetic force divided by the combined mass.

SOLUTION

(a) Since the magnetic field acts perpendicular to all parts of the wire, $\theta = 90°$ and the force on the voice coil is

$$F = ILB \sin \theta \tag{21.3}$$

$$F = (2.0 \text{ A})[55\pi(0.025 \text{ m})](0.10 \text{ T}) \sin 90° = \boxed{0.86 \text{ N}}$$

(b) The acceleration of the voice coil and cone is

$$a = \frac{F}{m} = \frac{0.86 \text{ N}}{0.020 \text{ kg}} = \boxed{43 \text{ m/s}^2} \tag{4.1}$$

This acceleration is more than four times the acceleration due to gravity.

THE PHYSICS OF . . .

a voice-coil positioner for a hard disk drive.

THE PHYSICS OF . . .

magnetohydrodynamic propulsion.

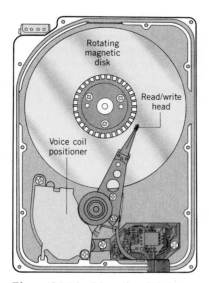

Figure 21.16 Many hard disk drives use a voice-coil positioner to move the read/write head to the appropriate location over the rotating disk.

The voice coil of a speaker moves when current is sent to it by a receiver. The same basic idea plays a role in some personal computer systems that incorporate a hard disk drive. In these drives the element that reads information from or writes information on the spinning disk is the read/write head (see Figure 21.16). Hard disk drives often use voice-coil positioners to move the read/write head to the proper location on the disk. In response to instructions from the user, current is sent to the voice-coil positioner, and a magnetic force causes the head to move across the surface of the disk to the appropriate location.

Magnetohydrodynamic (MHD) propulsion is a revolutionary type of propulsion system that uses a magnetic force on a current to power ships and submarines, without the need for propellers. This system uses a magnetic field and seawater, which conducts an electric current, to generate a magnetic force on water passing through a tube. The seawater is expelled as a water jet, which propels the vessel. This expulsion of water is analogous to how a jet engine uses air, taking it in the front of the engine and pushing it out the back to propel the plane forward. Figure 21.17*a* shows a side view of an experimental vessel with the MHD propulsion unit mounted underneath. Seawater enters the front of the unit and is expelled from the rear. The MHD propulsion system eliminates motors, drive shafts, gears, and propellers, so it promises to be a low-noise system with great reliability at relatively low cost.

Figure 21.17*b* shows an enlarged view of the propulsion unit. A pair of electrodes (metal plates), mounted on either side of the unit, is attached to a dc electrical generator. Because seawater is a conductor of electricity, electrical charge moves from one electrode to the other, passing through the water in between. This current is perpendicular to the motion of the water. A powerful magnetic field, generated by a magnet within the vessel, is oriented perpendicular to both the electric current and the motion of the seawater. The interaction of the magnetic field and the current produces a strong magnetic force on the current, and hence on the water. According to RHR-1, the direction of the magnetic force is perpendicular to both the current and the magnetic field, and, as the drawing shows, the water is forced out the back of the MHD unit. Since the MHD unit exerts a magnetic force on the water, the water exerts a force on the unit and the vessel attached to it. According to Newton's third law, this "reaction" force is equal in magnitude, but opposite in direction, and it is the reaction force that provides the thrust to drive the vessel.

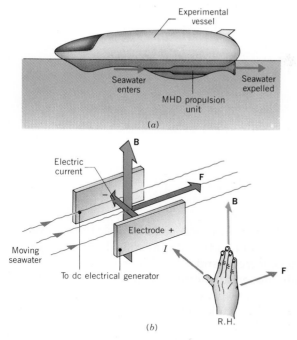

(a)

(b)

R.H.

Figure 21.17 (a) An experimental vessel that uses magnetohydrodynamic (MHD) propulsion. (b) In the MHD unit, the magnetic force exerted on the current forces water out the back.

21.7 THE TORQUE ON A CURRENT-CARRYING COIL

THE TORQUE

We have seen that a current-carrying wire can experience a force when placed in a magnetic field. If a loop of wire is suspended properly in a magnetic field, the magnetic force produces a torque that tends to rotate the loop. This torque is responsible for the operation of a number of useful devices, including galvanometers and electric motors.

Figure 21.18a shows a rectangular loop of wire attached to a vertical shaft. The shaft is mounted such that it is free to rotate in a uniform magnetic field. When

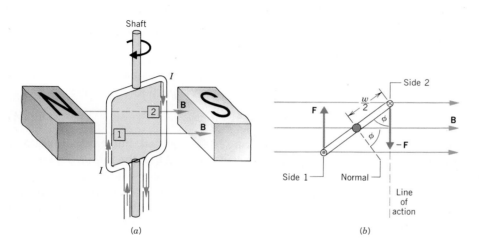

(a)

(b)

Figure 21.18 (a) A current-carrying loop of wire, which can rotate about a vertical shaft, is situated in a magnetic field. (b) A top view of the loop. The current in side 1 is directed out of the page (⊙), while the current in side 2 is directed into the page (⊗). The current in side 1 experiences a force **F** that is opposite in direction to the force exerted on side 2. The two forces produce a clockwise torque about the shaft.

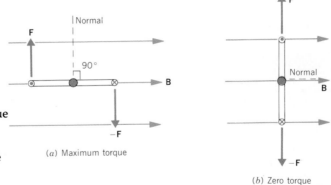

Figure 21.19 (a) Maximum torque occurs when the normal to the plane of the loop is perpendicular to the magnetic field, while (b) the torque is zero when the normal is parallel to the field.

there is a current in the loop, the loop rotates because a magnetic force is exerted on each of the two vertical sides, labeled 1 and 2 in the drawing. Part *b* shows a top view of the loop and the magnetic forces **F** and −**F** on the two sides. These two forces have the same magnitude, but an application of RHR-1 shows that they point in opposite directions, so the loop experiences no net force. The loop does, however, experience a net torque that tends to rotate the loop in a clockwise fashion about the vertical shaft. Figure 21.19*a* shows that the torque is maximum when the normal to the plane of the loop is perpendicular to the field. In contrast, part *b* shows that the torque is zero when the normal is parallel to the field. *When a current-carrying loop is placed in a magnetic field, the loop tends to rotate such that its normal becomes aligned with the magnetic field.* In this respect, a current loop behaves like a magnet (e.g., a compass needle) suspended in a magnetic field, since a magnet also rotates to line itself up with the magnetic field.

It is possible to determine the magnitude of the torque on the loop. From Equation 21.3 the magnetic force on each vertical side has a magnitude of $F = ILB \sin 90°$, where L is the length of side 1 or side 2, and $\theta = 90°$ because the current I always remains perpendicular to the magnetic field as the loop rotates. As Section 9.1 discusses, the torque produced by a force is the product of the force and the lever arm. In Figure 21.18*b* the lever arm is the perpendicular distance from the line of action of the force to the shaft. This distance is given by $(w/2) \sin \phi$, where w is the width of the loop, and ϕ is the angle between the normal to the plane of the loop and the direction of the magnetic field. The net torque is the sum of the torques on the two sides, so

$$\text{Net torque} = \tau = ILB(\tfrac{1}{2}w \sin \phi) + ILB(\tfrac{1}{2}w \sin \phi) = IAB \sin \phi$$

where the product Lw has been replaced by the area A of the loop. If the wire is wrapped so as to form a coil containing N loops, each of area A, the force on each side is N times larger, and the torque becomes proportionally greater:

$$\tau = NIAB \sin \phi \tag{21.4}$$

Equation 21.4 has been derived for a rectangular coil, but it is valid for any shape of flat coil, such as a circular coil. It is apparent that the torque depends on (1) the geometric properties of the coil itself and the current in it (NIA), (2) the magnitude B of the magnetic field, and (3) the orientation of the normal to the coil with respect to the direction of the field ($\sin \phi$). The quantity NIA is known as the ***magnetic moment*** of the coil, and its units are ampere·meter². The greater the magnetic moment of a current-carrying coil, the greater the torque that the coil experiences when placed in a magnetic field. Example 4 discusses the torque that a magnetic field applies to such a coil.

Example 4   The Torque Exerted on a Current-Carrying Coil

A coil of wire has an area of 2.0×10^{-4} m², consists of 100 loops or turns, and contains a current of 0.045 A. The coil is placed in a uniform magnetic field of magnitude 0.15 T. (a) Determine the magnetic moment of the coil. (b) Find the maximum torque that the magnetic field can exert on the coil.

REASONING AND SOLUTION

(a) The magnetic moment of the coil is

$$\text{Magnetic moment} = NIA$$

$$= (100)(0.045 \text{ A})(2.0 \times 10^{-4} \text{ m}^2) = \boxed{9.0 \times 10^{-4} \text{ A} \cdot \text{m}^2}$$

(b) According to Equation 21.4, the torque is the product of the magnetic moment NIA and $B \sin \phi$. However, the maximum torque occurs when $\phi = 90°$, so

$$\tau = (\text{Magnetic moment})(B \sin 90°)$$

$$\tau = (9.0 \times 10^{-4} \text{ A} \cdot \text{m}^2)(0.15 \text{ T}) = \boxed{1.4 \times 10^{-4} \text{ N} \cdot \text{m}}$$

THE GALVANOMETER

As we have noted in Section 20.11, the galvanometer is the basic component of nondigital ammeters and voltmeters. In measuring the current, a galvanometer relies on the fact that a current-carrying coil can rotate when placed in a magnetic field. Figure 21.20a shows the coil (only one turn is shown) of a galvanometer suspended in a magnetic field, the coil being able to rotate about a vertical shaft. Attached to the shaft is a pointer and a spring. When there is a current in the coil, the magnetic torque causes the coil to rotate. As the coil rotates, the spring winds

THE PHYSICS OF . . .

a galvanometer.

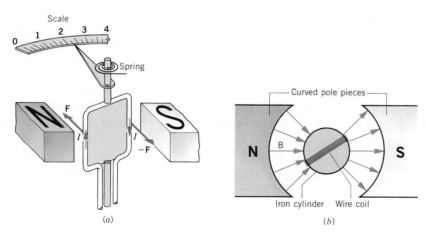

(a)   *(b)*

Figure 21.20   (a) The basic elements of a galvanometer. (b) Top view of a galvanometer mechanism showing the curved pole pieces, the iron cylinder, and the coil. For clarity, the scale, pointer, and spring have been omitted.

up and produces a countertorque. The coil comes to rest when the magnetic torque is counterbalanced by the spring torque. The greater the current, the greater the torque, and the further the coil rotates. In a properly designed instrument, the deflection of the coil and pointer is directly proportional to the current, so the measurement scale can be calibrated to indicate the magnitude of the current.

It is common practice to use a galvanometer coil that consists of many turns of wire, so as to generate a larger torque. The enhanced torque produces a greater deflection of the pointer for a given current, thus improving the ability of the instrument to detect a small current. Moreover, the permanent magnet is typically made with curved pole pieces, and the galvanometer coil surrounds a stationary iron cylinder, as part *b* of the drawing illustrates. The curved pole pieces and iron cylinder tend to orient the magnetic field so as to produce a uniform torque on the coil as it rotates.

THE DIRECT-CURRENT ELECTRIC MOTOR

The electric motor is found in many devices, such as tape decks, turntables, automobiles, washing machines, and air conditioners. Figure 21.21 shows the essential parts of a direct-current (dc) motor. The elements of a motor are similar to those of a galvanometer, except the spring is removed so the coil can rotate continuously in one direction. The coil of wire contains many turns and is wrapped around a movable iron cylinder, although these features have been omitted to simplify the drawing. The coil and iron cylinder assembly is known as the armature. Each end of the wire coil is attached to a metallic half-ring. Rubbing against each of the half-rings is a graphite contact called a brush. While the half-rings rotate with the coil, the graphite brushes remain stationary. The two half-rings and the associated brushes are referred to as a split-ring commutator, the purpose of which will be explained shortly.

The operation of a motor can be understood by considering Figure 21.22*a*. The current from the battery enters the coil through the left brush and half-ring, goes around the coil, and then leaves through the right half-ring and brush. According to RHR-1, the directions of the forces on the two sides of the coil are as shown in the drawing, and these forces produce the torque that turns the coil. The coil rotates until it reaches the position shown in part *b* of the drawing. In this position the half-rings momentarily lose electrical contact with the brushes, and, as a result, there is no current in the coil and no applied torque. However, like any moving object, the rotating coil does not stop immediately, for its rotational inertia carries it onward. When the half-rings reestablish contact with the brushes, there again is a current in the coil, and a magnetic torque again rotates the coil in the same direction. The split-ring commutator ensures that the current is always in the proper direction to yield a torque that produces a continuous rotation of the coil.

21.8 MAGNETIC FIELDS PRODUCED BY CURRENTS

INTRODUCTION

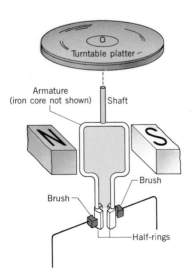

Figure 21.21 The basic components of a dc motor. The platter of a turntable is shown as it might be attached to the motor.

We have seen that a current-carrying wire can experience a magnetic force when placed in a magnetic field. The magnetic field is assumed to be produced by some external source, such as a permanent magnet. In this section we consider the

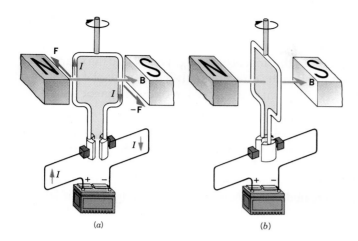

Figure 21.22 (*a*) When a current exists in the coil, the coil experiences a torque. (*b*) Because of its rotational inertia, the coil continues to rotate when there is no current.

phenomenon in which *a current-carrying wire produces a magnetic field.* Hans Christian Oersted (1777–1851) first discovered this effect in 1820 when he observed that a current-carrying wire influenced the orientation of a nearby compass needle. The compass needle aligns itself with the net magnetic field produced by the current and the magnetic field of the earth. Oersted's discovery, which linked the motion of electric charges with the creation of a magnetic field, marked the beginning of an important discipline called *electromagnetism.*

THE MAGNETIC FIELD PRODUCED BY A LONG, STRAIGHT, CURRENT-CARRYING WIRE

Figure 21.23*a* illustrates the essence of Oersted's discovery with a very long, straight wire. When a current is present, the compass needles are observed to point in a circular pattern about the wire. The pattern indicates that the magnetic field lines produced by the current are circles centered on the wire. If the direction of the current is reversed, the needles also reverse their directions, indicating that the direction of the magnetic field has reversed. The direction of the magnetic

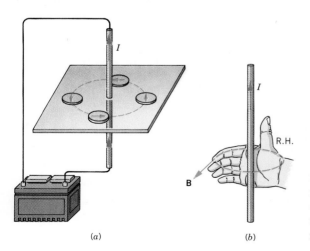

Figure 21.23 (*a*) A long, straight, current-carrying wire produces magnetic field lines that are circular about the wire. One such circular line is indicated by the compass needles. (*b*) If the thumb of the right hand (R.H.) is pointed in the direction of the current *I*, the curled fingers point in the direction of the magnetic field, according to RHR-2.

Figure 21.24 The magnetic field becomes stronger as the radial distance r decreases, so the field lines are closer together near the wire.

field can be obtained by using Right-Hand Rule No. 2 (RHR-2), as part b of the drawing indicates:

> *Right-Hand Rule No. 2.* Curl the fingers of the right hand into the shape of a half-circle. Point the thumb in the direction of the conventional current I, and the tips of the fingers will point in the direction of the magnetic field **B**.

With a Hall-effect probe, or some other device that measures the magnetic field, the magnitude of **B** can be measured as a function of the current I in the wire and the radial distance r from the wire. It is found that the magnitude of the field is directly proportional to the current and inversely proportional to the radial distance: $B \propto I/r$. The proportionality constant is written as $\mu_0/2\pi$. Thus, the magnitude of the magnetic field created by the current in a very long, straight wire is

$$\left[\begin{array}{c}\textbf{Long straight}\\ \textbf{wire}\end{array}\right] \qquad\qquad B = \frac{\mu_0 I}{2\pi r} \qquad\qquad (21.5)$$

The constant μ_0 is known as the *permeability of free space*, and its value is $\mu_0 = 4\pi \times 10^{-7}$ T·m/A. Since the magnetic field becomes stronger nearer the wire where r is smaller, the magnetic field lines near the wire are closer together than those located farther away, where the field is weaker. Figure 21.24 shows the pattern of field lines.

Many factories use industrial robots to carry materials or parts from one place to another. One type of robot follows a current-carrying cable buried in the floor. As Figure 21.25 suggests, the robot follows the cable by using special sensors to detect the magnetic field around the cable.

The magnetic field that a current-carrying wire produces can exert a force on a moving charge, as the next example illustrates.

THE PHYSICS OF . . .

an industrial robot.

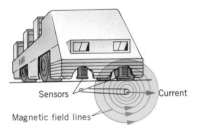

Figure 21.25 The robot follows the buried current-carrying cable by sensing the magnetic field that surrounds it.

Example 5   A Current Exerts a Magnetic Force on a Moving Charge

Figure 21.26 shows a long, straight wire carrying a current of $I = 3.0$ A. A particle of charge $q_0 = +6.5 \times 10^{-6}$ C is moving parallel to the wire at a distance of $r = 0.050$ m from it; the speed of the particle is $v = 280$ m/s. Determine the magnitude and direction of the magnetic force exerted on the moving charge by the current in the wire.

REASONING   The current generates a magnetic field in the space around the wire. A charge moving through this magnetic field experiences a magnetic force **F** whose magnitude is given by Equation 21.1 as $F = q_0 vB \sin\theta$, where θ is the angle between the magnetic field and the velocity of the charge. The magnitude of the magnetic field follows from Equation 21.5 as $B = \mu_0 I/(2\pi r)$. Thus, the magnitude of the magnetic force can be expressed as

$$F = q_0 vB \sin\theta = q_0 v\left(\frac{\mu_0 I}{2\pi r}\right)\sin\theta$$

The direction of the magnetic force is predicted by RHR-1.

SOLUTION   The drawing shows that the magnetic field **B** lies in the plane that is perpendicular to both the wire and velocity **v** of the particle. Thus, the angle between **B** and **v** is $\theta = 90°$ and the magnitude of the magnetic force is

$$F = q_0 v\left(\frac{\mu_0 I}{2\pi r}\right)\sin 90°$$

$$F = (6.5 \times 10^{-6} \text{ C})(280 \text{ m/s}) \left[\frac{(4\pi \times 10^{-7} \text{ T}\cdot\text{m/A})(3.0 \text{ A})}{2\pi(0.050 \text{ m})} \right] = \boxed{2.2 \times 10^{-8} \text{ N}}$$

The direction of the magnetic force is predicted by RHR-1 and, as the drawing shows, is radially inward toward the wire.

We have now seen that an electric current can create a magnetic field of its own. We have also seen earlier that an electric current can experience a force created by another magnetic field. Therefore, the magnetic field that a current creates can exert a force on another nearby current. Example 6 illustrates this magnetic interaction between two currents.

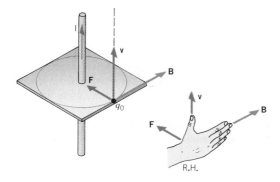

Figure 21.26 The moving charge experiences a magnetic force **F** because of the magnetic field **B** produced by the current in the wire.

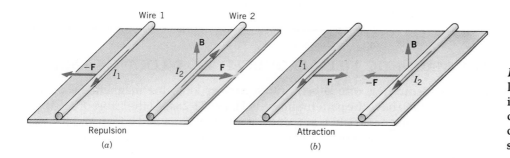

Figure 21.27 (a) Two long, parallel wires carrying currents I_1 and I_2 in opposite directions repel each other. (b) The wires attract each other when the currents are in the same direction.

Example 6 Two Current-Carrying Wires Exert Magnetic Forces on One Another

Figure 21.27 shows two parallel straight wires. The wires are separated by a distance of $r = 0.065$ m and carry currents of $I_1 = 15$ A and $I_2 = 7.0$ A. Find the magnitude and direction of the force that the magnetic field of wire 1 applies to a 1.5-m length of wire 2 when the currents are (a) in opposite directions and (b) in the same direction.

REASONING The current I_2 in wire 2 is situated in the magnetic field produced by the current in wire 1. The magnitude F of the magnetic force experienced by a length L of wire 2 is given by Equation 21.3 as $F = I_2 LB \sin \theta$. Here B is the magnitude of the magnetic field produced by wire 1 and is given by Equation 21.5 as $B = \mu_0 I_1/(2\pi r)$. The direction of the magnetic force can be determined by using RHR-1.

SOLUTION

(a) At wire 2, the magnitude of the magnetic field created by wire 1 is

$$B = \frac{\mu_0 I_1}{2\pi r} = \frac{(4\pi \times 10^{-7}\ \text{T}\cdot\text{m/A})(15\ \text{A})}{2\pi(0.065\ \text{m})} = 4.6 \times 10^{-5}\ \text{T} \tag{21.5}$$

The direction of this field is upward at the location of wire 2, as part *a* of the figure shows. The direction can be obtained using RHR-2 (thumb of right hand along I_1, curled fingers point upward at wire 2 and indicate the direction of **B**). The magnetic field is perpendicular to wire 2 ($\theta = 90°$), so the magnitude of the force on a 1.5-m length of wire 2 is

$$F = I_2 L B \sin\theta \tag{21.3}$$

$$F = (7.0\ \text{A})(1.5\ \text{m})(4.6 \times 10^{-5}\ \text{T}) \sin 90° = \boxed{4.8 \times 10^{-4}\ \text{N}}$$

The direction of the magnetic force on wire 2 is away from wire 1, as part *a* of the drawing indicates; the force direction is found by using RHR-1 (fingers of the right hand extended upward along **B**, thumb points along I_2, palm pushes in the direction of the force **F**).

In a like manner, the current in wire 2 also creates a magnetic field that produces a force on wire 1. Reasoning similar to that above shows that wire 1 is repelled from wire 2 with a force that also has a magnitude of 4.8×10^{-4} N. Thus, each wire generates a force on the other and, if the currents are in *opposite* directions, the wires *repel* each other. The fact that the two wires exert equal, but oppositely directed forces on each other is consistent with Newton's third law, the action–reaction law.

(b) If the current in wire 2 is reversed, as part *b* of the drawing indicates, wire 2 is attracted to wire 1, because the direction of the magnetic force is reversed. However, the magnitude of the force is the same as that calculated in part *a* above. Likewise, wire 1 is attracted to wire 2. Two parallel wires carrying currents in the *same* direction *attract* each other.

THE MAGNETIC FIELD PRODUCED BY A LOOP OF WIRE

If a current-carrying wire is bent into a circular loop, the magnetic field lines around the loop have the pattern shown in Figure 21.28*a*. At the *center* of a loop of radius R, the magnetic field is perpendicular to the plane of the loop and has the value $B = \mu_0 I/(2R)$, where I is the current in the loop. Often, the loop consists of N turns of wire that are wound sufficiently close together that they form a flat coil with a single radius. In this case, the magnetic fields of the individual turns add together to give a net field that is N times greater than that of a single loop. For such a coil the magnetic field at the center is

$$\begin{bmatrix} \textbf{Center of a} \\ \textbf{circular loop} \end{bmatrix} \qquad B = N\frac{\mu_0 I}{2R} \tag{21.6}$$

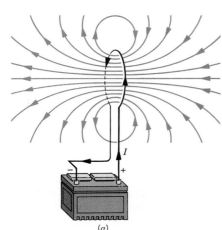

(a)

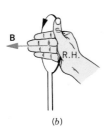

(b)

Figure 21.28 (*a*) The magnetic field lines in the vicinity of a current-carrying circular loop. (*b*) The direction of the magnetic field at the center of the loop is given by RHR-2.

The direction of the magnetic field at the center of the loop can be determined with the help of RHR-2. If the thumb of the right hand is pointed in the direction of the current and the curled fingers are placed at the center of the loop, as in Figure 21.28*b*, the fingers indicate that the magnetic field points from right to left.

Example 7 shows how the magnetic fields produced by the current in a loop of wire and the current in a long, straight wire combine to form a net magnetic field.

Example 7   Finding the Net Magnetic Field

A long, straight wire carries a current of 8.0 A. A portion of the wire is then bent into a circular loop (one turn) of radius 0.020 m, as Figure 21.29 illustrates. Find the magnitude and direction of the net magnetic field at the center C of the loop.

REASONING   The net magnetic field at the point C is the sum of two contributions: (1) the field that the circular loop produces at its center, and (2) the field that the long, straight wire generates. An application of RHR-2 shows that the magnetic field generated by the circular loop at C is directed out of the plane of the paper, toward the reader. Similarly, RHR-2 shows that the magnetic field created at C by the long, straight wire is directed into the plane of the paper. Therefore, the directions of the two magnetic field contributions are opposite.

SOLUTION   Taking the direction out of the paper as positive, the net magnetic field is

$$B = \underbrace{\frac{\mu_0 I}{2r}}_{\substack{\text{Center of} \\ \text{loop}}} - \underbrace{\frac{\mu_0 I}{2\pi r}}_{\substack{\text{Long} \\ \text{wire}}} = \frac{\mu_0 I}{2r}\left(1 - \frac{1}{\pi}\right)$$

$$B = \frac{(4\pi \times 10^{-7} \text{ T·m/A})(8.0 \text{ A})}{2\,(0.020 \text{ m})}\left(1 - \frac{1}{\pi}\right) = \boxed{1.7 \times 10^{-4} \text{ T}}$$

The net field is directed perpendicularly out of the plane of the paper.

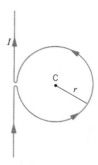

Figure 21.29   Part of a long, straight wire is bent into a circular loop that carries a current I.

PROBLEM SOLVING INSIGHT

Do not confuse the formula for the magnetic field produced at the center of a circular loop with that of a long, straight wire. The formulas are similar, differing only by a factor of π in the denominator.

A comparison of the magnetic field lines around the current loop in Figure 21.28a with those in the vicinity of the short bar magnet in Figure 21.30a shows that the two patterns are quite similar. Not only are the patterns similar, but the

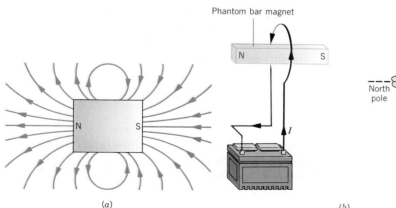

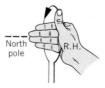

Figure 21.30   (a) The field lines around the bar magnet resemble those around the loop in Figure 21.28a. (b) The current loop can be imagined to be a phantom bar magnet with a north pole and a south pole.

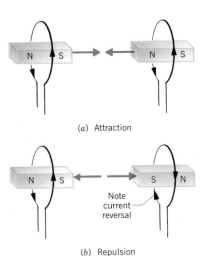

(a) Attraction

(b) Repulsion

Note current reversal

Figure 21.31 (a) The two current loops attract each other if the directions of the currents are the same and (b) repel each other if the directions of the currents are opposite. The "phantom" magnet included for each loop helps explain the attraction and repulsion between the loops.

loop itself behaves as a bar magnet with a "north pole" on one side and a "south pole" on the other side. Figure 21.30b emphasizes this point by including a "phantom" bar magnet at the center of the loop to symbolize that the loop may be imagined to be a bar magnet. The side of the loop that acts like a north pole can be determined with the aid of RHR-2; the fingers of the right hand not only point in the direction of **B**, but they also point toward the north pole.

Because a current-carrying loop acts like a bar magnet, two adjacent loops can be either attracted to or repelled from each other, depending on the relative directions of the currents. Figure 21.31 includes a "phantom" magnet for each loop and shows that the loops are attracted to each other when the currents are in the same direction and repelled from each other when the currents are in opposite directions. This behavior is analogous to that of the two long, straight wires discussed in Example 6.

THE SOLENOID

A solenoid is a long coil of wire wound in the shape of a helix (see Figure 21.32). If the wire is wound so the turns are packed close to each other and the solenoid is long compared to its diameter, the magnetic field lines have the appearance shown in the drawing. Notice that the field inside the solenoid and away from its ends is nearly constant in magnitude and directed parallel to the axis. The direction of the magnetic field inside the solenoid is given by RHR-2, just as it is for a circular current loop. The magnitude of the magnetic field in the interior of a long solenoid is

$$\left[\begin{array}{c}\textbf{Interior of a}\\\textbf{long solenoid}\end{array}\right] \qquad B = \mu_0 n I \qquad (21.7)$$

where n is the number of turns per unit length of the solenoid and I is the current. If, for example, the solenoid contains 100 turns and has a length of 0.05 m, the number of turns per unit length is $n = (100 \text{ turns})/(0.05 \text{ m}) = 2000 \text{ turns/m}$.

As with a single loop of wire, a solenoid can also be imagined to be a bar magnet, for the solenoid is just an array of connected current loops. And, as with a circular current loop, the location of the north pole can be determined with RHR-2. Figure 21.32 shows that the left end of the solenoid acts as a north pole,

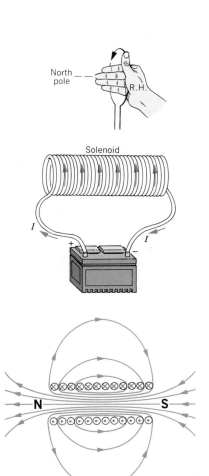

North pole

R.H.

Solenoid

I

I

N   S

Figure 21.32 A solenoid and a cross-sectional view of it, showing the magnetic field lines and the north and south poles.

and the right end behaves as a south pole. Solenoids are often referred to as *electromagnets,* and they have several advantages over permanent magnets. For one thing, the strength of the magnetic field can be altered by changing the current and/or the number of turns per unit length. Furthermore, the north and south poles of an electromagnet can be readily switched by reversing the current. An important application of electromagnetism is in tape recording, which we will discuss in the next section.

Television sets and computer display monitors use electromagnets to produce images by exerting magnetic forces on moving electrons. An evacuated glass tube, called a cathode-ray tube (CRT), contains an electron gun that sends a narrow beam of high-speed electrons toward the screen of the tube, as illustrated in Figure 21.33*a*. The inner surface of the screen is covered with a phosphor coating, and when the electrons strike it, they generate a spot of visible light. This spot is called a pixel (a contraction of "picture element").

To create a black-and-white picture, the electron beam is scanned rapidly from left to right across the screen. As the beam makes each horizontal scan, the intensity of the electrons striking the screen is changed by electronics controlling the electron gun, making the scan line brighter in some places and darker in others. When the beam reaches the right side of the screen, it is turned off and returned to the left side slightly below where it started (see part *b* of the figure).

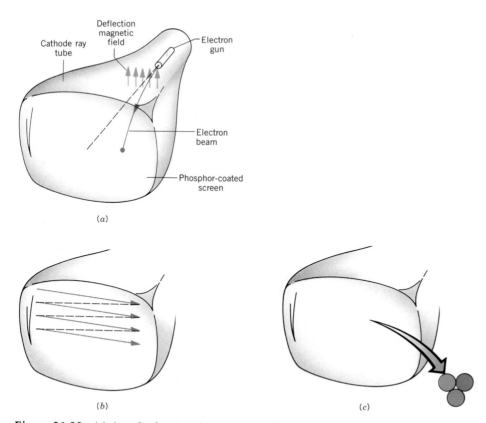

Figure 21.33   (*a*) A cathode-ray tube contains an electron gun, a magnetic field for deflecting the electron beam, and a phosphor-coated screen. (*b*) The image is formed by scanning the electron beam across the screen. (*c*) The red, green, and blue phosphors of a color TV.

An electromagnet, such as the one at the upper right, is being used to lift objects in this mechanized scrap yard in Boston, Massachusetts.

The beam is then scanned across the next line, and so on. In current TV sets, a complete picture consists of 525 scan lines from top to bottom and is formed in $\frac{1}{30}$ of a second. High-definition TV sets are being developed that use a greater number of scan lines.

The electron beam is deflected by a pair of electromagnets placed around the neck of the tube, between the electron gun and the screen. One electromagnet is responsible for producing the horizontal deflection of the beam and the other for the vertical deflection. For clarity, Figure 21.33a shows the net magnetic field at one instant generated by the electromagnets, and not the electromagnets themselves. The electric current in the electromagnets produces a net magnetic field that exerts a force on the moving electrons, causing their trajectories to bend and reach different points on the screen. Changing the current changes the field, so the electrons can be deflected to any point on the screen.

A color TV operates with three electron guns instead of one. And the single phosphor of a black-and-white TV is replaced by a large number of three-dot clusters of phosphors that glow red, green, and blue when struck by an electron beam, as indicated in Figure 21.33c. Each red, green, and blue color in a cluster is produced when electrons from one of the three guns strike the corresponding phosphor dot. The three dots are so close together that, from a normal viewing distance, they cannot be separately distinguished. Red, green, and blue are primary colors, so all other colors can be created by varying the intensities of the three beams focused on a cluster, thereby determining how much of each primary color goes into making a particular color.

21.9 MAGNETIC MATERIALS

FERROMAGNETISM

The similarity between the magnetic field lines in the neighborhood of a bar magnet and those around a current loop suggests that the magnetism in each case arises from a common cause. The field that surrounds the loop is created by the charges moving in the wire. The magnetic field around a bar magnet is also due to the motion of charges, but the motion is not that of a bulk current through the magnetic material. Instead, the motion responsible for the magnetism is that of the electrons within the atoms of the material.

The magnetism produced by electrons within an atom can arise from two motions. First, each electron orbiting the nucleus behaves like an atomic-sized loop of current that generates a small magnetic field; this situation is similar to the field created by the current loop in Figure 21.28. Second, each electron possesses a spin that also gives rise to a magnetic field. The net magnetic field created by the electrons within an atom is due to the combined fields created by their orbital and spin motions.

In most substances the magnetism produced at the atomic level tends to cancel out, with the result that the substance is nonmagnetic overall. However, there are some materials, known as *ferromagnetic materials,* in which the cancellation does not occur for groups of approximately 10^{16}–10^{19} neighboring atoms, because they have electron spins that are naturally aligned parallel to each other. This alignment results from a special type of quantum mechanical* interaction

* The branch of physics called quantum mechanics is mentioned in Section 29.5, although a detailed discussion of quantum mechanics is beyond the scope of this book.

between the spins. The result of the interaction is a small but highly magnetized region of about 0.01 to 0.1 mm in size, depending on the nature of the material; this region is called a *magnetic domain.* Each domain behaves as a small magnet with its own north and south poles.

Ferromagnetic materials are important technologically because some of them can be permanently magnetized and used, for example, as the magnetic medium in tape decks and computer disks. Common ferromagnetic materials are iron, nickel, cobalt, chromium dioxide, and alnico (an *aluminum–nickel–co*balt alloy).

INDUCED MAGNETISM

Often the magnetic domains in a ferromagnetic material are arranged randomly, as Figure 21.34*a* illustrates for a piece of iron. In such a situation, the magnetic fields of the domains cancel each other, so the iron displays little, if any, overall magnetism. However, an unmagnetized piece of iron can be magnetized by placing it in an external magnetic field provided by a permanent magnet or an electromagnet. The external magnetic field penetrates the unmagnetized iron and *induces* (or "brings about") a state of magnetism in the iron by causing two effects on the domains. Those domains whose magnetism is parallel or nearly parallel to the external magnetic field grow in size at the expense of other domains that are not so oriented. Part *b* of the drawing shows the growing domains in gold. In addition, the magnetic alignment of some domains may rotate and become more oriented in the direction of the external field. The resulting preferred alignment of the domains gives the iron an overall magnetism, so the iron behaves like a magnet with associated north and south poles. In some types of ferromagnetic materials, such as the chromium dioxide used in cassette tapes, the domains remain aligned for the most part when the external magnetic field is removed, and the material thus becomes permanently magnetized.

The magnetism induced in a ferromagnetic material can be surprisingly large, even in the presence of a weak external field. For instance, it is not unusual for the induced magnetic field to be a hundred to a thousand times stronger than the external field that causes the alignment. For this reason, high-field electromagnets are constructed by wrapping the current-carrying wire around a solid core made from iron or other ferromagnetic material.

Induced magnetism explains why a permanent magnet sticks to a refrigerator door and why an electromagnet can pick up scrap iron at a junkyard. Notice in

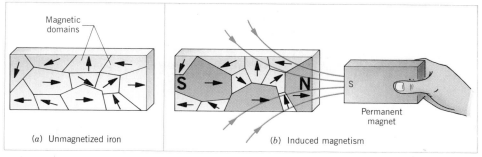

(a) Unmagnetized iron   (b) Induced magnetism

Figure 21.34 (*a*) Each magnetic domain is a highly magnetized region that behaves like a small magnet (represented by an arrow whose head indicates a north pole). An unmagnetized piece of iron consists of many domains that are randomly aligned. The size of each domain is exaggerated for clarity. (*b*) The external magnetic field of the permanent magnet causes those domains that are parallel or nearly parallel to the field to grow in size (shown in gold).

Figure 21.34*b* that there is a north pole at the end of the iron that is closest to the south pole of the permanent magnet. This north pole arises because the north poles of the magnetic domains within the iron tend to line up so they face the south pole of the permanent magnet. The net result is that the two opposite poles give rise to an attraction between the iron and the permanent magnet. Conversely, the north pole of the permanent magnet would also attract the piece of iron by inducing a south pole in the nearest side of the iron. In nonferromagnetic materials, such as aluminum and copper, the formation of magnetic domains does not occur, so magnetism cannot be induced into these substances. Consequently, magnets do not stick to aluminum cans or to copper pennies.

THE PHYSICS OF . . .

magnetic tape recording.

MAGNETIC TAPE RECORDING

The process of magnetic tape recording uses induced magnetism, as Figure 21.35 illustrates. The weak electrical signal from a microphone is routed to an amplifier where it is amplified. The current from the output of the amplifier is then sent to the recording head, which is a coil of wire wrapped around an iron core. The iron core has the approximate shape of a horseshoe with a small gap between the two ends. The ferromagnetic iron substantially enhances the magnetic field produced by the current in the wire.

When there is a current in the coil, the recording head becomes an electromagnet with a north pole at one end and a south pole at the other end. The magnetic field lines pass through the iron core and cross the gap. Within the gap, the lines are directed from the north pole to the south pole. Some of the field lines in the gap "bow outward," as Figure 21.35 indicates, the bowed region of magnetic field being called the *fringe field.* The fringe field penetrates the magnetic coating on the tape and induces magnetism in the coating. This induced magnetism is retained

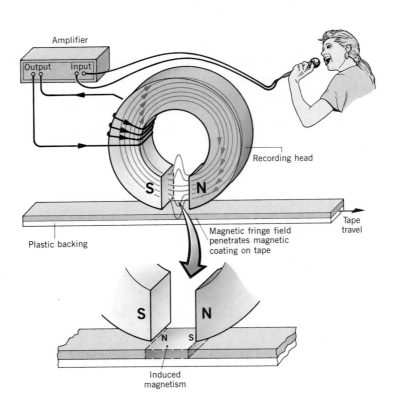

Figure 21.35 The magnetic fringe field of the recording head penetrates the magnetic coating on the tape and causes the coating to become magnetized.

when the tape leaves the vicinity of the recording head and, thus, provides a means for storing audio information. Audio information is retained, because at any instant in time the way in which the tape is magnetized depends on the amount and direction of current in the recording head. The current, in turn, depends on the sound intensity picked up by the microphone, so that changes in the sound intensity that occur from moment to moment are preserved as changes in the tape's induced magnetization.

MAGLEV TRAINS

A magnetically levitated train — or maglev, for short — uses forces that arise from induced magnetism to levitate or float above a guideway. Since it rides a few centimeters above the guideway, a maglev does not need wheels. Freed from friction with the guideway, the train can achieve significantly greater speeds than do conventional trains. For example, the Transrapid maglev in Figure 21.36*a* has achieved speeds of 110 m/s.

Figure 21.36*a* shows that the Transrapid maglev achieves levitation with electromagnets mounted on arms that extend around and under the guideway. When a current is sent to an electromagnet, the resulting magnetic field creates induced magnetism in a rail mounted in the guideway. The upward attractive force from the induced magnetism is balanced by the weight of the train, so the train moves without touching the rail or the guideway.

Magnetic levitation only lifts the train and does not move it forward. Figure 21.36*b* illustrates how magnetic propulsion is achieved. In addition to the levitation electromagnets, propulsion electromagnets are also placed underneath the train and along the guideway. By controlling the direction of the currents in the

THE PHYSICS OF . . .

a magnetically levitated train.

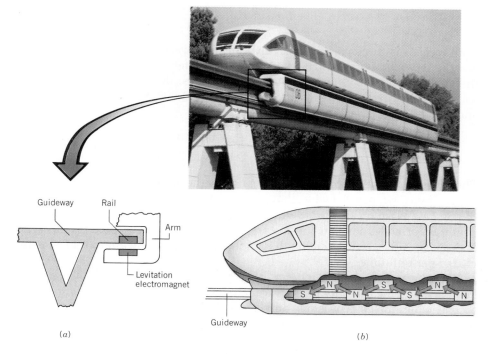

(a)

(b)

Figure 21.36 (*a*) The Transrapid maglev (a German train) has achieved speeds of 110 m/s (250 mph). The levitation electromagnets are drawn up toward the rail in the guideway, levitating the train. (*b*) The magnetic propulsion system.

train and guideway electromagnets, it is possible to create an unlike pole in the guideway just ahead of each electromagnet on the train, and a like pole just behind. Each electromagnet on the train is thus both pulled and pushed forward by electromagnets in the guideway. By adjusting the timing of the like and unlike poles in the guideway, the speed of the train can be adjusted. Reversing the poles in the guideway electromagnets relative to those in the train serves to brake the train.

21.10 OPERATIONAL DEFINITIONS OF THE AMPERE AND THE COULOMB

In Section 20.1 we defined current as the rate at which charge flows, or $I = q/t$ when the current is constant. Therefore, one way of measuring current is to determine the amount of charge q that flows in a time t. According to this procedure, one ampere of current exists when one coulomb of charge flows for one second. In practice, however, it is difficult to measure an ampere precisely by measuring the amount of charge flowing in a known time interval. It would be far superior if the ampere could be measured in terms of force and distance, quantities that can be measured with a high degree of precision. Such a measurement is possible in terms of the magnetic force that two current-carrying wires exert on each other.

Suppose the same current I is sent through two long, straight, parallel wires that are separated by a distance r. According to Equation 21.5, the magnetic field **B** produced at the location of one wire by the other wire has a magnitude of $B = \mu_0 I/(2\pi r)$. Since this magnetic field is perpendicular to the wire (see Figure 21.27), the field exerts a force **F** on a length L of the wire:

$$F = ILB \sin 90° = \frac{\mu_0 I^2 L}{2\pi r} \tag{21.3}$$

With special instruments, this force can be measured accurately. The wire length L and the separation r can also be determined accurately, and μ_0 has been assigned the value of $4\pi \times 10^{-7}$ T·m/A. With these values, the equation above can be solved for the current I. For instance, suppose $F = 2.000 \times 10^{-7}$ N, $r = 1.000$ m, and $L = 1.000$ m. The current is

$$I = \sqrt{\frac{2\pi r F}{\mu_0 L}} = \sqrt{\frac{2\pi(1.000 \text{ m})(2.000 \times 10^{-7} \text{ N})}{(4\pi \times 10^{-7} \text{ T·m/A})(1.000 \text{ m})}} = 1.000 \text{ A}$$

Therefore, one ampere of current is defined as the amount of electric current in each of two long, parallel wires that gives rise to a magnetic force per unit length of 2×10^{-7} N/m on each wire when the wires are separated by one meter. This definition provides a means for measuring current in terms of force and distance and obviates the need to define the ampere in terms of the amount of moving charge per unit time.

With the ampere defined in terms of force and distance, the coulomb can now be defined as the quantity of electrical charge that passes a given point in one second when the current is one ampere, or $1 \text{ C} = 1 \text{ A·s}$. This definition is preferred, since scientists can measure electric current and time more accurately than they can measure the amount of moving charge.

INTEGRATION OF CONCEPTS

FIELDS AND FORCES

The concept of an electric field is introduced in Chapter 18. An electric field is produced by one or more charged objects and exists in the region around them. Electric field lines are often drawn as an aid in visualizing the magnitude and direction of the electric field within the region. At any given location, the electric field exerts an electric force on a charged object placed there, the force being the product of the charge and the electric field at that point. The direction of the force is either parallel or antiparallel to the electric field, depending on whether the charge is positive or negative, respectively. In the present chapter, we see that a magnetic field is produced by permanent magnets or moving charges, such as an electric current, and exists in the region around them. Magnetic field lines are also drawn as an aid in visualizing the magnitude and direction of the magnetic field. As can an electric field, a magnetic field can exert a magnetic force on a charged object within it, but only if the object is moving and has a velocity component that is perpendicular to the magnetic field. The direction of the magnetic force is perpendicular to the plane defined by the velocity of the object and the magnetic field. Thus, the concept of a field is very useful, for it can be used to describe the electric and magnetic forces that are exerted on charged objects.

THE MAGNETIC FORCE AND NATURE'S FUNDAMENTAL FORCES

There are four fundamental forces in nature, fundamental in the sense that all other forces can be understood as manifestations of one or more of the four. Tension, friction, and the elastic force of a spring, for example, are not fundamental forces, but the gravitational force is. Another force that we have encountered is the force that one electrically charged particle exerts on another charged particle. This force is one part of a fundamental force called the electromagnetic force. The electromagnetic force contains two parts, an electric part and a magnetic part. Both parts, however, derive from the same source, the electric charge carried by the particles. Whether or not the particles are moving, they exert on each other the electric force specified in Chapter 18 by Coulomb's law. When the particles move, the other part of the electromagnetic force also appears, the part that we have called the magnetic force in the present chapter.

SUMMARY

A magnet has a north pole and a south pole. The north pole is the end that points toward the north magnetic pole of the earth when the magnet is freely suspended. **Like poles repel each other and unlike poles attract each other.**

A **magnetic field** exists in the space around a magnet.

The magnetic field is a vector whose direction at any point is the direction indicated by the north pole of a small compass needle placed at that point. The magnitude B of the magnetic field at any point in space is defined as $B = F/(q_0 v \sin \theta)$, where F is the magnitude of the magnetic force that acts on a charge q_0 whose velocity **v** makes an angle θ with respect to the magnetic field. The SI unit for the magnetic field is the tesla (T). The direction of the magnetic force is perpendicular to both **v** and **B**, and for a positive charge the direction can be determined with the aid of Right-Hand Rule No. 1 (RHR-1, see Section 21.2). The magnetic force on a moving negative charge is opposite to the force on a moving positive charge.

If a particle of charge q and mass m moves with speed v perpendicular to a uniform magnetic field **B**, the magnetic force causes the charge to move on a circular path of radius $r = mv/(qB)$. A constant magnetic force does no work on the charged particle, because the direction of the force is always perpendicular to the motion of the particle. Being unable to do work, the magnetic force cannot change the kinetic energy of the particle; however, the magnetic force does change the direction in which the particle moves.

A **mass spectrometer** is an instrument that can determine the masses of atoms and molecules. The **Hall emf** develops across a current-carrying metal or semiconductor that has been placed in a magnetic field, because the moving charges are deflected by the magnetic force. The polarity of the Hall emf indicates whether the charge carriers are positive or negative.

An electric current, being composed of moving charges, can experience a magnetic force when placed in a magnetic field. For a straight wire that has a length L and carries a current I, the magnetic force has a magnitude of $F = ILB \sin \theta$, where θ is the angle between the directions of I and **B**. The direction of the force is perpendicular to both I and **B** and is given by RHR-1.

Magnetic forces can exert a **torque** on a current-carrying loop of wire and thus cause the loop to rotate. If a current I exists in a coil of wire with N turns, each of area A, in the presence of a magnetic field **B**, the coil experiences a torque of magnitude $\tau = NIAB \sin \phi$, where ϕ is the angle between the direction of the magnetic field and the normal to the plane of the coil.

An electric current produces a magnetic field, with different current geometries giving rise to different field patterns. For a **long, straight wire,** the magnetic field lines are circles centered on the wire, and their direction is given by RHR-2 (see Section 21.8). The magnitude of the field at a radial distance r from the wire is $B = \mu_0 I/(2\pi r)$, where I is the current and μ_0 is the permeability of free space ($\mu_0 = 4\pi \times 10^{-7}$ T·m/A). The magnetic field at the center of a **flat circular coil** consisting of N turns, each of radius R, is $B = \mu_0 NI/(2R)$. The coil has associated with it a north pole on one side and a south pole on the other side. The side of the coil that behaves like a north pole can be predicted by using RHR-2. A **solenoid** is a coil of wire wound in the shape of a helix. Inside a long solenoid the magnetic field is nearly constant and has the value $B = \mu_0 nI$, where n is the number of turns per unit length of the solenoid. One end of a solenoid behaves like a north pole, and the other end like a south pole, as can be predicted by using RHR-2.

Ferromagnetic materials, such as iron, are made up of tiny regions called domains, each of which behaves as a small magnet. In an unmagnetized ferromagnetic material, the domains are randomly aligned. In a permanent magnet, many of the domains are aligned, and a high degree of magnetism results. An unmagnetized ferromagnetic material can be induced into becoming magnetized by placing it in an external magnetic field.

QUESTIONS

1. In all the drawings of magnetic field lines in this chapter there are no instances of two field lines crossing each other. Magnetic field lines, like electric field lines, never intersect. Suppose it were possible for two magnetic field lines to intersect at a point in space. Discuss what this would imply about the force(s) that act on a charge moving through such a point, thereby ruling out the possibility of field lines crossing.

2. Suppose you accidentally use your left hand, instead of your right hand, to determine the direction of the magnetic force on a positive charge moving in a magnetic field. Do you get the correct answer? If not, what direction do you get?

3. A charged particle, passing through a certain region of space, has a velocity whose magnitude and direction remain constant. (a) If it is known that the external magnetic field is zero everywhere in this region, can you conclude that the external electric field is also zero? (b) If it is known that the external electric field is zero everywhere, can you conclude that the external magnetic field is also zero? Explain.

4. A stationary charge is located between the poles of a horseshoe magnet. Is a magnetic force exerted on the charge? Why?

5. Three particles move through a constant magnetic field and follow the paths shown in the drawing. Determine whether each particle is positively charged, negatively charged, or neutral. Give a reason for each answer.

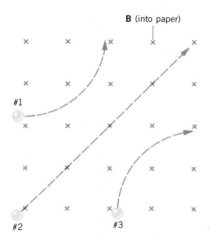

6. Three particles have identical charges and masses. They enter a constant magnetic field and follow the paths shown in the picture. Which particle is moving the fastest, and which is moving the slowest? Justify your answers.

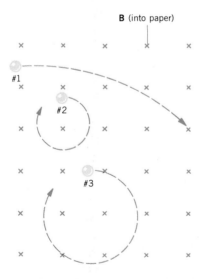

7. The drawing shows a top view of four interconnected chambers. A negative charge is fired into chamber 1. By turning on separate magnetic fields in each chamber, the charge can be made to exit from chamber 4. **(a)** Describe how the magnetic field in each chamber should be directed. **(b)** If the speed of the charge is v when it enters chamber 1, what is the speed of the charge when it exits chamber 4? Why?

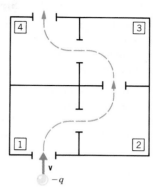

8. A positive charge moves along a circular path under the influence of a magnetic field. The magnetic field is perpendicular to the plane of the circle, as in Figure 21.9. If the velocity of the particle is reversed at some point along the path, will the particle retrace its path? If not, draw the new path. Explain.

9. A positively charged particle travels on a circular path in the presence of a magnetic field, as in Figure 21.9. A uniform electric field is then turned on. Draw the path of the particle when the electric field is directed parallel to the magnetic field.

10. The drawing shows a positive charge $+q$ located at the coordinate origin and a target located in the third quadrant. A magnetic field is directed perpendicularly into the plane of the paper. The charge can be projected in the plane of the paper only, along the positive or negative x or y axis. Thus, there are four possible initial directions along which the charge can be projected. The charge can be made to hit the target for only two of the four directions. Which two are they? Give your reasoning, along with the two paths that the charge can follow on its way to the target.

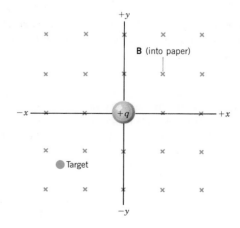

11. Refer to Figure 21.13. **(a)** What happens to the direction of the magnetic force if the current is reversed? (b) What happens to the direction of the force if *both* the current and the magnetic poles are reversed? Explain your answers.

12. Suppose that the magnet in a galvanometer has lost some of its magnetism due to age. Would the reading on the scale of the galvanometer be greater than or less than the value when the galvanometer was new? Account for your answer.

13. Suppose you are building a dc motor and use a wire of length *L* to make the coil. From the point of view of Equation 21.4, do you get more torque by winding the wire into a single circular loop or into a circular loop containing *N* turns? Give your reasoning.

14. The drawing shows an end-on view of three parallel wires that are perpendicular to the plane of the paper. In two of the wires the current is directed into the paper, while in the remaining wire the current is directed out of the paper. The two outermost wires are held rigidly in place. Which way will the middle wire move? Explain.

15. For each electromagnet at the left of the drawing, explain whether it will be attracted to or repelled from the adjacent magnet at the right.

16. Refer to Figure 21.5. If the earth's magnetism is assumed to originate from a large circular loop of current within the earth, how is the plane of this current loop oriented rela-

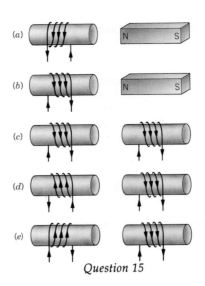

Question 15

tive to the magnetic axis, and what is the direction of the current around the loop?

17. Suppose you have two bars, one of which is a permanent magnet and the other of which is not a magnet, but is made from a ferromagnetic material like iron. The two bars look exactly alike. (a) Using a third bar, which is known to be a magnet, how can you determine which of the look-alike bars is the permanent magnet and which is not? (b) Can you determine the identities of the look-alike bars with the aid of a third bar that is not a magnet, but is made from a ferromagnetic material? Give a reason for your answer.

Section 21.1 Magnetic Fields, Section 21.2 The Force That a Magnetic Field Exerts on a Moving Charge

1. A charge of 12 μC, traveling with a speed of 9.0×10^6 m/s in a direction perpendicular to a magnetic field, experiences a magnetic force of 8.7×10^{-3} N. What is the magnitude of the field?

2. A particle with a charge of $+6.0 \, \mu$C and a speed of 25 m/s enters a uniform magnetic field whose magnitude is 0.15 T. For each of the cases in the drawing, find the magnitude and direction of the magnetic force on the charge.

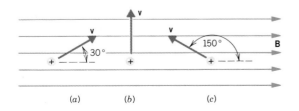

3. Due to friction with the air, an airplane has acquired a net charge of 1.7×10^{-5} C. The plane moves with a speed of 280 m/s at an angle θ with respect to the earth's magnetic field, the magnitude of which is 5.0×10^{-5} T. The magnetic force on the airplane has a magnitude of 2.3×10^{-7} N. Find the angle θ (there are two possible angles).

4. A proton, traveling with a velocity of 4.5×10^6 m/s due east, experiences a maximum magnetic force of 8.0×10^{-14} N due south. (a) What is the magnitude and direction of the magnetic field? (b) Answer part (a), assuming the proton is replaced by an electron.

5. A charged body, moving with a velocity of 8.0×10^4 m/s at an angle of 30.0° with respect to a magnetic field of 5.6×10^{-5} T, experiences a force of 2.0×10^{-4} N. What is the magnitude of the charge?

6. In New England, the horizontal component of the earth's magnetic field has a magnitude of 1.6×10^{-5} T. An electron is shot straight up from the ground with a speed of

2.1×10^6 m/s. What is the magnitude of the acceleration caused by the magnetic force?

*7. In a television set, electrons are accelerated from rest through a potential difference of 15 kV. The electrons then pass through a 0.35-T magnetic field that deflects them to the appropriate spot on the screen. Find the maximum force that an electron can experience.

Section 21.3 The Motion of a Charged Particle in a Magnetic Field, Section 21.4 The Mass Spectrometer, Section 21.5 The Hall Effect

8. An electron moves at a speed of 6.0×10^6 m/s perpendicular to a constant magnetic field. The path is a circle of radius 1.3×10^{-3} m. (a) Draw a sketch showing the magnetic field and the electron's path. (b) What is the magnitude of the field? (c) Find the magnitude of the electron's acceleration.

9. The solar wind is a thin hot gas given off by the sun. Charged particles in this gas enter the magnetic field of the earth and can experience a magnetic force. Suppose a charged particle traveling with a speed of 9.0×10^6 m/s encounters the earth's magnetic field at an altitude where the field has a magnitude of 1.2×10^{-7} T. Assuming the particle's velocity is perpendicular to the magnetic field, find the radius of the circular path on which the particle would move if it were (a) an electron and (b) a proton.

10. A beam of protons moves in a circle of radius 0.25 m. The beam moves perpendicular to a 0.30-T magnetic field. (a) What is the speed of each proton? (b) Determine the magnitude of the centripetal force that acts on each proton.

11. An ionized helium atom has a mass of 6.6×10^{-27} kg and a speed of 4.4×10^5 m/s. The atom moves perpendicular to a 0.75-T magnetic field on a circular path of radius 0.012 m. What is the magnitude of the charge of the helium atom?

12. A mass spectrometer uses a potential difference of 2.00 kV to accelerate a singly charged ion $(+e)$ to the proper speed. A 0.400-T magnetic field then bends the ion into a circular path of radius 0.226 m. What is the mass of the ion?

13. An ion source in a mass spectrometer produces deuterons (a deuteron is a particle that has twice the mass of a proton, but the same charge). Each deuteron is accelerated from rest through a potential difference of 2.00×10^3 V, after which it enters a 0.600-T magnetic field. Find the radius of its circular path.

14. Suppose that an ion source in a mass spectrometer produces *doubly* ionized gold ions (Au^{2+}), each with a mass of 3.27×10^{-25} kg. The ions are accelerated from rest through a potential difference of 1.00 kV. Then, a 0.500-T magnetic field causes the ions to follow a circular path. Determine the radius of the path.

15. Two isotopes of carbon, carbon-12 and carbon-13, have masses of 19.92×10^{-27} kg and 21.59×10^{-27} kg, re-

spectively. These two isotopes are singly ionized $(+e)$ and each is given a speed of 6.667×10^5 m/s. The ions then enter the bending region of a mass spectrometer where the magnetic field is 0.8500 T. Determine the spatial separation between the two isotopes after they have traveled through a half-circle. (*Hint: The spatial separation is the difference between the diameters of the trajectories.*)

16. A *velocity selector* is a device for measuring the speed of a charged particle. The drawing shows that a velocity selector consists of a cylindrical tube located within a constant magnetic field **B**. Inside the tube there is a parallel plate capacitor that produces an electric field **E**. The magnetic and electric fields are perpendicular to each other. A positive charge enters the left end of the tube and has a velocity that is perpendicular to both **B** and **E**. The charge experiences both a magnetic and an electric force. However, if the forces are adjusted so as to cancel each other, the net force acting on the charge is zero. The charge then moves down the tube at a constant speed v in a straight line and exits the right end of the tube. For such a situation, derive an expression for the speed of the particle in terms of B and E.

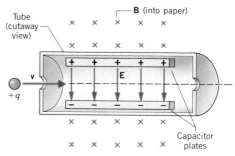

*17. Work problem 16 before attempting to solve this problem. A charged particle moves through a velocity selector at a constant speed in a straight line. The electric field of a velocity selector is 5.65×10^3 N/C, while the magnetic field is 0.114 T. When the electric field is turned off, the charged particle travels on a circular path whose radius is 2.90 cm. Find the charge-to-mass ratio of the particle.

*18. A proton with a speed of 2.2×10^6 m/s is shot into a region between two plates that are separated by a distance of 0.18 m. As the drawing shows, a magnetic field exists between the plates, and it is perpendicular to the velocity of the proton. What must be the magnitude of the magnetic field, so the proton just misses colliding with the opposite plate?

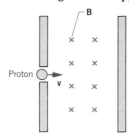

*19. A positively charged particle of mass 7.2×10^{-8} kg is traveling due east with a speed of 85 m/s. The particle enters a 0.31-T uniform magnetic field, and 2.2×10^{-3} s later the particle leaves the field one-quarter of a turn later, heading due south with a speed of 85 m/s. All during the motion the particle moves perpendicular to the magnetic field. (a) What is the magnitude of the magnetic force acting on the particle? (b) Determine the charge of the particle.

*20. An electron moves in a circular orbit of radius 1.7 m in a magnetic field of 2.2×10^{-5} T. The electron moves perpendicular to the magnetic field. Determine the kinetic energy of the electron.

*21. An α-particle is the nucleus of a helium atom; the orbiting electrons are missing. The α-particle contains two protons and two neutrons, and has a mass of 6.64×10^{-27} kg. Suppose an α-particle is accelerated from rest through a potential difference and then enters a region where its velocity is perpendicular to a 0.0210-T magnetic field. With what angular speed ω does the α-particle move on its circular path?

**22. Singly ionized atoms of neon-20 and neon-22 follow circular paths in the bending region of a mass spectrometer. (a) What is the ratio of the neon-22 radius to that of neon-20? (b) Repeat part (a) for a doubly ionized atom of neon-22.

Section 21.6 The Force on a Current in a Magnetic Field

23. The drawing shows wires of length L and current I, lying in a plane that is perpendicular to a magnetic field **B**. In all cases $B = 0.25$ T, $L = 0.60$ m, and $I = 15$ A. Find the magnitude and direction of the magnetic force on each wire.

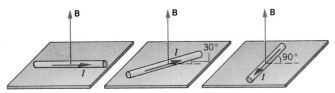

24. An electric power line carries a current of 1400 A in a location where the earth's magnetic field is 5.0×10^{-5} T. The line makes an angle of 75° with respect to the field. Determine the magnitude of the magnetic force on a 120-m length of line.

25. A square coil of wire containing a single turn is placed in a uniform 0.25-T magnetic field, as the drawing shows. Each side has a length of 0.32 m, and the current in the coil is 12 A. Determine the magnitude of the magnetic force on each of the four sides.

26. Near the equator in South America the earth's magnetic field has a strength of 3.0×10^{-5} T; the field is parallel to the surface of the earth and points due north. A straight wire, 25 m in length, has an east–west orientation and experiences a magnetic force of 0.041 N, directed vertically down (toward the earth). What is the magnitude and direction of the current in the wire?

27. A wire of length 0.655 m carries a current of 21.0 A. In the presence of a 0.470-T magnetic field, the wire experiences a force of 5.46 N. What is the angle between the wire and the magnetic field?

28. At New York City, the earth's magnetic field has a vertical (downward) component of 5.2×10^{-5} T and a horizontal component of 1.8×10^{-5} T that is directed toward geographic north. What is the magnitude of the magnetic force on a long, straight wire, 8.0 m in length, that carries a 35-A current due east?

*29. A 125-turn rectangular coil of wire is hung from one arm of a balance, as the drawing shows. With the magnetic field turned off, a mass M is added to the pan on the other arm to balance the mass of the coil. When a constant magnetic field of magnitude 0.200 T is turned on and there is a current of 8.50 A in the coil, how much *additional* mass m must be added to regain the balance?

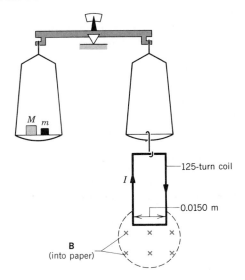

*30. A copper rod of length 0.85 m is lying on a frictionless table (see the drawing). Each end of the rod is attached to a fixed wire by an unstretched spring whose spring constant is $k = 75$ N/m. A magnetic field with a strength of 0.16 T is oriented perpendicular to the surface of the table. (a) What must be the direction of the current in the copper rod that causes the springs to stretch? (b) If the current is 12 A, by how much does each spring stretch?

**31. A 0.20-kg aluminum rod is lying on top of two conducting rails that are separated by 1.6 m. A 0.050-T magnetic

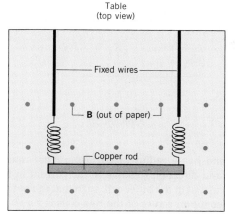

Table
(top view)

Fixed wires

B (out of paper)

Copper rod

Problem 30

field has the direction shown in the drawing. The coefficient of static friction between the rod and a rail is $\mu_s = 0.45$. (a) How much current must be sent through the rod before the rod begins to move? (b) In what direction will the rod move, toward the battery or away from it? Explain.

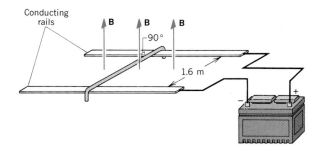

Conducting
rails

B   **B**   **B**

90°

1.6 m

** **32.** A horizontal wire of length 0.20 m and mass 0.080 kg is hung from the ceiling of a room by two massless strings. A 0.070-T magnetic field is directed from the ceiling to the floor. When a current of 42 A passes through the wire, the wire swings upward through an angle ϕ, as the drawing shows. Find (a) the angle ϕ and (b) the tension in each of the two strings.

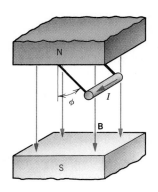

Section 21.7 The Torque on a Current-Carrying Coil

33. A circular coil of wire has a radius of 0.10 m. The coil has 50 turns and a current of 15 A, and is placed in a magnetic field whose magnitude is 0.20 T. (a) Determine the magnetic moment of the coil. (b) What is the maximum torque the coil can experience in this field?

34. The proton has an intrinsic magnetic moment of 1.4×10^{-26} A·m². If the magnetic moment makes an angle of $\phi = 64°$ with respect to a 0.65-T magnetic field, what is the torque exerted on the proton?

35. A coil carries a current and experiences a torque due to a magnetic field. The value of the torque is 80.0% of the maximum possible torque. (a) What is the smallest angle between the magnetic field and the normal to the plane of the coil? (b) Make a drawing, showing how this coil would be oriented relative to the magnetic field. Be sure to include the angle in the drawing.

36. The 1200-turn coil in a dc motor has an area of 1.1×10^{-2} m². The design for the motor specifies that the magnitude of the maximum torque is 5.8 N·m when the coil is placed in a 0.20-T magnetic field. What is the current in the coil?

37. A 0.50-m length of wire is formed into a single-turn, square loop in which there is a current of 12 A. The loop is placed in a magnetic field of 0.12 T, as in Figure 21.18a. What is the maximum torque that the loop can experience?

* **38.** In the model of the hydrogen atom due to Niels Bohr, the electron moves around the proton at a speed of 2.2×10^6 m/s in a circle of radius 5.3×10^{-11} m. Considering the orbiting electron to be a small current loop, determine the magnetic moment associated with this motion.

* **39.** A charge of 4.0×10^{-6} C is placed on a small conducting sphere that is located at the end of a thin insulating rod whose length is 0.20 m. The rod rotates with an angular speed of $\omega = 150$ rad/s about an axis that passes perpendicularly through its other end. Find the magnetic moment of the rotating charge.

Section 21.8 Magnetic Fields Produced by Currents

40. A long, straight wire produces a 4.5×10^{-5}-T magnetic field at a distance of 0.16 m from the wire. What is the current in the wire?

41. In a lightning bolt, 15 C of charge flows in a time of 1.5×10^{-3} s. Assuming that the lightning bolt can be represented as a long, straight line of current, what is the magnetic field at a distance of 25 m from the bolt?

42. A +6.00 μC charge is moving with a speed of 7.50×10^6 m/s parallel to a long, straight wire. The wire carries a current of 67.0 A in a direction opposite to that of the moving charge, and is 5.00 cm from the charge. Find the magnitude and direction of the force on the charge.

43. What must be the radius of a circular loop of wire so the magnetic field at its center is 1.4×10^{-4} T when the loop carries a current of 6.0 A?

44. A long solenoid consists of 1400 turns of wire and has a length of 0.65 m. There is a current of 4.7 A in the wire. What is the magnitude of the magnetic field within the solenoid?

45. Two long, straight wires are separated by 0.120 m. The wires carry currents of 8.0 A in opposite directions, as the drawing indicates. Find the magnitude of the net magnetic field at the points labeled A and B.

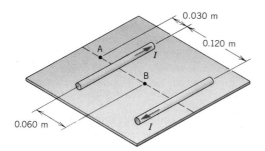

46. Two straight, parallel wires are separated by 0.15 m. The first wire carries a current of 125 A, and the magnetic field produced by this current exerts a force of 3.0×10^{-3} N on a 2.1-m length of the second wire. What is the current in the second wire?

47. Two rigid rods are oriented parallel to each other and to the ground. The rods carry the same current in the same direction. The length of each rod is 0.85 m, while the mass of each is 0.073 kg. One rod is held in place above the ground, and the other floats beneath it at a distance of 8.2×10^{-3} m. Determine the current in the rods.

*48. A piece of copper wire has a resistance per unit length of 5.90×10^{-3} Ω/m. The wire is wound into a thin, flat coil of many turns that has a radius of 0.140 m. The ends of the wire are connected to a 12.0-V battery. Find the magnetic field at the center of the coil.

*49. Two long, straight, parallel wires A and B are separated by a distance of one meter. They carry currents in opposite directions, and the current in wire A is one-third of that in wire B. On a line drawn perpendicular to the wires and passing through them, find the point where the net magnetic field is zero. Determine this point relative to wire A.

*50. A rectangular current loop is located near a long, straight wire that carries a current of 12 A (see the drawing). The current in the loop is 25 A. (a) Determine the net magnetic force that acts on the loop. (b) Is the loop attracted to or repelled from the straight wire? Why?

*51. Two circular coils are concentric and lie in the same plane. The inner coil contains 120 turns of wire, has a radius of 0.012 m, and carries a current of 6.0 A. The outer coil contains

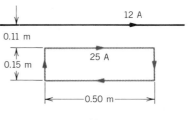

Problem 50

150 turns and has a radius of 0.017 m. What must be the magnitude and direction (relative to the current in the inner coil) of the current in the outer coil, such that the net magnetic field at the common center of the two coils is zero?

**52. The drawing shows two long, straight wires that are suspended from a ceiling. Each of the four strings suspending the wires has a length of 1.2 m. The mass per unit length of each wire is 0.050 kg/m. When the wires carry identical currents in opposite directions, the angle between the strings holding the two wires is 15°. What is the current in each wire?

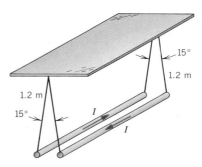

ADDITIONAL PROBLEMS

53. At a certain location, the horizontal component of the earth's magnetic field is 2.5×10^{-5} T, due north. A proton moves eastward with just the right speed, so the magnetic force on it balances its weight. Find the speed of the proton.

54. A charge $q_1 = 25.0$ μC moves with a speed of 4.50×10^3 m/s perpendicular to a uniform magnetic field. The charge experiences a magnetic force of 7.31×10^{-3} N. A second charge $q_2 = 5.00$ μC travels at an angle of 40.0° with respect to the same magnetic field and experiences a 1.90×10^{-3}-N force. Determine (a) the magnitude of the magnetic field and (b) the speed of q_2.

55. A circular loop of one turn is made from a wire of length 5.00×10^{-2} m. There is a current of 2.00 A in the wire. In the presence of a 1.50-T magnetic field, what is the largest torque that this loop can experience?

56. A long, straight wire carrying a current of 305 A is placed in a uniform magnetic field whose magnitude is

7.00 × 10⁻³ T. The wire is perpendicular to the field. Find a point in space where the net magnetic field is zero. Locate this point by specifying its perpendicular distance from the wire.

57. A long solenoid has 1400 turns per meter of length, and it carries a current of 3.5 A. A small circular coil of wire is placed inside the solenoid with the normal to the coil oriented at an angle of 90.0° with respect to the axis of the solenoid. The coil consists of 50 turns, has an area of 1.2×10^{-3} m², and carries a current of 0.50 A. Find the torque exerted on the coil.

58. A 45-m length of wire is stretched horizontally between two vertical posts. The wire carries a current of 75 A and experiences a magnetic force of 0.15 N. Find the magnitude of the earth's magnetic field at the location of the wire, assuming the field makes an angle of 60.0° with respect to the wire.

59. A charged particle with a charge-to-mass ratio of 5.7×10^8 C/kg travels on a circular path that is perpendicular to a magnetic field whose magnitude is 0.72 T. How much time does it take for the particle to complete one revolution?

60. A circular loop of wire and a long, straight wire each carry the same current, as the drawing shows. The loop and wire lie in the same plane. The net magnetic field at the center of the loop is zero. Find the distance H, expressing your answer in terms of R, the radius of the loop.

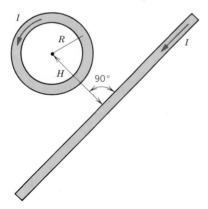

*61. The drawing shows a charge entering a 0.52-T magnetic field. The charge has a speed of 270 m/s and moves perpendicular to the magnetic field. Just as the particle enters the magnetic field, an electric field is turned on. What must be the magnitude and direction of the electric field such that the net force on the particle is twice the magnetic force?

*62. The electrons in the beam of a television tube have a kinetic energy of 2.40×10^{-15} J. Initially, the electrons move horizontally from west to east. The vertical component of the earth's magnetic field points down, toward the surface of the earth, and has a magnitude of 2.00×10^{-5} T. (a) In what direction are the electrons deflected by this field component? (b) What is the acceleration of an electron in part (a)?

*63. A charge of $+3.5 \times 10^{-5}$ C is distributed uniformly around a thin ring of insulating material. The ring has a radius of 0.25 m and rotates with an angular speed of $\omega = 6500$ rad/s about an axis perpendicular to the plane of the ring and passing through its center. Determine the magnitude of the magnetic field produced at the center of the ring.

*64. Two charged particles have the same *linear momentum*, but particle 1 has three times the charge of particle 2. Both travel perpendicular to a uniform magnetic field. What is the ratio r_1/r_2 of the radii of the circles on which they move?

*65. Suppose you have two identical lengths of thin insulated wire with which to make two planar coils. Coil A consists of a single circular turn, while coil B consists of two circular turns. The coils are located in the same magnetic field, and each carries the same current. What is the ratio τ_B/τ_A of the maximum torques that these coils can experience?

**66. The drawing shows two wires that carry the same current of $I = 85.0$ A and are oriented perpendicular to the plane of the paper. The current in one wire is directed out of the paper, while the current in the other is directed into the paper. Find the magnitude and direction of the net magnetic field at the point P.

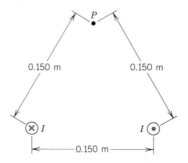

CHAPTER 22

ELECTROMAGNETIC INDUCTION

We have seen in Chapter 21 that an electric current produces a magnetic field. The interrelationship between electricity and magnetism goes even further than this, for we will now see that a magnetic field can be used to produce an electric current. The pickup in an electric guitar, for example, uses magnetic fields to generate the electrical signals that are ultimately converted into sound. The playback head in a stereo cassette deck works according to the same principle to reproduce the music on a tape. Bicycle computers that display speed and other information for cyclists also use a magnetic field to produce electricity. And on a much greater scale, generators at power plants utilize magnetic fields in providing the electricity for our homes and industries. The distribution of electricity to homes and industries depends on a device called a transformer, which is based on the fact that a magnetic field can be used to produce an electric current. The number and variety of such examples reveal that the use of a magnetic field to create an electric current is widespread.

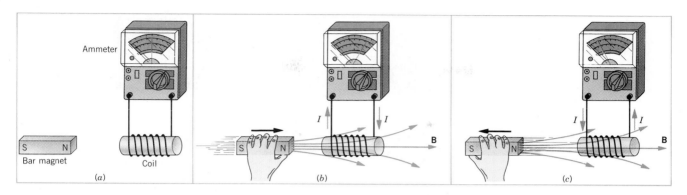

Figure 22.1 (*a*) When there is no relative motion between the coil of wire and the bar magnet, there is no current in the coil. (*b*) A current is created in the coil when the magnet moves toward the coil. (*c*) A current also exists when the magnet moves away from the coil, but the direction of the current is opposite to that in (*b*).

22.1 INDUCED EMF AND INDUCED CURRENT

There are a number of ways a magnetic field can be used to generate an electric current, and Figure 22.1 illustrates one of them. This drawing shows a bar magnet and a helical coil of wire to which an ammeter is connected. When there is no relative motion between the bar magnet and the coil, as in part *a* of the drawing, the ammeter reads zero, indicating there is no current. However, when the magnet moves toward the coil, as in part *b*, a current appears in the coil. As the magnet approaches, the magnetic field that it creates at the location of the coil becomes stronger and stronger, and it is this *changing* magnetic field that produces the current. When the magnet moves away from the coil, as in part *c*, a current also exists, but the direction of the current is reversed. Now the magnetic field at the coil becomes weaker as the magnet moves away. Once again, it is the *changing* magnetic field at the coil that generates the current.

A current would also be created in Figure 22.1 if the magnet were held stationary and the coil were moved, because the magnetic field at the coil would be changing as the coil approached or receded from the magnet. Only relative motion between the magnet and the coil is needed to generate a current; it does not matter which one moves.

The current in the coil is called an ***induced current,*** because the current is brought about (or "induced") by a changing magnetic field. Since a source of emf is always needed to produce a current, the coil itself behaves as if it were a source of emf. The emf is known as an ***induced emf.*** Thus, a changing magnetic field induces an emf in the coil, and the emf leads to an induced current.

Induced emf and induced current are used in bicycle computers, which are popular among cycling enthusiasts. This type of device consists of a small computer that attaches to the handlebar, a sensor that connects to the bike frame, and a magnet that attaches to the spokes (see Figure 22.2). The sensor contains a coil of wire in which an induced emf and induced current appear each time the magnet passes by as the wheel turns. The computer counts the pulses of current. Since the computer also includes an internal clock, it can determine the number of pulses per second. From this information and the radius of the wheel, the computer determines the speed and presents the result on a liquid crystal display (LCD).

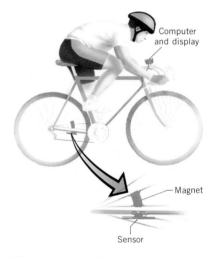

Figure 22.2 A bicycle computer uses the emf and current induced in a sensor coil to determine the speed of the bike.

THE PHYSICS OF . . .

bicycle computers.

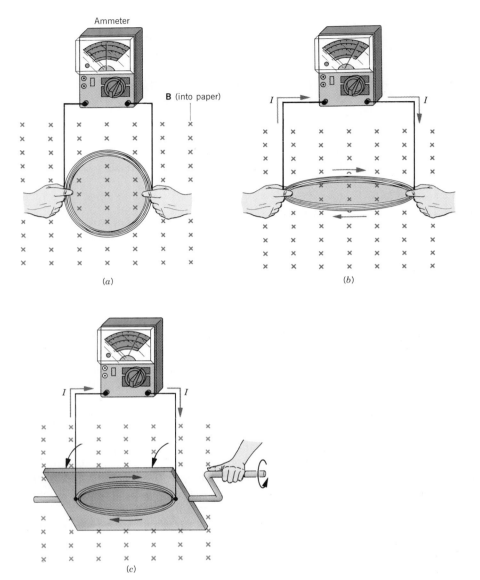

Figure 22.3 (*a*) No current exists in a coil of constant area (the shaded region) that is located in a constant magnetic field. (*b*) While the area of the coil is changing, an induced emf and current are generated. (*c*) An induced emf and current are also produced while the coil is rotating about an axis perpendicular to the magnetic field.

Figure 22.3 shows other ways to induce an emf and a current in a coil. Parts *a* and *b* of the drawing illustrate that an emf can be induced by *changing the area* of a coil in a constant magnetic field. Here the shape of the coil is being distorted so as to reduce the area. As long as the area is changing, an induced emf and current exist; they vanish when the area is no longer changing. If the distorted coil is returned to its original circular shape, thereby increasing the area, an oppositely directed current is generated while the area is changing.

Part *c* of Figure 22.3 indicates that an induced emf is also generated when a coil of constant area is rotated in a constant magnetic field and the *orientation* of the coil *changes* with respect to the field. When the rotation stops, the emf, and hence the current, vanishes.

In each of the examples above, both an emf and a current are induced in the coil because the coil is part of a complete, or closed, circuit. If the circuit were open —

perhaps because of an open switch — there would be no induced current. However, an emf would still be induced in the coil, whether the current exists or not.

Changing the magnetic field, changing the area of a coil, and changing the orientation of a coil are all methods that can be used to create an induced emf. The phenomenon of producing an induced emf with the aid of a magnetic field is called *electromagnetic induction.* The next section discusses how an induced emf arises when a conducting rod moves through a magnetic field.

22.2 MOTIONAL EMF

THE EMF INDUCED IN A MOVING CONDUCTOR

When a conducting rod moves through a constant magnetic field, an emf is induced in the rod. This special case of electromagnetic induction arises as a result of the magnetic force (see Section 21.2) that acts on a moving charge. Consider the metal rod of length L moving to the right in Figure 22.4a. The velocity $\mathbf{v}$ of the rod is constant and is perpendicular to a uniform magnetic field $\mathbf{B}$. Each charge within the rod also moves with a velocity $\mathbf{v}$ and experiences a magnetic force of magnitude $F = qvB$, according to Equation 21.1. By using RHR-1, it can be seen that the mobile, free electrons are driven to the bottom of the rod, leaving behind an equal amount of positive charge at the top. (Remember to reverse the direction of the

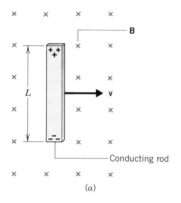

(a)

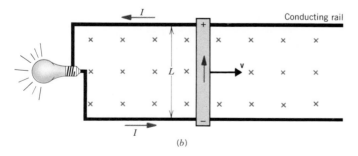

(b)

Figure 22.4 (*a*) When a conducting rod moves at right angles to a constant magnetic field, the magnetic force causes opposite charges to appear at the ends of the rod, giving rise to an induced emf. (*b*) The induced emf causes an induced current I to appear in the circuit.

force that RHR-1 predicts, since the electrons have a negative charge.) The positive and negative charges continue to accumulate, until the attractive electric force that the positive and negative charge groups exert on each other becomes equal in magnitude to the magnetic force. When the electric force balances the magnetic force, equilibrium is reached and no further charge separation occurs.

The separated charges on the ends of the moving conductor give rise to an induced emf, called a *motional emf*, because it originates from the motion of charges through a magnetic field. The motional emf exists as long as the rod moves. If the rod is brought to a halt, the magnetic force vanishes, with the result that the attractive electric force reunites the positive and negative charges and the emf disappears. The emf of the moving rod is similar to that between the terminals of a battery. The difference between a battery and the rod is that the emf of a battery is produced by chemical reactions, whereas the emf of the rod is created by the agent that moves the rod through the magnetic field.

The fact that the electric and magnetic forces balance at equilibrium in Figure 22.4a can be used to determine the magnitude of the motional emf $\mathscr{E}$. The electric force acting on the positive charge q at the top of the rod is Eq, where E is the magnitude of the electric field due to the separated charges. According to Equation 19.7,* the electric field magnitude is given by the voltage between the ends of the rod (the emf $\mathscr{E}$) divided by the length L of the rod. Thus, the electric force is $Eq = (\mathscr{E}/L)q$. The magnetic force is qvB, according to Equation 21.1, since the charge q moves perpendicular to the magnetic field. Since these two forces balance, it follows that $(\mathscr{E}/L)q = qvB$. The emf, then, is

$$\mathscr{E} = vBL \tag{22.1}$$

As expected, $\mathscr{E} = 0$ when $v = 0$, for no motional emf is developed in a stationary rod. Greater speeds and stronger fields lead to greater emfs for a given length L. As with batteries, $\mathscr{E}$ is expressed in volts. In Figure 22.4b the rod is sliding on conducting rails that form part of a closed circuit, and L is the length of the rod between the rails. Due to the emf, electrons flow in a clockwise direction around the circuit as long as the rod continues to move. Positive charge would flow in the direction opposite to the electron flow, so the conventional current I is drawn counterclockwise in the picture. Example 1 illustrates how to determine the electrical energy that the motional emf delivers to a device such as the light bulb in the drawing.

* Without the minus sign.

Example 1 Operating a Light Bulb with Motional Emf

Suppose the rod in Figure 22.4b is moving at a speed of 5.0 m/s in a direction perpendicular to a 0.80-T magnetic field. The rod has a length of 1.6 m and has negligible electrical resistance. The rails also have negligible resistance, and the light bulb has a resistance of 96 Ω. Find (a) the emf produced by the rod, (b) the induced current in the circuit, (c) the electrical power delivered to the bulb, and (d) the energy consumed by the bulb in 60.0 s.

REASONING The moving rod acts as a battery does in supplying an emf or voltage to the circuit. Ohm's law applies, and so does the expression of electrical power as current times voltage.

SOLUTION

(a) The motional emf is given by Equation 22.1 as

$$\mathcal{E} = vBL = (5.0 \text{ m/s})(0.80 \text{ T})(1.6 \text{ m}) = \boxed{6.4 \text{ V}}$$

(b) According to Ohm's law, the induced current is equal to the motional emf divided by the resistance of the circuit:

$$I = \frac{\mathcal{E}}{R} = \frac{6.4 \text{ V}}{96 \text{ }\Omega} = \boxed{0.067 \text{ A}} \tag{20.2}$$

(c) The electrical power P delivered to the light bulb is the product of the current I and the potential difference across the bulb:

$$P = I\mathcal{E} = (0.067 \text{ A})(6.4 \text{ V}) = \boxed{0.43 \text{ W}} \tag{20.6}$$

(d) Since power is energy per unit time, the energy E consumed in 60.0 s is the product of the power and the time:

$$E = Pt = (0.43 \text{ W})(60.0 \text{ s}) = \boxed{26 \text{ J}}$$

MOTIONAL EMF AND ELECTRICAL ENERGY

Motional emf arises because a magnetic force acts on the charges in a conductor that is moving through a magnetic field. Whenever this emf causes a current, a second magnetic force enters the picture. In Figure 22.4*b*, for instance, the second force arises because the current I in the rod is perpendicular to the magnetic field. The current, and hence the rod, experiences a magnetic force **F** whose magnitude is given by Equation 21.3 as $F = ILB \sin 90°$. Using the values of I, L, and B from Example 1, we see that $F = (0.067 \text{ A})(1.6 \text{ m})(0.80 \text{ T}) = 0.086 \text{ N}$. The direction of **F** is specified by RHR-1 and is *opposite* to the velocity **v** of the rod, as Figure 22.5 shows. Hence, **F** tends to *slow down* the rod, and here lies the crux of the matter. To keep the rod moving to the right with a constant velocity, a counterbalancing force must be applied to the rod by an external agent, such as somebody pushing it. The counterbalancing force must have a magnitude of 0.086 N and must be directed to the right in the drawing. If the counterbalancing force were removed, the rod would decelerate under the influence of the magnetic force **F** and eventually come to rest. During the deceleration, the motional emf would decrease and the light bulb would eventually go out.

In Example 1 the light bulb consumes 26 J of electrical energy in sixty seconds.

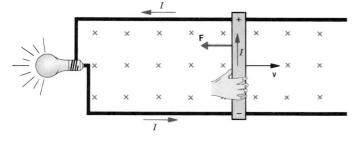

Figure 22.5 A magnetic force **F** is exerted on the current I in the moving rod and is opposite to the velocity **v**. The rod will slow down unless a counterbalancing force is applied by the hand.

We can now answer an important question — Who or what supplies this energy? The energy originates with the external agent that supplies the 0.086-N counterbalancing force needed to keep the rod moving. This agent does work, and Example 2 shows that the work done is equal to the electrical energy consumed by the light bulb.

Example 2   The Work Needed to Keep the Light Bulb Burning

In Example 1, an external agent (the hand in Figure 22.5) supplies the 0.086-N force that keeps the rod moving a constant speed of 5.0 m/s. Determine the work done in 60.0 s by the external agent.

REASONING AND SOLUTION   The work W done by the hand in Figure 22.5 is equal to the magnitude of the external force times the distance the rod moves, as given by Equation 6.1. The distance x traveled by the rod in a time t is $x = vt$, so the work is

$$W = Fx = F(vt) = (0.086 \text{ N})(5.0 \text{ m/s})(60.0 \text{ s}) = \boxed{26 \text{ J}}$$

The 26 J of work done on the rod by the external agent is the same as the 26 J of energy consumed by the light bulb. Hence, the moving rod converts mechanical work into electrical energy, much as a battery converts chemical energy into electrical energy.

It is interesting to speculate what would happen if the direction of the current in Figure 22.5 were *reversed*. If so, the magnetic force **F** would also be reversed and point in the same direction as **v**. Then the force **F** would cause the rod to accelerate, rather than decelerate. There would be no need for an external force to keep the rod moving. This hypothetical electric generator would create the energy that operates the light bulb out of nothing. Such a device cannot exist, because it would violate the principle of conservation of energy, which states that energy cannot be created or destroyed, but can only be converted from one form to another.

22.3 MAGNETIC FLUX

MOTIONAL EMF AND MAGNETIC FLUX

According to Equation 22.1, the emf induced in a rod moving perpendicular to a magnetic field is $\mathscr{E} = vBL$. This motional emf, as well as any other induced emf, can be described in terms of a concept called magnetic flux. To see how the expression for motional emf can be written in terms of this concept, look at Figure 22.6a, where a rod is shown moving through a magnetic field at a time $t = 0$ and a later time t_0. During this time interval, the rod moves a distance x_0 to the right. At an even later time t, the rod has moved an even greater distance x to the right, as part b of the drawing indicates. The speed v of the rod is the distance traveled divided by the elapsed time: $v = (x - x_0)/(t - t_0)$. Substituting this expression for v into $\mathscr{E} = vBL$ gives

$$\mathscr{E} = \left(\frac{x - x_0}{t - t_0}\right) BL = \left(\frac{xL - x_0 L}{t - t_0}\right) B$$

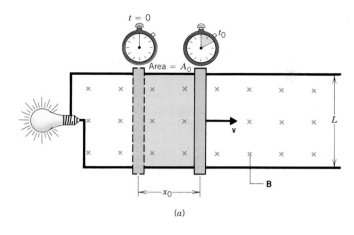

(a)

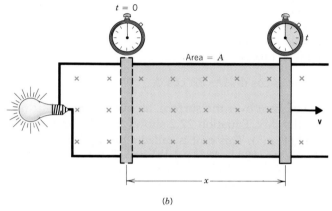

(b)

Figure 22.6   (a) In a time t_0, the moving rod sweeps out an area $A_0 = x_0 L$. (b) The area swept out in a time t is $A = xL$. In both parts of the figure the areas are shaded in color.

As the drawing indicates, the term $x_0 L$ is the area A_0 swept out by the rod in moving a distance x_0, while xL is the area A swept out in moving a distance x. In terms of these areas, the emf is

$$\mathscr{E} = \left(\frac{A - A_0}{t - t_0} \right) B = \frac{(BA) - (BA)_0}{t - t_0}$$

Notice how the product BA of the magnetic field strength and the area appears in the numerator of this expression. The quantity BA is given the name ***magnetic flux*** and is represented by the symbol Φ (Greek capital letter *phi*); thus, $\Phi = BA$. The magnitude of the induced emf is the *change* in flux $\Delta\Phi = \Phi - \Phi_0$ divided by the time interval $\Delta t = t - t_0$ during which the change occurs:

$$\mathscr{E} = \frac{\Phi - \Phi_0}{t - t_0} = \frac{\Delta\Phi}{\Delta t}$$

In other words, the induced emf equals the time rate of change of the magnetic flux.

    You will almost always see the equation above written with a minus sign, namely, $\mathscr{E} = -\Delta\Phi/\Delta t$. The minus sign is introduced for the following reason:

The direction of the current induced in the circuit is such that the magnetic force **F** acts on the rod to *oppose* its motion, thereby tending to slow down the rod (see Figure 22.5). The presence of the minus sign reminds us that the polarity of the induced emf sends the induced current in the proper direction so as to give rise to this opposing magnetic force.

The advantage of writing the induced emf as $\mathscr{E} = -\Delta\Phi/\Delta t$ is that this relation is far more general than our present discussion suggests. In Section 22.4 we will see that $\mathscr{E} = -\Delta\Phi/\Delta t$ can be applied to *all possible ways of generating induced emfs.*

A GENERAL EXPRESSION FOR MAGNETIC FLUX

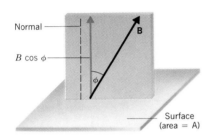

Figure 22.7 When computing the magnetic flux, the component of the magnetic field that is parallel to the normal to the surface must be used; this component is $B \cos \phi$.

In Figure 22.6 the direction of the magnetic field **B** is perpendicular to the surface swept out by the moving rod. In general, however, **B** may not be perpendicular to the surface. For instance, in Figure 22.7 the direction perpendicular to the surface is indicated by the normal to the surface, but the magnetic field is inclined at an angle ϕ with respect to this direction. In such a case the flux is computed using only the component of the field that is perpendicular to the surface, $B \cos \phi$. The general expression for magnetic flux is

$$\Phi = (B \cos \phi)A = BA \cos \phi \qquad (22.2)$$

If the magnetic field is not constant over the surface, an average magnetic field must be used in computing the flux. Equation 22.2 shows that the unit of magnetic flux is the tesla·meter² ($T \cdot m^2$). This unit is called a *weber* (Wb), after the German physicist Wilhelm Weber (1804–1891): 1 Wb = 1 T·m². Example 3 illustrates how to determine the magnetic flux for three different orientations of the surface of a coil relative to the magnetic field.

Example 3   Magnetic Flux

A rectangular coil of wire is situated in a constant magnetic field whose magnitude is 0.50 T. The coil has an area of 2.0 m². Determine the magnetic flux for the three orientations, $\phi = 0°$, 60.0°, and 90.0°, shown in Figure 22.8.

REASONING AND SOLUTION   The flux in the three cases can be computed using $\Phi = BA \cos \phi$:

$[\phi = 0°]$ $\qquad\qquad\qquad \Phi = (0.50 \text{ T})(2.0 \text{ m}^2) \cos 0° = \boxed{1.0 \text{ Wb}}$

$[\phi = 60.0°]$ $\qquad\qquad \Phi = (0.50 \text{ T})(2.0 \text{ m}^2) \cos 60.0° = \boxed{0.50 \text{ Wb}}$

$[\phi = 90.0°]$ $\qquad\qquad \Phi = (0.50 \text{ T})(2.0 \text{ m}^2) \cos 90.0° = \boxed{0}$

PROBLEM SOLVING INSIGHT

The magnetic flux Φ is determined by more than just the magnitude B of the magnetic field and the area A. It also depends on the angle ϕ (see Figure 22.7 and Equation 22.2).

GRAPHICAL INTERPRETATION OF MAGNETIC FLUX

It is possible to interpret the magnetic flux graphically by noting that *the flux is proportional to the number of magnetic field lines that passes through a surface.* This useful interpretation stems from the fact that the magnitude of the magnetic field **B** in any region of space is proportional to the number of magnetic field lines per unit area that passes through a surface perpendicular to the field

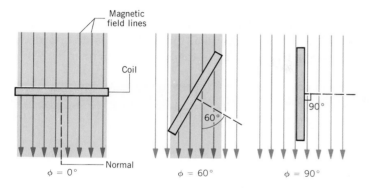

Figure 22.8 This picture shows three orientations of a rectangular coil (drawn as a side view), relative to the magnetic field lines. The magnetic field lines that pass through the coil are those in the regions shaded in blue.

lines (see Section 21.1). For instance, the magnitude of **B** in Figure 22.9a is three times larger than it is in part *b* of the drawing, since the number of field lines drawn through the identical surfaces is in the ratio of 3:1. Because Φ is directly proportional to B for a given area, the flux in part *a* is also three times larger than that in part *b*. Therefore, we can say that the magnetic flux is proportional to the number of magnetic field lines that passes through a surface.

The graphical interpretation of flux as being proportional to the number of magnetic field lines passing through a surface also applies when the surface is oriented at an angle with respect to **B**. For example, as the coil in Figure 22.8 is rotated from $\phi = 0°$ to 60° to 90°, the number of magnetic field lines passing through the surface (see the field lines in the regions shaded in blue) changes in the ratio of 8:4:0 or 2:1:0. The results of Example 3 show that the flux in the three orientations changes by the same ratio. Because the magnetic flux is proportional to the number of field lines passing through a surface, one often encounters phrases like "the flux that passes through a surface bounded by a loop of wire."

22.4 FARADAY'S LAW OF ELECTROMAGNETIC INDUCTION

Two scientists are given credit for the discovery of electromagnetic induction: the Englishman Michael Faraday (1791–1867) and the American Joseph Henry (1797–1878). Although Henry was the first to observe electromagnetic induction, Faraday investigated it in more detail and published his findings first. Consequently, the law that describes the phenomenon bears his name.

Faraday discovered that whenever there is a *change in flux* through a loop of wire, an emf is induced in the loop. A constant flux creates no emf. In fact, Faraday found that the magnitude of the induced emf is equal to the time rate of change of the magnetic flux. This is just the relation we obtained in Section 22.3 for the specific case of motional emf: $\mathscr{E} = -\Delta\Phi/\Delta t$.

Often the magnetic flux passes through a coil of wire containing more than one loop (or turn). If the coil consists of N loops, and if the same flux passes through each loop, it is found experimentally that the total induced emf is N times that

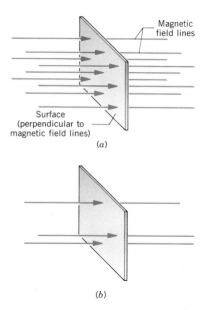

Figure 22.9 The magnitude of the magnetic field in (*a*) is three times greater than that in (*b*), because the number of magnetic field lines crossing the surfaces is in the ratio of 3:1.

induced in a single loop. An analogous situation occurs in a flashlight when two 1.5-V batteries are stacked in series on top of one another to give a total emf of 3.0 volts. For the general case of N loops, the total induced emf is described by *Faraday's law of electromagnetic induction* in the following manner.

Faraday's Law of Electromagnetic Induction

The average emf $\mathscr{E}$ induced in a coil of N loops is

$$\mathscr{E} = -N\left(\frac{\Phi - \Phi_0}{t - t_0}\right) = -N\frac{\Delta\Phi}{\Delta t} \tag{22.3}$$

where $\Delta\Phi$ is the average change in magnetic flux through one loop and Δt is the time interval during which the change occurs. The term $\Delta\Phi/\Delta t$ is the average time rate of change of the flux that passes through one loop.

SI Unit of Induced Emf: volt (V)

Faraday's law states that an emf is generated if the flux changes for any reason. Since the flux is given by Equation 22.2 as $\Phi = BA \cos \phi$, the flux depends on three factors, B, A, and ϕ, any of which may change. The examples in the remainder of this section illustrate how such changes can lead to an induced emf. Example 4 considers a changing magnetic field.

Example 4   The Emf Induced by a Changing Magnetic Field

A coil of wire consists of 20 turns, each of which has an area of 1.5×10^{-3} m². A magnetic field is perpendicular to the surface of each loop at all times, so that $\phi = \phi_0 = 0°$. At time $t_0 = 0$, the magnitude of the magnetic field at the location of the coil is $B_0 = 0.050$ T. At a later time of $t = 0.10$ s, the magnitude of the field at the coil has increased to $B = 0.060$ T. (a) Find the average emf induced in the coil during this time. (b) What would be the value of the induced emf if the magnitude of the magnetic field decreased from 0.060 T to 0.050 T in 0.10 s?

REASONING   To find the induced emf, we use Faraday's law of electromagnetic induction, combining it with the definition of magnetic flux from Equation 22.2.

SOLUTION
(a) Since $\phi = \phi_0$, the induced emf is

$$\mathscr{E} = -N\left(\frac{\Phi - \Phi_0}{t - t_0}\right) = -N\left(\frac{BA \cos \phi - B_0 A \cos \phi_0}{t - t_0}\right) = -NA \cos \phi \left(\frac{B - B_0}{t - t_0}\right)$$

We find that

$$\mathscr{E} = -(20)(1.5 \times 10^{-3}\text{ m}^2)(\cos 0°)\left(\frac{0.060\text{ T} - 0.050\text{ T}}{0.10\text{ s} - 0}\right) = \boxed{-3.0 \times 10^{-3}\text{ V}}$$

(b) The reasoning here is similar to that in part (a), except the initial and final values of B are interchanged. This interchange reverses the sign of the emf, so $\boxed{\mathscr{E} = +3.0 \times 10^{-3}\text{ V}}$. Because the polarity of the emf is reversed, the direction of the induced current is opposite to that in part (a).

PROBLEM SOLVING INSIGHT

The change in any quantity is the final value minus the initial value: e.g., the change in flux is $\Delta\Phi = \Phi - \Phi_0$ and the change in time is $\Delta t = t - t_0$.

The next example demonstrates that an emf can be created when a coil is rotated in a magnetic field.

Example 5   *Emf Induced in a Rotating Coil*

A flat coil of wire has an area of 0.020 m² and consists of 50 turns. At $t_0 = 0$ the coil is oriented so the normal to its surface is parallel ($\phi_0 = 0°$) to a constant magnetic field of magnitude 0.18 T. The coil is then rotated through an angle of $\phi = 30.0°$ in a time of 0.10 s (see Figure 22.3c). (a) Determine the average induced emf. (b) What would be the induced emf if the coil were returned to its initial orientation in the same time of 0.10 s?

REASONING AND SOLUTION
(a) Faraday's law yields

$$\mathscr{E} = -N\left(\frac{\Phi - \Phi_0}{t - t_0}\right) = -N\left(\frac{BA\cos\phi - BA\cos\phi_0}{t - t_0}\right) = -NBA\left(\frac{\cos\phi - \cos\phi_0}{t - t_0}\right)$$

$$\mathscr{E} = -(50)(0.18\text{ T})(0.020\text{ m}^2)\left(\frac{\cos 30.0° - \cos 0°}{0.10\text{ s} - 0}\right) = \boxed{+0.24\text{ V}}$$

(b) When the coil is rotated back to its initial orientation in a time of 0.10 s, the induced emf has the same magnitude, but opposite polarity, so $\boxed{\mathscr{E} = -0.24\text{ V}}$.

One application of Faraday's law that is found in the home is a safety device called a ground fault interrupter. This device protects against electric shock from an appliance, such as a clothes drier, and plugs directly into (or sometimes replaces) a wall socket, as in Figure 22.10. The interrupter consists of a circuit breaker that can be triggered to stop the current to the drier, depending on whether an induced voltage appears across a sensing coil. The sensing coil is wrapped around an iron ring, and the two wires carrying current to and from the drier pass through the ring. In the drawing, the current going to the drier is shown in red, while the returning current is shown in green. Each of the currents creates a

THE PHYSICS OF . . .

a ground fault interrupter.

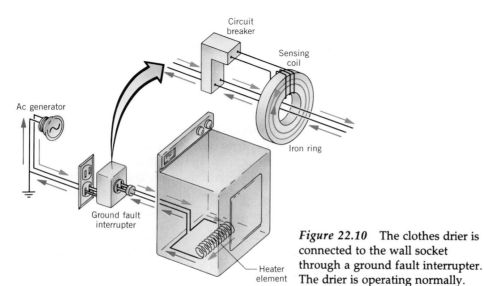

Figure 22.10   The clothes drier is connected to the wall socket through a ground fault interrupter. The drier is operating normally.

magnetic field that encircles the corresponding wire, according to RHR-2. However, the field lines have opposite directions since the currents have opposite directions. As the drawing shows, the iron ring guides the field lines through the sensing coil. Since the current is ac, the fields from the red and green current are changing. However, no induced voltage appears across the sensing coil, because the red and green field lines have opposite directions and the opposing fields cancel at all times. The net flux through the coil remains zero, and no emf is induced in the coil. Thus, when the drier operates normally, the circuit breaker is not triggered and does not shut down the current to the drier. The picture changes when the drier malfunctions, as when a wire inside the unit breaks and accidentally touches the metal case. When someone touches the case, some of the current begins to pass through the person's body and into the ground, returning to the source *without going through the return wire that passes through the ground fault interrupter.* As a result, the net magnetic field through the sensing coil is no longer zero and changes with time since the current is ac. The changing flux causes an induced voltage to appear across the sensing coil. This voltage triggers the circuit breaker to stop the current. Ground fault interrupters work very fast (in less than a millisecond) and can turn off the current before it reaches a dangerous level.

22.5 LENZ'S LAW

THE POLARITY OF THE INDUCED EMF

An induced emf drives current around a circuit just as the emf of a battery does. With a battery, conventional current is directed out of the positive terminal, through the attached device, and into the negative terminal. The same is true for an induced emf, although the location of the positive and negative terminals is generally not as obvious. Therefore, a method is needed for determining the polarity of the induced emf, so the terminals can be identified. As we discuss this method, it will be helpful to keep in mind that the net magnetic field penetrating a coil of wire is the combination of two contributions. One is the original magnetic field that produces the changing flux that leads to the induced emf. The other arises because of the induced current, which, like any current, creates its own magnetic field. The field created by the induced current is called the *induced magnetic field.*

To determine the polarity of the induced emf, we will use a method based on a discovery made by the Russian physicist Heinrich Lenz (1804–1865). He found that the polarity of an induced emf always leads to an induced current with the following characteristic: The direction of the induced current is such that the induced field either adds to or subtracts from the original field, whichever is necessary to help keep the flux from changing. This fact is known as *Lenz's law.*

Lenz's Law

The polarity of an induced emf is such that the induced current produces an induced magnetic field that opposes the change in flux causing the emf.

Lenz's law is best illustrated with examples. Each example is worked out according to the following procedure:

1. Determine whether the magnetic flux that penetrates a coil is increasing or decreasing.
2. Find what the direction of the induced magnetic field must be so that it can oppose the *change* in flux by adding to or subtracting from the original field.
3. Having found the direction of the induced magnetic field, use RHR-2 (see Section 21.8) to determine the direction of the induced current. Then the polarity of the induced emf can be assigned, because conventional current is directed out of the positive terminal, through the external circuit, and into the negative terminal.

Example 6   The Emf Produced by a Moving Magnet

Figure 22.11*a* shows a permanent magnet approaching a loop of wire. The external circuit attached to the loop consists of the resistance *R*, which could be the resistance of the filament in a light bulb, for instance. Find the direction of the induced current and the polarity of the induced emf.

REASONING   In solving this problem, we apply Lenz's law, the essence of which is that the change in magnetic flux must be opposed by the induced magnetic field.

SOLUTION   The magnetic flux through the loop is increasing, since the magnitude of the magnetic field at the loop is increasing as the magnet approaches. To oppose the increase in the flux, the direction of the induced magnetic field must be opposite to the field of the bar magnet. Since the field of the bar magnet passes through the loop from left to right in part *a* of the drawing, the induced field must pass through the loop from right to left, as in part *b*. To create such an induced field, the induced current must be directed *counterclockwise* around the loop, when viewed from the side nearest the magnet. (See the application of RHR-2 in the drawing.) The loop behaves as a source of emf, just like a battery, with the positive and negative terminals as shown in Figure 22.11*b*.

In Example 6 the direction of the induced magnetic field is opposite to the direction of the external field of the bar magnet. However, the induced field does not always have to be opposite to the external field, for Lenz's law requires only that it must oppose the *change* in the flux that generates the emf. Example 7 illustrates this point.

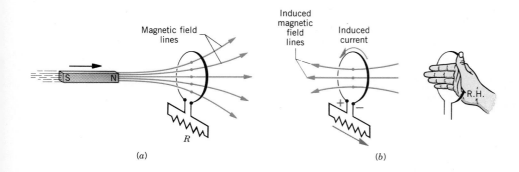

Figure 22.11 (*a*) As the magnet moves to the right, the magnetic flux through the loop increases. The external circuit attached to the loop has a resistance *R*. (*b*) The polarity of the induced emf is indicated by the + and − symbols.

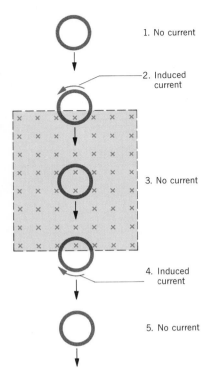

Figure 22.12 A copper ring passes through a rectangular region where a constant magnetic field is directed into the page. The picture shows that a current is induced in the ring at locations 2 and 4.

Example 7   *The Emf Produced by a Moving Copper Ring*

In Figure 22.12 there is a constant magnetic field in a rectangular region of space. This field is directed perpendicularly into the plane of the paper. Outside this region there is no magnetic field. A copper ring slides through the region, from position 1 to position 5. For each of the five positions, determine if an induced current exists in the ring and, if so, find the direction of the current.

REASONING   In applying Lenz's law here, we keep in mind that it does not require the induced magnetic field to be opposite to the external field. Rather, the induced field must oppose the change in flux, and sometimes this means that the induced field will reinforce the external field.

SOLUTION
Position 1: No flux passes through the ring, because the magnetic field is zero outside the rectangular region. Consequently, there is no change in flux and no induced emf or current.

Position 2: As the ring moves into the region of the magnetic field, the flux increases, and there is an induced emf and an induced current. To determine the direction of the current, we require that the induced magnetic field point opposite to the external field, so as to oppose the increase in the flux, in accord with Lenz's law. With the induced field pointing perpendicularly out of the plane of the paper, RHR-2 indicates that the direction of the induced current is counterclockwise, as the drawing indicates.

Position 3: Even though a flux passes through the moving ring, there is no induced emf or current, because the flux remains constant within the rectangular region. To induce an emf, it is not sufficient just to have a flux. The flux must *change* to generate an emf.

Position 4: As the ring leaves the magnetic field, the flux decreases. Once again, the induced magnetic field must be in such a direction so as to oppose this change. Since the change is a decrease in flux, the induced field must point in the *same* direction as the external field, namely, into the paper. With this orientation, the induced field increases the net magnetic field through the ring and thereby increases the flux. With the induced field pointing into the paper, RHR-2 indicates that the induced current is clockwise around the ring, opposite to what it was in position 2. By comparing the results for positions 2 and 4, it should be clear that the induced magnetic field does not always oppose the external field.

Position 5: As in position 1, there is no induced current, since the magnetic field is everywhere zero.

Lenz's law should not be thought of as an independent law, for it is a consequence of the law of conservation of energy. The connection between energy conservation and induced emf has already been discussed in Section 22.2 for the specific case of motional emf. However, the connection is valid for any type of induced emf. In fact, the polarity of the induced emf, as specified by Lenz's law, ensures that energy is conserved.

22.6 APPLICATIONS OF ELECTROMAGNETIC INDUCTION TO THE REPRODUCTION OF SOUND

Electromagnetic induction plays an important role in the technology used for the reproduction of sound. Figure 22.13 shows an audio system, to which an electric guitar, a cassette deck, and a microphone are connected. As we will now see, each

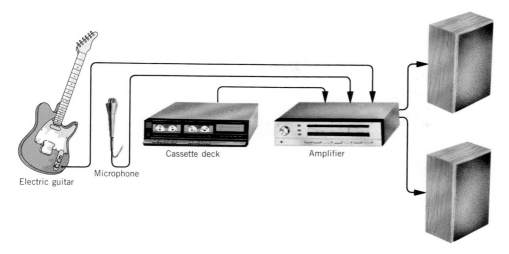

Figure 22.13   The operation of an electric guitar, a microphone, and a cassette deck is based on electromagnetic induction.

of these generates an induced emf. In general, however, the emf is rather small, so it is strengthened by an amplifier before being sent to the speakers.

THE ELECTRIC GUITAR PICKUP

Virtually all electric guitars use electromagnetic pickups in which an induced emf is generated in a coil of wire by a vibrating string. Most guitars have at least two pickups for each string and some, as Figure 22.14 illustrates, have three. These pickups are positioned at different locations under the string, so that each is sensitive to different harmonics produced by the vibrating string.

The guitar string is made from a magnetizable metal. The pickup itself consists of a coil of wire with a permanent magnet located inside the coil. The permanent magnet produces a magnetic field that penetrates the guitar string, causing it to become magnetized with north and south poles. When the magnetized string is plucked, it oscillates above the coil, thereby changing the magnetic flux that passes through the coil. The change in flux induces an emf in the coil. The polarity of the emf reverses with the vibratory motion of the string, so a string vibrating at 440 Hz, for example, induces a 440-Hz ac emf in the coil. This 440-Hz signal, after being amplified, is sent to the speakers, which produce a 440-Hz sound wave (concert A).

THE PHYSICS OF . . .

the electric guitar pickup.

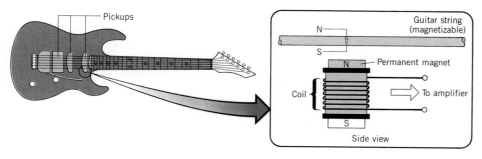

Figure 22.14   When the string of an electric guitar vibrates, an emf is induced in the coil of the pickup. The two ends of the coil are connected to the input of an amplifier.

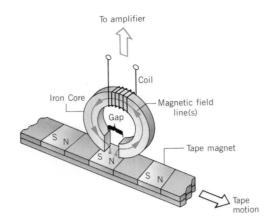

Figure 22.15 The magnetic playback head of a tape deck. As each tape magnet passes by the gap in the iron core, some of the magnetic field lines are routed through the core and the coil. The change in flux through the coil creates an induced emf. The width of the gap has been exaggerated.

THE PHYSICS OF . . .

the playback head of a tape deck.

THE PHYSICS OF . . .

the playback head of a tape deck.

THE PLAYBACK HEAD OF A TAPE DECK

The playback head of a cassette deck uses a moving tape to generate an emf in a coil of wire. Figure 22.15 shows a section of magnetized tape in which a series of "tape magnets" have been created in the magnetic layer of the tape during the recording process. The tape moves beneath the playback head, which consists of a coil of wire wrapped around an iron core. The iron core has the approximate shape of a horseshoe with a small gap between the two ends. The drawing shows an instant when a tape magnet is under the gap. Some of the magnetic field lines of the tape magnet are routed through the highly magnetizable iron core, and hence through the coil, as they proceed from the north pole to the south pole. Consequently, the flux through the coil changes as the tape moves past the gap. The change in flux leads to an ac emf, which is amplified and sent to the speakers, where the original sound is reproduced.

THE PHYSICS OF . . .

a moving coil and a moving magnet microphone.

THE MICROPHONE

There are a number of types of microphones, and Figure 22.16 illustrates the one known as a moving coil microphone. When a sound wave strikes the diaphragm of the microphone, the diaphragm vibrates back and forth, and the attached coil moves along with it. Nearby is a stationary magnet. As the coil alternately approaches and recedes from the magnet, the flux through the coil changes. Consequently, an ac emf is induced across the coil. This electrical signal is sent to an

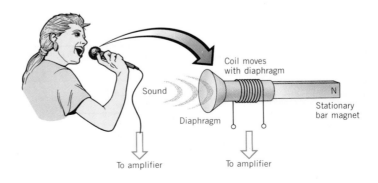

Figure 22.16 A moving coil microphone.

amplifier and then to the speakers. In the type of microphone called a moving magnet microphone, the magnet is attached to and moves along with the diaphragm. A nearby coil is fixed in place, and the flux through this coil changes as the magnet moves back and forth. Once again, an ac emf is induced, and sound is converted into electricity.

22.7 THE ELECTRIC GENERATOR

HOW A GENERATOR PRODUCES AN EMF

Electric generators are important because they produce virtually all the electrical energy consumed in the world. A generator produces electrical energy from mechanical work, just the opposite of what a motor does. In a motor, an *input* electric current causes a coil to rotate, thereby doing mechanical work on any object attached to the shaft of the motor. In a generator, the shaft is rotated by some mechanical means, such as by an engine or a turbine, and an emf is induced in a coil. If the generator is connected to an external circuit, an electric current is the *output* of the generator.

In its simplest form, an ac generator consists of a coil of wire that is rotated in a uniform magnetic field, as Figure 22.17a indicates. Although not shown in the picture, the wire is usually wound around an iron core, and, as in an electric

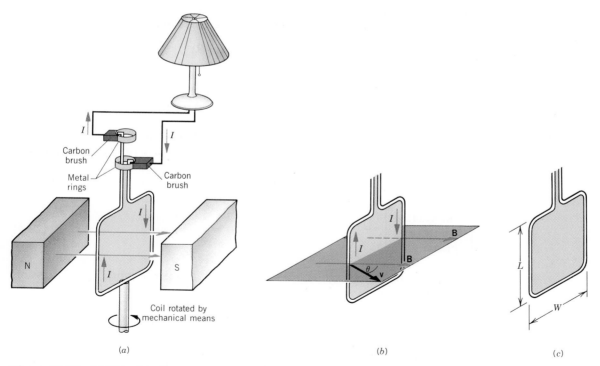

Figure 22.17 (a) This electric generator consists of a coil (only one loop is shown) of wire that is rotated in a magnetic field **B** by some mechanical means. (b) The current *I* arises because of the magnetic force exerted on the charges in the moving wire. (c) The dimensions of the coil.

motor, the coil/core combination is called the armature. Each end of the wire forming the coil is connected to the external circuit by means of a metal ring that rotates with the coil. Each ring slides against a stationary carbon brush, to which the external circuit (the lamp in the drawing) is connected.

To see how current is produced by the generator, consider the two vertical sides of the coil in Figure 22.17b. Since each is moving in a magnetic field **B,** the magnetic force exerted on the charges in the wire causes them to flow, thus creating a current. With the aid of RHR-1 (fingers of extended right hand point along **B,** thumb along the velocity **v,** palm pushes in the direction of the force on a positive charge), it can be seen that the direction of the current is from bottom to top in the left side and from top to bottom in the right side. Thus, charge flows around the loop. The upper and lower segments of the loop are also moving. However, the magnetic force on the charges in these segments can be ignored, because the force is toward the sides of the wire and not along the length. The emf generated in the coil results only from the magnetic force on the charges in the vertical sides.

The magnitude of the motional emf developed in a conductor moving through a magnetic field is given by Equation 22.1. To apply this expression to the left side of the coil, whose length is L (see Figure 22.17c), we need to use the velocity component $v_{\perp}$ that is perpendicular to **B.** Letting θ be the angle between **v** and **B,** it follows that $v_{\perp} = v \sin \theta$, and the emf can be written as

$$\mathscr{E} = BLv_{\perp} = BLv \sin \theta$$

The emf induced in the right side has the same magnitude and polarity as that for the left side, with the result that the emf for the complete loop is $\mathscr{E} = 2BLv \sin \theta$. If the coil consists of N loops, the net emf is N times greater than that of one loop, so

$$\mathscr{E} = N(2BLv \sin \theta)$$

It is convenient to express the variables v and θ in terms of the angular speed ω of the coil. Equation 8.2 shows that the angle θ is the product of the angular speed and the time, $\theta = \omega t$, if it is assumed that $\theta = 0$ when $t = 0$. Furthermore, any point on each vertical side moves on a circular path of radius $r = W/2$, where W is the width of the coil (see Figure 22.17c), so the tangential speed v of each side is related to the angular speed ω via Equation 8.9 as $v = r\omega = (W/2)\omega$. Substituting these expressions for θ and v in the equation above for $\mathscr{E}$, and recognizing that the product LW is the area A of the coil, we can write the induced emf as

$$\begin{bmatrix} \text{Emf induced} \\ \text{in a rotating} \\ \text{planar coil} \end{bmatrix} \qquad \mathscr{E} = NAB\omega \sin \omega t = \mathscr{E}_0 \sin \omega t \qquad (22.4)$$

In this equation, the angular speed ω is in radians per second. The expression shows that the emf varies sinusoidally with time. The peak, or maximum, emf $\mathscr{E}_0$ occurs when $\sin \omega t = 1$ and has the value $\mathscr{E}_0 = NAB\omega$. Although Equation 22.4 was derived for a rectangular coil, the result is valid for any planar shape of area A, such as a circle.

The emf of Equation 22.4 is plotted in Figure 22.18, which shows that the emf changes polarity as the coil rotates. This changing polarity is exactly the same as that discussed for an ac voltage in Section 20.5 and illustrated in Figure 20.10. If the external circuit connected to the generator is a closed circuit, an alternating current results that changes direction at the same rate as the emf changes polarity.

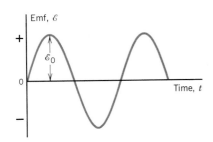

Figure 22.18 An ac generator produces an alternating emf $\mathscr{E}$ that varies as $\mathscr{E} = \mathscr{E}_0 \sin \omega t$.

Therefore, this electric generator is also called an *alternating current (ac) generator*. The next two examples show how Equation 22.4 is applied.

Example 8   An Ac Generator

In Figure 22.17 the armature of the ac generator rotates at a frequency of 60.0 Hz and develops an emf of 120 V(rms). The coil has an area of $A = 3.0 \times 10^{-3}$ m² and consists of $N = 500$ turns. Find the magnitude of the magnetic field in which the coil rotates.

REASONING   The magnetic field can be found from the relation $\mathcal{E}_0 = NAB\omega$. However, in using this equation we must remember that $\mathcal{E}_0$ is the peak emf, while the given quantity is not a peak value but an rms value of 120 V.

SOLUTION   The peak emf $\mathcal{E}_0$ is related to the rms emf by $\mathcal{E}_{rms} = \mathcal{E}_0/\sqrt{2}$ (Equation 20.13). Therefore, $\mathcal{E}_0 = \sqrt{2}\,\mathcal{E}_{rms} = \sqrt{2}\,(120\text{ V}) = 170$ V. Since one revolution corresponds to 2π radians, the angular speed of the coil is $\omega = 2\pi(60.0\text{ Hz}) = 377$ rad/s. The magnitude of the magnetic field is

$$B = \frac{\mathcal{E}_0}{NA\omega} = \frac{170\text{ V}}{(500)(3.0 \times 10^{-3}\text{ m}^2)(377\text{ rad/s})} = \boxed{0.30\text{ T}}$$

> **PROBLEM SOLVING INSIGHT**
>
> When using the equation $\mathcal{E}_0 = NAB\omega$, remember that the angular frequency ω must be expressed in rad/s, not in Hz. See Section 10.4 to review the relation between rad/s and Hz.

Example 9   A Bike Generator

A generator is mounted on a bicycle to power a headlight. A small wheel on the shaft of the generator is pressed against the bike tire and turns the armature 44 times for each revolution of the tire. The tire has a radius of 0.33 m. The armature has 75 turns, each with an area of 2.6×10^{-3} m², and rotates in a 0.10-T magnetic field. When the peak emf being generated is 6.0 V, what is the translational speed of the bike?

REASONING   The relation $\mathcal{E}_0 = NAB\omega$ provides a solution to this problem, if we take advantage of two additional facts. The first is that the angular speed ω of the armature is 44 times larger than the angular speed ω_{tire} of the bike tire. The second is that the tire is rolling, so that ω_{tire} is related to the translational speed v of the bike according to $\omega_{tire} = v/r$ (Equation 8.12), where r is the radius of the tire.

SOLUTION   Using $\omega = 44\omega_{tire} = 44(v/r)$, we find that the peak emf is $\mathcal{E}_0 = NAB \times 44(v/r)$. Therefore, the translational speed is

$$v = \frac{\mathcal{E}_0 r}{44NAB} = \frac{(6.0\text{ V})(0.33\text{ m})}{44(75)(2.6 \times 10^{-3}\text{ m}^2)(0.10\text{ T})} = \boxed{2.3\text{ m/s}}$$

> **THE PHYSICS OF . . .**
>
> a bike generator.

THE ELECTRICAL ENERGY DELIVERED BY A GENERATOR AND THE COUNTERTORQUE

Some power-generating stations burn fossil fuel (coal, gas, or oil) to heat water and produce pressurized steam for turning the blades of a turbine. Others use nuclear fuel or falling water as a source of energy. The shaft of the turbine is linked to that of the generator, so as the generator coil rotates, mechanical work is transformed into electrical energy.

The devices to which the generator supplies electricity are known collectively as the "load," because they place a burden or load on the generator by taking

Electrical generators, such as these supply electrical power to homes and industries. Faraday's law of electromagnetic induction explains how generators produce an induced emf.

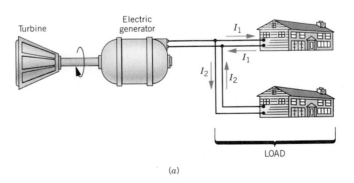

(a)

Figure 22.19 (a) Current is drawn when a load is connected to the generator. (b) The current $I = I_1 + I_2$ in the coil experiences a magnetic force **F** due to the magnetic field **B**. This magnetic force retards the motion of the coil.

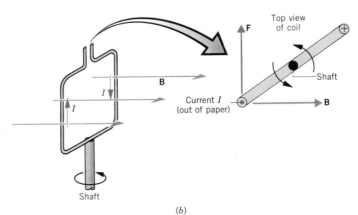

(b)

electrical energy from it. If all the devices are switched off, the generator runs under a no-load condition, because there is no current in the external circuit and the generator does not supply electrical energy. Then, the only work that the turbine does is to overcome friction and other mechanical losses within the generator itself, and fuel consumption is at a minimum.

Figure 22.19*a* illustrates a situation in which a load is connected to a generator. Because there is now a current $I = I_1 + I_2$ in the coil, and the coil is situated in a magnetic field, the current experiences a magnetic force **F**. Part *b* of the drawing shows the magnetic force acting on the left side of the coil, the direction of **F** being given by RHR-1. A force of equal magnitude but opposite direction acts on the right side of the coil, although this force is not shown in the drawing. The magnetic force **F** retards the motion of the coil. We encountered such a retarding force in Section 22.2 when discussing the motional emf of a rod sliding along two conducting rails (see Figure 22.5). In the ac generator, the magnetic force **F** gives rise to a *countertorque* that opposes the rotational motion. The greater the current drawn from the generator, the greater the countertorque, and the harder it is for the turbine to turn the coil. To compensate for this countertorque and to keep the coil rotating at a constant angular speed, work must be done on the coil by the turbine, which means more fuel must be burned. This is another example of the law of conservation of energy, since the electrical energy consumed by the load must ultimately come from the energy source used to drive the turbine.

THE BACK EMF GENERATED BY AN ELECTRIC MOTOR

A generator converts mechanical work into electrical energy; in contrast, an electric motor converts electrical energy into mechanical work. Both devices are similar and consist of a coil of wire that rotates in a magnetic field. In fact, as the armature of a motor rotates, the magnetic flux passing through the coil changes and an emf is induced in the coil. Thus, when a motor is operating, two sources of emf are present: (1) the applied emf V that supplies current to drive the motor (e.g., from a 120-V outlet), and (2) the emf $\mathscr{E}$ induced by the generator-like action of the rotating coil. The circuit diagram in Figure 22.20 shows these two emfs.

Consistent with Lenz's law, the induced emf $\mathscr{E}$ acts to oppose the applied emf V and is called the **back emf** or the **counter emf** of the motor. The greater the speed of the motor, the greater the flux change through the coil, and the greater is the back emf. Because V and $\mathscr{E}$ have opposite polarities, the net emf in the circuit is $V - \mathscr{E}$. If R in Figure 22.20 is the resistance of the wire in the coil, the current I

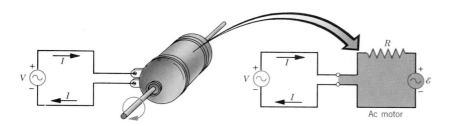

Figure 22.20 The applied emf V supplies the current I to drive the motor. The circuit shows V, along with the electrical equivalent of the motor, including the resistance R of its coil and the back emf $\mathscr{E}$.

drawn by the motor is determined from Ohm's law as the net emf divided by the resistance:

$$I = \frac{V - \mathcal{E}}{R} \tag{22.5}$$

The next example uses this result to illustrate that the current in a motor depends on both the applied emf V and the back emf $\mathcal{E}$.

Example 10 Operating a Motor

The coil of an ac motor has a resistance of $R = 4.1\ \Omega$. The motor is plugged into an outlet where $V = 120.0$ volts, and the coil develops a back emf of $\mathcal{E} = 118.0$ volts when rotating at normal speed. The motor is turning a wheel. Find (a) the current when the motor first starts up and (b) the current when the motor is operating at normal speed.

REASONING Once normal operating speed is attained, the motor need only work to compensate for frictional losses. But in bringing the wheel up to speed from rest, the motor must also do work to increase the wheel's rotational kinetic energy. Thus, bringing the wheel up to speed requires more work, and hence more current, than maintaining the normal operating speed. We expect our answers to parts (a) and (b) to reflect this fact.

SOLUTION
(a) When the motor just starts up, the coil is not rotating, so there is no back emf induced in the coil and $\mathcal{E} = 0$. The start-up current drawn by the motor is

$$I = \frac{V - \mathcal{E}}{R} = \frac{120.0\text{ V}}{4.1\ \Omega} = \boxed{29\text{ A}} \tag{22.5}$$

(b) At normal speed, the motor develops a back emf of $\mathcal{E} = 118.0$ volts, so the current is

$$I = \frac{V - \mathcal{E}}{R} = \frac{120.0\text{ V} - 118.0\text{ V}}{4.1\ \Omega} = \boxed{0.49\text{ A}}$$

PROBLEM SOLVING INSIGHT

The current in an electric motor depends on both the applied emf V and any back emf $\mathcal{E}$ developed because the coil of the motor is rotating.

Example 10 illustrates that when a motor is just starting, there is little back emf and, consequently, a relatively large current exists in the coil. As the motor speeds up, the back emf increases until it reaches a maximum value when the motor is rotating at normal speed. The back emf becomes almost equal to the applied emf, and the current is reduced to a relatively small value. This limiting value of the current is sufficient to provide the torque on the coil to drive the load (such as a fan) and to overcome frictional losses.

22.8 MUTUAL INDUCTANCE AND SELF-INDUCTANCE

MUTUAL INDUCTANCE

We have seen that an emf can be induced in a coil by keeping the coil stationary and moving a magnet nearby, or by moving the coil near a stationary magnet. Figure 22.21 illustrates another important method of inducing an emf. Here, two

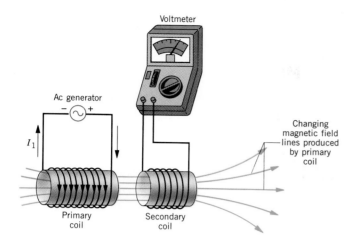

Voltmeter

Ac generator

I_1

Primary coil

Secondary coil

Changing magnetic field lines produced by primary coil

Figure 22.21 An alternating current I_1 in the primary coil creates an alternating magnetic field. This changing field induces an emf in the secondary coil.

coils of wire are placed close to each other. Coil 1 is connected to an ac generator that sends an alternating current I_1 through the coil. Coil 2 is not attached to a generator, although a voltmeter is connected between the ends of coil 2 to register any emf. It is customary to call coil 1 (the coil connected to the generator) the *primary coil* and coil 2 the *secondary coil.*

The current-carrying primary coil is an electromagnet and creates a magnetic field in the surrounding region. If the two coils are close to each other, a significant fraction of this magnetic field penetrates the secondary coil and produces a magnetic flux. The flux is changing in time, since the current in the primary coil and its associated magnetic field are changing in time. Because of the change in the flux, an emf is induced in the secondary coil.

The effect in which a changing current in one circuit induces an emf in another circuit is called *mutual induction.* According to Faraday's law of electromagnetic induction, the emf $\mathscr{E}_2$ induced in the secondary coil is proportional to the change in flux $\Delta\Phi_2$ passing through it. However, $\Delta\Phi_2$ is produced by the change in current ΔI_1 in the primary coil. Therefore, it is convenient to recast Faraday's law into a form that relates $\mathscr{E}_2$ to ΔI_1. To see how this recasting is accomplished, note that the net magnetic flux passing through the secondary coil is $N_2\Phi_2$, where N_2 is the number of loops in the secondary coil and Φ_2 is the flux through one loop (assumed to be the same for all loops). The net flux is proportional to the magnetic field, which, in turn, is proportional to the current I_1 in the primary. Thus, we can write $N_2\Phi_2 \propto I_1$. This proportionality can be converted into an equation in the usual manner by introducing a proportionality constant M, known as the *mutual inductance:*

$$N_2\Phi_2 = MI_1 \quad \text{or} \quad M = \frac{N_2\Phi_2}{I_1} \tag{22.6}$$

Substituting this equation into Faraday's law, we find that

$$\mathscr{E}_2 = -N_2\frac{\Delta\Phi_2}{\Delta t} = -\frac{\Delta(N_2\Phi_2)}{\Delta t} = -\frac{\Delta(MI_1)}{\Delta t} = -M\frac{\Delta I_1}{\Delta t}$$

$$\begin{bmatrix} \textbf{Emf due to} \\ \textbf{mutual} \\ \textbf{induction} \end{bmatrix} \qquad \mathscr{E}_2 = -M\frac{\Delta I_1}{\Delta t} \tag{22.7}$$

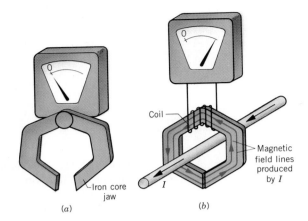

Figure 22.22 An induction ammeter with its iron-core jaw (*a*) open and (*b*) closed around a wire carrying an alternating current *I*. Some of the magnetic field lines that encircle the wire are routed through the coil by the iron core and lead to an induced emf. The meter detects the emf and is calibrated to display the amount of current in the wire.

Writing Faraday's law in this manner makes it clear that the emf $\mathscr{E}_2$ induced in the secondary coil is due to the change in the current ΔI_1 in the primary coil.

Equation 22.7 shows that the measurement unit for the mutual inductance M is V·s/A, which is called a henry (H) after Joseph Henry: 1 V·s/A = 1 H. The mutual inductance depends on the geometry of the coils and the nature of any ferromagnetic core material that is present. Although M can be calculated for some highly symmetrical arrangements, it is usually measured experimentally. In most situations, values of M are less than 1 H and are often on the order of millihenries (1 mH = 1×10^{-3} H) or microhenries (1 μH = 1×10^{-6} H).

Because mutual induction permits the transfer of electrical energy from one circuit to another without any physical contact between them, it has many applications. As an example, Figure 22.22*a* shows an induction ammeter, which is a device for measuring alternating current in situations where it would be too time-consuming or risky to disconnect a wire and insert a standard ammeter in the circuit. Suppose that you need to know whether a wire in a broken appliance carries a 60-Hz current, and, if so, how much. Part *b* of the figure indicates that the iron core "jaw" of the induction ammeter is slipped around the wire in question. The alternating current produces a changing magnetic field in the space around the wire. The iron core routes some of the field lines through a coil wrapped around the jaw of the ammeter, as part *b* of the drawing shows. The changing magnetic field induces an emf that is registered by the meter connected to the coil. Since the induced emf is proportional to the current in the appliance wire, the meter can be calibrated to read this current.

THE PHYSICS OF . . .

an induction ammeter.

SELF-INDUCTANCE

In all the examples of induced emfs presented so far, the magnetic field has been produced by an external source, such as a permanent magnet or an electromagnet. However, the magnetic field need not arise from an external source. An emf can be induced in a current-carrying coil by a change in the magnetic field that the current itself produces. For instance, Figure 22.23 shows a coil connected to an ac generator. The alternating current creates an alternating magnetic field that, in turn, creates a changing flux through the coil. The change in flux induces an emf in the coil, in accord with Faraday's law. The effect in which a changing current in a circuit induces an emf in the same circuit is referred to as *self-induction*.

When dealing with self-induction, as with mutual induction, it is customary to recast Faraday's law into a form in which the induced emf is proportional to the

Figure 22.23 The alternating current in the coil generates an alternating magnetic field that induces an emf in the coil.

change in current in the coil, rather than to the change in flux. If Φ is the magnetic flux that passes through one turn of the coil, then $N\Phi$ is the net flux through a coil of N turns. Since Φ is proportional to the magnetic field, and the magnetic field is proportional to the current I, it follows that $N\Phi \propto I$. By inserting a constant L, called the *self-inductance* or simply the *inductance* of the coil, we can convert this proportionality into Equation 22.8:

$$N\Phi = LI \quad \text{or} \quad L = \frac{N\Phi}{I} \tag{22.8}$$

Faraday's law of induction now gives the induced emf as

$$\mathscr{E} = -N\frac{\Delta\Phi}{\Delta t} = -\frac{\Delta(N\Phi)}{\Delta t} = -\frac{\Delta(LI)}{\Delta t} = -L\frac{\Delta I}{\Delta t}$$

$$\begin{bmatrix} \text{Emf due to} \\ \text{self-} \\ \text{induction} \end{bmatrix} \qquad \mathscr{E} = -L\frac{\Delta I}{\Delta t} \tag{22.9}$$

Like mutual inductance, L is measured in henries. The magnitude of L depends on the geometry of the coil and on the core material. By wrapping the coil around a ferromagnetic (iron) core, the magnetic flux—and therefore the inductance—can be increased substantially relative to that for an air core. Because of their self-inductance, coils are known as *inductors* and are widely used in electronics. Inductors come in all sizes, typically in the range between millihenries and microhenries. The next example shows how to determine the inductance of a solenoid.

Example 11   The Self-Inductance of a Solenoid

A solenoid of length $\ell = 8.0 \times 10^{-2}$ m and cross-sectional area $A = 5.0 \times 10^{-5}$ m^2 contains $n = 6500$ turns per unit length. (a) Find the self-inductance of the solenoid, assuming the core is air. (b) Determine the emf induced in the solenoid when the current increases from 0 to 1.5 A in a time of 0.20 s.

REASONING AND SOLUTION
(a) The self-inductance can be found by using Equation 22.8, $L = N\Phi/I$, provided the flux Φ can be determined. The flux is given by Equation 22.2 as $\Phi = BA \cos\phi$. In the case of a solenoid, the interior magnetic field is directed perpendicular to the plane of the loops, so $\phi = 0°$ and $\Phi = BA$. The magnetic field inside the solenoid has the value $B = \mu_0 nI$, according to Equation 21.7, where n is the number of turns per unit length.

$$L = \frac{N\Phi}{I} = \frac{N(BA)}{I} = \frac{N(\mu_0 nI)A}{I} = \mu_0 nNA = \mu_0 n^2 A\ell$$

where we have replaced N by $n\ell$. Substituting the given values into this result yields

$$L = \mu_0 n^2 A\ell = (4\pi \times 10^{-7} \text{ T·m/A})(6500 \text{ turns/m})^2$$

$$\times (5.0 \times 10^{-5} \text{ m}^2)(8.0 \times 10^{-2} \text{ m}) = \boxed{-2.1 \times 10^{-4} \text{ H}}$$

(b) The induced emf that results from the increasing current is

$$\mathscr{E} = -L\frac{\Delta I}{\Delta t} = -(2.1 \times 10^{-4} \text{ H})\left(\frac{1.5 \text{ A}}{0.20 \text{ s}}\right) = \boxed{-1.6 \times 10^{-3} \text{ V}} \tag{22.9}$$

The negative sign reminds us that the induced emf opposes the increasing current that induces the emf.

THE ENERGY STORED IN AN INDUCTOR

An inductor, like a capacitor, can store energy. This stored energy arises because a generator does work to establish a current in an inductor. Suppose an inductor is connected to a generator whose terminal voltage can be varied continuously from zero to some final value. As the voltage is increased, the current I in the circuit rises continuously from zero to its final value. While the current is rising, an induced emf $\mathscr{E} = -L(\Delta I/\Delta t)$ appears across the inductor. Conforming with Lenz's law, the polarity of the induced emf $\mathscr{E}$ is opposite to that of the generator voltage, so as to oppose the increase in the current. Thus, the generator must do work to push the charges through the inductor against this induced emf. The increment of work ΔW done by the generator in moving a small amount of charge ΔQ through the inductor is $\Delta W = (\Delta Q)\mathscr{E} = (\Delta Q)\,L\,(\Delta I/\Delta t)$, according to Equation 19.4. To ensure that the work done by the generator is positive, as it must be since the generator is driving charge against an opposing emf, the minus sign in front of the $L(\Delta I/\Delta t)$ term has been removed. Since $\Delta Q/\Delta t$ is the current I, the work done by the generator is

$$\Delta W = LI(\Delta I)$$

In this expression ΔW represents the work done by the generator to increase the current in the inductor by an amount ΔI. To determine the total work W done during the time interval when the current is changed from zero to its final value, all the small increments of work ΔW must be added together. This summation is left as an exercise at the end of this chapter (see problem 47). The result is $W = \frac{1}{2}LI^2$, where I represents the final current in the inductor. This work is stored as energy in the inductor, so that

$$\begin{bmatrix}\textbf{Energy stored} \\ \textbf{in an} \\ \textbf{inductor}\end{bmatrix} \qquad \text{Energy} = \tfrac{1}{2}\,LI^2 \qquad\qquad (22.10)$$

It is possible to regard the energy in an inductor as being stored in its magnetic field. For the special case of a long solenoid, Example 11 shows that the self-inductance is $L = \mu_0 n^2 A\ell$, where n is the number of turns per unit length, A is the cross-sectional area, and ℓ is the length of the solenoid. As a result, the energy stored in a solenoid is

$$\text{Energy} = \tfrac{1}{2}LI^2 = \tfrac{1}{2}\mu_0 n^2 A\ell I^2$$

Since $B = \mu_0 nI$ at the interior of a long solenoid (Equation 21.7), this energy can be expressed as

$$\text{Energy} = \frac{1}{2\mu_0}\, B^2 A\ell$$

The term $A\ell$ is the volume inside the solenoid in which the magnetic field exists, so the energy per unit volume or **energy density** is

$$\text{Energy density} = \frac{\text{Energy}}{\text{Volume}} = \frac{1}{2\mu_0}\, B^2 \qquad\qquad (22.11)$$

This result applies only to magnetic fields in air (or vacuum) or in nonmagnetic materials. Although it was obtained for the special case of a long solenoid, the relation is valid in general for any region of space where a magnetic field exists in air or vacuum. Thus, energy is stored in a magnetic field, just as it is in an electric field.

22.9 TRANSFORMERS

One of the most important applications of mutual induction and self-induction takes place in a transformer. A ***transformer*** is a device for increasing or decreasing an ac voltage. For example, whenever cordless appliances (e.g., a hand-held vacuum cleaner) are plugged into a wall receptacle to recharge the batteries, a transformer plays a role in reducing the 120-V ac voltage to a much smaller value. Typically, between 3 and 9 V are needed to energize batteries. In another instance, a picture tube in a television set needs about 15 000 V to accelerate the electron beam, and a transformer is used to obtain this high voltage from the 120 V provided at a wall socket.

Figure 22.24 shows a drawing of a transformer. The transformer consists of an iron core on which two coils are wound: a primary coil with N_p turns, and a secondary coil with N_s turns. The primary coil is connected to an ac generator. For the moment, suppose the switch in the secondary circuit is open, so there is no current in this circuit.

The alternating current in the primary coil establishes a changing magnetic field in the iron core. Because iron is easily magnetized, it greatly enhances the magnetic field relative to that in an air core and guides the field lines to the secondary coil. In a well-designed core, nearly all the magnetic flux Φ that passes through each turn of the primary also goes through each turn of the secondary. Since the magnetic field is changing, the flux through the primary and secondary coils is also changing, and consequently an emf is induced in both coils. In the secondary coil the induced emf $\mathscr{E}_s$ arises from mutual induction and is given by Faraday's law as

$$\mathscr{E}_s = -N_s \frac{\Delta \Phi}{\Delta t}$$

THE PHYSICS OF . . .

transformers.

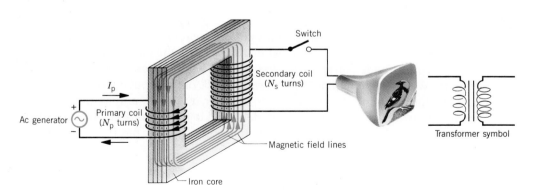

Figure 22.24 A transformer consists of a primary coil and a secondary coil, both wound on an iron core. The changing magnetic flux produced by the current in the primary coil induces an emf in the secondary coil. At the far right is the symbol for a transformer.

In the primary coil the induced emf $\mathscr{E}_p$ is due to self-induction and is specified by Faraday's law as

$$\mathscr{E}_p = -N_p \frac{\Delta\Phi}{\Delta t}$$

The term $\Delta\Phi/\Delta t$ is the same in both of these equations, since the same flux penetrates each turn of both coils. Dividing the two equations shows that

$$\frac{\mathscr{E}_s}{\mathscr{E}_p} = \frac{N_s}{N_p}$$

In a high-quality transformer the resistances of the coils are negligible, so the magnitudes of the emfs, $\mathscr{E}_s$ and $\mathscr{E}_p$, are nearly equal to the terminal voltages, V_s and V_p, across the coils (see Section 20.9 for a discussion of terminal voltage). The relation $\mathscr{E}_s/\mathscr{E}_p = N_s/N_p$ is called the *transformer equation* and is usually written in terms of the terminal voltages:

$$\begin{bmatrix} \textbf{Transformer} \\ \textbf{equation} \end{bmatrix} \qquad \frac{V_s}{V_p} = \frac{N_s}{N_p} \qquad (22.12)$$

Power distribution stations, such as this one at the World's Fair in Knoxville, Tennessee, use transformers (in red) to step-up or step-down voltages.

According to the transformer equation, if N_s is greater than N_p, the secondary (output) voltage is greater than the primary (input) voltage. In this case we have a *step-up* transformer. On the other hand, if N_s is less than N_p, the secondary voltage is less than the primary voltage, and we have a *step-down* transformer. The ratio N_s/N_p is referred to as the *turns ratio* of the transformer. A turns ratio of 8/1 (often written as 8 : 1) means, for example, that the secondary coil has eight times more turns than does the primary coil. Conversely, a turns ratio of 1 : 8 implies that the secondary coil has one-eighth as many turns as the primary coil.

A transformer operates with ac electricity and not with steady direct current. A steady direct current in the primary coil produces a flux that does not change, and thus no emf is induced in the secondary coil. The ease with which transformers can change voltages from one value to another is the principal reason why ac is preferred over dc.

If the switch in the secondary circuit of Figure 22.24 is closed, a current I_s exists in the circuit and electrical energy is fed to the TV tube. This energy comes from the ac generator connected to the primary coil. Although the secondary voltage V_s may be larger or smaller than the primary voltage V_p, energy is not being created or destroyed by the transformer. Energy conservation requires that the energy delivered to the secondary coil must be the same as the energy delivered to the primary coil, provided no energy is dissipated in heating these coils or is otherwise lost. In a well-designed transformer, less than 1% of the input energy is lost in the form of heat. Noting that power is energy per unit time, and assuming 100% energy transfer, the average power $\bar{P}_p$ delivered to the primary coil is equal to the average power $\bar{P}_s$ delivered to the secondary coil: $\bar{P}_p = \bar{P}_s$. But $P = IV$ (Equation 20.15a), so $I_p V_p = I_s V_s$, or

$$\frac{I_p}{I_s} = \frac{V_s}{V_p} = \frac{N_s}{N_p} \qquad (22.13)$$

Observe that V_s/V_p is equal to the turns ratio N_s/N_p, while I_s/I_p is equal to the inverse turns ratio N_p/N_s. Consequently, *a transformer that steps up the voltage simultaneously steps down the current, and a transformer that steps down the voltage steps up the current.* However, the power is neither stepped up nor stepped down, since $\overline{P}_p = \overline{P}_s$. Example 12 emphasizes this fact.

Example 12   A Step-Down Transformer

A step-down transformer inside a stereo receiver has 330 turns in the primary coil and 25 turns in the secondary coil. The plug connects the primary coil to a 120-V wall socket, and there is a current of 0.83 A in the primary coil while the receiver is turned on. Connected to the secondary coil are the transistor circuits of the receiver. Find (a) the voltage across the secondary coil, (b) the current in the secondary coil, and (c) the average electrical power delivered to the transistor circuits.

REASONING AND SOLUTION

(a) The voltage across the secondary coil can be found from the transformer equation:

$$V_s = V_p \frac{N_s}{N_p} = (120 \text{ V}) \left(\frac{25}{330} \right) = \boxed{9.1 \text{ V}}$$

(b) The current in the secondary coil follows from Equation 22.13 as

$$I_s = I_p \frac{N_p}{N_s} = (0.83 \text{ A}) \left(\frac{330}{25} \right) = \boxed{11 \text{ A}}$$

(c) The average power $\overline{P}_s$ delivered to the secondary is the product of I_s and V_s:

$$\overline{P}_s = I_s V_s = (11 \text{ A})(9.1 \text{ V}) = \boxed{1.0 \times 10^2 \text{ W}} \qquad (20.15a)$$

As a check on our calculation, we verify that the power delivered to the secondary coil is the same as that sent to the primary coil from the wall receptacle: $\overline{P}_p = I_p V_p = (0.83 \text{ A})(120 \text{ V}) = 1.0 \times 10^2 \text{ W}$.

Transformers play an important role in the transmission of power between electrical generating plants and the communities they serve. Whenever electricity is transmitted, there is always some loss of power in the transmission lines themselves due to resistive heating. Since the resistance of the wires is proportional to their length, the longer the wires the greater is the power loss. Power companies reduce this loss by using transformers that step up the voltage to high levels, while reducing the current. A smaller current means less power loss, since $P = I^2 R$, where R is the resistance of the transmission wires. Figure 22.25 shows one possible way of transmitting power. The power plant produces a voltage of 12 000 V. This voltage is then raised to 240 000 V by a 20 : 1 step-up transformer.

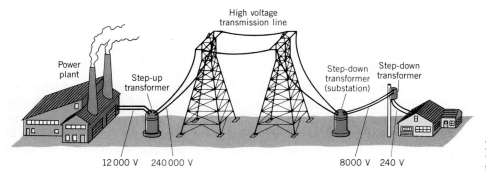

Figure 22.25   Transformers play a key role in the transmission of electric power.

The high-voltage power is sent over the long-distance transmission line. Upon arrival at the city, the voltage is reduced to about 8000 V at a substation using a 1 : 30 step-down transformer. However, before any domestic use, the voltage is further reduced to 240 V (or possibly 120 V) by another step-down transformer that is often mounted on a utility pole. The power is then distributed to consumers.

INTEGRATION OF CONCEPTS

ELECTROMAGNETIC INDUCTION AND ENERGY CONSERVATION

It could be said that electromagnetic induction is energy conservation in action. The conservation of energy principle stipulates that energy can only be transformed from one type to another; it cannot be created or destroyed. And energy transformation plays a large role in applications of electromagnetic induction. In an ac electrical generator, burning coal or oil, falling water, or nuclear fuel provides the energy to do the work of turning a coil in a magnetic field. As a result, electromagnetic induction converts the chemical energy of the coal or oil, the nuclear energy of uranium, or the gravitational potential energy of the water into electrical energy. In a small bicycle generator, the work of turning a coil in a magnetic field is done by the rider, and electromagnetic induction converts this work into electrical energy. In an electric guitar pickup, electromagnetic induction converts the energy of the vibrating guitar string into electrical energy, which is used to produce sound waves that carry acoustic energy. In a moving coil or moving magnet microphone, the energy in a sound wave causes a diaphragm to vibrate, and electromagnetic induction changes the acoustic energy into electrical energy. In a transformer, the electrical energy delivered to the primary coil is stored in the magnetic field of the coil. From there, it is transferred by electromagnetic induction to the secondary coil, where it appears again as electrical energy in the circuit connected to the secondary coil. Electromagnetic induction is indeed one of the most useful methods available for energy transformation.

SUMMARY

The **magnetic flux Φ** that passes through a surface is $\Phi = BA \cos \phi$, where A is the area of the surface, B is the magnitude of the magnetic field at the surface, and ϕ is the angle between $\mathbf{B}$ and the normal to the surface.

Electromagnetic induction is the phenomenon in which an emf is induced in a coil of wire by a change in the magnetic flux that passes through the coil.

Faraday's law of electromagnetic induction states that the average emf $\mathscr{E}$ induced in a coil of N loops is

$$\mathscr{E} = -N \left(\frac{\Phi - \Phi_0}{t - t_0} \right) = -N \frac{\Delta \Phi}{\Delta t}$$

where $\Delta \Phi$ is the average change in magnetic flux through one loop and Δt is the time interval during which the change occurs. For the special case of a conductor of length L moving with speed v perpendicular to a magnetic field $\mathbf{B}$, the induced emf is called motional emf and its value is given by $\mathscr{E} = vBL$.

Lenz's law provides a way to determine the polarity of an induced emf. Lenz's law states that the polarity of an induced emf is such that the induced current produces an induced magnetic field that opposes the change in flux causing the emf. Lenz's law is a consequence of the law of conservation of energy.

In its simplest form, an **electric generator** consists of a coil of N loops that rotates in a uniform magnetic field **B**. The emf produced by this generator is $\mathscr{E} = NAB\omega \times \sin \omega t = \mathscr{E}_0 \sin \omega t$, where A is the area of the coil, ω is the angular speed (in rad/s) of the coil, and $\mathscr{E}_0$ is the peak emf.

When an electric motor is running, it exhibits a generator-like behavior by producing an induced emf, called a **back emf.** The current I needed to keep the motor running at a constant speed is $I = (V - \mathscr{E})/R$, where V is the emf applied to the motor by an external source, $\mathscr{E}$ is the back emf, and R is the resistance of the coil.

Mutual induction is the effect in which a changing current in the primary coil induces an emf in the secondary coil. The emf $\mathscr{E}_2$ induced in the secondary coil by a change in current ΔI_1 in the primary coil is $\mathscr{E}_2 = -M(\Delta I_1/\Delta t)$, where Δt is the time interval during which the change occurs. The constant M is the **mutual inductance** between the two coils and is measured in henries (H).

Self-induction is the effect in which a change in current ΔI in a coil induces an emf $\mathscr{E} = -L(\Delta I/\Delta t)$ in the same coil. The constant L is the **self-inductance** or **inductance** of the coil and is measured in henries. To establish a current I in an inductor, work must be done by an external agent. This work is stored as energy in the inductor, the amount being Energy $= \frac{1}{2}LI^2$. The energy stored in an inductor can be regarded as being stored in its magnetic field. At any point in air or vacuum or in a nonmagnetic material where a magnetic field **B** exists, the **energy density,** or the energy stored per unit volume, is Energy density $= B^2/(2\mu_0)$.

A **transformer** consists of a primary coil of N_p turns and a secondary coil of N_s turns. When an emf $\mathscr{E}_p$ is applied to the primary, an emf $\mathscr{E}_s$ is induced in the secondary according to the relation $\mathscr{E}_s/\mathscr{E}_p = N_s/N_p$. A transformer functions with ac electricity, not with steady dc electricity. If the transformer is 100% efficient in transferring power from the primary coil to the secondary coil, the ratio of the primary current I_p to the secondary current I_s is $I_p/I_s = N_s/N_P$.

QUESTIONS

1. A uniform magnetic field points due east. A horizontal copper rod is perpendicular to this field and is oriented in the north–south direction. The rod falls freely to the earth. (a) Which end of the rod, north or south, becomes positively charged? (b) Which end of the rod becomes positively charged if the rod is initially oriented parallel to the magnetic field? Account for your answers.

2. In the discussion concerning Figure 22.5, we saw that a force of 0.086 N from an external agent was required to keep the rod moving at a constant speed. Suppose the light bulb in the figure is unscrewed from its socket. How much force would now be needed to keep the rod moving at a constant speed? Justify your answer.

3. Eddy currents are electric currents that can arise in a piece of metal when it moves through a region where the magnetic field is not the same everywhere. The picture shows, for example, a metal sheet moving to the right at a velocity **v** and a magnetic field **B** that is directed perpendicular to the sheet. At the instant represented, the magnetic field only extends over the left half of the sheet. An emf is induced that leads to the eddy current shown. Explain why this current causes the metal sheet to slow down. This action of eddy currents is used in various devices as a brake to damp out unwanted motion.

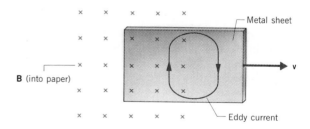

4. Suppose the magnetic flux through a 1-m^2 flat surface is known to be 2 Wb. From this data alone, is it possible to determine the average magnetic field at the surface? If it is not possible to determine the magnitude of the field, what can be ascertained about the field?

5. A square loop of wire is moving (but not rotating) through a uniform magnetic field. The normal to the loop is oriented parallel to the magnetic field. Is an emf induced in the loop? Give a reason for your answer.

6. Explain how a bolt of lightning can produce a current in the circuit of an electrical appliance, even when the lightning does not directly strike the appliance.

7. A robot is designed to move parallel to a cable hidden under the floor. The cable carries a steady direct current I. A sensor mounted on the robot consists of a coil of wire. The coil is near the floor and parallel to it. As long as the robot moves parallel to the cable, with the coil directly over it, no emf is induced in the coil, since the magnetic flux through the coil does not change. But when the robot deviates from the parallel path, an induced emf appears in the coil. The emf is sent to electronic circuits that bring the robot back to the path. Explain why an emf would be induced in the sensor coil.

8. In a car, the generator-like action of the alternator occurs while the engine is running and keeps the battery fully charged. The headlights would discharge an old and failing battery quickly if it were not for the alternator. Explain why the engine of a parked car runs more "quietly" with the headlights off than with them on when the battery is in bad shape.

9. In Figure 22.3b a coil of wire is being stretched. (a) Using Lenz's law, verify that the induced current in the coil has the direction shown in the drawing. (b) Deduce the direction of the induced current if the direction of the external magnetic field in the figure were reversed.

10. (a) When the switch in the circuit in the drawing is closed, a current is established in the coil and the metal ring

"jumps" upward. Explain this behavior. (b) Describe what would happen to the ring if the battery polarity were reversed.

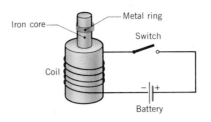

11. The string of an electric guitar vibrates in a standing wave pattern that consists of nodes and antinodes (Section 17.5 discusses standing waves). Where should an electromagnetic pickup be located in the standing wave pattern to produce a maximum emf, at a node or an antinode? Why?

12. An electric motor in a hair drier is running at normal speed and, thus, is drawing a relatively small current, as in part (b) of Example 10. What happens to the current drawn by the motor if the shaft is prevented from turning, so the back emf is suddenly reduced to zero? Remembering that the wire in the coil of the motor has some resistance, what happens to the temperature of the coil? Justify your answers.

13. Would a steady direct current in a wire register on the induction ammeter shown in Figure 22.22? Explain.

PROBLEMS

Section 22.2 Motional Emf

1. A spark can jump between two nontouching conductors if the potential difference between them is sufficiently large. Approximately, a potential difference of 940 V is required to produce a spark in an air gap of 1.0×10^{-4} m. Suppose the light bulb in Figure 22.4b is replaced by such a gap. How fast would a 1.6-m rod have to be moving in a magnetic field of 0.85 T to cause a spark to jump across the gap?

2. The wingspan (tip-to-tip) of a Boeing 747 jetliner is 59 m. The plane is flying horizontally at a speed of 220 m/s. The vertical component of the earth's magnetic field is 5.0×10^{-6} T. Find the emf induced between the wing tips.

3. Near San Francisco, where the vertically downward component of the earth's magnetic field is 4.8×10^{-5} T, a car is traveling at 25 m/s. An emf of 2.4×10^{-3} V is induced between the sides of the car. (a) Which side of the car is positive, the driver's side or the passenger's side? (b) What is the width of the car?

4. An emf of 0.35 V is generated between the ends of a metal bar moving through a magnetic field of 0.11 T, as in Figure 22.4a. What field strength would be needed to produce an emf of 1.5 V between the ends of the bar, assuming all other factors remain the same?

5. A metal rod (length = 0.75 m) moves perpendicular to a magnetic field of 0.15 T. An emf of 0.24 V exists between the ends of the rod. How far does the rod move in 7.0 s?

*6. Suppose the light bulb in Figure 22.4b is replaced by a 6.0-Ω electric heater that consumes 15 W of power. The conducting bar moves to the right at a constant speed, the field strength is 2.4 T, and the length of the bar between the rails is 1.2 m. (a) How fast is the bar moving? (b) What force must be applied to the bar to keep it moving to the right at the constant speed found in part (a)?

*7. Suppose the light bulb in Figure 22.4b is replaced with a short wire of zero resistance, and the resistance of the rails is negligible. The only resistance is from the moving rod, which is iron (resistivity = 9.7×10^{-8} Ω·m). The rod has a cross-sectional area of 3.1×10^{-6} m² and moves with a speed of 2.0 m/s. The magnetic field has a magnitude of 0.050 T. What is the current in the rod?

**8. A conducting rod slides down between two frictionless vertical copper tracks at a constant speed of 5.4 m/s perpendicular to a 0.30-T magnetic field (see the drawing). The resistance of the rod and tracks is negligible. The rod maintains electrical contact with the tracks at all times and has a length of 1.2 m. A 0.50-Ω resistor is attached between the tops of the tracks. (a) What is the mass of the rod? (b) Find the change in gravitational potential energy that occurs in a time of 0.20 s. (c) Find the electrical energy dissipated in the resistor in 0.20 s.

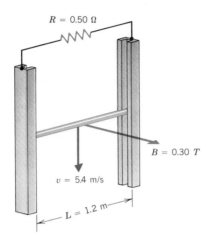

$R = 0.50\ \Omega$

$B = 0.30\ T$

$v = 5.4\ m/s$

$L = 1.2\ m$

Section 22.3 Magnetic Flux

9. A hand is held flat and placed in a uniform magnetic field of magnitude 0.35 T. The hand has an area of 0.0160 m² and negligible thickness. Determine the magnetic flux that passes through the hand when the normal to the hand is (a) parallel and (b) perpendicular to the magnetic field.

10. A magnetic field has a magnitude of 0.078 T and is uniform over a circular area that has a radius of 0.10 m. The field is oriented at an angle of $\phi = 25°$ with respect to the normal to the surface. What is the flux through the surface?

11. A rectangle (0.60 m × 0.30 m) lies in the xy plane. An identical rectangle lies in the xz plane. A uniform 0.17-T magnetic field points in the positive z direction. Find the flux through each rectangle.

12. A house has a floor area of 112 m² and an outside wall that has an area of 28 m². The earth's magnetic field here has a horizontal component of 2.6×10^{-5} T that points due north and a vertical component of 4.2×10^{-5} T that points straight down, toward the earth. Determine the magnetic flux through the wall if the wall faces (a) north and (b) east. (c) Calculate the magnetic flux that passes through the floor.

*13. A five-sided object, whose dimensions are shown in the drawing, is placed in a uniform magnetic field. The magnetic field has a magnitude of 0.25 T and points along the positive y direction. Determine the magnetic flux through each of the five sides.

Section 22.4 Faraday's Law of Electromagnetic Induction

14. A magnetic field is perpendicular to a 0.040-m × 0.060-m rectangular coil of wire that has one hundred turns.

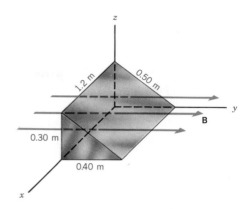

1.2 m   0.50 m

0.30 m

B

y

0.40 m

x

Problem 13

By how much must the field change so that a 1.5-V average emf is induced in the coil in a time of 0.050 s?

15. A circular loop of wire is placed in a uniform magnetic field that is parallel to the normal to the loop. The strength of the magnetic field is 3.0 T. The area of the loop begins shrinking at a constant rate of $\Delta A/\Delta t = 0.40$ m²/s. What is the magnitude of the emf induced in the loop while it is shrinking?

16. A circular coil (950 turns, radius = 0.060 m) is rotating in a uniform magnetic field. At $t = 0$, the normal to the coil is perpendicular to the magnetic field. At $t = 0.010$ s, the normal makes an angle of 45° with the field, because the coil has made one-eighth of a revolution. An average emf of magnitude 0.065 V is induced in the coil. Find the magnitude of the magnetic field at the location of the coil.

17. A 75-turn conducting coil has an area of 8.5×10^{-3} m² and the normal to the coil is parallel to a magnetic field $\mathbf{B}$. The coil has a resistance of 14 Ω. At what rate (in T/s) must the magnitude of $\mathbf{B}$ change for an induced current of 7.0 mA to exist in the coil?

18. Magnetic resonance imaging (MRI) is a medical technique for producing "pictures" of the body interior. The patient is placed within a strong magnetic field. One safety concern is what would happen to the positively and negatively charged particles in the body fluids if an equipment failure caused the magnetic field to be shut off suddenly. An induced emf could cause these particles to flow, producing an electric current within the body. Suppose the largest surface of the body through which flux passes has an area of 0.032 m² and a normal that is parallel to a magnetic field of 1.5 T. Determine the smallest time period during which the field can be allowed to vanish if the induced emf is to be kept less than 0.010 V.

19. A 1.8-m-long aluminum rod is rotating about an axis that is perpendicular to one end. A 0.27-T magnetic field is directed parallel to the axis. The rod rotates through one-fourth of a circle in 2.0 s. What is the magnitude of the average emf generated between the ends of the rod during this time?

*20. A copper rod is sliding on two conducting rails that

make an angle of 15° with respect to each other, as in the drawing. The rod is moving to the right with a constant speed of 0.40 m/s. A 0.42-T uniform magnetic field is perpendicular to the plane of the paper. Determine the magnitude of the average emf induced in the triangle *ABC* during the 5.0-s period after the rod has passed point *A*.

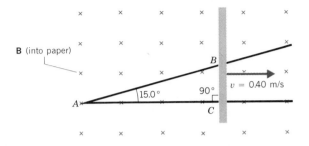

*21. A conducting coil of 1850 turns is connected to a galvanometer, and the total resistance of the circuit is 45.0 Ω. The area of each turn is 4.70×10^{-4} m². This coil is moved from a region where the magnetic field is zero into a region where it is nonzero, the normal to the coil being kept parallel to the magnetic field. The amount of charge that is induced to flow around the circuit is measured to be 8.87×10^{-3} C. Find the magnitude of the magnetic field. (Such a device can be used to measure the magnetic field strength and is called a *flux meter*.)

**22. Two 0.50-m-long conducting rods are rotating at the same speed in opposite directions, and both are perpendicular to a 4.0-T magnetic field. As the drawing shows, the ends of these rods come to within 1.0 mm of each other as they rotate. Moreover, the fixed ends about which the rods are rotating are connected by a wire, so these ends are at the same electric potential. If a potential difference of 4.5×10^3 V is required to cause a 1.0-mm spark in air, what is the angular speed (in rad/s) of the rods when a spark jumps across the gap?

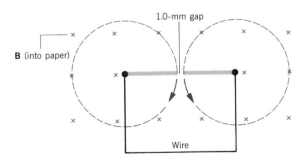

Section 22.5 Lenz's Law

23. In Figure 22.1, suppose the north and south poles of the magnet were interchanged. Determine the direction of the current through the ammeter in parts *b* and *c* of the picture (left-to-right or right-to-left). Give your rationale.

24. What is the direction of the induced current through *R*

in the drawing as the current *I* decreases to zero? Provide a reason for your answer.

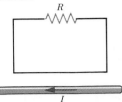

25. As the picture shows, a loop of copper wire is lying flat on a table and is attached to a battery via a switch. The current *I* in the loop establishes the magnetic field lines shown in color. There are also two smaller conducting loops *A* and *B* lying flat on the table, but not connected to batteries. Determine the direction of the induced current in loops *A* and *B* when the switch is (a) opened and (b) closed again. Specify the direction of the currents to be clockwise or counterclockwise when viewed from above the table.

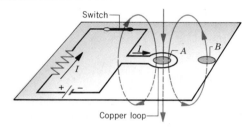

26. The drawing shows that a uniform magnetic field is directed perpendicularly out of the plane of the paper and fills the entire region to the left of the *y* axis. There is no magnetic field to the right of the *y* axis. A rigid right triangle *ABC* is made of copper wire. The triangle rotates counterclockwise about the origin at point *C*. What is the direction (clockwise or counterclockwise) of the induced current when the triangle is crossing (a) the $+y$ axis, (b) the $-x$ axis, (c) the $-y$ axis, and (d) the $+x$ axis? For each case, justify your answer.

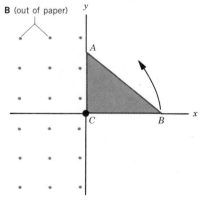

27. A long, straight wire lies on a table and carries a current *I*. As the drawing shows, a small circular loop of wire is pushed across the top of the table from position 1 to position 2. Determine the direction of the induced current, clockwise or counterclockwise, as the loop moves past each of the positions. Justify your answers.

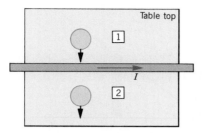

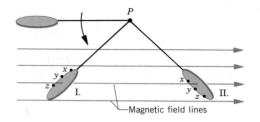

Magnetic field lines

*28. Indicate the direction of the electric field between the plates of the parallel plate capacitor shown in the drawing if the magnetic field is decreasing in time. Give your reasoning.

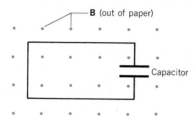

*29. The drawing shows a bar magnet falling through a metal ring. In part *a* the ring is solid all the way around, but in part *b* it has been cut through. (a) Explain why the motion of the magnet in part *a* is retarded when the magnet is above the ring and below the ring as well. Draw any induced currents that appear in the ring. (b) Explain why the motion of the magnet is unaffected by the ring in part *b*.

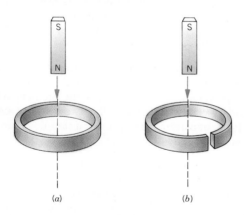

**30. A wire loop is suspended from a string that is attached to point *P* in the drawing. When released, the loop swings downward, from left to right, through a uniform magnetic field, with the plane of the loop remaining perpendicular to the plane of the paper at all times. (a) Determine the direction of the current induced in the loop as it swings past the locations labeled I and II. Specify the direction of the current in terms of the points *x*, *y*, and *z* on the loop (e.g., $x \rightarrow y \rightarrow z$ or $z \rightarrow y \rightarrow x$). The points *x*, *y*, and *z* lie behind the plane of the paper. (b) What is the direction of the induced current at the locations II and I when the loop swings back, from right to left? Provide reasons for your answers.

Section 22.7 The Electric Generator

31. A 200-turn rectangular coil has a cross-sectional area of 0.040 m^2. The coil is rotating at an angular speed of 15 rad/s about an axis that is perpendicular to a magnetic field of 1.5 T. Plot one cycle of the induced emf as a function of time, including on the graph numerical values for the maximum emf and the period.

32. You are requested to design a 60.0-Hz ac generator whose maximum emf is to be 5500 V. The generator is to contain a 150-turn coil whose area is 0.85 m^2. What should be the magnitude of the magnetic field in which the coil rotates?

33. The maximum strength of the earth's magnetic field is about 7.0×10^{-5} T near the south magnetic pole. In principle, this field could be used with a rotating coil to generate 60.0-Hz ac electricity. What is the minimum number of turns (area per turn = 0.016 m^2) that the coil must have so as to produce an rms voltage of 120 V?

34. A generator has a square coil consisting of 248 turns. The coil rotates at 79.1 rad/s in a 0.170-T magnetic field. The peak output of the generator is 75.0 V. What is the length of one side of the coil?

35. The current in the electric motor of a vacuum cleaner is 2.0 A when the cleaner is plugged into a 120.0-V receptacle and is running at normal speed. The coil resistance of the motor is 24 Ω. Find the back emf generated by the motor.

36. A generator produces a peak emf of 12.0 V when the armature rotates at 750 rev/min. What is the peak emf when the armature rotates at 2250 rev/min, assuming everything else remains the same?

*37. The coil of a generator has a radius of 0.14 m. When this coil is unwound, the wire from which it is made has a length of 5.7 m. The magnetic field of the generator is 0.20 T, and the coil rotates at an angular speed of 25 rad/s. What is the peak emf of this generator?

*38. At its normal operating speed, an electric fan motor draws a current of 1.2 A when plugged into a wall socket that provides 120.0 V. However, when the motor just begins to turn the fan blade, it draws a current of 6.0 A. What back emf does the motor generate at its normal operating speed?

**39. A motor is designed to operate on 117 V and draws a current of 12.2 A when it first starts up. At its normal operating speed, the motor draws a current of 2.30 A. Obtain (a) the resistance of the armature coil, (b) the back emf developed at normal speed, and (c) the current drawn by the motor at one-third normal speed.

Section 22.8 Mutual Inductance and Self-Inductance

40. The mutual inductance between two coils is $M = 8.0$ mH. The current in the primary coil changes at a constant rate from 2.0 to 5.5 A in 0.020 s. Determine the magnitude of the average emf induced in the secondary coil.

41. Mutual induction can be used as the basis for a metal detector. A typical setup uses two large coils that are parallel to each other and have a common axis. Because of mutual induction, the ac generator connected to the primary coil causes an emf of 0.46 V to be induced in the secondary coil. When someone without metal objects walks through the coils, the mutual inductance and, thus, the induced emf do not change much. But when a person carrying a handgun walks through, the mutual inductance increases. If the mutual inductance increases by a factor of three, find the new value of the induced emf. The change in emf can be used to trigger an alarm.

42. A coil consists of 275 turns and has a self-inductance of 0.0150 H. The coil carries a current of 0.0170 A. Obtain the magnetic flux through one turn of the coil.

43. Two coils have a mutual inductance of 4.0 mH. In the primary coil the current changes by 3.6 A in 0.030 s. The circuit that contains the secondary coil has a resistance of 1.5 Ω. Find the average current induced in the secondary coil.

44. How much energy is stored in a 0.085-H inductor that carries a current of 2.5 A?

45. The earth's magnetic field, like any magnetic field, stores energy. The maximum strength of the earth's field is about 7.0×10^{-5} T. Find the maximum magnetic energy stored in the space above a city if the space occupies an area of 5.0×10^{8} m^2 and has a height of 1500 m.

*46. A long, current-carrying solenoid with an air core has 1750 turns per meter of length and a radius of 0.0180 m. A coil of 125 turns is wrapped tightly around the outside of the solenoid. What is the mutual inductance of this system?

*47. The purpose of this problem is to show that the work W needed to establish a final current I_f in an inductor is $W = \frac{1}{2}LI_f^2$ (Equation 22.10). In Section 22.8 we saw that the amount of work ΔW needed to change the current through an inductor by an amount ΔI is $\Delta W = LI(\Delta I)$, where L is the inductance. The drawing shows a graph of LI versus I. Notice that $LI(\Delta I)$ is the area of the shaded vertical rectangle whose height is LI and whose width is ΔI. Use this fact to show that the total work W needed to establish a current I_f is $W = \frac{1}{2}LI_f^2$.

Section 22.9 Transformers

48. The batteries in a portable CD player are recharged by a unit that plugs into a wall socket. Inside the unit is a step-down

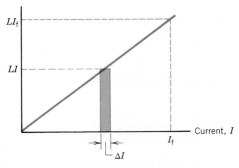

Problem 47

transformer with a turns ratio of 1:13. The wall socket provides 120 V. What voltage does the secondary coil of the transformer provide?

49. In some parts of the country, insect "zappers," with their blue lights, are a familiar sight on a summer's night. These devices use a high voltage to electrocute insects. One such device uses an ac voltage of 4150 V, which is obtained from a standard 120.0-V outlet by means of a transformer. If the primary coil has 17 turns, how many turns are in the secondary coil?

50. Electric doorbells found in many homes require 10.0 V to operate. To obtain this voltage from the standard 120-V supply, a transformer is used. Is a step-up or a step-down transformer needed, and what is its turns ratio?

51. A step-down transformer (turns ratio = 1:8) is used with an electric train to reduce the voltage from the wall receptacle to a value needed to operate the train. When the train is running, the current in the secondary coil is 3.4 A. What is the current in the primary coil?

52. The input to the primary coil of a transformer is 120 V, while the current in the secondary coil is 0.10 A. (a) When 60.0 W of power are being delivered to the circuit attached to the secondary coil, what is the voltage across the secondary coil? (b) Is the transformer a step-up or a step-down unit, and what is its turns ratio?

53. The secondary coil of a transformer provides the voltage that operates an electrostatic air filter. The turns ratio of the transformer is 43:1. The primary coil is plugged into a standard 120-V outlet. The current in the secondary coil is 1.5×10^{-3} A. Find the power consumed by the air filter.

*54. A generating station is producing 1.2×10^{6} W of power that is to be sent to a small town located 7.0 km away. Each of the two wires that comprise the transmission line has a resistance per unit of length of 5.0×10^{-2} Ω/km. (a) Find the power lost in heating the wires if the power is transmitted at 1200 V. (b) A 100:1 step-up transformer is used to raise the voltage before the power is transmitted. How much power is now lost in heating the wires?

**55. A generator is connected across the primary coil (N_p turns) of a transformer, while a resistance R_2 is connected across the secondary coil (N_s turns). This circuit is equivalent to a circuit in which a single resistance R_1 is connected directly across the generator, without the transformer. Show that $R_1 = (N_p/N_s)^2 R_2$, by starting with Ohm's law as applied to the secondary coil.

ADDITIONAL PROBLEMS

56. A 300-turn rectangular loop of wire has an area of 5.0×10^{-3} m². At $t_0 = 0$ a magnetic field is turned on, and its magnitude increases to 0.40 T when $t = 0.80$ s. The field is directed at an angle of 30.0° with respect to the normal of the loop. (a) Find the magnitude of the average emf induced in the loop. (b) If the loop is a closed circuit whose resistance is 6.0 Ω, determine the average induced current.

57. One generator uses a magnetic field of 0.10 T and has a coil area of 0.045 m². A second generator has a coil area of 0.015 m². The generator coils have the same number of turns and rotate at the same angular speed. What magnetic field should be used in the second generator, so that its peak emf is the same as that of the first generator?

58. Suppose in Figure 22.1 that the bar magnet is held stationary, but the coil of wire is free to move. Which way will current be directed through the ammeter, left-to-right or right-to-left, when the coil is moved (a) to the left and (b) to the right? Explain.

59. The coil of an electromagnet carries a steady direct current of 8.0 A and has a self-inductance of 0.150 H. Suddenly a switch is opened and the current decreases to zero in 7.0×10^{-3} s. Obtain the magnitude of the average emf induced in the coil during this time.

60. The resistances of the primary and secondary coils of a transformer are 56 and 14 Ω, respectively. Both coils are made from lengths of the same copper wire. The circular turns of each coil have the same diameter. Find the turns ratio N_s/N_p.

61. The back emf in a motor is 115 V when the motor is turning at 1800 rev/min. What is the back emf when the motor turns at 3600 rev/min, assuming all other factors remain the same?

*62. A magnetic field has a magnitude of 12 T. What is the magnitude of an electric field that stores the same energy per unit volume as this magnetic field?

*63. The armature of an electric drill motor has a resistance of 15.0 Ω. When connected to a 120.0-V outlet, the motor rotates at its normal speed and develops a back emf of 108 V. (a) What is the current through the motor? (b) If the armature "freezes up" due to a lack of lubrication in the bearings and can no longer rotate, what is the current in the stationary armature? (c) What is the current when the motor runs at only half speed?

*64. A large circular loop carries a current I. A much smaller circular loop is held above the center of the large loop, with the planes of the loops parallel. The small loop is released and falls downward through the large loop, all the while maintaining its parallel orientation. The center of the small loop remains in line with the center of the large loop at all times. Is the direction of the current induced in the small loop the same as I or opposite to I when (a) the small loop is above the large loop and (b) the small loop has fallen below the large loop? *(Hint: With the aid of Figure 21.28, first identify the direction of the magnetic field along the axis of the large loop.)* Justify your answers.

*65. A magnetic field is passing through a loop of wire whose area is 0.018 m². The direction of the magnetic field is parallel to the normal to the loop, and the magnitude of the field is increasing at the rate of 0.20 T/s. (a) Determine the magnitude of the emf induced in the loop. (b) Suppose the area of the loop can be enlarged or shrunk. If the magnetic field is increasing as in part (a), at what rate (in m²/s) should the area be changed at the instant when $B = 1.8$ T if the induced emf is to be zero? Explain whether the area is to be enlarged or shrunk.

**66. Coil 1 is a flat circular coil that has N_1 turns and a radius R_1. At its center is a much smaller flat, circular coil that has N_2 turns and radius R_2. The planes of the coils are parallel. Assume coil 2 is so small that the magnetic field at its location due to coil 1 is nearly constant. Determine an expression for the mutual inductance between these two coils in terms of μ_0, N_1, R_1, N_2, and R_2.

**67. The drawing shows a copper wire (negligible resistance) bent into a circular shape with a radius of 0.50 m. The radial section BC is fixed in place, while the copper bar AC sweeps around at an angular speed of 15 rad/s. The bar makes electrical contact with the wire at all times. A uniform magnetic field exists everywhere, is perpendicular to the plane of the circle, and has a magnitude of 3.8×10^{-3} T. Find the magnitude of the current induced in the loop ABC.

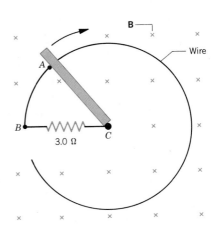

CHAPTER 23

ALTERNATING CURRENT CIRCUITS

The applications of alternating current (ac) circuits are so widespread that it is difficult to imagine living without them. The sound produced by the pianist in this photograph, for instance, is picked up by microphones placed around the piano. The ac voltage from each microphone is fed into the audio mixing console in the foreground of the picture. The console contains specialized circuits that mix the voltages in the desired proportions to produce a master recording of the performance. We use ac circuits in every room of our houses: lights, television, microwave oven, refrigerator, heating system, toaster, and hair dryer—the list goes on and on. In such applications, the basic elements of ac electricity are frequency, rms-voltage, rms-current, and power, as we have already discussed in Section 20.5. However, our discussion there focused on circuits that contain only resistors. In the present chapter, we deal with a number of important additional circuit components, including capacitors, inductors, diodes, and transistors, that make ac electricity vastly more useful.

23.1 CAPACITORS AND CAPACITIVE REACTANCE

Our experience with capacitors so far has been in dc circuits. As we have seen in Section 20.13, charge flows in a dc circuit only for the brief period after the battery voltage is applied across the capacitor. In other words, charge flows only while the capacitor is charging up. After the capacitor becomes fully charged, no more charge leaves the battery. However, suppose the battery connections to the fully charged capacitor were suddenly reversed, with the positive terminal being connected to the negative plate and the negative terminal being connected to the positive plate. Then charge would flow again, but in the reverse direction, until the battery recharged the capacitor according to the new connections. This is similar to what happens in an ac circuit, where the polarity of the voltage applied to the capacitor continually switches back and forth, and, in response, charges flow first one way around the circuit and then the other way. This flow of charge, surging back and forth, constitutes an alternating current. Thus, charge flows continuously in an ac circuit containing a capacitor.

To help set the stage for the present discussion, recall that $V_{rms} = I_{rms}R$ for a purely resistive ac circuit. The resistance R has the same value for any frequency. Figure 23.1 emphasizes this fact by showing that a graph of resistance versus frequency is a horizontal straight line.

For the rms-voltage across a capacitor the following expression applies, which is analogous to $V_{rms} = I_{rms}R$:

$$V_{rms} = I_{rms}X_C \qquad (23.1)$$

The term X_C appears in place of the resistance R and is called the **capacitive reactance.** The capacitive reactance, like resistance, is measured in *ohms* and determines how much rms-current exists in a capacitor in response to a given rms-voltage across the capacitor. It is found experimentally that the capacitive reactance X_C is inversely proportional to both the frequency f and the capacitance C, according to the following equation:

$$X_C = \frac{1}{2\pi f C} \qquad (23.2)$$

For a fixed value of the capacitance C, Figure 23.2 gives a plot of X_C versus frequency, according to Equation 23.2. A comparison of this drawing with Figure 23.1 reveals that a capacitor behaves differently than a resistor. As the frequency becomes very large, Figure 23.2 shows that X_C approaches zero, signifying that a capacitor offers only a negligibly small opposition to the alternating current. In contrast, in the limit of zero frequency (i.e., dc current), X_C becomes infinitely large, and a capacitor provides so much opposition to the motion of charges that there is no current. Example 1 illustrates how frequency and capacitance determine the amount of current in an ac circuit.

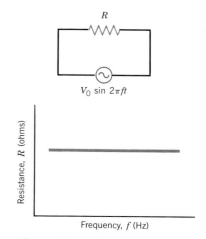

Figure 23.1 The resistance in a purely resistive circuit has the same value at all frequencies. The maximum emf of the generator is V_0.

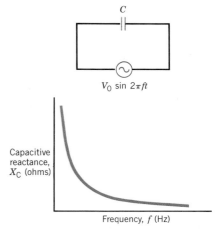

Figure 23.2 The capacitive reactance X_C is inversely proportional to the frequency f according to $X_C = 1/(2\pi f C)$.

Example 1   A Capacitor in an Ac Circuit

For the circuit in Figure 23.2, the capacitance of the capacitor is 1.50 μF and the rms-voltage of the generator is 25.0 V. What is the rms-current in the circuit when the frequency of the generator is (a) 1.00×10^2 Hz and (b) 5.00×10^3 Hz?

REASONING   The current can be found from $I_{rms} = V_{rms}/X_C$, once the capacitive reactance is determined. The values for the capacitive reactance will reflect the fact that the capacitor provides more opposition to the current when the frequency is smaller.

SOLUTION
(a) At a frequency of 1.00×10^2 Hz, we find

$$X_C = \frac{1}{2\pi f C} = \frac{1}{2\pi(1.00 \times 10^2 \text{ Hz})(1.50 \times 10^{-6} \text{ F})} = 1060 \ \Omega \qquad (23.2)$$

$$I_{rms} = \frac{V_{rms}}{X_C} = \frac{25.0 \text{ V}}{1060 \ \Omega} = \boxed{0.0236 \text{ A}} \qquad (23.1)$$

(b) When the frequency is 5.00×10^3 Hz, the calculations are similar:

$$X_C = \frac{1}{2\pi f C} = \frac{1}{2\pi(5.00 \times 10^3 \text{ Hz})(1.50 \times 10^{-6} \text{ F})} = 21.2 \ \Omega$$

$$I_{rms} = \frac{V_{rms}}{X_C} = \frac{25.0 \text{ V}}{21.2 \ \Omega} = \boxed{1.18 \text{ A}}$$

We now consider the behavior of the instantaneous (not rms) voltage and current. For comparison, Figure 23.3 shows graphs of voltage and current versus time in a resistive circuit. These graphs indicate that, when only resistance is present, the voltage and current are proportional to each other at every moment. For example, when the voltage increases from A to B on the graph, the current follows along in step, increasing from A' to B' during the same time. Likewise, when the voltage decreases from B to C, the current decreases from B' to C'. For this reason, the current in a resistance R is said to be *in phase* with the voltage across the resistance.

For a capacitor, this in-phase relation between instantaneous voltage and current does *not* exist. Figure 23.4 shows graphs of the ac voltage and current versus time for a circuit that contains only a capacitor. As the voltage increases from A to B, the charge on the capacitor increases and reaches its full value at B. The current, or rate of flow of charge, has a maximum positive value at the start of the charging process at A', when there is no charge on the capacitor and hence no capacitor voltage to oppose the generator voltage. When the capacitor is fully charged at B, the capacitor voltage has a magnitude equal to that of the generator and completely opposes the generator voltage. The result is that the current decreases to zero at B'. While the capacitor voltage decreases from B to C, the charges flow out of the capacitor in a direction opposite to that of the charging current, as indicated by the negative current from B' to C'. Thus, voltage and current are not in phase but are, in fact, one-quarter wave cycle out of step, or out of phase. More specifically, assuming the voltage fluctuates as $V_0 \sin(2\pi f t)$, the current varies as $I_0 \sin(2\pi f t + \pi/2) = I_0 \cos(2\pi f t)$. Since $\pi/2$ radians correspond to 90° and since the current reaches its maximum value *before* the voltage does, it is said that the current through a capacitor *leads* the voltage across the capacitor by a phase angle of 90°.

The fact that the current and voltage for a capacitor are 90° out of phase has an important consequence from the point of view of electric power, since power is the product of current and voltage. For the time interval between points A and B (or A' and B') in Figure 23.4, both current and voltage are positive. Therefore, the instantaneous power is also positive, meaning that the generator is delivering energy to the capacitor. However, during the period between B and C (or B' and C'), the current is negative while the voltage remains positive, and the power, being the product of the two, is negative. During this period, the capacitor is returning energy to the generator. Thus, the power alternates between positive and negative values for equal periods of time. In other words, the capacitor alternately absorbs and releases energy. Consequently, *on the average, the power is zero and a capacitor uses no energy in an ac circuit.*

It will prove useful later on to use a model for the voltage and current in ac circuits. In this model, voltage and current are represented by rotating arrows, often called *phasors*, whose lengths correspond to the maximum voltage V_0 and maximum current I_0, as Figure 23.5 indicates. These phasors rotate counterclockwise at a frequency f. For a resistor, the phasors are colinear as they rotate (see part

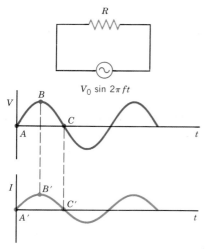

Figure 23.3 The instantaneous voltage V and current I in a resistive circuit are *in phase*, which means that they increase and decrease in step with one another.

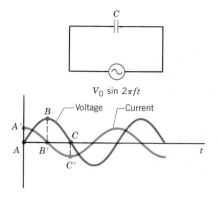

Figure 23.4 In a circuit containing only a capacitor, the instantaneous voltage and current are not in phase, as they are in a purely resistive circuit. Instead, the current *leads* the voltage by a phase angle of 90° (one-quarter of a cycle).

a of the drawing), because voltage and current are in phase. For a capacitor (see part *b*), the phasors remain perpendicular while rotating, because the phase angle between the current and the voltage is 90°. Since current leads voltage for a capacitor, the current phasor is ahead of the voltage phasor in the direction of rotation. In both cases in Figure 23.5, the instantaneous voltage and current are given by the vertical components of the phasors.

23.2 INDUCTORS AND INDUCTIVE REACTANCE

As Section 22.8 discusses, an inductor is usually a coil of wire, and the basis of its operation is Faraday's law of electromagnetic induction. According to Faraday's law, an inductor develops a voltage that opposes a change in the current. This voltage V is given by $V = -L(\Delta I/\Delta t)$ (see Equation 22.9*), where $\Delta I/\Delta t$ is the rate at which the current changes and L is the inductance of the inductor. In an ac circuit the current is always changing, and Faraday's law can be used to show that the rms-voltage across an inductor is

$$V_{\text{rms}} = I_{\text{rms}}X_{\text{L}} \qquad (23.3)$$

Equation 23.3 is analogous to $V_{\text{rms}} = I_{\text{rms}}R$, with the term X_{L} appearing in place of the resistance R and being called the **inductive reactance.** The inductive reactance is measured in ohms and determines how much rms-current exists in an inductor for a given rms-voltage across the inductor. It is found experimentally that the inductive reactance X_{L} is directly proportional to the frequency f and the inductance L, as indicated in the following equation:

$$X_{\text{L}} = 2\pi f L \qquad (23.4)$$

This relation indicates that the larger the inductance, the larger the inductive reactance. Note that the inductive reactance is directly proportional to the frequency ($X_{\text{L}} \propto f$), in contrast to the capacitive reactance, which is inversely proportional to the frequency ($X_{\text{C}} \propto 1/f$).

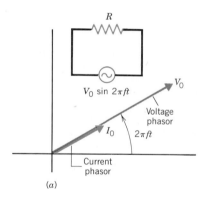

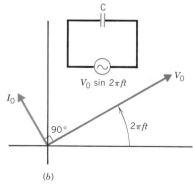

Figure 23.5 These rotating arrow models represent the voltage and the current in ac circuits that contain (*a*) only a resistor and (*b*) only a capacitor.

---

* When an inductor is used in a circuit, the notation is simplified if we designate the potential difference across the inductor as the voltage V, rather than the emf $\mathscr{E}$.

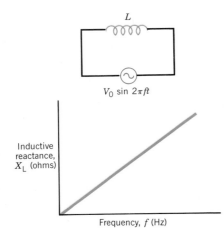

Figure 23.6 shows a graph of the inductive reactance versus frequency for a fixed value of the inductance, according to Equation 23.4. As frequency becomes very large, X_L also becomes very large. In such a situation, an inductor provides a large opposition to the alternating current. In the limit of zero frequency (i.e., direct current), X_L becomes zero, indicating that an inductor does not oppose direct current at all. The next example demonstrates the effect of inductive reactance on the current in an ac circuit.

Figure 23.6 In an ac circuit the inductive reactance X_L is directly proportional to the frequency f, according to $X_L = 2\pi f L$.

Example 2 An Inductor in an Ac Circuit

The circuit in Figure 23.6 contains a 3.60-mH inductor. The rms-voltage of the generator is 25.0 V. Find the rms-current in the circuit when the generator frequency is (a) 1.00×10^2 Hz and (b) 5.00×10^3 Hz.

REASONING The current can be calculated from $I_{rms} = V_{rms}/X_L$, provided the inductive reactance is obtained first. The inductor offers more opposition to the changing current when the frequency is larger, and the values for the inductive reactance will reflect this fact.

SOLUTION
(a) At a frequency of 1.00×10^2 Hz, we find

$$X_L = 2\pi f L = 2\pi(1.00 \times 10^2 \text{ Hz})(3.60 \times 10^{-3} \text{ H}) = 2.26 \ \Omega \qquad (23.4)$$

$$I_{rms} = \frac{V_{rms}}{X_L} = \frac{25.0 \text{ V}}{2.26 \ \Omega} = \boxed{11.1 \text{ A}} \qquad (23.3)$$

(b) The calculation is similar when the frequency is 5.00×10^3 Hz:

$$X_L = 2\pi f L = 2\pi(5.00 \times 10^3 \text{ Hz})(3.60 \times 10^{-3} \text{ H}) = 113 \ \Omega$$

$$I_{rms} = \frac{V_{rms}}{X_L} = \frac{25.0 \text{ V}}{113 \ \Omega} = \boxed{0.221 \text{ A}}$$

PROBLEM SOLVING INSIGHT

The inductive reactance X_L is directly proportional to the frequency f of the current. If the frequency increases by a factor of 50, for example, the inductive reactance increases by a factor of 50.

By virtue of its inductive reactance, an inductor affects the amount of current in an ac circuit. The inductor also influences the current in another way, as Figure 23.7 shows. This figure displays graphs of voltage and current versus time for a circuit containing only an inductor. At a maximum or minimum on the current graph, the current does not change much with time, so the voltage generated by the inductor to oppose a change in the current is zero. At the points on the current

Figure 23.7 The instantaneous voltage and current in a circuit containing only an inductor are not in phase. The current *lags behind* the voltage by a phase angle of 90° (one-quarter of a cycle).

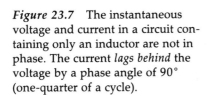

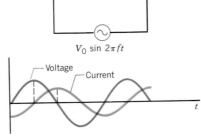

graph where the current is zero, the graph is at its steepest, and the current has the largest rate of increase or decrease. Correspondingly, the voltage generated by the inductor to oppose a change in the current has the largest positive or negative value. Thus, current and voltage are not in phase but are one-quarter of a wave cycle out of phase. If the voltage varies as $V_0 \sin (2\pi ft)$, the current fluctuates as $I_0 \sin (2\pi ft - \pi/2) = -I_0 \cos (2\pi ft)$. The current reaches its maximum *after* the voltage does, and it is said that the current *lags behind* the voltage by a phase angle of 90° ($\pi/2$ radians). In a purely capacitive circuit, in contrast, the current leads the voltage by 90°.

In an inductor the 90° phase difference between current and voltage leads to the same result for average power that it does in a capacitor. An inductor alternately absorbs and releases energy for equal periods of time, so *on the average, the power is zero and an inductor uses no energy in an ac circuit.*

As an alternative to the graphs in Figure 23.7, Figure 23.8 uses phasors to describe the instantaneous voltage and current in a circuit containing only an inductor. The voltage and current phasors remain perpendicular as they rotate, for there is a 90° phase angle between them. The current phasor lags behind the voltage phasor, relative to the direction of rotation, in contrast to the equivalent picture for a capacitor. Once again, the instantaneous values for voltage and current are given by the vertical components of the phasors.

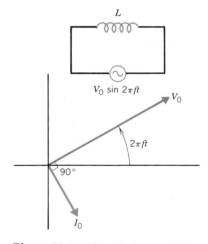

Figure 23.8 This phasor model represents the voltage and current in a circuit that contains only an inductor.

23.3 THE SERIES RCL-CIRCUIT

Capacitors and inductors can be combined along with resistors in a single circuit. The simplest combination is the series RCL-circuit, which contains a resistor, a capacitor, and an inductor, as Figure 23.9 shows. In a series RCL-circuit the total opposition to the flow of charge is called the ***impedance*** of the circuit and comes partially from (1) the resistance R, (2) the capacitive reactance X_C, and (3) the inductive reactance X_L. Figure 23.10 shows a graph of impedance versus frequency and emphasizes the frequency regions where each circuit component dominates. At low frequencies X_C becomes very large, and so does the impedance,

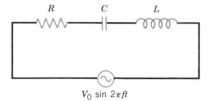

Figure 23.9 A series RCL-circuit contains a resistor, a capacitor, and an inductor.

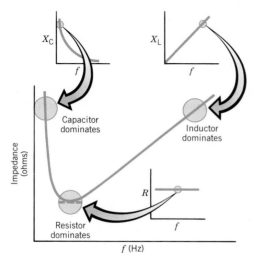

Figure 23.10 In a series RCL-circuit the impedance varies with frequency, as this graph shows.

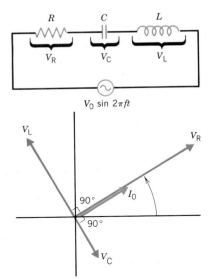

Figure 23.11 The relation between the three voltage phasors (V_R, V_C, and V_L) and the current phasor (I_0) for a series RCL-circuit.

with X_C making a much greater contribution than either X_L or R. At high frequencies X_L becomes very large, leading once again to a large impedance. However, in this case X_L dominates over X_C and R. At intermediate frequencies the impedance is smaller than it is at either extreme. In fact, we will see that there is a single frequency where the capacitive and inductive reactances cancel, leaving the frequency-independent resistance to dominate.

Because the resistor, the capacitor, and the inductor are wired in series, it is tempting to follow the analogy of a series combination of resistors and calculate the impedance by simply adding together R, X_C, and X_L. However, such a procedure is not correct. Instead, the phasors shown in Figure 23.11 must be used. The lengths of the voltage phasors in this drawing represent the maximum voltages V_R, V_C, and V_L across the resistor, the capacitor, and the inductor, respectively. The current is the same for each device, since the circuit is wired in series. The length of the current phasor represents the maximum current I_0. Notice that the drawing shows the current phasor to be (1) in phase with the voltage phasor for the resistor, (2) ahead of the voltage phasor for the capacitor by 90°, and (3) behind the voltage phasor for the inductor by 90°. These three facts are consistent with our earlier discussion in Sections 23.1 and 23.2.

The basis for dealing with the voltage phasors in Figure 23.11 is Kirchhoff's loop rule. In an ac circuit this rule applies to the *instantaneous* voltages across each circuit component and the generator. Therefore, it is necessary to take into account the fact that these voltages do not have the same phase, that is, the phasors V_R, V_C, and V_L point in different directions in the drawing. Kirchhoff's loop rule indicates that the phasors add together to give the total voltage V_0 that is supplied to the circuit by the generator. The addition, however, must be like a vector addition, to take into account the different directions. Since V_L and V_C point in opposite directions, they combine to give a resultant phasor of $V_L - V_C$, as Figure 23.12 shows. In this drawing the resultant $V_L - V_C$ is perpendicular to V_R and may be combined with it to give the total voltage V_0. Using the Pythagorean theorem, we find

$$V_0^2 = V_R^2 + (V_L - V_C)^2$$

In this equation each of the symbols stands for a maximum voltage and when divided by $\sqrt{2}$ gives the corresponding rms-voltage. Therefore, it is possible to divide both sides of the equation by $(\sqrt{2})^2$ and obtain a result for $V_{rms} = V_0/\sqrt{2}$. This result has exactly the same form as that above, but involves the rms-voltages $V_{R\text{-rms}}$, $V_{C\text{-rms}}$, and $V_{L\text{-rms}}$. However, to avoid such awkward symbols, we simply interpret V_R, V_C, and V_L as rms-quantities in the following expression:

$$V_{rms}^2 = V_R^2 + (V_L - V_C)^2 \tag{23.5}$$

The last step in determining the impedance of the circuit is to remember that $V_R = I_{rms}R$, $V_C = I_{rms}X_C$, and $V_L = I_{rms}X_L$. With these substitutions Equation 23.5 can be written as

$$V_{rms} = I_{rms}\sqrt{R^2 + (X_L - X_C)^2}$$

Therefore, for the entire RCL-circuit, it follows that

$$V_{rms} = I_{rms}Z \tag{23.6}$$

where the impedance Z of the circuit is

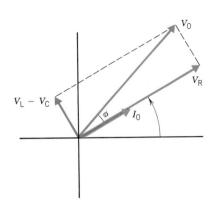

Figure 23.12 This simplified version of Figure 23.11 results when the phasors V_L and V_C, which point in opposite directions, are combined to give a resultant of $V_L - V_C$.

$$\left[\begin{array}{c} \textbf{Series} \\ \textbf{RCL-combination} \end{array}\right] \qquad Z = \sqrt{R^2 + (X_L - X_C)^2} \tag{23.7}$$

The impedance of the circuit, like R, X_C, and X_L, is measured in ohms. In Equation 23.7, $X_L = 2\pi fL$ and $X_C = 1/(2\pi fC)$, and a plot of Z versus frequency f gives the graph shown earlier in Figure 23.10. The minimum in the graph can now be seen to occur when $X_L = X_C$, so that at this point $Z = R$, the resistance in the circuit.

The phase angle between the current in and the voltage across a series RCL-combination is the angle ϕ between the current phasor I_0 and the voltage phasor V_0 in Figure 23.12. According to the drawing, the tangent of this angle is

$$\tan \phi = \frac{V_L - V_C}{V_R} = \frac{I_{rms}X_L - I_{rms}X_C}{I_{rms}R}$$

$$\left[\begin{array}{c} \textbf{Series} \\ \textbf{RCL-combination} \end{array}\right] \qquad \tan \phi = \frac{X_L - X_C}{R} \qquad (23.8)$$

The phase angle ϕ is important, because it has a major effect on the power dissipated by the circuit. Remember, on the average, only the resistance consumes power; that is, $\overline{P} = I_{rms}^2 R$ (Equation 20.15b). According to Figure 23.12, $\cos \phi = V_R/V_0 = (I_{rms}R)/(I_{rms}Z) = R/Z$, so that $R = Z \cos \phi$. Therefore,

$$\overline{P} = I_{rms}^2 Z \cos \phi = I_{rms}(I_{rms}Z) \cos \phi$$

$$\overline{P} = I_{rms}V_{rms} \cos \phi \qquad (23.9)$$

where V_{rms} is the rms-voltage of the generator. The term $\cos \phi$ is called the *power factor* of the circuit. As a check on the validity of Equation 23.9, note that if no resistance is present, $R = 0$, and $\cos \phi = R/Z = 0$. Consequently, $\overline{P} = I_{rms}V_{rms} \times \cos \phi = 0$, a result that is expected since neither a capacitor nor an inductor uses energy on the average. Conversely, if only resistance is present, $Z = \sqrt{R^2 + (X_L - X_C)^2} = R$, and $\cos \phi = R/Z = 1$. In this case, $\overline{P} = I_{rms}V_{rms} \times \cos \phi = I_{rms}V_{rms}$, which is the expression for the average power dissipated in a resistor. Example 3 deals with the current, voltages, and power for a series RCL-circuit.

This view of the cockpit of a Boeing 767 shows some of the plane's electronic controls and readout displays. The ac and dc circuits play a major role in the way these controls and displays work.

Example 3 Current, Voltages, and Power in a Series RCL-Circuit

A series RCL-circuit contains a 148-Ω resistor, a 1.50-μF capacitor, and a 35.7-mH inductor. The generator has a frequency of 512 Hz and an rms-voltage of 35.0 V. Obtain (a) the rms-voltage across each circuit element and (b) the electrical power consumed by the circuit.

REASONING The rms-voltages across each circuit element can be determined from $V_R = I_{rms}R$, $V_C = I_{rms}X_C$, and $V_L = I_{rms}X_L$, as soon as the rms-current and the reactances X_C and X_L are known. Since the rms-current can be found from $I_{rms} = V_{rms}/Z$, the first step in the solution is to find the impedance Z from the individual reactances. The power consumed is given by $\overline{P} = I_{rms}V_{rms} \cos \phi$, where the phase angle ϕ can be obtained from $\tan \phi = (X_L - X_C)/R$.

SOLUTION
(a) The individual reactances are

$$X_C = \frac{1}{2\pi fC} = \frac{1}{2\pi(512 \text{ Hz})(1.50 \times 10^{-6} \text{ F})} = 207 \ \Omega \qquad (23.2)$$

$$X_L = 2\pi fL = 2\pi(512 \text{ Hz})(35.7 \times 10^{-3} \text{ H}) = 115 \ \Omega \qquad (23.4)$$

The impedance of the circuit is

$$Z = \sqrt{R^2 + (X_L - X_C)^2} = \sqrt{(148\ \Omega)^2 + (115\ \Omega - 207\ \Omega)^2} = 174\ \Omega \quad (23.7)$$

The current through each circuit element is

$$I_{rms} = \frac{V_{rms}}{Z} = \frac{35.0\ V}{174\ \Omega} = 0.201\ A \quad (23.6)$$

The rms-voltages across each circuit element now follow immediately:

$$V_R = I_{rms}R = (0.201\ A)(148\ \Omega) = \boxed{29.7\ V} \quad (20.14)$$

$$V_C = I_{rms}X_C = (0.201\ A)(207\ \Omega) = \boxed{41.6\ V} \quad (23.1)$$

$$V_L = I_{rms}X_L = (0.201\ A)(115\ \Omega) = \boxed{23.1\ V} \quad (23.3)$$

Observe that these three rms-voltages do not add up to give the generator's rms-voltage, which is 35.0 V. Instead, the rms-voltages satisfy Equation 23.5. It is the sum of the *instantaneous* voltages across R, C, and L, rather than the sum of the rms-voltages, that add up to give the generator's *instantaneous* voltage, according to Kirchhoff's loop rule.

(b) The power consumed by the circuit is $\overline{P} = I_{rms}V_{rms}\cos\phi$. Therefore, a value for the phase angle ϕ is needed and can be obtained as follows:

$$\tan\phi = \frac{X_L - X_C}{R} = \frac{115\ \Omega - 207\ \Omega}{148\ \Omega} = -0.62 \quad (23.8)$$

$$\phi = \tan^{-1}(-0.62) = -32°$$

The phase angle is negative since the circuit is more capacitive than inductive (X_C is greater than X_L), and the current leads the voltage. The average power consumed is

$$\overline{P} = I_{rms}V_{rms}\cos\phi = (0.201\ A)(35.0\ V)\cos(-32°) = \boxed{6.0\ W} \quad (23.9)$$

PROBLEM SOLVING INSIGHT

In an RCL-series circuit, the rms-voltages across the resistor, capacitor, and inductor do not add up to equal the rms-voltage across the generator.

23.4 RESONANCE IN ELECTRIC CIRCUITS

The behavior of current and voltage in a series RCL-circuit can give rise to a condition of **resonance**. Resonance occurs when the frequency of a vibrating force exactly matches a natural frequency of the object to which the force is applied, as discussed in Section 10.8. In the electric case the vibrating force is provided by the oscillating electric field that is related to the voltage supplied by the generator.

Figure 23.13 helps us to understand why there is a resonance frequency for an ac circuit. This drawing presents an analogy between the electrical case (ignoring resistance) and the mechanical case of an object on a horizontal spring (ignoring friction). Part *a* shows a fully stretched spring that has just been released, with the initial speed *v* of the object being zero. All the energy is stored in the form of elastic potential energy. When the object begins to move, it gradually loses potential energy and picks up kinetic energy. In part *b*, the object moves with maximum kinetic energy through the position where the spring is unstretched (zero potential energy). Because of its inertia, the moving object coasts through this position and eventually comes to a halt in part *c* when the spring is fully compressed and all kinetic energy has been converted back into elastic potential energy. Part *d* of

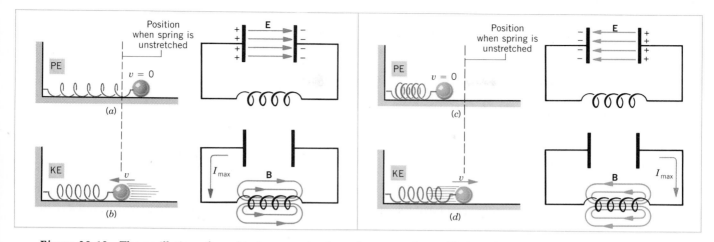

Figure 23.13 The oscillation of an object on a spring is analogous to the oscillation of the electric and magnetic fields that occur, respectively, in a capacitor and in an inductor.

the picture is like part b, except the direction of motion is reversed. The resonance frequency f_0 of the object on the spring is the natural frequency at which the object vibrates and is given as $f_0 = [1/(2\pi)]\sqrt{k/m}$ according to Equation 10.14. In this expression, m is the mass of the object, and k is the spring constant.

In the electrical case, Figure 23.13a begins with a fully charged capacitor that has just been connected to an inductor. At this instant the energy is stored in the electric field between the capacitor plates. As the capacitor discharges, the electric field between the plates decreases, while a magnetic field builds up around the inductor because of the increasing current in the circuit. The maximum current and the maximum magnetic field exist at the instant the capacitor is completely discharged, as in part b of the figure. Energy is now stored entirely in the magnetic field of the inductor. The voltage induced in the inductor keeps the charges flowing until the capacitor again becomes fully charged, but now with reverse polarity, as in part c. Once again, the energy is stored in the electric field between the plates and no energy resides in the magnetic field of the inductor. Part d of the cycle repeats part b, but with reversed directions of current and magnetic field. Thus, we see that an ac circuit can have a resonance frequency, because there is a natural tendency for energy to shuttle back and forth between the electric field of the capacitor and the magnetic field of the inductor.

We can determine the natural frequency at which energy shuttles back and forth between the capacitor and inductor by building on the analogy of the object on the spring. The larger the mass m, the greater is the inertia or tendency for the object to coast beyond the position where the spring is unstretched. Similarly, the larger the inductance L, the greater is the voltage induced in the coil, which keeps the charges flowing beyond the moment when the capacitor is uncharged. Now consider the spring constant k. Larger values of k mean that it is harder to compress the spring. In the electrical case, putting more charge on the capacitor plates is like compressing the spring. When a charge q is placed on the plates, the voltage V across the plates is $V = q/C$, according to Equation 19.8. Larger voltages result for larger values of $1/C$ and make it harder for more charge to be added to the plates. Therefore, larger values of $1/C$ are analogous to larger values of k. An expression for the natural frequency of the capacitor–inductor combination,

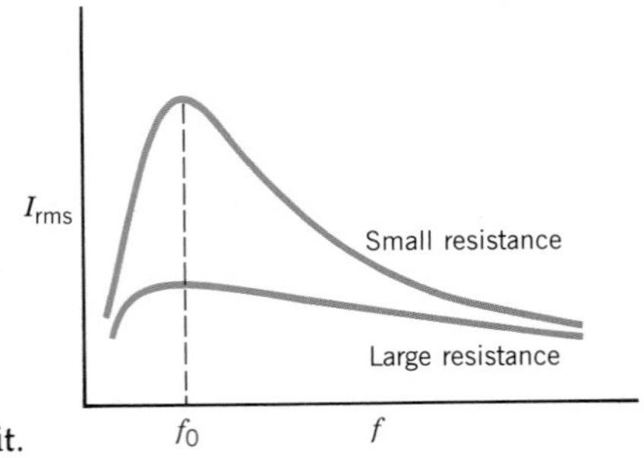

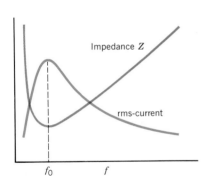

Figure 23.14 In a series RCL-circuit the impedance is a minimum, and the current is a maximum, when the frequency f equals the resonance frequency f_0 of the circuit.

then, can be obtained from $f_0 = [1/(2\pi)] \sqrt{k/m}$ by replacing m with L and k with $1/C$. The result is the resonance frequency f_0 for the RCL-circuit:

$$f_0 = \frac{1}{2\pi \sqrt{LC}} \qquad (23.10)$$

The resonant frequency is determined by the inductance and the capacitance, but not the resistance.

Equation 23.10 can be obtained in another way. In an RCL-circuit, the current is $I_{rms} = V_{rms}/Z$, where $Z = \sqrt{R^2 + (X_L - X_C)^2}$ is the impedance of the circuit. Thus, as Figure 23.14 illustrates, the rms-current is a maximum when the impedance Z is a minimum, assuming a given generator voltage. The minimum impedance occurs when the frequency is f_0, such that $X_L = X_C$ or $2\pi f_0 L = 1/(2\pi f_0 C)$, which can be solved to obtain Equation 23.10.

The effect of resistance on electrical resonance is to make the "sharpness" of the circuit response less pronounced, as Figure 23.15 indicates. When the resistance is small, the current-versus-frequency graph falls off suddenly on either side of the maximum current. When the resistance is large, the falloff is more gradual, and there is less current at the maximum.

The following example deals with one application of resonance in electrical circuits. In this example the focus is on the oscillation of energy between a capacitor and an inductor. Once a capacitor/inductor combination is energized, the energy will oscillate indefinitely as in Figure 23.13, provided there is some provision to replace any dissipative losses that occur because of resistance. Circuits that include this type of provision are called oscillator circuits.

Figure 23.15 The effect of resistance on the current in a series RCL-circuit.

THE PHYSICS OF . . .

a heterodyne metal detector.

Example 4   A Heterodyne Metal Detector

Figure 23.16 shows a heterodyne metal detector being used. As Figure 23.17 illustrates, this device utilizes two capacitor/inductor oscillator circuits, A and B. Each circuit produces its own resonance frequency, $f_{0A} = 1/(2\pi \sqrt{L_A C})$ and $f_{0B} = 1/(2\pi \sqrt{L_B C})$. Any difference between these two frequencies is detected through earphones as a beat frequency $f_{0B} - f_{0A}$, similar to the beat frequency that two musical tones produce. In the absence of any nearby metal object, the inductances L_A and L_B are the same, and f_{0A} and f_{0B} are identical. There is no beat frequency. When inductor B (the search coil) comes near a piece of metal, the inductance L_B decreases, the corresponding oscillator frequency f_{0B} increases, and a beat frequency is heard. Suppose that initially each inductor is adjusted so $L_B = L_A$, and each oscillator has a resonance frequency of 855.5 kHz. Assuming the inductance of search coil B decreases by 1.00% due to a nearby piece of metal, determine the beat frequency heard through the earphones.

REASONING   To find the beat frequency $f_{0B} - f_{0A}$, we need to determine the amount by which the resonance frequency f_{0B} changes because of a 1.00% decrease in the inductance L_B.

SOLUTION   We begin by obtaining the ratio of f_{0B} to f_{0A}:

$$\frac{f_{0B}}{f_{0A}} = \frac{\dfrac{1}{2\pi\sqrt{L_B C}}}{\dfrac{1}{2\pi\sqrt{L_A C}}} = \sqrt{\frac{L_A}{L_B}}$$

But due to the nearby piece of metal $L_B = 0.9900 L_A$, so that

$$\frac{f_{0B}}{f_{0A}} = \sqrt{\frac{L_A}{0.9900 L_A}} = 1.005$$

Therefore, the new value for f_{0B} is $f_{0B} = 1.005 f_{0A} = 1.005 \times (855.5 \text{ kHz}) = 859.8 \text{ kHz}$. As a result, the detected beat frequency is

$$f_{0B} - f_{0A} = 859.8 \text{ kHz} - 855.5 \text{ kHz} = \boxed{4.3 \text{ kHz}}$$

Figure 23.16   A heterodyne metal detector is used to locate buried metal objects.

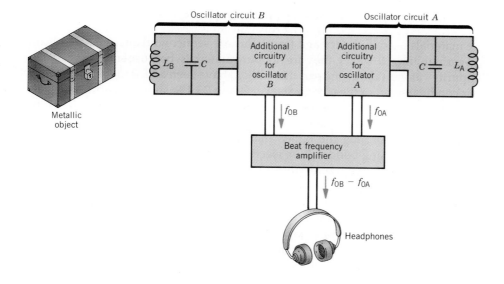

Metallic object

Figure 23.17   A heterodyne metal detector uses two electrical oscillators, *A* and *B*, in its operation. When the resonance frequency of oscillator *B* is changed due to the proximity of a metallic object, a beat frequency, whose value is $f_{0B} - f_{0A}$, is heard in the headphones.

23.5 SEMICONDUCTOR DEVICES

Semiconductor devices such as diodes and transistors are widely used in modern electronics, and Figure 23.18 illustrates one application. The drawing shows an audio system in which small ac voltages (originating in a compact disc player, etc.) are amplified so they can drive the speaker(s). The electric circuits that accomplish the amplification depend on the power provided by the power supply, which is simply a battery in portable units. In nonportable units, however, the power supply is a separate electric circuit containing diodes, along with other elements. As we will see, the diodes convert the 60-Hz ac voltage present at a wall outlet into

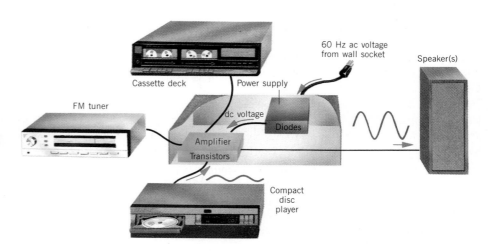

Figure 23.18 In a typical audio system, diodes are used in the power supply to create a dc voltage from the ac voltage present at the wall socket. This dc voltage is necessary so the transistors in the amplifier can perform their task of enlarging the small ac voltages originating in the compact disc player, etc.

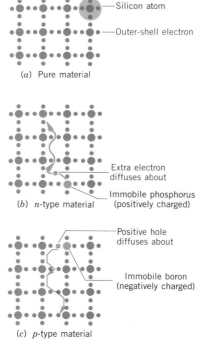

Figure 23.19 A silicon crystal that is (*a*) undoped or pure, (*b*) doped with phosphorus to produce an *n*-type material, and (*c*) doped with boron to produce a *p*-type material.

the dc voltage needed by the amplifier, which, in turn, performs its job of amplification with the aid of transistors.

n-TYPE AND *p*-TYPE SEMICONDUCTORS

The materials used in diodes and transistors are semiconductors, such as silicon and germanium. However, they are not pure materials, because small amounts of "impurity" atoms (about one part in a million) have been added to them to change their conductive properties. For instance, Figure 23.19*a* shows an array of atoms that symbolizes the crystal structure in pure silicon. Each silicon atom has four outer-shell* electrons, and each electron participates with electrons from neighboring atoms in forming the bonds that hold the crystal together. Since they participate in forming bonds, these electrons generally do not move throughout the crystal. Consequently, pure silicon and germanium are not good conductors of electricity. It is possible, however, to increase their conductivities by adding tiny amounts of impurity atoms, such as phosphorus or arsenic, whose atoms have five outer-shell electrons. For example, when a phosphorus atom replaces a silicon atom in the crystal, only four of the five outer-shell electrons of phosphorus fit into the crystal structure. The extra fifth electron does not fit in and is relatively free to diffuse throughout the crystal, as part *b* of the drawing suggests. A semiconductor containing small amounts of phosphorus can, therefore, be envisioned as containing immobile, positively charged phosphorus atoms and a pool of electrons that are free to wander throughout the material. These mobile electrons allow the semiconductor to conduct electricity.

The process of adding impurity atoms is called *doping.* A semiconductor doped with an impurity that contributes mobile electrons is called an ***n-type semiconductor,*** since the mobile charge carriers have a **n**egative charge. Note that an *n*-type semiconductor is overall electrically neutral, since it contains equal numbers of positive and negative charges.

It is also possible to dope a silicon crystal with an impurity whose atoms have only three outer-shell electrons (e.g., boron or gallium). Because of the missing fourth electron, there is a "hole" in the lattice structure at the boron atom, as

* Section 30.6 discusses the electronic structure of the atom in terms of "shells."

Figure 23.19c illustrates. An electron from a neighboring silicon atom can move into this hole, in which event the region around the boron atom, having acquired the electron, becomes negatively charged. Of course, when a nearby electron does move, it leaves behind a hole. This hole is positively charged, since it results from the removal of an electron from the vicinity of a neutral silicon atom. The vast majority of atoms in the lattice are silicon, so the hole is almost always next to another silicon atom. Consequently, an electron from one of these adjacent atoms can move into the hole, with the result that the hole moves to yet another location. In this fashion, a positively charged hole can wander through the crystal. This type of semiconductor can, therefore, be viewed as containing immobile, negatively charged boron atoms and an equal number of positively charged, mobile holes. Because of the mobile holes, the semiconductor can conduct electricity. In this case the charge carriers are positive, as can be verified by measuring the Hall emf (see Section 21.5). A semiconductor doped with an impurity that introduces mobile **positive** holes is called a *p-type semiconductor.*

A technician monitors the fabrication of a disc-shaped single wafer of crystalline silicon. This particular wafer consists of integrated circuit chips that will be used to control motors in computer disk drives.

THE SEMICONDUCTOR DIODE

A *p-n junction diode* is a device formed from a *p*-type semiconductor and an *n*-type semiconductor. The *p-n* junction between the two materials is of fundamental importance to the operation of diodes and transistors. Figure 23.20 shows separate *p*-type and *n*-type semiconductors, each electrically neutral. Figure 23.21a shows them joined together to form a diode. Mobile electrons from the *n*-type semiconductor and mobile holes from the *p*-type semiconductor flow across the junction and combine. This process leaves the *n*-type material with a positive charge layer and the *p*-type material with a negative charge layer, as part

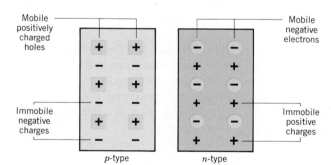

Figure 23.20   A *p*-type semiconductor and an *n*-type semiconductor.

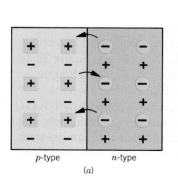
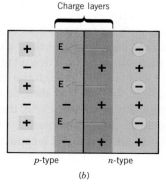

Figure 23.21   (*a*) At the junction between *n* and *p* materials, mobile electrons and holes combine and (*b*) create positive and negative charge layers. The electric field produced by the charge layers is **E**.

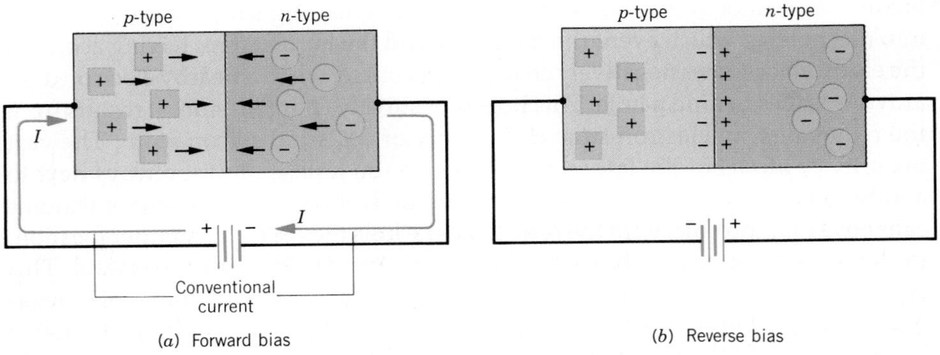

Figure 23.22 (*a*) There is an appreciable current through the diode when the diode is forward biased. (*b*) Under a reverse bias condition, there is almost no current through the diode.

b of the drawing indicates. The positive and negative charge layers on the two sides of the junction set up an electric field **E**, much like that in a parallel plate capacitor. This electric field tends to prevent any further movement of charge across the junction, and all charge flow quickly stops.

Suppose now that a battery is connected across the *p-n* junction, as in Figure 23.22*a*, where the negative terminal of the battery is attached to the *n*-material, and the positive terminal is attached to the *p*-material. In this situation the junction is said to be in a condition of *forward bias,* and as a result, there is a current in the circuit. The mobile electrons in the *n*-material are repelled by the negative terminal of the battery and move toward the junction. Likewise, the positive holes in the *p*-material are repelled by the positive terminal of the battery and also move toward the junction. At the junction the electrons fill the holes. In the meantime, the negative terminal of the battery provides a fresh supply of electrons to the *n*-material, and the positive terminal pulls off electrons from the *p*-material, forming new holes in the process. Consequently, a continual flow of charge, and hence a current, is maintained.

In Figure 23.22*b* the battery polarity has been reversed, and the *p-n* junction is in a condition known as *reverse bias.* The battery forces electrons in the *n*-material and holes in the *p*-material away from the junction. As a result, the potential across the junction builds up until it opposes the battery potential, and very little current can be sustained through the diode. The diode, then, is a unidirectional device, for it allows current to pass only in one direction.

The graph in Figure 23.23 shows the dependence of the current on the magnitude and polarity of the voltage applied across a *p-n* junction diode. The exact values of the current depend on the nature of the semiconductor and the extent of the doping. Also shown in the drawing is the symbol used for a diode (⟶▶⟶). The direction of the arrowhead in the symbol indicates the direction of the conventional current in the diode under a forward bias condition.

Because diodes are unidirectional devices, they are commonly used in *rectifier circuits,* which convert an ac voltage into a dc voltage. For instance, Figure 23.24 shows a circuit in which charges flow through the resistance *R* only while the ac generator biases the diode in the forward direction. Since current occurs only during one-half of every generator voltage cycle, the circuit is called a half-wave rectifier. A plot of the output voltage across the resistor reveals that only the positive halves of each cycle are present. If a capacitor is added in parallel with the resistor, as indicated in the drawing, the capacitor charges up and keeps the voltage from dropping to zero between each positive half-cycle. It is also possible

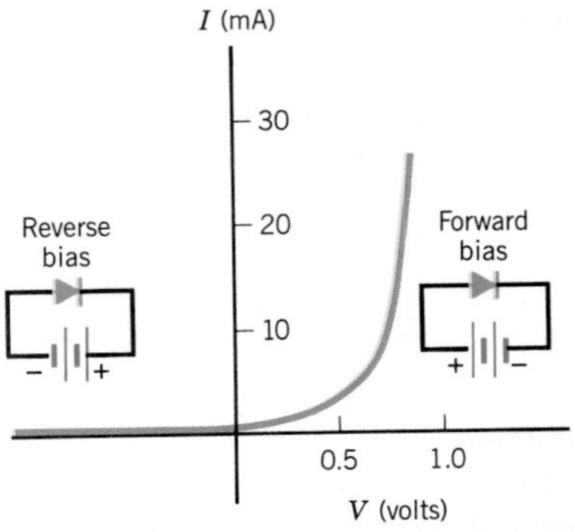

THE PHYSICS OF . . .

a semiconductor diode.

Figure 23.23 The current-versus voltage characteristics of a typical *p-n* junction diode.

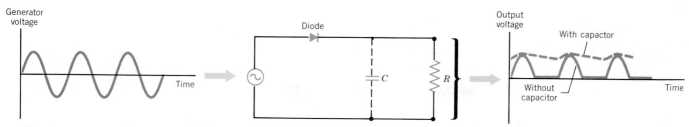

Figure 23.24 A half-wave rectifier circuit, together with a capacitor, constitutes a dc power supply, because the rectifier converts an ac voltage into a dc voltage.

to construct full-wave rectifier circuits, in which both halves of every cycle of the generator voltage drive current through the load resistor in the same direction (see question 8).

When a circuit such as that in Figure 23.24 includes a capacitor and also a transformer to establish the desired voltage level, the circuit is called a power supply. In the audio system in Figure 23.18, the power supply receives the 60-Hz ac voltage from a wall socket and produces a dc output voltage that is used for the transistors within the amplifier. Power supplies using diodes are also found in virtually all electronic appliances, such as televisions and microwave ovens.

SOLAR CELLS

Solar cells use *p-n* junctions to convert sunlight directly into electricity, as Figure 23.25 illustrates. The solar cell in this drawing consists of a *p*-type semiconductor surrounding an *n*-type semiconductor. As discussed earlier, charge layers form at the junction between the two types of semiconductors, leading an electric field **E** pointing from the *n*-type toward the *p*-type layer. The outer covering of *p*-type material is so thin that sunlight penetrates into the charge layers and ionizes some of the atoms there. In the process of ionization, the energy of the sunlight causes a negative electron to be ejected from the atom, leaving behind a positive hole. As the drawing indicates, the electric field in the charge layers causes the electron and the hole to move away from the junction. The electron moves into the *n*-type material, and the hole moves into the *p*-type material. As a result, the sunlight causes the solar cell to develop negative and positive terminals, much like the terminals of a battery. The current that a single solar cell can provide is small, so applications of solar cells often use many of them mounted to form large panels.

THE PHYSICS OF . . .

solar cells.

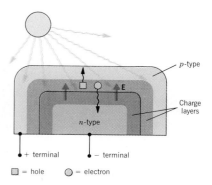

Figure 23.25 A solar cell formed from a *p-n* junction.

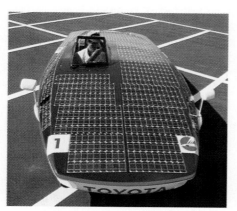

This electric car runs on energy derived from solar cells mounted on the top of the car.

TRANSISTORS

A number of different kinds of transistors are in use today. One type is the **bipolar junction transistor,** which consists of two *p-n* junctions formed by three layers of doped semiconductors. As Figure 23.26 indicates, there are *pnp* and *npn* transistors. In either case, the middle region is made very thin compared to the outer regions.

A transistor is useful because it can be used in circuits that amplify a smaller voltage into a larger one. A transistor plays the same kind of role in an amplifier circuit that a valve does when it controls the flow of water through a pipe. A small change in the valve setting produces a large change in the amount of water per second that flows through the pipe. In other words, a small change in the voltage applied as input to a transistor produces a large change in the output from the transistor.

Figure 23.27 shows a *pnp* transistor connected to two batteries, labeled V_E and V_C. The voltages V_E and V_C are applied in such a way that the *p-n* junction on the left has a forward bias, while the *p-n* junction on the right has a reverse bias. Moreover, the voltage V_C is usually much larger than V_E for a reason to be discussed shortly. The drawing also shows the standard symbol and nomenclature for the three sections of the transistor, namely, the *emitter*, the *base*, and the *collector*. The arrowhead in the symbol points in the direction of the conventional current through the emitter.

The positive terminal of V_E pushes the mobile positive holes in the *p*-type material of the emitter toward the emitter/base junction. And since this junction has a forward bias, the holes enter the base region readily. Once in the base region, the holes come under the strong influence of V_C and are attracted to its negative terminal. Since the base is so thin (about 10^{-6} m or so), approximately 98% of the holes are drawn through the base and into the collector. The remaining 2% of the holes combine with free electrons in the base region, thereby giving rise to a small base current I_B. As the drawing shows, the moving holes in the emitter and collector constitute currents that are labeled I_E and I_C, respectively. From Kirchhoff's junction rule it follows that $I_C = I_E - I_B$.

Because the base current I_B is small, the collector current is determined primarily by current from the emitter ($I_C = I_E - I_B \approx I_E$). This means that a change in I_E will cause a change in I_C of nearly the same amount. Furthermore, a substantial

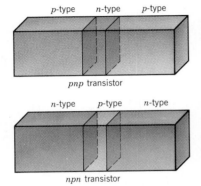

Figure 23.26 There are two kinds of bipolar junction transistors, *pnp* and *npn*.

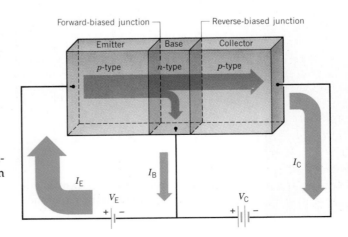

Figure 23.27 A *pnp* transistor, along with its bias voltages V_E and V_C. On the symbol for the *pnp* transistor, the emitter is marked with an arrowhead that denotes the direction of conventional current through the emitter.

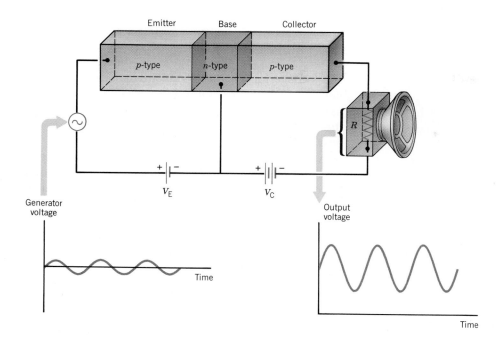

Figure 23.28 The basic *pnp* transistor amplifier in this drawing amplifies a small generator voltage to produce an enlarged voltage across the resistance *R*.

change in I_E can be caused by only a small change in the forward bias voltage V_E. To see that this is the case, look back at Figure 23.23 and notice how steep the current-versus-voltage curve is for a *p-n* junction; small changes in the forward bias voltage give rise to large changes in the current.

With the help of Figure 23.28 we can now appreciate what was meant by the earlier statement that a small change in the voltage applied as input to a transistor leads to a large change in the output. This picture shows an ac generator connected in series with the battery V_E, and a resistance R connected in series with the collector. The generator voltage could originate from many sources, such as an electric guitar pickup or a compact disc player, while the resistance R could represent a loudspeaker. The generator introduces small voltage changes in the forward bias across the emitter/base junction and, thus, causes large corresponding changes in the current I_C leaving the collector and passing through the resistance R. As a result, the output voltage across R is an enlarged or amplified version of the input voltage of the generator. The operation of an *npn* transistor is similar to that of a *pnp* transistor. The main difference is that the bias voltages (and current directions) are reversed.

It is important to realize that the increased power available at the output of a transistor amplifier does *not* come from the transistor itself. Rather, it comes from the power provided by the voltage source V_C. The transistor, acting like an automatic valve, merely allows the small, weak signals from the input generator to control the power taken from the source V_C and delivered to the resistance R.

Today it is possible to combine arrays of thousands of transistors, diodes, resistors, and capacitors on a tiny "chip" of silicon that usually measures less than a centimeter on a side. These arrays are called integrated circuits (ICs) and can be designed to perform almost any desired electronic function. Integrated circuits, such as the one in Figure 23.29, have revolutionized the electronics industry and lie at the heart of computers, hand-held calculators, digital watches, and programmable appliances.

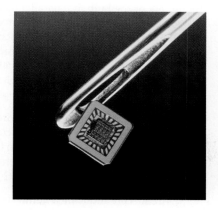

Figure 23.29 Because of their miniature size, integrated circuits such as this one are used in a wide variety of electronic devices.

SUMMARY

In an ac circuit the rms-voltage across a capacitor is related to the rms-current according to $V_{rms} = I_{rms}X_C$, where X_C is the **capacitive reactance.** The capacitive reactance is measured in ohms and is given by $X_C = 1/(2\pi fC)$ for a capacitance C and a frequency f. The current in a capacitor leads the voltage across the capacitor by a phase angle of $90°$, and as a result, a capacitor consumes no power, on the average.

For an inductor the rms-voltage and the rms-current are related by $V_{rms} = I_{rms}X_L$, where X_L is the **inductive reactance.** For an inductance L and frequency f the inductive reactance is given in ohms as $X_L = 2\pi fL$. The ac current in an inductor lags behind the voltage by a phase angle of $90°$. Consequently, an inductor, like a capacitor, consumes no power, on the average.

When a resistor, a capacitor, and an inductor are connected in series, the rms-voltage across the combination is related to the rms-current according to $V_{rms} = I_{rms}Z$, where Z is the **impedance** of the combination. The impedance (in ohms) for the series combination is $Z = \sqrt{R^2 + (X_L - X_C)^2}$. The phase angle ϕ between current and voltage for a series RCL-combination is given by $\tan\phi = (X_L - X_C)/R$. Only the resistor in the combination dissipates power on the average, according to the relation $\bar{P} = I_{rms}V_{rms}\cos\phi$. The term $\cos\phi$ is the **power factor** of the circuit.

A series RCL-circuit has a **resonance frequency** f_0 that is $f_0 = 1/(2\pi\sqrt{LC})$. At resonance the impedance of the circuit has a minimum value equal to the resistance R, and the rms-current has a maximum value.

In an *n*-type semiconductor, mobile negative electrons carry the current. An *n*-type material is produced by doping a semiconductor such as silicon with a small amount of impurity such as phosphorus. In a *p*-type **semiconductor,** mobile positive holes in the crystal structure carry the current. A *p*-type material is made by doping a semiconductor with an impurity such as boron. These two types of semiconductors are used in the *p-n* **junction diode** and in *pnp* **and** *npn* **bipolar junction transistors.**

QUESTIONS

1. A light bulb is connected directly to the 60-Hz ac voltage present at a wall outlet. (a) Describe what would happen to the brightness of the bulb, if a parallel plate capacitor (without a dielectric between the plates) is inserted in series between the light bulb and the wall outlet. (b) Describe what subsequently happens to the brightness when the capacitor is replaced with another capacitor, one that is identical, except the space between the plates is filled with a dielectric material. In both (a) and (b) explain your reasoning.

2. The ends of a long, straight wire are connected to the terminals of an ac generator, and the current is measured. The wire is then disconnected, wound into the shape of a multiple-turn coil, and reconnected to the generator. In which case does the generator deliver a larger current? Explain.

3. An air-core inductor is connected in series with a light bulb and this circuit is plugged into an electrical outlet. What happens to the brightness of the bulb when a piece of iron is inserted inside the inductor? Give a reason for your answer.

4. A light bulb is connected to an ac generator. When an inductor is added in series with the bulb, the brightness decreases, no matter what the value of the inductance is. However, if a capacitor is now added in series with the bulb and the inductor, the brightness may either increase or decrease, depending on the value of the capacitance. Why?

5. In a series circuit a resistor and an inductor are connected to a generator whose rms-voltage is 160 V. It is determined that the rms-voltage across the resistor is 110 V, while the rms-voltage across the inductor is also 110 V. Notice that 110 V + 110 V is greater than the 160 V provided by the generator. Does this situation involving rms-voltages violate Kirchhoff's loop rule? Explain.

6. An inductor and a capacitor are connected in parallel across the terminals of a generator. What happens to the current from the generator as the frequency becomes (a) very large and (b) very small?

7. Is it possible for two series RCL-circuits to have the same resonance frequencies and yet have (a) different R values and (b) different C and L values? Justify your answers.

8. The drawing shows a full-wave rectifier circuit (a so-called bridge-rectifier) that uses four diodes. The direction of the current in the load resistance R is the *same* for both positive and negative halves of the voltage cycle of the generator. (a) When the generator causes the potential at A to be positive relative to that at B, through which two diodes do charges pass, and what is the direction of the conventional current in the load resistance R? (b) Repeat part (a) when the potential at B becomes positive relative to that at A.

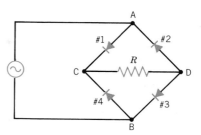

PROBLEMS

Note: For problems in this set, the ac current and voltage are rms-values and the power is an average value, unless indicated otherwise.

Section 23.1 Capacitors and Capacitive Reactance

1. At what frequency does a 7.50-μF capacitor have a reactance of 168 Ω?

2. What voltage is needed to create a current of 29.0 mA in a circuit containing only a 0.565-μF capacitor, when the frequency is 2.60 kHz?

3. Three capacitors are connected in parallel across the terminals of a 440-Hz generator that supplies a voltage of 17 V. The capacitances are 2.0, 4.0, and 7.0 μF. (a) Find the equivalent capacitance of these capacitors. (b) What is the total current supplied by the generator?

4. Two identical capacitors are connected in series to an ac generator that has a frequency of 620 Hz and produces a voltage of 24 V. The current in the circuit is 0.16 A. What is the capacitance of each capacitor?

5. A capacitor is attached to a 5.00-Hz generator. The current is observed to reach a maximum value at a certain time. What is the least amount of time that passes before the voltage across the capacitor reaches its maximum value?

*6. A capacitor consists of two square metal plates that are parallel, each having an area of 1.0×10^{-4} m². This capacitor is connected to a generator that has a frequency of 11 kHz and a voltage of 150 V. The current in the circuit is measured to be 9.4 μA. Assuming there is air between the plates, determine the distance between them.

Section 23.2 Inductors and Inductive Reactance

7. What is the inductance of an inductor that has a reactance of 1.8 kΩ at a frequency of 4.2 kHz?

8. The current in an inductor is 0.20 A, and the frequency is 750 Hz. If the inductance is 0.080 H, what is the voltage across the inductor?

9. The transformer for an electric toy train has a primary winding whose inductance is 2.4 H. A voltage of 120 V (frequency = 60.0 Hz) is applied to the primary coil when the transformer is plugged into an electrical outlet. Assuming the train is not connected to the secondary of the transformer, find the current in the primary.

10. A 0.047-H inductor is wired across the terminals of a generator that has a voltage of 2.1 V and supplies a current of 0.023 A. Find the frequency of the generator.

*11. A 0.313-H and a 0.127-H inductor are connected in parallel across the terminals of a generator. The generator has a voltage of 9.00 V and a frequency of 266 Hz. What is the total current that the generator delivers?

*12. The current in a solenoid is 0.036 A when the solenoid is connected to an 18-kHz generator. The solenoid has a cross-sectional area of 3.1×10^{-5} m² and a length of 2.5 cm. The solenoid has 135 turns. Determine the *peak voltage* of the generator.

Section 23.3 The Series RCL-Circuit

13. A series RCL-circuit includes a resistance of 275 Ω, an inductive reactance of 648 Ω, and a capacitive reactance of 415 Ω. The current in the circuit is 0.233 A. What is the voltage of the generator?

14. The purpose of this problem is to verify the shapes of the graphs of capacitive and inductive reactance versus frequency, which are shown in Figures 23.2 and 23.6. Plot these graphs for a 20.0-μF capacitor and a 5.00-mH inductor. Use a frequency of 10 Hz and five equally spaced frequencies between 100 and 1000 Hz.

15. An ac generator has a frequency of 5.60 kHz and produces a current of 0.0530 A in a series circuit that contains a 218-Ω resistor and a 0.100-μF capacitor. Obtain (a) the voltage of the generator and (b) the phase angle between the current and the voltage.

16. A series circuit has an impedance of 192 Ω, and the current leads the voltage by 75.0°. The circuit contains two different elements. (a) From the phase angle between current and voltage, decide which elements are present, R and C, R and L, or C and L. (b) Find values for the appropriate quantities, R and X_C, or R and X_L, or X_C and X_L.

17. A circuit consists of a 215-Ω resistor and a 0.200-H inductor. These two elements are connected in series across a generator that has a frequency of 106 Hz and a voltage of 234 V. (a) What is the current in the circuit? (b) Determine the phase angle between the current and the voltage.

18. A 2700-Ω resistor and a 1.1-μF capacitor are connected in series across a generator (60.0 Hz, 120 V). Determine the power dissipated in the circuit.

*19.** For the circuit shown in the drawing, find the current provided by the generator when the frequency is (a) very large and (b) very small.

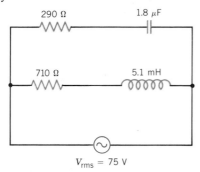

*20.** In reality, there is some resistance R in the wire from which an inductor is made. Therefore, an actual inductor should be represented as a resistor in series with an ideal (resistanceless) inductor. With this in mind, suppose the current in a 2.8-mH inductor is I_0 when the inductor is connected to a 12-V battery. However, when the battery is replaced with a 1500-Hz generator whose voltage is 12 V, the current is $I_0/3$. What is the resistance R of the wire?

*21.** A series circuit contains a resistor and an inductor. The voltage V of the generator is fixed. If $R = 16\ \Omega$ and $L = 4.0$ mH, find the frequency at which the current is one-half of its value at zero frequency.

22. When a resistor is connected by itself to an ac generator, the average power dissipated in the resistor is 1.000 W. When a capacitor is added in series with the resistor, the power dissipated is 0.500 W. When an inductor is added in series with the resistor (without the capacitor), the power dissipated is 0.250 W. Determine the power dissipated when both the capacitor and the inductor are added in series with the resistor.

Section 23.4 Resonance in Electric Circuits

23. A series RCL-circuit has a capacitance of 1.20 μF and an inductance of 2.00 mH. What is the resonance frequency of the circuit?

24. A series RCL-circuit has a resonance frequency of 690 kHz. If the value of the inductance is 26 μH, what is the value of the capacitance?

25. The resistor in a series RCL-circuit has a resistance of 92 Ω, while the voltage of the generator is 3.0 V. At resonance, what is the average power dissipated in the circuit?

26. A series RCL-circuit is at resonance and contains a variable resistor that is set to 175 Ω. The power dissipated in the circuit is 2.6 W. Assuming the voltage remains constant, how much power is dissipated when the variable resistor is set to 562 Ω?

27. Two series RCL-circuits have the same resonance frequency, yet circuit A has a capacitance of 2.3 μF while circuit B has a capacitance of 3.9 μF. Find the ratio L_A/L_B of the inductances in these circuits.

28. The power dissipated in a series RCL-circuit is 65.0 W, and the current is 0.530 A. The circuit is at resonance. Determine the voltage of the generator.

*29.** The ratio of the inductive reactance to the capacitive reactance is observed to be 5.36 in a series RCL-circuit. The resonance frequency of the circuit is 225 Hz. What is the frequency of the generator that is connected to the circuit?

*30.** Suppose you have a number of capacitors. Each of these capacitors is identical to the capacitor that is already in a series RCL-circuit. How many of these additional capacitors must be inserted in series in the circuit, so the resonance frequency triples?

*31.** The resonance frequency of a series RCL-circuit that contains an 18-μH inductor is 13 MHz. The capacitor is an empty parallel plate capacitor. What is the resonance frequency when the capacitor is filled with a material whose dielectric constant is 5.2?

32. In a series RCL-circuit the dissipated power drops by a factor of two when the frequency of the generator is changed from the resonance frequency to a nonresonance frequency. The peak voltage is held constant while this change is made. Determine the power factor of the circuit at the nonresonance frequency.

**33. When the frequency is twice the resonance frequency, the impedance of a series RCL-circuit is twice the value of the impedance at resonance. Obtain the ratios of the inductive and capacitive reactances to the resistance, that is, obtain X_L/R and X_C/R.

ADDITIONAL PROBLEMS

34. In a series circuit, a generator (1350 Hz, 15.0 V) is connected to a 16.0-Ω resistor, a 4.10-μF capacitor, and a 5.30-mH inductor. Find the voltage across each circuit element.

35. A circuit consists of a 3.00-μF and a 6.00-μF capacitor connected in series across the terminals of a 510-Hz generator. The voltage of the generator is 120 V. (a) Determine the equivalent capacitance of the two capacitors. (b) Find the current in the circuit.

36. A simple metal detector consists of a series circuit formed by a 1.70-mH inductor, a 3.00-μF capacitor, and a generator with a voltage of 9.00 V. The inductor has the shape of a large coil (see the drawing). The resistance of the wire in the coil is 3.50 Ω. The frequency of the generator is held constant at the resonance frequency that applies when there is no metal passing through the coil. When a person with a metal object walks through the coil, the inductance increases and, consequently, the current in the circuit changes. The change in current can be used to sound a warning. Determine the current in the circuit (a) when no metal is present and (b) when a metal object causes a 4.0% increase in the inductance. (c) Find the change in current. Is it an increase or a decrease?

37. An 8.2-mH inductor is connected to an ac generator (10.0 V, 620 Hz). Determine the *peak value* of the current supplied by the generator.

38. The resonance frequency of a series RCL-circuit is 7.8 kHz. The inductance and capacitance of the circuit are each doubled. What is the new resonance frequency?

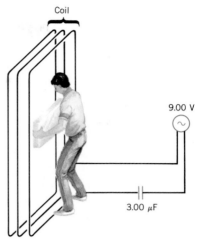

Coil

9.00 V

3.00 μF

Problem 36

*39. A series RCL-circuit contains a 5.10-μF capacitor and a generator whose voltage is 11.0 V. At a resonance frequency of 1.30 kHz the power dissipated in the circuit is 25.0 W. Find the values of (a) the inductance and (b) the resistance. (c) Calculate the power factor when the generator frequency is 2.31 kHz.

*40. The elements in a series RCL-circuit are a 106-Ω resistor, a 3.30-μF capacitor, and a 0.0310-H inductor. What is the impedance of the circuit and the phase angle between the current and the voltage when the frequency is 609 Hz?

**41. A 108-Ω resistor, a 0.200-μF capacitor, and a 5.42-mH inductor are connected in series to a generator whose voltage is 26.0 V. The current in the circuit is 0.141 A. Because of the shape of the current-versus-frequency graph (see Figure 23.14), there are two possible values for the frequency that correspond to this current. Obtain these two values.

Physics & the Environment

Since the beginning of the Industrial Revolution in the eighteenth century, factories and the consumers of the products they make have been throwing out waste products. Rivers were quickly polluted by the effluent from early mills, but the general assumption until recently was that the earth was so vast that the refuse would not affect it globally or permanently: "The solution to pollution is dilution," went the old adage. By the middle of this century the output of waste had increased to the point where its impact could no longer be ignored.

Imagine that you are an investigator for your state's Department of Environmental Protection, with the responsibility of monitoring water quality in drinking water reservoirs. Some days you spend scrambling around shorelines, or rowing out into deeper water, taking water samples, but more often than not you are in your laboratory analyzing and evaluating the samples that have been collected. How does basic physics affect your work?

Your lab is filled with instruments that analyze the chemicals contained in the water samples. All of these use the fundamental physics of matter to perform their functions. Spectrophotometers measure the unique frequencies of light absorbed by various elements and flu-orometers measure the fluorescent light emitted by the atoms and molecules. Today you are using a gas chromatograph. It indicates the chemical composition of the samples, using the fact that different elements and compounds have different masses, sizes, and boiling points. The time it takes different substances to get through the chromatograph varies with these factors, and the intensity of emission from the chromatograph tells how much there is of each substance in the water.

You compare the output of the chromatograph to a reference table, identifying decay products of herbicides and pesticides from agricultural activities upstream of the reservoir. Nitrates from fertilizer and phosphates from detergents and manufacturing are common pollutants. (In the Soviet Union, the nitrate pollution is so bad in the rivers feeding the Aral Sea that the water is undrinkable and lethal to fish. This area is considered to be one of the most polluted and dangerous bodies of water in the world.)

All the nitrate and phosphate readings are within Environmental Protection Agency (EPA) limits. All routine . . . no, what's that peak on the graph? It wasn't there last week. Checking the reference table makes it clear: dioxin! You look further, identifying three other industrial by-products banned by the EPA. Lethal wastes are leaching into the reservoir and a million and one-half people drink this water.

You study topological charts showing the water flowing into the reservoir. The water comes in from a river, let's call it the Darian, that in turn is fed by six smaller rivers and countless streams. To locate the dump site more water samples are needed, and fast. Three hours and several urgent phone calls later the first samples begin arriving. Twenty-five miles up river, beyond where the Acticala River joins the Darian, the samples show no dioxin. Below the Acticala, the dioxin is in the river. You direct your field teams to sample the water along the Acticala.

Here's where it gets tough. Samples show dioxin 8 miles up the Acticala, but the first industries along the river are 19 miles up river. Where the dioxin appears to be entering it, the river is lined with dairy farms and woodlands. Reconnaissance photos along the entire river reveal neither signs of illegal dumping nor the scruffy vegetation characteristic of areas containing pollutants. You conclude that the effluent is coming into the river through underground water, which seeps through porous rock (the aquifer).

If the pollutants are seeping into the aquifer and are being carried downhill by gravity into the Acticala, then the farther from their source they get, the more diluted they become. Maps show that there are plenty of wells for drinking and irrigation up from the river. You call for water samples from the wells. By noon the next day your map is covered with dioxin concentration numbers. As you hoped, the concentrations are pyramid-shaped, leading to an apex 3 miles up from the river.

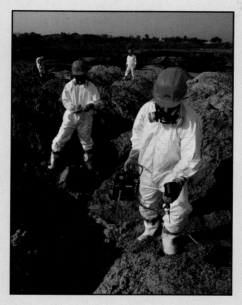

Protecting the environment against pollutants involves the coordinated efforts of many people. A number of the techniques used for detecting pollutants have their basis in physics principles.

You order a helicopter, load it with equipment, and head for that location. Hovering over the lightly wooded hillside, you see nothing suspicious. But somewhere there must be a buried, illegal dump. You point to a clearing one-half mile away and the 'copter sets down. You are wearing hot, confining, protective clothing, which makes you irritable and all the more eager to find the dump. A magnetometer is unloaded from the helicopter, which will measure the response of the ground around the suspected dump site to a magnetic field, similar to a metal detector for combing beaches. For an hour you systematically crisscross an acre of ground. Finally, you are rewarded by a signal from the magnetometer that indicates buried metal. Carefully, the crew begins digging and not 3 feet down they encounter the top of a 55-gallon drum. It is not the only one and several of them are leaking. While the cleanup is being done, you spend time back at your lab analyzing the contents of the drums for clues as to who manufactured it.

FOR INFORMATION ABOUT CAREERS RELATED TO THE ENVIRONMENT

Entrance into the environmental field in colleges and universities without specialized environmental studies programs is often through civil engineering or similar fields.

Encyclopedia of Careers and Vocational Guidance, *Vol. 3, 8th edition, 1990, William E. Hopke, ed., J. G. Ferguson, publisher, has an in-depth description of the working conditions groundwater professionals encounter.*

Career Information Center, *Vol. II, 4th edition, Glencoe/MacMillan, publishers, under the title "Environmentalists."*

Professional Careers Sourcebook, *K. M. Savage and C. A. Dorgan, eds., 1990, Gale Research, Inc., publisher, provides addresses and phone numbers for dozens of guides and other resources about careers in environment-related fields under the title "Civil Engineering."*

You can also write to the following, among many others listed in the books above:

■ *Association of Environmental Scientists and Engineers, 2718 Southwest Kelly, Portland, OR 97204*

■ *National Association of Environmental Professionals, Box 9400, Washington, DC 20016*

■ *American Society of Civil Engineers, 345 East 47th St., New York, NY 10017*

PART FIVE

LIGHT AND OPTICS

Light! We see by it, feel its warmth on a sunny day, and depend on it for our food supply. Plants convert sunlight into chemical energy by means of photosynthesis, and this energy sustains life for both plants and animals.

We have also learned how to use light in a number of ways, some of which are so commonplace that they are taken for granted. For example, mirrors are designed to reflect light, and they are commonplace items in homes and automobiles. Anyone who is nearsighted or farsighted uses eyeglasses or contact lenses to see better. These marvelous devices bend the light coming into the eye, so as to compensate for our vision problems. Microscopes, telescopes, and binoculars use a combination of mirrors and/or lenses to enlarge objects that would otherwise be too small to see. These instruments have immensely broadened our knowledge of the microscopic world and the ultra-macroscopic world of planets and stars. Cameras and video camcorders use lenses to record the good times with friends and family. Light is also used in the fields of communications and medicine. Many telephone cables are now being replaced by fiber optic bundles, which can carry an enormous number of high-quality telephone conversations in a very small space. And the light from lasers is used routinely in a number of surgical techniques. In Chapters 25 and 26 we will explore the physics of how light is reflected and bent and will discuss a number of optical devices.

Light is a wave and, therefore, interesting interference effects can arise when two or more light waves occupy the same space at the same time. We have already seen in Chapter 17 how interference between sound waves gives rise to the phenomena of diffraction, beats, and standing waves. In Chapter 27 we will investigate the interference of light, and show how it leads to colorful thin films (such as soap bubbles), to applications in compact disc technology, and to a limitation on the ability of the human eye to detect separately two objects that are close together.

CHAPTER 24

ELECTROMAGNETIC WAVES

*It was the great Scottish physicist James Clerk Maxwell (1831–1879) who showed that electric and magnetic fields fluctuating together can form a propagating wave, appropriately called an **electromagnetic wave.** We depend on electromagnetic waves in many important ways. For instance, virtually all our energy sources, such as petroleum, gas, and coal, are derived from sunlight, a mixture of electromagnetic waves having different wavelengths. And our sense of sight depends on us "seeing" a certain portion of this sunlight. Each wavelength in the light produces a different color sensation in our eyes, so the mixture of wavelengths can invoke the full spectrum of colors, from red to green to violet. The tropical reef fish in this photograph have beautiful colors, and yet the colors do not originate from the fish themselves. The colors are present in the sunlight that illuminates the fish. When the mixture of electromagnetic waves strikes the fish, different parts of the fish reflect certain wavelengths more strongly than other parts, giving each species its unique and colorful pattern. Thus, the colors of a fish "belong" to the electromagnetic waves that reflect from it. Most electromagnetic waves, however, are not visible to us, yet they are very important in our lives. Invisible electromagnetic waves are used each time we turn on a radio or TV, or when we use a cordless telephone. And these waves form the backbone of modern medicine, for they are used in a host of diagnostic tools, such as X-rays, magnetic resonance imaging, and gamma-ray treatment in cancer. This chapter is devoted to explaining the general properties of electromagnetic waves, and we will begin by showing one way to generate them.*

24.1 THE NATURE OF ELECTROMAGNETIC WAVES

Figure 24.1 illustrates one way to create an electromagnetic wave. The setup consists of two straight metal wires that are connected to the terminals of an ac generator and serve as an antenna. The potential difference between the terminals changes sinusoidally with time and has a period T. Part a shows the instant when there is no charge at the ends of either wire. Since there is no charge, there is no electric field at the point P just to the right of the antenna. As time passes, the top wire becomes positively charged and the bottom wire negatively charged, with the result that the electric field at P (drawn as a red arrow) points downward.* At one-quarter of a cycle later after part a ($t = \frac{1}{4}T$), the charges have attained their maximum values, as indicated in part b of the drawing. Correspondingly, the electric field at point P has increased to its maximum strength in the downward direction. This maximum field is represented by the red arrow in part b. Part b also indicates that the electric field created at earlier times (i.e., the black arrow in the picture) has not disappeared, but has moved to the right. Here lies the crux of the matter. At points far removed from the wire, the effect of the charges in creating the electric field is not felt immediately. Instead, the field is created first near the wires and then, like the effect of a pebble dropped into a pond, moves outward as a wave in all directions. Only the field moving to the right is shown in the picture for the sake of clarity. Eventually, after a time determined by the speed at which the wave travels, the field reaches distant points.

Parts c–e of Figure 24.1 show the creation of the electric field at point P (red arrow) at later times during the generator cycle. In each part, the fields produced earlier in the sequence (black arrows) continue propagating toward the right. Part d shows the charges on the wires when the polarity of the generator has reversed, so the top wire is negative and the bottom wire is positive. As a result, the electric field at P has reversed its direction and points upward. In part e of the sequence, a complete sine wave has been drawn through the tips of the electric field vectors to emphasize that the field changes sinusoidally.

So far, our focus has been on the electric field created by the charges on the antenna wires. However, a magnetic field is also created, because charges flowing in the antenna constitute an electric current, and an electric current creates a magnetic field. Figure 24.2 illustrates the magnetic field direction at point P at the instant when the current in the antenna wire is upward. With the aid of Right-Hand Rule No. 2 (thumb of right hand points along I, fingers curl in the direction of **B**), the magnetic field at P can be seen to point into the page. As the oscillating current changes in magnitude and direction, the magnetic field at P changes accordingly. As with electric fields, the magnetic fields created at earlier times propagate outward as a wave. Moreover, the magnetic field is zero when the electric field has its maximum positive or negative value, and the magnetic field has its maximum positive or negative value when the electric field is zero. In other words, the two fields are 90° out of phase with one another.

It is important to notice that the magnetic field in Figure 24.2 is perpendicular to the page, whereas the electric field in Figure 24.1 lies in the plane of the page. Thus, the electric and magnetic fields created by the antenna wires are mutually perpendicular, and they remain so as they move away from the antenna. More-

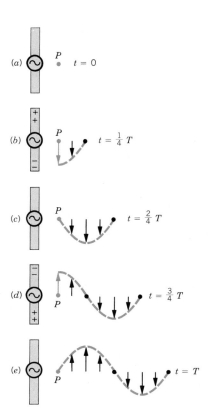

Figure 24.1 In each part of the drawing, the red arrow represents the electric field produced at point P by the oscillating charges on the antenna at the indicated time. The black arrows represent the electric fields created at earlier times. All the fields propagate to the right. Fields also are produced and propagate in other directions, but these fields are omitted for simplicity.

---

* The direction of the electric field can be obtained by imagining a positive test charge at P and determining the direction in which it would be pushed because of the charges on the wires.

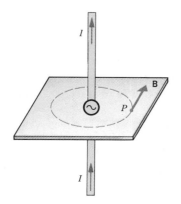

Figure 24.2 The oscillating current *I* in the antenna wires creates a magnetic field **B** at point *P*. The direction of **B** is tangent to a circle centered on the wire. When the current is upward, the magnetic field at point *P* is directed into the page. At a later time, when the current is downward, the field is directed out of the page.

over, both fields are perpendicular to the direction of travel. These perpendicular electric and magnetic fields, moving together, constitute an electromagnetic wave.

The electric and magnetic fields illustrated in the previous drawings decrease to zero rapidly with increasing distance from the antenna. Therefore, they exist mainly near the antenna and together are called the *near field.* Electric and magnetic fields do form an electromagnetic wave at large distances from the antenna, however. These fields arise from an effect that is different than that which produces the near field and are referred to as the *radiation field.* Faraday's law of electromagnetic induction provides part of the basis for the radiation field. As Section 22.4 discusses, this law describes the emf or potential difference produced by a changing magnetic field. And, as Section 19.4 explains, a potential difference can be related to an electric field. Thus, a changing magnetic field produces an electric field. Maxwell predicted that the reverse effect also occurs; namely, that a changing electric field produces a magnetic field. The radiation field arises, then, because the changing magnetic field creates an electric field that fluctuates in time and the changing electric field creates the magnetic field.

Figure 24.3 shows the electromagnetic wave of the radiation field far from the antenna. The picture shows only the part of the wave traveling along the +x axis. The part traveling in the other directions has been omitted for clarity. Note that for the radiation field, the electric and magnetic parts of the wave reach a maximum together. In other words, the fluctuating electric and magnetic fields are in phase, in contrast to the 90° phase relation for the near field.

It should be clear from Figure 24.3 that *an electromagnetic wave is a transverse wave,* because the electric and magnetic fields are both perpendicular to the direction in which the wave travels. Moreover, this kind of transverse wave, unlike a wave on a string, does not require a medium in which to propagate. *Electromagnetic waves can travel through a vacuum or a material substance,* since electric and magnetic fields can exist in either one.

Electromagnetic waves can be produced in situations that do not involve a wire antenna. In general, any electric charge that is accelerating emits an electromagnetic wave, whether the charge is inside a wire or not. In an alternating current, an electron oscillates in simple harmonic motion along the length of the wire and is one example of an accelerating charge.

All electromagnetic waves move through a vacuum at the same speed, and the symbol *c* is used to denote its value. This speed is called the *speed of light in a vacuum* and is $c = 3.00 \times 10^8$ m/s. In air, electromagnetic waves travel at nearly the same speed as they do in a vacuum, but, in general, they move through a substance such as glass at a speed that is less than *c*.

Figure 24.3 This picture shows the wave of the radiation field far away from the antenna. Observe that **E** and **B** are perpendicular to each other, and both are perpendicular to the direction of travel. The wave also travels outward in other directions, but for clarity only the part traveling along the positive *x* axis is shown.

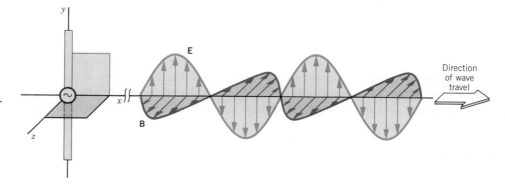

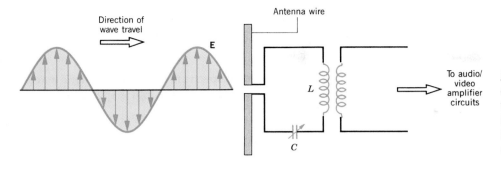

Figure 24.4 A radio wave can be detected with a receiving antenna wire that is parallel to the electric field of the wave. The magnetic field of the radio wave has been omitted for simplicity.

The frequency of an electromagnetic wave is determined by the vibration frequency of the electric charges at the source of the wave. In Figures 24.1–24.3 the wave frequency would equal the frequency of the ac generator. Suppose, for example, that the antenna is broadcasting the electromagnetic waves known as radio waves. The frequencies of AM radio waves lie between 545 and 1605 kHz, these numbers corresponding to the limits of the AM broadcast band on the dial. The frequencies of FM radio waves lie between 88 and 108 MHz on the dial. Television channels 2–6, on the other hand, utilize electromagnetic waves with frequencies between 54 and 88 MHz, while channels 7–13 use frequencies between 174 and 216 MHz.

Radio and television reception involves a process that is the reverse of that outlined earlier for the creation of electromagnetic waves. When broadcasted waves reach a receiving antenna, they interact with the electric charges in the antenna wires. Either the electric field or the magnetic field of the waves can be used. To take full advantage of the electric field, the wires of the receiving antenna must be parallel to the electric field, as Figure 24.4 indicates. The electric field acts on the electrons in the wire, forcing them to oscillate back and forth along the length of the wire. Consequently, an ac current exists in the antenna and the circuit connected to it. The variable-capacitor C and the inductor L in the circuit provide one way to select the frequency of the desired electromagnetic wave. By adjusting the value of the capacitance, it is possible to adjust the corresponding resonance frequency f_0 of the circuit [$f_0 = 1/(2\pi\sqrt{LC})$, Equation 23.10] to match the frequency of the wave. Under the condition of resonance there will be a maximum oscillating current in the inductor. Because of mutual inductance, this current creates a maximum voltage in the second coil in the drawing, and this voltage can then be amplified and processed by the remaining radio or television circuitry.

To detect the magnetic field of a broadcasted radio wave, a receiving antenna in the form of a loop can be used, as Figure 24.5 shows. For best reception, the normal to the plane of the wire loop is oriented parallel to the magnetic field. Then, as the wave sweeps by, the magnetic field penetrates the loop, and the

THE PHYSICS OF . . .

radio and television reception.

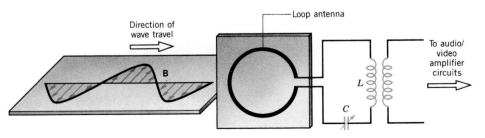

Figure 24.5 With a receiving antenna in the form of a loop, the magnetic field of a broadcasted radio wave can be detected. The normal to the plane of the loop should be parallel to the magnetic field for best reception. For clarity, the electric field of the radio wave has been omitted.

Both straight and loop antennas are being used on this cruise ship.

changing magnetic flux induces a voltage and a current in the loop, in accord with Faraday's law. Once again, the resonance frequency of a capacitor/inductor combination can be adjusted to match the frequency of the desired electromagnetic wave.

Radio waves are only one part of the broad spectrum of electromagnetic waves that has been discovered. The next section discusses the entire spectrum.

24.2 THE ELECTROMAGNETIC SPECTRUM

An electromagnetic wave, like any wave, has a frequency f and a wavelength λ that are related to the speed v of the wave by $v = f\lambda$ (Equation 16.1). For electromagnetic waves traveling through a vacuum or, to a good approximation, through air, the speed is $v = c$, so $c = f\lambda$.

As Figure 24.6 shows, electromagnetic waves exist with an enormous range of frequencies, from values less than 10^4 Hz to greater than 10^{22} Hz. The series of electromagnetic waves depicted in the drawing, arranged in order of their frequencies, is called the *electromagnetic spectrum.* Since all these waves travel through a vacuum at the same speed of $c = 3.00 \times 10^8$ m/s, Equation 16.1 can be used to find the correspondingly wide range of wavelengths that the picture also displays. Historically, regions of the electromagnetic spectrum have been given names such as radio waves and infrared waves. Although the boundary between two regions is drawn as a sharp line in the drawing, the boundary is not so well defined in practice, and the regions often overlap.

Beginning on the left in Figure 24.6, we find radio waves. The lower-frequency radio waves are generally produced by electric oscillator circuits, while the higher-frequency radio waves (called microwaves) are usually generated using electron tubes called klystrons. Infrared radiation, sometimes loosely called heat waves, originates with the vibration and rotation of molecules within a material. Visible light is emitted by hot objects, such as the sun, a burning log, or the

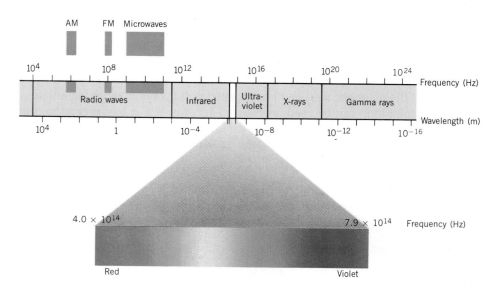

Figure 24.6 The electromagnetic spectrum.

filament of an incandescent light bulb, when the temperature is high enough to excite the electrons within an atom. Ultraviolet frequencies can be produced from the discharge of an electric arc. X-rays are produced by the sudden deceleration of high-speed electrons. And, finally, gamma rays are radiation from nuclear decay.

Of all the frequency ranges in the electromagnetic spectrum, the most familiar is that of visible light, although it is the smallest range indicated in Figure 24.6. Only waves with frequencies between about 4.0×10^{14} Hz and 7.9×10^{14} Hz are perceived by the human eye as visible light. Usually visible light is discussed in terms of wavelengths (in vacuum) rather than frequencies. As Example 1 indicates, the wavelengths of visible light are extremely small and, therefore, they are normally expressed in *nanometers* (nm); 1 nm $= 10^{-9}$ m. An obsolete (non-SI) unit occasionally used for wavelengths is the *angstrom* (Å); 1 Å $= 10^{-10}$ m.

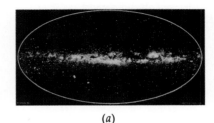

(a)

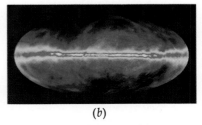

(b)

Two views of our galaxy (the Milky Way) as observed in two different regions of the electromagnetic spectrum. (a) The visible region and (b) the infrared region. (The infrared region is not visible to the eye, so the colors in this picture have been computer generated.)

Example 1   The Wavelengths of Visible Light

Find the range in wavelengths (in vacuum) for visible light in the frequency range between 4.0×10^{14} Hz (red light) and 7.9×10^{14} Hz (violet light). Express the answers in nanometers.

REASONING AND SOLUTION   The wavelength corresponding to a frequency of 4.0×10^{14} Hz is given by Equation 16.1 as

$$\lambda = \frac{c}{f} = \frac{3.00 \times 10^8 \text{ m/s}}{4.0 \times 10^{14} \text{ Hz}} = 7.5 \times 10^{-7} \text{ m}$$

Since 1 nm $= 10^{-9}$ m, it follows that $\boxed{\lambda = 750 \text{ nm}}$.

The calculation for a frequency of 7.9×10^{14} Hz is similar:

$$\lambda = \frac{c}{f} = \frac{3.00 \times 10^8 \text{ m/s}}{7.9 \times 10^{14} \text{ Hz}} = 3.8 \times 10^{-7} \text{ m} \quad \text{or} \quad \boxed{\lambda = 380 \text{ nm}}$$

The eye recognizes light of different wavelengths as different colors. A wavelength of 750 nm (in vacuum) is approximately the longest wavelength of red light, whereas 380 nm (in vacuum) is approximately the shortest wavelength of violet light. Between these limits are found the other familiar colors, as Figure 24.6 indicates.

The picture of light as a wave is supported by experiments that will be discussed in Chapter 27. However, there are also experiments indicating that light can behave as if it were composed of discrete particles, rather than waves. These experiments will be discussed in Chapter 29. Wave theories and particle theories of light have been around for hundreds of years, and it is now widely accepted that light, as well as other electromagnetic radiation, exhibits a dual nature. Either wave-like or particle-like behavior can be observed, depending on the kind of experiment being performed.

24.3 THE SPEED OF LIGHT

At a speed of 3.00×10^8 m/s, light travels from the earth to the moon in a little over a second, so the time required for light to travel between two places on earth is very short. In fact, early attempts at measuring the speed of light ran into just

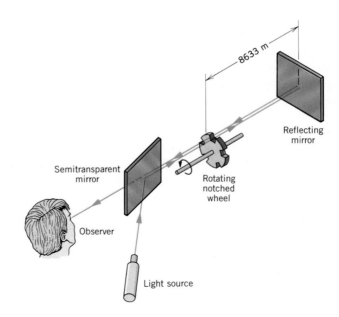

Figure 24.7 A version of the experiment that Fizeau carried out to measure the speed of light.

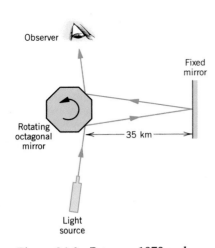

Figure 24.8 Between 1878 and 1931 Michelson used a rotating eight-sided mirror to measure the speed of light. The picture shown here presents a simplified version of the setup.

this problem. Galileo (1564–1642), for example, attempted to measure the speed of light at night by stationing a helper at a distant point. The helper was to uncover a lamp as soon as he saw the light from a lamp that Galileo had uncovered. Galileo hoped to measure the time it took for light to travel from his lamp to his assistant and for the reverse trip. Of course, the time was so short that the light appeared to move instantaneously from one point to the other, leaving Galileo only the conclusion that light travels very rapidly indeed.

In the first measurement of the speed of light that did not depend on astronomical observations, the French scientist Armand Fizeau (1819–1896) used a rotating notched wheel, as Figure 24.7 illustrates. The light from the source passes through one notch on its way to a mirror located some distance away (8633 m in Fizeau's experiment). After reflection, the light travels back and passes through another notch in the wheel only if the rotational speed of the wheel is appropriate. The rotational speed must be such that the time it takes for the light to travel the round-trip must match the time it takes for the second notch (or third or fourth, etc.) to rotate into the position previously occupied by the first notch. From a knowledge of the rotational speed of the wheel that permits the light to pass back through the second notch, Fizeau was able to determine the speed of light as 3.13×10^8 m/s.

More accurate measurements of the speed of light were performed later using a rotating mirror instead of a notched wheel. Figure 24.8 shows a simplified version of this setup. It was used first by the French scientist Jean Foucault (1819–1868) and later in a more refined version by the American physicist Albert Michelson (1852–1931), who obtained the value of $c = (2.997\ 96 \pm 0.000\ 04) \times 10^8$ m/s in 1926. The next example illustrates Michelson's method.

Example 2   The Angular Speed of Michelson's Rotating Mirror

If the angular speed of the rotating eight-sided mirror in Figure 24.8 is adjusted correctly, light reflected from one side travels to the fixed mirror, reflects, and can be detected after reflecting from another side that has rotated into place at just the right time. For one of his experiments, Michelson placed mirrors on Mt. San Antonio and Mt. Wilson in California, a distance of 35 km apart. Knowing that $c = 3.00 \times 10^8$ m/s, obtain the minimum angular speed for the eight-sided mirror in Michelson's experiment.

REASONING   The minimum angular speed is that at which one side of the mirror rotates one-eighth of a revolution during the time it takes for the light to make the round-trip between Mt. San Antonio and Mt. Wilson. The angular speed of the mirror can be determined by dividing its angular displacement ($\frac{1}{8}$ revolution) by the round-trip time. The round-trip time is the round-trip distance divided by the speed of light.

SOLUTION   The round-trip travel time is $t = 2(35 \times 10^3$ m$)/(3.00 \times 10^8$ m/s$) = 2.3 \times 10^{-4}$ s. The minimum angular speed ω is

$$\omega = \frac{\frac{1}{8}\text{ revolution}}{2.3 \times 10^{-4}\text{ s}} = \boxed{540\text{ rev/s}}$$

Today, the speed of light has been determined with such high accuracy that it is used to define the meter. As discussed in Section 1.2, the speed of light is now *defined* to be $c = 299\ 792\ 458$ m/s (although a value of 3.00×10^8 m/s is adequate for most calculations). The second is defined in terms of a cesium clock, and the meter is then defined as the distance light travels in a vacuum during a time of $1/(299\ 792\ 458)$ s.

In 1865 Maxwell determined theoretically that electromagnetic waves propagate through a vacuum at a speed given by

$$c = \frac{1}{\sqrt{\epsilon_0 \mu_0}} \tag{24.1}$$

where $\epsilon_0 = 8.85 \times 10^{-12}$ C^2/(N·m^2) is the (electric) permittivity of free space and $\mu_0 = 4\pi \times 10^{-7}$ T·m/A is the (magnetic) permeability of free space. Originally ϵ_0 was introduced in Section 18.5 as an alternative way of writing the proportionality constant k in Coulomb's law [$k = 1/(4\pi\epsilon_0)$] and, hence, plays a basic role in determining the strengths of the electric fields created by point charges. The role of μ_0 is similar for magnetic fields; it was introduced in Section 21.8 as part of a proportionality constant in the expression for the magnetic field created by the current in a long straight wire. Substituting the values for ϵ_0 and μ_0 into Equation 24.1 shows that

$$c = \frac{1}{\sqrt{[8.85 \times 10^{-12}\text{ C}^2/(\text{N·m}^2)][4\pi \times 10^{-7}\text{ T·m/A}]}} = 3.00 \times 10^8\text{ m/s}$$

The experimental and theoretical values for c agree. Maxwell's success in predicting c provided a basis for inferring that light behaves as a wave consisting of oscillating electric and magnetic fields.

24.4 THE ENERGY CARRIED BY ELECTROMAGNETIC WAVES

Electromagnetic waves, like water waves or sound waves, carry energy. The energy is carried by the electric and magnetic fields that comprise the wave. It is because of this energy, for example, that microwaves can cook a dinner. In a

THE PHYSICS OF . . .

a microwave oven.

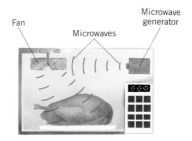

Fan

Microwaves

Microwave
generator

Figure 24.9 A microwave oven.
The rotating fan blades reflect the
microwaves to all parts of the oven.

microwave oven, microwaves pass directly through the food and deliver their energy to it in the process, as illustrated in Figure 24.9. The electric field of the microwaves is largely responsible for delivering the energy, and water molecules, one of the most abundant ingredients in food, readily absorb it. The reason water absorbs microwaves so readily is that water molecules have a permanent dipole moment; that is, one end of a water molecule has a slight positive charge and the other end a negative charge of equal magnitude. The electric field of the microwaves exerts forces on the positive and negative ends of the molecule, causing it to spin. Because the field is oscillating rapidly — about 2.4×10^9 times a second — the water molecules are kept spinning at a high rate. This produces "friction," which in turn converts the energy from the microwaves into heat, which is passed on to neighboring food molecules. Because microwaves penetrate all parts of the food, the heating is uniform as well as fast.

A measure of the energy stored in the electric field **E** of an electromagnetic wave, such as a microwave, is provided by the electric energy density. As we saw in Section 19.5, this density is the electric energy per unit volume of space in which the electric field exists:

$$\text{Electric energy density} = \frac{\text{Electric energy}}{\text{Volume}} = \tfrac{1}{2}\epsilon_0 E^2 \qquad (19.12)$$

where the dielectric constant κ has been set equal to unity, since we are dealing with an electric field in a vacuum (or air). From Section 22.8, the analogous expression for the magnetic energy density is

$$\text{Magnetic energy density} = \frac{\text{Magnetic energy}}{\text{Volume}} = \frac{1}{2\mu_0} B^2 \qquad (22.11)$$

The **total energy density** u of an electromagnetic wave in a vacuum is the sum of these two energy densities:

$$u = \frac{\text{Total energy}}{\text{Volume}} = \frac{1}{2}\epsilon_0 E^2 + \frac{1}{2\mu_0} B^2 \qquad (24.2)$$

In an electromagnetic wave propagating through a vacuum or air, the electric field and the magnetic field carry equal amounts of energy per unit volume of space. The fact that the two energy densities are equal means that the electric field is related to the magnetic field through the relation $E = cB$. To demonstrate this result, we set the electric energy density equal to the magnetic energy density:

$$\frac{1}{2}\epsilon_0 E^2 = \frac{1}{2\mu_0} B^2$$

Taking the square root of both sides of this equation and solving for E, we get $E = (1/\sqrt{\mu_0\epsilon_0})B$. But $c = 1/\sqrt{\mu_0\epsilon_0}$ according to Equation 24.1, so we have the result that

$$E = cB \qquad (24.3)$$

Since $\tfrac{1}{2}\epsilon_0 E^2 = (1/2\mu_0)B^2$, it is possible to rewrite Equation 24.2 for the total energy density in two additional, but equivalent, forms:

$$u = \frac{1}{2}\epsilon_0 E^2 + \frac{1}{2\mu_0} B^2 \qquad (24.2a)$$

$$u = \epsilon_0 E^2 \tag{24.2b}$$

$$u = \frac{1}{\mu_0} B^2 \tag{24.2c}$$

In an electromagnetic wave, the electric and magnetic fields fluctuate sinusoidally in time, so Equations 24.2 give the energy density of the wave at any instant in time. If an average value $\bar{u}$ for the total energy density is desired, average values are needed for E^2 and B^2. In Section 20.5 we faced a similar situation for alternating currents and voltages and introduced rms quantities. Using an analogous procedure here, it follows that the rms values for the electric and magnetic fields, E_{rms} and B_{rms}, are related to the maximum values of these fields, E_0 and B_0, by

$$E_{rms} = \frac{1}{\sqrt{2}} E_0 \quad \text{and} \quad B_{rms} = \frac{1}{\sqrt{2}} B_0$$

Equations 24.2a–c can now be interpreted as giving the average energy density $\bar{u}$, provided the symbols E and B are interpreted to mean the rms values given above. The average energy density of the sunlight reaching the earth is determined in the next example.

Example 3   The Average Energy Density of Sunlight

Sunlight enters the top of the earth's atmosphere with an electric field whose rms value is $E_{rms} = 720$ N/C. Find (a) the rms value of the sunlight's magnetic field and (b) the average total energy density of this electromagnetic wave.

REASONING   Since the magnitudes of the magnetic and electric fields are related according to Equation 24.3, the rms value of the magnetic field is $B_{rms} = E_{rms}/c$. The average energy density $\bar{u}$ in the sunlight can be obtained from Equation 24.2b, provided the rms value is used for the electric field.

SOLUTION
(a) The rms magnetic field is

$$B_{rms} = \frac{E_{rms}}{c} = \frac{720 \text{ N/C}}{3.0 \times 10^8 \text{ m/s}} = \boxed{2.4 \times 10^{-6} \text{ T}} \tag{24.3}$$

(b) The average energy density is

$$\bar{u} = \epsilon_0 E_{rms}^2 \tag{24.2b}$$

$$\bar{u} = (8.85 \times 10^{-12} \text{ C}^2/\text{N}\cdot\text{m}^2)(720 \text{ N/C})^2 = \boxed{4.6 \times 10^{-6} \text{ J/m}^3}$$

As an electromagnetic wave moves through space, it carries its energy from one region to another. This energy transport is characterized by the *intensity* of the wave. Recall that the intensity of a sound wave is the sound power that passes perpendicularly through a surface divided by the area of the surface (see Equation 16.8). In other words, intensity is power per unit area or energy per unit time per unit area. To help us apply this definition of intensity in the present situation, Figure 24.10 shows an electromagnetic wave traveling in a vacuum along the x axis. In a time t the wave travels the distance ct, passing through the surface of area A. Consequently, the volume of space through which the wave passes is ctA. The total (electric and magnetic) energy in this volume is

Total energy = (Total energy density) × Volume = $u(ctA)$

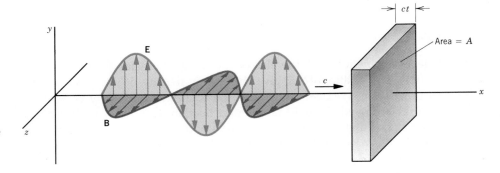

Figure 24.10   In a time t, an electromagnetic wave moves a distance ct along the x axis and passes through a surface of area A.

Using this expression, it can be seen that the intensity S and the total energy density u are related as follows:

$$S = \frac{\text{Total energy}}{\text{Time} \cdot \text{Area}} = \frac{uctA}{tA} = cu \qquad (24.4)$$

Substituting Equations 24.2a–c, one at a time, into Equation 24.4 shows that the intensity of an electromagnetic wave depends on the electric and magnetic fields according to the following equivalent relations:

$$S = cu = \frac{1}{2}\, c\epsilon_0 E^2 + \frac{c}{2\mu_0}\, B^2 \qquad (24.5a)$$

$$S = c\epsilon_0 E^2 \qquad (24.5b)$$

$$S = \frac{c}{\mu_0}\, B^2 \qquad (24.5c)$$

If the rms-values for the electric and magnetic fields are used in Equations 24.5a–c, the intensity becomes an average intensity $\overline{S}$, as Example 4 illustrates.

Example 4   A Neodymium-Glass Laser

A neodymium-glass laser emits short pulses of high-intensity electromagnetic waves. The electric field has an rms-value of $E_{\text{rms}} = 2.0 \times 10^9$ N/C. Find the average power of each pulse that passes through a 1.6×10^{-5}-m² surface that is perpendicular to the laser beam.

REASONING   Since the intensity of a wave is the power per unit area that passes through a surface perpendicular to the wave, the average power $\overline{P}$ of the wave is the product of the average intensity $\overline{S}$ and the area A.

SOLUTION   The average power of the wave is $\overline{P} = \overline{S}A$. But from Equation 24.5b, $\overline{S} = c\epsilon_0 E_{\text{rms}}^2$ so that

$$\overline{P} = c\epsilon_0 E_{\text{rms}}^2\, A$$

$$\overline{P} = (3.0 \times 10^8 \text{ m/s})(8.85 \times 10^{-12}\text{ C}^2/\text{N} \cdot \text{m}^2)(2.0 \times 10^9 \text{ N/C})^2(1.6 \times 10^{-5} \text{ m}^2)$$

$$= \boxed{1.7 \times 10^{11} \text{ W}}$$

24.5 POLARIZATION

POLARIZED ELECTROMAGNETIC WAVES

One of the essential features of electromagnetic waves is that they are transverse waves, and because of this feature they can be polarized. Figure 24.11 illustrates the idea of polarization by showing a transverse wave as it travels along a rope toward a slit. The wave is said to be *linearly polarized*, which means that its vibrations always occur along one direction. This direction is called the direction of polarization. In part *a* of the picture, the slit is oriented parallel to the direction of polarization, and the wave passes through easily. However, when the slit is turned perpendicular to the direction of polarization, as in part *b*, the wave cannot pass, because the slit prevents the rope from oscillating. Note that for longitudinal waves, such as sound waves, the notion of polarization has no meaning. In a longitudinal wave the direction of vibration is along the direction of travel, and, thus, the orientation of the slit would have no effect on the wave.

In an electromagnetic wave such as that in Figure 24.3, the electric field oscillates along the y axis. Similarly the magnetic field oscillates along the z axis. Therefore, the wave is linearly polarized, with the direction of polarization taken arbitrarily to be that along which the electric field oscillates. If the wave is a radio wave generated by a straight-wire antenna, the direction of polarization is determined by the orientation of the antenna. In comparison, the electromagnetic waves given off as light by an incandescent light bulb are completely unpolarized. In this case the light waves are emitted by a large number of atoms in the hot filament of the bulb. When an electron in an atom oscillates, the atom behaves as a miniature antenna that broadcasts light for brief periods of time, about 10^{-8} seconds. However, the directions of these atomic antennas change randomly, for the direction in which an electron oscillates changes randomly as a result of collisions. Unpolarized light, then, consists of many individual waves, emitted in short bursts by many "atomic antennas," each with its own direction of polarization. Figure 24.12 compares polarized and unpolarized light. In the unpolarized case, the arrows shown around the direction of wave travel symbolize the random directions of polarization of the individual waves that comprise the light.

Linearly polarized light can be produced from unpolarized light with the aid of certain materials. One commercially available material goes under the name of

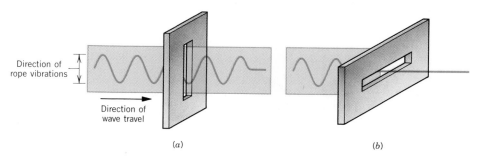

Direction of
rope vibrations

Direction of
wave travel

(a)   (b)

Figure 24.11   A transverse wave is linearly polarized when its vibrations always occur along one direction. (*a*) A linearly polarized wave on a rope can pass through a slit that is parallel to the direction of the rope vibrations, but (*b*) cannot pass through a slit that is perpendicular to the vibrations.

Figure 24.12 Polarized light consists of an electromagnetic wave in which the electric field fluctuates along a single direction. Unpolarized light consists of short bursts of electromagnetic waves emitted by many different atoms. The electric field directions of these bursts are perpendicular to the direction of wave travel but are distributed randomly about it.

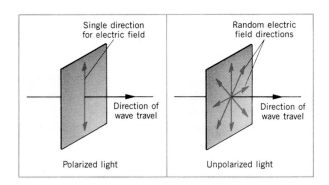

Polaroid. Such materials allow only the component of the electric field along one direction to pass through, while absorbing the field component perpendicular to this direction. As Figure 24.13 indicates, the direction of polarization that a polarizing material allows through is called the *transmission axis*. No matter how the transmission axis is oriented, the intensity of the transmitted polarized light is one-half that of the incident unpolarized light. The reason for this fact is that the unpolarized light contains all polarization directions to an equal extent. Moreover, the electric field for each direction can be resolved into components perpendicular and parallel to the transmission axis, with the result that the average components perpendicular and parallel to the transmission axis are equal. As a result, the polarizing material absorbs as much of the electric (and magnetic) field strength as it transmits, and the intensity of the transmitted polarized light is one-half the intensity of the incident unpolarized light.

MALUS' LAW

Once polarized light has been produced with a piece of polarizing material, it is possible to use a second piece to change the polarization direction and to adjust the intensity of the light. Figure 24.14 shows how. In this picture the first piece of polarizing material is called the *polarizer*, while the second piece is referred to as the *analyzer*. The transmission axis of the analyzer is oriented at an angle θ relative to the transmission axis of the polarizer. If the electric field strength of the polarized light incident on the analyzer is E, the field strength passing through is the component parallel to the transmission axis, $E \cos \theta$. According to Equation 24.5b, the intensity is proportional to the square of the electric field strength.

Figure 24.13 With the aid of a piece of polarizing material, polarized light may be produced from unpolarized light. The transmission axis of the material is the direction of polarization of the light that passes through the material.

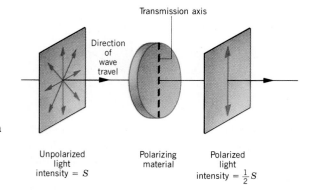

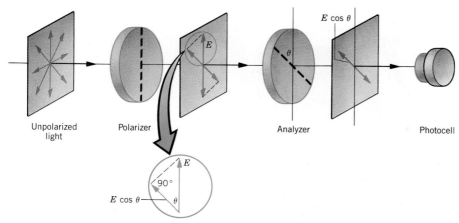

Figure 24.14 Two sheets of polarizing material, called the polarizer and the analyzer, may be used to adjust the polarization direction and intensity of the light reaching the photocell. This can be done by changing the angle θ between the transmission axes of the polarizer and analyzer.

Consequently, the average intensity of polarized light passing through the analyzer is proportional to $\cos^2 \theta$. Thus, both the polarization direction and the intensity of the light can be adjusted by rotating the transmission axis of the analyzer relative to that of the polarizer. The average intensity $\overline{S}$ of the light leaving the analyzer, then, is

[Malus' law]
$$\overline{S} = \overline{S}_0 \cos^2 \theta \qquad (24.6)$$

where $\overline{S}_0$ is the average intensity of the light entering the analyzer. Equation 24.6 is sometimes called *Malus' law*, for it was discovered by the French engineer Etienne-Louis Malus (1775–1812). Example 5 illustrates the use of Malus' law.

Example 5   Using Polarizers and Analyzers

What value of θ should be used in Figure 24.14, so the average intensity of the polarized light reaching the photocell is one-tenth the average intensity of the unpolarized light?

REASONING   Both the polarizer and the analyzer reduce the intensity of the light. The polarizer reduces the intensity by a factor of one-half, as discussed earlier. Therefore, if the average intensity of the unpolarized light is $\overline{I}$, the average intensity of the polarized light leaving the polarizer and striking the analyzer is $\overline{S}_0 = \overline{I}/2$. The angle θ must now be selected so the average intensity of the light leaving the analyzer is $\overline{S} = \overline{I}/10$.

SOLUTION   Using $\overline{S}_0 = \overline{I}/2$ and $\overline{S} = \overline{I}/10$ in Malus' law, we find that $\overline{I}/10 = (\overline{I}/2) \cos^2 \theta$. Solving this relation for $\cos \theta$ yields

$$\cos \theta = \sqrt{\tfrac{1}{5}} = 0.447 \quad \text{and} \quad \theta = \cos^{-1}(0.447) = \boxed{63.4°}$$

PROBLEM SOLVING INSIGHT

Remember that when unpolarized light strikes a polarizer, only one-half of the incident light is transmitted, the other half being absorbed by the polarizer.

When $\theta = 90°$ in Figure 24.14, the polarizer and analyzer are said to be *crossed,* and no light is transmitted by the polarizer/analyzer combination. As an

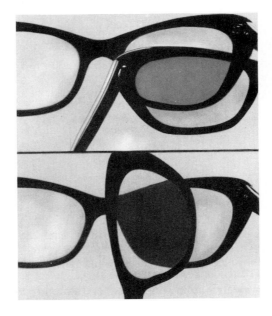

Figure 24.15   When Polaroid sunglasses are uncrossed (top photograph), the transmitted light is dimmed due to the extra thickness of tinted plastic. However, when they are crossed (lower photograph), the transmitted light is reduced to zero because of the effects of polarization.

illustration of this effect, Figure 24.15 shows two pairs of Polaroid sunglasses in uncrossed and crossed configurations.

An application of a crossed polarizer/analyzer combination occurs in one kind of liquid crystal display (LCD). LCDs are widely used in pocket calculators and digital watches. The display usually consists of blackened numbers and letters set against a light gray background. As Figure 24.16 indicates, each number or letter is formed from a combination of liquid crystal segments that have been "turned on" and appear black. Let us now see what it means for a liquid crystal to be turned on and how polarized light is used.

The liquid crystal part of an LCD segment consists of the liquid crystal material sandwiched between two transparent electrodes, as in Figure 24.17. When a voltage is applied between the electrodes, the liquid crystal is said to be "on." Part *a* of the picture shows that linearly polarized incident light passes through the "on" material without having its direction of polarization affected. When the voltage is removed, as in part *b*, the liquid crystal is said to be "off" and now rotates the direction of polarization by 90°.

T H E   P H Y S I C S   O F . . .

a liquid crystal display.

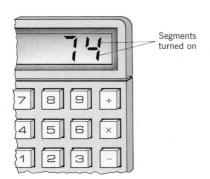

Figure 24.16   Liquid crystal displays use liquid crystal "segments" to form the numbers.

Segments turned on

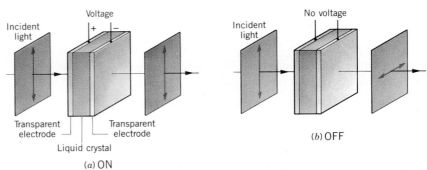

Incident light

Voltage

Transparent electrode

Transparent electrode

Liquid crystal

(*a*) ON

Incident light

No voltage

(*b*) OFF

Figure 24.17   A liquid crystal in its (*a*) "on" state and (*b*) "off" state.

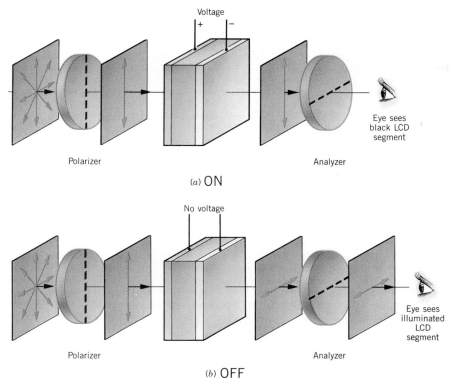

Figure 24.18 A liquid crystal display (LCD) incorporates a crossed polarizer/analyzer combination. (*a*) When the LCD segment is turned on, a voltage is applied to the electrodes, no light is transmitted through the analyzer, and the observer sees a black segment. (*b*) The LCD segment is turned off (no voltage), and light from the segment, which is the same color as the background light, reaches the observer. The segment is invisible, however, due to a lack of contrast between it and the background.

A complete LCD segment includes a crossed polarizer/analyzer combination, as Figure 24.18 illustrates. The polarizer, analyzer, electrodes, and liquid crystal material are packaged as a single unit. The polarizer produces polarized light from incident unpolarized light. With the display segment turned on, as in part *a*, the polarized light emerges from the liquid crystal only to be absorbed by the analyzer, since the light is polarized perpendicular to the transmission axis of the analyzer. Since no light emerges from the analyzer, an observer sees a black segment against a light gray background, as in Figure 24.16. On the other hand, the segment is turned off when the voltage is removed, in which case the liquid crystal rotates the direction of polarization by 90° to coincide with the axis of the analyzer, as in Figure 24.18*b*. The light now passes through the analyzer and enters the eye of the observer. However, the light coming from the segment has been designed to have the same color and shade (light gray) as the background of the display, so the segment becomes indistinguishable from the background.

THE OCCURRENCE OF POLARIZED LIGHT IN NATURE

Polaroid is a familiar material because of its widespread use in sunglasses. Such sunglasses are designed so that the axis of the Polaroid is oriented vertically when the glasses are worn in the usual fashion. Thus, the glasses prevent any light that

THE PHYSICS OF . . .

Polaroid sunglasses.

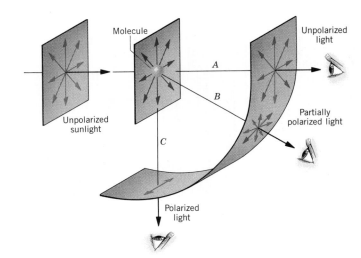

Figure 24.19 In the process of being scattered from atmospheric molecules, unpolarized light from the sun becomes polarized.

"Fido" learns about polarizing sunglasses.

is polarized horizontally from reaching the eye. Light from the sun is unpolarized, but a considerable amount of horizontally polarized sunlight originates by reflection from horizontal surfaces such as that of a lake. Section 26.4 discusses this effect. Polaroid sunglasses reduce glare by preventing the horizontally polarized reflected light from reaching the eyes.

Polarized sunlight also originates from the scattering of light by molecules in the atmosphere. Figure 24.19 shows light being scattered by a single atmospheric molecule. The electric fields in the unpolarized sunlight cause the electrons in the molecule to vibrate perpendicular to the direction in which the light is traveling. The electrons, in turn, reradiate the electromagnetic waves in different directions, as the drawing illustrates. The light radiated straight ahead in direction *A* is unpolarized, just like the incident light. But light radiated perpendicular to the incident light in direction *C* is polarized. Light radiated in the intermediate direction *B* is partially polarized.

INTEGRATION OF CONCEPTS

ELECTROMAGNETIC WAVES AND OTHER KINDS OF WAVES

A wave is a disturbance that travels from place to place, carrying energy as it goes. The waves that we have discussed so far include water waves, waves on a string, and sound waves. To this list we now add electromagnetic waves. As an electromagnetic wave passes by, the electric and magnetic fields of the wave can disturb electric charges and currents by exerting forces on them. Electromagnetic waves share with all waves the ability to transport energy. To describe electromagnetic waves, we use the same concepts that apply to other waves, namely, wavelength, frequency, amplitude, speed, and intensity. As is the case for other waves, the product of the wavelength and frequency of an electromagnetic wave equals the wave speed. Like a wave on a string, an electromagnetic wave is a transverse wave, for its electric and magnetic fields are perpendicular to the direction in which the wave travels. Because of its transverse nature, an electromagnetic wave can be polarized. In contrast, a sound wave is a longitudinal wave and, for such waves, the idea of polarization has no meaning. Electromagnetic waves also have some characteristics that distinguish them from other kinds of waves. Most significantly, they are the only waves that can travel through a vacuum. Compared to other kinds of waves, electromagnetic waves travel very fast (3.00×10^8 m/s, in a vacuum). And they encompass a remarkably wide range of frequencies, from less than 10^4 Hz to greater than 10^{22} Hz. These distinguishing characteristics have led to the widespread use of electromagnetic waves to carry information in the field of telecommunications.

SUMMARY

An **electromagnetic wave** in a vacuum consists of mutually perpendicular and oscillating electric and magnetic fields. The wave is a transverse wave, since the fields are perpendicular to the direction in which the wave travels. All electromagnetic waves, regardless of their frequency, travel through a vacuum at the same speed, the speed of light c ($c = 3.00 \times 10^8$ m/s).

The frequency f and wavelength λ of an electromagnetic wave in a vacuum are related to its speed through the relation $c = f\lambda$. The series of electromagnetic waves, arranged in order of their frequencies, is called the **electromagnetic spectrum**. The electromagnetic spectrum is composed of groups of waves that are known as radio waves, infrared radiation, visible light, ultraviolet radiation, X-rays, and gamma rays. **Visible light** has frequencies between about 4.0×10^{14} Hz and 7.9×10^{14} Hz. The human eye and brain perceive different frequencies as different colors.

Maxwell calculated that the speed of light in a vacuum is $c = 1/\sqrt{\epsilon_0 \mu_0}$, where ϵ_0 is the (electric) permittivity of free space and μ_0 is the (magnetic) permeability of free space.

The **total energy density** u of an electromagnetic wave is the total energy per unit volume of the wave and, in a vacuum, is given by $u = \frac{1}{2}\epsilon_0 E^2 + B^2/(2\mu_0)$, where E and B are the magnitudes of the electric and magnetic fields. In a vacuum, E and B are related by $E = cB$, and the electric and magnetic parts of the total energy density are equal.

The **intensity** of an electromagnetic wave is the power that the wave carries perpendicularly through a surface divided by the area of the surface. In a vacuum, the intensity S is related to the total energy density u according to $S = cu$.

A **linearly polarized** electromagnetic wave is one in which all oscillations of the electric field occur along one direction, which is taken to be the direction of polarization. In **unpolarized light** the direction of polarization does not remain fixed, but fluctuates randomly in time.

Polarizing materials allow only the component of the wave's electric field along one direction to pass through them. The preferred transmission direction for the electric field is called the **transmission axis** of the material. When unpolarized light is incident on a piece of polarizing material, the transmitted polarized light has an intensity that is one-half that of the incident light. When two pieces of polarizing material are used one after the other, the first one is called the polarizer, while the second one is referred to as the analyzer. If the average intensity of polarized light falling on an analyzer is $\overline{S}_0$, the average intensity $\overline{S}$ of the light leaving the analyzer is given by **Malus' law** as $\overline{S} = \overline{S}_0 \cos^2 \theta$, where θ is the angle between the transmission axes of the polarizer and analyzer. When $\theta = 90°$, the polarizer and the analyzer are said to be "crossed," and no light passes through the analyzer.

QUESTIONS

1. Compare the properties of electromagnetic waves and sound waves by answering "yes" or "no" to the questions in the table below.

Property	Electromagnetic Wave	Sound Wave
Transverse?		
Longitudinal?		
Can be polarized?		
Can travel through a vacuum?		
Can travel through a material such as glass?		
Involves electric and magnetic fields?		
Involves pressure oscillations?		

2. A transmitting antenna is located at the origin of an x, y, z axis system and broadcasts an electromagnetic wave whose electric field oscillates along the y axis. The wave travels along the $+x$ axis. There are three possible wire loops that can be used with an LC-turned circuit to detect this wave: One loop lies in the xy plane, another in the xz plane, and the third in the yz plane. Which of the loops will detect the wave? Why?

3. In Section 17.3 we discussed the diffraction of sound waves, that is, the ability of the waves to bend around obstacles. Electromagnetic waves also have the same ability. On the basis of this earlier discussion and the fact that AM radio waves have larger wavelengths than FM radio waves, would you expect AM or FM radio waves to exhibit a greater ability to bend around an obstacle such as a building? Explain.

4. The speed of sound is about one-fifth of a mile per second, and there is a rule of thumb based on this fact. The rule specifies that if you divide the number of seconds between a lightning flash and the associated thunder clap by five, you get the approximate distance (in miles) to the lightning. Explain why this rule works, using what you know about the speed of sound compared to the speed of light. Would this rule be useful if the speed of light were nearly equal to the speed of sound? Why?

5. Refer to Figure 24.8 and Example 2 in the text. Would light be detected by the observer if the eight-sided mirror were to rotate at an angular speed of 1080 rev/s, instead of 540 rev/s? Explain.

6. Why is it said that astronomers looking at distant stars are "looking back in time"?

7. Malus' law applies to the setup in Figure 24.14, which shows the analyzer rotated through an angle θ and the polarizer held fixed. Does Malus' law apply when the analyzer is held fixed and the polarizer is rotated? Give your reasoning.

8. Light is incident from the left on two pieces of polarizing material, 1 and 2. As part a of the drawing illustrates, the transmission axis of material 1 is along the vertical direction, while that of material 2 makes an angle of θ with respect to the vertical. In part b of the drawing the two polarizing materials are interchanged. (a) Assume the incident light is unpolarized and determine whether the intensity of the transmitted light in part a is greater than, equal to, or less than that in part b. (b) Repeat part (a), assuming the incident light is linearly polarized along the vertical direction. Justify your answers to both parts (a) and (b).

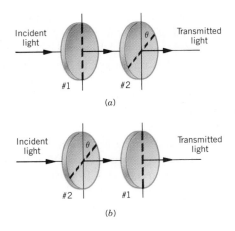

PROBLEMS

Section 24.1 The Nature of Electromagnetic Waves

1. The mean distance from the sun to the earth is 1.50×10^{11} m. How long does it take for sunlight to reach the earth?

2. In astronomy, distances are often expressed in light-years. One light-year is the distance traveled by light in one year. The distance to Alpha Centauri, the closest star to us other than our own sun, is 4.3 light-years. Express this distance in meters.

3. In Figure 24.4 the value of the inductance is $2.6 \times$

10^{-4} H. For an AM radio station broadcasting at a frequency of 1200 kHz, find the value to which the capacitor must be adjusted.

4. For an FM radio station broadcasting at a frequency of 88.0 MHz, the capacitance in Figure 24.4 must be adjusted to a value of 23.0×10^{-12} F. Assuming the inductance does not change, determine the value of the capacitance for an FM station broadcasting at 108.0 MHz.

*** 5.** Equation 16.3, $y = A \sin (2\pi ft - 2\pi x/\lambda)$, gives the mathematical representation of a wave oscillating in the y direction and traveling in the positive x direction. Let y in this equation equal the electric field of an electromagnetic wave traveling in a vacuum. Assuming that the maximum electric field is $A = 156$ N/C and that the frequency is $f = 1.50 \times 10^8$ Hz, plot a graph of the electric field strength versus position, using for x the following values: 0, 0.50, 1.00, 1.50, and 2.00 m. Plot this graph for (a) a time $t = 0$ and (b) a time t that is one-fourth of the wave's period.

Section 24.2 The Electromagnetic Spectrum

6. Some of the X-rays produced in an X-ray machine have a wavelength of 2.1 nm. What is the frequency of these electromagnetic waves?

7. A truck driver is broadcasting at a frequency of 26.965 MHz with a CB (citizen's band) radio. Determine the wavelength of the electromagnetic wave being used.

8. "Microwave" is the term applied to electromagnetic waves that have shorter wavelengths than radio waves, but longer wavelengths than infrared and visible waves. What is the ratio of the wavelength of microwaves ($f = 2.4 \times 10^9$ Hz) used in a microwave oven to radio waves ($f = 6.0 \times 10^5$ Hz) of an AM radio station?

9. A radio station broadcasts a radio wave whose wavelength is 274 m. (a) What is the frequency of the wave? (b) Is this radio wave AM or FM? (See Figure 24.6.)

10. Obtain the wavelengths in vacuum for blue light with a frequency of 6.34×10^{14} Hz and for orange light with a frequency of 4.95×10^{14} Hz. Express your answers in nanometers (nm).

*** 11.** Sections 17.5 and 17.6 deal with standing waves. Electromagnetic waves also can form standing waves. In a standing wave pattern formed from microwaves, the distance between a node and an adjacent antinode is 0.50 cm. What is the microwave frequency?

Section 24.3 The Speed of Light

12. Ghost images are formed in a TV picture when the electromagnetic wave from the broadcasting antenna reflects from a building or other large object and arrives at the TV set shortly after the wave coming directly from the broadcasting

antenna. If the reflected wave arrives 4.0×10^{-6} s after the direct wave, what is the difference in distances traveled by the two waves?

13. (a) Neil A. Armstrong was the first person to walk on the moon. The distance between the earth and the moon is 3.85×10^8 m. Find the time it took for his voice to reach earth via radio waves. (b) Determine the communication time for the first person who will some day walk on Mars, which is 5.6×10^{10} m from earth at the point of closest approach.

14. In Fizeau's experiment (see Figure 24.7) the rotating wheel contained 720 notches. Knowing that the speed of light in air is 3.00×10^8 m/s, obtain the angular speed (in rev/s) of the wheel if the incident light is to pass through one notch and the reflected light is to pass through an adjacent notch.

15. The distance between earth and the moon can be determined from the time it takes for a laser beam to travel from earth to a reflector on the moon and back. If the round-trip time can be measured to an accuracy of one-tenth of a nanosecond (1 ns = 10^{-9} s), what is the corresponding error in the earth–moon distance?

*** 16.** A mirror faces a cliff, located some distance away. Mounted on the cliff is a second mirror, directly opposite the first mirror and facing toward it. A gun is fired very close to the first mirror. The speed of sound is 343 m/s. How many times does the flash of the gunshot travel the round-trip distance between the mirrors before the echo of the gunshot is heard?

*** 17.** The President of the United States holds a press conference, which is televised live. A television viewer hears the sound picked up by a microphone directly in front of the president. This viewer is seated 2.0 m from the television set. A reporter at the press conference is located 5.0 m from the stage and hears the president's words directly *at the very same instant* that the television viewer hears them. Using a value of 343 m/s for the speed of sound, determine the maximum distance between the television viewer and the president.

Section 24.4 The Energy Carried by Electromagnetic Waves

18. A laser emits a narrow beam of light. The radius of the beam is 1.0×10^{-3} m, and the power is 1.2×10^{-3} W. What is the intensity of the laser beam?

19. Suppose the electric field in an electromagnetic wave has a maximum strength of 2140 N/C. What is the maximum strength of the magnetic field of the wave?

20. The rms-value of the electric field in an electromagnetic wave is 123 N/C. The wave passes perpendicularly through a surface of area 0.350 m². How much energy does this wave carry across the surface in one minute?

21. A future space station in orbit about the earth is being powered by an electromagnetic beam from the earth. The beam has a cross-sectional area of 135 m² and transmits an average power of 1.20×10^4 W. What are the rms-values of the (a) electric and (b) magnetic fields?

22. Show that, in addition to Equations 24.2a–24.2c, the total energy density for an electromagnetic wave can be expressed as $u = (\sqrt{\epsilon_0/\mu_0})EB$.

* **23.** An argon-ion laser produces a cylindrical beam of light whose average power is 0.750 W. How much energy is contained in a 2.50-m length of the beam?

* **24.** In an electromagnetic wave the magnetic field strength has an rms-value of 9.11×10^{-8} T. Find (a) the rms-value of the electric field strength, (b) the average total energy density of the wave, and (c) the average intensity of the wave.

** **25.** The mean distance between earth and the sun is 1.50×10^{11} m. The average intensity of solar radiation incident on the upper atmosphere of the earth is 1390 W/m². Assuming the sun emits radiation uniformly in all directions, determine the total power radiated by the sun.

Section 24.5 Polarization

26. Linearly polarized light is incident on a piece of polarizing material. What is the ratio of the transmitted light intensity to the incident light intensity when the angle between the transmission axis and the incident electric field is (a) 25° and (b) 65°?

27. Unpolarized light whose intensity is 1.10 W/m² is incident on the polarizer in Figure 24.14. (a) What is the intensity of the light leaving the polarizer? (b) If the analyzer is set at an angle of $\theta = 75°$ with respect to the polarizer, what is the intensity of the light that reaches the photocell?

28. What should be the angle between the transmission axes of the polarizer and the analyzer in Figure 24.14, so the polarized light reaching the photocell has an intensity that is (a) one-half and (b) one-fourth of the intensity of the incident unpolarized light?

29. In the polarizer/analyzer combination in Figure 24.14, 90.0% of the light intensity falling on the analyzer is absorbed. Determine the angle between the transmission axes of the polarizer and the analyzer.

* **30.** The orientation of the transmission axis for each of the three sheets of polarizing material in the drawing is labeled relative to the vertical. A beam of light, polarized in the vertical direction, is incident on the first sheet. The intensity of the incident beam is 1550 W/m². Obtain the intensity of the beam transmitted through the three sheets when: (a) $\theta_1 = 0°$, $\theta_2 = 40.0°$, $\theta_3 = 75.0°$; (b) $\theta_1 = 30.0°$, $\theta_2 = 30.0°$, $\theta_3 = 70.0°$.

* **31.** The polarizer and the analyzer in Figure 24.14 are crossed ($\theta = 90.0°$), and no light falls on the photocell. Then, a third piece of polarizing material is put between the polarizer

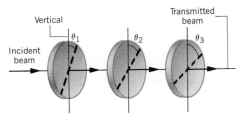

Problem 30

and the analyzer, with its transmission axis oriented at 45.0° relative to the transmission axes of the polarizer and the analyzer. If the unpolarized light intensity incident on the polarizer is I, what fraction of I now falls on the photocell?

** **32.** More than one analyzer can be used in a setup like that in Figure 24.14, each analyzer following the previous one. Suppose that the transmission axis of the first analyzer is rotated 27° relative to the transmission axis of the polarizer, and that the transmission axis of each additional analyzer is rotated 27° relative to the transmission axis of the previous one. What is the minimum number of analyzers needed, so the light reaching the photocell has an intensity that is reduced by at least a factor of one hundred relative to that striking the first analyzer?

ADDITIONAL PROBLEMS

33. Determine the range of wavelengths for FM radio waves with frequencies between 88.0 and 108.0 MHz.

34. An industrial laser is used to burn a hole through a piece of metal. The average intensity of the light is $\bar{S} = 1.23 \times 10^9$ W/m². What is the rms-value of (a) the electric field and (b) the magnetic field in the electromagnetic wave emitted by the laser?

35. In 1980 and 1981, two Voyager spacecraft sent back beautiful photographs of Saturn via radio transmission. If the distance between earth and Saturn was 1.277×10^{12} m, how much time (in minutes) was required for the transmission?

36. The average intensity of sunlight at the top of the earth's atmosphere is 1390 W/m². What is the maximum energy that a 25-m × 45-m solar panel could collect in one hour in this sunlight?

37. TV channel 3 (VHF) broadcasts at a frequency of 63.0 MHz. TV channel 23 (UHF) broadcasts at a frequency of 527 MHz. Find the ratio (VHF/UHF) of the wavelengths for these channels.

38. Polarized light strikes a piece of polarizing material. The incident light is polarized at an angle of 30.0° relative to the transmission axis of the material. What percentage of the light intensity is transmitted?

*39. As Section 24.3 discusses, Maxwell calculated that the speed of light in a vacuum is given by $c = 1/\sqrt{\epsilon_0\mu_0}$. The unit for ϵ_0 is $C^2/(N \cdot m^2)$ and the unit for μ_0 is $T \cdot m/A$. Show that the unit for $1/\sqrt{\epsilon_0\mu_0}$ is meters per second.

*40. In experiment 1, unpolarized light falls on the polarizer in Figure 24.14, with $\theta = 60.0°$. In experiment 2, the unpolarized light is replaced with light of the same intensity, but which is polarized along the direction of the polarizer transmission axis. By how many *additional* degrees and in what direction must the analyzer be rotated, so the light falling on the photocell has the same intensity as it did in experiment 1?

*41. A flat coil of wire is used with an LC-tuned circuit as a receiving antenna. The coil has a radius of 0.25 m and consists of 450 turns. The transmitted radio wave has a frequency of 1.2 MHz. The magnetic field of the wave is parallel to the normal to the coil and has a maximum value of 2.0×10^{-13} T. Using Faraday's law of electromagnetic induction and the fact that the magnetic field changes from zero to its maximum value in one-quarter of a wave period, find the magnitude of the average emf induced in the antenna during this time.

*42. The average intensity of sunlight reaching the earth is 1390 W/m^2. A charge of 2.6×10^{-8} C is placed in the path of this electromagnetic wave. (a) What is the maximum electric force that the charge experiences? (b) If the charge is moving at a speed of 3.7×10^4 m/s, what is the maximum magnetic force that the charge could experience?

**43. A tiny source of light emits light uniformly in all directions. The average power emitted is 60.0 W. For a point located 8.00 m away from this source, determine the rms electric and magnetic field strengths in the light waves.

**44. Suppose that the light falling on the polarizer in Figure 24.14 is partially polarized (average intensity = $\bar{S}_P$) and partially unpolarized (average intensity = $\bar{S}_U$). The total incident intensity is $\bar{S}_P + \bar{S}_U$, and the percentage polarization is $100\bar{S}_P/(\bar{S}_P + \bar{S}_U)$. When the polarizer is rotated in such a situation, the intensity reaching the photocell varies between a minimum value of $\bar{S}_{min}$ and a maximum value of $\bar{S}_{max}$. Show that the percentage polarization can be expressed as $100(\bar{S}_{max} - \bar{S}_{min})/(\bar{S}_{max} + \bar{S}_{min})$.

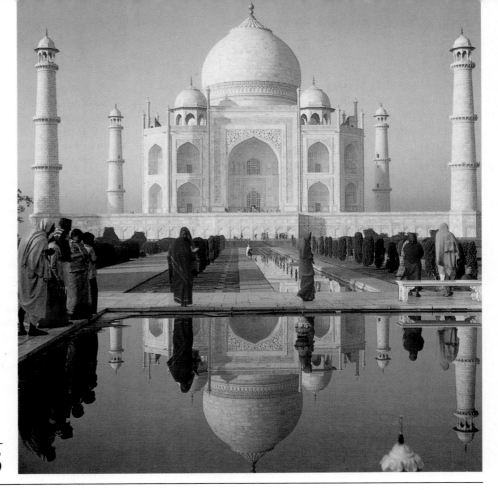

CHAPTER 25

THE REFLECTION OF LIGHT: MIRRORS

The reflection of light can produce dramatic and beautiful results, as it does in the photograph above, where the smooth surface of the water acts like a mirror and creates the stunning reflection of the Taj Mahal, near New Delhi, India. The reflection of light does more than produce pretty pictures, however. Without reflection, for example, our sense of vision would not be nearly as useful as it is. When we "see" an object, light from it enters our eyes and evokes the sensation of vision. Some objects themselves produce the light that we see, like the sun, a flame, or a light bulb. Most objects, however, reflect into our eyes light that originates elsewhere. Reflection also plays a major part, as we will learn, in some of the techniques being used to harness solar energy as an alternate energy source. And we have already seen in Section 18.9 that the reflection of light plays a role in the operation of laser printers. We begin our discussion of reflection with the fundamental concepts of a wave front and a ray of light.

25.1 WAVE FRONTS AND RAYS

To introduce the concepts of a wave front and a ray, we take advantage of a topic that we have studied before, sound waves. Both sound and light are kinds of waves and therefore have some similarities. In particular, the ideas of a wave front and a ray apply to both.

Consider a small spherical object whose surface is pulsating in simple harmonic motion. A sound wave is emitted that moves spherically outward from the object at a constant speed. To represent this wave, we draw surfaces through all points of the wave that are in the same phase of motion. These surfaces of constant phase are called *wave fronts*. Figure 25.1 shows a two-dimensional view of the wave fronts. In this view the wave fronts appear as concentric circles about the vibrating object. If the wave fronts are drawn through the condensations, or crests, of the sound wave, as they are in the picture, the distance between adjacent wave fronts equals the wavelength λ. The radial lines pointing outward from the source and perpendicular to the wave fronts are called *rays*. The rays point in the direction of the velocity of the wave.

Figure 25.2*a* shows a small section of two adjacent spherical wave fronts. At large distances from the source, the wave fronts become less and less curved and approach the shape of flat surfaces, as part *b* of the drawing shows. Waves whose wave fronts are flat surfaces (i.e., planes) are known as *plane waves* and are

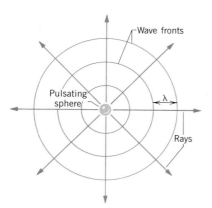

Figure 25.1   A cross-sectional view of a sound wave emitted by a pulsating sphere. The wave fronts are drawn through the condensations of the wave, so the distance between two successive wave fronts is the wavelength λ. The rays are perpendicular to the wave fronts and point in the direction of the velocity of the wave.

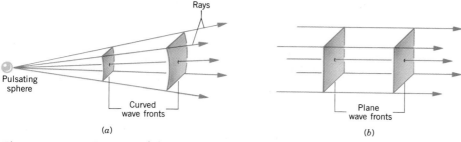

Figure 25.2   (*a*) Portions of the spherical wave fronts are shown. The rays are perpendicular to the wave fronts and diverge. (*b*) For a plane wave, the wave fronts are flat surfaces, and the rays are parallel to each other.

important in understanding the properties of mirrors and lenses. Since rays are perpendicular to the wave fronts, the rays for a plane wave are parallel to each other.

The concepts of wave fronts and rays can also be applied to light waves. For light waves, the ray concept is particularly convenient for showing the path taken by the light. We will make frequent use of light rays, and they can be regarded essentially as narrow beams of light.

25.2 THE REFLECTION OF LIGHT

Most objects reflect a certain portion of the light falling on them. Suppose a ray of light is incident on a flat, shiny surface, such as the mirror in Figure 25.3. As the drawing shows, the *angle of incidence* θ_i is the angle that the incident ray makes with respect to the normal, which is a line drawn perpendicular to the surface at the point of incidence. The *angle of reflection* θ_r is the angle that the reflected ray makes with the normal. The **law of reflection** describes the behavior of the incident and reflected rays.

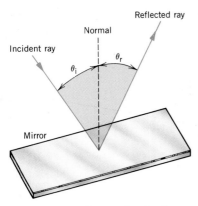

Figure 25.3 The angle of reflection θ_r equals the angle of incidence θ_i. These angles are measured with respect to the normal, which is a line drawn perpendicular to the surface of the mirror at the point of incidence.

Law of Reflection

The incident ray, the reflected ray, and the normal to the surface all lie in the same plane, and the angle of reflection θ_r equals the angle of incidence θ_i:

$$\theta_r = \theta_i$$

When parallel light rays strike a smooth, plane surface, such as that in Figure 25.4a, the reflected rays are parallel to each other. This type of reflection is known as *specular reflection* and is important in determining the properties of mirrors. Most surfaces, however, are not perfectly smooth, for they contain irregularities the sizes of which are equal to or greater than the wavelength of light. The irregular surface reflects the light rays in various directions, as part b of the drawing suggests. This type of reflection is known as *diffuse reflection*. Common surfaces that give rise to diffuse reflection are most papers, wood, nonpolished metals, and walls covered with a "flat" (nongloss) paint.

Figure 25.4 (a) The drawing shows specular reflection from a polished plane surface, such as a mirror. The reflected rays are parallel to each other. (b) The rough surface reflects the light rays in all directions; this type of reflection is known as diffuse reflection.

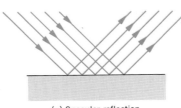

(a) Specular reflection

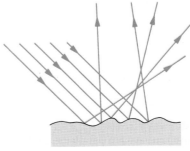

(b) Diffuse reflection

25.3 THE FORMATION OF IMAGES BY A PLANE MIRROR

When you look into a plane (flat) mirror, you see an image of yourself that has four properties:

1. The image is upright.
2. The image is the same size as you are.
3. The image is located as far behind the mirror as you are in front of it.
4. The image has left–right reversal. That is, if you wave your *right* hand, it is the *left* hand of the image that waves back, as Figure 25.5a illustrates. Therefore, letters and words held up to a mirror are reversed. Ambulances and other emergency vehicles are often lettered in reverse, as in part *b* of the drawing, so the letters will appear normal when seen in the rearview mirror of a car.

To illustrate how an image appears to originate from behind a plane mirror, Figure 25.6a shows a light ray leaving the top of an object. This ray reflects from the mirror (angle of reflection equals angle of incidence) and enters the eye. To the eye, it appears that the ray originates from behind the mirror, somewhere back along the dashed line. Actually, rays going in all directions leave each point on the object. But only a small bundle of such rays is intercepted by the eye. Part *b* of the figure shows a bundle of two rays leaving the top of the object and a similar bundle leaving the bottom. All the rays that leave a given point on the object, no matter what angle θ they have when they strike the mirror, appear to originate from a corresponding point on the image behind the mirror (see the dashed lines in part *b*). For each point on the object, there is a single corresponding point on the image, and it is this fact that makes the image in a plane mirror a sharp and undistorted one.

Although rays of light *seem* to come from the image, it is evident from Figure 25.6*b* that no light emanates from behind the plane mirror where the image appears to be. Because the rays of light do not actually emanate from the image, it is called a **virtual image**. In this text the parts of the light rays that appear to come from a virtual image are represented by dashed lines. *Curved* mirrors, on the other

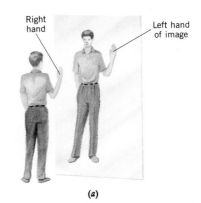

(a)

(b)

Figure 25.5 (*a*) The person's right hand becomes the image's left hand when viewed in a plane mirror. (*b*) Many emergency vehicles are reverse-lettered so the lettering appears normal when viewed through the rearview mirror of a car.

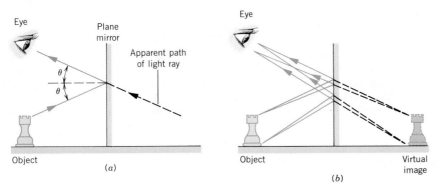

Figure 25.6 (*a*) A ray of light from the top of the chess piece reflects from the mirror. To the eye, the ray seems to come from behind the mirror. (*b*) Two bundles of rays from the object appear to originate from the image behind the mirror.

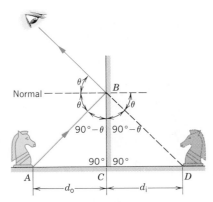

Figure 25.7 This drawing illustrates the geometry used to show that for a plane mirror the image distance d_i equals the object distance d_o.

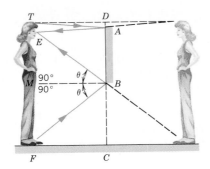

Figure 25.8 For the woman to see her full-sized image, only a half-sized mirror AB is needed.

Mirrors work equally well in and out of water.

hand, can produce images from which light rays actually do emanate. Such images are known as *real images,* and are discussed in later sections.

With the aid of the law of reflection, it is possible to show that the image is located as far behind a plane mirror as the object is in front of it. In Figure 25.7 the object distance is d_o and the image distance is d_i. A ray of light leaves the base of the object, strikes the mirror at an angle of incidence θ, and is reflected at the same angle. To the eye, this ray appears to come from the base of the image. In the drawing there are two right triangles, ABC and DBC, that share a common side BC. The angles at the top of each triangle are both equal to $90° - \theta$, as the drawing shows. Since the triangles share a common side BC and have two identical angles, $90°$ and $90° - \theta$, the triangles are identical (congruent). It then follows that the object distance d_o equals the image distance d_i.

By starting with a light ray from the top of the object, rather than from the bottom, we can extend the line of reasoning given above to show that the height of the image also equals the height of the object.

Example 1 discusses an interesting feature of plane mirrors.

Example 1 Looking into a Plane Mirror

A woman stands in front of a plane mirror to see her full height. She is 1.68 m tall, and her eyes are 0.08 m below the top of her head. Show that the mirror need only be *half as tall* as the woman.

REASONING Many "full-length" mirrors are sold today. But parts of these mirrors are useless. They are useless because light rays coming from a person's body and striking them do not reflect into the person's eyes. By using the law of reflection, we will find that only "half-size" mirrors are really needed.

SOLUTION In Figure 25.8 the mirror is labeled AB. A ray from the woman's foot F strikes the bottom of the mirror at B, with an angle of incidence equal to θ. The ray is reflected with an angle of reflection equal to θ and proceeds to the woman's eye E. The two right triangles EBM and FBM are identical, since they share the common side MB and have two angles, θ and $90°$, that are identical. Therefore,

$$EM = MF = \tfrac{1}{2}EF = \tfrac{1}{2}(1.68 \text{ m} - 0.08 \text{ m}) = 0.80 \text{ m}$$

which is also the distance BC. Similarly, a ray from the top of the woman's head T strikes the top of the mirror at A and proceeds to her eye. The same line of reasoning as above leads us to the conclusion that

$$DA = \tfrac{1}{2}TE = \tfrac{1}{2}(0.08 \text{ m}) = 0.04 \text{ m}$$

Thus, the length AB of the mirror is given by $AB = DC - BC - DA = 1.68$ m $- 0.80$ m $- 0.04$ m $= 0.84$ m. The mirror, then, need only be *half as tall* as the woman, if its bottom edge is located 0.80 m off the ground. Note that the conclusions here are valid regardless of how far the person stands from the mirror.

25.4 SPHERICAL MIRRORS

The most common type of curved mirror is a spherical mirror. As Figure 25.9 shows, a spherical mirror has the shape of a section from the surface of a sphere. If the inside or concave surface of the mirror is polished, it is a *concave mirror.* If the outside or convex surface is polished, it is a *convex mirror.* The drawing shows both types of mirrors, with a light ray reflecting from the polished surface. The law of reflection applies, just as it does for a plane mirror. But for either type of spherical mirror, the normal is drawn perpendicular to the mirror at the point of incidence. For each type, the center of curvature is located at point C, and the radius of curvature is R. The *principal axis* of the mirror is a straight line drawn through C and the midpoint of the mirror.

Figure 25.10 shows a point on a tree from which light rays are emanating. This point lies on the principal axis of the mirror and is beyond the center of curvature C. Those rays that are near the principal axis are reflected from the concave mirror and cross the axis at a point called the image point. The rays continue to diverge from the image point as if there were an object there. Since light rays actually come from the image point, the image is a real image.

If the tree in Figure 25.10 is infinitely far from the mirror, the rays are parallel to each other and to the principal axis as they approach the mirror. Figure 25.11 shows that rays near and parallel to the principal axis are reflected from the mirror and pass through an image point. In this special case the image point is referred to as the *focal point F* of the mirror. Therefore, an object infinitely far away on the principal axis gives rise to an image at the focal point of the mirror. The distance between the focal point and the middle of the mirror is the *focal length f* of the mirror.

We can show that the focal point F lies halfway between the center of curvature C and the middle of the mirror. In Figure 25.12, a light ray parallel to the principal axis strikes the concave mirror at point A. The line CA is the radius of the mirror and therefore is the normal to the spherical surface at the point of incidence. The

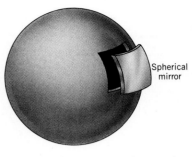

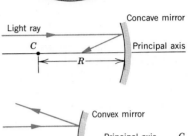

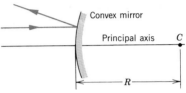

Figure 25.9 A spherical mirror has the shape of a segment of a spherical surface. The center of curvature is point C and the radius is R. For a concave mirror, the reflecting surface is the inner one, while for a convex mirror it is the outer one.

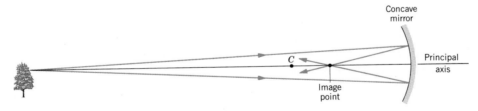

Figure 25.10 A point on the tree lies on the principal axis of the concave mirror. Rays from this point that are near the principal axis are reflected from the mirror and cross the axis at the image point.

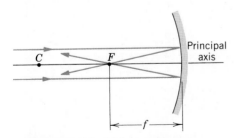

Figure 25.11 Light rays that are near and parallel to the principal axis are reflected from a concave mirror and converge at the focal point F. The focal length f of the mirror is the distance between F and the middle of the mirror.

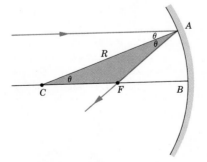

Figure 25.12 This drawing is used to show that the focal point F of a concave mirror is halfway between the center of curvature C and the mirror at point B.

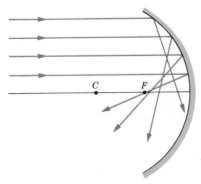

Figure 25.13 Rays that are farthest from the principal axis have the greatest angle of incidence and miss the focal point *F* after reflection from the mirror.

THE PHYSICS OF . . .

capturing solar energy with mirrors.

ray reflects from the mirror such that the angle of reflection θ equals the angle of incidence. Furthermore, the angle *ACF* is also θ, because the radial line *CA* is a transversal of two parallel lines. Since two of its angles are equal, the triangle *CAF* is an isosceles triangle; thus, sides *CF* and *FA* are equal. But when the incoming ray lies close to the principal axis, the angle of incidence θ is small, and the distance *FA* does not differ appreciably from the distance *FB*. Therefore, in the limit that θ is small, *CF = FA = FB*, and the focal point *F* lies halfway between the center of curvature and the mirror. In other words, the focal length f is one-half of the radius *R*:

$$\left[\begin{array}{l}\textbf{Focal length of}\\\textbf{a concave mirror}\end{array}\right] \qquad f = \tfrac{1}{2}R \qquad\qquad (25.1)$$

Rays that lie close to the principal axis are known as *paraxial rays,** and Equation 25.1 is valid only for such rays. Rays that are far from the principal axis do not converge to a single point after reflection from the mirror, as Figure 25.13 shows. The result is a blurred image. The fact that a spherical mirror does not bring all rays parallel to the axis to a single image point is known as *spherical aberration.* Spherical aberration can be minimized by using a mirror whose height is small compared to the radius of curvature.

A sharp image point can be obtained with a large mirror, if the mirror is parabolic in shape instead of spherical. The shape of a parabolic mirror is such that all light rays parallel to the principal axis, regardless of their distance from the axis, are reflected through a single image point. However, parabolic mirrors are costly to manufacture and are used where the sharpest images are required, as in telescopes used for scientific purposes. Parabolic mirrors are also used in one method of capturing solar energy for commercial purposes. Figure 25.14 shows a solar energy "farm" consisting of long rows of concave parabolic mirrors that reflect the sun's rays to the focal point. Located at the focal point and running the length of each row is an oil-filled pipe. The focused rays of the sun heat the oil, the heat from which is used to generate steam. The steam, in turn, drives a turbine connected to an electric generator.

Figure 25.14 This solar-thermal electric plant in the Mojave Desert uses long rows of parabolic mirrors to focus the sun's rays on an oil-filled pipe, which is located at the focal point of each mirror.

* Paraxial rays are close to the principal axis, but not necessarily parallel to it.

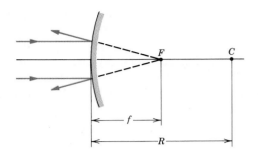

Figure 25.15 When paraxial light rays that are parallel to the principal axis strike a convex mirror, the reflected rays appear to originate from the focal point *F*. The radius of curvature is *R* and the focal length is *f*.

Not all mirrors are concave. Figure 25.15 shows parallel rays incident on a convex mirror. Clearly, the rays diverge after being reflected. If the incident parallel rays are paraxial, the reflected rays seem to come from a single point *F* behind the mirror. This point is the focal point of the convex mirror, and its distance from the midpoint of the mirror is the focal length *f*. The focal length of a convex mirror is also one-half of the radius of curvature, just as it is for a concave mirror. However, we assign the focal length of a convex mirror a negative value, because it will be convenient later on to do so:

$$\begin{bmatrix} \textbf{Focal length of} \\ \textbf{a convex mirror} \end{bmatrix} \qquad f = -\tfrac{1}{2}R \qquad\qquad (25.2)$$

25.5 THE FORMATION OF IMAGES BY SPHERICAL MIRRORS

IMAGE FORMATION BY A CONCAVE MIRROR

As we have seen, some of the light rays emitted from an object in front of a mirror strike the mirror, reflect from it, and form an image. For a concave mirror, three paraxial rays are particularly helpful in determining the location and size of the image. Figure 25.16 shows these rays—labeled 1, 2, and 3—leaving a point on the top of an object:

> **Ray 1.** This ray is initially parallel to the principal axis and therefore passes through the focal point *F* after reflection from the mirror.
>
> **Ray 2.** This ray passes through the focal point *F* and is reflected parallel to the principal axis. Ray 2 is analogous to ray 1, except the order of the incident and reflected rays is interchanged.
>
> **Ray 3.** This ray travels along a line that passes through the center of curvature *C* and follows a radius of the spherical mirror; as a result, the ray strikes the mirror perpendicularly and reflects back on itself.

If rays 1, 2, and 3 are superimposed on a scale drawing, they converge at a point on the top of the image, as can be seen in Figure 25.17a.* Although three rays have been used here to locate the image, only two are really needed; the third ray is

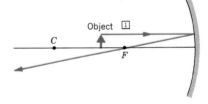

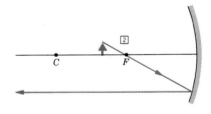

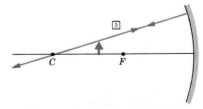

Figure 25.16 The rays labeled 1, 2, and 3 are useful in locating the image of an object that stands in front of a convex spherical mirror. The object is represented as a vertical arrow.

---

* In the drawings that follow, we assume the rays are paraxial, although the distance between the rays and the principal axis is often exaggerated for clarity.

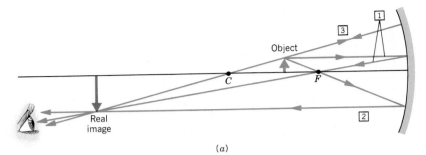

(a)

Figure 25.17 (a) When an object is placed between the focal point F and the center of curvature C of a concave mirror, a real image is formed. The image is enlarged and inverted with respect to the object. (b) When the object is located beyond the center of curvature C, a real image is created that is reduced in size and inverted with respect to the object.

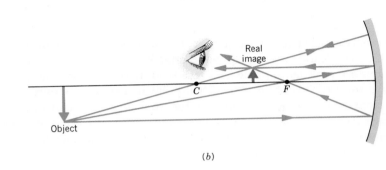

(b)

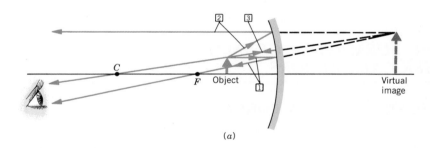

(a)

(b)

Figure 25.18 (a) When an object is located between the focal point F and a concave mirror, an enlarged, upright, virtual image is produced. (b) A makeup mirror is concave, and normally the image is formed when the object is within the focal point of the mirror.

usually drawn to serve as a check. In a similar fashion, rays from all other points on the object locate corresponding points on the image, and the mirror forms a complete image of the object. If you place your eye as shown in the drawing, you will see an image that is *larger* and *inverted* relative to the object. The image is real, because the light rays actually pass through the image.

If the object and image in Figure 25.17a are interchanged, the situation in part *b* of the drawing results. The three rays in part *b* are the same as those in part *a*, except the directions are reversed. These drawings illustrate the *principle of reversibility*, which states that *if the direction of a light ray is reversed, the light retraces its original path.* This principle is quite general and is not restricted just to reflection from mirrors.

When the object is placed between the focal point F and the mirror, as in Figure 25.18a, three rays can again be drawn to find the image. But now ray 2 does not go through the focal point on its way to the mirror, since the object is inside the focal point. However, when projected backward, ray 2 appears to come from the focal point. Therefore, after ray 2 is reflected it is directed parallel to the principal axis. In this case the three reflected rays diverge from each other and do not converge to a common point. However, when projected behind the mirror, the three rays appear to come from a point behind the mirror; thus, a virtual image is formed. This virtual image is larger than the object and upright. Makeup and shaving mirrors are concave mirrors. When you place your face between the mirror and its focal point, you see an enlarged virtual image of yourself, as part *b* of the drawing shows.

IMAGE FORMATION BY A CONVEX MIRROR

The procedure for determining the location and size of an image in a convex mirror is similar to that for a concave mirror. The same three rays are used. However, the focal point and center of curvature of a convex mirror lie behind the mirror, not in front of it. Figure 25.19a shows the three rays that are summarized below:

Ray 1. This ray is initially parallel to the principal axis and therefore appears to originate from the focal point F after reflection from the mirror.

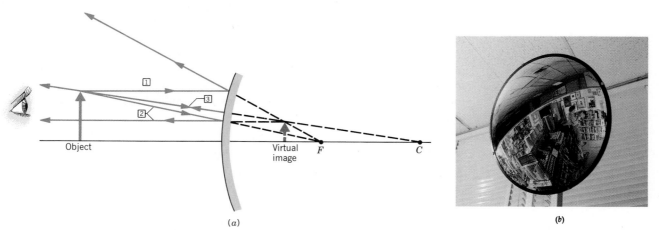

(a)   (b)

Figure 25.19   (a) An object placed in front of a convex mirror produces a virtual image behind the mirror. The virtual image is reduced in size and upright. (b) Convex mirrors are often used for security purposes.

Ray 2. This ray heads toward *F*, emerging parallel to the principal axis after reflection. Ray 2 is analogous to ray 1, except the order of the incident and reflected rays is interchanged.

Ray 3. This ray travels toward the center of curvature *C*; as a result, the ray strikes the mirror perpendicularly and reflects back on itself.

These three rays appear to come from a point on a virtual image that is behind the mirror. The virtual image is diminished in size and upright, relative to the object. A convex mirror *always* forms a virtual image of the object, no matter where in front of the mirror the object is placed. Because of the shape of convex mirrors, they give a wider field of view than do other types of mirrors. Therefore, they are often used for security purposes, as in Figure 25.19*b*. A mirror with a wide field of view is also needed to give a driver a good rear view. Thus, the outside mirror on the passenger side is often a convex mirror. Printed on such a mirror is usually the warning "VEHICLES IN MIRROR ARE CLOSER THAN THEY APPEAR." The reason for the warning is that the virtual image in Figure 25.19*a* is reduced in size and therefore looks smaller, just as a distant object would look in a plane mirror. An unwary driver, thinking that the side-view mirror is a plane mirror, might incorrectly deduce from the small size of the image that the car behind is far enough away to be ignored.

THE PHYSICS OF . . .

passenger-side automobile mirrors.

25.6 THE MIRROR EQUATION AND THE MAGNIFICATION EQUATION

CONCAVE MIRRORS

Ray diagrams drawn to scale are useful for determining the location and size of the image formed by a mirror. However, for an accurate description of the image, a more analytical technique is needed. It is possible to derive an equation, called the mirror equation, that gives the image distance if the object distance and the focal length of the mirror are known. The image distance is the distance between the image and the mirror, while the object distance is the distance between the object and the mirror. In Figure 25.20 the image and object distances are labeled d_i and d_o, respectively. The height of the image is h_i, and the height of the object is h_o. Part *a* of the drawing shows a ray leaving the top of the object and striking the mirror at the point where the principal axis intersects the mirror. Since the principal axis is perpendicular to the mirror, it is also the normal at the point of incidence. Therefore, the ray reflects at an equal angle and passes through the image. The pink and the yellow triangles are similar because they have equal angles, so

$$\frac{h_o}{h_i} = \frac{d_o}{d_i}$$

In part *b* another ray leaves the top of the object, this time passing through the focal point *F*, reflecting parallel to the principal axis, and then passing through the image. Provided the ray remains close to the axis, the pink triangle and the yellow area can be considered to be similar triangles, with the result that

$$\frac{h_o}{h_i} = \frac{d_o - f}{f}$$

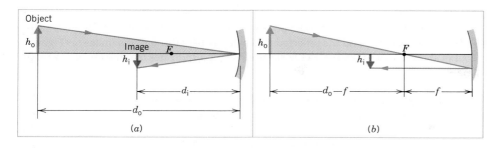

Figure 25.20 These diagrams are used to derive the mirror equation and the magnification equation. (*a*) The triangle shaded in pink is similar to the triangle shaded in yellow. (*b*) If the ray is close to the principal axis, the two shaded regions are almost similar triangles.

Setting the two equations above equal to each other yields $d_o/d_i = (d_o - f)/f$. Rearranging this result gives the *mirror equation:*

$$\boxed{\begin{array}{c}\textbf{Mirror}\\\textbf{equation}\end{array}} \qquad\qquad \frac{1}{d_o} + \frac{1}{d_i} = \frac{1}{f} \qquad\qquad (25.3)$$

We have derived this equation for a real image formed in front of a concave mirror. In this case, the image distance is a positive quantity, as are the object distance and the focal length. However, we have seen in the last section that a concave mirror can also form a virtual image, if the object is located between the focal point and the mirror. Equation 25.3 can also be applied to such a situation, provided that we adopt the convention that d_i is negative for an image behind the mirror, as it is for a virtual image. *When using the mirror equation under any circumstances, it is useful to construct a ray diagram to guide your thinking and check on your calculation.*

The *magnification m* of a mirror is defined as the ratio of the image height to the object height: $m = h_i/h_o$. If the image height is less than the object height, m is less than one. Conversely, if the image is larger than the object, m is greater than one. Since $h_i/h_o = d_i/d_o$, it follows that

$$\boxed{\begin{array}{c}\textbf{Magnification}\\\textbf{equation}\end{array}} \qquad m = \frac{\text{Image height}}{\text{Object height}} = -\frac{d_i}{d_o} \qquad (25.4)$$

The minus sign is included in Equation 25.4 because of the convention that d_i is negative for images behind the mirror. As Examples 2 and 3 show, the value of m is positive if the image is upright and negative if the image is inverted.

Example 2   A Real Image Formed by a Concave Mirror

A 2.0-cm-high object is placed 7.10 cm from a concave mirror whose radius of curvature is 10.20 cm. Find (a) the location of the image and (b) its size.

REASONING   Since $f = \frac{1}{2}R = (10.20 \text{ cm})/2 = 5.10$ cm, the object is located between the focal point F and the center of curvature C of the mirror, as illustrated in Figure 25.17*a*. Thus, we expect that relative to the object, the image is real, further away from the mirror, inverted, and larger.

SOLUTION
(a) With $d_o = 7.10$ cm and $f = 5.10$ cm, the mirror equation can be used to find the image distance:

$$\frac{1}{d_i} = \frac{1}{f} - \frac{1}{d_o} = \frac{1}{5.10 \text{ cm}} - \frac{1}{7.10 \text{ cm}} = 0.055 \text{ cm}^{-1} \quad \text{or} \quad \boxed{d_i = 18 \text{ cm}}$$

In this calculation, f and d_o are positive numbers, indicating that the focal point and the object are in front of the mirror. The positive answer for d_i means that the image is also in front of the mirror, and the reflected rays actually pass through the image, as Figure 25.17a shows. In other words, the positive value for d_i indicates that the image is a real image.

(b) The height of the image can be determined once the magnification m of the mirror is known. The magnification equation can be used to find m:

$$m = -\frac{d_i}{d_o} = -\frac{18 \text{ cm}}{7.10 \text{ cm}} = -2.5$$

The image height is $h_i = mh_o = (-2.5)(2.0 \text{ cm}) = \boxed{-5.0 \text{ cm}}$. The image is 2.5 times larger than the object, the negative values for m and h_i indicating that the image is inverted with respect to the object, as in Figure 25.17a.

Example 3 *A Virtual Image Formed by a Concave Mirror*

An object is placed 6.00 cm in front of a concave mirror that has a 10.0-cm focal length. (a) Determine the location of the image. (b) If the object is 1.2 cm high, find the image height.

REASONING The object is located between the focal point and the mirror, as in Figure 25.18a. The setup is analogous to a person using a shaving or makeup mirror. Therefore, we expect that, relative to the object, the image is virtual, upright, and larger.

SOLUTION
(a) Using the mirror equation with $d_o = 6.00$ cm and $f = 10.0$ cm, we have

$$\frac{1}{d_i} = \frac{1}{f} - \frac{1}{d_o} = \frac{1}{10.0 \text{ cm}} - \frac{1}{6.00 \text{ cm}} = -0.067 \text{ cm}^{-1} \quad \text{or} \quad \boxed{d_i = -15 \text{ cm}}$$

The answer for d_i is negative, indicating that the image is *behind* the mirror. Thus, as expected, the image is a virtual image.

(b) The image height h_i can be found from the magnification and the object height h_o:

$$m = -\frac{d_i}{d_o} = -\frac{(-15 \text{ cm})}{6.00 \text{ cm}} = 2.5$$

The image height is $h_i = mh_o = (2.5)(1.2 \text{ cm}) = \boxed{3.0 \text{ cm}}$. The image is larger than the object, and the positive values for m and h_i indicate that the image is upright (see Figure 25.18a).

CONVEX MIRRORS

The mirror equation and the magnification equation can also be used with convex mirrors, provided the focal length f is taken to be a *negative number*, as indicated earlier in Equation 25.2. One way to remember this is to recall that the focal point of a convex mirror lies *behind* the mirror. Example 4 deals with a convex mirror.

Example 4 A Virtual Image Formed by a Convex Mirror

A convex mirror is used to reflect light from an object placed 66 cm in front of the mirror. The focal length of the mirror is $f = -46$ cm (note the minus sign). Find (a) the location of the image and (b) its size relative to the object.

REASONING It is evident in Figure 25.19*a* that the image lies behind the mirror (a virtual image) and is smaller than the object. These characteristics should also result from our analysis here.

SOLUTION
(a) With $d_o = 66$ cm and $f = -46$ cm, the mirror equation gives

$$\frac{1}{d_i} = \frac{1}{f} - \frac{1}{d_o} = \frac{1}{-46\ \text{cm}} - \frac{1}{66\ \text{cm}} = -0.037\ \text{cm}^{-1} \quad \text{or} \quad \boxed{d_i = -27\ \text{cm}}$$

The negative sign for d_i indicates that the image is behind the mirror and, therefore, is a virtual image.

(b) The size of the image relative to the object is given by the magnification equation:

$$m = -\frac{d_i}{d_o} = -\frac{(-27\ \text{cm})}{66\ \text{cm}} = \boxed{0.41}$$

The image is smaller (m is less than one) and upright (m is positive) with respect to the object.

This enormous sphere (called La Geode) is in Paris, France. Its polished surface acts like a convex mirror.

Convex mirrors, like plane (flat) mirrors, always produce virtual images behind the mirror. However, the virtual image in a convex mirror is closer to the mirror than it would be if the mirror were planar, as Example 5 illustrates.

Example 5 A Convex Versus a Plane Mirror

An object is placed 9.00 cm in front of a mirror. The image is 3.00 cm closer to the mirror when the mirror is convex than when it is planar. Find the focal length of the convex mirror.

REASONING When the mirror is planar, the image is located the same distance behind the mirror as the object is in front of the mirror. Thus, the image would be 9.00 cm behind a plane mirror. If the image in a convex mirror is 3.00 cm closer than this, the image must be located 6.00 cm behind the convex mirror. In other words, when the object distance is $d_o = 9.00$ cm, the image distance for the convex mirror is $d_i = -6.00$ cm (negative, because the image is virtual). The mirror equation can be used to find the focal length of the mirror.

SOLUTION According to the mirror equation the reciprocal of the focal length is

$$\frac{1}{f} = \frac{1}{d_o} + \frac{1}{d_i} = \frac{1}{9.00\ \text{cm}} + \frac{1}{(-6.00\ \text{cm})} = -0.0556\ \text{cm}^{-1} \quad \text{or} \quad \boxed{f = -18.0\ \text{cm}}$$

SUMMARY OF SIGN CONVENTIONS

We conclude this section by summarizing the sign conventions that are used with the mirror equation and the magnification equation. These conventions apply to both concave and convex mirrors:

Object distance

d_o is + if the object is in front of the mirror (real object).

d_o is − if the object is behind the mirror (virtual object).*

Image distance

d_i is + if the image is in front of the mirror (real image).

d_i is − if the image is behind the mirror (virtual image).

Focal length

f is + for a concave mirror.

f is − for a convex mirror.

Magnification

m is + for an image that is upright with respect to the object.

m is − for an image that is inverted with respect to the object.

* Sometimes optical systems use two (or more) mirrors, and the image formed by the first mirror serves as the object for the second mirror. Occasionally, such an object falls *behind* the second mirror. In this case the object distance is negative, and the object is said to be a virtual object.

SUMMARY

Wave fronts are surfaces on which all points of a wave are in the same phase of motion. If the wave fronts are flat surfaces, the wave is called a **plane wave. Rays** are lines that are perpendicular to the wave fronts and point in the direction of the velocity of the wave.

When light reflects from a smooth surface, the reflected light obeys the **law of reflection,** which states that (a) the incident ray, the reflected ray, and the normal to the surface all lie in the same plane, and (b) the angle of reflection equals the angle of incidence. The law of reflection explains how mirrors form images. A **virtual image** is one from which rays of light do not actually come, but only appear to do so. A **real image** is one from which rays of light actually emanate. A **plane mirror** forms an upright, virtual image that is located as far behind the mirror as the object is in front of the mirror. In addition, the heights of the image and the object are equal.

A **spherical mirror** has the shape of a section from the surface of a sphere. The **principal axis** of a mirror is a straight line drawn through the center of curvature and the middle of the mirror's surface. Rays that lie close to the principal axis are known as **paraxial rays.** The

radius of curvature R of the mirror is the distance from the center of curvature to the mirror. The **focal point** of a concave spherical mirror is a point on the principal axis, in front of the mirror. Incident paraxial rays that are parallel to the principal axis converge to the focal point after being reflected from the concave mirror. The focal point of a convex spherical mirror is a point on the principal axis behind the mirror. For a convex mirror, paraxial rays that are parallel to the principal axis seem to diverge from the focal point after reflecting from the mirror. The **focal length** f of a mirror is the distance from the focal point to the middle of the mirror. The focal length and the radius of curvature are related by $f = \frac{1}{2}R$ for a concave mirror and $f = -\frac{1}{2}R$ for a convex mirror.

The image produced by a concave mirror can be located by the **ray technique,** using the three rays shown in Figure 25.16. Similarly, the image produced by a convex mirror can be found using the three rays shown in Figure 25.19.

The **mirror equation** can be used with either concave or convex mirrors and specifies the relation between the image distance d_i, the object distance d_o, and the focal length f of the mirror: $1/d_o + 1/d_i = 1/f$. The **magnifi-**

cation m of a mirror is the ratio of the image height h_i to the object height h_o. The magnification is also related to d_i and d_o by the **magnification equation:** $m = h_i/h_o = -d_i/d_o$. The algebraic sign conventions for the variables appearing in these equations are summarized at the end of Section 25.6.

QUESTIONS

1. A sign painted on a store window is reversed when viewed from inside the store. If a person inside the store views the reversed sign in a plane mirror, does the sign appear as it would when viewed from outside the store? (Try it by writing some letters on a transparent sheet of paper and then holding the back side of the paper up to a mirror.)

2. Which kind of spherical mirror, concave or convex, can be used to start a fire with sunlight? For the best results, how far from the mirror should the paper to be ignited be placed? Explain.

3. Why is your image distorted when you look at yourself in a small shiny sphere, such as a Christmas tree ornament?

4. (a) Can the image formed by a concave mirror ever be projected directly onto a screen, without the help of other mirrors or lenses? If so, specify where the object should be placed relative to the mirror. (b) Repeat part (a) assuming the mirror is convex.

5. When you look at the back side of a shiny teaspoon, held at arm's length, you see yourself upright. When you look at the other side of the spoon, you see yourself upside down. Why?

6. Suppose you wish to design a searchlight that produces a parallel beam of light. The searchlight consists of a light bulb in front of a concave spherical mirror. Where should the bulb be positioned along the principal axis of the mirror? Give your reasoning.

7. Sometimes news personnel covering an event use a microphone arrangement that is designed to increase the ability of the mike to pick up weak sounds. The drawing shows that the arrangement consists of a "hollowed-out" shell behind the mike. The shell acts like a mirror for sound waves. Explain how this arrangement enables the mike to detect weak sounds.

8. When you see the image of yourself formed by a mirror, it is because (1) light rays actually coming from a real image enter your eyes or (2) light rays appearing to come from a virtual image enter your eyes. If light rays from the image do not enter your eyes, you do not see yourself. Are there any places on the principal axis where you cannot see yourself when you are in front of a mirror that is (a) convex and (b) concave? If so, where are these places?

9. Suppose you stand in front of a spherical mirror (concave or convex). Is it possible for your image to be (a) real and upright or (b) virtual and inverted? Justify your answers.

PROBLEMS

Section 25.2 The Reflection of Light, Section 25.3 The Formation of Images by a Plane Mirror

1. Two plane mirrors are separated by 120°, as the drawing illustrates. If a ray strikes mirror M_1 at a 65° angle of incidence, at what angle θ does it leave mirror M_2?

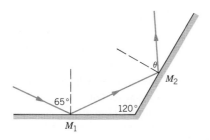

2. A person stands 3.6 m in front of a wall that is covered floor-to-ceiling with a plane mirror. His eyes are 1.8 m above the floor. He holds a flashlight between his feet and manages to point it at the mirror. At what angle of incidence must the light strike the mirror so the light will reach his eyes?

3. A person whose eyes are 1.50 m above the floor stands in front of a plane mirror. The top of his head is 0.10 m above his eyes. (a) What is the height of the shortest mirror in which he can see his entire image? (b) How far above the floor should the bottom edge of the mirror be placed?

4. Two diverging light rays, originating from the same point, have an angle of 10° between them. After the rays reflect from a plane mirror, what is the angle between them?

Construct one possible ray diagram that supports your answer.

5. Suppose you walk with a speed of 0.90 m/s toward a plane mirror. What is the speed of your image *relative to you*, when your velocity is (a) perpendicular to the mirror and (b) at an angle of 50.0° with respect to the normal to the mirror?

*6. A person, trying on a new suit in a clothing store, stands in front of two mirrors to examine the "fit." The mirrors intersect at a 90° angle. When the person looks into the mirrors, he can see three images of himself. Draw the rays and show where the images are located.

*7. The drawing shows a top view of a square room. One wall is missing, and the other three are each mirrors. From point *P* in the center of the open side, a laser is fired, with the intent of hitting a small target located at the center of one wall. Identify five directions in which the laser can be fired and score a hit, assuming the light does not strike any mirror more than once. Draw the rays to confirm your choices.

**8. A ray of light strikes a plane mirror at a 45° angle of incidence. The mirror is then rotated by 15° into the position shown in red in the drawing, while the incident ray is kept fixed. (a) Through what angle ϕ does the reflected ray rotate? (b) What is the answer to part (a) if the angle of incidence is 60°?

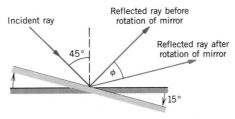

Section 25.4 Spherical Mirrors, Section 25.5 The Formation of Images by Spherical Mirrors

9. A 2.0-cm-high object is situated 15.0 cm in front of a concave mirror that has a radius of curvature of 10.0 cm. Using a ray diagram drawn to scale on a piece of paper, measure the location and the height of the image. The radius of curvature must be drawn to scale, and the scale chosen for the horizontal and vertical directions must be the same.

10. Repeat problem 9 for a concave mirror with a focal length of 20.0 cm, an object distance of 12.0 cm, and a 2.0-cm-high object.

11. Repeat problem 9 for a convex mirror with a radius of curvature of 1.00×10^2 cm, an object distance of 25 cm, and a 10.0-cm-high object.

12. Repeat problem 9 for a concave mirror with a focal length of 7.50 cm, an object distance of 11.0 cm, and a 1.0-cm-high object.

Section 25.6 The Mirror Equation and the Magnification Equation

13. The focal length of a concave mirror is 17 cm. An object is located 38 cm in front of this mirror. Where is the image located?

14. An object is 14 cm in front of a convex mirror that has a focal length of −23 cm. Determine the location of the image.

15. A concave mirror has a focal length of 42 cm. The image formed by this mirror is 97 cm in front of the mirror. What is the object distance?

16. A clown is using a concave makeup mirror to get ready for a show and is 27 cm in front of the mirror. The image is 65 cm *behind* the mirror. Find (a) the focal length of the mirror and (b) its magnification.

17. The image behind a convex mirror (radius of curvature = 68 cm) is located 22 cm from the mirror. (a) Where is the object located and (b) what is the magnification of the mirror? State whether the image is (c) upright or inverted and (d) larger or smaller than the object.

18. The image of a very distant car is located 12 cm behind a convex mirror. (a) What is the radius of curvature of the mirror? (b) Draw a ray diagram to scale showing this situation.

19. Convex mirrors are being used to monitor the aisles in a store. The mirrors have a radius of curvature of 4.0 m. (a) What is the image distance if a customer is 15 m in front of the mirror? (b) Is the image real or virtual? (c) If a customer is 1.6 m tall, how tall is the image?

20. A concave mirror (*R* = 64.0 cm) is used to project a transparent slide onto a wall. The slide is located at a distance of 38.0 cm from the mirror, and a small flashlight shines light through the slide and onto the mirror. (a) How far from the wall should the mirror be located? (b) The height of the object on the slide is 1.20 cm. What is the height of the image? (c) Is the image upright or inverted relative to the object?

21. A dentist's mirror is placed 1.5 cm from a tooth. The *enlarged* image is located 4.3 cm behind the mirror. (a) What kind of mirror (plane, concave, or convex) is being used? (b) Determine the focal length of the mirror. (c) What is the magnification? (d) How is the image oriented relative to the object?

22. A small postage stamp is placed in front of a concave mirror (radius = R), such that the image distance equals the object distance. (a) In terms of R, what is the object distance? (b) What is the magnification of the mirror? (c) State whether the image is upright or inverted relative to the object. Draw a ray diagram to guide your thinking.

*23. A concave shaving mirror is designed so the virtual image is twice the size of the object, when the distance between the object and the mirror is 15 cm. (a) Determine the radius of curvature of the mirror. (b) Draw a ray diagram to scale showing this situation.

*24. A gemstone is placed 20.0 cm in front of a concave mirror and is within the focal point. When the concave mirror is replaced with a plane mirror, the image moves 15.0 cm toward the mirror. Find the focal length of the concave mirror.

*25. The radius of curvature of a plane mirror is infinite ($R = \infty$). With this in mind, show that the mirror equation and the magnification equation correctly predict the location, magnification, orientation (upright or inverted), and nature (real or virtual) of an image formed by a plane mirror.

*26. Show that to produce an image with magnification m, using a mirror whose focal length is f, the object must be placed at a distance d_o from the mirror, where $d_o = (m - 1)f/m$.

**27. A spherical mirror is polished on both sides. When used as a convex mirror, the magnification is $+1/4$. What is the magnification when used as a concave mirror, the object remaining the same distance from the mirror?

**28. Using the mirror equation and the magnification equation, show that for a convex mirror the image is always (a) virtual (i.e., d_i is always negative) and (b) upright and smaller, relative to the object (i.e., m is positive and less than one).

ADDITIONAL PROBLEMS

29. The image formed by a convex mirror is located a distance of 22 cm behind the mirror, when the object is 34 cm in front of the mirror. What is the focal length of the mirror?

30. A coin is placed 8.0 cm in front of a concave mirror. The mirror produces a real image that has a diameter 4.0 times larger than that of the coin. What is the image distance?

31. The intent of this problem is to demonstrate the phenomenon of spherical aberration. Draw a semicircle with a radius of 10 cm to represent a concave spherical mirror. Recall that a radial line drawn between the center of curvature and any point on the arc is perpendicular to the arc and, hence, is a normal to the mirror. (a) Draw a ray parallel to the principal axis at a distance of 5 cm from the axis. Where the ray strikes the mirror, draw the normal and the reflected ray, such that the angle of reflection equals the angle of incidence. Extend the reflected ray until it intersects the principal axis (this ray should cross the axis just inside the focal point). (b) Repeat (a) with a ray drawn at a distance of 7.5 cm from the principal axis and note how much farther from the focal point this ray crosses the axis.

32. The image produced by a concave mirror is located 26 cm in front of the mirror. The focal length of the mirror is 12 cm. How far in front of the mirror is the object located?

*33. Two plane mirrors are facing each other. They are parallel, 3.00 cm apart, and 17.0 cm in length, as the picture indicates. A laser beam is directed at the top mirror, from the left edge of the bottom mirror. What is (a) the smallest and (b) largest angle of incidence θ that allows the beam to hit each mirror once or less?

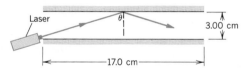

*34. (a) Where should a diamond ring be placed in front of a concave mirror, such that the image is twice the size of the ring? There are two answers, depending on whether the image is upright or inverted. Express your answers in terms of the radius of curvature R. (b) Draw ray diagrams to confirm your answers.

**35. A concave mirror has a focal length of 30.0 cm. The distance between an object and its image is 45.0 cm. Find the object and image distances assuming that (a) the object lies beyond the center of curvature and (b) the object lies within the focal point.

**36. In the drawing for problem 7, a laser is fired from point P in the center of the open side of the square room. The laser is pointed at the mirrored wall on the right. At what angle of incidence must the light strike the right-hand wall, so that after being reflected, the light hits the left corner of the back wall?

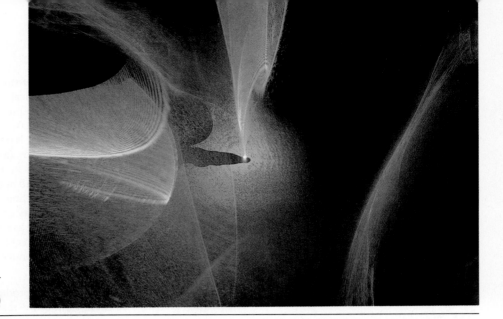

CHAPTER 26

THE REFRACTION OF LIGHT: LENSES AND OPTICAL INSTRUMENTS

Light can travel through many materials, although it does so at different speeds. Glass, for instance, is transparent to light. In this chapter, we will see that when light passes from a medium such as air into another medium such as glass, the difference in speeds leads to a change in the direction of travel. This directional change lies at the heart of some remarkable effects, depending on the nature of the materials and their shapes. The time lapse photograph above, for example, was made by shining laser light into a rotating piece of shaped glass that causes the direction of the outgoing beam to change continually. The change in direction is also responsible for rainbows and the sparkle of diamonds. It is the basis of the important field of fiber optics, which has revolutionized telecommunications and medical technology. The lenses used to correct vision problems and to construct microscopes and telescopes depend on the changes in travel direction that occur when light enters and leaves the lenses. To see what happens when light passes from one medium into another, let us begin with a discussion of how the speed of light changes in different materials.

26.1 THE INDEX OF REFRACTION

Many materials, such as air, water, and glass are transparent to light. As it goes from one transparent material into another, a ray of light will deviate from its incident direction (see Figure 26.1*a*), unless it enters perpendicular to the surface between the materials (normal incidence). This change in direction as light passes from one medium into another is called *refraction*. Refraction plays a central role

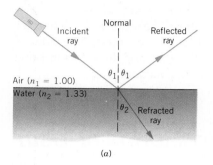

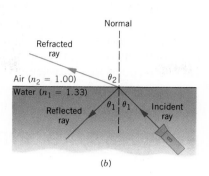

Figure 26.1 (*a*) When a ray of light is directed from air into water, part of the light is reflected at the surface and the remainder is refracted into the water. The refracted ray is bent *toward* the normal $(\theta_2 < \theta_1)$. (*b*) When a ray of light is directed from water into air, the refracted ray in air is bent *away* from the normal $(\theta_2 > \theta_1)$.

in determining the properties of the lenses used in a wide variety of optical instruments, including eyeglasses, cameras, microscopes, telescopes, and even the human eye itself.

As we will see, refraction depends on the speed of light in a material, and when light travels through a solid, a liquid, or a gas, its speed is different than that in a vacuum. The *index of refraction* (or *refractive index*) n is the ratio of the speed of light c in a vacuum to the speed of light v in the material:

$$\begin{bmatrix} \text{Index of} \\ \text{refraction} \end{bmatrix} \qquad n = \frac{\text{Speed of light in a vacuum}}{\text{Speed of light in the material}} = \frac{c}{v} \qquad (26.1)$$

Table 26.1 lists the refractive indices for some common substances. As the table indicates, the values of n are greater than unity, so the speed of light in a material medium is less than it is in a vacuum. For example, the index of refraction for

Table 26.1   Index of Refraction[a] for Various Substances

Substance	Index of Refraction, n
Solids at 20 °C	
Diamond	2.419
Glass, crown	1.52
Ice (0 °C)	1.309
Sodium chloride	1.544
Quartz	
Crystalline	1.544
Fused	1.458
Liquids at 20 °C	
Benzene	1.501
Carbon tetrachloride	1.461
Ethyl alcohol	1.362
Water	1.333
Gases at 0 °C, 1 atm	
Air	1.000 293
Carbon dioxide	1.000 45
Oxygen, O_2	1.000 271
Hydrogen, H_2	1.000 139

[a] Measured with light whose wavelength in a vacuum is 589 nm.

diamond is $n = 2.42$, so the speed of light in diamond is $v = c/n = (3.00 \times 10^8 \text{ m/s})/2.42 = 1.24 \times 10^8 \text{ m/s}$. In contrast, the index of refraction for air and other gases is so close to unity that $n_{\text{air}} = 1$ for most purposes. The index of refraction depends slightly on the wavelength of the light, and the values in Table 26.1 correspond to a wavelength of $\lambda = 589$ nm in a vacuum.

26.2 SNELL'S LAW AND THE REFRACTION OF LIGHT

SNELL'S LAW

When light strikes the interface between two transparent materials, such as air and water, the light generally divides into two parts, as Figure 26.1a illustrates. Part of the light is reflected, with the angle of reflection equaling the angle of incidence. The remainder of the light is transmitted across the interface. If the incident ray does not strike the interface at normal incidence, the transmitted ray has a different direction than the incident ray. The ray that enters the second material is said to be refracted.

In Figure 26.1a the light travels from a medium where the refractive index is smaller (air) into a medium where it is larger (water), and the refracted ray is bent *toward* the normal. Both the incident and refracted rays obey the principle of reversibility, so their directions can be reversed to give a situation like that in part b of the drawing. Here light travels from a material with a greater refractive index (water) into one with a smaller refractive index (air), and the refracted ray is bent *away* from the normal. In this case the reflected ray lies in the water, rather than in the air. In both parts of the drawing the angles of incidence, refraction, and reflection are measured relative to the normal. The index of refraction of air is labeled n_1 in part a, while it is labeled n_2 in part b, because *we label all variables associated with the incident (and reflected) ray with a subscript 1 and all variables associated with the refracted ray with a subscript 2.*

The angle of refraction θ_2 depends on the angle of incidence θ_1 and on the indices of refraction, n_2 and n_1, of the two media. The relation between these quantities is known as *Snell's law of refraction,* after the Dutch mathematician Willebrord Snell (1591–1626) who discovered it experimentally. A proof of Snell's law is presented at the end of this section.

Snell's Law of Refraction

When light travels from a material with refractive index n_1 into a material with refractive index n_2, the refracted ray, the incident ray, and the normal to the surface all lie in the same plane. The angle of refraction θ_2 is related to the angle of incidence θ_1 by

$$n_1 \sin \theta_1 = n_2 \sin \theta_2 \qquad (26.2)$$

Example 1 illustrates Snell's law.

Example 1 Determining the Angle of Refraction

A light ray strikes an air/water surface at an angle of 46° with respect to the normal. Find the angle of refraction when the direction of the ray is (a) from air to water and (b) from water to air.

REASONING AND SOLUTION

(a) The incident ray is in air, so $\theta_1 = 46°$ and $n_1 = 1.00$. The refracted ray is in water, so $n_2 = 1.33$. Snell's law can be used to find the angle of refraction:

$$\sin \theta_2 = \frac{n_1 \sin \theta_1}{n_2} = \frac{(1.00) \sin 46°}{1.33} = 0.54 \qquad (26.2)$$

$$\theta_2 = \sin^{-1}(0.54) = \boxed{33°}$$

The refracted ray is bent *toward* the normal, since θ_2 is less than θ_1, as Figure 26.1a shows.

(b) Now the incident ray propagates in water ($\theta_1 = 46°$, $n_1 = 1.33$), and the refracted ray propagates in air ($n_2 = 1.00$). Snell's law yields

$$\sin \theta_2 = \frac{n_1 \sin \theta_1}{n_2} = \frac{(1.33) \sin 46°}{1.00} = 0.96$$

$$\theta_2 = \sin^{-1}(0.96) = \boxed{74°}$$

Since θ_2 is greater than θ_1, the refracted ray is bent *away* from the normal, as Figure 26.1b indicates.

PROBLEM SOLVING INSIGHT

The angle of incidence θ_1 and the angle of refraction θ_2 that appear in Snell's law are measured with respect to the normal to the surface. They are not measured with respect to the surface itself.

The simultaneous reflection and refraction of light at an interface has applications in a number of devices. For instance, many cars come equipped with an interior rearview mirror that has an adjustment lever. One position of the lever sets the mirror for day driving, while another position sets it for night driving. The night setting is useful for reducing glare from the headlights of the car behind. As Figure 26.2a indicates, this kind of mirror is a glass wedge, the back side of which is silvered and highly reflecting. Part b of the picture shows the day setting. Light from the car behind follows the path ABCD in reaching the driver's eye. At points A and C, where the light strikes the air–glass surface, there are both reflected and

THE PHYSICS OF . . .

rearview mirrors that have a day–night adjustment.

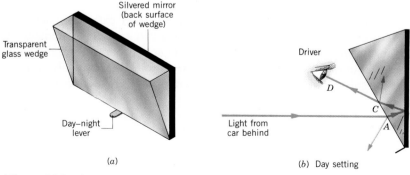
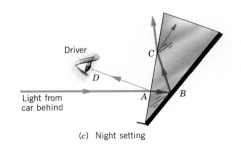

(a)   (b) Day setting   (c) Night setting

Figure 26.2 An interior rearview mirror with a day–night adjustment lever.

refracted rays. The reflected rays are drawn as thin lines, the thinness denoting that only a small percentage (about 10%) of the daylight is reflected. The weak reflected rays at A and C do not reach the driver's eye. In contrast, almost all the light reaching the silvered back surface at B is reflected toward the driver. Since most of the light follows the path ABCD, the driver sees a bright image of the car behind during the day.

During the night, the adjustment lever can be used to rotate the mirror clockwise (see part c of the drawing), away from the driver. Now, most of the light from the headlights behind follows the path ABC and does not reach the driver. Only the light that is weakly reflected from the front surface along path AD is seen. As a result, there is significantly less glare.

APPARENT DEPTH

One interesting consequence of refraction is that an object lying under water appears to be closer to the surface than it actually is. Example 2 sets the stage for explaining why, by showing what must be done to shine a light on such an object.

PROBLEM SOLVING INSIGHT

Remember that the refractive indices are written as n_1 for the medium in which the incident light travels and n_2 for the medium in which the refracted light travels.

Example 2   Finding a Sunken Chest

A searchlight on a yacht is being used at night to illuminate a sunken chest, as in Figure 26.3. At what angle of incidence θ_1 should the light be aimed?

REASONING AND SOLUTION   The angle of incidence θ_1 can be determined from Snell's law, provided the angle of refraction θ_2 can be found. From the data in the drawing it follows that $\tan \theta_2 = (2.0\ \text{m})/(3.3\ \text{m})$, so $\theta_2 = 31°$. With $n_1 = 1.00$ for air and $n_2 = 1.33$ for water, Snell's law gives

$$\sin \theta_1 = \frac{n_2 \sin \theta_2}{n_1} = \frac{(1.33) \sin 31°}{1.00} = 0.69$$

$$\theta_1 = \sin^{-1}(0.69) = \boxed{44°}$$

When the sunken chest in Example 2 is viewed from the boat (Figure 26.4), light rays from the chest pass upward through the water, refract away from the normal when they enter the air, and then travel to the observer. This picture is similar to Figure 26.3, except the direction of the rays is reversed and the searchlight is replaced by an observer. The rays entering the air are extended back into the water (see dashed lines) and indicate that the observer sees a virtual image of the chest at an *apparent depth* that is less than the actual depth. The image is virtual, because light rays do not actually pass through it. When the observer is *directly above* the submerged object, the apparent depth d' is related to the actual depth d by

$$\begin{bmatrix} \text{Apparent depth with} \\ \text{observer directly} \\ \text{above object} \end{bmatrix} \qquad d' = d\left(\frac{n_2}{n_1}\right) \qquad (26.3)$$

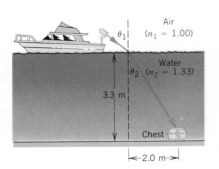

Figure 26.3 The beam from the searchlight is refracted when it enters the water.

In this result, n_1 is the refractive index of the medium associated with the incident ray (the medium in which the object is located), while n_2 refers to the medium associated with the refracted ray (the medium in which the observer is situated). The proof of Equation 26.3 is left as problem 19 at the end of the chapter. Example 3 illustrates that the effect of apparent depth is quite noticeable in water.

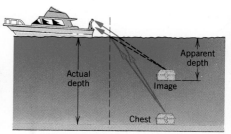

Example 3   The Apparent Depth of a Swimming Pool

A swimmer is treading water (with her head above the water) at the surface of a 3.00-m-deep pool. She sees a coin on the bottom directly below. How deep does the coin appear to be?

REASONING   Equation 26.3 may be used to find the apparent depth, provided we remember that the light rays travel from the coin to the swimmer. Therefore, the incident ray is coming from the coin under the water ($n_1 = 1.33$), while the refracted ray is in the air ($n_2 = 1.00$).

SOLUTION   The apparent depth d' of the coin is related to its actual depth d by Equation 26.3:

$$d' = d\left(\frac{n_2}{n_1}\right) = (3.00\text{ m})\left(\frac{1.00}{1.33}\right) = \boxed{2.26\text{ m}}$$

The coin appears to be closer than it actually is.

Figure 26.4   Because light from the chest is refracted away from the normal when the light enters the air, the apparent depth of the image is less than the actual depth.

THE DISPLACEMENT OF LIGHT BY A TRANSPARENT SLAB OF MATERIAL

A common use of a transparent material, such as glass, is for windows. A window pane consists of a plate of glass with parallel surfaces. When a ray of light passes through the glass, the emergent ray is parallel to the incident ray, but displaced from it, as Figure 26.5 shows. This result can be verified by applying Snell's law to each of the two glass surfaces, with the result that $n_1 \sin\theta_1 = n_2 \sin\theta_2 = n_3 \sin\theta_3$. Since air surrounds the glass, $n_1 = n_3$, and it follows that $\sin\theta_1 = \sin\theta_3$. Therefore, $\theta_1 = \theta_3$, and the emergent ray is parallel to the incident ray. However, the drawing shows that the emergent ray is displaced laterally relative to the incident ray. The extent of the lateral displacement depends on the angle of incidence and on the thickness and refractive index of the glass.

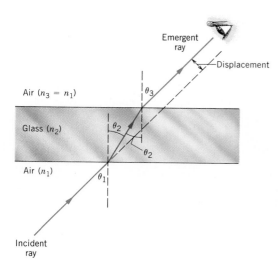

Figure 26.5   When a ray of light passes through a pane of glass that has parallel surfaces and is surrounded by air, the emergent ray is parallel to the incident ray ($\theta_3 = \theta_1$), but is displaced from it.

DERIVATION OF SNELL'S LAW

Snell's law can be derived by considering what happens to the wave fronts when the light passes from one medium into another. Figure 26.6a shows light propagating from medium 1, where the speed is relatively large, into medium 2, where the speed is smaller; therefore, n_1 is less than n_2. The plane wave fronts in this picture are drawn perpendicular to the incident and refracted rays. Since the part of each wave front that penetrates medium 2 slows down first, the wave fronts in medium 2 are rotated clockwise relative to those in medium 1. Correspondingly, the refracted ray in medium 2 is bent toward the normal, as the drawing shows.

Although the incident and refracted waves have different speeds, *they have the same frequency f.* Each wave front crosses the boundary between the two media. There are no "pileups." Therefore, the number of wave fronts per second arriving at the boundary equals the number of wave fronts per second leaving the boundary, so the frequencies of the incident and refracted waves are the same.

The distance between successive wave fronts in Figure 26.6a has been chosen to be the wavelength λ. Since the frequencies are the same in both media, but the speeds are different, it follows from Equation 16.1 that $\lambda_1 = v_1/f$ and $\lambda_2 = v_2/f$. Since v_1 is assumed to be larger than v_2, λ_1 is larger than λ_2, and the wave fronts are farther apart in medium 1.

Figure 26.6b shows an enlarged view of the incident and refracted wave fronts at the surface. The angles θ_1 and θ_2 within the colored right triangles are, respectively, the angles of incidence and refraction. In addition, the triangles share the same hypotenuse h. Therefore,

$$\sin\theta_1 = \frac{\lambda_1}{h} = \frac{(v_1/f)}{h} = \frac{v_1}{hf} \quad \text{and} \quad \sin\theta_2 = \frac{\lambda_2}{h} = \frac{(v_2/f)}{h} = \frac{v_2}{hf}$$

Combining these two equations into a single equation by eliminating the common term hf gives

$$\frac{\sin\theta_1}{v_1} = \frac{\sin\theta_2}{v_2}$$

By multiplying each side of this result by c, the speed of light in a vacuum, and

Figure 26.6 (a) The wave fronts are refracted as the light ray passes from medium 1 into medium 2. (b) An enlarged view of the incident and refracted wave fronts at the surface.

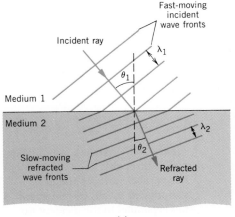

(a)

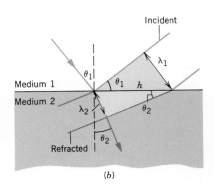

(b)

recognizing that the ratio c/v is the index of refraction n, we arrive at Snell's law of refraction: $n_1 \sin \theta_1 = n_2 \sin \theta_2$.

26.3 TOTAL INTERNAL REFLECTION

THE CRITICAL ANGLE AND TOTAL INTERNAL REFLECTION

When light passes from a medium of larger refractive index into one of smaller refractive index — for example, from water to air — the refracted ray bends *away* from the normal, as in Figure 26.7a. As the angle of incidence increases, the angle of refraction also increases. When the angle of incidence reaches a certain value, called the **critical angle** θ_c, the angle of refraction is 90°. Then the refracted ray points along the surface. Part b illustrates what happens at the critical angle. When the angle of incidence exceeds the critical angle, as in part c of the drawing, there is no refracted light. All the incident light is reflected back into the medium from which it came, a phenomenon called **total internal reflection.**

Total internal reflection occurs only when light travels from a higher-index medium toward a lower-index medium. Total internal reflection does not occur when light propagates in the reverse direction — for example, from air to water. In this situation, the refracted ray bends toward the normal, rather than away from it, so there is always a refracted ray, regardless of the angle of incidence.

An expression for the critical angle θ_c can be obtained from Snell's law by setting $\theta_1 = \theta_c$ and $\theta_2 = 90°$:

$$\begin{bmatrix} \text{Critical} \\ \text{angle} \end{bmatrix} \qquad \sin \theta_c = \frac{n_2 \sin 90°}{n_1} = \frac{n_2}{n_1} \qquad (n_1 > n_2) \qquad (26.4)$$

For example, the critical angle for light traveling from water ($n_1 = 1.33$) to air ($n_2 = 1.00$) is $\theta_c = \sin^{-1}(1.00/1.33) = 48.8°$. For incident angles greater than 48.8°, Snell's law predicts that $\sin \theta_2$ is greater than unity, a value that is not possible. Thus, for all light rays with incident angles exceeding 48.8° there is no refracted light, and the light is totally reflected back into the water. The next example illustrates how the critical angle changes when the indices of refraction change.

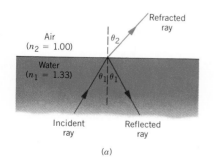
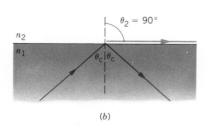
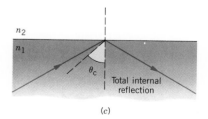

Figure 26.7 (a) When light travels from a higher-index medium (water) into a lower-index medium (air), the refracted ray is bent away from the normal. (b) When the angle of incidence is equal to the critical angle θ_c, the angle of refraction is 90°. (c) If θ_1 is greater than θ_c, there is no refracted ray, and total internal reflection occurs.

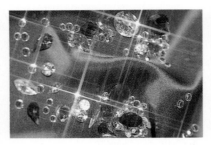

A collection of diamonds and other precious gems. The sparkle of many jewels is related to the total internal reflection of light.

Example 4 **Total Internal Reflection**

A beam of light is propagating through diamond ($n_1 = 2.42$) and strikes a diamond–air interface at an angle of incidence of 28°. (a) Will part of the beam enter the air ($n_2 = 1.00$) or will the beam be totally reflected at the interface? (b) Repeat part (a), assuming the diamond is surrounded by water ($n_2 = 1.33$).

REASONING AND SOLUTION

(a) The critical angle θ_c for total internal reflection at the diamond–air interface is given by Equation 26.4 as

$$\theta_c = \sin^{-1}\left(\frac{n_2}{n_1}\right) = \sin^{-1}\left(\frac{1.00}{2.42}\right) = 24.4°$$

Because the angle of incidence of 28° is greater than the critical angle, there is no refraction, and the light is totally reflected back into the diamond.

(b) If water, rather than air, surrounds the diamond, the critical angle for total internal reflection becomes larger:

$$\theta_c = \sin^{-1}\left(\frac{n_2}{n_1}\right) = \sin^{-1}\left(\frac{1.33}{2.42}\right) = 33.3°$$

Now a ray of light that has an angle of incidence of 28° (less than the critical angle) at the diamond–water interface is refracted into the water.

THE PHYSICS OF . . .

why a diamond sparkles.

The critical angle influences how well a diamond sparkles. For instance, Figure 26.8 shows the critical angle for light incident on a diamond–air interface. Light with a large range of incident angles strikes a bottom facet. Rays with angles of incidence exceeding the critical angle are totally reflected back into the diamond, eventually exiting out the top to give the diamond its sparkle. If the diamond is placed in water, the critical angle increases. Therefore, more rays now strike the bottom facet at an angle less than the critical angle, and some of this light escapes from the diamond. As a result, the gem sparkles less in water.

PRISMS AND TOTAL INTERNAL REFLECTION

Many optical instruments, such as binoculars, periscopes, and telescopes, use glass prisms to turn a beam of light through 90° or 180°. Figure 26.9a shows a light ray striking a 45°–45°–90° glass prism ($n_1 = 1.5$). Most of the light enters

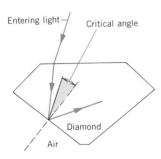

Figure 26.8 When a light ray strikes the bottom facet of a diamond and the angle of incidence is greater than the critical angle, the light is totally reflected.

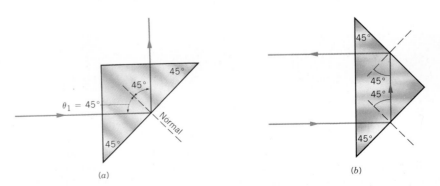

Figure 26.9 Total internal reflection at a glass–air interface can be used to turn a ray of light through an angle of (a) 90° or (b) 180°.

Figure 26.10   Two prisms, each reflecting the light by total internal reflection, are sometimes used in binoculars.

the prism and is directed toward the hypotenuse of the prism with a 45° angle of incidence. The critical angle for a glass–air interface is $\theta_c = \sin^{-1}(n_2/n_1) = \sin^{-1}(1.0/1.5) = 42°$. Since the angle of incidence is greater than the critical angle, the light is totally reflected at the hypotenuse and is directed vertically upward in the drawing, having been turned through an angle of 90°. Part *b* of the picture shows how the same prism can turn the beam through 180° when total internal reflection occurs twice. Prisms can also be used in tandem to produce a lateral displacement of a light beam, while leaving its initial direction unaltered. Figure 26.10 illustrates such an application in binoculars.

FIBER OPTICS

Another application of total internal reflection is in fiber optics, where hair-thin threads of glass or plastic, called optical fibers, "pipe" light from one place to another. Figure 26.11 shows that an optical fiber consists of a cylindrical inner *core* that carries the light and an outer concentric shell, the *cladding*. The core is made from transparent glass or plastic that has a relatively high index of refraction. The cladding, also made of glass, has a relatively low index of refraction. Light enters one end of the core, strikes the core/cladding surface at an angle of incidence greater than the critical angle, and, therefore, is reflected back into the core. Light, then, travels inside the optical fiber along a zigzag path. In a well-designed fiber, little light is lost as a result of absorption by the core, so light can travel many kilometers before its intensity diminishes appreciably. Fibers are often bundled together to produce cables that usually contain 72 fibers. Because the fibers themselves are so thin, the cables are relatively small and flexible, and they can fit into places inaccessible to larger wire cables.

Optical fibers are revolutionizing video, telephone, and computer-data communications, because a light beam can carry information through an optical fiber

THE PHYSICS OF . . .

fiber optics.

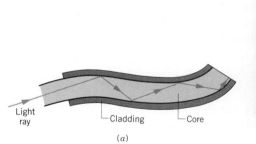

Light ray — └Cladding └Core

(a)

(b)

Figure 26.11   (*a*) Light can travel with little loss in a curved optical fiber, because the light is totally reflected whenever it strikes the core–cladding interface and because the absorption of light by the glass core itself is small. (*b*) The thickness of the optical fiber (core plus cladding) is about the thickness of a human hair.

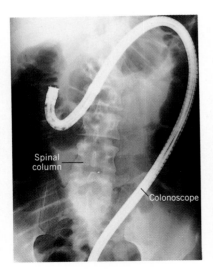

Fiber optics has led to the development of the colonoscope, which is used to examine the interior of the colon. The colonoscope provides one of the best ways to detect colon cancer in its early stages.

just as electricity carries information through copper wires and radio waves carry information through space. The information-carrying capacity of light, however, is thousands of times greater than that of electricity or radio waves. A laser beam traveling through a single optical fiber can carry tens of thousands of telephone conversations and several TV programs simultaneously. Optical fiber cables are the medium of choice for high-quality telecommunications, because the cables are relatively immune to external electrical interference.

Flexible, fiber optic cables are also used in medicine. For instance, such a cable can be passed through the esophagus into the stomach to search for ulcers and other abnormalities. Light is carried into the stomach by the outer fibers of the cable, reflected back by the stomach wall, and transmitted out of the stomach by the inner fibers of the same cable. The image can be displayed on a TV monitor or recorded on film. In arthroscopic surgery, a small surgical instrument, several millimeters in diameter, is mounted at the end of an optical fiber cable. The surgeon can insert the instrument and cable into a joint, such as the knee, with only a tiny incision and minimal damage to the surrounding tissue.

26.4 POLARIZATION AND THE REFLECTION AND REFRACTION OF LIGHT

For incident angles other than $0°$, unpolarized light becomes partially polarized in reflecting from a nonmetallic surface, such as water. To demonstrate this fact, rotate a pair of Polaroid sunglasses in the sunlight reflected from a lake. You will see that the light intensity transmitted through the glasses is a minimum when the glasses are oriented as they are normally worn. Since sunglasses are built with the transmission axis aligned in the vertical direction, it follows that the light reflected from the lake is partially polarized in the horizontal direction.

There is one special angle of incidence at which the reflected light is completely polarized parallel to the surface, the refracted ray being only partially polarized. This angle is called the *Brewster angle* θ_B. Figure 26.12 summarizes what happens when unpolarized light strikes a nonmetallic surface at the Brewster angle. The value of θ_B is given by *Brewster's law*, in which n_1 and n_2 are, respectively, the refractive indices of the materials in which the incident and refracted rays propagate:

$$\begin{bmatrix} \textbf{Brewster's} \\ \textbf{law} \end{bmatrix} \qquad \tan \theta_B = \frac{n_2}{n_1} \qquad (26.5)$$

This relation is named after the Scotsman David Brewster (1781–1868) who discovered it.

Figure 26.12 also indicates that the reflected and refracted rays are perpendicular to each other when light strikes the surface at the Brewster angle. This result follows directly from Snell's law and Brewster's law:

$$\sin \theta_B = \left(\frac{n_2}{n_1} \right) \sin \theta_2 = \tan \theta_B \sin \theta_2 = \left(\frac{\sin \theta_B}{\cos \theta_B} \right) \sin \theta_2$$

In other words, $\cos \theta_B = \sin \theta_2$. Since $\sin \theta_2 = \cos(90° - \theta_2)$, we have that $\cos \theta_B = \sin \theta_2 = \cos(90° - \theta_2)$. Thus, $\theta_B = 90° - \theta_2$, so $\theta_B + \theta_2 = 90°$, and the reflected and refracted rays are perpendicular.

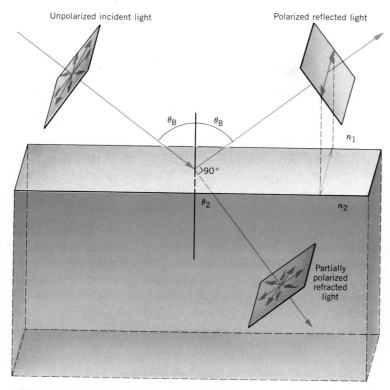

Unpolarized incident light

Polarized reflected light

θ_B θ_B

n_1

90°

θ_2 n_2

Partially
polarized
refracted
light

Figure 26.12 When unpolarized light is incident on a nonmetallic surface at the Brewster angle θ_B, the reflected light is 100% polarized in a direction parallel to the surface. The angle between the reflected and refracted rays is 90°.

26.5 THE DISPERSION OF LIGHT: PRISMS AND RAINBOWS

Figure 26.13*a* shows a ray of light passing through a glass prism. When the light enters the prism at the left face, the refracted ray is bent toward the normal, for the refractive index of glass is greater than that of air. Conversely, when the light leaves the prism at the right face and enters the air, the light is refracted away from the normal. Thus, the net effect of the prism is to change the direction of the

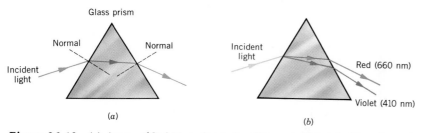

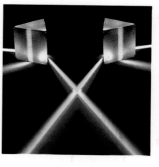

Glass prism

Normal Normal

Incident
light

(a)

Incident
light

Red (660 nm)

Violet (410 nm)

(b)

(c)

Figure 26.13 (a) A ray of light is refracted as it passes through the prism. (b) Two different colors are refracted by different amounts. For clarity, the amount of refraction has been exaggerated. (c) The prism disperses sunlight into its color components.

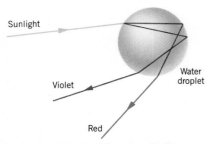

Figure 26.14 When sunlight emerges from a water droplet, the light is dispersed into its constituent colors, of which only two are shown.

Table 26.2 Indices of Refraction *n* of Selected Materials at Various Wavelengths

Approximate Color	Wavelength in Vacuum (nm)	Crown Glass	Flint Glass	Diamond
Red	660	1.520	1.662	2.410
Orange	610	1.522	1.665	2.415
Yellow	580	1.523	1.667	2.417
Green	550	1.526	1.674	2.426
Blue	470	1.531	1.684	2.444
Violet	410	1.538	1.698	2.458

THE PHYSICS OF . . .

rainbows.

ray. Because the refractive index of the glass depends on wavelength (see Table 26.2), the rays corresponding to different colors are bent by different amounts by the prism and depart traveling in different directions. The greater the index of refraction for a given color, the greater the bending, and part *b* of the drawing shows the refractions for the colors red and violet, which are at opposite ends of the visible spectrum. If a beam of sunlight, which contains all colors, is sent through the prism, the sunlight is separated into the spectrum of colors, as part *c* shows. The spreading of light into its color components is called *dispersion.*

Another example of dispersion occurs in rainbows, in which refraction by water droplets gives rise to the colors. Rainbows are often seen just as a storm is leaving, if we look at the departing rain with the sun at our backs. When light from the sun enters a spherical raindrop, as in Figure 26.14, light of each color is refracted or bent by an amount that depends on the refractive index of water for that wavelength. After reflection from the back surface of the droplet, the different colors are again refracted as they reenter the air. Although each droplet disperses the light into its full spectrum of colors, the observer in Figure 26.15 sees only one color of light coming from any given droplet, since only one color travels in the right direction to reach the observer's eyes. However, all colors are visible in

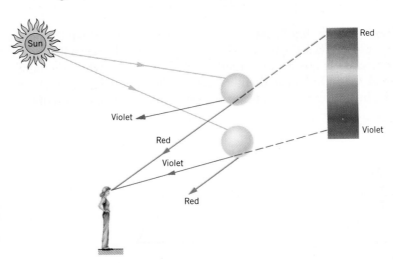

Figure 26.15 The different colors seen in a rainbow originate from water droplets at different angles of elevation.

a rainbow, because each color originates from different droplets at different angles of elevation.

26.6 LENSES

CONVERGING LENSES

The lenses used in optical instruments, such as eyeglasses, cameras, and telescopes, are made from transparent materials that refract light. They refract the light in such a way that an image of the source of the light is formed. Figure 26.16a shows a crude lens formed from two glass prisms. Suppose an object, centered on the principal axis, is infinitely far from the lens so the rays from the object are parallel to the principal axis. In passing through the prisms, these rays are bent toward the axis because of refraction. Unfortunately, the rays do not all cross the axis at the same place, and, therefore, such a crude lens gives rise to a "blurred" image of the object.

A better lens can be constructed from a single piece of transparent material with properly curved surfaces, often spherical, as in part b of the drawing. With this improved lens, rays that are near the principal axis (paraxial rays) and parallel to it converge to a single point on the axis after emerging from the lens. This point is called the *focal point F* of the lens. Thus, an object located infinitely far away on the principal axis leads to an image at the focal point of the lens. The distance between the focal point and the lens is the *focal length f*. In what follows, we assume the lens is sufficiently thin compared to f that it makes no difference whether f is measured between the focal point and either surface of the lens or the center of the lens. The type of lens in Figure 26.16b is known as a *converging lens*, because it causes incident parallel rays to converge at the focal point.

DIVERGING LENSES

Another type of lens found in optical instruments is a *diverging lens*, which causes incident parallel rays to diverge after exiting the lens. Two prisms can also be used to form a crude diverging lens, as in Figure 26.17a. In a properly designed diverging lens, such as that in part b of the picture, paraxial rays that are parallel to the principal axis appear to originate from a single point on the axis after passing

The appearance of a rainbow is a nice bonus after a summer shower. Water droplets in the air disperse the sunlight into its spectrum of colors.

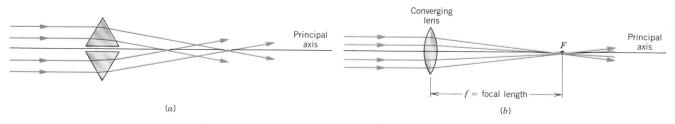

Figure 26.16 (a) These two prisms cause rays of light that are parallel to the principal axis to change direction and cross the axis at different points. (b) With a converging lens, paraxial rays that are parallel to the principal axis converge to the focal point F after passing through the lens.

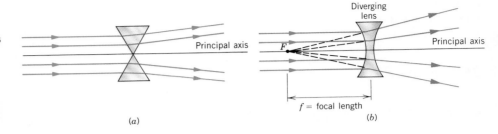

Figure 26.17 (a) These two prisms cause parallel rays to diverge. (b) With a diverging lens, paraxial rays that are parallel to the principal axis appear to originate from the focal point *F* after passing through the lens.

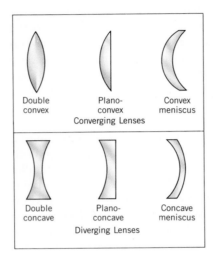

Figure 26.18 Converging and diverging lenses come in a variety of shapes.

through the lens. This point is the focal point *F* of the diverging lens, and its distance *f* from the lens is the focal length. Again, we assume that the lens is thin compared to the focal length.

Converging and diverging lenses come in a variety of shapes, as Figure 26.18 illustrates. Observe that converging lenses are thicker at the center than at the edges, whereas diverging lenses are thinner at the center.

26.7 THE FORMATION OF IMAGES BY LENSES

RAY DIAGRAMS

Each point on an object emits light rays in all directions, and when some of these rays pass through a lens, they form an image. As with mirrors, ray diagrams can be drawn to determine the location and size of the image. Lenses differ from mirrors, however, in that light can pass through a lens from left to right or from right to left. Therefore, when constructing ray diagrams, begin by locating a focal point *F* on *each side of the lens;* each point lies on the principal axis at the same distance *f* from the lens. The lens is assumed to be a thin lens, in that its thickness is small compared with the focal length and the distances of the object and the image from the lens. For convenience, it is also assumed that the object is located to the left of the lens and is oriented perpendicular to the principal axis. There are three paraxial rays that leave a point on the top of the object and are especially

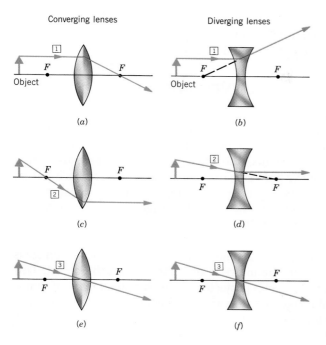

Converging lenses   Diverging lenses

(a)   (b)

(c)   (d)

(e)   (f)

Figure 26.19 The rays shown here are useful in determining the nature of the images formed by converging and diverging lenses.

helpful in drawing ray diagrams. They are labeled 1, 2, and 3 in Figure 26.19 and are as follows:

Converging Lens	Diverging Lens
Ray 1	
This ray initially travels parallel to the principal axis. In passing through a converging lens, the ray is refracted toward the axis and travels through the focal point on the right side of the lens, as Figure 26.19a shows.	This ray initially travels parallel to the principal axis. In passing through a diverging lens, the ray is refracted away from the axis, and *appears* to have originated from the focal point on the left of the lens. The dashed line in Figure 26.19b represents the apparent path of the ray.
Ray 2	
This ray first passes through the focal point on the left and then is refracted by the lens in such a way that it leaves traveling parallel to the axis, as in part c.	This ray leaves the object and moves toward the focal point on the right of the lens. Before reaching the focal point, however, the ray is refracted by the lens so as to exit parallel to the axis. See part d, where the dashed line indicates the ray's path in the absence of the lens.
Ray 3	
This ray travels directly through the center of the thin lens without any appreciable bending, as in part e.	This ray travels directly through the center of the thin lens without any appreciable bending, as in part f.

Ray 3 does not bend as it proceeds through the lens, because the front and back surfaces of the lens are nearly parallel at the center. Thus, the lens behaves as a transparent slab. As Figure 26.5 shows, the rays incident on and exiting from a slab travel in the same direction with only a lateral displacement. If the lens is sufficiently thin, the displacement is negligibly small.

IMAGE FORMATION BY A CONVERGING LENS

Figure 26.20a illustrates the formation of a real image by a converging lens. Here the object is located at a distance from the lens that is greater than twice the focal length (beyond the point labeled 2F). To locate the image, any two of the three special rays can be drawn from the tip of the object, although all three are shown in the drawing. The point on the right side of the lens where these rays intersect locates the tip of the image. The ray diagram indicates that the image is real, inverted, and smaller than the object. This optical arrangement is similar to that used in a camera, where a piece of film records the image (see part b of the drawing).

THE PHYSICS OF . . .

a camera.

When the object is placed between 2F and F, as in Figure 26.21a, the image is still real and inverted; however, the image is now larger than the object. This optical system is used in a slide or film projector in which a small piece of film is the object and the enlarged image falls on a screen. However, to obtain an image that is right-side up, the film must be placed in the projector upside down.

THE PHYSICS OF . . .

a slide or film projector.

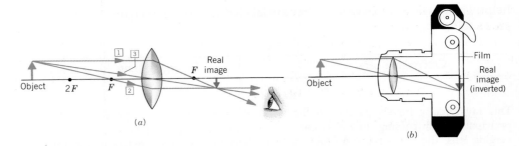

(a)

(b)

Figure 26.20 (a) When the object is placed to the left of the point labeled 2F, a real, inverted, and smaller image is formed. (b) The arrangement in part a is like that used in a camera.

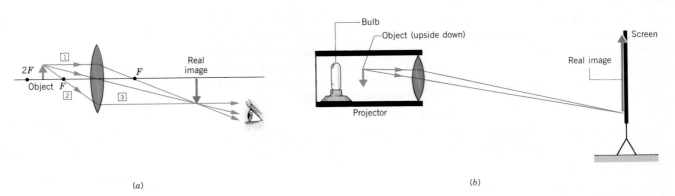

(a)

(b)

Figure 26.21 (a) When the object is placed between 2F and F, the image is real, inverted, and larger than the object. (b) This arrangement is found in projectors.

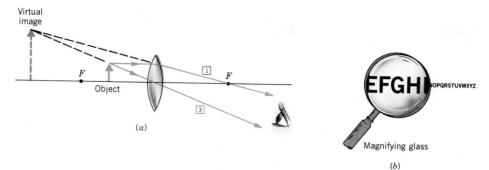

Figure 26.22   (*a*) When an object is placed inside the focal point *F* of a converging lens, an upright, enlarged, and virtual image is created. (*b*) Such an image is seen when looking through a magnifying glass.

When the object is located between the focal point and the lens, as in Figure 26.22, the rays diverge after leaving the lens. To a person viewing the diverging rays, they appear to come from an image behind and to the left of the lens. Because the rays do not actually come from the image, it is a virtual image. The ray diagram shows that the virtual image is upright and enlarged. A magnifying glass uses this arrangement.

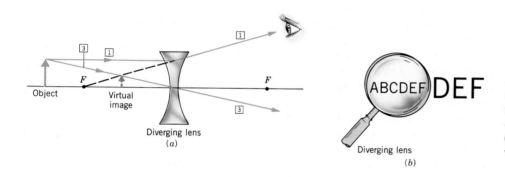

Figure 26.23   (*a*) A diverging lens always forms a virtual image of a real object. The image is upright and smaller relative to the object. (*b*) The image seen through a diverging lens.

IMAGE FORMATION BY A DIVERGING LENS

Light rays diverge upon leaving a diverging lens, as Figure 26.23 shows, and the ray diagram indicates that a virtual image is formed on the left side of the lens. In fact, regardless of the position of a real object, a diverging lens always forms a virtual image that is upright and smaller relative to the object.

26.8 THE THIN-LENS EQUATION AND THE MAGNIFICATION EQUATION

For an object in front of a mirror, it is possible to determine the location, size, and nature of the image with the mirror equation and the magnification equation. A similar analysis can be carried out for thin lenses, and the resulting equations are identical to those used with mirrors. These equations are

$$\begin{bmatrix} \textbf{Thin-lens} \\ \textbf{equation} \end{bmatrix} \qquad \frac{1}{d_o} + \frac{1}{d_i} = \frac{1}{f} \qquad\qquad (26.6)$$

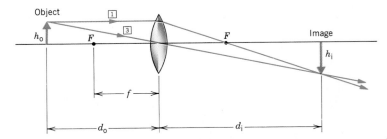

Figure 26.24 The drawing shows the focal length f, the object distance d_o, and the image distance d_i, for a converging lens. The object and image heights are, respectively, h_o and h_i.

$$\left[\begin{array}{c}\textbf{Magnification}\\\textbf{equation}\end{array}\right] \qquad m = \frac{\text{Image height}}{\text{Object height}} = \frac{h_i}{h_o} = -\frac{d_i}{d_o} \qquad (26.7)$$

Figure 26.24 defines the symbols in these expressions with the aid of a thin converging lens, but the expressions also apply to a diverging lens, if it is thin. The derivations of these equations are presented at the end of this section.

Certain sign conventions accompany the use of the thin-lens and magnification equations, and the conventions are similar to those used with mirrors in Section 25.6. These conventions allow the equations to convey information about whether the image is real or virtual, upright or inverted, and enlarged or reduced with respect to the object. The issue of real-versus-virtual images, however, is slightly different with lenses than with mirrors. With a mirror, a real image is formed on the *same side* of the mirror as the object, in which case the image distance d_i is a positive number (see Figure 25.17). With a lens, a positive value for d_i also means the image is real. But, starting with an actual object, a real image is formed on the *opposite side* of the lens as the object (see Figure 26.24). The **sign conventions** listed below apply to light rays traveling from left to right from a real object.

Object distance

d_o is $+$ if the object is to the left of the lens (real object), as is usually the case.

d_o is $-$ if the object is to the right of the lens (virtual object).*

Image distance

d_i is $+$ for an image (real) formed to the right of the lens by a real object.

d_i is $-$ for an image (virtual) formed to the left of the lens by a real object.

Focal length

f is $+$ for a converging lens.

f is $-$ for a diverging lens.

Magnification

m is $+$ for an image that is upright with respect to the object.

m is $-$ for an image that is inverted with respect to the object.

Examples 5 and 6 illustrate the use of the thin-lens and magnification equations.

---

* This situation arises in systems containing more than one lens, where the image formed by the first lens becomes the object for the second lens. In such a case, the object of the second lens may lie to the right of that lens, in which event d_o is assigned a negative value and the object is called a virtual object.

Example 5   The Real Image Formed by A Camera Lens

A 1.70-m-tall person is standing 2.50 m in front of a camera. The camera uses a converging lens whose focal length is 0.0500 m. (a) Find the image distance (the distance between the lens and the film) and determine whether the image is real or virtual. (b) Find the magnification and the height of the image on the film.

REASONING   This optical arrangement is similar to that in Figure 26.20, where the object distance is greater than twice the focal length of the lens. Therefore, we expect the image to be real, inverted, and smaller than the object.

SOLUTION
(a) To find the image distance d_i we use the thin-lens equation with $d_o = 2.50$ m and $f = 0.0500$ m:

$$\frac{1}{d_i} = \frac{1}{f} - \frac{1}{d_o} = \frac{1}{0.0500 \text{ m}} - \frac{1}{2.50 \text{ m}} = 19.6 \text{ m}^{-1} \quad \text{or} \quad \boxed{d_i = 0.0510 \text{ m}}$$

Since the image distance is a positive number, a $\boxed{\text{real image}}$ is formed on the film.

(b) The magnification follows from Equation 26.7:

$$m = -\frac{d_i}{d_o} = -\frac{0.0510 \text{ m}}{2.50 \text{ m}} = \boxed{-0.0204}$$

The image is 0.0204 times as large as the object, and it is inverted since m is negative. Since the object height is $h_o = 1.70$ m, the image height is

$$h_i = mh_o = (-0.0204)(1.70 \text{ m}) = \boxed{-0.0347 \text{ m}}$$

PROBLEM SOLVING INSIGHT

According to the thin-lens equation, the image distance d_i has a reciprocal given by $d_i^{-1} = f^{-1} - d_o^{-1}$, where f is the focal length and d_o is the object distance. After combining the reciprocals f^{-1} and d_o^{-1}, do not forget to take the reciprocal of the result to find d_i.

Example 6   The Virtual Image Formed by a Diverging Lens

An object is placed 7.10 cm to the left of a diverging lens whose focal length is $f = -5.08$ cm (a diverging lens has a negative focal length). (a) Find the image distance and determine whether the image is real or virtual. (b) Obtain the magnification.

REASONING   This situation is similar to that in Figure 26.23. The ray diagram shows that the image is virtual, erect, and smaller than the object.

SOLUTION
(a) The thin-lens equation can be used to find the image distance d_i:

$$\frac{1}{d_i} = \frac{1}{f} - \frac{1}{d_o} = \frac{1}{-5.08 \text{ cm}} - \frac{1}{7.10 \text{ cm}} = -0.338 \text{ cm}^{-1} \quad \text{or} \quad \boxed{d_i = -2.96 \text{ cm}}$$

The image distance is negative, indicating that the image is $\boxed{\text{virtual}}$ and located to the left of the lens.

(b) Since d_i and d_o are known, the magnification can be determined:

$$m = -\frac{d_i}{d_o} = -\frac{(-2.96 \text{ cm})}{7.10 \text{ cm}} = \boxed{0.417}$$

The image is upright (m is +) and smaller ($m < 1$) than the object.

The thin-lens and magnification equations can be derived by considering rays 1 and 3 in Figure 26.25a. Ray 1 is shown separately in part b of the drawing, where the angle θ is the same in each of the two pink triangles. Thus, $\tan \theta$ is the same for each triangle, so

$$\tan \theta = \frac{h_o}{f} = \frac{h_i}{d_i - f}$$

Ray 3 is shown separately in part c of the drawing, where the angles θ' are the same. Therefore,

$$\tan \theta' = \frac{h_o}{d_o} = \frac{h_i}{d_i}$$

The first equation gives $h_i/h_o = (d_i - f)/f$, while the second equation yields $h_i/h_o = d_i/d_o$. Equating these two expressions for h_i/h_o and rearranging the result produces the thin-lens equation, $1/d_o + 1/d_i = 1/f$.

The magnification equation follows directly from the equation $h_i/h_o = d_i/d_o$, if we recognize that h_i/h_o is the magnification m of the lens. As with mirrors, a minus sign is inserted in front of the ratio d_i/d_o, so a positive value for m indicates an image that is upright relative to the object, while a negative value indicates the opposite.

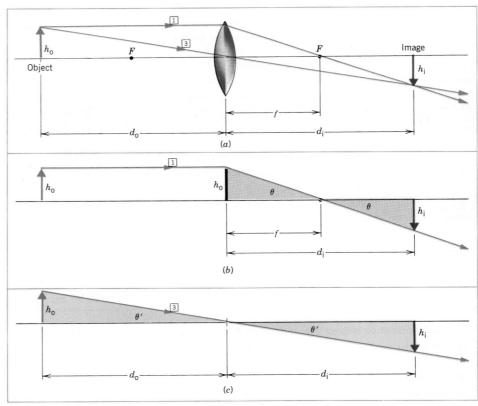

Figure 26.25 These ray diagrams are used for deriving the thin-lens and magnification equations.

26.9 LENSES IN COMBINATION

Many optical instruments, such as microscopes and telescopes, use a number of lenses together to produce an image. Among other things, a multiple-lens system can produce an image that is magnified more than is possible with a single lens. For instance, Figure 26.26a shows a two-lens system used in a microscope. The first lens, the lens closest to the object, is referred to as the *objective*. The second lens is known as the *eyepiece* (or *ocular*). The object is placed just outside the focal point F_o of the objective. The image formed by the objective—called the "first image" in the drawing—is real, inverted, and enlarged compared to the object. This first image then serves as the object for the eyepiece. Since the first image falls between the eyepiece and its focal point F_e, the eyepiece forms an enlarged, virtual, final image, which is what the observer sees.

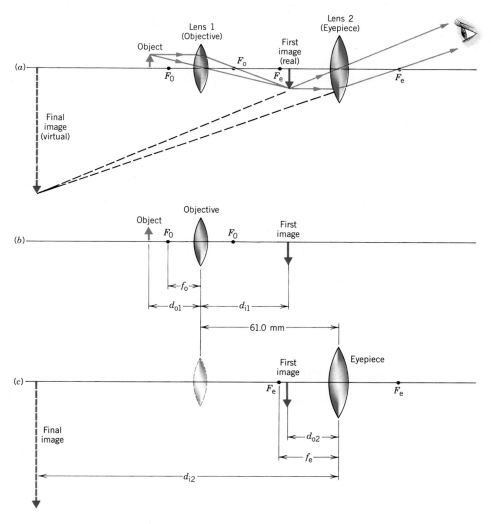

Figure 26.26 (a) This two-lens system can be used as a compound microscope to produce a virtual, enlarged, and inverted final image. (b) The objective forms the first image and (c) the eyepiece forms the final image.

PROBLEM SOLVING INSIGHT

The location of the final image in a multiple-lens system can be determined by applying the thin-lens equation to each lens separately. The key point to remember in such situations is that *the image produced by one lens serves as the object for the next lens,* as the next example illustrates.

Example 7   A Microscope — Two Lenses in Combination

The objective and eyepiece of the compound microscope in Figure 26.26 are both converging lenses and have focal lengths of $f_o = 15.0$ mm and $f_e = 25.5$ mm. A distance of 61.0 mm separates the lenses. The microscope is being used to examine an object placed $d_{o1} = 24.1$ mm in front of the objective. Find the final image distance.

REASONING   We begin by using the thin-lens equation to find the location of the image produced by the first lens, the objective. This image then becomes the object for the second lens, the eyepiece. The thin-lens equation can be used again to locate the final image produced by the eyepiece.

SOLUTION   The "first image" distance d_{i1} can be determined using the thin-lens equation with $d_{o1} = 24.1$ mm and $f_o = 15.0$ mm (see Figure 26.26b):

$$\frac{1}{d_{i1}} = \frac{1}{f_o} - \frac{1}{d_{o1}} = \frac{1}{15.0 \text{ mm}} - \frac{1}{24.1 \text{ mm}} = 0.0252 \text{ mm}^{-1} \quad \text{or} \quad d_{i1} = 39.7 \text{ mm}$$

The first image now becomes the object for the eyepiece (see part c of the drawing). Since the distance between the lenses is 61.0 mm, the object distance for the eyepiece is $d_{o2} = 61.0 \text{ mm} - d_{i1} = 61.0 \text{ mm} - 39.7 \text{ mm} = 21.3 \text{ mm}$. Noting that the focal length of the eyepiece is $f_e = 25.5$ mm, we can determine the final image distance with the aid of the thin-lens equation:

$$\frac{1}{d_{i2}} = \frac{1}{f_e} - \frac{1}{d_{o2}} = \frac{1}{25.5 \text{ mm}} - \frac{1}{21.3 \text{ mm}} = -0.0077 \text{ mm}^{-1} \quad \text{or} \quad \boxed{d_{i2} = -130 \text{ mm}}$$

The fact that d_{i2} is negative indicates that the final image is virtual and lies to the left of the eyepiece, as the drawing shows.

26.10 THE HUMAN EYE

THE ANATOMY OF THE EYE

Without doubt, the human eye is the most remarkable of all optical devices. Figure 26.27 shows some of its main anatomical features. The eyeball is approximately spherical with a diameter of about 25 mm. Light enters the eye through a transparent membrane (the *cornea*). This membrane covers a clear liquid region (the *aqueous humor*), behind which is a diaphragm (the *iris*), the *lens,* a region filled with a jelly-like substance (the *vitreous humor*), and, finally, the *retina.* The retina is the light-sensitive part of the eye, consisting of millions of structures called *rods* and *cones.* When stimulated by light, these structures send electrical impulses via the *optic nerve* to the brain, which interprets the image on the retina.

The iris is the colored portion of the eye and controls the amount of light reaching the retina. The iris acts as a controller, because it is a muscular diaphragm with a variable opening at its center, through which the light passes. The opening

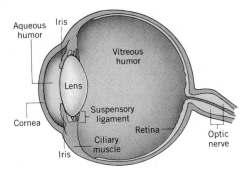

Figure 26.27 A cross-sectional view of the human eye.

is called the *pupil*. The diameter of the pupil varies from about 2 to 7 mm, decreasing in bright light and increasing (dilating) in dim light.

Of prime importance to the operation of the eye is the fact that the lens is flexible, and its shape can be altered by the action of the *ciliary muscle*. The lens is connected to the ciliary muscle by the *suspensory ligaments* (see the drawing). We will see shortly how the shape-changing ability of the lens affects the focusing property of the eye.

THE OPTICS OF THE EYE

Optically, the eye and the camera are similar; both have a lens system and a diaphragm with a variable opening or aperture at its center. Moreover, the retina of the eye and the film in the camera serve similar functions, for both record the image formed by the lens system. In the eye, the image formed on the retina is real, inverted, and smaller than the object, just as it is in a camera. Although the image on the retina is inverted, it is interpreted by the brain as being right-side up.

For clear vision, the eye must refract the incoming light rays, so as to form a sharp image on the retina. In reaching the retina, the light travels through five different media, each with a different index of refraction n: air ($n = 1.00$), the cornea ($n = 1.38$), the aqueous humor ($n = 1.33$), the lens ($n = 1.40$, on the average), and the vitreous humor ($n = 1.34$). Each time light passes from one medium into another, it is refracted at the boundary. Collectively, all the boundaries participate in refracting the light to form the image on the retina. However, the greatest amount of refraction, about 70% or so, occurs at the air/cornea boundary. According to Snell's law, the large refraction at this interface occurs primarily because the refractive index of air ($n = 1.00$) is so different from that of the cornea ($n = 1.38$). The refraction at all the other boundaries is relatively small, because the indices of refraction on either side of these boundaries are nearly equal. The lens itself contributes only about 20–25% of the total refraction, since the surrounding aqueous and vitreous humors have indices of refraction that are nearly the same as that of the lens.

Even though the lens contributes only a quarter of the total refraction or less, the function of the lens is an important one. The eye has a fixed image distance, that is, the distance between the lens and the retina is constant. Therefore, the only way for images to be produced on the retina for objects located at different distances is for the focal length of the lens to be adjustable. And it is the ciliary muscle that adjusts the focal length. When the eye looks at a very distant object, the ciliary muscle is not tensed. The lens has its least curvature and, consequently,

THE PHYSICS OF . . .

the human eye.

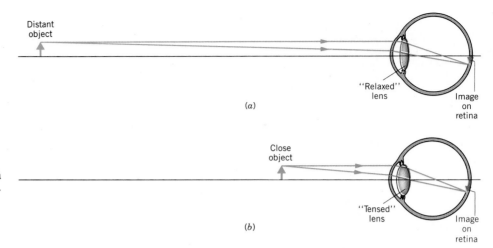

Figure 26.28 (a) When fully relaxed, the lens of the eye has its longest focal length, and an image of a very distant object is formed on the retina. (b) When the ciliary muscle is tensed, the lens is thicker and has a shorter focal length. Consequently, an image of a closer object is also formed on the retina.

its longest focal length. Under this condition the eye is said to be "fully relaxed," and the rays form a sharp image on the retina, as in Figure 26.28a. When the object moves closer to the eye, the ciliary muscle automatically tenses, thereby increasing the curvature of the lens, shortening the focal length, and permitting a sharp image to form again on the retina (Figure 26.28b). When a sharp image of an object is formed on the retina, we say the eye is "focused" on the object. The process in which the lens changes its focal length to focus on objects at different distances is called *accommodation* and occurs so swiftly that we are usually unaware of it.

When you hold a book too close, the print is blurred because the lens cannot adjust enough to bring the book into focus. The point nearest the eye at which an object can be placed and still produce a sharp image on the retina is called the *near point* of the eye. The ciliary muscle is fully tensed when an object is placed at the near point. For people in their early twenties with normal vision, the near point is located about 25 cm from the eye. It increases to about 50 cm at age 40 and to roughly 500 cm at age 60. Since most reading material is held at a distance of 45 cm or so from the eye, older adults typically need eyeglasses to overcome the loss of accommodation. The *far point* of the eye is the location of the farthest object on which the fully relaxed eye can focus. A person with normal eyesight can see objects very far away, such as the planets and stars, and thus has a far point located nearly at infinity.

NEARSIGHTEDNESS

A person who is *nearsighted (myopic)* can focus on nearby objects but cannot clearly see objects far away. For such a person, the far point of the eye is not at infinity and may even be as close to the eye as three or four meters. When a nearsighted eye tries to focus on a distant object, the eye is fully relaxed, like a normal eye. However, the nearsighted eye has a focal length that is shorter than it should be, so rays from the distant object form a sharp image in front of the retina, as Figure 26.29a shows, and blurred vision results.

The nearsighted eye can be corrected with glasses or contacts that use *diverging* lenses, as Figure 26.29b suggests. The rays from the object diverge after leaving the eyeglass lens. Therefore, when they are subsequently refracted toward the principal axis by the eye, a sharp image is formed farther back and falls on the

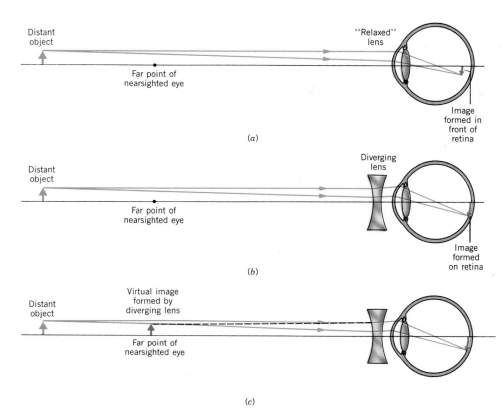

(a)

(b)

(c)

Figure 26.29 (a) When a near-sighted person views a distant object, the image is formed in front of the retina. The result is blurred vision. (b) With a diverging lens in front of the eye, the image is moved onto the retina and clear vision results. (c) The diverging lens is designed to form a virtual image at the far point of the nearsighted eye.

retina. Since the relaxed (but nearsighted) eye can focus on an object at the eye's far point—but not on objects farther away—the diverging lens is designed to transform a very distant object into an image located at the far point. Part *c* of the drawing shows this transformation, and the next example illustrates how to determine the focal length of the diverging lens.

Example 8   Eyeglasses for the Nearsighted Person

A nearsighted person has a far point located only 521 cm from the eye. Assuming that eyeglasses are to be worn 2 cm in front of the eye, find the focal length needed for the diverging lenses of the glasses so the person can see distant objects.

REASONING   In Figure 26.29c the far point is 521 cm away from the eye. Since the glasses are worn 2 cm from the eye, the far point is 519 cm to the left of the diverging lens. The image distance, then, is −519 cm, the negative sign indicating that the image is a virtual image formed to the left of the lens. The object is assumed to be infinitely far from the diverging lens. The thin-lens equation can be used to find the focal length of the eyeglasses.

SOLUTION   With $d_i = -519$ cm and $d_o = \infty$, the focal length can be found as follows:

$$\frac{1}{f} = \frac{1}{d_o} + \frac{1}{d_i} = \frac{1}{\infty} + \frac{1}{-519 \text{ cm}} \quad \text{or} \quad \boxed{f = -519 \text{ cm}} \qquad (26.6)$$

The value for f is negative, indicating the lens is a diverging lens.

THE PHYSICS OF . . .

eyeglasses.

PROBLEM SOLVING INSIGHT

Eyeglasses are worn about 2 cm from the eyes. Be sure, if necessary, to take this 2 cm into account when determining the object and image distances (d_o and d_i) that are used in the thin-lens equation.

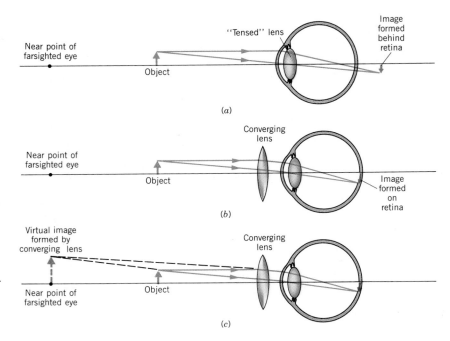

Figure 26.30 (*a*) When a farsighted person views an object located inside the near point, an image is formed behind the retina, causing blurred vision. (*b*) With a converging lens in front of the eye, the image is moved onto the retina and clear vision results. (*c*) The converging lens is designed to form a virtual image at the near point of the farsighted eye.

FARSIGHTEDNESS

A *farsighted (hyperopic)* person can usually see distant objects clearly, but cannot focus on those nearby. Whereas the near point of a "normal" eye is located about 25 cm from the eye, the near point of a farsighted eye may be considerably farther away than that, perhaps as far as several hundred centimeters. When a farsighted eye tries to focus on a book held closer than the near point, the eye accommodates and shortens its focal length as much as it can. However, even at its shortest, the focal length of a farsighted eye is longer than it should be. Therefore, the light rays from the book would form a sharp image behind the retina, as Figure 26.30*a* indicates, leading to blurred vision.

Figure 26.30*b* shows that farsightedness can be corrected by placing a *converging* lens in front of the eye. The lens refracts the light rays more toward the principal axis before they enter the eye. Consequently, when the rays are refracted even more by the eye, they converge to form an image on the retina. Part *c* of the figure illustrates what the eye sees when it looks through the converging lens. The lens is designed so that the eye perceives the light to be coming from a virtual image located at the near point. Example 9 shows how the focal length of the converging lens is determined to correct for farsightedness.

THE PHYSICS OF . . .

contact lenses.

Example 9 Contact Lenses for the Farsighted Person

A farsighted person has a near point located 210 cm from the eyes. Obtain the focal length of the converging lenses in a pair of contacts that can be used to read a book held 25.0 cm from the eyes.

<u>*REASONING*</u> A contact lens is placed directly against the eye. Thus, the object distance, which is the distance from the book to the lens, is 25.0 cm. The lens forms an

image of the book at the near point of the eye, so the image distance is -210 cm. The minus sign indicates that the image is a virtual image formed to the left of the lens, as in Figure 26.30c. The focal length can be obtained from the thin-lens equation.

SOLUTION   With $d_o = 25.0$ cm and $d_i = -210$ cm, the focal length can be determined as follows:

$$\frac{1}{f} = \frac{1}{d_o} + \frac{1}{d_i} = \frac{1}{25.0 \text{ cm}} + \frac{1}{-210 \text{ cm}} = 0.0352 \text{ cm}^{-1} \quad \text{or} \quad \boxed{f = 28.4 \text{ cm}} \quad (26.6)$$

In addition to improving your vision, contact lenses can now be used to coordinate your eye color with your wardrobe.

THE REFRACTIVE POWER OF A LENS—THE DIOPTER

The extent to which rays of light are refracted by a lens depends on its focal length. However, the optometrists who prescribe correctional lenses and the opticians who make the lenses do not specify the focal length directly in prescriptions. Instead, they use the concept of *refractive power* to decribe the extent to which a lens refracts light:

$$\begin{array}{c} \text{Refractive power} \\ \text{of a lens} \\ \text{(in diopters)} \end{array} = \frac{1}{f \text{ (in meters)}} \quad (26.8)$$

The refractive power is measured in units of *diopters*. One diopter is 1 m^{-1}.

Equation 26.8 shows that a converging lens has a refractive power of 1 diopter if it focuses parallel light rays to a focal point 1 m beyond the lens. If a lens refracts parallel rays even more and converges them to a focal point only 0.25 m beyond the lens, the lens has four times more refractive power, or 4 diopters. Since a converging lens has a positive focal length and a diverging lens has a negative focal length, the refractive power of a converging lens is positive while that of a diverging lens is negative. For instance, the eyeglasses in Example 8 would be described in a prescription from an optometrist in the following way: Refractive power $= 1/(-5.19 \text{ m}) = -0.193$ diopters. The contact lenses in Example 9 would be described in a similar fashion: Refractive power $= 1/(0.284 \text{ m}) = 3.52$ diopters.

26.11 ANGULAR MAGNIFICATION AND THE MAGNIFYING GLASS

If you hold a penny at arm's length, the penny looks larger than the moon. The reason is that the penny, being so close, forms a larger image on the retina of the eye than does the more distant moon. The brain interprets the larger image of the penny as arising from a larger object. As far as the brain is concerned, then, the size of the image on the retina determines how large an object appears to be. However, the size of the image on the retina is difficult to measure. Alternatively, the angle θ subtended by the image can be used as an indication of the image size. Figure 26.31 shows this alternative, which has the advantage that θ is also the angle subtended by the object and, hence, can be measured easily. The angle θ is called the *angular size* of both the image and the object. The larger the angular size, the larger the image on the retina, and the larger the object appears to be.

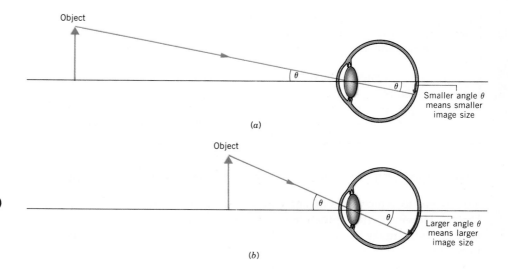

Figure 26.31 In both (*a*) and (*b*) the object is the same size, but in (*b*) the image on the retina is larger because the object is closer to the eye. The angle θ is the angular size of both the image and the object.

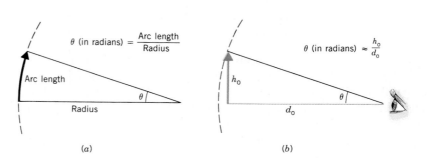

Figure 26.32 (*a*) The angle θ, measured in radians, is the arc length divided by the radius. (*b*) For small angles (less than 9°), θ is approximately equal to h_o/d_o, where h_o and d_o are the object height and distance.

According to Equation 8.1, the angle θ (measured in radians) is the length of the circular arc subtended by the angle divided by the radius of the arc, as Figure 26.32*a* indicates. Part *b* of the drawing shows the situation when we view an object of height h_o at a distance d_o from the eye. Comparing part *a* with part *b*, it is evident that when θ is small, h_o is approximately equal to the arc length and d_o is nearly equal to the radius. Thus, for small angles, θ can be expressed in terms of the measurable quantities h_o and d_o:

$$\theta \text{ (in radians)} = \text{Angular size} \approx \frac{h_o}{d_o}$$

This approximation is good to within one percent for angles of 9° or smaller. In the next example the angular size of the penny is compared with that of the moon.

Example 10   A Penny and the Moon

Compare the angular size of a penny ($h_o = 1.9$ cm) held at arm's length ($d_o = 71$ cm) with that of the moon ($h_o = 3.5 \times 10^6$ m, and $d_o = 3.9 \times 10^8$ m).

__REASONING AND SOLUTION__   The angular size θ of an object is given by its height h_o divided by its distance d_o from the eye, $\theta = h_o/d_o$:

[Penny]	$\theta = \dfrac{h_o}{d_o} = \dfrac{1.9 \text{ cm}}{71 \text{ cm}} =$	$\boxed{0.027 \text{ rad } (1.5°)}$
[Moon]	$\theta = \dfrac{h_o}{d_o} = \dfrac{3.5 \times 10^6 \text{ m}}{3.9 \times 10^8 \text{ m}} =$	$\boxed{0.0090 \text{ rad } (0.52°)}$

The penny thus appears to be about three times larger than the moon.

An optical instrument, such as a magnifying glass, allows us to view small or distant objects, because it produces a larger image on the retina than would be possible otherwise. In other words, an optical instrument magnifies the angular size of the object. The *angular magnification* (or *magnifying power*) M is the angular size θ' of the final image produced by the instrument divided by a reference angular size θ. The reference angular size is the angular size of the object when seen without the instrument.

$$\begin{bmatrix} \text{Angular} \\ \text{magnification} \end{bmatrix} \quad M = \dfrac{\begin{matrix}\text{Angular size of} \\ \text{final image produced} \\ \text{by optical instrument}\end{matrix}}{\begin{matrix}\text{Reference angular size} \\ \text{of object seen without} \\ \text{optical instrument}\end{matrix}} = \dfrac{\theta'}{\theta} \qquad (26.9)$$

A magnifying glass is the simplest device that provides angular magnification. To find its angular magnification, we first determine the reference angular size θ. In this case, θ is chosen to be the angular size of the object when placed at the near point of the eye and seen without the magnifying glass. Since an object cannot be brought closer than the near point of the eye and still produce a sharp image on the retina, θ represents the largest angular size obtainable without the magnifying glass. Figure 26.33*a* indicates that the reference angular size is $\theta \approx h_o/N$, where N is the distance from the eye to the near point. To compute θ', recall from Section 26.7 and Figure 26.22 that a magnifying glass is usually a single converging lens, with the object located inside the focal point. In this situation, Figure 26.33*b* indicates that the lens produces a virtual image that is enlarged and upright with respect to the object. Assuming the eye is next to the magnifying glass, the angular size θ' seen by the eye is $\theta' \approx h_o/d_o$, where d_o is the object distance. The angular magnification is

$$M = \dfrac{\theta'}{\theta} \approx \dfrac{h_o/d_o}{h_o/N} = \dfrac{N}{d_o}$$

According to the thin-lens equation, d_o is related to the image distance d_i and the focal length f of the lens by

$$\dfrac{1}{d_o} = \dfrac{1}{f} - \dfrac{1}{d_i}$$

Substituting this expression for $1/d_o$ into the expression above for M leads to the following result:

$$\begin{bmatrix} \text{Angular magni-} \\ \text{fication of a} \\ \text{magnifying glass} \end{bmatrix} \quad M = \dfrac{\theta'}{\theta} \approx \left(\dfrac{1}{f} - \dfrac{1}{d_i} \right) N \qquad (26.10)$$

Two special cases of this result are of interest, depending on whether the image

THE PHYSICS OF . . .

a magnifying glass.

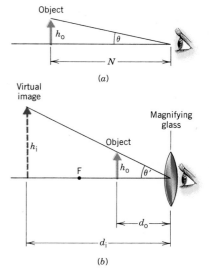

Figure 26.33 (*a*) Without a magnifying glass, the largest angular size θ occurs when the object is placed at the near point of the eye. The distance from the near point to the eye is N. (*b*) A magnifying glass produces an enlarged, virtual image of an object placed inside the focal point F of the lens. The angular size of both the image and the object is θ'.

is located as close to the eye as possible or as far away as possible. To be seen clearly, the closest the image can be relative to the eye is at the near point; for this situation, $d_i = -N$. The minus sign indicates that the image lies to the left of the lens and is virtual. Under this condition, Equation 26.10 becomes $M \approx N/f + 1$. The farthest the image can be from the eye is at infinity ($d_i = -\infty$); this occurs when the object is placed at the focal point of the lens. When the image is at infinity, Equation 26.10 simplifies to $M \approx N/f$. Clearly, the angular magnification is greater when the image is at the near point of the eye rather than at infinity. In either case, however, the greatest magnification is achieved by using a magnifying glass with the shortest possible focal length.

Example 11   Examining a Diamond Ring with a Magnifying Glass

A jeweler, whose near point is 40.0 cm from his eye, is using a small magnifying glass (called a loupe) to examine a diamond ring. The lens of the magnifying glass has a focal length of 5.00 cm, and the image of the ring is -185 cm from the lens. The image distance is negative because the image is virtual and is formed on the same side of the lens as the object. (a) Determine the angular magnification of the magnifying glass. (b) Where should the image be located so the jeweler's eye is fully relaxed and has the least strain? What is the angular magnification under this "least strain" condition?

REASONING   The angular magnification of the magnifying glass can be determined from Equation 26.10. In part (a) the image distance is -185 cm. In part (b) the ciliary muscle of the jeweler's eye is fully relaxed when examining the ring, so the image must be infinitely far from the eye, as Section 26.10 discusses.

SOLUTION
(a) With $f = 5.00$ cm, $d_i = -185$ cm, and $N = 40.0$ cm, the angular magnification of the magnifying glass is

$$M = \left(\frac{1}{f} - \frac{1}{d_i}\right)N = \left(\frac{1}{5.00 \text{ cm}} - \frac{1}{-185 \text{ cm}}\right)(40.0 \text{ cm}) = \boxed{8.22}$$

(b) When the jeweler's eye is fully relaxed, the image distance is $d_i = -\infty$, so $1/d_i = 1/(-\infty) = 0$. The angular magnification is

$$M = \left(\frac{1}{f} - \frac{1}{d_i}\right)N = \left(\frac{1}{5.00 \text{ cm}} - \frac{1}{-\infty}\right)(40.0 \text{ cm}) = \boxed{8.00}$$

Jewelers often prefer to minimize eye strain when viewing objects, even though it means a slight reduction in angular magnification.

THE PHYSICS OF . . .

the compound microscope.

26.12 THE COMPOUND MICROSCOPE

To increase the angular magnification beyond that possible with a magnifying glass, an additional converging lens can be included to "premagnify" the object before the magnifying glass comes into play. The result is an optical instrument known as the *compound microscope* (Figure 26.34). As discussed in Section 26.9, the magnifying glass is called the eyepiece and the additional lens is called the objective. The object is placed just outside the focal point F_o of the objective, as in Figure 26.35a.

In obtaining the angular magnification M of the compound microscope, we follow the same approach used in the last section and begin with $M = \theta'/\theta$ (Equation 26.9), where θ' is the angular size of the final image and θ is the reference angular size. As with the magnifying glass in Figure 26.33, the reference angular size is determined by the height h_o of the object when the object is located at the near point of the unaided eye: $\theta \approx h_o/N$, where N is the distance between the eye and the near point.

To find the angular size θ' of the final image produced by the microscope, note from Figure 26.35a that the objective produces a "first image," which then serves as the object for the eyepiece. The eyepiece, in turn, produces the final image. Part b of the drawing shows that the angular size θ' of the final image is equal to that of the first image, so we need only to determine θ' from the first image. The first image is inverted and its height is mh_o, where m is the magnification of the objective and h_o is the height of the object. The magnification equation (Equation 26.7) specifies m in terms of the object distance d_o and the image distance d_i, according to $m = -d_i/d_o$. If, as in part a of the drawing, the object is placed just outside the focal point F_o of the objective to achieve a large magnification, then $d_o \approx f_o$, where f_o is the focal length of the objective. Furthermore, if the microscope is designed so the eye is fully relaxed when viewing the final image, the final image must be very far from the eyepiece, or near "infinity." This location of the final image implies that the first image must fall just inside the focal point F_e of the eyepiece, as part b of the drawing indicates. Therefore, $d_i \approx L - f_e$, where L is the separation between the two lenses and f_e is the focal length of the eyepiece.

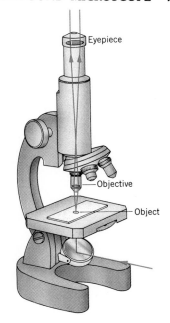

Figure 26.34   A compound microscope.

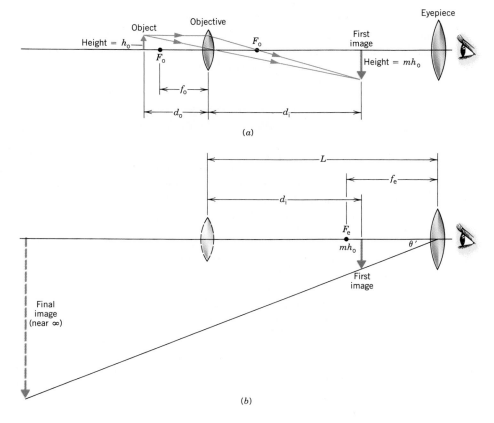

Figure 26.35   (a) In a two-lens compound microscope, the objective forms a "first image" of the object. This first image is enlarged, real, and inverted. (b) The first image becomes the object for the eyepiece, which produces an enlarged, virtual, final image near infinity.

With these approximations for d_o and d_i, the magnification of the objective becomes

$$m = -\frac{d_i}{d_o} \approx -\frac{(L - f_e)}{f_o}$$

Since the distance between the first image and the eyepiece is nearly equal to f_e, the angular size θ' of the first image (and also of the final image) is

$$\theta' \approx \frac{\text{Height of first image}}{\text{Distance of first image from eyepiece}} \approx \frac{mh_o}{f_e} \approx -\left(\frac{L - f_e}{f_o}\right)\frac{h_o}{f_e}$$

Using the expressions above for θ and θ', we find that the angular magnification of the compound microscope is

$$M = \frac{\theta'}{\theta} \approx \frac{-\left(\dfrac{L - f_e}{f_o}\right)\dfrac{h_o}{f_e}}{\dfrac{h_o}{N}}$$

$$\begin{bmatrix}\text{Angular magni-}\\ \text{fication of a}\\ \text{compound}\\ \text{microscope}\end{bmatrix} \qquad M \approx -\frac{(L - f_e)N}{f_o f_e} \qquad (26.11)$$

Equation 26.11 shows that the angular magnification is greatest when f_o and f_e are as small as possible (since they are in the denominator) and when the distance L between the lenses is as large as possible. Furthermore, L must be greater than the sum of f_o and f_e for this equation to be valid. Example 12 deals with the angular magnification of a compound microscope.

Example 12   The Angular Magnification of a Compound Microscope

The objective of a compound microscope has a focal length of $f_o = 0.40$ cm, while that of the eyepiece is $f_e = 3.0$ cm. The two lenses are separated by a distance of $L = 20.0$ cm. A person with normal eyes ($N = 25$ cm) is using the microscope. (a) Determine the angular magnification of the microscope. (b) Compare the answer in part (a) with the largest angular magnification obtainable by using the eyepiece alone as a magnifying glass.

REASONING   The angular magnification of the compound microscope can be obtained directly from Equation 26.11, since all the variables are known. When the eyepiece is used alone as a magnifying glass, as in Figure 26.33b, the largest angular magnification occurs when the image seen through the eyepiece is as close as possible to the eye. The image in this case is at the near point, and according to Equation 26.10, the angular magnification is $M \approx (N/f_e) + 1$.

SOLUTION
(a) The angular magnification of the compound microscope is

$$M \approx -\frac{(L - f_e)N}{f_o f_e} = -\frac{(20.0\text{ cm} - 3.0\text{ cm})(25\text{ cm})}{(0.40\text{ cm})(3.0\text{ cm})} = \boxed{-350}$$

The minus sign indicates that the final image is inverted relative to the initial object.

(b) The maximum angular magnification of the eyepiece by itself is

$$M \approx \frac{N}{f_e} + 1 = \frac{25 \text{ cm}}{3.0 \text{ cm}} + 1 = \boxed{9.3}$$

The effect of the objective is to increase the angular magnification of the compound microscope by a factor of $350/9.3 = 38$ compared to that of a magnifying glass.

26.13 THE TELESCOPE

THE PHYSICS OF . . .

the telescope.

A telescope is an instrument for magnifying distant objects, such as stars and planets. Like a microscope, a telescope consists of an objective and an eyepiece (also called the ocular). Since the object is usually far away, the light rays striking the telescope are nearly parallel, and the "first image" is formed just beyond the focal point F_o of the objective, as Figure 26.36a illustrates. The first image is real and inverted. Unlike that in the compound microscope, however, this image is *smaller* than the object. If, as in part *b* of the drawing, the telescope is constructed so the first image lies just inside the focal point F_e of the eyepiece, the eyepiece acts like a magnifying glass. It forms a final image that is greatly enlarged, virtual, and located near infinity. This final image can then be viewed with a fully relaxed eye.

The angular magnification M of a telescope, like that of a magnifying glass or a microscope, is the angular size θ' subtended by the final image of the telescope

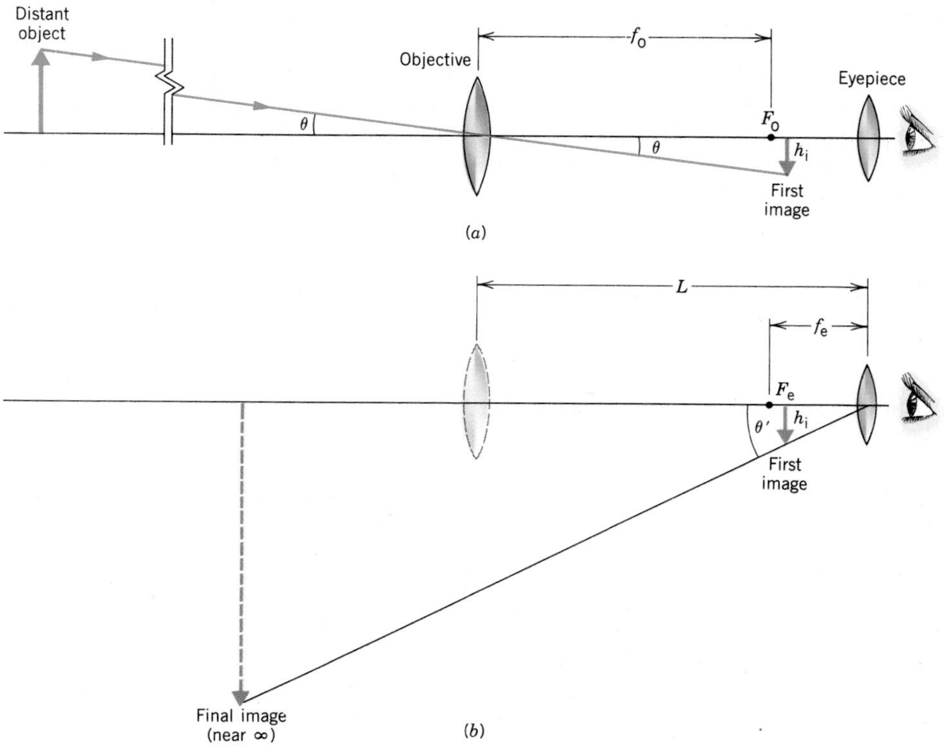

(a)

(b)

Figure 26.36 (a) An astronomical telescope is used to view distant objects (note the "break" in the principal axis, between the object and the objective). The objective produces a real, inverted, first image. (b) The eyepiece magnifies the first image to produce the final image near infinity.

divided by the reference angular size θ of the object as seen without the telescope. For an astronomical object, such as a planet, it is convenient to use as a reference the angular size of the object seen in the sky with the unaided eye. Since the object is far away, the angular size seen by the unaided eye is nearly the same as the angle θ subtended at the objective of the telescope in Figure 26.36a. Moreover, θ is also the angle subtended by the first image, so $\theta \approx h_i/f_o$, where h_i is the height of the first image and f_o is the focal length of the objective. To obtain an expression for θ', note in part b of the figure that the first image is located very near the focal point F_e of the eyepiece, so $\theta' \approx -h_i/f_e$, where f_e is the focal length of the eyepiece lens. The minus sign is present because the first image is inverted. The angular magnification of the telescope is, then,

$$\left[\begin{array}{c} \textbf{Angular} \\ \textbf{magnification of} \\ \textbf{an astronomical} \\ \textbf{telescope} \end{array}\right] \qquad M = \frac{\theta'}{\theta} \approx \frac{-h_i/f_e}{h_i/f_o} = -\frac{f_o}{f_e} \qquad (26.12)$$

The angular magnification is determined by the ratio of the focal length of the objective to the focal length of the eyepiece. For large angular magnifications, the objective should have a long focal length and the eyepiece a short one. Some of the design features of a telescope are the topic of the next example.

Figure 26.37 An astronomical telescope. The viewfinder is a separate small telescope with low magnification and serves as an aid in locating the object. Once the object has been found, the viewer looks through the eyepiece to obtain the full magnification of the telescope.

Example 13   The Angular Magnification of an Astronomical Telescope

The telescope shown in Figure 26.37 has the following specifications: $f_o = 985$ mm and $f_e = 5.00$ mm. From these data, find (a) the angular magnification of the telescope and (b) the approximate length of the telescope.

REASONING   The angular magnification of the telescope follows directly from Equation 26.12, since the focal lengths of the objective and eyepiece are known. We can find the length of the telescope by noting that it is equal to the distance L between the objective and eyepiece. Figure 26.36a shows that the first image is located just beyond the focal point F_o of the objective, and just inside the focal point F_e of the eyepiece. These two focal points are, therefore, very close together, so the distance L is approximately the sum of the two focal lengths: $L \approx f_o + f_e$.

SOLUTION
(a) The angular magnification is

$$M = -\frac{f_o}{f_e} = -\frac{985 \text{ mm}}{5.00 \text{ mm}} = \boxed{-197}$$

(b) The approximate length of the telescope is

$$L \approx f_o + f_e = 985 \text{ mm} + 5.00 \text{ mm} = \boxed{990 \text{ mm}}$$

Two types of astronomical telescopes are used today, the **refracting telescope** and the **reflecting telescope**. Both include a lens for the eyepiece. The refracting telescope also uses a lens for the objective and is the type illustrated in Figure 26.36. The reflecting telescope, however, uses a concave mirror for the objective, as Figure 26.38 indicates. The incoming light rays reflect from a concave mirror at the right end. A small plane mirror intercepts the reflected, converging rays and

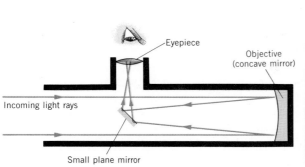

Figure 26.38   A reflecting telescope with a Newtonian focus.

directs them toward the eyepiece. This particular design was developed by Isaac Newton and is said to have a Newtonian focus.

Because the light from distant stars is so faint, telescopes need a great light-gathering ability, so their objectives have large diameters. Since a lens can be supported only at its edge, a large lens begins to sag under its own weight and produce distorted images. Consequently, large concave mirrors are preferable, because they can be supported over the entire back surface. Also, a mirror has only one surface to be ground and polished rather than two, and a mirror is free from the chromatic aberration that exists in all lenses.* All the larger astronomical telescopes in the world are reflectors.

26.14 LENS ABERRATIONS

Rather than forming a sharp image, a single lens typically forms an image that is slightly out of focus. This lack of sharpness arises because the rays originating from a single point on the object are not focused to a single point on the image. As a result, each point on the image becomes a small "blur." The lack of point-to-point correspondence between object and image is called an aberration.

One common type of aberration is *spherical aberration,* and it occurs with converging and diverging lenses made with spherical surfaces. Figure 26.39a shows how spherical aberration arises with a converging lens. Ideally, all rays traveling parallel to the principal axis are refracted so they cross the axis at the same point after passing through the lens. However, rays far from the principal axis are refracted more by the lens than those closer in. Consequently, the outer rays cross the axis closer to the lens than do the inner rays, so a lens with spherical aberration does not have a unique focal point. Instead, as the drawing suggests, there is a location along the principal axis where the light converges to the smallest cross-sectional area. This area is circular and is known as the *circle of least confusion.* The circle of least confusion is where the most satisfactory image can be formed by the lens.

Spherical aberration can be reduced substantially by using a variable-aperture diaphragm to allow only those rays close to the principal axis to pass through the lens. Figure 26.39b indicates that a reasonably sharp focal point can be achieved

* Section 26.14 discusses chromatic aberration.

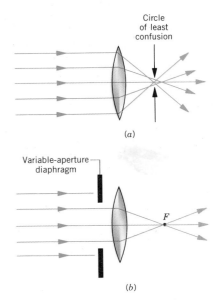

Figure 26.39   (a) In a converging lens, spherical aberration prevents light rays parallel to the principal axis from converging to a common point. (b) Spherical aberration can be reduced by allowing only rays near the principal axis to pass through the lens. The refracted rays now converge more nearly to a single focal point F.

Figure 26.40 (a) Chromatic aberration arises when different colors are focused at different points along the principal axis: F_V = focal point for violet light, F_R = focal point for red light. (b) A converging and a diverging lens in tandem can be designed to bring different colors more nearly to the same focal point F.

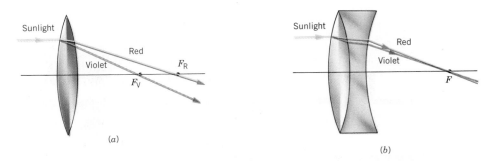

(a)   (b)

by this method, although less light now passes through the lens. Lenses with parabolic surfaces are also used to reduce this type of aberration, but they are difficult and expensive to make.

Chromatic aberration also causes blurred images. It arises because the index of refraction of the material from which the lens is made varies with wavelength. Section 26.5 discusses how this variation leads to the phenomenon of dispersion, whereby different colors refract by different amounts. Figure 26.40a shows sunlight incident on a converging lens, in which the light spreads into its color spectrum because of dispersion. For clarity, however, the picture shows only the colors at the opposite ends of the visible spectrum—red and violet. Violet is refracted more than red, so the violet ray crosses the principal axis closer to the lens than does the red ray. Thus, the focal length of the lens is shorter for violet than for red, with intermediate values of the focal length corresponding to the colors in between. As a result of chromatic aberration, an undesirable color fringe surrounds the image.

Chromatic aberration can be greatly reduced by using a compound lens, such as the combination of a converging lens and a diverging lens shown in Figure 26.40b. Each lens is made from a different type of glass. With this lens combination the red and violet rays almost come to a common focus and, thus, chromatic aberration is reduced. A lens combination designed to reduce chromatic aberration is called an *achromatic lens* (from the Greek "achromatos," meaning "without color"). All high-quality cameras use achromatic lenses.

INTEGRATION OF CONCEPTS

THE REFLECTION AND REFRACTION OF LIGHT AND ENERGY CONSERVATION

We first encountered reflection in Chapter 25 in association with mirrors and found that the law of reflection is obeyed. This law is also obeyed when light reflects from a surface between two transparent materials (e.g., air and water). In the current chapter, we have seen that refraction also occurs at such a surface. In refraction, light penetrates into the second material and travels in a direction that differs from the incident direction, as specified by Snell's law. Thus, reflection and refraction of light waves occur simultaneously at a surface between two transparent materials. It is important to keep in mind that light waves are electromagnetic in nature. They are composed of electric and magnetic fields, which carry energy. As a result, we can apply the principle of conservation of energy (see Chapter 6) to the phenomena of reflection and refraction. When only reflection and refraction occur at a surface, energy conservation indicates that the energy reflected plus the energy refracted must add up to equal the energy carried by the incident light. The percentage of incident energy that appears as reflected versus refracted light depends on the angle of incidence and the refractive indices of the materials on either side of the surface. For instance, when light travels from air toward water at perpendicular incidence, most of the light energy is refracted and little is reflected. But when the angle of incidence is nearly 90° and the light barely grazes the water surface, most of the light energy is reflected, with only a small amount refracted into the water. On a rainy night, you probably have experienced the annoying glare that results when light from an oncoming car just grazes the wet road. Under such conditions, most of the light energy reflects into your eyes.

LENSES AND MIRRORS

In passing through a lens, light is refracted in such a way that an image of the source of the light (the object) is formed. In this chapter, we have concentrated on lenses that are so thin that their thicknesses may be ignored. Mirrors also use light, but they depend on reflection, not refraction. In Chapter 25 we dealt with spherical mirrors. In spite of the fact that thin lenses and spherical mirrors use light in fundamentally different ways, they have much in common as optical devices. For example, they both have focal lengths. The thin-lens equation and the mirror equation show that the focal length is related to the object distance and the image distance in the same mathematical way for both kinds of devices. Both can produce real or virtual images and can produce upright or inverted images. Both can also create images that are larger or smaller than the original objects, an ability described by the same magnification equation in each case. It is convenient that thin lenses and spherical mirrors function as optical devices in so many similar ways.

SUMMARY

When light strikes the interface between two media, part of the light is reflected and the remainder is transmitted across the interface. The change in the direction of travel as light passes from one medium into another is called **refraction.** The **index of refraction** n of a material is the ratio of the speed of light c in a vacuum to the speed of light v in the material: $n = c/v$. **Snell's law of refraction** states that (1) the refracted ray, the incident ray, and the normal to the interface all lie in the same plane, and (2) the angle of refraction θ_2 is related to the angle of incidence θ_1 by $n_1 \sin \theta_1 = n_2 \sin \theta_2$, where n_1 and n_2 are the indices of refraction of the incident and

refracting media, respectively. The angles are measured relative to the normal.

Because of refraction, a submerged object has an **apparent depth** that is different than its actual depth. If the observer is directly above the object, the apparent depth d' is related to the actual depth d by $d' = d(n_2/n_1)$, where n_1 and n_2 are the refractive indices of the media in which the object and the observer, respectively, are located.

When light passes from a medium of larger refractive index n_1 into one of smaller refractive index n_2, the refracted ray is bent away from the normal. If the incident ray is at the **critical angle** θ_c, the angle of refraction is $90°$. The critical angle is determined from Snell's law and is given by $\sin \theta_c = n_2/n_1$. When the angle of incidence exceeds the critical angle, all the incident light is reflected back into the medium from which it came, a phenomenon known as **total internal reflection.**

When light is incident on a nonmetallic surface at the **Brewster angle** θ_B, the reflected light is completely polarized parallel to the surface. The Brewster angle is given by $\tan \theta_B = n_2/n_1$, where n_1 and n_2 are the refractive indices of the incident and refracting media, respectively. When light is incident at the Brewster angle, the reflected and refracted rays are perpendicular to each other.

A glass prism can spread a beam of sunlight into a spectrum of colors, because the index of refraction of the glass depends on the wavelength of the light. The spreading of light into its color components is known as **dispersion.**

Converging lenses and **diverging lenses** depend on the phenomenon of refraction in forming an image. With a converging lens, paraxial rays that are parallel to the principal axis are focused to a point on the axis by the lens. This point is called the **focal point** of the lens, and its distance from the lens is the **focal length** f. Paraxial light rays that are parallel to the principal axis of a diverging lens appear to originate from its focal point after passing through the lens. The image produced by a converging or a diverging lens can be located with the help of a **ray diagram,** which can be constructed using the rays shown in Figure 26.19.

The **thin-lens equation** can be used with either converging or diverging lenses that are thin, and it relates the object distance d_o, the image distance d_i, and the focal length f of the lens: $1/d_o + 1/d_i = 1/f$. The magnification m of a lens is the ratio of the image height h_i to the object height h_o. The magnification is also related to d_i and d_o by the **magnification equation:** $m = -(d_i/d_o)$. The algebraic sign conventions for the variables appearing in the thin-lens and magnification equations are summarized in Section 26.8. When two or more lenses are used in combination, the image produced by one lens serves as the object for the next lens.

In the **human eye,** a real, inverted image is formed on a light-sensitive surface, called the retina. **Accommodation** is the process by which the focal length of the eye is automatically adjusted, so that objects at different distances produce focused images on the retina. The **near point** of the eye is the point nearest the eye at which an object can be placed and still produce a sharp image on the retina. The **far point** of the eye is the location of the farthest object on which the fully relaxed eye can focus.

For a normal eye, the near point is located 25 cm from the eye and the far point is located at infinity.

A **nearsighted (myopic)** eye is one that can focus on nearby objects, but not on distant ones. Nearsightedness can be corrected by wearing eyeglasses or contacts made from diverging lenses. A **farsighted (hyperopic)** eye can see distant objects clearly, but not those close up. Farsightedness can be corrected by using converging lenses.

The **refractive power** of a lens is measured in diopters and is given by $1/f$, where f is the focal length of the lens in meters. A converging lens has a positive refractive power, while a diverging lens has a negative refractive power.

The **angular size** of an object is the angle that it subtends at the eye of the viewer. For small angles, the angular size in radians is $\theta \approx h_o/d_o$, where h_o is the height of the object and d_o is the object distance. The **angular magnification** M of an optical instrument is the angular size θ' of the final image produced by the instrument divided by the reference angular size θ of the object, which is that seen without the instrument: $M = \theta'/\theta$.

A **magnifying glass** is usually a single converging lens that forms an enlarged, upright, and virtual image of an object placed at or inside the focal point of the lens. For a magnifying glass held close to the eye, the angular magnification is approximately $M \approx (1/f - 1/d_i)N$, where f is the focal length of the lens, d_i is the image distance, and N is the distance of the viewer's near point from the eye.

A **compound microscope** usually consists of two lenses, an objective and an eyepiece. The final image is enlarged, inverted, and virtual. The angular magnification of such a microscope is $M \approx -(L - f_e)N/(f_o f_e)$, where f_o and f_e are, respectively, the focal lengths of the objective and eyepiece, L is the distance between the two lenses, and N is the distance of the viewer's near point from the eye.

An **astronomical telescope** magnifies distant objects with the aid of an objective and eyepiece, and it produces a final image that is inverted and virtual. The angular magnification of a telescope is $M \approx -f_o/f_e$, where f_o and f_e are the focal lengths of the objective and eyepiece.

Lens aberrations limit the formation of perfectly focused or sharp images by optical instruments. **Spherical aberration** occurs because rays that pass through the outer edge of a lens with spherical surfaces are not focused at the same point as those that pass through the center of the lens. **Chromatic aberration** arises because a lens focuses different colors at different points.

QUESTIONS

1. Two slabs with parallel faces are made from different types of glass. A ray of light enters each slab at the same angle of incidence, as the drawing shows. Which slab has the greater index of refraction? Why?

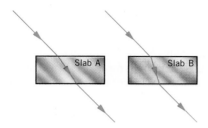

2. Two blocks, made from the same transparent material, are immersed in different liquids. A ray of light strikes each block at the same angle of incidence. From the drawing, determine which liquid has the greater index of refraction. Justify your answer.

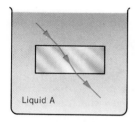
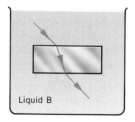

3. When an observer peers over the edge of a deep empty bowl, he does not see the entire bottom surface, so a small object lying on the bottom remains hidden from view. However, when the bowl is filled with water, the object can be seen. Explain this effect.

4. Two identical containers, one filled with water ($n = 1.33$) and the other filled with ethyl alcohol ($n = 1.36$) are viewed from directly above. Which container (if either) appears to have a greater depth of fluid? Why?

5. To a pearl diver swimming under water, does an object suspended in the air above the water appear to be at the actual height above the surface (assuming the diver is not wearing goggles)? If not, does the object appear to be higher or lower than the actual height? Why?

6. When you look through an aquarium window at a fish, is the fish as close as it appears? Explain.

7. At night, when it's dark outside and you are standing in a brightly lit room, it is easy to see your reflection in a window. During the day it is not so easy. Account for these facts.

8. A man is fishing from a dock. (a) If he is using a bow and arrow, should he aim above the fish, at the fish, or below the fish, to strike it? (b) How would he aim if he were using a laser gun? Give your reasoning.

9. Two rays of light converge to a point on a screen. A plane-parallel plate of glass is placed in the path of this converging light, and the glass plate is parallel to the screen. Will the point of convergence remain on the screen? If not, will the point move toward the glass or away from it? Justify your answer by drawing a diagram and showing how the rays are affected by the glass.

10. A person sitting at the beach is wearing a pair of Polaroid sunglasses and notices little discomfort due to the glare from the water on a bright sunny day. When she lies on her side, however, she notices that the glare increases. Why?

11. Linearly polarized light is incident on a surface at the Brewster's angle. The incident light is polarized such that its electric field has *no component* parallel to the surface. Is any light reflected from the surface? Explain.

12. Suppose a narrow beam of sunlight passes through a plate of glass with parallel sides, as in Figure 26.5. When the light leaves the glass, is the light dispersed into colors, as when light leaves a glass prism? Justify your answer, commenting on whether the separation of the colors (if they are separated) would be larger or smaller with thicker glass plates.

13. Figure 26.13c illustrates the dispersion of sunlight by prisms. Would the dispersion be enhanced or reduced if the prisms were immersed in water? Explain.

14. Suppose you want to make a rainbow by spraying water from a garden hose into the air. (a) Where must you stand relative to the water and the sun to see the rainbow? (b) Why can't you ever walk under the rainbow?

15. A beam of blue light is propagating in glass. When the light reaches the boundary between the glass and the surrounding air, the beam is totally reflected back into the glass. However, red light with the same angle of incidence is not totally reflected, and some of the light is refracted into the air. Explain the difference in the behavior of these colors.

16. A beacon in a lighthouse is to produce a parallel beam of light. The beacon consists of a bulb and a converging lens. Should the bulb be placed outside the focal point, at the focal point, or inside the focal point of the lens? State your reason.

17. A lens can be used in bright sunlight to start a fire. (a) Should a converging or a diverging lens be used? Why? (b) Relative to the lens, where should a piece of paper be placed so the fire can be started as quickly as possible?

18. A spherical mirror and a lens are immersed in water. Compared to the way they work in air, which one do you expect will be more affected by the water? Why?

19. In a TV mystery program, a photographic negative is introduced as evidence in a court trial. The negative shows an image of a house (now burned down) that was the scene of the crime. At the trial the defendant's acquittal depends on knowing exactly how far above the ground a window was. An expert called by the defense claims that this height can be calculated from only two pieces of information: (1) the measured height on the film, and (2) the focal length of the camera lens. Explain whether the expert is making sense, using the thin-lens and magnification equations to guide your thinking.

20. Suppose two people who wear glasses are camping. One of them is nearsighted and the other is farsighted. Whose glasses may be useful in starting a fire with the sun's rays? Give your reasoning.

21. Suppose that a 21-year-old with normal vision (near point = 25 cm) is standing in front of a plane mirror. How close can he stand to the mirror and still see himself in focus?

22. If we read for a long time, our eyes become "tired." When this happens, it helps to stop reading and look at a distant object. From the point of view of the ciliary muscle, why does this refresh the eyes?

23. To a swimmer under water, objects look blurred and out of focus. However, when the swimmer wears goggles that keep the water away from the eyes, the objects appear sharp and in focus. Why do goggles improve a swimmer's underwater vision?

24. The refractive power of the lens of the eye is 15 diopters when surrounded by the aqueous and vitreous humors. If this lens is removed from the eye and surrounded by air, its refractive power increases to about 150 diopters. Why is the refractive power of the lens so much greater outside the eye?

25. Two lenses have refractive powers of 1 and 4 diopters. Draw each of the lenses and locate their focal points to scale. Parallel rays of light are incident on each lens. (a) Draw the refracted rays as they pass through the focal point of each lens.

(b) From your drawings, decide which lens bends the rays to the greatest extent.

26. By means of a ray diagram, show that the eyes of a person wearing glasses appear to be (a) smaller when the glasses use diverging lenses to correct for nearsightedness and (b) larger when the glasses use converging lenses to correct for farsightedness.

27. Can a diverging lens be used as a magnifying glass? Justify your answer with a ray diagram.

28. Who benefits more from using a magnifying glass, a person whose near point is located 25 cm away from the eyes or a person whose near point is located 75 cm away from the eyes? Provide a reason for your answer.

29. Two lenses, whose focal lengths are 3.0 and 45 cm, are used to build a telescope. Which lens should be the objective? Why?

30. Two refracting telescopes have identical eyepieces, although one telescope is twice as long as the other. Which telescope has the greater angular magnification? Provide a reason for your answer.

31. Suppose a well-designed optical instrument is composed of two converging lenses separated by 14 cm. The focal lengths of the lenses are 0.60 and 4.5 cm. Is the instrument a microscope or a telescope? Why?

32. It is often thought that virtual images are somehow less important than real images. To show that this is not true, identify which of the following instruments normally produce final images that are virtual: (a) a projector, (b) a camera, (c) a magnifying glass, (d) eyeglasses (e) a compound microscope, and (f) an astronomical telescope.

33. Why does chromatic aberration occur in lenses, but not in mirrors?

PROBLEMS

Unless specified otherwise, use the values given in Table 26.1 for the refractive index.

Section 26.1 The Index of Refraction

1. What is the speed of light in benzene?

2. Find the ratio of the speed of light in diamond to the speed of light in ice.

3. Light travels at a speed of 2.201×10^8 m/s in a certain substance. What substance in Table 26.1 could this be? Use 2.998×10^8 m/s for the speed of light in a vacuum.

4. A light wave has a frequency of 5.09×10^{14} Hz in water and diamond. What is the wavelength of this light in each medium?

5. A glass window ($n = 1.5$) has a thickness of 4.0×10^{-3} m. How long does it take light to pass perpendicularly through the plate?

6. The speed of light is fifty percent larger in material A than it is in material B. Determine the ratio n_A/n_B of the refractive indices of these materials.

***7.** In a certain time, light travels 3.50 km in a vacuum. During the same time, light travels only 2.50 km in a liquid. What is the refractive index of the liquid?

***8.** When light enters a medium whose index of refraction is n, the frequency of the light does not change, but the wavelength and speed do. (a) Show that the wavelength λ' in the medium is $\lambda' = \lambda/n$, where λ is the wavelength of the light in a vacuum.

Section 26.2 Snell's Law and the Refraction of Light

9. A light ray in air is incident on a water surface at a 43° angle of incidence. Find (a) the angle of reflection and (b) the angle of refraction.

10. A layer of oil ($n = 1.45$) floats on an unknown liquid. A ray of light shines from the oil into the unknown liquid. The angles of incidence and refraction are, respectively, 65.0° and 53.0°. What is the index of refraction of the unknown liquid?

11. A ray of light is propagating in water and strikes a plate of fused quartz. The angle of refraction in the quartz is measured to be 36.7°. What is the angle of incidence?

12. A beam of light is traveling in air and strikes a material. The angles of incidence and refraction are 50.0° and 30.3°, respectively. Obtain the speed of light in the material.

13. A block of crown glass is placed on top of a printed page. The block is 6.00 cm thick. When viewed directly from above, how far *above* the page does the printing appear to be?

14. A ray of sunlight hits a frozen lake at a 45° angle of incidence. At what angle of refraction does the ray penetrate (a) the ice and (b) the water beneath the ice?

15. A spotlight on a boat is 2.5 m above the water, and the light strikes the water at a point that is 8.0 m horizontally displaced from the spotlight (see the drawing). The depth of the water is 4.0 m. Determine the distance d, which locates the point where the light strikes the bottom.

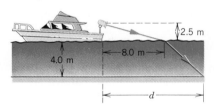

*16. A prism is made from ice and is surrounded by air. The cross section of the prism is an isosceles right triangle. As the drawing indicates, a ray of light hits the prism. Once inside the prism, the ray travels parallel to the hypotenuse of the right triangle. Find the angle of incidence θ_1 of the entering ray and the angle of refraction θ_2 of the exiting ray.

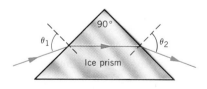

*17. A silver medallion is sealed within a transparent block of plastic. An observer in air, viewing the medallion from directly above, sees the medallion at an apparent depth of 1.6 cm beneath the top surface of the block. How far below the top surface would the medallion appear if the observer (not wearing goggles) and the block were under water?

**18. A small logo is embedded in a thick block of crown glass ($n = 1.52$), 3.20 cm beneath the top surface of the glass. The block is put under water, so there is 1.50 cm of water above the top surface of the block. The logo is viewed from directly above by an observer in air. How far beneath the top surface of the water does the logo appear to be?

**19. Refer to Figure 26.4 and assume the observer is nearly above the submerged object. For this situation, derive the expression for the apparent depth: $d' = d(n_2/n_1)$, Equation 26.3. (*Hint: Use Snell's law of refraction and the fact that the angles of incidence and refraction are small, so* $\tan \theta \approx \sin \theta$.)

**20. The back wall of a home aquarium is a mirror that is 30.0 cm away from the front wall. The walls of the tank are negligibly thin. A fish is swimming midway between the front and back walls. (a) How far from the front wall does the fish seem to be located? (b) An image of the fish appears behind the mirror. How far is this image from the front wall of the aquarium? (c) Would the refractive index of the liquid have to be larger or smaller in order for the image of the fish to appear in *front* of the mirror, rather than behind it? Why?

Section 26.3 Total Internal Reflection

21. What is the critical angle for light emerging from ice into air?

22. Light is propagating from diamond into crown glass. (a) Find the critical angle. (b) Is there a critical angle for light propagating from crown glass into diamond? If so, find its value.

23. One method of determining the refractive index of a transparent solid is to measure the critical angle when the solid is in air. If θ_c is found to be 40.5°, what is the index of refraction of the solid?

24. A point source of light is submerged 2.2 m below the surface of a lake and emits rays in all directions. On the surface of the lake, directly above the source, the area illuminated is a circle. What is the maximum radius that this circle could have?

25. The drawing shows a crown glass slab with a rectangular cross section. As illustrated, a laser beam strikes the upper surface at an angle of 60.0°. After reflecting from the upper surface, the beam reflects from the side and bottom surfaces. (a) If the glass is surrounded by air, determine where part of the beam first exits the glass, at point A, B, or C. (b) Repeat part (a), assuming the glass is surrounded by water.

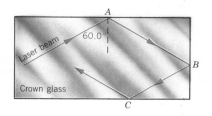

26. A swimmer, under water ($n = 1.333$), looks upward toward the surface and sees a honey bee. The swimmer's line of sight makes an angle of 48.6° with respect to the normal to the surface. Is the honey bee in the air above the water or floating on the surface of the water?

*27. A glass block ($n = 1.60$) is immersed in a liquid. A ray of light within the glass hits a glass–liquid surface at a 65.0° angle of incidence. Some of the light enters the liquid. What is the smallest possible refractive index for the liquid?

*28. Three materials, A, B, and C, have refractive indices n_A, n_B, and n_C. The materials are in the form of parallel plates and are stacked on top of one another with A on the bottom and B in the middle. A ray of light originates in material A and strikes the A–B surface with an angle of incidence θ_A. It is observed that the light penetrates into material B only when θ_A is less than 50.0° and penetrates into material C only when θ_A is less than 30.0°. Find n_B/n_A and n_B/n_C.

Section 26.4 Polarization and the Reflection and Refraction of Light

29. Light is reflected from a glass coffee table. When the angle of incidence is 56.7°, the reflected light is completely polarized parallel to the surface of the glass. What is the index of refraction of the glass?

30. At what angle of incidence is sunlight completely polarized upon being reflected from the surface of a lake (a) in the summer and (b) in the winter when the water is frozen?

31. Find Brewster's angle when light is reflected off a piece of glass ($n = 1.530$) submerged in ethyl alcohol ($n = 1.362$).

32. Light is incident from air onto a beaker of benzene. If the reflected light is 100% polarized, what is the angle of refraction of the light that penetrates into the benzene?

*33. When light strikes the surface between two materials from above, the Brewster angle is 65.0°. What is the Brewster angle when the light encounters the same surface from below?

**34. For a surface between two nonconducting materials, prove that the Brewster angle is never larger than the critical angle, so that light can never be 100% reflected from such a surface and simultaneously be 100% polarized.

Section 26.5 The Dispersion of Light: Prisms and Rainbows

35. A beam of sunlight encounters a plate of crown glass at a 45.00° angle of incidence. Using the data in Table 26.2, find the angle between the violet ray and the red ray in the glass.

36. Red light is incident on a block of plastic. The index of refraction for the red light is 1.67 and the angle of refraction is 22.6°. Green light is also incident on the plastic with the same angle of incidence as the red light. The green light has an angle of refraction of 21.9°. What is the index of refraction for the green light?

37. Yellow light strikes a diamond at a 45.0° angle of incidence and is refracted when it enters the diamond. Blue light strikes a piece of flint glass and has the same angle of refraction as does the yellow light in the diamond. See Table 26.2 for data. What is the angle of incidence of the blue light?

38. Horizontal rays of red light ($\lambda = 660$ nm, in vacuum) and violet light ($\lambda = 410$ nm, in vacuum) are incident on the flint-glass prism shown in the drawing. See Table 26.2 for any necessary data. What is the angle of refraction for each ray as it emerges from the prism?

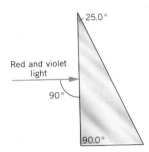

*39. This problem relates to Figure 26.13b, which illustrates the dispersion of light by a prism. The prism is made from flint glass (see Table 26.2), and its cross section is an equilateral triangle. The angle of incidence for both the red and violet light is 60.0°. Find the angles of refraction at which the red and violet rays emerge into the air from the prism.

Section 26.6 Lenses, Section 26.7 The Formation of Images by Lenses, Section 26.8 The Thin-Lens Equation and the Magnification Equation

40. An object is located 9.0 cm in front of a converging lens ($f = 6.0$ cm). Using an accurately drawn ray diagram, determine where the image is located.

41. A figurine is placed 15.0 cm in front of a converging lens ($f = 40.0$ cm). Using a ray diagram drawn to scale, find (a) the image distance and (b) the magnification.

42. A macroscopic (or macro) lens for a camera is usually a lens of normal focal length built into a lens barrel that can be adjusted to provide the additional lens-to-film distance needed when focusing at very close range. Suppose that a macro lens ($f = 50.0$ mm) has a maximum lens-to-film distance of 275 mm. How close can the object be located in front of the lens?

43. A converging lens ($f = 12.0$ cm) is held 8.00 cm in front of a newspaper. Find (a) the image distance and (b) the magnification.

44. A diverging lens has a focal length of -25 cm. (a) Find the image distance when an object is placed 38 cm from the lens. (b) Is the image real or virtual?

45. An object, 0.50 cm high, is placed 8.6 cm in front of a diverging lens whose focal length is -7.5 cm. Find the height of the image.

46. (a) For a diverging lens ($f = -20.0$ cm), construct a ray diagram to scale and find the image distance for an object that is 20.0 cm from the lens. (b) Determine the magnification of the lens from the diagram.

47. A diverging lens has a focal length of -38 cm. An object is placed 28 cm in front of this lens. Calculate (a) the image distance and (b) the magnification. Is the image (c) real or virtual, (d) upright or inverted, and (e) enlarged or reduced in size?

48. A magnifying glass uses a converging lens whose focal length is 15 cm. The magnifying glass produces a virtual and upright image that is 3.0 times larger than the object. (a) How far is the object from the lens? (b) What is the image distance?

*49. From a distance of sixty meters, a photographer uses a telephoto lens ($f = 500.0$ mm) to take a picture of a charging rhinoceros. How far from the rhinoceros would the photographer have to be to record an image of the same size using a lens whose focal length is 50.0 mm?

*50. On a roll of movie film, each picture has a width of 70.0 mm. The projector lens has a focal length of 305 mm. If the screen in a theater is 60.0 m from the projector lens, what is the width of the image projected on the screen?

*51. The moon's diameter is 3.48×10^6 m and its mean distance from the earth is 3.85×10^8 m. The moon is being photographed by a camera whose lens has a focal length of 50.0 mm. (a) Find the diameter of the moon's image on the slide film. (b) When the slide is projected onto a screen that is 15.0 m from the lens of the projector ($f = 110.0$ mm), what is the diameter of the moon's image on the screen?

**52. An object is 20.0 cm from a converging lens, and the image falls on a screen. When the object is moved 4.00 cm closer to the lens, the screen must be moved 2.70 cm farther away from the lens to register a sharp image. Determine the focal length of the lens.

**53. A converging lens ($f = 25.0$ cm) is used to project an image of an object onto a screen. The object and the screen are 125 cm apart, and between them the lens can be placed at either of two locations. Find the two object distances.

Section 26.9 Lenses in Combination

54. A converging lens ($f = 12.0$ cm) is located 30.0 cm to the left of a diverging lens ($f = -6.00$ cm). A postage stamp is placed 36.0 cm to the left of the converging lens. (a) Locate the final image of the stamp relative to the diverging lens. (b) Find the overall magnification. (c) Is the final image real or virtual? With respect to the original object, is the final image (d) upright or inverted, and is it (e) larger or smaller?

55. An object, 0.75 cm tall, is placed 12.0 cm to the left of a diverging lens ($f = -8.00$ cm). A converging lens is placed 8.00 cm to the right of the diverging lens. The final image is virtual and is 29.0 cm to the left of the diverging lens. Determine (a) the focal length of the converging lens and (b) the height of the final image.

56. A converging lens ($f = 12.0$ cm) is 28.0 cm to the left of a diverging lens ($f = -14.0$ cm). An object is located 6.00 cm to the left of the converging lens. Draw an accurate ray diagram and from it find (a) the final image distance, measured from the diverging lens, and the overall magnification. (b) Confirm your answers to part (a) by using the thin-lens and magnification equations.

*57. A coin is located 15.00 cm to the left of a converging lens ($f = 10.00$ cm). A second, identical lens is placed to the right of the first lens, such that the image formed by the combination has the same size and orientation as the original coin. Find the separation between the lenses.

**58. Two converging lenses ($f_1 = 9.00$ cm and $f_2 = 6.00$ cm) are separated by 18.0 cm. The lens on the left has the longer focal length. An object stands 12.0 cm to the left of the combination. (a) Locate the final image relative to the lens on the right. (b) Obtain the overall magnification. (c) Is the final image real or virtual? With respect to the original object, is the final image (d) upright or inverted and is it (e) larger or smaller?

Section 26.10 The Human Eye

59. A farsighted person cannot see clearly objects that are closer to the eye than 73 cm. Determine the focal length of contact lenses that will enable this person to read a magazine at a distance of 25 cm.

60. A nearsighted person has a far point located only 220 cm from his eyes. Determine the focal length of contact lenses that will enable him to see distant objects clearly.

61. Suppose your friend wears contact lenses that have a focal length of 35.1 cm. The lenses are designed so she can read a magazine held as close as 25.0 cm. Where is the near point of her unaided eyes?

62. A person has far points of 5.0 m from the right eye and 6.5 m from the left eye. Write a prescription for the refractive power of each corrective contact lens.

63. A farsighted person cannot focus clearly on objects that are less than 145 cm from his eyes. To correct this problem, the person wears eyeglasses that are located 2.0 cm in front of his eyes. Determine the focal length that will permit this person to read a newspaper at a distance of 32.0 cm from his eyes.

64. A person holds a book 25 cm in front of the effective lens of her eye; the print in the book is 2.0 mm high. If the effective lens of the eye is located 1.7 cm from the retina, what is the size of the print image on the retina?

*65. A nearsighted person wears contacts to correct for a far point that is only 3.62 m from his eyes. The near point of his unaided eyes is 25.0 cm from his eyes. If he does not remove the lenses when reading, how close can he hold a book and see it clearly?

**66. The far point of a nearsighted person is 4.37 m from her eyes, and she wears contacts that enable her to see distant objects clearly. A tree is 12.0 m away and 3.00 m high. (a) When she looks through the contacts at the tree, what is its image distance? (b) How high is the image formed by the contacts?

**67. The contacts worn by a farsighted person allow her to see objects clearly that are as close as 25.0 cm, even though her uncorrected near point is 79.0 cm from her eyes. When she is looking at a poster, the contacts form an image of the poster at a distance of 217 cm from her eyes. (a) How far away is the poster actually located? (b) If the poster is 0.350 m tall, how tall is the image formed by the contacts?

Section 26.11 Angular Magnification and the Magnifying Glass

68. A spectator, seated in the left field stands, is watching a 1.9-m-tall baseball player who is 75 m away. On a TV screen, the same player has a 0.12-m image. To a viewer located 3.0 m from the screen, does the ball player appear to be larger or smaller than what the spectator sees? Give your reasoning.

69. An engraver uses a magnifying glass ($f = 9.5$ cm) to examine some work. The image he sees is located 25 cm from his eye, which is his near point. (a) What is the distance between the work and the magnifying glass? (b) What is the angular magnification of the magnifying glass?

70. A magnifying glass is held above a magazine such that the image is located at the near point of the eye. The near point is 0.30 m away from the eye, and the angular magnification is 3.4. Find the focal length of the magnifying glass.

71. A butterfly collector is examining a rare specimen and uses a magnifying glass with a refractive power of 10.0 diopters. The magnifying glass is held close to the eye, and the butterfly-to-lens distance is adjusted so a virtual image is formed at infinity. The angular magnification is 4.0. What is the distance between the collector's eyes and the near point?

*72. A stamp collector is viewing a special stamp. The collector's near point is 25 cm from his eyes. (a) What is the refracting power of a magnifying glass that has an angular magnification of 6.0 when the image of the stamp is located at the near point of the eye? (b) What is the angular magnification when the image of the stamp is 45 cm from the eye?

**73. A farsighted person can read printing as close as 25.0 cm when she wears contacts that have a focal length of 45.4 cm. One day, however, she forgets her contacts and uses a magnifying glass, which has a maximum angular magnification of 7.50 for a young person with a normal near point of 25.0 cm. What is the maximum angular magnification that the magnifying glass can provide for her?

Section 26.12 The Compound Microscope

74. An insect subtends an angle of only 4.0×10^{-3} rad at the unaided eye when placed at the near point. What is the angular size (magnitude only) when the insect is viewed through a microscope whose angular magnification has a magnitude of 160?

75. A microscope for viewing blood cells has an objective with a focal length of 0.620 cm and an eyepiece with a focal length of 4.40 cm. The distance between the objective and eyepiece is 23.0 cm. If a blood cell subtends an angle of 2.40×10^{-5} rad when viewed with the naked eye at a near point of 25.0 cm, what angle (magnitude only) does it subtend when viewed through the microscope?

76. A compound microscope has a barrel whose length is 16.0 cm and an eyepiece whose focal length is 1.4 cm. The viewer has a near point located 25 cm from his eyes. What focal length must the objective have so the angular magnification of the microscope is -320?

77. An anatomist is viewing heart muscle cells with a microscope that has two selectable objectives with refracting powers of 100 and 300 diopters. When she uses the 100-diopter objective, the image of a cell subtends an angle of 3×10^{-3} rad with the eye. What angle is subtended when she uses the 300-diopter objective?

*78. The maximum angular magnification of a magnifying glass is 12.0 when a person uses it who has a near point that is 25.0 cm from his eyes. The same person finds that a microscope, using this magnifying glass as the eyepiece, has an angular magnification of -525. The separation between the eyepiece and the objective of the microscope is 23.0 cm. Obtain the focal length of the objective.

*79. It is possible to interchange the eyepiece and the objective of a microscope without changing the angular magnification of the instrument, provided the separation L between the two lenses is suitably adjusted. Derive an equation that gives the new separation L' in terms of the original separation L, the focal length f_o of the original objective, and the focal length f_e of the original eyepiece.

Section 26.13 The Telescope

80. An astronomical telescope has an objective with a focal length of 96 cm and interchangeable eyepieces whose focal lengths are 3.0 and 0.80 cm. What angular magnifications are possible?

81. A refracting telescope has an objective and an eyepiece

that have refractive powers of 1.25 diopters and 250 diopters, respectively. Find the angular magnification of the telescope.

82. The moon subtends an angle of 9.0×10^{-3} rad at the unaided eye. An astronomical telescope uses an eyepiece with a focal length of 0.42 m and an objective with a focal length of 1.8 m. When viewed through this telescope, what angle (magnitude only) does the moon subtend?

83. A refracting astronomical telescope for hobbyists has an angular magnification of -155. The eyepiece has a focal length of 5.00 mm. (a) Determine the focal length of the objective. (b) About how long is the telescope?

84. An astronomical telescope has an angular magnification of -184 and uses an objective with a focal length of 48.0 cm. What is the focal length of the eyepiece?

*85. A refracting telescope has an angular magnification of -109. The length of the barrel is 1.10 m. What are the focal lengths of the objective and the eyepiece?

*86. The refracting telescope at Yerkes Observatory in Wisconsin has an objective whose focal length is 19.4 m. Its eyepiece has a focal length of 10.0 cm. (a) What is the angular magnification of the telescope? (b) If the telescope is used to look at a lunar crater (diameter = 1500 m), what is the size of the first image, assuming the moon is 3.85×10^8 m from the earth? (c) How close does the crater appear to be when seen through the telescope?

**87. An astronomical telescope is being used to examine a relatively close object that is only 114 m away from the objective of the telescope. The objective and eyepiece have focal lengths of 1.500 and 0.070 m, respectively. Noting that the expression $M \approx -f_o/f_e$ is no longer applicable because the object is so close, use the thin-lens and magnification equations to find the angular magnification of the telescope. (Hint: See Figure 26.36 and note that the focal points F_o and F_e are so close together that the distance between them may be ignored.)

ADDITIONAL PROBLEMS

88. When a diverging lens is held 13 cm above a line of print, as in Figure 26.23, the image is 5.0 cm beneath the lens. What is the focal length of the lens?

89. The distance between the lenses in a microscope is 18 cm. The focal length of the objective is 1.5 cm. If the microscope is to provide an angular magnification of -83 when used by a person with a normal near point (25 cm from the eye), what must be the focal length of the eyepiece?

90. A beam of light impinges from air onto a block of ice at a 60.0° angle of incidence. Assuming this angle remains the same, find the percentage by which the angle of refraction changes when the ice turns to water, and state whether the change is an increase or a decrease.

91. A woman can read the large print in a newspaper only when it is at a distance of 65 cm or more from her eyes. (a) Is she myopic or hyperopic? (b) What should be the refractive power of her glasses (worn 2.0 cm from the eyes), so she can read the newspaper at a distance of 25 cm from the eyes?

92. A person looks at a scene through a diverging lens with a focal length of -12.5 cm. The lens forms a virtual image 5.00 cm from the lens. Find the magnification.

93. A camera is supplied with two interchangeable lenses, whose focal lengths are 35.0 and 150.0 mm. A woman whose height is 1.80 m stands 8.00 m in front of the camera. What is the height (including sign) of her image that each lens produces on the film?

94. A jeweler whose near point is 65 cm from his eye uses a small magnifying glass to examine a watch held 5.0 cm from the lens. Find the angular magnification of the magnifying glass.

95. An amateur astronomer decides to build a telescope from a discarded pair of eyeglasses. One of the lenses has a refractive power of 11 diopters, while the other has a refractive power of 1.3 diopters. (a) Which lens should be the objective? (b) How far apart should the lenses be separated? (c) What is the angular magnification of the telescope?

96. A ray of light in air enters a liquid at an angle of incidence of 47.00°. The angle of refraction is 29.16°. Identify the liquid on the basis of the information in Table 26.1.

97. An optometrist prescribes contact lenses that have a focal length of 55.0 cm. (a) Are the lenses converging or diverging, and (b) is the person who wears them nearsighted or farsighted? (c) Where is the unaided near point of the person located, if the lenses are designed so that objects no closer than 35.0 cm can be seen clearly?

98. To focus a camera on objects at different distances, the converging lens is moved toward or away from the film, so a sharp image always falls on the film. A camera with a telephoto lens ($f = 200.0$ mm) is to be focused on an object located first at a distance of 3.5 m and then at 50.0 m. Over what distance must the lens be movable?

99. Amber ($n = 1.546$) is a transparent brown-yellow fossil resin. An insect, trapped and preserved within the amber, appears to be 2.5 cm beneath the surface, when viewed directly from above. How far below the surface is the insect actually located?

100. An object is located 30.0 cm to the left of a converging lens whose focal length is 50.0 cm. (a) Draw a ray diagram to scale and from it determine the image distance and the magnification. (b) Use the thin-lens and magnification equations to verify your answers to part (a).

*101. A person using a magnifying glass observes that for clear vision its maximum angular magnification is 25% larger than its minimum angular magnification. Assuming that the person has a near point located 25 cm from her eye, what is the focal length of the magnifying glass?

*102. An office copier uses a lens to place an image of a document onto a rotating drum. The copy is made from this image. (a) What kind of lens is used? (b) If the document and its copy are to have the same size, but are inverted with respect to one another, how far from the document is the lens located and how far from the lens is the image located? Express your answers in terms of the focal length f of the lens.

*103. A reflecting telescope uses a concave mirror whose radius of curvature is 2.4 m. If the angular magnification of the telescope is -360, what is the focal length of the eyepiece?

*104. When a converging lens is used in a camera (as in Figure 26.20), the film must be placed at a distance of 0.210 m from the lens to record an image of an object that is 4.00 m from the lens. The same lens is then used in a projector (see Figure 26.21), with the screen 0.500 m from the lens. How far from the projector lens should the film be placed?

*105. At age forty, a man requires contact lenses ($f = 65.0$ cm) to read a book held 25.0 cm from his eyes. At age forty-five, he finds that while wearing these contacts he must now hold a book 29.0 cm from his eyes. (a) By what distance has his near point changed? (b) What focal length lenses does he require at age forty-five to read a book at 25.0 cm?

**106. The angular magnification of a refracting telescope is 32 800 times larger when you look through the correct end of the telescope than when you look through the wrong end. What is the angular magnification of the telescope?

**107. Bill is farsighted and has a near point located 125 cm from his eyes. Anne is also farsighted, but her near point is 75.0 cm from her eyes. Both have glasses that correct their vision to a normal near point (25.0 cm from the eyes), and both wear the glasses 2.0 cm from the eyes. Relative to the eyes, what is the closest object that can be seen clearly (a) by Anne when she wears Bill's glasses and (b) by Bill when he wears Anne's glasses?

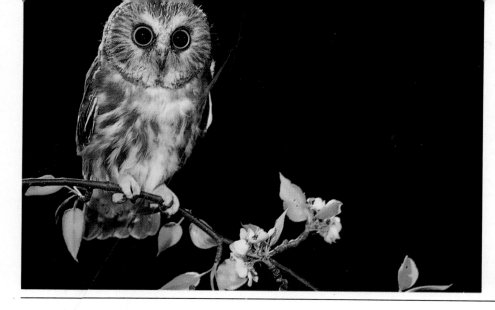

CHAPTER 27

INTERFERENCE AND THE WAVE NATURE OF LIGHT

*Saw-whet owls such as this one are famous for their striking eyes, which have such large pupils. For a number of reasons, owl eyes have the ability to distinguish between two objects that are close together. One of the most important reasons involves the large pupils and the wave nature of light, as we will discuss. Being a wave, light can exhibit effects due to interference and diffraction, which influence vision quality. The study of the interference and diffraction of light is referred to as **wave optics** or **physical optics**, to distinguish it from the material that we studied in Chapters 25 and 26. Those two chapters dealt with situations in which light traveled through a uniform medium in a straight line and changed direction only when it reflected or refracted at a boundary between two different media. The study of the straight-line motion of light and reflection and refraction is called **geometrical optics.** Geometrical optics cannot explain interference and diffraction, which are the same phenomena that sound waves exhibit. To explain them, we will use the principle of linear superposition, as we did in Chapter 17. We will see that interference and diffraction are important when the dimensions of the system under study become small enough that they begin to approach the wavelength of light. A number of areas in addition to vision will be discussed, including lenses, compact disc technology, and the production of computer chips.*

27.1 THE PRINCIPLE OF LINEAR SUPERPOSITION

The principle of linear superposition states that when two or more waves are present simultaneously at the same place, the resultant wave is the sum of the individual waves. Light is an electromagnetic wave; thus, the electric fields of two

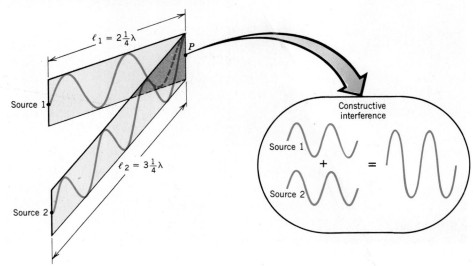

Figure 27.1 The waves emitted by source 1 and source 2 arrive at point P in phase, leading to constructive interference at that point.

light waves passing through a given point combine to give the total electric field at that point. And the square of the electric field strength is proportional to the light intensity, which is the light energy per second per square meter. The intensity, in turn, is related to the brightness of the light.

Figure 27.1 illustrates what happens when two identical waves arrive at the point P in phase, that is, crest to crest and trough to trough. According to the principle of linear superposition, the waves reinforce each other and *constructive interference* occurs. The resulting total wave at P has an amplitude that is twice the amplitude of either individual wave, and in the case of light waves, the brightness at P is greater than that of either wave alone. The waves have the same wavelength λ and are in phase at P, because this spot is located away from the sources of the waves at distances ℓ_1 and ℓ_2 that differ by one wavelength. In Figure 27.1, the distances are $\ell_1 = 2\frac{1}{4}$ wavelengths and $\ell_2 = 3\frac{1}{4}$ wavelengths. In general, constructive interference will result at P whenever the distances are the same or differ by any integer number of wavelengths; in other words, assuming ℓ_2 is the larger distance, whenever $\ell_2 - \ell_1 = m\lambda$, where $m = 0, 1, 2, 3, \ldots$.

Figure 27.2 shows what occurs when two identical waves arrive at the point P

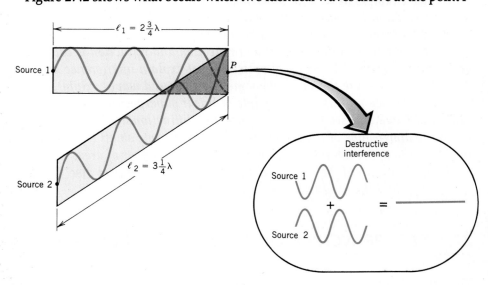

Figure 27.2 The waves emitted by the two sources arrive at point P out of phase, and destructive interference occurs at P.

out of phase with one another, or crest to trough. Now the waves mutually cancel, according to the principle of linear superposition, and *destructive interference* results. With light waves this would mean that there is no brightness. The waves are out of phase at P, because the distances through which they travel in reaching this spot differ by one-half of a wavelength ($\ell_1 = 2\frac{3}{4}\lambda$ and $\ell_2 = 3\frac{1}{4}\lambda$ in the drawing). Destructive interference will take place at P whenever the distances differ by any odd integer number of half-wavelengths, that is, whenever $\ell_2 - \ell_1 = (m + \frac{1}{2})\lambda$, where $m = 0, 1, 2, 3, \ldots$, and ℓ_2 is the larger distance.

If constructive or destructive interference is to continue occurring at a point, the sources of the waves must be *coherent sources.* Two sources are coherent if the waves they emit maintain a constant phase relation. Effectively, this means that the waves do not shift relative to one another as time passes. For instance, suppose that the wave pattern of source 1 in Figure 27.2 shifted forward or backward by random amounts at random moments. Then, on average, neither constructive nor destructive interference would be observed at point P, because there would be no stable relation between the two wave patterns. Lasers are coherent sources of light, while incandescent light bulbs and fluorescent lamps are incoherent sources.

Many people have observed the effects of interference between two electromagnetic waves while watching TV. Television programming is carried by electromagnetic waves that are emitted by a transmitting antenna and detected by a receiving antenna connected to the TV set, as Figure 27.3 suggests. Sometimes these waves reflect from an airplane, and then the receiving antenna may detect waves coming from two sources. Source 1 is the transmitting antenna of the station, which is located at a fixed distance ℓ_1 from your house. The airplane acts as source 2, and its distance ℓ_2 from the house is changing as the plane moves. As a result, the difference between ℓ_1 and ℓ_2 is also changing, and the conditions of constructive and destructive interference come and go at the receiving antenna. Correspondingly, the receiving antenna delivers to the TV set a signal that increases and decreases, causing the picture to flutter as the plane passes overhead.

THE PHYSICS OF . . .

how airplanes overhead cause TV pictures to "flutter."

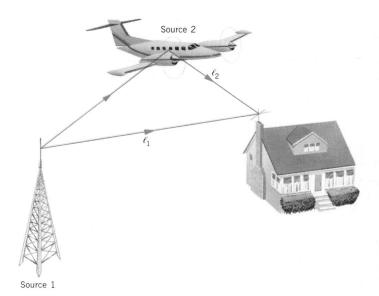

Figure 27.3 Source 1 is a TV transmitting antenna that emits electromagnetic waves. One wave reaches the house directly, while another arrives after being reflected from a passing airplane, which acts as wave source 2. As the plane flies by, conditions of constructive and destructive interference come and go at the receiving antenna, and the TV picture flutters.

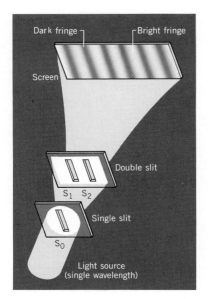

Figure 27.4 In Young's double-slit experiment, two slits S_1 and S_2 act as coherent sources of light. Light waves from these slits interfere constructively and destructively on the screen to produce, respectively, the bright and dark fringes. The slit widths and the distance between the slits have been exaggerated for clarity.

27.2 YOUNG'S DOUBLE-SLIT EXPERIMENT

In 1801 the English scientist Thomas Young (1773–1829) performed a historic experiment that demonstrated the wave nature of light by showing that two overlapping light waves interfered with each other. His experiment was particularly important, because he was also able to determine the wavelength of the light from his measurements, the first such determination of this important property. Figure 27.4 shows one arrangement of Young's experiment, in which light of a single wavelength (monochromatic light) passes through a single narrow slit S_0 and falls on two closely spaced, narrow slits S_1 and S_2. These two slits act as coherent sources of light waves that interfere constructively and destructively at different points on the screen to produce a pattern of alternating bright and dark fringes. The purpose of the single slit S_0 is to ensure that only light from one direction falls on the double slit. Without it, light coming from different points on the light source would strike the double slit from different directions and cause the pattern on the screen to be washed out. The slits S_1 and S_2 act as coherent sources of light waves, because the light from each originates from the same primary source, namely, the single slit S_0.

To help explain the origin of the bright and dark fringes, Figure 27.5 presents three top views of the double slit and the screen. Part a illustrates how a bright fringe arises directly opposite the midpoint between the two slits. At this location on the screen, the distances ℓ_1 and ℓ_2 to the slits are equal, each containing the same number of wavelengths. Therefore, constructive interference results, leading to the bright fringe. Part b indicates that constructive interference produces another bright fringe on one side of the midpoint when the distance ℓ_2 is larger than ℓ_1 by exactly one wavelength. A bright fringe also occurs symmetrically on the other side of the midpoint when the distance ℓ_1 exceeds ℓ_2 by one wavelength; for clarity, however, this bright fringe is not shown. Constructive interference produces additional bright fringes on both sides of the middle wherever the difference between ℓ_1 and ℓ_2 is an integer number of wavelengths. Part c shows how the first dark fringe arises. Here the distance ℓ_2 is larger than ℓ_1 by exactly one-half a wavelength, so the waves interfere destructively, giving rise to the dark fringe. Destructive interference creates additional dark fringes on both sides of

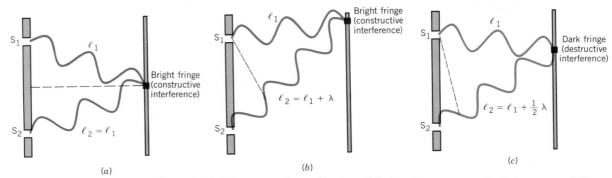

Figure 27.5 The waves from slits S_1 and S_2 interfere constructively (parts a and b) or destructively (part c) on the screen, depending on the difference in distances between the slits and the screen. The slit widths and the distance between the slits have been exaggerated for clarity.

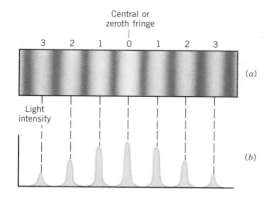

Central or
zeroth fringe

3 2 1 0 1 2 3

Light
intensity

(a)

(b)

Figure 27.6 The results of Young's double-slit experiment, showing (a) the light pattern formed and (b) a graph of the light intensity. The central fringe is brightest (greatest intensity). The intensities of other bright fringes decrease to either side of the central bright fringe, as the graph indicates.

the center wherever the difference between ℓ_1 and ℓ_2 equals an odd integer number of half-wavelengths.

The brightness of the fringes in Young's experiment varies. As an indication of the brightness, Figure 27.6 gives a graph of the light intensity for the fringe pattern. The central fringe is labeled with a zero, while the other bright fringes are numbered in ascending order on either side of center. It can be seen that the central fringe has the greatest intensity. To either side of center, the intensities of the other fringes decrease in a way that depends on how small the slit widths are relative to the wavelength of the light. Figure 27.7 shows a photograph of the fringe pattern observed in a typical Young's experiment.

The position of the fringes observed on the screen in Young's experiment can be calculated with the aid of Figure 27.8. If the screen is located very far away compared with the separation d of the slits, then the lines labeled ℓ_1 and ℓ_2 in part

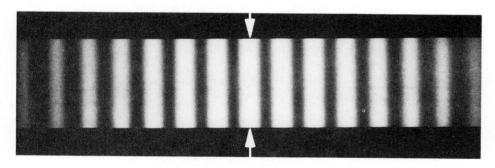

Figure 27.7 This photograph shows the fringe pattern obtained in a typical Young's double-slit experiment. The arrows indicate the central bright fringe.

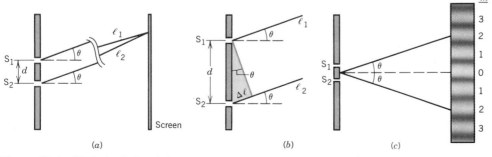

(a)

(b)

(c)

Figure 27.8 With the help of these pictures, Equation 27.1 is derived in the text. This equation gives the angles θ at which the bright fringes occur.

a are nearly parallel. Being nearly parallel, these lines make approximately equal angles θ with the horizontal. The distances ℓ_1 and ℓ_2 differ by an amount $\Delta\ell$, which is the length of the short side of the colored triangle in part *b* of the drawing. Since the triangle is a right triangle, it follows that $\Delta\ell = d\sin\theta$. Constructive interference occurs when the distances differ by an integer number *m* of wavelengths λ, or $\Delta\ell = d\sin\theta = m\lambda$. Therefore, the angle θ for the interference maxima can be determined from the following expression:

$$\begin{bmatrix}\textbf{Bright fringes}\\ \textbf{of a double}\\ \textbf{slit}\end{bmatrix} \qquad \sin\theta = m\frac{\lambda}{d} \qquad m = 0, 1, 2, 3, \ldots \qquad (27.1)$$

The value of *m* specifies the *order* of the fringe. Thus, $m = 2$ identifies the "second-order" bright fringe. Part *c* of the drawing stresses that the angle θ given by Equation 27.1 locates bright fringes on either side of the midpoint between the slits. A similar line of reasoning leads to the conclusion that the dark fringes, which lie between the bright fringes, are located according to

$$\begin{bmatrix}\textbf{Dark fringes}\\ \textbf{of a double}\\ \textbf{slit}\end{bmatrix} \qquad \sin\theta = (m+\tfrac{1}{2})\frac{\lambda}{d} \qquad m = 0, 1, 2, 3, \ldots \qquad (27.2)$$

Example 1 illustrates the application of these expressions and shows how to determine the distance of a higher-order bright fringe from the central bright fringe.

Example 1 Young's Double-Slit Experiment

Red light ($\lambda = 713$ nm in vacuum) is used in Young's experiment with the slits separated by a distance $d = 1.20 \times 10^{-4}$ m. The screen is located at a distance from the slits given by $L = 2.75$ m. Find the distance *y* on the screen between the central bright fringe and the third-order bright fringe (see Figure 27.9).

REASONING This problem can be solved by first using Equation 27.1 to determine the value of θ that locates the third-order bright fringe. Then trigonometry can be used to obtain the distance *y*.

SOLUTION According to Equation 27.1, we find

$$\sin\theta = m\frac{\lambda}{d} = 3\left(\frac{713 \times 10^{-9}\text{ m}}{1.20 \times 10^{-4}\text{ m}}\right) = 1.78 \times 10^{-2}$$
$$\theta = \sin^{-1}(1.78 \times 10^{-2}) = 1.02°$$

According to Figure 27.9, the distance *y* can be calculated from $\tan\theta = y/L$:

$$y = L\tan\theta = (2.75\text{ m})\tan 1.02° = \boxed{0.0490\text{ m}}$$

Historically, Young's experiment provided strong evidence that light has a wavelike character. If light behaved only as a stream of "tiny particles," as others believed at the time,* then the two slits would deliver the light energy into only

* It is now known that the particle or corpuscular theory of light, which Isaac Newton promoted, does indeed explain some experiments that the wave theory cannot explain. Today, light is regarded as having both particle and wave characteristics. Chapter 29 discusses this dual nature of light.

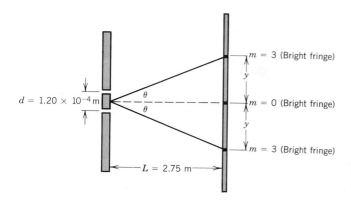

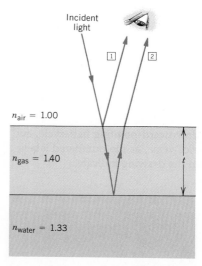

Figure 27.9 The third-order bright fringe ($m = 3$) is observed on the screen at a distance y from the central bright fringe ($m = 0$).

two bright fringes located directly opposite the slits on the screen. Instead, Young's experiment shows that wave interference redistributes the energy from the two slits into many bright fringes.

27.3 THIN FILM INTERFERENCE

Young's double-slit experiment is one example of interference between light waves. Interference also occurs in more common circumstances. For instance, Figure 27.10 shows a thin film, such as gasoline floating on water. The film is assumed to have a constant thickness. Consider what happens when monochromatic light (a single wavelength) strikes the film nearly perpendicularly. At the top surface of the film reflection occurs and produces the light wave represented by ray 1. However, refraction also occurs, and some light enters the film. Part of this light reflects from the bottom surface and passes back up through the film, eventually reentering the air. Thus, a second light wave, which is represented by ray 2, also exists. Moreover, this wave, having traversed the film twice, has traveled farther than wave 1. Because of the extra travel distance, there can be interference between the two waves. If constructive interference occurs, an observer, whose eyes detect the superposition of waves 1 and 2, would see a uniformly bright film. If destructive interference occurs, an observer would see a

Figure 27.10 Because of reflection and refraction, two light waves, represented by rays 1 and 2, enter the eye when light shines on a thin film of gasoline floating on water. Interference occurs between these waves.

The phenomenon of thin-film interference of light gives this unusual soap "bubble" its colors.

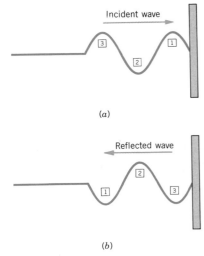

(a)

(b)

Figure 27.11 When a wave on a string reflects from a wall, the wave undergoes a phase change. Thus, after reflection an upward-pointing half-cycle of the wave becomes a downward-pointing half-cycle, and vice versa, as the numbered labels in the drawing indicate.

The beautiful iridescent colors of many birds, like the blues and greens in the peacock here, are due to the thin-film interference of light that occurs because of the layer of oil covering the feathers.

uniformly dark film. The controlling factor is whether the extra distance for wave 2 is an integer number of whole wavelengths or an odd integer number of half-wavelengths.

In Figure 27.10, the difference in path lengths between waves 1 and 2 occurs inside the thin film. Therefore, *the wavelength that is important for thin film interference is the wavelength within the film*, not the wavelength in vacuum. The wavelength within the film can be calculated from the wavelength in vacuum by using the index of refraction n for the film, since $n = c/v = (c/f)/(v/f) = \lambda_{\text{vacuum}}/\lambda_{\text{film}}$. In other words,

$$\lambda_{\text{film}} = \frac{\lambda_{\text{vacuum}}}{n} \qquad (27.3)$$

In explaining the interference that can occur in Figure 27.10, we need to add one more important part to the story. Whenever waves reflect at a boundary, it is possible for them to change phase. Figure 27.11, for example, shows that a wave on a string is inverted when it reflects from the end that is tied to a wall (see also Section 17.5). This inversion is equivalent to a half-cycle of the wave, as if the wave had traveled an additional distance of one-half of a wavelength. In contrast, a phase change does not occur when a wave on a string reflects from the end of a string that is hanging free. When light waves undergo reflection, similar phase changes occur as follows:

1. When light travels through a material with a smaller refractive index toward a material with a larger refractive index (e.g., air to gasoline), reflection at the boundary occurs along with a phase change that is equivalent to one-half of a wavelength.

2. When light travels from a larger toward a smaller refractive index, there is no phase change upon reflection at the boundary.

The next example indicates how the phase change that can accompany reflection is taken into account when dealing with thin film interference.

Example 2   Colored Thin Films of Gasoline

(a) A thin film of gasoline floats on a puddle of water. Sunlight falls almost perpendicularly on the film and reflects into your eyes. Although sunlight is white, since it contains all colors, the film has a yellow hue, because destructive interference eliminates the color of blue ($\lambda_{\text{vacuum}} = 469$ nm) from the reflected light. If the refractive indices of the blue light in gasoline and water are 1.40 and 1.33, respectively, determine the minimum nonzero thickness t of the film. (b) Repeat part (a) assuming the gasoline is on glass ($n_{\text{glass}} = 1.52$) instead of water.

REASONING   To solve this problem, we must express the condition for destructive interference in terms of the film thickness t. And we must take into account any phase changes that occur upon reflection. These phase changes will be different in parts (a) and (b), because different indices of refraction are involved.

SOLUTION
(a) In Figure 27.10, the phase change for wave 1 is equivalent to one-half of a wavelength, since this light travels from a smaller refractive index ($n_{\text{air}} = 1.00$) toward a larger refractive index ($n_{\text{gas}} = 1.40$). In contrast, there is no phase change when wave 2 reflects from the bottom surface of the film, since this light travels from a larger

refractive index (n_{gas} = 1.40) toward a smaller one (n_{water} = 1.33). The net phase change between waves 1 and 2 due to reflection is, thus, equivalent to one-half of a wavelength, $\frac{1}{2}\lambda_{film}$. This half-wavelength must be combined with the extra travel distance for wave 2, to determine the condition for destructive interference. For destructive interference, the combined total must be an odd integer number of half-wavelengths. Since wave 2 travels back and forth through the film and since light strikes the film nearly perpendicularly, the extra travel distance is twice the film thickness, or 2t. Thus, the condition for destructive interference is $2t + \frac{1}{2}\lambda_{film} = \frac{1}{2}\lambda_{film}$, $1\frac{1}{2}\lambda_{film}$, and so forth. This condition is satisfied when

$$2t = m\lambda_{film} \qquad m = 0, 1, 2, 3, \ldots$$

With m = 1, the expression above gives the minimum nonzero film thickness for which the blue color is missing in the reflected light: $t = \frac{1}{2}\lambda_{film}$. Equation 27.3 gives the wavelength of blue light in the film as λ_{film} = (469 nm)/1.40 = 335 nm. Therefore, the minimum film thickness is

$$t = \frac{1}{2}\lambda_{film} = \frac{1}{2}(335 \text{ nm}) = \boxed{168 \text{ nm}}$$

(b) When the water in Figure 27.10 is replaced by glass, the phase change that accompanies the reflection of wave 2 from the bottom surface of the film is no longer zero. Instead, the phase change is equivalent to one-half of a wavelength, because the wave now travels from a smaller refractive index (n_{gas} = 1.40) toward a larger refractive index (n_{glass} = 1.52). In other words, wave 2 now behaves exactly as wave 1. Consequently, there is no net phase change between the waves due to reflection. As a result, destructive interference occurs when the extra distance traveled by wave 2 is an odd integer number of half-wavelengths, which is a different condition than that in part (a):

$$2t = (m + \tfrac{1}{2})\lambda_{film} \qquad m = 0, 1, 2, 3, \ldots$$

The minimum nonzero thickness for destructive interference corresponds to m = 0 in this equation, so

$$t = \frac{(m + \frac{1}{2})\lambda_{film}}{2} = \frac{\lambda_{film}}{4} = \frac{335 \text{ nm}}{4} = \boxed{83.8 \text{ nm}}$$

Under natural conditions a thin film does not have a uniform thickness. Consequently, destructive interference eliminates different colors from the light reflected at different points on the film, depending on the thickness. Thus, a gasoline film floating on water looks multicolored, as does a soap bubble. The colors also depend on the viewing angle. At an oblique angle, the light corresponding to ray 2 in Figure 27.10 would travel a greater distance within the film than it does at nearly perpendicular incidence. The greater distance would lead to destructive interference for a different wavelength.

Thin film interference can be beneficial in optical instruments. For example, many cameras may contain up to six or more lenses. Reflections from all the lens surfaces can reduce considerably the amount of light directly reaching the film. In addition, multiple reflections from the lenses often reach the film indirectly and produce a background "haze" that degrades the quality of the image. To minimize such unwanted reflections, high-quality lenses are often covered with a thin nonreflective coating of magnesium fluoride (n = 1.38). This situation is like that in part (b) of Example 2, except the thickness is usually chosen to ensure that destructive interference eliminates the reflection of green light, which is in the middle of the visible spectrum. It should be pointed out that the absence of any reflected light does not mean that it has been destroyed by the nonreflective

THE PHYSICS OF . . .

nonreflecting coatings for high-quality lenses.

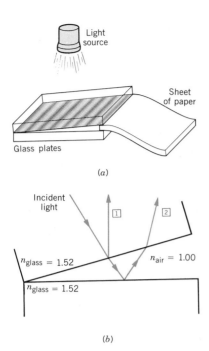

Figure 27.12 (a) The wedge of air formed between two flat glass plates causes an interference pattern of alternating dark and bright fringes to appear in reflected light. (b) A side view of the glass plates and the air wedge.

PROBLEM SOLVING INSIGHT

When light reflects from a boundary between two materials, the reflected light will experience a phase change only if the light travels from the material with a smaller refractive index toward the material with a larger refractive index. Be sure to take such a phase change into account when analyzing thin-film interference phenomena.

---

coating. Rather, the "missing" light has been transmitted into the coating and the lens it covers.

Another interesting illustration of thin film interference is the air wedge. As Figure 27.12a shows, an air wedge is formed when two flat plates of glass are separated along one side, perhaps by a thin sheet of paper. The thickness of this film of air varies between zero, where the plates touch, and the thickness of the paper. When monochromatic light reflects from this arrangement, alternate bright and dark fringes are formed by constructive and destructive interference, as the picture indicates. Example 3 deals with the interference caused by an air wedge.

Example 3   An Air Wedge

(a) Assuming that green light ($\lambda_{vacuum} = 552$ nm) strikes the glass plates nearly perpendicularly in Figure 27.12, determine the number of bright fringes that occurs between the place where the plates touch and the edge of the sheet of paper (thickness = 4.10×10^{-5} m). (b) Explain why there is a dark fringe where the plates touch.

REASONING   A bright fringe occurs wherever constructive interference occurs, as determined by any phase changes due to reflection and the effects of the thickness of the air wedge. We examine the phase changes and the effects of the thickness separately.

SOLUTION
(a) There is no phase change upon reflection for wave 1, since this light travels from a larger (glass) toward a smaller (air) refractive index. In contrast, there is a half-wavelength phase change for wave 2, since the ordering of the refractive indices is reversed at the lower air/glass boundary where reflection occurs. The net phase change due to reflection for waves 1 and 2, then, is equivalent to a half wavelength. Now we combine any extra distance traveled by ray 2 with this half-wavelength and determine the condition for the constructive interference that creates the bright fringes. Constructive interference occurs whenever the *combination* yields an integer number of wavelengths. Therefore, the extra travel distance alone must be an odd integer number of half-wavelengths if constructive interference is to occur. At nearly perpendicular incidence, the extra travel distance is approximately twice the thickness t of the wedge at any point, so the condition for constructive interference is

$$2t = (m + \tfrac{1}{2})\lambda_{film} \qquad m = 0, 1, 2, 3, \ldots$$

In this expression, note that the "film" is a film of air. Since the refractive index of air is nearly one, λ_{film} is virtually the same as that in vacuum, so $\lambda_{film} = 552$ nm. When t equals the thickness of the paper holding the plates apart, the corresponding value of m can be obtained from the equation above:

$$m = \frac{2t}{\lambda_{film}} - \frac{1}{2} = \frac{2(4.10 \times 10^{-5} \text{ m})}{552 \times 10^{-9} \text{ m}} - \frac{1}{2} = 148$$

Since the first bright fringe occurs when $m = 0$, the number of bright fringes is $m + 1 = \boxed{149}$.

(b) Where the plates touch, there is a dark fringe because of destructive interference between the light waves represented by rays 1 and 2. Destructive interference occurs, since the thickness of the wedge is zero here and the only difference between the rays is the half-wavelength phase change due to reflection from the lower plate.

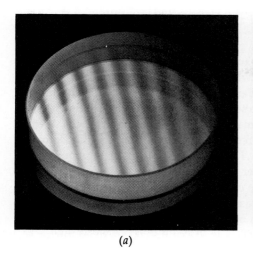

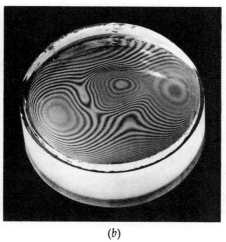

(a)   (b)

Figure 27.13 In reflected light a pattern of interference fringes can be observed due to the air wedge between two glass plates. (*a*) When the plates are ultraflat, or optically flat, the pattern consists of straight fringes. (*b*) When the plates are not flat, the pattern is wavy.

Figure 27.13*a* shows a photograph of the fringes observed for an air wedge between two ultraflat, or optically flat, plates (plates that are flat to within a fraction of a wavelength of the light). Part *b* shows the fringe pattern observed when the plates are not optically flat. In fact, one way to test a plate for flatness is to use it to form an air wedge with a second reference plate that is known to be flat. When the fringes are straight as in part *a*, rather than wavy as in part *b*, no further polishing is needed to flatten the plate being tested.

Another type of air wedge can also be used to determine the degree to which the surface of a lens or mirror is spherical. When an accurate spherical surface is put in contact with an optically flat plate, as in Figure 27.14*a*, the circular interference fringes shown in part *b* of the figure can be observed. The circular fringes are called *Newton's rings*. They arise in the same way that the straight fringes arise in Figure 27.12. When the curved surface is irregular, as in Figure 27.14*c*, the interference fringes are no longer circular.

THE PHYSICS OF . . .

testing a surface for flatness using thin film interference.

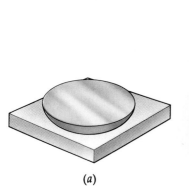

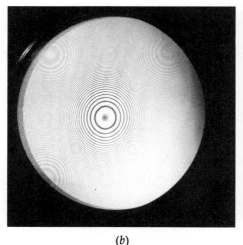

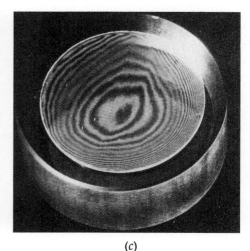

(a)   (b)   (c)

Figure 27.14 (*a*) The air wedge between an accurate spherical glass surface and an optically flat plate leads to (*b*) a pattern of circular interference fringes that is known as Newton's rings. (*c*) When the curved surface is irregular, the interference fringes are not circular.

27.4 THE MICHELSON INTERFEROMETER

An interferometer is an apparatus that can be used to measure the wavelength of light by utilizing interference between two light waves. One particularly famous interferometer is that developed by Albert A. Michelson (1852–1931). The Michelson interferometer uses reflection to set up conditions where two light waves interfere. Figure 27.15 presents a schematic drawing of the instrument. Waves emitted by the monochromatic light source strike a *beam splitter,* so called because it splits the beam of light into two parts. The beam splitter is a glass plate, the far side of which is coated with a thin layer of silver that reflects part of the beam upward as wave A in the drawing. The coating is so thin, however, that it also allows the remainder of the beam to pass directly through as wave F. Wave A strikes an adjustable mirror and reflects back on itself. It again crosses the beam splitter and then enters the viewing telescope. Wave F strikes a fixed mirror M_F and returns, to be partly reflected into the viewing telescope by the beam splitter. Note that wave A passes through the glass plate of the beam splitter three times in reaching the viewing scope, while wave F passes through it only once. The compensating plate in the path of wave F has the same thickness as the beam splitter plate and ensures that wave F also passes three times through the same thickness of glass on the way to the viewing scope. Thus, an observer who views the combination of waves A and F through the telescope sees constructive or destructive interference, depending only on the difference in path lengths D_A and D_F traveled by the two waves.

Now suppose the mirrors are perpendicular to each other, the beam splitter makes a 45° angle with each, and the distances D_A and D_F are equal. The waves A and F travel the same distance, and the field of view in the telescope is uniformly bright due to constructive interference. However, if the adjustable mirror were

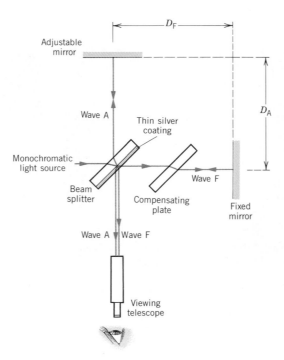

Figure 27.15 A schematic drawing of a Michelson interferometer.

moved a distance of $\frac{1}{4}\lambda$, one wave would travel back and forth by an amount that is twice this distance, leading to an extra distance of $\frac{1}{2}\lambda$. Then, the two waves would be out of phase when they reached the viewing scope, destructive interference would occur, and the viewer would see a dark field. If the adjustable mirror were moved further, brightness would return as soon as the waves were in phase and interfered constructively. The in-phase condition would occur when one of the waves travels a total extra distance of λ relative to the other. Thus, as the mirror is continuously moved, the viewer sees the field of view change from bright to dark, then back to bright, and so on. The amount by which D_A has been changed can be measured and related to the wavelength of the light, since a bright field changes into a dark field and back again each time D_A is changed by a half wavelength. (The back-and-forth change in distance is λ.) If a sufficiently large number of wavelengths are counted in this manner, the Michelson interferometer can be used to obtain a very accurate value for the wavelength from the measured changes in D_A.

27.5 DIFFRACTION

As Section 17.3 discusses, *diffraction* is a bending of waves around obstacles or the edges of an opening. In Figure 27.16, for example, sound waves are leaving a room through an open doorway. Because the exiting sound waves bend, or diffract, around the edges of the opening, a listener outside the room can hear the sound even when standing around the corner from the doorway.

Diffraction is an interference effect, and the Dutch scientist Christian Huygens (1629–1695) developed a principle that is useful in explaining why diffraction arises. *Huygens' principle* describes how a wave front that exists at one instant gives rise to the wave front that exists later on. The principle states that *every point on a wave front acts as a source of tiny wavelets that move forward with the same speed as the wave; the wave front at a later instant is the surface that is tangent to the wavelets.*

We begin by using Huygens' principle to explain the diffraction of sound waves in Figure 27.16. The drawing shows the top view of a plane wave front of sound approaching a doorway and identifies five points on the wave front just as it is leaving the opening. According to Huygens' principle, each of these points acts as a source of wavelets, which are shown as red circular arcs at some moment after they are emitted. The tangent to the wavelets from points 2, 3, and 4 indicates that in front of the doorway the wave front is flat and is moving straight ahead at this later instant. But at the edges, points 1 and 5 are the last points that produce wavelets. Huygens' principle suggests that in conforming to the curved shape of the wavelets emitted near the edges, the new wave front moves into regions that it would not reach otherwise. The sound wave, then, bends or diffracts around the edges of the doorway.

Huygens' principle applies not only to sound waves, but to all kinds of waves. For instance, light has a wavelike nature and, consequently, exhibits diffraction. Therefore, you may ask, "Since I can hear around the edges of a doorway, why can't I also see around them?" As a matter of fact, light waves do bend around the edges of a doorway. However, the degree of bending is extremely small, so the diffraction of light is not enough to allow you to see around the corner.

As we will learn, the extent to which a wave bends around the edges of an

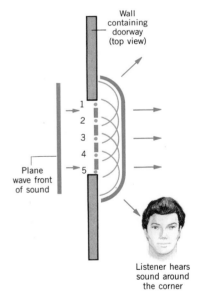

Figure 27.16 Sound bends or diffracts around the edges of a doorway, so even a person who is not standing directly in front of the opening can hear the sound. The five red points within the doorway act as sources and emit the five Huygens wavelets shown in red.

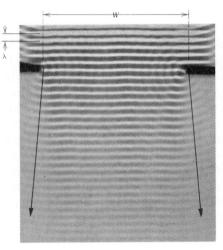
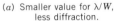
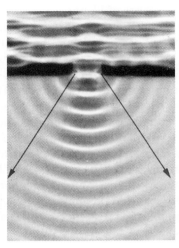

Figure 27.17 These photographs show the plane wave fronts (horizontal lines) of water waves approaching an opening whose width *W* is smaller in (*b*) than it is in (*a*). In addition, the wavelength λ of the waves is larger in (*b*) than in (*a*). Therefore, the ratio λ/W increases from (*a*) to (*b*) and so does the extent of the diffraction. The waves bend around the edges of the opening more in (*b*) than in (*a*).

(*a*) Smaller value for λ/W, less diffraction.

(*b*) Larger value for λ/W, more diffraction.

opening is determined by the ratio λ/W, where λ is the wavelength of the wave and *W* is the width of the opening. The photographs in Figure 27.17 illustrate the effect of this ratio on the diffraction of water waves. In part *a*, the ratio λ/W is small, because the wavelength (as indicated by the distance between the wave fronts) is small relative to the width of the opening. The wave fronts move through the opening with little bending or diffraction into the regions around the corners of the edges. In part *b*, the wavelength is larger and the width of the opening is smaller. As a result, the ratio λ/W is larger, and the degree of bending becomes more pronounced, with the wave fronts penetrating more into the regions around the corners of the edges.

Based on the pictures in Figure 27.17, we might expect that light waves of wavelength λ will bend or diffract appreciably when they pass through an opening whose width *W* is small enough to make the ratio λ/W sufficiently large. This is indeed the case, as Figure 27.18 illustrates. In this picture, it is assumed that parallel rays (or plane wave fronts) of light fall on a very narrow slit and illuminate a viewing screen that is located far from the slit. Part *a* of the drawing shows what would happen if there were no diffraction. Light would pass through the slit

Figure 27.18 (*a*) If light were to pass through a very narrow slit *without* diffraction, only the region on the screen directly opposite the slit would be illuminated. (*b*) Diffraction causes the light to bend around the edges of the slit into regions it would not otherwise reach, forming a pattern of alternating bright and dark fringes on the screen. The slit width has been exaggerated for clarity.

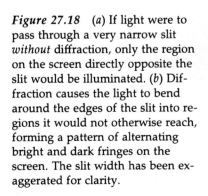
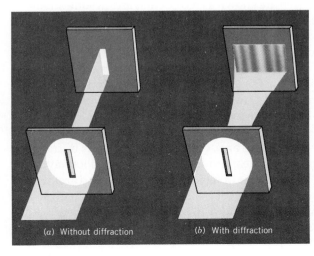

(*a*) Without diffraction

(*b*) With diffraction

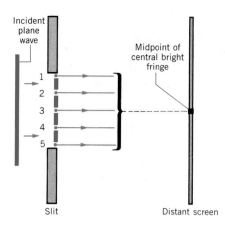

Figure 27.19   A plane wave front
is incident on a single slit. This top
view of the slit shows five sources
of Huygens wavelets. The wavelets
travel toward the midpoint of the
central bright fringe on the screen,
as the rays indicate. The screen is
very far from the slit.

without bending around the edges and produce an image of the slit on the screen.
Part *b* shows what actually happens. The light diffracts around the edges of the
slit and brightens regions on the screen that are not directly opposite the slit. The
diffraction pattern on the screen consists of a bright central band, accompanied by
a series of narrower faint fringes that are parallel to the slit itself.

To help explain how the pattern of diffraction fringes arises, Figure 27.19
shows a top view of a plane wave front approaching the slit and singles out five
sources of Huygens wavelets. Consider how the light from these five sources
reaches the midpoint on the screen. To simplify things, the screen is assumed to be
so far from the slit that the rays from each Huygens source are nearly parallel.*
Then, all the wavelets travel the same distance to the midpoint, arriving there in
phase. As a result, constructive interference creates a bright central fringe on the
screen, directly opposite the slit.

The wavelets emitted by the Huygens sources in the slit can also interfere
destructively on the screen, as Figure 27.20 illustrates. Part *a* shows the light
traveling from each source toward the first dark fringe. The angle θ gives the
position of this dark fringe relative to the line between the midpoint of the slit and

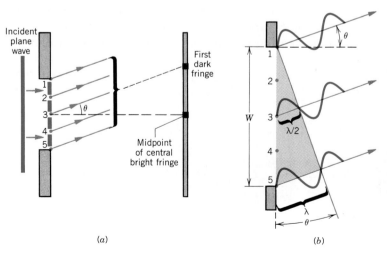

(a)

(b)

Figure 27.20   These drawings pertain to single-slit diffraction and
show how destructive interference
leads to the first dark fringe on either side of the central bright
fringe. For clarity, only one of the
dark fringes is shown. The screen
is very far from the slit.

* When the rays are parallel, the diffraction is called Fraunhofer diffraction in tribute to the German
optician Josef Fraunhofer (1787–1826). When the rays are not parallel, the diffraction is referred to as
Fresnel diffraction, named for the French physicist Augustin Jean Fresnel (1788–1827).

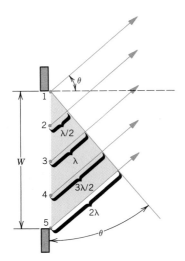

Figure 27.21 In a single-slit diffraction pattern, multiple dark fringes occur on either side of the central bright fringe. This drawing shows how destructive interference creates the second dark fringe on a very distant screen.

the central bright fringe. Since the screen is very far from the slit, the rays of light from each Huygens source are nearly parallel and oriented at nearly the same angle θ, as in part b of the drawing. The wavelet from source 1 travels the shortest distance to the screen, while that from source 5 travels the farthest. Destructive interference creates the first dark fringe when the extra distance traveled by the wavelet from source 5 is exactly one wavelength, as the colored right triangle in the drawing indicates. Under this condition, the extra distance traveled by the wavelet from source 3 at the center of the slit is exactly one-half of a wavelength. Therefore, wavelets from sources 1 and 3 are exactly out of phase and interfere destructively when they reach the screen. Similarly, a wavelet that originates slightly below source 1 cancels a wavelet that originates the same distance below source 3. Thus, each wavelet from the upper half of the slit cancels a corresponding wavelet from the lower half, and no light reaches the screen. As can be seen from the colored right triangle, the angle θ locating the first dark fringe is given by $\sin \theta = \lambda/W$, where W is the width of the slit.

Figure 27.21 shows the condition that leads to destructive interference at the second dark fringe on either side of the midpoint on the screen. In reaching the screen, the light from source 5 now travels a distance of two wavelengths farther than the light from source 1. Under this condition, the wavelet from source 5 travels one wavelength farther than the wavelet from source 3, and the wavelet from source 3 travels one wavelength farther than the wavelet from source 1. Therefore, each half of the slit can be treated as the entire slit was in the previous paragraph; all the wavelets from the top half interfere destructively with each other, and all the wavelets in the bottom half do likewise. As a result, no light from either half reaches the screen, and another dark fringe occurs. The colored triangle in the drawing shows that this second dark fringe occurs when $\sin \theta = 2\lambda/W$. Similar arguments hold for the third- and higher-order dark fringes, with the general result being

$$\begin{bmatrix} \text{Dark fringes} \\ \text{for single-slit} \\ \text{diffraction} \end{bmatrix} \qquad \sin \theta = m\,\frac{\lambda}{W} \qquad m = 1, 2, 3, \ldots \qquad (27.4)$$

Between each pair of dark fringes there is a bright fringe due to constructive interference. The brightness of the fringes is related to the light intensity, just as loudness is related to sound intensity. The intensity of the light at any location on the screen is the amount of light energy per second per unit area that strikes the screen there. Figure 27.22 gives a graph of the light intensity along with a photo-

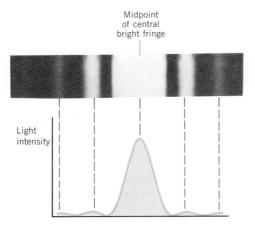

Figure 27.22 The photograph shows a single-slit diffraction pattern, with a bright and wide central fringe. The higher-order bright fringes are much less intense than the central fringe, as the graph indicates.

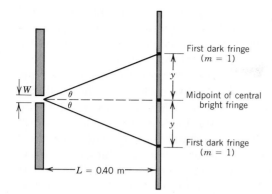

Figure 27.23   The distance $2y$ is the width of the central bright fringe.

graph of a single-slit diffraction pattern. The central bright fringe, which is approximately twice as wide as the other bright fringes, has by far the greatest intensity. The width of the central fringe provides one indication of the extent of diffraction, as Example 4 illustrates.

Example 4   Single-Slit Diffraction

Light passes through a slit and shines on a flat screen that is located $L = 0.40$ m away (see Figure 27.23). The width of the slit is $W = 4.0 \times 10^{-6}$ m. The distance between the middle of the central bright fringe and the first dark fringe is y. Determine the width $2y$ of the central bright fringe when the wavelength of the light is (a) $\lambda = 690$ nm in vacuum (red) and (b) $\lambda = 410$ nm in vacuum (violet).

REASONING   The width of the central bright fringe is determined by two factors. One is the angle θ that locates the first dark fringe on either side of the midpoint. The other is the distance L between the screen and the slit. Larger values for θ and L lead to a wider central bright fringe.

SOLUTION
(a) The angle θ in Equation 27.4 locates the first dark fringe when $m = 1$; $\sin \theta = (1)\lambda/W$. Therefore,

$$\theta = \sin^{-1}\left(\frac{\lambda}{W}\right) = \sin^{-1}\left(\frac{690 \times 10^{-9} \text{ m}}{4.0 \times 10^{-6} \text{ m}}\right) = 9.9°$$

According to Figure 27.23, $\tan \theta = y/L$, so

$$y = L \tan \theta = (0.40 \text{ m}) \tan 9.9° = 0.070 \text{ m}$$

The width of the central fringe, then, is $\boxed{2y = 0.14 \text{ m}}$.

(b) Repeating the calculation above with $\lambda = 410$ nm shows that $\theta = 5.9°$ and $\boxed{2y = 0.083 \text{ m}}$. Notice that the central fringe is narrower when the wavelength is smaller. This occurs because diffraction always becomes less when the ratio λ/W becomes smaller.

In the production of computer chips it is important to minimize the effects of diffraction. As Figure 23.29 illustrates, such chips are very small and yet contain enormous numbers of electronic components. Such miniaturization is achieved using the techniques of photolithography. The patterns on the chip are created first on a so-called "mask," which is similar to a photographic negative. Light is then directed through the mask onto silicon wafers that have been coated with a photosensitive material. The light-activated parts of the coating can be removed chemically, to leave the ultra-thin lines that form the miniature patterns on the chip. As the light passes through the narrow slit-like patterns on the mask, the light spreads out due to diffraction. If excessive diffraction occurs, the light spreads out so much that sharp patterns are not formed on the photosensitive material coating the silicon wafer. Ultra-miniaturization of the patterns requires the absolute minimum of diffraction, and currently this is achieved by using ultraviolet light, which has a wavelength shorter than that of visible light. The shorter the wavelength λ, the smaller the ratio λ/W, and the less the diffraction, as illustrated in Example 4. Intensive research programs are under way to develop the technique of X-ray lithography. The wavelengths of X-rays are even shorter than those of ultraviolet light and, thus, will reduce diffraction even more, allowing further miniaturization.

Another example of diffraction can be seen when light from a point source falls on an opaque disk, such as a coin (Figure 27.24). The effects of diffraction modify the dark shadow cast by the disk in several ways. First, the light waves diffracted around the circular edge of the disk interfere constructively at the center of the shadow to produce a small bright spot. There are also circular bright fringes (not visible in the drawing) in the shadow area. In addition, the boundary between the circular shadow and the lighted screen is not sharply defined, but consists of concentric bright and dark fringes. The various fringes are analogous to those produced by a single slit and are due to interference between Huygens wavelets that originate from different points near the edge of the disk.

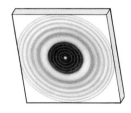

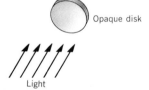

Light

Figure 27.24 The diffraction pattern formed by an opaque disk consists of a small bright spot in the middle of the shadow, circular bright fringes (not visible in the drawing) within the shadow, and concentric bright and dark fringes surrounding the shadow.

27.6 RESOLVING POWER

The *resolving power* is the ability of an optical instrument to distinguish between two closely spaced objects. In a fashion similar to that for a single slit, diffraction also occurs when light passes through circular or nearly circular openings, such as those that admit light into cameras, microscopes, telescopes, or human eyes. The resulting diffraction pattern places a natural limit on the resolving power of such optical instruments.

Figure 27.25 shows the diffraction pattern created by a small circular opening when the viewing screen is far from the opening. The pattern consists of a central bright circular region, surrounded by alternating bright and dark circular fringes. These fringes are analogous to the rectangular fringes that a single slit produces. The angle θ in the picture locates the first circular dark fringe relative to the central bright region and is given by

$$\sin \theta = 1.22 \frac{\lambda}{D} \tag{27.5}$$

where λ is the wavelength of the light and D is the diameter of the circular opening. This expression is similar to Equation 27.4 for a slit ($\sin \theta = \lambda/D$, when $m = 1$) and is valid when the distance to the screen is much larger than the diameter of the aperture.

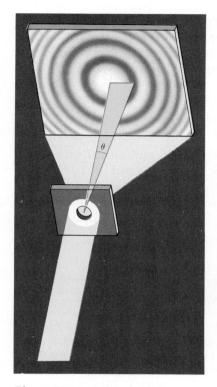

Figure 27.25 When light passes through a small circular opening, a circular diffraction pattern is observed on a viewing screen. The angle θ locates the first dark fringe relative to the central bright region. The intensities of the bright fringes, as well as the diameter of the opening, have been exaggerated in the interest of clarity.

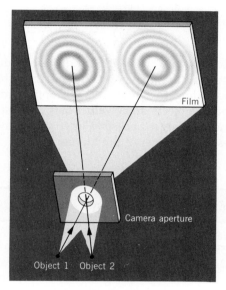

Figure 27.26 When light from two point objects passes through the circular aperture of a camera, two circular diffraction patterns are formed as images on the film. The images here are completely separated or resolved, because the objects themselves are widely separated.

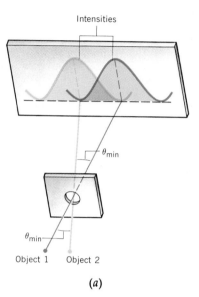

(a)

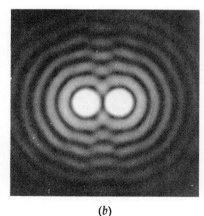

(b)

Figure 27.27 (a) According to the Rayleigh criterion, two point objects are just resolved when the first dark fringe (zero intensity) of one of the images falls on the central bright fringe (maximum intensity) of the other. (b) This photograph shows two overlapping but still resolvable diffraction patterns.

An optical instrument with the ability to resolve two closely spaced objects can produce images of them that can be identified separately. For instance, think about the images on the film when light from two widely separated point objects passes through the circular aperture of a camera. As Figure 27.26 illustrates, each image is a circular diffraction pattern, but the two patterns do not overlap and are completely resolved. On the other hand, if the objects are sufficiently close together, the intensity patterns created by the diffraction overlap, as Figure 27.27a suggests. In fact, if the overlap is extensive it may no longer be possible to distinguish the patterns separately. In such a case, the picture from a camera would show a single blurred object instead of two separate objects. In Figure 27.27b the diffraction patterns overlap, but not enough to prevent us from seeing that two objects are present. Ultimately, then, diffraction limits the ability of an optical instrument to produce distinguishable images of objects that are close together.

It is useful to have a criterion for judging whether two closely spaced objects are resolved by an optical instrument. Figure 27.27*a* presents the *Rayleigh criterion* for resolution, first proposed by Lord Rayleigh (1842–1919): *Two point objects are just resolved when the first dark fringe in the diffraction pattern of one falls directly on the central bright fringe in the diffraction pattern of the other.* The minimum angle θ_{min} between the two objects in the drawing is that given by Equation 27.5. If θ_{min} is small and is expressed in radians, $\sin\theta_{min} \approx \theta_{min}$. Then, Equation 27.5 can be rewritten as

$$\theta_{min} \approx 1.22\,\frac{\lambda}{D} \qquad (\theta_{min}\ \text{in radians}) \qquad (27.6)$$

For a given wavelength λ and aperture diameter D, this result specifies the smallest angle that two point objects can subtend at the aperture and still be resolved. According to Equation 27.6, optical instruments designed to resolve closely spaced objects (small values of θ_{min}) must utilize the smallest possible wavelength and the largest possible aperture diameter. Examples 5 and 6 deal with the resolving power of the human eye.

Example 5   The Human Eye Versus the Eagle's Eye

(a) A hang glider is flying at an altitude of $H = 120$ m. Green light (wavelength = 555 nm in vacuum) enters the pilot's eye through a pupil that has a diameter $D = 2.5$ mm. The average index of refraction of the material in the eye is approximately $n = 1.36$. Determine how far apart two point objects must be on the ground if the pilot is to have any hope of distinguishing between them (see Figure 27.28). (b) An eagle's eye has a pupil with a diameter of $D = 6.2$ mm and has about the same refractive index as does a human eye. Repeat part (a) for an eagle flying at the same altitude as the glider.

REASONING   According to the Rayleigh criterion, the two objects must be separated by a distance s sufficient to subtend an angle $\theta_{min} \approx 1.22\lambda/D$ at the pupil of the pilot's eye. This expression gives the angle in radians, and from this value and the altitude H, we can find s. In applying the expression for θ_{min}, we must keep in mind that λ is the wavelength within the eye, which is where the diffraction occurs that limits the resolution.

SOLUTION
(a) The wavelength λ within the eye takes into account the refractive index of the eye and is given by Equation 27.3 as $\lambda = \lambda_{vacuum}/n = (555\ \text{nm})/1.36 = 408$ nm. Therefore,

$$\theta_{min} \approx 1.22\,\frac{\lambda}{D} = 1.22\left(\frac{408 \times 10^{-9}\ \text{m}}{2.5 \times 10^{-3}\ \text{m}}\right) = 2.0 \times 10^{-4}\ \text{rad} \qquad (27.6)$$

According to Equation 8.1, θ_{min} in radians is $\theta_{min} \approx s/H$, so

$$s \approx \theta_{min}H = (2.0 \times 10^{-4}\ \text{rad})(120\ \text{m}) = \boxed{0.024\ \text{m}}$$

(b) Since the pupil of an eagle's eye is larger than that of a human eye, diffraction creates less of a limitation for the eagle. A calculation like that above, using $D = 6.2$ mm, reveals that the diffraction limit for the eagle is $\boxed{s = 0.0096\ \text{m}}$.

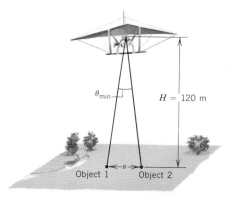

Figure 27.28 The Rayleigh criterion can be used to estimate the smallest distance *s* that can separate two objects on the ground, if a person on a hang glider is to have any hope of distinguishing one object from the other.

Example 6   *The Great Wall of China*

The Great Wall of China is an enormous structure, its width approaching 7.0 m in places. Some people even claim that it is possible to identify the separate sides of the Great Wall from the moon with the unaided eye. Evaluate this claim in the following way. Assume that green light (wavelength = 555 nm in vacuum) reflects from the Great Wall and enters the eye (pupil diameter $D = 2.5$ mm, refractive index $n = 1.36$) of an astronaut. Determine the maximum distance this astronaut could be from the earth and still resolve the two sides of the Great Wall at their widest point. Compare this maximum distance to the distance between the earth and the moon.

REASONING   This problem is similar to that in Example 5, except that the astronaut takes the place of the hang glider pilot and the two sides of the Great Wall serve as object 1 and object 2 in Figure 27.28. From Example 5, we know, then, that the sides of the Great Wall must subtend an angle of at least $\theta_{min} \approx 2.0 \times 10^{-4}$ rad at the astronaut's eye if they are to be resolvable. Using this value for θ_{min} and the fact that $s = 7.0$ m, we can find H, the distance for the astronaut.

SOLUTION   Referring to Figure 27.28 and using Equation 8.1, we can express θ_{min} in radians as $\theta_{min} \approx s/H$. Therefore, the maximum distance for the astronaut is

$$H \approx \frac{s}{\theta_{min}} = \frac{7.0 \text{ m}}{2.0 \times 10^{-4} \text{ rad}} = \boxed{3.5 \times 10^4 \text{ m}}$$

In comparison, the moon is 3.85×10^8 m from the earth. Thus, it is not possible to see the two sides of the Great Wall from the moon.

27.7 THE DIFFRACTION GRATING

Diffraction patterns of bright and dark fringes occur when monochromatic light passes through a single or double slit. Fringe patterns also result when light passes through more than two slits, and an arrangement consisting of a large number of parallel, closely spaced slits is called a ***diffraction grating***. Gratings with as many as 40 000 slits per centimeter can be made, depending on the production method.

THE PHYSICS OF . . .

the diffraction grating.

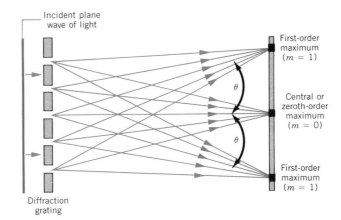

Figure 27.29   When light passes through a diffraction grating, a central bright fringe ($m = 0$) and higher-order bright fringes ($m = 1, 2, \ldots$) form when the light falls on a distant viewing screen.

In one method a diamond-tipped cutting tool is used to inscribe closely spaced parallel lines on a glass plate, the spaces between the lines serving as the slits. In fact, the number of slits per centimeter is often quoted as the number of lines per centimeter.

Figure 27.29 illustrates how light travels to a distant viewing screen from each of five slits in a grating and forms the central bright fringe and the first-order bright fringes on either side. Higher-order bright fringes are also formed but are not shown in the drawing. Each bright fringe is located by an angle θ relative to the central fringe. These bright fringes are sometimes called the *principal fringes* or *principal maxima,* since they are places where the light intensity is a maximum. The term "principal" distinguishes them from other, much less bright fringes that are referred to as secondary fringes or secondary maxima.

Constructive interference creates the principal fringes. To show how, we assume the screen is far from the grating, so that the rays remain nearly parallel while the light travels toward the screen, as in Figure 27.30. In reaching the place on the screen where the first-order maximum is located, light from slit 2 travels a distance of one wavelength farther than light from slit 1. Similarly, light from slit 3 travels one wavelength farther than light from slit 2, and so forth, as emphasized by the colored right triangles. For the first-order maximum, the enlarged

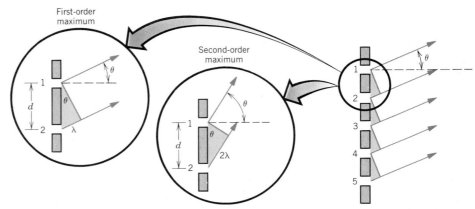

Figure 27.30   The conditions shown here lead to the first- and second-order intensity maxima in the diffraction pattern produced by a diffraction grating.

view of one of these right triangles shows that constructive interference occurs when $\sin\theta = \lambda/d$, where d is the separation between the slits. The second-order maximum forms when the extra distance traveled by light from adjacent slits is two wavelengths, so that $\sin\theta = 2\lambda/d$. The general result is

$$\begin{bmatrix} \textbf{Principal maxima} \\ \textbf{of a diffraction} \\ \textbf{grating} \end{bmatrix} \qquad \sin\theta = m\,\frac{\lambda}{d} \qquad m = 0,\,1,\,2,\,3,\,\ldots \qquad (27.7)$$

The separation d between the slits can be calculated from the number of slits per centimeter of grating; for instance, a grating with 2500 slits per centimeter has a slit separation of $d = 1/2500$ cm $= 4.0 \times 10^{-4}$ cm.

Equation 27.7 is identical to Equation 27.1 for the double slit. A grating, however, produces bright fringes that are much *narrower* or *sharper* than those from a double slit, as the intensity patterns in Figure 27.31 reveal. Consider a grating with 100 slits/cm, for instance. The extra distance traveled by light from adjacent slits in forming the first-order maximum is exactly one wavelength λ. For an extra distance slightly larger than one wavelength, the light reaches the screen at a point slightly displaced from the maximum. If the extra distance is $\lambda + \lambda/100$, this slightly displaced point is already a place where complete destructive interference occurs, according to the following reasoning. The extra distance traveled by light from slit 51 compared to that from slit 1 is $50(\lambda + \lambda/100) = 50.5\lambda$. The additional half wavelength means that crests and troughs from these two slits combine to create destructive interference. The same result applies to light from slits 52 and 2, 53 and 3, and so on—in other words, to all the light from the grating. If a grating contains 10 000 slits/cm instead of 100 slits/cm, then destructive interference occurs when the extra distance traveled by light from adjacent slits is $\lambda + \lambda/10\,000$, instead of $\lambda + \lambda/100$. Thus, for a greater number of slits per centimeter, a smaller displacement from the maximum is required to produce the adjacent point of destructive interference on the screen, with the result that the principal fringes are even narrower. Note in Figure 27.31 that between the principal fringes there are secondary maxima with much smaller intensities.

The next example illustrates the ability of a grating to separate the components in a mixture of colors.

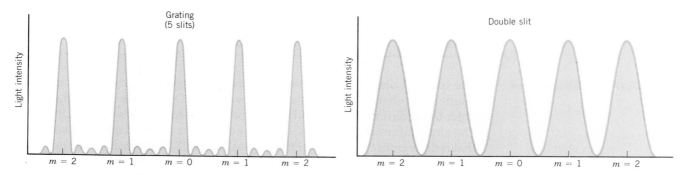

Figure 27.31 The bright fringes produced by a diffraction grating are much narrower than those produced by a double slit. Note the three small secondary bright fringes between the principal bright fringes of the grating. For a large number of slits, these secondary fringes become very small.

Example 7 Separating Colors with a Diffraction Grating

A mixture of violet light ($\lambda = 410$ nm in vacuum) and red light ($\lambda = 660$ nm in vacuum) falls on a grating that contains 1.0×10^4 lines/cm. For each wavelength, find the angle θ that locates the first-order maxima.

REASONING AND SOLUTION   Before Equation 27.7 can be used here, a value for the separation d between the slits is needed: $d = 1/(1.0 \times 10^4 \text{ lines/cm}) = 1.0 \times 10^{-4}$ cm, or 1.0×10^{-6} m. For violet light, the angle θ_{violet} for the first-order maxima ($m = 1$) is given by $\sin\theta_{\text{violet}} = m\lambda/d = \lambda/d$. Consequently,

$$\theta_{\text{violet}} = \sin^{-1}\frac{\lambda}{d} = \sin^{-1}\left(\frac{410 \times 10^{-9}\text{ m}}{1.0 \times 10^{-6}\text{ m}}\right) = \boxed{24°}$$

For red light, a similar calculation with $\lambda = 660 \times 10^{-9}$ m shows that $\boxed{\theta_{\text{red}} = 41°}$.

Because θ_{violet} and θ_{red} are different, separate first-order bright fringes are seen for violet and red light on a viewing screen.

If the light in Example 7 had been sunlight, the angles for the first-order maxima would cover all values in the range between 24° and 41°, since sunlight contains all colors or wavelengths between violet and red. Consequently, a rainbow-like dispersion of the colors would be observed to either side of the central fringe on a screen. The central bright fringe would be white, however, since all the colors overlap there.

An instrument designed to measure the angles at which the principal maxima of a grating occur is called a grating spectroscope. With a measured value of the angle, calculations such as those in Example 7 can be turned around to provide the corresponding value of the wavelength. As we will point out in Chapter 30, the atoms in a hot gas emit discrete wavelengths, and determining the values of these wavelengths is one important technique used to identify the atoms. Figure 27.32 contains a sketch of a grating spectroscope. The slit that admits light from the source (e.g., a hot gas) is located at the focal point of the collimating lens, so the light rays striking the grating are parallel. The telescope is used to detect the bright fringes and, hence, to measure the angle θ.

THE PHYSICS OF . . .

a grating spectroscope.

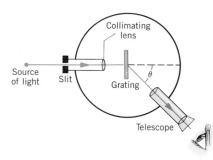

Figure 27.32 A grating spectroscope.

27.8 COMPACT DISCS AND THE USE OF INTERFERENCE

The compact disc (CD) has revolutionized stereo sound reproduction and uses interference effects in some interesting ways. A CD contains a spiral track that holds the audio information and is analogous to the spiral groove on an LP record. The audio information on the CD track, however, is detected using a laser beam that reflects from the bottom of the disc, as Figure 27.33 illustrates. The information is encoded in the form of raised areas on the bottom of the disc. These raised areas appear as "pits" when viewed from the *top* or labeled side of the CD. They are separated by flat areas called "land." The pits and land are covered with a transparent plastic coating, which has been omitted from the drawing for simplicity.

As the CD rotates, the laser beam reflects off the disc and into a detector. The

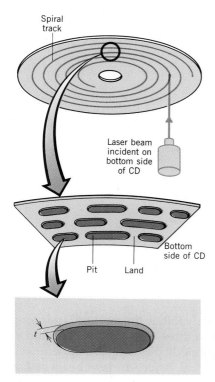

Figure 27.33 The bottom surface of a compact disc (CD) carries the audio information in the form of raised areas ("pits") and land along a spiral track. A CD is played by using a laser beam that strikes the bottom surface of the disc and reflects from it.

reflected light intensity fluctuates as the pits and land areas pass by, and the fluctuations convey the audio information as a series of binary numbers (zeros and ones). To make the fluctuations easier to detect, the pit thickness t (see Figure 27.33) is chosen with destructive interference in mind. As the laser beam overlaps the edges of a pit, part of the beam is reflected from the raised pit surface and part from the land. The part that reflects from the land travels an additional distance of $2t$. The thickness of the pits is chosen so that $2t$ is one-half of a wavelength of the laser beam in the plastic coating. With this choice, destructive interference occurs when the two parts of the reflected beam combine. As a result, there is markedly less reflected intensity when the laser beam passes over a pit edge than when it passes over land alone. Thus, the fluctuations in reflected light that occur while the disc rotates are large enough to detect because of the effects of destructive interference. Example 8 determines the theoretical thickness of the pits on a compact disc. In reality, a value slightly less than that obtained in the example is used for technical reasons that are not pertinent here.

THE PHYSICS OF . . .

retrieving information from compact discs.

Example 8 *Pit Thickness on a Compact Disc*

The laser in a CD player has a wavelength of 790 nm in a vacuum. The plastic coating over the pits has an index of refraction of $n = 1.5$. Find the thickness of the pits on a CD.

REASONING AND SOLUTION As we have discussed, the thickness t is chosen so that $2t = \frac{1}{2}\lambda_{coating}$ in order to achieve destructive interference. Equation 27.3 gives the wavelength in the plastic coating as $\lambda_{coating} = \lambda_{vacuum}/n$. The thickness, then, is

$$t = \frac{\lambda_{coating}}{4} = \frac{\lambda_{vacuum}}{4n} = \frac{790 \times 10^{-9} \text{ m}}{4(1.5)} = \boxed{1.3 \times 10^{-7} \text{ m}}$$

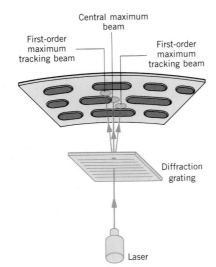

Figure 27.34 A three-beam tracking method is often used in CD players to ensure that the laser follows the spiral track correctly. The three beams are derived from a single laser beam by using a diffraction grating.

THE PHYSICS OF . . .

the three-beam tracking method for compact discs.

As a CD rotates, the laser beam must accurately follow the pits and land along the spiral track. One type of CD player utilizes a three-beam tracking method to ensure that the laser beam follows the spiral properly. Figure 27.34 shows that a diffraction grating is the key element in this method. Before the laser beam strikes the CD, the beam passes through a grating that produces a central maximum and two first-order maxima, one on either side. As the picture indicates, the central maximum beam falls directly on the spiral track. This beam reflects into a detector, and the reflected light intensity fluctuates as the pits and land areas pass by, the fluctuations conveying the audio information. The two first-order maxima beams are called tracking beams. They hit the CD between the arms of the spiral and also reflect into detectors of their own. Under perfect conditions, the intensities of the two reflected tracking beams do not fluctuate, since they originate from the smooth surface between the arms of the spiral where there are no pits. As a result, each tracking beam detector puts out the same constant electrical signal. However, if the tracking drifts to either side, the reflected intensity of each tracking beam changes because of the pits. In response, the tracking beam detectors produce different electrical signals. The difference between the signals is used in a "feedback" circuit to correct for the drift and put the three beams back into their proper positions.

THE PHYSICS OF . . .

X-ray diffraction.

27.9 X-RAY DIFFRACTION

Not all diffraction gratings are commercially made. Nature also creates diffraction gratings, although these gratings do not look like an array of closely spaced slits. Instead, nature's gratings are the arrays of regularly spaced atoms that exist in crystalline solids. For example, Figure 27.35 shows a crystal of ordinary salt (NaCl). Typically, the atoms in a crystalline solid are separated by distances of about 1.0×10^{-10} m, so we might expect a crystalline array of atoms to act like a grating with roughly this "slit" spacing for electromagnetic waves of the appropriate wavelength. Assuming that $\sin \theta = 0.5$ and that $m = 1$ in Equation 27.7, it

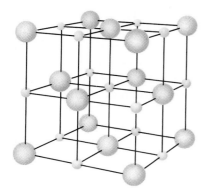

Figure 27.35 In this drawing of the crystalline structure of sodium chloride, the small red spheres represent positive sodium ions, while the large blue spheres represent negative chloride ions.

Figure 27.36 The X-ray diffraction pattern from crystalline NaCl.

follows that $0.5 = \lambda/d$. A value of $d = 1.0 \times 10^{-10}$ m in this equation gives a wavelength of $\lambda = 0.5 \times 10^{-10}$ m. This wavelength is much shorter than that of visible light and falls in the X-ray region of the electromagnetic spectrum. (See Figure 24.6.)

A diffraction pattern does indeed result when X-rays are directed onto a crystalline material, as Figure 27.36 illustrates for a crystal of NaCl. The pattern consists of a complicated arrangement of spots, because a crystal has a complex three-dimensional structure. It is from patterns such as these that the spacing between atoms and the nature of the crystal structure can be determined.

SUMMARY

The **principle of linear superposition** states that when two or more waves are present simultaneously in the same region of space, the resultant wave is the sum of the individual waves. According to this principle, two or more light waves can interfere constructively or destructively when they exist at the same place at the same time, provided they originate from **coherent sources.** Two sources are coherent if they emit waves that have a constant phase relationship.

In **Young's double-slit experiment,** light passes through a pair of closely spaced narrow slits and produces a pattern of alternating bright and dark fringes on a viewing screen. The fringes arise because of constructive and destructive interference. The angle θ for the mth higher-order bright fringes on either side of the central bright fringe is given by $\sin \theta = m\lambda/d$, where d is the spacing between the narrow slits, λ is the wavelength of the light, and $m = 0, 1, 2, 3, \ldots$. Similarly, the angle for the dark fringes is given by $\sin \theta = (m + \frac{1}{2})\lambda/d$.

Constructive and destructive interference of light waves can occur with **thin films** of transparent materials. The interference occurs between light waves that reflect from the top and bottom surfaces of the film. One important factor in thin film interference is the thickness of the film relative to the wavelength of the light within the film. The wavelength within the film is $\lambda_{\text{film}} = \lambda_{\text{vacuum}}/n$, where n is the refractive index of the film. A second important factor is the phase change that can occur when light undergoes reflection at each surface of the film; the exact nature of the phase change is discussed in Section 27.3.

Diffraction is a bending of waves around obstacles or the edges of an opening. Diffraction is an interference effect that can be explained with the aid of **Huygens' principle.** This principle states that every point on a wave front acts as a source of tiny wavelets that move forward with the same speed as the wave; the wave front at a later instant is the surface that is tangent to the wavelets. When light passes through a single narrow slit and falls on a viewing screen, a pattern of bright and dark fringes is formed because of the superposition of such wavelets. The angle θ for the mth dark fringe on either side of the central bright fringe is given by $\sin \theta = m\lambda/W$, where W is the slit width and $m = 1, 2, 3, \ldots$.

The **resolving power** of an optical instrument is the

ability of the instrument to distinguish between two closely spaced objects. Resolving power is limited by the diffraction that occurs when light waves enter an instrument, often through a circular opening. Consideration of the diffraction fringes leads to the **Rayleigh criterion** for resolution. This criterion specifies that two point objects are just resolved when the first dark fringe in the diffraction pattern of one falls directly on the central bright fringe in the diffraction pattern of the other. According to this specification, the minimum angle that two point objects can subtend at an aperture

of diameter D and still be resolved as separate objects is $\theta_{min} \approx 1.22\lambda/D$, where λ is the wavelength of the light.

A **diffraction grating** is a device consisting of a large number of parallel, closely spaced slits. When light passes through a diffraction grating and falls on a viewing screen, the light forms a pattern of bright and dark fringes. The bright fringes are referred to as principal maxima and are found at an angle θ, such that $\sin \theta = m\lambda/d$, where d is the separation between two successive slits and $m = 0, 1, 2, 3, \ldots$.

QUESTIONS

1. A Young's double-slit experiment is performed using sunlight. The zeroth-order or central bright fringe on the screen is observed to be white, while rainbow-like patterns of color are observed on either side of the central fringe. Explain, using Equation 27.1 to guide your thinking.

2. (a) How would the pattern of bright and dark fringes produced in a Young's double-slit experiment change if the light waves coming from *both* of the slits had their phases shifted by a half wavelength? (b) How would the pattern change, if the light coming from *only one* of the slits had its phase shifted by a half wavelength?

3. Replace slits S_1 and S_2 in Figure 27.4 with identical loudspeakers and use the same ac electrical signal to drive them. The two sound waves produced will then be identical and you will have the audio equivalent of Young's double-slit experiment. In terms of loudness and softness, describe what you would hear as you walked along the screen, starting from the center and going to either end.

4. A camera lens is covered with a nonreflective coating that eliminates the reflection of perpendicularly incident green light. Recalling Snell's law of refraction, would you expect the reflected green light to be eliminated if it were incident on the lens at an angle of 45° rather than perpendicularly? Justify your answer.

5. When white light reflects from a soap bubble, the bubble appears multicolored, because destructive interference removes different wavelengths from the light reflected at different places, depending on the thickness of the soap film. As a bubble becomes thinner and thinner, it looks darker and darker, appearing black just before it bursts. The blackness means that destructive interference removes *all* wavelengths from the reflected light when the soap film is very thin. Explain why.

6. In Figure 27.14*b* there is a dark spot at the center of the pattern of Newton's rings. By considering the phase changes

that occur when light reflects from the upper curved surface and the lower flat surface, account for the dark spot.

7. A transparent coating is deposited on a glass plate and has a refractive index that is *larger than that of the glass*, not smaller, as it is for a typical nonreflective coating. For a certain wavelength within the coating, the thickness of the coating is a quarter wavelength. The coating *enhances* the reflection of the light corresponding to this wavelength. Explain why, referring to part (a) of Example 2 in the text to guide your thinking.

8. Do any dark fringes appear in a single-slit diffraction pattern (see Figure 27.18*b*) when the wavelength of the light is greater than the width of the slit? Why? (*Hint: See Equation 27.4.*)

9. Account for the fact that a sound wave diffracts much more than a light wave does when the two pass through the same doorway.

10. The French postimpressionist artist Georges Seurat developed a technique of painting in which dots of color are placed close together on the canvas. From sufficiently far away the individual dots are no longer distinguishable, and the images in the picture take on a more normal appearance. Explain why this is so, utilizing what you have learned about the resolving power of the eye.

11. Four light bulbs are arranged at the corners of a rectangle that is three times longer than it is wide. You look at this arrangement perpendicular to the plane of the rectangle. From very far away, your eyes cannot resolve the individual bulbs and you see a single "smear" of light. From close in, you see the individual bulbs. Between these two extremes, what do you see? Draw two pictures to illustrate the possibilities that exist, depending on how far away you are. Explain your drawings.

12. Suppose the pupil of your eye were elliptical instead of circular in shape, with the long axis of the ellipse oriented in the vertical direction. (a) Would the resolving power of your eye be the same in the horizontal and vertical directions? (b) In which direction would the resolving power be greatest? Justify your answers by discussing how the diffraction of light waves would differ in the two directions.

13. The cameras used for taking photographs of the earth from satellites have large-diameter lenses. From the point of view of photographing two objects that are close together, account for this feature. (*Hint: See Equation 27.6.*)

14. Suppose you were designing an eye and could select the size of the pupil and the wavelength of the electromagnetic waves to which the eye is sensitive. As far as the limitation created by diffraction is concerned, rank the following design choices in order of decreasing resolving power (greatest first): (a) large pupil and ultraviolet wavelengths, (b) small pupil and infrared wavelengths, and (c) small pupil and ultraviolet wavelengths.

15. What would happen to the distance between the bright fringes produced by a diffraction grating if the entire interference apparatus (light source, grating, and screen) were immersed in water? Why?

PROBLEMS

Section 27.1 The Principle of Linear Superposition, Section 27.2 Young's Double-Slit Experiment

1. Two sources are in phase and emit waves that have a wavelength of 0.44 m. Determine whether constructive or destructive interference occurs at a point whose distances from the two sources are as follows: (a) 1.32 and 3.08 m; (b) 2.67 and 3.33 m; (c) 2.20 and 3.74 m; (d) 1.10 and 4.18 m.

2. A Young's double-slit experiment is performed using light that has a wavelength of 630 nm. The separation between the slits is 5.3×10^{-5} m. Find the angles that locate the first-, second-, and third-order bright fringes on the screen.

3. In a Young's double-slit experiment, the angle that locates the first dark fringe on either side of the central bright fringe is 1.6°. Find the ratio of the slit separation d to the wavelength λ of the light.

4. A flat observation screen is placed at a distance of 4.5 m from a pair of slits. The separation on the screen between the central bright fringe and the first-order bright fringe is 0.037 m. The light illuminating the slits has a wavelength of 490 nm. Determine the slit separation.

5. In a Young's double-slit experiment, the seventh dark fringe is located 0.025 m to the side of the central bright fringe on a flat screen, which is 1.1 m away from the slits. The slits are 1.4×10^{-4} m apart. What is the wavelength of the light being used?

*6. At most, how many bright fringes can be formed on either side of the central bright fringe when light of wavelength 625 nm falls on a double slit whose slit separation is 3.76×10^{-6} m?

*7. In a Young's double-slit experiment the separation y between the first-order bright fringe and the central bright fringe on a flat screen is 0.0240 m, when light is used that has a wavelength of 475 nm. Assume the angles that locate the fringes on the screen are small enough so that $\sin \theta \approx \tan \theta$. Find the separation y when the light has a wavelength of 611 nm.

*8. Two parallel slits are illuminated by light composed of two wavelengths, one of which is 645 nm. On a viewing screen, the light whose wavelength is known produces its third dark fringe at the same place where the light whose wavelength is unknown produces its fourth-order bright fringe. The fringes are counted relative to the central or zeroth-order bright fringe. What is the unknown wavelength?

Section 27.3 Thin Film Interference

9. A nonreflective coating of magnesium fluoride ($n = 1.38$) covers the glass ($n = 1.52$) of a camera lens. Assuming the coating prevents reflection of yellow-green light (wavelength in vacuum = 565 nm), determine the minimum nonzero thickness that the coating can have.

10. A transparent film ($n = 1.43$) is deposited on a glass plate ($n = 1.52$) to form a nonreflective coating. The film has a thickness of 1.07×10^{-7} m. What is the longest possible wavelength of light (in vacuum) for which this film has been designed?

11. A layer of transparent plastic ($n = 1.61$) on glass ($n = 1.52$) looks dark when illuminated by light whose wavelength is 589 nm in vacuum. Find the two smallest possible nonzero values for the thickness of the layer.

12. Example 3(a) in the text deals with the air wedge formed between two plates of glass ($n = 1.52$). Repeat this example, assuming the wedge of air is replaced by water ($n = 1.33$).

13. A mixture of red light ($\lambda_{\text{vacuum}} = 661$ nm) and blue light ($\lambda_{\text{vacuum}} = 472$ nm) shines perpendicularly on a thin layer of gasoline ($n_{\text{gas}} = 1.40$) lying on water ($n_{\text{water}} = 1.33$). The gasoline layer has a uniform thickness of 2.36×10^{-7} m. Destructive interference removes one of the colors from the reflected light. By means of suitable calculations decide whether the gasoline looks red or blue in reflected light.

*14. A film of oil lies on wet pavement. The refractive index of the oil exceeds that of the water. The film has the minimum nonzero thickness such that it appears dark due to destructive interference when viewed in red light (wavelength = 660 nm in vacuum). Assuming the visible spectrum extends from 380 to 750 nm, what are the visible wavelength(s) (in vacuum) for which the film will appear bright due to constructive interference?

*15. A layer of glycerol ($n = 1.47$) on glass ($n = 1.52$) has a thickness of 1.02×10^{-7} m. This is the minimum nonzero thickness for which the layer will look dark in a certain monochromatic light. What is the next largest thickness that the layer could have and still look dark in the same light?

**16. A uniform layer of water ($n = 1.33$) lies on a glass plate ($n = 1.52$). Light shines perpendicularly on the layer, which looks maximally bright when the wavelength of the light is 432 nm in vacuum and when it is 648 nm in vacuum. (a) Obtain the minimum thickness of the film. (b) Assuming that the film has the minimum thickness and that the visible spectrum extends from 380 to 750 nm, determine the visible wavelength(s) (in vacuum) for which the film appears completely dark.

Section 27.5 Diffraction

17. Light shines through a single slit whose width is 5.6×10^{-4} m. A diffraction pattern is formed on a flat screen located 4.0 m away. The distance between the middle of the central bright fringe and the first dark fringe is 3.5 mm. What is the wavelength of the light?

18. A diffraction pattern forms when light passes through a single slit. The wavelength of the light is 675 nm. Determine the angle that locates the first dark fringe when the width of the slit is (a) 1.8×10^{-4} m and (b) 1.8×10^{-6} m.

19. A slit whose width is 4.50×10^{-5} m is located 1.23 m from a flat screen. Light shines through the slit and falls on the screen. Find the width of the central fringe of the diffraction pattern when the wavelength of the light is 580 nm.

20. A doorway is 0.91 m wide. (a) Obtain the angle that locates the first dark fringe in the Fraunhofer diffraction pattern formed when red light (wavelength = 660 nm) passes through the doorway. (b) Repeat part (a) for a 440-Hz tone (concert A), assuming the speed of sound is 343 m/s.

21. A single slit (width = 4.85×10^{-7} m) produces *no* dark fringes on an observation screen, even though light shines on the slit. Determine a minimum value for the wavelength of the light being used.

*22. A loudspeaker produces an 1100-Hz tone and a 3100-Hz tone. The sound waves are emitted through a vertically oriented slit. The angle locating the first diffraction minimum, or audio "dark fringe," is 15° for the 3100-Hz tone. What is the corresponding angle for the 1100-Hz tone?

*23. In a single-slit diffraction pattern, the central fringe is 450 times as wide as the slit. The screen is 18 000 times as far from the slit as the slit is wide. What is the ratio λ/W, where λ is the wavelength of the light shining through the slit and W is the width of the slit? Assume the angle that locates a dark fringe on the screen is small, so that $\sin \theta \approx \tan \theta$.

Section 27.6 Resolving Power

24. In a dot matrix printer, an array of dots is used to form the printed characters. If the dots are close enough together, they cannot be resolved individually by the eye and, therefore, appear to form solid lines. Suppose that the pupil of the eye has a diameter of 2.0 mm in bright yellow-green light (wavelength = 563 nm in vacuum), that the material in the eye has an average refractive index of $n = 1.36$, and that the printed page is to be read at a distance of 0.31 m. Considering the limit created by diffraction, find the smallest separation between the dots that the eye can see.

25. Late one night on a highway, a car speeds by you and fades into the distance. Under these conditions the pupils of your eyes (average refractive index = 1.36) have diameters of about 7.0 mm. The taillights of this car are separated by a distance of 1.2 m and emit red light (wavelength = 660 nm in vacuum). How far away from you is this car when its taillights appear to merge into a single spot of light because of the effects of diffraction?

26. You are looking down at the earth from inside a commercial jetliner flying at an altitude of 8690 m, and the pupil of your eye has a diameter of 2.00 mm. The average refractive index of the material in the eye is 1.36. Determine how far apart two cars must be on the ground if you are to have any hope of distinguishing between them in (a) red light (wavelength = 665 nm in vacuum) and (b) violet light (wavelength = 405 nm in vacuum).

27. Two stars are 3.7×10^{11} m apart and are equally distant from the earth. A telescope has an objective lens with a diameter of 1.02 m and detects these stars as separate objects. Assume that light of wavelength 550 nm is being observed. Diffraction effects, rather than atmospheric turbulence, limit the resolving power of the telescope. Find the maximum distance that these stars could be from the earth.

28. The largest refracting telescope in the world is at the Yerkes Observatory in Williams Bay, Wisconsin. The objective of the telescope has a diameter of 1.02 m. With light whose wavelength is 565 nm, it is said that this telescope can resolve two objects separated 0.0254 m apart at a distance of 3.75×10^4 m. Verify this statement with a calculation of your own.

*29. In an experiment, red light from a ruby laser (wavelength = 694.3 nm) is passed through a telescope in reverse and is sent on its way to the moon. At the surface of the moon, which is 3.77×10^8 m away, the light strikes a reflector left there by astronauts. The reflected light returns to the earth, where it is detected. When it leaves the telescope, the circular beam of light has a diameter of about 0.20 m, and diffraction causes the beam to spread as the light travels to the moon. In effect, the first circular dark fringe in the diffraction pattern defines the size of the central bright spot on the moon. Determine the diameter (not the radius) of the central bright spot on the moon.

*30. You are using a microscope to examine a blood sample. Recall from Section 26.12 that the sample should be placed just outside the focal point of the objective lens of the microscope. (a) If the specimen is being illuminated with light of wavelength λ and the diameter of the objective equals its focal length, determine the closest distance between two blood cells that can just be resolved. Express your answer in terms of λ. (b) Based on your answer to (a), should you use light with a longer wavelength or a shorter wavelength if you wish to resolve two blood cells that are even closer together?

*31. The pupil of an eagle's eye has a diameter of 6.0 mm. The refractive index in the eye is 1.36. Two field mice are separated by 0.010 m. From a distance of 175 m, the eagle sees them as one unresolved object and dives toward them at a speed of 17 m/s. Assume that the eagle's eye detects light that has a wavelength of 550 nm in a vacuum. How much time passes until the eagle sees the mice as separate objects?

Section 27.7 The Diffraction Grating, Section 27.8 Compact Discs and the Use of Interference

32. The diffraction gratings discussed in the text are transmission gratings, because light *passes through* them. There are also gratings in which the light *reflects from* the grating to form a pattern of fringes. Equation 27.7 also applies to a reflection grating with straight parallel lines when the incident light shines perpendicularly on the grating. The surface of a compact disc (CD) has a multicolored appearance because it acts like a reflection grating and spreads sunlight into its colors. The arms of the spiral track on the CD are separated by 1.1×10^{-6} m. Using Equation 27.7, estimate the angle that corresponds to the first-order maximum for a wavelength of (a) 660 nm (red) and (b) 410 nm (violet).

33. Light of wavelength 490 nm falls on a diffraction grating. The third-order maximum occurs at an angle of 25°. Find the separation between the slits of the grating.

34. A diffraction grating produces a first-order bright fringe that is 0.0894 m away from the central bright fringe on a flat screen. The separation between the slits of the grating is 4.17×10^{-6} m, and the distance between the grating and the screen is 0.625 m. What is the wavelength of the light shining on the grating?

35. When a grating is used with light that has a wavelength of 575 nm, a second-order maximum is formed at an angle of 11.2°. How many lines per centimeter does this grating have?

36. The wavelength of the laser beam used in a compact

disc player is 790 nm. Suppose that a diffraction grating produces first-order tracking beams that are 1.2 mm apart at a distance of 3.0 mm from the grating. Estimate the spacing between the slits of the grating.

37. Monochromatic light shines on a diffraction grating. When the light source and the grating are in air, the first-order maximum occurs at an angle of 33°. At what angle does the first-order maximum occur when the source and the grating are immersed in water ($n = 1.33$)?

*38. A diffraction grating contains 4820 lines/cm and is used with blue light (wavelength = 470 nm). What is the highest-order bright fringe that can be seen with this grating?

*39. There are 5620 lines per centimeter in a grating that is used with light whose wavelength is 471 nm. A flat observation screen is located at a distance of 0.750 m from the grating. What is the minimum width that the screen must have so the *centers* of all the principal maxima formed on either side of the central maximum fall on the screen?

**40. The separation between the slits of a grating is 2.2×10^{-6} m. This grating is used with light that contains all wavelengths between 410 and 660 nm. Rainbow-like spectra form on a screen 3.2 m away. How wide (in meters) is (a) the first-order spectrum and (b) the second-order spectrum?

ADDITIONAL PROBLEMS

41. A single slit has a width of 2.1×10^{-6} m and is used to form a diffraction pattern. Find the angle that locates the second dark fringe when the wavelength of the light is (a) 430 nm and (b) 660 nm.

42. A rock concert is being held in an open field. Two loudspeakers are separated by 7.00 m. As an aid in arranging the seating, a test is conducted in which both speakers vibrate in phase and produce an 80.0-Hz bass tone simultaneously. The speed of sound is 343 m/s. A reference line is marked out in front of the speakers, perpendicular to the midpoint of the line between the speakers. Relative to either side of this reference line, what is the smallest angle that locates the places where destructive interference occurs? People seated in these places would have trouble hearing the 80.0-Hz bass tone.

43. The first-order maximum produced by a grating is located by an angle of $\theta = 28°$. What is the angle for the second-order maximum with the same light?

44. In a Young's double-slit experiment, the angle that locates the second-order bright fringe is 2.0°. The slit separation is 3.8×10^{-5} m. What is the wavelength of the light?

45. A mixture of yellow light (wavelength = 580 nm in vacuum) and violet light (wavelength = 410 nm in vacuum) falls perpendicularly on a film of gasoline that is floating on a

puddle of water. For both wavelengths, the refractive index of gasoline is $n = 1.40$ and of water is $n = 1.33$. What is the minimum nonzero thickness of the film in a spot that looks (a) yellow and (b) violet?

46. It is claimed that some professional baseball players can see which way the ball is spinning as it travels toward home plate. One way to judge this claim is to estimate the distance at which a batter can first hope to resolve two points on opposite sides of a baseball, which has a diameter of 0.0738 m. (a) Estimate this distance, assuming the pupil of the eye has a diameter of 2.0 mm, the material within the eye has a refractive index of 1.36, and the wavelength of the light is 550 nm in vacuum. (b) Considering that the distance between the pitcher's mound and home plate is 18.4 m, can you rule out the claim based on your answer to part (a)?

47. A telescope is being used to view two objects that are separated by 480 m on the moon's surface. The surface of the moon is 3.77×10^8 m away from the surface of the earth. Assume that diffraction effects, rather than atmospheric turbulence, limits the resolving power of the telescope and that the wavelength of the light being used is 550 nm. Determine the diameter that the objective lens of this telescope must have if the two objects on the moon are to be resolved.

*48. The same diffraction grating is used with two different wavelengths of light, λ_A and λ_B. The fourth-order principal maximum of light A falls exactly on top of the third-order principal maximum of light B. Find the ratio λ_A / λ_B.

*49. A film of gasoline ($n = 1.40$) floats on water ($n = 1.33$). Yellow light (wavelength = 580 nm in vacuum) shines perpendicularly on this film. (a) Determine the minimum nonzero thickness of the film, such that the film appears bright yellow due to constructive interference. (b) Repeat part (a), assuming the gasoline film is on glass ($n = 1.52$) instead of water.

*50. Violet light (wavelength = 410 nm) and red light (wavelength = 660 nm) lie at opposite ends of the visible spectrum. (a) For each wavelength, find the angle θ that locates the first-order maximum produced by a grating with 3300 lines/cm. This grating converts a mixture of all colors between violet and red into a rainbow-like dispersion between the two angles. Repeat the calculation above for (b) the second-order maximum and (c) the third-order maximum. (d) From your results, decide whether there is any overlap between any of the "rainbows" and specify which orders overlap.

*51. Two slits are 0.158 mm apart. A mixture of red light (wavelength = 665 nm) and yellow-green light (wavelength = 565 nm) falls on the slits. A flat observation screen is located 2.24 m away. What is the distance on the screen between the third-order red fringe and the third-order yellow-green fringe?

**52. Two gratings A and B have slit separations d_A and d_B, respectively. They are used with the same light and the same observation screen. When grating A is replaced with grating B, it is observed that the first-order maximum of A is exactly replaced by the second-order maximum of B. (a) Determine the ratio d_B / d_A of the spacings between the slits of the gratings. (b) Find the next two principal maxima of grating A and the principal maxima of B that exactly replace them when the gratings are switched. Identify these maxima by their order numbers.

**53. A piece of curved glass has a radius of curvature of 10.0 m and is used to form Newton's rings, as in Figure 27.14. Not counting the dark spot at the center of the pattern, there are one hundred dark fringes, the last one being at the outer edge of the curved piece of glass. The light being used has a wavelength of 654 nm in vacuum. What is the radius of the outermost dark ring in the pattern?

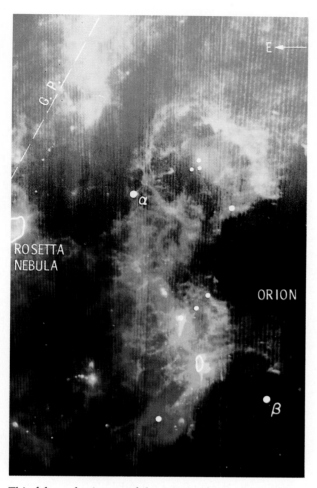

This false-color image of the region of sky around the constellation Orion was produced from data from the Infrared Astronomical Satellite (IRAS), and shows a much different view than that seen from optical telescopes. The intensity of infrared radiation is represented by colors: red indicates strong 100-micron-wavelength radiation, green indicates strong 60-micron-wavelength radiation, and blue shows strong 12-micron-wavelength radiation. Note that 1 micron $= 10^{-6}$ m. Well-known regions of star formation are apparent, such as the Orion molecular cloud (large feature dominating lower half of picture), located in and surrounding the sword of Orion. Part of the Milky Way crosses the upper left corner. Extended infrared cirrus clouds associated with the galaxy and the solar system are also seen throughout the image.

Physics & Space Science

Our ability to put probes into deep space and satellites into earth orbit provides the human race with a great opportunity to understand both the earth and our astronomical environment. By using orbiting telescopes and cameras that are sensitive to visible light, and to infrared and ultraviolet radiation, scientists are able to gather tremendous amounts of information about the earth's surface. By using other telescopes in orbit that are sensitive to X-rays and gamma rays, astronomers are able to learn many things about the stars and interstellar matter that are otherwise invisible to us on earth.

Imagine that you are a Space Shuttle astronaut. Consider the physics involved in putting you in orbit, and the physics that you use once you are there. The Space Shuttle rocket engines exert 6.3 million pounds of thrust. By firing tremendous streams of hot gases out the bottom of the Shuttle's rockets, you are forced upward into space. The Shuttle has only enough fuel to put you into an orbit 300 hundred miles above the earth's surface; it cannot carry enough fuel to take you to the moon.

The acceleration you feel getting into space is determined by Newton's second law, $\Sigma F = ma$. To maintain the height you reach, the Shuttle is put in orbit; otherwise, like a rock thrown straight up in the air, it would fall right back to the surface. As you move around the earth while in orbit, your potential and kinetic energies change. However, the sum of the two energies

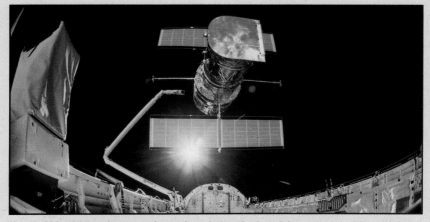

The Hubble space telescope, shown here being deployed from the cargo bay of the Space Shuttle, is one of the many artificial satellites that orbit the earth. Such satellites gather unique and valuable information about our planet, solar system, and galaxy.

remains constant throughout the orbit. (Actually, your Shuttle continually loses energy owing to its colliding with air molecules even at that altitude, but the effect is negligible over the length of time you will be up there.)

The crew has three assignments: To examine the effects on agriculture of the prolonged drought that has occurred in the western United States, to study the sun's X-ray emissions, and to salvage a spy satellite.

Your primary diagnostic instrument for studying crops is an infrared camera. Everything stores, and re-emits, energy received from the sun, and living things also generate and emit their own energy. Some of this energy is emitted as infrared radiation. The health of trees and crops can be determined by the infrared radiation they are generating and emitting. When the infrared-sensitive film is developed on earth, scientists will be able to see which forests and crop regions are not thriving. With that information in hand, they will be able to provide vital information on how to reallocate water.

As your Shuttle passes over the Caribbean, you observe the characteristic spiral shape of a developing tropical storm. The spiral shape results from effects of the earth's rotation on regions of different pressure in the atmosphere, an application of basic physics principles. You photograph the storm and radio news of it to Houston.

Now one of your colleagues uses the Shuttle's mechanical arm to pull an X-ray telescope out of the cargo bay and to aim it at the sun. Al-

though it appears yellow to our eyes, the sun not only emits all visible colors of light but also emits all kinds of nonvisible radiation, including radio waves, infrared radiation, ultraviolet radiation, X-rays and gamma rays. To a telescope sensitive to X-rays, the sun does not appear at all like the uniform disk we normally see. It is splotchy as seen by means of X-rays because this radiation leaks out through holes created in the sun's surface by magnetic fields. The distribution of X-rays you record will give astronomers important insights into the sun's internal activity.

A piece of space debris, pulled by the earth's gravity, flashes past the Shuttle. The earth's atmosphere creates enormous friction on the surface of the space rock (called a meteoroid) as it is pulled earthward. The friction generates heat, which causes the meteoroid to begin melting and shedding its outer layer. Staring out a window, you are treated to the sight of the burning meteor, as the meteoroid is called while it glows in our atmosphere. Without the protection of the heat tiles on the bottom of the Shuttle, the same thing would happen to your ship as you reentered the atmosphere. Material scientists developed the heat tiles by using the principles of basic physics and chemistry, and the perfection of the tiles was essential to reusable spacecraft.

Ahead is the rapidly rotating satellite you are to retrieve. Its gyroscopes have failed and now it is useless because its antennae don't remain aimed at the earth; it cannot reliably receive commands sent to it. Retrieving and repairing it will save millions of dollars, compared

with replacing it. You prepare for a space walk. You and a colleague will go out and spin it down so that it can be safely retrieved in the Shuttle's cargo bay. The conservation of angular momentum makes this challenging, both in grabbing the satellite and in spinning it down. Any ideas how you'll do it?

Returning to earth, the Shuttle becomes one of the most expensive gliders in history. There is no room for mistakes, since the Shuttle's engines cannot be used to bring your ship around for a second try. The Shuttle's aerodynamics, state of the art navigation and communication, and outstanding piloting bring you down safely.

FOR INFORMATION ABOUT CAREERS IN SPACE SCIENCE

Professional Careers Sourcebook, *K. M. Savage and C. A. Dorgan, eds., 1990, Gale Research, Inc., publisher, provides addresses and phone numbers for dozens of guides and other resources about careers in many space and related earth science fields.*

Encyclopedia of Careers and Vocational Guidance, *Vol. 2, 8th edition, William E. Hopke, ed., 1990, J. G. Ferguson, publisher, has an in-depth description of the working conditions astronauts, and space and related earth scientists encounter.*

You can also write to the following, among many others listed in the books above:

■ *American Institute of Physics, 335 East 45th St., New York, NY 10017*
■ *American Astronomical Society, 2000 Florida Ave., NW, No. 300, Washington, DC 20009*
■ *NASA, Office of Educational Programs and Services, 400 Maryland Ave., Washington, DC 20025*
■ *American Institute of Aeronautics and Astronautics, 370 L'Enfant Promenade, SW, Washington, DC 20024*
■ *American Society of Civil Engineers, 345 East 47th St., New York, NY 10017*

PART SIX

MODERN PHYSICS

Near the turn of the century, scientists believed they had a firm understanding of the physical aspect of our world. For more than two hundred years, Newton's laws had accurately described the motion of all types of objects, from projectiles to orbiting planets. The theories of heat and sound had been successfully established by using a kinetic theory of matter based on Newton's laws. And physicists had attained a clear picture of electricity and magnetism, as well as considerable insight into the nature of electromagnetic waves.

Yet all was not well, for there remained important phenomena and new experimental data that could not be explained with these so-called classical theories. The shortcomings of the classical theories turned out to be so serious that, during the first part of the twentieth century, physicists were forced to develop entirely new views of the universe. First came Einstein's theory of special relativity, which describes how the world behaves when viewed at speeds near the speed of light. Then came quantum mechanics, which offers extraordinary insight into the microscopic realm of the atomic world, a world that consists of electrons, protons,

neutrons, atoms, and molecules. The special theory of relativity and quantum mechanics form the foundation of what is now called "modern physics."

The first chapter in our discussion of modern physics deals with Einstein's theory of special relativity. We will see that this theory radically alters our notions about space and time. And special relativity predicts a surprising union between mass and energy, as expressed in Einstein's famous equation $E_0 = mc^2$, where E_0 is the rest energy, m is the mass, and c is the speed of light. In the following two chapters, we will encounter some of the experiments that led up to the development of quantum mechanics. Then the "wave-particle duality" of nature will be discussed, according to which a wave can exhibit particle-like characteristics and a particle can exhibit wave-like characteristics. In addition, we will see how quantum mechanics describes the nature of the atom, how X rays are produced, and how lasers operate. Finally, the remaining two chapters treat some of the more recent developments in modern physics, including nuclear physics and radioactivity, the biological effects of ionizing radiation, and elementary particles.

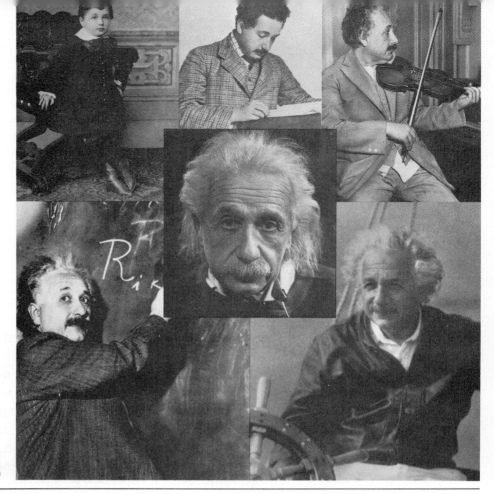

CHAPTER 28

SPECIAL RELATIVITY

Ask anyone to name the most famous twentieth century scientist, and you are likely to get Albert Einstein (1879–1955) for an answer. Few scientists have become as well known as he. The number, diversity, and fundamental significance of his scientific accomplishments are remarkable. He is best known for the theory of special relativity, published in 1905 when he was 26 years old. Even today, this theory is a rich source of amazement, for it alters in radical ways so many of our basic ideas about the physical world. We will see that in the aftermath of Einstein's theory, time no longer has a unique meaning. And neither does distance. In addition, we will find that mass (inertia) and energy are not independent ideas as classical physics assumes. Instead, they are equivalent, with the result that the energy E_0 of a stationary object with mass m is $E_0 = mc^2$, where c is the speed of light in a vacuum. Another surprising result is that objects with mass cannot be accelerated to speeds at or beyond the speed of light, no matter how much force is applied. All of these features of the theory of special relativity will be discussed in this chapter. But first we need to explain the idea of an inertial reference frame, since this idea plays such a fundamental role in the theory.

28.1 EVENTS AND INERTIAL REFERENCE FRAMES

The theory of special relativity deals with the way that an event is measured by observers who are moving relative to the event. An *event*, such as the launching of the space shuttle in Figure 28.1, is a physical "happening" that occurs at a certain place and time. In this drawing two observers are watching the lift-off, one standing on the earth and one seated in an airplane that is flying at a constant velocity relative to the earth. To record the event, each observer uses a *reference frame* that consists of a set of x, y, z axes (called a *coordinate system*) and a clock. The coordinate system is used to establish where the event occurs, and the clock is used to specify when it happens. Each observer is at rest relative to his own reference frame. Since the earth-based observer and the airborne observer are moving relative to each other, their respective reference frames are also in relative motion.

The theory of special relativity deals with a "special" kind of reference frame, called an *inertial reference frame.* As Section 4.2 discusses, an inertial reference frame is one in which Newton's law of inertia is valid. That is, if the net force acting on a body is zero, the body either remains at rest or moves at a constant velocity. In other words, the acceleration of such a body is zero when measured in an inertial reference frame. Rotating and otherwise accelerating reference frames are not inertial reference frames. The earth-based reference frame in Figure 28.1 is not quite an inertial frame, because it is subjected to centripetal accelerations as the earth spins on its axis and revolves around the sun. In most situations, however, the effects of these accelerations are small, so we can consider an earth-based reference frame to be an inertial one. To the extent that the earth-based reference frame is an inertial frame, so is the plane-based reference frame, for the plane moves at a constant velocity relative to the earth. The next section discusses why inertial reference frames are important in relativity.

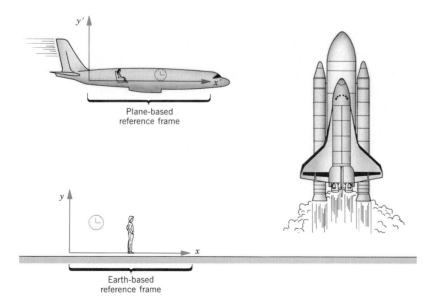

Figure 28.1 Using an earth-based reference frame, an observer standing on the earth records the location and time of an event (the lift-off). Likewise, an observer in the airplane uses a plane-based reference frame to describe the event.

28.2 THE POSTULATES OF SPECIAL RELATIVITY

Einstein built his theory of special relativity on two fundamental assumptions or postulates about the way nature behaves.

The Postulates of Special Relativity

1. **The Relativity Postulate.** The laws of physics are the same in every inertial reference frame.
2. **The Speed of Light Postulate.** The speed of light in a vacuum, measured in any inertial reference frame, always has the same value of c, no matter how fast the source of light and the observer are moving relative to each other.

It is not difficult to accept the relativity postulate. For instance, in Figure 28.1 each observer, using his own inertial reference frame, can make measurements on the motion of the space shuttle. The relativity postulate asserts that both observers find their data to be consistent with Newton's laws of motion. Similarly, both observers find that the behavior of the electronics on board the space shuttle is described by the laws of electromagnetism, such as Faraday's law of electromagnetic induction. According to the relativity postulate, *any inertial reference frame is as good as any other for expressing the laws of physics, because the laws are the same in all such frames.* In other words, with regard to inertial reference frames, nature does not play favorites.

Since the laws of physics are the same in all inertial reference frames, there is no experiment that can distinguish between an inertial frame that is at rest and one that is moving at a constant velocity. When you are seated on the aircraft in Figure 28.1, for instance, it is just as valid to say that you are at rest and the earth is moving as it is to say the reverse. It is not possible to single out one particular inertial reference frame as being at "absolute rest." Consequently, it is meaningless to talk about the "absolute velocity" of an object—that is, its velocity measured relative to a reference frame at "absolute rest." Thus, the earth moves relative to the sun, which itself moves relative to the center of our galaxy. And the galaxy moves relative to other galaxies, and so on. According to Einstein, only the relative velocity between objects, not their absolute velocities, can be measured and is physically meaningful.

While the relativity postulate is not too difficult to accept, the speed of light postulate seems to defy common sense. For instance, Figure 28.2 illustrates a person standing on the bed of a truck that is moving at a constant speed of 15 m/s relative to the ground. Now, suppose you are standing on the ground and the

Figure 28.2 Both the person on the truck and the observer on the earth measure the speed of the light to be c, regardless of the speed of the truck.

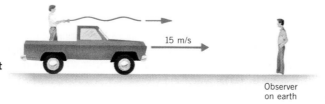

15 m/s

Observer
on earth

person on the truck shines a flashlight at you. The person on the truck observes the speed of light to be c. What do you measure for the speed of light? You might guess that the speed of light would be $c + 15$ m/s. However, this guess is inconsistent with the speed of light postulate, which states that all observers in inertial reference frames measure the speed of light to be c—nothing more, nothing less. Therefore, you must also measure the speed of light to be c, the same as that measured by the person on the truck. According to the speed of light postulate, the fact that the flashlight is moving toward you has no influence whatsoever on the speed of the light approaching you. This property of light, although surprising, has been verified many times by experiment.

Since waves, such as water waves and sound waves, require a medium through which to propagate, it was natural for scientists before Einstein to assume that light did too. This hypothetical medium was called the *luminiferous ether* and was assumed to fill all of space. Furthermore, it was believed that light traveled only at the speed c when measured with respect to the ether. According to this view, an observer moving relative to the ether would measure a speed for light that was slower or faster than c, depending on whether the observer moved with or against the light. During the years 1883–1887, however, the American scientists A. A. Michelson and E. W. Morley carried out a series of famous experiments whose results were not consistent with the ether theory. Their results indicated that the speed of light is indeed the same in all inertial reference frames and does not depend on the motion of the observer relative to the source of the light. These experiments, and others, led eventually to the demise of the ether theory and the acceptance of the theory of special relativity.

28.3 THE RELATIVITY OF TIME: TIME DILATION

TIME DILATION

Common experience indicates that time passes just as quickly for a person standing on the ground as it does for an astronaut in a spacecraft. In contrast, the theory of special relativity reveals that the person on the ground measures time passing more slowly for the astronaut than for himself. We can see how this curious effect arises with the help of the clock illustrated in Figure 28.3. This clock uses a pulse of light to mark time. A short pulse of light is emitted by a light source, reflects from a mirror, and then strikes a detector that is situated next to the source. Each time a pulse reaches the detector, a "tick" registers on the chart recorder, another short pulse of light is emitted, and the cycle repeats. Thus, the time interval between successive "ticks" is marked by a beginning event (the firing of the light source) and an ending event (the pulse striking the detector). The source and detector are so close to each other that the two events can be considered to occur at the same location.

Suppose two identical clocks are built. One clock is kept on earth, and the other is placed aboard a spacecraft that travels at a constant velocity relative to the earth. The astronaut is at rest with respect to the clock on the spacecraft and, therefore, sees the light pulse move on the up/down path shown in Figure 28.4a. According to the astronaut, the time interval Δt_0 required for the light to follow this path is the distance $2D$ divided by the speed of light c; $\Delta t_0 = 2D/c$. To the astronaut, Δt_0 is the time interval between the "ticks" of the spacecraft clock, that

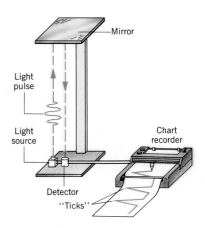

Figure 28.3 A light clock.

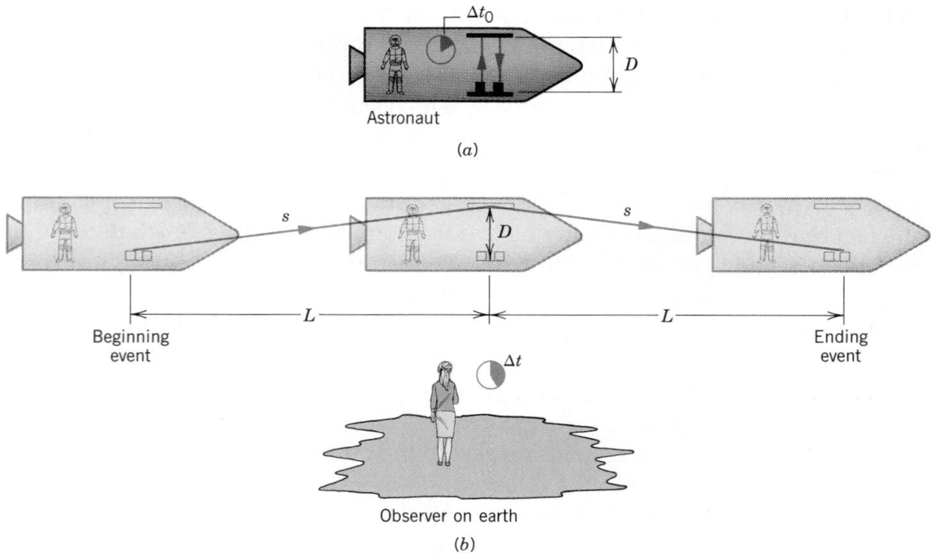

Figure 28.4 (a) The astronaut measures the time interval Δt_0 between successive "ticks" of his light clock. (b) An observer on earth watches the astronaut's clock and sees the light pulse travel a greater distance between "ticks" than it does in part a. Consequently, the earth-based observer measures a time interval Δt between "ticks" that is greater than Δt_0.

is, the time interval between the beginning and ending events of the clock. An earth-based observer, however, does *not* measure Δt_0 as the time interval between these two events. Since the spacecraft is moving, the earth-based observer sees the light pulse follow the diagonal path shown in part b of the drawing. This diagonal path is longer than the up/down path seen by the astronaut. But light travels at the *same speed c* for both observers, in accord with the speed of light postulate. Therefore, the earth-based observer measures a time interval Δt between the two events that is *greater* than the time interval Δt_0 measured by the astronaut. In other words, the earth-based observer, using his own earth-based clock to measure the performance of the astronaut's clock, finds that the astronaut's clock runs slowly. This result of the theory of special relativity is known as *time dilation*. (To *dilate* means to expand, and the time interval Δt is "expanded" relative to Δt_0.)

The time interval Δt that the earth-based observer measures in Figure 28.4b can be determined as follows. While the light pulse travels from the source to the detector, the spacecraft moves a distance $2L = v \, \Delta t$ to the right, where v is the speed of the spacecraft relative to the earth. From the drawing it can be seen that the light pulse travels a total diagonal distance of $2s$ during the time interval Δt. Applying the Pythagorean theorem, we find that

$$2s = 2\sqrt{D^2 + L^2} = 2\sqrt{D^2 + \left(\frac{v \, \Delta t}{2}\right)^2}$$

But the distance $2s$ is also equal to the speed of light times the time interval Δt, so $2s = c \, \Delta t$. Therefore,

$$c \, \Delta t = 2\sqrt{D^2 + \left(\frac{v \, \Delta t}{2}\right)^2}$$

Squaring this result and solving for Δt gives

$$\Delta t = \frac{2D}{c} \frac{1}{\sqrt{1 - \dfrac{v^2}{c^2}}}$$

But $2D/c = \Delta t_0$, the time interval between successive "ticks" of the spacecraft's clock as measured by the astronaut. With this substitution, the equation above can be expressed as

$$\left[\begin{array}{c} \textbf{Time} \\ \textbf{dilation} \end{array} \right] \qquad \Delta t = \frac{\Delta t_0}{\sqrt{1 - \dfrac{v^2}{c^2}}} \qquad \text{(28.1)}$$

The symbols in this formula are summarized below:

Δt_0 = time interval between two events, as measured by an observer who is at rest with respect to the events and who views the events as occurring *at the same place*

Δt = time interval measured by an observer who is in motion with respect to the events and who views the events as occurring at *different places*

v = relative speed between the two observers

c = speed of light in a vacuum

For a speed v that is less than c, the term $\sqrt{1 - v^2/c^2}$ in Equation 28.1 is less than 1, and the dilated time interval Δt is greater than Δt_0. Example 1 illustrates this time dilation effect.

The fact that each of these clocks has a different time is allowed by special relativity, but only if each one moves at a different speed relative to the others.

Example 1   Time Dilation

The spacecraft in Figure 28.4 is moving past the earth at a constant speed v that is 0.92 times the speed of light. Thus, $v = (0.92)(3.0 \times 10^8 \text{ m/s})$, which is often written as $v = 0.92c$. The astronaut measures the time interval between successive "ticks" of the spacecraft clock to be $\Delta t_0 = 1.0$ s. What is the time interval Δt that an earth observer measures between "ticks" of the astronaut's clock?

REASONING   Since the clock on the spacecraft is moving relative to the earth observer, the earth observer measures a greater time interval Δt between "ticks" than does the astronaut, who is at rest relative to the clock. The dilated time interval Δt can be determined from the time dilation relation, Equation 28.1.

SOLUTION   The dilated time interval is

$$\Delta t = \frac{\Delta t_0}{\sqrt{1 - \dfrac{v^2}{c^2}}} = \frac{1.0 \text{ s}}{\sqrt{1 - \left(\dfrac{0.92c}{c}\right)^2}} = \boxed{2.6 \text{ s}}$$

From the point of view of the earth-based observer, the astronaut is using a clock that is running slowly, for the earth-based observer measures a time between "ticks" that is longer (2.6 s) than what the astronaut measures (1.0 s). The earth observer measures the clock on the spacecraft to "lose" 1.6 s every second.

Example 1 shows that time dilation is appreciable when the speed v is comparable to the speed of light. The speeds we experience in everyday life, however, are far too small for time dilation to be noticeable. For instance, for a clock aboard a jetliner traveling at $v = 0.000\ 000\ 75c$ (about 500 miles per hour), the time intervals Δt and Δt_0 in Example 1 would differ by only 2.8×10^{-13} s. This small difference in time means that 110 000 years would have to pass on an earth-based clock before the clock on the jetliner would lose 1 second.

PROPER TIME INTERVAL

In Figure 28.4 both the astronaut and the person standing on the earth are measuring the time interval between a beginning event (the firing of the light source) and an ending event (the light pulse striking the detector). For the astronaut, who is at rest with respect to the light clock, the two events occur at the same location. Being at rest with respect to a clock is the usual or "proper" situation, so the time interval Δt_0 measured by the astronaut is called the *proper time interval.* In general, the proper time interval Δt_0 between two events is the time interval measured by an observer who is at rest relative to the events and sees the events at the *same location* in space. On the other hand, the earth-based observer does not see the two events occurring at the same location in space, since the spacecraft is in motion. The time interval Δt that the earth-based observer measures is, therefore, not a proper time interval in the sense that we have defined it.

To understand situations involving time dilation, it is essential to distinguish between Δt_0 and Δt. In such situations it is helpful if one first identifies the two events that define the time interval. These events may be something other than the firing of a light source and the light pulse striking a detector. Then determine the reference frame in which the two events occur at the same place. For an observer at rest in this reference frame, the time interval is the proper time interval Δt_0.

SPACE TRAVEL

One of the intriguing aspects of time dilation occurs in conjunction with space travel. Since enormous distances are involved, travel to even the closest star outside our solar system would take a long time. However, as the following example shows, the time for such a trip can be considerably less for the passengers than one might guess.

Example 2   Space Travel

The star closest to our solar system is Alpha Centauri, which is 4.3 light-years away. This means that, as measured by a person on earth, it would take light 4.3 years to reach this star. If a rocket leaves for Alpha Centauri at a speed of $v = 0.95c$ relative to the earth, by how much will the passengers have aged, according to their own clock, when they reach their destination?

REASONING   The two events in this problem are the departure from earth and the arrival at Alpha Centauri. At departure, earth is just outside the spaceship. Upon arrival at the destination, Alpha Centauri is just outside. Therefore, relative to the passengers, the two events occur at the same place, namely, just outside the spaceship. Thus, the passengers measure the proper time interval Δt_0 on their clock, and it is this interval that we must find. For a person left behind on earth, the events occur at *different places,* so such a person measures the dilated time interval Δt rather than the proper time interval. Since it takes 4.3 years for light to traverse the distance and the rocket is moving at $v = 0.95c$ relative to the earth, a person on earth measures the dilated time interval to be $\Delta t = (4.3 \text{ years})/0.95 = 4.5$ years. This value can be used with the time-dilation equation to find the proper time interval Δt_0.

SOLUTION   Using the time-dilation equation, we find that the proper time interval by which the passengers judge their own aging is

$$\Delta t_0 = \Delta t \sqrt{1 - \frac{v^2}{c^2}} = (4.5 \text{ years}) \sqrt{1 - \left(\frac{0.95c}{c}\right)^2} = \boxed{1.4 \text{ years}}$$

Thus, the people aboard the rocket have aged by only 1.4 years when they reach Alpha Centauri, and not the 4.5 years an earthbound observer has calculated.

VERIFICATION OF TIME DILATION

A striking confirmation of time dilation was achieved in 1971 by an experiment carried out by J. C. Hafele and R. E. Keating.* They transported very precise cesium-beam atomic clocks around the world on commercial jets. Since the speed of a jet plane is considerably less than c, the time-dilation effect is extremely small. However, the atomic clocks were accurate to about $\pm 10^{-9}$ s, so the effect could be measured. The clocks were in the air for 45 hours, and their times were compared to reference atomic clocks kept on earth. The experimental results revealed that, within experimental error, the readings on the clocks on board the planes were different than those on earth by an amount that agreed with the prediction of relativity.

Time dilation has also been confirmed with experiments using subatomic particles called *muons*. These particles are created high in the atmosphere, at altitudes of about 10 000 m. When at rest, muons are short-lived, existing for a time of about 2.2×10^{-6} s before disintegrating into other particles. With such a short lifetime, these particles could never make it down to the earth's surface, even if they traveled close to the speed of light. However, *a large number of muons do reach the earth*. The only way they can do so is to live longer because of time dilation, as Example 3 illustrates.

* J. C. Hafele and R. E. Keating, Around the world atomic clocks: Relativistic time gains observed. *Science* 168 (July 14, 1972).

Example 3   *The Lifetime of a Muon*

The average lifetime of a muon at rest is 2.2×10^{-6} s. A muon created in the upper atmosphere travels toward the earth at a speed of $v = 0.998c$. Find, on the average, (a) how long a muon lives according to an observer on earth, and (b) how far the muon travels before disintegrating.

REASONING   The two events of interest are the generation and subsequent disintegration of the muon. When the muon is at rest, these events occur at the same place, so the muon's average (at rest) lifetime of 2.2×10^{-6} s is a proper time interval Δt_0. When the muon moves at a speed $v = 0.998c$ relative to the earth, an observer on the earth measures a dilated lifetime Δt that is given by Equation 28.1. The average distance x traveled by a muon, as measured by an earth observer, is equal to the muon's speed times the dilated time interval.

SOLUTION

(a) The observer on earth measures a dilated lifetime given by

$$\Delta t = \frac{\Delta t_0}{\sqrt{1 - \frac{v^2}{c^2}}} = \frac{2.2 \times 10^{-6} \text{ s}}{\sqrt{1 - \left(\frac{0.998c}{c}\right)^2}} = \boxed{35 \times 10^{-6} \text{ s}} \qquad (28.1)$$

PROBLEM SOLVING INSIGHT

The proper time interval Δt_0 is always shorter than the dilated time interval Δt.

(b) The distance traveled by the muon before it disintegrates is

$$x = v \, \Delta t = (0.998)(3.00 \times 10^8 \text{ m/s})(35 \times 10^{-6} \text{ s}) = \boxed{1.0 \times 10^4 \text{ m}}$$

Thus, the dilated, or extended, lifetime provides sufficient time for the muon to reach the surface of the earth. If its lifetime were only 2.2×10^{-6} s, a muon would travel only 660 m before disintegrating and could never reach the earth.

28.4 THE RELATIVITY OF LENGTH: LENGTH CONTRACTION

Because of time dilation, observers moving at a constant velocity relative to each other measure different time intervals between two events. For instance, Example 2 in the previous section illustrates that a trip from earth to Alpha Centauri at a speed of $v = 0.95c$ takes 4.5 years according to a clock on earth, but only 1.4 years according to a clock in the rocket. These two times differ by the factor $\sqrt{1 - v^2/c^2}$. Since the times for the trip are different, one might ask if the observers measure different distances between earth and Alpha Centauri. The answer, according to special relativity, is yes. After all, both the earth-based observer and the rocket passenger agree that the relative speed between the rocket and earth is $v = 0.95c$. Since speed is distance divided by time and the time is different for the two observers, it follows that the distances must also be different, if the relative speed is to be the same for both individuals. Thus, the earth observer determines the distance to Alpha Centauri to be $L_0 = v \, \Delta t = (0.95c)(4.5 \text{ years}) = 4.3$ light-years. On the other hand, a passenger aboard the rocket finds the distance is only $L = v \, \Delta t_0 = (0.95c)(1.4 \text{ years}) = 1.3$ light-years. The passenger, measuring the shorter time, also measures the shorter distance. This shortening of the distance between two points is one example of a phenomenon known as *length contraction.*

To space travelers heading toward a distant galaxy, such as the Spiral Galaxy shown here, the distance is not as great as observers on Earth measure it to be, because of the effect of length contraction in special relativity.

The relation between the distances measured by two observers in relative motion at a constant velocity can be obtained with the aid of Figure 28.5. Part *a* of the drawing shows the situation from the point of view of the earth-based observer. This person measures the time of the trip to be Δt, the distance to be L_0, and the relative speed of the rocket to be $v = L_0/\Delta t$. Part *b* of the drawing presents the point of view of the passenger, for whom the rocket is at rest, and the earth and Alpha Centauri appear to move by at a speed v. The passenger determines the distance of the trip to be L, the time to be Δt_0, and the relative speed to be $v = L/\Delta t_0$. Since the relative speed computed by the passenger equals that computed by the earth-based observer, it follows that $v = L/\Delta t_0 = L_0/\Delta t$. Using this result and the time-dilation equation, Equation 28.1, we obtain the following relation between L and L_0:

$$\begin{bmatrix} \textbf{Length} \\ \textbf{contraction} \end{bmatrix} \qquad L = L_0 \sqrt{1 - \frac{v^2}{c^2}} \qquad\qquad (28.2)$$

The length L_0 is called the ***proper length;*** it is the length (or distance) between two points *as measured by an observer at rest with respect to them.* Since v is less than c, the term $\sqrt{1 - v^2/c^2}$ is less than 1, and L is less than L_0. It is important to note that this length contraction occurs only along the direction of the motion. Those dimensions that are perpendicular to the motion are not shortened, as the next example discusses.

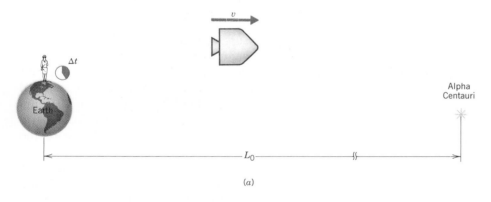

(a)

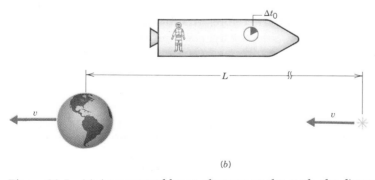

(b)

Figure 28.5 (*a*) As measured by an observer on the earth, the distance to Alpha Centauri is L_0 and the time required to make the trip is Δt. (*b*) According to the passenger on the spacecraft, the earth and Alpha Centauri move with speed v relative to the craft. The passenger measures the distance and time of the trip to be L and Δt_0, respectively, both quantities less than those in part *a*.

Example 4   The Contraction of a Meter Stick

An astronaut, using a meter stick that is at rest relative to a cylindrical spacecraft, measures the length and diameter of the spacecraft to be 82 and 21 m, respectively. The spacecraft moves with a constant speed of $v = 0.95c$ relative to the earth, as in Figure 28.5. What are the dimensions of the spacecraft, as measured by an observer on earth?

REASONING AND SOLUTION   The length of 82 m is a proper length L_0, since it is measured using a meter stick that is at rest relative to the spacecraft. The length L measured by the observer on earth can be determined from the length-contraction formula:

$$L = L_0 \sqrt{1 - \frac{v^2}{c^2}} = (82 \text{ m}) \sqrt{1 - \left(\frac{0.95c}{c}\right)^2} = \boxed{26 \text{ m}}$$

The diametric dimension is perpendicular to the motion, so the earth observer does not measure any change in the diameter: $\boxed{\text{Diameter} = 21 \text{ m}}$. Figure 28.5a shows the size of the spacecraft as measured by the earth observer, while part b shows the size measured by the astronaut.

PROBLEM SOLVING INSIGHT

The proper length L_0 is always larger than the contracted length L.

When dealing with relativistic effects we need to distinguish carefully between the criteria for the proper time interval and the proper length. The proper time interval Δt_0 between two events is the time interval measured by an observer who is at rest relative to the events and who sees them occurring at the *same place.* All other moving inertial observers will measure a larger value for this time interval. The proper length L_0 of an object is the length measured by an observer who is *at rest* with respect to the object. All other moving inertial observers will measure a shorter value for this length. The observer who measures the proper time interval may not be the same one who measures the proper length. For instance, Figure 28.5 shows that the astronaut measures the proper time interval for the trip between earth and Alpha Centauri, while the earth-based observer measures the proper length (or distance) for the trip.

It should be emphasized that the word "proper" in the phrases proper time and proper length does *not* mean that these quantities are the correct or preferred quantities. If this were so, the observer measuring these quantities would be using a preferred reference frame for making the measurement, a situation that is prohibited by the relativity postulate. According to this postulate, all inertial reference frames are equivalent. Therefore, when two observers are moving relative to each other at a constant velocity, each measures the other person's clock to run more slowly than his own, and each measures the other person's length to be contracted.

28.5 RELATIVISTIC MOMENTUM

Thus far we have discussed how time intervals and distances between two events are measured by observers moving at a constant velocity relative to each other. The theory of special relativity also alters our ideas about momentum and energy.

Recall from Chapter 7 that the conservation of linear momentum states that when two or more objects interact, the total linear momentum of an isolated system of objects remains constant at all times (an isolated system is one in which the sum of the external forces acting on the objects is zero). The conservation of linear momentum is a law of physics and, in accord with the relativity postulate, is valid in all inertial reference frames. That is, when the total linear momentum is conserved in one inertial reference frame, it is conserved in all inertial reference frames.

As an example of momentum conservation, suppose several people are watching two billiard balls collide on a pool table. One person is standing next to the pool table and the other is moving past the table with a constant velocity. Since the two balls constitute an isolated system, the relativity postulate requires that both observers must find the total linear momentum of the two-ball system to be the same before, during, and after the collision. For this kind of situation in Chapter 7 (see Example 5 there), we used the definition of the linear momentum **p** of an object to be the product of its mass m and velocity **v**. As a result, the magnitude of the momentum was $p = mv$. As long as the speed of an object is considerably smaller than the speed of light, this definition is adequate. However, when the speed approaches the speed of light, an analysis of the collision shows that the total linear momentum is not conserved in all inertial reference frames if one defines linear momentum as the product of mass and velocity. In order to preserve the conservation of linear momentum, it is necessary to modify this definition as shown below in Equation 28.3. The momentum given by this expression, which is valid for all speeds, is called the *relativistic momentum* and is conserved in all inertial reference frames.

[Relativistic momentum]
$$p = \frac{mv}{\sqrt{1 - \dfrac{v^2}{c^2}}}$$
(28.3)

We see that the relativistic momentum differs from the nonrelativistic momentum (mv) by the same factor $\sqrt{1 - v^2/c^2}$ that also occurs in the time-dilation and length-contraction equations. Since this factor is always less than 1 and occurs in the denominator of Equation 28.3, the relativistic momentum is always larger than the nonrelativistic momentum. To illustrate how the two momenta differ as the speed of an object increases, Figure 28.6 shows a plot of the ratio of the relativistic momentum to the nonrelativistic momentum as a function of

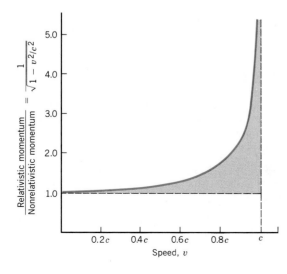

Figure 28.6 This graph shows how the ratio of the relativistic momentum to the nonrelativistic momentum increases as the speed of an object approaches the speed of light.

speed. According to Equation 28.3, this momentum ratio is just $1/\sqrt{1 - v^2/c^2}$. The graph shows that for speeds attained by ordinary objects, such as cars and planes, the relativistic and nonrelativistic momenta are almost equal because their ratio is nearly 1. Thus, at speeds much less than the speed of light, either the nonrelativistic momentum or the relativistic momentum can be used to describe collisions. On the other hand, when the speed of the object becomes comparable to the speed of light, the relativistic momentum becomes significantly greater than the nonrelativistic momentum, and the relativistic momentum must be used. Example 5 deals with the relativistic momentum of an electron traveling close to the speed of light.

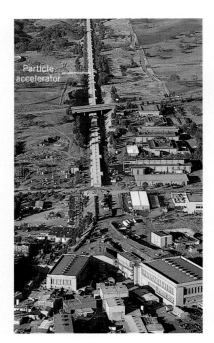

Particle accelerator

Figure 28.7 The Stanford three-kilometer linear accelerator accelerates electrons to nearly the speed of light.

Example 5   The Relativistic Momentum of a High-Speed Electron

The particle accelerator at Stanford University (Figure 28.7) is three kilometers long and accelerates electrons to a speed of 0.999 999 999 7c, which is very nearly equal to the speed of light. Find the relativistic momentum of an electron emerging from the accelerator and compare its value with the nonrelativistic value.

REASONING AND SOLUTION   The relativistic momentum of the electron can be obtained from Equation 28.3 if we recall that the mass of an electron is $m = 9.11 \times 10^{-31}$ kg:

$$p = \frac{mv}{\sqrt{1 - \dfrac{v^2}{c^2}}} = \frac{(9.11 \times 10^{-31} \text{ kg})(0.999\ 999\ 999\ 7c)}{\sqrt{1 - \dfrac{(0.999\ 999\ 999\ 7c)^2}{c^2}}} = \boxed{1 \times 10^{-17} \text{ kg} \cdot \text{m/s}}$$

This value for the momentum agrees with that measured experimentally when the electrons are deflected by a magnetic field as they emerge from the accelerator. The relativistic momentum is greater than the nonrelativistic momentum by a factor of

$$\frac{1}{\sqrt{1 - \dfrac{v^2}{c^2}}} = \frac{1}{\sqrt{1 - \dfrac{(0.999\ 999\ 999\ 7c)^2}{c^2}}} = \boxed{4 \times 10^4}$$

28.6 THE EQUIVALENCE OF MASS AND ENERGY

THE TOTAL ENERGY OF AN OBJECT

One of the most astonishing results of special relativity is that mass and energy are equivalent, in the sense that a gain or loss of mass can be regarded equally well as a gain or loss of energy. Consider, for example, an object of mass m traveling at a speed v. Einstein showed that the *total energy* E of the moving object is related to its mass and speed by the following relation:

$$\begin{bmatrix} \text{Total energy} \\ \text{of an object} \end{bmatrix} \qquad E = \frac{mc^2}{\sqrt{1 - \dfrac{v^2}{c^2}}} \qquad (28.4)$$

To gain some understanding of Equation 28.4, consider the special case when the object is at rest. When $v = 0$, the total energy is called the *rest energy E_0*, and Equation 28.4 reduces to Einstein's now-famous equation:

$$\begin{bmatrix} \text{Rest energy} \\ \text{of an object} \\ \text{when } v = 0 \end{bmatrix} \qquad E_0 = mc^2 \qquad (28.5)$$

The rest energy represents the energy equivalent of the mass of an object at rest. As Example 6 shows, even a small mass is equivalent to an enormous amount of energy.

Example 6   The Energy Equivalent of a Golf Ball

A 0.046-kg golf ball is lying on the green. (a) Find the rest energy of the golf ball. (b) If this rest energy were used to operate a 75-W light bulb, for how many years could the bulb stay on?

REASONING   The rest energy E_0 that is equivalent to the mass m of the golf ball is found from the relation $E_0 = mc^2$. The 75-W light bulb consumes 75 J of energy per second. If the entire rest energy is available to keep the light bulb burning, the bulb could stay on for a time equal to the rest energy divided by the power of the light bulb.

SOLUTION
(a) The rest energy of the golf ball is

$$E_0 = mc^2 = (0.046 \text{ kg})(3.0 \times 10^8 \text{ m/s})^2 = \boxed{4.1 \times 10^{15} \text{ J}} \qquad (28.5)$$

(b) This rest energy can keep the light bulb burning for a time t given by

$$t = \frac{\text{Rest energy}}{\text{Power}} = \frac{4.1 \times 10^{15} \text{ J}}{75 \text{ W}} = 5.5 \times 10^{13} \text{ s}$$

Since one year contains 3.2×10^7 s, we find $\boxed{t = 1.7 \times 10^6 \text{ yr}}$, or 1.7 million years!

When an object is accelerated from rest to a speed v, the object acquires kinetic energy in addition to its rest energy. The total energy E is the sum of the rest energy E_0 and the kinetic energy KE, or $E = E_0 + \text{KE}$. Using Equation 28.4, we can write the kinetic energy as

$$\text{KE} = E - E_0 = mc^2 \left(\frac{1}{\sqrt{1 - \frac{v^2}{c^2}}} - 1 \right) \qquad (28.6)$$

This equation is the relativistically correct expression for the kinetic energy of an object of mass m moving at speed v; the kinetic energy is the difference between the object's total energy E and its rest energy E_0.

Equation 28.6 looks nothing like the kinetic energy expression introduced in Chapter 6, namely, $\text{KE} = \frac{1}{2}mv^2$. However, for speeds much less than the speed of light ($v \ll c$), the relativistic equation for the kinetic energy reduces to KE =

$\frac{1}{2}mv^2$, as can be seen by using the binomial expansion* to represent the square root term in Equation 28.6:

$$\frac{1}{\sqrt{1 - \frac{v^2}{c^2}}} = 1 + \frac{1}{2}\left(\frac{v^2}{c^2}\right) + \frac{3}{8}\left(\frac{v^2}{c^2}\right) + \cdots$$

Suppose v is much smaller than c, say $v = 0.01c$. The second term in the binomial expansion has the value $\frac{1}{2}(v^2/c^2) = 5.0 \times 10^{-5}$, while the third term has the much smaller value $\frac{3}{8}(v^2/c^2)^2 = 3.8 \times 10^{-9}$. The additional terms are even smaller than the third term, so if $v \ll c$, we can neglect the third and additional terms in comparison with the first and second terms. Substituting the first two terms of the binomial expansion into Equation 28.6 gives

$$KE \approx mc^2 \left(1 + \frac{1}{2}\frac{v^2}{c^2} - 1\right) = \frac{1}{2}mv^2$$

which is the familiar form for the kinetic energy. However, Equation 28.6 gives the correct kinetic energy for all speeds and must be used for speeds near the speed of light, as in Example 7.

Example 7   A High-Speed Electron

An electron ($m = 9.109 \times 10^{-31}$ kg) is accelerated from rest to a speed of $v = 0.9995c$ in a particle accelerator. Determine the electron's (a) rest energy, (b) total energy, and (c) kinetic energy.

REASONING AND SOLUTION

(a) The electron's rest energy is

$$E_0 = mc^2 = (9.109 \times 10^{-31} \text{ kg})(2.998 \times 10^8 \text{ m/s})^2 = \boxed{8.187 \times 10^{-14} \text{ J}}$$

Energy is often expressed in units of electron volts (eV). Since 1 eV = 1.602×10^{-19} J, the electron's rest energy is

$$8.187 \times 10^{-14} \text{ J} \left(\frac{1 \text{ eV}}{1.602 \times 10^{-19} \text{ J}}\right) = \boxed{5.11 \times 10^5 \text{ eV} \quad \text{or} \quad 0.511 \text{ MeV}}$$

(b) The total energy of an electron traveling at a speed of $v = 0.9995c$ is

$$E = \frac{mc^2}{\sqrt{1 - \frac{v^2}{c^2}}} = \frac{(9.109 \times 10^{-31} \text{ kg})(2.998 \times 10^8 \text{ m/s})^2}{\sqrt{1 - \left(\frac{0.9995c}{c}\right)^2}}$$

$$= \boxed{2.59 \times 10^{-12} \text{ J} \quad \text{or} \quad 16.2 \text{ MeV}} \tag{28.4}$$

(c) The kinetic energy is the difference between the total energy and the rest energy:

$$KE = E - E_0 = 2.59 \times 10^{-12} \text{ J} - 8.2 \times 10^{-14} \text{ J}$$

$$= \boxed{2.51 \times 10^{-12} \text{ J} \quad \text{or} \quad 15.7 \text{ MeV}} \tag{28.6}$$

---

* The binomial expansion states that $(1 - x)^n = 1 - nx + n(n - 1)x^2/2 + \cdots$. In our case, $x = v^2/c^2$ and $n = -1/2$.

Since mass and energy are equivalent, any change in one is accompanied by a corresponding change in the other. For instance, life on earth is dependent on electromagnetic energy (light) from the sun. Because this energy is leaving the sun, there is a decrease in the sun's mass. Example 8 illustrates how to determine this decrease.

Example 8   The Sun Is Losing Mass

The sun radiates electromagnetic energy at the rate of 3.92×10^{26} W. (a) What is the change in the sun's mass during each second that it is radiating energy? (b) The mass of the sun is 1.99×10^{30} kg. What fraction of the sun's mass is lost during a human lifetime of 75 years?

REASONING   Since power is energy per unit time, the amount of electromagnetic energy radiated during each second is 3.92×10^{26} J. Thus, during each second, the sun's rest energy decreases by this amount. The change ΔE_0 in the sun's rest energy is related to the change Δm in its mass by $\Delta E_0 = (\Delta m)c^2$, according to Equation 28.5.

SOLUTION
(a) For each second that the sun radiates energy, the change in its mass is

$$\Delta m = \frac{\Delta E_0}{c^2} = \frac{3.92 \times 10^{26}\text{ J}}{(3.00 \times 10^8 \text{ m/s})^2} = \boxed{4.36 \times 10^9 \text{ kg}}$$

Over 4 billion kilograms of mass are lost by the sun during each second.

(b) The amount of mass lost by the sun in 75 years is

$$\Delta m = (4.36 \times 10^9 \text{ kg/s}) \left(\frac{3.16 \times 10^7 \text{ s}}{1 \text{ year}}\right)(75 \text{ years}) = 1.0 \times 10^{19} \text{ kg}$$

While this is an enormous amount of mass, it represents only a tiny fraction of the sun's mass:

$$\frac{\Delta m}{m_{sun}} = \frac{1.0 \times 10^{19} \text{ kg}}{1.99 \times 10^{30} \text{ kg}} = \boxed{5.0 \times 10^{-12}}$$

Any change in the rest energy of a system causes a change in the mass of the system according to $\Delta E_0 = (\Delta m)c^2$. It does not matter whether the change in energy is due to a change in electromagnetic energy, potential energy, thermal energy, or so on. While any change in energy gives rise to a change in mass, in most instances the change in mass is too small to be detected. For instance, if 4186 J of heat is used to raise the temperature of 1 kg of water by 1 C°, the mass changes by only $\Delta m = \Delta E_0/c^2 = (4186 \text{ J})/(3.00 \times 10^8 \text{ m/s})^2 = 4.7 \times 10^{-14}$ kg.

It is also possible to transform matter itself into other forms of energy, just as potential energy can be transformed into kinetic energy and vice versa. For example, the positron (see Section 31.4), created in high-energy accelerators, has the same mass as an electron, but an opposite electrical charge. If these two particles of matter collide, they are completely annihilated, and a burst of high-energy electromagnetic waves is produced. Thus, matter is transformed into electromagnetic waves, the energy of the electromagnetic waves being equal to the rest energies of the two colliding particles.

The transformation of electromagnetic waves into matter also happens. In one experiment, an extremely high-energy electromagnetic wave, called a gamma ray (see Section 31.4) passes close to the nucleus of an atom. If the gamma ray has sufficient energy, it can create an electron and a positron. The gamma ray disappears and the two particles of matter appear in its place. Except for picking up some momentum, the nearby nucleus remains unchanged. The process in which the gamma ray is transformed into two antiparticles is known as *pair production.*

THE SPEED OF LIGHT IS THE ULTIMATE SPEED

One of the important consequences of the theory of special relativity is that objects with mass cannot reach the speed of light. Thus, the speed of light represents the ultimate speed. To see that this speed limitation is a consequence of special relativity, consider Equation 28.6, which gives the kinetic energy of a moving object. As v approaches the speed of light c, the $\sqrt{1 - v^2/c^2}$ term in the denominator approaches zero. Hence, the kinetic energy becomes infinitely large. However, the work–energy theorem (Chapter 6) tells us that an infinite amount of work would have to be done to give the object an infinite kinetic energy. Since an infinite amount of work is not available, we are left with the conclusion that particles with mass cannot attain the speed of light.

28.7 THE RELATIVISTIC ADDITION OF VELOCITIES

The velocity of an object relative to an observer plays a central role in special relativity, for the effects on time, length, momentum, and energy depend on how fast the relative motion is, compared to the speed of light. To determine the velocity of an object relative to one or more observers, it is sometimes necessary to add two or more velocities together. For instance, Figure 28.8 illustrates a truck moving at a constant velocity of $v = 15$ m/s toward an observer standing on the earth. Suppose someone on the truck throws a baseball toward the observer at a velocity of $u' = 8.0$ m/s relative to the truck. We might conclude that the observer on earth sees the ball approaching at a velocity of $u = u' + v = 23$ m/s. Although this conclusion seems reasonable, careful measurements would show that this is

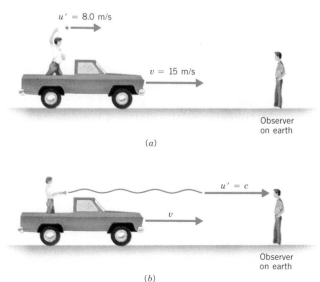

Figure 28.8   (*a*) The truck is approaching the earth-based observer at a relative velocity of $v = 15$ m/s. The velocity of the baseball relative to the truck is $u' = 8.0$ m/s. (*b*) The speed of the light emitted by the flashlight is c relative to both the truck and the observer on earth.

not quite right. The equation $u = u' + v$ is not valid, because if the velocity of the truck were close to the speed of light, the equation would predict that the observer on earth would see the baseball moving at a velocity greater than the speed of light. This is an impossibility, since no object with a finite mass can move faster than the speed of light.

For the case where the truck and ball are moving along the same direction, the theory of special relativity states that the velocities are related according to the *velocity-addition formula:*

$$\begin{bmatrix} \text{Velocity} \\ \text{addition} \end{bmatrix} \qquad u = \frac{u' + v}{1 + \dfrac{u'v}{c^2}} \qquad (28.7)$$

In this equation the symbols have the following meanings:

u = the velocity of the object as measured by the observer on the earth

u' = the velocity of the object measured by the person on the truck, which itself is moving at a velocity v relative to the earth

When the motion occurs along a straight line, the velocities in Equation 28.7 can have either positive or negative values, depending on whether they are directed along the positive or negative direction. For instance, in Figure 28.8a, $u' = 8.0$ m/s and $v = 15$ m/s, assuming that the direction to the right is positive. Equation 28.7 differs from the nonrelativistic formula ($u = u' + v$) by the presence of the $u'v/c^2$ term in the denominator. When u' and v are small compared to c, the $u'v/c^2$ term is small compared to 1, so the velocity-addition formula reduces to $u \approx u' + v$. However, when either u' or v is comparable to c, the results can be quite different, as Example 9 illustrates.

Example 9 The Relativistic Addition of Velocities

Imagine a hypothetical situation in which the truck in Figure 28.8a is approaching the observer on the earth at a relative velocity of $v = 0.8c$. A person riding on the truck throws a baseball toward the observer at a velocity of $u' = 0.5c$ relative to the truck. At what velocity does the observer on earth see the ball approaching?

REASONING The observer on earth does *not* see the baseball approaching at $u = 0.5c + 0.8c = 1.3c$. This cannot be, because the velocity of the ball would then exceed the speed of light. The velocity-addition formula gives the correct velocity, which is less than the speed of light.

SOLUTION The earth-based observer sees the ball approaching with a velocity of

$$u = \frac{u' + v}{1 + \dfrac{u'v}{c^2}} = \frac{0.5c + 0.8c}{1 + \dfrac{(0.5c)(0.8c)}{c^2}} = \frac{1.3c}{1 + 0.4} = \boxed{0.93c}$$

The velocity-addition formula is consistent with the speed of light postulate, which states that all observers in inertial reference frames measure the speed of light to be c. Consider Figure 28.8b, which shows the person riding on the truck and holding a flashlight. The speed of the light, as measured by this person, is

$u' = c$. According to the observer standing on the earth, the speed of this light is given by the velocity-addition formula as

$$u = \frac{u' + v}{1 + \dfrac{u'v}{c^2}} = \frac{c + v}{1 + \dfrac{cv}{c^2}} = \frac{(c + v)c}{(c + v)} = c$$

Thus, the velocity-addition formula indicates that the observer on earth and the person on the truck both measure the speed of light to be c, independent of the relative velocity v between them.

INTEGRATION OF CONCEPTS

SPECIAL RELATIVITY AND NEWTONIAN MECHANICS

Einstein's theory of special relativity reveals a number of startling results that conflict with our traditional ideas about space and time. Among the most famous revelations of special relativity are that moving clocks run slow, moving objects appear shortened, mass and energy are equivalent, and the speed of light is the ultimate speed for an object with mass. However surprising these predictions are, a large number of experiments are in complete agreement with the special theory of relativity, and today scientists accept it.

But what about our traditional ideas concerning space and time? We should remember that these ideas come from centuries of experience and experiments that support the concepts developed by Galileo, Newton, and others. These traditional concepts have been discussed in Part I of this text under the title of "Mechanics," or "Newtonian Mechanics," as it is often called. Newtonian mechanics, in contrast to the special theory of relativity, presumes that time and length are not different in different inertial reference frames. And Newtonian mechanics does not remotely hint that mass and energy are equivalent, or that the speed of light is the ultimate speed.

Is the theory provided by Newton and others wrong, then, because it fails to predict those things for which special relativity has become famous? Newtonian mechanics is certainly not wrong. It is just more limited in scope than is special relativity, which applies to all speeds between zero and the speed of light. The Newtonian view, in contrast, is valid only for speeds much smaller than the speed of light. In fact, in the limit of small speeds, both the Newtonian and the relativistic pictures of reality are in agreement. Thus, special relativity does not contradict the results of Newtonian mechanics, but generalizes them. Suppose, for example, that your car could attain speeds that are near the speed of light. As you accelerated toward the speed of light, you would observe the world around you change gradually. At the start, you would see the familiar Newtonian picture, where, for example, length is not contracted and time is not dilated. As you accelerated, this picture would change, until you saw a relativistic picture in which the lengths of objects speeding by the car are contracted and the moving clocks run slow.

SUMMARY

The special theory of relativity is based on two postulates. The **relativity postulate** states that the laws of physics are the same in every inertial reference frame. The **speed of light postulate** says that the speed of light in a vacuum, measured in any inertial reference frame, always has the same value of c, no matter how fast the source of light and the observer are moving relative to each other.

The **proper time interval** Δt_0 between two events is the time interval measured by an observer who is at rest relative to the events and views them occurring at the same location. A moving observer who does *not* see the two events occurring at the same location measures a dilated time interval Δt. The dilated time interval is greater than the proper time interval, according to the **time-dilation equation:** $\Delta t = \Delta t_0 / \sqrt{1 - v^2/c^2}$. In this expression, v is the relative speed between the observer who measures Δt_0 and the observer who measures Δt.

The **proper length** L_0 between two points is the length measured by an observer who is at rest relative to the points. An observer moving with a relative speed v parallel to the line between the two points does not measure the proper length. Instead, such an observer measures a contracted length L given by the **length-contraction formula:** $L = L_0 \sqrt{1 - v^2/c^2}$.

An object of mass m, moving with speed v, has a **relativistic momentum** given by $p = mv / \sqrt{1 - v^2/c^2}$.

Energy and mass are equivalent. The total energy E of an object of mass m, moving at speed v, is $E = mc^2 / \sqrt{1 - v^2/c^2}$. The total energy of an object is the sum of its rest energy, $E_0 = mc^2$, and its kinetic energy KE: $E = E_0 + \text{KE}$. The kinetic energy is, therefore, $\text{KE} = E - E_0$. The speed of an object with mass cannot equal the speed of light, which is the **ultimate speed** for such an object.

When an object is moving with respect to a reference frame that itself is moving relative to an observer, the **velocity-addition formula** (Equation 28.7) gives the velocity of the object as measured by the observer.

QUESTIONS

1. A baseball player at home plate hits a pop fly straight up (the beginning event) that is caught by the catcher at home plate (the ending event). Which of the following observers record the proper time interval between the two events: (a) a spectator sitting in the stands, (b) a spectator watching the game on TV, and (c) the third baseman running in to cover the play? Explain your answers.

2. Suppose you are standing at a railroad crossing, watching a train go by. (a) Both you and a passenger in the train are looking at a clock on the train. Which of you measures the proper time interval? (b) Who measures the proper length of the train car? (c) Who measures the proper distance between the railroad ties under the track? Justify your answers.

3. The speed limit on many interstate highways is 65 miles per hour. If the speed of light were 65 miles per hour, would you be able to drive at the speed limit? Give your reasoning.

4. There are tables that list data for the various particles of matter that physicists have discovered. Often, such tables list the masses of the particles in units of energy, such as in MeV (million electron volts), rather than in kilograms. Why is this possible?

5. Does a compressed spring with elastic potential energy have more mass than a noncompressed spring (assume the spring is not vibrating)? Explain.

6. Do two positive, electric charges separated by a finite distance have more mass than when they are infinitely far apart (assume the charges remain stationary)? Provide a reason for your answer.

7. A person is approaching you in a truck that is traveling very close to the speed of light. This person throws a baseball toward you. Relative to the truck, the ball is thrown with a speed nearly equal to the speed of light, so the person on the truck sees the baseball move away from the truck at a very high speed. Yet you see the baseball move away from the truck very slowly. Why? Use the velocity-addition formula to guide your thinking.

8. Which of the following quantities will two observers always measure to be the *same*, regardless of the relative velocity between the observers: (a) the time interval between two events; (b) the length of an object; (c) the speed of light; (d) the relative speed between the observers. In each case, give a reason for your answer.

9. If the speed of light were infinitely large instead of 3.0×10^8 m/s, would the effects of time dilation and length contraction be observable? Explain, using the equations presented in the text to support your reasoning.

PROBLEMS

Before doing any calculations involving time dilation and length contraction, it is useful to identify which observer measures the proper time interval Δt_0 or the proper length L_0.

Section 28.3 The Relativity of Time: Time Dilation

1. A law enforcement officer in an intergalactic "police car" turns on a red flashing light and sees it generate a flash every 1.5 s. A person on earth measures that the time between flashes is 2.5 s. How fast is the "police car" moving relative to the earth?

2. Suppose that you are on board a spacecraft moving toward the earth at a speed of $0.960c$. You have just finished exercising, and your heart is beating at a rate of 155 beats per minute. Your pulse rate is also being monitored by a clock on earth. What is your pulse rate according to the clock on earth?

3. In 1986, Kristin Otto set a world's record for the 100-m freestyle. Suppose that this race had been monitored from a spaceship traveling at a speed of $0.900c$ relative to the earth and that the space travelers measured the time interval of the race to be 125.6 s. What was the time recorded on earth?

4. A spacecraft is passing through the solar system at a speed of $0.850c$ relative to the earth. What does the captain measure for the number of hours in an earth day if the spacecraft is (a) moving toward the earth or (b) away from the earth?

*5. An astronaut travels at a speed of 7800 m/s relative to the earth, a speed that is very small compared to c. According to a clock on the earth, the trip lasts 15 days. Determine the *difference* (in seconds) between the time recorded by the earth clock and the astronaut's clock. *Hint: When $v \ll c$, the following approximation is valid:* $\sqrt{1 - v^2/c^2} \approx 1 - \tfrac{1}{2}(v^2/c^2)$.

*6. A 5.00-kg object oscillates back and forth at the end of a spring whose spring constant is 49.3 N/m. An observer is traveling at a speed of 2.80×10^8 m/s relative to the fixed end of the spring. What does this observer measure for the period of oscillation?

**7. A certain type of bacteria is known to double in number every 24.0 hours. Two cultures of these bacteria are prepared, each consisting initially of one bacterium. One culture is left on earth and the other placed on a rocket that travels at a speed of $0.866c$ relative to the earth. At a time when the earthbound culture has grown to 256 bacteria, how many bacteria are in the culture on the rocket?

Section 28.4 The Relativity of Length: Length Contraction

8. The mean distance between earth and Jupiter is 6.29×10^{11} m. How fast would you have to travel in a spacecraft so the distance has decreased to 2.00×10^{11} m?

9. A land speed record for a jet-propelled car was set by Craig Breedlove when his car attained an average speed of 274 m/s (613 mi/h) over a distance of 604 m. If the speed of light were 355 m/s, what distance would Breedlove have measured while driving the car?

10. Suppose the straight-line distance between New York and San Francisco is 4.2×10^6 m (neglecting the curvature of the earth). A UFO is flying between these two cities at a speed of $0.70c$ relative to the earth. What do the voyagers aboard the UFO measure for this distance?

11. Suppose you are traveling in space and pass a rectangular landing pad on a planet. Your spacecraft has a speed of $0.85c$ relative to the planet and moves in a direction parallel to the length of the pad. While moving, you measure the length to be 1800 m and the width to be 1500 m. What are the dimensions of the landing pad according to the engineer who built it?

*12. As the drawing shows, a carpenter on a space station has constructed a $30.0°$ ramp. A rocket moves past the space station with a relative speed of $0.850c$ in a direction parallel to side x. What does a person aboard the rocket measure for the angle of the ramp?

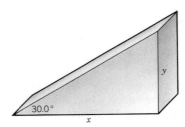

**13. A rectangle has the dimensions of 3.0 m × 2.0 m when viewed by someone at rest with respect to it. When you move past the rectangle along one of its sides, the rectangle looks like a square. What dimensions do you observe when you move at the same speed along the adjacent side of the rectangle?

Section 28.5 Relativistic Momentum

14. A small meteor, moving through the solar system at a speed of $0.70c$ relative to the earth, has a mass of 10.2 kg. What is the relativistic momentum of the meteor?

15. A car, whose mass is 1550 kg, is traveling at 15.0 m/s. If the speed of light were 25.0 m/s, what would be the momentum of the car as measured by a person standing on the ground?

16. At what speed is the relativistic momentum of a particle three times its nonrelativistic momentum?

*17. Starting from rest, two skaters "push off" against each other on smooth level ice, where friction is negligible. One is a woman and one is a man. The woman moves away with a velocity of $+2.5$ m/s. The mass of the woman is 54 kg, and the mass of the man is 88 kg. Assuming that the speed of light is 3.0 m/s, so that the relativistic momentum must be used, find the recoil velocity of the man. *(Hint: This problem is similar to Example 4 in Chapter 7.)*

Section 28.6 The Equivalence of Mass and Energy

18. Radium is a radioactive element whose nucleus emits an α particle (a helium nucleus) that has a kinetic energy of about 7.8×10^{-13} J (4.9 MeV). To what amount of mass is this energy equivalent?

19. The total amount of energy consumed in the United States during 1988 is estimated to have been about 8.4×10^{19} J. One penny has a mass of 2.9×10^{-3} kg. How many pennies have the equivalent of this amount of energy?

20. The amount of heat required to melt 1 kg of ice at 0 °C is 3.35×10^5 J. What is the difference between the mass of the water and that of the ice? Which has the greater mass?

21. A nuclear power reactor generates 3.0×10^9 W of power. In one year, what is the change in the mass of the nuclear fuel due to the energy being taken from the reactor?

22. An elementary particle called a pion has been observed to decay completely into electromagnetic radiation. The pion has a mass of 2.4×10^{-28} kg. (a) What is the kinetic energy of the pion at a speed of $0.850c$? (b) How much energy in the form of electromagnetic radiation is released when the high-speed pion decays?

***23.** In a TV picture tube, an electron is accelerated from rest through a potential difference of 2.40×10^4 V before striking the screen. (a) What is the kinetic energy (in joules) of the electron just before the electron hits the screen? (b) What is the speed of the electron?

Section 28.7 The Relativistic Addition of Velocities

24. A rocket ship is moving directly toward the earth with a velocity of $0.80c$ relative to the earth. The ship sends out a pulse of light that is aimed at the earth. What is the velocity that a person on earth sees for the approaching pulse?

25. An observer on the earth sees a spaceship approaching at a velocity of $0.50c$. The spacecraft then launches an exploration vehicle that, according to the earth observer, approaches at $0.70c$. What is the velocity of the exploration vehicle relative to the spaceship?

26. It has been proposed that spaceships of the future will be powered by ion propulsion engines. In one such engine the ions are to be ejected with a speed of $0.80c$ relative to the engine. If the ship were traveling away from the earth with a velocity of $0.70c$, what would be the velocity of the ions rela-

tive to the earth? (Be sure to assign the correct plus or minus signs to the velocities.)

***27.** An intergalactic cruiser has two types of guns: a photon cannon that fires a beam of laser light, and an ion gun that shoots atomic ions at a velocity of $0.950c$ relative to the cruiser. The cruiser closes in on an alien spacecraft at a velocity of $0.800c$ relative to this spacecraft. The captain fires both types of guns. At what velocity do the aliens see (a) the laser light and (b) the ions approach them? At what velocity do the aliens see (c) the laser light and (d) the ions move away from the cruiser?

***28.** A person on earth notices a rocket approaching from the right at a speed of $0.75c$ and another rocket approaching from the left at $0.65c$. What is the relative velocity between the two rockets, as measured by a passenger on one of them?

****29.** Two atomic particles approach each other in a head-on collision. Each particle has a mass of 2.16×10^{-25} kg and a speed of 2.40×10^8 m/s when measured by an observer standing in the laboratory. (a) What is the speed of one particle as seen by the other particle? (b) Determine the relativistic momentum of one particle, as would be observed by the other.

ADDITIONAL PROBLEMS

30. An electron and a positron each have a mass of 9.11×10^{-31} kg. They collide and annihilate each other, with only electromagnetic radiation appearing after the collision. If each particle is moving at a speed of $0.20c$ relative to the laboratory before the collision, determine the energy of the electromagnetic radiation. *(Hint: For each particle, use the expression for its total energy.)*

31. A particle known as a pion lives, on average, for a proper time of 2.6×10^{-8} s before breaking apart into other particles. How long does this particle live according to a laboratory observer if the particle moves past the observer at a speed of $0.67c$?

32. How fast must a meter stick be moving if its length is observed to shrink to one-half a meter?

33. A woman is 1.7 m tall and has a mass of 49 kg. She moves past an observer with the direction of the motion parallel to her height. The observer measures her relativistic momentum to be 3.0×10^{10} kg·m/s. What does the observer measure for her height?

34. How much work must be done on an electron to accelerate it from rest to a speed of $0.99c$?

***35.** A rocket is moving away from the earth with a speed of $0.75c$. An escape pod of length 45 m (as measured by the rocket crew) is launched from the rocket toward the earth with a speed of $0.55c$ relative to the rocket. What is the length of the escape pod as determined by an observer on earth?

*36. Four kilograms of water are heated from 20.0 °C to 60.0 °C. (a) How much heat is required to produce this change in temperature? [The specific heat capacity of water is 4186 J/(kg·C°).] (b) By how much does the mass of the water increase?

**37. Twins who are 19.0 years of age leave the earth and travel to a distant planet 12.0 light-years away. Assume the planet and earth are at rest with respect to each other. The twins depart at the same time on different spaceships. One twin travels at a speed of 0.900c, while the other twin travels at 0.500c. (a) According to the theory of special relativity, what is the difference between their ages when they meet again at the earliest possible time? (b) Which twin is older?

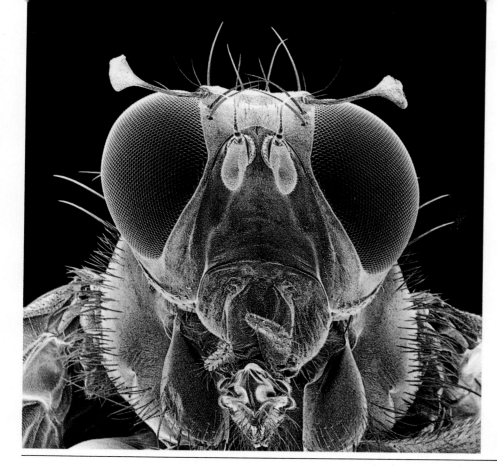

CHAPTER 29

PARTICLES AND WAVES

This highly magnified view of the head of a Mediterranean fruit fly was made with a scanning electron microscope (SEM). Unlike an optical microscope, which operates with light waves, an electron microscope uses electrons. The impressive resolution of fine detail that can be obtained in a SEM image is a result of the fact that particles of matter, such as the electron, behave as waves. The experimental and theoretical basis for this surprising behavior will be discussed in this chapter. We will find that it is even possible to determine a wavelength for a particle. For the electrons in an electron microscope, the wavelength is very small, and the high resolution of a SEM image is a direct result of the small wavelength. The wave nature of matter also plays the central role in one of the most unusual scientific principles, the Heisenberg uncertainty principle. As we will see, this principle places a limit on our knowledge of certain aspects of the physical world. We turn now to the wave nature of matter and the wave-particle duality.

29.1 THE WAVE–PARTICLE DUALITY

The ability to exhibit interference effects is an essential characteristic of waves. For instance, Section 27.2 discusses Young's famous experiment in which light passes through two closely spaced slits and produces a pattern of bright and dark

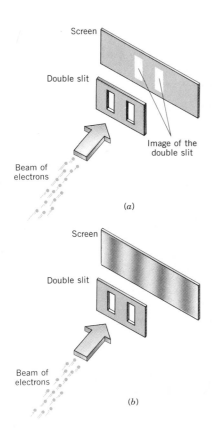

Figure 29.1 (*a*) If electrons behaved as discrete particles with no wave properties, they would pass through one or the other of the two slits and strike the screen, causing it to glow and produce exact images of the slits. (*b*) In reality, the screen reveals a pattern of bright and dark fringes, similar to the pattern produced when a beam of light is used and interference occurs between the light waves coming from each slit.

fringes on a screen (see Figure 27.4). The fringe pattern is a direct indication that interference is occurring between the light waves coming from each slit.

One of the most incredible discoveries of twentieth-century physics is that particles can also behave like waves and exhibit interference effects. For instance, Figure 29.1 shows a version of Young's experiment performed by directing *a beam of electrons* onto a double slit. In this experiment, the screen is like a television screen and glows wherever an electron strikes it. Part *a* of the picture indicates the pattern that would be seen on the screen if each electron, behaving strictly as a particle, were to pass through one slit or the other and strike the screen. The pattern would consist of an image of each slit. Part *b* shows the pattern actually observed, which consists of bright and dark fringes, reminiscent of that obtained when light waves pass through the double slit. The fringe pattern indicates that the electrons are exhibiting interference effects that are associated with waves.

But how can electrons behave like waves in the experiment shown in Figure 29.1*b*? And what kind of waves are they? The answers to these profound questions will be discussed later in this chapter. For the moment, we intend only to emphasize that the picture of an electron as a tiny discrete particle of matter does not account for the fact that the electron can behave as a wave in some circumstances. In other words, the electron exhibits a dual nature, with both particle-like characteristics and wave-like characteristics.

There is another interesting question: If a particle can exhibit wave-like properties, can waves exhibit particle-like behavior? As the next three sections reveal, the answer is yes. In fact, experiments that demonstrated the particle-like behavior of waves were performed near the beginning of the twentieth century, before the experiments that demonstrated the wave-like properties of the electron. In any event, scientists now accept the *wave–particle duality* as an essential part of nature: *Waves can exhibit particle-like characteristics and particles can exhibit wave-like characteristics.*

Section 29.2 begins the remarkable story of the wave-particle duality by discussing the electromagnetic waves that are radiated by a perfect blackbody. It is appropriate to begin with blackbody radiation, because it provided the first link in the chain of experimental evidence leading to our present understanding of the wave–particle duality.

29.2 BLACKBODY RADIATION AND PLANCK'S CONSTANT

All bodies, no matter how hot or cold, continuously radiate electromagnetic waves. For instance, we see the glow of very hot objects, because they emit electromagnetic waves in the visible region of the spectrum. A temperature of about 1700 K produces the white-hot appearance of the filament in an incandescent light bulb, while a temperature near 1000 K creates the characteristic cherry red color of burning charcoal. However, at relatively low temperatures, cooler objects emit visible light waves only weakly and, as a result, do not appear to be glowing. Certainly the human body, at only 310 K, does not emit enough visible light to be seen in the dark with the unaided eye. But the body does emit electromagnetic waves in the infrared region of the spectrum, and these can be detected with infrared sensitive detectors.

At a given temperature, the intensities of the electromagnetic waves emitted by an object vary from wavelength to wavelength throughout the visible, the infra-red, and other regions of the spectrum. Figure 29.2 illustrates how the intensity per unit wavelength depends on wavelength for a perfect blackbody emitter. As Section 13.3 discusses, a perfect blackbody at a constant temperature absorbs and reemits all the electromagnetic radiation that falls on it. The two curves in the drawing show that at a higher temperature the maximum emitted intensity in-creases and shifts toward shorter wavelengths. In accounting for the shape of these curves, the German physicist Max Planck (1858–1947) took the first step toward our present understanding of the wave–particle duality.

In 1900 Planck calculated the blackbody radiation curves, using a model that represents a blackbody as a large number of atomic oscillators, each of which emits and absorbs electromagnetic waves. To obtain agreement between the theoretical and experimental curves, Planck assumed that the energy E of an atomic oscillator could have only the discrete values of $E = 0$, hf, $2hf$, $3hf$, and so on. In other words, he assumed that

$$E = nhf \qquad n = 0, 1, 2, 3, \ldots \qquad (29.1)$$

where n is a positive integer, f is the frequency of vibration (in hertz), and h is a constant now called **Planck's constant**.* Experiment has shown that Planck's constant has a value of

$$h = 6.626\ 0755 \times 10^{-34}\ \text{J·s}$$

The radical feature of Planck's assumption was that the energy of an atomic oscillator could have only discrete values (hf, $2hf$, $3hf$, etc.), with energies in between these values being forbidden. Whenever the energy of a system can have only certain definite values, and nothing in between, the energy is said to be *quantized*. This quantization of the energy was unexpected on the basis of the traditional physics of the time. However, it was soon realized that energy quanti-zation had wide-ranging implications.

Conservation of energy requires that the energy carried off by the electromag-netic waves must equal the energy lost by the atomic oscillators. Suppose, for example, that an oscillator with an energy of $3hf$ emits an electromagnetic wave. According to Equation 29.1, the next smallest allowed value for the energy of the oscillator is $2hf$. In such a case, the energy carried off by the electromagnetic wave would have the value of hf, equaling the amount of energy lost by the oscillator. Thus, Planck's model for blackbody radiation sets the stage for the idea that electromagnetic energy occurs as a collection of discrete amounts or packets of energy, the energy of a packet being equal to hf. As the next section discusses, it was Einstein who made the specific proposal that light consists of such energy packets.

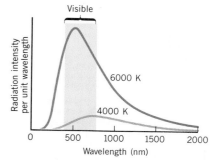

Figure 29.2 The electromagnetic radiation emitted by a perfect blackbody has an intensity per unit wavelength that varies from wave-length to wavelength, as each curve indicates. At the higher tempera-ture, the intensity per unit wave-length is greater and the maximum occurs at a shorter wavelength.

29.3 PHOTONS AND THE PHOTOELECTRIC EFFECT

Einstein proposed that light consists of energy packets in connection with a phenomenon called the *photoelectric effect*. Figure 29.3 illustrates the effect. If light with a sufficiently high frequency shines on a metal plate, electrons are

---

* It is now known that the energy of a harmonic oscillator is $E = (n + \frac{1}{2})hf$, the extra term of $\frac{1}{2}$ being unimportant to the present discussion.

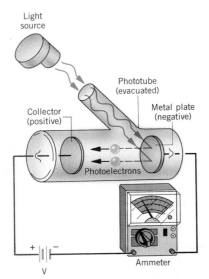

Figure 29.3 In the photoelectric effect, light shines on a metal surface, and if the frequency of the light is sufficiently high, electrons are ejected from the surface. These photoelectrons, as they are called, are drawn to the positive collector, thus producing a current.

emitted from the plate. The emitted electrons move toward a positive electrode called the collector and cause a current to register on the ammeter. Because the electrons are ejected with the aid of light, they are called *photoelectrons.* As will be discussed shortly, a number of features of the photoelectric effect could not be explained solely with the ideas of classical physics.

In 1905 Einstein presented an explanation of the photoelectric effect that took advantage of Planck's work concerning blackbody radiation. It was primarily for his theory of the photoelectric effect that he was awarded the Nobel prize in physics in 1921. In his photoelectric theory, Einstein proposed that light of frequency f could be regarded as a collection of discrete packets of energy, each packet containing an amount of energy E given by

$$\left[\begin{array}{l}\textbf{Energy of}\\ \textbf{a photon}\end{array}\right] \qquad\qquad E = hf \qquad\qquad (29.2)$$

where h is Planck's constant. Today these energy packets are called *photons.* The light energy given off by a light bulb, for instance, is carried by photons. The brighter the light shining on a given area, the greater is the number of photons per second that strike the area. Example 1 estimates the number of photons emitted per second by a typical light bulb.

Example 1   *Photons from a Light Bulb*

In converting electrical energy into light energy, a sixty-watt incandescent light bulb operates at about 2.1% efficiency. Assuming that all the light is green light (vacuum wavelength = 555 nm), determine the number of photons given off per second by the bulb.

REASONING   The number of photons emitted per second can be found by dividing the amount of light energy emitted per second by the energy E of one photon. The energy of a single photon is $E = hf$, according to Equation 29.2. The frequency f of the photon is related to its wavelength λ by Equation 16.1 as $f = c/\lambda$.

SOLUTION   At an efficiency of 2.1%, the number of joules of light energy emitted per second by a sixty-watt bulb is $(0.021)(60.0 \text{ J/s}) = 1.3 \text{ J/s}$. The energy of a single photon is

$$E = hf = \frac{hc}{\lambda} = \frac{(6.63 \times 10^{-34} \text{ J·s})(3.00 \times 10^{8} \text{ m/s})}{555 \times 10^{-9} \text{ m}} = 3.58 \times 10^{-19} \text{ J}$$

Therefore,

$$\begin{array}{l}\text{Number of}\\ \text{photons emitted} \\ \text{per second}\end{array} = \frac{1.3 \text{ J/s}}{3.58 \times 10^{-19} \text{ J/photon}} = \boxed{3.6 \times 10^{18} \text{ photons/s}}$$

According to Einstein, when light shines on a metal, a photon can give up its energy to an electron in the metal. If the photon has enough energy to do the work of removing the electron from the metal, the electron can be ejected. The work required depends on how strongly the electron is held. For the *least strongly* held electrons, the necessary work has a minimum value W_0 and is called the *work function* of the metal. If a photon has energy in excess of the work needed to remove an electron, the excess energy appears as kinetic energy of the ejected electron. Thus, the least strongly held electrons are ejected with the maximum kinetic energy KE_{max}. Einstein applied the conservation of energy principle and proposed the following relation to describe the photoelectric effect:

$$\underbrace{hf}_{\substack{\text{Photon}\\\text{energy}}} = \underbrace{KE_{max}}_{\substack{\text{Maximum}\\\text{kinetic energy}\\\text{of ejected}\\\text{electron}}} + \underbrace{W_0}_{\substack{\text{Minimum}\\\text{work needed to}\\\text{eject electron}}} \qquad (29.3)$$

According to this equation, $KE_{max} = hf - W_0$, which is plotted in Figure 29.4, with KE_{max} along the ordinate and f along the abscissa. The graph is a straight line that crosses the abscissa at $f = f_0$. At this frequency, the electron leaves the metal with no kinetic energy ($KE_{max} = 0$). According to Equation 29.3, when $KE_{max} = 0$ the energy hf_0 of the incident photon is equal to the work function W_0 of the metal: $hf_0 = W_0$.

The photon picture provides an explanation for a number of features of the photoelectric experiment that are difficult to explain without using the concept of photons. It is known, for instance, that only light with a frequency above a certain minimum value f_0 will eject electrons. If the frequency of the light is below this value, no electrons are ejected, regardless of how intense the light is. The next example illustrates how Einstein's theory accounts for this minimum value of the frequency.

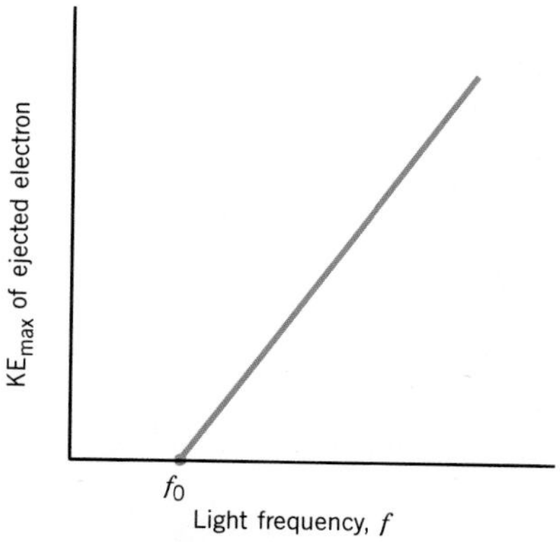

Figure 29.4 Photons of light can eject electrons from a metal when the light frequency is above a minimum value f_0. For frequencies above this minimum value, the ejected electrons have a maximum kinetic energy KE_{max} that is linearly related to the frequency of the light, as the graph shows.

Example 2   The Photoelectric Effect for a Silver Surface

The work function for a silver surface is $W_0 = 4.73$ eV. Find the minimum frequency that light must have to eject electrons from this surface.

REASONING   The minimum frequency f_0 is that frequency at which the photon energy equals the work function W_0 of the metal, so the electron is ejected with zero kinetic energy. Since 1 eV $= 1.60 \times 10^{-19}$ J, the work function expressed in joules is $W_0 = (4.73 \text{ eV})(1.60 \times 10^{-19} \text{ J/1 eV}) = 7.57 \times 10^{-19}$ J. Using Equation 29.3, we find that the minimum frequency needed to eject an electron from the metal is

$$hf_0 = \underbrace{KE_{max}}_{=0} + W_0 \quad \text{or} \quad f_0 = \frac{W_0}{h}$$

SOLUTION   The minimum frequency f_0 is

$$f_0 = \frac{W_0}{h} = \frac{7.57 \times 10^{-19} \text{ J}}{6.63 \times 10^{-34} \text{ J·s}} = \boxed{1.14 \times 10^{15} \text{ Hz}}$$

Photons with frequencies less than f_0 do not have enough energy to eject electrons from a silver surface. Since $\lambda_0 = c/f_0$, the wavelength of this light is $\lambda_0 = 263$ nm, which is in the ultraviolet region of the electromagnetic spectrum.

PROBLEM SOLVING INSIGHT

The work function of a metal is the minimum energy needed to eject an electron from the metal. An electron that has received this minimum energy has no kinetic energy once outside the metal.

Another significant feature of the photoelectric effect is that the maximum kinetic energy of the ejected electrons remains the same when the intensity of the light increases, provided the light frequency remains the same. As the light intensity increases, more photons per second strike the metal, and consequently more electrons per second are ejected. However, since the frequency is the same for each photon, the energy of each photon is also the same. Thus, the ejected electrons always have the same maximum kinetic energy.

Whereas the photon model of light explains the photoelectric effect satisfactorily, the electromagnetic wave picture of light does not. Certainly, it is possible to imagine that the electric field of an electromagnetic wave would cause electrons in the metal to oscillate and tear free from the surface when the amplitude of oscillation becomes large enough. However, were this the case, higher intensity light would eject electrons with a greater maximum kinetic energy, a fact that experiment does not confirm. Moreover, in the electromagnetic wave picture, a relatively long time would be required with low-intensity light before the electrons would build up a sufficiently large oscillation amplitude to tear free. Instead, experiment shows that even the weakest light intensity causes electrons to be ejected almost instantaneously, provided the frequency of the light is above the minimum value f_0. The failure of the electromagnetic wave picture to explain the photoelectric effect does not mean that the wave model should be abandoned. But we must recognize that the wave picture does not account for all the characteristics of light. The photon model also makes an important contribution to our understanding of the way light behaves when it interacts with matter.

Because a photon has energy, the photon can eject an electron from a metal surface when it interacts with the electron. However, a photon is different than a normal particle. A normal particle has a mass and can travel at speeds up to, but not equal to, the speed of light. A photon, on the other hand, travels at the speed of light in a vacuum and does not exist as an object at rest. The energy of a photon is entirely kinetic in nature, for it has no rest energy and no mass. To show that a photon has no mass, we rewrite Equation 28.4 for the total energy E as

$$E \sqrt{1 - \frac{v^2}{c^2}} = mc^2$$

The term $\sqrt{1 - (v^2/c^2)}$ is zero because a photon travels at the speed of light, $v = c$. Since the energy E of the photon is finite, the left side of the equation above is zero. Thus, the right side must also be zero, so $m = 0$ and the photon has no mass.

There are a number of interesting applications of the photoelectric effect. These applications depend on the fact that the moving photoelectrons in Figure 29.3 constitute a current, a current that changes as the intensity of the light changes. For example, one type of burglar alarm uses a beam of light that passes across a room before striking the metal surface within the phototube. Ultraviolet light is often used because it is invisible to the naked eye. When an intruder passes through the beam, the light intensity drops momentarily. The corresponding drop in the current of photoelectrons is sensed by electronic circuitry and activates an alarm.

Another application of the photoelectric effect occurs in motion pictures. The sound produced by most motion pictures is contained in the optical soundtrack of the film. The soundtrack is printed alongside the picture frames and consists of a pattern of light and dark regions, as Figure 29.5 illustrates. During the showing of a film, a beam of light in the projector passes through the soundtrack portion of the film and onto a phototube located behind the film, as in part b of the drawing. As the film moves past the beam, the light and dark regions of the soundtrack vary the intensity of the light reaching the phototube, producing a fluctuating current. The fluctuations in the current are a replica of the sound encoded on the soundtrack. The current is then sent to an amplifier, which drives the speakers.

THE PHYSICS OF . . .

a photoelectric burglar alarm.

THE PHYSICS OF . . .

an optical soundtrack in a motion picture.

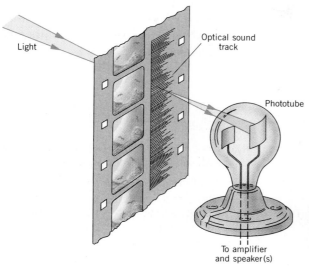

Figure 29.5 The optical sound-track is adjacent to the picture frames, and it varies the intensity of the light reaching the phototube.

29.4 THE MOMENTUM OF A PHOTON AND THE COMPTON EFFECT

Although Einstein presented his photon model for the photoelectric effect in 1905, it was not until 1923 that the photon picture began to achieve widespread acceptance. It was then that the American physicist Arthur H. Compton (1892–1962) used the photon model to explain his research on the scattering of X-rays by the electrons in graphite. X-rays are high-frequency electromagnetic waves and, like light, they are composed of photons.

Figure 29.6 illustrates what happens when an X-ray photon strikes an electron in a piece of graphite. Like two billiard balls colliding on a pool table, the X-ray photon scatters in one direction and the electron recoils in another direction after the collision. Compton observed that the scattered photon has a frequency f' that is smaller than the frequency f of the incident photon, indicating that the photon loses energy during the collision. In addition, he found that the difference between the two frequencies depends on the angle θ at which the scattered photon leaves the collision. The phenomenon in which an X-ray photon is scattered from an electron, the scattered photon having a smaller frequency than the incident photon, is called the *Compton effect.*

In Section 7.3 the collision between two objects is analyzed using the fact that the total kinetic energy and the total linear momentum of the objects are the same before and after the collision. Similar analysis can be applied to the collision between a photon and an electron. The electron is assumed to be initially at rest and essentially free, that is, not bound to the atoms of the material. According to the principle of conservation of energy,

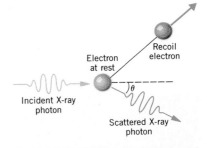

Figure 29.6 In an experiment performed by Arthur H. Compton, an X-ray photon collides with a stationary electron. The scattered photon and the recoil electron depart the collision in different directions.

$$\underbrace{hf}_{\substack{\text{Energy of} \\ \text{incident} \\ \text{photon}}} = \underbrace{hf'}_{\substack{\text{Energy of} \\ \text{scattered} \\ \text{photon}}} + \underbrace{\text{KE}}_{\substack{\text{Kinetic energy} \\ \text{of recoil} \\ \text{electron}}} \qquad (29.4)$$

where the relation $E = hf$ has been used for the photon energies. It follows, then,

844 CHAPTER 29/PARTICLES AND WAVES

that $hf' = hf - KE$, which shows that the energy and corresponding frequency f' of the scattered photon are less than the energy and frequency of the incident photon, just as Compton observed. Since $\lambda' = c/f'$, the wavelength of the scattered X-rays is larger than that of the incident X-rays.

For an initially stationary electron, conservation of total linear momentum requires that

$$\begin{array}{c}\text{Momentum of} \\ \text{incident photon}\end{array} = \begin{array}{c}\text{Momentum of} \\ \text{scattered photon}\end{array} + \begin{array}{c}\text{Momentum of} \\ \text{recoil electron}\end{array} \qquad (29.5)$$

To find an expression for the magnitude p of the photon's momentum, we use Equations 28.3 and 28.4. According to these equations, the momentum of any particle is $p = mv/\sqrt{1 - (v^2/c^2)}$ and its total energy is $E = mc^2/\sqrt{1 - (v^2/c^2)}$. Dividing these two equations, we find that $p/E = v/c^2$. Since a photon travels at the speed of light, $v = c$ and $p/E = 1/c$. Therefore, the momentum of a photon is $p = E/c$. But the energy of a photon is $E = hf$, while the wavelength is $\lambda = c/f$. Therefore, the magnitude of the momentum is

$$p = \frac{hf}{c} = \frac{h}{\lambda} \qquad (29.6)$$

Using Equations 29.4, 29.5, and 29.6, Compton showed that the difference between the wavelength λ' of the scattered photon and the wavelength λ of the incident photon is related to the scattering angle θ by

$$\lambda' - \lambda = \frac{h}{mc}(1 - \cos\theta) \qquad (29.7)$$

In this equation m is the mass of the electron. The quantity h/mc is referred to as the **Compton wavelength of the electron**, and has the value $h/mc = 2.43 \times 10^{-12}$ m. Since $\cos\theta$ varies between $+1$ and -1, the shift $\lambda' - \lambda$ in the wavelength can vary between zero and $2h/mc$, depending on the value of θ, a fact observed by Compton. (The reason why photons are more affected by scattering from electrons than from more massive particles, such as atoms or molecules, is the subject of question 5 at the end of this chapter.) The photoelectric effect and the Compton effect provided compelling evidence that electromagnetic waves can exhibit particle-like characteristics attributable to energy packets called photons.

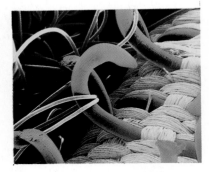

The scanning electron microscope takes advantage of the wave nature of the electron to produce highly magnified pictures containing great detail, such as this one of the common fastener Velcro. Velcro is a nylon material manufactured in two separate pieces, one with a hooked surface (right) and the other with a loop-covered surface (left). When the two surfaces are pressed together, the hooks catch in the loops to form a strong bond.

29.5 THE DE BROGLIE WAVELENGTH AND THE WAVE NATURE OF MATTER

As a graduate student in 1923, Louis de Broglie (1892–1987) made the astounding suggestion that since light waves could exhibit particle-like behavior, particles of matter should exhibit wave-like behavior. De Broglie proposed that the wavelength λ of a particle is given by the same relation (Equation 29.6) that applies to a photon:

$$\begin{bmatrix} \textbf{De Broglie} \\ \textbf{wavelength} \end{bmatrix} \qquad \lambda = \frac{h}{p} \qquad (29.8)$$

where h is Planck's constant and p is the magnitude of the relativistic momentum of the particle. Today, λ is known as the *de Broglie wavelength* of the particle.

Confirmation of de Broglie's suggestion came in 1927 from the experiments of the American physicists Clinton J. Davisson (1881–1958) and Lester H. Germer (1896–1971) and, independently, the English physicist George P. Thomson (1882–1975). Davisson and Germer directed a beam of electrons onto a crystal of nickel and observed that the electrons exhibited a diffraction behavior, analogous to that seen when X-rays are diffracted by a crystal (see Section 27.9 for a discussion of X-ray diffraction). The wavelength of the electrons revealed by the diffraction pattern matched that predicted by de Broglie's hypothesis, $\lambda = h/p$. More recently, Young's double-slit experiment has been performed with electrons, and they exhibit the effects of wave interference illustrated in Figure 29.1.

Particles other than electrons can also exhibit wave-like properties. For instance, neutrons are sometimes used in diffraction studies of crystal structure. Figure 29.7 compares the neutron diffraction pattern and the X-ray diffraction pattern caused by a crystal of rock salt (NaCl).

Although all moving particles have a de Broglie wavelength, the effects of this wavelength are observable only for particles whose masses are very small, on the order of the mass of an electron or a neutron, for instance. Example 3 illustrates why.

(a)

Example 3   The de Broglie Wavelength of an Electron and a Baseball

Determine the de Broglie wavelength for (a) an electron (mass $= 9.1 \times 10^{-31}$ kg) moving at a speed of 6.0×10^6 m/s and (b) a baseball (mass $= 0.15$ kg) moving at a speed of 13 m/s.

REASONING AND SOLUTION

(a) Since the speed of the electron is small compared with the speed of light, we can ignore relativistic effects, and the magnitude of the electron's momentum is the product of its mass and speed:

$$p = mv = (9.1 \times 10^{-31} \text{ kg})(6.0 \times 10^6 \text{ m/s}) = 5.5 \times 10^{-24} \text{ kg·m/s}$$

The de Broglie wavelength of the electron is

$$\lambda = \frac{h}{p} = \frac{6.63 \times 10^{-34} \text{ J·s}}{5.5 \times 10^{-24} \text{ kg·m/s}} = \boxed{1.2 \times 10^{-10} \text{ m}} \qquad (29.8)$$

A de Broglie wavelength of 1.2×10^{-10} m is about the size of the interatomic spacing in a solid, such as the nickel crystal used by Davisson and Germer, and, therefore, leads to the observed diffraction effects.

(b) Calculations similar to those in part (a) show that the magnitude of the baseball's momentum and its de Broglie wavelength are $p = 2.0$ kg·m/s and $\boxed{\lambda = 3.3 \times 10^{-34} \text{ m}}$. This wavelength is incredibly small, even by comparison with the size of an atom (10^{-10} m) or a nucleus (10^{-14} m). A wavelength of 3.3×10^{-34} m is so small that the wave characteristics of a baseball cannot be observed.

(b)

Figure 29.7 (*a*) The neutron diffraction pattern and (*b*) the X-ray diffraction pattern for a crystal of sodium chloride (NaCl).

The de Broglie equation for particle wavelength provides no hint as to what kind of wave is associated with a particle of matter. To gain some insight into the nature of this wave, we now turn our attention to Figure 29.8. This picture shows

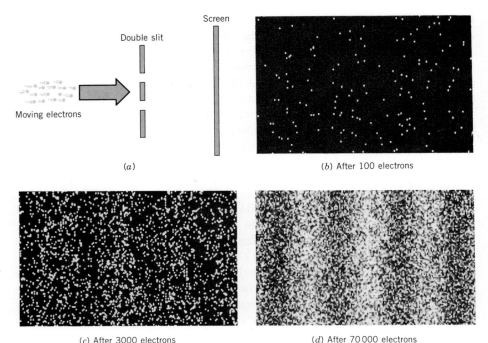

(a)

(b) After 100 electrons

(c) After 3000 electrons

(d) After 70 000 electrons

Figure 29.8 This electron version of Young's double-slit experiment was produced by Tonomura et al. The characteristic fringe pattern becomes recognizable only after a sufficient number of electrons have struck the screen. (*Source:* A. Tonomura, J. Endo, T. Matsuda, and T. Kawasaki, *Am. J. Phys.* 57(2): 117, Feb. 1989.)

how the fringe pattern emerges on the screen when electrons are used in a version of Young's double-slit experiment. The bright fringes occur in places on the screen where particle waves coming from each slit interfere constructively, while the dark fringes occur in places where the waves interfere destructively.

When an electron passes through the double-slit arrangement and strikes a spot on the screen, the screen glows at that spot, and Figure 29.8 illustrates how the spots accumulate in time. As more and more electrons strike the screen, the spots eventually form the fringe pattern that is evident in part *d* of the drawing. Bright fringes occur where there is a high probability of electrons striking the screen, and dark fringes occur where there is a low probability. Here lies the key to understanding particle waves. *Particle waves are waves of probability,* waves whose magnitude at a point in space gives an indication of the probability that the particle will be found at that point. At the place where the screen is located, the pattern of probabilities conveyed by the particle waves causes the fringe pattern to emerge. The fact that no fringe pattern is apparent in part *b* of the picture does not mean that there are no probability waves present; it just means that too few electrons have struck the screen for the fringe pattern to be recognizable.

The pattern of probabilities that leads to the fringes in Figure 29.8 is analogous to the pattern of light intensities that creates the fringes in Young's original experiment with light waves (see Figure 27.4). Section 24.4 discusses the fact that the intensity of the light is proportional to either the square of the electric field strength or the square of the magnetic field strength of the wave. In an analogous fashion in the case of particle waves, the probability is proportional to the square of the magnitude Ψ (Greek letter psi) of the wave. Ψ is referred to as the *wave function* of the particle.

In 1925 the Austrian physicist Erwin Schrödinger (1887–1961) and the German physicist Werner Heisenberg (1901–1976) independently developed theoretical frameworks for determining the wave function. In so doing, they established a new branch of physics called *quantum mechanics.* The word "quantum" refers to the fact that in the world of the atom, where particle waves must be considered, the particle energy is quantized, so only certain energies are allowed. To understand the structure of the atom and the phenomena related to it, quantum mechanics is essential, and the Schrödinger equation for calculating the wave function is now widely used. A discussion of the Schrödinger equation is beyond the scope of this text. But in the next chapter, we will explore the structure of the atom based on the ideas of quantum mechanics.

29.6 THE HEISENBERG UNCERTAINTY PRINCIPLE

As the previous section discusses, the bright fringes in Figure 29.8 indicate the places where there is a high probability of an electron striking the screen. And since there are a number of bright fringes, there is more than one place where each electron has some probability of hitting. Yet, any given electron can strike the screen in only one place after passing through the double slit. As a result, it is not possible to specify in advance exactly where on the screen an individual electron will fall. All we can do is speak of the probability that the electron may end up in a number of different places. No longer is it possible to say, as Newton's laws would suggest, that a single electron, fired through the double slit, will travel directly forward in a straight line and strike the screen. This simple picture just does not apply when a particle as small as an electron passes through a pair of closely spaced narrow slits. Because the wave nature of particles is important in such circumstances, we lose the ability to predict with 100% certainty the path that a single particle will follow. Instead, only the average behavior of large numbers of particles is predictable, and the behavior of any individual particle is uncertain.

To see more clearly into the nature of the uncertainty, consider electrons passing through the single slit in Figure 29.9. After a sufficient number of electrons strike the screen, a diffraction pattern emerges. The electron diffraction pattern consists of alternating bright and dark fringes and is analogous to that for light waves shown in Figure 27.22. Figure 29.9 shows the slit and locates the first

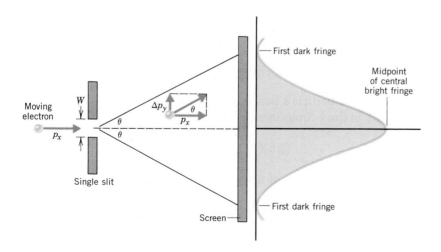

Figure 29.9 When a sufficient number of electrons pass through a single slit and strike the screen, a diffraction pattern of bright and dark fringes emerges (only the central bright fringe is shown). This pattern is due to the wave nature of the electrons and is analogous to that produced by light waves.

dark fringe on either side of the central bright fringe. The central fringe is bright because electrons strike the screen over the entire region between the dark fringes. If the electrons striking the screen outside the central bright fringe can be neglected, the extent to which the electrons are diffracted is given by the angle θ in the drawing. To reach locations within the central bright fringe, some electrons must have acquired momentum in the y direction, despite the fact that they enter the slit traveling along the x direction, and thus have no momentum in the y direction to start with. The figure illustrates that the y component of the momentum may be as large as Δp_y. The notation Δp_y indicates the difference between the maximum value of the y component of the momentum after the electron passes through the slit and its value of zero before the electron passes through the slit. Δp_y represents the *uncertainty* in the y component of the momentum, in that a diffracted electron may have any value from zero to Δp_y.

It is possible to relate Δp_y to the width W of the slit. To do this, we assume that Equation 27.4, which applies to light waves, also applies to particle waves whose de Broglie wavelength is λ. This equation, $\sin \theta = \lambda/W$, specifies the angle θ that locates the first dark fringe. If θ is small, $\sin \theta \approx \tan \theta$. Moreover, Figure 29.9 indicates that $\tan \theta = \Delta p_y/p_x$, where p_x is the x component of the momentum of the electron. Therefore, $\Delta p_y/p_x \approx \lambda/W$. But $p_x = h/\lambda$ according to de Broglie's equation, so that

$$\frac{\Delta p_y}{h/\lambda} \approx \frac{\lambda}{W}$$

As a result,

$$\Delta p_y \approx \frac{h}{W} \tag{29.9}$$

which indicates that a smaller slit width leads to a larger uncertainty in the y component of the electron's momentum.

It was Heisenberg who first suggested that the uncertainty Δp_y in the y component of the momentum is related to the uncertainty in the y position of the electron as the electron passes through the slit. Since the electron can pass through anywhere over the width W, the uncertainty in the y position of the electron is $\Delta y = W$. Substituting Δy for W in Equation 29.9 shows that $\Delta p_y \approx h/\Delta y$ or $(\Delta p_y)(\Delta y) \approx h$. The result of Heisenberg's more complete analysis is given below in Equation 29.10 and is known as the **Heisenberg uncertainty principle.**

The Heisenberg Uncertainty Principle

$$(\Delta p_y)(\Delta y) \geq \frac{h}{2\pi} \tag{29.10}$$

Δy = uncertainty in a particle's position along the y direction
Δp_y = uncertainty in the y component of the linear momentum of the particle

$$(\Delta E)(\Delta t) \geq \frac{h}{2\pi} \tag{29.11}$$

ΔE = uncertainty in the energy of a particle when the particle is in a certain state
Δt = time interval during which the particle is in the state

The Heisenberg uncertainty principle places limits on the accuracy with which the momentum and position of a particle can be simultaneously specified, and these limits are not just limits due to faulty measuring techniques. They are fundamental limits imposed by nature, in the sense that the second law of thermodynamics places a natural limit on the efficiency of a heat engine. There are no ways to circumvent such limits. Equation 29.10 indicates that Δp_y and Δy cannot both be arbitrarily small at the same time. If one is small, then the other must be large, so that their product equals or exceeds Planck's constant divided by 2π. For example, if the position of a particle is known exactly, so that Δy is zero, then Δp_y is an infinitely large number, and the momentum of the particle is completely uncertain. Conversely, if we assume that Δp_y is zero, then Δy is an infinitely large number, and the position of the particle is completely uncertain. In other words, the Heisenberg uncertainty principle states that it is impossible to specify precisely both the momentum and position of a particle at the same time.

There is also an uncertainty principle that deals with energy and time, as expressed by Equation 29.11. The product of the uncertainty ΔE in the energy of a particle and the time interval Δt during which the particle remains in a given energy state is greater than or equal to Planck's constant divided by 2π. Therefore, the shorter the lifetime of a particle in a given energy state, the greater is the uncertainty in the energy of that state.

Example 4 shows that the uncertainty principle has significant consequences for the motion of tiny particles such as electrons but has little effect on the motion of macroscopic objects, even those with as little mass as a Ping-Pong ball.

Example 4   The Heisenberg Uncertainty Principle

Assume that the position of an object is known so precisely that the uncertainty in the position is only $\Delta y = 1.5 \times 10^{-11}$ m. (a) Determine the minimum uncertainty in the momentum of the object. Find the corresponding minimum uncertainty in the speed of the object, if the object is (b) an electron (mass = 9.1×10^{-31} kg) and (c) a Ping-Pong ball (mass = 2.2×10^{-3} kg).

REASONING   The minimum uncertainty Δp_y in the y component of the momentum is given by the Heisenberg uncertainty principle as $\Delta p_y = h/(2\pi\, \Delta y)$, where Δy is the uncertainty in the position of the object. Both the electron and the Ping-Pong ball have the same uncertainty in their momenta, because they have the same uncertainty in their positions. However, these objects have very different masses. As a result, we will find that the uncertainty in the speeds of these objects is very different.

SOLUTION
(a) The minimum uncertainty in the y component of the momentum is

$$\Delta p_y = \frac{h}{2\pi\, \Delta y} = \frac{6.63 \times 10^{-34}\ \text{J·s}}{2\pi(1.5 \times 10^{-11}\ \text{m})} = \boxed{7.0 \times 10^{-24}\ \text{kg·m/s}} \qquad (29.10)$$

(b) Since $\Delta p_y = m\, \Delta v_y$, the minimum uncertainty in the speed of the electron is

$$\Delta v_y = \frac{\Delta p_y}{m} = \frac{7.0 \times 10^{-24}\ \text{kg·m/s}}{9.1 \times 10^{-31}\ \text{kg}} = \boxed{7.7 \times 10^{6}\ \text{m/s}}$$

Thus, the small uncertainty in the y position of the electron gives rise to a large uncertainty in the speed of the electron.

PROBLEM SOLVING INSIGHT

The Heisenberg uncertainty principle states that the product of Δp_y and Δy is greater than or equal to $h/2\pi$. The minimum uncertainty occurs when the product is equal to $h/2\pi$.

(c) The uncertainty in the speed of the Ping-Pong ball is

$$\Delta v_y = \frac{\Delta p_y}{m} = \frac{7.0 \times 10^{-24} \text{ kg·m/s}}{2.2 \times 10^{-3} \text{ kg}} = \boxed{3.2 \times 10^{-21} \text{ m/s}}$$

Because the mass of the Ping-Pong ball is relatively large compared to that of the electron, the uncertainty in the speed of the ball is unobservable.

INTEGRATION OF CONCEPTS

PHOTONS AND THE CONSERVATION PRINCIPLES FOR ENERGY AND MOMENTUM

The conservation of energy and the conservation of momentum are two of the most firmly established principles in physics. A large body of experimental evidence, accumulated over many years, indicates that these principles are fundamental ones. The word "fundamental" means that other phenomena and concepts must obey these principles. If a newly observed phenomena or proposed idea seems to conflict with energy or momentum conservation, scientists immediately suspect that an important feature of the story has been left out. In this sense, then, these conservation principles serve as a test that new ideas must pass. The photon picture of light, for instance, indicates that light consists of packets of energy. Each photon has a discrete energy given by the product of Planck's constant and the frequency of the light. And each photon has a momentum given by Planck's constant divided by the wavelength of the light. This picture was not widely accepted until Einstein and Compton showed that these relationships for photon energy and momentum could be used with the principles of energy and momentum conservation to explain the photoelectric effect and the Compton effect. Scientists now routinely incorporate photon energy and momentum when they apply the conservation principles for energy and momentum. Thus, the newer physics of the twentieth century combines with the older classical physics to give a more complete view of the physical world.

SUMMARY

The **wave–particle duality** refers to the fact that a wave can exhibit particle-like characteristics and a particle can exhibit wave-like characteristics.

At a constant temperature, a perfect blackbody absorbs and reemits all the electromagnetic radiation that falls on it. Max Planck calculated the emitted radiation intensity per unit wavelength as a function of wavelength. In his theory, Planck assumed that a blackbody consists of atomic oscillators that can have only quantized energies. Planck's quantized energies are given by $E = nhf$, where $n = 0, 1, 2, 3, \ldots$, h is **Planck's con-** stant (6.63×10^{-34} J·s), and f is the vibration frequency.

All electromagnetic radiation consists of **photons,** which are packets of energy. The energy of a photon is $E = hf$, where h is Planck's constant and f is the frequency of the light. A photon in a vacuum always travels at the speed of light c and has no mass. The **photoelectric effect** is the phenomenon in which light shining on a metal surface causes electrons to be ejected from the surface. The **work function** W_0 of a metal is the minimum work that must be done to eject an electron from the metal. In accordance with the conservation of energy, the electrons ejected from a metal have a maximum kinetic energy KE_{max} that is related to the energy hf of the incident photon by $hf = KE_{max} + W_0$.

The **Compton effect** is the scattering of a photon by an electron in a material, the scattered photon having a smaller frequency than the incident photon. The difference between the wavelength λ' of the scattered photon and the wavelength λ of the incident photon is related to the scattering angle θ by $\lambda' - \lambda = (h/mc)(1 - \cos\theta)$, where m is the mass of the electron and the quantity h/mc is known as the Compton wavelength of the electron.

The **de Broglie wavelength** of a particle is $\lambda = h/p$, where p is the magnitude of the relativistic momentum of the particle. Because of its de Broglie wavelength, a particle can exhibit wave-like characteristics. The wave associated with a particle is a wave of probability.

The **Heisenberg uncertainty principle** places limits on our knowledge about the behavior of a particle. The uncertainty principle indicates that $(\Delta p_y)(\Delta y) \geq h/2\pi$, where Δy and Δp_y are, respectively, the uncertainties in the position and momentum of the particle. The uncertainty principle also states that $(\Delta E)(\Delta t) \geq h/2\pi$, where ΔE is the uncertainty in the energy of a particle when the particle is in a certain state and Δt is the time interval during which the particle is in the state.

QUESTIONS

1. Radiation of a given wavelength causes electrons to be emitted from the surface of one metal but not from the surface of another metal. Explain why this could be.

2. Which of the colored lights (red, orange, yellow, green, or blue) on a Christmas tree emits photons with (a) the least energy and (b) the greatest energy? Account for your answers.

3. When a sufficient number of visible light photons strike a piece of photographic film, the film becomes exposed. An X-ray photon is more energetic than a visible light photon. Yet, most photographic films are not exposed by the X-ray machines used at airport security checkpoints. Explain what these observations imply about the number of photons emitted by the X-ray machines.

4. In a Compton scattering experiment, an electron is accelerated straight ahead in the same direction as that of the incident X-ray photon. Which way does the scattered photon move? Explain your reasoning, using the principle of conservation of momentum.

5. Photons can undergo Compton scattering from a molecule such as nitrogen, just as they do from an electron. However, the change in photon wavelength is much less than when an electron is scattered. Explain why, using Equation 29.7 for a nitrogen molecule instead of an electron.

6. In Section 14.3 the impulse-momentum theorem is used to analyze how gas molecules exert a force on a wall. (a) Use similar reasoning to discuss why a beam of light exerts a force on a surface it strikes. (b) Do you think a beam of light exerts more force on a mirror that reflects the light or on a black surface that absorbs the light? In your discussion assume the beam is perpendicular to both the mirror and the black surface.

7. A stone is dropped from the top of a building. Explain what happens to the de Broglie wavelength of the stone as the stone falls.

8. A bullet leaving the barrel of a gun is analogous to an electron passing through the single slit in Figure 29.9. With this analogy in mind, explain whether the uncertainty principle is likely to have any effect on your success as a hunter.

PROBLEMS

In working these problems, ignore relativistic effects.

Section 29.3 Photons and the Photoelectric Effect

1. Ultraviolet light is responsible for sun tanning. Find the wavelength of an ultraviolet photon whose energy is 6.4×10^{-19} J.

2. The wavelengths (in vacuum) of visible light occur between 380 and 750 nm. Determine the range of photon energies (in joules) to which this range of wavelengths corresponds.

3. An FM radio station broadcasts at a frequency of 98.1 MHz. The power radiated from the antenna is 5.0×10^4 W. How many photons per second does the antenna emit?

4. The work function for a sodium surface is 2.28 eV. What is the maximum wavelength that an electromagnetic wave can have and still eject electrons from this surface?

5. Light is shining perpendicularly on the surface of the earth with an intensity of 680 W/m². Assuming all the photons in the light have a wavelength of 730 nm, determine the number of photons per second per square meter that reach the earth.

6. A magnesium surface has a work function of 3.68 eV. Electromagnetic waves with a wavelength of 215 nm strike the surface and eject electrons. Find the maximum kinetic energy of the ejected electrons. Express your answer in electron volts.

7. An AM radio station broadcasts an electromagnetic wave at a frequency of 665 kHz, while an FM station broadcasts at 91.9 MHz. How many AM photons are needed to have a total energy equal to that of one FM photon?

*8. An owl has good night vision because its eyes can detect a light intensity as small as 5.0×10^{-13} W/m². What is the minimum number of photons per second that an owl eye can detect if its pupil has a diameter of 8.5 mm and the light has a wavelength of 510 nm?

*9. Radiation with a wavelength of 281 nm shines on a metal surface and ejects electrons that have a maximum speed of 3.48×10^5 m/s. Which one of the following metals is present, the values in parentheses being the work functions: potassium (2.24 eV), calcium (2.71 eV), uranium (3.63 eV), aluminum (4.08 eV), and gold (4.82 eV)?

*10. The maximum wavelength for which an electromagnetic wave can eject electrons from a platinum surface is 196 nm. When radiation with a wavelength of 141 nm shines on the surface, what is the maximum speed of the ejected electrons?

*11. At night, approximately 530 photons per second must enter an unaided human eye for an object to be seen, assuming the light is green. The light bulb in Example 1 in the text emits green light uniformly in all directions and the diameter of the pupil of the eye is 7.0 mm. What is the maximum distance from which the bulb could be seen?

**12. (a) How many photons (wavelength = 620 nm) must be absorbed to melt a 2.0-kg block of ice at 0 °C into water at 0 °C? (b) On the average, how many H_2O molecules does one photon convert from the ice phase to the water phase?

**13. A laser emits 1.30×10^{18} photons per second in a beam of light that has a diameter of 2.00 mm and a wavelength of 514.5 nm. Determine (a) the average electric field strength and (b) the average magnetic field strength for the electromagnetic wave that constitutes the beam.

Section 29.4 The Momentum of a Photon and the Compton Effect

14. The microwaves used in a microwave oven have a wavelength of about 0.13 m. What is the momentum of a microwave photon?

15. A photon has the same momentum as an electron moving with a speed of 2.0×10^5 m/s. What is the wavelength of the photon?

16. Determine the *change* in the photon's wavelength that occurs when an electron scatters an X-ray photon (a) straight back at an angle of $\theta = 180.0°$ and (b) at an angle of $\theta = 30.0°$. All angles are measured as in Figure 29.6.

17. In a Compton scattering experiment, the incident X-rays have a wavelength of 0.2685 nm, while the scattered X-rays have a wavelength of 0.2702 nm. At what angle θ in Figure 29.6 are the X-rays scattered?

*18. The X-rays detected at a scattering angle of $\theta = 163°$ in Figure 29.6 have a wavelength of 0.1867 nm. Find (a) the wavelength of an incident photon, (b) the energy of an incident photon, (c) the energy of a scattered photon, and (d) the kinetic energy of the recoil electron. (For accuracy, use $h = 6.626 \times 10^{-34}$ J·s and $c = 2.998 \times 10^8$ m/s.)

Section 29.5 The de Broglie Wavelength and the Wave Nature of Matter

19. A honeybee (mass = 1.3×10^{-4} kg) is crawling at a speed of 0.020 m/s. What is the de Broglie wavelength of the bee?

20. A particle has a speed of 1.2×10^6 m/s. Its de Broglie wavelength is 8.4×10^{-14} m. What is the mass of the particle?

21. The de Broglie wavelength of a proton in a particle accelerator is 1.30×10^{-14} m. Determine the kinetic energy of the proton.

22. How fast does a proton have to be moving to have the same de Broglie wavelength as an electron does when the electron moves at 4.5×10^6 m/s?

23. Recall from Section 14.3 that the average kinetic energy of an atom in an ideal gas is given by $\overline{KE} = \frac{3}{2}kT$, where $k = 1.38 \times 10^{-23}$ J/K and T is the Kelvin temperature of the gas. Determine the de Broglie wavelength of a helium atom (mass = 6.65×10^{-27} kg) that has the average kinetic energy at room temperature (293 K).

*24. In a Young's double-slit experiment performed with electrons, the two slits are separated by a distance of 2.0×10^{-6} m. The first-order bright fringes are located on the observation screen at a position given by $\theta = 1.6 \times 10^{-4}$ degrees in Equation 27.1. Find (a) the wavelength, (b) the momentum, and (c) the kinetic energy of the electrons.

*25. In a television picture tube, electrons are accelerated from rest through a potential difference of 21 000 V. What is the de Broglie wavelength of the electrons?

**26. The kinetic energy of a particle is equal to the energy of a photon. The particle moves at 5.0% of the speed of light. Find the ratio of the photon wavelength to the de Broglie wavelength of the particle.

Section 29.6 The Heisenberg Uncertainty Principle

27. The speed of a golf ball (mass = 0.045 kg) and of an electron is 71 m/s. If the uncertainty in the speed is 1.0%, estimate the minimum uncertainty in the position of each object.

28. An electron is trapped within a sphere whose diameter is 2.0×10^{-15} m. What is the minimum uncertainty in the electron's momentum?

29. A prisoner (mass = 75 kg) paces back and forth in his cell, and the uncertainty in his speed is 0.10 m/s. (a) Use the uncertainty principle to estimate the minimum uncertainty in his position. (b) Repeat part (a), assuming Planck's constant has a value of 663 J·s instead of 6.63×10^{-34} J·s.

30. Suppose the minimum uncertainty in the position of a particle is equal to its de Broglie wavelength. If the particle has an average speed of 4.5×10^5 m/s, what is the minimum uncertainty in its speed?

31. When electrons pass through a single slit, as in Figure 29.9, they form a diffraction pattern. As Section 29.6 discusses, the central bright fringe extends to either side of the midpoint, according to an angle θ given by $\sin \theta = \lambda/W$, where λ is the de Broglie wavelength of the electron and W is the width of the slit. When λ is the same size as W, $\theta = 90°$, and the central fringe fills the entire observation screen. In this case, an electron passing through the slit has roughly the same probability of hitting the screen either straight ahead or anywhere off to one side or the other. Now, imagine yourself in a world where Planck's constant is large enough so you exhibit similar effects when you walk through a 0.90-m-wide doorway. If your mass is 82 kg and you walk at a speed of 0.50 m/s, how large would Planck's constant have to be in this hypothetical world?

ADDITIONAL PROBLEMS

32. What is the wavelength of (a) a 1.0-eV photon and (b) a 1.0-eV electron?

33. Two photons have energies of 3.3×10^{-16} J and 1.3×10^{-20} J. Using Figure 24.6, identify the appropriate region in the electromagnetic spectrum for each of these photons.

34. As Section 27.5 discusses, sound waves diffract or bend around the edges of a doorway. Larger wavelengths diffract more than smaller wavelengths. (a) The speed of sound is 343 m/s. With what speed would a 55.0-kg person have to move through a doorway to diffract to the same extent as a 128-Hz bass tone? (b) At the speed calculated in part (a), how long (in years) would it take the person to move a distance of one meter?

35. Incident X-rays have a wavelength of 0.3365 nm and are scattered by the "free" electrons in carbon. The scattering angle in Figure 29.6 is $\theta = 125°$. What is the magnitude of the momentum of (a) the incident photon and (b) the scattered photon? (For accuracy, use $h = 6.626 \times 10^{-34}$ J·s and $c = 2.998 \times 10^8$ m/s.)

36. Radiation of a certain wavelength causes electrons with a maximum kinetic energy of 0.68 eV to be ejected from a metal whose work function is 2.75 eV. What will be the maximum kinetic energy (in eV) with which this same radiation ejects electrons from another metal whose work function is 2.17 eV?

*37. The width of the central bright fringe in a diffraction pattern on a screen is identical when either electrons or red light (vacuum wavelength = 661 nm) pass through a single slit. The distance between the screen and the slit is the same in each case and is large compared to the slit width. (a) How fast are the electrons moving? (b) To judge whether the speed in (a) is fast or slow for an electron, determine the speed acquired by an electron in accelerating from rest through a potential difference of one volt.

*38. Example 1 in the text calculates the number of photons per second given off by a sixty-watt incandescent light bulb. The photons are emitted uniformly in all directions. From a distance of 3.1 m you glance at this bulb for 0.10 s. The light from the bulb travels directly to your eye and does not reflect from anything. The pupil of the eye has a diameter of 2.0 mm. How many photons enter your eye?

*39. An electron and a proton have the same kinetic energy. Determine the ratio of the de Broglie wavelength of the electron to that of the proton.

**40. A beam of visible light has a wavelength of 395 nm and shines perpendicularly on a surface. As a result, there are 3.0×10^{18} photons per second striking the surface. By using the impulse-momentum theorem (Section 7.1), obtain the average force that this beam applies to the surface when (a) the surface is a mirror, so the momentum of each photon is reversed after reflection, and (b) the surface is black, so each photon is absorbed and the momentum of the photon is reduced to zero in the process.

CHAPTER 30

THE NATURE OF THE ATOM

The idea that all matter is composed of atoms is fundamental to our modern view of the world. Compared to the recorded history of man, the atomic viewpoint is a relative newcomer, for only in the twentieth century has the existence of the atom become universally accepted. Atomic theory has proved successful because it has explained phenomena that cannot otherwise be understood, and it has given us a firm basis for understanding the properties of solids, liquids, and gases. This understanding has led to a host of useful devices, such as transistors, lasers, and superconducting magnets. For example, in the rock concert above, the laser beams arise because atoms generate light as they undergo transitions from one atomic energy level to another. Our venture into the atomic world begins with the concept of the nuclear atom and continues with the first successful model of the hydrogen atom, the Bohr model. This model introduces many basic atomic features, including the notions of discrete energy states and quantum numbers. Next we will see that quantum mechanics has displaced the Bohr model and provides a more complete description of the atom. Following an overview of quantum mechanics, the Pauli exclusion principle will be discussed. This principle is important for explaining how electrons are arranged in complex atoms and how the elements are ordered in the periodic table. We will then apply our knowledge of atomic structure to describe how X-rays and laser light are produced.

30.1 RUTHERFORD SCATTERING AND THE NUCLEAR ATOM

An atom contains a small, positively charged nucleus (radius $\approx 10^{-15}$ m), which is surrounded at relatively large distances (radius $\approx 10^{-10}$ m) by a number of electrons, as Figure 30.1a illustrates. In the natural state, an atom is electrically

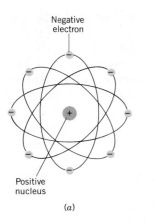

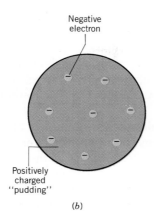

Negative
electron

Positive
nucleus

(a)

Negative
electron

Positively
charged
"pudding"

(b)

Figure 30.1   (*a*) The nuclear atom.
(*b*) The "plum pudding" model of
the atom (now discredited).

neutral, because the nucleus contains a number of protons (each with a charge of
+*e*) that equals the number of electrons (each with a charge of −*e*). This model of
the atom is universally accepted now and is referred to as the "nuclear atom."

The nuclear atom is a relatively recent idea. In the early part of the twentieth
century a widely accepted model, due to the English physicist Joseph J. Thomson
(1856–1940), pictured the atom very differently. In Thomson's view there was no
nucleus at the center of an atom. Instead, the positive charge was assumed to be
spread throughout the atom, forming a kind of "paste" or "pudding," in which
the negative electrons were suspended like "plums." Figure 30.1 compares this
"plum-pudding" model with the currently accepted view of the atom.

The "plum-pudding" model was discredited in 1911 when the New Zealand
physicist Ernest Rutherford (1871–1937) published experimental results that the
model could not explain. As Figure 30.2 indicates, Rutherford and his co-workers
directed a beam of alpha particles (α particles) at a thin metal foil made of gold.
Alpha particles are positively charged particles (the nuclei of helium atoms,
although this was not recognized at the time) emitted by some radioactive materials. If the "plum-pudding" model were correct, the α particles would be expected
to pass nearly straight through the foil. After all, there is nothing in this model to
deflect the relatively massive α particles, since the electrons have a comparatively
small mass and the positive charge is spread out in a diluted "pudding." Using a
zinc sulfide screen, which flashed briefly when struck by an α particle, Ruther-

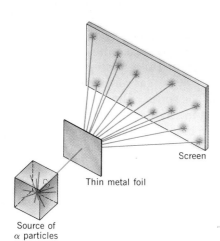

Screen

Thin metal foil

Source of
α particles

Figure 30.2   An illustration of a
Rutherford scattering experiment in
which α particles are scattered by
a thin metal foil. The entire apparatus is placed in a vacuum
chamber (not shown).

ford and co-workers were able to determine that not all the α particles passed straight through the foil. Instead, some were deflected at large angles, even backward. Rutherford himself said, "It was almost as incredible as if you had fired a fifteen inch shell at a piece of tissue and it came back and hit you." Rutherford concluded that the positive charge, instead of being distributed thinly and uniformly throughout the atom, was concentrated in a small region called the nucleus.

But how could the electrons in a nuclear atom remain separated from the positively charged nucleus? If the electrons were stationary, they would be pulled inward by the attractive electric force of the nuclear charge. Therefore, it was realized that the electrons had to be moving around the nucleus in some fashion, like the planets revolving around the sun. Such a planetary model of the atom has its own difficulties, however. For instance, an electron moving on a curved path has a centripetal acceleration, as Section 5.2 discusses. And when an electron is accelerating, it radiates electromagnetic waves. The difficulty is that the waves carry away energy, which decreases the energy of the electrons. With their energy constantly being depleted, the electrons would spiral inward and eventually collapse into the nucleus. Since matter is stable, such a collapse does not occur. Thus, the planetary model, while providing a more realistic picture of the atom than the "plum-pudding" model, must be telling only part of the story. The full story of atomic structure is fascinating, and the next section describes another aspect of it.

30.2 LINE SPECTRA

We have seen in Sections 13.3 and 29.2 that all objects emit electromagnetic waves, and we will see in Section 30.3 how this radiation arises. For a solid object, such as the hot filament of a light bulb, these waves have a continuous range of wavelengths, some of which are in the visible region of the spectrum. The continuous range of wavelengths is characteristic of the entire collection of atoms that make up the solid. In contrast, individual atoms, free from the strong interatomic interactions that are present in a solid, emit only certain specific wavelengths, rather than a continuous range. These wavelengths are characteristic of the atom and provide important clues about its structure. To study the behavior of individual atoms, low-pressure gases are used, in which the atoms are relatively far apart.

A low-pressure gas contained within a sealed tube can be made to emit electromagnetic waves by applying a sufficiently large potential difference between two electrodes located within the tube. With a grating spectroscope like that in Figure 27.32, the individual wavelengths emitted by the gas can be separated and identified as a series of bright fringes or lines. The series of lines is called a *line spectrum*. The simplest line spectrum is that of the hydrogen atom (H).* Figure 30.3 shows the visible part of the line spectrum of atomic hydrogen, along with the visible parts of the line spectra of more complicated atoms such as neon and mercury. The specific visible wavelengths emitted by neon and mercury are

THE PHYSICS OF . . .

neon signs and mercury vapor street lamps.

* Molecular hydrogen (H_2) consists of two hydrogen atoms bonded together and has a more complicated line spectrum than that of atomic hydrogen.

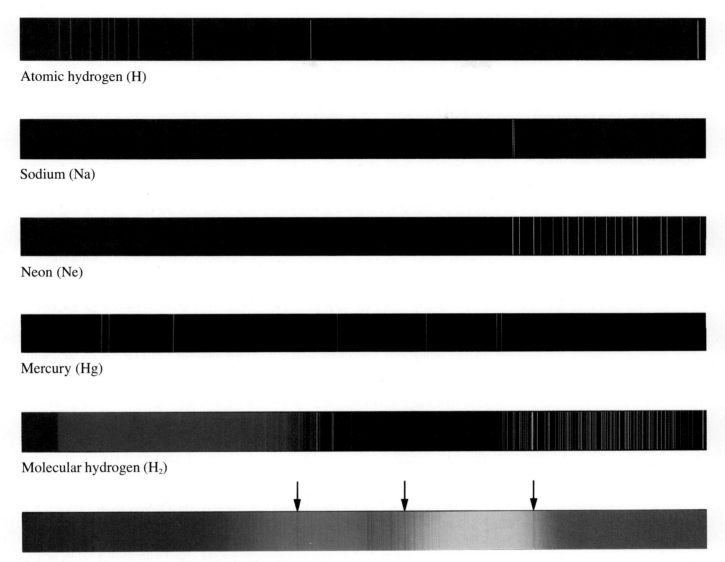

Atomic hydrogen (H)

Sodium (Na)

Neon (Ne)

Mercury (Hg)

Molecular hydrogen (H₂)

Solar absorption spectrum (Fraunhofer lines)

Figure 30.3 Line spectra for various atoms and molecules, along with the continuous spectrum of the sun. The dark lines in the sun's spectrum are called Fraunhofer lines, three of which are marked by arrows.

familiar, because they give neon signs and mercury vapor street lamps their characteristic colors.

Much effort has been devoted to understanding the pattern of wavelengths observed in the line spectrum of atomic hydrogen. In addition to the series of lines found in the visible region, analogous series have been found in shorter and longer wavelength regions (nonvisible) of the electromagnetic spectrum. In schematic form, Figure 30.4 illustrates some of the series of lines for atomic hydrogen. The group of lines in the visible region is known as the *Balmer series*, in recognition of Johann J. Balmer (1825–1898), a Swiss schoolteacher who found an empirical equation that gave the values for the observed wavelengths. This equa-

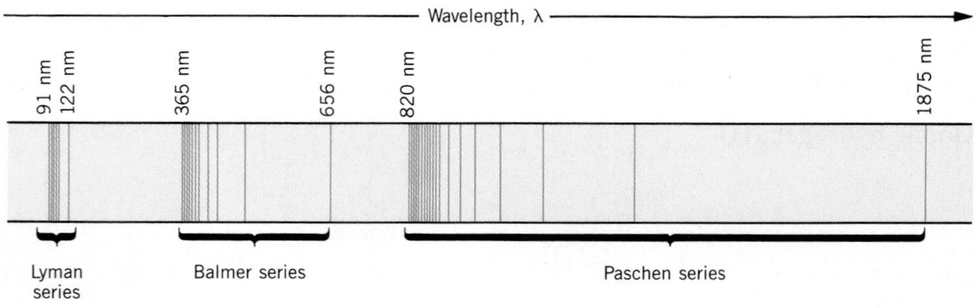

Figure 30.4 Line spectrum of atomic hydrogen. Only the Balmer series lies in the visible region of the electromagnetic spectrum.

tion is given below, along with similar equations that apply to the *Lyman series* and *Paschen series*, which are also shown in the drawing:

[Lyman series] $\qquad \dfrac{1}{\lambda} = R\left(\dfrac{1}{1^2} - \dfrac{1}{n^2}\right) \qquad n = 2, 3, 4, \ldots$ $\qquad$ (30.1)

[Balmer series] $\qquad \dfrac{1}{\lambda} = R\left(\dfrac{1}{2^2} - \dfrac{1}{n^2}\right) \qquad n = 3, 4, 5, \ldots$ $\qquad$ (30.2)

[Paschen series] $\qquad \dfrac{1}{\lambda} = R\left(\dfrac{1}{3^2} - \dfrac{1}{n^2}\right) \qquad n = 4, 5, 6, \ldots$ $\qquad$ (30.3)

In these equations, the constant term R has the value of $R = 1.097 \times 10^7 \text{ m}^{-1}$ and is called the *Rydberg constant.* An essential feature of each group of lines is that there is a long and a short wavelength limit, with the lines being increasingly crowded together toward the short wavelength limit. Figure 30.4 also gives the wavelength limits for each of the three series, and Example 1 determines them for the Balmer series.

Example 1   The Balmer Series

Find (a) the longest and (b) the shortest wavelengths of the Balmer series.

REASONING $\quad$ Each wavelength in the series corresponds to one value for the integer n in Equation 30.2. Longer wavelengths are associated with smaller values of n.

SOLUTION

(a) The longest wavelength of the series occurs when n has its smallest value ($n = 3$) in Equation 30.2:

$$\frac{1}{\lambda} = R\left(\frac{1}{2^2} - \frac{1}{n^2}\right) = (1.097 \times 10^7 \text{ m}^{-1})\left(\frac{1}{2^2} - \frac{1}{3^2}\right)$$

$$= 1.524 \times 10^6 \text{ m}^{-1} \quad \text{or} \quad \boxed{\lambda = 656 \text{ nm}}$$

(b) The shortest wavelength in the series arises when the integer n has a very large value, so that $1/n^2$ is essentially zero:

$$\frac{1}{\lambda} = (1.097 \times 10^7 \text{ m}^{-1})\left(\frac{1}{2^2} - 0\right) = 2.743 \times 10^6 \text{ m}^{-1} \quad \text{or} \quad \boxed{\lambda = 365 \text{ nm}}$$

Equations 30.1–30.3 are useful, because they reproduce the wavelengths that hydrogen atoms radiate. However, these equations are empirical, and they provide no insight as to *why* certain wavelengths are radiated and others are not. It was the Danish physicist Niels Bohr (1885 – 1962) who provided the first model of the atom that predicted the discrete wavelengths emitted by atomic hydrogen. Bohr's model started us on the way toward understanding how the structure of the atom restricts the radiated wavelengths to certain values.

30.3 THE BOHR MODEL OF THE HYDROGEN ATOM

THE MODEL

In 1913 Bohr presented a model that led to equations such as Balmer's for predicting the specific wavelengths that the hydrogen atom radiates. Bohr's theory begins with Rutherford's picture of an atom as a nucleus surrounded by electrons moving in circular orbits. In analyzing this picture, Bohr made a number of assumptions in order to combine the new quantum ideas of Planck and Einstein with the traditional description of a particle in uniform circular motion.

Adopting Planck's idea of quantized energy levels, Bohr hypothesized that in a hydrogen atom there can be only certain values of the total energy (electron kinetic energy plus potential energy). These allowed energy levels correspond to different orbits for the electron as it moves around the nucleus, the larger orbits being associated with larger total energies. Figure 30.5 illustrates two of the orbits. In addition, Bohr assumed that an electron in one of these orbits *does not* radiate electromagnetic waves. For this reason, the orbits are called **stationary orbits** or **stationary states**. Bohr recognized that radiationless orbits violated the laws of physics, as they were then known. But the assumption of such orbits was necessary, because the traditional laws indicated that an electron radiates electromagnetic waves as it accelerates around a circular path, and the loss of the energy carried by the waves would lead to the collapse of the orbit.

To incorporate Einstein's photon concept, Bohr theorized that a photon is emitted only when the electron *changes* orbits from a larger one with a higher energy to a smaller one with a lower energy, as Figure 30.5 indicates. But how do electrons get into the higher-energy orbits in the first place? They get there by picking up energy when atoms collide, which happens more often when a gas is heated, or by acquiring energy when a high voltage is applied to a gas.

When an electron in an initial orbit with a larger energy E_i changes to a final orbit with a smaller energy E_f, the emitted photon has an energy of $E_i - E_f$, consistent with the law of conservation of energy. But according to Einstein, the energy of a photon is hf, where f is its frequency and h is Planck's constant. Thus,

$$E_i - E_f = hf \tag{30.4}$$

Since the frequency of an electromagnetic wave is related to the wavelength by $f = c/\lambda$, Bohr could use Equation 30.4 to determine the wavelengths radiated by a hydrogen atom. First, however, he had to derive expressions for the energies E_i and E_f.

THE ENERGIES AND RADII OF THE BOHR ORBITS

For an electron of mass m and speed v in an orbit of radius r, the total energy is the kinetic energy (KE $= \frac{1}{2}mv^2$) of the electron plus the electric potential energy. The potential energy is the product of the charge $(-e)$ on the electron and the electric

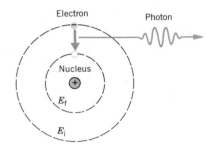

Figure 30.5 In the Bohr model of the hydrogen atom, a photon is emitted when the electron drops from a larger, higher energy orbit (energy $= E_i$) to a smaller, lower energy orbit (energy $= E_f$).

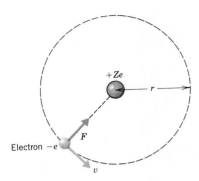

Figure 30.6 In the Bohr model, the electron is in uniform circular motion around the nucleus. The centripetal force F is the electrostatic force of attraction that the positive nuclear charge exerts on the electron.

potential produced by the positive nuclear charge, in accord with Equation 19.3. We assume that the nucleus contains Z protons,* for a total nuclear charge of $+Ze$. The electric potential at a distance r from a point charge of $+Ze$ is given as $+kZe/r$ by Equation 19.6, where $k = 8.99 \times 10^9$ N·m²/C². The electric potential energy is, then, EPE $= (-e)(+kZe/r)$. Consequently, the total energy E of the atom is

$$E = \text{KE} + \text{EPE} = \tfrac{1}{2}mv^2 - \frac{kZe^2}{r} \tag{30.5}$$

But a centripetal force of magnitude mv^2/r (Equation 5.3) must act on a particle in uniform circular motion. As Figure 30.6 indicates, the centripetal force is provided by the electrostatic force of attraction that the protons in the nucleus exert on the electron. According to Coulomb's law, the magnitude of the electrostatic force is $F = kZe^2/r^2$. Therefore, $mv^2/r = kZe^2/r^2$, or

$$mv^2 = \frac{kZe^2}{r} \tag{30.6}$$

We can use this relation to eliminate the term mv^2 from Equation 30.5, with the result that

$$E = \frac{1}{2}\left(\frac{kZe^2}{r}\right) - \frac{kZe^2}{r} = -\frac{kZe^2}{2r} \tag{30.7}$$

The total energy of the atom is negative, because the negative electric potential energy is larger in magnitude than the positive kinetic energy.

A value for the radius r is needed, if Equation 30.7 is to be useful. To determine r, Bohr made an assumption about the angular momentum of the electron. The angular momentum L is given by Equation 9.10 as $L = I\omega$, where $I = mr^2$ is the moment of inertia of the electron moving on its circular path and $\omega = v/r$ is the angular speed of the electron in radians per second. Thus, the angular momentum is $L = (mr^2)(v/r) = mvr$. Bohr conjectured that the angular momentum of the electron can assume only certain discrete values; in other words, L is quantized. He postulated that the allowed values are integer multiples of Planck's constant divided by 2π:

$$L_n = mv_n r_n = n\frac{h}{2\pi} \qquad n = 1, 2, 3, \ldots \tag{30.8}$$

Solving this equation for v_n and substituting the result into Equation 30.6 leads to the following expression for the radius r_n of the nth Bohr orbit:

$$r_n = \left(\frac{h^2}{4\pi^2 mke^2}\right)\frac{n^2}{Z} \qquad n = 1, 2, 3, \ldots \tag{30.9}$$

With $h = 6.626 \times 10^{-34}$ J·s, $m = 9.109 \times 10^{-31}$ kg, $k = 8.988 \times 10^9$ N·m²/C², and $e = 1.602 \times 10^{-19}$ C, this expression reveals that

$$\begin{bmatrix} \textbf{Radii for} \\ \textbf{Bohr orbits} \end{bmatrix} \quad r_n = (5.29 \times 10^{-11} \text{ m})\frac{n^2}{Z} \qquad n = 1, 2, 3, \ldots \tag{30.10}$$

Therefore, in the hydrogen atom ($Z = 1$) the smallest Bohr orbit ($n = 1$) has a

---

* For hydrogen, $Z = 1$, but we also wish to consider situations in which Z is greater than 1.

radius of $r_1 = 5.29 \times 10^{-11}$ m. This particular value is called the **Bohr radius.** Figure 30.7 shows the first three Bohr orbits for the hydrogen atom.

The expression for the radius of a Bohr orbit can be substituted into Equation 30.7 to show that the corresponding total energy for the nth orbit is

$$E_n = -\left(\frac{2\pi^2 m k^2 e^4}{h^2}\right)\frac{Z^2}{n^2} \qquad n = 1, 2, 3, \ldots \qquad (30.11)$$

Substituting values for h, m, k, and e into this expression yields

$$\begin{bmatrix} \text{Bohr energy} \\ \text{levels in} \\ \text{joules} \end{bmatrix} \quad E_n = -(2.18 \times 10^{-18} \text{ J})\frac{Z^2}{n^2} \qquad n = 1, 2, 3, \ldots \qquad (30.12)$$

Often, atomic energies are expressed in electron volts rather than joules. Since 1.60×10^{-19} J $= 1$ eV, the result above can be rewritten as

$$\begin{bmatrix} \text{Bohr energy} \\ \text{levels in} \\ \text{electron volts} \end{bmatrix} \quad E_n = -(13.6 \text{ eV})\frac{Z^2}{n^2} \qquad n = 1, 2, 3, \ldots \qquad (30.13)$$

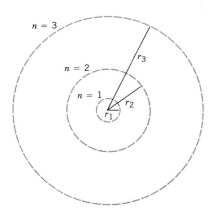

Figure 30.7 The first Bohr orbit in the hydrogen atom has a radius $r_1 = 5.29 \times 10^{-11}$ m. The second and third Bohr orbits have radii $r_2 = 4r_1$ and $r_3 = 9r_1$, respectively.

ENERGY LEVEL DIAGRAMS

It is useful to represent the energy values given by Equation 30.13 on an *energy level diagram,* as in Figure 30.8. In this diagram, which applies to the hydrogen atom ($Z = 1$), the highest energy level corresponds to $n = \infty$ in Equation 30.13 and

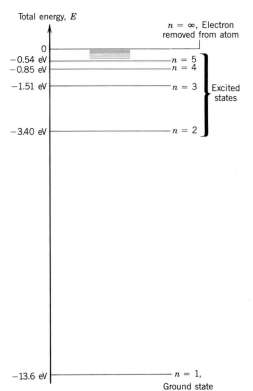

Figure 30.8 Energy level diagram for the hydrogen atom.

has an energy of 0 eV. This is the energy of the atom when the electron is completely removed ($r = \infty$) from the nucleus and is at rest. In contrast, the lowest energy level corresponds to $n = 1$ and has a value of -13.6 eV. The lowest energy level is called the *ground state*, to distinguish it from the higher levels, which are called *excited states.* Observe how the energies of the excited states come closer and closer together as n increases.

The electron in a hydrogen atom at room temperature spends most of its time in the ground state. To raise the electron from the ground state ($n = 1$) to the highest possible excited state ($n = \infty$), 13.6 eV of energy must be supplied. Supplying this amount of energy removes the electron from the atom, producing the positive hydrogen ion H^+. The energy needed to remove the electron is called the *ionization energy.* Thus, the Bohr model predicts that the ionization energy of atomic hydrogen is 13.6 eV, in excellent agreement with the experimental value. In Example 2 the Bohr model is applied to doubly ionized lithium.

Example 2   The Ionization Energy of Li^{2+}

The Bohr model does not apply when more than one electron orbits the nucleus, because the model does not account for the electrostatic forces that one electron exerts on another. For instance, an electrically neutral lithium atom (Li) contains three electrons in orbit around a nucleus that includes three protons ($Z = 3$), and Bohr's analysis is not applicable. However, the Bohr model can be used for the doubly charged positive ion of lithium (Li^{2+}) that results when two electrons are removed from the neutral atom, leaving only one electron to orbit the nucleus. Obtain the ionization energy that is needed to remove the remaining electron from Li^{2+}.

REASONING   The lithium ion Li^{2+} contains three times the positive nuclear charge as that of the hydrogen atom. Therefore, the orbiting electron is attracted more strongly to the nucleus in Li^{2+} than in the hydrogen atom. As a result, we expect that more energy is required to ionize Li^{2+} than the 13.6 eV required for atomic hydrogen.

SOLUTION   The Bohr energy levels for Li^{2+} are given by Equation 30.13 with $Z = 3$; $E_n = -(13.6 \text{ eV})(3^2/n^2)$. Therefore, the ground state ($n = 1$) energy is

$$E_1 = -(13.6 \text{ eV})\frac{3^2}{1^2} = -122 \text{ eV}$$

To remove the electron from Li^{2+}, 122 eV of energy must be supplied: Ionization energy $= 122$ eV . This value for the ionization energy agrees well with the experimental value of 122.4 eV.

THE LINE SPECTRA OF THE HYDROGEN ATOM

To determine the wavelengths radiated by the hydrogen atom, Bohr substituted Equation 30.11 for the energies into Equation 30.4 and used $f = c/\lambda$. He obtained the following result:

$$\frac{1}{\lambda} = \frac{2\pi^2 mk^2 e^4}{h^3 c} (Z^2)\left(\frac{1}{n_f^2} - \frac{1}{n_i^2}\right) \tag{30.14}$$

$$n_i, n_f = 1, 2, 3, \ldots \quad \text{and} \quad n_i > n_f$$

With the known values for h, m, k, e, and c, it can be seen that $2\pi^2 mk^2 e^4/(h^3 c) =$

$1.097 \times 10^7 \text{ m}^{-1}$, in agreement with the Rydberg constant R that appears in Equations 30.1–30.3. The agreement between the theoretical and experimental values of the Rydberg constant was a major accomplishment of Bohr's theory.

With $Z = 1$ and $n_f = 1$, Equation 30.14 reproduces Equation 30.1 for the Lyman series. Thus, Bohr's model shows that the Lyman series of lines occurs when electrons make transitions from higher energy levels with $n_i = 2, 3, 4, \ldots$ to the first energy level where $n_f = 1$. Figure 30.9 shows these transitions. Notice that when an electron makes a transition from $n_i = 2$ to $n_f = 1$, the longest wavelength photon in the Lyman series is emitted, since the energy change is the smallest possible. When an electron makes a transition from the highest level where $n_i = \infty$ to the lowest level where $n_f = 1$, the shortest wavelength is emitted, since the energy change is the largest possible. Since the higher energy levels are increasingly close together, the lines in the series become more and more crowded together toward the short wavelength limit, as observed. Figure 30.9 also shows the energy level transitions for the Balmer and Paschen series. In the Balmer series $n_i = 3, 4, 5, \ldots$, while $n_f = 2$. In the Paschen series $n_i = 4, 5, 6, \ldots$, while $n_f = 3$. The next example deals further with the line spectrum of the hydrogen atom.

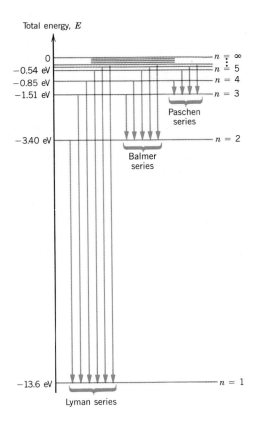

Figure 30.9 The Lyman, Balmer, and Paschen series of lines in the hydrogen atom spectrum correspond to transitions that the electron makes between higher and lower energy levels, as indicated here.

Example 3   The Brackett Series for Atomic Hydrogen

In the line spectrum of atomic hydrogen there is also a group of lines known as the Brackett series. These lines are produced when electrons, excited to high energy levels, make transitions to the $n = 4$ level. Determine (a) the longest wavelength in this series and (b) the wavelength that corresponds to the transition from $n_i = 6$ to $n_f = 4$. (c) Refer to Figure 24.6 and identify the spectral region in which these lines are found.

REASONING AND SOLUTION

(a) The longest wavelength corresponds to the transition that has the smallest energy change. This would be between the $n_i = 5$ and $n_f = 4$ levels in Figure 30.9. Using Equation 30.14 with $Z = 1$, we find that

$$\frac{1}{\lambda} = (1.097 \times 10^7 \text{ m}^{-1})(1^2)\left(\frac{1}{4^2} - \frac{1}{5^2}\right)$$

$$= 2.468 \times 10^5 \text{ m}^{-1} \quad \text{or} \quad \boxed{\lambda = 4051 \text{ nm}}$$

(b) The calculation here is similar to that above:

$$\frac{1}{\lambda} = (1.097 \times 10^7 \text{ m}^{-1})(1^2)\left(\frac{1}{4^2} - \frac{1}{6^2}\right)$$

$$= 3.809 \times 10^5 \text{ m}^{-1} \quad \text{or} \quad \boxed{\lambda = 2625 \text{ nm}}$$

(c) According to Figure 24.6, these lines lie in the infrared region of the spectrum.

The various lines in the hydrogen atom spectrum are produced when electrons change from higher to lower energy levels. During these energy-level transitions photons are emitted, and, consequently, the spectral lines are called *emission lines*. Electrons can also make transitions from lower energy levels to higher energy levels, in a process known as absorption. In this case, an atom absorbs a photon that has precisely the energy needed to produce a transition from a lower energy level to a higher energy level. Thus, if photons with a continuous range of wavelengths pass through a gas and then are analyzed with a grating spectroscope, a series of dark *absorption lines* appear in the continuous spectrum. Such absorption lines can be seen in Figure 30.3 in the spectrum of the sun, where they are called Fraunhofer lines, after their discoverer. They are due to atoms, located in the outer and cooler layers of the sun, that absorb radiation coming from within the sun. The inner and hotter portion of the sun emits a continuous spectrum of wavelengths, since it is too hot for individual atoms to retain their structures.

THE PHYSICS OF . . .

absorption lines in the sun's spectrum.

The Bohr model provides a great deal of insight into atomic structure. However, this model is now known to be oversimplified and has been superseded by a more detailed picture provided by quantum mechanics and the Schrödinger equation (see Section 30.5).

30.4 DE BROGLIE'S EXPLANATION OF BOHR'S ASSUMPTION ABOUT ANGULAR MOMENTUM

Of all the assumptions Bohr made in his model of the hydrogen atom, perhaps the most puzzling is the one about the angular momentum of the electron ($L_n = mv_n r_n = nh/2\pi$; $n = 1, 2, 3, \ldots$). Why should the angular momentum have only those values that are integer multiples of Planck's constant divided by 2π? In 1923, ten years after Bohr's work, de Broglie pointed out that his own theory for the wavelength of a moving particle could provide an answer to this question.

In de Broglie's way of thinking, the electron in its circular Bohr orbit must be pictured as a particle wave. And like waves traveling on a string, particle waves can lead to standing waves under resonance conditions. Section 17.5 discusses these conditions for a string. Standing waves form when the total distance traveled by a wave down the string and back is one wavelength, two wavelengths, or any integer number of wavelengths. The total distance around a Bohr orbit of radius r is the circumference of the orbit or $2\pi r$. By the same reasoning, then, the condition for standing particle waves for the electron in a Bohr orbit would be

$$2\pi r = n\lambda \qquad n = 1, 2, 3, \ldots$$

where n is the number of whole wavelengths that fit into the circumference of the circle. But according to Equation 29.8 the de Broglie wavelength of the electron is $\lambda = h/p$, where p is the magnitude of the electron's momentum. If the speed of the electron is much less than the speed of light, the momentum is $p = mv$, and the condition for standing particle waves becomes $2\pi r = nh/(mv)$. A rearrangement of this result gives

$$mvr = n\frac{h}{2\pi} \qquad n = 1, 2, 3, \ldots$$

which is just what Bohr assumed for the angular momentum of the electron. As an example, Figure 30.10 illustrates the standing particle wave on a Bohr orbit for which $2\pi r = 4\lambda$.

De Broglie's explanation of Bohr's assumption about angular momentum emphasizes an important fact, namely, that particle waves play a central role in the structure of the atom. Moreover, the theoretical framework of quantum mechanics provides the basis for determining the wave function Ψ (Greek letter *psi*) that represents a particle wave. The next section deals with the picture that quantum mechanics gives for atomic structure, a picture that supersedes the Bohr model. In any case, the Bohr model can be applied when a single electron orbits the nucleus, while the theoretical framework of quantum mechanics can be applied, in principle, to atoms that contain an arbitrary number of electrons.

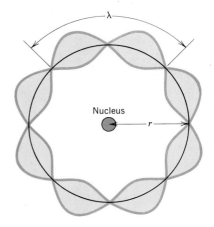

Figure 30.10   De Broglie suggested standing particle waves as an explanation for Bohr's angular momentum assumption. Here, a standing particle wave is illustrated on a Bohr orbit where four de Broglie wavelengths fit into the circumference of the orbit.

30.5 THE QUANTUM MECHANICAL PICTURE OF THE HYDROGEN ATOM

QUANTUM NUMBERS

The picture of the hydrogen atom that quantum mechanics and the Schrödinger equation provide differs in a number of ways from the Bohr model. The Bohr model uses a single integer number n to identify the various electron orbits and the associated energies. Because this number can have only discrete values, rather than a continuous range of values, n is called a *quantum number*. In contrast, quantum mechanics reveals that four different quantum numbers are needed to describe each state of the hydrogen atom. These four are described below.

1. **The principal quantum number n.** As in the Bohr model, this number determines the total energy of the atom and can have only integer values: $n = 1, 2, 3, \ldots$. In fact, the Schrödinger equation predicts* that the energy of the hydrogen atom is identical to that obtained from the Bohr model: $E_n = -(13.6 \text{ eV})Z^2/n^2$.

* This prediction requires that small relativistic effects and small interactions within the atom be ignored, and assumes that the hydrogen atom is not located in an external magnetic field.

2. **The orbital quantum number ℓ.** This number determines the angular momentum of the electron due to its orbital motion. The values that ℓ can have depend on the value of n, and only the following integers are allowed:

$$\ell = 0, 1, 2, \ldots, (n-1)$$

For instance, if $n = 1$, the orbital quantum number can have only the value $\ell = 0$, but if $n = 4$, the values $\ell = 0, 1, 2,$ and 3 are possible. The magnitude L of the angular momentum of the electron is

$$L = \sqrt{\ell(\ell+1)} \frac{h}{2\pi} \qquad (30.15)$$

3. **The magnetic quantum number m_ℓ.** The word "magnetic" is used here because an externally applied magnetic field influences the energy of the atom, and this quantum number is used in describing the effect. The effect was discovered by the Dutch physicist Pieter Zeeman (1865–1943) and, hence, is known as the *Zeeman effect*. When there is no external magnetic field, m_ℓ plays no role in determining the energy. In either event, the magnetic quantum number determines the component of the angular momentum along a specific direction, which is called the z direction by convention. The values that m_ℓ can have depend on the value of ℓ, with only the following positive and negative integers being permitted:

$$m_\ell = 0, \pm1, \pm2, \ldots, \pm\ell$$

For example, if the orbital quantum number is $\ell = 2$, then the magnetic quantum number can have the values $m_\ell = -2, -1, 0, +1,$ and $+2$. The component L_z of the angular momentum in the z direction is

$$L_z = m_\ell \frac{h}{2\pi} \qquad (30.16)$$

4. **The spin quantum number m_s.** This number is needed because the electron has an intrinsic property called spin angular momentum. Loosely speaking, we can view the electron as spinning while it orbits the nucleus, analogous to the way the earth spins as it moves around the sun. There are two possible values for the spin quantum number of the electron:

$$m_s = +\tfrac{1}{2} \quad \text{or} \quad m_s = -\tfrac{1}{2}$$

Sometimes the phrases "spin up" and "spin down" are used to refer to the directions of the spin angular momentum associated with the values for m_s.

Table 30.1 summarizes the four quantum numbers that are needed to describe each state of the hydrogen atom. One set of values for n, ℓ, m_ℓ, and m_s corre-

Table 30.1 Quantum Numbers for the Hydrogen Atom

Name	Symbol	Allowed Values
Principal quantum number	n	1, 2, 3, . . .
Orbital quantum number	ℓ	0, 1, 2, . . . , $(n-1)$
Magnetic quantum number	m_ℓ	0, ±1, ±2, . . . , $\pm\ell$
Spin quantum number	m_s	$\pm\tfrac{1}{2}$

sponds to one state. As the principal quantum number n increases, the number of possible combinations of the four quantum numbers rises rapidly, as Example 4 illustrates.

Example 4   States of the Hydrogen Atom

Determine the number of possible states for the hydrogen atom when the principal quantum number is (a) $n = 1$ and (b) $n = 2$.

REASONING   Each different combination of the four quantum numbers summarized in Table 30.1 corresponds to a different state. We begin with the value for n and find the allowed values for ℓ. Then, for each ℓ value we find the possibilities for m_ℓ. Finally, m_s may be $+\frac{1}{2}$ or $-\frac{1}{2}$ for each group of values for n, ℓ, and m_ℓ.

SOLUTION

(a) The diagram below shows the possibilities for ℓ, m_ℓ, and m_s when $n = 1$:

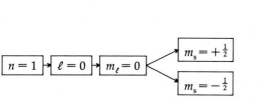

	State		
n	ℓ	m_ℓ	m_s
1	0	0	$+\frac{1}{2}$
1	0	0	$-\frac{1}{2}$

Thus, there are two different states for the hydrogen atom. These two states have the same energy, since they have the same value of n.

(b) When $n = 2$, there are eight possible combinations for the values of n, ℓ, m_ℓ, and m_s, as the diagram below indicates:

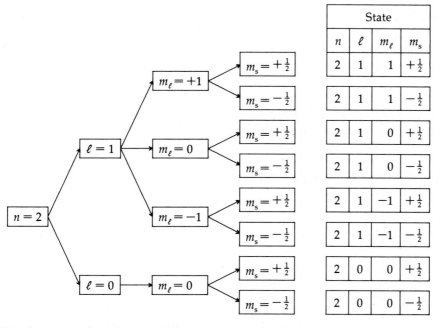

	State		
n	ℓ	m_ℓ	m_s
2	1	1	$+\frac{1}{2}$
2	1	1	$-\frac{1}{2}$
2	1	0	$+\frac{1}{2}$
2	1	0	$-\frac{1}{2}$
2	1	-1	$+\frac{1}{2}$
2	1	-1	$-\frac{1}{2}$
2	0	0	$+\frac{1}{2}$
2	0	0	$-\frac{1}{2}$

With the same value of $n = 2$, all eight states have the same energy.

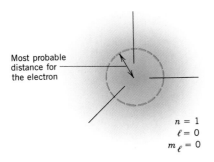

Figure 30.11 The electron probability cloud for the ground state ($n = 1$, $\ell = 0$, $m_\ell = 0$) of the hydrogen atom.

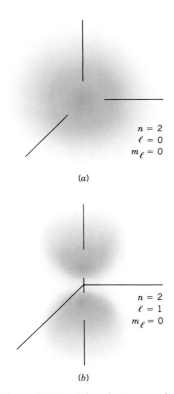

Figure 30.12 The electron probability clouds for the hydrogen atom when (a) $n = 2$, $\ell = 0$, $m_\ell = 0$ and (b) $n = 2$, $\ell = 1$, $m_\ell = 0$.

ELECTRON PROBABILITY CLOUDS

According to the Bohr model, the nth orbit is a circle of radius r_n, and every time the position of the electron in this orbit is measured, the electron is found exactly a distance r_n away from the nucleus. The quantum mechanical picture of the atom is quite different than that of the Bohr model. Suppose the electron is in a quantum mechanical state for which $n = 1$, and we imagine making a number of measurements of the electron's position with respect to the nucleus. We would find that the position of the electron is uncertain, in the sense that even in a state for which $n = 1$, there is a probability of finding the electron sometimes very near the nucleus, sometimes very far from the nucleus, and sometimes at intermediate locations. The probability is determined by the wave function Ψ, as Section 29.5 discusses. We can make a three-dimensional picture of our findings by marking a dot at each location where the electron is found. After a sufficient number of measurements is made, a picture of the quantum mechanical state emerges. A greater number of dots occur at places where the probability of finding the electron is higher. Figure 30.11 shows the spatial distribution for an electron in a state for which $n = 1$, $\ell = 0$, and $m_\ell = 0$. This picture is constructed from so many measurements that the individual dots are no longer visible, but have merged to form a kind of probability "cloud" whose density changes gradually from place to place. The dense regions indicate places where the probability of finding the electron is higher, while the less dense regions indicate places where the probability is lower. Also indicated in Figure 30.11 is the radius where quantum mechanics predicts the greatest probability per unit radial distance of finding the electron in the $n = 1$ state. This radius matches exactly the radius of 5.29×10^{-11} m found for the first Bohr orbit.

For a principal quantum number of $n = 2$, the probability clouds are different than for $n = 1$. In fact, more than one cloud shape is possible, because with $n = 2$ the orbital quantum number can be either $\ell = 0$ or $\ell = 1$. While the value of ℓ does not affect the energy of the hydrogen atom, the value does have a significant effect on the shape of the probability clouds. Figure 30.12a shows the cloud for $n = 2$, $\ell = 0$, and $m_\ell = 0$. Part b of the drawing shows that when $n = 2$, $\ell = 1$, and $m_\ell = 0$, the cloud has a two-lobe shape with the nucleus at the center between the lobes. For larger values of n, the probability clouds become increasingly complex and are spread out over larger volumes of space.

30.6 THE PAULI EXCLUSION PRINCIPLE AND THE PERIODIC TABLE OF THE ELEMENTS

MULTIPLE-ELECTRON ATOMS

Except for hydrogen, all electrically neutral atoms contain more than one electron, the number being given by the atomic number Z of the element. In addition to being attracted by the nucleus, the electrons repel each other. This repulsion contributes to the total energy of a multiple-electron atom. As a result, the one-electron energy expression for hydrogen [$E_n = -(13.6 \text{ eV})Z^2/n^2$] provided by the Bohr model and also by quantum mechanics does not apply to other neutral atoms. However, the simplest approach for dealing with a multiple-electron atom still uses the four quantum numbers n, ℓ, m_ℓ, and m_s.

Detailed quantum mechanical calculations reveal that the energy level of each

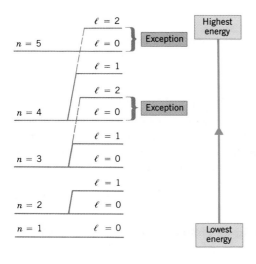

Figure 30.13 When there is more than one electron in an atom, the total energy of a given state depends on the principal quantum number n and the orbital quantum number ℓ. Generally, the energy increases with increasing n and, for a fixed n, with increasing ℓ. There are exceptions to the general rule, however, as indicated here. For clarity, levels for $n = 6$ and higher are not shown.

state of a multiple-electron atom depends on both the principal quantum number n and the orbital quantum number ℓ. Figure 30.13 illustrates that the energy generally increases as n increases. Furthermore, for a given n, the energy also increases as ℓ increases, but there are some exceptions, as the drawing indicates.

In a multiple-electron atom, all electrons with the same value of n are said to be in the same *shell*. Electrons with $n = 1$ are in a single shell (sometimes called the K shell), electrons with $n = 2$ are in another shell (the L shell), those with $n = 3$ are in a third shell (the M shell), and so on. Those electrons with the same values for both n and ℓ are often referred to as being in the same *subshell*. The $n = 1$ shell consists of a single $\ell = 0$ subshell. The $n = 2$ shell has two subshells, one with $\ell = 0$ and one with $\ell = 1$. Similarly, the $n = 3$ shell has three subshells.

In the hydrogen atom near room temperature, the electron spends most of its time in the lowest energy level or ground state, namely, in the $n = 1$ shell. Similarly, when an atom contains more than one electron and is near room temperature, the electrons spend most of their time in the lowest energy levels possible. The lowest energy state for an atom is called the *ground state.* However, when a multiple-electron atom is in its ground state, not every electron is crowded into the $n = 1$ shell. The reason the electrons are not all in the same shell is that they obey a principle discovered by the Austrian physicist Wolfgang Pauli (1900–1958).

The Pauli Exclusion Principle

No two electrons in an atom can have the same set of values for the four quantum numbers n, ℓ, m_ℓ, and m_s.

For instance, suppose two electrons in an atom have three quantum numbers that are identical: $n = 3$, $m_\ell = 1$, and $m_s = -\frac{1}{2}$. According to the exclusion principle, it is not possible for each to have $\ell = 2$, for example, since each would then have the same four quantum numbers. Each electron must have a different value for ℓ ($\ell = 1$ and $\ell = 2$, for instance) and, consequently, be in a different subshell. With the aid of the Pauli exclusion principle, we can determine which energy levels are

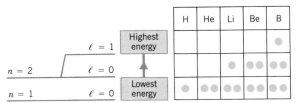

Figure 30.14 The electrons (●) in the ground state of an atom fill the available energy levels "from the bottom up," that is, from the lowest to the highest energy, consistent with the Pauli exclusion principle. The ranking of the energy levels in this figure is meant to apply for a given atom, not between one atom and another.

occupied by the electrons in an atom in its ground state, as the next example demonstrates.

Example 5   Ground States of Atoms

Determine which of the energy levels in Figure 30.13 are occupied by the electrons in the ground state of hydrogen (1 electron), helium (2 electrons), lithium (3 electrons), beryllium (4 electrons), and boron (5 electrons).

REASONING AND SOLUTION   As the colored dot in Figure 30.14 indicates, the electron in the hydrogen atom (H) is in the $n = 1$, $\ell = 0$ subshell, which has the lowest possible energy. A second electron is present in the helium atom (He) and both electrons can have the quantum numbers $n = 1$, $\ell = 0$, and $m_\ell = 0$. However, in accord with the Pauli exclusion principle, each electron must have a different spin quantum number, $m_s = +\frac{1}{2}$ for one electron and $m_s = -\frac{1}{2}$ for the other. Thus, the drawing shows both electrons in the lowest energy level.

The third electron that is present in the lithium atom (Li) would violate the exclusion principle if it were also in the $n = 1$, $\ell = 0$ subshell, no matter what the value for m_s. Thus, the $n = 1$, $\ell = 0$ subshell is filled when occupied by two electrons. With this level filled, the $n = 2$, $\ell = 0$ subshell becomes the next lowest energy level available and is where the third electron of lithium is found (see Figure 30.14). In the beryllium atom (Be), the fourth electron is in the $n = 2$, $\ell = 0$ subshell, along with the third electron. This is possible, since the third and fourth electrons can have different values for m_s.

With the first four electrons in place as discussed above, the fifth electron in the boron atom (B) cannot fit into the $n = 2$, $\ell = 0$ subshell without violating the exclusion principle. Therefore, the fifth electron is found in the $n = 2$, $\ell = 1$ subshell, which is the next available energy level with the lowest energy, as Figure 30.14 indicates. For this electron, m_ℓ can be -1, 0, or $+1$ and m_s can be $+\frac{1}{2}$ or $-\frac{1}{2}$ in each case. However, in the absence of an external magnetic field, all of these possibilities correspond to the same energy.

Because of the Pauli exclusion principle, there is a maximum number of electrons that can fit into an energy level or subshell. Example 5 shows that the $n = 1$, $\ell = 0$ subshell can hold at most two electrons. The $n = 2$, $\ell = 1$ subshell, however, can hold six electrons, because with $\ell = 1$, there are three possibilities for m_ℓ (-1, 0, and $+1$), and for each of these, the value of m_s can be $+\frac{1}{2}$ or $-\frac{1}{2}$. In general, m_ℓ can have the values $0, \pm 1, \pm 2, \ldots, \pm\ell$, for $2\ell + 1$ possibilities. Since each

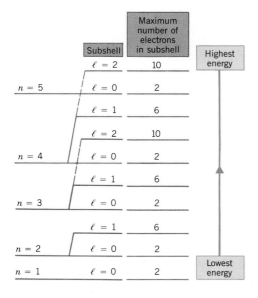

Figure 30.15   The maximum number of electrons that the ℓth subshell can hold is $2(2\ell + 1)$.

of these can be combined with two possibilities for m_s, the total number of different combinations for m_ℓ and m_s is $2(2\ell + 1)$. This, then, is the maximum number of electrons the ℓth subshell can hold, as Figure 30.15 summarizes.

SHORTHAND NOTATION FOR THE ELECTRONIC CONFIGURATION OF THE ATOM

For historical reasons, there is a widely used convention in which each subshell of an atom is referred to by a letter, rather than by the value of its orbital quantum number ℓ. For instance, an $\ell = 0$ subshell is called an s subshell. An $\ell = 1$ subshell and an $\ell = 2$ subshell are known as p and d subshells, respectively. The higher values of $\ell = 3, 4$, etc., are referred to as f, g, etc., in alphabetical sequence, as Table 30.2 indicates.

This convention of letters is used in a shorthand notation that is convenient for simultaneously indicating the principal quantum number n, the orbital quantum number ℓ, and the number of electrons in the n, ℓ subshell. The notation is as follows:

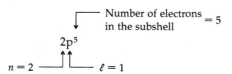

Table 30.2   The Convention of Letters Used to Refer to the Orbital Quantum Number

Orbital Quantum Number ℓ	Letter
0	s
1	p
2	d
3	f
4	g
5	h

With this notation, the arrangement or configuration of the electrons in an atom can be specified efficiently. For instance, in Example 5, we found that the electron configuration for boron has two electrons in the $n = 1, \ell = 0$ subshell, two in the $n = 2, \ell = 0$ subshell, and one in the $n = 2, \ell = 1$ subshell. In shorthand notation this arrangement is expressed as $1s^2\ 2s^2\ 2p^1$. Table 30.3 gives the ground state electron configurations written in this fashion for elements containing up to eighteen electrons. The first five entries are those worked out in Example 5.

Table 30.3 Ground State Electronic
Configurations of Atoms

Element	Number of Electrons	Configuration of the Electrons
Hydrogen (H)	1	$1s^1$
Helium (He)	2	$1s^2$
Lithium (Li)	3	$1s^2 \, 2s^1$
Beryllium (Be)	4	$1s^2 \, 2s^2$
Boron (B)	5	$1s^2 \, 2s^2 \, 2p^1$
Carbon (C)	6	$1s^2 \, 2s^2 \, 2p^2$
Nitrogen (N)	7	$1s^2 \, 2s^2 \, 2p^3$
Oxygen (O)	8	$1s^2 \, 2s^2 \, 2p^4$
Fluorine (F)	9	$1s^2 \, 2s^2 \, 2p^5$
Neon (Ne)	10	$1s^2 \, 2s^2 \, 2p^6$
Sodium (Na)	11	$1s^2 \, 2s^2 \, 2p^6 \, 3s^1$
Magnesium (Mg)	12	$1s^2 \, 2s^2 \, 2p^6 \, 3s^2$
Aluminum (Al)	13	$1s^2 \, 2s^2 \, 2p^6 \, 3s^2 \, 3p^1$
Silicon (Si)	14	$1s^2 \, 2s^2 \, 2p^6 \, 3s^2 \, 3p^2$
Phosphorus (P)	15	$1s^2 \, 2s^2 \, 2p^6 \, 3s^2 \, 3p^3$
Sulfur (S)	16	$1s^2 \, 2s^2 \, 2p^6 \, 3s^2 \, 3p^4$
Chlorine (Cl)	17	$1s^2 \, 2s^2 \, 2p^6 \, 3s^2 \, 3p^5$
Argon (Ar)	18	$1s^2 \, 2s^2 \, 2p^6 \, 3s^2 \, 3p^6$

THE PHYSICS OF . . .

the periodic table of the elements.

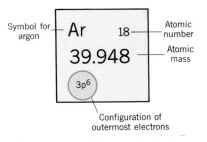

Symbol for argon — Ar   18 — Atomic number
39.948 — Atomic mass
3p⁶
Configuration of outermost electrons

Figure 30.16 The entries in the periodic table of the elements often include the ground state configuration of the outermost electrons.

THE PERIODIC TABLE

Each entry in the periodic table of the elements often includes the ground state electronic configuration, as Figure 30.16 illustrates for argon. To save space, only the configuration of the outermost electrons and unfilled subshells is specified.

Originally the periodic table was developed by the Russian chemist Dmitri Mendeleev (1834–1907) on the basis that certain groups of elements exhibit similar chemical properties. There are eight of these groups, plus the transition elements in the middle of the table, the lanthanide series, and the actinide series. The similar chemical properties within a group can be explained on the basis of the configurations of the outer electrons of the elements in the group. Thus, quantum mechanics and the Pauli exclusion principle offer an explanation for the chemical behavior of the atoms.

The full periodic table can be found on the inside of the back cover. Group 0, the last column of elements on the right side of the table, consists of the noble gases, such as helium (He), neon (Ne), and argon (Ar). Chemically, these elements are relatively inert, because their outermost electrons form a shell or subshell that is completely full. Being full, the shell or subshell is very stable, not readily forming chemical bonds by accepting electrons from or donating electrons to other elements.

Group I is made up of the alkali metals. Sodium (Na) and potassium (K) are familiar members of this group. These elements are chemically reactive, because they have only a single electron in an outermost s subshell. This electron can be easily lost to other elements in a chemical reaction. Thus, elements in group I often form singly charged positive ions, such as the sodium ion Na^+.

Group VII consists of the halogens and includes fluorine (F), chlorine (Cl), and bromine (Br). These elements have outermost electrons in a p subshell that is only one electron shy of being full. The halogens are highly reactive. Their chemistry is characterized by reactions in which they accept a single electron from other elements to form a stable, filled subshell. For this reason, the halogens readily form ions such as the chloride ion Cl⁻, which carry a single negative charge.

By looking at the other groups (II–VI) in the periodic table, you can see that elements within a group have outermost electrons in either s or p subshells. Within a group, the subshells are filled to the same extent. The transition elements are elements formed when electrons fill out primarily 3d, 4d, 5d, and 6d subshells. The lanthanide series involves completing mainly the 5d and 4f subshells. And finally, the actinide series corresponds to filling out primarily the 5f and 6d subshells.

30.7 X-RAYS

THE PHYSICS OF . . .

X-rays.

X-rays were discovered by Wilhelm K. Roentgen (1845–1923), a Dutch physicist who performed much of his work in Germany. X-rays can be produced when electrons, accelerated through a large potential difference, collide with a metal target made from molybdenum or platinum, for example. The target is contained within an evacuated glass tube, as Figure 30.17 shows. A plot of X-ray intensity per unit wavelength versus the wavelength looks similar to Figure 30.18 and consists of sharp peaks or lines superimposed on a broad continuous spectrum.

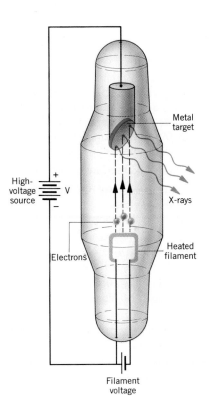

Figure 30.17 In an X-ray tube, electrons are emitted by a heated filament, accelerate through a large potential difference *V*, and strike a metal target. The X-rays originate from the interaction between the electrons and the metal target.

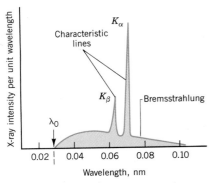

Figure 30.18 When a molybdenum target is bombarded with electrons that have been accelerated from rest through a potential difference of 45 000 V, the X-ray spectrum shown here is produced. The vertical axis is not drawn to scale.

PROBLEM SOLVING INSIGHT

Equation 30.13 for the Bohr energy levels [$E_n = -(13.6 \text{ eV})Z^2/n^2$, $n = 1$] can be used in rough calculations of the energy levels involved in the production of K_α X-rays. In such calculations, the value of the atomic number Z must be reduced by one, to account approximately for the shielding of one K-shell electron by the other K-shell electron.

The sharp peaks are called characteristic lines or ***characteristic X-rays***, because they are characteristic of the target material. The broad continuous spectrum is referred to as ***Bremsstrahlung*** (German for "braking radiation"). Bremsstrahlung X-rays are emitted when the electrons decelerate or "brake" upon hitting the target.

In Figure 30.18 the characteristic lines are marked K_α and K_β, because they involve the $n = 1$ or K shell of a metal atom. If an electron with enough energy strikes the target, one of the K-shell electrons can be knocked entirely out of a target atom. An electron in one of the outer shells can then fall into the K shell, and an X-ray photon is emitted in the process. Example 6 shows that a large potential difference is needed to operate an X-ray tube, so the electrons impinging on the metal target have sufficient energy to generate the characteristic X-rays.

Example 6   The Voltage Needed to Operate an X-Ray Tube

Strictly speaking, the Bohr model does not apply to multiple-electron atoms, but it can be used to make estimates. Use the Bohr model to estimate the minimum energy that an incoming electron must have to knock a K-shell electron entirely out of an atom in a platinum ($Z = 78$) target of an X-ray tube.

REASONING   According to the Bohr model, the energy of a K-shell electron is given by Equation 30.13 with $n = 1$: $E_n = -(13.6 \text{ eV})Z^2/n^2$. When striking a platinum target, an incoming electron must have at least enough energy to raise the K-shell electron from this low energy level up to the 0-eV level that corresponds to a very large distance from the nucleus. Only then will the incoming electron knock the K-shell electron out of the platinum atom.

SOLUTION   The energy of the Bohr $n = 1$ level is

$$E_1 = -(13.6 \text{ eV})\frac{Z^2}{n^2} = -(13.6 \text{ eV})\frac{77^2}{1^2} = -8.1 \times 10^4 \text{ eV}$$

In this calculation we have used 77 rather than 78 for the value of Z. In so doing, we account approximately for the fact that each of the two K-shell electrons applies a repulsive force to the other. This repulsive force balances the attractive force of one nuclear proton. In effect, one electron shields the other from the force of that proton. Therefore, to raise the K-shell electron up to the 0-eV level, the minimum energy for an incoming electron is $\boxed{8.1 \times 10^4 \text{ eV}}$. One electron volt is the kinetic energy acquired when an electron accelerates from rest through a potential difference of one volt. Thus, a potential difference of 81 000 V must be applied to the X-ray tube.

The K_α line in Figure 30.18 arises when an electron in the $n = 2$ level falls into the vacancy that the impinging electron has created in the $n = 1$ level. Similarly, the K_β line arises when an electron in the $n = 3$ level falls to the $n = 1$ level. Example 7 determines an estimate for the K_α wavelength of platinum.

Example 7   The K_α Characteristic X-Ray for Platinum

Use the Bohr model to estimate the wavelength of the K_α line in the X-ray spectrum of platinum.

REASONING AND SOLUTION   This example is very similar to Example 3, which deals with the emission line spectrum of the hydrogen atom. As in that example, we use Equation 30.14, this time with the initial value of n being $n_i = 2$ and the final value being $n_f = 1$. As in Example 6, a value of 77 rather than 78 is used for Z, to account approximately for the shielding effect of the single K-shell electron in canceling out the attraction of one nuclear proton:

$$\frac{1}{\lambda} = (1.097 \times 10^7 \text{ m}^{-1})(77^2)\left(\frac{1}{1^2} - \frac{1}{2^2}\right)$$

$$= 4.9 \times 10^{10} \text{ m}^{-1} \quad \text{or} \quad \boxed{\lambda = 2.0 \times 10^{-11} \text{ m}}$$

This answer is close to an experimental value of 1.9×10^{-11} m.

Another interesting feature of the X-ray spectrum in Figure 30.18 is the sharp cutoff that occurs at a wavelength of λ_0 on the short wavelength side of the Bremsstrahlung. This cutoff wavelength is independent of the target material, but depends on the energy of the impinging electrons. An impinging electron cannot give up any more than all of its kinetic energy when decelerated by the metal target in an X-ray tube. Thus, at most, an emitted X-ray photon can have an energy equal to the kinetic energy KE of the electron and a frequency given by Equation 29.2 as $f = (\text{KE})/h$, where h is Planck's constant. But the kinetic energy acquired by an electron in accelerating from rest through a potential difference V is eV, according to earlier discussions in Section 19.2; V is the potential difference applied across the X-ray tube. Thus, the maximum photon frequency is $f_0 = (eV)/h$. Since $f_0 = c/\lambda_0$, a maximum frequency corresponds to a minimum wavelength, which is the cutoff wavelength λ_0:

$$\lambda_0 = \frac{hc}{eV} \tag{30.17}$$

Figure 30.18, for instance, assumes a potential difference of 45 000 V, which corresponds to a cutoff wavelength of

$$\lambda_0 = \frac{(6.63 \times 10^{-34} \text{ J·s})(3.00 \times 10^8 \text{ m/s})}{(1.60 \times 10^{-19} \text{ C})(45\,000 \text{ V})} = 2.8 \times 10^{-11} \text{ m}$$

30.8 THE LASER

THE PHYSICS OF . . .

the laser.

The laser is certainly one of the most useful inventions of the twentieth century. Today, there are many types of lasers, and most of them work in a way that depends directly on the quantum mechanical structure of the atom.

When an electron makes a transition from a higher energy state to a lower energy state, a photon is emitted. The emission process can be one of two types, spontaneous emission or stimulated emission. In *spontaneous emission* (see Figure 30.19a), the photon is emitted spontaneously, in a random direction, without external provocation. In *stimulated emission* (see Figure 30.19b), an incoming photon induces or stimulates the electron to change energy levels. To produce stimulated emission, however, the incoming photon must have an energy that exactly matches the difference between the energies of the two levels, namely, $E_i - E_f$. Stimulated emission is similar to a resonance process, in which the in-

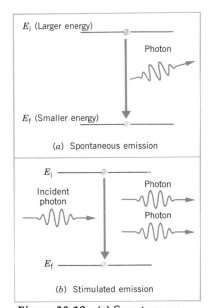

(a) Spontaneous emission

(b) Stimulated emission

Figure 30.19 (a) Spontaneous emission of a photon occurs when the electron (●) makes an unprovoked transition from a higher to a lower energy level, the photon departing in a random direction. (b) Stimulated emission of a photon occurs when an incoming photon with the correct energy induces an electron to change energy levels, the emitted photon traveling in the same direction as the incoming photon.

coming photon "jiggles" the electron at just the frequency to which it is particularly sensitive and, thus, causes the change between levels. This frequency is given by Equation 30.4 as $f = (E_i - E_f)/h$. The operation of lasers depends on stimulated emission.

Stimulated emission has three important features. First, one photon goes in and two photons come out. In this sense, the process amplifies the number of photons. In fact, this is the origin of the word "laser," which is an acronym for **l**ight **a**mplification by the **s**timulated **e**mission of **r**adiation. Second, the emitted photon travels in the same direction as the incoming photon. And third, the emitted photon is exactly in step with or has the same phase as the incoming photon. In other words, the two electromagnetic waves that these two photons represent are coherent and are locked in step with one another. In contrast, two photons emitted by the filament of an incandescent light bulb are emitted independently. They are not coherent, since one does not stimulate the emission of the other.

While stimulated emission plays a pivotal role in a laser, other factors are also important. For instance, an external source of energy must be provided to excite electrons into higher energy levels. The energy can be provided in a number of ways, including intense flashes of ordinary light and high-voltage discharges. If sufficient energy is delivered to the atoms, more electrons will be excited to a higher energy level than remain in a lower energy level, a condition known as a *population inversion.* Figure 30.20 compares a normal energy level population with a population inversion. The population inversions used in lasers involve a higher energy state that is *metastable,* in the sense that electrons remain in it for a much longer period of time than they do in an ordinary excited state (10^{-3} s versus 10^{-8} s, for example). The requirement of a metastable higher energy state is essential, so that there is more time to enhance the population inversion.

Figure 30.21 shows the widely used helium/neon laser. To sustain the necessary population inversion a high voltage is discharged across a low-pressure mixture of 15% helium and 85% neon contained in a glass tube. The laser process begins when an atom, via spontaneous emission, emits a photon parallel to the axis of the tube. This photon, via stimulated emission, causes another atom to emit two photons parallel to the tube axis. These two photons, in turn, stimulate two more atoms, yielding four photons. Four yield eight, and so on, in a kind of avalanche effect. To ensure that more and more photons are created by stimulated emission, both ends of the tube are silvered to form mirrors that reflect the photons back and forth through the helium/neon mixture. One end is only

Figure 30.20 (a) In a normal situation at room temperature, most of the electrons in atoms are found in a lower or ground state energy level. (b) If an external energy source is provided to excite electrons into a higher energy level, a population inversion can be created, in which more electrons are in the higher level than in the lower level.

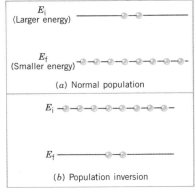

(a) Normal population

(b) Population inversion

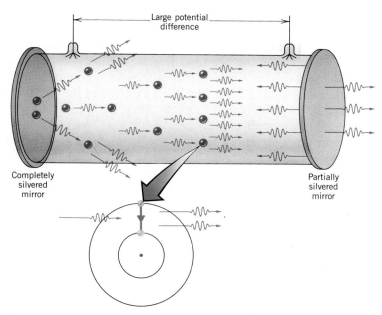

Figure 30.21 A schematic drawing of a helium/neon laser.

partially silvered, however, so that some of the photons can escape from the tube to form the laser beam. When the stimulated emission in a laser involves only a single pair of energy levels, the output beam has a single frequency or wavelength; that is, the radiation in the laser beam is monochromatic.

A laser beam is also exceptionally narrow. The width is determined by the size of the opening through which the beam exits, and very little spreading-out occurs, except that due to diffraction around the edges of the opening. A laser beam does not spread much, because any photons emitted at an angle with respect to the tube axis are quickly reflected out the sides of the tube by the silvered ends (see Figure 30.21). These ends are carefully arranged to be perpendicular to the tube axis. Since all the power in a laser beam can be confined to a narrow region, the intensity, or power per unit area, can be quite large.

Figure 30.22 shows the pertinent energy levels for a helium/neon laser. By coincidence, helium and neon have nearly identical metastable higher energy states, respectively located 20.61 and 20.66 eV above the ground state. The high-voltage discharge across the gaseous mixture excites electrons in helium atoms to the 20.61-eV state. Then, when an excited helium atom collides inelastically with

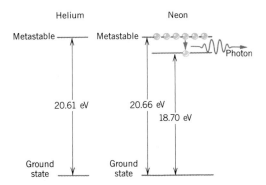

Figure 30.22 These energy levels are involved in the operation of a helium/neon laser.

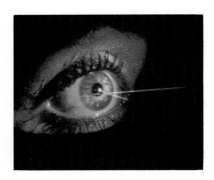

This helium-neon laser beam is being used to diagnose diseases of the eye. The use of laser technology in the field of ophthalmology is widespread.

a neon atom, the 20.61 eV of energy is given to an electron in the neon atom, along with 0.05 eV of kinetic energy from the moving atoms. As a result, the electron in the neon atom is raised to the 20.66-eV state. In this fashion, a population inversion is sustained in the neon, relative to an energy level that is 18.70 eV above the ground state. In producing the laser beam, stimulated emission causes electrons in neon to drop from the 20.66-eV level to the 18.70-eV level. The energy change of 1.96 eV corresponds to a wavelength of 633 nm, which is in the red region of the visible spectrum.

The helium/neon laser is not the only kind of laser. There are many different types, including the ruby laser, the argon-ion laser, the carbon dioxide laser, the gallium arsenide solid-state laser, and chemical dye lasers. Depending on the type and whether the laser operates continuously or in pulses, the available beam power ranges from milliwatts to megawatts. Since lasers provide coherent monochromatic electromagnetic radiation that can be confined to an intense narrow beam, they are useful in a wide variety of situations. Today they are used to reproduce music in compact disc players, to weld parts of automobile frames together, to measure distances accurately in surveying, to transmit telephone conversations and other forms of communication over long distances, and to study molecular structure. Medical applications of lasers include delicate eye surgery, removal of kidney stones, and removal of tooth and gum decay. These are only a few of the uses for lasers that have been found since the laser was invented in 1960.

INTEGRATION OF CONCEPTS

THE BOHR MODEL AND PHYSICS PRINCIPLES

Physicists often use models to help them understand how nature works. Most models, like the Bohr model of the hydrogen atom, are theoretical and exist as a collection of interconnected relations that express fundamental principles of nature. The principles are put together in a way that, hopefully, allows the physicist to predict the results of experiments. To the extent that the predictions of the model match experimental results, the model is a successful one. The Bohr model was the first to succeed in predicting the measured line spectra of the hydrogen atom.

The Bohr model brings together a remarkably wide array of physics principles to create a picture of atomic structure. The model is based on the ideas of uniform circular motion and centripetal force, as expressed by Newton's second law of motion. It incorporates the concepts of kinetic energy and potential energy. It uses Coulomb's law for the electrostatic force that one charged particle exerts on another. It uses the electrostatic potential energy that is related to the conservative nature of the electrostatic force. Bohr's model stresses the importance of angular momentum. And it takes advantage of the relation between frequency, wavelength, and the speed of an electromagnetic wave. All of the ideas above are the older and traditional ones that have become known as classical physics.

The Bohr model not only draws on the ideas of classical physics. It also combines them with many of the newer concepts of physics that emerged during the early twentieth century and are known as modern physics. It uses

Planck's idea of quantized energy levels and Einstein's idea of photons. And it introduces the ideas of radiationless orbits and quantized angular momentum.

Today, the Schrödinger equation provides us with the quantum mechanical picture of the atom, which supersedes the Bohr model. Nevertheless, we should not lose sight of the fact that the Bohr model is a *tour de force* of integrating the principles of classical and modern physics.

SUMMARY

The idea of a **nuclear atom** originated in 1911, as a result of experiments by Ernest Rutherford, in which α particles were scattered by a thin metal foil. The phrase "nuclear atom" refers to the fact that an atom consists of a small, positively charged nucleus surrounded at relatively large distances by a number of electrons, whose negative charge equals the positive nuclear charge when the atom is electrically neutral.

A **line spectrum** is a series of discrete electromagnetic wavelengths emitted by the atoms of a low-pressure gas that is subjected to a sufficiently high potential difference. Certain groups of discrete wavelengths are referred to as "series." The line spectrum of atomic hydrogen includes the **Lyman series,** the **Balmer series,** and the **Paschen series** of wavelengths.

The Bohr model applies to atoms that have only a single electron orbiting a nucleus containing Z protons. This model assumes that the electron exists in circular orbits that are called **stationary orbits,** because the electron does not radiate electromagnetic waves while in them. According to this model, a photon is emitted only when an electron changes from a higher energy orbit to a lower energy orbit. The model also assumes that the orbital angular momentum L_n of the electron can only have values that are integer multiples of Planck's constant divided by 2π: $L_n = n(h/2\pi)$; $n = 1, 2, 3, \ldots$. With the assumptions above, it can be shown that the nth Bohr orbit has a radius of $r_n = (5.29 \times 10^{-11} \text{ m})(n^2/Z)$ and that the total energy associated with this orbit is $E_n = -(13.6 \text{ eV})(Z^2/n^2)$. The **ionization energy** is the energy needed to remove an electron completely from an atom. The Bohr model predicts that the wavelengths comprising the line spectrum emitted by an atom are given by Equation 30.14.

Quantum mechanics describes the hydrogen atom in terms of four quantum numbers: (1) **the principal quantum number** n, which can have the integer values $n = 1, 2, 3, \ldots$; (2) **the orbital quantum number** ℓ, which can have the integer values $\ell = 0, 1, 2, \ldots$, $(n - 1)$; (3) **the magnetic quantum number** m_ℓ, which can have the positive and negative integer values $m_\ell = 0, \pm 1, \pm 2, \ldots, \pm \ell$; and (4) **the spin quantum number** m_s, which, for an electron, can be either $m_s = +\frac{1}{2}$ or $m_s = -\frac{1}{2}$. According to quantum mechanics, an electron does not reside in a circular orbit but, rather, has some probability of being found at various distances from the nucleus.

The Pauli exclusion principle states that no two electrons in an atom can have the same set of values for the four quantum numbers n, ℓ, m_ℓ, and m_s. This principle determines the way in which the electrons in multiple-electron atoms are distributed into shells (defined by the value of n) and subshells (defined by the values of n and ℓ). Table 30.2 summarizes the conventional notation for atomic subshells. The arrangement of the periodic table of the elements is related to the exclusion principle.

X-rays are electromagnetic waves emitted when high-energy electrons strike a metal target contained within an evacuated glass tube. The emitted X-ray spectrum of wavelengths consists of sharp "peaks" or "lines," called **characteristic X-rays,** superimposed on a broad continuous range of wavelengths called **Bremsstrahlung.** The minimum wavelength, or cutoff wavelength, of the Bremsstrahlung is determined by the kinetic energy of the electrons striking the target in the X-ray tube.

A **laser** is a device that generates electromagnetic waves via a process known as **stimulated emission.** In this process, one photon stimulates the production of another photon, by causing an electron in an atom to fall from a higher energy level to a lower energy level. Because of this mechanism of photon production, the electromagnetic waves generated by a laser are coherent and may be confined to a very narrow beam.

QUESTIONS

1. At room temperature, most of the atoms of atomic hydrogen contain electrons that are in the ground state or $n = 1$ energy level. A tube is filled with atomic hydrogen. Electromagnetic radiation with a continuous spectrum of wavelengths, including those in the Lyman, Balmer, and Paschen series, enters one end of this tube and leaves the other end. The exiting radiation is found to contain absorption lines. To which one (or more) of the series do the wavelengths of these absorption lines correspond? Explain.

2. When the outermost electron in an atom is in an excited state, the atom is more easily ionized than when the outermost electron is in the ground state. Why?

3. For a principal quantum number of $n = 2$, is it possible for the electron to have zero orbital angular momentum L in (a) the Bohr model and (b) quantum mechanics? Give your reasoning in each case.

4. In the Bohr model for the hydrogen atom, the closer the electron is to the nucleus, the smaller is the total energy of the atom. Is this also true in the quantum mechanical picture of the hydrogen atom? Justify your answer.

5. Consider two different hydrogen atoms. The electron in each atom is in an excited state. Is it possible for the two electrons to have different energies but the same orbital angular momentum L (a) according to the Bohr model and (b) according to quantum mechanics? Is it possible for the two electrons to have different orbital angular momenta but the same energy (c) according to the Bohr model and (d) according to quantum mechanics? Account for your answer in each case.

6. Can a 5g subshell contain (a) 22 electrons and (b) 17 electrons? Why?

7. Explain why you would not expect hydrogen and helium atoms in their ground state to emit characteristic X-rays.

8. The drawing shows the X-ray spectra produced by an X-ray tube when the tube is operated at two different potential differences. Explain why the characteristic lines occur at the same wavelengths in the two spectra, while the cutoff wavelength λ_0 shifts to the right when a smaller potential difference is used to operate the tube.

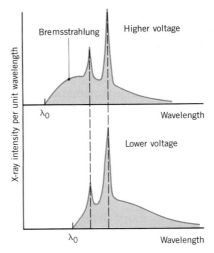

9. The short wavelength side of X-ray spectra ends abruptly at a cutoff wavelength λ_0 (see Figure 30.18). Does this cutoff wavelength depend on the target material used in the X-ray tube? Give your reasoning.

10. Explain why a laser beam focused to a small spot can cut through a piece of metal.

PROBLEMS

In working these problems, ignore relativistic effects.

Section 30.1 Rutherford Scattering and the Nuclear Atom

1. The electron in the hydrogen atom is in orbit around the nucleus. The radius of the orbit is about 53 000 times the radius of the nucleus. The earth is similarly in orbit around the sun. The sun has a radius of 6.96×10^8 m. If the scale of the earth–sun system were the same as that in the hydrogen atom, what would be the radius of the earth's orbit? For comparison, Pluto is the most distant planet and has a mean distance from the sun of about 5.9×10^{12} m.

2. The nucleus of the hydrogen atom has a radius of about 1×10^{-15} m. The electron is normally at a distance of about 5.3×10^{-11} m from the nucleus. Assuming the hydrogen atom is a sphere with a radius of 5.3×10^{-11} m, find (a) the volume of the atom, (b) the volume of the nucleus, and (c) the percentage of the volume of the atom that is occupied by the nucleus.

3. The mass of an α particle is 6.64×10^{-27} kg. An α particle used in a scattering experiment has a kinetic energy of 7.00×10^{-13} J. What is the de Broglie wavelength of the particle?

4. There are Z protons in the nucleus of an atom, where Z is the atomic number of the element. An α particle carries a charge of $+2e$. In a scattering experiment, an α particle, heading directly toward a nucleus in a metal foil, will come to a halt when all the particle's kinetic energy is converted to electric potential energy. In such a situation, how close will an α particle with a kinetic energy of 5.0×10^{-13} J come to a gold nucleus ($Z = 79$)?

*5. The nucleus of an aluminum atom contains 13 protons and has a radius of 3.6×10^{-15} m. How much work (in electron volts) must be done to bring a proton from infinity, where it is at rest, to the "surface" of an aluminum nucleus?

Section 30.2 Line Spectra, Section 30.3 The Bohr Model of the Hydrogen Atom

6. If the line with the longest wavelength in the Balmer series for atomic hydrogen is counted as the first line, what is the wavelength of the third line?

7. On a piece of graph paper, make a copy of Figure 30.8. This figure shows the energy level diagram for the hydrogen atom ($Z = 1$), as determined from Equation 30.13. On the same piece of graph paper, alongside the hydrogen atom diagram, draw to scale the energy level diagram that Equation 30.13 predicts for singly ionized helium He⁺ ($Z = 2$). In these drawings include only the first two energy levels for hydrogen and the first four levels for He⁺. Which levels have the same energy in both diagrams?

8. In the line spectrum of atomic hydrogen there is also a group of lines known as the Pfund series. These lines are produced when electrons, excited to high energy levels, make transitions to the $n = 5$ level. Determine (a) the longest wavelength and (b) the shortest wavelength in this series. (c) Refer to Figure 24.6 and identify the region of the electromagnetic spectrum in which these lines are found.

9. Determine the ionization energy (in electron volts) that is needed to remove the remaining electron from a singly ionized helium atom He⁺ ($Z = 2$).

10. Find the energy (in joules) of the photon that is emitted when the electron in a hydrogen atom undergoes a transition from the $n = 7$ energy level to produce a line in the Paschen series.

11. What is the radius for the $n = 5$ Bohr orbit in a doubly ionized lithium atom Li²⁺ ($Z = 3$)?

12. Using the Bohr model, compare the nth orbit of a triply ionized beryllium atom Be³⁺ ($Z = 4$) to the nth orbit of a hydrogen atom (H) by calculating the ratio (Be³⁺/H) of the following quantities: (a) the energies and (b) the radii.

13. The electron in a hydrogen atom is in the first excited state, when the electron acquires an additional 2.86 eV of energy. What is the quantum number of the state into which the electron moves?

*14. For atomic hydrogen, the Paschen series of lines occurs when $n_f = 3$, while the Brackett series occurs when $n_f = 4$ in

Equation 30.14. Using this equation, show that the ranges of wavelengths in these two series overlap.

*15. In the Bohr model, Equation 30.12 or 30.13 gives the total energy (kinetic plus potential). In a certain Bohr orbit, the total energy is -4.90 eV. For this orbit, determine the kinetic energy and the electric potential energy of the electron.

*16. In an unidentified ionized atom, only one electron moves about the nucleus. The radius of the $n = 3$ Bohr orbit is 2.38×10^{-10} m. What is the energy (in electron volts) of the $n = 7$ orbit for this atom?

*17. The Bohr model can be applied to singly ionized helium He⁺ ($Z = 2$). Using this model, consider the series of lines that is produced when the electron makes a transition from higher energy levels into the $n_f = 4$ level. Some of the lines in this series lie in the visible region of the spectrum (380–750 nm). What are the values of n_i for the energy levels from which the electron makes the transitions corresponding to these lines?

**18. (a) Derive an expression for the velocity of the electron in the nth Bohr orbit, in terms of Z, n, and the constants k, e, and h. For the hydrogen atom, determine the velocity in (b) the $n = 1$ orbit and (c) the $n = 2$ orbit. (d) Generally, when speeds are less than one-tenth the speed of light, the effects of special relativity can be ignored. Do the speeds found in (b) and (c) justify ignoring relativistic effects in the Bohr model?

**19. A diffraction grating is used in the first order to separate the wavelengths in the Balmer series of atomic hydrogen (Section 27.7 discusses diffraction gratings). The grating and an observation screen are separated by a distance of 75.0 cm, as Figure 27.29 illustrates. You may assume that θ is small, so $\sin \theta \approx \theta$ when radian measure is used for θ. How many lines per centimeter should the grating have, so the longest and the next-to-the-longest wavelengths in the series are separated by 5.00 cm on the screen?

Section 30.5 The Quantum Mechanical Picture of the Hydrogen Atom

20. Suppose the value of the principle quantum number is $n = 4$. What are the possible values for the magnetic quantum number m_ℓ?

21. Write down the eighteen possible sets of the four quantum numbers when the principal quantum number is $n = 3$.

22. The principal quantum number for an electron in an atom is $n = 6$, while the magnetic quantum number is $m_\ell = 2$. What possible values for the orbital quantum number ℓ could this electron have?

23. An electron in an atom has a value for the magnetic quantum number of $m_\ell = 4$. What are the *minimum* values that (a) the orbital quantum number and (b) the principal quantum number can have?

*24. For an electron in a hydrogen atom, the z component of the angular momentum has a *maximum* value of $L_z = 2.11 \times 10^{-34}$ J·s. Find the three smallest possible values (algebraically) for the total energy (in electron volts) that this atom could have.

*25. For the hydrogen atom, the Bohr model and quantum mechanics both give the same value for the energy of the nth state. However, they do not give the same value for the orbital angular momentum L. (a) For $n = 1$, determine the values of L (in units of $h/2\pi$) predicted by the Bohr model and quantum mechanics. (b) Repeat part (a) for $n = 2$, noting that quantum mechanics permits more than one value of ℓ when the electron is in the $n = 2$ state.

Section 30.6 The Pauli Exclusion Principle and the Periodic Table of the Elements

26. In the style indicated in Table 30.3, write down the ground state electronic configuration of calcium Ca ($Z = 20$).

27. In the style shown in Table 30.3, write down the ground state electronic configuration for arsenic As ($Z = 33$). Note from Figure 30.15 that the 4s subshell fills before the 3d subshell.

28. Figure 30.15 was constructed using the Pauli exclusion principle and indicates that the $n = 1$ shell holds 2 electrons, the $n = 2$ shell holds 8 electrons, and the $n = 3$ shell holds 18 electrons. These numbers can be obtained by adding the numbers given in the figure for the subshells contained within a given shell. How many electrons can be put into (a) the $n = 4$ shell and (b) the $n = 5$ shell, neither of which is completely shown in the figure?

29. When an electron makes a transition between energy levels of an atom, there are no restrictions on the initial and final values of the principal quantum number n. According to quantum mechanics, however, there is a rule that restricts the initial and final values of the orbital quantum number ℓ. This rule is called a *selection rule* and states that $\Delta \ell = \pm 1$. In other words, when an electron makes a transition between energy levels, the value of ℓ can only increase or decrease by one. The value of ℓ may not remain the same or increase or decrease by more than one. According to this rule, which of the following energy level transitions are allowed: (a) 2s → 1s, (b) 2p → 1s, (c) 4p → 2p, (d) 4s → 2p, and (e) 3d → 3s?

Section 30.7 X-Rays

30. Suppose an X-ray machine in a doctor's office uses a potential difference of 61 kV to operate the X-ray tube. What is the shortest X-ray wavelength emitted by this machine?

31. By using the Bohr model to estimate the atomic number

Z, decide which element is likely to emit a K_α X-ray with a wavelength of 4.5×10^{-9} m.

32. Molybdenum has an atomic number of $Z = 42$. Using the Bohr model, estimate the wavelength of the K_α X-ray.

33. An X-ray tube is being operated at a potential difference of 45.0 kV. What is the Bremsstrahlung wavelength that corresponds to 25.0% of the kinetic energy with which an electron collides with the metal target in the tube?

*34. Metal A emits a K_α X-ray that has a wavelength of 4.7×10^{-11} m. Metal B emits a K_α X-ray that has a wavelength of 1.8×10^{-11} m. The atomic number of metal A is $Z_A = 51$. Use the Bohr model to find an estimate for Z_B, the atomic number of metal B.

Section 30.8 The Laser

35. A laser used in the removal of tooth decay has a wavelength of 193 nm. What is the energy of a photon that this laser produces?

36. In the helium/neon laser, there is an energy difference of 1.96 eV between the levels that participate in stimulated emission. Verify, by performing a calculation, that the laser produces a wavelength of 633 nm.

37. A carbon dioxide laser produces a wavelength of 1.06×10^{-5} m. A semiconductor laser produces a wavelength of 7.90×10^{-7} m. (a) Which laser produces the more energetic photons? Explain. (b) By what factor are they more energetic?

38. A laser is used in eye surgery to weld a detached retina back into place. The wavelength of the laser beam is 514 nm, while the power is 2.0 W. During surgery, the laser beam is turned on for 0.10 s. During this time, how many photons are emitted by the laser?

ADDITIONAL PROBLEMS

39. Following the style used in Table 30.3, determine the electronic configuration of the ground state for yttrium Y ($Z = 39$). Refer to Figure 30.15 to see the order in which the subshells fill.

40. What is the minimum potential difference that must be applied to an X-ray tube to knock a K-shell electron completely out of an atom in a copper ($Z = 29$) target? Use the Bohr model as needed.

41. It is possible to use electromagnetic radiation to ionize atoms. To do so, the atoms must absorb the radiation, the photons of which must have enough energy to remove an electron from an atom. What is the longest radiation wave-

length that can be used to ionize the ground state hydrogen atom?

42. Write down the fourteen sets of the four quantum numbers that correspond to the electrons in a completely filled 4f subshell.

43. The K_β characteristic X-ray line for tungsten has a wavelength of 1.84×10^{-11} m. What is the difference in energy between the two energy levels that give rise to this line? Express the answer in joules and in electron volts.

44. A 790-kg synchronous communications satellite has a period of one day. The radius of the orbit is 4.23×10^7 m. Suppose that Bohr's assumption about the angular momentum of the electron in orbit about the nucleus applies to this satellite. What is the quantum number n for the orbit of the satellite?

*45. The total orbital angular momentum of the electron in a hydrogen atom has a magnitude of $L = 3.66 \times 10^{-34}$ J·s. What values can the angular momentum component L_z have?

*46. In the line spectrum emitted by doubly ionized lithium atoms Li^{2+} ($Z = 3$), the shortest wavelength in one series of lines is 162.1 nm. This line is produced when an electron makes a transition to a final energy level whose principal quantum number is n_f. What is the value of n_f?

*47. Consider a particle of mass m that can exist only between $x = 0$ and $x = +L$ on the x axis. We could say that this particle is confined to a "box" of length L. In this situation, imagine the standing de Broglie waves that can fit into the box. For example, the drawing shows the first three possibilities. Note in this picture that there are either one, two, or three half-wavelengths that fit into the distance L. Use Equation 29.8 for the de Broglie wavelength of a particle and derive an expression for the allowed energies (only kinetic energy) that the particle can have. This expression involves m, L, Planck's

constant, and a quantum number n that can have only the values 1, 2, 3, ...

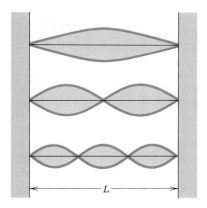

*48. A wavelength of 410.2 nm is emitted by the hydrogen atoms in a high-voltage discharge tube. What are the initial and final values of the quantum number n for the energy level transition that produces this wavelength? (*Hint: Identify the region of the electromagnetic spectrum in which the given wavelength is to be found.*)

**49. (a) Derive an expression for the time it takes the electron in the nth Bohr orbit to make one complete revolution around the nucleus. Express your answer in terms of Z, n, and the constants k, e, h, and m. For a hydrogen atom, determine this time for (b) the $n = 1$ orbit and (c) the $n = 2$ orbit.

**50. A certain species of ionized atoms produces an emission line spectrum according to the Bohr model, but the number of protons Z in the nucleus is unknown. A group of lines in the spectrum forms a series in which the shortest wavelength is 22.79 nm and the longest wavelength is 41.02 nm. Find the next-to-the-longest wavelength in the series of lines.

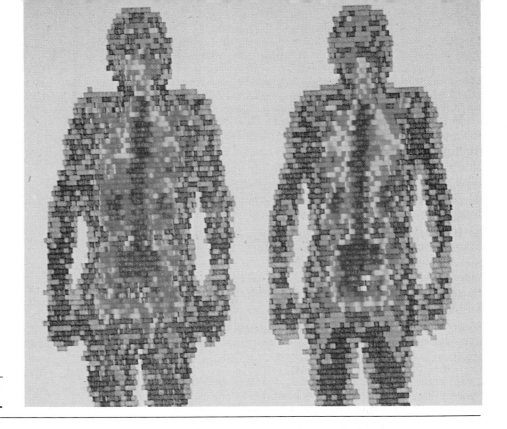

CHAPTER 31

NUCLEAR PHYSICS AND RADIOACTIVITY

The nucleus of an atom is incredibly small, for if the diameter of the atom were enlarged to the size of a football field, the nucleus would only be the size of a small BB. The nucleus, however, contains more than 99.9% of the mass of the atom, because each of its constituents has a mass that is about 1800 times greater than that of the electron. We will see that many nuclei, particularly those of the heavier elements, are unstable, and they spontaneously break apart, or disintegrate, into other nuclei. This spontaneous disintegration is called radioactive decay, and it is accompanied by the emission of certain types of particles and high-energy photons called gamma rays. There are many beneficial uses of radioactive decay in medicine. For example, the photograph above shows a bone scan of a healthy person who has been injected with a radioactive material or isotope, which tends to concentrate in the spine, ribs, and pelvis. The front of the person is on the left and the back is on the right. This scan was made by a gamma camera that detects the gamma rays emitted by the isotope. The scans are color-coded according to the intensity of the gamma rays, ranging from blue for the lowest emission, through green, yellow, orange, and brown for the highest emission. Such scans can reveal whether cancer has invaded the bone structure. Another application of radioactive decay is found in archeology, where radioactive dating is used to determine the age of ancient artifacts. In this chapter, we will examine some of the principles of nuclear physics and radioactive decay.

31.1 NUCLEAR STRUCTURE

Atoms consist of electrons in orbit about a central nucleus. As we have seen in Chapter 30, the electron orbits are quantum mechanical in nature and have interesting characteristics. In our previous discussion, however, little has been said about the nucleus itself. But the nucleus is fascinating in its own right, and now we consider it in greater detail.

The nucleus of an atom consists of neutrons and protons, collectively referred to as *nucleons.* The *neutron,* discovered in 1932 by the English physicist James Chadwick (1891–1974) carries no electrical charge and has a mass slightly larger than that of a proton (see Table 31.1).

The number of protons in the nucleus is different in different elements and is given by the *atomic number Z.* In an electrically neutral atom, the number of nuclear protons equals the number of electrons in orbit around the nucleus. The number of neutrons in the nucleus is N. The total number of protons and neutrons is referred to as the *atomic mass number A,* because the total nuclear mass is *approximately* equal to A times the mass of a single nucleon:

$$A = Z + N \qquad (31.1)$$

Sometimes, A is also called the *nucleon number.* A shorthand notation is often used to specify Z and A along with the chemical symbol for the element. For instance, the nuclei of all naturally occurring aluminum atoms have $A = 27$, and the atomic number for aluminum is $Z = 13$. In shorthand notation, then, the aluminum nucleus is specified as $^{27}_{13}\text{Al}$. The number of neutrons in an aluminum nucleus is $N = 14$. In general, for an element whose chemical symbol is X, the symbol for the nucleus is

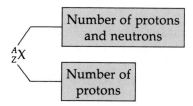

For a proton the symbol is ^1_1H, since the proton is the nucleus of a hydrogen atom. A neutron is denoted by ^1_0n. In the case of an electron we use $^{\ 0}_{-1}\text{e}$, where $A = 0$ because an electron has no nucleus and $Z = -1$ because the electron has a negative charge.

Nuclei that contain the same number of protons, but a different number of neutrons are known as *isotopes.* Carbon, for example, occurs in nature in two stable forms. In most carbon atoms (98.90%), the nucleus is the $^{12}_6\text{C}$ isotope,

Table 31.1   Properties of Particles in the Atom

Particle	Electric Charge (C)	Mass	
		Kilograms (kg)	Atomic Mass Units (u)
Electron	-1.60×10^{-19}	$9.109\ 390 \times 10^{-31}$	$5.485\ 799 \times 10^{-4}$
Proton	$+1.60 \times 10^{-19}$	$1.672\ 623 \times 10^{-27}$	$1.007\ 276$
Neutron	0	$1.674\ 929 \times 10^{-27}$	$1.008\ 665$

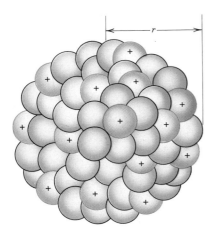

Figure 31.1 The nucleus is approximately spherical (radius = r) and contains protons (⊕) clustered closely together with neutrons (◯).

consisting of six protons and six neutrons. A small fraction (1.10%), however, contains nuclei that have six protons and seven neutrons, namely, the $^{13}_{6}$C isotope. The percentages given above are the natural abundances of the isotopes. The atomic masses in the periodic table are average atomic masses, taking into account the abundances of the various isotopes.

The protons and neutrons in the nucleus are clustered together to form an approximately spherical region, as Figure 31.1 illustrates. Experiment shows that the radius r of the nucleus depends on the atomic mass number A and is given approximately in meters by

$$r \approx (1.2 \times 10^{-15} \text{ m})A^{1/3} \tag{31.2}$$

The radius of the aluminum nucleus ($A = 27$), for example, is $r \approx (1.2 \times 10^{-15} \text{ m})27^{1/3} = 3.6 \times 10^{-15}$ m.

We can interpret Equation 31.2 to mean that the nuclear density, or the nuclear mass per unit volume, is approximately the same for all nuclei. The reasoning is as follows. Equation 31.2 indicates that r^3 is proportional to the nucleon number A. But $\frac{4}{3}\pi r^3$ is the volume of a sphere, so the volume of the nucleus is proportional to the number of nucleons it contains, the nucleons being clustered together as incompressible pieces of matter. The nucleon number, in turn, is nearly proportional to the total nuclear mass, since all nucleons have roughly the same mass. Thus, the volume and mass of a nucleus are nearly proportional, and the nuclear mass per unit volume is approximately the same for all nuclei.

31.2 THE STRONG NUCLEAR FORCE AND THE STABILITY OF THE NUCLEUS

Two positive charges that are as close together as they are in a nucleus repel one another with a very strong electrostatic force. What, then, keeps the nucleus from flying apart? Clearly, some kind of attractive force must hold the nucleus together, since many kinds of naturally occurring atoms contain stable nuclei. The gravitational force of attraction between nucleons is too weak to counteract the repulsive electric force, so it must be that a different type of force exists within the nucleus. This force is the *strong nuclear force* and is one of only four fundamental forces that have been discovered, fundamental in the sense that all forces in nature can be explained in terms of these four. We have already encountered two other fundamental forces, the gravitational force and the electromagnetic force. The remaining one will be mentioned in Section 31.5.

Many features of the strong nuclear force are well known. The strong nuclear force is independent of electric charge. At a given separation distance, the same nuclear force of attraction exists between two protons, two neutrons, or between a proton and a neutron. The range of action of the strong nuclear force is extremely short, with the force of attraction being very strong when two nucleons are as close as 10^{-15} m and essentially zero at larger distances. In contrast, the electric force between two protons decreases to zero only gradually as the separation distance increases to large values and, therefore, has a relatively long range of action.

The limited range of action of the strong nuclear force plays an important role in the stability of the nucleus. For a nucleus to be stable, the electrostatic repulsion between the protons must be balanced by the attraction between the nucleons

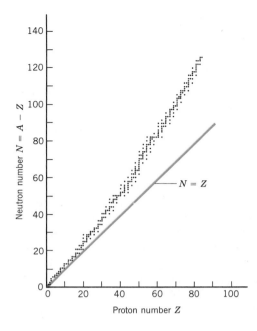

Figure 31.2 With few exceptions, the naturally occurring stable nuclei have a number N of neutrons that equals or exceeds the number Z of protons.

due to the strong nuclear force. But one proton repels all other protons within the nucleus, since the electrostatic force has such a long range of action. In contrast, a proton or a neutron attracts only its nearest neighbors via the strong nuclear force, because the range of the strong nuclear force is very limited. As the number Z of protons in the nucleus increases under these conditions, the number N of neutrons has to increase even more, if stability is to be maintained. Figure 31.2 shows a plot of N versus Z for naturally occurring elements that have stable nuclei. For reference, the plot also includes the straight line that represents the condition $N = Z$. With few exceptions, the points representing stable nuclei fall above this reference line, reflecting the fact that the number of neutrons becomes greater than the number of protons as the atomic number Z increases.

As more and more protons occur in a nucleus, there comes a point when a balance of repulsive and attractive forces cannot be achieved by an increased number of neutrons. Eventually, the limited range of action of the strong nuclear force prevents extra neutrons from balancing the long-range electric repulsion of extra protons. The stable nucleus with the largest number of protons ($Z = 83$) is that of bismuth, $^{209}_{83}\text{Bi}$, which contains 126 neutrons. All nuclei with more than 83 protons (e.g., uranium ($Z = 92$)) are unstable and spontaneously break apart or rearrange their internal structures as time passes. This spontaneous disintegration or rearrangement of internal structure is called *radioactivity*, first discovered in 1896 by the French physicist Henri Becquerel (1852–1908).

When an unstable nucleus disintegrates, certain kinds of particles and/or high-energy photons are released. These particles and photons are collectively called "rays." Section 31.4 will discuss three kinds of rays produced by naturally occurring radioactivity, namely, *α rays*, *β rays*, and *γ rays.* They are named according to the first three letters of the Greek alphabet, alpha (α), beta (β), and gamma (γ), to indicate the extent of their ability to penetrate matter. α rays are the least penetrating, being blocked by a thin (≈ 0.01-mm) sheet of lead, while β rays penetrate into lead much farther (≈ 0.1 mm). γ rays are the most penetrating and can pass through an appreciable thickness (≈ 100 mm) of lead.

31.3 THE MASS DEFECT OF THE NUCLEUS AND NUCLEAR BINDING ENERGY

Because of the strong nuclear force, the nucleons in a stable nucleus are held tightly together. Therefore, energy is required to separate a stable nucleus into its constituent protons and neutrons, as Figure 31.3 illustrates. The more stable the nucleus is, the greater the amount of energy needed to break it apart. The required energy is called the *binding energy* of the nucleus.

In Einstein's theory of special relativity, energy and mass are equivalent. A change Δm in the mass of a system is equivalent to a change ΔE_0 in the total rest energy of the system by an amount $\Delta E_0 = (\Delta m)c^2$, where c is the speed of light in a vacuum. Thus, in Figure 31.3, the binding energy used to disassemble the nucleus appears as extra mass of the separated nucleons. In other words, the sum of the individual masses of the separated protons and neutrons is greater by an amount Δm than the mass of the stable nucleus. The difference in mass Δm is known as the *mass defect* of the nucleus. As Example 1 shows, the binding energy of a nucleus can be determined from the mass defect according to Equation 31.3:

$$\text{Binding energy} = (\text{Mass defect})c^2 = (\Delta m)c^2 \qquad (31.3)$$

Figure 31.3 Energy must be supplied to break the nucleus apart into its constituent protons and neutrons. Each of the separated nucleons is at rest and out of the range of the forces of the other nucleons.

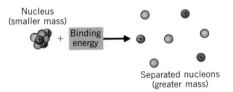

Nucleus (smaller mass)

+ Binding energy →

Separated nucleons (greater mass)

Example 1   The Binding Energy of the Helium Nucleus

The most abundant isotope of helium has a $_2^4$He nucleus whose mass is 6.6447×10^{-27} kg. For this nucleus, find (a) the mass defect and (b) the binding energy.

REASONING   The symbol $_2^4$He indicates that the helium nucleus contains $Z = 2$ protons and $N = 4 - 2 = 2$ neutrons. To obtain the mass defect Δm, we first determine the sum of the individual masses of the separated protons and neutrons. Then we subtract from this sum the mass of the $_2^4$He nucleus. Finally, we use Equation 31.3 to calculate the binding energy from the value for Δm.

SOLUTION
(a) Using data from Table 31.1, we find that the sum of the individual masses of the nucleons is

$$\underbrace{2(1.6726 \times 10^{-27}\,\text{kg})}_{\text{Two protons}} + \underbrace{2(1.6749 \times 10^{-27}\,\text{kg})}_{\text{Two neutrons}} = 6.6950 \times 10^{-27}\,\text{kg}$$

This value is greater than the mass of the intact $_2^4$He nucleus, and the mass defect is

$$\Delta m = 6.6950 \times 10^{-27}\,\text{kg} - 6.6447 \times 10^{-27}\,\text{kg} = \boxed{0.0503 \times 10^{-27}\,\text{kg}}$$

(b) According to Equation 31.3, the binding energy is

$$\frac{\text{Binding}}{\text{energy}} = (\Delta m)c^2 = (0.0503 \times 10^{-27} \text{ kg})(3.00 \times 10^8 \text{ m/s})^2 = 4.53 \times 10^{-12} \text{ J}$$

Usually, binding energies are expressed in energy units of electron volts instead of joules (1 eV = 1.60×10^{-19} J):

$$\frac{\text{Binding}}{\text{energy}} = (4.53 \times 10^{-12} \text{ J})\left(\frac{1 \text{ eV}}{1.60 \times 10^{-19} \text{ J}}\right) = 2.83 \times 10^7 \text{ eV} = \boxed{28.3 \text{ MeV}}$$

This value is more than two million times greater than the energy required to remove an orbital electron from an atom.

In calculations such as that in Example 1, it is customary to use the *atomic mass unit* (u) instead of the kilogram. As introduced in Section 14.1, the atomic mass unit is one-twelfth of the mass of a $^{12}_{6}$C atom of carbon. In terms of this unit, the mass of a $^{12}_{6}$C atom is exactly 12 u. Table 31.1 also gives the masses of the electron, the proton, and the neutron in atomic mass units. For future use, the energy equivalent of one atomic mass unit can be calculated by observing that the mass of a proton is 1.6726×10^{-27} kg or 1.0073 u, so that

$$1 \text{ u} = (1 \text{ u})\left(\frac{1.6726 \times 10^{-27} \text{ kg}}{1.0073 \text{ u}}\right) = 1.6605 \times 10^{-27} \text{ kg}$$

and

$$\Delta E_0 = (\Delta m)c^2 = (1.6605 \times 10^{-27} \text{ kg})(2.9979 \times 10^8 \text{ m/s})^2 = 1.4924 \times 10^{-10} \text{ J}$$

In electron volts, therefore, one atomic mass unit is equivalent to

$$1 \text{ u} = (1.4924 \times 10^{-10} \text{ J})\left(\frac{1 \text{ eV}}{1.6022 \times 10^{-19} \text{ J}}\right) = 9.315 \times 10^8 \text{ eV} = 931.5 \text{ MeV}$$

A table of the isotopes, such as that in Appendix F, gives masses in atomic mass units. Typically, however, the given masses are not nuclear masses. They are *atomic masses*, that is, the masses of neutral atoms, including the mass of the orbital electrons. Example 2 deals again with the $^{4}_{2}$He nucleus and shows how to take into account the effect of the orbital electrons when using data from a table of isotopes to determine energies.

Example 2   The Binding Energy of the Helium Nucleus, Revisited

Using atomic mass units instead of kilograms, obtain the binding energy of the $^{4}_{2}$He nucleus.

REASONING   To determine the binding energy, we calculate the mass defect in atomic mass units and then use the fact that one atomic mass unit is equivalent to 931.5 MeV of energy. The table in Appendix F gives a mass of 4.0026 u for $^{4}_{2}$He, *which includes the mass of the two electrons in the neutral helium atom.* To calculate the mass defect, we must subtract 4.0026 u from the sum of the individual masses of the nucleons, including the mass of the electrons. As Figure 31.4 illustrates, the electron mass will be included if the masses of two hydrogen atoms are used in the calculation instead of the masses of two protons. The mass of a $^{1}_{1}$H hydrogen atom is 1.0078 u according to Appendix F, and the mass of a neutron is given in Table 31.1 as 1.0087 u.

<u>SOLUTION</u>   The sum of the individual masses is

$$\underbrace{2(1.0078\ u)}_{\substack{\text{Two hydrogen}\\\text{atoms}}} + \underbrace{2(1.0087\ u)}_{\text{Two neutrons}} = 4.0330\ u$$

The mass defect is $\Delta m = 4.0330\ u - 4.0026\ u = 0.0304\ u$. Since 1 u is equivalent to 931.5 MeV, the binding energy is $\boxed{\text{Binding energy} = 28.3\ \text{MeV}}$, which matches that obtained in Example 1.

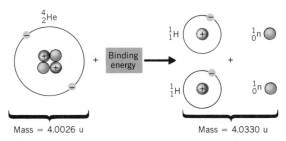

Mass = 4.0026 u   Mass = 4.0330 u

Figure 31.4   Tables of isotopes usually give the mass of the neutral atom (including the orbital electrons), rather than the mass of the nucleus. When using data from such tables to determine the mass defect of a nucleus, the mass of the orbital electrons must be taken into account, as this drawing illustrates for the ^4_2He isotope of helium.

To see how the nuclear binding energy varies from nucleus to nucleus, it is necessary to compare the binding energy for each nucleus on a per-nucleon basis. Figure 31.5 shows a graph in which the binding energy divided by the nucleon number A is plotted against the nucleon number itself. In the graph, the peak for the ^4_2He isotope of helium indicates that the ^4_2He nucleus is particularly stable. The binding energy per nucleon increases rapidly for nuclei with small masses and reaches a maximum of approximately 8.7 MeV/nucleon for a nucleon number of about $A = 60$. For greater nucleon numbers, the binding energy per nucleon decreases gradually. Eventually, the binding energy per nucleon decreases

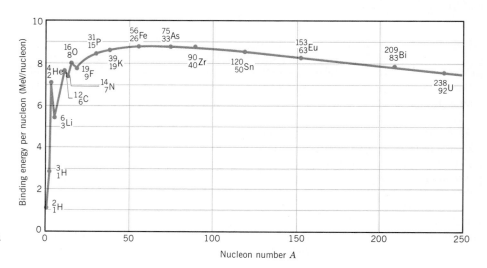

Figure 31.5   A plot of binding energy per nucleon versus the nucleon number A.

enough so there is insufficient binding energy to hold the nucleus together. Nuclei more massive than the $^{209}_{83}$Bi nucleus of bismuth are unstable and hence radioactive.

31.4 RADIOACTIVITY

When an unstable (radioactive) nucleus disintegrates spontaneously and produces α, β, or γ rays, the process must obey the laws of physics that are summarized below:

1. The conservation of *mass/energy* (see Sections 6.8 and 28.6)
2. The conservation of *electric charge* (see Section 18.2)
3. The conservation of *linear momentum* (see Section 7.2)
4. The conservation of *angular momentum* (See Section 9.6)
5. The conservation of *nucleon number*

Except for the conservation of nucleon number, these laws have been discussed earlier. No reaction has ever been observed in which the number of nucleons present before the reaction has differed from the number of nucleons after the reaction. Therefore, the number of nucleons is conserved during a nuclear disintegration. As applied to the disintegration of a nucleus, the laws require that the energy, electric charge, linear momentum, angular momentum, and nucleon number that a nucleus possesses must remain unchanged when the nucleus disintegrates into nuclear fragments and accompanying α, β, or γ rays.

The three types of radioactivity that occur naturally can be observed in a relatively simple experiment. A piece of radioactive material is placed at the bottom of a narrow hole in a lead cylinder. The cylinder is located within an evacuated chamber, as Figure 31.6 illustrates. A magnetic field is directed perpendicular to the plane of the paper, and a photographic plate is positioned above the hole. Three spots appear on the developed plate, which are associated with the radioactivity of the nuclei in the material. Since moving particles are deflected by a magnetic field only when they are electrically charged, an analysis of this experiment reveals that two types of radioactivity (α and β rays, as it turns out) consist of charged particles, while the third type (γ rays) does not.

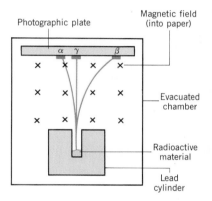

Figure 31.6 α and β rays are deflected by a magnetic field and, therefore, consist of moving charged particles. γ rays are not deflected by a magnetic field and, consequently, must be uncharged.

α DECAY

When a nucleus disintegrates and produces α rays, it is said to undergo α *decay*. Experimental evidence shows that α rays consist of positively charged particles, each particle being the $^4_2$He nucleus of helium. Thus, an α particle has a charge of $+2e$ and a nucleon number of $A = 4$. Since the grouping of 2 protons and 2 neutrons in a $^4_2$He nucleus is particularly stable, as we have seen in connection with Figure 31.5, it is not surprising that an α particle can be ejected as a unit from a more massive unstable nucleus.

Figure 31.7 shows the disintegration process for one example of α decay:

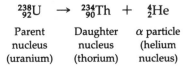

$$^{238}_{92}\text{U} \;\rightarrow\; ^{234}_{90}\text{Th} \;+\; ^4_2\text{He}$$

| Parent nucleus (uranium) | Daughter nucleus (thorium) | α particle (helium nucleus) |

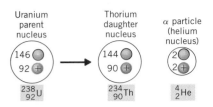

Figure 31.7 α decay occurs when an unstable parent nucleus emits an α particle and in the process is converted into a different or daughter nucleus.

The original nucleus is referred to as the parent nucleus (P), and the nucleus remaining after disintegration is called the daughter nucleus (D). Upon emission of an α particle, the uranium $^{238}_{92}\text{U}$ parent is converted into the $^{234}_{90}\text{Th}$ daughter, which is an isotope of thorium. The parent and daughter nuclei are different, so α decay converts one element into another, a process known as **transmutation.**

Electric charge is conserved during α decay. In Figure 31.7, for instance, 90 of the 92 protons in the uranium nucleus end up in the thorium nucleus, and the remaining 2 protons are carried off by the α particle. The total number of 92, however, is the same before and after disintegration. α decay also conserves the number of nucleons, for the number is the same before (238) and after (234 + 4) disintegration. Consistent with the conservation of electric charge and nucleon number, the general form for α decay is

$$[\alpha \text{ decay}] \qquad {}^{A}_{Z}\text{P} \;\longrightarrow\; {}^{A-4}_{Z-2}\text{D} \;+\; {}^{4}_{2}\text{He}$$

$$\underset{\substack{\text{Parent}\\\text{nucleus}}}{} \qquad \underset{\substack{\text{Daughter}\\\text{nucleus}}}{} \qquad \underset{\substack{\alpha \text{ particle}\\(\text{helium nucleus})}}{}$$

When a nucleus releases an α particle, the nucleus also releases energy. In fact, the energy released by radioactive decay is responsible, in part, for keeping the interior of the earth hot and, in some places, even molten. The following example shows how the conservation of mass/energy can be used to determine the amount of energy released in α decay.

Example 3 α Decay and the Release of Energy

Determine the energy released when α decay converts $^{238}_{92}\text{U}$ into $^{234}_{90}\text{Th}$.

REASONING   Energy is released during the α decay, because the combined mass of the $^{234}_{90}\text{Th}$ daughter nucleus and the α particle is less than the mass of the $^{238}_{92}\text{U}$ parent nucleus. The difference in mass is equivalent to the energy released. We determine the difference in mass in atomic mass units and then use the fact that 1 u is equivalent to 931.5 MeV.

SOLUTION   Appendix F gives the masses shown below:

$$^{238}_{92}\text{U} \;\longrightarrow\; ^{234}_{90}\text{Th} \;+\; ^{4}_{2}\text{He}$$

$$238.0508 \text{ u} \qquad\quad \underbrace{234.0436 \text{ u} \qquad 4.0026 \text{ u}}_{238.0462 \text{ u}}$$

The decrease in mass is $238.0508 \text{ u} - 238.0462 \text{ u} = 0.0046 \text{ u}$. As usual, the masses from Appendix F are atomic masses and include the mass of the orbital electrons. But this causes no error here, because the same total number of electrons is included for $^{238}_{92}\text{U}$, on the one hand, and for $^{234}_{90}\text{Th}$ plus $^{4}_{2}\text{He}$, on the other. Since 1 u is equivalent to 931.5 MeV, the released energy is $\boxed{4.3 \text{ MeV}}$. Except for a small portion carried away as a γ ray, this energy appears as kinetic energy of the α particle and the recoiling $^{234}_{90}\text{Th}$ nucleus. However, $^{234}_{90}\text{Th}$ is much more massive than the α particle, so that $^{234}_{90}\text{Th}$ recoils with only a small velocity and correspondingly small kinetic energy. The law of conservation of momentum can be applied (see problem 24) to determine the individual velocities and, hence, kinetic energies.

THE PHYSICS OF . . .

radioactivity and smoke detectors.

One widely used application of α decay is in smoke detectors. Figure 31.8 illustrates how a smoke detector operates. Two small and parallel metal plates are

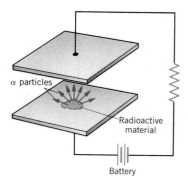

Figure 31.8 A smoke detector.

separated by a distance of about one centimeter. A tiny amount of radioactive material at the center of one of the plates emits α particles, which collide with air molecules. During the collisions, the air molecules are ionized to form positive and negative ions. The voltage from a battery causes one plate to be positive and the other negative, so that each plate attracts ions of opposite charges. As a result there is a current in the circuit attached to the plates. The presence of smoke particles between the plates reduces the current, since the ions that collide with a smoke particle are usually neutralized. The drop in current that smoke particles cause is used to trigger an alarm.

β DECAY

The β rays in Figure 31.6 are deflected by the magnetic field in a direction opposite to that of the positively charged α rays. Consequently, these β rays, which are the most common kind, consist of negatively charged particles or β^- particles. Experiment shows that β^- particles are electrons. As an illustration of β^- decay, consider the thorium $^{234}_{90}$Th nucleus, which decays by emitting a β^- particle, as in Figure 31.9:

$$^{234}_{90}\text{Th} \rightarrow {}^{234}_{91}\text{Pa} + {}^{0}_{-1}\text{e}$$

| Parent nucleus (thorium) | Daughter nucleus (protactinium) | β^- particle (electron) |

β^- decay, like α decay, causes a transmutation of one element into another. In this case, thorium $^{234}_{90}$Th is converted into protactinium $^{234}_{91}$Pa. The law of conservation of charge is obeyed, since the net number of positive charges is the same before (90) and after (91 − 1) the β^- emission. The law of conservation of nucleon number is obeyed, since the nucleon number remains at $A = 234$. The general form for β^- decay is

[β^- decay]
$$^{A}_{Z}\text{P} \rightarrow {}^{A}_{z+1}\text{D} + {}^{0}_{-1}\text{e}$$

| Parent nucleus | Daughter nucleus | β^- particle (electron) |

The electron emitted in β^- decay does *not* actually exist within the parent nucleus and is *not* one of the orbital electrons. Instead, the electron is created when a neutron decays into a proton and an electron; when this occurs, the proton number of the parent nucleus increases from Z to Z + 1 and the nucleon number

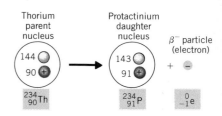

Figure 31.9 β decay occurs when a neutron in an unstable parent nucleus decays into a proton and an electron, the electron being emitted as the β^- particle. In the process, the parent nucleus is transformed into the daughter nucleus.

remains unchanged. The electron is usually fast-moving and escapes from the atom, leaving behind a positively charged atom.

Example 4 illustrates that energy is released during β^- decay, just as it is during α decay, and that the conservation of mass/energy applies.

Example 4 β^- Decay and the Release of Energy

Find the energy released when β^- decay changes $^{234}_{90}\text{Th}$ into $^{234}_{91}\text{Pa}$.

REASONING AND SOLUTION To find the energy released, we follow the usual procedure of determining how much the mass has decreased because of the decay and then calculating the equivalent energy. The masses (see Appendix F) are shown below:

$$^{234}_{90}\text{Th} \longrightarrow {}^{234}_{91}\text{Pa} + {}^{0}_{-1}\text{e}$$

$$234.043\ 59\ \text{u} \qquad \underbrace{\qquad\qquad\qquad}_{234.043\ 30\ \text{u}}$$

When the $^{234}_{90}\text{Th}$ nucleus of a thorium atom is converted into a $^{234}_{91}\text{Pa}$ nucleus, the number of orbital electrons remains the same, so the resulting protactinium atom is missing one orbital electron. However, the mass taken from Appendix F includes all 91 electrons of a neutral protactinium atom. In effect, then, the value of 234.043 30 u for $^{234}_{91}\text{Pa}$ already includes the mass of the β^- particle. The mass decrease that accompanies the β^- decay is 234.043 59 u − 234.043 30 u = 0.000 29 u. The equivalent energy (1 u = 931.5 MeV) is $\boxed{0.27\ \text{MeV}}$. This is the maximum kinetic energy that the emitted electron can have.

PROBLEM SOLVING INSIGHT

In β^- decay, be careful not to include the mass of the electron (${}_{-1}^{0}$e) twice. As discussed here for the daughter atom ($^{234}_{91}$Pa), the masses given in Appendix F already include the mass of the emitted electron.

A second kind of β decay sometimes occurs.† In this process the particle emitted by the nucleus is a *positron*, rather than an electron. A positron, also called a β^+ particle, has the same mass as an electron, but carries a charge of $+e$ instead of $-e$. The disintegration process for β^+ decay is

[β^+ decay] $\qquad {}^{A}_{Z}\text{P} \rightarrow {}^{A}_{Z-1}\text{D} + {}^{0}_{1}\text{e}$

Parent nucleus · Daughter nucleus · β^+ particle (positron)

The emitted positron does *not* exist within the nucleus but, rather, is created when a nuclear proton is transformed into a neutron. In the process, the proton number of the parent nucleus decreases from Z to $Z-1$, and the nucleon number remains the same. As with β^- decay, the laws of conservation of charge and nucleon number are obeyed, and there is a transmutation of one element into another.

γ DECAY

The nucleus, like the orbital electrons, exists only in discrete energy states or levels. When a nucleus changes from an excited energy state (denoted by an asterisk *) to a lower energy state, a photon is emitted. The process is similar to the one discussed in Section 30.3 for the photon emission that leads to the hydrogen

† A third kind of β decay also occurs, in which a nucleus pulls in or captures one of the orbital electrons from outside the nucleus. The process is called *electron capture*, or *K capture*, since the electron normally comes from the innermost or K shell.

atom line spectrum. With nuclear energy levels, however, the photon has a much greater energy and is called a γ ray. The γ decay process is written as follows:

[γ decay]
$$_Z^A P^* \rightarrow \ _Z^A P \ + \ \gamma$$

Excited · Lower · γ ray
energy state · energy state

γ decay does *not* cause a transmutation of one element into another. In the next example the wavelength of a γ ray photon is determined.

Example 5   The Wavelength of the Photon Emitted in γ Decay

What is the wavelength of the 0.186 MeV γ ray photon emitted by radium $_{88}^{226}$Ra?

REASONING   The photon energy is the difference between two nuclear energy levels. Equation 30.4 gives the relation between the energy level separation ΔE and the frequency f of the photon as $\Delta E = hf$. Since $f\lambda = c$, the wavelength of the photon is $\lambda = hc/\Delta E$.

SOLUTION   First we must convert the photon energy into joules:

$$\Delta E = (0.186 \times 10^6 \text{ eV}) \left(\frac{1.60 \times 10^{-19} \text{ J}}{1 \text{ eV}} \right) = 2.98 \times 10^{-14} \text{ J}$$

The wavelength of the photon is

$$\lambda = \frac{hc}{\Delta E} = \frac{(6.63 \times 10^{-34} \text{ J} \cdot \text{s})(3.00 \times 10^8 \text{ m/s})}{2.98 \times 10^{-14} \text{ J}} = \boxed{6.67 \times 10^{-12} \text{ m}}$$

PROBLEM SOLVING INSIGHT

The energy ΔE of a γ ray photon, like that of photons in other regions of the electromagnetic spectrum (visible, infrared, microwave, etc.) is equal to the product of Planck's constant h and the frequency f of the photon: $\Delta E = hf$.

31.5 THE NEUTRINO

When a β particle is emitted by a radioactive nucleus, energy is simultaneously released, as Example 4 illustrates. Experimentally, however, it is found that most β particles do not have enough kinetic energy to account for all the energy released. If a β particle carries away only part of the energy, where does the remainder go? The question puzzled physicists until 1930, when Wolfgang Pauli proposed that part of the energy is carried away by another particle that is emitted along with the β particle. This additional particle is called the *neutrino,* and its existence was verified experimentally in 1956. The Greek letter nu (ν) is used to symbolize the neutrino. For instance, the β^- decay of thorium $_{90}^{234}$Th (see Section 31.4) is more correctly written as

$$_{90}^{234}\text{Th} \rightarrow \ _{91}^{234}\text{Pa} + \ _{-1}^{0}e + \bar{\nu}$$

The bar above the ν is included, because the neutrino emitted in this particular decay process is an antimatter neutrino or antineutrino. A normal neutrino (ν without the bar) is emitted when β^+ decay occurs.

The neutrino has zero electrical charge. Moreover, at present there is no convincing experimental evidence to indicate that the neutrino has any mass. A particle with zero mass, like a photon, travels at the speed of light. The neutrino,

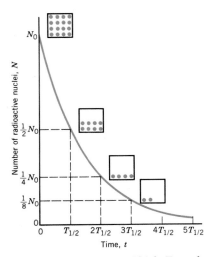

Figure 31.10 The half-life $T_{1/2}$ of a radioactive decay is the time in which one-half of the radioactive nuclei disintegrate.

therefore, travels near or at the speed of light. The emission of β particles and neutrinos involves a fundamental force that has not been mentioned before in this text. This force is much weaker than the strong nuclear force and weaker than the electromagnetic force; hence, it is referred to as the *weak nuclear force*. Since this force is so weak, it plays a negligible role in determining nuclear structure, but is nonetheless responsible for β decay.

31.6 RADIOACTIVE DECAY AND ACTIVITY

The question of which radioactive nucleus in a group disintegrates at a given instant is decided like the drawing of numbers in a state lottery; individual disintegrations occur randomly. As time passes, the number N of parent nuclei decreases, as Figure 31.10 shows. This graph of N versus time indicates that the decrease occurs in a smooth fashion, with N approaching zero after enough time has passed. To help describe the graph, it is useful to define the *half-life $T_{1/2}$* of a radioactive isotope as the time required for one-half of the nuclei present to disintegrate. For example, radium $^{226}_{88}$Ra has a half-life of 1600 years, for it takes this amount of time for one-half of a given quantity of this isotope to disintegrate into radon $^{222}_{86}$Rn. In another 1600 years, one-half of the remaining radium atoms will disintegrate, leaving only one-fourth of the original number intact. In Figure 31.10, the number of nuclei present at time $t = 0$ is $N = N_0$, while the number present at $t = T_{1/2}$ is $N = \frac{1}{2}N_0$. The number present at $t = 2T_{1/2}$ is $N = \frac{1}{4}N_0$, and so forth. The value of the half-life depends on the nature of the radioactive nucleus. Values ranging from a fraction of a second to billions of years have been found, as Table 31.2 indicates. Example 6 deals with the half-life of radon $^{222}_{86}$Rn.

Table 31.2 Some Half-lives for Radioactive Decay

Isotope		Half-life	Decay Mode
Polonium	$^{214}_{84}$Po	1.64×10^{-4} s	α, γ
Krypton	$^{89}_{36}$Kr	3.16 min	β^-, γ
Radon	$^{222}_{86}$Rn	3.83 days	α, γ
Strontium	$^{90}_{38}$Sr	28.5 yr	β^-
Radium	$^{226}_{88}$Ra	1.6×10^3 yr	α, γ
Carbon	$^{14}_{6}$C	5.73×10^3 yr	β^-
Uranium	$^{238}_{92}$U	4.47×10^9 yr	α, γ
Indium	$^{115}_{49}$In	4.41×10^{14} yr	β^-

THE PHYSICS OF . . .

radioactive radon gas in houses.

Example 6   The Radioactive Decay of Radon Gas

Radon $^{222}_{86}$Rn is a radioactive gas produced when radium $^{226}_{88}$Ra undergoes α decay. There is growing concern about radon as a health hazard, because it can become trapped in houses, entering primarily through cracks in walls and floors and in the drinking water. Suppose 3.0×10^7 radon atoms are trapped in a basement at the time the basement is sealed against further entry of the gas. The half-life of radon is 3.83 days. How many radon atoms remain after 31 days?

REASONING   During each half-life, the number of radon atoms is reduced by a factor of two. Thus, we determine the number of half-lives there are in a period of 31 days and reduce the number of radon atoms by a factor of two for each one.

SOLUTION   In a period of 31 days there are 31 days/3.83 days = 8.1 half-lives. In 8 half-lives the number of radon atoms is reduced by a factor of $2^8 = 256$. Ignoring the difference between 8 and 8.1 half-lives, we find that the number of atoms remaining is $3.0 \times 10^7/256 = \boxed{1.2 \times 10^5}$.

The *activity* of a radioactive sample is the number of disintegrations per second that occur. Each time a disintegration occurs, the number N of radioactive nuclei decreases. As a result, the activity can be obtained by dividing ΔN, the change in the number of nuclei, by Δt, the time interval during which the change takes place; the average activity over the time interval Δt is the magnitude of $\Delta N/\Delta t$. Since the decay of any individual nucleus is completely random, the number of disintegrations per second that occur in a sample is proportional to the number of radioactive nuclei present, so that

$$\frac{\Delta N}{\Delta t} = -\lambda N \tag{31.4}$$

where λ is a proportionality constant referred to as the *decay constant*. The minus sign is present in this equation because each disintegration decreases the number N of nuclei originally present.

The SI unit for activity is the *becquerel* (Bq); one becquerel equals one disintegration per second. Activity is also measured in terms of a unit called the *curie* (Ci), in honor of Marie (1867–1934) and Pierre (1859–1906) Curie, the discoverers of radium and polonium. Historically, the curie was chosen as a unit because it is roughly the activity of one gram of pure radium. In terms of becquerels,

$$1 \text{ Ci} = 3.70 \times 10^{10} \text{ Bq}$$

The activity of the radium put into the dial of a watch to make it glow in the dark is about 4×10^4 Bq, and the activity used in radiation therapy for cancer treatment is approximately 4×10^{13} Bq.

The mathematical expression for the graph of N versus t shown in Figure 31.10 can be obtained from Equation 31.4 with the aid of calculus. The result for the number N of radioactive nuclei present at time t is

$$N = N_0\, e^{-\lambda t} \tag{31.5}$$

assuming that the number at $t = 0$ is N_0. The exponential e has the value $e = 2.718 \ldots$, and many calculators provide the value of e^x. By substituting $N = \frac{1}{2}N_0$ and $t = T_{1/2}$ into Equation 31.5, we find that $\frac{1}{2} = e^{-\lambda T_{1/2}}$. Taking the natural logarithm of both sides of this equation reveals that

$$T_{1/2} = \frac{\ln 2}{\lambda} = \frac{0.693}{\lambda} \tag{31.6}$$

The following example illustrates the use of Equations 31.5 and 31.6.

Example 7   The Activity of Radon $^{222}_{86}Rn$

As in Example 6, suppose there are 3.0×10^7 radon atoms ($T_{1/2} = 3.83$ days or 3.31×10^5 s) trapped in a basement. (a) How many radon atoms remain after 31 days? Find the activity (b) just after the basement is sealed against further entry of radon and (c) 31 days later.

REASONING AND SOLUTION

(a) The answer can be obtained directly from Equation 31.5, provided the decay constant is first determined from the half-life:

$$\lambda = \frac{0.693}{T_{1/2}} = \frac{0.693}{3.83 \text{ days}} = 0.181 \text{ days}^{-1}$$

$$N = N_0\, e^{-\lambda t} = (3.0 \times 10^7)e^{-(0.181 \text{ days}^{-1})(31 \text{ days})} = \boxed{1.1 \times 10^5}$$

This value is slightly less than that found in Example 6, because there we ignored the difference between 8.0 and 8.1 half-lives.

(b) The activity can be obtained from Equation 31.4, provided the decay constant is expressed in reciprocal seconds: $\lambda = 0.693/(3.31 \times 10^5 \text{ s}) = 2.09 \times 10^{-6} \text{ s}^{-1}$. According to Equation 31.4,

$$\frac{\Delta N}{\Delta t} = -\lambda N = -(2.09 \times 10^{-6} \text{ s}^{-1})(3.0 \times 10^7) = -63 \text{ disintegrations/s}$$

The activity is the magnitude of $\Delta N/\Delta t$, so initially $\boxed{\text{Activity} = 63 \text{ Bq}}$.

(c) From part (a), the number of radioactive nuclei remaining at the end of 31 days is $N = 1.1 \times 10^5$, and reasoning similar to that in part (b) reveals that $\boxed{\text{Activity} = 0.23 \text{ Bq}}$.

THE PHYSICS OF . . .

radioactive dating.

31.7 RADIOACTIVE DATING

One important application of radioactivity is the determination of the age of archeological or geological samples. If an object contains radioactive nuclei when it is formed, then the decay of these nuclei marks the passage of time like a clock, half of the nuclei disintegrating during each half-life. If the half-life is known, a measurement of the number of nuclei present today relative to the number present initially can give the age of the sample. According to Equation 31.4, the activity of a sample is proportional to the number of radioactive nuclei, so one way to obtain the age is to compare present activity with initial activity. A more accurate way is to determine the present number of radioactive nuclei with the aid of a mass spectrometer.

The present activity of a sample can be measured, but how is it possible to know what the original activity was, perhaps thousands of years ago? Radioactive dating methods entail certain assumptions that make it possible to estimate the original activity. For instance, the radiocarbon technique utilizes the $^{14}_{6}C$ isotope of carbon, which undergoes β^- decay with a half-life of 5730 yr. This isotope is present in the earth's atmosphere at an equilibrium concentration of about one atom for every 8.3×10^{11} atoms of normal carbon $^{12}_{6}C$. It is often assumed* that this value has remained constant over the years, because $^{14}_{6}C$ is created when cosmic rays interact with the earth's upper atmosphere, a production method that offsets the loss via β^- decay. Moreover, nearly all living organisms ingest the equilibrium concentration of $^{14}_{6}C$. However, once an organism dies, metabolism no longer sustains the input of $^{14}_{6}C$, and β^- decay causes half of the $^{14}_{6}C$ nuclei to disintegrate every 5730 yr.

It is possible to calculate the $^{14}_{6}C$ activity of one gram of carbon in a living organism. One gram of carbon (atomic mass = 12 u) is 1.0/12 mol, and since there are 6.02×10^{23} atoms per mole (Avogadro's number), the number of $^{14}_{6}C$ atoms present is

$$\left(\frac{1.0}{12}\ mol\right)\left(6.02 \times 10^{23}\ \frac{atoms}{mol}\right)\left(\frac{1}{8.3 \times 10^{11}}\right) = 6.0 \times 10^{10}\ atoms$$

Since the half-life is 5730 yr (1.81×10^{11} s), the decay constant of $^{14}_{6}C$ is $\lambda = 0.693/T_{1/2} = 0.693/(1.81 \times 10^{11}\ s) = 3.83 \times 10^{-12}\ s^{-1}$. Therefore, Equation 31.4 indicates that the activity, or the magnitude of $\Delta N/\Delta t$, is

Activity of one
gram of carbon in $= \lambda N = (3.83 \times 10^{-12}\ s^{-1})(6.0 \times 10^{10}) = 0.23\ Bq$
a living organism

An organism that lived thousands of years ago presumably had an activity of about 0.23 Bq per gram of carbon. When the organism died, the activity began decreasing. From a sample of the remains, the current activity per gram of carbon can be measured and compared to the value of 0.23 Bq to determine the time that has transpired since death. This procedure is illustrated in Example 8.

Radioactive dating is used to determine the age of artifacts found in archaeological digs, such as this Late Classic Maya grave in Caracol, Belize.

* The assumption that the $^{14}_{6}C$ concentration has always been at its present equilibrium value has been evaluated by comparing $^{14}_{6}C$ ages with ages determined by counting tree rings. More recently, ages determined using the radioactive decay of uranium $^{238}_{92}U$ have been used for comparison. These comparisons indicate that the equilibrium value of the $^{14}_{6}C$ concentration has indeed remained constant for the past 1000 years. However, from there back about 30 000 years, it appears that the $^{14}_{6}C$ concentration in the atmosphere was larger than its present value by up to 40%. In this text, as a first approximation we ignore such discrepancies.

Example 8   The Dead Sea Scrolls

The Dead Sea Scrolls are famous ancient manuscripts, discovered in 1947. They were dated by applying the radiocarbon method to a sample of the linen in which they were wrapped. Linen is made from the flax plant. A $^{14}_{6}C$ activity of about 0.18 Bq per gram of carbon was measured. Determine the age of the scrolls.

REASONING   According to Equation 31.5, the number of nuclei remaining at time t is $N = N_0 e^{-\lambda t}$. Multiplying both sides of this expression by the decay constant λ and recognizing that the product of λ and N is the activity A, we find that $A = A_0 e^{-\lambda t}$, where $A_0 = 0.23$ Bq is the activity at time $t = 0$ for one gram of carbon. The decay constant λ can be determined from the value of 5730 yr for the half-life of $^{14}_{6}C$, using Equation 31.6. With known values for A_0 and λ, the given activity of $A = 0.18$ Bq per gram of carbon can be used to find the age t of the Dead Sea Scrolls.

> **SOLUTION**   For $^{14}_{6}C$, the decay constant is $\lambda = 0.693/T_{1/2} = 0.693/(5730 \text{ yr}) = 1.21 \times 10^{-4} \text{ yr}^{-1}$. Since $A = 0.18$ Bq and $A_0 = 0.23$ Bq, the age can be determined from
>
> $$A = 0.18 \text{ Bq} = (0.23 \text{ Bq})e^{-(1.21 \times 10^{-4} \text{ yr}^{-1})t}$$
>
> Taking the natural logarithm of both sides of this result gives
>
> $$\ln\left(\frac{0.18 \text{ Bq}}{0.23 \text{ Bq}}\right) = -(1.21 \times 10^{-4} \text{ yr}^{-1})t$$
>
> The age of the sample is $\boxed{t = 2.0 \times 10^3 \text{ yr}}$.

Radiocarbon dating is not the only radioactive dating method. For example, other methods utilize uranium $^{238}_{92}U$, potassium $^{40}_{19}K$, and lead $^{210}_{82}Pb$. And the related technique of thermoluminescence is being increasingly used.

31.8 RADIOACTIVE DECAY SERIES

When an unstable parent nucleus decays, the resulting daughter nucleus is sometimes also unstable. If so, the daughter then decays and produces its own daughter, and so on, until a completely stable nucleus is produced. This sequential decay

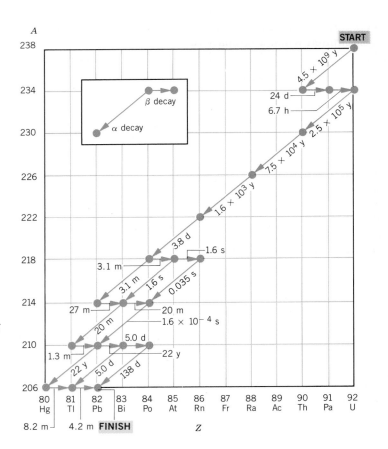

Figure 31.11   A radioactive decay series that begins with uranium $^{238}_{92}U$ and ends with lead $^{206}_{82}Pb$. The half-lives are given in seconds (s), minutes (m), hours (h), days (d), or years (y). The insert in the upper left corner of the graph identifies the type of decay that each nucleus undergoes.

of one nucleus after another is called a *radioactive decay series.* Examples 3 and 4 discuss the first two steps of a series that begins with uranium $^{238}_{92}$U:

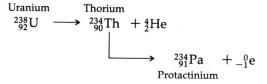

$$\text{Uranium} \qquad \text{Thorium}$$
$$^{238}_{92}\text{U} \longrightarrow {}^{234}_{90}\text{Th} + {}^{4}_{2}\text{He}$$
$$\longrightarrow {}^{234}_{91}\text{Pa} + {}^{0}_{-1}\text{e}$$
$$\text{Protactinium}$$

Furthermore, Example 6 deals with radon $^{222}_{86}$Rn, which is formed down the line in the $^{238}_{92}$U radioactive decay series. Figure 31.11 shows the entire series. At several points in the series, branches occur, because more than one kind of decay is possible for an intermediate species. Ultimately, however, the series ends with lead $^{206}_{82}$Pb, which is stable.

The $^{238}_{92}$U series and other such series are the only sources of some of the radioactive elements found in nature. Radium $^{226}_{88}$Ra, for instance, has a half-life of 1600 yr, which is short enough that all the $^{226}_{88}$Ra created when the earth was formed billions of years ago has now disappeared. The $^{238}_{92}$U series provides a continuing supply of $^{226}_{88}$Ra, however.

31.9 DETECTORS OF RADIATION

THE PHYSICS OF . . .

detectors of radiation.

There are a number of devices that can be used to detect the effects of the particles and photons (γ rays) emitted when a radioactive nucleus decays. Such devices detect the ionization that these particles and photons cause as they pass through matter.

The most familiar detector is the *Geiger counter,* which Figure 31.12 illustrates.

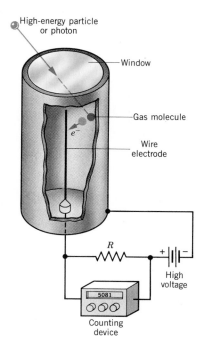

Figure 31.12   A Geiger counter.

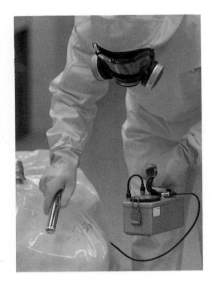

Workers in the nuclear industry monitor levels of radioactivity as part of routine safety inspections.

The Geiger counter consists of a gas-filled metal cylinder. The α, β, or γ rays enter the cylinder through a thin window at one end. γ rays can also penetrate directly through the metal. A wire electrode runs along the center of the tube and is kept at a high positive voltage (1000–3000 V) relative to the outer cylinder. When a high-energy particle or photon enters the cylinder, it collides with and ionizes a gas molecule. The electron produced from the gas molecule accelerates toward the positive wire, ionizing other molecules in its path. Additional electrons are formed, and an avalanche of electrons rushes toward the wire, leading to a pulse of current through the resistor R. This pulse can be counted or made to produce a ''click'' in a loudspeaker. The number of counts or clicks is related to the number of high-energy particles or photons present, or equivalently, to the number of disintegrations that produced the particles or photons.

The *scintillation counter* is another important radiation detector. As Figure 31.13 indicates, this device consists of a scintillator mounted on a photomultiplier tube. Often the scintillator is a crystal (e.g., cesium iodide) containing a small amount of impurity (thallium), but plastic, liquid, and gaseous scintillators are also used. In response to ionizing radiation, the scintillator emits a flash of visible light. The photons of the flash then strike the photocathode of the photomultiplier tube. The photocathode is made of a material that emits electrons because of the photoelectric effect. These photoelectrons are then attracted to a special electrode kept at a voltage of about +100 V relative to the photocathode. The electrode is coated with a substance that emits several additional electrons for

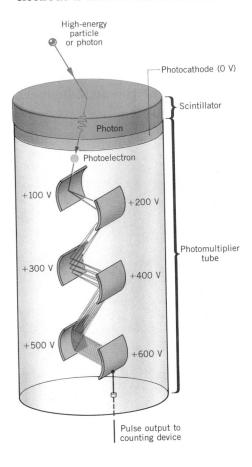

Figure 31.13   A scintillation counter.

every electron striking it. The additional electrons are attracted to a second similar electrode (voltage = +200 V) where they generate even more electrons. Commercial photomultiplier tubes contain as many as 15 of these special electrodes, so photoelectrons resulting from the light flash of the scintillator lead to a cascade of electrons and a pulse of current. As in a Geiger tube, the current pulses can be counted.

Ionizing radiation can also be detected with several types of *semiconductor detectors.* Such devices utilize *n*- and *p*-type materials, and their operation depends on the electrons and holes formed in the materials as a result of the radiation. One of the main advantages of semiconductor detectors is their ability to discriminate between two particles with only slightly different energies.

A number of instruments provide a pictorial representation of the path that high-energy particles follow after they are emitted from unstable nuclei. In a *cloud chamber,* a gas is cooled just to the point where it will condense into droplets, provided nucleating agents are available on which the droplets can form. When a high-energy particle, such as an α particle or a β particle, passes through the gas, the ions it leaves behind serve as nucleating agents, and droplets form along the path of the particle. A *bubble chamber* works in a similar fashion, except it contains a liquid that is just at the point of boiling. Tiny bubbles form along the trail of a high-energy particle passing through the liquid. The paths revealed in a cloud or bubble chamber can be photographed to provide a permanent record of the event. Figure 31.14 shows a photograph of the tracks in a bubble chamber. A *photographic emulsion* also can be used directly to produce a record of the path taken by a particle of ionizing radiation. The ions formed as the particle passes through the emulsion cause silver to be deposited along the track when the emulsion is developed.

Figure 31.14 Particle tracks in a bubble chamber.

SUMMARY

The nucleus of an atom consists of protons and neutrons, which are collectively referred to as **nucleons.** A **neutron** is an electrically neutral particle whose mass is slightly larger than that of a proton. The **atomic number** Z is the number of protons in the nucleus. The **atomic mass number** or **nucleon number** A is the total number of protons and neutrons in the nucleus: $A = Z + N$, where N is the number of neutrons. Nuclei that contain the same number of protons, but a different number of neutrons, are called **isotopes.**

The **strong nuclear force** is the force of attraction between nucleons and is one of the four fundamental forces of nature. This force balances the electrostatic force of repulsion between protons and holds the nucleus together. The strong nuclear force has a very short range of action and does not depend on electric charge.

The **binding energy** of a nucleus is the energy required to separate the nucleus into its constituent protons and neutrons. The binding energy is equal to

$(\Delta m)c^2$, where c is the speed of light in a vacuum and Δm is the mass defect of the nucleus. The **mass defect** is the amount by which the sum of the individual masses of the protons and neutrons exceeds the mass of the parent nucleus.

When specifying nuclear masses, it is customary to use the **atomic mass unit** (u), which is one-twelfth of the mass of a $^{12}_{6}C$ atom. One atomic mass unit is equivalent to an energy of 931.5 MeV.

Unstable nuclei spontaneously decay by breaking apart or rearranging their internal structures in a process called **radioactivity.** Naturally occurring radioactivity produces α rays, β rays, and γ rays. α **rays** consist of positively charged particles, each particle being the $^{4}_{2}He$ nucleus of helium. The most common kind of β **ray** consists of negatively charged particles, or β^- particles, which are electrons. Another kind of β ray consists of positively charged particles, or β^+ particles. A β^+ particle, also called a **positron,** has the same mass as an electron but carries a charge of $+e$ instead of $-e$. γ **rays** are high-energy photons. If a radioactive parent nucleus disintegrates into a daughter nucleus that has a different atomic number, one element has been converted into another element, the conversion being referred to as a **transmutation.**

The **neutrino** is an electrically neutral particle that has near zero or zero mass. The neutrino travels near or at the speed of light and is emitted along with β particles.

The **half-life** $T_{1/2}$ of a radioactive isotope is the time required for one-half of the nuclei present to disintegrate or decay. The **activity** is the number of disintegrations per second that occur. In other words, the activity is the magnitude of $\Delta N / \Delta t$, where ΔN is the change in the number N of radioactive nuclei and Δt is the time interval during which the change occurs. The SI unit for activity is the becquerel (Bq), one becquerel being one disintegration per second. Radioactive decay obeys the following relation: $\Delta N / \Delta t = -\lambda N$, where λ is the **decay constant.** This equation can be solved to show that $N = N_0 e^{-\lambda t}$, where N_0 is the number of nuclei present initially. The decay constant is related to the half-life according to $T_{1/2} = 0.693/\lambda$.

The sequential decay of one nucleus after another is called a **radioactive decay series,** and Figure 31.11 illustrates one such series.

QUESTIONS

1. A material is known to be an isotope of lead, although the particular isotope is not known. From such limited information, which of the following quantities can you specify: (a) its atomic number, (b) its neutron number, and (c) its atomic mass number? Explain.

2. Two nuclei have different nucleon numbers A_1 and A_2. Are the two nuclei necessarily isotopes of the same element? Give your reasoning and support it with an example from Appendix F.

3. The density of an atom as a whole is much less than the density of the nucleus of the atom. Considering what Section 30.1 discusses about the structure of the atom, explain why.

4. Using Figure 31.5, rank the following nuclei in ascending order according to the binding energy per nucleon (smallest first): phosphorus $^{31}_{15}\text{P}$, cobalt $^{59}_{27}\text{Co}$, tungsten $^{184}_{74}\text{W}$, and thorium $^{232}_{90}\text{Th}$.

5. Describe qualitatively how the radius of the daughter nucleus compares to that of the parent nucleus for (a) α decay and (b) β decay. Justify your answers in terms of Equation 31.2.

6. Uranium $^{238}_{92}\text{U}$ decays into thorium $^{234}_{90}\text{Th}$ by means of α decay, as Example 3 in the text discusses. A reasonable question to ask is, Why doesn't the $^{238}_{92}\text{U}$ nucleus just emit a single proton, instead of an α particle? This hypothetical decay scheme is shown below, along with the pertinent atomic masses:

$$^{238}_{92}\text{U} \quad \rightarrow \quad ^{237}_{91}\text{Pa} \quad + \quad ^{1}_{1}\text{H}$$

Uranium	Protactinium	Proton
238.050 78 u	237.051 14 u	1.007 83 u

For a decay to be possible, it must bring the parent nucleus toward a more stable state by allowing the release of energy. Compare the total mass of the products of this hypothetical decay with the mass of $^{238}_{92}\text{U}$ and decide whether the emission of a single proton is possible for $^{238}_{92}\text{U}$. Explain.

7. Explain why unstable nuclei with short half-lives typically have only a small or zero natural abundance.

8. On the basis of the half-lives given in the isotope table in Appendix F, decide which isotope of hydrogen or oxygen might be of use to date the pure H_2O in a sealed bottle. The water is thought to be about 5–15 yr old. Account for your choice.

9. To which of the following objects, each about 1000 yr old, can the radiocarbon dating technique *not* be applied: a glass vial, a wooden box, a gold statue, and a skeleton? Explain.

10. Suppose there were a greater number of carbon $^{14}_{6}\text{C}$ atoms in an animal living 5000 yr ago than is currently believed. When the bones of this animal are tested today using radiocarbon dating, is the age obtained too small or too large? Give your reasoning.

11. Tritium is an isotope of hydrogen and undergoes β^- decay with a half-life of 12.33 yr. Like carbon $^{14}_{6}\text{C}$, tritium is produced in the atmosphere because of cosmic rays and can be used in a radioactive dating technique. In any such technique, there must be a sufficient number of radioactive nuclei left in a sample to detect if the technique is to be useful. Can tritium dating be used to determine a reliable date for a sample that is about 700 yr old? Account for your answer.

PROBLEMS

The data given for atomic masses in these problems include the mass of the electrons orbiting the nucleus of the electrically neutral atom.

Section 31.1 Nuclear Structure, Section 31.2 The Strong Nuclear Force and the Stability of the Nucleus

1. How many protons and neutrons are there in the nucleus of (a) oxygen $^{18}_{8}O$ and (b) tin $^{120}_{50}Sn$?

2. What is the radius of a nucleus of uranium $^{238}_{92}U$?

3. Two isotopes of chlorine occur in nature. The $^{35}_{17}Cl$ isotope has an atomic mass of 34.968 85 u and a natural abundance of 75.77%. The $^{37}_{17}Cl$ isotope has an atomic mass of 36.965 90 u and a natural abundance of 24.23%. By a calculation of your own, verify that the value of 35.45 u listed in the periodic table is a weighted average of the individual atomic masses.

4. By what factor does the nucleon number of a nucleus have to increase in order for the nuclear radius to double?

5. In the nucleus of gold $^{197}_{79}Au$, what is the electrostatic force of repulsion between two protons, assuming the centers of the protons are located at opposite ends of a diameter of the gold nucleus?

*6. Two naturally occurring isotopes of carbon are $^{12}_{6}C$ (atomic mass = 12.000 000 u) and $^{13}_{6}C$ (atomic mass = 13.003 355 u). In one gram of each of these isotopes there are different numbers of atoms. Which contains more atoms, and how many more?

*7. One isotope (X) contains an equal number of protons and neutrons. Another isotope (Y) of the same element has twice the number of neutrons as the first isotope does. Determine the ratio r_Y/r_X of the nuclear radii of the isotopes.

**8. (a) Determine an approximate value for the density (in kg/m³) of the nucleus. (b) If a BB (radius = 2.3 mm) from an air rifle had a density equal to the nuclear density, what mass would the BB have? (c) Assuming the mass of a supertanker is about 1.5×10^8 kg, how many "supertankers" of mass would this hypothetical BB have?

Section 31.3 The Mass Defect of the Nucleus and Nuclear Binding Energy

9. Determine the mass defect of the nucleus for cobalt $^{59}_{27}Co$, which has an atomic mass of 58.933 198 u. Express your answer in (a) atomic mass units and (b) kilograms.

10. Find the binding energy (in MeV) for lithium $^{7}_{3}Li$ (atomic mass = 7.016 003 u).

11. What is the binding energy (in MeV) for oxygen $^{16}_{8}O$ (atomic mass = 15.994 915 u)?

12. For radium $^{226}_{88}Ra$ (atomic mass = 226.025 402 u) obtain (a) the mass defect in atomic mass units, (b) the binding energy in MeV, and (c) the binding energy per nucleon.

*13. (a) Energy is required to separate a nucleus into its constituent nucleons, as Figure 31.3 indicates; this energy is the *total* binding energy of the nucleus. In a similar way one can speak of the energy that binds a single nucleon to the remainder of the nucleus. For example, separating nitrogen $^{14}_{7}N$ into nitrogen $^{13}_{7}N$ and a neutron takes energy equal to the binding energy of the neutron, as shown below:

$$^{14}_{7}N + \text{Energy} \rightarrow ^{13}_{7}N + ^{1}_{0}n$$

Find the energy that binds the neutron to the $^{14}_{7}N$ nucleus by considering the mass of $^{13}_{7}N$ and the mass of $^{1}_{0}n$, as compared to the mass of $^{14}_{7}N$ (see Appendix F for masses). (b) Similarly, one can speak of the energy that binds a single proton to the $^{14}_{7}N$ nucleus:

$$^{14}_{7}N + \text{Energy} \rightarrow ^{13}_{6}C + ^{1}_{1}H$$

Following the procedure outlined in part (a), determine the energy that binds the proton to the $^{14}_{7}N$ nucleus. (c) Which nucleon is more tightly bound, the neutron or the proton?

Section 31.4 Radioactivity

14. α decay occurs for each of the nuclei given below. Write the decay process for each, including the chemical symbols and values for Z and A for the daughter nuclei: (a) $^{228}_{90}Th$ and (b) $^{231}_{91}Pa$.

15. Write the α decay process for each of the nuclei given below, including the chemical symbols and values for Z and A for the daughter nuclei: (a) $^{235}_{92}U$ and (b) $^{239}_{94}Pu$.

16. For the following nuclei, each undergoing β^- decay, write the decay process, identifying each daughter nucleus with its chemical symbol and values for Z and A: (a) $^{14}_{6}C$ and (b) $^{212}_{82}Pb$.

17. Write the β^- decay process for $^{60}_{27}Co$, including the chemical symbol and values for Z and A.

18. Carbon $^{14}_{6}C$ (atomic mass = 14.003 241 u) is converted into nitrogen $^{14}_{7}N$ (atomic mass = 14.003 074 u) via β^- decay. (a) Write this process in symbolic form, giving Z and A for the parent and daughter nuclei and the β^- particle. (b) Determine the energy (in MeV) released.

19. What is the wavelength of the 0.510-MeV γ ray that is emitted by radon $^{222}_{86}Rn$?

20. In the form $_Z^A X$, identify the daughter nucleus that results when (a) plutonium $_{94}^{242}Pu$ undergoes α decay, (b) sodium $_{11}^{24}Na$ undergoes β^- decay, and (c) nitrogen $_7^{13}N$ undergoes β^+ decay.

*21. Determine the symbol $_Z^A X$ for the parent nucleus whose α decay produces the same daughter as the β^- decay of thallium $_{81}^{208}Tl$.

*22. Find the energy (in MeV) released when β^+ decay converts sodium $_{11}^{22}Na$ (atomic mass = 21.994 434 u) into neon $_{10}^{22}Ne$ (atomic mass = 21.991 383 u). Notice that the atomic mass for $_{11}^{22}Na$ includes the mass of 11 electrons, whereas the atomic mass for $_{10}^{22}Ne$ includes the mass of only 10 electrons.

*23. Thorium $_{90}^{232}Th$ undergoes α decay to produce a daughter nucleus that itself undergoes β^- decay. In the form $_Z^A X$, identify the nucleus that ultimately results.

**24. Example 3 in the text deals with the α decay of uranium $_{92}^{238}U$, which produces thorium $_{90}^{234}Th$ (atomic mass = 234.0436 u). The energy released in the decay is determined in this example to be 4.3 MeV. Use the conservation of linear momentum (see Example 4 in Chapter 7), and determine how much of this energy is carried away by the recoiling $_{90}^{234}Th$ daughter nucleus and how much by the α particle. Assume the energy of each particle is kinetic energy, and ignore the small amount of energy carried away by the γ ray that is also emitted. In addition, ignore relativistic effects.

**25. Sodium $_{11}^{24}Na$ emits a γ ray that has an energy of 0.423 MeV. Assuming the $_{11}^{24}Na$ nucleus is initially at rest, use the conservation of linear momentum to find the speed with which the nucleus recoils. Ignore relativistic effects.

Section 31.6 Radioactive Decay and Activity

26. In 9.0 days the number of radioactive nuclei decreases to one-eighth the number present initially. What is the half-life (in days) of the material?

27. The number of radioactive nuclei present at the start of an experiment is 4.60×10^{15}. The number present twenty days later is 8.14×10^{14}. What is the half-life (in days) of the nuclei?

28. The $_1^3H$ isotope of hydrogen is called tritium and has a half-life of 12.33 yr. What is its decay constant in units of s^{-1}?

29. Strontium $_{38}^{90}Sr$ has a half-life of 28.5 yr. It is chemically similar to calcium, enters the body through the food chain, and collects in the bones. Consequently, $_{38}^{90}Sr$ is a particularly serious health hazard. How long (in years) will it take for 99.9900% of the $_{38}^{90}Sr$ released in a nuclear reactor accident to disappear?

30. To make the dial of a watch glow in the dark, 1.00×10^{-9} kg of radium $_{88}^{226}Ra$ is used. The half-life of this isotope is 1.6×10^3 yr. How many kilograms of radium *disappear* while the watch is in use for fifty years?

*31. A device used in radiation therapy for cancer contains 0.50 g of cobalt $_{27}^{60}Co$ (59.933 819 u). The half-life of $_{27}^{60}Co$ is 5.27 yr. Determine the activity of the radioactive material.

*32. Two waste products from nuclear reactors are strontium $_{38}^{90}Sr$ and cesium $_{55}^{134}Cs$. The half-life of $_{38}^{90}Sr$ is 28.5 yr, while that of $_{55}^{134}Cs$ is 2.06 yr. If these two species are initially present in a ratio of $Sr/Cs = 7.80 \times 10^{-3}$, what is this ratio fifteen years later?

*33. If the activity of a radioactive substance is initially 398 disintegrations/min and two days later it is 285 disintegrations/min, what is the activity four days later still? Give your answer in disintegrations/min.

*34. A sample of ore containing a radioactive element has an activity of 4.0×10^4 Bq. How many grams of the element are in the sample, assuming the element is (a) radium $_{88}^{226}Ra$ ($T_{1/2} = 1.6 \times 10^3$ yr) and (b) uranium $_{92}^{238}U$ ($T_{1/2} = 4.47 \times 10^9$ yr)?

**35. Outside the nucleus, the neutron decays into a proton, an electron, and an antineutrino. The half-life for the neutron is 10.4 min. Over what distance will a beam of 5.00-eV neutrons travel before the number of neutrons per unit volume of the beam decreases to 75.0% of its initial value? Ignore relativistic effects.

Section 31.7 Radioactive Dating, Section 31.8 Radioactive Decay Series

36. Bones of the woolly mammoth have been found in North America. The youngest of these bones has a $_6^{14}C$ activity per gram of carbon that is about 21% of what was present in the live animal. How long ago (in years) did this animal disappear from North America?

37. The practical limit to ages that can be determined by radiocarbon dating is about 41 000 yr. In a 41 000-yr-old sample, what percentage of the original $_6^{14}C$ atoms remains?

38. The half-life for the α decay of uranium $_{92}^{238}U$ is 4.47×10^9 yr. Determine the age of a rock that contains sixty percent of its original $_{92}^{238}U$ atoms.

39. A sample of fossilized bones has a $_6^{14}C$ activity of 0.0061 Bq per gram of carbon. (a) Find the age of the sample, assuming that the activity per gram of carbon in a living organism has been constant at a value of 0.23 Bq. (b) Evidence suggests that the value of 0.23 Bq might have been as much as 40% larger. Repeat part (a), taking into account this 40% increase.

*40. Using the isotope table in Appendix F, construct a plot like that in Figure 31.11, showing the radioactive series that begins with thorium $_{90}^{232}Th$ and ends with lead $_{82}^{208}Pb$. You need not include half-lives.

**41. When any radioactive dating method is used, experimental error in the measurement of the sample's activity leads to error in the estimated age. In an application of the radiocarbon dating technique to certain fossils, an activity of 0.10 Bq per gram of carbon is measured to within an accuracy of ± ten percent. Find the age of the fossils and the maximum error (in years) in the value obtained. Assume that there is no error in the 5730-year half-life of $^{14}_{6}C$.

ADDITIONAL PROBLEMS

42. The $^{208}_{82}Pb$ isotope of lead has an atomic mass of 207.976 627 u. Obtain the binding energy per nucleon (in MeV).

43. According to the periodic table on the inside of the back cover, what element does each symbol "X" represent: $^{195}_{78}X$, $^{32}_{16}X$, $^{63}_{29}X$, $^{11}_{5}X$, and $^{239}_{94}X$?

44. Find the energy (in MeV) released when α decay converts radium $^{226}_{88}Ra$ (atomic mass = 226.025 40 u) into radon $^{222}_{86}Rn$ (atomic mass = 222.017 57 u).

45. How many half-lives are required for the number of radioactive nuclei to decrease to one one-millionth of the initial number?

46. Write the β^+ decay process for each of the following nuclei, being careful to include Z and A and the proper chemical symbol for each daughter nucleus: (a) $^{18}_{9}F$ and (b) $^{15}_{8}O$.

47. Two isotopes of a certain element have binding energies that differ by 5.03 MeV. The isotope with the larger binding energy contains one more neutron than the other isotope. Find the difference in atomic mass between the two isotopes.

*48. The photomultiplier tube in a commercial scintillator counter contains 15 of the special electrodes or dynodes. Each dynode produces 3 electrons for every electron that strikes it. One photoelectron strikes the first dynode. What is the maximum number of electrons that strikes the 15th dynode?

*49. To see why one curie of activity was chosen to be 3.7×10^{10} Bq, determine the activity (in disintegrations per second) of one gram of radium $^{226}_{88}Ra$ ($T_{1/2} = 1.6 \times 10^3$ yr).

*50. Plutonium $^{239}_{94}Pu$ (atomic mass = 239.052 16 u) undergoes α decay. Assuming all the released energy is in the form of kinetic energy of the α particle and ignoring the recoil of the daughter nucleus, find the speed of the α particle. Ignore relativistic effects.

**51. Both gold $^{198}_{79}Au$ ($T_{1/2} = 2.69$ days) and iodine $^{131}_{53}I$ ($T_{1/2} = 8.04$ days) are used in diagnostic medicine related to the liver. At the time laboratory supplies are monitored, the activity of the gold is observed to be five times greater than the activity of the iodine. How many days later will the two activities be equal?

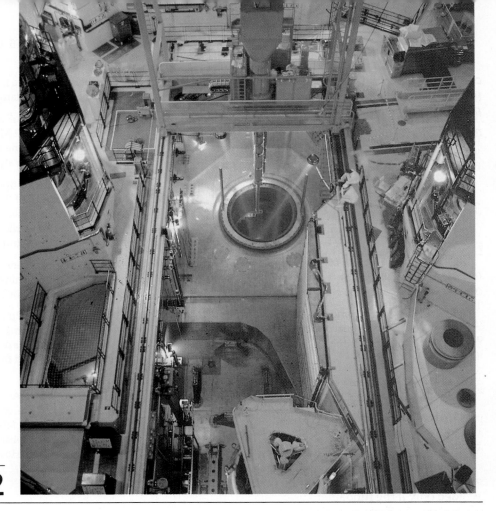

CHAPTER 32

IONIZING RADIATION, NUCLEAR ENERGY, AND ELEMENTARY PARTICLES

Ionizing radiation is present everywhere in our environment, whether from natural sources or from man-made sources like the nuclear reactor in the photograph above. Since ionizing radiation can have a serious effect on our health, we will examine the ways in which it is measured, especially with regard to its effects on living organisms. Particular attention will be paid to the sources of radiation in our environment, and to the amounts of radiation they produce. Starting with Section 32.2, we will discuss two ways in which energy is obtained from nuclear reactions. The most well-known method is by fission, whereby a heavy nucleus splits into two smaller nuclei with the release of a relatively large amount of energy. Controlled fission is the means by which nuclear reactors ultimately generate electrical energy. The other energy-releasing process, fusion, occurs when two lighter nuclei combine to form a heavier nucleus. Stars such as our sun produce energy by this method. Finally, we turn to the elementary particles that have been found within the nucleus. We will discuss the current theory of particle physics that suggests that protons and neutrons, as well as other particles, are made of smaller, indivisible particles called quarks.

32.1 BIOLOGICAL EFFECTS OF IONIZING RADIATION

IONIZING RADIATION

Ionizing radiation consists of photons and/or moving particles that have sufficient energy to knock an electron out of an atom or molecule, thus forming an ion. The photons usually lie in the ultraviolet, X-ray, or γ-ray regions of the electromagnetic spectrum, while the moving particles can be the α and β particles emitted during radioactive decay. An energy of roughly 1 to 35 eV is needed to ionize atoms or molecules, and the particles and γ rays emitted during nuclear disintegration often have energies of several million eV. Therefore, a single α particle, β particle, or γ ray can ionize thousands of molecules.

Nuclear radiation is potentially harmful to humans, because the ionization it produces can significantly alter the structure of molecules within a living cell. The alterations cause the cell to malfunction and, if severe enough, can lead to the death of the cell and even the organism itself. Despite the potential hazards, ionizing radiation can be used in medicine for diagnostic and therapeutic purposes, such as locating bone fractures and treating cancer. The hazards can be avoided only if the fundamentals of radiation exposure, including dose units and the biological effects of radiation, are understood.

Exposure is a measure of the ionization produced in air by X-rays or γ rays, and it is defined in the following manner. A beam of X-rays or γ rays is sent through a mass m of dry air at standard temperature and pressure (STP: 0 °C, 1 atm pressure). In passing through the air, the beam produces positive ions whose total charge is q. Exposure is defined as the total charge per unit mass of air: exposure $= q/m$. The SI unit of exposure is coulombs per kilogram (C/kg). However, the first radiation unit to be defined was the *roentgen* (R), and it is still used today. The exposure in roentgens is given by

$$\text{Exposure (in roentgens)} = \left(\frac{1}{2.58 \times 10^{-4}}\right)\frac{q}{m} \qquad (32.1)$$

Thus, when X-rays or γ rays produce an exposure of one roentgen, $q = 2.58 \times 10^{-4}$ C of positive charge are produced in $m = 1$ kg of dry air:

$$1 \text{ R} = 2.58 \times 10^{-4} \text{ C/kg} \qquad \text{(dry air, at STP)}$$

Since the concept of exposure is defined in terms of the ionizing abilities of X-rays and γ rays in air, it does not specify the effect of radiation on living tissue. For biological purposes, the ***absorbed dose*** is a more suitable quantity, because it is the energy absorbed from the radiation per unit mass of absorbing material:

$$\text{Absorbed dose} = \frac{\text{Energy absorbed}}{\text{Mass of absorbing material}} \qquad (32.2)$$

The SI unit of absorbed dose is the *gray* (Gy), which is a unit of energy divided by a unit of mass: 1 Gy = 1 J/kg. Equation 32.2 is applicable to all types of radiation and absorbing media.

Another unit is often used for absorbed dose, namely, the *rad* (rd). The word rad is an acronym for **r**adiation **a**bsorbed **d**ose. The rad and the gray are related by 1 rad = 0.01 gray.

The amount of biological damage produced by ionizing radiation is different for different kinds of radiation. For instance, a 1-rad dose of neutrons is far more effective in producing eye cataracts than a 1-rad dose of X-rays. To compare the damage caused by different types of radiation, the ***relative biological effectiveness*** (RBE) is used.* The relative biological effectiveness of a particular type of radiation compares the dose of that radiation needed to produce a certain biological effect to the dose of 200-keV X-rays needed to produce the same biological effect:

$$\text{Relative biological effectiveness (RBE)} = \frac{\text{The dose of 200-keV X-rays that produces a certain biological effect}}{\text{The dose of radiation that produces the same biological effect}} \quad (32.3)$$

The RBE depends on the nature of the ionizing radiation and its energy, as well as the type of tissue being irradiated. Table 32.1 lists some typical RBE values for different kinds of radiation, assuming an "average" biological tissue is being irradiated. The values of RBE = 1 indicate that γ rays and β^- particles produce the same biological damage as do 200-keV X-rays. The larger RBE values indicate that protons, α particles, and fast neutrons cause substantially more damage. The RBE is often used in conjunction with the absorbed dose to reflect the damage-producing character of the radiation on tissue. The product of the absorbed dose in rads (not in grays) and the RBE is the ***biologically equivalent dose:***

$$\text{Biologically equivalent dose (in rem)} = \text{Absorbed dose (in rad)} \times \text{RBE} \quad (32.4)$$

The unit for the biologically equivalent dose is the *rem*, short for **r**oentgen **e**quivalent, **m**an. Example 1 illustrates the use of the biologically equivalent dose.

*The RBE is sometimes called the *quality factor* (QF).

Table 32.1 Relative Biological Effectiveness (RBE) for Various Types of Radiation

Type of Radiation	RBE
200-keV X-rays	1
γ rays	1
β^- particles (electrons)	1
Protons	10
α particles	10–20
Neutrons	
Slow	2
Fast	10

Example 1 Comparing Absorbed Doses of γ Rays and Neutrons

A biological tissue is irradiated with γ rays that have an RBE of 0.70. The absorbed dose of γ rays is 850 rd. The tissue is then exposed to neutrons whose RBE is 3.5. The biologically equivalent dose of the neutrons is the same as that of the γ rays. What is the absorbed dose of neutrons?

REASONING The biologically equivalent doses of the neutrons and the γ rays are the same. Therefore, the tissue damage produced in each case is the same. However, the RBE of the neutrons is larger than the RBE of the γ rays by a factor of 3.5/0.70 = 5.0. Consequently, we will find that the absorbed dose of the neutrons is only one-fifth as great as that of the γ rays.

SOLUTION The biologically equivalent dose of the γ rays is the product of the absorbed dose (in rads) and the RBE:

$$\text{Biologically equivalent dose of } \gamma \text{ rays} = (850 \text{ rad})(0.70) = 6.0 \times 10^2 \text{ rem} \qquad (32.4)$$

For the neutrons (RBE = 3.5), the biologically equivalent dose is the same. Therefore, 6.0×10^2 rem = (Absorbed dose of neutrons)(3.5) and

$$\frac{\text{Absorbed dose of neutrons}}{} = \frac{6.0 \times 10^2 \text{ rem}}{3.5} = \boxed{170 \text{ rd}}$$

THE EFFECTS OF IONIZING RADIATION ON HUMANS

Everyone is continually exposed to background radiation from natural sources, such as cosmic rays (high-energy particles that come from outside the solar system), radioactive materials in the environment, radioactive nuclei—primarily carbon $^{14}_{6}$C and potassium $^{40}_{19}$K—within our own bodies, and radon. Table 32.2 lists the average biologically equivalent doses received from these sources by a person in the United States. According to this table, radon is a major contributor to the natural background radiation. Radon is an odorless radioactive gas and poses a health hazard because, when inhaled, it can damage the lungs and cause cancer. Radon is found in soil and rocks and enters houses through cracks and crevices in the foundation. The amount of radon in the soil varies greatly throughout the country, with some localities having significant amounts and others having virtually none. Accordingly, the dose that any individual receives can vary widely from the average value of 200 mrem/yr given in Table 32.2 (1 mrem = 10^{-3} rem). In many houses, the entry of radon can be reduced significantly by sealing the foundation against entry of the gas and providing good ventilation so it does not accumulate.

Table 32.2 Average Biologically Equivalent Doses of Radiation Received by a U. S. Resident[a]

Source of Radiation	Biologically Equivalent Dose (mrem/yr)[b]
Natural background radiation	
Cosmic rays	28
Radioactive earth and air	28
Internal radioactive nuclei	39
Inhaled radon	≈ 200
Man-made radiation	
Consumer products	10
Medical/dental diagnostics	39
Nuclear medicine	14
Rounded total:	360

[a] National Council on Radiation Protection and Measurement, Report No. 93, "Ionizing Radiation Exposure of the Population of the United States," 1987.
[b] 1 mrem = 10^{-3} rem.

To the natural background of radiation, a significant amount of man-made radiation has been added, mostly from medical/dental diagnostic X-rays. Table 32.2 indicates an average total dose of 360 mrem/yr from all sources.

The effects of radiation on humans can be grouped into two categories, according to the time span between initial exposure and the appearance of physiological effects: (1) short-term or acute effects that appear within a matter of minutes, days, or weeks, and (2) long-term or latent effects that appear years, decades, or even generations later.

Radiation sickness is the general term applied to the acute effects of radiation. Depending on the severity of the dose, a person with radiation sickness can exhibit nausea, vomiting, fever, diarrhea, and loss of hair. Ultimately, death can occur. The severity of radiation sickness is related to the dose received, and in the following discussion the biologically equivalent doses quoted are whole-body, single doses. A dose less than 50 rem causes no short-term, ill effects. A dose between 50 and 300 rem brings on radiation sickness, the severity increasing with increasing dosage. A whole-body dose in the range of 400–500 rem is classified as an LD_{50} dose, meaning that it is a lethal dose (LD) for about 50% of the people so exposed; death occurs within a few months. Whole-body doses greater than 600 rem result in death for almost all individuals.

Long-term or latent effects of radiation may appear as a result of high-level, brief exposure or low-level exposure over a long period of time. Some long-term effects are loss of hair, eye cataracts, and various kinds of cancer. In addition, genetic defects caused by mutated genes may be passed on from one generation to the next.

Because of the hazards of radiation, the federal government has established dose limits. The permissible dose for an individual is defined as the dose, accumulated over a long period of time or resulting from a single exposure, that carries negligible probability of a severe health hazard. Current federal standards (1991) state that an individual in the general population should not receive more than 500 mrem of man-made radiation each year, *exclusive* of medical sources. A person exposed to radiation in the workplace (e.g., a radiation therapist) should not receive more than 5 rem per year from work-related sources.

32.2 INDUCED NUCLEAR REACTIONS

Section 31.4 discusses how a radioactive parent nucleus disintegrates spontaneously into a daughter nucleus. It is also possible to bring about or "induce" the disintegration of a stable nucleus by striking it with another nucleus, an atomic or subatomic particle, or a γ-ray photon. A **nuclear reaction** is said to occur whenever the incident nucleus, particle, or photon causes a change to occur in a target nucleus.

In 1919 Ernest Rutherford observed that when an α particle strikes a nitrogen nucleus, an oxygen nucleus and a proton are produced. This nuclear reaction is written as

$$\underbrace{^{4}_{2}\text{He}}_{\substack{\text{Incident}\\ \alpha \text{ particle}}} + \underbrace{^{14}_{7}\text{N}}_{\substack{\text{Nitrogen}\\ \text{(target)}}} \longrightarrow \underbrace{^{17}_{8}\text{O}}_{\text{Oxygen}} + \underbrace{^{1}_{1}\text{H}}_{\text{Proton}}$$

Because the incident α particle induces the transmutation of nitrogen into oxygen, this reaction is an example of an ***induced nuclear transmutation.***

Nuclear reactions are often written in a shorthand form. For example, the reaction above is designated by $^{14}_{7}N(\alpha, p)^{17}_{8}O$. The first and last symbols represent the initial and final nuclei, respectively. The symbols inside the parentheses denote the incident α particle (on the left) and the small emitted particle or proton (on the right). Some other induced nuclear transmutations are listed below, together with the corresponding shorthand notations:

Nuclear Reaction	Notation
$^{1}_{0}n + {}^{10}_{5}B \rightarrow {}^{7}_{3}Li + {}^{4}_{2}He$	$^{10}_{5}B(n, \alpha)^{7}_{3}Li$
$\gamma + {}^{25}_{12}Mg \rightarrow {}^{24}_{11}Na + {}^{1}_{1}H$	$^{25}_{12}Mg(\gamma, p)^{24}_{11}Na$
$^{1}_{1}H + {}^{13}_{6}C \rightarrow {}^{14}_{7}N + \gamma$	$^{13}_{6}C(p, \gamma)^{14}_{7}N$

In any nuclear reaction, both the total electric charge of the nucleons and the total number of nucleons are conserved during the process, as discussed in Section 31.4. The fact that these quantities are conserved makes it possible to identify the nucleus produced in a nuclear reaction, as the next example illustrates.

Example 2   An Induced Nuclear Transmutation

An α particle strikes an aluminum $^{27}_{13}Al$ nucleus, and a nucleus $^{A}_{Z}X$ and a neutron are produced:

$$^{4}_{2}He + {}^{27}_{13}Al \rightarrow {}^{A}_{Z}X + {}^{1}_{0}n$$

Identify the nucleus produced, including its atomic number Z (the number of protons) and its atomic mass number A (the number of nucleons).

REASONING AND SOLUTION   Since the total electric charge of the nucleons and the total number of nucleons are conserved, it is possible to write the equations listed below:

Conserved Quantity	Before Reaction		After Reaction
Total electric charge (number of protons)	2 + 13	=	Z + 0
Total number of nucleons	4 + 27	=	A + 1

Solving these equations for Z and A gives $Z = 15$ and $A = 30$. Since $Z = 15$ identifies the element as phosphorus, the nucleus produced is $\boxed{^{30}_{15}P}$.

Induced nuclear transmutations can be used to produce isotopes that are not found naturally. In 1934, Enrico Fermi suggested a method for producing elements with a higher atomic number than uranium ($Z = 92$). These elements — neptunium ($Z = 93$), plutonium ($Z = 94$), americium ($Z = 95$), and so on — are known as *transuranium elements.* None of the transuranium elements occurs naturally. They are created in a nuclear reaction between a suitably chosen lighter

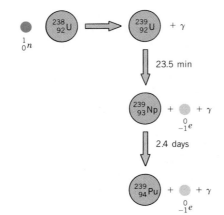

Figure 32.1 An induced nuclear reaction in which $^{238}_{92}U$ is transmuted into the transuranium element plutonium $^{239}_{94}Pu$.

element and a small incident particle, usually a neutron or an α particle. For example, Figure 32.1 shows a reaction that produces plutonium from uranium. A neutron is captured by a uranium $^{238}_{92}U$ nucleus, producing $^{239}_{92}U$ and a γ ray. The $^{239}_{92}U$ nucleus is radioactive and decays with a half-life of 23.5 min into neptunium $^{239}_{93}Np$. Neptunium is also radioactive and disintegrates with a half-life of 2.4 days into plutonium $^{239}_{94}Pu$. Plutonium is the final product and has a half-life of 24 100 yr.

The neutrons that participate in nuclear reactions can have kinetic energies that cover a wide range. In particular, those that have a kinetic energy of about 0.04 eV or less are called ***thermal neutrons.*** The name derives from the fact that such a relatively small kinetic energy is comparable to the average translational kinetic energy of a molecule at room temperature. Thermal neutrons are used in one type of bomb detection system that can expose hidden explosives. As Figure 32.2 illustrates, the system bathes luggage suspected of containing a bomb in low doses of thermal neutrons. Some of the neutrons are captured by the nuclei of the luggage and its contents, including explosives. These nuclei subsequently emit γ rays, the energies of which are unique to the nuclei that emit them. By analyzing the γ rays, it is possible to determine the chemical natures of the materials in the luggage. In particular, the system looks for certain nitrogen compounds that signal the presence of a bomb.

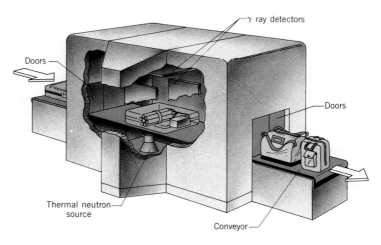

Figure 32.2 A bomb detection system that uses thermal neutrons.

32.3 NUCLEAR FISSION

THE FISSION PROCESS

In 1939 four German scientists, Otto Hahn, Lise Meitner, Fritz Strassmann, and Otto Frisch, made an important discovery that ushered in the atomic age. They found that a uranium nucleus, after absorbing a neutron, splits into two fragments, each with a smaller mass than the original nucleus. The splitting of a massive nucleus into two less-massive fragments is known as *nuclear fission.*

Figure 32.3 shows a fission reaction in which a uranium $^{235}_{92}U$ nucleus is split into barium $^{141}_{56}Ba$ and krypton $^{92}_{36}Kr$ nuclei. The reaction begins when $^{235}_{92}U$ absorbs a slowly moving neutron, creating a "compound nucleus," $^{236}_{92}U$. The compound nucleus disintegrates quickly into $^{141}_{56}Ba$, $^{92}_{36}Kr$, and three neutrons according to the following reaction:

$$^{1}_{0}n + {}^{235}_{92}U \longrightarrow \underbrace{{}^{236}_{92}U}_{\substack{\text{Compound} \\ \text{nucleus} \\ \text{(unstable)}}} \longrightarrow \underbrace{{}^{141}_{56}Ba}_{\text{Barium}} + \underbrace{{}^{92}_{36}Kr}_{\text{Krypton}} + \underbrace{3{}^{1}_{0}n}_{\text{3 neutrons}}$$

This reaction is only one of the many possible reactions that can occur when uranium fissions. For example, another reaction is

$$^{1}_{0}n + {}^{235}_{92}U \longrightarrow \underbrace{{}^{236}_{92}U}_{\substack{\text{Compound} \\ \text{nucleus} \\ \text{(unstable)}}} \longrightarrow \underbrace{{}^{140}_{54}Xe}_{\text{Xenon}} + \underbrace{{}^{94}_{38}Sr}_{\text{Strontium}} + \underbrace{2{}^{1}_{0}n}_{\text{2 neutrons}}$$

Some reactions produce as many as 5 neutrons; however, the average number produced per fission is 2.5.

When a neutron collides with and is absorbed by a uranium nucleus, the uranium nucleus begins to vibrate and becomes distorted. The vibration continues until the distortion becomes so severe that the attractive strong nuclear force

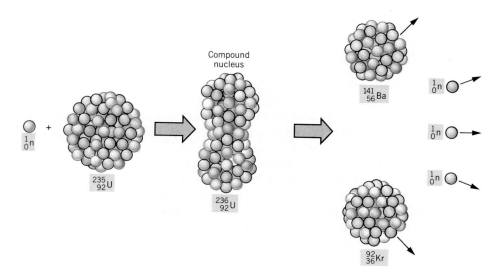

Figure 32.3 The slow neutron causes the uranium nucleus $^{235}_{92}U$ to fission into barium $^{141}_{56}Ba$, krypton $^{92}_{36}Kr$, and three neutrons.

can no longer balance the electrostatic repulsion between the nuclear protons. At this point, the nucleus bursts apart into fragments, which carry off energy, primarily in the form of kinetic energy. The energy carried off by the fragments is enormous and was stored in the original nucleus primarily in the form of electric potential energy. An average of roughly 200 MeV of energy is released per fission. This energy is approximately 10^8 times greater than the energy released per molecule in an ordinary chemical reaction, such as the combustion of gasoline or coal. Example 3 demonstrates how to estimate the energy released during the fission of a nucleus.

Example 3 The Energy Released During Nuclear Fission

Estimate the amount of energy released when a massive nucleus ($A = 240$) fissions.

REASONING   Figure 31.5 shows that the binding energy of a nucleus with $A = 240$ is about 7.6 MeV per nucleon. We assume that this nucleus fissions into two fragments, each with $A \approx 120$. According to Figure 31.5, the binding energy of the fragments increases to about 8.5 MeV per nucleon. Consequently, when a massive nucleus fissions, there is a release of about 8.5 MeV − 7.6 MeV = 0.9 MeV of energy per nucleon.

SOLUTION   Since there are 240 nucleons involved in the fission process, the total energy released per fission is approximately (0.9 MeV/nucleon)(240 nucleons) ≈ $\boxed{200 \text{ MeV}}$.

Virtually all naturally occurring uranium is composed of two isotopes. These isotopes and their natural abundances are $^{238}_{92}\text{U}$ (99.275%) and $^{235}_{92}\text{U}$ (0.720%). Although $^{238}_{92}\text{U}$ is by far the most abundant isotope, the probability that it will capture a neutron and fission is very small. For this reason, $^{238}_{92}\text{U}$ is not the isotope of choice for generating nuclear energy. In contrast, the isotope $^{235}_{92}\text{U}$ readily captures a neutron and fissions, *provided the neutron is a thermal neutron* (kinetic energy ≈ 0.04 eV or less). The probability of a thermal neutron causing $^{235}_{92}\text{U}$ to fission is about five hundred times greater than a neutron whose energy is relatively high, say 1 MeV. Thermal neutrons can also be used to fission other nuclei, such as plutonium $^{239}_{94}\text{Pu}$.

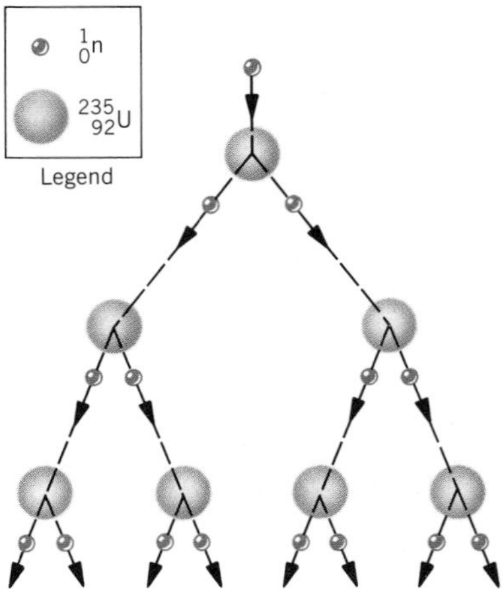

CHAIN REACTION

The fact that the uranium fission reaction releases 2.5 neutrons, on the average, makes it possible for a self-sustaining series of fissions to occur. As Figure 32.4 illustrates, each neutron released by a fission can initiate another fission process, resulting in the emission of still more neutrons, followed by more fissions, and so on. A *chain reaction* is a series of nuclear fissions whereby some of the neutrons produced by each fission cause additional fissions. During an uncontrolled chain reaction, it would not be unusual for the number of fissions to increase a thousandfold within a few millionths of a second. With an average energy of about 200 MeV being released per fission, an uncontrolled chain reaction can generate an incredible amount of energy in a very short time, as happens in an atomic bomb (which is actually a *nuclear* bomb).

By limiting the number of neutrons in the environment of the fissile nuclei, it is

Figure 32.4   A chain reaction. For clarity, it is assumed that each fission generates two neutrons (2.5 neutrons are actually liberated on the average). The fission fragments are not shown for clarity.

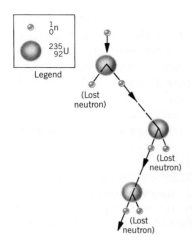

Legend

(Lost neutron)

(Lost neutron)

(Lost neutron)

Figure 32.5 In a controlled chain reaction, only one neutron, on average, from each fission event causes another nucleus to fission. As a result, energy is released at a steady or controlled rate.

possible to establish a condition whereby each fission event contributes, on average, only *one neutron* that fissions another nucleus (see Figure 32.5). In this manner, the chain reaction and the rate of energy production are *controlled*. The controlled fission chain reaction is the principle behind the nuclear reactors used in the commercial generation of electric power.

32.4 NUCLEAR REACTORS

BASIC COMPONENTS

A nuclear reactor is a type of furnace in which energy is generated by a controlled fission chain reaction. The first nuclear reactor was built by Enrico Fermi in 1942, on the floor of a squash court under the west stands of Stagg Field at the University of Chicago. Today, there are many kinds and sizes of reactors, but they all have three basic components: fuel elements, a neutron moderator, and control rods. Figure 32.6 illustrates these components.

The *fuel elements* contain the fissile fuel and, for example, may be in the shape of thin rods about 1 cm in diameter. In a large power reactor there may be thousands of fuel elements placed close together, the entire region of fuel elements being known as the *reactor core.*

Uranium $^{235}_{92}$U is a common reactor fuel. Since the natural abundance of this isotope is only about 0.7%, there are special uranium-enrichment plants to increase the percentage. Most commercial reactors use uranium in which the amount of $^{235}_{92}$U has been enriched to about 3%.

While neutrons with energies of about 0.04 eV (or less) readily fission $^{235}_{92}$U, the neutrons released during the fission process have significantly greater energies of several MeV or so. Consequently, a nuclear reactor must contain some type of material that will decrease or moderate the speed of such energetic neutrons so they can readily fission additional $^{235}_{92}$U nuclei. The material that slows down the neutrons is called a *moderator*. One commonly used moderator is water. When an energetic neutron leaves a fuel element, the neutron enters the surrounding water and collides with water molecules. With each collision, the neutron loses an

THE PHYSICS OF . . .

nuclear reactors.

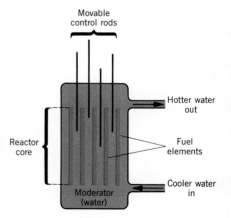

Figure 32.6 A nuclear reactor consists of fuel elements, control rods, and a moderator (in this case, water).

appreciable fraction of its energy and slows down. Once slowed down to thermal energy by the moderator, a process that takes less than 10^{-3} s, the neutron is capable of initiating a fission event upon reentering a fuel element.

If the output power from a reactor is to remain constant, only one neutron from each fission event must trigger a new fission, as Figure 32.5 suggests. When each fission leads to one additional fission—no more or no less—the reactor is said to be *critical*. A reactor normally operates in a critical condition, for then it produces a steady output of energy. The reactor is *subcritical* when, on average, the neutrons from each fission trigger *less than one* subsequent fission. In a subcritical reactor, the chain reaction is not self-sustaining and eventually dies out. When the neutrons from each fission trigger *more than one* additional fission, the reactor is *supercritical*. During a supercritical condition, the energy released by a reactor increases. If left unchecked, the increasing energy can lead to a partial or total meltdown of the reactor core, with the possible release of radioactive material into the environment.

Clearly, a control mechanism is needed to keep the reactor in its normal or critical state. This control is accomplished by a number of **control rods** that can be moved into and out of the reactor core (see Figure 32.6). The control rods contain an element, such as boron or cadmium, that readily absorbs neutrons without fissioning. If the reactor becomes supercritical, the control rods are automatically moved farther into the core to absorb the excess neutrons causing the condition. In response, the reactor returns to its critical state. Conversely, if the reactor becomes subcritical, the control rods are partially withdrawn from the core, so fewer neutrons are absorbed. Thus, more neutrons are available for fission, and the reactor returns to its critical state.

THE PRESSURIZED WATER REACTOR

Figure 32.7 illustrates a pressurized water reactor. In such a reactor, the heat generated within the fuel rods is carried away by water that surrounds the rods. To remove as much heat as possible, the water is heated to a high temperature (above 300 °C). To prevent boiling, which occurs at 100 °C at 1 atmosphere of pressure, the water is pressurized in excess of 150 atmospheres. The hot water is pumped through a heat exchanger, where heat is transferred to water flowing in a

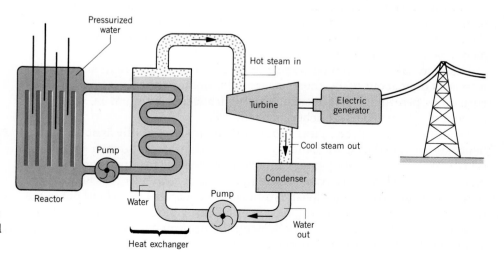

Figure 32.7 Diagram of a nuclear power plant that uses a pressurized water reactor.

second, closed system. The heat transferred to the second system produces steam that drives a turbine. The turbine is coupled to an electric generator, whose output electrical power is delivered to consumers via high-voltage transmission lines. After exiting the turbine, the steam is condensed back into water that is returned to the heat exchanger.

32.5 NUCLEAR FUSION

In Example 3 of Section 32.3, the binding-energy-per-nucleon curve is used to estimate the amount of energy released in the fission process. As summarized in Figure 32.8, the massive nuclei at the right end of the curve have a binding energy of about 7.6 MeV per nucleon. The less-massive fission fragments are near the center of the curve and have a binding energy of approximately 8.5 MeV per nucleon. The energy released per nucleon by fission is the difference between these two values, or about 0.9 MeV per nucleon.

A glance at the far left end of the diagram suggests another means of generating energy. Two very-low-mass nuclei with relatively small binding energies per nucleon, could be combined or "fused" into a single, more massive nucleus that has a greater binding energy per nucleon. This process is called *nuclear fusion.* A substantial amount of energy can be released during a fusion reaction, as Example 4 shows.

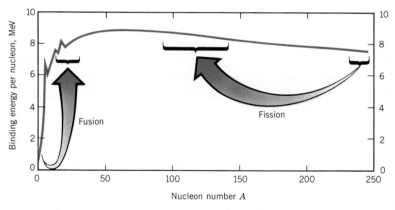

Figure 32.8 When fission occurs, a massive nucleus divides into two fragments whose binding energy per nucleon is greater than that of the original nucleus. When fusion occurs, two low-mass nuclei combine to form a more massive nucleus whose average binding energy per nucleon is greater than that of the original nuclei.

Example 4   *The Energy Released During Nuclear Fusion*

Two isotopes of hydrogen, $^{2}_{1}$H (deuterium, D) and $^{3}_{1}$H (tritium, T), fuse to form $^{4}_{2}$He and a neutron according to the following reaction:

$$^{2}_{1}\text{H} + {}^{3}_{1}\text{H} \rightarrow {}^{4}_{2}\text{He} + {}^{1}_{0}\text{n}$$

Determine the energy released by this fusion reaction.

This nuclear explosion on May 8, 1951, was part of Operation Greenhouse on the South Pacific atoll of Eniwetok.

REASONING AND SOLUTION   The masses of the initial and final nuclei in this reaction, as well as that of the neutron, are given in the isotope table in Appendix F:

Initial Masses		Final Masses	
$^2_1$H	2.014 u	$^4_2$He	4.003 u
$^3_1$H	3.016 u	$^1_0$n	1.009 u
Total:	5.030 u	Total:	5.012 u

The mass defect is $\Delta m = (5.030\ u - 5.012\ u) = 0.018\ u$. Since 1 u is equivalent to 931.5 MeV, the energy released is $\boxed{17\ \text{MeV}}$.

There are five nucleons that participate in the fusion, so the energy released per nucleon is about 3.4 MeV. This energy per nucleon is greater than that released in a fission process (≈ 0.9 MeV per nucleon). Thus, for a given mass of fuel, a fusion reaction yields more energy than a fission reaction.

Because fusion reactions release substantial amounts of energy, there is considerable interest in fusion reactors, although to date no commercial units have been constructed. The difficulties in building a fusion reactor arise mainly because the two low-mass nuclei must be brought sufficiently near each other so that the short-range strong nuclear force can pull them together, leading to fusion. But each nucleus has a positive charge and repels the other electrically. For the nuclei to get sufficiently close in the presence of the repulsive electric force, they must have large kinetic energies to start with. For example, to start a deuterium–deuterium fusion reaction, it has been estimated that each nucleus needs an initial kinetic energy of about 0.25 MeV.

In Chapter 14, we saw that the average translational kinetic energy $\overline{KE}$ of an atom in an ideal gas is directly proportional to the Kelvin temperature T of the gas according to $\overline{KE} = \frac{3}{2}kT$ (Equation 14.6), where $k = 1.38 \times 10^{-23}$ J/K is the Boltzmann constant. The kinetic energy of 0.25 MeV (4.0×10^{-14} J) needed to start a fusion reaction corresponds to a gas temperature of

$$T = \frac{2(\overline{KE})}{3k} = \frac{2(4.0 \times 10^{-14}\ \text{J})}{3(1.38 \times 10^{-23}\ \text{J/K})} = 2 \times 10^9\ \text{K}$$

This is two billion kelvins! Actually, the temperature of the deuterium gas need not be quite this high, because this temperature corresponds to the average kinetic energy of the nuclei. Some nuclei within the gas have energies substantially greater than the average energy, and these higher-energy nuclei can fuse together to produce a net outflow of energy. Typically, the temperature needed to start a deuterium–deuterium fusion reaction is about 4×10^8 K. Nevertheless, it is no trivial task to create such a temperature in the laboratory. Reactions that require such extremely high temperatures are called *thermonuclear reactions*. The most important thermonuclear reactions occur in stars, such as our own sun. The energy radiated by the sun comes from such reactions deep within its core, where the temperature is high enough to initiate the fusion process. One group of reactions thought to occur in the sun is the *proton–proton* cycle. This cycle is a series of reactions whereby six protons form a helium nucleus, two positrons, two γ rays, and two protons. The energy released by the proton–proton cycle is about 27 MeV.

Man-made fusion reactions have been carried out in a fusion-type nuclear bomb—commonly called a hydrogen bomb. In a hydrogen bomb, the fusion reaction is ignited by a fission bomb using uranium or plutonium. The temperature produced by the fission bomb is sufficiently high to initiate a thermonuclear reaction where, for example, hydrogen isotopes are fused into helium, releasing even more energy.

For fusion to be useful as a commercial energy source, the energy must be released in a steady, controlled manner—unlike the uncontrolled energy released by a hydrogen bomb. To date, scientists have not succeeded in constructing a fusion device that produces more energy on a continual basis than is expended in operating the device. A fusion device uses a high temperature to start a reaction, and under such a condition, all the atoms are completely ionized to form a *plasma* (a gas composed of charged particles, like ${}^2_1H^+$ and e^-). The problem is to confine the hot plasma for a long enough time so that collisions among the ions can lead to fusion.

One ingenious method of confining the plasma, called *magnetic confinement,* uses a magnetic field. Charges moving in a magnetic field are subject to magnetic forces, and it is hoped that the plasma can be confined to a region of space by these forces. The problem of magnetic confinement is a difficult one, although steady progress is being made. Figure 32.9 shows the Tokamak Fusion Test Reactor, which uses magnetic confinement.

Another type of confinement scheme, known as *inertial confinement,* is also being developed. Tiny, solid pellets of fuel are dropped into a container. As each pellet reaches the center of the container, a number of high-intensity lasers or electron beams strike the pellet simultaneously. The heating causes almost instantaneous vaporization of the pellet. However, the inertia of the vaporized atoms keeps them from expanding outward as fast as the vapor is being formed.

THE PHYSICS OF . . .

nuclear fusion using magnetic confinement.

THE PHYSICS OF . . .

nuclear fusion using inertial confinement.

Figure 32.9 The Tokamak Fusion Test Reactor at Princeton University uses the method of magnetic confinement to contain the hot plasma during nuclear fusion.

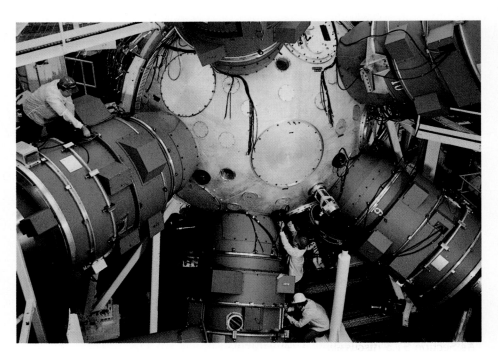

Figure 32.10 The NOVA Laser Facility at Lawrence Livermore National Laboratory at Lawrence, California. Short pulses of laser light travel simultaneously through each of the cylindrical tubes. Inside the spherical chamber, the pulses strike a small pellet containing deuterium and tritium. The laser pulses heat the pellet to an extremely high temperature, initiating fusion.

As a result, high pressures, high densities, and high temperatures are achieved at the center of the pellet, thus causing fusion. Figure 32.10 shows the inertial confinement facility at the Lawrence Livermore National Laboratory.

When compared to fission, fusion has some attractive features as an energy source. As we have seen in Example 4, fusion yields more energy than fission, for a given mass of fuel. Moreover, one type of fuel, $^2_1$H (deuterium), is found in the waters of the oceans and is plentiful, cheap, and easy to separate from the common $^1_1$H isotope of hydrogen. Fissile materials like naturally occurring uranium $^{235}_{92}$U are much less available and supplies could be depleted within a century or two. Unfortunately, the commercial use of fusion to provide cheap energy remains in the future.

32.6 ELEMENTARY PARTICLES

SETTING THE STAGE

By 1932 the electron, the proton, and the neutron had been discovered and were thought to be nature's three *elementary particles*, in the sense that they were the basic building blocks from which all matter is constructed. Experimental evidence obtained since then, however, shows that several hundred additional particles exist, and scientists no longer believe that the proton and the neutron are elementary particles.

Most of these new particles have masses greater than the electron's mass, and many are more massive than protons or neutrons. Virtually all the new particles are unstable and decay with times between about 10^{-6} and 10^{-23} s.

Often, new particles are produced by accelerating protons or electrons to high energies and letting them collide with a target nucleus. For example, Figure 32.11

shows a collision between an energetic proton and a stationary proton. If the incoming proton has sufficient energy, the collision produces an entirely new particle, the *neutral pion* (π^0). The π^0 particle lives for only about 0.8×10^{-16} s before it decays into two γ-ray photons. Since the pion did not exist before the collision, the pion was created from part of the incident proton's energy. Because a new particle such as the neutral pion is often created from energy, it is customary to report the mass of the particle in terms of its equivalent *rest energy*. Often energy units of MeV are used. For instance, detailed analyses of experiments reveal that the mass of the π^0 particle is equivalent to a rest energy of 135.0 MeV. For comparison, the more massive proton has a rest energy of 938.3 MeV. Analyses of experiments also provide the electric charge and other properties of particles created in high-energy collisions.

In the limited space available here, it is not possible to describe all the new particles that have been found. However, we will highlight some of the more significant discoveries.

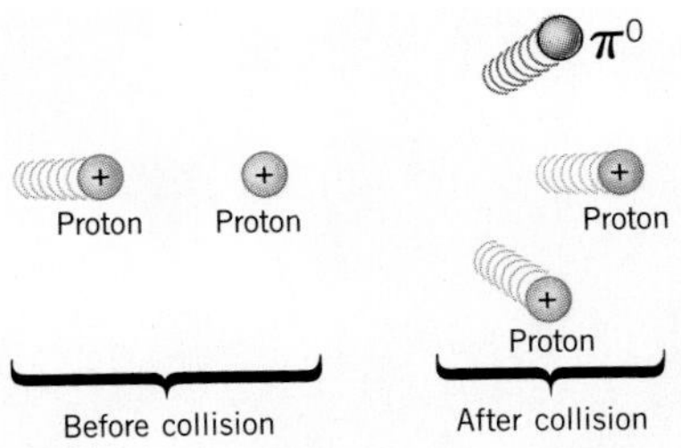

Figure 32.11 When an energetic proton collides with a stationary proton, a neutral pion (π^0) is produced. Part of the energy of the incident proton goes into creating the pion.

NEUTRINOS

In 1930 Wolfgang Pauli suggested that a particle called the **neutrino** (now known as the electron neutrino) should accompany the β decay of a radioactive nucleus. As Section 31.5 discusses, the neutrino has no electric charge, has a very small (possibly zero) mass, and travels near or at the speed of light. Neutrinos were finally discovered in 1956. Today, neutrinos are created in abundance in nuclear reactors and particle accelerators and are thought to be plentiful in the universe.

POSITRONS AND ANTIPARTICLES

The year 1932 saw the discovery of the *positron* (a contraction for "positive electron"). The positron has the same mass as the electron, but carries an opposite charge of $+e$. A collision between a positron and an electron is likely to annihilate both particles, converting them into electromagnetic energy in the form of γ rays. For this reason, positrons never coexist with ordinary matter for any appreciable length of time.

The positron is an example of an antiparticle, and after its discovery, scientists came to realize that for every particle there is an antiparticle. The antiparticle is a form of matter that has the same mass as the particle, but carries an opposite electric charge (e.g., the electron–positron pair) or a magnetic moment that is oriented in an opposite direction relative to the spin (e.g., the neutrino–antineutrino pair). A few electrically neutral particles, like the photon and the neutral pion (π^0), are their own antiparticles.

Antimatter would consist of positrons and antinucleons, such as antiprotons and antineutrons. While antimatter cannot coexist with matter, there is speculation that regions of the universe might consist entirely of antimatter. Presently, however, there is no experimental evidence to support such speculation.

MUONS AND PIONS

In 1937 the American physicists S. H. Neddermeyer and C. D. Anderson discovered a new charged particle whose mass was about 207 times greater than the mass of the electron. The particle is designated by the Greek letter μ (mu) and is

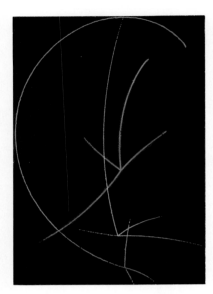

Subatomic particles resulting from induced nuclear transmutation produce these red, blue, and green tracks in a cloud chamber. The colors are computer generated.

known as a *muon*. There are two muons that have the same mass but opposite charge: the particle μ^- and its antiparticle μ^+. The μ^- muon has the same charge as the electron, while the μ^+ muon has the same charge as the positron. Both muons are unstable, with a lifetime of 2.2×10^{-6} s. The μ^- muon decays into an electron (β^-), a muon neutrino (ν_μ), and an electron antineutrino ($\bar{\nu}_e$), according to the following reaction:

$$\mu^- \rightarrow \beta^- + \nu_\mu + \bar{\nu}_e$$

The μ^+ muon decays into a positron (β^+), a muon antineutrino ($\bar{\nu}_\mu$), and an electron neutrino (ν_e):

$$\mu^+ \rightarrow \beta^+ + \bar{\nu}_\mu + \nu_e$$

Muons interact with protons and neutrons via the weak nuclear force.

The Japanese physicist Hidekei Yukawa (1907–1981) predicted in 1935 that *pions* exist, but they were not discovered until 1947. Pions came in three varieties: one that is positively charged, the negatively charged antiparticle with the same mass, and the neutral pion, mentioned earlier, which is its own antiparticle. The symbols for these pions are, respectively, π^+, π^-, and π^0. The charged pions are unstable and have a lifetime of 2.6×10^{-8} s. The decay of a charged pion almost always produces a muon:

$$\pi^- \rightarrow \mu^- + \bar{\nu}_\mu$$
$$\pi^+ \rightarrow \mu^+ + \nu_\mu$$

As mentioned earlier, the neutral pion π^0 is also unstable and decays into two γ-ray photons, the lifetime being 0.8×10^{-16} s. The pions are of great interest because, unlike the muons, the pions interact with protons and neutrons via the strong nuclear force.

CLASSIFICATION OF PARTICLES

It is useful to group the known particles into three families, the photons, the leptons, and the hadrons, as Table 32.3 summarizes. This grouping is made according to the nature of the force by which a particle interacts with other particles. The *photon family*, for instance, has only one member, the photon. The photon interacts only with charged particles, and the interaction is only via the *electromagnetic force*. No other particle behaves in this manner.

The *lepton family* consists of particles that interact by means of the *weak nuclear force*. Leptons can also exert gravitational and electromagnetic (if the leptons are charged) forces on other particles. The four better-known leptons are the electron, the muon, the electron neutrino ν_e, and the muon neutrino ν_μ. Table 32.3 lists these particles together with their antiparticles. Recently, two other leptons have been discovered, the tau particle (τ) and its neutrino (ν_τ), bringing the number of particles in the lepton family to six.

The *hadron family* contains the particles that interact by means of the *strong nuclear force*. Hadrons can also interact by gravitational and electromagnetic forces, but at short distances ($\leq 10^{-15}$ m) the strong nuclear force dominates. Among the hadrons are the proton, the neutron, and the pions. As Table 32.3 indicates, most hadrons are short-lived. The hadrons are subdivided into two groups, the *mesons* and the *baryons*, for a reason that will be discussed in connection with the idea of quarks.

Table 32.3 Some Elementary Particles

Family	Particle	Particle Symbol	Antiparticle Symbol	Rest Energy (MeV)	Lifetime (s)
Photon	Photon	γ	Self[a]	0	Stable
Lepton	Electron	e^- (or β^-)	e^+ (or β^+)	0.511	Stable
	Muon	μ^-	μ^+	105.7	2.2×10^{-6}
	Tau	τ^-	τ^+	1784	10^{-13}
	Electron neutrino	ν_e	$\bar{\nu}_e$	≈ 0	Stable
	Muon neutrino	ν_μ	$\bar{\nu}_\mu$	≈ 0	Stable
	Tau neutrino	ν_τ	$\bar{\nu}_\tau$	≈ 0	Stable
Hadron *Mesons*					
	Pion	π^+	π^-	139.6	2.6×10^{-8}
		π^0	Self[a]	135.0	0.8×10^{-16}
	Kaon	K^+	K^-	493.7	1.2×10^{-8}
		K_S^0	$\bar{K}_S^0$	497.7	0.9×10^{-10}
		K_L^0	$\bar{K}_L^0$	497.7	5.2×10^{-8}
	Eta	η^0	Self[a]	548.8	$< 10^{-18}$
	. . .				
	Plus other mesons				
Baryons	Proton	p	$\bar{p}$	938.3	Stable
	Neutron	n	$\bar{n}$	939.6	900
	Lambda	Λ^0	$\bar{\Lambda}^0$	1116	2.6×10^{-10}
	Sigma	Σ^+	$\bar{\Sigma}^-$	1189	0.8×10^{-10}
		Σ^0	$\bar{\Sigma}^0$	1192	6×10^{-20}
		Σ^-	$\bar{\Sigma}^+$	1197	1.5×10^{-10}
	Omega	Ω^-	Ω^+	1672	0.82×10^{-10}
	. . .				
	Plus other baryons				

[a] The particle is its own antiparticle.

QUARKS

As more and more hadrons were discovered, it became clear that they were not all elementary particles. The suggestion was made that the hadrons are made up of smaller, more elementary particles called *quarks*. In 1963 a quark theory was advanced independently by M. Gell-Mann (1929–) and G. Zweig (1937–). The theory proposed that there are three quarks and three corresponding antiquarks, and that hadrons are constructed from combinations of these. Thus, the quarks are elevated to the status of elementary particles for the hadron family. The particles in the photon and lepton families are considered to be elementary, and as such they are not composed of quarks.

The three quarks were named *up* (*u*), *down* (*d*), and *strange* (*s*), and were assumed to have, respectively, fractional charges of $+\frac{2}{3}e$, $-\frac{1}{3}e$, and $-\frac{1}{3}e$. In other words, a quark possesses a charge smaller than the charge of an electron. Table

Table 32.4   Quarks and Antiquarks

Name	Quarks Symbol	Quarks Charge	Antiquarks Symbol	Antiquarks Charge
Up	u	$+\frac{2}{3}e$	$\bar{u}$	$-\frac{2}{3}e$
Down	d	$-\frac{1}{3}e$	$\bar{d}$	$+\frac{1}{3}e$
Strange	s	$-\frac{1}{3}e$	$\bar{s}$	$+\frac{1}{3}e$
Charm	c	$+\frac{2}{3}e$	$\bar{c}$	$-\frac{2}{3}e$
Top	t	$+\frac{2}{3}e$	$\bar{t}$	$-\frac{2}{3}e$
Bottom	b	$-\frac{1}{3}e$	$\bar{b}$	$+\frac{1}{3}e$

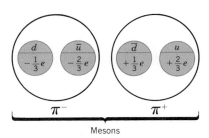

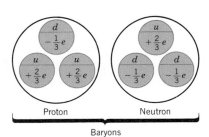

Figure 32.12   According to the original quark model of hadrons, all mesons consist of a quark and an antiquark, while baryons contain three quarks.

32.4 lists the symbols and electric charges of these quarks. Experimentally, quarks should be recognizable by their fractional charges, but in spite of an extensive search for them, free quarks have never been found.

According to the original quark theory, the mesons are different from the baryons, for each meson consists of only two quarks — a quark and an antiquark — while a baryon contains three quarks. For instance, the π^- pion (a meson) is composed of a d quark and a $\bar{u}$ antiquark, $\pi^- = d + \bar{u}$, as Figure 32.12 shows. These two quarks combine to give the π^- pion a net charge of $-e$. Similarly, the π^+ pion is a combination of the $\bar{d}$ and u quarks, $\pi^+ = \bar{d} + u$. In contrast, protons and neutrons, being baryons, consist of three quarks. A proton contains the combination $d + u + u$, while a neutron contains the combination $d + d + u$ (see Figure 32.12). These groups of three quarks give the correct charges for the proton and neutron.

The original quark model was extremely successful in predicting not only the correct charges for the hadrons, but other properties as well. Except for the fact that no one had succeeded in isolating a quark, the quark theory was a phenomenal success. However, in 1974 a new particle, the J/ψ meson, was discovered. This meson has a rest energy of 3100 MeV, far higher than other known mesons. The existence of the J/ψ meson could be explained only if a new quark–antiquark pair existed; this new quark was named *charm* (*c*). With the discovery of more and more particles, it has been necessary to postulate a fifth and a sixth quark; their names are *top* (*t*) and *bottom* (*b*), although some scientists prefer to call these quarks *truth* and *beauty*. Today, there is firm evidence for the five quarks called up, down, strange, charm, and bottom, and there is provisional evidence for the sixth or top quark. Each quark has a corresponding antiquark. All of the hundreds of the known hadrons can be accounted for in terms of these six quarks and their antiquarks. Whether the story is complete, however, remains to be seen.

SUMMARY

Ionizing radiation consists of photons and/or moving particles that have enough energy to ionize an atom or molecule. **Exposure** is a measure of the ionization produced in air by X-rays or γ rays. An exposure of one roentgen (R) produces 2.58×10^{-4} coulombs of positive charge in 1 kg of dry air at STP conditions.

The **absorbed dose** is the amount of energy absorbed from the radiation per unit mass of absorbing material. The SI unit of absorbed dose is the gray (Gy); 1 Gy = J/kg. However, the rad (rd) is another unit that is often used; 1 rd = 0.01 Gy.

The amount of biological damage produced by ionizing radiation is different for different types of radiation. The **relative biological effectiveness** (RBE) is the dose of 200-keV X-rays required to produce a certain biological effect divided by the dose of a given type of radiation to produce the same biological effect. The **biologically equivalent dose** is the product of the absorbed dose (in rads) and the RBE. The unit for the biologically equivalent dose is the rem.

An **induced nuclear transmutation** is the process whereby an incident particle or photon strikes a nucleus and causes the production of a new element.

Nuclear fission occurs when a massive nucleus splits into two less-massive fragments. Fission can be induced by the absorption of a thermal neutron. When a massive nucleus fissions, energy is released, because the binding energy per nucleon is greater for the fragments than for the original nucleus. Neutrons are also released during nuclear fission. These neutrons can, in turn, induce

other nuclei to fission and lead to a process known as a **chain reaction**. A **fission reactor** is a device that generates energy by a controlled chain reaction.

In a **fusion** process, two nuclei with smaller masses combine to form a single nucleus with a larger mass. Energy is released by fusion when the binding energy per nucleon is greater for the larger nucleus than for the smaller nuclei.

Subatomic particles are divided into three families: the **photon** family, the **lepton** family, and the **hadron** family. **Elementary particles** are the basic building blocks of matter. It is believed that all members of the photon and lepton families are elementary particles. The quark theory proposes that the hadrons are not elementary particles, but are composed of elementary particles called **quarks**. Currently, the hundreds of hadron particles can be accounted for in terms of six quarks and their antiquarks.

QUESTIONS

1. When a dentist X-rays your teeth, a lead apron is placed over your chest and lower body. What is the purpose of this apron?

2. State whether the two quantities in each of the following cases are related and, if so, give the relation between them: (a) rads and grays, and (b) rads and roentgens.

3. Explain why the following reactions are *not* allowed: (a) $^{60}_{28}\text{Ni}(\alpha, p)^{62}_{29}\text{Cu}$, (b) $^{27}_{13}\text{Al}(n, n)^{28}_{13}\text{Al}$, (c) $^{39}_{19}\text{K}(p, \alpha)^{36}_{17}\text{Cl}$.

4. Why is it possible for a thermal neutron (i.e., one with a small kinetic energy) to penetrate a nucleus, whereas a proton or an α particle would need a large amount of energy to penetrate the same nucleus?

5. Would a release of energy accompany the fission of a

nucleus of mass number 25 into two fragments of about equal mass? Using the curve in Figure 32.8, account for your answer.

6. In the fission of $^{235}_{92}\text{U}$ there are, on the average, 2.5 neutrons released per fission. Suppose a *different* element is being fissioned and, on the average, only 1.0 neutron is released per fission. If a small fraction of the thermal neutrons absorbed by the nuclei does *not* produce a fission, can a self-sustaining chain reaction be produced using this element? Explain.

7. The mass of coal consumed in a coal-burning electric power plant is about two million times greater than the mass of $^{235}_{92}\text{U}$ used to fuel a comparable nuclear power plant. Why?

8. Explain the difference between fission and fusion and why each process produces energy.

PROBLEMS

Section 32.1 Biological Effects of Ionizing Radiation

1. A beam of γ rays passes through 4.0×10^{-3} kg of dry air and generates 1.7×10^{12} ions, each with a charge of $+e$. What is the exposure (in roentgens)?

2. A film badge worn by a radiologist indicates that she has received an absorbed dose of 2.5×10^{-3} Gy. The mass of the radiologist is 65 kg. How much energy has she absorbed?

3. What absorbed dose (in rads) of α particles (RBE = 20)

causes as much biological damage as a 60-rad dose of protons (RBE = 10)?

4. A person who receives a 500-rem dose of proton radiation (RBE = 10) has a 50% chance of dying within a few months or so. What is the absorbed dose (in rads) of this radiation?

5. Someone stands near a radioactive source and receives doses of the following types of radiation: γ rays (20 mrad, RBE = 1), electrons (30 mrad, RBE = 1), protons (4 mrad,

RBE = 10), and slow neutrons (5 mrad, RBE = 2). What is the total biologically equivalent dose (in mrem) received?

6. A beam of α particles is directed at a 0.015-kg tumor. There are 1.6×10^{10} particles per second reaching the tumor, and the energy of each particle is 4.0 MeV. The RBE for the radiation is 14. Find the biologically equivalent dose given to the tumor in 25 s.

*7. A water sample receives a 750-rad dose of radiation. Find the rise in the water temperature.

*8. A 2.0-kg tumor is being irradiated by a radioactive source. The tumor receives an absorbed dose of 12 Gy in a time of 850 s. Each particle in the radiation delivers an energy of 0.40 MeV to the tumor. What is the activity $\Delta N/\Delta t$ (see Section 31.6) of the radioactive source?

Section 32.2 Induced Nuclear Reactions

9. What is the nucleon number A in the reaction $^{27}_{13}\text{Al}$ $(\alpha, n)^A_{15}\text{P}$?

10. Write the reactions below in the shorthand form discussed in the text.

(a) $^{27}_{13}\text{Al} + ^1_0\text{n} \rightarrow ^{27}_{12}\text{Mg} + ^1_1\text{H}$
(b) $^{40}_{18}\text{Ar} + ^4_2\text{He} \rightarrow ^{43}_{19}\text{K} + ^1_1\text{H}$

11. Write the equation for the reaction $^{17}_8\text{O}(\gamma, \alpha n)^{12}_6\text{C}$. The notation "$\alpha n$" means that an α particle and a neutron are produced by the reaction.

12. Complete the following nuclear reactions, assuming the unknown quantity signified by the question mark is a single entity:

(a) $^{43}_{20}\text{Ca}(\alpha, ?)^{46}_{21}\text{Sc}$ (d) $?(\alpha, p)^{17}_8\text{O}$
(b) $^9_4\text{Be}(?, n)^{12}_6\text{C}$ (e) $^{55}_{25}\text{Mn}(n, \gamma)?$
(c) $^9_4\text{Be}(p, \alpha)?$

*13. During a nuclear reaction, an unknown particle is absorbed by a copper $^{63}_{29}\text{Cu}$ nucleus, and the reaction products are $^{62}_{29}\text{Cu}$, a neutron, and a proton. What is the name, atomic number, and nucleon number of the *compound nucleus*?

Section 32.3 Nuclear Fission, Section 32.4 Nuclear Reactors

14. Determine the number of neutrons released during the following fission reaction: $^1_0\text{n} + ^{235}_{92}\text{U} \rightarrow ^{133}_{51}\text{Sb} + ^{99}_{41}\text{Nb} +$ neutrons.

15. $^{235}_{92}\text{U}$ absorbs a thermal neutron and fissions into rubidium $^{93}_{37}\text{Rb}$ and cesium $^{141}_{55}\text{Cs}$. What other *nucleons* are produced by the fission, and how many are there?

16. During an underground nuclear test, an atomic bomb is detonated. The bomb produces an amount of energy equivalent to 36 kilotons of TNT (1.0 kiloton of TNT releases about 5.0×10^{12} J of energy). To what amount of mass is this energy equivalent?

17. When a $^{235}_{92}\text{U}$ nucleus fissions, about 200 MeV of energy is released. What is the ratio of this energy to the rest energy of the uranium nucleus?

18. Uranium $^{235}_{92}\text{U}$ fissions into two fragments plus three neutrons: $^1_0\text{n} + ^{235}_{92}\text{U} \rightarrow$ (2 fragments) $+ 3^1_0\text{n}$. The mass of a neutron is 1.008 665 u and that of $^{235}_{92}\text{U}$ is 235.043 924 u. If 225.0 MeV of energy is released during the fission, what is the combined mass of the two fragments?

19. A particular fission reaction produces an energy of 210 MeV per fission. How many fissions occur per second if a reactor is generating 130 MW of power?

20. Neutrons released by a fission reaction must be slowed by collisions with the moderator nuclei before the neutrons can cause further fissions. Suppose a 1.5-MeV neutron leaves each collision with 65% of its incident energy. How many collisions are required to reduce the neutron's energy to at least 0.040 eV, which is the energy of a thermal neutron?

*21. When 1.0 kg of coal is burned, about 3.0×10^7 J of energy is released. If the energy released per $^{235}_{92}\text{U}$ fission is 2.0×10^2 MeV, how many kilograms of coal must be burned to produce the same energy as 1.0 kg of $^{235}_{92}\text{U}$?

*22. (a) If each fission of a $^{235}_{92}\text{U}$ nucleus releases about 2.0×10^2 MeV of energy, determine the energy (in joules) released by the complete fissioning of 1.0 gram of $^{235}_{92}\text{U}$. (b) How many grams of $^{235}_{92}\text{U}$ are consumed in one year, in order to supply the energy needs of a household that uses 30.0 kWh of energy per day, on the average?

*23. The water that cools a reactor core enters the reactor at 216 °C and leaves at 287 °C. (The water is pressurized, so it does not turn to steam.) The core is generating 5.6×10^9 W of power. Assume the specific heat of water is 4420 J/(kg · C°) over the temperature range stated above, and find the mass of water that passes through the core each second.

**24. A 20.0 kiloton atomic bomb releases as much energy as 20.0 kilotons of TNT (1.0 kiloton of TNT releases about 5.0×10^{12} J of energy). Recall that about 2.0×10^2 MeV of energy is released when each $^{235}_{92}\text{U}$ nucleus fissions. (a) How many $^{235}_{92}\text{U}$ nuclei are fissioned to produce the bomb's energy? (b) How many grams of uranium are fissioned? (c) What is the equivalent mass (in grams) of the bomb's energy?

**25. A nuclear power plant is 25% efficient, meaning that 25% of the power it generates goes into producing usable electricity. The remaining 75% is wasted as heat. The plant generates 8.0×10^8 watts of usable electric power. If each fission releases 2.0×10^2 MeV of energy, how many kilograms of $^{235}_{92}\text{U}$ are fissioned per year?

Section 32.5 Nuclear Fusion

26. Two deuterium (^2_1H) nuclei fuse and form ^3_2He and a neutron. The atomic masses are ^2_1H (2.0141 u), ^3_2He (3.0160 u), and ^1_0n (1.0087 u). Find the energy (in MeV) released.

27. In one type of fusion reaction a proton fuses with a neutron to form a deuterium nucleus: $^1_1\text{H} + ^1_0\text{n} \rightarrow ^2_1\text{H}$. The masses are ^1_1H (1.0078 u), ^1_0n (1.0087 u), and ^2_1H (2.0141 u). How much energy (in MeV) is released by this reaction?

***28.** Imagine your car is powered by a fusion engine in which the following reaction occurs: $3\,^2_1\text{H} \rightarrow ^4_2\text{He} + ^1_1\text{H} + ^1_0\text{n}$. The masses are ^2_1H (2.0141 u), ^4_2He (4.0026 u), ^1_1H (1.0078 u), and ^1_0n (1.0087 u). The engine uses 6.1×10^{-6} kg of deuterium ^2_1H fuel. If one gallon of gasoline produces 2.1×10^9 J of energy, how many gallons of gasoline would have to be burned to equal the energy released by all the deuterium fuel?

***29.** Deuterium (^2_1H) is an attractive fuel for fusion reactions because it is abundant in the waters of the oceans. In the oceans, deuterium makes up about 0.015% of the hydrogen in the water (H_2O). (a) How many deuterium atoms are there in one kilogram of water? (b) If each deuterium nucleus produces about 7.5 MeV in a fusion reaction, how many kilograms of ocean water would be needed to supply the energy needs of the United States for one year, estimated to be 8.4×10^{19} J?

Section 32.6 Elementary Particles

30. A high-energy proton collides with a stationary proton, and the reaction $p + p \rightarrow n + p + \pi^+$ occurs. The rest energy of the π^+ pion is 139.6 MeV. Ignore momentum conservation and find the minimum energy (in MeV) the incident proton must have.

31. A collision between two protons produces three new particles: $p + p \rightarrow p + \pi^+ + \Lambda^0 + K^0$. The rest energies of the new particles are π^+ (139.6 MeV), Λ^0 (1116 MeV), and K^0 (497.7 MeV). Note that one proton disappears during the reaction. How much of the protons' incident energy (in MeV) is transformed into matter during this reaction?

32. An electron and its antiparticle annihilate each other, producing two γ-ray photons. The kinetic energies of the particles are negligible. For each photon, determine its (a) energy (in MeV), and (b) wavelength.

***33.** Suppose a neutrino is created and has an energy of 35 MeV. (a) If the neutrino, like the photon, has no mass and travels at the speed of light, find the momentum of the neutrino. (b) Determine the de Broglie wavelength of the neutrino.

***34.** An energetic proton is fired at a stationary proton. For the reaction to produce new particles, the two protons must approach each other to within a distance of about 8.0×10^{-15} m. The moving proton must have a sufficient speed to overcome the repulsive Coulomb force. What must be the minimum initial kinetic energy (in MeV) of the proton?

ADDITIONAL PROBLEMS

35. During an X-ray examination, a person is exposed to radiation at a rate of 3.1×10^{-5} grays per second. The exposure time is 0.10 s, and the mass of the exposed tissue is 1.2 kg. Determine the energy absorbed.

36. A Σ^+ particle (see Table 32.3) decays into a π^0 particle and a proton: $\Sigma^+ \rightarrow \pi^0 + p$. Ignore the kinetic energy of the Σ^+ particle, and determine how much energy (in MeV) is released in the process.

37. A nitrogen $^{14}_7\text{N}$ nucleus absorbs a deuterium ^2_1H nucleus during a nuclear reaction. What is the name, atomic number, and nucleon number of the compound nucleus?

38. What energy (in MeV) is liberated by the following fission reaction?

$$^1_0\text{n} + ^{235}_{92}\text{U} \longrightarrow ^{141}_{56}\text{Ba} + ^{92}_{36}\text{Kr} + 3\,^1_0\text{n}$$
$$1.009\,\text{u} \quad 235.044\,\text{u} \qquad 140.914\,\text{u} \quad 91.926\,\text{u} \quad 3(1.009\,\text{u})$$

39. Within the core of a nuclear reactor there are 3.0×10^{19} nuclei fissioning each second. The energy released by each fission is about 2.0×10^2 MeV. Determine the power (in watts) being generated.

***40.** During an X-ray examination of the chest, a person receives an exposure of 0.015 R. How many singly charged ions would be produced if the X-rays passed through 2.0 m³ of dry air at STP conditions (density of air = 1.29 kg/m³)?

***41.** One proposed fusion reaction combines lithium ^6_3Li (6.015 u) with deuterium ^2_1H (2.014 u) to give helium ^4_2He (4.003 u): $^2_1\text{H} + ^6_3\text{Li} \rightarrow 2\,^4_2\text{He}$. How many kilograms of lithium would be needed to supply the energy needs of one household for a year, estimated to be 3.8×10^{10} J?

***42.** The energy consumed in one year in the United States is about 8.4×10^{19} J. When each $^{235}_{92}\text{U}$ nucleus fissions, about 2.0×10^2 MeV of energy is released. How many kilograms of $^{235}_{92}\text{U}$ would be needed to generate this energy if all the nuclei fissioned?

***43.** Suppose the $^{239}_{94}\text{Pu}$ nucleus fissions into two fragments whose mass ratio is 0.32 : 0.68. With the aid of Figure 32.8, estimate the energy (in MeV) released during this fission.

****44.** One kilogram of dry air at STP conditions is exposed to 1.0 R of X-rays. One roentgen is defined by Equation 32.1. An equivalent definition can be based on the fact that an exposure of one roentgen deposits 8.3×10^{-3} J of energy per kilogram of dry air. Using the two definitions, determine the average energy (in eV) needed to produce a single ion in air.

Physics & Medicine

Medicine has made more advances in the last one hundred years than it had during the previous 10 thousand years. Are the phenomenal achievements in healing occurring on their own, or are they accompanied, perhaps driven, by other scientific progress? In particular, just how much does medicine benefit from the insights physicists have made into the nature of fundamental physical processes over the past century?

Consider the physics that is involved in helping you in a medical emergency. Suppose, for example, that you were involved in a serious accident in which you received head and abdominal wounds. You're transported to the hospital in an ambulance powered by an internal combustion engine moving on nearly frictionless bearings. Your vital signs are called into the hospital over a static-free FM radio. Once in the bright, clean, emergency room your wounds are X-rayed and the films are developed in a matter of minutes. The doctor examines the negative through a screen that is back-lit by fluorescent lights. Meanwhile, you are connected to a cardiac monitor, and its output is displayed on a cathode-ray tube (CRT) monitor.

The skull X-rays are inconclusive, but the abdominal X-rays show clear signs of internal damage, and the doctor determines that immediate surgery is necessary. She operates on you with a laser, as well as knives of surgical stainless steel. Your blood pressure is monitored continuously and automatically. Your internal wounds are repaired, the incisions sutured, and you are rolled down the hall to the magnetic resonance imager (MRI) for a brain scan to see if your brain was injured. Happily, it was not.

With all the exposure television and personal experience have given you, you are probably not surprised by any of the details of this incident. But if you lived in the middle of the last century, say, you'd consider virtually every step nothing short of miraculous. In fact, the fundamental physics discovered over the past century plays crucial roles in every step of this scenario. Consider: The engine in the ambulance has been optimized by using the physics of combustion and thermodynamics; the vehicle's bearings are created from special, scientifically derived compounds for smooth running and long life; the FM (frequency modulation) radio is a result of decades of refinement of electromagnetic theory; the air in the emergency room is electrostatically cleaned to prevent infection; the X-ray machine and X-ray sensitive, rapid-development film are results of fundamental research in physics done since the end of the last century. The heart monitor measures changes in your body's internal resistance, and the CRT, a simplified television monitor, is based on the atomic theory of the atom, Ohm's law of electrical circuits, and electromagnetism. The automatic blood pressure monitor uses a piezoelectric crystal that converts changes in pressure that the blood pressure cuff senses as blood pulses through your veins into electrical impulses.

One of the crowning applications of physics in medicine is the series of internal imaging devices (Computerized Axial Tomography—CAT, Magnetic Resonance Imaging—MRI, and Positron Emission Tomography—PET). These apply the theories of electromagnetism, atomic and nuclear physics, and computer technology to allow physicians to see details of the inner workings of the human body.

If you become a medical professional, virtually every test you make, and every procedure you do, will be steeped in applications of fundamental physics. And new advances in physics are continually being applied to medicine. For example, during the past decade, surgeons have moved beyond metal scalpels to using lasers and ultrasonics in some applications. The laser was first built in 1960, based on a detailed study of atomic physics. Ultrasonics, the generation, detection, and application of ultra-high-frequency sound waves, evolved in the 1940s and was refined over the next several decades. By understanding the basic physics of compressional waves (sound), physicists have learned how to focus them for use in pinpoint surgery.

Another interesting application of physics to medicine is radioactivity. Exposure to unrestricted X-radiation is lethal to life. Among other things, it modifies DNA, allowing for unre-

stricted cell growth, which we call cancer. The impact of nuclear weapons use and tests, and the radiation leak from the Chernobyl nuclear reactor, clearly demonstrate this. On the other hand, restricted, focused radiation is successfully used in medicine to combat cancer.

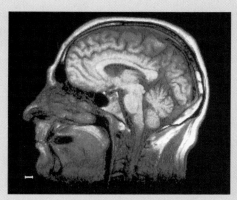

This image of a human brain was obtained by using magnetic resonance imaging (MRI). MRI scans use magnetic fields that enable nuclei within the body to generate radio waves. These waves can be detected and reveal, with remarkable clarity, images of the interior of the body. MRI scans have become an invaluable diagnostic tool.

A well-collimated (focused) beam of X-rays can be aimed to kill just cancer cells, leaving nearby healthy cells intact.

The profound advances in our understanding of physics during this century have enabled medical researchers to understand many of the human body's functions, right down to the molecular level. As a result, medicine now provides us with cures for many of the most crippling and lethal diseases that, until now, have claimed thousands, and even millions of lives, every year. Even as recently as the early 1950s, before the vaccine for polio was developed, most people were petrified to be out and about during the summer, when polio spread like

wildfire. And it is at the level of basic physics and chemistry that researchers are learning about, and searching for the cure and prevention of cancer, AIDS, and other viral diseases.

If the advances in physics made during this past century had not come along until now, most of the devices described in this essay would *still* not exist. This process of applying basic physics to medicine is not over yet. To this very day, new discoveries in science are finding important and immediate applications in medicine. And it's fair to say that without the present understanding of the basic principles of physics, medicine would still be a medieval art, and the quality of our lives would be much lower.

FOR INFORMATION ABOUT CAREERS IN MEDICINE

Professional Careers Sourcebook, *K. M. Savage and C. A. Dorgan, eds., 1990, Gale Research, Inc., publisher, provides addresses and phone numbers for dozens of guides and other resources about careers in many medically related fields.*

Encyclopedia of Careers and Vocational Guidance, *Vol. 2, 8th edition, William E. Hopke, ed., 1990, J. G. Ferguson, publisher, has an in-depth description of the working conditions physicians encounter.*

You can also write to the following, among many others listed in the books above:

■ *American Institute of Physics, 335 East 45th St., New York, NY 10017*

■ *American Medical Association, 535 North Dearborn St., Chicago, IL 60610*

■ *Association of American Medical Colleges, 1 Dupont Circle, NW, Washington, DC 20036*

■ *American College of Physicians, 4200 Pine St., Philadelphia, PA 19104*

POWERS OF TEN AND SCIENTIFIC NOTATION

In science, very large and very small decimal numbers are conveniently expressed in terms of powers of ten, some of which are listed below:

$$10^3 = 10 \times 10 \times 10 = 1000 \qquad 10^{-3} = \frac{1}{10 \times 10 \times 10} = 0.001$$

$$10^2 = 10 \times 10 = 100 \qquad 10^{-2} = \frac{1}{10 \times 10} = 0.01$$

$$10^1 = 10 \qquad 10^{-1} = \frac{1}{10} = 0.1$$

$$10^0 = 1$$

Using powers of ten, we can write the radius of the earth in the following way, for example:

$$\text{Earth radius} = 6\ 380\ 000 \text{ m} = 6.38 \times 10^6 \text{ m}$$

The factor of ten raised to the sixth power is ten multiplied by itself six times, or one million, so the earth's radius is 6.38 million meters. Alternatively, the factor of ten raised to the sixth power indicates that the decimal point in the term 6.38 is to be moved six places *to the right* to obtain the radius as a number without powers of ten.

For numbers less than one, negative powers of ten are used. For instance, the Bohr radius of the hydrogen atom is

$$\text{Bohr radius} = 0.000\ 000\ 000\ 0529 \text{ m} = 5.29 \times 10^{-11} \text{ m}$$

The factor of ten raised to the minus eleventh power indicates that the decimal point in the term 5.29 is to be moved eleven places *to the left* to obtain the radius as a number without powers of ten. Numbers expressed with the aid of powers of ten are said to be in *scientific notation.*

Calculations that involve the multiplication and division of powers of ten are carried out as in the following examples:

$$(2.0 \times 10^6)(3.5 \times 10^3) = (2.0 \times 3.5) \times 10^{6+3} = 7.0 \times 10^9$$

$$\frac{9.0 \times 10^7}{2.0 \times 10^4} = \left(\frac{9.0}{2.0}\right) \times 10^7 \times 10^{-4} = \left(\frac{9.0}{2.0}\right) \times 10^{7-4} = 4.5 \times 10^3$$

The general rules for such calculations are:

$$\frac{1}{10^n} = 10^{-n} \qquad\qquad\qquad (A\text{-}1)$$

$$10^n \times 10^m = 10^{n+m} \qquad \text{(Exponents added)} \qquad (A\text{-}2)$$

$$\frac{10^n}{10^m} = 10^{n-m} \quad \text{(Exponents subtracted)} \quad \text{(A-3)}$$

where n and m are any positive or negative number.

Scientific notation is convenient because of the ease with which it can be used in calculations. Moreover, scientific notation provides a convenient way to express the significant figures in a number, as Appendix B discusses.

SIGNIFICANT FIGURES

The number of *significant figures* in a number is the number of digits whose values are known with certainty. For instance, a person's height is measured to be 1.78 m, with the measurement error being in the third decimal place. All three digits are known with certainty, so that the number contains three significant figures. If a zero is given as the last digit to the right of the decimal point, the zero is presumed to be significant. Thus, the number 1.780 m contains four significant figures. As another example, consider a distance of 1500 m. This number contains only two significant figures, the one and the five. The zeros immediately to the left of the unexpressed decimal point are not counted as significant figures. However, zeros located between significant figures are significant, so a distance of 1502 m contains four significant figures.

Scientific notation is particularly convenient from the point of view of significant figures. Suppose it is known that a certain distance is fifteen hundred meters, to four significant figures. Writing the number as 1500 m presents a problem, because it implies that only two significant figures are known. In contrast, the scientific notation of 1.500×10^3 m has the advantage of indicating that the distance is known to four significant figures.

When two or more numbers are used in a calculation, the number of significant figures in the answer is limited by the number of significant figures in the original data. For instance, a rectangular garden with sides of 9.8 m and 17.1 m has an area of (9.8 m)(17.1 m). A calculator gives 167.58 m² for this product. However, one of the original lengths is known only to two significant figures, so the final answer is limited to only two significant figures and should be rounded off to 170 m². In general, *when numbers are multiplied or divided, the final answer has a number of significant figures that equals the smallest number of significant figures in any of the original factors.*

The number of significant figures in the answer to an addition or subtraction is also limited by the original data. Consider the total distance along a biker's trail that consists of three segments with the distances shown below:

$$
\begin{array}{rl}
 & 2.5 \text{ km} \\
 & 11 \quad\;\; \text{km} \\
 & \underline{5.26} \text{ km} \\
\text{Total} & 18.76 \text{ km}
\end{array}
$$

The distance of 11 km contains no significant figures to the right of the decimal point. Therefore, neither does the sum of the three distances, and the total distance should not be reported as 18.76 km. Instead, the answer is rounded off to 19 km. In general, *when numbers are added or subtracted, the last significant figure in the answer occurs in the last column (counting from left to right) containing a number that results from a combination of digits that are all significant.* In the answer of 18.76 km, the eight is the sum of $2 + 1 + 5$, each digit being significant. However, the seven is the sum of $5 + 0 + 2$, and the zero is not significant, since it comes from the 11-km distance, which contains no significant figures to the right of the decimal point.

ALGEBRA

C1 PROPORTIONS AND EQUATIONS

Physics deals with physical variables and the relations between them. Typically, variables are represented by the letters of the English and Greek alphabets. Sometimes, the relation between variables is expressed as a proportion or inverse proportion. Other times, however, it is more convenient or necessary to express the relation by means of an equation, which is governed by the rules of algebra.

If two variables are *directly proportional* and one of them doubles, then the other variable also doubles. Similarly, if one variable is reduced to one-half its original value, then the other is also reduced to one-half its original value. In general, if x is directly proportional to y, then increasing or decreasing one variable by a given factor causes the other variable to change in the same way by the same factor. This kind of relation is expressed as $x \propto y$, where the symbol $\propto$ means "is proportional to."

Since the proportional variables x and y always increase and decrease by the same factor, the ratio of x to y must have a constant value, or $x/y = k$, where k is a constant, independent of the values for x and y. Consequently, a proportionality such as $x \propto y$ can also be expressed in the form of an equation: $x = ky$. The constant k is referred to as a *proportionality constant.*

If two variables are *inversely proportional* and one of them increases by a given factor, then the other decreases by the same factor. An inverse proportion is written as $x \propto 1/y$. This kind of proportionality is equivalent to the following equation: $xy = k$, where k is a proportionality constant, independent of x and y.

C2 SOLVING EQUATIONS

Some of the variables in an equation typically have known values, and some do not. It is often necessary to solve the equation so that a variable whose value is unknown is expressed in terms of the known quantities. *In the process of solving an equation, it is permissible to manipulate the equation in any way, as long as a change made on one side of the equals sign is also made on the other side.* For example, consider the equation $v = v_0 + at$. Suppose values for v, v_0, and a are available, and the value of t is required. To solve the equation for t, we begin by subtracting v_0 from *both* sides:

$$v \quad = v_0 + at$$

$$\frac{-v_0 \qquad -v_0}{v - v_0 = \quad at}$$

Next, we divide *both* sides of $v - v_0 = at$ by the quantity a:

$$\frac{v - v_0}{a} = \frac{at}{a} = (1)t$$

On the right side, the a in the numerator divided by the a in the denominator equals one, so that

$$t = \frac{v - v_0}{a}$$

It is always possible to check the correctness of the algebraic manipulations performed in solving an equation by substituting the answer back into the original equation. In the previous example, we substitute into $v = v_0 + at$:

$$v = v_0 + a\left(\frac{v - v_0}{a}\right) = v_0 + (v - v_0) = v$$

The result $v = v$ implies that our algebraic manipulations were done correctly.

Algebraic manipulations other than addition, subtraction, multiplication, and division may play a role in solving an equation. The same basic rule applies, however: Whatever is done to the left side of an equation must also be done to the right side. As another example, suppose it is necessary to express v_0 in terms of v, a, and s, where $v^2 = v_0^2 + 2as$. By subtracting $2as$ from both sides, we isolate v_0^2 on the right:

$$
\begin{array}{rcr}
v^2 & = & v_0^2 + 2as \\
-2as & & -2as \\
\hline
v^2 - 2as & = & v_0^2
\end{array}
$$

To solve for v_0, we take the positive or negative square root of *both* sides of $v^2 - 2as = v_0^2$:

$$v_0 = \pm\sqrt{v^2 - 2as}$$

C3 SIMULTANEOUS EQUATIONS

When more than one variable in a single equation is unknown, additional equations are needed if solutions are to be found for all of the unknown quantities. Thus, the equation $3x + 2y = 7$ cannot be solved by itself to give unique values for both x and y. However, if x and y simultaneously obey the equation $x - 3y = 6$, then both unknowns can be found.

There are a number of methods by which such simultaneous equations can be solved. One method is to solve one equation for x in terms of y and substitute the result into the other equation to obtain an expression containing only the single unknown variable y. The equation $x - 3y = 6$, for instance, can be solved for x by adding $3y$ to each side, with the result that $x = 6 + 3y$. The substitution into the equation $3x + 2y = 7$ is shown below:

$$3x + 2y = 7$$
$$3(6 + 3y) + 2y = 7$$
$$18 + 9y + 2y = 7$$

We find, then, that $18 + 11y = 7$, a result that can be solved for y:

$$
\begin{array}{rl}
18 + 11y = & 7 \\
\underline{-18} & \underline{-18} \\
11y = & -11
\end{array}
$$

Dividing both sides of this result by 11 shows that $y = -1$. The value of $y = -1$ can be substituted in either of the original equations to obtain a value for x:

$$
\begin{array}{rcl}
x - 3y & = & 6 \\
x - 3(-1) & = & 6 \\
\\
x + 3 & = & 6 \\
\underline{-3} & & \underline{-3} \\
x & = & 3
\end{array}
$$

C4 THE QUADRATIC FORMULA

Equations occur in physics that include the square of a variable. Such equations are said to be *quadratic* in that variable and often can be put into the following form:

$$ ax^2 + bx + c = 0 \tag{C-1} $$

where a, b, and c are constants independent of x. This equation can be solved to give the **quadratic formula**, which is

$$ x = \frac{-b \pm \sqrt{b^2 - 4ac}}{2a} \tag{C-2} $$

The $\pm$ in the quadratic formula indicates that there are two solutions. For instance, if $2x^2 - 5x + 3 = 0$, then $a = 2$, $b = -5$, and $c = 3$. The quadratic formula gives the two solutions as follows:

$$
\left[\begin{array}{l} \text{Solution 1:} \\ \text{Plus sign} \end{array} \right] \qquad\qquad \left[\begin{array}{l} \text{Solution 2:} \\ \text{Minus sign} \end{array} \right]
$$

$$
\begin{aligned}
x &= \frac{-b + \sqrt{b^2 - 4ac}}{2a} \\
&= \frac{-(-5) + \sqrt{(-5)^2 - 4(2)(3)}}{2(2)} \\
&= \frac{+5 + \sqrt{1}}{4} = \frac{3}{2}
\end{aligned}
\qquad
\begin{aligned}
x &= \frac{-b - \sqrt{b^2 - 4ac}}{2a} \\
&= \frac{-(-5) - \sqrt{(-5)^2 - 4(2)(3)}}{2(2)} \\
&= \frac{+5 - \sqrt{1}}{4} = 1
\end{aligned}
$$

EXPONENTS AND LOGARITHMS

Appendix A discusses powers of ten such as 10^3, which means ten multiplied by itself three times, or $10 \times 10 \times 10$. The three is referred to as an *exponent.* The use of exponents extends beyond powers of ten. In general, the term y^n means the factor y is multiplied by itself n times. For example, y^2, or y squared, is familiar and means $y \times y$. Similarly, y^5 means $y \times y \times y \times y \times y$.

The rules that govern algebraic manipulations of exponents are the same as those given in Appendix A (see Equations A-1, A-2, and A-3) for powers of ten:

$$\frac{1}{y^n} = y^{-n} \tag{D-1}$$

$$y^n y^m = y^{n+m} \qquad \text{(Exponents added)} \tag{D-2}$$

$$\frac{y^n}{y^m} = y^{n-m} \qquad \text{(Exponents subtracted)} \tag{D-3}$$

To the three rules above we add two more that are useful. One of these is

$$y^n z^n = (yz)^n \tag{D-4}$$

The following example helps to clarify the reasoning behind this rule:

$$3^2 5^2 = (3 \times 3)(5 \times 5) = (3 \times 5)(3 \times 5) = (3 \times 5)^2$$

The other additional rule is

$$(y^n)^m = y^{nm} \qquad \text{(Exponents multiplied)} \tag{D-5}$$

To see why this rule applies, consider the following example:

$$(5^2)^3 = (5^2)(5^2)(5^2) = 5^{2+2+2} = 5^{2 \times 3}$$

Roots, such as a square root or a cube root, can be represented with fractional exponents. For instance,

$$\sqrt{y} = y^{1/2} \quad \text{and} \quad \sqrt[3]{y} = y^{1/3}$$

In general, the nth root of y is given by

$$\sqrt[n]{y} = y^{1/n} \tag{D-6}$$

The rationale for Equation D-6 can be explained using the fact that $(y^n)^m = y^{nm}$. For instance, the fifth root of y is the number that, when multiplied by itself five times, gives back y. As shown below, the term $y^{1/5}$ satisfies this definition:

$$(y^{1/5})(y^{1/5})(y^{1/5})(y^{1/5})(y^{1/5}) = (y^{1/5})^5 = y^{(1/5) \times 5} = y$$

Logarithms are closely related to exponents. To see the connection between the two, note that it is possible to express any number y as another number B raised to the exponent x. In other words,

$$y = B^x \qquad (D\text{-}7)$$

The exponent x is called the **_logarithm_** of the number y. The number B is called the **base number.** One of two choices for the base number is usually used. If $B = 10$, the logarithm is known as the *common logarithm,* for which the notation "log" applies:

[Common logarithm] $\qquad\qquad y = 10^x \quad \text{or} \quad x = \log y \qquad (D\text{-}8)$

If $B = e = 2.718 \ldots$, the logarithm is referred to as the *natural logarithm,* and the notation "ln" is used:

[Natural logarithm] $\qquad\qquad y = e^z \quad \text{or} \quad z = \ln y \qquad (D\text{-}9)$

The two kinds of logarithms are related by

$$\ln y = 2.3026 \log y \qquad (D\text{-}10)$$

Both kinds of logarithms are often given on calculators.

The logarithm of the product or quotient of two numbers A and C can be obtained from the logarithms of the individual numbers according to the rules below. These rules are illustrated here for natural logarithms, but they are the same for any kind of logarithm.

$$\ln AC = \ln A + \ln C \qquad (D\text{-}11)$$

$$\ln \left(\frac{A}{C}\right) = \ln A - \ln C \qquad (D\text{-}12)$$

Thus, the logarithm of the product of two numbers is the sum of the individual logarithms, and the logarithm of the quotient of two numbers is the difference between the individual logarithms. Another useful rule concerns the logarithm of a number A raised to an exponent n:

$$\ln A^n = n \ln A \qquad (D\text{-}13)$$

Rules D-11, D-12, and D-13 can be derived from the definition of the logarithm and the rules governing exponents.

GEOMETRY AND TRIGONOMETRY

E1 GEOMETRY

ANGLES

Two angles are equal if

1. They are vertical angles (see Figure E1).
2. Their sides are parallel (see Figure E2).
3. Their sides are mutually perpendicular (see Figure E3).

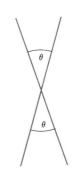

Figure E1

TRIANGLES

1. The *sum of the angles* of any triangle is 180° (see Figure E4).
2. A *right triangle* has one angle that is 90°.
3. An *isosceles triangle* has two sides that are equal.
4. An *equilateral triangle* has three sides that are equal. Each angle of an equilateral triangle is 60°.
5. Two triangles are *similar* if two of their angles are equal (see Figure E5). The corresponding sides of similar triangles are proportional to each other:

$$\frac{a_1}{a_2} = \frac{b_1}{b_2} = \frac{c_1}{c_2}$$

6. Two similar triangles are *congruent* if they can be placed on top of one another to make an exact fit.

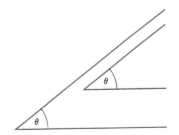

Figure E2

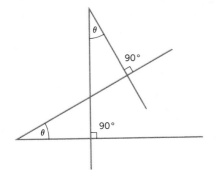

Figure E3

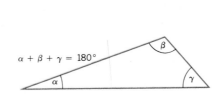

$\alpha + \beta + \gamma = 180°$

Figure E4

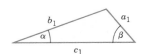

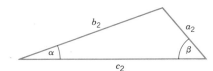

Figure E5

CIRCUMFERENCES, AREAS, AND VOLUMES OF SOME COMMON SHAPES

1. Triangle of base b and altitude h (see Figure E6): Area $= \frac{1}{2}bh$.
2. Circle of radius r: Circumference $= 2\pi r$, Area $= \pi r^2$.
3. Sphere of radius r: Surface area $= 4\pi r^2$, Volume $= \frac{4}{3}\pi r^3$.
4. Right circular cylinder of radius r and height h (see Figure E7):

$$\text{Surface area} = 2\pi r^2 + 2\pi rh, \quad \text{Volume} = \pi r^2 h.$$

E2 TRIGONOMETRY

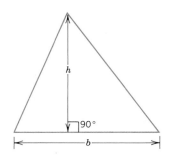

Figure E6

Figure E7

BASIC TRIGONOMETRIC FUNCTIONS

1. For a right triangle, the sine, cosine, and tangent of an angle θ are as follows (see Figure E8):

$$\sin\theta = \frac{\text{Side opposite }\theta}{\text{Hypotenuse}} = \frac{h_o}{h}$$

$$\cos\theta = \frac{\text{Side adjacent to }\theta}{\text{Hypotenuse}} = \frac{h_a}{h}$$

$$\tan\theta = \frac{\text{Side opposite }\theta}{\text{Side adjacent to }\theta} = \frac{h_o}{h_a}$$

2. The secant, cosecant, and cotangent of an angle θ are defined as follows:

$$\sec\theta = \frac{1}{\cos\theta} \qquad \csc\theta = \frac{1}{\sin\theta} \qquad \cot\theta = \frac{1}{\tan\theta}$$

TRIANGLES AND TRIGONOMETRY

1. The *Pythagorean theorem* states that the square of the hypotenuse of a right triangle is equal to the sum of the squares of the other two sides (see Figure E8):

$$h^2 = h_o{}^2 + h_a{}^2$$

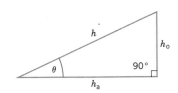

Figure E8

2. The *law of cosines* and the *law of sines* apply to any triangle, not just a right triangle, and they relate the angles and the lengths of the sides (see Figure E9):

[Law of cosines] $\qquad c^2 = a^2 + b^2 - 2ab\cos\gamma$

[Law of sines] $\qquad \dfrac{a}{\sin\alpha} = \dfrac{b}{\sin\beta} = \dfrac{c}{\sin\gamma}$

OTHER TRIGONOMETRIC IDENTITIES

1. $\sin\theta/\cos\theta = \tan\theta$.
2. $\sin^2\theta + \cos^2\theta = 1$.

3. $\sin(\alpha \pm \beta) = \sin\alpha\cos\beta \pm \cos\alpha\sin\beta$.

If $\alpha = 90°$, $\sin(90° \pm \beta) = \cos\beta$.

If $\alpha = \beta$, $\sin 2\beta = 2\sin\beta\cos\beta$.

4. $\cos(\alpha \pm \beta) = \cos\alpha\cos\beta \mp \sin\alpha\sin\beta$.

If $\alpha = 90°$, $\cos(90° \pm \beta) = \mp\sin\beta$.

If $\alpha = \beta$, $\cos 2\beta = \cos^2\beta - \sin^2\beta = 1 - 2\sin^2\beta$.

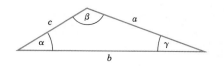

Figure E9

SELECTED ISOTOPES[a]

Atomic Number Z	Element	Symbol	Atomic Mass Number A	Atomic Mass u	Percent Abundance, or Decay Mode if Radioactive	Half-life (If Radioactive)
0	(Neutron)	n	1	1.008 665	β^-	10.37 min
1	Hydrogen	H	1	1.007 825	99.985	
	Deuterium	D	2	2.014 102	0.015	
	Tritium	T	3	3.016 050	β^-	12.33 yr
2	Helium	He	3	3.016 030	0.000 138	
			4	4.002 603	≈ 100	
3	Lithium	Li	6	6.015 121	7.5	
			7	7.016 003	92.5	
4	Beryllium	Be	7	7.016 928	EC, γ	53.29 days
			9	9.012 182	100	
5	Boron	B	10	10.012 937	19.9	
			11	11.009 305	80.1	
6	Carbon	C	11	11.011 432	β^+, EC	20.39 min
			12	12.000 000	98.90	
			13	13.003 355	1.10	
			14	14.003 241	β^-	5730 yr
7	Nitrogen	N	13	13.005 738	β^+, EC	9.965 min
			14	14.003 074	99.634	
			15	15.000 108	0.366	
8	Oxygen	O	15	15.003 065	β^+, EC	122.2 s
			16	15.994 915	99.762	
			18	17.999 160	0.200	
9	Fluorine	F	18	18.000 937	EC, β^+	1.8295 h
			19	18.998 403	100	
10	Neon	Ne	20	19.992 435	90.51	
			22	21.991 383	9.22	
11	Sodium	Na	22	21.994 434	β^+, EC, γ	2.602 yr
			23	22.989 767	100	
			24	23.990 961	β^-, γ	14.659 h
12	Magnesium	Mg	24	23.985 042	78.99	
13	Aluminum	Al	27	26.981 539	100	
14	Silicon	Si	28	27.976 927	92.23	
			31	30.975 362	β^-, γ	2.622 h
15	Phosphorus	P	31	30.973 762	100	
			32	31.973 907	β^-	14.282 days

[a] Data for atomic masses are taken from *Handbook of Chemistry and Physics*, 66th ed., CRC Press, Boca Raton, FL. The masses are those for the neutral atom, including the Z electrons. Data for percent abundance, decay mode, and half-life are taken from E. Browne and R. Firestone, *Table of Radioactive Isotopes*, V. Shirley, Ed., Wiley, New York, 1986. α = alpha particle emission, β^- = negative beta emission, β^+ = positron emission, γ = γ-ray emission, EC = electron capture.

APPENDIX F: Selected Isotopes — *Continued*

Atomic Number Z	Element	Symbol	Atomic Mass Number A	Atomic Mass u	Percent Abundance, or Decay Mode if Radioactive	Half-life (If Radioactive)
16	Sulfur	S	32	31.972 070	95.02	
			35	34.969 031	β^-	87.51 days
17	Chlorine	Cl	35	34.968 852	75.77	
			37	36.965 903	24.23	
18	Argon	Ar	40	39.962 384	99.600	
19	Potassium	K	39	38.963 707	93.2581	
			40	39.963 999	β^-, EC, γ	1.277×10^9 yr
20	Calcium	Ca	40	39.962 591	96.941	
21	Scandium	Sc	45	44.955 910	100	
22	Titanium	Ti	48	47.947 947	73.8	
23	Vanadium	V	51	50.943 962	99.750	
24	Chromium	Cr	52	51.940 509	83.789	
25	Manganese	Mn	55	54.938 047	100	
26	Iron	Fe	56	55.934 939	91.72	
27	Cobalt	Co	59	58.933 198	100	
			60	59.933 819	β^-, γ	5.271 yr
28	Nickel	Ni	58	57.935 346	68.27	
			60	59.930 788	26.10	
29	Copper	Cu	63	62.939 598	69.17	
			65	64.927 793	30.83	
30	Zinc	Zn	64	63.929 145	48.6	
			66	65.926 034	27.9	
31	Gallium	Ga	69	68.925 580	60.1	
32	Germanium	Ge	72	71.922 079	27.4	
			74	73.921 177	36.5	
33	Arsenic	As	75	74.921 594	100	
34	Selenium	Se	80	79.916 520	49.7	
35	Bromine	Br	79	78.918 336	50.69	
36	Krypton	Kr	84	83.911 507	57.0	
			89	88.917 640	β^-, γ	3.16 min
			92	91.926 270	β^-, γ	1.840 s
37	Rubidium	Rb	85	84.911 794	72.165	
38	Strontium	Sr	86	85.909 267	9.86	
			88	87.905 619	82.58	
			90	89.907 738	β^-	28.5 yr
			94	93.915 367	β^-, γ	1.235 s
39	Yttrium	Y	89	88.905 849	100	
40	Zirconium	Zr	90	89.904 703	51.45	
41	Niobium	Nb	93	92.906 377	100	
42	Molybdenum	Mo	98	97.905 406	24.13	
43	Technecium	Tc	98	97.907 215	β^-, γ	4.2×10^6 yr
44	Ruthenium	Ru	102	101.904 348	31.6	
45	Rhodium	Rh	103	102.905 500	100	

APPENDIX F: Selected Isotopes—*Continued*

Atomic Number Z	Element	Symbol	Atomic Mass Number A	Atomic Mass u	Percent Abundance, or Decay Mode if Radioactive	Half-life (If Radioactive)
46	Palladium	Pd	106	105.903 478	27.33	
47	Silver	Ag	107	106.905 092	51.839	
			109	108.904 757	48.161	
48	Cadmium	Cd	114	113.903 357	28.73	
49	Indium	In	115	114.903 880	95.7; β^-	4.41×10^{14} yr
50	Tin	Sn	120	119.902 200	32.59	
51	Antimony	Sb	121	120.903 821	57.3	
52	Tellurium	Te	130	129.906 229	33.8; β^-	2.5×10^{21} yr
53	Iodine	I	127	126.904 473	100	
			131	130.906 114	β^-, γ	8.040 days
54	Xenon	Xe	132	131.904 144	26.9	
			136	135.907 214	8.9	
			140	139.921 620	β^-, γ	13.6 s
55	Cesium	Cs	133	132.905 429	100	
			134	133.906 696	β^-, EC, γ	2.062 yr
56	Barium	Ba	137	136.905 812	11.23	
			138	137.905 232	71.70	
			141	140.914 363	β^-, γ	18.27 min
57	Lanthanum	La	139	138.906 346	99.91	
58	Cerium	Ce	140	139.905 433	88.48	
59	Praseody-mium	Pr	141	140.907 647	100	
60	Neodymium	Nd	142	141.907 719	27.13	
61	Promethium	Pm	145	144.912 743	EC, α, γ	17.7 yr
62	Samarium	Sm	152	151.919 729	26.7	
63	Europium	Eu	153	152.921 225	52.2	
64	Gadolinium	Gd	158	157.924 099	24.84	
65	Terbium	Tb	159	158.925 342	100	
66	Dysprosium	Dy	164	163.929 171	28.2	
67	Holmium	Ho	165	164.930 319	100	
68	Erbium	Er	166	165.930 290	33.6	
69	Thulium	Tm	169	168.934 212	100	
70	Ytterbium	Yb	174	173.938 859	31.8	
71	Lutetium	Lu	175	174.940 770	97.41	
72	Hafnium	Hf	180	179.946 545	35.100	
73	Tantalum	Ta	181	180.947 992	99.988	
74	Tungsten (wolfram)	W	184	183.950 928	30.67	
75	Rhenium	Re	187	186.955 744	62.60; β^-	4.6×10^{10} yr
76	Osmium	Os	191	190.960 920	β^-, γ	15.4 days
			192	191.961 467	41.0	
77	Iridium	Ir	191	190.960 584	37.3	
			193	192.962 917	62.7	

APPENDIX F: Selected Isotopes—*Continued*

Atomic Number Z	Element	Symbol	Atomic Mass Number A	Atomic Mass u	Percent Abundance, or Decay Mode if Radioactive	Half-life (If Radioactive)
78	Platinum	Pt	195	194.964 766	33.8	
79	Gold	Au	197	196.966 543	100	
			198	197.968 217	β^-, γ	2.6935 days
80	Mercury	Hg	202	201.970 617	29.80	
81	Thallium	Tl	205	204.974 401	70.476	
			208	207.981 988	β^-, γ	3.053 min
82	Lead	Pb	206	205.974 440	24.1	
			207	206.975 872	22.1	
			208	207.976 627	52.4	
			210	209.984 163	α, β^-, γ	22.3 yr
			211	210.988 735	β^-, γ	36.1 min
			212	211.991 871	β^-, γ	10.64 h
			214	213.999 798	β^-, γ	26.8 min
83	Bismuth	Bi	209	208.980 374	100	
			211	210.987 255	α, β^-, γ	2.14 min
			212	211.991 255	β^-, α, γ	1.0092 h
84	Polonium	Po	210	209.982 848	α, γ	138.376 days
			212	211.988 842	α, γ	45.1 s
			214	213.995 176	α, γ	163.69 μs
			216	216.001 889	α, γ	150 ms
85	Astatine	At	218	218.008 684	α, β^-	1.6 s
86	Radon	Rn	220	220.011 368	α, γ	55.6 s
			222	222.017 570	α, γ	3.825 days
87	Francium	Fr	223	223.019 733	α, β^-, γ	21.8 min
88	Radium	Ra	224	224.020 186	α, γ	3.66 days
			226	226.025 402	α, γ	1.6×10^3 yr
			228	228.031 064	β^-, γ	5.75 yr
89	Actinium	Ac	227	227.027 750	α, β^-, γ	21.77 yr
			228	228.031 015	β^-, γ	6.13 h
90	Thorium	Th	228	228.028 715	α, γ	1.913 yr
			231	231.036 298	β^-, γ	1.0633 days
			232	232.038 054	100; α, γ	1.405×10^{10} yr
			234	234.043 593	β^-, γ	24.10 days
91	Protactinium	Pa	231	231.035 880	α, γ	3.276×10^4 yr
			234	234.043 303	β^-, γ	6.70 h
			237	237.051 140	β^-, γ	8.7 min
92	Uranium	U	232	232.037 130	α, γ	68.9 yr
			233	233.039 628	α, γ	1.592×10^5 yr
			235	235.043 924	0.7200; α, γ	7.037×10^8 yr
			236	236.045 562	α, γ	2.342×10^7 yr
			238	238.050 784	99.2745; α, γ	4.468×10^9 yr
			239	239.054 289	β^-, γ	23.47 min
93	Neptunium	Np	239	239.052 933	β^-, γ	2.355 days
94	Plutonium	Pu	239	239.052 157	α, γ	2.411×10^4 yr
			242	242.058 737	α, γ	3.763×10^5 yr

APPENDIX F: Selected Isotopes—*Continued*

Atomic Number Z	Element	Symbol	Atomic Mass Number A	Atomic Mass u	Percent Abundance, or Decay Mode if Radioactive	Half-life (If Radioactive)
95	Americium	Am	243	243.061 375	α, γ	7.380×10^3 yr
96	Curium	Cm	245	245.065 483	α, γ	8.5×10^3 yr
97	Berkelium	Bk	247	247.070 300	α, γ	1.38×10^3 yr
98	Californium	Cf	249	249.074 844	α, γ	350.6 yr
99	Einsteinium	Es	254	254.088 019	α, γ, β^-	275.7 days
100	Fermium	Fm	253	253.085 173	EC, α, γ	3.00 days
101	Mendelevium	Md	255	255.091 081	EC, α	27 min
102	Nobelium	No	255	255.093 260	EC, α	3.1 min
103	Lawrencium	Lr	257	257.099 480	α, EC	646 ms
104	Rutherfordium	Rf	261	261.108 690	α	1.08 min
105	Hahnium	Ha	262	262.113 760	α	34 s

ANSWERS TO ODD-NUMBERED PROBLEMS

CHAPTER 1

1. (a) 5700 s   (b) 86 400 s   3. 27.4 km   5. 4048 m²
7. 1.1×10^{-2} m/s²
9. 80.1 km, at 25.9° south of west   11. 0.707 m
13. 340 m   15. 35.3°
17. 7.80 km, at 35.0° north of west
19. 5.70×10^2 N, at 33.6° south of west
21. (a) 5.31 km, due south   (b) 5.31 km, due north
23. 11.7 km, at 13.1° south of east
25. (a) 6.00 units   (b) 36.9° north of west
(c) 6.00 units   (d) 36.9° south of west
27. (a) −128 m   (b) 68.1 m
29. 147 km (east), and 47.9 km (north)   31. 150 m/s
33. (a) $A_x = 650$ units   (b) $A_{x'} = 570$ units
$A_y = 380$ units $A_{y'} = -480$ units
35. 4.80 km, at 24.0° north of east
37. 268 km, at 38.5° north of east
39. 46 paces north, 88 paces west
41. (a) 192 units   (b) 49.7 units
43. 6.88 km, −26.9°   45. 13.5 m   47. 115 m³
49. 3.00 m, at 42.8° above the negative x axis
51. (a) 6.43 m, at 70.0° above the negative x axis
(b) 7.66 m, at 20.0° below the negative x axis
53. (a) 130 N, at 53° south of east
(b) 130 N, at 53° north of west

CHAPTER 2

1. 4.69 s   3. 7.2 m/s   5. 60 s   7. 6.25 m/s, down
9. 3.16 m/s², south   11. 3.0 m/s²
13. 3.44 m/s, due west   15. 8.0 m/s, motorcycle A
17. 13 m/s   19. 3.1 m/s², southward
21. 5.0×10^1 m   23. (a) 2.0×10^1 s   (b) 580 m
25. 17.7 m   27. 15 s   29. 14 s   31. 91.5 m/s
33. 15.4 m/s   35. 1.3 m   37. 153 m/s, upward
39. 39 m/s, upward   41. 5.0 m   43. 35.6 s
45. 10.6 m   47. 8.18 m/s
49. $a_A = 2$ m/s², $a_B = 0$, $a_C = 3$ m/s²
51. The answer is in graphical form.
53. (a) 1.5 m/s²   (b) 1.5 m/s²,   (c) Yes, 76 m
55. 1.1 s   57. 2.5 m/s
59. (a) 26 700 m   (b) 6.74 m/s   61. 35 m/s
63. −4.6 m/s²   65. 19 m   67. 1.4 m/s²

CHAPTER 3

1. 41 m   3. 8600 m   5. 16.9 m/s   7. 4.7 m/s
9. (a) 3.3 km, at 87° above the horizontal
(b) 3.0 m/s, at 87° above the horizontal
11. 4.42 s   13. (a) 93.6 m   (b) 4.37 s
15. (a) 0.57 s   (b) 630 m   17. 10.3 m   19. 39 m/s
21. 14°   23. 2.40 m   25. 48 m   27. 58 m
29. 74.9 m   31. 42°
33. 4.52 m/s, 59.4° above the horizontal
35. 55.0 m/s, 35.1° above the horizontal
37. 5.8 m/s   39. 0.141° and 89.859°
41. The answer is a proof.   43. 24 m/s, due south
45. (a) 41 m/s, due east   (b) 41 m/s, due west
47. 24.2 m/s, 21.1° south of east
49. 7.50 m/s, 23.1° north of east
51. (a) 19.5°   (b) 955 s   53. 515 m/s
55. 26 m/s, 36° west of north   57. 1.7 s
59. 54.7°   61. The answer is a proof.
63. 2.44×10^6 m

CHAPTER 4

1. 36 500 N   3. 6500 N, due north
5. 2900 N   7. 2
9. $a_{man} = 0.55$ m/s² due east, $a_{woman} = 0.94$ m/s² due west
11. $a_{left} = 14.4$ m/s² at 56.3° above the x axis, $a_{center} = 18.5$ m/s² at 27.2° above the x axis, $a_{right} = 20.0$ m/s² along the x axis
13. 51.8 N, directed 88° above the $+x$ axis   15. 7×10^{28} N
17. 1.70×10^{-5} N   19. (a) 1130 N   (b) zero
21. 22.2 N   23. $g_{mountain}/g_{sea\ level} = 0.997$
25. 645 N   27. 730 N   29. 3.46×10^8 m
31. 0.19 m/s², up   33. (a) 447 N   (b) 241 N
35. 267 N   37. (a) 550 N   (b) 7.2 m/s
39. 0.235   41. 58.8 N
43. (a) 57 600 N   (b) 20 600 N
45. 9.70 N   47. 4300 N   49. 620 N
51. 7260 N   53. (a) 79.0 N   (b) 219 N
55. 220 N, pointing 64° north of east
57. 1730 N, due west
59. (a) 914 N   (b) 822 N   61. 29 400 N
63. 1.30 m/s²   65. (a) 5.97 m/s²   (b) 101 N
67. 8.17 s   69. 0.788 m/s²   71. 3.9 m/s²

73. (a) 3.68 m/s^2 (b) 11.8 m/s^2
(c) There is more mass (inertia) in part (a).
75. (a) 1.00×10^2 N   (b) 41.7 N
77. (a) 1710 N   (b) zero   **79.** 929 N
81. 1.00×10^2 N, directed $53.1°$ S of E   **83.** 1.5 m/s^2
85. (a) 0.0640 m/s^2 (b) 1.58×10^7 m
87. (a) 3.56 m/s^2 (b) 281 N   **89.** 4.7 kg
91. $T_{\text{right}} = 317$ N, $T_{\text{left}} = 249$ N   **93.** 0.265 m
95. (a) $a = g \tan \theta$ (b) 1.73 m/s^2,   (c) $0°$
97. 0.665

CHAPTER 5

1. 0.016 s
3. 0.479 m/s^2, toward the center of the takeup reel
5. 4   **7.** 0.79 m/s^2 **9.** $10\,600$ rev/min
11. 0.68 m/s
13. (a) 0.189 N
(b) No, the centripetal force increases by a factor of four.
15. 0.75 **17.** 0.186 **19.** $23°$
21. (a) $\tan \theta = v^2/(rg)$ (b) $25.2°$ **23.** 4.20×10^4 m/s
25. 1.54×10^9 m   **27.** 3070 m/s
29. 0.611 yr or 223 days   **31.** 4720 m   **33.** 14.0 m/s
35. 2.9×10^4 N   **37.** 426 N
39. (a) 3.00×10^4 m/s   (b) 2.02×10^{30} kg
41. 1.43×10^4 s   **43.** 66 m   **45.** 105 m
47. (a) The centripetal force is provided by the normal force exerted by the wall on the rider.
(b) 1670 N   (c) 0.323

CHAPTER 6

1. (a) 2980 J   (b) 3290 J   **3.** 270 J
5. (a) $24\,300$ J   (b) $-10\,400$ J   **7.** -5800 J
9. 203 N   **11.** 39 m/s   **13.** 3180 J
15. 4030 N, opposite the car's motion
17. (a) 38 J   (b) 3800 N
19. 3.0×10^1 m/s   **21.** 1.0×10^3 N
23. (a) -44 J   (b) 44 J   **25.** 5.24×10^5 J   **27.** 6.6 m/s

29.

Height (m)	KE(J)	PE(J)	E(J)
20.0	0	392	392
15.0	98	294	392
10.0	196	196	392
5.00	294	98	392
0	392	0	392

31. 15.0 m/s   **33.** 1.4 m   **35.** 4.43 m/s   **37.** 40.8 kg
39. -4.51×10^4 J
41. (a) -1086 J   (b) -2.01 m, skateboarder is below starting point.

43. 4130 N   **45.** 4.17 m/s   **47.** 3.6×10^6 J
49. (a) 0.14 hp   (b) 0.71 hp   **51.** (a) 58 hp   (b) 77 hp
53. (a) 4.00×10^3 W   (b) 3.35×10^4 W
55. (a) 22.2 m/s   (b) 22.2 m/s   (c) 22.2 m/s
57. 6.4×10^5 J   **59.** 55.8 m/s   **61.** 3.0 m/s
63. 3.1 m/s   **65.** 0.327 m

CHAPTER 7

1. 14 N · s   **3.** (a) 1.79×10^{29} kg · m/s
(b) No. The earth's linear momentum changes direction as the earth travels in a nearly circular orbit about the sun. The gravitational force of the sun causes the earth's linear momentum to change.
5. 4.24 kg · m/s, at $45.0°$ south of east
7. (a) 1.3 kg · m/s, parallel to the velocity of the ball.
(b) 220 N, parallel to the velocity of the ball.
9. 0.53 kg · m/s, down   **11.** $v_2 = 9.28$ m/s, $v_3 = 19.4$ m/s
13. -1.5×10^{-4} m/s   **15.** 5.7 kg   **17.** 4500 m/s
19. 0.66 m   **21.** $+3.2$ m/s   **23.** (a) $+5.25$ m/s
(b) 1.41 m
25. $+9.3$ m/s
27. (a) $v_{f1} = -0.400$ m/s, $v_{f2} = +1.60$ m/s
(b) $v_f = +0.800$ m/s
29. $m_1 = 1.00$ kg, $m_2 = 1.00$ kg
31. (a) $\theta = 73.0°$ (b) $v_f = 4.28$ m/s
33. 8 bounces   **35.** 2.2 kg/s
37. -0.14 m/s (opposite to the velocity of the bullet)
39. (a) $+6.71$ m/s   (b) 0.559 **41.** $+69$ N
43. $+7.1 \times 10^5$ m/s
45. -0.330 m/s (opposite to the horizontal component of the velocity of the stone)
47. 77 N
49. (a) 5.56 m/s
(b) -2.83 m/s (1.50-kg ball), $+2.73$ m/s (4.60-kg ball)
(c) 0.409 m (1.50-kg ball), 0.380 m (4.60-kg ball)

CHAPTER 8

1. (a) 0.79 rad   (b) 3.1 rad   (c) 6.3 rad
3. $\theta_{\text{moon}} = 9.04 \times 10^{-3}$ rad, $\theta_{\text{sun}} = 9.27 \times 10^{-3}$ rad
5. (a) 9.4×10^{-4} s   (b) 0.13 m   **7.** 1.15 rad/s^2
9. 157.3 rad/s   **11.** 825 m   **13.** 2.00×10^{-2} s
15. (a) 40.0 rad   (b) 15.0 rad/s   **17.** (a) 4.60×10^3 rad
(b) 2.00×10^2 rad/s^2
19. (a) 1.2×10^4 rad   (b) 110 s
21. -0.504 rad/s^2 **23.** 3.20×10^4 rad   **25.** 466 m/s
27. 4.7 rad/s   **29.** 0.817 m   **31.** $18.6°$
33. 380 m/s^2 **35.** (a) 1.60 m/s   (b) 50.1 m/s^2
37. 1.73 m   **39.** 0.213 s   **41.** 1450 rad

43. (a) 7.50 rad/s
(b) -1.73×10^{-3} rad/s², indicating that the angular velocity is decreasing.
45. 0.300 m/s   **47.** 0.267 rad/s²   **49.** (a) 15.0 rad/s
(b) 2.55 m/s
51. (a) 0.583 m/s²   (b) 31.0°   **53.** 492 rad/s
55. 28.0 rad/s   **57.** $L_1/L_2 = 1/\sqrt{3}$ **59.** 0.611 rad

CHAPTER 9

1. 1.70×10^3 N · m   **3.** 1.5 N · m   **5.** 0.667 m
7. 0.100 m (distance from right edge), 0.300 m (distance from bottom edge)   **9.** 0.90 m   **11.** 1.20×10^3 N
13. T = 56.4 N, down, F = 70.6 N, up   **15.** 1.60×10^5 N (front wheel), 4.20×10^5 N (each rear wheel)
17. $F_N = 212$ N, the horizontal and vertical components of the force exerted on her shoes are 212 and 5.00×10^2 N, respectively.
19. (a) 2310 N   (b) The horizontal and vertical components of the force that the wall exerts on the beam are 1630 and 1550 N, respectively.
21. 51.4 N   **23.** 1.1 kg · m²/s²
25. (a) -7.0 N · m   (b) -12 rad/s²
27. -3.32×10^{-2} N · m   **29.** 1.25 kg · m²
31. 460 N   **33.** 0.34 N   **35.** 1.33 m, 1.67 m
37. 2.6×10^{29} J   **39.** (a) 3.80×10^5 J   (b) 6.22×10^3 J
(c) 3.86×10^5 J   **41.** 2/5   **43.** 3/4
45. (a) 0.877 m/s   (b) 175 rad/s   **47.** 7.11 rad/s
49. 0.34 rad/s   **51.** 0.573 m   **53.** 0.25 m
55. 1.83 rad/s   **57.** 3   **59.** (a) 2.52×10^4 N
(b) 2.35×10^4 N
61. (a) 0.500 rad/s, the disk rotates in a direction that is opposite to the motion of the person.
(b) 2.99 s
63. 69 N

CHAPTER 10

1. 23 m   **3.** 7.7×10^{-5} m   **5.** 1.4×10^{-6}
7. (a) 1.4×10^{-3}
(b) Figure 10.9 shows that the strain calculated from Hooke's law is less than the actual strain that the material experiences.
9. -3.6×10^{-4} **11.** (a) 6.3×10^{-2} m
(b) 7.3×10^{-2} m
13. 4.6×10^{-4} **15.** (a) 710 N/m²
(b) 3.5×10^{-8} (c) 3.5×10^{-10} m
17. 3.6×10^{-4} m (tungsten), 6.4×10^{-4} m (steel)
19. (a) 7.44 N   (b) 7.44 N   **21.** 0.015 m   **23.** 237 N
25. 0.92   **27.** (a) 1.00×10^3 N/m   (b) 0.340
29. (a) 0.152 m   (b) 1.85 m/s²   **31.** 2.04 m/s
33. 0.069 m   **35.** 9.93×10^{-3} m   **37.** 4.3 kg

39. (a) 46.9 J   (b) 55.9 m/s   **41.** 18 m/s
43.

h (meters)	KE	PE (gravity)	PE (elastic)	E
0	0	0	8.76 J	8.76 J
0.100	0.75 J	1.96 J	6.05 J	8.76 J
0.200	1.00 J	3.92 J	3.84 J	8.76 J
0.300	0.75 J	5.88 J	2.13 J	8.76 J
0.400	0	7.84 J	0.92 J	8.76 J

45. 16 m/s
47. 0.556 m/s (29.2-kg block), 1.11 m/s (14.6-kg block)
49. 2.76×10^3 N/m   **51.** 921 m/s   **53.** 0.995 m
55. 12.5 m   **57.** 0.816   **59.** (a) 5.08×10^{-2} m
(b) 5.00×10^{-2} s   (c) 6.40 m/s
61. 174 N/m   **63.** 66 Hz   **65.** 6.9×10^{-2} m
67. (a) 3.59×10^{-2} m and 4.24 Hz
(b) 5.08×10^{-2} m and 4.24 Hz
69. 0.240 m   **71.** 2.08 m/s

CHAPTER 11

1. 8750 N, the water bed should not be purchased.
3. 3400 N   **5.** 0.13 m   **7.** 63%   **9.** (a) 4.50×10^2 N
(b) 53.2 N
11. 41 N   **13.** 0.11 m   **15.** 5.66×10^3 Pa
17. 6.0×10^4 Pa   **19.** (a) 9.98×10^4 Pa
(b) Worse, because atmospheric pressure is less, so there is less force to push the water up.
21. 46.2 mm Hg
23. 0.741 m (mercury), 0.259 m (water)
25. $R_1 = 13.3$ m, $R_2 = 16.2$ m   **27.** 2.1×10^4 N
29. 4   **31.** (a) 72.8 N   (b) 74.7 N   **33.** 3.3×10^{-3} m³
35. 2.2×10^3 kg   **37.** 2.04×10^{-3} m³
39. 6.34×10^{-3} kg   **41.** (a) $r_1 = 5.28 \times 10^{-2}$ m
(b) $r_2 = 6.20 \times 10^{-2}$ m
43. 4.5×10^{-5} kg/s   **45.** 8.12×10^6 gal
47. (a) 2.2×10^5 kg   (b) 16 m/s   **49.** (a) 150 Pa
(b) The pressure inside the roof is greater than the pressure outside the roof, so there is a net outward force.
51. (a) 22.2 m/s   (b) 3.47×10^5 Pa   **53.** (a) 7.00 m/s
(b) 1.21×10^5 Pa
55. 0.19 m   **57.** (a) 32.8 m/s   (b) 54.9 m
59. (a) The larger hole is near the top of the tank.
(b) 1.19
61. 20 Pa   **63.** 0.5 m/s   **65.** (a) 5.7×10^{-5} N
(b) 5.2 m/s
67. 20.6 m   **69.** 4.89 m   **71.** 2.1 m/s   **73.** 317 m²
75. (a) 7.5×10^{-2} N   (b) 6.5 N   **77.** (a) 1.6×10^{-4} m³/s
(b) 2.0×10^1 m/s
79. 20 logs   **81.** 1.57 kg   **83.** 60.3%
85. The answer is a proof.

CHAPTER 12

1. (a) $-12.2°$ and $41°C$ (b) 261 K and 314 K
3. (a) $75°F$ (b) 297 K   5. $-459.67°F$ 7. $18°C$
9. 0.084 m   11. $33°C$ 13. 9.3×10^{-6} $(C°)^{-1}$
15. $110°C$ 17. $41°C$
19. (a) Since the ruler shrinks as the temperature decreases, a tension is needed to stretch the ruler.
(b) 9.6×10^7 N/m^2
21. (a) Conservation of angular momentum requires that $I\omega$ remains constant. As temperature increases, I increases, so ω decreases.
(b) -0.90%
23. 1.1×10^{-2} m^3 25. 1.8×10^{-4} m^3
27. 1.2×10^{-5} m^3 29. 4.5×10^{-3} m
31. (a) The sphere at $25°C$ weighs more, because it is subjected to less buoyant force.   (b) 18 N
33. 6.9   35. 1.6×10^8 J   37. 76 kcal or 3.2×10^5 J
39. 250 s   41. $230   43. 2.3×10^5 N
45. 1.1×10^3 N   47. 0.13 kg   49. (a) 4.52×10^6 J
(b) 5.36×10^6 J
51. 1.79×10^5 J   53. 1.49×10^{14} J
55. (a) 7.1×10^7 J (water)   (b) 4.4×10^8 J (G. salt)
(c) 5.0×10^7 J (water) and 3.6×10^7 J (G. salt)
57. 1.9×10^4 J/kg   59. 0.237 kg   61. Vapor phase
63. 4800 Pa   65. $10°C$ 67. 28%   69. 3.9×10^5 J
71. $940°C$ 73. 5.8 m   75. 690 J/(kg · C°)
77. 210×10^{-6} $(C°)^{-1}$ 79. 4.0×10^{-4} $(C°)^{-1}$
81. $79°C$ 83. $26°C$

CHAPTER 13

1. 12 J   3. (a) 1.0×10^4 J   (b) 3.0×10^{-2} kg
5. 4.8   7. (a) 78 W
(b) Virtually all the power is conducted through the copper bar because of its much larger thermal conductivity.
9. 17   11. (a) $130°C$ (b) 830 J   (c) $237°C$
13. 0.11 mm   15. 4.29×10^2 W/m^2 17. 1.7
19. $831°C$ 21. 14.5 days   23. (a) 9×10^6 m
(b) 7×10^8 kg/m^3
25. $558°C$ 27. 2.6×10^{-5} m^2
29. Aluminum, copper, silver   31. 1.67   33. 732 K

CHAPTER 14

1. (a) 16.043   (b) 2.6641×10^{-26} kg
3. 1.2×10^{24} 5. 141 g   7. (a) 7.65×10^{-26} kg
(b) 2.11×10^{25} molecules
9. 0.550 kg   11. 1050 K   13. 67.0 m^3
15. (a) 8×10^{-18} mol/m^3 (b) 2×10^{-16} Pa
17. 0.140 m   19. 10.3 m   21. 0.205
23. 5.61×10^5 Pa   25. 1.2×10^4 m/s

27. (a) 1.23   (b) 1.11   29. 1.01×10^4 K   31. 399 J
33. (a) 120 N   (b) 120 N   (c) 4.0×10^5 Pa
35. 7.0×10^{-3} m/s   37. 0.0919 kg/m^4
39. (a) The answer is a proof.   (b) 31 s
41. 2.3×10^{-2} mol   43. 11 s   45. 0.27 m
47. 4.0×10^1 Pa   49. 11.7

CHAPTER 15

1. -4.3×10^5 J, heat is given off.
3. -210 K, a decrease in temperature
5. (a) -3.42×10^5 J   (b) 81.7 nutritional calories
7. 1.3×10^5 Pa   9. (a) 3.0×10^3 J
(b) Work is done by the system, so the work is positive.
11. 1.3×10^4 J   13. $W = \frac{1}{2}(P_A + P_B)(V_B - V_A)$
15. 2.9 J   17. -8.0×10^4 J, heat flows out of the system.
19. 1.29×10^4 J   21. 3.0   23. 3.17   25. 18.0:1
27. $T_f = 327$ K, $V_f = 0.132$ m^3 29. 2400 J
31. 310 J   33. $\frac{5}{2}$ 35. 75 K   37. 2.38×10^4 J
39. 1.82×10^4 J   41. 0.631   43. 0.31   45. 0.050
47. The greatest improvement is made by lowering the temperature of the cold reservoir.
49. (a) 0.360   (b) 1.3×10^{13} J   51. 5.7 C°
53. 890 J   55. 5.86×10^5 J   57. 1.4   59. 0.30 cents
61. (a) The entropy change of the environment must be greater than or equal to 25 J/K, depending on whether the process is irreversible or reversible.
(b) The disorder of the environment increases.
63. (a) Reversible   (b) -125 J/K
65. (a) $+541$ J/K
(b) The entropy of the universe increases.
67. -3700 J, heat flows out of the gas.
69. 3.0×10^5 Pa   71. (a) $+3.68 \times 10^3$ J/K
(b) $+1.82 \times 10^4$ J/K
(c) The conversion of a liquid to a vapor creates more disorder.
73. 9.03   75. (a) -3.1×10^3 J   (b) The work is negative.
77. (a) 0.50   (b) 1.5   (c) 0.67
79. $\frac{2}{3}$ (Engine A), $\frac{1}{3}$ (Engine B)   81. 23%

CHAPTER 16

1. 0.083 Hz   3. 4.5×10^{14} Hz   5. 8.19×10^{-2} m
7. 1.7 m/s   9. 45 m/s
11. (a) 1.33 m/s   (b) 5.33 m   13. 64 N
15. 6.1×10^{-4} kg   17. 1.2 m/s^2 19. 3.3×10^{-4} kg/m
21. (a) $+x$ direction   (b) -0.080 m
23. (a) $A = 0.45$ m, $f = 4.0$ Hz, $\lambda = 2.0$ m, $v = 8.0$ m/s
(b) $-x$ direction
25. $y = (2.00$ mm$) \sin(314t - 8.38)$
27. (a) 20 m (at $f = 20$ Hz), 0.02 m (at $f = 20$ kHz)
(b) The statement is incorrect.

29. (a) 4.31×10^2 m/s (b) 3.22×10^2 m/s
31. 2.8×10^{-4} s   33. 2.06   35. 690 rad/s
37. Tungsten   39. 61 m   41. 0.404 m
43. 6.7×10^{-9} W   45. 2.4×10^{-5} W/m²
47. 1.98%   49. 1.0×10^{-6} W/m²   51. 63.1 W/m²
53. 66 dB   55. (a) 7.4 dB
(b) Since an increase of 10 dB is needed to double the loudness, it will not sound twice as loud.
57. 6.3   59. $r_1 = 3.9$ m and $r_2 = 4.9$ m
61. 838 Hz   63. 0.340 m   65. 31 m/s
67. 0.26 m   69. 3.9×10^{-3} m   71. 5.9 m/s
73. 3.9×10^{-3} W   75. $y = (0.15$ m$) \sin(160t - 22x)$
77. 0.25 m   79. $m_1 = 28.6$ kg and $m_2 = 14.3$ kg
81. 21.2 m/s   83. 76.8 dB   85. 6

CHAPTER 17

1. The answer is a drawing.
3. The answer is a drawing.   5. 107 Hz
7. (a) Destructive interference   (b) Constructive interference
9. (a) 44° (for $f = 2$ kHz), 13° (for $f = 6$ kHz)   (b) 0.10 m
11. 3.4°   13. 0.5 s   15. 8 Hz
17. 3 Hz or 7 Hz   19. 88 m/s   21. 171 N
23. 5.30   25. 3.92×10^{-3} kg/m   27. 0.49
29. (a) 100 Hz   (b) The tube is open at only one end.
31. 0.50 m   33. 0.30 m   35. 35.3 Hz
37. $f_n = n\left(\dfrac{v}{2v}\right)$ $n = 1, 2, 3, \ldots$ 39. 1.2 m/s
41. 4   43. 5.06 m   45. 570 m/s   47. $n = 66$
49. 1.68×10^5 Pa

CHAPTER 18

1. 2.4×10^{13} 3. 3.1×10^{13}
5. (a) $+4q$ (b) $+4q$ 7. (a) 0.83 N
(b) The force is attractive, because the spheres have charges of different polarities.
9. 0.14 N   11. (a) 1.4 N in the $-x$ direction,
(b) 1.4 N in the $+x$ direction, because q_2 has the same magnitude but opposite sign.
13. Placing the -9.0 μC charge on the north corner, the $+8.0$ μC charge on the west corner, and the $+2.0$ μC charge on the east corner, the answer is 6.8 N pointing 66°N of E.
15. $-2\sqrt{2}q$ 17. 92 N/m   19. 0.37 N
21. Assuming the negative charge is on the left, the answers are -5.58 μC and 0.957 μC or -0.957 μC and 5.58 μC.
23. 0.45 N, due east   25. -2.5×10^{-5} C
27. (a) 4.10×10^{12} N/C (radially outward)
(b) 6.56×10^{-7} N (radially inward)   (c) 4.37×10^6 m/s

29. (a) 2.9×10^5 N/C, in the $+x$ direction
(b) 2.3 N, in the $-x$ direction
31. 2.8×10^5 N/C, in the $-x$ direction   33. 0.364
35. (a) The charges have like polarities.   (b) 27.0
37. 120 N   39. 9
41. 0.38 N, at 49° below the $-x$ axis (third quadrant)
43. (a) The charges have like polarities.
(b) 1.7×10^{-16} C
45. (a) The charges have like polarities.
(b) 8.4×10^{-6} C
47. (a) 15.4°   (b) 0.813 N

CHAPTER 19

1. 1.5×10^{-20} J   3. (a) 5.80×10^{-3} J
(b) 32.2 V   (c) Point B is at the higher potential.
5. 9.4×10^7 m/s   7. 15 m   9. 42.8
11. 3.6×10^{-9} C, the charge is positive.   13. 41 V
15. 0.61 J   17. 1.2 m on either side of the negative charge.
19. 2.4 J   21. (a) 8.68×10^{-18} J   (b) 54.3 eV
23. (a) $-\frac{2}{3}q$ (b) $-2q$ 25. 0.15 m
27. 1.3×10^7 V/m   29. 9.0×10^3 V
31. (a) 2.54 m and 3.54 m
(b) 3.54 V (for $r = 2.54$ m), 2.54 V (for $r = 3.54$ m)
33. 7.0×10^{13} 35. 2×10^{-8} F
37. (a) 1×10^{-12} C   (b) 6×10^6
39. (a) 2.5×10^6 V/m   (b) 1.2×10^6 V/m
41. 52 V   43. 1.1×10^{-2} m   45. 2.18×10^{-5} m
47. 5.3   49. 8.0×10^{-5} C   51. 1.0×10^{-4} C
53. 2.77×10^6 m/s

CHAPTER 20

1. 0.025 A   3. 1.0×10^6 J   5. 22 A
7. (a) 7.9×10^5 C   (b) 350 A   9. 0.58 Ω
11. 1.64   13. 9.9×10^{-3} m   15. 189 Ω
17. 140° C   19. 228 W   21. 1.0×10^3 W
23. (a) 4.4 Ω   (b) 2.8 A   25. 1.3   27. 1.77 A
29. 1.8 A   31. (a) 9.0×10^2 W   (b) 1.8×10^3 W
33. (a) 2.0×10^3 W   (b) 450 s
35. (a) 145 Ω   (b) 74 V   37. 21.7 V   39. 32 Ω
41. (a) 28.9 V   (b) 16.7 W   43. 446 Ω   45. 64
47. (a) 0.75 A   (b) 160 Ω   49. 5.24 Ω and 0.76 Ω
51. 1.0×10^2 Ω   53. 6.76 Ω   55. 4.6 Ω
57. 600 Ω   59. 0.054 Ω   61. 0.449 Ω   63. 30
65. (a) 12 V   (b) 24 V   67. 33 A
69. 0.75 V, the left end is at the higher potential.
71. 1.43 V, point A is at the higher potential.
73. 0.667 Ω   75. 7.7%
77. (a) 30.0 V   (b) 28.1 V   79. 9.23 μF

81. 25 V   **83.** 1.88   **85.** 55 V   **87.** 1.80 V
89. 0.15 s   **91.** 0.0089 s   **93.** 4.68×10^{-4} C
95. 3.5×10^4 A   **97.** 0.21 A   **99.** 3.4×10^4 s
101. 6.00 Ω, 0.545 Ω, 3.67 Ω, 2.75 Ω, 2.20 Ω, 1.50 Ω, 1.33 Ω, 0.833 Ω
103. $92   **105.** 0.054
107. $R_1 = 13.0$ Ω, $R_3 = 4.0$ Ω, 9.8 V
109. $L_{tungsten}/L_{carbon} = 70$

CHAPTER 21

1. 8.1×10^{-5} T   **3.** 75° or 105°   **5.** 8.9×10^{-5} C
7. 4.1×10^{-12} N
9. (a) 4.3×10^2 m   (b) 7.8×10^5 m
11. 3.2×10^{-19} C   **13.** 1.52×10^{-2} m
15. 8.2×10^{-3} m   **17.** 1.50×10^7 C/kg
19. (a) 4.3×10^{-3} N   (b) 1.6×10^{-4} C
21. 1.01×10^6 rad/s   **23.** 2.3 N (for each wire)
25. 0.96 N (top and bottom sides), 0 (left and right sides)
27. 57.6°   **29.** 0.325 kg   **31.** (a) 11 A
(b) Away from the battery.
33. (a) 24 A · m²   (b) 4.8 N · m
35. (a) 53.1°   (b) The answer is a drawing.
37. 2.3×10^{-2} N · m   **39.** 1.2×10^{-5} A · m²
41. 8.0×10^{-5} T   **43.** 2.7×10^{-2} m
45. A: 4.3×10^{-5} T, B: 5.3×10^{-5} T   **47.** 190 A
49. 0.50 m, on the side of wire A that is away from wire B.
51. 6.8 A, opposite to the direction of the current in the inner coil.
53. 4.1×10^{-3} m/s   **55.** 5.97×10^{-4} N · m
57. 1.9×10^{-4} N · m   **59.** 1.5×10^{-8} s
61. 140 V/m, directed toward the bottom of the page.
63. 9.1×10^{-8} T   **65.** $\frac{1}{2}$

CHAPTER 22

1. 690 m/s   **3.** (a) The driver's side   (b) 2.0 m
5. 15 m   **7.** 3.2 A   **9.** (a) 5.6×10^{-3} Wb,   (b) 0 Wb
11. 0.031 Wb for x, y plane and 0 Wb for x, z plane.
13. 0.090 Wb for the 1.2 m × 0.30 m side, 0 Wb for the triangular ends and the bottom side, 0.090 Wb for the 1.2 m × 0.50 m side.
15. 1.2 V   **17.** 0.15 T/s   **19.** 0.34 V   **21.** 0.459 T
23. Right to left in part (b) and left to right in part (c).
25. (a) Clockwise in loop A and counterclockwise in loop B
(b) Counterclockwise in loop A and clockwise in loop B.
27. Clockwise at both positions 1 and 2.
29. (a) Retardation occurs because of an induced current that flows counterclockwise in the ring while the magnet approaches from above and clockwise while the magnet moves away from the bottom.
(b) No induced current can flow in the cut ring.
31. $\mathcal{E}_0 = 180$ V, T = 0.42 s   **33.** 4.0×10^5
35. 72 V   **37.** 2.0 V   **39.** (a) 9.59 Ω
(b) 95 V   (c) 8.9 A
41. 1.4 V   **43.** The magnitude of the current is 0.32 A.
45. 1.5×10^9 J   **47.** The answer is a proof.
49. 588   **51.** 0.43 A   **53.** 7.7 W
55. The answer is a proof.   **57.** 0.30 T   **59.** 170 V
61. 230 V   **63.** (a) 0.80 A   (b) 8.00 A   (c) 4.40 A
65. (a) 3.6×10^{-3} V   (b) 2.0×10^{-3} m²/s (The area must shrink.)
67. 2.4×10^{-3} A

CHAPTER 23

1. 126 Hz   **3.** (a) 13.0 μF   (b) 0.61 A
5. 5.00×10^{-2} s   **7.** 6.8×10^{-2} H   **9.** 0.13 A
11. 5.96×10^{-2} A   **13.** 84.0 V
15. (a) 19.0 V   (b) −52.5° (The minus sign means that the current leads the voltage.)
17. (a) 0.925 A   (b) 31.7°
19. (a) 0.26 A   (b) 0.11 A   **21.** 1100 Hz
23. 3250 Hz   **25.** 9.8×10^{-2} W   **27.** 1.7
29. 521 Hz   **31.** 5.7 MHz
33. $X_L/R = 4/\sqrt{3}$, $X_C/R = 1/\sqrt{3}$
35. (a) 2.00 μF   (b) 0.77 A   **37.** 0.44 A
39. (a) 2.94×10^{-3} H   (b) 4.84 Ω   (c) 0.164
41. 3.11×10^3 Hz, 7.50×10^3 Hz

CHAPTER 24

1. 5.00×10^2 s   **3.** 6.8×10^{-11} F
5. The answers are in graphical form.   **7.** 11.118 m
9. (a) 1090 kHz   (b) AM   **11.** 1.5×10^{10} Hz
13. (a) 1.28 s   (b) 1.9×10^2 s   **15.** 0.015 m
17. 2.6×10^6 m   **19.** 7.13×10^{-6} T
21. (a) 1.83×10^2 N/C   (b) 6.10×10^{-7} T
23. 6.25×10^{-9} J   **25.** 3.93×10^{26} W
27. (a) 0.550 W/m²   (b) 3.7×10^{-2} W/m²
29. 71.6°   **31.** 0.125   **33.** 2.78 m to 3.41 m
35. 70.9 min   **37.** 8.37
39. The answer is obtained by using the definitions of the tesla and the ampere.
41. 8.5×10^{-5} V   **43.** 5.30 N/C and 1.77×10^{-8} T

CHAPTER 25

1. 55°   **3.** (a) 0.80 m   (b) 0.75 m
5. (a) 1.8 m/s   (b) 1.2 m/s

7. The answer is a drawing.

9. A ray diagram indicates that the image distance is 7.5 cm; the image height is 1.0 cm and the image is inverted relative to the object.

11. A ray diagram indicates that the image is 17 cm behind the mirror; the image height is 6.8 cm and the image is upright relative to the object.

13. 31 cm   15. 74 cm   17. (a) 62 cm
(b) +0.35   (c) Upright   (d) Smaller

19. (a) −1.8 m   (b) Virtual   (c) 0.19 m

21. (a) Concave   (b) 2.3 cm   (c) +2.9   (d) Upright

23. (a) 6.0×10^1 cm   (b) The answer is a drawing.

25. The answer is a proof.   27. $-\frac{1}{2}$ 29. −62 cm

31. The answers are drawings.

33. (a) 62.1°   (b) 80.0°

35. (a) $d_o = 9.0 \times 10^1$ cm, $d_i = 45$ cm
(b) $d_o = 15$ cm, $d_i = -3.0 \times 10^1$ cm

CHAPTER 26

1. 2.00×10^8 m/s   3. Ethyl alcohol

5. 2.0×10^{-11} s   7. 1.40   9. (a) 43°   (b) 31°

11. 41.0°   13. 2.05 cm   15. 12.1 m

17. 2.1 cm   19. The answer is a derivation.

21. 49.81°   23. 1.54   25. (a) Point B   (b) Point A

27. 1.45   29. 1.52   31. 48.32°   33. 25.0°

35. 0.35°   37. 29.5°   39. 52.7° (red), 56.2° (violet)

41. (a) −24 cm   (b) 1.6   43. (a) −24 cm   (b) 3.0

45. 0.23 cm   47. (a) −16 cm   (b) 0.57
(c) Virtual   (d) Upright   (e) Reduced

49. 6.00 m   51. (a) 4.52×10^{-4} m
(b) 6.10×10^{-2} m

53. 35 cm and 90.5 cm   55. (a) 19.6 cm   (b) 0.87 cm

57. 60.0 cm   59. 38 cm   61. 87.0 cm

63. 38.0 cm   65. 26.9 cm

67. (a) 31.3 cm   (b) 2.43 m

69. (a) 6.9 cm   (b) 3.6   71. 0.40 m   73. 15.4

75. 4.09×10^{-3} rad   77. 9×10^{-3} rad

79. $L' = L - f_e + f_o$ 81. -2.0×10^2

83. (a) 0.775 m   (b) 7.80×10^{-1} m

85. $f_o = 1.09$ m and $f_e = 1.00 \times 10^{-2}$ m   87. −31

89. 3.0 cm   91. (a) Hyperopic   (b) 2.8 diopters

93. -7.90×10^{-3} m (35.0 mm lens) and -3.44×10^{-2} m (150.0 mm lens)

95. (a) 1.3-diopter lens   (b) 0.86 m,   (c) −8.5

97. (a) Converging   (b) Farsighted   (c) 96.3 cm

99. 3.9 cm   101. 6.3 cm   103. 3.3×10^{-3} m

105. (a) 11.8 cm   (b) 47.8 cm

107. (a) 22.4 cm   (b) 28.4 cm

CHAPTER 27

1. (a) Constructive   (b) Destructive   (c) Destructive
(d) Constructive

3. 18   5. 490 nm   7. 0.0309 m   9. 102 nm

11. 183 nm, 366 nm   13. Blue

15. 3.06×10^{-7} m   17. 490 nm   19. 0.32 m

21. 485 nm   23. 0.013   25. 14 000 m

27. 5.6×10^{17} m   29. 3.2×10^3 m   31. 3.2 s

33. 3.5×10^{-6} m   35. 1690 lines/cm   37. 24°

39. 1.95 m   41. (a) 24°   (b) 39°

43. 7.0×10^1 degrees

45. (a) 1.0×10^{-7} m   (b) 7.3×10^{-8} m

47. 0.52 m   49. (a) 1.0×10^{-7} m   (b) 2.1×10^{-7} m

51. 0.43 cm   53. 0.0256 m

CHAPTER 28

1. 2.4×10^8 m/s   3. 54.7 s   5. 4.4×10^{-4} s

7. 16   9. 384 m   11. 3400 m (length), 1500 m (width)

13. 3.0 m × 1.3 m   15. 2.91×10^4 kg · m/s

17. −2.0 m/s   19. 3.2×10^5 21. 1.1 kg

23. (a) 3.8×10^{-15} J   (b) $0.30\,c$ 25. $0.31\,c$

27. (a) c (b) $0.994\,c$ (c) $0.200\,c$ (d) $0.194\,c$

29. (a) 2.93×10^8 m/s   (b) 2.9×10^{-16} kg · m/s

31. 3.5×10^{-8} s   33. 0.74 m   35. 42 m

37. (a) 4.3 years
(b) The twin who travels at $0.500c$ is older.

CHAPTER 29

1. 310 nm   3. 7.7×10^{29} photons/s

5. 2.5×10^{21} photons/(m² · s)   7. 138

9. Aluminum (4.08 eV)   11. 1.4×10^5 m

13. (a) 7760 N/C   (b) 2.59×10^{-5} T

15. 3.6×10^{-9} m   17. 73°   19. 2.6×10^{-28} m

21. 7.79×10^{-13} J   23. 7.38×10^{-11} m

25. 8.5×10^{-12} m

27. 1.6×10^{-4} m (electron), 3.3×10^{-33} m (golf ball)

29. (a) 1.4×10^{-35} m   (b) 14 m   31. 37 J · s

33. X-ray and infrared   35. (a) 1.97×10^{-24} kg · m/s
(b) 1.95×10^{-24} kg · m/s

37. (a) 1.10×10^3 m/s   (b) 5.93×10^5 m/s

39. 42.8

CHAPTER 30

1. 3.7×10^{13} m   3. 6.88×10^{-15} m

5. 5.2×10^6 eV   7. $E_{1,H} = E_{2,He}$ and $E_{2,H} = E_{4,He}$

9. 54.4 eV   11. 4.41×10^{-10} m   13. 5

15. KE = 4.90 eV, EPE = −9.80 eV   17. $6 \le n_i \le 19$

19. 3920 lines/cm

21.

n	ℓ	m_ℓ	m_s
3	0	0	$\frac{1}{2}$
3	0	0	$-\frac{1}{2}$
3	1	1	$\frac{1}{2}$
3	1	1	$-\frac{1}{2}$
3	1	0	$\frac{1}{2}$
3	1	0	$-\frac{1}{2}$
3	1	-1	$\frac{1}{2}$
3	1	-1	$-\frac{1}{2}$
3	2	2	$\frac{1}{2}$
3	2	2	$-\frac{1}{2}$
3	2	1	$\frac{1}{2}$
3	2	1	$-\frac{1}{2}$
3	2	0	$\frac{1}{2}$
3	2	0	$-\frac{1}{2}$
3	2	-1	$\frac{1}{2}$
3	2	-1	$-\frac{1}{2}$
3	2	-2	$\frac{1}{2}$
3	2	-2	$-\frac{1}{2}$

23. (a) $\ell = 4$ (b) $n = 5$

25. (a) $L = h/2\pi$ (Bohr model), $L = 0$ (quantum mechanics)
(b) $L = 2h/2\pi$ (Bohr model)
   for $\ell = 0$, $L = 0$ $\}$ (quantum
   for $\ell = 1$, $L = \sqrt{2}h/2\pi$ $\}$ mechanics)

27. $1s^2 2s^2 2p^6 3s^2 3p^6 4s^2 3d^{10} 4p^3$

29. (a) Not allowed   (b) Allowed   (c) Not allowed
(d) Allowed   (e) Not allowed

31. $Z = 6.2$ (carbon)   **33.** 1.11×10^{-10} m

35. 1.03×10^{-18} J

37. (a) The semiconductor laser   (b) 13.4

39. $1s^2 2s^2 2p^6 3s^2 3p^6 4s^2 3d^{10} 4p^6 5s^2 4d^1$ **41.** 91.2 nm

43. 1.08×10^{-14} J or 6.75×10^4 eV

45. $0, \pm 1.05 \times 10^{-14}$ J $\cdot$ s, $\pm 2.11 \times 10^{-14}$ J $\cdot$ s, $\pm 3.16 \times 10^{-14}$ J $\cdot$ s

47. KE $= n^2 h^2/(8\,mL^2)$ **49.** (a) $T_n = h^3 n^3/(4\pi^2 mk^2 e^4 Z^2)$
(b) 1.53×10^{-16} s   (c) 1.22×10^{-15} s

CHAPTER 31

1. (a) 8 protons, 10 neutrons   (b) 50 protons, 70 neutrons

3. 35.45 u   **5.** 1.2 N   **7.** 1.14

9. (a) 0.555 357 u   (b) 9.2217×10^{-28} kg

11. 127.6 MeV   **13.** (a) 10.55 MeV
(b) 7.55 MeV   (c) The neutron

15. (a) $^{231}_{90}$Th   (b) $^{235}_{92}$U   **17.** $^{60}_{27}$Co $\rightarrow$ $^{60}_{28}$Ni $+$ $^{0}_{-1}$e

19. 2.44×10^{-12} m   **21.** $^{212}_{84}$Po   **23.** $^{228}_{89}$Ac

25. 5.67×10^3 m/s   **27.** 8.00 da   **29.** 379 yrs

31. 2.1×10^{13} Bq   **33.** 146 disintegrations/min

35. 8.00×10^6 m   **37.** 0.70%

39. (a) 3.0×10^4 yr   (b) 3.3×10^4 yr

41. 6900 yr, maximum error is 900 yr   **43.** Platinum (Pt), sulfur (S), copper (Cu), boron (B), plutonium (Pu)

45. 19.9   **47.** 1.003 26 u   **49.** 3.7×10^{10} Bq

51. 9.37 da

CHAPTER 32

1. 0.26 R   **3.** 30 rd   **5.** 100 mrem

7. 1.8×10^{-3} C°   **9.** $A = 30$

11. $^{17}_{8}$O $+ \gamma \rightarrow$ $^{12}_{6}$C $+$ $^{4}_{2}$He $+$ $^{1}_{0}$n   **13.** Zinc, $Z = 30$, $A = 64$

15. Two neutrons   **17.** 9×10^{-4}

19. 3.9×10^{18} fissions/s   **21.** 2.7×10^6 kg

23. 1.8×10^4 kg/s   **25.** 1200 kg   **27.** 2.2 MeV

29. 7.0×10^9 kg   **31.** 815 MeV

33. (a) 1.9×10^{-20} kg $\cdot$ m/s   (b) 3.6×10^{-14} m

35. 3.7×10^{-6} J   **37.** Oxygen, $Z = 8$, $A = 16$

39. 9.6×10^8 W   **41.** 1.1×10^{-4} kg   **43.** 160 MeV

Photo Credits

PART IV *Opener:* Agence Vandystadt/Allsport.

CHAPTER 18 *Opener:* Warren Faidley/Weatherstock. *Page 504:* Leif Skoogfors/Woodfin Camp & Associates. *Page 506:* Courtesy BMW of North America. *Page 509:* Ron & Valerie Taylor/Bruce Coleman, Inc.

CHAPTER 19 *Opener:* Dan Carroll/The Image Bank. *Page 530:* Mark Antman/The Image Works. *Page 544 (top):* Adam Hart-Davis/Photo Researchers. *Page 544 (bottom):* Animals, Animals/Earth Scenes.

CHAPTER 20 *Opener:* Tom Mareschal/The Image Bank. *Page 556:* Jim Sulley/The Image Works. *Figure 20.5:* Courtesy Macintosh. *Page 561:* Courtesy Kinetic Corporation, Louisville, KY. *Page 563:* William Bacon III/Photo Researchers.

CHAPTER 21 *Opener:* Everett Collection, Inc. *Figure 21.4a:* Kathy Bendo. *Page 598:* Science Photo Library/Photo Researchers. *Page 620:* Tom Hollyman/Photo Researchers. *Figure 21.36a:* Courtesy Deutsche Bundesbahn/Transrapid.

CHAPTER 22 *Opener:* Mike Guastella/Star File. *Page 654:* Robert Frerck/Woodfin Camp & Associates. *Page 662:* Will McIntyre/Photo Researchers.

CHAPTER 23 *Opener:* Jon Love/The Image Bank. *Page 679:* J. F. Towers/The Stock Market. *Figure 23.16:* Matt Bradley/Southern Stock Photos. *Page 685:* Science Photo Library/Photo Researchers. *Page 687:* Fuji Fotos/The Image Works. *Figure 23.29:* Uniphoto, Inc.

INTERSECTION, *Page 693b:* Phil Degginger/FPSA/Tony Stone World Wide/Chicago, Ltd.

PART V *Opener:* David Lawrence/The Stock Market.

CHAPTER 24 *Opener:* E. R. Degginger/FPSA. *Page 698:* Jurgen Vogt/The Image Bank. *Page 699:* Astronomical Society of the Pacific. *Figure 24.15:* Courtesy Bausch & Lomb. *Page 710:* Hans Neleman/The Image Bank.

CHAPTER 25 *Opener:* Ted Mahieu/The Stock Market. *Figure 25.5b:* Gary Bistram/The Image Bank. *Page 720:* Jim and Cathy Church/FPG International. *Figure 25.14:* Science Source/Photo Researchers. *Figure 25.18b:* Ken Karp Photography. *Figure 25.19b:* Paul Silverman/Fundamental Photos. *Page 729:* S. Achernar/The Image Bank.

CHAPTER 26 *Opener:* King Dexter. *Page 742:* John Michael/International Stock Photo. *Figure 26.11b:* T. J. Florian/Rainbow.

Page 744: Courtesy The Johns Hopkins University School of Medicine. *Page 745:* Science Source/Photo Researchers. *Page 747:* Michael Giannechini/Photo Researchers. *Page 761:* Courtesy Bausch and Lomb. *Figure 26.37:* Courtesy Tasco. *Figure 26.38:* Courtesy Celestron, Inc.

CHAPTER 27 *Opener:* Joe McDonald/Bruce Coleman, Inc. *Figure 27.7:* From *Atlas of Optical Phenomena,* Michael Cagnet, Springer-Verlag, Berlin. *Page 787:* Mickey Pfleger. *Page 788:* Priscilla Connell/Photo-NATS. *Figures 27.13 and 27.14:* Courtesy Bausch & Lomb. *Figure 27.17:* Education Development Center. *Figures 27.22 and 27.27b:* From *Atlas of Optical Phenomena,* Michael Cagnet, Springer-Verlag, Berlin. *Figure 27.36:* From *Fundamentals of College Physics,* by W. Wallace McCormick.

INTERSECTION, *Page 813a:* Courtesy NASA.

PART VI *Opener:* Agence Vandystadt/Allsport.

CHAPTER 28 *Opener (top left):* AIP Neils Bohr Library; *(top center):* Lotte Jacobi; *(top right):* New York Times Pictures; *(bottom right):* AIP Neils Bohr Library; *(bottom center):* Philippe Halsman; *(bottom left):* UPI/Bettmann Newsphotos. *Page 819:* Mark Antman/The Image Works. *Page 822:* David Malin/Anglo Australian Telescope Board. *Figure 28.7:* Bill W. Marsh/Photo Researchers.

CHAPTER 29 *Opener:* David Scharf/Peter Arnold, Inc. *Page 844:* Science Photo Library/Photo Researchers. *Figure 29.7a:* From "Physics Review" 73, 527 (1948), by Wollan, Shull, and Marney. *Figure 29.7b:* From *Fundamentals of College Physics,* by W. Wallace McCormick. *Figure 29.8:* Akira Tonomura, J. Endo, T. Matsuda, T. Kawasaki, *American Journal of Physics,* February 1989.

CHAPTER 30 *Opener:* Todd Kaplan/Star File. *Figure 30.3:* Courtesy Bausch & Lomb. *Page 878:* Alexander Tsiaras/Science Source/Photo Researchers.

CHAPTER 31 *Opener:* Science Source/Photo Researchers. *Page 899:* Thomas Ives. *Page 902:* William Rivelli/The Image Bank. *Figure 31.14:* Dan McCoy/Rainbow.

CHAPTER 32 *Opener:* Y. Arthus-Bertrand/Photo Researchers. *Page 920:* U.S. Department of Energy. *Figure 32.9:* Lawrence Livermore Laboratory. *Figure 32.10:* Science Photo Library/Photo Researchers. *Page 924:* Lawrence Berkeley Laboratory/Science Photo Library/Photo Researchers.

INTERSECTION, *Page 929b:* Science Photo Library/Photo Researchers.

INDEX

SI UNITS

Quantity	Name of Unit	Symbol	Expression in Terms of Other SI Units	Quantity	Name of Unit	Symbol	Expression in Terms of Other SI Units
Length	meter	m	Base unit	Viscosity	—	—	Pa·s
Mass	kilogram	kg	Base unit	Electric charge	coulomb	C	A·s
Time	second	s	Base unit	Electric field	—	—	N/C
Electric current	ampere	A	Base unit	Electric potential	volt	V	J/C
Temperature	kelvin	K	Base unit	Resistance	ohm	Ω	V/A
Amount of substance	mole	mol	Base unit	Capacitance	farad	F	C/V
Velocity	—	—	m/s	Inductance	henry	H	V·s/A
Acceleration	—	—	m/s²	Magnetic field	tesla	T	N·s/(C·m)
Force	newton	N	kg·m/s²	Magnetic flux	weber	Wb	T·m²
Work, energy	joule	J	N·m	Specific heat capacity	—	—	J/(kg·K) or J/(kg·C°)
Power	watt	W	J/s	Thermal conductivity	—	—	J/(s·m·K) or J/(s·m·C°)
Impulse, momentum	—	—	kg·m/s	Entropy	—	—	J/K
Plane angle	radian	rad	m/m	Radioactive activity	becquerel	Bq	s⁻¹
Angular velocity	—	—	rad/s	Absorbed dose	gray	Gy	J/kg
Angular acceleration	—	—	rad/s²	Exposure	—	—	C/kg
Torque	—	—	N·m				
Frequency	hertz	Hz	s⁻¹				
Density	—	—	kg/m³				
Pressure, stress	pascal	Pa	N/m²				

THE GREEK ALPHABET

Alpha	A	α	Iota	I	ι	Rho	P	ρ
Beta	B	β	Kappa	K	κ	Sigma	Σ	σ
Gamma	Γ	γ	Lambda	Λ	λ	Tau	T	τ
Delta	Δ	δ	Mu	M	μ	Upsilon	Y	υ
Epsilon	E	ϵ	Nu	N	ν	Phi	Φ	ϕ
Zeta	Z	ζ	Xi	Ξ	ξ	Chi	X	χ
Eta	H	η	Omicron	O	o	Psi	Ψ	ψ
Theta	Θ	θ	Pi	Π	π	Omega	Ω	ω